Geometry

Formulas for Area (A), Perimeter (P), Circumference (C), and Volume (V)

Square

$A = s^2$

$P = 4s$

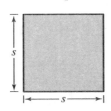

Rectangle

$A = lw$

$P = 2l + 2w$

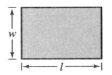

Triangle

$A = \frac{1}{2}bh$

$P = a + b + c$

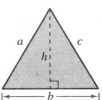

Circle

$A = \pi r^2$

$C = 2\pi r$

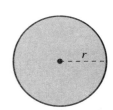

Trapezoid

$A = \frac{1}{2}h(b_1 + b_2)$

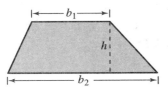

Parallelogram

$A = bh$

$P = 2a + 2b$

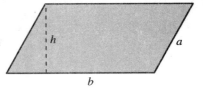

Cube

$V = s^3$

Rectangular Solid

$V = lwh$

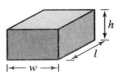

Circular Cylinder

$V = \pi r^2 h$

Sphere

$V = \frac{4}{3}\pi r^3$

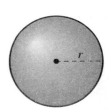

Special Triangles

Equilateral Triangle

Isosceles Triangle

Pythagorean Theorem

$a^2 + b^2 = c^2$

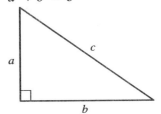

Right Triangle

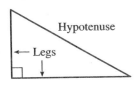

Hypotenuse

Legs

Similar Triangles

$\frac{a}{d} = \frac{b}{e} = \frac{c}{f}$

Common Formulas

Distance

$d = rt$

d = distance traveled
t = time
r = rate

Temperature

$F = \dfrac{9}{5}C + 32$

F = degrees Fahrenheit
C = degrees Celsius

Simple Interest

$I = Prt$

I = interest
P = principal
r = annual interest rate
t = time in years

Compound Interest

$A = P\left(1 + \dfrac{r}{n}\right)^{nt}$

A = balance
P = principal
r = annual interest rate
n = compoundings per year
t = time in years

Coordinate Plane: Midpoint Formula

Midpoint of line segment
joining (x_1, y_1) and (x_2, y_2)

$\left(\dfrac{x_1 + x_2}{2}, \dfrac{y_1 + y_2}{2}\right)$

Coordinate Plane: Distance Formula

d = distance between points (x_1, y_1) and (x_2, y_2)

$d = \sqrt{(x_2 - x_1)^2 + (y_2 - y_1)^2}$

Quadratic Formula

Solutions of $ax^2 + bx + c = 0$

$x = \dfrac{-b \pm \sqrt{b^2 - 4ac}}{2a}$

Rules of Exponents

(Assume $a \neq 0$ and $b \neq 0$.)

$a^0 = 1$

$a^m \cdot a^n = a^{m+n}$

$(ab)^m = a^m \cdot b^m$

$(a^m)^n = a^{mn}$

$\dfrac{a^m}{a^n} = a^{m-n}$

$\left(\dfrac{a}{b}\right)^m = \dfrac{a^m}{b^m}$

$a^{-n} = \dfrac{1}{a^n}$

$\left(\dfrac{a}{b}\right)^{-n} = \dfrac{b^n}{a^n}$

Basic Rules of Algebra

Commutative Property of Addition

$a + b = b + a$

Commutative Property of Multiplication

$ab = ba$

Associative Property of Addition

$(a + b) + c = a + (b + c)$

Associative Property of Multiplication

$(ab)c = a(bc)$

Left Distributive Property

$a(b + c) = ab + ac$

Right Distributive Property

$(a + b)c = ac + bc$

Additive Identity Property

$a + 0 = 0 + a = a$

Multiplicative Identity Property

$a \cdot 1 = 1 \cdot a = a$

Additive Inverse Property

$a + (-a) = 0$

Multiplicative Inverse Property

$a \cdot \dfrac{1}{a} = 1, \quad a \neq 0$

Properties of Equality

Addition Property of Equality

If $a = b$, then $a + c = b + c$.

Multiplication Property of Equality

If $a = b$, then $ac = bc$.

Cancellation Property of Addition

If $a + c = b + c$, then $a = b$.

Cancellation Property of Multiplication

If $ac = bc$, and $c \neq 0$, then $a = b$.

Zero Factor Property

If $ab = 0$, then $a = 0$ or $b = 0$.

College Prep Algebra

Ron Larson
The Pennsylvania State University
The Behrend College

With the assistance of Kimberly Nolting
Hillsborough Community College

BROOKS/COLE
CENGAGE Learning™

ralia • Brazil • Japan • Korea • Mexico • Singapore • Spain • United Kingdom • United States

College Prep Algebra

Ron Larson

Senior Publisher: Charlie Van Wagner

Senior Acquiring Sponsoring Editor: Marc Bove

Associate Development Editor: Stefanie Beeck

Assistant Editor: Lauren Crosby

Senior Editorial Assistant: Jennifer Cordoba

Media Editor: Bryon Spencer

Senior Market Development Manager: Danae April

Content Project Manager: Jill Quinn

Manufacturing Planner: Doug Bertke

VP, Director, Advanced and Elective Products Program:
    Alison Zetterquist

Editorial Coordinator, Advanced and Elective Products Program:
    Jean Woy

Rights Acquisition Specialist: Shalice Shah-Caldwell

Text and Cover Designer: Larson Texts, Inc.

Cover Image: front cover, ©iStockphoto.com/Christopher Futcher;
back cover, wavebreakmedia/Shutterstock.com

Compositor: Larson Texts, Inc.

For product information and technology assistance, contact us at **Cengage Learning Custome & Sales Support, 1-800-354-9706.**

For permission to use materl from this text or product, submit all requests online at **www.cngage.com/permissions.** Further permissions quetons can be emailed to **permissionrecest@cengage.com.**

Library of Congress Control Numr: 2012955392

Student Edition:
ISBN-13: 978-1-285-18262-9
ISBN-10: 1-285-18262-6

Brooks/Cole
20 Channel Center Street
Boston, MA 02210
USA

Cengage Learning products are resented in high schools by Holt McDougal, a division of Hoıton Mifflin Harcourt.

Cengage Learning is a leading prıer of customized learning solutions with office locations ard the globe, including Singapore, the United Kingdom, Australia, Λco, Brazil and Japan. Locate your local office at **International.cenp.com/region**

Cengage Learning products are rısented in Canada by Nelson Education, Ltd.

For your course and learning solus, visit **www.cengage.com**.

Printed in the United States of America

5 6 7 16

Contents

1 ▶ THE REAL NUMBER SYSTEM   **1**

1.1	Real Numbers	2
1.2	Adding and Subtracting Integers	10
1.3	Multiplying and Dividing Integers	18
	Mid-Chapter Quiz	28
1.4	Operations with Rational Numbers	30
1.5	Exponents and Properties of Real Numbers	40
	Chapter Summary	48
	Review Exercises	50
	Chapter Test	54

2 ▶ FUNDAMENTALS OF ALGEBRA   **55**

2.1	Writing and Evaluating Algebraic Expressions	56
2.2	Simplifying Algebraic Expressions	64
	Mid-Chapter Quiz	74
2.3	Algebra and Problem Solving	76
2.4	Introduction to Equations	86
	Chapter Summary	94
	Review Exercises	96
	Chapter Test	100

3 ▶ EQUATIONS, INEQUALITIES, AND PROBLEM SOLVING   **101**

3.1	Solving Linear Equations	102
3.2	Equations that Reduce to Linear Form	110
3.3	Problem Solving with Percents	118
3.4	Ratios and Proportions	126
	Mid-Chapter Quiz	134
3.5	Geometric and Scientific Applications	136
3.6	Linear Inequalities	144
3.7	Absolute Value Equations and Inequalities	152
	Chapter Summary	160
	Review Exercises	162
	Chapter Test	165
	Cumulative Test: Chapters 1−3	166

4 ▶ GRAPHS AND FUNCTIONS   **167**

4.1	Ordered Pairs and Graphs	168
4.2	Graphs of Equations in Two Variables	176
4.3	Relations, Functions, and Graphs	184
	Mid-Chapter Quiz	192
4.4	Slope and Graphs of Linear Equations	194
4.5	Equations of Lines	202
4.6	Graphs of Linear Inequalities	210
	Chapter Summary	218
	Review Exercises	220
	Chapter Test	224

5 ▶ EXPONENTS AND POLYNOMIALS   225

5.1   Integer Exponents and Scientific Notation   226
5.2   Adding and Subtracting Polynomials   234
      Mid-Chapter Quiz   242
5.3   Multiplying Polynomials: Special Products   244
5.4   Dividing Polynomials and Synthetic Division   254
      Chapter Summary   262
      Review Exercises   264
      Chapter Test   268

6 ▶ FACTORING AND SOLVING EQUATIONS   269

6.1   Factoring Polynomials with Common Factors   270
6.2   Factoring Trinomials   278
6.3   More About Factoring Trinomials   286
      Mid-Chapter Quiz   294
6.4   Factoring Polynomials with Special Forms   296
6.5   Solving Polynomial Equations by Factoring   304
      Chapter Summary   312
      Review Exercises   314
      Chapter Test   317
      Cumulative Test: Chapters 4−6   318

7 ▶ RATIONAL EXPRESSIONS, EQUATIONS, AND FUNCTIONS   319

7.1   Rational Expressions and Functions   320
7.2   Multiplying and Dividing Rational Expressions   328
7.3   Adding and Subtracting Rational Expressions   336
      Mid-Chapter Quiz   344
7.4   Complex Fractions   346
7.5   Solving Rational Equations   354
7.6   Applications and Variation   362
      Chapter Summary   370
      Review Exercises   372
      Chapter Test   376

8 ▶ SYSTEMS OF EQUATIONS AND INEQUALITIES   377

8.1   Solving Systems of Equations by Graphing and Substitution   378
8.2   Solving Systems of Equations by Elimination   388
8.3   Linear Systems in Three Variables   396
      Mid-Chapter Quiz   404
8.4   Matrices and Linear Systems   406
8.5   Determinants and Linear Systems   416
8.6   Systems of Linear Inequalities   426
      Chapter Summary   434
      Review Exercises   436
      Chapter Test   440

9 ▶ **RADICALS AND COMPLEX NUMBERS**   **441**

9.1   Radicals and Rational Exponents   442
9.2   Simplifying Radical Expressions   450
9.3   Adding and Subtracting Radical Expressions   458
      Mid-Chapter Quiz   466
9.4   Multiplying and Dividing Radical Expressions   468
9.5   Radical Equations and Applications   476
9.6   Complex Numbers   484
      Chapter Summary   492
      Review Exercises   494
      Chapter Test   498
      Cumulative Test: Chapters 7–9   499

10 ▶ **QUADRATIC EQUATIONS, FUNCTIONS, AND INEQUALITIES**   **501**

10.1   Solving Quadratic Equations   502
10.2   Completing the Square   510
10.3   The Quadratic Formula   518
       Mid-Chapter Quiz   526
10.4   Graphs of Quadratic Functions   528
10.5   Applications of Quadratic Equations   536
10.6   Quadratic and Rational Inequalities   544
       Chapter Summary   552
       Review Exercises   554
       Chapter Test   558

11 ▶ **EXPONENTIAL AND LOGARITHMIC FUNCTIONS**   **559**

11.1   Exponential Functions   560
11.2   Composite and Inverse Functions   570
11.3   Logarithmic Functions   578
       Mid-Chapter Quiz   586
11.4   Properties of Logarithms   588
11.5   Solving Exponential and Logarithmic Equations   596
11.6   Applications   604
       Chapter Summary   612
       Review Exercises   614
       Chapter Test   618

12 ► **CONICS**   **619**

   12.1   Circles and Parabolas   620
   12.2   Ellipses   628
          Mid-Chapter Quiz   636
   12.3   Hyperbolas   638
   12.4   Solving Nonlinear Systems of Equations   646
          Chapter Summary   654
          Review Exercises   656
          Chapter Test   660
          Cumulative Test: Chapters 10−12   661

13 ► **SEQUENCES, SERIES, AND THE
          BINOMIAL THEOREM**   **663**

   13.1   Sequences and Series   664
   13.2   Arithmetic Sequences   672
          Mid-Chapter Quiz   680
   13.3   Geometric Sequences   682
   13.4   The Binomial Theorem   690
          Chapter Summary   698
          Review Exercises   700
          Chapter Test   703

APPENDICES

Appendix A   Review of Elementary Algebra Topics   A1

   A.1   The Real Number System   A1
   A.2   Fundamentals of Algebra   A6
   A.3   Equations, Inequalities, and Problem Solving   A9
   A.4   Graphs and Functions   A16
   A.5   Exponents and Polynomials   A24
   A.6   Factoring and Solving Equations   A32

Appendix B   Introduction to Graphing Calculators (web)*

Appendix C   Further Concepts in Geometry (web)*
   C.1   Exploring Congruence and Similarity
   C.2   Angles

Appendix D   Further Concepts in Statistics (web)*

Appendix E   Introduction to Logic (web)*
   E.1   Statements and Truth Tables
   E.2   Implications, Quantifiers, and Venn Diagrams
   E.3   Logical Arguments

Appendix F   Counting Principles (web)*

Appendix G   Probability (web)*

Answers to Odd-Numbered Exercises   A41
Index of Applications   A89
Index   A92

*Available at the text-specific website *www.cengagebrain.com*

Preface

Welcome to *College Prep Algebra*. I am proud to present a pedagogically sound, mathematically precise, and comprehensive textbook that will prepare you for college.

I'm very excited about this textbook. As I was writing, I kept one thought in mind—provide students what they need to learn algebra within reach. As you study from this book, you should notice right away that something is different. I've structured the book so that examples and exercises are on the same page—within reach. I am also offering a companion website at **CollegePrepAlgebra.com**. This site offers many resources that will help you as you study algebra. All of these resources are just a click away—within reach.

My goal for this textbook is to provide students with the tools that they need to master algebra. I hope that you find *College Prep Algebra*, together with CollegePrepAlgebra.com, will accomplish just that.

Textbook Features

Exercises Within Reach

The exercise sets have been carefully and extensively reviewed to ensure they are relevant and cover all topics suggested by our users. Additionally, the exercises have been completely restructured. Exercises now appear on the *same* page and immediately follow a corresponding example. There is no need to flip back and forth from example to exercise. The end-of-section exercises focus on mastery of conceptual understanding. View and listen to worked-out solutions at CollegePrepAlgebra.com.

Data Spreadsheets

Download editable spreadsheets from CollegePrepAlgebra.com, and use this data to solve exercises.

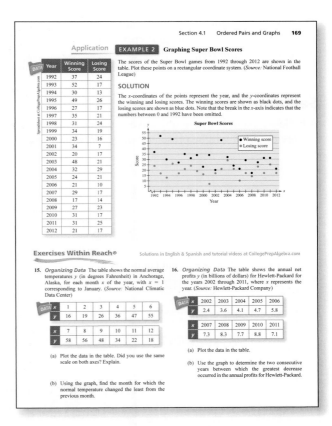

CollegePrepAlgebra.com

This companion website offers multiple tools and resources to supplement your learning. Access to these features is free. View and listen to worked-out solutions of thousands of exercises in English or Spanish, download data sets, take diagnostic tests, watch lesson videos and much more.

Concept Summary

This simple review of important concepts appears at the end of every section. Each Concept Summary reviews *What*, *How*, and *Why*—what concepts you studied, how to apply the concepts, and why the concepts are important. The Concept Summary includes four exercises to check your understanding.

Math Helps

Additional instruction is available for every example and many exercises at CollegePrepAlgebra.com. Just click on *Math Help*.

Section Objectives

A bulleted list of learning objectives provides you the opportunity to preview what will be presented in the upcoming section.

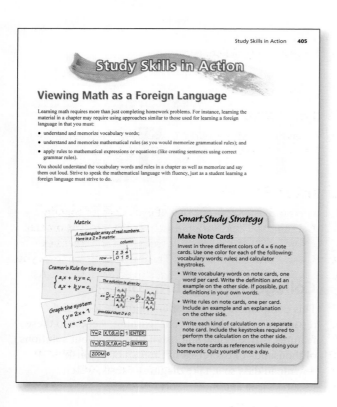

Study Skills in Action

Each chapter presents a study skill essential to success in mathematics. Read and apply these throughout the course. Print them out at CollegePrepAlgebra.com to keep them as reminders to develop strong study skills.

Applications

A wide variety of real-life applications are integrated throughout the text in examples and exercises. These applications demonstrate the relevance of algebra in the real world. Many of these applications use current, real data.

Chapter Summaries

The *Chapter Summary* now includes explanations and examples of the objectives taught in the chapter. Review exercises that cover these objectives are listed to check your understanding of the material.

Examples

Each example has been carefully chosen to illustrate a particular mathematical concept or problem-solving technique. The examples cover a wide variety of problems and are titled for easy reference. Many examples include detailed, step-by-step solutions with side comments, which explain the key steps of the solution process.

Study Tips

Study Tips offer students specific point-of-use suggestions for studying algebra, as well as pointing out common errors and discussing alternative solution methods. They appear in the margins.

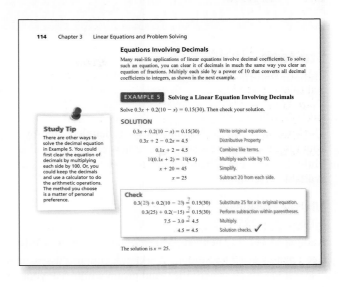

Technology Tips

Point-of-use instructions for using graphing calculators or software appear in the margins as *Technology Tips*. These features encourage the use of graphing technology as a tool for visualization of mathematical concepts, for verification of other solution methods, and for help with computations.

Cumulative Review

Each exercise set (except in Chapter 1) is followed by *Cumulative Review* exercises that cover concepts from previous sections. This serves as a review and also a way to connect old concepts with new concepts.

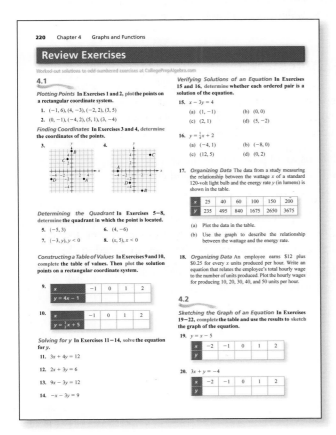

Mid-Chapter Quiz

Each chapter contains a *Mid-Chapter Quiz*. View and listen to worked-out solutions at CollegePrepAlgebra.com.

Chapter Review

The *Review Exercises* at the end of each chapter contain skill-building and application exercises that are first ordered by section, and then grouped according to the objectives stated at the start of the section. This organization allows you to easily identify the appropriate sections and concepts for study and review.

Chapter Test

Each chapter ends with a *Chapter Test*. View and listen to worked-out solutions at CollegePrepAlgebra.com.

Cumulative Test

The *Cumulative Tests* that follow Chapters 3, 6, 9, and 12 provide a comprehensive self-assessment tool that helps you check your mastery of previously covered material. View and listen to worked-out solutions at CollegePrepAlgebra.com.

Supplements

Student

Student Solutions Manual

ISBN 978-1-285-18275-9

Author: Ron Larson

The Student Solutions Manual provides detailed, step-by-step solutions to all odd-numbered problems in both the section exercise sets and review exercises. It also contains detailed, step-by-step solutions to all Mid-Chapter Quiz, Chapter Test, and Cumulative Test questions. Ask your sales representative about available discounts.

Student Workbook

ISBN 978-1-285-18277-3

Author: Maria H. Andersen, Muskegon Community College

Get a head start! The Student Workbook contains assessments, activities and worksheets for classroom discussions, in-class activities, and group work. Ask your sales representative about available discounts.

Printed Access Card: 978-1-285-18395-4

Instant Access Code: 978-1-285-18394-7

CourseMate is a perfect study tool for bringing concepts to life with interactive learning, studying, and exam preparation tools that support the printed textbook. CourseMate includes an interactive eBook, videos, quizzes, and more. Ask your sales representative about available discounts.

Test Prep Guide

ISBN 978-1-285-18287-2

The Test Prep Guide provides students with testing tips and advice, additional practice, as well as study tools designed to help them excel on various college mathematics placement tests. Ask your sales representative about available discounts.

Teacher

Complete Solutions Manual

ISBN 978-1-285-18270-4

Author: Ron Larson

The Complete Solutions Manual provides detailed step-by-step solutions to all problems in the text. It contains Chapter and Final Exam test forms with answer keys as well as individual test items and answers for Chapters 1–13.

Instructor's Resource Binder

ISBN 978-0-538-73675-6

Author: Maria H. Andersen, Muskegon Community College

The Instructor's Resource Binder contains uniquely designed Teaching Guides, which include instruction tips, examples, activities, worksheets, overheads, and assessments with answers to accompany them.

Printed Access Card: 978-1-285-18395-4

Instant Access Code: 978-1-285-18394-7

CourseMate is a perfect study tool for students, and requires no setup from you. CourseMate brings concepts to life with interactive learning, study, and exam preparation tools that support the printed textbook. CourseMate includes an interactive eBook, videos, quizzes, and more. For teachers, CourseMate includes Engagement Tracker, a first-of-its- kind tool that monitors student engagement. Ask your sales representative about available discounts for student access.

Lesson Plans

ISBN: 978-1-285-45752-9

Author: Melodie Cleckler Carr, Parkview High School

Keyed to the text by section, these Lesson Plans provide comprehensive coverage of the course, with section objectives, examples, and tips for the teacher.

PowerLecture with Examview

ISBN: 978-1-285-18286-5

Author: Ron Larson

This supplement provides the teacher with dynamic media tools for teaching. Create, deliver, and customize tests (both printed and online) in minutes with *Examview*® *Computerized Testing Featuring Algorithmic Equations*. Easily build solution sets for homework or exams using *Solution Builder*'s online solution manual. Microsoft® Powerpoint® lecture slides including *all* examples from the text, figures from the book, and easy-to-use PDF testbanks, in electronic format, are also included on this DVD-ROM.

Solution Builder

This online teacher database offers complete worked-out solutions to all exercises in the text, allowing you to create customized, secure solutions printouts (in PDF format) matched exactly to the problems you assign in class. For more information, visit www.cengage.com/solutionbuilder.

Acknowledgements

My thanks to Robert Hostetler, The Behrend College, The Pennsylvania State University, David Heyd, The Behrend College, The Pennsylvania State University, and Patrick Kelly, Mercyhurst University, for their significant contributions to previous editions of this text.

I would also like to thank the staff of Larson Texts, Inc., who assisted in preparing the manuscript, rendering the art package, and typesetting and proofreading the pages and the supplements.

On a personal level, I am grateful to my spouse, Deanna Gilbert Larson, for her love, patience, and support. Also, a special thanks goes to R. Scott O'Neil.

If you have suggestions for improving this text, please feel free to write to me. Over the past two decades I have received many useful comments from both instructors and students, and I value these comments very much.

Ron Larson
Professor of Mathematics
Penn State University
www.RonLarson.com

1 The Real Number System

1.1   Real Numbers

1.2   Adding and Subtracting Integers

1.3   Multiplying and Dividing Integers

1.4   Operations with Rational Numbers

1.5   Exponents and Properties of Real Numbers

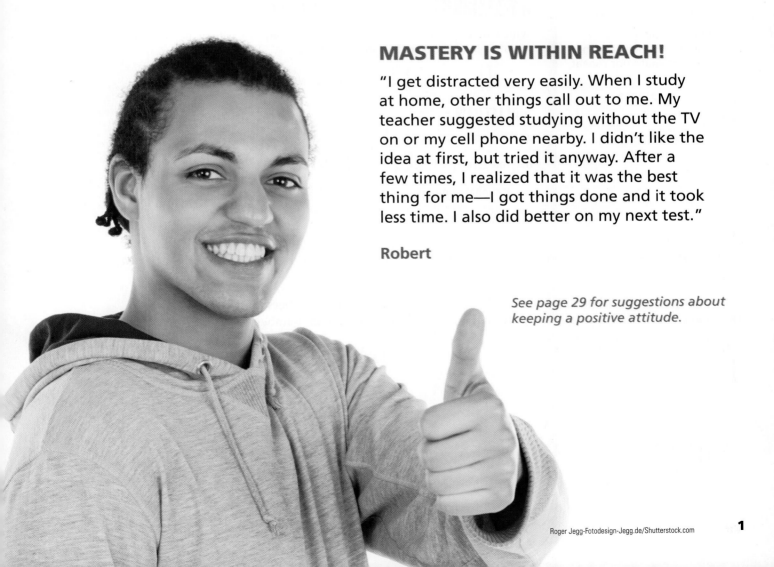

MASTERY IS WITHIN REACH!

"I get distracted very easily. When I study at home, other things call out to me. My teacher suggested studying without the TV on or my cell phone nearby. I didn't like the idea at first, but tried it anyway. After a few times, I realized that it was the best thing for me—I got things done and it took less time. I also did better on my next test."

Robert

See page 29 for suggestions about keeping a positive attitude.

1.1 Real Numbers

▶ Classify numbers and plot them on a real number line.

▶ Use the real number line and inequality symbols to order real numbers.

▶ Find the absolute value of a number.

Classifying Real Numbers and the Real Number Line

The numbers you use in everyday life are called **real numbers**. They are classified into different categories, as shown at the right.

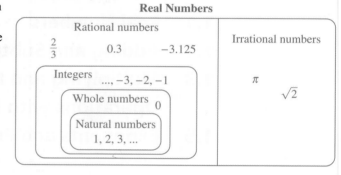

Study Tip

In *decimal form*, you can recognize rational numbers as decimals that terminate

$$\frac{1}{2} = 0.5 \quad \text{or} \quad \frac{3}{8} = 0.375$$

or repeat

$$\frac{4}{3} = 1.\overline{3} \quad \text{or} \quad \frac{2}{11} = 0.\overline{18}.$$

Irrational numbers are represented by decimals that neither terminate nor repeat, as in

$$\sqrt{2} = 1.414213562 \ldots$$

or

$$\pi = 3.141592654 \ldots$$

EXAMPLE 1 Classifying Real Numbers

Which of the numbers in the following set are (a) natural numbers, (b) integers, (c) rational numbers, and (d) irrational numbers?

$$\left\{ \frac{1}{2}, -1, 0, 4, -\frac{5}{8}, \frac{4}{2}, -\frac{3}{1}, 0.86, \sqrt{2}, \sqrt{9} \right\}$$

SOLUTION

a. Natural numbers: $\left\{ 4, \frac{4}{2} = 2, \sqrt{9} = 3 \right\}$

b. Integers: $\left\{ -1, 0, 4, \frac{4}{2} = 2, -\frac{3}{1} = -3, \sqrt{9} = 3 \right\}$

c. Rational numbers: $\left\{ \frac{1}{2}, -1, 0, 4, -\frac{5}{8}, \frac{4}{2}, -\frac{3}{1}, 0.86, \sqrt{9} = 3 \right\}$

d. Irrational number: $\left\{ \sqrt{2} \right\}$

Exercises Within Reach ®

Solutions in English & Spanish and tutorial videos at CollegePrepAlgebra.com

Classifying Real Numbers In Exercises 1−4, determine which of the numbers in the set are (a) natural numbers, (b) integers, (c) rational numbers, and (d) irrational numbers.

1. $\left\{ -3, 20, \pi, -\frac{3}{2}, \frac{9}{3}, 4.5, -\sqrt{3} \right\}$

2. $\left\{ \frac{1}{5}, \sqrt{5}, -\frac{24}{3}, -42, -4.5, 10, -\pi \right\}$

3. $\left\{ \sqrt{7}, -\sqrt{25}, -\frac{5}{1}, 9.4, 0, -12, \frac{7}{14} \right\}$

4. $\left\{ \frac{6}{1}, -\frac{6}{18}, -\sqrt{11}, \sqrt{36}, -1, -9.98, 22 \right\}$

The diagram used to represent real numbers is called the **real number line**.

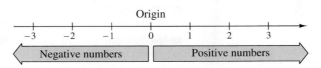

Each point on the real number line corresponds to exactly one real number, and each real number corresponds to exactly one point on the real number line.

EXAMPLE 2   Plotting Real Numbers

Plot each number on the real number line.

a. $-\dfrac{1}{2}$ b. 2   c. $-\dfrac{3}{2}$ d. 1

SOLUTION

a.

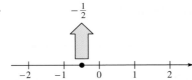

b.

c.

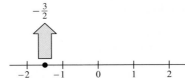

d.

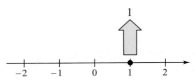

Exercises Within Reach®

Identifying the Number In Exercises 5–10, identify the real number that corresponds to the point plotted on the real number line.

5.

6.

7.

8.

9.

10.

Plotting Real Numbers In Exercises 11–16, plot the numbers on the real number line.

11. $\dfrac{5}{2}$, 5

12. $-\dfrac{3}{2}$, -3

13. $-6, -7, -3$

14. 9, 6, 10

15. $-0.8, 1.2, 1.8$

16. $-1.4, -0.3, 0.8$

Study Tip

The symbols < and > are called **inequality symbols**.

Ordering Real Numbers

The real number line provides you with a way of comparing any two real numbers. If a is to the left of b, then a is **less than** b, which is written as $a < b$. You can also describe this relationship by saying that b is **greater than** a, or $b > a$.

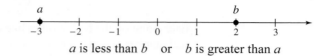

a is less than b or b is greater than a

EXAMPLE 3   **Ordering Real Numbers**

a. **Ordering Integers**

$3 < 5$ (3 is less than 5.) $-5 < -3$ (-5 is less than -3.)

b. **Ordering Decimals**

$-3.1 < 2.8$ (-3.1 is less than 2.8.) $-1.90 < -1.09$ (-1.90 is less than -1.09.)

c. **Ordering Fractions**

To order two fractions, you can write both fractions with the same denominator, or you can rewrite both fractions in decimal form. Here are two examples.

$$\frac{1}{3} = \frac{4}{12} \quad \text{and} \quad \frac{1}{4} = \frac{3}{12} \quad \Longrightarrow \quad \frac{1}{3} > \frac{1}{4}$$

$$\frac{11}{131} \approx 0.084 \quad \text{and} \quad \frac{19}{209} \approx 0.091 \quad \Longrightarrow \quad \frac{11}{131} < \frac{19}{209}$$

The symbol $\approx$ means "is approximately equal to."

Exercises Within Reach®

Solutions in English & Spanish and tutorial videos at CollegePrepAlgebra.com

Ordering Real Numbers In Exercises 17−24, **plot each real number as a point on the real number line and place the correct inequality symbol (< or >) between the real numbers.**

17. 3 [ ] -4

18. 6 [ ] -2

19. 3 [ ] 9

20. -4 [ ] -7

21. -4.6 [ ] 1.5

22. 28.60 [ ] -3.75

23. -6.58 [ ] -7.66

24. 20.156 [ ] 54.235

Ordering Fractions In Exercises 25−28, **order the fractions by (a) writing both fractions with the same denominator and (b) rewriting both fractions in decimal form.**

25. $\frac{9}{16}$ [ ] $\frac{5}{8}$

26. $-\frac{3}{8}$ [ ] $-\frac{5}{4}$

27. $-\frac{7}{3}$ [ ] $-\frac{5}{2}$

28. $\frac{3}{4}$ [ ] $\frac{5}{6}$

Application   EXAMPLE 4   **Comparing Profits**

The bar graph shows the profits for a company from 2008 through 2013.

a. Compare the profit for 2008 with the profit for 2009.

b. Write a statement that summarizes the trend in the company's profits.

Company Profits

A negative profit is called a *loss*. However, when comparing profits and losses, it is better to list all the figures as profits. This makes comparisons easier.

SOLUTION

a. The profit for 2008 was greater than the profit for 2009.

$$\$42.1 > -\$15.2$$

Profit for 2008 > Profit for 2009

b. Here is one possible summary statement.

"The company's profits had a big fall from 2008 to 2009. After that, profits have been increasing, but as of 2013, they have still not reached the level of profit in 2008."

Exercises Within Reach®   Solutions in English & Spanish and tutorial videos at CollegePrepAlgebra.com

29. *Miniature Golf* The table shows your scores relative to par in six consecutive rounds of miniature golf.

  (a) In golf, the lowest score wins. Which was your best score?

  (b) Describe the trend in your scores.

Round	Score
1	4
2	1
3	0
4	−2
5	−3
6	−5

30. *Temperature* The line graph shows the temperature (in degrees Celsius) of a cut of meat moved from a freezer to a refrigerator to thaw.

  (a) Compare the temperatures at hour 0 and hour 1.

  (b) Write a statement that describes the trend in the temperature over the 6-hour period.

Thawing Meat

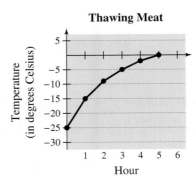

Study Tip

Two real numbers are **opposites** of each other if they lie the same distance from, but on opposite sides of, zero. For example, −2 is the opposite of 2, and 4 is the opposite of −4.

Finding Absolute Value

For any real number, the distance between the number and 0 on the real number line is its **absolute value**. A pair of vertical bars, | |, is used to denote absolute value. For instance, the absolute value of −8 is 8.

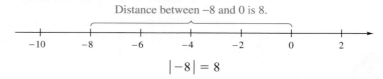

$$|-8| = 8$$

EXAMPLE 5 Evaluating an Absolute Value

a. $|-10| = 10$, because the distance between -10 and 0 is 10.

b. $\left|\frac{3}{4}\right| = \frac{3}{4}$, because the distance between $\frac{3}{4}$ and 0 is $\frac{3}{4}$.

c. $|-3.2| = 3.2$, because the distance between -3.2 and 0 is 3.2.

d. $-|-6| = -(6) = -6$

EXAMPLE 6 Comparing Real Numbers

Place the correct symbol ($<$, $>$, or $=$) between the real numbers.

a. $|-9|$ ▢ $|9|$ b. $|-3|$ ▢ 5 c. 0 ▢ $|-7|$

d. -4 ▢ $-|-4|$ e. $|12|$ ▢ $|-15|$ f. 2 ▢ $-|-2|$

SOLUTION

a. $|-9| = |9|$, because $|-9| = 9$ and $|9| = 9$.

b. $|-3| < 5$, because $|-3| = 3$ and 3 is less than 5.

c. $0 < |-7|$, because $|-7| = 7$ and 0 is less than 7.

d. $-4 = -|-4|$, because $-|-4| = -4$ and -4 is equal to -4.

e. $|12| < |-15|$, because $|12| = 12$, $|-15| = 15$, and 12 is less than 15.

f. $2 > -|-2|$, because $-|-2| = -2$ and 2 is greater than -2.

Exercises Within Reach ®

Solutions in English & Spanish and tutorial videos at CollegePrepAlgebra.com

Using Absolute Value In Exercises 31−34, **find** the distance between a and zero on the real number line.

31. $a = 2$ **32.** $a = 5$ **33.** $a = -8$ **34.** $a = -17$

Evaluating an Absolute Value In Exercises 35−46, **evaluate** the expression.

35. $|10|$ **36.** $|1|$ **37.** $|-3|$ **38.** $|-19|$

39. $|-3.4|$ **40.** $|-16.2|$ **41.** $\left|-\frac{7}{2}\right|$ **42.** $\left|-\frac{9}{16}\right|$

43. $-|-23.6|$ **44.** $-|-0.08|$ **45.** $|0|$ **46.** $|\pi|$

Comparing Real Numbers In Exercises 47−52, **place** the correct symbol ($<$, $>$, or $=$) between the real numbers.

47. $|-16|$ ▢ $|16|$ **48.** $|525|$ ▢ $|-525|$ **49.** $|-4|$ ▢ $|3|$

50. $|16|$ ▢ $|-25|$ **51.** $-|-48.5|$ ▢ $|-48.5|$ **52.** $|-\pi|$ ▢ $-|-2\pi|$

Application   EXAMPLE 7   **Explaining Elevation**

You are visiting Death Valley, California. Your GPS receiver lists your elevation as −282 feet. However, an outdoor sign states that the elevation is +282 feet below sea level. How can you explain the difference in signs?

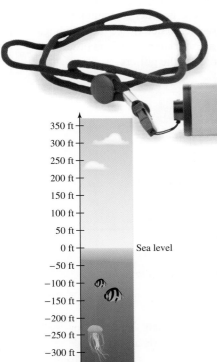

350 ft
300 ft
250 ft
200 ft
150 ft
100 ft
50 ft
0 ft   Sea level
−50 ft
−100 ft
−150 ft
−200 ft
−250 ft
−300 ft
−350 ft

SOLUTION

GPS receivers usually display elevations above sea level as positive numbers and elevations below sea level as negative numbers. If you want to talk about the elevation "below sea level," take the absolute value of the actual elevation.

$$|-282| = 282$$

GPS elevation   Elevation *below* sea level

Exercises Within Reach ®

Solutions in English & Spanish and tutorial videos at CollegePrepAlgebra.com

53. *Oceanography* A whale is hunting at an elevation of −456 meters relative to sea level. How many meters below sea level is the whale?

54. *Falcon* A soaring falcon dives to the surface of a lake to catch a fish. The change in the falcon's elevation is −192 feet. What was the falcon's elevation relative to the lake before the dive?

55. *Checking Account* What payment amount should you enter for check number 143? Explain.

Check number	Date	Transaction description	Payment amount	Deposit amount	Account balance
142	1/11	Art text	78.99		-78.99 23.24
	1/12	Deposit		100.00	+100.00 123.24
143	1/14	Art supplies			-49.12 74.12
	1/16	Deposit			+150.00 224.12

56. *Checking Account* What deposit amount should you enter for January 16th? To find the amount, can you use the same method you used in Exercise 55? Do you *need* to use the same method you used in Exercise 55? Explain.

Concept Summary: Ordering Real Numbers

What

When you are asked to order two **real numbers**, the goal is to determine which of the two numbers is greater.

EXAMPLE

Order $-\dfrac{6}{3}$ and $-\dfrac{5}{2}$.

How

You can use the **real number line** to order two real numbers. For example, to order two **fractions**, rewrite them with the same denominator, or rewrite them as decimals. Then plot each number on a number line.

EXAMPLE

$$-\frac{6}{3} = -2, \quad -\frac{5}{2} = -2.5$$

$$-\frac{6}{3} > -\frac{5}{2}$$

Why

There are many situations in which you need to order real numbers. For instance, to determine the standings at a golf tournament, you order the scores of the golfers.

Exercises Within Reach ®

Worked-out solutions to odd-numbered exercises at CollegePrepAlgebra.com

Concept Summary Check

57. *Using a Number Line* Explain how the number line above shows that $-2 > -2.5$.

58. *Using a Number Line* Describe the meaning of the expression $-\frac{6}{3} > -\frac{5}{2}$.

59. *Ordering Methods* Which method for ordering fractions is shown in the solution above?

60. *Rewriting Fractions* Describe another way to rewrite and order $-\frac{6}{3}$ and $-\frac{5}{2}$.

Extra Practice

Plotting Real Numbers In Exercises 61−64, **plot** the numbers on the real number line.

61. $\frac{5}{2}, \pi, -1, -|-3|$

62. $3.7, \frac{16}{3}, -|-1.9|, -\frac{1}{2}$

63. $-5, \frac{7}{3}, |-3|, 0, -|4.5|$

64. $|-2.3|, 3.2, -2.3, -|3.2|$

Distance on the Real Number Line In Exercises 65−70, **find all real numbers whose distance from *a* is given by *d*.**

65. $a = 8, d = 12$

66. $a = 6, d = 7$

67. $a = 21.3, d = 6$

68. $a = 42.5, d = 7$

69. $a = -2, d = 3.5$

70. $a = -7, d = 7.2$

Identifying Real Numbers In Exercises 71−76, **give three examples that satisfy the given conditions.**

71. A real number that is not a rational number

72. A real number that is not an irrational number

73. An integer that is a rational number

74. A rational number that is not a negative number

75. A real number that is not a positive rational number

76. An integer that is not a whole number

77. *Determining a Solution Set* Describe the real numbers n that satisfy the equation $n + |-n| = 2n$.

78. *Determining a Solution Set* Describe the real numbers n that satisfy the equation $n + |-n| = 0$.

79. *Volcanoes* The *summit elevation* of a volcano is the elevation of the top of the volcano relative to sea level. The table shows the summit elevations of several volcanoes.

Volcano	Summit Elevation
Kīlauea	1277 m
Lo`ihi	−969 m
Mauna Loa	4170 m
Ruby	−230 m
Anatahan	790 m

(a) Which of the volcanoes have summits below sea level?

(b) Which volcano summit is closest to sea level?

(c) Which volcano summit is farthest from sea level?

80. *Scuba Diving* The positions relative to sea level of two scuba divers are shown.

(a) Which position is represented by the greater integer?

(b) Which integer has the greater absolute value?

(c) What does the absolute value of each integer represent in the context of the problem?

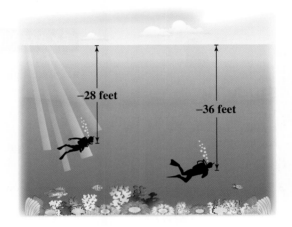

Explaining Concepts

81. Explain the difference between plotting the numbers 4 and −4 on the real number line.

82. How many numbers are three units from 0 on the real number line? Explain your answer.

83. *Writing* Explain why $\frac{8}{4}$ is a natural number, but $\frac{7}{4}$ is not.

84. *Number Sense* Which real number lies farther from 0 on the real number line, −15 or 10? Explain.

85. *Number Sense* Which real number lies farther from −4 on the real number line, 3 or −10? Explain.

86. *Precision* Which real number is smaller, $\frac{3}{8}$ or 0.37? Explain.

True or False? **In Exercises 87−92, decide whether the statement is true or false. Justify your answer.**

87. The absolute value of any real number is always positive.

88. The absolute value of a number is equal to the absolute value of its opposite.

89. The absolute value of a rational number is a rational number.

90. A given real number corresponds to exactly one point on the real number line.

91. The opposite of a positive number is a negative number.

92. Every rational number is an integer.

1.2 Adding and Subtracting Integers

▶ Add integers using a number line.

▶ Add integers with like signs and with unlike signs.

▶ Subtract integers with like signs and with unlike signs.

Adding Integers Using a Number Line

To find the sum $a + b$ using a number line, start at 0. Move right or left a units depending on whether a is positive or negative. From that position, move right or left b units depending on whether b is positive or negative. The final position is the **sum**.

Study Tip

The sum of a number and its opposite is 0. For instance,

$$-4 + 4 = 0.$$

EXAMPLE 1   **Adding Integers Using a Number Line**

a. Like signs: $5 + 2 = 7$

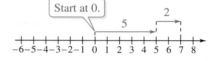

b. Like signs: $-3 + (-5) = -8$

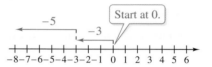

c. Unlike signs: $-5 + 2 = -3$

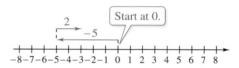

d. Unlike signs: $7 + (-3) = 4$

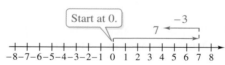

Exercises Within Reach ®

Solutions in English & Spanish and tutorial videos at CollegePrepAlgebra.com

Adding Integers Using a Number Line In Exercises 1−12, **find the sum and demonstrate the addition on the real number line.**

1. $2 + 7$

2. $3 + 9$

3. $-8 + (-3)$

4. $-4 + (-7)$

5. $10 + (-3)$

6. $14 + (-8)$

7. $-6 + 4$

8. $-12 + 5$

9. $3 + (-9)$

10. $-2 + 7$

11. $-5 + 9$

12. $5 + (-8)$

Application   EXAMPLE 2   **Finding the Total Yardage**

The yards gained or lost by a football team on its four downs are shown.

1st Down: +3 yards
2nd Down: −4 yards
3rd Down: +7 yards
4th Down: +4 yards

Did the team earn a new first down?
(*Note:* In football, a team earns a
new first down when it gains 10
or more yards in 4 downs.)

SOLUTION

You can use a number line to find the total yards gained.

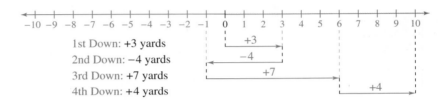

1st Down: +3 yards
2nd Down: −4 yards
3rd Down: +7 yards
4th Down: +4 yards

The team gained 10 yards total. So, it did earn a new 1st down.

$$3 + (-4) + 7 + 4 = 10$$

Exercises Within Reach®   Solutions in English & Spanish and tutorial videos at CollegePrepAlgebra.com

13. *Highway Speed* While driving on a highway, you set the cruise control at the speed limit. You change your speed by the following amounts for traffic.

−7 miles per hour
+6 miles per hour
−3 miles per hour
+8 miles per hour

Are you traveling within 5 miles per hour of the speed limit after these changes?

14. *Game Score* Your team's score in a game is the sum of the scores of the four players on your team.

Player 1:   +7 points
Player 2:   −1 point
Player 3:   +5 points
Player 4:   −4 points

Is your team's score greater than the opposing team's score of 8 points?

15. *Temperature Change* When you left for class in the morning, the temperature was 25°C. By the time class ended, the temperature had increased by 4°. While you studied, the temperature increased by 3°. During your soccer practice, the temperature decreased by 9°. What was the temperature after your soccer practice?

16. *Temperature Change* When you left for class in the morning, the temperature was 40°F. By the time class ended, the temperature had increased by 13°. While you studied, the temperature decreased by 5°. During your club meeting, the temperature decreased by 6°. What was the temperature after your club meeting?

Adding Integers Algebraically

1. **To add two integers with like signs,** add their absolute values and attach the common sign to the result.

2. **To add two integers with unlike signs,** subtract the lesser absolute value from the greater absolute value and attach the sign of the integer with the greater absolute value.

> **EXAMPLE 3** **Adding Integers**

a. Like Signs: $-18 + (-62) = -\left(|-18| + |-62|\right)$

$$= -(18 + 62)$$

$$= -80$$

b. Unlike Signs: $22 + (-17) = |22| - |-17|$

$$= 22 - 17$$

$$= 5$$

c. Unlike Signs: $-84 + 14 = -\left(|-84| - |14|\right)$

$$= -(84 - 14)$$

$$= -70$$

There are different ways to add three or more integers. You can use the carrying algorithm with a vertical format with nonnegative integers, as shown below, or you can add them two at a time, as illustrated in Example 4.

$$
\begin{array}{r}
1\ 1\\
1\ 4\ 8\\
6\ 2\\
+\ 5\ 3\ 6\\
\hline
7\ 4\ 6
\end{array}
$$

Vertical carrying algorithm

$148 + 62 + 536 = 746$

Exercises Within Reach®

Solutions in English & Spanish and tutorial videos at CollegePrepAlgebra.com

Adding Integers In Exercises 17−46, **find the sum.**

17. $6 + 10$

18. $8 + 3$

19. $14 + (-14)$

20. $10 + (-10)$

21. $-45 + 45$

22. $-23 + 23$

23. $-14 + (-13)$

24. $-20 + (-19)$

25. $-23 + (-4)$

26. $-32 + (-16)$

27. $18 + (-12)$

28. $34 + (-16)$

29. $-75 + 100$

30. $-54 + 68$

31. $9 + (-14)$

32. $18 + (-26)$

33.
$$
\begin{array}{r}
110\\
45\\
+\ 208\\
\hline
\end{array}
$$

34.
$$
\begin{array}{r}
44\\
115\\
+\ 380\\
\hline
\end{array}
$$

35.
$$
\begin{array}{r}
250\\
354\\
+\ 122\\
\hline
\end{array}
$$

36.
$$
\begin{array}{r}
275\\
416\\
+\ 316\\
\hline
\end{array}
$$

37. $10 + (-6) + 34$

38. $7 + (-4) + 1$

39. $-15 + (-3) + 8$

40. $-82 + (-36) + 82$

41. $9 + (-18) + 4$

42. $2 + (-51) + 13$

43. $803 + (-104) + (-613) + 214$

44. $4365 + (-2145) + (-1873) + 40{,}084$

45. $312 + (-564) + 119 + (-100)$

46. $1200 + (-1300) + 62 + (-275)$

Application   EXAMPLE 4   **Finding Your Account Balance**

At the beginning of a month, your account balance was $28. During the month, you deposited $60 and withdrew $40. What was your balance at the end of the month?

ANY BANK
12345 Main Street
Anytown, NY 01234

CHECKING ACCOUNT STATEMENT

	Statement period	Account No.
	01/01/2013 - 02/01/2013	00005-123-456-7

Date	Description	Ref.	Withdrawals	Deposits	Balance
01/01	Beginning Balance				$28.00
01/01	Deposit			$60.00	?
01/01	Withdrawal		$40.00		?
01/01	Ending Balance				?

SOLUTION

$$\$28 + \$60 + (-\$40) = (\$28 + \$60) + (-\$40)$$
$$= \$88 + (-\$40)$$
$$= \$48 \qquad \text{Balance}$$

Your balance at the end of the month was $48.

Exercises Within Reach®

Solutions in English & Spanish and tutorial videos at CollegePrepAlgebra.com

47. *Finding Your Account Balance* At the beginning of a month, your account balance was $46. During the month, you deposited $552 and wrote a check for $489. What was your balance at the end of the month?

Date	Description	Withdrawals	Deposits	Balance
09/01	Beginning Balance			$46.00
09/08	Deposit		$552.00	?
09/18	Check #321	$489.00		?
09/30	Ending Balance			?

48. *Finding Your Account Balance* At the beginning of a month, your checking account balance was $89. During the month, you deposited $120 and withdrew $108. What was your balance at the end of the month?

Date	Description	Withdrawals	Deposits	Balance
10/01	Beginning Balance			$89.00
10/09	Deposit		$120.00	?
10/22	Withdrawal	$108.00		?
10/31	Ending Balance			?

49. *Fishing Depth* A fisherman drops a line 56 feet below the surface of the water. The fisherman raises the line 18 feet, then 16 feet more. Finally, the fisherman drops the line 45 feet and catches a fish. How far below the surface does the fisherman catch the fish?

50. *Forensics* A forensic archaeologist finds a human skull 220 centimeters below ground. The archaeologist had already found a leg bone 75 centimeters above the skull and a hand bone 36 centimeters below the leg bone. At what depth was the hand bone?

Subtracting Integers Algebraically

To subtract one integer from another, add the opposite of the integer being subtracted to the other integer. The result is called the **difference** of the two integers.

> **EXAMPLE 5**   **Subtracting Integers**

a. $3 - 8 = 3 + (-8) = -5$ Add the opposite of 8.

b. $10 - (-13) = 10 + 13 = 23$ Add the opposite of -13.

c. $-5 - 12 = -5 + (-12) = -17$ Add the opposite of 12.

For subtraction problems involving two nonnegative integers, you can use the borrowing algorithm shown below.

$$\begin{array}{r} \overset{3\;\;10\;\;15}{\cancel{4}\;\;\cancel{1}\;\;\cancel{5}} \\ -2\;\;7\;\;6 \\ \hline 1\;\;3\;\;9 \end{array}$$

Vertical borrowing algorithm

$415 - 276 = 139$

> **EXAMPLE 6**   **Evaluating Expressions**

a. $-13 - 7 + 11 - (-4) = -13 + (-7) + 11 + 4$ Add opposites.

                $= -20 + 11 + 4$ Add -13 and -7.

                $= -9 + 4$ Add -20 and 11.

                $= -5$ Add.

b. $-1 - 3 - 4 + 6 = -1 + (-3) + (-4) + 6$ Add opposites.

                $= -4 + (-4) + 6$ Add -1 and -3.

                $= -8 + 6$ Add -4 and -4.

                $= -2$ Add.

Exercises Within Reach ®

Solutions in English & Spanish and tutorial videos at CollegePrepAlgebra.com

Subtracting Integers In Exercises 51−82, **find the difference.**

51. $21 - 18$ **52.** $47 - 12$ **53.** $51 - 25$ **54.** $37 - 37$

55. $1 - (-4)$ **56.** $7 - (-8)$ **57.** $15 - (-10)$ **58.** $8 - (-31)$

59. $18 - (-18)$ **60.** $62 - (-28)$ **61.** $27 - 57$ **62.** $18 - 32$

63. $61 - 85$ **64.** $53 - 74$ **65.** $22 - 131$ **66.** $48 - 222$

67. $2 - 11$ **68.** $3 - 15$ **69.** $13 - 24$ **70.** $26 - 34$

71. $-135 - (-114)$ **72.** $-63 - (-8)$ **73.** $-4 - (-4)$ **74.** $-942 - (-942)$

75. $-10 - (-4)$ **76.** $-12 - (-7)$ **77.** $-71 - 32$ **78.** $-84 - 55$

79. $-210 - 400$ **80.** $-120 - 142$ **81.** $-110 - (-30)$ **82.** $-2500 - (-600)$

Evaluating an Expression In Exercises 83−88, **evaluate the expression.**

83. $-3 + 2 - 20 + 9$ **84.** $-1 + 3 - (-4) + 10$ **85.** $12 - 6 + 3 - (-8)$

86. $6 + 7 - 12 - 5$ **87.** $-(-5) + 7 - 18 + 4$ **88.** $-15 - (-2) + 4 - 6$

<u>Application</u>   EXAMPLE 7   **Finding a Temperature**

Minnehaha Falls in Minneapolis has a 53-foot drop. It is located near the confluence of Minnehaha Creek and the Mississippi.

The temperature in Minneapolis, Minnesota, at 4 P.M. was 15°F. By midnight, the temperature had decreased by 18°. What was the temperature in Minneapolis at midnight?

SOLUTION

To find the temperature at midnight, subtract 18 from 15.

$$15 - 18 = 15 + (-18)$$
$$= -3$$

The temperature in Minneapolis at midnight was -3°F.

EXAMPLE 8   **Using a Calculator**

Evaluate each expression using a calculator.

a. $-4 - 5$ 　　　　　　　　　　　　**b.** $2 - (-3) + 9$

SOLUTION

	Keystrokes	*Display*	
a.	4 [+/−] [−] 5 [=]	−9	Scientific
	[(−)] 4 [−] 5 [ENTER]	−9	Graphing

	Keystrokes	*Display*	
b.	2 [−] [(] 3 [+/−] [)] [+] 9 [=]	14	Scientific
	2 [−] [(] [(−)] 3 [)] [+] 9 [ENTER]	14	Graphing

Exercises Within Reach ®   Solutions in English & Spanish and tutorial videos at CollegePrepAlgebra.com

89. *Altitude* An airplane flying at a cruising altitude of 31,000 feet is instructed to descend as shown in the diagram below. How many feet must the airplane descend?

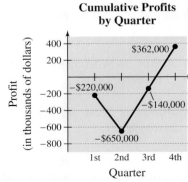

90. *Profit* A telephone company lost $650,000 during the first half of the year. By the end of the year, the company had an overall profit of $362,000. What was the company's profit during the second half of the year?

🔲 *Using a Calculator* In Exercises 91−96, write the keystrokes used to evaluate the expression with a calculator (either scientific or graphing). Then evaluate the expression.

91. $-3 - 7$ 　　　　　　　　　**92.** $9 - (-2)$ 　　　　　　　　　**93.** $6 + 5 - (-7)$

94. $4 - 3 - (-9)$ 　　　　　　　**95.** $-6 + (-2) - 5$ 　　　　　　　**96.** $-3 + (-7) - 9$

Concept Summary: *Adding and Subtracting Integers*

What

When you are asked to find the **sum** of two integers, you add the integers.

EXAMPLE

Evaluate the equation
$-6 + 3$.

How

To find the sum of two integers,

- use a number line.

- add the integers algebraically.

EXAMPLE

$$-6 + 3 = -\big(|-6| - |3|\big)$$
$$= -(6 - 3)$$
$$= -3$$

Why

When you know how to add any two integers, you can subtract any two integers by adding the **opposite**.

Exercises Within Reach ®

Worked-out solutions to odd-numbered exercises at CollegePrepAlgebra.com

Concept Summary Check

97. *Explaining the Process* In the first step of the solution above, why is subtraction used between the absolute value expressions? Why is there a negative sign in front of the expression in parentheses?

98. *Using a Number Line* Describe the process for adding -6 and 3 using a number line.

99. *Adding Integers* Explain how to add two integers with like signs.

100. *Subtracting Integers* State the rule for subtracting integers.

Extra Practice

101. *Think About It* What number must be added to 10 to obtain -5?

102. *Think About It* What number must be added to 36 to obtain -12?

103. *Think About It* What number must be subtracted from -12 to obtain 24?

104. *Think About It* What number must be subtracted from -20 to obtain 15?

105. *Trail Elevation* The North Bass Trail in the Grand Canyon starts at Swamp Point and descends 244 meters to Muav Saddle. The trail continues on to Shinumo Creek, which is 1433 meters lower than Swamp Point, and then descends another 182 meters to the Colorado River. How much lower than Muav Saddle is the Colorado River?

106. *Trail Elevation* The North Kaibab Trail in the Grand Canyon descends 439 meters from the North Kaibab trailhead to the Supai Tunnel. The trail continues on to Roaring Springs, which is 921 meters lower than the North Kaibab trailhead, and then descends another 457 meters to Ribbon Falls. How much lower than the Supai Tunnel is Ribbon Falls?

The Grand Canyon, located in Arizona, is 277 river miles long and 1 mile deep on average.

107. *Stock Values* On Monday, you purchased $500 worth of stock. The values of the stock during the remainder of the week are shown in the bar graph.

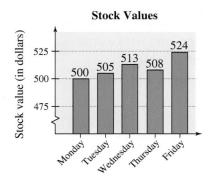

Stock Values

(a) Use the graph to complete the table.

Day	Daily Gain or Loss
Tuesday	
Wednesday	
Thursday	
Friday	

(b) What was the total change in the value of the stock from trading on Thursday and Friday?

(c) Find the sum of the daily gains and losses. Interpret the result in the context of the problem. How could you determine this sum from the graph?

108. *Profits* The bar graph shows the yearly revenues and expenses of a company for 2010 through 2013.

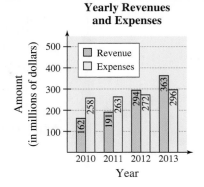

Yearly Revenues and Expenses

(a) Use the graph to complete the table for the yearly profit, using Profit = Revenue − Expenses.

Year	Profit
2010	
2011	
2012	
2013	

(b) Find the total profit for the years 2010 through 2013. How is the total profit related to the total revenue and total expenses for these years?

(c) From 2010 through 2013, between which two consecutive years was the change in profits the greatest?

Explaining Concepts

Adding Integers Using a Number Line In Exercises 109 and 110, an addition problem is shown visually on the real number line. (a) **Write** the addition problem and find the sum. (b) **State** the rule for adding integers that is demonstrated.

109.

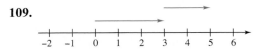

110.

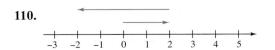

111. *Writing* Explain why the sum of two negative integers is a negative integer.

112. *Writing* When is the sum of a positive integer and a negative integer a positive integer?

113. *Writing* Is is possible that the sum of two positive integers is a negative integer? Explain.

114. *Writing* Is it possible that the difference of two negative integers is a positive integer? Explain.

1.3 Multiplying and Dividing Integers

▶ Multiplying integers with like signs and with unlike signs.

▶ Divide integers with like signs and with unlike signs.

▶ Find factors and prime factors of an integer.

▶ Represent the definitions and rules of arithmetic symbolically.

Multiplying Integers

Multiplication of two integers can be described as repeated addition or subtraction. The result of multiplying one number by another is called a **product**.

1. The product of an integer and zero is 0.

2. The product of two integers with like signs is positive.

3. The product of two integers with unlike signs is negative.

Study Tip

To find the product of more than two numbers, first find the product of their absolute values. If the number of negative factors is even, then the product is positive. If the number of negative factors is odd, then the product is negative.

> **EXAMPLE 1**   **Multiplying Integers**

a. $4(10) = 40$ (Positive) • (positive) = positive

b. $-6 \cdot 9 = -54$ (Negative) • (positive) = negative

c. $-5(-7) = 35$ (Negative) • (negative) = positive

d. $3(-12) = -36$ (Positive) • (negative) = negative

e. $-12 \cdot 0 = 0$ (Negative) • (zero) = zero

f. $-2(8)(-3)(-1) = -(2 \cdot 8 \cdot 3 \cdot 1)$ Odd number of negative factors
        $= -48$ Answer is negative.

To multiply two integers having two or more digits, use the vertical multiplication algorithm shown below. The sign of the product is determined by the usual multiplication rule.

Vertical multiplication algorithm

$47(23) = 1081$

$$
\begin{array}{r}
47 \\
\times\ 23 \\
\hline
141 \\
94 \\
\hline
1081
\end{array}
$$

Multiply 3 times 47.
Multiply 2 times 47.
Add columns.

Exercises Within Reach ®

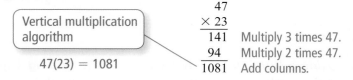

Solutions in English & Spanish and tutorial videos at CollegePrepAlgebra.com

Multiplying Integers **In Exercises 1−24, find the product.**

1. 5×7
2. 8×3
3. $0 \cdot 4$
4. $-12 \cdot 0$
5. $2(-16)$
6. $8(-7)$
7. $-9(4)$
8. $-6(5)$
9. $230(-3)$
10. $175(-2)$
11. $-7(-13)$
12. $-40(-4)$
13. $-200(-8)$
14. $-150(-4)$
15. $3(-5)(6)$
16. $4(2)(-6)$
17. $7(-3)(-1)$
18. $-2(5)(-3)$
19. $-2(-3)(-5)$
20. $-10(-4)(-2)$
21. $|(-3)4|$
22. $|8(-9)|$
23. $|3(-5)(6)|$
24. $|8(-3)(5)|$

Multiplying Integers Vertically **In Exercises 25−32, use the vertical multiplication algorithm to find the product.**

25. 26×13
26. -14×24
27. $75(-63)$
28. $-72(866)$
29. $-13(-20)$
30. $-11(-24)$
31. $-21(-429)$
32. $-14(-585)$

Application   EXAMPLE 2   Geometry: **Finding the Volume of a Box**

Find the volume of the rectangular box.

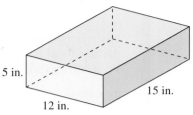

5 in.

12 in.

15 in.

Jets typically provide about 30 cubic feet of air per passenger in coach and about 50 cubic feet per passenger in first class.

SOLUTION

To find the volume, multiply the length, width, and height of the box.

Volume = (Length) • (Width) • (Height)

  = (15 inches) • (12 inches) • (5 inches)

  = 900 cubic inches

So, the box has a volume of 900 cubic inches.

Exercises Within Reach®

33. *Geometry* Find the area of the football field.

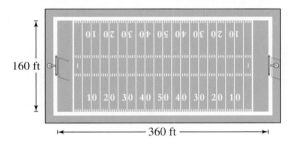

160 ft

360 ft

34. *Geometry* Find the area of the city park.

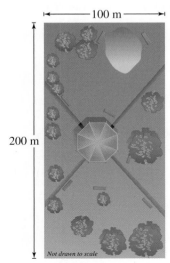

100 m

200 m

Not drawn to scale

35. *Geometry* Find the volume of the rectangular shipping box.

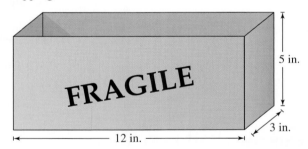

FRAGILE

5 in.

12 in.

3 in.

36. *Geometry* Find the volume of the rectangular shipping box.

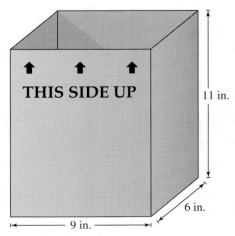

THIS SIDE UP

11 in.

9 in.

6 in.

Dividing Integers

The result of dividing one integer by another is called the **quotient** of the integers. Division is denoted by the symbol ÷, or by /, or by a horizontal line. For instance,

$$30 \div 6, \quad 30/6 \quad \text{and} \quad \frac{30}{6}$$

all denote the quotient of 30 and 6, which is 5. Using the form $30 \div 6$, 30 is called the **dividend** and 6 is the **divisor**. In the forms 30/6 and $\frac{30}{6}$, 30 is the **numerator** and 6 is the **denominator**.

1. Zero divided by a nonzero integer is 0, whereas a nonzero integer divided by zero is undefined.

2. The quotient of two nonzero integers with like signs is positive.

3. The quotient of two nonzero integers with unlike signs is negative.

EXAMPLE 3 **Dividing Integers**

a. $\dfrac{-42}{-6} = 7$ because $-42 = 7(-6)$.

b. $36 \div (-9) = -4$ because $36 = (-4)(-9)$.

c. $0 \div (-13) = 0$ because $0 = (0)(-13)$.

d. $-105 \div 7 = -15$ because $-105 = (-15)(7)$.

e. $-97 \div 0$ is undefined.

When dividing large numbers, the long division algorithm can be used. For instance, the long division algorithm below shows that $351 \div 13 = 27$.

$$
\begin{array}{r}
27 \\
13\overline{)351} \\
26 \\
\hline
91 \\
91 \\
\hline
0
\end{array}
$$

Long division algorithm

$351 \div 13 = 27$

Exercises Within Reach®

Solutions in English & Spanish and tutorial videos at CollegePrepAlgebra.com

Dividing Integers In Exercises 37–52, perform the division, if possible. If not possible, state the reason.

37. $27 \div 9$
38. $35 \div 7$
39. $72 \div (-12)$
40. $54 \div (-9)$
41. $-28 \div 4$
42. $-108 \div 9$
43. $-56 \div (-8)$
44. $-68 \div (-4)$
45. $8 \div 0$
46. $17 \div 0$
47. $0 \div 8$
48. $0 \div 17$
49. $\dfrac{-81}{-3}$
50. $\dfrac{-125}{-25}$
51. $\dfrac{-28}{4}$
52. $\dfrac{72}{-12}$

Using Long Division In Exercises 53–60, use the long division algorithm to find the quotient.

53. $1440 \div 45$
54. $936 \div 52$
55. $1440 \div (-9)$
56. $936 \div (-8)$
57. $-1312 \div 16$
58. $-5152 \div 23$
59. $-9268 \div (-28)$
60. $-6804 \div (-36)$

Using a Calculator In Exercises 61–64, use a calculator to find the quotient.

61. $\dfrac{44,290}{515}$
62. $\dfrac{33,511}{47}$
63. $\dfrac{169,290}{-162}$
64. $\dfrac{-1,027,500}{250}$

Application   EXAMPLE 4   **Finding an Average Gain in Stock Prices**

On Monday, you bought $500 worth of stock in a company. During the rest of the week, you recorded the gains and losses in your stock's value.

Tuesday	Wednesday	Thursday	Friday
Gained $15	Lost $18	Lost $23	Gained $10

Several million shares of stock are traded on the New York Stock Exchange each business day.

a. What was the value of the stock at the close of Wednesday?

b. What was the value of the stock at the end of the week?

c. What would the total loss have been if Thursday's loss had occurred on each of the four days?

d. What was the average daily gain (or loss) for the four days recorded?

SOLUTION

a. The value at the close of Wednesday was

$$500 + 15 - 18 = \$497.$$

b. The value of the stock at the end of the week was

$$500 + 15 - 18 - 23 + 10 = \$484.$$

c. The loss on Thursday was $23. If this loss had occurred each day, the total loss would have been

$$4(23) = \$92.$$

d. To find the average daily gain (or loss), add the gains and losses of the four days and divide by 4.

$$\text{Average} = \frac{15 + (-18) + (-23) + 10}{4} = \frac{-16}{4} = -4$$

This means that during the four days, the stock had an average loss of $4 per day.

Exercises Within Reach ®

Solutions in English & Spanish and tutorial videos at CollegePrepAlgebra.com

65. *Average Speed* A space shuttle orbiting Earth travels about 45 miles in 9 seconds. What is the average speed of the space shuttle in miles per second?

66. *Sports* A football team gains a total of 20 yards after four plays.

(a) What is the average number of yards gained per play?

(b) The gains on the four plays are 8 yards, 4 yards, 2 yards, and 6 yards. Plot each of the gains and the average gain on the real number line.

(c) Find the difference between each gain and the average gain. Find the sum of these differences and give a possible explanation of the result.

67. *Exam Scores* A student has a total of 328 points after four 100-point exams.

(a) What is the average number of points scored per exam?

(b) The scores on the four exams are 87, 73, 77, and 91. Plot each of the scores and the average score on the real number line.

(c) Find the difference between each score and the average score. Find the sum of these differences and give a possible explanation of the result.

68. *Average Speed* A hiker jogs a mountain trail that is 6 miles long in 54 minutes. How many minutes does the hiker average per mile?

Factors and Prime Numbers

If a and b are positive integers, then a is a **factor** (or divisor) of b if and only if a divides evenly into b. For instance, 1, 2, 3, and 6 are all factors of 6.

The concept of factors allows you to classify positive integers into three groups: prime numbers, composite numbers, and the number 1.

1. An integer greater than 1 with no factors other than itself and 1 is called a **prime number**, or simply a prime.

2. An integer greater than 1 with more than two factors is called a **composite number**, or simply a composite.

Every composite number can be expressed as a unique product of prime factors.

Study Tip

A tree diagram is a nice way to record your work when you are factoring a composite number into prime numbers. For instance, the following tree diagram shows that $45 = 3 \cdot 3 \cdot 5$.

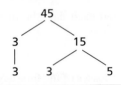

EXAMPLE 5 **Prime Factorization**

Write the prime factorization of each number.

a. 84   **b.** 78   **c.** 133   **d.** 43

SOLUTION

a. 2 is a recognized divisor of 84. So,

$$84 = 2 \cdot 42 = 2 \cdot 2 \cdot 21 = 2 \cdot 2 \cdot 3 \cdot 7.$$

b. 2 is a recognized divisor of 78. So,

$$78 = 2 \cdot 39 = 2 \cdot 3 \cdot 13.$$

c. If you do not recognize a divisor of 133, you can start by dividing any of the prime numbers 2, 3, 5, 7, 11, 13, etc., into 133. You will find 7 to be the first prime to divide into 133. So,

$$133 = 7 \cdot 19.$$

d. In this case, none of the primes less than 43 divides 43. So, 43 is prime.

Exercises Within Reach ®

Solutions in English & Spanish and tutorial videos at CollegePrepAlgebra.com

Classifying an Integer In Exercises 69−80, decide whether the number is prime or composite.

69. 2   **70.** 3   **71.** 4   **72.** 5

73. 7   **74.** 9   **75.** 12   **76.** 35

77. 240   **78.** 533   **79.** 643   **80.** 257

Prime Factorization In Exercises 81−98, write the prime factorization of the number.

81. 11   **82.** 13   **83.** 4

84. 6   **85.** 16   **86.** 27

87. 37   **88.** 29   **89.** 12

90. 52   **91.** 561   **92.** 245

93. 210   **94.** 525   **95.** 192

96. 264   **97.** 2535   **98.** 1521

Application   EXAMPLE 6   **Finding Factors of Note Frequencies**

From A220 to A440 on the piano, there are 12 semitones.

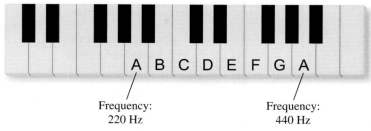

A, A#, B, C, C#, D, D#, E, F, F#, G, G#, A

Frequency:
220 Hz

Frequency:
440 Hz

Explain why there is no simple standard way for piano tuners to set the frequencies of the notes between A220 and A440.

SOLUTION

The reason for this is a bit complicated. However, you can see the problem when you consider that piano tuners have two conflicting goals.

Bartolomeo Cristofori (1655–1731) of Padua, Italy, is credited with the invention of the modern piano. He was employed by the famous Medici family.

1. One goal is that from each note to the next higher note, you should have the same multiple. For instance, you want a number a such that

(Frequency of A)(a) = (Frequency of A#)
(Frequency of A#)(a) = (Frequency of B), . . . and so on.

The number that works is $a \approx 1.0594$.

2. A second goal is that you want the frequencies of the 11 notes between A220 and A440 to have as many common factors with 220 as possible. Notes whose frequencies have common factors harmonize.

When you try satisfying these two goals, you will see that they are conflicting.

Exercises Within Reach ®

Solutions in English & Spanish and tutorial videos at CollegePrepAlgebra.com

99. *Packaging* You are designing a shipping carton to hold 144 tomatoes. The carton will contain multiple layers of tomatoes arranged in rows and columns. One possible arrangement has 4 layers of 4 rows and 9 columns, as shown. Use the factors of 144 to determine the possible numbers of rows and columns when the carton has 6 layers.

100. *Inventory* You need to arrange 72 boxes on a skid with the same number of rows and columns in each layer. One possible arrangement has 6 layers of 3 rows and 4 columns, as shown. Use the factors of 72 to determine the possible numbers of rows and columns in an arrangement of 4 layers. Which of these arrangements do you prefer? Explain your reasoning.

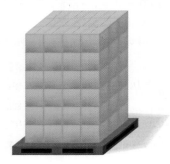

Summary of Rules and Definitions

At its simplest level, algebra is a symbolic form of arithmetic. This arithmetic-algebra connection can be illustrated in the following way.

Arithmetic   *Algebra*

Verbal rules and definitions ┐
                             ├─── Symbolic rules and definitions
Specific examples of ────────┘
rules and definitions

Arithmetic Summary

Definitions: Let a, b, and c be integers.

Definition	*Example*
1. Subtraction	
$a - b = a + (-b)$	$5 - 7 = 5 + (-7)$
2. Multiplication: (a is a positive integer)	
$a \cdot b = \underbrace{b + b + \cdots + b}_{a \text{ terms}}$	$3 \cdot 5 = 5 + 5 + 5$
3. Division: ($b \neq 0$)	
$a \div b = c$ if and only if $a = c \cdot b$.	$12 \div 4 = 3$ because $12 = 3 \cdot 4$.
4. Less than:	
$a < b$ if there is a positive real number c such that $a + c = b$.	$-2 < 1$ because $-2 + 3 = 1$.
5. Absolute value: $|a| = \begin{cases} a, & \text{if } a \geq 0 \\ -a, & \text{if } a < 0 \end{cases}$	$|-3| = -(-3) = 3$
6. Divisor:	
a is a divisor of b if and only if there is an integer c such that $a \cdot c = b$.	7 is a divisor of 21 because $7 \cdot 3 = 21$.

Rules: Let a and b be integers.

Rule	*Example*
1. Addition	
(a) To add two integers with *like* signs, add their absolute values and attach the common sign to the result.	$3 + 7 = |3| + |7| = 10$
(b) To add two integers with *unlike* signs, subtract the smaller absolute value from the larger absolute value and attach the sign of the integer with the larger absolute value.	$-5 + 8 = |8| - |-5|$ $= 8 - 5$ $= 3$
2. Multiplication:	
(a) $a \cdot 0 = 0 = 0 \cdot a$	$3 \cdot 0 = 0 = 0 \cdot 3$
(b) Like signs: $a \cdot b > 0$	$(-2)(-5) = 10$
(c) Unlike signs: $a \cdot b < 0$	$(2)(-5) = -10$
3. Division:	
(a) $\dfrac{0}{a} = 0, a \neq 0$	$\dfrac{0}{4} = 0$
(b) $\dfrac{a}{0}$ is undefined.	$\dfrac{6}{0}$ is undefined.
(c) Like signs: $\dfrac{a}{b} > 0, b \neq 0$	$\dfrac{-2}{-3} = \dfrac{2}{3}$
(d) Unlike signs: $\dfrac{a}{b} < 0, b \neq 0$	$\dfrac{-5}{7} = -\dfrac{5}{7}$

The word *algebra* comes from the Arabic language and much of its methods from Arabic/Islamic mathematics.

EXAMPLE 7   **Using Rules and Definitions**

a. Use the definition of subtraction to complete the statement.

$$4 - 9 = \boxed{}$$

b. Use the definition of multiplication to complete the statement.

$$6 + 6 + 6 + 6 = \boxed{}$$

c. Use the definition of absolute value to complete the statement.

$$|-9| = \boxed{}$$

d. Use the rule for adding integers with unlike signs to complete the statement.

$$-7 + 3 = \boxed{}$$

e. Use the rule for multiplying integers with unlike signs to complete the statement.

$$-9 \times 2 = \boxed{}$$

SOLUTION

a. $4 - 9 = 4 + (-9) = -5$

b. $6 + 6 + 6 + 6 = 4 \cdot 6 = 24$

c. $|-9| = -(-9) = 9$

d. $-7 + 3 = -(|-7| - |3|) = -4$

e. $-9 \times 2 = -18$

Exercises Within Reach ®

Solutions in English & Spanish and tutorial videos at CollegePrepAlgebra.com

Using a Rule or Definition **In Exercises 101–104, complete the statement using the indicated definition or rule.**

101. Definition of division: $12 \div 4 = \boxed{}$

102. Definition of absolute value: $|-8| = \boxed{}$

103. Rule for multiplying integers by 0:

$$6 \cdot 0 = \boxed{} = 0 \cdot 6$$

104. Rule for dividing integers with unlike signs:

$$\frac{30}{-10} = \boxed{}$$

Analyzing a Rule **In Exercises 105 and 106, write an example and an algebraic description of the arithmetic rule.**

105. The product of 1 and any real number is the real number itself.

106. Any nonzero real number divided by itself is 1.

Finding a Pattern **In Exercises 107–110, complete the pattern. Decide which rule the pattern demonstrates.**

107. $2(0) = 0$
$1(0) = 0$
$-1(0) = \boxed{}$
$-2(0) = \boxed{}$

108. $0 \div 2 = 0$
$0 \div 1 = 0$
$0 \div (-1) = \boxed{}$
$0 \div (-2) = \boxed{}$

109. $|2| = 2$
$|1| = 1$
$|0| = \boxed{}$
$|-1| = \boxed{}$
$|-2| = \boxed{}$

110. When $a + 2 = b, a < b.$
When $a + 1 = b, a < b.$
When $a + 0 = b, a \boxed{} b.$
When $a + (-1) = b, a \boxed{} b.$
When $a + (-2) = b, a \boxed{} b.$

Concept Summary: *Multiplying and Dividing Integers*

What

When you are asked to multiply or divide two integers a and b, the goal is to determine the correct sign of the **product** or **quotient** of the absolute values of the integers.

EXAMPLE

Find the product $-5 \cdot 8$.

Find the quotient $\dfrac{-40}{-8}$.

How

To determine the correct sign, use the rules for multiplying and dividing (nonzero) integers.

Like signs: $a \cdot b > 0, \dfrac{a}{b} > 0$

Unlike signs: $a \cdot b < 0, \dfrac{a}{b} < 0$

EXAMPLE

$$-5 \cdot 8 = -40$$

$$\frac{-40}{-8} = 5$$

Why

The sign is as important as the number in applications involving multiplying and dividing integers. In accounting, for instance, an incorrect sign could make a loss appear to be a profit.

Exercises Within Reach ®

Worked-out solutions to odd-numbered exercises at CollegePrepAlgebra.com

Concept Summary Check

111. *The Signs of the Factors* Do the integers multiplied in the example above have like signs or unlike signs?

112. *The Signs in a Quotient* Do the integers divided in the example above have like signs or unlike signs?

113. *A Sign of a Product* Is the product of -6 and -4 positive or negative? Explain.

114. *The Sign of a Quotient* Is the quotient of 12 and -4 positive or negative? Explain.

Extra Practice

Multiplying Integers In Exercises 115–126, find the product.

115. $-5(2)(-3)(-4)$

116. $-2(6)(-2)(-1)$

117. $4(-5)(2)(-1)$

118. $2(-3)(4)(-9)$

119. $-7(-2)(-5)(-3)(4)$

120. $-6(-2)(-3)(-4)(5)$

121. $-10(3)(7)(-2)(-6)$

122. $-10(4)(5)(-8)(-2)$

123. $4(-2)(3)(-4)(2)(-1)$

124. $5(-3)(2)(-2)(-1)(4)$

125. $8(-3)(2)(-4)(5)(0)$

126. $7(-2)(-1)(5)(0)(-9)$

Relatively Prime Numbers Two or more numbers that have no common prime factors are called *relatively prime*. In Exercises 127–138, **decide** whether the numbers are relatively prime.

127. 15, 28

128. 18, 49

129. 64, 232

130. 27, 36

131. 495, 784

132. 621, 1496

133. 63, 1375

134. 403, 899

135. 51, 85, 119

136. 24, 65, 161

137. 21, 20, 143

138. 84, 289, 325

139. *Temperature Change* The temperature measured by a weather balloon is decreasing approximately 3° for each 1000-foot increase in altitude. The balloon rises 8000 feet. What is the total temperature change?

140. *Savings Plan* A homeowner saves $250 per month for home improvements. After 2 years, how much money has the homeowner saved?

141. *Stock Price* The price per share of a technology stock drops $0.29 on each of four consecutive days, and then increases $0.32 on the fifth day. What is the total price change per share during the five days?

142. *Loss Leaders* To attract customers, a grocery store runs a sale on bananas. The bananas are *loss leaders*, which means the store loses money on the bananas but hopes to make it up on other items. The store sells 800 pounds of bananas at a loss of $0.26 per pound, and 1200 pounds of potatoes at a profit of $0.22 per pound. What is the store's profit?

143. *Geometry* The rectangular prism shown below has a volume of 4095 cubic feet. What is the height of the prism?

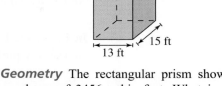

144. *Geometry* The rectangular prism shown below has a volume of 3456 cubic feet. What is the height of the prism?

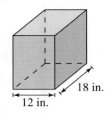

Explaining Concepts

145. *Reasoning* What is the only even prime number? Explain why there are no other even prime numbers.

146. *Logic* The number 1997 is not divisible by a prime number that is less than 45. Explain why this implies that 1997 is a prime number.

147. *Number Sense* Explain why the product of an even integer and any other integer is even. What can you conclude about the product of two odd integers?

148. *Reasoning* Explain how to check the result of a division problem.

149. *Think About It* A nonzero product has 25 factors, 17 of which are negative. What is the sign of the product? Explain.

150. *Think About It* A nonzero product has 25 factors, 16 of which are negative. What is the sign of the product? Explain.

151. *Investigation* Twin primes are prime numbers that differ by 2. For instance, 3 and 5 are twin primes. What are the other twin primes less than 100?

152. *Investigation* The **proper factors** of a number are all its factors less than the number itself. A number is **perfect** if the sum of its proper factors is equal to the number. A number is **abundant** if the sum of its proper factors is greater than the number. Which numbers less than 25 are perfect? Which are abundant? Try to find the first perfect number greater than 25.

153. *Investigation* The numbers 14, 15, and 16 are an example of three consecutive composite numbers. Is it possible to find ten consecutive composite numbers? If so, list an example. If not, explain why.

154. *Think About It* An integer *n* is divided by 2 and the quotient is an even integer. What does this tell you about *n*? Give an example.

155. *The Sieve of Eratosthenes* Write the integers from 1 through 100 in 10 lines of 10 numbers each.

(a) Cross out the number 1. Cross out all multiples of 2 other than 2 itself. Do the same for 3, 5, and 7.

(b) Of what type are the remaining numbers? Explain why this is the only type of number left.

Mid-Chapter Quiz: Sections 1.1–1.3

Solutions in English & Spanish and tutorial videos at CollegePrepAlgebra.com

Take this quiz as you would take a quiz in class. After you are done, check your work against the answers in the back of the book.

In Exercises 1 and 2, plot the numbers on the real number line and place the correct inequality symbol (< or >) between the numbers.

1. -4.5 ⬛ -6

2. $\frac{3}{4}$ ⬛ $\frac{3}{2}$

In Exercises 3 and 4, find the distance between the real numbers.

3. -15 and 7

4. -8.75 and -2.25

In Exercises 5 and 6, evaluate the expression.

5. $\left|-7.6\right|$

6. $-\left|9.8\right|$

In Exercises 7–16, evaluate the expression. Write fractions in simplest form.

7. $32 + (-18)$

8. $-12 - (-17)$

9. $\frac{3}{4} + \frac{7}{4}$

10. $\frac{2}{3} - \frac{1}{6}$

11. $(-3)(2)(-10)$

12. $\left(-\frac{4}{5}\right)\left(\frac{15}{32}\right)$

13. $\frac{7}{12} \div \frac{5}{6}$

14. $\left(-\frac{3}{2}\right)^3$

15. $3 - 2^2 + 25 \div 5$

16. $\dfrac{18 - 2(3 + 4)}{6^2 - (12 \cdot 2 + 10)}$

In Exercises 17 and 18, identify the property of real numbers illustrated by each statement.

17. (a) $8(u - 5) = 8 \cdot u - 8 \cdot 5$ (b) $10x - 10x = 0$

18. (a) $(7 + y) - z = 7 + (y - z)$ (b) $2x \cdot 1 = 2x$

Applications

19. During one month, you made the following transactions in your non-interest-bearing checking account. Find the balance at the end of the month.

NUMBER OR CODE	DATE	TRANSACTION DESCRIPTION	PAYMENT AMOUNT	✓	FEE	DEPOSIT AMOUNT	BALANCE $1406.98
2103	1/5	Car Payment	$375 03				
2104	1/7	Phone	$ 59 20				
	1/8	Withdrawal	$225 00				
	1/12	Deposit				$320 45	

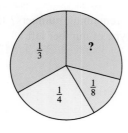

20. You deposit $45 in a retirement account twice each month. How much will you deposit in the account in 8 years?

21. Find the unknown fractional part of the circle graph at the left. Explain how you were able to make this determination.

Study Skills in Action

Keeping a Positive Attitude

A student's experiences during the first three weeks in a math class often determine whether the student sticks with it or not. You can get yourself off to a good start by immediately acquiring a positive attitude and the study behaviors to support it.

Using Study Strategies

In each *Study Skills in Action* feature, you will learn a new study strategy that will help you progress through the class. Each strategy will help you do the following.

- Set up good study habits
- Organize information into smaller pieces
- Create review tools
- Memorize important definitions and rules
- Learn the math at hand

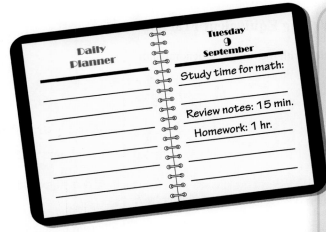

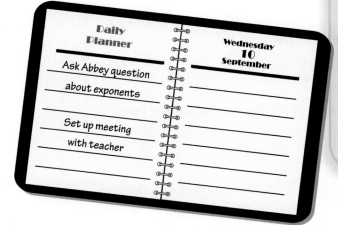

Smart Study Strategy

Create a Positive Study Environment

1 ▶ After the first math class, set aside time for reviewing your notes and the textbook, and completing homework.

2 ▶ Find a productive study environment. Set up a place for studying at home that is comfortable, but not too comfortable. It needs to be away from all potential distractions.

3 ▶ Study with a friend. Students who study well together can help each other out when someone gets sick and keep each other's attitudes positive.

4 ▶ Meet with your teacher at least once during the first two weeks. Ask the teacher what he or she advises for study strategies in the class. This will help you and let the teacher know that you really want to do well.

1.4 Operations with Rational Numbers

▶ Rewrite fractions as equivalent fractions.
▶ Add and subtract fractions.
▶ Multiply and divide fractions.
▶ Add, subtract, multiply, and divide decimals.

Rewriting Fractions

To write a fraction in simplest form, divide both the numerator and denominator by their greatest common factor (GCF).

EXAMPLE 1   Writing Fractions in Simplest Form

a. $\dfrac{18}{24} = \dfrac{\overset{1}{\cancel{2}} \cdot \overset{1}{\cancel{3}} \cdot 3}{2 \cdot 2 \cdot \underset{1}{\cancel{2}} \cdot \underset{1}{\cancel{3}}} = \dfrac{3}{4}$ **b.** $\dfrac{35}{21} = \dfrac{5 \cdot \overset{1}{\cancel{7}}}{3 \cdot \underset{1}{\cancel{7}}} = \dfrac{5}{3}$ **c.** $\dfrac{24}{72} = \dfrac{\overset{1}{\cancel{2}} \cdot \overset{1}{\cancel{2}} \cdot \overset{1}{\cancel{2}} \cdot \overset{1}{\cancel{3}}}{\underset{1}{\cancel{2}} \cdot \underset{1}{\cancel{2}} \cdot \underset{1}{\cancel{2}} \cdot \underset{1}{\cancel{3}} \cdot 3} = \dfrac{1}{3}$

Divide out GCF of 6.   Divide out GCF of 7.   Divide out GCF of 24.

You can obtain an **equivalent fraction** by multiplying or dividing the numerator and denominator by the same nonzero number. For instance, 9/12 and 3/4 are equivalent.

EXAMPLE 2   Writing Equivalent Fractions

Write an equivalent fraction with the indicated denominator.

a. $\dfrac{2}{3} = \dfrac{}{15}$ **b.** $\dfrac{9}{15} = \dfrac{}{35}$

SOLUTION

a. $\dfrac{2}{3} = \dfrac{2 \cdot 5}{3 \cdot 5} = \dfrac{10}{15}$ Multiply numerator and denominator by 5.

b. $\dfrac{9}{15} = \dfrac{\cancel{3} \cdot 3}{\cancel{3} \cdot 5} = \dfrac{3 \cdot 7}{5 \cdot 7} = \dfrac{21}{35}$ Reduce first, then multiply numerator and denominator by 7.

$\dfrac{9}{12}$

$\dfrac{3}{4}$

$\dfrac{9}{12} = \dfrac{\overset{1}{\cancel{3}} \cdot 3}{\underset{1}{\cancel{3}} \cdot 4}$

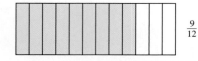

$\dfrac{9}{12}$ and $\dfrac{3}{4}$ are equivalent.

Exercises Within Reach ®

Solutions in English & Spanish and tutorial videos at CollegePrepAlgebra.com

Finding the GCF In Exercises 1–8, find the greatest common factor.

1. 5, 10   **2.** 3, 9   **3.** 20, 45   **4.** 48, 64
5. 45, 90   **6.** 27, 54   **7.** 18, 84, 90   **8.** 84, 98, 192

Writing a Fraction in Simplest Form In Exercises 9–16, write the fraction in simplest form.

9. $\dfrac{2}{4}$ **10.** $\dfrac{4}{16}$ **11.** $\dfrac{12}{15}$ **12.** $\dfrac{14}{35}$
13. $\dfrac{60}{192}$ **14.** $\dfrac{90}{225}$ **15.** $\dfrac{28}{350}$ **16.** $\dfrac{88}{154}$

Writing an Equivalent Fraction In Exercises 17–20, write an equivalent fraction with the indicated denominator.

17. $\dfrac{3}{8} = \dfrac{}{16}$ **18.** $\dfrac{4}{5} = \dfrac{}{15}$ **19.** $\dfrac{6}{15} = \dfrac{}{25}$ **20.** $\dfrac{21}{49} = \dfrac{}{28}$

Application   EXAMPLE 3   **Comparing Fraction and Decimal Systems**

Why do humans continue to use fractions to represent numbers? Why don't they simply write all numbers in decimal form?

SOLUTION

The simple answer is that in the decimal (base 10) system, most fractions cannot be written in exact form using a finite number of digits. For instance, the decimal form of the fraction $\frac{1}{3}$ has infinitely many digits.

$$\frac{1}{3} = 0.333333\ldots$$

Rather than using base 10, if you use a base that has more factors, then it is possible to write more fractions in exact form using a finite number of digits. For instance, the ancient Babylonians used base 60 for their number system. This is still present today in the measurement of time—there are 60 minutes in an hour.

The ancient Babylonians used a base 60 number system.

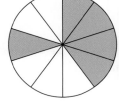

Exercises Within Reach®

Solutions in English & Spanish and tutorial videos at CollegePrepAlgebra.com

Writing the Decimal Form of a Fraction In Exercises 21–24, each figure is divided into regions of equal area. **Write** the fraction in simplest form that represents the shaded portion of the figure. Then use a calculator to **find** the decimal form of the fraction. Can the fraction be written in decimal form using a finite number of digits?

21.

22.

23.

24.

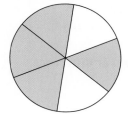

Adding and Subtracting Fractions

Let a, b, and c be integers with $c \neq 0$. To add or subtract the fractions a/c and b/c, use the following rules.

1. **With like denominators:**

$$\frac{a}{c} + \frac{b}{c} = \frac{a+b}{c} \quad \text{or} \quad \frac{a}{c} - \frac{b}{c} = \frac{a-b}{c}$$

2. **With unlike denominators:** Rewrite the fractions so that they have like denominators. Then use the rule for adding and subtracting fractions with like denominators.

Study Tip

To find a like denominator for two or more fractions, find the least common multiple (or LCM) of their denominators. For instance, the LCM of the denominators of 3/8 and 5/12 is 24.

EXAMPLE 4   **Adding and Subtracting with Like Denominators**

a. $\dfrac{3}{12} + \dfrac{4}{12} = \dfrac{3+4}{12} = \dfrac{7}{12}$ **b.** $\dfrac{7}{9} - \dfrac{2}{9} = \dfrac{7-2}{9} = \dfrac{5}{9}$

EXAMPLE 5   **Adding and Subtracting with Unlike Denominators**

a. $\dfrac{4}{5} + \dfrac{11}{15} = \dfrac{4(3)}{5(3)} + \dfrac{11}{15}$ LCM of 5 and 15 is 15.

$= \dfrac{12}{15} + \dfrac{11}{15}$ Rewrite with like denominators.

$= \dfrac{23}{15}$ Add numerators.

b. $1\dfrac{7}{9} - \dfrac{11}{12} = \dfrac{16}{9} - \dfrac{11}{12}$ Rewrite $1\frac{7}{9}$ as $\frac{16}{9}$.

$= \dfrac{16(4)}{9(4)} - \dfrac{11(3)}{12(3)}$ LCM of 9 and 12 is 36.

$= \dfrac{64}{36} - \dfrac{33}{36}$ Rewrite with like denominators.

$= \dfrac{31}{36}$ Subtract numerators.

Exercises Within Reach®

Solutions in English & Spanish and tutorial videos at CollegePrepAlgebra.com

Adding and Subtracting with Like Denominators In Exercises 25−32, **find the sum or difference. Write the result in simplest form.**

25. $\dfrac{7}{15} + \dfrac{1}{15}$ **26.** $\dfrac{13}{35} + \dfrac{5}{35}$ **27.** $\dfrac{3}{2} + \dfrac{5}{2}$ **28.** $\dfrac{5}{6} + \dfrac{13}{6}$

29. $\dfrac{9}{16} - \dfrac{3}{16}$ **30.** $\dfrac{15}{32} - \dfrac{7}{32}$ **31.** $\dfrac{3}{4} - \dfrac{5}{4}$ **32.** $\dfrac{7}{8} - \dfrac{9}{8}$

Adding and Subtracting with Unlike Denominators In Exercises 33−48, **find the sum or difference. Write the result in simplest form.**

33. $\dfrac{1}{2} + \dfrac{1}{3}$ **34.** $\dfrac{3}{5} + \dfrac{1}{2}$ **35.** $\dfrac{3}{16} + \dfrac{3}{8}$ **36.** $\dfrac{2}{3} + \dfrac{4}{9}$

37. $\dfrac{5}{6} - \dfrac{1}{3}$ **38.** $\dfrac{2}{3} - \dfrac{1}{6}$ **39.** $\dfrac{3}{4} - \dfrac{2}{5}$ **40.** $\dfrac{5}{6} - \dfrac{2}{7}$

41. $3\dfrac{1}{2} + 5\dfrac{2}{3}$ **42.** $5\dfrac{3}{4} + 8\dfrac{1}{10}$ **43.** $1\dfrac{3}{16} - 2\dfrac{1}{4}$ **44.** $5\dfrac{7}{8} - 2\dfrac{1}{2}$

45. $15 - 20\dfrac{1}{4}$ **46.** $6 - 3\dfrac{5}{8}$ **47.** $-5\dfrac{1}{3} - 4\dfrac{5}{12}$ **48.** $-2\dfrac{3}{4} - 3\dfrac{1}{5}$

Application   EXAMPLE 6   **Finding the Yardage for a Clothing Design**

A designer uses $3\frac{1}{6}$ yards of material to make a skirt and $2\frac{3}{4}$ yards to make a shirt. Find the total amount of material required.

SOLUTION

$$3\frac{1}{6} + 2\frac{3}{4} = \frac{19}{6} + \frac{11}{4}$$ Write mixed numbers as fractions.

$$= \frac{19(2)}{6(2)} + \frac{11(3)}{4(3)}$$ LCM of 6 and 4 is 12.

$$= \frac{38}{12} + \frac{33}{12}$$ Rewrite with like denominators.

$$= \frac{71}{12}$$ Add numerators.

$$= 5\frac{11}{12}$$ Write fraction as mixed number.

The designer needs $5\frac{11}{12}$ (about 6) yards of material.

Exercises Within Reach ®

Solutions in English & Spanish and tutorial videos at CollegePrepAlgebra.com

49. **Construction Project** A sign near a construction site indicates what fraction of the work has been completed. At the beginnings of May and June, the fractions of work completed were $\frac{5}{16}$ and $\frac{2}{3}$, respectively. What fraction of the work was completed during the month of May?

50. **Fund Drive** During a fund drive, a charity has a display showing how close it is to reaching its goal. At the end of the first week, the display shows $\frac{1}{8}$ of the goal. At the end of the second week, the display shows $\frac{3}{5}$ of the goal. What fraction of the goal was gained during the second week?

51. **Geometry** Determine the unknown fractional part of the circle graph.

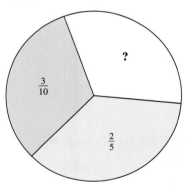

52. **Geometry** Determine the unknown fractional part of the circle graph.

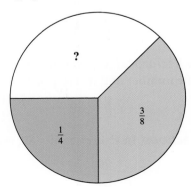

Multiplying and Dividing Fractions

Let a, b, c, and d be integers with $b \neq 0$ and $d \neq 0$. Then the product of $\frac{a}{b}$ and $\frac{c}{d}$ is

$$\frac{a}{b} \cdot \frac{c}{d} = \frac{a \cdot c}{b \cdot d}.$$ Multiply numerators and denominators.

EXAMPLE 7 **Multiplying Fractions**

$$\frac{5}{8} \cdot \frac{3}{2} = \frac{5(3)}{8(2)}$$ Multiply numerators and denominators.

$$= \frac{15}{16}$$ Simplify.

Let a, b, c, and d be integers with $b \neq 0$, $c \neq 0$, and $d \neq 0$. Then the quotient of $\frac{a}{b}$ and $\frac{c}{d}$ is

$$\frac{a}{b} \div \frac{c}{d} = \frac{a}{b} \cdot \frac{d}{c}.$$ Invert divisor and multiply.

EXAMPLE 8 **Dividing Fractions**

$$\frac{5}{8} \div \frac{20}{12} = \frac{5}{8} \cdot \frac{12}{20}$$ Invert divisor and multiply.

$$= \frac{(5)(12)}{(8)(20)}$$ Multiply numerators and denominators.

$$= \frac{(5)(3)(4)}{(8)(4)(5)}$$ Divide out common factors.

$$= \frac{3}{8}$$ Write in simplest form.

Exercises Within Reach®

Solutions in English & Spanish and tutorial videos at CollegePrepAlgebra.com

Multiplying Fractions In Exercises 53−64, **evaluate the expression. Write the result in simplest form.**

53. $\frac{1}{2} \cdot \frac{3}{4}$　　**54.** $\frac{3}{5} \cdot \frac{1}{2}$　　**55.** $-\frac{2}{3} \cdot \frac{5}{7}$　　**56.** $-\frac{5}{6} \cdot \frac{1}{2}$

57. $\frac{2}{3}\left(-\frac{9}{16}\right)$　　**58.** $\frac{5}{3}\left(-\frac{3}{5}\right)$　　**59.** $-\frac{3}{4}\left(-\frac{4}{9}\right)$　　**60.** $-\frac{15}{16}\left(-\frac{12}{5}\right)$

61. $\frac{11}{12}\left(-\frac{9}{44}\right)$　　**62.** $\frac{5}{12}\left(-\frac{6}{25}\right)$　　**63.** $9\left(\frac{4}{15}\right)$　　**64.** $24\left(\frac{7}{18}\right)$

Finding the Reciprocal In Exercises 65−68, **find the reciprocal of the number. Show that the product of the number and its reciprocal is 1.**

65. 7　　**66.** 14　　**67.** $\frac{4}{7}$　　**68.** $-\frac{5}{9}$

Dividing Fractions In Exercises 69−80, **evaluate the expression and write the result in simplest form.**

69. $\frac{3}{8} \div \frac{3}{4}$　　**70.** $\frac{5}{16} \div \frac{25}{8}$　　**71.** $-\frac{5}{12} \div \frac{45}{32}$　　**72.** $-\frac{16}{21} \div \frac{12}{27}$

73. $\frac{8}{3} \div \frac{8}{3}$　　**74.** $\frac{5}{7} \div \frac{5}{7}$　　**75.** $\frac{3}{5} \div \frac{7}{5}$　　**76.** $\frac{7}{8} \div \frac{3}{8}$

77. $-\frac{5}{6} \div \left(-\frac{8}{10}\right)$　　**78.** $-\frac{14}{15} \div \left(-\frac{24}{25}\right)$　　**79.** $-10 \div \frac{1}{9}$　　**80.** $-6 \div \frac{1}{3}$

Application   EXAMPLE 9   **Finding the Number of Calories Burned**

You decide to take a tennis class. You burn about 400 calories per hour playing tennis. In one week, you played tennis for $\frac{3}{4}$ hour on Tuesday, 2 hours on Wednesday, and $1\frac{1}{2}$ hours on Thursday. How many total calories did you burn playing tennis during that week? What was your average number of calories burned per day playing tennis?

SOLUTION

The total number of calories you burned playing tennis during the week was

$$400\left(\frac{3}{4}\right) + 400(2) + 400\left(1\frac{1}{2}\right) = 300 + 800 + 600$$
$$= 1700 \text{ calories.}$$

The average number of calories burned per day was

$$\frac{1700 \text{ calories}}{3 \text{ days}} = 566\frac{2}{3} \text{ calories per day.}$$

Exercises Within Reach ®

Solutions in English & Spanish and tutorial videos at CollegePrepAlgebra.com

81. **Basketball** You burn about 600 calories per hour playing basketball. You played for $1\frac{1}{4}$ hours on Monday, $1\frac{1}{2}$ hours on Tuesday, and $\frac{5}{6}$ hour on Wednesday. How many total calories did you burn playing basketball on the three days? What was the average number of calories you burned per day playing basketball?

82. **Racewalking** You burn about 240 calories per hour racewalking. You racewalked for $\frac{3}{4}$ hour on Thursday, $1\frac{1}{6}$ hours on Friday, and $1\frac{4}{5}$ hours on Saturday. How many total calories did you burn racewalking on the three days? What was the average number of calories you burned per day racewalking?

83. **Skiing Time** You cross-country ski on a trail that is $5\frac{3}{5}$ miles long at an average rate of $9\frac{1}{3}$ miles per hour. How long does it take you to ski the entire trail?

84. **Walking Time** Your apartment is $\frac{3}{4}$ mile from the subway. You walk at the rate of $3\frac{1}{4}$ miles per hour. How long does it take you to walk to the subway?

Operations with Decimals

To round a decimal, use the following rules.

1. Determine the number of digits of accuracy you wish to keep. The digit in the last position you keep is called the **rounding digit**, and the digit in the first position you discard is called the **decision digit**.

2. If the decision digit is 5 or greater, round up by adding 1 to the rounding digit.

3. If the decision digit is 4 or less, round down by leaving the rounding digit unchanged.

EXAMPLE 10   **Adding and Multiplying Decimals**

a.
```
   1 1
  0.583
  1.06
+ 2.9104
  4.5534
```

b.
```
     −3.57     Two decimal places
  × 0.032      Three decimal places
      714
     1071
  −0.11424     Five decimal places
```

EXAMPLE 11   **Dividing Decimals**

$$\frac{1.483}{0.56}$$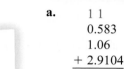

```
        2.648
56)148.300
    112
    36 3
    336
    270
    224
    460
    448
```

Rounded to two decimal places, the answer is 2.65. This can be written as

$$\frac{1.483}{0.56} \approx 2.65.$$

Study Tip

Rational numbers can be represented as terminating or repeating decimals. Here are some examples.

Terminating Decimals

$\frac{1}{4} = 0.25$ $\frac{3}{8} = 0.375$

Repeating Decimals

$\frac{1}{6} = 0.1666 \ldots$ or $0.1\overline{6}$

$\frac{8}{33} = 0.2424 \ldots$ or $0.\overline{24}$

Exercises Within Reach®

Solutions in English & Spanish and tutorial videos at CollegePrepAlgebra.com

Adding and Subtracting Decimals   In Exercises 85−96, evaluate the expression.

85. $12.33 + 14.76$

86. $6.983 + 241.5$

87. $0.287 + 1.65 + 2.1932$

88. $2.013 + 0.1145 + 1.12$

89. $132.1 + (−25.45)$

90. $408.9 + (−13.12)$

91. $8.673 − 2.55$

92. $5.1146 − 1.028$

93. $4.54 − 6.668$

94. $4.25 − 7.998$

95. $1.21 + 4.06 − 3.00$

96. $3.4 + 1.062 − 5.13$

Multiplying Decimals   In Exercises 97−100, evaluate the expression. Round your answer to two decimal places.

97. $−6.3(9.05)$

98. $3.7(−14.8)$

99. $−0.05(−85.95)$

100. $−0.09(−0.45)$

Dividing Decimals   In Exercises 101−104, evaluate the expression. Round your answer to two decimal places.

101. $4.69 \div 0.12$

102. $7.14 \div 0.94$

103. $1.062 \div (−2.1)$

104. $2.011 \div (−3.3)$

EXAMPLE 12   **Finding a Cell Phone Charge**

A cellular phone company charges $5.35 for the first 200 text messages per month and $0.10 for each additional text message.

a. Find the cost of 263 text messages.

b. Can you save money by switching to a plan that allows unlimited text messages for $10 per month?

SOLUTION

a. You sent or received 63 text messages above 200. The cost of these is

$$(\$0.10)(63) = \$6.30.$$

So, your total charge for the month is

$$\$5.35 + \$6.30 = \$11.65.$$

b. If you continue to send and receive this number of text messages each month, you can save money by switching to a plan that allows unlimited texts messages for $10 per month.

Exercises Within Reach®

105. *Consumer Awareness* At a convenience store, you buy two gallons of milk at $3.94 per gallon and three loaves of bread at $2.47 per loaf. You give the clerk a 20-dollar bill. How much change will you receive? (Assume there is no sales tax.)

106. *Stock Purchase* You buy 200 shares of stock at $23.63 per share and 300 shares at $86.25 per share.

(a) Estimate the total cost of the stock.

(b) Use a calculator to find the total cost of the stock.

107. *Fuel Efficiency* The sticker on a new car gives the fuel efficiency as 22.3 miles per gallon. The average cost of fuel is $3.479 per gallon. You expect to drive the car 12,000 miles per year. Find the expected annual fuel cost for the car.

108. *Consumer Awareness* The prices per gallon of regular unleaded gasoline at three service stations are $3.439, $3.479, and $3.589, respectively. Find the average price per gallon.

Concept Summary: Operations with Rational Numbers

What

As you study mathematics, you will be asked to add, subtract, multiply, or divide two fractions or two decimals.

EXAMPLE

Evaluate each expression.

a. $\frac{1}{2} + \frac{2}{3}$ **b.** $\frac{1}{2} \div \frac{2}{3}$

c. $2.2 + 10.03$ **d.** $0.2 \cdot 6.33$

How

To perform such operations with rational numbers, you can use the rules for fractions and the rules for decimals.

EXAMPLE

a. $\frac{1}{2} + \frac{2}{3} = \frac{3}{6} + \frac{4}{6} = \frac{7}{6}$

b. $\frac{1}{2} \div \frac{2}{3} = \frac{1}{2} \cdot \frac{3}{2} = \frac{3}{4}$

c.
$$\begin{array}{r} 2.2 \\ + 10.03 \\ \hline 12.23 \end{array}$$

d.
$$\begin{array}{r} 6.33 \\ \times 0.2 \\ \hline 1.266 \end{array}$$

Why

When you know the rules for the operations with rational numbers, you will be able to add, subtract, multiply, or divide any two rational numbers. For instance, to determine the gas mileage for your car, you will be able to divide the number of miles you drove by the number of gallons you used.

Exercises Within Reach ®

Worked-out solutions to odd-numbered exercises at CollegePrepAlgebra.com

Concept Summary Check

109. *Using the Least Common Denominator* Explain how to use the least common denominator to add two fractions.

110. *Dividing Fractions* Describe how to divide two fractions.

111. *Adding Decimals* Describe how to add two decimals using a vertical format.

112. *Multiplying Decimals* When you multiply two decimals, how do you determine where to place the decimal point in the product?

Extra Practice

Geometry In Exercises 113 and 114, **determine the unknown fractional part of the circle graph.**

113.

114.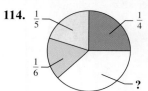

Operations with Fractions In Exercises 115−134, **evaluate the expression and write the result in simplest form. If it is not possible, explain why.**

115. $-\frac{23}{11} + \frac{12}{11}$

116. $-\frac{39}{23} + \frac{11}{23}$

117. $\frac{7}{10} + \left(-\frac{3}{10}\right)$

118. $\frac{11}{15} + \left(-\frac{2}{15}\right)$

119. $-\frac{1}{8} - \frac{1}{6}$

120. $-\frac{13}{8} - \frac{3}{4}$

121. $\frac{2}{5} + \frac{4}{5} + \frac{1}{5}$

122. $\frac{2}{9} + \frac{4}{9} + \frac{1}{9}$

123. $2\frac{3}{4} \cdot 3\frac{2}{3}$

124. $-8\frac{1}{2} \cdot 3\frac{2}{5}$

125. $-5\frac{2}{3} \cdot 4\frac{1}{2}$

126. $2\frac{4}{5} \cdot 6\frac{2}{3}$

127. $-\frac{3}{2}\left(-\frac{15}{16}\right)\left(\frac{12}{25}\right)$

128. $\frac{1}{2}\left(-\frac{4}{15}\right)\left(-\frac{5}{24}\right)$

129. $6\left(\frac{3}{4}\right)\left(\frac{2}{9}\right)$

130. $8\left(\frac{5}{12}\right)\left(\frac{3}{10}\right)$

131. $\frac{3}{5} \div 0$

132. $\frac{11}{13} \div 0$

133. $3\frac{3}{4} \div 1\frac{1}{2}$

134. $2\frac{4}{9} \div 5\frac{1}{3}$

Operations with Decimals In Exercises 135−138, **evaluate the expression. Round your answer to two decimal places.**

135. $-0.0005 - 2.01 + 0.111$

136. $-1.0012 - 3.25 + 0.2$

137. $-2.54(3.8)(6.55)$

138. $7.8(12.32)(-0.95)$

139. *Consumer Awareness* The prices of a 16-ounce bottle of soda at three different convenience stores are $1.09, $1.25, and $1.10, respectively. Find the average price for the bottle of soda.

140. *Cooking* You make 60 ounces of dough for breadsticks. Each breadstick requires $\frac{5}{4}$ ounces of dough. How many breadsticks can you make?

Explaining Concepts

141. *Number Sense* Is it true that the sum of two fractions of like signs is positive? If not, give an example that shows the statement is false.

142. *Structure* Does $\frac{2}{3} + \frac{3}{2} = \frac{(2+3)}{(3+2)} = 1$? Explain your answer.

143. *Writing* In your own words, describe the rule for determining the sign of the product of two fractions.

144. *Precision* Is it true that $\frac{2}{3} = 0.67$? Explain your answer.

145. *Modeling* Use the figure to determine how many one-fourths are in 3. Explain how to obtain the same result by division.

146. *Modeling* Use the figure to determine how many one-sixths are in $\frac{2}{3}$. Explain how to obtain the same result by division.

147. *Investigation* When using a calculator to perform operations with decimals, you should try to get in the habit of rounding your answers *only* after all the calculations are done. By rounding the answer at a preliminary stage, you can introduce unnecessary roundoff error. The dimensions of a box are $l = 5.24$, $w = 3.03$, and $h = 2.749$. Find the volume $l \cdot w \cdot h$, by multiplying the numbers and then rounding the answer to one decimal place. Now use a second method, first rounding each dimension to one decimal place and then multiplying the numbers. Compare your answers, and explain which of these techniques produces the more accurate answer.

True or False? In Exercises 148–153, decide whether the statement is true or false. Justify your answer.

148. The reciprocal of every nonzero integer is an integer.

149. The reciprocal of every nonzero rational number is a rational number.

150. The product of two nonzero rational numbers is a rational number.

151. The product of two positive rational numbers is greater than either factor.

152. If $u > v$, then $u - v > 0$.

153. If $u > 0$ and $v > 0$, then $u - v > 0$.

154. *Estimation* Use mental math to determine whether $\left(5\frac{3}{4}\right) \times \left(4\frac{1}{8}\right)$ is less than 20. Explain your reasoning.

155. *Think About It* Determine the placement of the 3, 4, 5, and 6 in the following addition problem so that you obtain the specified sum. Use each number only once.

$$\frac{\boxed{}}{\boxed{}} + \frac{\boxed{}}{\boxed{}} = \frac{13}{10}$$

156. *Think About It* If the fractions represented by the points P and R are multiplied, what point on the number line best represents their product: M, S, N, P, or T? (*Source:* National Council of Teachers of Mathematics)

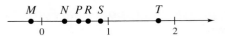

1.5 Exponents and Properties of Real Numbers

▶ Write expressions in exponential form and evaluate exponential expressions.

▶ Evaluate expressions using the order of operations.

▶ Identify and use the properties of real numbers.

Exponents

Repeated multiplication can be described in exponential form.

Repeated Multiplication	Exponential Form
$\underbrace{7 \cdot 7 \cdot 7 \cdot 7}_{\text{4 factors of 7}}$	7^4

In the exponential form, 7 is the **base** and it specifies the repeated factor. The number 4 is the **exponent** and it indicates how many times the base occurs as a factor. When you write the exponential form, you can say that you are raising 7 to the fourth **power**.

Study Tip

Keep in mind that an exponent applies only to the factor (number) directly preceding it. Parentheses are needed to include a negative sign or other factors as part of the base. Here is an example.

$(-5)^2 = (-5)(-5) = 25$

$-5^2 = -(5 \cdot 5) = -25$

EXAMPLE 1   Evaluating Exponential Expressions

a. $2^5 = 2 \cdot 2 \cdot 2 \cdot 2 \cdot 2$ Rewrite expression as a product.

$= 32$ Simplify.

b. $\left(\dfrac{2}{3}\right)^4 = \dfrac{2}{3} \cdot \dfrac{2}{3} \cdot \dfrac{2}{3} \cdot \dfrac{2}{3}$ Rewrite expression as a product.

$= \dfrac{2 \cdot 2 \cdot 2 \cdot 2}{3 \cdot 3 \cdot 3 \cdot 3}$ Multiply fractions.

$= \dfrac{16}{81}$ Simplify.

EXAMPLE 2   Evaluating Exponential Expressions

a. $(-4)^3 = (-4)(-4)(-4)$ Rewrite expression as a product.

$= -64$ Simplify.

b. $(-3)^4 = (-3)(-3)(-3)(-3)$ Rewrite expression as a product.

$= 81$ Simplify.

c. $-3^4 = -(3 \cdot 3 \cdot 3 \cdot 3)$ Rewrite expression as a product.

$= -81$ Simplify.

Exercises Within Reach ®

Solutions in English & Spanish and tutorial videos at CollegePrepAlgebra.com

Writing an Expression in Exponential Form In Exercises 1 and 2, rewrite the expression in exponential form.

1. $2 \cdot 2 \cdot 2 \cdot 2 \cdot 2 \cdot 2$

2. $4 \cdot 4 \cdot 4 \cdot 4 \cdot 4 \cdot 4$

Evaluating an Exponential Expression In Exercises 3−14, evaluate the expression.

3. 3^2

4. 4^3

5. 2^6

6. 5^3

7. $\left(\dfrac{1}{4}\right)^3$

8. $\left(\dfrac{4}{5}\right)^4$

9. $(-5)^3$

10. $(-4)^2$

11. -4^2

12. $-(-6)^3$

13. $(-1.2)^3$

14. $(-1.5)^4$

Application   EXAMPLE 3   **Finding the Amount a Truck Can Transport**

A truck can transport a load of motor oil that is 6 cases high, 6 cases wide, and 6 cases long. Each case contains 6 quarts of motor oil. How many quarts can the truck transport?

SOLUTION

A sketch can help you solve this problem.

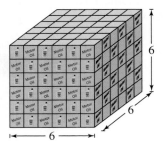

There are $6 \cdot 6 \cdot 6$ cases of motor oil, and each case contains 6 quarts. You can see that 6 occurs as a factor 4 times, which implies that the total number of quarts is

$$(6 \cdot 6 \cdot 6) \cdot 6 = 6^4 = 1296.$$

So, the truck can transport 1296 quarts of oil.

Exercises Within Reach®

Solutions in English & Spanish and tutorial videos at CollegePrepAlgebra.com

15. *Cereal* A grocery store has a cereal display that is 8 boxes high, 8 boxes wide, and 8 boxes long. How many cereal boxes are in the display?

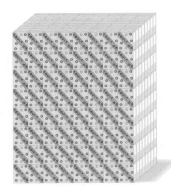

16. *Propane* A truck can transport a load of propane tanks that is 4 cases high, 4 cases wide, and 4 cases long. Each case contains 4 propane tanks. How many tanks can the truck transport?

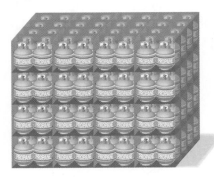

17. *Exponential Communication* You send an email to 5 people. Each person forwards the message to 5 other people, and so on. Complete the table to show the results of the first 6 stages. After what stage have more than 3900 people seen the message?

Stage	Emails sent, as a power	Emails sent
1	5^1	5
2	5^2	25
3		
4		
5		
6		

18. *Exponential Communication* You send an email to 3 people. Each person forwards the message to 3 other people, and so on. How many stages does it take for the total number of people who see the message to exceed the total number after Stage 4 in Exercise 17? Explain.

Order of Operations

The accepted priorities for the **order of operations** are summarized below.

1. Perform operations inside symbols of grouping: **P**arentheses, brackets, or absolute value symbols, starting with the innermost symbols.
2. Evaluate all **E**xponential expressions.
3. Perform all **M**ultiplications and **D**ivisions from left to right.
4. Perform all **A**dditions and **S**ubtractions from left to right.

EXAMPLE 4 Using Order of Operations

a. $7 - [(5 \cdot 3) + 2^3] = 7 - [15 + 2^3]$ Multiply inside the parentheses.

$\qquad\qquad\qquad\quad = 7 - [15 + 8]$ Evaluate exponential expression.

$\qquad\qquad\qquad\quad = 7 - 23$ Add inside the brackets.

$\qquad\qquad\qquad\quad = -16$ Subtract.

b. $36 \div (3^2 \cdot 2) - 6 = 36 \div (9 \cdot 2) - 6$ Evaluate exponential expression.

$\qquad\qquad\qquad\quad = 36 \div 18 - 6$ Multiply inside the parentheses.

$\qquad\qquad\qquad\quad = 2 - 6$ Divide.

$\qquad\qquad\qquad\quad = -4$ Subtract.

c. $\dfrac{8}{3}\left(\dfrac{1}{6} + \dfrac{1}{4}\right) = \dfrac{8}{3}\left(\dfrac{2}{12} + \dfrac{3}{12}\right)$ Find common denominator.

$\qquad\qquad\quad = \dfrac{8}{3}\left(\dfrac{5}{12}\right)$ Add inside the parentheses.

$\qquad\qquad\quad = \dfrac{40}{36}$ Multiply fractions.

$\qquad\qquad\quad = \dfrac{10}{9}$ Simplify.

Exercises Within Reach®

Solutions in English & Spanish and tutorial videos at CollegePrepAlgebra.com

Using Order of Operations In Exercises 19−58, evaluate the expression. If it is not possible, state the reason. Write fractional answers in simplest form.

19. $4 - 6 + 10$ **20.** $8 + 9 - 12$ **21.** $5 - (8 - 15)$ **22.** $13 - (12 - 3)$

23. $15 + 3 \cdot 4$ **24.** $9 - 5 \cdot 2$ **25.** $25 - 32 \div 4$ **26.** $16 + 24 \div 8$

27. $(45 \div 10) \cdot 2$ **28.** $(38 \div 5) \cdot 4$ **29.** $(16 - 5) \div (3 - 5)$ **30.** $(19 - 4) \div (7 - 2)$

31. $(10 - 16) \cdot (20 - 26)$ **32.** $(14 - 17) \cdot (13 - 19)$ **33.** $17 - |2 - (6 + 5)|$

34. $125 - |10 - (25 - 3)|$ **35.** $[360 - (8 + 12)] \div 5$ **36.** $[127 - (13 + 4)] \div 10$

37. $5 + (2^2 \cdot 3)$ **38.** $181 - (13 \cdot 3^2)$ **39.** $(-6)^2 - (48 \div 4^2)$ **40.** $(-3)^3 + (12 \div 2^2)$

41. $\left(3 \cdot \dfrac{5}{9}\right) + 1 - \dfrac{1}{3}$ **42.** $\dfrac{2}{3}\left(\dfrac{3}{4}\right) + 2 - \dfrac{3}{2}$ **43.** $18\left(\dfrac{1}{2} + \dfrac{2}{3}\right)$ **44.** $4\left(-\dfrac{2}{3} + \dfrac{4}{3}\right)$

45. $\dfrac{7}{25}\left(\dfrac{7}{16} - \dfrac{1}{8}\right)$ **46.** $\dfrac{3}{2}\left(\dfrac{2}{3} + \dfrac{1}{6}\right)$ **47.** $\dfrac{7}{3}\left(\dfrac{2}{3}\right) \div \dfrac{28}{15}$ **48.** $\dfrac{3}{8}\left(\dfrac{1}{5}\right) \div \dfrac{25}{32}$

49. $\dfrac{3 + [15 \div (-3)]}{16}$ **50.** $\dfrac{5 + [(-12) \div 4]}{24}$ **51.** $\dfrac{1 - 3^2}{-2}$ **52.** $\dfrac{2^2 + 4^2}{5}$

53. $\dfrac{7^2 - 4^2}{0}$ **54.** $\dfrac{0}{3^2 - 1^2}$ **55.** $\dfrac{0}{6^2 + 1}$

56. $\dfrac{3^3 + 1}{0}$ **57.** $\dfrac{5^2 + 12^2}{13}$ **58.** $\dfrac{8^2 - 2^3}{4}$

Properties of Real Numbers

You are now ready for the symbolic versions of the properties that are true about operations with real numbers. These properties are referred to as **properties of real numbers**. The table shows a verbal description and an illustrative example for each property. Keep in mind that the letters a, b, and c represent real numbers, even though only rational numbers have been used to this point.

Properties of Real Numbers

Let a, b, and c be real numbers.

Property	*Example*
1. *Commutative Property of Addition:*	
Two real numbers can be added in either order.	
$a + b = b + a$	$3 + 5 = 5 + 3$
2. *Commutative Property of Multiplication:*	
Two real numbers can be multiplied in either order.	
$ab = ba$	$4 \cdot (-7) = -7 \cdot 4$
3. *Associative Property of Addition:*	
When three real numbers are added, it makes no difference which two are added first.	
$(a + b) + c = a + (b + c)$	$(2 + 6) + 5 = 2 + (6 + 5)$
4. *Associative Property of Multiplication:*	
When three real numbers are multiplied, it makes no difference which two are multiplied first.	
$(ab)c = a(bc)$	$(3 \cdot 5) \cdot 2 = 3 \cdot (5 \cdot 2)$
5. *Distributive Property:*	
Multiplication distributes over addition.	
$a(b + c) = ab + ac$	$3(8 + 5) = 3 \cdot 8 + 3 \cdot 5$
$(a + b)c = ac + bc$	$(3 + 8)5 = 3 \cdot 5 + 8 \cdot 5$
6. *Additive Identity Property:*	
The sum of zero and a real number equals the number itself.	
$a + 0 = 0 + a = a$	$3 + 0 = 0 + 3 = 3$
7. *Multiplicative Identity Property:*	
The product of 1 and a real number equals the number itself.	
$a \cdot 1 = 1 \cdot a = a$	$4 \cdot 1 = 1 \cdot 4 = 4$
8. *Additive Inverse Property:*	
The sum of a real number and its opposite is zero.	
$a + (-a) = 0$	$3 + (-3) = 0$
9. *Multiplicative Inverse Property:*	
The product of a nonzero real number and its reciprocal is 1.	
$a \cdot \dfrac{1}{a} = 1, a \neq 0$	$8 \cdot \dfrac{1}{8} = 1$

EXAMPLE 5 Identifying Properties of Real Numbers

Identify the property of real numbers illustrated by each statement.

a. $3(6 + 2) = 3 \cdot 6 + 3 \cdot 2$ **b.** $5 \cdot \frac{1}{5} = 1$

c. $7 + (5 + 4) = (7 + 5) + 4$ **d.** $(12 + 3) + 0 = 12 + 3$

SOLUTION

a. This statement illustrates the Distributive Property.

b. This statement illustrates the Multiplicative Inverse Property.

c. This statement illustrates the Associative Property of Addition.

d. This statement illustrates the Additive Identity Property.

EXAMPLE 6 Using Properties of Real Numbers

Complete each statement using the specified property of real numbers.

a. Commutative Property of Addition: $5 + 9 = $ ▢

b. Associative Property of Multiplication: $6(5 \cdot 13) = $ ▢

c. Distributive Property: $4 \cdot 3 + 4 \cdot 7 = $ ▢

SOLUTION

a. By the Commutative Property of Addition, you can write

$$5 + 9 = 9 + 5.$$

b. By the Associative Property of Multiplication, you can write

$$6(5 \cdot 13) = (6 \cdot 5)13.$$

c. By the Distributive Property, you can write

$$4 \cdot 3 + 4 \cdot 7 = 4(3 + 7).$$

Exercises Within Reach ®

Identifying a Property of Real Numbers **In Exercises 59−70, identify the property of real numbers illustrated by the statement.**

59. $(10 + 3) + 2 = 10 + (3 + 2)$ **60.** $(32 + 8) + 5 = 32 + (8 + 5)$ **61.** $6(-3) = -3(6)$

62. $16 + 10 = 10 + 16$ **63.** $5 + 10 = 10 + 5$ **64.** $-2(8) = 8(-2)$

65. $6(3 + 13) = 6 \cdot 3 + 6 \cdot 13$ **66.** $(14 + 2)3 = 14 \cdot 3 + 2 \cdot 3$ **67.** $7\left(\frac{1}{7}\right) = 1$

68. $1 \cdot 4 = 4$ **69.** $0 + 15 = 15$ **70.** $25 + (-25) = 0$

Using a Property of Real Numbers **In Exercises 71−74, complete the statement using the specified property of real numbers.**

71. Commutative Property of Multiplication: $10(-3) = $ ▢

72. Distributive Property: $6(19 + 2) = $ ▢

73. Associative Property of Addition: $18 + (12 + 9) = $ ▢

74. Associative Property of Multiplication: $12(3 \cdot 4) = $ ▢

Application   **EXAMPLE 7**   **Finding the Area of a Billboard**

You measure the width of a billboard and find that it is 60 feet. You are told that its height is 22 feet less than its width.

a.   Write an expression for the area of the billboard.

b.   Use the Distributive Property to rewrite the expression.

c.   Find the area of the billboard.

SOLUTION

(60 − 22) ft

60 ft

a.   Begin by drawing and labeling a diagram. To find an expression for the area of the billboard, multiply the width by the height.

$$\text{Area} = \text{Width} \times \text{Height}$$
$$= 60(60 - 22)$$

b.   To rewrite the expression $60(60 - 22)$ using the Distributive Property, distribute 60 over the subtraction.

$$60(60 - 22) = 60(60) - 60(22)$$

c.   To find the area of the billboard, evaluate the expression in part (b) as follows.

$$60(60) - 60(22) = 3600 - 1320 \qquad \text{Multiply.}$$
$$= 2280 \qquad \text{Subtract.}$$

So, the area of the billboard is 2280 square feet.

Exercises Within Reach ®

Solutions in English & Spanish and tutorial videos at CollegePrepAlgebra.com

75. *Movie Screen* The width of a movie screen is 30 feet and its height is 8 feet less than its width.

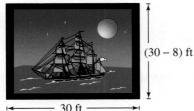

(30 − 8) ft

30 ft

(a) Write an expression for the area of the movie screen.

(b) Use the Distributive Property to rewrite the expression.

(c) Find the area of the movie screen.

76. *Picture Frame* A picture frame is 36 inches wide and its height is 9 inches less than its width.

(a) Write an expression for the area of the picture frame.

(b) Use the Distributive Property to rewrite the expression.

(c) Find the area of the picture frame.

77. *Geometry* Consider the rectangle shown in the figure.

(a) Find the area of the rectangle by adding the areas of Regions I and II.

(b) Find the area of the rectangle by multiplying its length by its width.

(c) Explain how the results of parts (a) and (b) relate to the Distributive Property.

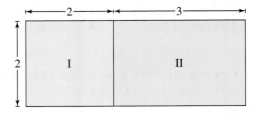

78. *Sales Tax* You purchase a sweater for $35.95. There is a 6% sales tax, which means that the total amount you must pay is $35.95 + 0.06(35.95)$.

(a) Use the Distributive Property to rewrite the expression.

(b) How much must you pay for the sweater including sales tax?

Concept Summary: Using the Order of Operations

What

To evaluate expressions consisting of grouping symbols, **exponents**, or more than one operation correctly, you need to use the established **order of operations**.

EXAMPLE

Evaluate

$(7 + 3)2^4 + 5$.

How

You can apply the order of operations as follows.

1. **P**arentheses
2. **E**xponents
3. **M**ultiplication and **D**ivision
4. **A**ddition and **S**ubtraction

EXAMPLE

$$(7 + 3)2^4 + 5 = 10 \cdot 2^4 + 5$$
$$= 10 \cdot 16 + 5$$
$$= 160 + 5$$
$$= 165$$

Why

When you use the order of operations, along with the **properties of real numbers**, you will be able to evaluate expressions correctly and efficiently.

Exercises Within Reach ®

Worked-out solutions to odd-numbered exercises at CollegePrepAlgebra.com

Concept Summary Check

79. *Understanding Exponents* Which part of the exponential expression 2^4 is the base? Which part is the exponent?

80. *Order of Operations* Explain why addition is the first operation performed in the solution steps above.

81. *Order of Operations* Explain how each step in the solution above relates to PEMDAS.

82. *Equivalent Expressions* Explain why the expressions $10 \cdot 16 + 5$ and $(10 \cdot 16) + 5$ are equivalent.

Extra Practice

Using Order of Operations **In Exercises 83–86, evaluate the expression. If it is not possible, state the reason. Write fractional answers in simplest form.**

83. $\dfrac{3 \cdot 6 - 4 \cdot 6}{5 + 1}$

84. $\dfrac{5 \cdot 3 + 5 \cdot 6}{7 - 2}$

85. $7 - \dfrac{4 + 6}{2^2 + 1} + 5$

86. $11 - \dfrac{3^3 - 30}{8 + 1} + 1$

Using a Calculator **In Exercises 87–90, use a calculator to evaluate the expression. Round your answer to two decimal places.**

87. $300\left(1 + \dfrac{0.1}{12}\right)^{24}$

88. $1000 \div \left(1 + \dfrac{0.09}{4}\right)^8$

89. $\dfrac{1.32 + 4(3.68)}{1.5}$

90. $\dfrac{4.19 - 7(2.27)}{14.8}$

91. *Finding Inverses* Find (a) the additive inverse and (b) the multiplicative inverse of 50.

92. *Finding Inverses* Find (a) the additive inverse and (b) the multiplicative inverse of -8.

93. *Geometry* Write and evaluate an expression for the perimeter of the triangle.

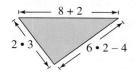

94. *Geometry* Find the area of the region.

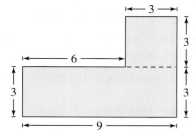

Explaining Concepts

95. *Writing* Are -6^2 and $(-6)^2$ equal? Explain.

96. *Writing* Are $2 \cdot 5^2$ and 10^2 equal? Explain.

Writing **In Exercises 97–106, explain why the statement is true.**

97. $4 \cdot 6^2 \neq 24^2$

98. $-3^2 \neq (-3)(-3)$

99. $4 - (6 - 2) \neq 4 - 6 - 2$

100. $\dfrac{8 - 6}{2} \neq 4 - 6$

101. $100 \div 2 \times 50 \neq 1$

102. $\frac{16}{2} \cdot 2 \neq 4$

103. $5(7 + 3) \neq 5(7) + 3$

104. $-7(5 - 2) \neq -7(5) - 7(2)$

105. $\frac{8}{0} \neq 0$

106. $5\left(\frac{1}{5}\right) \neq 0$

107. *Error Analysis* Describe and correct the error.

$$-9 + \frac{9 + 20}{3(5)} - (-3) = -9 + \frac{9}{3} + \frac{20}{5} - (-3)$$
$$= -9 + 3 + 4 - (-3)$$
$$= 1$$

108. *Error Analysis* Describe and correct the error.

$$7 - 3(8 + 1) - 15 = 4(8 + 1) - 15$$
$$= 4(9) - 15$$
$$= 36 - 15$$
$$= 21$$

109. *Matching* Match each expression in the first column with its value in the second column.

Expression	Value
$(6 + 2) \cdot (5 + 3)$	19
$(6 + 2) \cdot 5 + 3$	22
$6 + 2 \cdot 5 + 3$	64
$6 + 2 \cdot (5 + 3)$	43

110. *Determining Order of Operations* Using the established order of operations, which of the following expressions has a value of 72? For those that do not, decide whether you can insert parentheses into the expression so that its value is 72.

(a) $4 + 2^3 - 7$ (b) $4 + 8 \cdot 6$

(c) $93 - 25 - 4$ (d) $70 + 10 \div 5$

(e) $60 + 20 \div 2 + 32$ (f) $35 \cdot 2 + 2$

Geometry **In Exercises 111 and 112, find the area of the shaded rectangle in two ways. Explain how the results are related to the Distributive Property.**

111.

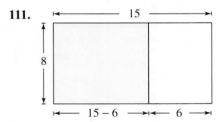

112.

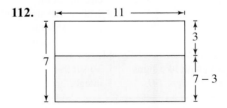

Think About It **In Exercises 113 and 114, determine whether the order in which the two activities are performed is "commutative." That is, do you obtain the same result regardless of which activity is performed first?**

113. (a) "Put on your socks."

(b) "Put on your shoes."

114. (a) "Weed the flower beds."

(b) "Mow the lawn."

1 Chapter Summary

What did you learn?	Explanation and Examples	Review Exercises
1.1 Classify numbers and plot them on a real number line *(p. 2)*.	The real numbers include rational numbers, irrational numbers, integers, whole numbers, and natural numbers. Each point on the real number line corresponds to exactly one real number. Origin −3 −2 −1 0 1 2 3 Negative numbers Positive numbers The Real Number Line	1−10
Use the real number line and inequality symbols to order real numbers *(p. 4)*.	a b −3 −2 −1 0 1 2 3 a is less than b or b is greater than a	11−16
Find the absolute value of a number *(p. 6)*.	The absolute value of a number is its distance from 0 on the real number line. Distance between −8 and 0 is 8. −10 −8 −6 −4 −2 0 2 $\lvert -8 \rvert = 8$	17−38
1.2 Add integers using a number line *(p. 10)*.	To find $a + b$, start at 0. Move a units along the number line. From that position, move b units. The final position is the sum. $-5 + 2 = -3$ 2 −5 Start at 0. −8−7−6−5−4−3−2−1 0 1 2 3 4 5 6 7 8	39−42
Add integers with like signs and with unlike signs *(p. 12)*.	To add two integers with like signs, add their absolute values and attach the common sign to the result. To add two integers with unlike signs, subtract the lesser absolute value from the greater absolute value and attach the sign of the integer with the greater absolute value.	43−54
Subtract integers with like signs and with unlike signs *(p. 14)*.	To subtract one integer from another, add the opposite of the integer being subtracted to the other integer.	55−66
1.3 Multiply integers with like signs and with unlike signs *(p. 18)*.	1. The product of an integer and zero is 0. 2. The product of two integers with like signs is positive. 3. The product of two integers with unlike signs is negative.	67−80
Divide integers with like signs and with unlike signs *(p. 20)*.	1. Zero divided by a nonzero integer is 0, whereas a nonzero integer divided by zero is undefined. 2. The quotient of two nonzero integers with like signs is positive. 3. The quotient of two nonzero integers with unlike signs is negative.	81−94
Find factors and prime factors of an integer *(p. 22)*.	If a and b are positive integers, then a is a factor of b if and only if a divides evenly into b. The prime factorization of b is its representation as the product of all its prime factors.	95−106

What did you learn?	Explanation and Examples	Review Exercises
1.3 Represent the definitions and rules of arithmetic symbolically *(p. 24).*	*Arithmetic* → *Algebra* Verbal rules and definitions Specific examples of rules and definitions → Symbolic rules and definitions	107–110
Rewrite fractions as equivalent fractions *(p. 30).*	To write a fraction in simplest form, divide both the numerator and denominator by the greatest common factor. You can obtain an equivalent fraction by multiplying the numerator and denominator by the same nonzero number.	111–124
Add and subtract fractions *(p. 32).*	**1. With like denominators:** $$\frac{a}{c} + \frac{b}{c} = \frac{a+b}{c} \quad \text{or} \quad \frac{a}{c} - \frac{b}{c} = \frac{a-b}{c}$$ **2. With unlike denominators:** Rewrite the fractions so that they have like denominators. Then use the rule for adding and subtracting fractions with like denominators.	125–140
1.4 Multiply and divide fractions *(p. 34).*	$\frac{a}{b} \cdot \frac{c}{d} = \frac{a \cdot c}{b \cdot d}$, where a, b, c, and d are integers with $b \neq 0$ and $d \neq 0$. $\frac{a}{b} \div \frac{c}{d} = \frac{a}{b} \cdot \frac{d}{c}$, where a, b, c, and d are integers with $b \neq 0$, $c \neq 0$, and $d \neq 0$.	141–156
Add, subtract, multiply, and divide decimals *(p. 36).*	**Add:** $\begin{array}{r} 0.43 \\ + 1.1 \\ \hline 1.53 \end{array}$ **Subtract:** $\begin{array}{r} 8.40 \\ -1.38 \\ \hline 7.02 \end{array}$ **Multiply:** $\begin{array}{r} 2.32 \\ \times 0.4 \\ \hline 0.928 \end{array}$ **Divide:** $12\overline{)28.8}$ gives 2.4	157–166
Write expressions in exponential form and evaluate exponential expressions *(p. 40).*	*Repeated Multiplication* *Exponential Form* $7 \cdot 7 \cdot 7 \cdot 7$ (4 factors of 7) 7^4	167–176
Evaluate expressions using the order of operations *(p. 42).*	1. Perform operations inside symbols of grouping: **P**arentheses, brackets, or absolute value symbols, starting with the innermost symbols. 2. Evaluate all **E**xponential expressions. 3. Perform all **M**ultiplications and **D**ivisions from left to right. 4. Perform all **A**dditions and **S**ubtractions from left to right.	177–202
1.5 Identify and use the properties of real numbers *(p. 43).*	Let a, b, and c be real numbers. *Commutative Property of Addition:* $a + b = b + a$ *Commutative Property of Multiplication:* $ab = ba$ *Associative Property of Addition:* $(a + b) + c = a + (b + c)$ *Associative Property of Multiplication:* $(ab)c = a(bc)$ *Distributive Property:* $a(b + c) = ab + ac \qquad (a + b)c = ac + bc$ *Additive Identity Property:* $a + 0 = 0 + a = a$ *Multiplicative Identity Property:* $a \cdot 1 = 1 \cdot a = a$ *Additive Inverse Property:* $a + (-a) = 0$ *Multiplicative Inverse Property:* $a \cdot \frac{1}{a} = 1, \, a \neq 0$	203–215

Review Exercises

Worked-out solutions to odd-numbered exercises at CollegePrepAlgebra.com

1.1

Classifying Real Numbers In Exercises 1−4, determine which of the numbers in the set are (a) natural numbers, (b) integers, (c) rational numbers, and (d) irrational numbers.

1. $\left\{-1, 4.5\frac{2}{5}, -\frac{1}{7}, \sqrt{4}, \sqrt{5}\right\}$

2. $\left\{10, -3, \frac{4}{5}, \pi, -3.16, -\frac{19}{11}\right\}$

3. $\left\{\frac{30}{2}, 2, -\sqrt{3}, 1.5, -\pi, -\frac{10}{7}\right\}$

4. $\left\{3.75, 33, \frac{2}{3}, \frac{1}{10}, -92, -\frac{\pi}{4}\right\}$

Plotting Real Numbers In Exercises 5−10, plot the numbers on the real number line.

5. $-3, 5$
6. $-8, 11$
7. $-6, \frac{5}{4}$
8. $-\frac{7}{2}, 9$
9. $-1, 0, \frac{1}{2}$
10. $-2, -\frac{1}{3}, 5$

Ordering Real Numbers In Exercises 11−16, plot each real number as a point on the real number line and place the correct inequality symbol (< or >) between the real numbers.

11. $-\frac{1}{10}$ ⬜ 4
12. $\frac{25}{3}$ ⬜ $\frac{5}{3}$
13. -3 ⬜ -7
14. 10.6 ⬜ -3.5
15. 5 ⬜ $\frac{7}{2}$
16. $\frac{3}{8}$ ⬜ $\frac{4}{9}$

Using Absolute Value In Exercises 17−20, find the distance between a and zero on the real number line.

17. $a = 152$
18. $a = -10.4$
19. $a = -\frac{7}{3}$
20. $a = \frac{2}{3}$

Evaluating an Absolute Value In Exercises 21−28, evaluate the expression.

21. $|-8.5|$
22. $|-9.6|$
23. $|3.4|$
24. $|5.98|$
25. $-|-6.2|$
26. $-\left|-\frac{7}{9}\right|$
27. $-\left|\frac{8}{5}\right|$
28. $-|4|$

Comparing Real Numbers In Exercises 29−34, place the correct symbol (<, >, or =) between the real numbers.

29. $|-84|$ ⬜ $|84|$
30. $|-10|$ ⬜ $|4|$
31. $\left|\frac{5}{2}\right|$ ⬜ $\left|\frac{8}{9}\right|$
32. $-|-1.8|$ ⬜ $|5.7|$
33. $\left|\frac{3}{10}\right|$ ⬜ $-\left|\frac{4}{5}\right|$
34. $|2.3|$ ⬜ $-|2.3|$

Distance on the Real Number Line In Exercises 35−38, find all real numbers whose distance from a is given by d.

35. $a = 5, d = 7$
36. $a = -1, d = 4$
37. $a = 2.6, d = 5$
38. $a = -3, d = 6.5$

1.2

Adding Integers Using a Number Line In Exercises 39−42, find the sum and demonstrate the addition on the real number line.

39. $4 + 3$
40. $15 + (-6)$
41. $-1 + (-4)$
42. $-6 + (-2)$

Adding Integers In Exercises 43−52, find the sum.

43. $16 + (-5)$
44. $25 + (-10)$
45. $-125 + 30$
46. $-54 + 12$
47. $-13 + (-76)$
48. $-24 + (-25)$
49. $-10 + 21 + (-6)$
50. $-23 + 4 + (-11)$
51. $-17 + (-3) + (-9)$
52. $-16 + (-2) + (-8)$

53. **Profit** A small software company had a profit of $95,000 in January, a loss of $64,400 in February, and a profit of $51,800 in March. What was the company's overall profit (or loss) for the three months?

54. *Account Balance* At the beginning of a month, your account balance was $3090. During the month, you withdrew $870 and $465, deposited $109, and earned $10.05 in interest. What was your balance at the end of the month?

Subtracting Integers In Exercises 55−64, **find the difference.**

55. $28 - 7$

56. $43 - 12$

57. $8 - 15$

58. $17 - 26$

59. $14 - (-19)$

60. $28 - (-4)$

61. $-18 - 4$

62. $-37 - 14$

63. $-12 - (-7) - 4$

64. $-26 - (-8) - (-10)$

65. *Account Balance* At the beginning of a month, your account balance was $1560. During the month, you withdrew $50, $255, and $490. What was your balance at the end of the month?

66. *Gasoline Prices* At the beginning of a month, gas cost $4.14 per gallon. During the month, the price increased by $0.05 and $0.02, decreased by $0.10, and then increased again by $0.07. How much did gas cost at the end of the month?

1.3

Multiplying Integers In Exercises 67−78, **find the product.**

67. $15 \cdot 3$

68. $21 \cdot 4$

69. $-3 \cdot 24$

70. $-2 \cdot 44$

71. $6(-8)$

72. $12(-5)$

73. $-5(-9)$

74. $-10(-81)$

75. $3(-6)(3)$

76. $15(-2)(7)$

77. $-4(-5)(-2)$

78. $-12(-2)(-6)$

79. *Savings Plan* You save $150 per month for 2 years. What is the total amount you have saved?

80. *Average Speed* A truck drives 65 miles per hour for 5 hours. How far has the truck traveled?

Dividing Integers In Exercises 81−92, **perform** the **division, if possible. If not possible, state the reason.**

81. $72 \div 8$

82. $63 \div 9$

83. $\dfrac{-72}{6}$

84. $\dfrac{-162}{9}$

85. $75 \div (-5)$

86. $48 \div (-4)$

87. $\dfrac{-52}{-4}$

88. $\dfrac{-64}{-4}$

89. $0 \div 815$

90. $0 \div 25$

91. $135 \div 0$

92. $26 \div 0$

93. *Average Speed* A commuter train travels 195 miles between two cities in 3 hours. What is the average speed of the train in miles per hour?

94. *Unit Price* At an auction, you buy a box of six glass canisters for a total of $78. All the canisters are of equal value. How much is each one worth?

Classifying an Integer In Exercises 95−100, decide **whether the number is prime or composite.**

95. 137

96. 296

97. 839

98. 909

99. 1764

100. 1847

Prime Factorization In Exercises 101−106, write the **prime factorization of the number.**

101. 264

102. 195

103. 378

104. 858

105. 1612

106. 1787

Using a Rule or Definition In Exercises 107−110, complete **the statement using the indicated definition or rule.**

107. Rule for multiplying integers with unlike signs:

$12 \times (-3) = $ ▢

108. Definition of multiplication:

$(-4) + (-4) + (-4) = $ ▢

109. Definition of absolute value:

$|-7| = $ ▢

110. Rule for adding integers with unlike signs:

$-9 + 5 = $ ▢

1.4

Finding the GCF In Exercises 111−116, **find the greatest common factor.**

111. 54, 90

112. 154, 220

113. 2, 6, 9

114. 8, 12, 24

115. 63, 84, 441

116. 99, 132, 253

Writing a Fraction in Simplest Form In Exercises 117−120, write the fraction in simplest form.

117. $\frac{3}{12}$

118. $\frac{15}{25}$

119. $\frac{30}{48}$

120. $\frac{126}{162}$

Writing an Equivalent Fraction In Exercises 121−124, write an equivalent fraction with the indicated denominator.

121. $\frac{2}{3} = \frac{\boxed{}}{15}$

122. $\frac{3}{7} = \frac{\boxed{}}{28}$

123. $\frac{6}{10} = \frac{\boxed{}}{25}$

124. $\frac{9}{12} = \frac{\boxed{}}{16}$

Adding and Subtracting Fractions In Exercises 125−138, find the sum or difference. Write the result in simplest form.

125. $\frac{3}{25} + \frac{7}{25}$

126. $\frac{9}{64} + \frac{7}{64}$

127. $\frac{27}{16} - \frac{15}{16}$

128. $-\frac{5}{12} + \frac{1}{12}$

129. $\frac{3}{8} + \frac{1}{2}$

130. $\frac{7}{12} + \frac{5}{18}$

131. $-\frac{5}{9} + \frac{2}{3}$

132. $\frac{7}{15} - \frac{2}{25}$

133. $-\frac{25}{32} + \left(-\frac{7}{24}\right)$

134. $-\frac{7}{8} - \frac{11}{12}$

135. $5 - \frac{15}{4}$

136. $\frac{12}{5} - 3$

137. $5\frac{3}{4} - 3\frac{5}{8}$

138. $-3\frac{7}{10} + 1\frac{1}{20}$

139. **Meteorology** The table shows the daily amounts of rainfall (in inches) during a five-day period. What was the total amount of rainfall for the five days?

Day	Mon	Tue	Wed	Thu	Fri
Rainfall (in inches)	$\frac{3}{8}$	$\frac{1}{2}$	$\frac{1}{8}$	$1\frac{1}{4}$	$\frac{1}{2}$

140. **Fuel Consumption** The morning and evening readings of the fuel gauge on a car were 7/8 and 1/3, respectively. What fraction of the tank of fuel was used that day?

Multiplying and Dividing Fractions In Exercises 141−154, evaluate the expression and write the result in simplest form. If it is not possible, explain why.

141. $\frac{5}{8} \cdot \frac{-2}{15}$

142. $\frac{3}{32} \cdot \frac{32}{3}$

143. $35\left(\frac{1}{35}\right)$

144. $-6\left(\frac{5}{36}\right)$

145. $\frac{3}{8}\left(-\frac{2}{27}\right)$

146. $-\frac{5}{12}\left(-\frac{4}{25}\right)$

147. $\frac{5}{14} \div \frac{15}{28}$

148. $-\frac{7}{10} \div \frac{4}{15}$

149. $-\frac{3}{4} \div \left(-\frac{7}{8}\right)$

150. $\frac{15}{32} \div \left(-\frac{5}{4}\right)$

151. $-\frac{5}{9} \div 0$

152. $0 \div \frac{1}{12}$

153. $-5 \cdot 0$

154. $0 \cdot \frac{1}{2}$

155. **Meteorology** During an eight-hour period, $6\frac{3}{4}$ inches of snow fell. What was the average rate of snowfall per hour?

156. **Sports** In three strokes on a golf course, you hit your ball a total distance of $64\frac{7}{8}$ meters. What is your average distance per stroke?

Operations with Decimals In Exercises 157−164, evaluate the expression. Round your answer to two decimal places.

157. $4.89 + 0.76$

158. $1.29 + 0.44$

159. $3.815 - 5.19$

160. $7.234 - 8.16$

161. $1.49(-0.5)$

162. $2.34(-1.2)$

163. $5.25 \div 0.25$

164. $10.18 \div 1.6$

165. **Consumer Awareness** An engagement ring is advertised for $299.99 plus $26.99 per month for 24 months. Find the total cost of the engagement ring.

166. **Consumer Awareness** A plasma television costs $599.99 plus $32.96 per month for 18 months. Find the total cost of the television.

1.5

Writing an Expression in Exponential Form In Exercises 167−170, rewrite the expression in exponential form.

167. $6 \cdot 6 \cdot 6 \cdot 6 \cdot 6$

168. $(-3) \cdot (-3) \cdot (-3)$

169. $\left(\frac{6}{7}\right) \cdot \left(\frac{6}{7}\right) \cdot \left(\frac{6}{7}\right) \cdot \left(\frac{6}{7}\right)$

170. $-[(3.3) \cdot (3.3)]$

Evaluating an Exponential Expression In Exercises 171−176, evaluate the expression.

171. 2^4

172. $(-6)^2$

173. $\left(-\frac{3}{4}\right)^3$

174. $\left(\frac{2}{3}\right)^2$

175. -7^2

176. $-(-3)^3$

Using Order of Operations In Exercises 177−196, evaluate the expression. Write fractional answers in simplest form.

177. $12 - 2 \cdot 3$

178. $1 + 7 \cdot 3 - 10$

179. $18 \div 6 \cdot 7$

180. $3^2 \cdot 4 \div 2$

181. $20 + (8^2 \div 2)$

182. $(8 - 3) \div 15$

183. $240 - (4^2 \cdot 5)$

184. $5^2 - (625 \cdot 5^2)$

185. $3^2(5 - 2)^2$

186. $-5(10 - 7)^3$

187. $\frac{3}{4}\left(\frac{5}{6}\right) + 4$

188. $75 - 24 \div 2^3$

189. $122 - [45 - (32 + 8) - 23]$

190. $-58 - (48 - 12) - (-30 - 4)$

191. $\dfrac{6 \cdot 4 - 36}{4}$

192. $\dfrac{144}{2 \cdot 3 \cdot 3}$

193. $\dfrac{54 - 4 \cdot 3}{6}$

194. $\dfrac{3 \cdot 5 + 125}{10}$

195. $\dfrac{78 - |-78|}{5}$

196. $\dfrac{300}{15 - |-15|}$

⊞ *Using a Calculator* In Exercises 197−200, use a calculator to evaluate the expression. Round your answer to two decimal places.

197. $(5.8)^4 - (3.2)^5$

198. $\dfrac{(15.8)^3}{(2.3)^8}$

199. $\dfrac{3000}{(1.05)^{10}}$

200. $500\left(1 + \dfrac{0.07}{4}\right)^{40}$

201. *Depreciation* After 3 years, the value of a $25,000 car is given by $25,000\left(\frac{3}{4}\right)^3$.

    (a) What is the value of the car after 3 years?

    (b) How much has the car depreciated during the 3 years?

202. *Geometry* The volume of water in a hot tub is given by $V = 6^2 \cdot 3$ (see figure). How many cubic feet of water will the hot tub hold? Find the total weight of the water in the tub. (Use the fact that 1 cubic foot of water weighs 62.4 pounds.)

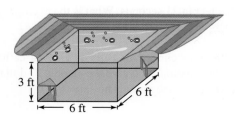

3 ft

6 ft

6 ft

Identifying a Property of Real Numbers In Exercises 203−210, identify the property of real numbers illustrated by the statement.

203. $123 - 123 = 0$

204. $9 \cdot \frac{1}{9} = 1$

205. $14(3) = 3(14)$

206. $5(3 \cdot 8) = (5 \cdot 3)8$

207. $17 \cdot 1 = 17$

208. $10 + 6 = 6 + 10$

209. $-2(7 + 12) = (-2)7 + (-2)12$

210. $2 + (3 + 19) = (2 + 3) + 19$

Using a Property of Real Numbers In Exercises 211−214, complete the statement using the specified property of real numbers.

211. Additive Identity Property:

    $-16 + 0 =$ ▢

212. Distributive Property:

    $8(7 + 2) =$ ▢

213. Commutative Property of Addition:

    $24 + 1 =$ ▢

214. Associative Property of Multiplication:

    $8(5 \cdot 7) =$ ▢

215. *Geometry* Find the area of the shaded rectangle in two ways. Explain how the results are related to the Distributive Property.

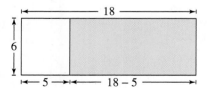

18

6

5

18 − 5

Chapter Test

Solutions in English & Spanish and tutorial videos at CollegePrepAlgebra.com

Take this test as you would take a test in class. After you are done, check your work against the answers in the back of the book.

1. Which of the following are (a) natural numbers, (b) integers, (c) rational numbers, and (d) irrational numbers?

   $$\left\{4, -6, \tfrac{1}{2}, 0, \pi, \tfrac{7}{9}\right\}$$

2. Place the correct inequality symbol ($<$ or $>$) between the real numbers.

   $$-\frac{3}{5} \quad \boxed{} \quad -\lvert -2 \rvert$$

In Exercises 3−20, evaluate the expression. Write fractional answers in simplest form.

3. $\lvert -13 \rvert$

4. $-\lvert -6.8 \rvert$

5. $16 + (-20)$

6. $-50 - (-60)$

7. $7 + \lvert -3 \rvert$

8. $64 - (25 - 8)$

9. $-5(32)$

10. $\dfrac{-72}{-9}$

11. $\dfrac{15(-6)}{3}$

12. $-\dfrac{(-2)(5)}{10}$

13. $\dfrac{5}{6} - \dfrac{1}{8}$

14. $-\dfrac{9}{50}\left(-\dfrac{20}{27}\right)$

15. $\dfrac{7}{16} \div \dfrac{21}{28}$

16. $\dfrac{-8.1}{0.3}$

17. $-(0.8)^2$

18. $35 - (50 \div 5^2)$

19. $5(3 + 4)^2 - 10$

20. $18 - 7 \cdot 4 + 2^3$

In Exercises 21−24, identify the property of real numbers illustrated by the statement.

21. $3(4 + 6) = 3 \cdot 4 + 3 \cdot 6$

22. $5 \cdot \tfrac{1}{5} = 1$

23. $3 + (4 + 8) = (3 + 4) + 8$

24. $3(7 + 2) = (7 + 2)3$

25. Write the fraction $\frac{36}{162}$ in simplest form.

26. Write the prime factorization of 216.

27. An electric train travels 1218 feet in 21 seconds. What is the average speed of the train in feet per second?

28. At the grocery store, you buy five cartons of eggs at $1.49 a carton and two gallons of orange juice at $3.06 a gallon. You give the clerk a 20-dollar bill. How much change will you receive? (Assume there is no sales tax.)

2 Fundamentals of Algebra

2.1 Writing and Evaluating Algebraic Expressions

2.2 Simplifying Algebraic Expressions

2.3 Algebra and Problem Solving

2.4 Introduction to Equations

MASTERY IS WITHIN REACH!

"When I am in math class, I struggle with keeping all my notes neat. After class, I rewrite them so that I can review what we have covered and so I have a neat set for future reference. A friend of mine does this too."

Abbey

See page 75 for suggestions about reworking your notes.

2.1 Writing and Evaluating Algebraic Expressions

▶ Define and identify terms, variables, and coefficients of algebraic expressions.
▶ Define exponential form and interpret exponential expressions.
▶ Evaluate algebraic expressions using real numbers.

Study Tip

The terms of an algebraic expression depend on the way the expression is written. Rewriting the expression can (and, in fact, usually does) change its terms. For instance, the expression $4 + x + 2$ has three terms, but the equivalent expression $6 + x$ has only two terms.

Variables and Algebraic Expressions

A collection of letters (**variables**) and real numbers (**constants**) combined by using addition, subtraction, multiplication, or division is an **algebraic expression**.
The **terms** of an algebraic expression are those parts that are separated by *addition*. For example, the expression $x^2 - 4x + 5$ has three terms: x^2, $-4x$, and 5. Note that $-4x$, rather than $4x$, is a term of $x^2 - 4x + 5$ because

$$x^2 - 4x + 5 = x^2 + (-4x) + 5. \qquad \text{To subtract, add the opposite.}$$

For terms such as x^2, $-4x$, and 5, the numerical factor is called the **coefficient** of the term. Here, the coefficients are 1, -4, and 5.

EXAMPLE 1 **Identifying the Terms of Expressions**

Algebraic Expression	Terms
a. $x + 2$	$x, 2$
b. $3x + \dfrac{1}{2}$	$3x, \dfrac{1}{2}$
c. $2y - 5x - 7$	$2y, -5x, -7$

EXAMPLE 2 **Identifying the Coefficients of Terms**

Term	Coefficient	Comment
a. $-5x^2$	-5	Note that $-5x^2 = (-5)x^2$.
b. x^3	1	Note that $x^3 = 1 \cdot x^3$.
c. $-\dfrac{2x}{3}$	$-\dfrac{2}{3}$	Note that $-\dfrac{2x}{3} = -\dfrac{2}{3}(x)$.

Exercises Within Reach ®

Solutions in English & Spanish and tutorial videos at CollegePrepAlgebra.com

Identifying Variables In Exercises 1−4, identify the variable(s) in the expression.

1. $x + 3$ **2.** $y - 1$ **3.** $m + n$ **4.** $a + b$

Identifying Terms In Exercises 5−10, identify the terms of the expression.

5. $4x + 3$ **6.** $5 - 3t^2$ **7.** $\dfrac{5}{3} - 3y^3$

8. $6x + \dfrac{2}{3}$ **9.** $a^2 + 4ab + b^2$ **10.** $x^2 + 18xy + y^2$

Identifying the Coefficient In Exercises 11−18, identify the coefficient of the term.

11. $14x$ **12.** $25y$ **13.** $-\dfrac{1}{3}y$ **14.** $\dfrac{2}{3}n$

15. $\dfrac{2x}{5}$ **16.** $-\dfrac{3x}{4}$ **17.** $2\pi x^2$ **18.** πt^4

Exponential Form

In general, for any positive integer n and any real number a, you have

$$a^n = \underbrace{a \cdot a \cdot a \cdots a}_{n \text{ factors}}.$$

This rule applies to factors that are variables as well as factors that are *algebraic expressions*.

Study Tip

Be sure you understand the difference between repeated addition

$$\underbrace{x + x + x + x}_{4 \text{ terms}} = 4x$$

and repeated multiplication

$$\underbrace{x \cdot x \cdot x \cdot x}_{4 \text{ factors}} = x^4.$$

EXAMPLE 3   **Interpreting Exponential Expressions**

a. $3^4 = 3 \cdot 3 \cdot 3 \cdot 3$

b. $3x^4 = 3 \cdot x \cdot x \cdot x \cdot x$

c. $(-3x)^4 = (-3x)(-3x)(-3x)(-3x)$
$$= (-3)(-3)(-3)(-3) \cdot x \cdot x \cdot x \cdot x$$

d. $(y + 2)^3 = (y + 2)(y + 2)(y + 2)$

e. $(5x)^2 y^3 = (5x)(5x) \cdot y \cdot y \cdot y$
$$= 5 \cdot 5 \cdot x \cdot x \cdot y \cdot y \cdot y$$

Be sure you understand the priorities for order of operations involving exponents. Here are some examples that tend to cause problems.

Expression	Correct Evaluation	Incorrect Evaluation
-3^2	$-(3 \cdot 3) = -9$	$\cancel{(-3)(-3) = 9}$
$(-3)^2$	$(-3)(-3) = 9$	$\cancel{-(3 \cdot 3) = -9}$
$3x^2$	$3 \cdot x \cdot x$	$\cancel{(3x)(3x)}$
$-3x^2$	$-3 \cdot x \cdot x$	$\cancel{-(3x)(3x)}$
$(-3x)^2$	$(-3x)(-3x)$	$\cancel{-(3x)(3x)}$

Exercises Within Reach®

Solutions in English & Spanish and tutorial videos at CollegePrepAlgebra.com

Interpreting an Exponential Expression **In Exercises 19−36, expand the expression as a product of factors.**

19. y^5

20. $(-x)^6$

21. $2^2 x^4$

22. $(-5)^3 x^2$

23. $4y^2 z^3$

24. $3uv^4$

25. $\left(a^2\right)^3$

26. $\left(z^3\right)^3$

27. $-4x^3 \cdot x^4$

28. $a^2 y^2 \cdot y^3$

29. $-9(ab)^3$

30. $2(xz)^4$

31. $(x + y)^2$

32. $(s - t)^5$

33. $\left(\dfrac{a}{3s}\right)^4$

34. $\left(-\dfrac{2}{5x}\right)^3$

35. $[2(a - b)^3][2(a - b)^2]$

36. $[3(r + s)^2][3(r + s)]^2$

Evaluating Algebraic Expressions

In applications of algebra, you are often required to **evaluate** an algebraic expression. This means you are to find the value of an expression when its variables are replaced by real numbers.

EXAMPLE 4   **Evaluating Algebraic Expressions**

Evaluate each expression when $x = -3$ and $y = 5$.

a. $-x$ **b.** $x - y$ **c.** $3x + 2y$

d. $y - 2(x + y)$ **e.** $y^2 - 3y$

SOLUTION

a. When $x = -3$, the value of $-x$ is

$$-x = -(-3)$$ Substitute -3 for x.

$$= 3.$$ Simplify.

b. When $x = -3$ and $y = 5$, the value of $x - y$ is

$$x - y = (-3) - 5$$ Substitute -3 for x and 5 for y.

$$= -8.$$ Simplify.

c. When $x = -3$ and $y = 5$, the value of $3x + 2y$ is

$$3x + 2y = 3(-3) + 2(5)$$ Substitute -3 for x and 5 for y.

$$= -9 + 10$$ Multiply.

$$= 1.$$ Add.

d. When $x = -3$ and $y = 5$, the value of $y - 2(x + y)$ is

$$y - 2(x + y) = 5 - 2[(-3) + 5]$$ Substitute -3 for x and 5 for y.

$$= 5 - 2(2)$$ Add.

$$= 1.$$ Simplify.

e. When $y = 5$, the value of $y^2 - 3y$ is

$$y^2 - 3y = (5)^2 - 3(5)$$ Substitute 5 for y.

$$= 25 - 15$$ Simplify.

$$= 10.$$ Subtract.

> **Study Tip**
>
> As shown in parts (a), (b), (c), and (d) of Example 4, it is a good idea to use parentheses when substituting a negative number for a variable.

Exercises Within Reach®

Solutions in English & Spanish and tutorial videos at CollegePrepAlgebra.com

Evaluating an Algebraic Expression In Exercises 37–44, evaluate the algebraic expression for the given values of the variable(s).

	Expression	*Values*			*Expression*	*Values*	
37.	$2x - 1$	(a) $x = \frac{1}{2}$	(b) $x = -4$	**38.**	$3x - 2$	(a) $x = \frac{4}{3}$	(b) $x = -1$
39.	$2x^2 - 5$	(a) $x = 2$	(b) $x = 3$	**40.**	$64 - 16t^2$	(a) $t = 2$	(b) $t = 3$
41.	$3x - 2y$	(a) $x = 4, y = 3$		**42.**	$10u - 3v$	(a) $u = 3, v = 10$	
		(b) $x = \frac{2}{3}, y = -1$				(b) $u = -2, v = \frac{4}{7}$	
43.	$x - 3(x - y)$	(a) $x = 3, y = 3$		**44.**	$-3x + 2(x + y)$	(a) $x = -2, y = 2$	
		(b) $x = 4, y = -4$				(b) $x = 0, y = 5$	

EXAMPLE 5   **Evaluating Algebraic Expressions**

a. When $y = -6$, the value of y^2 is

$$y^2 = (-6)^2 = 36.$$

b. When $y = -6$, the value of $-y^2$ is

$$-y^2 = -(y^2) = -(-6)^2 = -36.$$

c. When $x = 4$ and $y = -6$, the value of $y - x$ is

$$y - x = (-6) - 4 = -10.$$

d. When $x = 4$ and $y = -6$, the value of $|y - x|$ is

$$|y - x| = |(-6) - 4| = |-10| = 10.$$

e. When $x = 4$ and $y = -6$, the value of $|x - y|$ is

$$|x - y| = |4 - (-6)| = |4 + 6| = |10| = 10.$$

EXAMPLE 6   **Evaluating an Algebraic Expression**

When $x = -5$, $y = -2$, and $z = 3$, the value of $\dfrac{y + 2z}{5y - xz}$ is

$$\frac{y + 2z}{5y - xz} = \frac{-2 + 2(3)}{5(-2) - (-5)(3)} \qquad \text{Substitute for } x, y, \text{ and } z.$$

$$= \frac{-2 + 6}{-10 - (-15)} \qquad \text{Multiply.}$$

$$= \frac{-2 + 6}{-10 + 15} \qquad \text{Add the opposite of } -15.$$

$$= \frac{4}{5}. \qquad \text{Simplify.}$$

Exercises Within Reach ®

Evaluating an Algebraic Expression In Exercises 45−54, evaluate the algebraic expression for the given values of the variables. If it is not possible, state the reason.

Expression	*Values*	*Expression*	*Values*
45. $b^2 - 4ab$	(a) $a = 2, b = -3$	**46.** $a^2 + 2ab$	(a) $a = -2, b = 3$
	(b) $a = 6, b = -4$		(b) $a = 4, b = -2$
47. $\lvert 2x - 3y \rvert$	(a) $x = 2, y = 3$	**48.** $y - \lvert -3x + y \rvert$	(a) $x = -2, y = -1$
	(b) $x = -1, y = 4$		(b) $x = 7, y = 3$
49. $\dfrac{x - 2y}{x + 2y}$	(a) $x = 4, y = 2$	**50.** $\dfrac{5x}{y - 3}$	(a) $x = 2, y = 4$
	(b) $x = 4, y = -2$		(b) $x = 2, y = 3$
51. $\dfrac{-y}{x^2 + y^2}$	(a) $x = 0, y = 5$	**52.** $\dfrac{2x - y}{y^2 + 1}$	(a) $x = 1, y = 2$
	(b) $x = 1, y = -3$		(b) $x = 1, y = 3$
53. $(x + 2y)(-3x - z)$	(a) $x = 2, y = -1, z = -1$	**54.** $\dfrac{yz - 3}{x + 2z}$	(a) $x = 0, y = -7, z = 3$
	(b) $x = -3, y = 2, z = -2$		(b) $x = -2, y = -3, z = 3$

Application EXAMPLE 7 **Converting Temperatures**

There is one temperature that has the same degree measure in the Fahrenheit and Celsius scales. Use the formula $F = \frac{9}{5}C + 32$ to find this temperature.

SOLUTION

A spreadsheet is useful to answer this type of question. Enter several Celsius temperatures into Column A. Then program Column B to calculate corresponding Fahrenheit temperatures.

	A	B	
1	**Celsius**	**Fahrenheit**	
2	100	212	(9/5)*A2+32
3	90	194	
4	80	176	
5	70	158	
6	60	140	
7	50	122	
8	40	104	
9	30	86	
10	20	68	
11	10	50	
12	0	32	
13	−10	14	
14	−20	−4	
15	−30	−22	
16	−40	−40	
17	−50	−58	
18			

Spreadsheet at CollegePrepAlgebra.com

Inuvik is a town in the Northwest Territories of Canada. The population is about 3500. The average low temperature in January is about −32°C.

From the result, you can see that the temperature −40 degrees is the same on both scales.

Exercises Within Reach®

Solutions in English & Spanish and tutorial videos at CollegePrepAlgebra.com

55. *Savings* The interest a savings account earns is given by $I = 850(0.095)t$, where I is the interest the account earns after t years. Use a spreadsheet to determine the interest the account earns after 8 years.

	A	B	
1	**Years**	**Interest**	
2	1		
3	2		
4	3		
5	4		
6	5		
7	6		
8	7		
9	8		
10			

Spreadsheet at CollegePrepAlgebra.com

56. *Traveling* The distance a car travels is given by $d = 63t$, where d is the distance (in miles) the car travels after t hours. Use a spreadsheet to determine the distance the car travels after 7 hours.

	A	B	
1	**Hours**	**Distance**	
2	1		
3	2		
4	3		
5	4		
6	5		
7	6		
8	7		
9			

Spreadsheet at CollegePrepAlgebra.com

Application   EXAMPLE 8   **Writing an Algebraic Expression**

You accept a part-time job for $9 per hour. The job offer states that you will be expected to work between 15 and 30 hours a week. Because you do not know how many hours you will work during a week, your total income for a week is unknown. Moreover, your income will probably *vary* from week to week. By representing the variable quantity (the number of hours worked) by the letter x, you can represent the weekly income by the following algebraic expression.

$9 per hour   Number of hours worked

$9x$

In the product $9x$, the number 9 is a *constant* and the letter x is a *variable*.

Application   EXAMPLE 9   Geometry: **Writing an Algebraic Expression**

Write an expression for the area of the rectangle shown at the left. Then evaluate the expression to find the area of the rectangle when $x = 7$.

SOLUTION

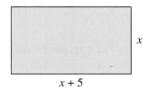

$x + 5$

Area of a rectangle = Length • Width

$$= (x + 5) \cdot x \qquad \text{Substitute.}$$

To find the area of the rectangle when $x = 7$, substitute 7 for x in the expression for the area.

$$(x + 5) \cdot x = (7 + 5) \cdot 7 \qquad \text{Substitute 7 for } x.$$
$$= 12 \cdot 7 \qquad \text{Add.}$$
$$= 84 \qquad \text{Multiply.}$$

So, the area of the rectangle is 84 square units.

Exercises Within Reach ®

Solutions in English & Spanish and tutorial videos at CollegePrepAlgebra.com

Writing an Algebraic Expression In Exercises 57–60, write an algebraic expression for the statement.

57. The income earned at $7.55 per hour for w hours

58. The cost for a family of n people to see a movie when the cost per person is $8.25

59. The cost of m pounds of meat when the cost per pound is $3.79

60. The total weight of x bags of fertilizer when each bag weighs 50 pounds

Geometry In Exercises 61–64, write an expression for the area of the figure. Then evaluate the expression for the given value(s) of the variable(s).

61. $n = 8$ 62. $x = 10, y = 3$ 63. $a = 5, b = 4$ 64. $x = 9$

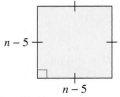

$n - 5$

$n - 5$

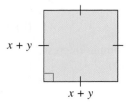

$x + y$

$x + y$

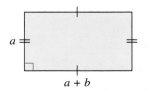

a

$a + b$

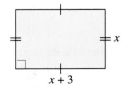

x

$x + 3$

Concept Summary: *Writing and Evaluating Algebraic Expressions*

What

When solving a word problem, the goal is to translate the words and phrases into an **algebraic expression**. Then **evaluate** the expression.

EXAMPLE

You can earn $12 every hour you work. Write an expression that represents the amount of money you earn working x hours. How much money do you earn after working 8 hours?

How

When translating a word problem into an algebraic expression, represent unknown quantities with **variables** and known quantities with **constants**. Then to evaluate the expression, replace the variables with real numbers, and simplify.

EXAMPLE

Write: $12x$

Evaluate: $12x = 12(8) = 96$

Why

When given a real-life problem with unknown quantities, you can use algebra to write an expression to represent the situation.

Then you can find the value of the expression for different values of the variable. For example, you can use the expression $12x$ to find the amount of money you earn for any number of hours you work.

Exercises Within Reach ®

Worked-out solutions to odd-numbered exercises at CollegePrepAlgebra.com

Concept Summary Check

65. *Interpreting Expressions* Are the expressions -3^2 and $(-3)^2$ equivalent? Explain.

66. *Writing Algebraic Expressions* Do you always have to use x to represent an unknown value when writing an algebraic expression? Give an example of when you may want to use another letter.

67. *Writing Algebraic Expressions* Name four mathematical operations you can use to write an algebraic expression.

68. *Number Sense* What value of y would cause $3y + 2$ to equal 8? Explain.

Extra Practice

Identifying Variables **In Exercises 69 and 70,** identify the variable(s) in the expression.

69. $2^3 - k$

70. $3^2 + z$

Identifying Terms **In Exercises 71 and 72,** identify the terms of the expression.

71. $3(x + 5) + 10$

72. $\dfrac{6}{t} - 22$

Rewriting in Exponential Form **In Exercises 73−76,** rewrite the product in exponential form.

73. $-2 \cdot u \cdot u \cdot u \cdot u$

74. $\frac{1}{3} \cdot x \cdot x \cdot x \cdot x \cdot x$

75. $-3 \cdot (x - y) \cdot (x - y) \cdot (-3) \cdot (-3)$

76. $(u - v) \cdot (u - v) \cdot 8 \cdot 8 \cdot 8 \cdot (u - v)$

Evaluating an Algebraic Expression **In Exercises 77−80,** evaluate the algebraic expression for the given values of the variable(s).

77. Area of a Triangle
$\frac{1}{2}bh$
    (a) $b = 3, h = 5$
    (b) $b = 2, h = 10$

78. Distance Traveled
rt
    (a) $r = 50, t = 3.5$
    (b) $r = 35, t = 4$

79. Volume of a Rectangular Prism
lwh
    (a) $l = 4, w = 2, h = 9$
    (b) $l = 100, w = 0.8, h = 4$

80. Simple Interest
Prt
    (a) $P = 1000, r = 0.08, t = 3$
    (b) $P = 500, r = 0.07, t = 5$

81. *Advertising* An advertisement for a new pair of basketball shoes claims that the shoes will help you jump 6 inches higher than without shoes.

 (a) Let x represent the height (in inches) jumped without shoes. Write an expression that represents the height of a jump while wearing the new shoes.

 (b) You can jump 23 inches without shoes. How high can you jump while wearing the new shoes?

 (c) Your friend can jump 20.5 inches without shoes. How high can she jump while wearing the new shoes?

82. *Distance* You are driving 60 miles per hour on the highway.

 (a) Write an expression that represents the distance you travel in t hours.

 (b) How far will you travel in 2.75 hours?

83. *Exploration* For any natural number n, the sum of the numbers 1, 2, 3, . . . , n is equal to

$$\frac{n(n + 1)}{2}, \quad n \geq 1.$$

Verify the formula for (a) $n = 3$, (b) $n = 6$, and (c) $n = 10$.

84. *Exploration* A convex polygon with n sides has

$$\frac{n(n - 3)}{2}, \quad n \geq 4$$

diagonals. Verify the formula for (a) a square (two diagonals), (b) a pentagon (five diagonals), and (c) a hexagon (nine diagonals).

Explaining Concepts

85. *Identifying Terms* Is $3x$ a term of $4 - 3x$? Explain.

86. *Number Sense* Is it possible to evaluate the expression

$$\frac{x + 2}{y - 3}$$

when $x = 5$ and $y = 3$? Explain.

87. *Logic* Explain why the formulas in Exercises 83 and 84 will always yield natural numbers.

88. *Error Analysis* Describe and correct the error in evaluating $y - 2(x - y)$ for $x = 2$ and $y = -4$.

$$y - 2(x - y) = -4 - 2(2 - 4)$$
$$= -4 - 2(-2)$$
$$= -4 + 4$$
$$= 0$$

Cumulative Review

In Exercises 89−96, evaluate the expression.

89. $10 - (-7)$

90. $6 - 10 - (-12) + 3$

91. $-5 + 10 - (-9) - 4$

92. $-(-8) + 6 - 4 - 2$

93. $(-6)(-4)$

94. $\dfrac{-56}{7}$

95. $\dfrac{-144}{-12}$

96. $5(-7)$

In Exercises 97−100, identify the property of real numbers illustrated by the statement.

97. $3(4) = 4(3)$

98. $10 - 10 = 0$

99. $3(6 + 2) = 3 \cdot 6 + 3 \cdot 2$

100. $7 + (8 + 5) = (7 + 8) + 5$

2.2 Simplifying Algebraic Expressions

▶ Use the properties of algebra.

▶ Combine like terms of an algebraic expression.

▶ Simplify an algebraic expression by rewriting the terms.

▶ Use the Distributive Property to remove symbols of grouping.

Properties of Algebra

You can rewrite and simplify algebraic expressions using the properties of real numbers listed on page 43.

EXAMPLE 1 Applying Properties of Real Numbers

Use the indicated rule to complete each statement.

a. Additive Identity Property: $(x - 2) + \boxed{} = x - 2$

b. Commutative Property of Multiplication: $5(y + 6) = \boxed{}$

c. Associative Property of Addition: $(x^2 + 3) + 7 = \boxed{}$

d. Additive Inverse Property: $\boxed{} + 4x = 0$

SOLUTION

a. $(x - 2) + 0 = x - 2$

b. $5(y + 6) = (y + 6)5$

c. $(x^2 + 3) + 7 = x^2 + (3 + 7)$

d. $-4x + 4x = 0$

EXAMPLE 2 Using the Distributive Property

Use the Distributive Property to expand each expression.

a. $2(7 - x)$ b. $(10 - 2y)3$ c. $2x(x + 4y)$ d. $-(1 - 2y)$

SOLUTION

a. $2(7 - x) = 2 \cdot 7 - 2 \cdot x$
$= 14 - 2x$

b. $(10 - 2y)3 = 10(3) - 2y(3)$
$= 30 - 6y$

c. $2x(x + 4y) = 2x(x) + 2x(4y)$
$= 2x^2 + 8xy$

d. $-(1 - 2y) = (-1)(1 - 2y)$
$= (-1)(1) - (-1)(2y)$
$= -1 + 2y$

Exercises Within Reach ®

Solutions in English & Spanish and tutorial videos at CollegePrepAlgebra.com

Applying Properties of Real Numbers In Exercises 1−4, complete the statement. Then state the property of algebra that you used.

1. $v \cdot 2 = \boxed{}$

2. $(2x - y)(-3) = -3\boxed{}$

3. $5(t - 2) = 5\left(\boxed{}\right) + 5\left(\boxed{}\right)$

4. $x(y + 4) = x\left(\boxed{}\right) + x\left(\boxed{}\right)$

Using the Distributive Property In Exercises 5−12, use the Distributive Property to expand the expression.

5. $2(16 + 8z)$

6. $3(7 - 4a)$

7. $-5(2x - y)$

8. $-3(11y - 6)$

9. $(x + 1)8$

10. $(r + 10)2$

11. $-6s(6s - 1)$

12. $-(u - v)$

Application   **EXAMPLE 3**   Geometry: **Visualizing the Distributive Property**

Write the area of each component of each figure. Then demonstrate the Distributive Property by writing the total area of each figure in two ways.

a.

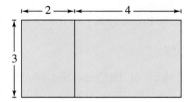

b.

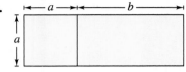

SOLUTION

a.

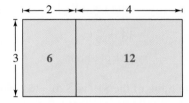

The total area is $3(2 + 4) = 3 \cdot 2 + 3 \cdot 4$
$$= 6 + 12$$
$$= 18.$$

b.

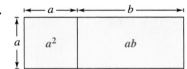

The total area is $a(a + b) = a \cdot a + a \cdot b$
$$= a^2 + ab.$$

Exercises Within Reach®   Solutions in English & Spanish and tutorial videos at CollegePrepAlgebra.com

Visualizing the Distributive Property In Exercises 13–16, write the area of each component of the figure. Then demonstrate the Distributive Property by writing the total area of each figure in two ways.

13.

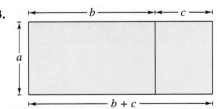

14.

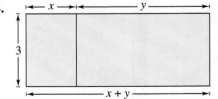

15.

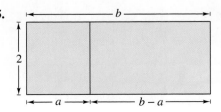

16.

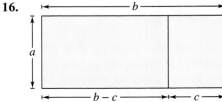

Application **EXAMPLE 4** **Using Mental Math in Everyday Applications**

a. You earn $14 per hour and time-and-a-half for overtime. Show how you can use the Distributive Property to find your overtime wage mentally.

b. You are buying 15 potted plants that cost $19 each. Show how you can use the Distributive Property to find the total cost mentally.

SOLUTION

a. You can think of the "time-and-a-half" as $1\frac{1}{2}$. So, using the Distributive Property, you can think of the following *without* writing anything on paper.

$$14\left(1\frac{1}{2}\right) = 14\left(1 + \frac{1}{2}\right)$$
$$= 14(1) + 14\left(\frac{1}{2}\right)$$
$$= 14 + 7$$
$$= \$21 \text{ per hour}$$

b. You can think of the $19 as $20 − $1. So, using the Distributive Property, you can think of the following *without* writing anything on paper.

$$15(19) = 15(20 − 1)$$
$$= 15(20) − 15(1)$$
$$= 300 − 15$$
$$= \$285$$

Exercises Within Reach®

Solutions in English & Spanish and tutorial videos at CollegePrepAlgebra.com

17. *Notebooks* You are shopping for school supplies. You want to buy 8 notebooks for $1.25 each. Show how you can use the Distributive Property to find the total of the notebooks mentally.

18. *Work* You earn $12 per hour working at a grocery store. You receive a 10% raise in pay. Show how you can use the Distributive Property to find your new hourly pay rate.

19. *Lighting* You are installing solar lighting units along the driveway to your house. Each unit is on sale for $5. You purchase 18 units. Show how you can use the Distributive Property to find the total cost of the lighting units mentally.

20. *Movies* You are shopping for DVDs. You want to buy 7 DVDs for $19.99 each. Show how you can use the Distributive Property to find the total cost of the DVDs mentally.

Combining Like Terms

In an algebraic expression, two terms are said to be **like terms** if they are both constant terms or if they have the same variable factor(s). Factors such as x in $5x$ and ab in $6ab$ are called **variable factors**.

EXAMPLE 5   Identifying Like Terms

	Expression	*Like Terms*
a.	$5xy + 1 - xy$	$5xy$ and $-xy$
b.	$12 - x^2 + 3x - 5$	12 and -5
c.	$7x - 3 - 2x + 5$	$7x$ and $-2x$, -3 and 5

EXAMPLE 6   Combining Like Terms

a. $5x + 2x - 4 = (5 + 2)x - 4$ Distributive Property
$$= 7x - 4 \qquad \text{Add.}$$

b. $-5 + 8 + 7y - 5y = (-5 + 8) + (7 - 5)y$ Distributive Property
$$= 3 + 2y \qquad \text{Simplify.}$$

c. $2y - 3x - 4x = 2y - x(3 + 4)$ Distributive Property
$$= 2y - x(7) \qquad \text{Add.}$$
$$= 2y - 7x \qquad \text{Simplify.}$$

d. $7x + 3y - 4x = 3y + 7x - 4x$ Commutative Property
$$= 3y + (7x - 4x) \qquad \text{Associative Property}$$
$$= 3y + (7 - 4)x \qquad \text{Distributive Property}$$
$$= 3y + 3x \qquad \text{Subtract.}$$

e. $12a - 5 - 3a + 7 = 12a - 3a - 5 + 7$ Commutative Property
$$= (12a - 3a) + (-5 + 7) \qquad \text{Associative Property}$$
$$= (12 - 3)a + (-5 + 7) \qquad \text{Distributive Property}$$
$$= 9a + 2 \qquad \text{Simplify.}$$

Exercises Within Reach ®

Solutions in English & Spanish and tutorial videos at CollegePrepAlgebra.com

Identifying Like Terms In Exercises 21−24, identify the like terms.

21. $16t^3 + 4t - 5t + 3t^3$

22. $-\frac{1}{4}x^2 - 3x + \frac{3}{4}x^2 + x$

23. $4rs^2 - 5 - 2r^2s + 12rs^2 + 1$

24. $3 + 6x^2y + 2xy - 2 - 4x^2y$

Combining Like Terms In Exercises 25−32, simplify the expression by combining like terms.

25. $3y - 5y$

26. $-16x + 25x$

27. $x + 5 - 3x$

28. $7s + 3 - 3s$

29. $2x + 9x + 4$

30. $10x - 6 - 5x$

31. $5r + 6 - 2r + 1$

32. $2t - 4 + 8t + 9$

Simplifying Algebraic Expressions

To **simplify an algebraic expression** generally means to remove symbols of grouping and combine like terms.

> **EXAMPLE 7** **Simplifying Algebraic Expressions**
>
> Simplify each expression.
>
> **a.** $-3(-5x)$ **b.** $7(-x)$ **c.** $\dfrac{5x}{3} \cdot \dfrac{3}{5}$
>
> **d.** $x^2(-2x^3)$ **e.** $(-2x)(4x)$ **f.** $(2rs)(r^2s)$

SOLUTION

a. $\begin{aligned} -3(-5x) &= (-3)(-5)x && \text{Associative Property} \\ &= 15x && \text{Multiply.} \end{aligned}$

b. $\begin{aligned} 7(-x) &= 7(-1)(x) && \text{Coefficient of } -x \text{ is } -1. \\ &= -7x && \text{Multiply.} \end{aligned}$

c. $\begin{aligned} \dfrac{5x}{3} \cdot \dfrac{3}{5} &= \left(\dfrac{5}{3} \cdot x\right) \cdot \dfrac{3}{5} && \text{Coefficient of } \dfrac{5x}{3} \text{ is } \dfrac{5}{3}. \\ &= \left(\dfrac{5}{3} \cdot \dfrac{3}{5}\right) \cdot x && \text{Commutative and Associative Properties} \\ &= 1 \cdot x && \text{Multiplicative Inverse} \\ &= x && \text{Multiplicative Identity} \end{aligned}$

d. $\begin{aligned} x^2(-2x^3) &= (-2)(x^2 \cdot x^3) && \text{Commutative and Associative Properties} \\ &= -2 \cdot x \cdot x \cdot x \cdot x \cdot x && \text{Repeated multiplication} \\ &= -2x^5 && \text{Exponential form} \end{aligned}$

e. $\begin{aligned} (-2x)(4x) &= (-2 \cdot 4)(x \cdot x) && \text{Commutative and Associative Properties} \\ &= -8x^2 && \text{Exponential form} \end{aligned}$

f. $\begin{aligned} (2rs)(r^2s) &= 2(r \cdot r^2)(s \cdot s) && \text{Commutative and Associative Properties} \\ &= 2 \cdot r \cdot r \cdot r \cdot s \cdot s && \text{Repeated multiplication} \\ &= 2r^3s^2 && \text{Exponential form} \end{aligned}$

Exercises Within Reach®

Solutions in English & Spanish and tutorial videos at CollegePrepAlgebra.com

Simplifying an Algebraic Expression In Exercises 33–46, simplify the expression.

33. $2(6x)$

34. $-7(5a)$

35. $-(4x)$

36. $-(5t)$

37. $(-2x)(-3x)$

38. $(-3y)(-4y)$

39. $(-5z)(2z^2)$

40. $(10t)(-4t^2)$

41. $\dfrac{18a}{5} \cdot \dfrac{15}{6}$

42. $\dfrac{5x}{8} \cdot \dfrac{16}{5}$

43. $\left(-\dfrac{3x^2}{2}\right)\left(\dfrac{4x}{18}\right)$

44. $\left(\dfrac{4x}{3}\right)\left(\dfrac{3x}{16}\right)$

45. $(12xy^2)(-2x^3y^2)$

46. $(7r^2s^3)(3rs)$

Symbols of Grouping

EXAMPLE 8   **Removing Symbols of Grouping**

Simplify each expression.

a. $-(3y + 5)$ **b.** $5x + (x - 7)2$ **c.** $-2(4x - 1) + 3x$

d. $5x - 2[4x + 3(x - 1)]$ **e.** $-7y + 3[2y - (3 - 2y)] - 5y + 4$

SOLUTION

a. $-(3y + 5) = -3y - 5$ Distributive Property

b. $5x + (x - 7)2 = 5x + 2x - 14$ Distributive Property

$= 7x - 14$ Combine like terms.

c. $-2(4x - 1) + 3x = -8x + 2 + 3x$ Distributive Property

$= -8x + 3x + 2$ Commutative Property

$= -5x + 2$ Combine like terms.

d. $5x - 2[4x + 3(x - 1)]$

$= 5x - 2[4x + 3x - 3]$ Distributive Property

$= 5x - 2[7x - 3]$ Combine like terms.

$= 5x - 14x + 6$ Distributive Property

$= -9x + 6$ Combine like terms.

e. $-7y + 3[2y - (3 - 2y)] - 5y + 4$

$= -7y + 3[2y - 3 + 2y] - 5y + 4$ Distributive Property

$= -7y + 3[4y - 3] - 5y + 4$ Combine like terms.

$= -7y + 12y - 9 - 5y + 4$ Distributive Property

$= (-7y + 12y - 5y) + (-9 + 4)$ Group like terms.

$= -5$ Combine like terms.

Study Tip

When a parenthetical expression is preceded by a *plus* sign, you can remove the parentheses without changing the signs of the terms inside.

$3y + (-2y + 7)$
$= 3y - 2y + 7$

When a parenthetical expression is preceded by a *minus* sign, however, you must change the sign of each term to remove the parentheses.

$3y - (2y - 7)$
$= 3y - 2y + 7$

Remember that $-(2y - 7)$ is equal to $(-1)(2y - 7)$, and the Distributive Property can be used to "distribute the minus sign" to obtain $-2y + 7$.

Exercises Within Reach®

Solutions in English & Spanish and tutorial videos at CollegePrepAlgebra.com

Removing Symbols of Grouping   In Exercises 47−66, simplify the expression by removing symbols of grouping and combining like terms.

47. $2(x - 2) + 4$ **48.** $3(x - 5) - 2$

49. $6(2s - 1) + s + 4$ **50.** $(2x - 1)2 + x + 9$

51. $m - 3(m - 7)$ **52.** $8l - (3l - 7)$

53. $-6(2 - 3x) + 10(5 - x)$ **54.** $3(r - 2s) - 5(3r - 5s)$

55. $\frac{2}{3}(12x + 15) + 16$ **56.** $\frac{3}{8}(4 - y) - \frac{5}{2} + 10$

57. $3 - 2[6 + (4 - x)]$ **58.** $10x + 5[6 - (2x + 3)]$

59. $7x(2 - x) - 4x$ **60.** $-6x(x - 1) + x^2$

61. $4x^2 + x(5 - x) - 3$ **62.** $-z(z - 2) + 3z^2 + 5$

63. $-3t(4 - t) + t(t + 1)$ **64.** $-2x(x - 1) + x(3x - 2)$

65. $3t[4 - (t - 3)] + t(t + 5)$ **66.** $4y[5 - (y + 1)] + 3y(y + 1)$

<u>Application</u> **EXAMPLE 9** **Geometry: Writing and Simplifying a Formula**

Write and simplify an expression for (a) the perimeter and (b) the area of the triangle.

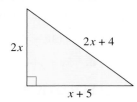

SOLUTION

a. Perimeter of a Triangle = Sum of the Three Sides

$= 2x + (2x + 4) + (x + 5)$ Substitute.

$= (2x + 2x + x) + (4 + 5)$ Group like terms.

$= 5x + 9$ Combine like terms.

b. Area of a Triangle $= \frac{1}{2} \cdot$ Base $\cdot$ Height

$= \frac{1}{2}(x + 5)(2x)$ Substitute.

$= \frac{1}{2}(2x)(x + 5)$ Commutative Property

$= x(x + 5)$ Multiply.

$= x^2 + 5x$ Distributive Property

Exercises Within Reach ®   Solutions in English & Spanish and tutorial videos at CollegePrepAlgebra.com

Geometry In Exercises 67 and 68, write and simplify expressions for (a) the perimeter and (b) the area of the rectangular sandboxes.

67.

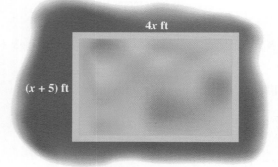

68.

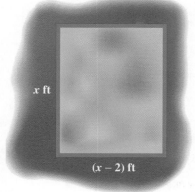

<u>Application</u>   EXAMPLE 10   **Geometry: Writing and Simplifying a Formula**

The formula for the area of a trapezoid is $A = \frac{1}{2}h(b_1 + b_2)$. Use this formula to write and simplify an expression for the area of the proposed trapezoidal park.

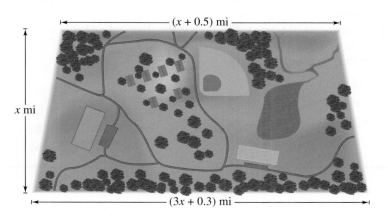

SOLUTION

Begin by assigning the following values.

$$h = x \text{ mi} \qquad b_1 = (x + 0.5) \text{ mi} \qquad b_2 = (3x + 0.3) \text{ mi}$$

Then use the formula to write an expression.

$$\frac{1}{2}h(b_1 + b_2) = \frac{1}{2}x[(x + 0.5) + (3x + 0.3)]$$
$$= \frac{1}{2}x[x + 0.5 + 3x + 0.3]$$
$$= \frac{1}{2}x[4x + 0.8]$$
$$= 2x^2 + 0.4x$$

Exercises Within Reach ®   Solutions in English & Spanish and tutorial videos at CollegePrepAlgebra.com

Geometry **In Exercises 69 and 70, use the formula for the area of a trapezoid to write and simplify an expression for the area of the trapezoidal house lot and park.**

69.

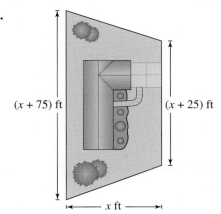

$(x + 75)$ ft $(x + 25)$ ft

x ft

70.

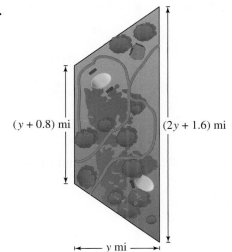

$(y + 0.8)$ mi $(2y + 1.6)$ mi

y mi

Concept Summary: *Simplifying Algebraic Expressions*

What

You can use properties of algebra to **simplify algebraic expressions**. To simplify these types of expressions usually means to remove symbols of grouping and combine **like terms**.

EXAMPLE

Simplify the expression
$3x + 2(x + 4)$.

How

The main tool for removing symbols of grouping and combining like terms is the Distributive Property.

EXAMPLE

$3x + 2(4 + x)$

$\begin{aligned} &= 3x + 8 + 2x &&\text{Distributive Prop.} \\ &= 3x + 2x + 8 &&\text{Comm. Prop.} \\ &= x(3 + 2) + 8 &&\text{Distributive Prop.} \\ &= 5x + 8 &&\text{Simplify.} \end{aligned}$

Why

Simplifying an algebraic expression into a more usable form is one of the most frequently used skills in algebra.

Exercises Within Reach ®

Worked-out solutions to odd-numbered exercises at CollegePrepAlgebra.com

Concept Summary Check

71. *Describing Like Terms* In your own words, state the definition of like terms. Give an example of like terms and an example of unlike terms.

72. *Combining Like Terms* Describe how to combine like terms. Give an example of an expression that can be simplified by combining like terms.

73. *Writing* In your own words, describe the procedure for removing symbols of grouping.

74. *Writing* What does it mean to simplify an algebraic expression?

Extra Practice

Simplifying an Algebraic Expression In Exercises 75–86, simplify the expression.

75. $x^2 - 2xy + 4 + xy$

76. $r^2 + 3rs - 6 - rs$

77. $5z - 5 + 10z + 2z + 16$

78. $7x - 4x + 8 + 3x - 6$

79. $(7y^2)(-3y)$

80. $(-2t^3)(4t^2)$

81. $\left(\dfrac{2x}{5}\right)\left(\dfrac{4x}{8}\right)$

82. $\left(-\dfrac{6y^2}{7}\right)\left(-\dfrac{y}{6}\right)$

83. $-4(2 - 5x) + 3(x + 6)$

84. $5(x + 9) - 2(30 + 4x)$

85. $7 - 3[7 - (3 + x)]$

86. $2x[1 - (x - 4)] + x(x - 3)$

Geometry In Exercises 87 and 88, write and simplify an expression for the perimeter of the triangle.

87.

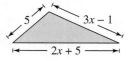

88.

Geometry **In Exercises 89 and 90, write and simplify expressions for (a) the perimeter and (b) the area of the rectangle.**

89.

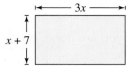

90.

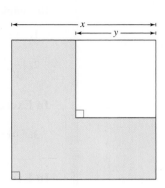

91. *Geometry* The remaining area of a square with side length x after a smaller square with side length y has been removed (see figure) is $(x + y)(x - y)$.

(a) Show that the remaining area can also be expressed as $x(x - y) + y(x - y)$, and give a geometric explanation for the area represented by each term in this expression.

(b) Find the remaining area of a square with side length 9 after a square with side length 5 has been removed.

Explaining Concepts

Writing **In Exercises 92 and 93, explain why the two expressions are not like terms.**

92. $\frac{1}{2}x^2y, \frac{5}{2}xy^2$

93. $-16x^2y^3, 7x^2y$

94. *Using Order of Operations* Does the expression $[x - (3 \cdot 4)] \div 5$ change when the parentheses are removed? Does it change when the brackets are removed? Explain.

95. *Writing* Discuss the difference between $(6x)^4$ and $6x^4$.

96. *Error Analysis* Describe and correct the error.

$$4x - 3(x - 1) = 4x - 3(x) - 3(1)$$
$$= 4x - 3x - 3$$
$$= x - 3$$

Cumulative Review

In Exercises 97–102, evaluate the expression.

97. $0 - (-12)$

98. $5 - 4 \div 2 + 6$

99. $-12 - 2 + |-3|$

100. $6 + 3(4 + 2)$

101. $\frac{5}{16} - \frac{3}{10}$

102. $\frac{9}{16} + 2\frac{3}{12}$

In Exercises 103 and 104, evaluate the algebraic expression for the given values of the variable.

103. $3x - 2$

(a) $x = 2$

(b) $x = -1$

104. $2x^2 + 3$

(a) $x = 3$

(b) $x = -4$

Mid-Chapter Quiz: Sections 2.1–2.2

Solutions in English & Spanish and tutorial videos at CollegePrepAlgebra.com

Take this quiz as you would take a quiz in class. After you are done, check your work against the answers in the back of the book.

In Exercises 1 and 2, evaluate the algebraic expression for the given values of the variable(s). If it is not possible, state the reason.

1. $x^2 - 3x$ (a) $x = 3$ (b) $x = -2$

2. $\dfrac{x}{y-3}$ (a) $x = 5, y = 3$ (b) $x = 0, y = -1$

In Exercises 3 and 4, identify the terms of the expression and their coefficients.

3. $4x^2 - 2x$ 4. $5x + 3y - z$

In Exercises 5 and 6, rewrite the product in exponential form.

5. $(-3y)(-3y)(-3y)(-3y)$ 6. $2 \cdot (x - 3) \cdot (x - 3) \cdot 2 \cdot 2$

In Exercises 7–10, identify the property of algebra illustrated by the statement.

7. $-3(2y) = (-3 \cdot 2)y$ 8. $(x + 2)y = xy + 2y$

9. $3y \cdot \dfrac{1}{3y} = 1, y \neq 0$ 10. $x - x^2 + 2 = -x^2 + x + 2$

In Exercises 11 and 12, use the Distributive Property to expand the expression.

11. $2x(3x - 1)$ 12. $-6(2y + 3y^2 - 6)$

In Exercises 13–20, simplify the expression.

13. $-4(-5y^2)$ 14. $\dfrac{x}{3}\left(-\dfrac{3x}{5}\right)$

15. $(-3y)^2 y^3$ 16. $\dfrac{2z^2}{3y} \cdot \dfrac{5z}{7}$

17. $y^2 - 3xy + y + 7xy$ 18. $(2x + 2)3 + x - 10$

19. $5(a - 2b) + 3(a + b)$ 20. $4x + 3[2 - 4(x + 6)]$

Applications

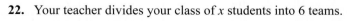

21. Write and simplify an expression for the perimeter of the triangle (see figure).

22. Your teacher divides your class of x students into 6 teams.

    (a) Write an expression representing the number of students on each team.

    (b) There are 30 students in your class. How many students are on each team?

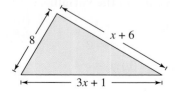

Study Skills in Action

Absorbing Details Sequentially

Math is a sequential subject. Learning new math concepts successfully depends on how well you understand all the previous concepts. So, it is important to learn and remember concepts as they are encountered. One way to work through a section sequentially is by following these steps.

1 ▶ Work through an example. If you have trouble, consult your notes or seek help from a classmate or teacher.

2 ▶ Complete the exercises following the example.

3 ▶ If you get the exercises correct, move on to the next example. If not, make sure you understand your mistake(s) before you move on.

4 ▶ When you have finished working through all the examples in the section, take a short break of 5 to 10 minutes. This will give your brain time to process everything.

5 ▶ Start the exercises that follow the lesson.

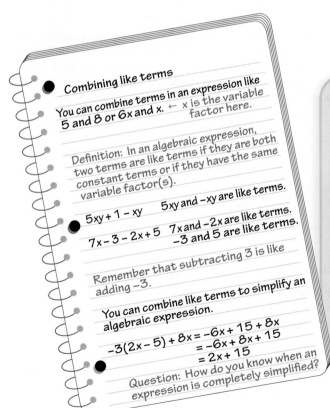

Combining like terms

You can combine terms in an expression like 5 and 8 or 6x and x. ← x is the variable factor here.

Definition: In an algebraic expression, two terms are like terms if they are both constant terms or if they have the same variable factor(s).

$5xy + 1 - xy$ $5xy$ and $-xy$ are like terms.

$7x - 3 - 2x + 5$ $7x$ and $-2x$ are like terms.
    -3 and 5 are like terms.

Remember that subtracting 3 is like adding −3.

You can combine like terms to simplify an algebraic expression.

$-3(2x - 5) + 8x = -6x + 15 + 8x$
$= -6x + 8x + 15$
$= 2x + 15$

Question: How do you know when an expression is completely simplified?

Smart Study Strategy

Rework Your Notes

It is almost impossible to write down in your notes all the detailed information you are taught in class. A good way to reinforce the concepts and put them into your long-term memory is to rework your notes. When you take notes, leave extra space on the pages. You can go back after class and fill in:

- important definitions and rules.

- additional examples.

- questions you have about the material.

2.3 Algebra and Problem Solving

▶ Construct verbal mathematical models from written statements.
▶ Translate verbal phrases into algebraic expressions.
▶ Identify hidden operations when writing algebraic expressions.
▶ Use problem-solving strategies to solve application problems.

Constructing Verbal Models

Algebra is a problem-solving language that is used to solve real-life problems. It has four basic components, which tend to nest within each other.

1. Rules of arithmetic

2. Algebraic expressions: rewriting into equivalent forms

3. Algebraic equations: creating and solving

4. Functions and graphs: relationships among variables

In the first two sections of this chapter, you studied techniques for rewriting and simplifying algebraic expressions. In this section, you will study ways to write algebraic expressions from written statements by first constructing a **verbal mathematical model**.

Take another look at Example 8 in Section 2.1 (page 61). In that example, you are paid $9 per hour and your weekly pay can be represented by the verbal model

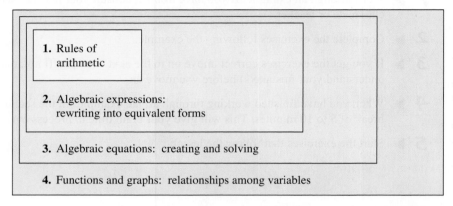

$$\boxed{\text{Pay per hour}} \cdot \boxed{\text{Number of hours}} = 9 \text{ dollars} \cdot x \text{ hours} = 9x.$$

Verbal model

Algebraic expression

Exercises Within Reach ®

Solutions in English & Spanish and tutorial videos at CollegePrepAlgebra.com

Constructing a Verbal Model In Exercises 1−6, construct a verbal model for the given situation.

1. You earn $10 per hour. How much do you earn for working x hours?

2. You bought x CDs online and y CDs from a store. How many total CDs did you buy?

3. You have 20 coupons. You use c coupons when you pay the bill. How many coupons do you have left?

4. You want to evenly divide 15 tickets between p people. How many tickets will each person receive?

5. A carton of eggs costs $2.89. How much will it cost to buy m cartons?

6. You have x dollars. How much money will you have after loaning $5 to a friend?

Application **EXAMPLE 1** **Writing an Algebraic Expression**

You are paid 5 cents for each aluminum soda can and 3 cents for each plastic soda bottle you collect. Write an algebraic expression that represents the total weekly income (in cents) for this recycling activity.

Study Tip

In Example 1, notice that c is used to represent the number of cans and b is used to represent the number of bottles. When writing algebraic expressions, it is convenient to choose variables that can be identified with the unknown quantities.

SOLUTION

Before writing an algebraic expression for the weekly income, it is helpful to construct an informal verbal model. For instance, the following verbal model could be used.

$$\boxed{\text{Pay per can}} \cdot \boxed{\text{Number of cans}} + \boxed{\text{Pay per bottle}} \cdot \boxed{\text{Number of bottles}} \qquad \text{Verbal model}$$

Note that the word *and* in the problem indicates addition. Because both the number of cans and the number of bottles can vary from week to week, you can use the two variables c and b, respectively, to write the following algebraic expression.

$$\boxed{5 \text{ cents}} \cdot \boxed{c \text{ cans}} + \boxed{3 \text{ cents}} \cdot \boxed{b \text{ bottles}} = 5c + 3b \qquad \text{Algebraic expression}$$

Exercises Within Reach®

Solutions in English & Spanish and tutorial videos at CollegePrepAlgebra.com

7. **Money** A cash register contains d dimes. Write an algebraic expression that represents the total amount of money (in dollars).

8. **Fruit** A bag of apples costs $4.99. Write an algebraic expression that represents the total cost of b bags of apples.

9. **Money** A cash register contains d dimes and q quarters. Write an algebraic expression that represents the total amount of money (in dollars).

10. **Fruit** Apples cost $3.29 per pound and oranges cost $2.99 per pound. Write an algebraic expression that represents the total cost of buying a pounds of apples and g pounds of oranges.

Translating Phrases

Translating Phrases into Algebraic Expressions

Key Words and Phrases	Verbal Description	Expression
Addition:		
Sum, plus, greater than, increased by, more than, exceeds, total of	The sum of 6 and x	$6 + x$
	Eight more than y	$y + 8$
Subtraction:		
Difference, minus, less than, decreased by, subtracted from, reduced by	Five decreased by a	$5 - a$
	Four less than z	$z - 4$
Multiplication:		
Product, multiplied by, twice, times, percent of	Seven times x	$7x$
Division:		
Quotient, divided by, ratio, per	The ratio of x and 3	$\dfrac{x}{3}$

Study Tip

Order is important when writing subtraction and division expressions. For instance, *three less than m* means $m - 3$, not $3 - m$, and the *quotient of n and 7* means $\dfrac{n}{7}$, not $\dfrac{7}{n}$.

EXAMPLE 2 Translating Verbal Phrases into Algebraic Expressions

a. Three less than m

$\qquad m - 3$ "Less than" indicates subtraction.

b. y decreased by 10

$\qquad y - 10$ "Decreased by" indicates subtraction.

c. The product of 5 and x

$\qquad 5x$ "Product" indicates multiplication.

d. The quotient of n and 7

$\qquad \dfrac{n}{7}$ "Quotient" indicates division.

Exercises Within Reach®

Solutions in English & Spanish and tutorial videos at CollegePrepAlgebra.com

Translating a Verbal Phrase In Exercises 11−22, translate the verbal phrase into an algebraic expression.

11. x increased by 5

12. 17 more than y

13. b decreased by 25

14. k decreased by 7

15. Six less than g

16. Ten more than x

17. Twice h

18. The product of 30 and c

19. w divided by 3

20. d divided by 100

21. The ratio of x and 50

22. One-half of y

EXAMPLE 3   **Translating Algebraic Expressions into Verbal Phrases**

Without using a variable, write a verbal description for each expression.

a. $x - 12$

b. $7(x + 12)$

c. $5 + \dfrac{x}{2}$

d. $\dfrac{5 + x}{2}$

SOLUTION

a. *Algebraic expression:* $x - 12$

    Operation: Subtraction

    Key phrase: Less than

    Verbal description: Twelve less than a number

b. *Algebraic expression:* $7(x + 12)$

    Operations: Multiplication, addition

    Key words: Times, sum

    Verbal description: Seven times the sum of a number and 12

c. *Algebraic expression:* $5 + \dfrac{x}{2}$

    Operations: Addition, division

    Key words: Plus, quotient

    Verbal description: Five plus the quotient of a number and 2

d. *Algebraic expression:* $\dfrac{5 + x}{2}$

    Operations: Addition, division

    Key words: Sum, divided by

    Verbal description: The sum of 5 and a number, all divided by 2

Exercises Within Reach ®

Solutions in English & Spanish and tutorial videos at CollegePrepAlgebra.com

Translating an Algebraic Expression In Exercises 23−36, write a verbal description of the algebraic expression, without using a variable. (There is more than one correct answer.)

23. $x - 10$

24. $x + 9$

25. $3x + 2$

26. $4 - 7x$

27. $\dfrac{1}{2}x - 6$

28. $9 - \dfrac{1}{4}x$

29. $3(2 - x)$

30. $-10(t - 6)$

31. $\dfrac{t + 1}{2}$

32. $\dfrac{y - 3}{4}$

33. $\dfrac{1}{2} - \dfrac{t}{5}$

34. $\dfrac{1}{4} + \dfrac{y}{8}$

35. $x^2 + 5$

36. $x^3 - 1$

Verbal Models with Hidden Operations

Application   **EXAMPLE 4**   **Discovering Hidden Operations**

A cash register contains *n* nickels and *d* dimes. Write an algebraic expression for this amount of money in cents.

SOLUTION

The amount of money is a sum of products.

Verbal Model:	Value of nickel	·	Number of nickels	+	Value of dime	·	Number of dimes

Labels:	Value of nickel = 5	(cents)
	Number of nickels = *n*	(nickels)
	Value of dime = 10	(cents)
	Number of dimes = *d*	(dimes)

Expression:	$5n + 10d$	(cents)

EXAMPLE 5   **Discovering Hidden Operations**

A person riding a bicycle travels at a constant rate of 12 miles per hour. Write an algebraic expression showing how far the person can ride in *t* hours.

SOLUTION

The distance traveled is a product.

Verbal Model:	Rate of travel · Time traveled	$(d = r \cdot t)$
Labels:	Rate of travel = 12	(miles per hour)
	Time traveled = *t*	(hours)
Expression:	$12t$	(miles)

Study Tip

In Example 5, the final answer is listed in terms of miles. This unit is found as follows.

$$12 \frac{\text{miles}}{\text{hour}} \cdot t \text{ hours}$$

Note that the hours "divide out," leaving miles as the unit of measure. This technique is called *unit analysis* and can be very helpful in determining the final unit of measure.

Exercises Within Reach ®   Solutions in English & Spanish and tutorial videos at CollegePrepAlgebra.com

37. *Sales Tax* The sales tax on a purchase of *L* dollars is 6%. Write an algebraic expression that represents the total amount of sales tax. (*Hint:* Use the decimal form of 6%.)

38. *Income Tax* The state income tax on a gross income of *I* dollars in Pennsylvania is 3.07%. Write an algebraic expression that represents the total amount of income tax. (*Hint:* Use the decimal form of 3.07%.)

39. *Rentals* A movie rental costs $3 per day. A video game rental costs $4 per day. Write an algebraic expression that represents the total cost of renting *m* movies and *v* video games per day.

40. *Frame* The height of a rectangular picture frame is 1.5 times the width *w*. Write an algebraic expression that represents the perimeter of the picture frame.

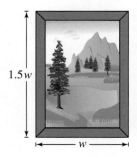

1.5*w*

w

Application EXAMPLE 6 **Discovering Hidden Operations**

A person paid x dollars plus 6% sales tax for an automobile. Write an algebraic expression for the total cost of the automobile.

SOLUTION

The total cost is a sum.

Verbal Model:	Cost of automobile	+	Sales tax rate	·	Cost of automobile

Labels: Sales tax rate $= 0.06$ (decimal form)
Cost of automobile $= x$ (dollars)

Expression: $x + 0.06x = (1 + 0.06)x$ (dollars)
$= 1.06x$

Application EXAMPLE 7 **Discovering Hidden Operations**

A truck travels 100 miles at an average speed of r miles per hour. Write an expression that represents the total travel time.

SOLUTION

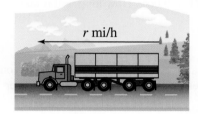

r mi/h

Verbal Model:	$\dfrac{\text{Distance}}{\text{Rate}}$	$\left(t = \dfrac{d}{r} \right)$

Labels: Distance $= 100$ (miles)
Rate $= r$ (miles per hour)

Expression: $\dfrac{100}{r}$ (hours)

Exercises Within Reach ®

Solutions in English & Spanish and tutorial videos at CollegePrepAlgebra.com

41. *Hourly Wage* An employee's hourly wage is $12.50 per hour plus $0.75 for each of the q units produced during the hour. Write an algebraic expression that represents the employee's total hourly earnings.

42. *Camping* A campground charges $15 for adults and $2 for children. Write an algebraic expression that represents the total camping fee for m adults and n children.

43. *Mobile Device* Applications for a cellular phone cost $0.99 each. Ringtones cost $1.99 each. Write an algebraic expression that represents the total cost of buying a applications and r ringtones.

44. *Tickets* You buy t tickets to a baseball game for a total of $45. Write an algebraic expression that represents the cost of each ticket.

Additional Problem-Solving Strategies

Summary of Additional Problem-Solving Strategies

1. **Guess, Check, and Revise** Guess a reasonable solution based on the given data. Check the guess, and revise it, if necessary. Continue guessing, checking, and revising until a correct solution is found.

2. **Make a Table/Look for a Pattern** Make a table using the data in the problem. Look for a number pattern. Then use the pattern to complete the table or find a solution.

3. **Draw a Diagram** Draw a diagram that shows the facts of the problem. Use the diagram to visualize the action of the problem. Use algebra to find a solution. Then check the solution against the facts.

4. **Solve a Simpler Problem** Construct a simpler problem that is similar to the original problem. Solve the simpler problem. Then use the same procedure to solve the original problem.

Application

EXAMPLE 8   Guess, Check, and Revise

You deposit $500 in an account that earns 6% interest compounded annually. The balance A in the account after t years is $A = 500(1 + 0.06)t$. How long will it take for your investment to double?

SOLUTION

You can solve this problem using a guess, check, and revise strategy. For instance, you might guess that it will take 10 years for your investment to double. The balance after 10 years is

$$A = 500(1 + 0.06)^{10} \approx \$895.42$$

Because the amount has not yet doubled, you increase your guess to 15 years.

$$A = 500(1 + 0.06)^{15} \approx \$1198.28$$

Because this amount is greater than double the investment, your next guess should be a number between 10 and 15. After trying several more numbers, you can determine that your balance will double in about 11.9 years.

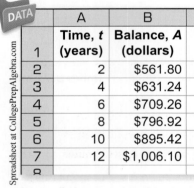

	A	B
	Time, t (years)	Balance, A (dollars)
1		
2	2	$561.80
3	4	$631.24
4	6	$709.26
5	8	$796.92
6	10	$895.42
7	12	$1,006.10
8		

Spreadsheet at CollegePrepAlgebra.com

Another way to solve this problem is to use a spreadsheet to make a table.

Exercises Within Reach ®

Solutions in English & Spanish and tutorial videos at CollegePrepAlgebra.com

Guess, Check, and Revise **In Exercises 45−50, an expression for the balance in an account is given. Use a guess, check, and revise strategy to determine the time (in years) necessary for the investment of $1000 to double.**

45. Interest rate: 7%

$1000(1 + 0.07)^t$

46. Interest rate: 5%

$1000(1 + 0.05)^t$

47. Interest rate: 6%

$1000(1 + 0.06)^t$

48. Interest rate: 8%

$1000(1 + 0.08)^t$

49. Interest rate: 6.5%

$1000(1 + 0.065)^t$

50. Interest rate: 7.5%

$1000(1 + 0.075)^t$

Application   EXAMPLE 9   **Draw a Diagram**

The outer dimensions of a rectangular apartment are 25 feet by 40 feet. The combination living room, dining room, and kitchen areas occupy two-fifths of the apartment's area. Find the total area of the remaining rooms.

SOLUTION

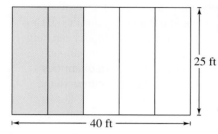

For this problem, it helps to draw a diagram, as shown at the left. From the figure, you can see that the total area of the apartment is

$$\text{Area} = (\text{Length})(\text{Width})$$
$$= (40)(25)$$
$$= 1000 \text{ square feet.}$$

The area occupied by the living room, dining room, and kitchen is

$$\frac{2}{5}(1000) = 400 \text{ square feet.}$$

This implies that the remaining rooms must have a total area of

$$1000 - 400 = 600 \text{ square feet.}$$

Application   EXAMPLE 10   **Solve a Simpler Problem**

You are driving on an interstate highway at an average speed of 60 miles per hour. How far will you travel in $12\frac{1}{2}$ hours?

SOLUTION

One way to solve this problem is to use the formula that relates distance, rate, and time. Suppose, however, that you have forgotten the formula. To help you remember, you could solve some simpler problems.

- If you travel 60 miles per hour for 1 hour, you will travel 60 miles.
- If you travel 60 miles per hour for 2 hours, you will travel 120 miles.
- If you travel 60 miles per hour for 3 hours, you will travel 180 miles.

From the examples, it appears that you can find the total miles traveled by multiplying the rate by the time. So, if you travel 60 miles per hour for $12\frac{1}{2}$ hours, you will travel a distance of

$$(60)(12.5) = 750 \text{ miles.}$$

Exercises Within Reach ®

Solutions in English & Spanish and tutorial videos at CollegePrepAlgebra.com

Drawing a Diagram In Exercises 51 and 52, draw figures satisfying the specified conditions.

51. The sides of a square have lengths of *a* centimeters. Draw the square. Draw the rectangle obtained by extending two parallel sides of the square 6 centimeters. Find expressions for the perimeter and area of each figure.

52. The dimensions of a rectangular lawn are 150 feet by 250 feet. The property owner buys a rectangular strip of land *x* feet wide along one 250-foot side of the lawn. Draw diagrams representing the lawn before and after the purchase. Write an expression for the area of each.

Solving a Simpler Problem In Exercises 53 and 54, solve the problem.

53. A bubble rises through water at a rate of about 1.15 feet per second. How far will the bubble rise in 5 seconds?

54. A train travels at an average speed of 50 miles per hour. How long will it take the train to travel 350 miles?

Concept Summary: Using Verbal Models to Write Algebraic Expressions

What

You can construct algebraic expressions from written statements by constructing **verbal mathematical models**.

EXAMPLE

A gallon of milk costs $3.29. Write an algebraic expression that represents the total cost of buying g gallons of milk.

How

Determine any known and unknown quantities from the written statement. Then construct a verbal model. Use the model to write an algebraic expression.

EXAMPLE

| Cost per gallon | $\cdot$ | Number of gallons | = |

$$\$3.29 \cdot g \text{ gallons} =$$

$$3.29 \cdot g = 3.29g$$

Note that nowhere in the written statement does it say to multiply 3.29 by g. It is *implied* in the statement.

Why

Algebra is a problem-solving language that you can use to solve real-life problems. Some of these problems can consist of several known and unknown quantities. Using a verbal mathematical model can help you organize your thoughts about the known and unknown quantities, as well as the overall solution to the real-life problem.

Exercises Within Reach ®

Worked-out solutions to odd-numbered exercises at CollegePrepAlgebra.com

Concept Summary Check

55. *Describing Strategies* Describe other problem-solving strategies besides constructing verbal models that you can use to solve exercises in this chapter.

56. *Assigning Variables* Two unknown quantities in a verbal model are "Number of cherries" and "Number of strawberries." What variables would you use to represent these quantities. Explain.

57. *Identifying Operations* When constructing a verbal model from a written statement, what are some key words and phrases that indicate the four operations of arithmetic?

58. *Writing* What is a *hidden operation* in a verbal phrase? Explain how to identify hidden operations.

Extra Practice

Translating an Algebraic Expression In Exercises 59−62, write a verbal description of the algebraic expression, without using a variable. (There is more than one correct answer.)

59. $t(t + 16)$

60. $\frac{4}{5}(w + 10)$

61. $\dfrac{4}{x - 2}$

62. $x^2 - (x + 17)$

Translating a Verbal Phrase In Exercises 63−68, translate the verbal phrase into an algebraic expression. Simplify the expression.

63. x times the sum of x and 3

64. n times the difference of 6 and n

65. x minus the sum of 25 and x

66. The sum of 4 and x added to the sum of x and -8

67. The square of x decreased by the product of x and $2x$

68. The square of x added to the product of x and $x + 1$

69. *Supplies* Pens cost $0.25 each. Pencils cost $0.10 each. Write an algebraic expression that represents the total cost of buying p pens and n pencils.

70. *Gasoline* A consumer buys g gallons of gasoline for a total of d dollars. Write an algebraic expression that represents the price per gallon.

71. *Finding a Pattern* Describe the pattern and use your description to find the value of the expression when $n = 20$.

n	0	1	2	3	4	5
Value of expression	-1	1	3	5	7	9

72. *Finding a Pattern* Find values for a and b such that the expression $an + b$ yields the values in the table.

n	0	1	2	3	4	5
$an + b$	4	9	14	19	24	29

Explaining Concepts

75. *Reasoning* Determine which verbal phrase(s) is (are) equivalent to the expression $n + 4$.

  (a) 4 more than n (b) the sum of n and 4

  (c) n less than 4   (d) the ratio of n to 4

  (e) the total of 4 and n

76. *Number Sense* Determine whether order is important when translating each verbal phrase into an algebraic expression. Explain.

  (a) x increased by 10

  (b) 10 decreased by x

  (c) The product of x and 10

  (d) The quotient of x and 10

Finding Additional Information **In Exercises 73 and 74,** describe **what additional information is needed to solve the problem. (Do not solve the problem.)**

73. A family taking a Sunday drive through the country travels at an average speed of 45 miles per hour. How far have they traveled by 3:00 P.M.?

74. You paint a rectangular room that is twice as long as it is wide. One gallon of paint covers 100 square feet. How much money do you spend on paint?

77. *Writing* Give two interpretations of "the quotient of 5 and a number times 3." Explain why $\dfrac{3n}{5}$ is not a possible interpretation.

78. *Writing* Give two interpretations of "the difference of 6 and a number divided by 3." Explain why $\dfrac{n - 6}{3}$ is not a possible interpretation.

Cumulative Review

In Exercises 79–84, evaluate the expression.

79. $(-6)(-13)$ **80.** $|4(-6)(5)|$

81. $\left(-\frac{4}{3}\right)\left(-\frac{9}{16}\right)$ **82.** $\frac{7}{8} \div \frac{3}{16}$

83. $\left|-\frac{5}{9}\right| + 2$ **84.** $-7\frac{3}{5} - 3\frac{1}{2}$

In Exercises 85–88, identify the property of algebra illustrated by the statement.

85. $2a + b = b + 2a$

86. $-4x(1) = -4x$

87. $2(c - d) = 2c - 2d$

88. $-3y^3 + 3y^3 = 0$

2.4 Introduction to Equations

▶ Check whether a given value is a solution of an equation.
▶ Use properties of equality to form equivalent equations.
▶ Use a verbal model to write an algebraic equation.

Checking Solutions of Equations

An **equation** is a statement that two algebraic expressions are equal. For instance,

$$x = 3, \quad 5x - 2 = 8, \quad \frac{x}{4} = 7, \quad \text{and} \quad x^2 - 9 = 0$$

are equations. To **solve** an equation involving the variable x means to find all values of x for which the equation is true. Such values are called **solutions**. For instance, $x = 2$ is a solution of the equation $5x - 2 = 8$ because $5(2) - 2 = 8$ is a true statement. The solutions of an equation are said to **satisfy** the equation.

EXAMPLE 1 Checking Solutions of an Equation

Determine whether (a) $x = -2$ and (b) $x = 2$ are solutions of $x^2 - 5 = 4x + 7$.

SOLUTION

a.

$x^2 - 5 = 4x + 7$	Write original equation.
$(-2)^2 - 5 \stackrel{?}{=} 4(-2) + 7$	Substitute -2 for x.
$4 - 5 \stackrel{?}{=} -8 + 7$	Simplify.
$-1 = -1$	Solution checks. ✔

b.

$x^2 - 5 = 4x + 7$	Write original equation.
$(2)^2 - 5 \stackrel{?}{=} 4(2) + 7$	Substitute 2 for x.
$4 - 5 \stackrel{?}{=} 8 + 7$	Simplify.
$-1 \neq 15$	Solution does not check. ✗

So, $x = -2$ is a solution, but $x = 2$ is not a solution.

Exercises Within Reach ®

Solutions in English & Spanish and tutorial videos at CollegePrepAlgebra.com

Checking Solutions of an Equation **In Exercises 1−10, determine whether each value of x is a solution of the equation.**

	Equation	*Values*			*Equation*	*Values*	
1.	$2x - 18 = 0$	(a) $x = 0$	(b) $x = 9$	**2.**	$3x - 3 = 0$	(a) $x = 4$	(b) $x = 1$
3.	$6x + 1 = -11$	(a) $x = 2$	(b) $x = -2$	**4.**	$2x + 5 = -15$	(b) $x = -10$	(b) $x = 5$
5.	$x + 5 = 2x$	(a) $x = -1$	(b) $x = 5$	**6.**	$15 - 2x = 3x$	(a) $x = 3$	(b) $x = 5$
7.	$7x + 1 = 4(x - 2)$	(a) $x = 1$	(b) $x = 12$	**8.**	$5x - 1 = 3(x + 5)$	(a) $x = 8$	(b) $x = -2$
9.	$2x + 10 = 7(x + 1)$	(a) $x = \frac{3}{5}$	(b) $x = -\frac{2}{3}$	**10.**	$3(3x + 2) = 9 - x$	(a) $x = -\frac{3}{4}$	(b) $x = \frac{3}{10}$

EXAMPLE 2   **Comparing Equations and Expressions**

Make a table that compares algebraic expressions and algebraic equations.

SOLUTION

Algebraic Expression	Algebraic Equation
• Example: $4(x - 1)$	• Example: $4(x - 1) = 12$
• Contains no equal sign	• Contains an equal sign and is true for only certain values of the variable
• Can be evaluated for any real number for which the expression is defined	• Solution is found by forming equivalent equations using the properties of equality:
• Can sometimes be simplified to an equivalent form: $4(x - 1)$ simplifies to $4x - 4$	$4(x - 1) = 12$ $4x - 4 = 12$ $4x = 16$ $x = 4$

Exercises Within Reach®

Solutions in English & Spanish and tutorial videos at CollegePrepAlgebra.com

Comparing Equations and Expressions In Exercises 11−16, determine whether the statement describes an algebraic expression or an algebraic equation.

11. Contains an equal sign

12. Contains no equal sign

13. Can be evaluated for any real number for which it is defined

14. Is true for only certain values of the variable

15. The solution is found by using the properties of equality

16. A statement of equality

Comparing Equations and Expressions In Exercises 17−26, determine whether an algebraic expression or an algebraic equation is given.

17. $3x$

18. $\frac{1}{2}x$

19. $-7x - 8 = 0$

20. $7 = 9 - x$

21. $\frac{5}{6}x + 1$

22. $x - 4$

23. $2x + 1 = 4(x - 10)$

24. $\frac{1}{9}(4 + 8x) = -15$

25. $x^2 + 2x + 1$

26. $x^2 - 4$

Forming Equivalent Equations

Forming Equivalent Equations: Properties of Equality

An equation can be transformed into an **equivalent equation** using one or more of the following procedures.

		Original Equation	*Equivalent Equation(s)*
1.	*Simplify either side:* Remove symbols of grouping, combine like terms, or simplify fractions on one or both sides of the equation.	$3x - x = 8$	$2x = 8$
2.	*Apply the Addition Property of Equality:* Add (or subtract) the same quantity to (from) *each* side of the equation.	$x - 2 = 5$	$x - 2 + 2 = 5 + 2$ $x = 7$
3.	*Apply the Multiplication Property of Equality:* Multiply (or divide) each side of the equation by the same *nonzero* quantity.	$3x = 9$	$\dfrac{3x}{3} = \dfrac{9}{3}$ $x = 3$
4.	*Interchange the two sides of the equation.*	$7 = x$	$x = 7$

EXAMPLE 3 **Forming Equivalent Equations**

The second and third procedures in the above list can be used to eliminate terms or factors of an equation. For example, to solve the equation $x - 5 = 1$, you need to eliminate the term -5 on the left side. This is accomplished by adding its opposite, 5, to each side.

$$x - 5 = 1 \qquad \text{Write original equation.}$$
$$x - 5 + 5 = 1 + 5 \qquad \text{Add 5 to each side.}$$
$$x + 0 = 6 \qquad \text{Combine like terms.}$$
$$x = 6 \qquad \text{Solution}$$

These four equations are equivalent, and they are called the *steps* of the solution.

Exercises Within Reach ®

Solutions in English & Spanish and tutorial videos at CollegePrepAlgebra.com

Solving an Equation In Exercises 27–32, solve the equation.

27. $x - 8 = 5$

28. $4 = x - 11$

29. $x + 3 = 19$

30. $10 + x = 3$

31. $3x = 30$

32. $\frac{3}{2}x = 9$

EXAMPLE 4   **Identifying Properties of Equality**

Identify the property of equality used to solve each equation.

a. $x - 5 = 0$ Original equation

$x - 5 + 5 = 0 + 5$ Add 5 to each side.

$x = 5$ Solution

b. $\dfrac{x}{5} = -2$ Original equation

$\dfrac{x}{5}(5) = -2(5)$ Multiply each side by 5.

$x = -10$ Solution

c. $4x = 9$ Original equation

$\dfrac{4x}{4} = \dfrac{9}{4}$ Divide each side by 4.

$x = \dfrac{9}{4}$ Solution

SOLUTION

a.   The Addition Property of Equality is used to add 5 to each side of the equation in the second step. Adding 5 eliminates the term -5 from the left side of the equation.

b.   The Multiplication Property of Equality is used to multiply each side of the equation by 5 in the second step. Multiplying by 5 eliminates the denominator from the left side of the equation.

c.   The Multiplication Property of Equality is used to divide each side of the equation by 4 (or multiply each side by 1/4) in the second step. Dividing by 4 eliminates the coefficient from the left side of the equation.

> **Study Tip**
>
> In Example 4(c), each side of the equation is divided by 4 to eliminate the coefficient 4 on the left side. You could just as easily *multiply* each side by 1/4. Both techniques are legitimate—which one you decide to use is a matter of personal preference.

Exercises Within Reach ®

Solutions in English & Spanish and tutorial videos at CollegePrepAlgebra.com

Identifying Properties of Equality In Exercises 33−38, justify each step of the equation. Then identify any properties of equality used to solve the equation.

33. $x - 8 = 3$

$x - 8 + 8 = 3 + 8$

$x = 11$

34. $x + 4 = 16$

$x + 4 - 4 = 16 - 4$

$x = 12$

35. $\dfrac{2}{3}x = 12$

$\dfrac{3}{2}\left(\dfrac{2}{3}x\right) = \dfrac{3}{2}(12)$

$x = 18$

36. $\dfrac{4}{5}x = -28$

$\dfrac{5}{4}\left(\dfrac{4}{5}x\right) = \dfrac{5}{4}(-28)$

$x = -35$

37. $5x + 12 = 22$

$5x + 12 - 12 = 22 - 12$

$5x = 10$

$\dfrac{5x}{5} = \dfrac{10}{5}$

$x = 2$

38. $14 - 3x = 5$

$14 - 3x - 14 = 5 - 14$

$14 - 14 - 3x = 5 - 14$

$-3x = -9$

$\dfrac{-3x}{-3} = \dfrac{-9}{-3}$

$x = 3$

Writing Equations

It is helpful to use two phases in writing equations that model real-life situations, as shown below.

| Value description | ⇒ | Verbal model | ⇒ | Assign labels | ⇒ | Algebraic equation |

Phase 1   Phase 2

In the first phase, you translate the verbal description into a verbal model. In the second phase, you assign labels and translate the verbal model into a mathematical model or an algebraic equation.

Application   **EXAMPLE 5**   **Using a Verbal Model to Write an Equation**

Write an algebraic equation for the following problem.

> The total income that an employee received in a year was $40,950. How much was the employee paid each week? Assume that each weekly paycheck contained the same amount, and that the year consisted of 52 weeks.

SOLUTION

Verbal Model:

$$\text{Income for year} = \text{Number of weeks in a year} \cdot \text{Weekly pay}$$

Labels:

Income for year = 40,950   (dollars)
Weekly pay = x (dollars per week)
Number of weeks = 52   (weeks)

Expression: $40,950 = 52x$

> ### Study Tip
>
> When you write an equation, be sure to check that both sides of the equation represent the *same* unit of measure. For instance, in Example 5, both sides of the equation $40,950 = 52x$ represent dollar amounts.

Exercises Within Reach®

Solutions in English & Spanish and tutorial videos at CollegePrepAlgebra.com

Writing an Algebraic Equation In Exercises 39−42, write an algebraic equation. Do *not* solve the equation.

39. After your instructor added 6 points to each student's test score, your score is 94. What was your original score?

40. With the 1.2-inch rainfall today, the total for the month is 4.5 inches. How much had been recorded for the month before today's rainfall?

41. During a football game, a running back carried the ball 18 times and his average number of yards per carry was 4.5. How many yards did the running back gain for the game?

42. The total cost of admission for 6 adults at an aquarium is $132. What is the cost per adult?

Application   EXAMPLE 6   **Using a Verbal Model to Write an Equation**

Write an algebraic equation for the following problem.

Returning to college after spring break, you travel 3 hours and stop for lunch. You know that it takes 45 minutes to complete the last 36 miles of the 180-mile trip. What was the average speed during the first 3 hours of the trip?

SOLUTION

Verbal Model: Distance = Rate · Time

Labels:	Distance = $180 - 36 = 144$	(miles)
	Rate = r	(miles per hour)
	Time = 3	(hours)

Expression: $144 = 3r$

Study Tip

In Example 6, the information that it takes 45 minutes to complete the last part of the trip is unnecessary information. This type of unnecessary information in an applied problem is sometimes called a *red herring*.

Application   EXAMPLE 7   **Using a Verbal Model to Write an Equation**

Write an algebraic equation for the following problem.

Tickets for a concert cost $175 for each floor seat and $95 for each stadium seat. There were 2500 seats on the main floor, and these were sold out. The total revenue from ticket sales was $865,000. How many stadium seats were sold?

SOLUTION

Verbal Model: Total revenue = Revenue from floor seats + Revenue from stadium seats

Labels:	Total revenue = 865,000	(dollars)
	Price per floor seat = 175	(dollars per seat)
	Number of floor seats = 2500	(seats)
	Price per stadium seat = 95	(dollars per seat)
	Number of stadium seats = x	(seats)

Expression: $865,000 = 175(2500) + 95x$

Exercises Within Reach ®

Solutions in English & Spanish and tutorial videos at CollegePrepAlgebra.com

Writing an Algebraic Equation In Exercises 43 and 44, **write** an algebraic equation. Do *not* solve the equation.

43. You want to volunteer at a soup kitchen for 150 hours over a 15-week period. After 8 weeks, you have volunteered for 72 hours. How many hours will you have to work per week over the remaining 7 weeks to reach your goal?

44. A textile corporation buys equipment with an initial purchase price of $750,000. It is estimated that its useful life will be 3 years and at that time its value will be $75,000. The total depreciation is divided equally among the three years. (Depreciation is the difference between the initial price of an item and its current value.) What is the total amount of depreciation declared each year?

Concept Summary: *Writing and Solving Equations*

What

You can use verbal models to write **equations** and algebra to **solve** equations.

EXAMPLE

The sale price of a coat is $250. The discount is $25. What is the original price?

How

To write an equation that models a real-life situation, first translate the verbal description into a verbal model. Then assign labels and translate the verbal model into an algebraic equation. You can solve the equation by creating **equivalent equations**.

EXAMPLE

Verbal Model:

$$\boxed{\text{Original price}} - \boxed{\text{Discount}} = \boxed{\text{Sale price}}$$

Labels: Original price $= x$
Discount $= 25$
Sale price $= 250$

Equation: $x - 25 = 250$

Why

You can use verbal models to write algebraic equations that model many real-life situations. Knowing how to use properties of equality will help you solve these equations.

Exercises Within Reach®

Worked-out solutions to odd-numbered exercises at CollegePrepAlgebra.com

Concept Summary Check

45. *Vocabulary* Is there more than one way to write a verbal model? Explain.

46. *Writing* Describe the steps that can be used to transform an equation into an equivalent equation.

47. *Number Sense* When dividing each side of an equation by the same quantity, why must the quantity be nonzero?

48. *Writing* Describe how to solve $x - 25 = 250$.

Extra Practice

Identifying Properties of Equality **In Exercises 49 and 50, justify each step of the equation. Then identify any properties of equality used to solve the equation.**

49.
$$\frac{x}{3} = x + 1$$
$$3\left(\frac{x}{3}\right) = 3(x + 1)$$
$$x = 3x + 3$$
$$x - 3x = 3x + 3 - 3x$$
$$x - 3x = 3x - 3x + 3$$
$$-2x = 3$$
$$\frac{-2x}{-2} = \frac{3}{-2}$$
$$x = -\frac{3}{2}$$

50.
$$\frac{4}{5}x = 4x - 16$$
$$\frac{5}{4}\left(\frac{4}{5}x\right) = \frac{5}{4}(4x - 16)$$
$$x = 5x - 20$$
$$x - 5x = 5x - 20 - 5x$$
$$x - 5x = 5x - 5x - 20$$
$$-4x = -20$$
$$\frac{-4x}{-4} = \frac{-20}{-4}$$
$$x = 5$$

Checking Solutions of an Equation **In Exercises 51–54, determine whether each value of x is a solution of the equation.**

Equation	Values

51. $\dfrac{2}{x} - \dfrac{1}{x} = 1$ (a) $x = 0$ (b) $x = \dfrac{1}{3}$

52. $\dfrac{4}{x} + \dfrac{2}{x} = 1$ (a) $x = 0$ (b) $x = 6$

53. $\dfrac{5}{x-1} + \dfrac{1}{x} = 5$ (a) $x = 3$ (b) $x = \dfrac{1}{6}$

54. $\dfrac{3}{x-2} = x$ (a) $x = -1$ (b) $x = 3$

Writing an Algebraic Equation **In Exercises 55–58, write an algebraic equation. Do *not* solve the equation.**

55. A student has n quarters and seven \$1 bills totaling \$8.75. How many quarters does the student have?

56. A school science club conducts a car wash to raise money. The club spends \$12 on supplies and charges \$5 per car. After the car wash, the club has a profit of \$113. How many cars did the members of the science club wash?

57. A high school earned \$986 in revenue for a play. Tickets for the play cost \$10 for adults and \$6 for students. The number of students attending the play was $\frac{3}{4}$ the number of adults attending the play. How many adults and student attended the play?

58. An ice show earns a revenue of \$11,041 one night. Tickets for the ice show cost \$18 for adults and \$13 for children. The number of adults attending the ice show was 33 more than the number of children attending the show. How many adults and children attended the show?

Explaining Concepts

59. *Number Sense* Are there any equations of the form $ax = b$ $(a \neq 0)$ that are true for more than one value of x? Explain.

60. *Structure* Determine which equations are equivalent to $14 = x + 8$.

  (a) $x + 8 = 14$ (b) $8x = 14$

  (c) $x - 8 = 14$ (d) $8 + x = 14$

  (e) $2(x + 4) - x = 14$

  (f) $3(x + 6) - 2x + 5 = 14$

61. *Modeling* Describe a real-life problem that uses the following verbal model.

$$\begin{array}{|c|} \hline \text{Revenue} \\ \text{of \$840} \\ \hline \end{array} = \begin{array}{|c|} \hline \text{\$35 per} \\ \text{case} \\ \hline \end{array} + \begin{array}{|c|} \hline \text{Number} \\ \text{of cases} \\ \hline \end{array}$$

62. *Writing* Explain the difference between simplifying an expression and solving an equation. Give an example of each.

Cumulative Review

In Exercises 63–68, simplify the expression.

63. $t^2 \cdot t^5$

64. $(-3y^3)y^2$

65. $6x + 9x$

66. $4 - 3t + t$

67. $-(-8b)$

68. $7(-10x)$

In Exercises 69–72, translate the phrase into an algebraic expression. Let x represent the real number.

69. 23 more than x

70. c divided by 6

71. Seven more than 4 times y

72. Nine times the difference of h and 3

2 Chapter Summary

	What did you learn?	*Explanation and Examples*	*Review Exercises*
2.1	Define and identify terms, variables, and coefficients of algebraic expressions *(p. 56)*.	In the expression $4x + 3y$, $4x$ and $3y$ are the terms of the expression, x and y are the variables, and 4 and 3 are the coefficients.	$1-10$
	Define exponential form and interpret exponential expressions *(p. 57)*.	Repeated multiplication can be expressed in exponential form using a base a and an exponent n, where a is a real number, variable, or algebraic expression, and n is a positive integer. $$a^n = \underbrace{a \cdot a \cdot a \cdot a \cdots a}_{n \text{ factors}}$$	$11-16$
	Evaluate algebraic expressions using real numbers *(p. 58)*.	To evaluate an algebraic expression, replace every occurrence of the specified variable in the expression with the appropriate real number, and perform the operation(s).	$17-24$
2.2	Use the properties of algebra *(p. 64)*.	Let a, b, and c represent real numbers, variables, or algebraic expressions. Commutative Property of Addition $\qquad a + b = b + a$ Commutative Property of Multiplication $\qquad ab = ba$ Associative Property of Addition $\quad (a + b) + c = a + (b + c)$ Associative Property of Multiplication $\quad (ab)c = a(bc)$ Distributive Property $\quad a(b + c) = ab + ac \qquad a(b - c) = ab - ac$ $\quad (a + b)c = ac + bc \qquad (a - b)c = ac - bc$ Additive Identity Property $\qquad a + 0 = 0 + a = a$ Multiplicative Identity Property $\qquad a \cdot 1 = 1 \cdot a = a$ Additive Inverse Property $\qquad a + (-a) = 0$ Multiplicative Inverse Property $\qquad a \cdot \dfrac{1}{a} = 1,\, a \neq 0$	$25-30$
	Combine like terms of an algebraic expression *(p. 67)*.	To combine like terms in an algebraic expression, add their respective coefficients and attach the common variable factor(s).	$31-42$
	Simplify an algebraic expression by rewriting the terms *(p. 68)*.	To simplify an algebraic expression, remove symbols of grouping and combine like terms.	$43-54$
	Use the Distributive Property to remove symbols of grouping *(p. 69)*.	The main tool for removing symbols of grouping is the Distributive Property.	$55-66$
2.3	Construct verbal mathematical models from written statements *(p. 76)*.	A person riding a scooter travels at a constant rate of 32 miles per hour. Write an expression showing how far the person can ride in t hours. Verbal Model: $\quad$ Rate of travel $\cdot$ Time traveled Expression: $\quad 32t$	67, 68

What did you learn?	*Explanation and Examples*	*Review Exercises*
2.3 Translate verbal phrases into algebraic expressions *(p. 78).*	When translating verbal phrases into algebraic expressions, look for key words and phrases that indicate the four different operations of arithmetic. **Addition:** Sum, plus, greater than, increased by, more than, exceeds, total of **Subtraction:** Difference, minus, less than, decreased by, subtracted from, reduced by **Multiplication:** Product, multiplied by, twice, times, percent of **Division:** Quotient, divided by, ratio, per	69–78
Identify hidden operations when writing algebraic expressions *(p. 80).*	Most real-life problems do not contain verbal expressions that clearly identify all the arithmetic operations involved. You need to rely on past experience and the physical nature of the problem in order to identify the operations hidden in the problem statement. Multiplication is the operation most commonly hidden in real-life applications.	79–82
Use problem-solving strategies to solve application problems *(p. 82).*	1. **Guess, Check, and Revise** Guess a reasonable solution based on the given data. Check the guess, and revise it, if necessary. Continue guessing, checking, and revising until a correct solution is found. 2. **Make a Table/Look for a Pattern** Make a table using the data in the problem. Look for a number pattern. Then use the pattern to complete the table or find a solution. 3. **Draw a Diagram** Draw a diagram that shows the facts of the problem. Use the diagram to visualize the action of the problem. Use algebra to find a solution. Then check the solution against the facts. 4. **Solve a Simpler Problem** Construct a simpler problem that is similar to the original problem. Solve the simpler problem. Then use the same procedure to solve the original problem.	83, 84
2.4 Check whether a given value is a solution of an equation *(p. 86).*	To check whether a given value is a solution of an equation, substitute the value into the original equation. If the substitution results in a true statement, then the value is a solution of the equation. If the substitution results in a false statement, then the value is not a solution of the equation.	85–94
Use properties of equality to form equivalent equations *(p. 88).*	An equation can be transformed into an equivalent equation using one or more of the following procedures. 1. *Simplify either side:* Remove symbols of grouping, combine like terms, or simplify fractions on one or both sides of the equation. 2. *Apply the Addition Property of Equality:* Add (or subtract) the same quantity to (from) *each* side of the equation. 3. *Apply the Multiplication Property of Equality:* Multiply (or divide) each side of the equation by the same *nonzero* quantity. 4. *Interchange the two sides of the equation.*	95–98
Use a verbal model to write an algebraic equation *(p. 90).*	First, construct a verbal model. Then, assign labels to the known and unknown quantities and translate the verbal model into an algebraic equation.	99–102

Review Exercises

Worked-out solutions to odd-numbered exercises at CollegePrepAlgebra.com

2.1

Identifying Variables In Exercises 1−4, identify the variable(s) in the expression.

1. $15 - x$

2. $t - 5^2$

3. $a - 3b$

4. $y + z$

Identifying Terms and Coefficients In Exercises 5−10, identify the terms of the expression and their coefficients.

5. $12y + y^2$

6. $4x - \frac{1}{2}x^3$

7. $5x^2 - 3xy + 10y^2$

8. $y^2 - 10yz + \frac{2}{3}z^2$

9. $\frac{2y}{3} - \frac{4x}{y}$

10. $-\frac{4b}{9} + \frac{11a}{b}$

Rewriting in Exponential Form In Exercises 11−16, rewrite the product in exponential form.

11. $5z \cdot 5z \cdot 5z$

12. $\frac{3}{8}y \cdot \frac{3}{8}y \cdot \frac{3}{8}y \cdot \frac{3}{8}y$

13. $(-3x) \cdot (-3x) \cdot (-3x) \cdot (-3x) \cdot (-3x)$

14. $\left(-\frac{2}{7}\right) \cdot \left(-\frac{2}{7}\right) \cdot \left(-\frac{2}{7}\right)$

15. $(b - c) \cdot (b - c) \cdot 6 \cdot 6$

16. $2 \cdot (a + b) \cdot 2 \cdot (a + b) \cdot 2$

Evaluating an Algebraic Expression In Exercises 17−24, evaluate the algebraic expression at the given values of the variable(s).

	Expression		*Values*
17.	$x^2 - 2x + 5$	(a) $x = 0$	(b) $x = 2$
18.	$x^3 - 8$	(a) $x = 2$	(b) $x = 4$
19.	$x^2 - x(y + 1)$	(a) $x = 2, y = -1$	
		(b) $x = 1, y = 2$	
20.	$2r + r(t^2 - 3)$	(a) $r = 3, t = -2$	
		(b) $r = -2, t = 3$	
21.	$\frac{x + 5}{y}$	(a) $x = -5, y = 3$	
		(b) $x = 2, y = -1$	
22.	$\frac{a - 9}{2b}$	(a) $a = 7, b = -3$	
		(b) $a = -4, b = 5$	
23.	$x^2 - 2y + z$	(a) $x = 1, y = 2, z = 0$	
		(b) $x = 2, y = -3, z = -4$	
24.	$\frac{m + 2n}{-p}$	(a) $m = -1, n = 2, p = 3$	
		(b) $m = 4, n = -2, p = -7$	

2.2

Identifying Properties In Exercises 25−30, identify the property of algebra illustrated by the statement.

25. $xy \cdot \frac{1}{xy} = 1$

26. $u(vw) = (uv)w$

27. $(x - y)(2) = 2(x - y)$

28. $(a + b) + 0 = a + b$

29. $2x + (3y - z) = (2x + 3y) - z$

30. $x(y + z) = xy + xz$

Combining Like Terms In Exercises 31−40, simplify the expression by combining like terms.

31. $3a - 5a$

32. $6c - 2c$

33. $3p - 4q + q + 8p$

34. $10x - 4y - 25x + 6y$

35. $\frac{1}{4}s - 6t + \frac{7}{2}s + t$

36. $\frac{2}{3}a + \frac{3}{5}a - \frac{1}{2}b + \frac{2}{3}b$

37. $x^2 + 3xy - xy + 4$

38. $uv^2 + 10 - 2uv^2 + 2$

39. $5x - 5y + 3xy - 2x + 2y$

40. $y^3 + 2y^2 + 2y^3 - 3y^2 + 1$

41. *Number Sense* Simplify the algebraic expression that represents the sum of three consecutive odd integers, $2n - 1$, $2n + 1$, and $2n + 3$.

42. *Number Sense* Simplify the algebraic expression that represents the sum of three consecutive even integers, $2n$, $2n + 2$, and $2n + 4$.

Simplifying an Algebraic Expression **In Exercises 43–50, simplify the expression.**

43. $12(4t)$

44. $8(7x)$

45. $-5(-9x^2)$

46. $-10(-3b^3)$

47. $(-6x)(2x^2)$

48. $(-3y^2)(15y)$

49. $\dfrac{12x}{5} \cdot \dfrac{10}{3}$

50. $\dfrac{4z}{15} \cdot \dfrac{9}{2}$

51. *Geometry* Write and simplify an expression for (a) the perimeter and (b) the area of the rectangle.

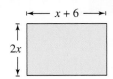

52. *Geometry* Write and simplify an expression for the area of the triangle.

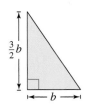

53. *Geometry* The face of a DVD player has the dimensions shown in the figure. Write an algebraic expression that represents the area of the face of the DVD player excluding the compartment holding the disc.

54. *Geometry* Write an expression for the perimeter of the figure. Use the rules of algebra to simplify the expression.

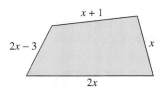

Removing Symbols of Grouping **In Exercises 55–66, simplify the expression by removing symbols of grouping and combining like terms.**

55. $5(u - 4) + 10$

56. $16 - 3(v + 2)$

57. $3s - (r - 2s)$

58. $50x - (30x + 100)$

59. $-3(1 - 10z) + 2(1 - 10z)$

60. $8(15 - 3y) - 5(15 - 3y)$

61. $\frac{1}{3}(42 - 18z) - 2(8 - 4z)$

62. $\frac{1}{4}(100 + 36s) - (15 - 4s)$

63. $10 - [8(5 - x) + 2]$

64. $3[2(4x - 5) + 4] - 3$

65. $2[x + 2(y - x)]$

66. $2t[4 - (3 - t)] + 5t$

2.3

Writing an Algebraic Expression **In Exercises 67 and 68,** construct **a verbal model and then** write **an algebraic expression that represents the specified quantity.**

67. The total hourly wage for an employee who earns $8.25 per hour and an additional $0.60 for each unit produced per hour.

68. The total cost for a family to stay one night at a campground when the charge is $18 for the parents plus $3 for each of the children.

Translating a Verbal Phrase **In Exercises 69–78,** translate **the verbal phrase into an algebraic expression.**

69. The sum of two-thirds of x and 5

70. One hundred decreased by the product of 5 and b

71. Ten less than twice y

72. The ratio of c and 10

73. Fifty increased by the product of 7 and z

74. Ten decreased by the quotient of a and 2

75. The sum of s and 10, all divided by 8

76. The product of 15 and d, all decreased by 2

77. The sum of the square of g and 64

78. The absolute value of the sum of t and -10

79. *Commission* A salesperson earns 5% commission on his total weekly sales x. Write an algebraic expression that represents the amount of commission that the salesperson earns in a week.

80. *Sale Price* A cordless phone is advertised for 20% off the list price of L dollars. Write an algebraic expression that represents the sale price of the phone.

81. *Rent* The monthly rent for your apartment is $625. Write an algebraic expression that represents the total rent for n months.

82. *Distance* A car travels for 10 hours at an average speed of s miles per hour. Write an algebraic expression that represents the total distance traveled by the car.

83. *Finding a Pattern* Describe the pattern, and use your description to find the value of the expression when $n = 20$.

n	0	1	2	3	4	5
Value of expression	4	7	10	13	16	19

84. *Finding a Pattern* Find values of a and b such that the expression $an + b$ yields the values in the table.

n	0	1	2	3	4	5
$an + b$	4	9	14	19	24	29

2.4

Checking Solutions of an Equation **In Exercises 85–94,** determine **whether each value of x is a solution of the equation.**

	Equation	*Values*
85.	$5x + 6 = 36$	(a) $x = 3$ (b) $x = 6$
86.	$17 - 3x = 8$	(a) $x = 3$ (b) $x = -3$
87.	$3x - 12 = x$	(a) $x = -1$ (b) $x = 6$

	Equation	Values	
88.	$8x + 24 = 2x$	(a) $x = 0$	(b) $x = -4$
89.	$4(2 - x) = 3(2 + x)$	(a) $x = \frac{2}{7}$	(b) $x = -\frac{2}{3}$
90.	$5x + 2 = 3(x + 10)$	(a) $x = 14$	(b) $x = -10$
91.	$\frac{4}{x} - \frac{2}{x} = 5$	(a) $x = -1$	(b) $x = \frac{2}{5}$
92.	$\frac{x}{3} + \frac{x}{6} = 1$	(a) $x = \frac{2}{9}$	(b) $x = -\frac{2}{9}$
93.	$x(x - 7) = -12$	(a) $x = 3$	(b) $x = 4$
94.	$x(x + 1) = 2$	(a) $x = 1$	(b) $x = -2$

Identifying Properties of Equality In Exercises
95–98, **justify each step of the equation. Then identify
any properties of equality used to solve the equation.**

95.
$$-7x + 20 = -1$$
$$-7x + 20 - 20 = -1 - 20$$
$$-7x = -21$$
$$\frac{-7x}{-7} = \frac{-21}{-7}$$
$$x = 3$$

96.
$$3(x - 2) = x + 2$$
$$3x - 6 = x + 2$$
$$3x - 6 - x = x + 2 - x$$
$$3x - x - 6 = x - x + 2$$
$$2x - 6 = 2$$
$$2x - 6 + 6 = 2 + 6$$
$$2x = 8$$
$$\frac{2x}{2} = \frac{8}{2}$$
$$x = 4$$

97.
$$x = -(x - 14)$$
$$x = -x + 14$$
$$x + x = -x + 14 + x$$
$$x + x = -x + x + 14$$
$$2x = 14$$
$$\frac{2x}{2} = \frac{14}{2}$$
$$x = 7$$

98.
$$\frac{x}{4} = x - 2$$
$$4\left(\frac{x}{4}\right) = 4(x - 2)$$
$$x = 4x - 8$$
$$x - 4x = 4x - 8 - 4x$$
$$x - 4x = 4x - 4x - 8$$
$$-3x = -8$$
$$\frac{-3x}{-3} = \frac{-8}{-3}$$
$$x = \frac{8}{3}$$

Writing an Algebraic Equation In Exercises 99–102,
write an algebraic equation. Do *not* solve the equation.

99. The sum of a number and its reciprocal is $\frac{37}{6}$. What is
the number?

100. A car travels 135 miles in t hours with an average
speed of 45 miles per hour (see figure). How many
hours did the car travel?

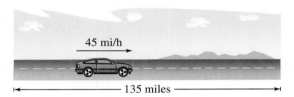

45 mi/h

135 miles

101. The area of the shaded region in the figure is 24 square
inches. What is the length of the rectangle?

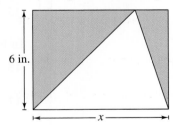

6 in.

x

102. The perimeter of the face of a rectangular traffic light
is 72 inches (see figure). What are the dimensions of
the traffic light?

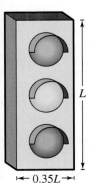

L

$0.35L$

Chapter Test

Solutions in English & Spanish and tutorial videos at CollegePrepAlgebra.com

Take this test as you would take a test in class. After you are done, check your work against the answers in the back of the book.

1. Identify the terms of the expression and their coefficients.

   $2x^2 - 7xy + 3y^3$

2. Rewrite the product in exponential form.

   $x \cdot (x + y) \cdot x \cdot (x + y) \cdot x$

In Exercises 3–6, identify the property of algebra illustrated by the statement.

3. $(5x)y = 5(xy)$

4. $2 + (x - y) = (x - y) + 2$

5. $7xy + 0 = 7xy$

6. $(x + 5) \cdot \dfrac{1}{(x + 5)} = 1$

In Exercises 7–10, use the Distributive Property to expand the expression.

7. $3(x + 8)$

8. $5(4r - s)$

9. $-y(3 - 2y)$

10. $-9(4 - 2x + x^2)$

In Exercises 11–14, simplify the expression.

11. $3b - 2a + a - 10b$

12. $15(u - v) - 7(u - v)$

13. $3z - (4 - z)$

14. $2[10 - (t + 1)]$

In Exercises 15 and 16, evaluate the expression for $x = 2$ and $y = -10$.

15. $x^3 - 2$

16. $x^2 + 4(y + 2)$

17. Explain why it is not possible to evaluate $\dfrac{a + 2b}{3a - b}$ when $a = 2$ and $b = 6$.

18. Translate the phrase "four less than one-third of n" into an algebraic expression.

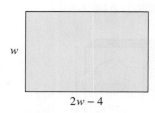

w

$2w - 4$

19. (a) Write expressions for the perimeter and area of the rectangle at the left. Simplify each expression.

    (b) Evaluate each expression for $w = 7$.

20. The prices of concert tickets for adults and children are $25 and $20, respectively.

    (a) Write an algebraic expression that represents the total income from the concert for m adults and n children.

    (b) How much will it cost two adults and three children to attend the concert?

21. Determine whether the values of x are solutions of $6(3 - x) - 5(2x - 1) = 7$.

    (a) $x = -2$

    (b) $x = 1$

3

Equations, Inequalities, and Problem Solving

3.1   Solving Linear Equations

3.2   Equations that Reduce to Linear Form

3.3   Problem Solving with Percents

3.4   Ratios and Proportions

3.5   Geometric and Scientific Applications

3.6   Linear Inequalities

3.7   Absolute Value Equations and Inequalities

MASTERY IS WITHIN REACH!

"When I study with a tutor, I always understand better and quicker when I do the problem on the board. I just see it better and it clicks. It turns out that I am a visual and kinesthetic learner when I study math. I spend a lot of time in the tutoring center where there are whiteboards."

Cynthia

See page 135 for suggestions about using your preferred learning modality.

3.1 Solving Linear Equations

▶ Solve a linear equation in standard form.

▶ Solve a linear equation in nonstandard form.

Solving Linear Equations in Standard Form

A **linear equation in one variable** x is an equation that can be written in the standard form

$$ax + b = 0.$$ a and b are real numbers with $a \neq 0$.

EXAMPLE 1   **Solving Linear Equations in Standard Form**

Solve $3x - 15 = 0$. Then check the solution.

SOLUTION

$3x - 15 = 0$	Write original equation.
$3x - 15 + 15 = 0 + 15$	Add 15 to each side.
$3x = 15$	Combine like terms.
$\dfrac{3x}{3} = \dfrac{15}{3}$	Divide each side by 3.
$x = 5$	Simplify.

The solution is $x = 5$. You can check this as shown at the right.

Study Tip

Remember that to *solve* an equation involving x means to find all values of x that make the equation true.

Check

$3x - 15 = 0$	Write original equation.
$3(5) - 15 \overset{?}{=} 0$	Substitute 5 for x.
$15 - 15 \overset{?}{=} 0$	Multiply.
$0 = 0$	Solution checks. ✔

Exercises Within Reach ®

Solutions in English & Spanish and tutorial videos at CollegePrepAlgebra.com

Mental Math In Exercises 1−4, solve **the equation mentally.**

1. $x - 9 = 0$ **2.** $u - 3 = 0$ **3.** $x + 6 = 0$ **4.** $a + 5 = 0$

Deciding How to Start In Exercises 5−12, decide **which operation you would use first to solve the equation.**

5. $x + 8 = 0$ **6.** $p - 1 = 0$ **7.** $\dfrac{y}{2} = 0$ **8.** $3z = 0$

9. $2x + 4 = 0$ **10.** $2d + 10 = 0$ **11.** $3x - 6 = 0$ **12.** $4x - 20 = 0$

Justifying Steps In Exercises 13 and 14, justify **each step of the solution.**

13.
$$5x + 15 = 0$$
$$5x + 15 - 15 = 0 - 15$$
$$5x = -15$$
$$\frac{5x}{5} = \frac{-15}{5}$$
$$x = -3$$

14.
$$-2x - 8 = 0$$
$$-2x - 8 + 8 = 0 + 8$$
$$-2x = 8$$
$$\frac{-2x}{-2} = \frac{8}{-2}$$
$$x = -4$$

EXAMPLE 2   **Solving Linear Equations in Standard Form**

a.
$$2x + 18 = 0$$ Original equation

$$2x + 18 - 18 = 0 - 18$$ Subtract 18 from each side.

$$2x = -18$$ Combine like terms.

$$\frac{2x}{2} = -\frac{18}{2}$$ Divide each side by 2.

$$x = -9$$ Simplify. Check in original equation.

b.
$$5x - 12 = 0$$ Original equation

$$5x - 12 + 12 = 0 + 12$$ Add 12 to each side.

$$5x = 12$$ Combine like terms.

$$\frac{5x}{5} = \frac{12}{5}$$ Divide each side by 5.

$$x = \frac{12}{5}$$ Simplify. Check in original equation.

c.
$$\frac{x}{3} + 3 = 0$$ Original equation

$$\frac{x}{3} + 3 - 3 = 0 - 3$$ Subtract 3 from each side.

$$\frac{x}{3} = -3$$ Combine like terms.

$$3\left(\frac{x}{3}\right) = 3(-3)$$ Multiply each side by 3.

$$x = -9$$ Simplify. Check in original equation.

Study Tip

As you gain experience, you will find that you can perform some of the solution steps in your head. For instance, you might solve part (c) by writing only the following steps.

$$\frac{x}{3} + 3 = 0$$

$$\frac{x}{3} = -3$$

$$x = -9$$

Exercises Within Reach ®

Solutions in English & Spanish and tutorial videos at CollegePrepAlgebra.com

Solving a Linear Equation In Exercises 15−32, solve the equation and check your solution.

15. $10x + 10 = 0$

16. $8f + 8 = 0$

17. $9g - 18 = 0$

18. $8x - 16 = 0$

19. $4x - 24 = 0$

20. $7x - 21 = 0$

21. $2x + 52 = 0$

22. $3x + 21 = 0$

23. $8x - 2 = 0$

24. $6x - 4 = 0$

25. $\frac{y}{4} + 7 = 0$

26. $\frac{x}{2} + 1 = 0$

27. $-x + 9 = 0$

28. $-q + 1 = 0$

29. $-5x - 15 = 0$

30. $-11y - 44 = 0$

31. $-3p + 1 = 0$

32. $-9w + 6 = 0$

Solving Linear Equations in Nonstandard Form

EXAMPLE 3   **Solving a Linear Equation in Nonstandard Form**

$3y + 8 - 5y = 4$	Original equation
$3y - 5y + 8 = 4$	Group like terms.
$-2y + 8 = 4$	Combine like terms.
$-2y + 8 - 8 = 4 - 8$	Subtract 8 from each side.
$-2y = -4$	Combine like terms.
$\dfrac{-2y}{-2} = \dfrac{-4}{-2}$	Divide each side by -2.
$y = 2$	Simplify. Check in original equation.

EXAMPLE 4   **Solving Linear Equations: Special Cases**

a.

$2x + 3 = 2(x + 4)$	Original equation
$2x + 3 = 2x + 8$	Apply Distributive Property.
$2x - 2x + 3 = 2x - 2x + 8$	Subtract $2x$ from each side.
$3 \neq 8$	Simplify.

Because 3 does not equal 8, the original equation has no solution.

b.

$4(x + 3) = 4x + 12$	Original equation
$4x + 12 = 4x + 12$	Apply Distributive Property.
$4x - 4x + 12 = 4x - 4x + 12$	Subtract $4x$ from each side.
$12 = 12$	Simplify.

Because the last equation is true for any value of x, the original equation has infinitely many solutions.

Study Tip

Equations like the one in Example 4(b) that are true for all values of x are called **identities**.

Exercises Within Reach ®   Solutions in English & Spanish and tutorial videos at CollegePrepAlgebra.com

Solving a Linear Equation in Nonstandard Form In Exercises 33–50, **solve the equation and check your solution. (Some of the equations have no solution.)**

33. $3y - 2 = 2y$

34. $2s - 13 = 28s$

35. $4 - 7x = 5x$

36. $24 - 5x = x$

37. $4 - 5t = 16 + t$

38. $3x + 4 = x + 10$

39. $-3t + 5 = -3t$

40. $4z + 2 = 4z$

41. $4x - 2 = 3x + 1$

42. $7x + 9 = 3x + 1$

43. $4x - 6 = 4x - 6$

44. $5 - 3x = 5 - 3x$

45. $2x + 4 = -3(x - 2)$

46. $4(y + 1) = -y + 5$

47. $5(3 - x) = x - 12$

48. $12 - w = -2(3w - 1)$

49. $2x = -3x$

50. $6t = 9t$

Application   EXAMPLE 5   **Geometry: Finding Dimensions**

You have 96 feet of fencing to enclose a rectangular exercise area for your Border Collie. The exercise area is to be three times as long as it is wide. Find the dimensions of the exercise area.

SOLUTION

Begin by drawing and labeling a diagram.

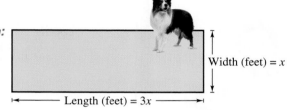

Labels & Diagram:

Width (feet) = x

$\longmapsto$ — Length (feet) = $3x$ — $\longrightarrow$

The perimeter of a rectangle is the sum of twice its length and twice its width.

Verbal Model: Perimeter = 2 • Length + 2 • Width
Equation: $96 = 2(3x) + 2x$

You can solve this equation as follows.

$96 = 6x + 2x$	Multiply.
$96 = 8x$	Combine like terms.
$\dfrac{96}{8} = \dfrac{8x}{8}$	Divide each side by 8.
$12 = x$	Simplify.

So, the width is 12 feet and the length is $3(12) = 36$ feet.

The Border Collie is a herding dog breed developed in the English-Scottish border region for herding livestock, especially sheep. Typically energetic and athletic, they often compete with success in dog sports.

Exercises Within Reach ®

Solutions in English & Spanish and tutorial videos at CollegePrepAlgebra.com

51. *Geometry* The sides of a yield sign all have the same length (see figure). The perimeter of a roadway yield sign is 225 centimeters. Find the length of each side.

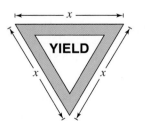

Figure for 51   Figure for 52

52. *Geometry* The length of a tennis court is 6 feet more than twice the width (see figure). Find the width of the court when the length is 78 feet.

53. *Geometry* The perimeter of the Jamaican flag is 120 inches. Its length is twice its width. Find the dimensions of the flag.

54. *Geometry* You are asked to cut a 12-foot board into 3 pieces. Two pieces are to have the same length and the third is to be twice as long as the others. How long are the pieces?

Application    EXAMPLE 6   **Finding the Number of Stadium Seats Sold**

Tickets for a concert cost $175 for each floor seat and $95 for each stadium seat. There were 2500 seats on the main floor, and these were sold out. The total revenue from ticket sales was $865,000. How many stadium seats were sold?

SOLUTION

Verbal Model: | Total revenue | = | Revenue from floor seats | + | Revenue from stadium seats |

Labels:
Total revenue = 865,000   (dollars)
Price per floor seat = 175   (dollars per seat)
Number of floor seats = 2500   (seats)
Price per stadium seat = 95   (dollars per seat)
Number of stadium seats = x (seats)

Equation: $865,000 = 175(2500) + 95x$ (See page 91.)

You can solve this equation as follows.

$$865,000 = 175(2500) + 95x$$ Write equation.
$$865,000 = 437,500 + 95x$$ Simplify.
$$865,000 - 437,500 = 437,500 - 437,500 + 95x$$ Subtract 437,500 from each side.
$$427,500 = 95x$$ Combine like terms.
$$\frac{427,500}{95} = \frac{95x}{95}$$ Divide each side by 95.
$$4500 = x$$ Simplify.

There were 4500 stadium seats sold. To check this, go back to the original statement of the problem.

Exercises Within Reach®

Solutions in English & Spanish and tutorial videos at CollegePrepAlgebra.com

55. *Ticket Sales* Tickets for a community theater cost $10 for each main floor seat and $8 for each balcony seat. There are 400 seats on the main floor, and these seats were sold out for the evening performance. The total revenue from ticket sales was $5200. How many balcony seats were sold?

56. *Ticket Sales* Tickets for a drumline competition cost $5 at the gate and $3 in advance. Eight hundred tickets were sold at the gate. The total revenue from tickets sales was $5500. How many advance tickets were sold?

57. *Car Repair* The bill (including parts and labor) for the repair of your car is shown. Some of the bill is unreadable. From what is given, can you determine how many hours were spent on labor? Explain.

Parts	$285.00
Labor ($44 per hour)	$
Total	**$384.00**

58. *Car Repair* The bill for the repair of your car was $553. The cost for parts was $265. The cost for labor was $48 per hour. How many hours did the repair work take?

Application   EXAMPLE 7   **Finding Your Gross Pay per Paycheck**

Write an algebraic equation that represents the following problem. Then solve the equation and answer the question.

> *You have accepted a job offer at an annual salary of $40,830. This salary includes a year-end bonus of $750. You are paid twice a month. What will your gross pay be for each paycheck?*

SOLUTION

You will receive 24 paychecks and 1 bonus check during the year.

Verbal Model: $\boxed{\text{Income for year}} = 24 \times \boxed{\text{Amount of each paycheck}} + \boxed{\text{Bonus}}$

Labels: Income for year $= 40{,}830$ (dollars)
Amount of each paycheck $= x$ (dollars)
Bonus $= 750$ (dollars)

Equation:

$40{,}830 = 24x + 750$	Write equation.
$40{,}080 = 24x$	Subtract 750 from each side.
$\dfrac{40{,}080}{24} = \dfrac{24x}{24}$	Divide each side by 24.
$1670 = x$	Simplify.

Each paycheck will be $1670. Check this in the original statement of the problem.

Exercises Within Reach ®

Solutions in English & Spanish and tutorial videos at CollegePrepAlgebra.com

59. *Job Offer* You have accepted a job offer at an annual salary of $37,120. This salary includes a year-end bonus of $2800. You are paid twice per month. What will your gross pay be for each paycheck?

60. *Assembly Line Production* You have a job on an assembly line for which you earn $10 per hour plus $0.75 for each unit you produce. Your earnings for an 8-hour day are $146. Find the number of units you produced.

61. *Internship* An internship pays $320 per week plus an additional $75 for a training session. The total pay for the internship and training is $2635. How many weeks long is the internship?

62. *Sales Position* You have a job as a salesperson for which you are paid $6 per hour plus $1.25 for each sale made. Your earnings for an 8-hour day are $88. Find the number of sales you made during the day.

Concept Summary: *Solving Linear Equations*

What

When given a **linear equation in one variable**, the goal is usually to solve the equation by isolating the variable on one side of the equation.

EXAMPLE

Solve the equation

$3x + 2 = 11$.

How

To isolate the variable, you "get rid" of terms and factors by using inverse operations.

EXAMPLE

$$3x + 2 = 11$$

$$3x + 2 - 2 = 11 - 2 \qquad \text{Subtract.}$$

$$3x = 9$$

$$\frac{3x}{3} = \frac{9}{3} \qquad \text{Divide.}$$

$$x = 3$$

The solution is $x = 3$. ✓

Why

An equation is like a scale. To keep the equation balanced, you must do the same thing to each side of the equation. The resulting equation is said to be equivalent to the original equation.

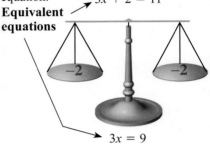

Equivalent equations

$3x + 2 = 11$

$3x = 9$

Exercises Within Reach ®   Worked-out solutions to odd-numbered exercises at CollegePrepAlgebra.com

Concept Summary Check

63. *Checking a Solution* Explain how to check a solution of an equation. Then illustrate your explanation using the equation $3x + 2 = 11$.

64. *Isolating the Variable* Is it okay to isolate the variable on the right side of the equation? Illustrate your answer using the equation $11 = 3x + 2$.

65. *Inverse Operations* In the solution above, explain how you know to subtract 2 from each side of the equation. What operation is "subtract 2" the inverse of?

66. *Equivalent Equations* Two equations are called *equivalent* if they have exactly the same solutions. Are each of the five equations in the solution steps above equivalent? Justify your answer.

Extra Practice

Solving a Linear Equation In Exercises 67–78, **solve** the equation and check your solution. **(Some of the equations have no solution.)**

67. $10x = 50$

68. $-3x = 21$

69. $\dfrac{x}{3} = 10$

70. $-\dfrac{x}{2} = 3$

71. $15x - 3 = 15 - 3x$

72. $2x - 5 = 7x + 10$

73. $2x - 5 + 10x = 3$

74. $-4x + 10 + 10x = 4$

75. $5t - 4 + 3t = 8t - 4$

76. $7z - 5z - 8 = 2z - 8$

77. $3(2 - 7x) = 3(4 - 7x)$

78. $2(5 + 6x) = 4(3x - 1)$

79. *Geometry* The perimeter of a rectangle is 260 meters. The length is 30 meters greater than the width. Find the dimensions of the rectangle.

80. *Geometry* A 10-foot board is cut so that 1 piece is 4 times as long as the other. Find the length of each piece.

81. *Hourly Wage* Your hourly wage is $8.30 per hour plus $0.60 for each unit you produce. How many units must you produce in 1 hour so that your hourly wage is $15.50?

82. *Labor Cost* The total cost for a new deck (including materials and labor) was $1830. The materials cost $1500 and the cost of labor was $55 per hour. How many hours did it take to build the deck?

|← 10 ft →|

83. *Summer Jobs* During the summer, you work 30 hours per week at a gas station and earn $8.75 per hour. You also work as a landscaper for $11.00 an hour and can work as many hours as you want. You want to earn a total of $400 per week. How many hours must you work at the landscaping job?

84. *Summer Jobs* During the summer, you work 40 hours per week at a coffee shop and earn $9.25 per hour. You also tutor for $10.00 per hour and can work as many hours as you want. You want to earn a total of $425 per week. How many hours must you tutor?

Explaining Concepts

85. *Reasoning* The scale below is balanced. Each blue box weighs 1 ounce. If you remove three blue boxes from each side, would the scale still balance? What property of equality does this illustrate? How much does each red box weigh?

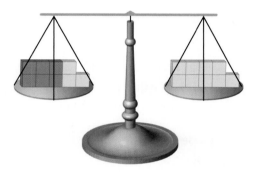

True or False? **In Exercises 86–89, determine whether the statement is true or false. Justify your answer.**

86. Subtracting 0 from each side of an equation yields an equivalent equation.

87. Multiplying each side of an equation by 0 yields an equivalent equation.

88. The sum of two odd integers is even.

89. The sum of an odd integer and an even integer is even.

90. *Finding a Pattern* The length of a rectangle is t times its width (see figure). The perimeter of the rectangle is 1200 meters, which implies that $2w + 2(tw) = 1200$, where w is the width of the rectangle.

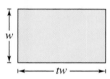

(a) Complete the table.

t	Width	Length	Area
1			
1.5			
2			
3			
4			
5			

(b) Use the completed table to write a conclusion about the area of a rectangle of fixed perimeter as its length increases relative to its width.

Cumulative Review

In Exercises 91–94, plot the numbers on the real number line.

91. $2, -3$

92. $-2.5, 0$

93. $\frac{3}{2}, 1, -1$

94. $4, -\frac{1}{2}, 2.6$

In Exercises 95–98, determine whether (a) $x = -1$ or (b) $x = 2$ is a solution of the equation.

95. $x - 8 = -9$

96. $x + 1.5 = 3.5$

97. $x + 4 = 2x$

98. $-2(x - 1) = 2 - 2x$

3.2 Equations that Reduce to Linear Form

▶ Solve linear equations containing symbols of grouping.
▶ Solve linear equations involving fractions.
▶ Solve linear equations involving decimals.

Equations Containing Symbols of Grouping

To solve a linear equation that contains symbols of grouping, use the following guidelines.

1. Remove symbols of grouping from each side by using the Distributive Property.
2. Combine like terms.
3. Isolate the variable using properties of equality.
4. Check your solution in the original equation.

| EXAMPLE 1 | Solving Linear Equations Involving Parentheses

a.

$4(x - 3) = 8$	Original equation
$4 \cdot x - 4 \cdot 3 = 8$	Distributive Property
$4x - 12 = 8$	Simplify.
$4x - 12 + 12 = 8 + 12$	Add 12 to each side.
$4x = 20$	Combine like terms.
$\dfrac{4x}{4} = \dfrac{20}{4}$	Divide each side by 4.
$x = 5$	Simplify.

b.

$3(2x - 1) + x = 11$	Original equation
$6x - 3 + x = 11$	Distributive Property
$6x + x - 3 = 11$	Group like terms.
$7x - 3 = 11$	Combine like terms.
$7x - 3 + 3 = 11 + 3$	Add 3 to each side.
$7x = 14$	Combine like terms.
$x = 2$	Divide each side by 7.

Exercises Within Reach ®

Solutions in English & Spanish and tutorial videos at CollegePrepAlgebra.com

Solving a Linear Equation Involving Parentheses In Exercises 1−14, solve the equation and check your solution.

1. $2(y - 4) = 0$
2. $9(y - 7) = 0$
3. $12(x - 3) = 0$
4. $4(z - 8) = 0$

5. $7(x + 5) = 49$
6. $4(x + 1) = 24$
7. $-5(t + 3) = 10$
8. $-3(x + 1) = 18$

9. $15(x + 1) - 8x = 29$
10. $7x - 2(x - 2) = 12$
11. $4 - (z + 6) = 8$

12. $25 - (y + 3) = 10$
13. $3 - (2x - 18) = 3$
14. $16 - (3x - 10) = 5$

EXAMPLE 2   **Equations Involving Symbols of Grouping**

a.

$5(x + 2) = 2(x - 1)$	Original equation
$5x + 10 = 2x - 2$	Distributive Property
$5x - 2x + 10 = 2x - 2x - 2$	Subtract $2x$ from each side.
$3x + 10 = -2$	Combine like terms.
$3x + 10 - 10 = -2 - 10$	Subtract 10 from each side.
$3x = -12$	Combine like terms.
$x = -4$	Divide each side by 3.

b.

$2(x - 7) - 3(x + 4) = 4 - (5x - 2)$	Original equation
$2x - 14 - 3x - 12 = 4 - 5x + 2$	Distributive Property
$-x - 26 = -5x + 6$	Combine like terms.
$-x + 5x - 26 = -5x + 5x + 6$	Add $5x$ to each side.
$4x - 26 = 6$	Combine like terms.
$4x - 26 + 26 = 6 + 26$	Add 26 to each side.
$4x = 32$	Combine like terms.
$x = 8$	Divide each side by 4.

c.

$5x - 2[4x + 3(x - 1)] = 8 - 3x$	Original equation
$5x - 2[4x + 3x - 3] = 8 - 3x$	Distributive Property
$5x - 2[7x - 3] = 8 - 3x$	Combine like terms inside brackets.
$5x - 14x + 6 = 8 - 3x$	Distributive Property
$-9x + 6 = 8 - 3x$	Combine like terms.
$-9x + 3x + 6 = 8 - 3x + 3x$	Add $3x$ to each side.
$-6x + 6 = 8$	Combine like terms.
$-6x + 6 - 6 = 8 - 6$	Subtract 6 from each side.
$-6x = 2$	Combine like terms.
$x = -\frac{2}{6} = -\frac{1}{3}$	Divide each side by -6.

Technology Tip

Try using your calculator to check the solution found in Example 2(b). This will give you practice working with parentheses on a calculator.

Exercises Within Reach ®   Solutions in English & Spanish and tutorial videos at CollegePrepAlgebra.com

Symbols of Grouping In Exercises 15−30, solve the equation and check your solution. (Some of the equations have no solution.)

15. $5(x - 4) = 2(2x + 5)$

16. $-(4x + 10) = 6(x + 2)$

17. $3(2x - 1) = 3(2x + 5)$

18. $4(z - 2) = 2(2z - 4)$

19. $-3(x + 4) = 4(x + 4)$

20. $-8(x - 6) = 3(x - 6)$

21. $-6(3 + x) + 2(3x + 5) = 0$

22. $-3(5x + 2) + 5(1 + 3x) = 0$

23. $7 = 3(x + 2) - 3(x - 5)$

24. $24 = 12(z + 1) - 3(4z - 2)$

25. $4 - (y - 3) = 3(y + 1) - 4(1 - y)$

26. $12 - 2(y + 3) = 4(y - 6) - (y - 1)$

27. $2[(3x + 5) - 7] = 3(4x - 3)$

28. $3[(5x + 1) - 4] = 4(2x - 3)$

29. $4x + 3[x - 2(2x - 1)] = 4 - 3x$

30. $16 + 4[5x - 4(x + 2)] = 7 - 2x$

Equations Involving Fractions

To solve a linear equation that contains one or more fractions, it is usually best to first clear the equation of fractions by multiplying each side by the least common multiple (LCM) of the denominators.

EXAMPLE 3 **Solving Linear Equations Involving Fractions**

a.

$$\frac{3x}{2} - \frac{1}{3} = 2$$ Original equation

$$6\left(\frac{3x}{2} - \frac{1}{3}\right) = 6 \cdot 2$$ Multiply each side by LCM 6.

$$6 \cdot \frac{3x}{2} - 6 \cdot \frac{1}{3} = 12$$ Distributive Property

$$9x - 2 = 12$$ Simplify.

$$9x = 14$$ Add 2 to each side.

$$x = \frac{14}{9}$$ Divide each side by 9.

b.

$$\frac{x}{5} + \frac{3x}{4} = 19$$ Original equation

$$20\left(\frac{x}{5}\right) + 20\left(\frac{3x}{4}\right) = 20(19)$$ Multiply each side by LCM 20.

$$4x + 15x = 380$$ Simplify.

$$19x = 380$$ Combine like terms.

$$x = 20$$ Divide each side by 19.

c.

$$\frac{2}{3}\left(x + \frac{1}{4}\right) = \frac{1}{2}$$ Original equation

$$\frac{2}{3}x + \frac{2}{12} = \frac{1}{2}$$ Distributive Property

$$12 \cdot \frac{2}{3}x + 12 \cdot \frac{2}{12} = 12 \cdot \frac{1}{2}$$ Multiply each side by LCM 12.

$$8x + 2 = 6$$ Simplify.

$$8x = 4$$ Subtract 2 from each side.

$$x = \frac{4}{8} = \frac{1}{2}$$ Divide each side by 8.

Study Tip

Notice in Example 3(c) that to clear all fractions in the equation, you multiply by 12, which is the LCM of 3, 4, and 2.

Exercises Within Reach ®

Solutions in English & Spanish and tutorial videos at CollegePrepAlgebra.com

Solving a Linear Equation Involving Fractions In Exercises 31–42, solve the equation and check your solution.

31. $\frac{6x}{25} = \frac{3}{5}$

32. $\frac{8x}{9} = \frac{2}{3}$

33. $\frac{5x}{4} + \frac{1}{2} = 0$

34. $\frac{3z}{7} + \frac{5}{14} = 0$

35. $\frac{x}{5} - \frac{1}{2} = 3$

36. $\frac{y}{6} - \frac{5}{8} = 2$

37. $\frac{x}{5} - \frac{x}{2} = 1$

38. $\frac{x}{3} + \frac{x}{4} = 1$

39. $\frac{2}{3}\left(x - \frac{5}{4}\right) = -\frac{1}{3}$

40. $\frac{1}{2}\left(1 - \frac{4}{3}x\right) = \frac{1}{4}$

41. $3x + \frac{1}{4} = \frac{3}{4}$

42. $2x - \frac{3}{8} = \frac{5}{8}$

Application   EXAMPLE 4   **Finding a Test Score**

To get an A in a course, you must have an average of at least 90 points for 4 tests of 100 points each. For the first 3 tests, your scores are 87, 92, and 84. What must you score on the fourth test to earn a 90% average for the course?

SOLUTION

Verbal Model: $\dfrac{\text{Sum of 4 tests}}{4} = 90$

Labels: Score on 4th test $= x$ (points)

Scores on first 3 tests: 87, 92, 84   (points)

Equation: $\dfrac{87 + 92 + 84 + x}{4} = 90$

You can solve this equation by multiplying each side by 4.

$\dfrac{87 + 92 + 84 + x}{4} = 90$ Write equation.

$4\left(\dfrac{87 + 92 + 84 + x}{4}\right) = 4(90)$ Multiply each side by LCM 4.

$87 + 92 + 84 + x = 360$ Simplify.

$263 + x = 360$ Combine like terms.

$x = 97$ Subtract 263 from each side.

You need to score 97 on the fourth test to earn a 90% average.

Exercises Within Reach ®

Solutions in English & Spanish and tutorial videos at CollegePrepAlgebra.com

43. *Time to Complete a Task* Two people can complete 80% of a task in t hours, where t must satisfy the equation $\dfrac{t}{10} + \dfrac{t}{15} = 0.8$. How long will it take for the two people to complete 80% of the task?

44. *Time to Complete a Task* Two machines can complete a task in t hours, where t must satisfy the equation $\dfrac{t}{10} + \dfrac{t}{15} = 1$. How long will it take for the two machines to complete the task?

45. *Course Grade* To get a B in a course, you must have an average of at least 80 points for 4 tests of 100 points each. For the first 3 tests, your scores are 79, 83, and 81. What must you score on the fourth test to earn an 80% average for the course?

46. *Weight Loss* You want to lose an average of 4 pounds per month on a new weight loss program. In the first 3 months, you lost 3 pounds, 7 pounds, and 4 pounds. How much weight must you lose during the fourth month to maintain an average weight loss of 4 pounds per month?

Equations Involving Decimals

Many real-life applications of linear equations involve decimal coefficients. To solve such an equation, you can clear it of decimals in much the same way you clear an equation of fractions. Multiply each side by a power of 10 that converts all decimal coefficients to integers, as shown in the next example.

EXAMPLE 5 **Solving a Linear Equation Involving Decimals**

Solve $0.3x + 0.2(10 - x) = 0.15(30)$. Then check your solution.

SOLUTION

$0.3x + 0.2(10 - x) = 0.15(30)$	Write original equation.
$0.3x + 2 - 0.2x = 4.5$	Distributive Property
$0.1x + 2 = 4.5$	Combine like terms.
$10(0.1x + 2) = 10(4.5)$	Multiply each side by 10.
$x + 20 = 45$	Simplify.
$x = 25$	Subtract 20 from each side.

Study Tip

There are other ways to solve the decimal equation in Example 5. You could first clear the equation of decimals by multiplying each side by 100. Or, you could keep the decimals and use a calculator to do the arithmetic operations. The method you choose is a matter of personal preference.

Check

$0.3(25) + 0.2(10 - 25) \stackrel{?}{=} 0.15(30)$	Substitute 25 for x in original equation.
$0.3(25) + 0.2(-15) \stackrel{?}{=} 0.15(30)$	Perform subtraction within parentheses.
$7.5 - 3.0 \stackrel{?}{=} 4.5$	Multiply.
$4.5 = 4.5$	Solution checks. ✓

The solution is $x = 25$.

Exercises Within Reach ®

Solutions in English & Spanish and tutorial videos at CollegePrepAlgebra.com

Solving a Linear Equation Involving Decimals In Exercises 47−64, solve the equation. Round your answer to two decimal places.

47. $0.2x + 5 = 6$

48. $4 - 0.3x = 1$

49. $5.6 = 1.1x - 1.2$

50. $7.2x - 4.7 = 62.3$

51. $1.2x - 4.3 = 1.7$

52. $16 - 2.4x = -8$

53. $0.234x + 1 = 2.805$

54. $2.75x - 3.13 = 5.12$

55. $3 + 0.03x = 5$

56. $0.4x - 0.1 = 2$

57. $1.205x - 0.003 = 0.5$

58. $5.225 + 3.001x = 10.275$

59. $0.42x - 0.4(x + 2.4) = 0.3(5)$

60. $1.6x + 0.25(12 - x) = 0.43(-12)$

61. $\dfrac{x}{3.25} + 1 = 2.08$

62. $\dfrac{x}{4.08} + 7.2 = 5.14$

63. $\dfrac{x}{3.155} = 2.850$

64. $\dfrac{3x}{4.5} = \dfrac{1}{8}$

Application   EXAMPLE 6   **Analyzing Postsecondary Enrollment**

The enrollment y (in millions) at postsecondary schools from 2000 through 2009 can be approximated by the linear model $y = 0.35t + 9.1$, where t represents the year, with $t = 0$ corresponding to 2000. Use the model to predict the year in which the enrollment will be 14 million students. (*Source:* U.S. Department of Education)

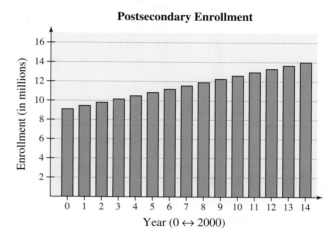

Postsecondary Enrollment

SOLUTION

To find the year in which the enrollment will be 14 million students, substitute 14 for y in the original equation and solve the equation for t.

$14 = 0.35t + 9.1$	Substitute 14 for y in original equation.
$4.9 = 0.35t$	Subtract 9.1 from each side.
$14 = t$	Divide each side by 0.35.

Because $t = 0$ corresponds to 2000, the enrollment at postsecondary schools will be 14 million during 2014. Check this in the original statement of the problem.

Exercises Within Reach ®

Solutions in English & Spanish and tutorial videos at CollegePrepAlgebra.com

65. *Data Analysis* The table shows the projected numbers N (in millions) of people living in the United States. (*Source:* U.S. Census Bureau)

Year	N
2020	341.4
2025	357.5
2030	373.5
2035	389.5

A model for the data is $N = 3.21t + 277.3$, where t represents time in years, with $t = 20$ corresponding to the year 2020. According to the model, in what year will the population exceed 450 million?

66. *Data Analysis* The table shows the sales S (in billions) of Coach for the years 2005 through 2010. (*Source:* Coach, Inc.)

Year	S
2005	1.71
2006	2.11
2007	2.61
2008	3.18
2009	3.23
2010	3.61

A model for the data is $S = 0.384t - 0.14$, where t represents time in years, with $t = 5$ corresponding to the year 2005. According to the model, in what year will the sales exceed 6 billion?

Concept Summary: Solving Equations

What

Some linear equations involve symbols of grouping, fractions, and decimals. To solve such equations, you must first reduce the equation to linear form.

EXAMPLE

Solve $\dfrac{2x}{4} = \dfrac{1-2x}{3}$.

How

To solve a linear equation involving fractions, first clear the equation of fractions. To do this, multiply each side of the equation by the least common multiple of the denominators.

EXAMPLE

$12 \cdot \dfrac{2x}{4} = 12 \cdot \dfrac{1-2x}{3}$ Multiply by LCM 12.

$6x = 4 - 8x$ Simplify.

Why

Many real-life applications are modeled by equations involving symbols of grouping, fractions, and decimals. Knowing how to reduce such equations to linear form will help you solve these equations.

Exercises Within Reach ®

Worked-out solutions to odd-numbered exercises at CollegePrepAlgebra.com

Concept Summary Check

67. *Least Common Multiple* What is the least common multiple of the denominators of two or more fractions?

68. *Writing* Discuss one method for finding the least common multiple of the denominators of two fractions.

69. *Structure* When solving an equation that contains fractions, explain what is accomplished by multiplying each side of the equation by the least common multiple of the denominators.

70. *Number Sense* What is the LCM of the denominators in the equation $\dfrac{2x}{7} = \dfrac{3x}{2}$?

Extra Practice

Solving a Linear Equation In Exercises 71–82, solve the equation and check your solution. (Some of the equations have no solution.)

71. $5z - 2 = 2(3z - 4)$

72. $3 - 4x = 5(x - 3)$

73. $7(y + 7) = 5y + 59$

74. $40 + 14k = 2(-4k - 13)$

75. $3.7y + 7 = 8.1y - 19.4$

76. $5(1.2x + 6) = 7.1x + 34.4$

77. $2s + \dfrac{3}{2} = 2s + 2$

78. $\dfrac{3}{4} + 5s = -2 + 5s$

79. $\dfrac{x}{4} = \dfrac{1-2x}{3}$

80. $\dfrac{x+1}{6} = \dfrac{3x}{10}$

81. $\dfrac{100-4u}{3} = \dfrac{5u+6}{4} + 6$

82. $\dfrac{8-3x}{2} - 4 = \dfrac{x}{6}$

Geometry In Exercises 83 and 84, the perimeter of the figure is 15. Find the value of x.

83.

$\frac{2}{5}x - 1$ $\frac{1}{2}x$

$\frac{1}{10}x + 6$

84.

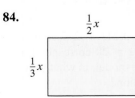

$\frac{1}{2}x$

$\frac{1}{3}x$

Solving a Mixture Problem **In Exercises 85 and 86, use the equation and solve for x.**

$$p_1x + p_2(a - x) = p_3a$$

85. *Mixture Problem* Determine the number of quarts of a 10% solution that must be mixed with a 30% solution to obtain 100 quarts of a 25% solution. ($p_1 = 0.1$, $p_2 = 0.3$, $p_3 = 0.25$, and $a = 100$.)

86. *Mixture Problem* Determine the number of gallons of a 25% solution that must be mixed with a 50% solution to obtain 5 gallons of a 30% solution. ($p_1 = 0.25$, $p_2 = 0.5$, $p_3 = 0.3$, and $a = 5$.)

Fireplace Construction **In Exercises 87 and 88, use the following information. A fireplace is 93 inches wide. Each brick in the fireplace has a length of 8 inches, and there is $\frac{1}{2}$ inch of mortar between adjoining bricks (see figure). Let *n* be the number of bricks per row.**

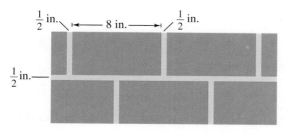

87. Explain why the number of bricks per row is the solution of the equation $8n + \frac{1}{2}(n - 1) = 93$.

88. Find the number of bricks per row in the fireplace.

Explaining Concepts

89. *Structure* You could solve $3(x - 7) = 15$ by applying the Distributive Property as the first step. However, there is another way to begin. What is it?

90. *Error Analysis* Describe and correct the error.

$$\frac{1}{4}x - \frac{2}{3} = 4$$
$$12\left(\frac{1}{4}x\right) - 12\left(\frac{2}{3}\right) = 4$$
$$3x - 8 = 4$$
$$3x = 12$$
$$x = 4$$

91. *Logic* Explain what happens when you divide each side of a linear equation by a variable factor.

92. *Writing* When simplifying an algebraic *expression* involving fractions, why can't you simplify the expression by multiplying by the least common multiple of the denominators?

Cumulative Review

In Exercises 93–100, simplify the expression.

93. $(-2x)^2x^4$

94. $-y^2(-2y)^3$

95. $5z^3(z^2)$

96. $a^2 + 3a + 4 - 2a - 6$

97. $\frac{5x}{3} - \frac{2x}{3} - 4$

98. $2x^2 - 4 + 5 - 3x^2$

99. $-y^2(y^2 + 4) + 6y^2$

100. $5t(2 - t) + t^2$

In Exercises 101–104, solve the equation and check your solution.

101. $3x - 5 = 12$

102. $-5x + 9 = 9$

103. $4 - 2x = 22$

104. $x - 3 + 7x = 29$

3.3 Problem Solving with Percents

▶ Convert percents to decimals and fractions, and vice versa
▶ Solve linear equations involving percents.
▶ Solve problems involving markups and discounts.

Percents

"Cent" implies 100, as in the word *century*. A **percent** is the number of parts per one hundred.

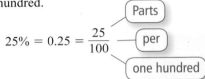

$$25\% = 0.25 = \frac{25}{100}$$

Parts
per
one hundred

Study Tip

In Examples 1 and 2, there is a quick way to convert between percent form and decimal form.

• To convert from percent form to decimal form, move the decimal point two places to the left. For instance,

3.5% = 0.035.

• To convert from decimal form to percent form, move the decimal point two places to the right. For instance,

1.20 = 120%

EXAMPLE 1 **Converting Decimals and Fractions to Percents**

a. $1.20 = \frac{120}{100} = 120\%$ **b.** $\frac{3}{5} = \frac{3(20)}{5(20)} = \frac{60}{100} = 60\%$

EXAMPLE 2 **Converting Percents to Decimals and Fractions**

a. Convert 3.5% to a decimal. **b.** Convert 55% to a fraction.

SOLUTION

a. $3.5\% = \frac{3.5}{100}$

$= \frac{3.5(10)}{100(10)}$

$= \frac{35}{1000}$

$= 0.035$

b. $55\% = \frac{55}{100}$

$= \frac{5(11)}{5(20)}$

$= \frac{11}{20}$

Exercises Within Reach ®

Solutions in English & Spanish and tutorial videos at CollegePrepAlgebra.com

Converting to a Percent In Exercises 1−8, convert the decimal or fraction to a percent.

1. 0.62 **2.** 0.57 **3.** 0.075 **4.** 0.005

5. $\frac{4}{5}$ **6.** $\frac{1}{4}$ **7.** $\frac{5}{4}$ **8.** $\frac{6}{5}$

Converting to a Decimal In Exercises 9−16, convert the percent to a decimal.

9. 12% **10.** 95% **11.** 125% **12.** 250%

13. 8.5% **14.** 0.3% **15.** $\frac{3}{4}\%$ **16.** $4\frac{4}{5}\%$

Converting to a Fraction In Exercises 17−24, convert the percent to a fraction.

17. 30% **18.** 85% **19.** 130% **20.** 350%

21. 1.4% **22.** 0.7% **23.** $\frac{1}{2}\%$ **24.** $2\frac{3}{10}\%$

The Percent Equation

The primary use of percents is to compare two numbers. For example, 2 is 50% of 4, and 5 is 25% of 20. The following model is helpful.

Verbal Model: $a = p$ percent of b

Labels: $b =$ base number
$p =$ percent (in decimal form)
$a =$ number being compared to b

Equation: $a = p \cdot b$

Application   **EXAMPLE 3**   **Solving Percent Equations for *a***

a.   What number is 30% of 70?

b.   A union negotiates for a cost-of-living raise of 7%. What is the raise for a union member whose salary is $40,240? What is this person's new salary?

SOLUTION

a.   *Verbal Model:*   What number $=$ 30% of 70

Labels: $a =$ unknown number

Equation: $a = (0.3)(70) = 21$

So, 21 is 30% of 70.

b.   *Verbal Model:*   Raise $= \dfrac{\text{Percent}}{\text{(in decimal form)}} \cdot$ Salary

Labels: Raise $= a$ (dollars)
Percent $= 7\% = 0.07$ (decimal form)
Salary $= 40{,}240$ (dollars)

Equation: $a = 0.07(40{,}240) = 2816.80$

So, the raise is $2816.80 and the new salary is $40{,}240.00 + 2816.80 = \$43{,}056.80$.

Exercises Within Reach ®

Solutions in English & Spanish and tutorial videos at CollegePrepAlgebra.com

Solving a Percent Equation In Exercises 25–28, solve the percent equation.

25. What number is 30% of 150?

26. What number is 60% of 820?

27. What number is 0.75% of 56?

28. What number is 325% of 450?

29. *Salary Raise* You accept a job with a salary of $35,600. After 6 months, you receive a 5% raise. What is your new salary?

30. *Salary Raise* A union negotiates for a cost-of-living raise of 4.5%. What is the raise for a union member whose salary is $37,380? What is this person's new salary?

EXAMPLE 4 **Solving Percent Equations for *b***

a. 14 is 25% of what number?

b. You missed an A in your chemistry course by only 3 points. Your point total for the course was 402. How many points were possible in the course? (Assume that you needed 90% of the course total for an A.)

SOLUTION

Study Tip

It may help to think of *a* as a "new" amount and *b* as the "original" amount.

a. *Verbal Model:* $\boxed{14} = \boxed{25\% \text{ of what number}}$

   Labels: $b = \text{unknown number}$

   Equation: $14 = 0.25b$

   $$\frac{14}{0.25} = b$$

   $$56 = b$$

So, 14 is 25% of 56.

b. *Verbal Model:* $\boxed{\begin{array}{c}\text{Your}\\\text{points}\end{array}} + \boxed{\begin{array}{c}3\\\text{points}\end{array}} = \boxed{\begin{array}{c}\text{Percent}\\\text{(in decimal form)}\end{array}} \cdot \boxed{\begin{array}{c}\text{Total}\\\text{points}\end{array}}$

   Labels: Your points = 402 (points)
   Percent = 90% = 0.9 (decimal form)
   Total points for course = b (points)

   Equation: $402 + 3 = 0.9b$ Write equation.

   $405 = 0.9b$ Add.

   $450 = b$ Divide each side by 0.9

So, there were 450 possible points in the course.

Exercises Within Reach ®

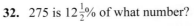

Solutions in English & Spanish and tutorial videos at CollegePrepAlgebra.com

Solving a Percent Equation In Exercises 31–34, solve the percent equation.

31. 903 is 43% of what number?

32. 275 is $12\frac{1}{2}\%$ of what number?

33. 594 is 450% of what number?

34. 51.2 is 0.08% of what number?

35. *Course Grade* You missed an A in your art course by only 5 points. Your point total for the course was 382. How many points were possible in the course? (Assume that you needed 90% of the course total for an A.)

36. *Course Grade* You were 6 points shy of a B in your mathematics course. Your point total for the course was 394. How many points were possible in the course? (Assume that you needed 80% of the course total for a B.)

Application   **EXAMPLE 5**   **Solving Percent Equations for *p***

a.   135 is what percent of 27?

b.   A real estate agency receives a commission of $8092.50 for the sale of a $124,500 house. What percent commission is this?

SOLUTION

a.   *Verbal Model:* $\boxed{135}$ $=$ $\boxed{\text{What percent of 27}}$

Labels: p = unknown percent (in decimal form)

Equation: $135 = p(27)$

$$\frac{135}{27} = p$$

$$5 = p$$

So, 135 is 500% of 27.

b.   *Verbal Model:* $\boxed{\text{Commission}}$ $=$ $\boxed{\begin{matrix}\text{Percent}\\ \text{(in decimal form)}\end{matrix}}$ $\cdot$ $\boxed{\text{Sale price}}$

Labels: Commission = 8092.50   (dollars)
            Percent = p (decimal form)
            Sale Price = 124,500   (dollars)

Equation: $8092.50 = p(124,500)$ Write equation.

$$\frac{8092.50}{124,500} = p$$ Divide each side by 124,500.

$$0.065 = p$$ Simplify.

So, the real estate agency receives a commission of 6.5%.

In 2010, approximately 321,000 new houses were sold, with a median price of $221,900. (*Source:* U.S. Census Bureau)

Exercises Within Reach ®

Solutions in English & Spanish and tutorial videos at CollegePrepAlgebra.com

Solving a Percent Equation In Exercises 37–40, solve the percent equation.

37. 576 is what percent of 800?

38. 38 is what percent of 5700?

39. 22 is what percent of 800?

40. 148.8 is what percent of 960?

41. ***Commission*** A real estate agency receives a commission of $14,506.50 for the sale of a $152,700 house. What percent commission is this?

42. ***Commission*** A car salesman receives a commission of $1145 for the sale of a $45,800 car. What percent commission is this?

Markups and Discounts

Retail stores generally sell items for more than what they paid for the items. The difference is called **markup**.

$$\boxed{\text{Selling price}} = \boxed{\text{Cost}} + \boxed{\text{Markup}}$$

In markup problems, the markup may be known or it may be expressed as a percent of the cost. This percent is called the **markup rate**.

$$\boxed{\text{Markup}} = \boxed{\text{Markup rate}} \cdot \boxed{\text{Cost}}$$

Study Tip

The cost of an item is sometimes called the wholesale cost.

EXAMPLE 6 Solving Markup Problems

a. The cost is $45. The markup rate is 55%. What is the selling price?
b. The selling price is $98. The markup rate is 60%. What is the cost?
c. The selling price is $60. The cost is $24. What is the markup rate?

SOLUTION

a. $\boxed{\text{Selling price}} = \boxed{\text{Cost}} + \boxed{\text{Markup}}$

$= 45 + (0.55)(45)$	Substitute known values.
$= 45 + 24.75$	Multiply.
$= \$69.75$	Add.

b. $\boxed{\text{Selling price}} = \boxed{\text{Cost}} + \boxed{\text{Markup}}$

$98 = C + 0.6C$	Substitute known values.
$98 = 1.6C$	Combine like terms.
$\dfrac{98}{1.6} = C$	Divide each side by 1.6.
$\$61.25 = C$	Simplify.

c. $\boxed{\text{Selling price}} = \boxed{\text{Cost}} + \boxed{\text{Markup rate}} \cdot \boxed{\text{Cost}}$

$60 = 24 + p\,(24)$	Substitute known values.
$36 = 24p$	Subtract 24 from each side.
$\dfrac{36}{24} = p$	Divide each side by 24.
$1.5 = p$	Simplify.

So, the markup rate is 150%.

Exercises Within Reach ® Solutions in English & Spanish and tutorial videos at CollegePrepAlgebra.com

Solving a Markup Problem In Exercises 43−50, find the missing quantities.

	Cost	Selling Price	Markup	Markup Rate		Cost	Selling Price	Markup	Markup Rate
43.	$26.97	$49.95			**44.**	$71.97	$119.95		
45.		$74.38		81.5%	**46.**		$69.99		55.5%
47.		$125.98	$56.69		**48.**		$350.00	$80.77	
49.	$13,250.00			20%	**50.**	$107.97			85.2%

Retail stores also sometimes sell items at a **discount**.

$$\boxed{\text{Sale price}} = \boxed{\text{Original price}} - \boxed{\text{Discount}}$$

The discount is given in dollars, and the **discount rate** is given as a percent of the original price.

$$\boxed{\text{Discount}} = \boxed{\text{Discount rate}} \cdot \boxed{\text{Original price}}$$

Application   **EXAMPLE 7**   **Solving Discount Problems**

a. The original price of a lawn mower was $199.95. During a midsummer sale, the lawn mower is on sale for $139.95. What is the discount rate?

b. A drug store advertises 40% off the prices of all summer tanning products. The original price of a bottle of suntan oil is $3.49. What is the sale price?

SOLUTION

a. *Verbal Model:* $\boxed{\text{Discount}} = \boxed{\text{Discount rate}} \cdot \boxed{\text{Original price}}$

    Labels: Discount = 199.95 − 139.95 = 60   (dollars)
               Original price = 199.95   (dollars)
               Discount rate = p (decimal form)

    Equation: $60 = p(199.95)$ Write equation.

                  $0.30 \approx p$ Divide each side by 199.95.

So, the discount rate is 30%.

b. *Verbal Model:* $\boxed{\text{Sale price}} = \boxed{\text{Original price}} - \boxed{\text{Discount}}$

    Labels: Original price = 3.49   (dollars)
               Discount rate = 0.4   (decimal form)
               Discount = 0.4(3.49) = 1.396   (dollars)
               Sale Price = x (dollars)

    Equation: $x = 3.49 - 1.396 = \$2.09$

Exercises Within Reach ®

Solutions in English & Spanish and tutorial videos at CollegePrepAlgebra.com

Solving a Discount Problem **In Exercises 51−60, find the missing quantities.**

	Original Price	Sale Price	Discount	Discount Rate		Original Price	Sale Price	Discount	Discount Rate
51.	$39.95	$29.95			**52.**	$50.99	$45.99		
53.		$18.95		20%	**54.**		$189.00		40%
55.	$189.99		$30.00		**56.**	$18.95		$8.00	
57.	$119.96			50%	**58.**	$84.95			65%
59.		$695.00	$300.00		**60.**		$259.97	$135.00	

Concept Summary: Using the Percent Equation

What

The primary use of **percents** is to compare two numbers. You can use the percent equation to solve three basic types of percent problems.

EXAMPLE

1. What number is p percent of b?

2. a is p percent of what number?

3. a is what percent of b?

How

To solve these types of problems, substitute the known quantities into the percent equation and solve for the unknown quantity.

Percent Equation

Verbal Model:

$$a = p \text{ percent of } b$$

Equation: $a = p \cdot b$

Why

When you know how to use the percent equation, you can apply it in real-life situations. For instance, you can use the percent equation to find **markups** and **discounts**.

Exercises Within Reach ®

Worked-out solutions to odd-numbered exercises at CollegePrepAlgebra.com

Concept Summary Check

61. *Writing* Explain what is meant by the word *percent*.

62. *Number Sense* Can any positive terminating decimal be written as a percent? Explain.

63. *Structure* Write an equation that can be used to find the number x that is 25% of a number y.

64. *Writing* In your own words, explain what each variable in the percent equation represents.

Extra Practice

Finding Equivalent Forms of a Percent **In Exercises 65–74, complete** the table showing the equivalent forms of the percent.

	Percent	Parts out of 100	Decimal	Fraction		Percent	Parts out of 100	Decimal	Fraction
65.	40%				**66.**	16%			
67.	7.5%				**68.**	75%			
69.		63			**70.**		10.5		
71.			0.155		**72.**			0.80	
73.				$\frac{3}{5}$	**74.**				$\frac{3}{20}$

Finding a Percent **In Exercises 75–78, determine** the percent of the figure that is shaded.
(There are 360° in a circle.)

75.

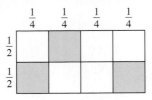

76.

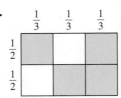

77.

78.

79. Lawn Tractor You purchase a lawn tractor for $3750, and 1 year later you note that the price has increased to $3900. Find the percent increase in the price of the lawn tractor.

80. Membership Drive Because of a membership drive for a public television station, the current membership is 125% of what it was a year ago. The current number of members is 7815. How many members did the station have last year?

81. Geometry A rectangular plot of land measures 650 feet by 825 feet. A square garage with side lengths of 24 feet is built on the plot of land. What percent of the plot of land is occupied by the garage?

82. Geometry A circular target is attached to a rectangular board, as shown in the figure. The radius of the circle is $4\frac{1}{2}$ inches, and the measurements of the board are 12 inches by 15 inches. What percent of the board is covered by the target? (The area of a circle is $A = \pi r^2$, where r is the radius of the circle.)

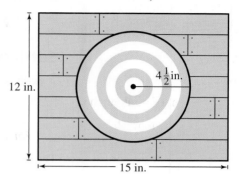

Explaining Concepts

83. Number Sense The fraction $\frac{a}{b}(a > 0, b > 0)$ is converted to a percent. For what values of a and b is the percent greater than 100%? Less than 100%? Equal to 100%? Explain.

84. Reasoning Would you rather receive a 3% raise followed by a 9% raise or a 9% raise followed by a 3% raise? Explain.

True or False? In Exercises 85–88, decide whether the statement is true or false. Justify your answer.

85. $1 = 1\%$

86. Every percent can be written as a fraction.

87. The question "What is 68% of 50?" can be answered by solving the equation $a = 68(50)$.

88. $\frac{1}{2}\% = 50\%$

Cumulative Review

In Exercises 89 and 90, evaluate the expression.

89. $8 - |-7 + 11| + (-4)$

90. $34 - [54 - (-16 + 4) + 6]$

In Exercises 91 and 92, evaluate the algebraic expression for the specified values of the variables.

91. $x^2 - y^2$

    (a) $x = 4, y = 3$ (b) $x = -5, y = 3$

92. $\dfrac{z^2 + 2}{x^2 - 1}$

    (a) $x = 1, z = 2$ (b) $x = 2, z = 1$

In Exercises 93 and 94, use the Distributive Property to expand the expression.

93. $4(2x - 5)$ **94.** $-z(xz - 2y^2)$

In Exercises 95–98, solve the equation and check your solution.

95. $4(x + 3) = 0$

96. $-3(y - 2) = 21$

97. $22 - (z + 1) = 33$

98. $\dfrac{w}{3} = \dfrac{8}{12}$

3.4 Ratios and Proportions

▶ Compare relative sizes using ratios.
▶ Find the unit price of a consumer item.
▶ Solve proportions that equate two ratios.

Setting Up Ratios

A **ratio** is a comparison of one number with another by division. The ratio of the real number a to the real number b is given by

$$\frac{a}{b}.$$ Ratio of a to b

The ratio of a to b is sometimes written as $a:b$. Note the order implied by a ratio. The ratio of a to b means $a:b$, whereas the ratio of b to a means $b:a$.

EXAMPLE 1   **Writing Ratios in Fractional Form**

a. The ratio of 7 to 5 is given by $\frac{7}{5}$.

b. The ratio of 12 to 8 is given by $\frac{12}{8} = \frac{3}{2}$.

Note that the fraction $\frac{12}{8}$ can be written in simplest form as $\frac{3}{2}$.

c. The ratio of 10 to 2 is given by $\frac{10}{2} = \frac{5}{1}$.

d. The ratio of $3\frac{1}{2}$ to $5\frac{1}{4}$ is given by

$$\frac{3\frac{1}{2}}{5\frac{1}{4}} = \frac{\frac{7}{2}}{\frac{21}{4}}$$ Rewrite mixed numbers as fractions.

$$= \frac{7}{2} \cdot \frac{4}{21}$$ Invert divisor and multiply.

$$= \frac{2}{3}$$ Simplify.

Study Tip

There are many applications of ratios. For instance, ratios are used to describe opinion surveys (for/against), populations (male/female, unemployed/employed), and mixtures (oil/gasoline, water/alcohol).

Exercises Within Reach ®

Solutions in English & Spanish and tutorial videos at CollegePrepAlgebra.com

Writing a Ratio in Fractional Form In Exercises 1−12, write the ratio as a fraction in simplest form.

1. 36 to 9

2. 45 to 15

3. 27 to 54

4. 27 to 63

5. $5\frac{2}{3}$ to $1\frac{1}{3}$

6. $2\frac{1}{4}$ to $3\frac{3}{8}$

7. $14:21$

8. $12:30$

9. $144:16$

10. $60:45$

11. $3\frac{1}{5}:5\frac{3}{10}$

12. $1\frac{2}{7}:\frac{1}{2}$

When comparing two measurements by a ratio, you should use the *same unit of measurement* in both the numerator and the denominator. For example, to find the ratio of 4 feet to 8 inches, you could write either of the following.

$$\frac{4 \text{ feet}}{8 \text{ inches}} = \frac{48 \text{ inches}}{8 \text{ inches}} = \frac{48}{8} = \frac{6}{1}$$ Convert feet to inches.

$$\frac{4 \text{ feet}}{8 \text{ inches}} = \frac{4 \text{ feet}}{\frac{8}{12} \text{ foot}} = 4 \cdot \frac{12}{8} = \frac{6}{1}$$ Convert inches to feet.

EXAMPLE 2   **Comparing Measurements**

Find ratios to compare the relative sizes of the following.

a. 5 gallons to 7 gallons   **b.** 3 meters to 40 centimeters

c. 200 cents to 3 dollars   **d.** 30 months to $1\frac{1}{2}$ years

SOLUTION

a. Because the units of measurement are the same, the ratio is $\frac{5}{7}$.

b. Because the units of measurement are different, begin by converting meters to centimeters *or* centimeters to meters. Here, it is easier to convert meters to centimeters by multiplying by 100.

$$\frac{3 \text{ meters}}{40 \text{ centimeters}} = \frac{3(100) \text{ centimeters}}{40 \text{ centimeters}}$$ Convert meters to centimeters.

$$= \frac{300}{40}$$ Multiply in numerator.

$$= \frac{15}{2}$$ Simplify.

c. Because 200 cents is the same as 2 dollars, the ratio is

$$\frac{200 \text{ cents}}{3 \text{ dollars}} = \frac{2 \text{ dollars}}{3 \text{ dollars}} = \frac{2}{3}.$$

d. Because $1\frac{1}{2}$ years $= 18$ months, the ratio is

$$\frac{30 \text{ months}}{1\frac{1}{2} \text{ years}} = \frac{30 \text{ months}}{18 \text{ months}} = \frac{30}{18} = \frac{5}{3}.$$

3 meters = 300 centimeters

40 cm

Exercises Within Reach ®

Solutions in English & Spanish and tutorial videos at CollegePrepAlgebra.com

Comparing Measurements In Exercises 13−24, **find a ratio that compares the relative sizes of the quantities. (Use the same units of measurement for both quantities.)**

13. 42 inches to 21 inches

14. 81 feet to 27 feet

15. $40 to $60

16. 24 pounds to 30 pounds

17. 60 milliliters to 1 liter

18. 3 inches to 2 feet

19. 7 nickels to 3 quarters

20. 24 ounces to 3 pounds

21. 3 hours to 90 minutes

22. 21 feet to 35 yards

23. 75 centimeters to 2 meters

24. 2 weeks to 7 days

Unit Prices

The **unit price** of an item is given by the ratio of the total price to the total number of units.

$$\text{Unit price} = \frac{\text{Total price}}{\text{Total units}}$$

Study Tip

The word *per* is used to state unit prices. For instance, the unit price for a particular brand of coffee might be $4.69 per pound.

EXAMPLE 3 **Finding a Unit Price**

Find the unit price (in dollars per ounce) for a 5-pound, 4-ounce box of detergent that sells for $7.14.

SOLUTION

$$\text{Unit price} = \frac{\text{Total price}}{\text{Total units}} = \frac{\$7.14}{84 \text{ ounces}} = \$0.085 \text{ per ounce}$$

$5(16 \text{ oz}) + 4 \text{ oz}$

EXAMPLE 4 **Comparing Unit Prices**

Which has the lower unit price: a 12-ounce box of breakfast cereal for $2.79 or a 16-ounce box of the same cereal for $3.49?

SOLUTION

The unit price for the smaller box is

$$\text{Unit price} = \frac{\text{Total price}}{\text{Total units}} = \frac{\$2.79}{12 \text{ ounces}} = \$0.23 \text{ per ounce.}$$

The unit price for the larger box is

$$\text{Unit price} = \frac{\text{Total price}}{\text{Total units}} = \frac{\$3.49}{16 \text{ ounces}} = \$0.22 \text{ per ounce.}$$

16 oz: $3.49

12 oz: $2.79

So, the larger box has a slightly lower unit price.

Exercises Within Reach ®

Solutions in English & Spanish and tutorial videos at CollegePrepAlgebra.com

Finding a Unit Price In Exercises 25–28, find the unit price (in dollars per ounce).

25. A 20-ounce can of pineapple for $0.98

26. An 18-ounce box of cereal for $4.29

27. A 1-pound, 4-ounce loaf of bread for $1.46

28. A 1-pound package of cheese for $3.08

Comparing Unit Prices In Exercises 29–32, determine which product has the lower unit price.

29. (a) An 18-ounce jar of peanut butter for $1.92
    (b) A 28-ounce jar of peanut butter for $3.18

30. (a) A 16-ounce bag of chocolates for $1.99
    (b) An 18-ounce bag of chocolates for $2.29

31. (a) A 4-pound bag of sugar for $1.89
    (b) A 10-pound bag of sugar for $4.49

32. (a) A gallon of orange juice for $3.49
    (b) A half-gallon of orange juice for $1.70

Solving Proportions

A **proportion** is a statement that equates two ratios. For example, if the ratio of a to b is the same as the ratio of c to d, you can write the proportion as

$$\frac{a}{b} = \frac{c}{d}.$$

In typical applications, you know three of the values and are required to find the fourth. To solve a proportion, you can use *cross multiplication*. If $\frac{a}{b} = \frac{c}{d}$, then $ad = bc$.

EXAMPLE 5   **Solving Proportions**

a.

$\dfrac{50}{x} = \dfrac{2}{28}$	Original proportion
$50(28) = 2x$	Cross-multiply.
$\dfrac{1400}{2} = x$	Divide each side by 2.
$700 = x$	Simplify.

So, the ratio of 50 to 700 is the same as the ratio of 2 to 28.

b.

$\dfrac{x-2}{5} = \dfrac{4}{3}$	Original proportion
$3(x-2) = 20$	Cross-multiply.
$3x - 6 = 20$	Distributive Property
$3x = 26$	Add 6 to each side.
$x = \dfrac{26}{3}$	Divide each side by 3.

Exercises Within Reach ®

Solutions in English & Spanish and tutorial videos at CollegePrepAlgebra.com

Solving a Proportion In Exercises 33−44, solve the proportion.

33. $\dfrac{5}{3} = \dfrac{20}{y}$

34. $\dfrac{9}{x} = \dfrac{18}{5}$

35. $\dfrac{5}{x} = \dfrac{3}{2}$

36. $\dfrac{4}{t} = \dfrac{2}{25}$

37. $\dfrac{z}{35} = \dfrac{5}{8}$

38. $\dfrac{y}{25} = \dfrac{12}{10}$

39. $\dfrac{0.5}{0.8} = \dfrac{n}{0.3}$

40. $\dfrac{2}{4.5} = \dfrac{t}{0.5}$

41. $\dfrac{x+1}{5} = \dfrac{3}{10}$

42. $\dfrac{z-3}{8} = \dfrac{3}{10}$

43. $\dfrac{x+6}{3} = \dfrac{x-5}{2}$

44. $\dfrac{x-2}{4} = \dfrac{x+10}{10}$

Application EXAMPLE 6 Geometry: **Using Similar Triangles**

A triangular lot has perpendicular sides with lengths of 100 feet and 210 feet. You are making a proportional sketch of this lot using 8 inches as the length of the shorter side. How long should you make the longer side?

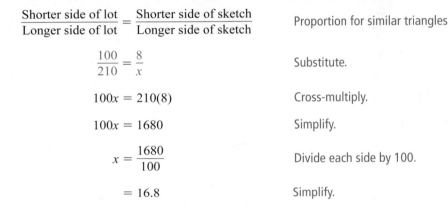

Triangular lot   *Sketch*

SOLUTION

This is a case of similar triangles in which the ratios of the corresponding sides are equal.

$$\frac{\text{Shorter side of lot}}{\text{Longer side of lot}} = \frac{\text{Shorter side of sketch}}{\text{Longer side of sketch}}$$ Proportion for similar triangles

$$\frac{100}{210} = \frac{8}{x}$$ Substitute.

$$100x = 210(8)$$ Cross-multiply.

$$100x = 1680$$ Simplify.

$$x = \frac{1680}{100}$$ Divide each side by 100.

$$= 16.8$$ Simplify.

So, the length of the longer side of the sketch should be 16.8 inches.

Exercises Within Reach ®

Solutions in English & Spanish and tutorial videos at CollegePrepAlgebra.com

Geometry In Exercises 45–48, **find the length** *x* **of the side of the larger triangle. (Assume that the two triangles are similar, and use the fact that corresponding sides of similar triangles are proportional.)**

45.

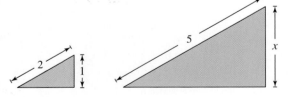

46.

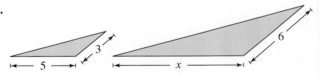

47.

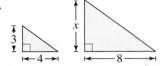

48.

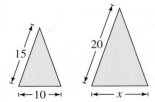

Application   **EXAMPLE 7**   **Finding Gasoline Cost**

You are driving from New York to Phoenix, a trip of 2450 miles. You begin the trip with a full tank of gas, and after traveling 424 miles, you refill the tank for $58. Assuming gas prices will be the same for the duration of your trip, how much should you plan to spend on gasoline for the entire trip?

SOLUTION

Verbal Model: $\dfrac{\text{Cost for entire trip}}{\text{Cost for one tank}} = \dfrac{\text{Miles for entire trip}}{\text{Miles for one tank}}$

Labels:

Cost of gas for entire trip $= x$	(dollars)
Cost of gas for one tank $= 58$	(dollars)
Miles for entire trip $= 2450$	(miles)
Miles for one tank $= 424$	(miles)

Equation:

$\dfrac{x}{58} = \dfrac{2450}{424}$ Write proportion.

$x = 58\left(\dfrac{2450}{424}\right)$ Multiply each side by 58.

$x \approx 335.14$ Simplify.

You should plan to spend approximately $335 for gasoline on the trip. Check this in the original statement of the problem.

Exercises Within Reach ®

Solutions in English & Spanish and tutorial videos at CollegePrepAlgebra.com

49. *Amount of Fuel* A car uses 20 gallons of gasoline for a trip of 500 miles. How many gallons would be used on a trip of 400 miles?

50. *Amount of Fuel* A tractor requires 4 gallons of diesel fuel to plow for 90 minutes. How many gallons of fuel would be required to plow for 8 hours?

51. *Polling Results* In a poll, 624 people from a sample of 1100 indicated they would vote for the Republican candidate. How many votes can the candidate expect to receive from 40,000 votes cast?

52. *Quality Control* A quality control engineer found two defective units in a sample of 50. At this rate, what is the expected number of defective units in a shipment of 10,000 units?

53. *Building Material* One hundred cement blocks are required to build a 16-foot wall. How many blocks are needed to build a 40-foot wall?

54. *Force on a Spring* A force of 50 pounds stretches a spring 4 inches. How much force is required to stretch the spring 6 inches?

Concept Summary: *Solving Proportions*

What

A **proportion** is a statement that equates two **ratios**. When solving a proportion, you usually know three of the values and are asked to find the fourth.

EXAMPLE

Solve $\frac{6}{18} = \frac{x}{6}$.

How

One way to solve a proportion is to use cross-multiplication.

EXAMPLE

$\frac{6}{18} = \frac{x}{6}$ Write proportion.

$6(6) = 18x$ Cross-multiply.

$\frac{36}{18} = x$ Divide each side by 18.

$2 = x$ Simplify.

Why

There are many real-life applications involving ratios and proportions. For example, knowing how to use ratios will help you identify **unit prices**.

Exercises Within Reach ®

Worked-out solutions to odd-numbered exercises at CollegePrepAlgebra.com

Concept Summary Check

55. *Creating an Example* Give an example of a real-life problem that you can represent with a ratio.

56. *Writing* In your own words, explain what the following proportion represents.

$$\frac{1 \text{ gallon of milk}}{\$3.89} = \frac{3 \text{ gallons of milk}}{\$11.67}$$

57. *Structure* Explain how to solve a proportion.

58. *Number Sense* Determine whether the following statement is a proportion. Explain.

$$\frac{5}{25} \stackrel{?}{=} \frac{6}{36}$$

Extra Practice

Comparing Measurements **In Exercises 59–64, find a ratio that compares the relative sizes of the quantities. (Use the same units of measurement for both quantities.)**

59. 1 quart to 1 gallon

60. 3 miles to 2000 feet

61. 2 kilometers to 2500 meters

62. $5\frac{1}{2}$ pints to 2 quarts

63. 3000 pounds to 5 tons

64. 4 days to 30 hours

Comparing Unit Prices **In Exercises 65 and 66, determine which product has the lower unit price.**

65. (a) A 2 liter bottle (67.6 ounces) of soft drink for $1.09
(b) Six 12-ounce cans of soft drink for $1.69

66. (a) A 1-quart container of oil for $2.12
(b) A 2.5-gallon container of oil for $19.99

Writing a Ratio **In Exercises 67 and 68, express the statement as a ratio in simplest form.**

67. You study 4 hours per day and are in class 6 hours per day. Find the ratio of the number of study hours to class hours.

68. You have $22 of state tax withheld from your paycheck per week when your gross pay is $750. Find the ratio of tax to gross pay.

69. *Map Scale* On a map, $1\frac{1}{4}$ inches represents 80 miles. Estimate the distance between two cities that are 6 inches apart on the map.

70. *Map Scale* On a map, $1\frac{1}{2}$ inches represents 40 miles. Estimate the distance between two cities that are 4 inches apart on the map.

71. *Geometry* Find the length of the shadow of the man shown in the figure. (*Hint:* Use similar triangles to create a proportion.)

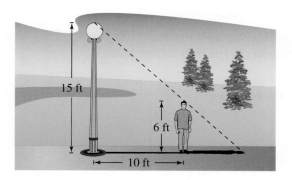

72. *Geometry* Find the height of a tree shown in the figure. (*Hint:* Use similar triangles to create a proportion.)

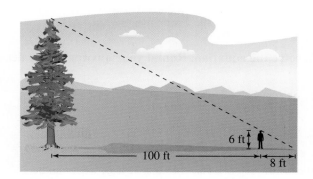

73. *Pumping Time* A pump can fill a 750-gallon tank in 35 minutes. How long will it take to fill a 1000-gallon tank with this pump?

74. *Recipe* Two cups of flour are required to make one batch of cookies. How many cups are required for $2\frac{1}{2}$ batches?

75. *Salt Water* The fresh water to salt ratio for a mixture is 25 to 1. How much fresh water is required to produce a mixture that contains one-half pound of salt?

76. *Building Material* The ratio of cement to sand in an 80-pound bag of dry mix is 1 to 4. Find the number of pounds of sand in the bag. (*Note:* Dry mix is composed of only cement and sand).

77. *Resizing a Picture* You have an 8-by-10-inch photo of a soccer player that must be reduced to a size of 1.6 inches by 2 inches for the school newsletter. What percent does the photo need to be reduced to in order for it to fit in the allotted space?

78. *Resizing a Picture* You have an 7-by-5-inch photo of the math club that must be reduced to a size of 5.6 inches by 4 inches for the school yearbook. What percent does the photo need to be reduced to in order for it to fit in the allotted space?

Explaining Concepts

79. *Writing* You are told that the ratio of men to women in a class is 2 to 1. Does this information tell you the total number of people in the class? Explain.

80. *Writing* Explain the following statement. "When setting up a ratio, be sure you are comparing apples to apples and not apples to oranges."

81. *Creating a Problem* Create a proportion problem. Exchange problems with another student and solve the problem you receive.

82. *Writing* Explain how to find the unit price of an item.

Cumulative Review

In Exercises 83−88, evaluate the expression.

83. $3^2 - (-4)$

84. $(-5)^3 + 3$

85. 9.3×10^6

86. $\dfrac{-|7 + 3^2|}{4}$

87. $(-4)^2 - (30 \div 50)$

88. $(8 \cdot 9) + (-4)^3$

In Exercises 89−92, solve the percent equation.

89. What number is 25% of 250?

90. What number is 45% of 90?

91. 150 is 250% of what number?

92. 465 is what percent of 500?

Mid-Chapter Quiz: Sections 3.1–3.4

Solutions in English & Spanish and tutorial videos at CollegePrepAlgebra.com

Take this quiz as you would take a quiz in class. After you are done, check your work against the answers in the back of the book.

In Exercises 1–10, solve the equation.

1. $74 - 12x = 2$

2. $10(y - 8) = 0$

3. $3x + 1 = x + 20$

4. $6x + 8 = 8 - 2x$

5. $-10x + \dfrac{2}{3} = \dfrac{7}{3} - 5x$

6. $\dfrac{x}{5} + \dfrac{x}{7} = 1$

7. $\dfrac{9 + x}{3} = 15$

8. $3 - 5(4 - x) = -6$

9. $\dfrac{x + 3}{6} = \dfrac{4}{3}$

10. $\dfrac{x + 7}{5} = \dfrac{x + 9}{7}$

In Exercises 11 and 12, solve the equation. Round your answer to two decimal places.

11. $32.86 - 10.5x = 11.25$

12. $\dfrac{x}{5.45} + 3.2 = 12.6$

13. What number is 62% of 25?

14. What number is $\frac{1}{2}$% of 8400?

15. 300 is what percent of 150?

16. 145.6 is 32% of what number?

Applications

17. You work 40 hours per week at a candy store and earn $7.50 per hour. You also earn $7.00 per hour babysitting and can work as many hours as you want. You want to earn $370 a week. How many hours must you babysit?

18. A region has an area of 42 square meters. It must be divided into the three subregions so that the second has twice the area of the first, and the third has twice the area of the second. Find the area of each subregion.

19. To get an A in a psychology course, you must have an average of at least 90 points for 3 tests of 100 points each. For the first 2 tests, your scores are 84 and 93. What must you score on the third test to earn a 90% average for the course?

20. You budget 30% of your annual after-tax income for housing. Your after-tax income is $38,500. What amount can you spend on housing?

21. Two people can paint a room in t hours, where t must satisfy the equation $\dfrac{t}{4} + \dfrac{t}{12} = 1$. How long will it take for the two people to paint the room?

22. A large round pizza has a radius of $r = 15$ inches, and a small round pizza has a radius of $r = 8$ inches. Find the ratio of the area of the large pizza to the area of the small pizza. (*Hint:* The area of a circle is $A = \pi r^2$.)

23. A car uses 30 gallons of gasoline for a trip of 800 miles. How many gallons would be used on a trip of 700 miles?

Study Skills in Action

Knowing Your Preferred Learning Modality

Math is a specific system of rules, properties, and calculations used to solve problems. However, you can take different approaches to learning this specific system based on learning modalities. A learning modality is a preferred way of taking in information that is then transferred into the brain for processing. The three modalities are *visual, auditory,* and *kinesthetic*. The following are brief descriptions of these modalities.

- **Visual** You take in information more productively if you can see the information.

- **Auditory** You take in information more productively when you listen to an explanation and talk about it.

- **Kinesthetic** You take in information more productively if you can experience it or use physical activity in studying.

You may find that one approach, or even a combination of approaches, works best for you.

Smart Study Strategy

Use Your Preferred Learning Modality

Visual *Draw a picture of a word problem.*

- Draw a picture of a word problem before writing a verbal model. You do not have to be an artist.

- When making a review card for a word problem, include a picture. This will help recall the information while taking a test.

- Make sure your notes are visually neat for easy recall.

Auditory *Talk about a word problem.*

- Explain how to do a word problem to another student. This is a form of thinking out loud. Write the instructions down on a review card.

- Find several students as serious as you are about math and form a study group.

- Teach the material to an imaginary person when studying alone.

Kinesthetic *Incorporate a physical activity.*

- Act out a word problem as much as possible. Use props when you can.

- Solve a word problem on a large whiteboard—the physical action of writing is more kinesthetic when the writing is larger and you can move around while doing it.

- Make a review card.

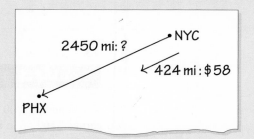

"It takes $58 worth of gas to travel 424 miles. To find the cost of traveling 2450 miles, I can set up and solve a proportion."

3.5 Geometric and Scientific Applications

▶ Use common formulas to solve application problems.
▶ Solve mixture problems.
▶ Solve work-rate problems.

Using Formulas

Miscellaneous Common Formulas

Temperature: F = degrees Fahrenheit, C = degrees Celsius

$$F = \frac{9}{5}C + 32$$

Simple Interest: I = interest, P = principal, r = interest rate (decimal form), t = time (years)

$$I = Prt$$

Distance: d = distance traveled, r = rate, t = time

$$d = rt$$

Application   **EXAMPLE 1**   **Using the Simple Interest Formula**

You deposit $5000 in an account paying simple interest. After 6 months, the account has earned $162.50 in interest. What is the annual interest rate?

SOLUTION

$I = Prt$	Simple interest formula
$162.50 = 5000(r)(0.5)$	Substitute for I, P, and t.
$162.50 = 2500r$	Simplify.
$\dfrac{162.50}{2500} = r$	Divide each side by 2500.
$0.065 = r$	Simplify.

The annual interest rate is $r = 0.065$ (or 6.5%). Check this solution in the original statement of the problem.

Exercises Within Reach ®   Solutions in English & Spanish and tutorial videos at CollegePrepAlgebra.com

Solving a Simple Interest Problem In Exercises 1−4, find the missing interest, principal, interest rate, or time.

1. $I =$ ▮
    $P = \$870$
    $r = 3.8\%$
    $t = 18$ months

2. $I = \$180$
    $P =$ ▮
    $r = 4.5\%$
    $t = 3$ years

3. $I = \$54$
    $P = \$450$
    $r =$ ▮
    $t = 2$ years

4. $I = \$97.30$
    $P = \$1200$
    $r = 4.5\%$
    $t =$ ▮

Common Formulas for Area, Perimeter, and Volume

Square

$A = s^2$

$P = 4s$

Rectangle

$A = lw$

$P = 2l + 2w$

Circle

$A = \pi r^2$

$C = 2\pi r$

Triangle

$A = \frac{1}{2}bh$

$P = a + b + c$

Cube

$V = s^3$

Rectangular Solid

$V = lwh$

Circular Cylinder

$V = \pi r^2 h$

Sphere

$V = \frac{4}{3}\pi r^3$

Application   **EXAMPLE 2**   **Using a Geometric Formula**

You own a rectangular lot that is 500 feet deep and has an area of 100,000 square feet. To pay for a new sewer system, you are assessed $5.50 per foot of lot frontage. (a) Find the frontage of your lot. (b) How much will you be assessed for the new sewer system?

SOLUTION

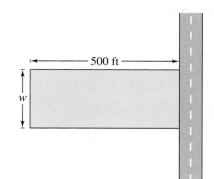

a.

$A = lw$	Area of a rectangle
$100,000 = 500(w)$	Substitute 100,000 for A and 500 for l.
$200 = w$	Divide each side by 500.

The frontage of the rectangular lot is 200 feet.

b. If each foot of frontage costs $5.50, then your total assessment will be $200(5.50) = \$1100$.

Exercises Within Reach ®

Solutions in English & Spanish and tutorial videos at CollegePrepAlgebra.com

5. **Geometry** A triangular piece of stained glass has an area of 6 square inches and a height of 3 inches. What is the length of the base?

6. **Geometry** A dime has a circumference of about 56.27 millimeters. What is the radius of a dime? Round your answer to two decimal places.

7. **Geometry** An Olympic-size swimming pool in the shape of a rectangular solid has a volume of 3125 cubic meters, a length of 50 meters, and a width of 25 meters. What is the depth of the pool?

8. **Geometry** A cylindrical bass drum has a volume of about 3054 cubic inches and a radius of 9 inches. What is the height of the drum? Round your answer to one decimal place.

Solving Mixture Problems

Many real-life problems involve combinations of two or more quantities that make up a new or different quantity. Such problems are called **mixture problems**.

First component   Second component   Final mixture

$$\left[\begin{array}{c}\text{First}\\\text{rate}\end{array}\right] \cdot \left[\text{Amount}\right] + \left[\begin{array}{c}\text{Second}\\\text{rate}\end{array}\right] \cdot \left[\text{Amount}\right] = \left[\begin{array}{c}\text{Final}\\\text{rate}\end{array}\right] \cdot \left[\begin{array}{c}\text{Final}\\\text{Amount}\end{array}\right]$$

Application

EXAMPLE 3   **Solving an Investment Mixture Problem**

You invested a total of \$10,000 in 2 funds earning $4\frac{1}{2}\%$ and $5\frac{1}{2}\%$ simple interest. During 1 year, the 2 funds earned a total of \$508.75 in interest. How much did you invest in each fund?

SOLUTION

Verbal Model:

$$\left[\begin{array}{c}\text{Interest earned}\\\text{at } 4\frac{1}{2}\%\end{array}\right] + \left[\begin{array}{c}\text{Interest earned}\\\text{at } 5\frac{1}{2}\%\end{array}\right] = \left[\begin{array}{c}\text{Total interest}\\\text{earned}\end{array}\right]$$

Labels:

Amount invested at $4\frac{1}{2}\% = x$	(dollars)
Amount invested at $5\frac{1}{2}\% = 10{,}000 - x$	(dollars)
Interest earned at $4\frac{1}{2}\% = (x)(0.045)(1)$	(dollars)
Interest earned at $5\frac{1}{2}\% = (10{,}000 - x)(0.055)(1)$	(dollars)
Total interest earned $= 508.75$	(dollars)

Equation:

$0.045x + 0.055(10{,}000 - x) = 508.75$	Write equation.
$0.045x + 550 - 0.055x = 508.75$	Distributive Property
$550 - 0.01x = 508.75$	Simplify.
$-0.01x = -41.25$	Subtract 550 from each side.
$x = 4125$	Divide each side by -0.01.

So, you invested \$4125 at $4\frac{1}{2}\%$ and $10{,}000 - x = 10{,}000 - 4125 = \$5875$ at $5\frac{1}{2}\%$. Check this in the original statement of the problem.

Exercises Within Reach ®

Solutions in English & Spanish and tutorial videos at CollegePrepAlgebra.com

9. **Investment Mixture** You invested a total of \$6000 in 2 funds earning 7% and 9% simple interest. During 1 year, the 2 funds earned a total of \$500 in interest. How much did you invest in each fund?

10. **Investment Mixture** You invested a total of \$30,000 in 2 funds earning 8.5% and 10% simple interest. During 1 year, the 2 funds earned a total of \$2700 in interest. How much did you invest in each fund?

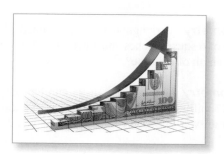

Application   EXAMPLE 4   **Solving a Solution Mixture Problem**

A pharmacist needs to strengthen a 15% alcohol solution with a pure alcohol solution to obtain a 32% solution. How much pure alcohol should be added to 100 milliliters of the 15% solution?

SOLUTION

In this problem, the rates are the alcohol *percents* of the solutions.

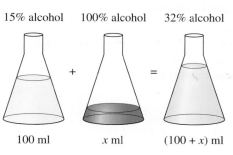

15% alcohol   100% alcohol   32% alcohol

100 ml   *x* ml   (100 + *x*) ml

Verbal Model:

Amount of alcohol in 15% alcohol solution	+	Amount of alcohol in 100% alcohol solution	=	Amount of alcohol in final alcohol solution

Labels:

15% solution: Percent alcohol = 0.15   (decimal form)
Amount of alcohol solution = 100   (milliliters)

100% solution: Percent alcohol = 1.00   (decimal form)
Amount of alcohol solution = *x*   (milliliters)

32% solution: Percent alcohol = 0.32   (decimal form)
Amount of alcohol solution = 100 + *x*   (milliliters)

Equation:

$0.15(100) + 1.00(x) = 0.32(100 + x)$	Write equation.
$15 + x = 32 + 0.32x$	Simplify.
$0.68x = 17$	Simplify.
$x = \dfrac{17}{0.68}$	Divide each side by 0.68.
$= 25$	Simplify.

So, the pharmacist should add 25 milliliters of pure alcohol to the 15% solution. This will result in $100 + x = 100 + 25 = 125$ milliliters of the 32% solution.

Exercises Within Reach ®

Solutions in English & Spanish and tutorial videos at CollegePrepAlgebra.com

11. **Mixture** You need to strengthen a 19% alcohol solution with a pure alcohol solution to obtain a 40% solution. How much pure alcohol should you add to 100 milliliters of the 19% solution?

12. **Chemistry** You need 175 milliliters of a 6% hydrochloric acid solution for an experiment. Your chemistry lab has a bottle of 3% hydrochloric acid solution and a bottle of 10% hydrochloric acid solution. How many milliliters of each solution should you mix together?

Solving Work-Rate Problems

Although not generally referred to as such, **work-rate problems** are actually mixture problems because they involve two or more rates.

$$\text{First rate} \cdot \text{Time} + \text{Second rate} \cdot \text{Time} = \begin{array}{l}\text{1 (one whole}\\\text{job completed)}\end{array}$$

Application   **EXAMPLE 5**   **Solving a Work-Rate Problem**

Consider two machines in a paper manufacturing plant. Machine 1 can complete one job in 3 hours. Machine 2 is newer and can complete one job in $2\frac{1}{2}$ hours. How long will it take the two machines working together to complete one job?

SOLUTION

Verbal Model: $\boxed{\begin{array}{l}\text{Portion done}\\\text{by Machine 1}\end{array}} + \boxed{\begin{array}{l}\text{Portion done}\\\text{by Machine 2}\end{array}} = \boxed{\begin{array}{l}\text{1 (one whole}\\\text{job completed)}\end{array}}$

Labels:

One whole job completed $= 1$	(job)
Rate (Machine 1) $= \frac{1}{3}$	(job per hour)
Time (Machine 1) $= t$	(hours)
Rate (Machine 2) $= \frac{2}{5}$	(job per hour)
Time (Machine 2) $= t$	(hours)

Equation:

$\left(\frac{1}{3}\right)t + \left(\frac{2}{5}\right)t = 1$ Write equation.

$\left(\frac{11}{15}\right)t = 1$ Combine like terms.

$t = \frac{15}{11}$ Multiply each side by $\frac{15}{11}$.

It will take $\frac{15}{11}$ hours (or about 1.36 hours) for the machines to complete the job working together. Check this solution in the original statement of the problem.

Exercises Within Reach®

Solutions in English & Spanish and tutorial videos at CollegePrepAlgebra.com

13. *Mowing a Lawn* You can mow a lawn in 2 hours using a riding mower, and your friend can mow the same lawn in 3 hours using a push mower. Using both machines together, how long will it take you and your friend to mow the lawn?

14. *Typing Project* One person can complete a typing project in 6 hours, and another can complete the same project in 8 hours. How long will it take the two people working together to complete the project?

15. *Work Rate* One worker can complete a task in m minutes while a second can complete the task in $9m$ minutes. Show that by working together they can complete the task in $t = \frac{9}{10}m$ minutes.

16. *Work Rate* One worker can complete a task in h hours while a second can complete the task in $3h$ hours. Show that by working together they can complete the task in $t = \frac{3}{4}h$ hours.

Application   EXAMPLE 6   Solving a Fluid-Rate Problem

An above ground swimming pool has a capacity of 15,600 gallons. A drain pipe can empty the pool in $6\frac{1}{2}$ hours. At what rate (in gallons per minute) does the water flow through the drain pipe?

15,600 gallons

Drain pipe

SOLUTION

To begin, change the time from hours to minutes by multiplying by 60. That is, $6\frac{1}{2}$ hours is equal to $(6.5)(60)$ or 390 minutes.

Verbal Model: $\boxed{\text{Volume of pool}} = \boxed{\text{Rate}} \cdot \boxed{\text{Time}}$

Labels: Volume = 15,600   (gallons)
            Rate = r (gallons per minute)
            Time = 390   (minutes)

Equation: $15,600 = r(390)$ Write equation.
              $40 = r$ Divide each side by 390.

The water is flowing through the drain pipe at a rate of 40 gallons per minute. Check this solution in the original statement of the problem.

Exercises Within Reach ®

Solutions in English & Spanish and tutorial videos at CollegePrepAlgebra.com

17. *Intravenous Bag* A 1000-milliliter intravenous bag is attached to a patient with a tube and is empty after 8 hours. At what rate does the solution flow through the tube?

18. *Swimming Pool* A swimming pool has a capacity of 10,800 gallons. A drain pipe empties the pool at a rate of 12 gallons per minute. How long, in hours, will it take for the pool to empty?

19. *Flower Order* A floral shop receives a $384 order for roses and carnations. The order contains twice as many roses as carnations. The prices per dozen for the roses and carnations are $18 and $12, respectively. How many of each type of flower are in the order?

20. *Ticket Sales* Ticket sales for a play totaled $1700. The number of tickets sold to adults was three times the number sold to children. The prices of the tickets for adults and children were $5 and $2, respectively. How many of each type were sold?

Concept Summary: Solving Geometric and Scientific Applications

What

Many real-life problems involve geometric applications such as perimeter, area, and volume. Other problems might involve distance, temperature, or interest.

EXAMPLE

You jog at an average rate of 8 kilometers per hour. How long will it take you to jog 14 kilometers?

How

To solve such problems, you can use formulas. For example, to solve a problem involving distance, rate, and time, you can use the distance formula.

EXAMPLE

$$\text{Distance} = \text{Rate} \cdot \text{Time}$$

$d = rt$	Distance formula
$14 = 8(t)$	Substitute for d and r.
$\dfrac{14}{8} = t$	Divide each side by 8.
$1.75 = t$	Simplify.

It will take you 1.75 hours.

Why

You can use formulas to solve many real-life applications. Some formulas occur so frequently that it is to your benefit to memorize them.

Exercises Within Reach ®

Worked-out solutions to odd-numbered exercises at CollegePrepAlgebra.com

Concept Summary Check

21. **Formulas** What is the formula for the volume of a circular cylinder?

22. **Create an Example** Give an example in which you need to find the perimeter of a real-life object.

23. **Writing** In your own words, describe the units of measure used for perimeter, area, and volume. Give examples of each.

24. **Structure** Rewrite the formula for simple interest by solving for P.

Extra Practice

Using the Distance Formula In Exercises 25−28, find the missing distance, rate, or time.

	Distance, d	Rate, r	Time, t		Distance, d	Rate, r	Time, t
25.		4 m/min	12 min	26.		62 mi/hr	$2\frac{1}{2}$ hr
27.	210 mi	50 mi/hr		28.	2054 m		18 sec

Solving for a Variable In Exercises 29 and 30, solve for the specified variable.

29. Solve for h: $A = \frac{1}{2}bh$

30. Solve for r: $A = P + Prt$

Mixture Problem In Exercises 31−34, determine the numbers of units of solution 2 required to obtain the desired percent alcohol concentration of the final solution. Then find the amount of the final solution.

	Concentration Solution 1	Amount of Solution 1	Concentration Solution 2	Concentration Final Solution		Concentration Solution 1	Amount of Solution 1	Concentration Solution 2	Concentration Final Solution
31.	10%	25 gal	30%	25%	32.	25%	4 L	50%	30%
33.	15%	5 qt	45%	30%	34.	70%	18.75 gal	90%	75%

35. **Interest Rate** Find the annual interest rate on a certificate of deposit that earned $128.98 interest in 1 year on a principal of $1500.

36. **Interest** How long must $700 be invested at an annual interest rate of 6.25% to earn $460 interest?

37. **Geometry** Two sides of a triangle have the same length. The third side is 7 meters less than 4 times that length. The perimeter is 83 meters. What are the lengths of the three sides of the triangle?

38. **Geometry** The longest side of a triangle is 3 times the length of the shortest side. The third side of the triangle is 4 inches longer than the shortest side. The perimeter is 49 inches. What are the lengths of the three sides of the triangle?

39. **Distance** Two cars start at a given point and travel in the same direction at average speeds of 45 miles per hour and 52 miles per hour (see figure). How far apart will they be in 4 hours?

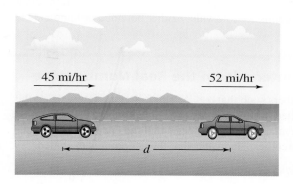

45 mi/hr   52 mi/hr

d

40. **Distance** Two planes leave Orlando International Airport approximately the same time and fly in opposite directions (see figure). Their speeds are 510 miles per hour and 600 miles per hour. How far apart will the planes be after $1\frac{1}{2}$ hours?

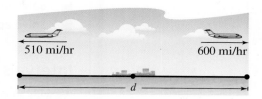

510 mi/hr   600 mi/hr

d

41. **Travel Time** On the first part of a 225-mile automobile trip, you averaged 55 miles per hour. On the last part of the trip, you averaged 48 miles per hour because of increased traffic congestion. The total trip took 4 hours and 15 minutes. Find the travel time for each part of the trip.

42. **Time** A jogger leaves a point on a fitness trail running at a rate of 4 miles per hour. Ten minutes later, a second jogger leaves from the same location running at 5 miles per hour. How long will it take the second jogger to overtake the first? How far will each have run at that point?

Explaining Concepts

43. **Mental Math** It takes you 4 hours to drive 180 miles. Explain how to use mental math to find your average speed. Then explain how your method is related to the formula $d = rt$.

44. **Error Analysis** A student solves the equation $S = 2lw + 2lh + 2wh$ for w and his answer is

$$w = \frac{S - 2lw - 2lh}{2h}.$$

Describe and correct the student's error.

45. **Structure** Write three equations that are equivalent to $A = \frac{1}{2}(x + y)h$ by solving for each variable, where A is the area, h is the height, and x and y are the bases of a trapezoid. Explain when you would use each equation.

46. **Think About It** When the height of a triangle doubles, does the area of the triangle double? Explain.

47. **Think About It** When the radius of a circle doubles, does its circumference double? Does its area double? Explain.

Cumulative Review

In Exercises 48 and 49, determine which of the numbers in the set are (a) natural numbers, (b) integers, (c) rational numbers, and (d) irrational numbers.

48. $\left\{-6, \frac{7}{4}, 2.1, \sqrt{49}, -8, \frac{4}{3}\right\}$

49. $\left\{1.8, \frac{1}{10}, 7, -2.75, 1, -3\right\}$

In Exercises 50 – 55, solve the proportion.

50. $\dfrac{x}{3} = \dfrac{28}{12}$

51. $\dfrac{1}{4} = \dfrac{y}{36}$

52. $\dfrac{z}{18} = \dfrac{8}{12}$

53. $\dfrac{3}{2} = \dfrac{9}{x}$

54. $\dfrac{5}{t} = \dfrac{75}{165}$

55. $\dfrac{34}{x} = \dfrac{102}{48}$

3.6 Linear Inequalities

▶ Sketch the graphs of inequalities.
▶ Solve linear inequalities.
▶ Solve application problems involving inequalities.

Intervals on the Real Number Line

As with an equation, you **solve an inequality** in the variable x by finding all values of x for which the inequality is true. Such values are called **solutions** and are said to satisfy the inequality. The set of all solutions of the inequality is the **solution set** of the inequality. The **graph** of an inequality is obtained by plotting its solution set on the real number line. Often, these graphs are intervals—either bounded or unbounded.

Study Tip

The **length** of an interval is the distance between its endpoints.

Bounded Intervals on the Real Number Line

Let a and b be real numbers such that $a < b$. The following intervals on the real number line are called **bounded intervals**. The numbers a and b are the **endpoints** of each interval. A bracket indicates that the endpoint is included in the interval, and a parenthesis indicates that the endpoint is excluded.

Notation	Interval Type	Inequality	Graph
$[a, b]$	Closed	$a \le x \le b$	
(a, b)	Open	$a < x < b$	
$[a, b)$		$a \le x < b$	
$(a, b]$		$a < x \le b$	

EXAMPLE 1 **Finding the Length of an Interval**

Find the length of the interval $[-1, 1]$.

SOLUTION

The length of the interval $[-1, 1]$ is
$$|1 - (-1)| = 2.$$

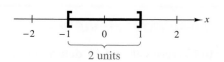

2 units

Exercises Within Reach ®   Solutions in English & Spanish and tutorial videos at CollegePrepAlgebra.com

Finding the Length of an Interval In Exercises 1−6, find the length of the interval.

1. $[-3, 5]$ 2. $[4, 10]$ 3. $(-9, 2]$

4. $[5, 13)$ 5. $(-3, 0)$ 6. $(0, 7)$

Study Tip

The symbols ∞ (**positive infinity**) and −∞ (**negative infinity**) do not represent real numbers. They are simply convenient symbols used to describe the unboundedness of an interval such as (5, ∞).

Unbounded Intervals on the Real Number Line

Let a and b be real numbers. The following intervals on the real number line are called **unbounded intervals**.

Notation	Interval Type	Inequality	Graph
$[a, \infty)$		$x \geq a$	
(a, ∞)	Open	$x > a$	
$(-\infty, b]$		$x \leq b$	
$(-\infty, b)$	Open	$x < b$	
$(-\infty, \infty)$	Entire real line		

EXAMPLE 2   **Graphing Inequalities**

Sketch the graph of each inequality.

a. $-3 < x \leq 1$ **b.** $0 < x < 2$

c. $-3 < x$ **d.** $x \leq 2$

SOLUTION

a. The graph of $-3 < x \leq 1$ is a bounded interval.

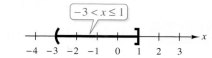

b. The graph of $0 < x < 2$ is a bounded interval.

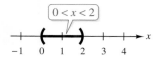

c. The graph of $-3 < x$ is an unbounded interval.

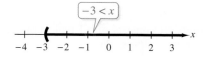

d. The graph of $x \leq 2$ is an unbounded interval.

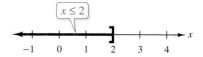

Exercises Within Reach®

Solutions in English & Spanish and tutorial videos at CollegePrepAlgebra.com

Graphing an Inequality In Exercises 7−20, sketch the graph of the inequality.

7. $x \leq 4$

8. $x > -6$

9. $x > 3.5$

10. $x \leq -2.5$

11. $x \geq \frac{1}{2}$

12. $x < \frac{1}{4}$

13. $-5 < x \leq 3$

14. $-1 < x \leq 5$

15. $4 > x \geq 1$

16. $9 \geq x \geq 3$

17. $\frac{3}{2} \geq x > 0$

18. $-\frac{15}{4} < x < -\frac{5}{2}$

19. $3.5 < x \leq 4.5$

20. $6.5 \geq x > -2.5$

Solving Linear Inequalities

Properties of Inequalities

1. *Addition and Subtraction Properties*

   Adding the same quantity to, or subtracting the same quantity from, each side of an inequality produces an equivalent inequality.

   If $a < b$, then $a + c < b + c$.

   If $a < b$, then $a - c < b - c$.

2. *Multiplication and Division Properties: Positive Quantities*

   Multiplying or dividing each side of an inequality by a positive quantity produces an equivalent inequality.

   If $a < b$ and c is positive, then $ac < bc$.

   If $a < b$ and c is positive, then $\dfrac{a}{c} < \dfrac{b}{c}$.

3. *Multiplication and Division Properties: Negative Quantities*

   Multiplying or dividing each side of an inequality by a negative quantity produces an equivalent inequality in which the inequality symbol is reversed.

   If $a < b$ and c is negative, then $ac > bc$.   Reverse inequality.

   If $a < b$ and c is negative, then $\dfrac{a}{c} > \dfrac{b}{c}$.   Reverse inequality.

4. *Transitive Property*

   Consider three quantities for which the first quantity is less than the second, and the second is less than the third. It follows that the first quantity must be less than the third quantity.

   If $a < b$ and $b < c$, then $a < c$.

EXAMPLE 3 **Solving a Linear Inequality**

$$x + 6 < 9 \qquad \text{Original inequality}$$
$$x + 6 - 6 < 9 - 6 \qquad \text{Subtract 6 from each side.}$$
$$x < 3 \qquad \text{Combine like terms.}$$

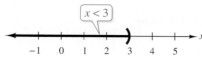

The solution set consists of all real numbers that are less than 3. The solution set in interval notation is $(-\infty, 3)$.

Exercises Within Reach ®

Solutions in English & Spanish and tutorial videos at CollegePrepAlgebra.com

Identifying Equivalent Inequalities In Exercises 21−24, determine whether the inequalities are equivalent.

21. $3x - 2 < 12, \quad 3x < 10$

22. $6x + 7 \geq 11, \quad 6x \geq 18$

23. $7x - 6 \leq 3x + 12, \quad 4x \leq 18$

24. $11 - 3x \geq 7x + 1, \quad 10 \geq 10x$

Solving an Inequality In Exercises 25−30, solve the inequality and sketch the solution on the real number line.

25. $x - 4 \geq 0$

26. $x + 1 < 0$

27. $x + 7 \leq 9$

28. $z - 5 > 0$

29. $2x < 8$

30. $3x \geq 12$

EXAMPLE 4   **Solving a Linear Inequality**

$8 - 3x \leq 20$	Original inequality
$8 - 8 - 3x \leq 20 - 8$	Subtract 8 from each side.
$-3x \leq 12$	Combine like terms.
$\dfrac{-3x}{-3} \geq \dfrac{12}{-3}$	Divide each side by -3 and reverse the inequality symbol.
$x \geq -4$	Simplify.

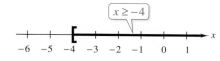

The solution set consists of all real numbers that are greater than or equal to -4. The solution set in interval notation is $[-4, \infty)$.

EXAMPLE 5   **Solving a Linear Inequality**

$7x - 3 > 3(x + 1)$	Original inequality
$7x - 3 > 3x + 3$	Distributive Property
$7x - 3x - 3 > 3x - 3x + 3$	Subtract $3x$ from each side.
$4x - 3 > 3$	Combine like terms.
$4x - 3 + 3 > 3 + 3$	Add 3 to each side.
$4x > 6$	Combine like terms.
$\dfrac{4x}{4} > \dfrac{6}{4}$	Divide each side by 4.
$x > \dfrac{3}{2}$	Simplify.

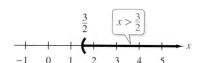

The solution set consists of all real numbers that are greater than $\frac{3}{2}$. The solution set in interval notation is $\left(\frac{3}{2}, \infty\right)$.

Exercises Within Reach ®

Solutions in English & Spanish and tutorial videos at CollegePrepAlgebra.com

Solving an Inequality In Exercises 31–46, solve the inequality and sketch the solution on the real number line.

31. $-9x \geq 36$

32. $-6x \leq 24$

33. $-\frac{3}{4}x < -6$

34. $-\frac{1}{5}x > -2$

35. $5 - x \leq -2$

36. $1 - y \geq -5$

37. $2x - 5.3 > 9.8$

38. $1.6x + 4 \leq 12.4$

39. $5 - 3x < 7$

40. $12 - 5x > 5$

41. $3x - 11 > -x + 7$

42. $21x - 11 \leq 6x + 19$

43. $-3x + 7 < 8x - 13$

44. $6x - 1 > 3x - 11$

45. $-3(y + 10) \geq 4(y + 10)$

46. $2(4 - z) \geq 8(1 + z)$

Two inequalities joined by the word *and* or the word *or* constitute a **compound inequality**. When two inequalities are joined by the word *and*, the solution set consists of all real numbers that satisfy *both* inequalities. The solution set for the compound inequality $-4 \le 5x - 2$ *and* $5x - 2 < 7$ can be written more simply as the **double inequality**

$$-4 \le 5x - 2 < 7.$$

EXAMPLE 6 Solving a Double Inequality

$-7 \le 5x - 2 < 8$	Original inequality
$-7 + 2 \le 5x - 2 + 2 < 8 + 2$	Add 2 to all three parts.
$-5 \le 5x < 10$	Combine like terms.
$\dfrac{-5}{5} \le \dfrac{5x}{5} < \dfrac{10}{5}$	Divide each part by 5.
$-1 \le x < 2$	Simplify.

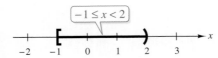

The solution set consists of all real numbers that are greater than or equal to -1 and less than 2. The solution set in interval notation is $[-1, 2)$.

EXAMPLE 7 Solving a Compound Inequality

$-3x + 6 \le 2$	or	$-3x + 6 \ge 7$	Original inequality
$-3x + 6 - 6 \le 2 - 6$		$-3x + 6 - 6 \ge 7 - 6$	Subtract 6 from all parts.
$-3x \le -4$		$-3x \ge 1$	Combine like terms.
$\dfrac{-3x}{-3} \ge \dfrac{-4}{-3}$		$\dfrac{-3x}{-3} \le \dfrac{1}{-3}$	Divide all parts by -3 and reverse both inequality symbols.
$x \ge \dfrac{4}{3}$		$x \le -\dfrac{1}{3}$	Solution set

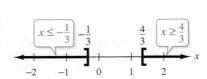

The solution set consists of all real numbers that are less than or equal to $-\frac{1}{3}$ or greater than or equal to $\frac{4}{3}$.

Exercises Within Reach ®

Solutions in English & Spanish and tutorial videos at CollegePrepAlgebra.com

Solving an Inequality In Exercises 47–60, solve the inequality and sketch the solution on the real number line. (Some inequalities have no solution.)

47. $0 < 2x - 5 < 9$

48. $-6 \le 3x - 9 < 0$

49. $8 < 6 - 2x \le 12$

50. $-10 \le 4 - 7x < 10$

51. $-1 < -0.2x < 1$

52. $-2 < -0.5s \le 0$

53. $2x - 4 \le 4$ and $2x + 8 > 6$

54. $7 + 4x < -5 + x$ and $2x + 10 \le -2$

55. $8 - 3x > 5$ and $x - 5 \ge 10$

56. $9 - x \le 3 + 2x$ and $3x - 7 \le -22$

57. $3x + 11 < 3$ or $4x - 1 \ge 9$

58. $4x + 10 \le -6$ or $-2x + 5 < -4$

59. $7.2 - 1.1x > 1$ or $1.2x - 4 > 2.7$

60. $0.4x - 3 \le 8.1$ or $4.2 - 1.6x \le 3$

Application

Application   **EXAMPLE 8**   Finding the Maximun Width of a Package

An overnight delivery service will not accept any package with a combined length and girth (perimeter of a cross section perpendicular to the length) exceeding 132 inches. Consider a rectangular box that is 68 inches long and has square cross sections. What is the maximum acceptable width of such a box?

SOLUTION

First make a sketch as shown below. The length of the box is 68 inches, and because a cross section is square, the width and height are each x inches.

Verbal Model:	Length + Girth ≤ 132 inches	
Labels:	Width of a side $= x$	(inches)
	Length $= 68$	(inches)
	Girth $= 4x$	(inches)
Inequality:	$68 + 4x \leq 132$	
	$4x \leq 64$	
	$x \leq 16$	

The width of the box can be at most 16 inches.

Exercises Within Reach ®

Solutions in English & Spanish and tutorial videos at CollegePrepAlgebra.com

61. *Budget* A student group has $4500 budgeted for a field trip. The cost of transportation for the trip is $1900. To stay within the budget, all other costs C must be no more than what amount?

62. *Budget* You have budgeted $1800 per month for your total expenses. The cost of rent per month is $600 and the cost of food is $350. To stay within your budget, all other costs C must be no more than what amount?

63. *Geometry* The width of a rectangle is 22 meters. The perimeter of the rectangle must be at least 90 meters and not more than 120 meters. Find the interval for the length x.

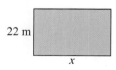

64. *Geometry* The length of a rectangle is 12 centimeters. The perimeter of the rectangle must be at least 30 centimeters and not more than 42 centimeters. Find the interval for the width x.

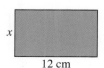

Concept Summary: Solving Linear Inequalities

What

When you **solve an inequality**, you find all the values for which the inequality is true. The procedures for solving inequalities are similar to those for solving equations.

EXAMPLE

Solve the inequality
$-2x + 4 > 8$.

How

To solve an inequality, isolate the variable by using the properties of inequalities.

EXAMPLE

$$-2x + 4 > 8 \qquad \text{Write inequality.}$$
$$-2x + 4 - 4 > 8 - 4 \quad \text{Subtract 4.}$$
$$-2x > 4 \qquad \text{Combine terms.}$$
$$\frac{-2x}{-2} < \frac{4}{-2} \qquad \begin{array}{l}\text{Divide by } -2 \text{ and} \\ \text{reverse inequality.}\end{array}$$
$$x < -2 \qquad \text{Solution set}$$

Why

Many real-life applications involve phrases like "at least" or "no more than." Using inequalities, you will be able to solve such problems.

Exercises Within Reach ®

Worked-out solutions to odd-numbered exercises at CollegePrepAlgebra.com

Concept Summary Check

65. *Writing* Does the graph of $x < -2$ contain a parenthesis or a square bracket? Explain.

66. *Structure* State whether each inequality is equivalent to $x > 3$. Explain your reasoning in each case.

    (a) $x < 3$ (b) $3 < x$

    (c) $-x < -3$ (d) $-3 < x$

67. *Reasoning* Is dividing each side of an inequality by 5 the same as multiplying each side by $\frac{1}{5}$? Explain.

68. *Writing* Describe two types of situations in which you must reverse the inequality symbol of an inequality.

Extra Practice

Matching In Exercises 69–74, **match the inequality with its graph.**

(a)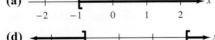

(b)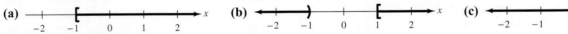

(c)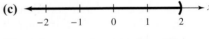

(d)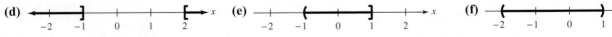

(e)

(f)

69. $x \geq -1$ **70.** $-1 < x \leq 1$ **71.** $x \leq -1$ or $x \geq 2$

72. $x < -1$ or $x \geq 1$ **73.** $-2 < x < 1$ **74.** $x < 2$

Solving an Inequality In Exercises 75–86, **solve the inequality and sketch the solution on the real number line.**

75. $\dfrac{x}{4} > 2 - \dfrac{x}{2}$

76. $\dfrac{x}{6} - 1 \leq \dfrac{x}{4}$

77. $\dfrac{x-4}{3} + 3 \leq \dfrac{x}{8}$

78. $\dfrac{x+3}{6} + \dfrac{x}{8} \geq 1$

79. $\dfrac{3x}{5} - 4 < \dfrac{2x}{3} - 3$

80. $\dfrac{4x}{7} + 1 > \dfrac{x}{2} + \dfrac{5}{7}$

81. $-4 \leq 2 - 3(x + 2) < 11$

82. $16 < 4(y + 2) - 5(2 - y) \leq 24$

83. $-3 < \dfrac{2x-3}{2} < 3$

84. $0 \leq \dfrac{x-5}{2} < 4$

85. $1 > \dfrac{x-4}{-3} > -2$

86. $-\dfrac{2}{3} < \dfrac{x-4}{-6} \leq \dfrac{1}{3}$

87. *Meteorology* Miami's average temperature is greater than the average temperature in Washington, D.C., and the average temperature in Washington, D.C., is greater than the average temperature in New York City. How does the average temperature in Miami compare with the average temperature in New York City?

88. *Elevation* The elevation (above sea level) of San Francisco is less than the elevation of Dallas, and the elevation of Dallas is less than the elevation of Denver. How does the elevation of San Francisco compare with the elevation of Denver?

89. *Operating Costs* A utility company has a fleet of vans. The annual operating cost per van is $C = 0.35m + 2900$, where m is the number of miles traveled by a van in a year. What is the maximum number of miles that will yield an annual operating cost that is no more than $12,000?

90. *Operating Costs* A fuel company has a fleet of trucks. The annual operating cost per truck is $C = 0.58m + 7800$, where m is the number of miles traveled by a truck in a year. What is the maximum number of miles that will yield an annual operating cost that is less than $25,000?

91. *Hourly Wage* Your company requires you to select one of two payment plans. One plan pays a straight $12.50 per hour. The second plan pays $8.00 per hour plus $0.75 per unit produced per hour. Write an inequality for the number of units that must be produced per hour so that the second option yields the greater hourly wage. Solve the inequality.

92. *Monthly Wage* Your company requires you to select one of two payment plans. One plan pays a straight $3000 per month. The second plan pays $1000 per month plus a commission of 4% of your gross sales. Write an inequality for the gross sales per month for which the second option yields the greater monthly wage. Solve the inequality.

Explaining Concepts

93. *Writing* Describe any differences between properties of equalities and properties of inequalities.

94. *Precision* If $-3 \leq x \leq 10$, then $-x$ must be in what interval? Explain.

95. *Logic* Discuss whether the solution set of a linear inequality is a *bounded* interval or an *unbounded* interval.

96. *Logic* Two linear inequalities are joined by the word *or* to form a compound inequality. Discuss whether the solution set is a bounded interval.

Writing a Compound Inequality **In Exercises 97−100, let a and b be real numbers such that $a < b$. Use a and b to write a compound algebraic inequality in x with the given type of solution. Explain your reasoning.**

97. A bounded interval

98. Two unbounded intervals

99. The set of all real numbers

100. No solution

Cumulative Review

In Exercises 101−104, place the correct symbol ($<$, $>$, or $=$) between the real numbers.

101. $|4|$ ____ $|-5|$ 102. $|-4|$ ____ $|-6|$

103. $|-7|$ ____ $|7|$ 104. $-|5|$ ____ $-(5)$

In Exercises 105−108, determine whether each value of the variable is a solution of the equation.

105. $3x = 27$; $x = 6$, $x = 9$

106. $x - 14 = 8$; $x = 6$, $x = 22$

107. $7x - 5 = 7 + x$; $x = 2$, $x = 6$

108. $2 + 5x = 8x - 13$; $x = 3$, $x = 5$

In Exercises 109−112, solve the equation.

109. $2x - 17 = 0$

110. $x - 17 = 4$

111. $32x = -8$

112. $14x + 5 = 2 - x$

3.7 Absolute Value Equations and Inequalities

▶ Solve absolute value equations.

▶ Solve absolute value inequalities.

▶ Solve real-life problems involving absolute value.

Solving Equations Involving Absolute Value

> ### Solving an Absolute Value Equation
>
> Let x be a variable or an algebraic expression and let a be a real number such that $a \geq 0$. The solutions of the equation $|x| = a$ are given by $x = -a$ and $x = a$. That is,
>
> $$|x| = a \implies x = -a \quad \text{or} \quad x = a.$$

EXAMPLE 1 Solving Absolute Value Equations

Solve each absolute value equation.

a. $|x| = 10$ **b.** $|x| = 0$ **c.** $|y| = -1$

SOLUTION

a. This equation is equivalent to the two linear equations

$$x = -10 \quad \text{and} \quad x = 10. \qquad \text{Equivalent linear equations}$$

So, the absolute value equation has two solutions: $x = -10$ and $x = 10$.

b. This equation is equivalent to the two linear equations

$$x = -(0) = 0 \quad \text{and} \quad x = 0. \qquad \text{Equivalent linear equations}$$

Because both equations are the same, you can conclude that the absolute value equation has only one solution: $x = 0$.

c. This absolute value equation has *no solution* because it is not possible for the absolute value of a real number to be negative.

> **Study Tip**
>
> The strategy for solving an absolute value equation is to *rewrite* the equation in *equivalent forms* that can be solved by previously learned methods. This is a common strategy in mathematics. That is, when you encounter a new type of problem, you try to rewrite the problem so that it can be solved by techniques you already know.

Exercises Within Reach ®

Solutions in English & Spanish and tutorial videos at CollegePrepAlgebra.com

Checking a Solution In Exercises 1−4, determine whether the value of the variable is a solution of the equation.

	Equation	*Value*		*Equation*	*Value*
1.	$|4x + 5| = 10$	$x = -3$	**2.**	$|2x - 16| = 10$	$x = 3$
3.	$|6 - 2w| = 2$	$w = 4$	**4.**	$\left|\frac{1}{2}t + 4\right| = 8$	$t = 6$

Solving an Equation In Exercises 5−10, solve the equation. (Some equations have no solution.)

5. $|x| = 4$ **6.** $|x| = 3$ **7.** $|t| = -45$

8. $|s| = 16$ **9.** $|h| = 0$ **10.** $|x| = -82$

EXAMPLE 2   **Solving an Absolute Value Equation**

Solve $|3x + 4| = 10$.

SOLUTION

$|3x + 4| = 10$		Write original equation.
$3x + 4 = -10$ or $3x + 4 = 10$		Equivalent equations
$3x + 4 - 4 = -10 - 4$ $3x + 4 - 4 = 10 - 4$		Subtract 4 from each side.
$3x = -14$ $3x = 6$		Combine like terms.
$x = -\dfrac{14}{3}$ $x = 2$		Divide each side by 3.

Check

$$|3x + 4| = 10$$
$$\left|3\left(-\tfrac{14}{3}\right) + 4\right| \overset{?}{=} 10$$
$$|-14 + 4| \overset{?}{=} 10$$
$$|-10| = 10 ✔$$

$$|3x + 4| = 10$$
$$|3(2) + 4| \overset{?}{=} 10$$
$$|6 + 4| \overset{?}{=} 10$$
$$|10| = 10 ✔$$

EXAMPLE 3   **Solving an Absolute Value Equation**

Solve $|2x - 1| + 3 = 8$.

SOLUTION

$|2x - 1| + 3 = 8$		Write original equation.
$|2x - 1| = 5$		Write in standard form.
$2x - 1 = -5$ or $2x - 1 = 5$		Equivalent equations
$2x = -4$ $2x = 6$		Add 1 to each side.
$x = -2$ $x = 3$		Divide each side by 2.

The solutions are $x = -2$ and $x = 3$. Check these in the original equation.

Exercises Within Reach®   Solutions in English & Spanish and tutorial videos at CollegePrepAlgebra.com

Solving an Equation In Exercises 11−24, solve the equation. (Some equations have no solution.)

11. $|5x| = 15$

12. $\left|\tfrac{1}{3}x\right| = 2$

13. $|x + 1| = 5$

14. $|x + 5| = 7$

15. $|4 - 3x| = 0$

16. $|3x - 2| = -5$

17. $\left|\dfrac{2x + 3}{5}\right| = 5$

18. $\left|\dfrac{7a + 6}{4}\right| = 2$

19. $|5 - 2x| + 10 = 6$

20. $|5x - 3| + 8 = 22$

21. $\left|\dfrac{x - 2}{3}\right| + 6 = 6$

22. $\left|\dfrac{x - 2}{5}\right| + 4 = 4$

23. $3|2x - 5| + 4 = 7$

24. $2|4 - 3x| - 6 = -2$

| EXAMPLE 4 | Solving an Equation Involving Two Absolute Values |

Solve $|3x - 4| = |7x - 16|$.

SOLUTION

$$|3x - 4| = |7x - 16| \qquad \text{Write original equation.}$$

$$3x - 4 = 7x - 16 \quad \text{or} \quad 3x - 4 = -(7x - 16) \qquad \text{Equivalent equations}$$

$$-4x - 4 = -16 \qquad\qquad 3x - 4 = -7x + 16$$

$$-4x = -12 \qquad\qquad\qquad 10x = 20$$

$$x = 3 \qquad\qquad\qquad\quad x = 2 \qquad \text{Solutions}$$

The solutions are $x = 3$ and $x = 2$. Check these in the original equation.

| EXAMPLE 5 | Solving an Equation Involving Two Absolute Values |

Solve $|x + 5| = |x + 11|$.

SOLUTION

By equating the expression $(x + 5)$ to the opposite of $(x + 11)$, you obtain

$$x + 5 = -(x + 11) \qquad \text{Equivalent equation}$$

$$x + 5 = -x - 11 \qquad \text{Distributive Property}$$

$$2x + 5 = -11 \qquad \text{Add } x \text{ to each side.}$$

$$2x = -16 \qquad \text{Subtract 5 from each side.}$$

$$x = -8. \qquad \text{Divide each side by 2.}$$

However, by setting the two expressions equal to each other, you obtain

$$x + 5 = x + 11 \qquad \text{Equivalent equation}$$

$$x = x + 6 \qquad \text{Subtract 5 from each side.}$$

$$0 = 6 \qquad \text{Subtract } x \text{ from each side.}$$

which is a false statement. So, the original equation has only one solution: $x = -8$. Check this solution in the original equation.

Study Tip

When solving an equation of the form

$$|ax + b| = |cx + d|$$

it is possible that one of the resulting equations will not have a solution. Note this occurrence in Example 5.

Exercises Within Reach ®

Solutions in English & Spanish and tutorial videos at CollegePrepAlgebra.com

Solving an Equation In Exercises 25−32, solve the equation.

25. $|2x + 1| = |x - 4|$

26. $|2x - 5| = |x + 10|$

27. $|x + 8| = |2x + 1|$

28. $|10 - 3x| = |x + 7|$

29. $|3x + 1| = |3x - 3|$

30. $|2x + 7| = |2x + 9|$

31. $|4x - 10| = 2|2x + 3|$

32. $3|2 - 3x| = |9x + 21|$

Solving Inequalities Involving Absolute Value

Solving an Absolute Value Inequality

Let x be a variable or an algebraic expression and let a be a real number such that $a > 0$.

1. The solutions of $|x| < a$ are all values of x that lie between $-a$ and a. That is,

$$|x| < a \quad \text{if and only if} \quad -a < x < a.$$

2. The solutions of $|x| > a$ are all values of x that are less than $-a$ or greater than a. That is,

$$|x| > a \quad \text{if and only if} \quad x < -a \text{ or } x > a.$$

These rules are also valid if $<$ is replaced by $\leq$ and $>$ is replaced by $\geq$.

EXAMPLE 6   Solving an Absolute Value Inequality

Solve $|x - 5| < 2$.

SOLUTION

$\|x - 5\| < 2$	Write original inequality.
$-2 < x - 5 < 2$	Equivalent double inequality
$-2 + 5 < x - 5 + 5 < 2 + 5$	Add 5 to all three parts.
$3 < x < 7$	Combine like terms.

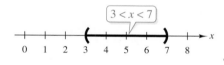

The solution set consists of all real numbers that are greater than 3 and less than 7. The solution set in interval notation is (3, 7).

Exercises Within Reach ®

Solutions in English & Spanish and tutorial videos at CollegePrepAlgebra.com

Checking a Solution In Exercises 33−36, determine **whether the x-value is a solution of the inequality.**

	Inequality	*Value*		*Inequality*	*Value*
33.	$\|x\| < 3$	$x = 2$	**34.**	$\|x\| \leq 5$	$x = -7$
35.	$\|x - 7\| \geq 3$	$x = 9$	**36.**	$\|x - 3\| > 5$	$x = 16$

Solving an Inequality In Exercises 37−46, solve the inequality.

37. $|y| < 4$ **38.** $|x| < 6$

39. $|x| \geq 6$ **40.** $|y| \geq 4$

41. $|x + 6| > 10$ **42.** $|y - 2| \leq 4$

43. $|2x| < 14$ **44.** $|4z| \leq 9$

45. $\left|\dfrac{y}{3}\right| \leq \dfrac{1}{3}$ **46.** $\left|\dfrac{t}{5}\right| < \dfrac{3}{5}$

EXAMPLE 7 **Solving an Absolute Value Inequality**

$$|3x - 4| \geq 5 \qquad \text{Original inequality}$$

$$3x - 4 \leq -5 \quad \text{or} \quad 3x - 4 \geq 5 \qquad \text{Equivalent inequalities}$$

$$3x - 4 + 4 \leq -5 + 4 \qquad 3x - 4 + 4 \geq 5 + 4 \qquad \text{Add 4 to all parts.}$$

$$3x \leq -1 \qquad 3x \geq 9 \qquad \text{Combine like terms.}$$

$$\frac{3x}{3} \leq \frac{-1}{3} \qquad \frac{3x}{3} \geq \frac{9}{3} \qquad \text{Divide each side by 3.}$$

$$x \leq -\frac{1}{3} \qquad x \geq 3 \qquad \text{Simplify.}$$

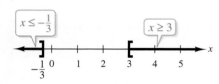

The solution set consists of all real numbers that are less than or equal to $-\frac{1}{3}$ or greater than or equal to 3.

EXAMPLE 8 **Solving an Absolute Value Inequality**

$$\left|2 - \frac{x}{3}\right| \leq 0.01 \qquad \text{Original inequality}$$

$$-0.01 \leq 2 - \frac{x}{3} \leq 0.01 \qquad \text{Equivalent double inequality}$$

$$-0.01 - 2 \leq 2 - 2 - \frac{x}{3} \leq 0.01 - 2 \qquad \text{Subtract 2 from all three parts.}$$

$$-2.01 \leq -\frac{x}{3} \leq -1.99 \qquad \text{Combine like terms.}$$

$$-2.01(-3) \geq -\frac{x}{3}(-3) \geq -1.99(-3) \qquad \text{Multiply all three parts by } -3 \text{ and reverse both inequality symbols.}$$

$$6.03 \geq x \geq 5.97 \qquad \text{Simplify.}$$

$$5.97 \leq x \leq 6.03 \qquad \text{Solution set in standard form}$$

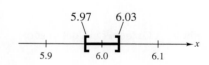

The solution set consists of all real numbers that are greater than or equal to 5.97 and less than or equal to 6.03. The solution set in interval notation is [5.97, 6.03].

Exercises Within Reach ®

Solving an Inequality In Exercises 47–56, solve the inequality. (Some inequalities have no solution.)

47. $|2x + 3| > 9$

48. $|7r - 3| > 11$

49. $|2x - 1| \leq 7$

50. $|6t + 15| \geq 30$

51. $|3x + 10| < -1$

52. $|4x - 5| > -3$

53. $\dfrac{|a + 6|}{2} \geq 16$

54. $\dfrac{|y - 16|}{4} < 30$

55. $|0.2x - 3| < 4$

56. $|1.5t - 8| \leq 16$

Applications

Application   **EXAMPLE 9**   **Oil Production**

The estimated daily production at an oil refinery is given by the absolute value inequality $|x - 200{,}000| \le 25{,}000$, where x is measured in barrels of oil. Solve the inequality to determine the maximum and minimum production levels.

SOLUTION

$\|x - 200{,}000\| \le 25{,}000$	Write original inequality.
$-25{,}000 \le x - 200{,}000 \le 25{,}000$	Equivalent double inequality
$175{,}000 \le x \le 225{,}000$	Add 200,000 to all three parts.

So, the oil refinery produces a maximum of 225,000 barrels of oil a minimum of 175,000 barrels of oil per day.

Application   **EXAMPLE 10**   **Creating a Model**

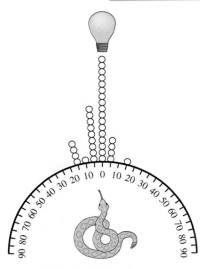

To test the accuracy of a rattlesnake's "pit-organ sensory system," a biologist blindfolds a rattlesnake and presents the snake with a warm "target." Of 36 strikes, the snake is on target 17 times. Let A represent the number of degrees by which the snake is off target. Then $A = 0$ represents a strike that is aimed directly at the target. Positive values of A represent strikes to the right of the target, and negative values of A represent strikes to the left of the target. Use the diagram shown to write an absolute value inequality that describes the interval in which the 36 strikes occurred.

SOLUTION

From the diagram, you can see that in the 36 strikes, the snake is never off by more than 15 degrees in either direction. As a compound inequality, this can be represented by $-15 \le A \le 15$. As an absolute value inequality, this interval can be represented by $|A| \le 15$.

Exercises Within Reach ®

Solutions in English & Spanish and tutorial videos at CollegePrepAlgebra.com

57. *Speed Skating* Each skater in a 500-meter short track speed skating final had a time that satisfied the inequality $|t - 42.238| \le 0.412$, where t is the time in seconds. Sketch the graph of the solution of the inequality. What were the fastest and slowest possible times?

58. *Time Study* A time study was conducted to determine the length of time required to perform a task in a manufacturing process. The times required by approximately two-thirds of the workers in the study satisfied the inequality

$$\left| \frac{t - 15.6}{1.9} \right| \le 1$$

where t is time in minutes. Sketch the graph of the solution of the inequality. What were the maximum and minimum possible times?

Concept Summary: Solving Absolute Value Equations

What

The **absolute value equation** $|x| = a$, $a \geq 0$, has exactly two solutions:

$x = -a$ and $x = a$.

EXAMPLE

Solve $|x + 1| = 12$.

How

Use these steps to solve an absolute value equation.

1. Write the equation in **standard form**, if necessary.

2. Rewrite the equation in equivalent forms and solve using previously learned methods.

EXAMPLE

$|x + 1| = 12$

$x + 1 = 12 \quad$ or $\quad x + 1 = -12$

$x = 11 \quad$ or $\quad x = -13$

Why

Understanding how to solve absolute value equations will help you when you learn how to solve absolute value inequalities.

Exercises Within Reach®

Worked-out solutions to odd-numbered exercises at CollegePrepAlgebra.com

Concept Summary Check

59. *Solving an Equation* In your own words, explain how to solve an absolute value equation. Illustrate your explanation with an example.

60. *Number of Solutions* In the equation $|x| = b$, b is a positive real number. How many solutions does this equation have? Explain.

61. *Writing an Equation* Write an absolute value equation that represents the verbal statement.

"The distance between x and zero is a."

62. *True or False?* The solutions of $|x| = -2$ are $x = -2$ and $x = 2$. Justify your answer.

Extra Practice

Solving an Equation **In Exercises 63 and 64, solve the equation.**

63. $|4x + 1| = \frac{1}{2}$

64. $\frac{1}{4}|3x + 1| = 4$

Think About It **In Exercises 65 and 66, write an absolute value equation that represents the verbal statement.**

65. The distance between x and 4 is 9.

66. The distance between -3 and t is 5.

Writing an Inequality **In Exercises 67−72, write an absolute value inequality that represents the interval.**

67.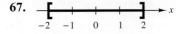

68.

69.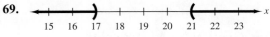

70.

71.

72.

Solving an Inequality **In Exercises 73 and 74, solve the inequality.**

73. $\left|\dfrac{3x - 2}{4}\right| + 5 \geq 5$

74. $\left|\dfrac{2x - 4}{5}\right| - 9 \leq 3$

Writing an Inequality In Exercises 75−78, write an absolute value inequality that represents the verbal statement.

75. The set of all real numbers x whose distance from 0 is less than 3.

76. The set of all real numbers x whose distance from 0 is more than 2.

77. The set of all real numbers x for which the distance from 0 to 3 less than twice x is more than 5.

78. The set of all real numbers x for which the distance from 0 to 5 more than half of x is less than 13.

79. *Accuracy of Measurements* In woodshop class, you must cut several pieces of wood to within $\frac{3}{16}$ inch of the teacher's specifications. Let $(s - x)$ represent the difference between the specification s and the measured length x of a cut piece.

(a) Write an absolute value inequality that describes the values of x that are within specifications.

(b) The length of one piece of wood is specified to be $5\frac{1}{8}$ inches. Describe the acceptable lengths for this piece.

80. *Body Temperature* Physicians generally consider an adult's body temperature x to be normal if it is within $1°F$ of the temperature $98.6°F$.

(a) Write an absolute value inequality that describes the values of x that are considered normal.

(b) Describe the range of body temperatures that are considered normal.

81. *Height* The heights h of two-thirds of the members of a population satisfy the inequality

$$\left| \frac{h - 68.5}{2.7} \right| \le 1$$

where h is measured in inches. Determine the interval on the real number line in which these heights lie.

82. *Geometry* The side of a square is measured as 10.4 inches with a possible error of $\frac{1}{16}$ inch. Using these measurements, determine the interval containing the possible areas of the square.

Explaining Concepts

83. *Reasoning* The graph of the inequality $|x - 3| < 2$ can be described as *all real numbers that are within two units of 3*. Give a similar description of $|x - 4| < 1$.

84. *Precision* Write an absolute value inequality to represent all the real numbers that are more than $|a|$ units from b. Then write an example showing the solution of the inequality for sample values of a and b.

85. *Reasoning* Complete $|2x - 6| \le$ ▨ so that the solution is $0 \le x \le 6$.

86. *Writing* Describe and correct the error. Explain how you can recognize that the solution is wrong without solving the inequality.

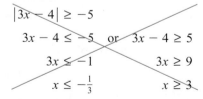

$$|3x - 4| \ge -5$$
$$3x - 4 \le -5 \quad \text{or} \quad 3x - 4 \ge 5$$
$$3x \le -1 \qquad\qquad 3x \ge 9$$
$$x \le -\tfrac{1}{3} \qquad\qquad x \ge 3$$

Cumulative Review

In Exercises 87 and 88, translate the verbal phrase into an algebraic expression.

87. Four times the sum of a number n and 3

88. Eight less than two times a number n

In Exercises 89−92, solve the inequality.

89. $x - 7 > 13$

90. $x + 7 \le 13$

91. $4x + 11 \ge 27$

92. $-4 < x + 2 < 12$

3 Chapter Summary

	What did you learn?	Explanation and Examples	Review Exercises
3.1	Solve a linear equation in standard form *(p. 102)*.	To solve a linear equation, use inverse operations to isolate the variable.	1–4
	Solve a linear equation in nonstandard form *(p. 104)*.	To solve a linear equation in nonstandard form, simplify each side of the equation before using inverse operations to isolate the variable.	5–14
3.2	Solve linear equations containing symbols of grouping *(p. 110)*.	To solve a linear equation that contains symbols of grouping, remove symbols of grouping from each side using the Distributive Property, combine like terms, isolate the variable using properties of equality, and check your solution in the original equation.	15–20
	Solve linear equations involving fractions *(p. 112)*.	To clear an equation of fractions, multiply each side by the least common multiple (LCM) of the denominators.	21–26, 31, 32
	Solve linear equations involving decimals *(p. 114)*.	To clear an equation of decimals, multiply each side by a power of 10 that converts all decimal coefficients to integers.	27–30
3.3	Convert percents to decimals and fractions, and vice versa *(p. 118)*.	To convert from percent form to decimal form, move the decimal point two places to the left. To convert from decimal form to percent form, move the decimal point two places to the right.	33–40
	Solve linear equations involving percents *(p. 119)*.	Use the percent equation. The percent equation $a = p \cdot b$ compares two numbers. b = base number p = percent (in decimal form) a = number being compared to b	41–46
	Solve problems involving markups and discounts *(p. 122)*.	**1.** Write a verbal model that describes the problem. Selling price = Cost + Markup Markup = Market rate · Cost Sale price = Original price − Discount Discount = Discount rate · Original price **2.** Assign labels to fixed quantities and variable quantities. **3.** Rewrite the verbal model as an algebraic equation using the assigned labels. **4.** Solve the resulting algebraic equation. **5.** Check to see that your solution satisfies the original problem as stated.	47, 48
3.4	Compare relative sizes using ratios *(p. 126)*.	When comparing the relative sizes of two quantities, be sure to use the same units of measurement for both quantities.	49–52
	Find the unit price of a consumer item *(p. 128)*.	The unit price of an item is given by the ratio of the total price to the total number of units.	53, 54

What did you learn?	Explanation and Examples	Review Exercises
3.4 Solve proportions that equate two ratios *(p. 129)*.	A proportion equates two ratios. If $\dfrac{a}{b} = \dfrac{c}{d}$, then $ad = bc$.	55−62
3.5 Use common formulas to solve application problems *(p. 136)*.	Temperature: F = degrees Fahrenheit $\quad\quad\quad\quad\;\; C$ = degrees Celsius $$F = \tfrac{9}{5}C + 32$$ Simple Interest: I = interest $\quad\quad\quad\quad\quad\;\; P$ = principal $\quad\quad\quad\quad\quad\;\; r$ = interest rate (decimal form) $\quad\quad\quad\quad\quad\;\; t$ = time (years) $$I = Prt$$ Distance: d = distance traveled $\quad\quad\quad\;\; r$ = rate $\quad\quad\quad\;\; t$ = time $$d = rt$$ Also see page 137 for common formulas for area, perimeter, and volume.	63−72
Solve mixture problems *(p. 138)*.	Mixture problems involve combinations of two or more quantities that make up a new or different quantity.	73, 74
Solve work-rate problems *(p. 140)*.	Work-rate problems are actually mixture problems because they involve two or more rates.	75, 76
3.6 Sketch the graphs of inequalities *(p. 144)*.	See examples of bounded and unbounded intervals on pages 144 and 145.	77−80
Solve linear inequalities *(p. 146)*.	**1.** Addition and subtraction: $\quad$ If $a < b$, then $a + c < b + c$. $\quad$ If $a < b$, then $a - c < b - c$. **2.** Multiplication and division (c is positive): $\quad$ If $a < b$, then $ac < bc$. If $a < b$, then $\dfrac{a}{c} < \dfrac{b}{c}$. **3.** Multiplication and division (c is negative): $\quad$ If $a < b$, then $ac > bc$. If $a < b$, then $\dfrac{a}{c} > \dfrac{b}{c}$. **4.** Transitive property: $\quad$ If $a < b$ and $b < c$, then $a < c$.	81−98
Solve application problems involving inequalities *(p. 149)*.	Use a verbal model to write an inequality and then solve the inequality.	99, 100
3.7 Solve absolute value equations *(p. 152)*.	$|x| = a \Rightarrow x = -a \;$ or $\; x = a$.	101−108
Solve absolute value inequalities *(p. 155)*.	**1.** $|x| < a \;$ if and only if $-a < x < a$. **2.** $|x| > a \;$ if and only if $x < -a$ or $x > a$.	109−118
Solve real-life problems involving absolute value *(p. 157)*.	Use the information given in the problem to write an absolute value equation or an absolute value inequality. Then solve the equation or inequality.	119, 120

Review Exercises

Worked-out solutions to odd-numbered exercises at CollegePrepAlgebra.com

3.1

Solving a Linear Equation In Exercises 1–12, solve the equation and check your solution.

1. $2x - 10 = 0$
2. $12y + 72 = 0$
3. $5x - 3 = 0$
4. $-8x + 6 = 0$
5. $x + 10 = 13$
6. $x - 3 = 8$
7. $10x = 50$
8. $-3x = 21$
9. $8x + 7 = 39$
10. $12x - 5 = 43$
11. $\frac{x}{5} = 4$
12. $-\frac{x}{14} = \frac{1}{2}$

13. *Earnings* Your hourly wage is $9.75 per hour plus $0.80 for each unit you produce. How many units must you produce in 1 hour so that your hourly wage is $19.35?

14. *Geometry* A 12-foot board is cut so that one piece is 5 times as long as the other. Find the length of each piece.

3.2

Solving a Linear Equation In Exercises 15–26, solve the equation and check your solution.

15. $3x - 2(x + 5) = 10$
16. $4x + 2(7 - x) = 5$
17. $2(x + 3) = 6(x - 3)$
18. $8(x - 2) = 3(x + 2)$
19. $7 - [2(3x + 4) - 5] = x - 3$
20. $14 + [3(6x - 15) + 4] = 5x - 1$
21. $\frac{2}{3}x - \frac{1}{6} = \frac{9}{2}$
22. $\frac{1}{8}x + \frac{3}{4} = \frac{5}{2}$
23. $\frac{u}{10} + \frac{u}{5} = 6$
24. $\frac{x}{3} + \frac{x}{5} = 1$
25. $\frac{x + 3}{5} = \frac{x + 7}{12}$
26. $\frac{y - 2}{6} = \frac{y + 1}{15}$

Solving a Linear Equation In Exercises 27–30, solve the equation. Round your answer to two decimal places.

27. $5.16x - 87.5 = 32.5$
28. $2.825x + 3.125 = 12.5$
29. $\frac{x}{4.625} = 48.5$
30. $5x + \frac{1}{4.5} = 18.125$

31. *Time to Complete a Task* Two people can complete 50% of a task in t hours, where t must satisfy the equation $\frac{t}{10} + \frac{t}{15} = 0.5$. How long will it take for the two people to complete 50% of the task?

32. *Course Grade* To get an A in a course, you must have an average of at least 90 points for 4 tests of 100 points each. For the first 3 tests, your scores are 85, 96, and 89. What must you score on the fourth test to earn a 90% average for the course?

3.3

Finding Equivalent Forms of a Percent In Exercises 33–40, complete the table showing the equivalent forms of the percent.

	Percent	Parts out of 100	Decimal	Fraction
33.	60%			
34.	35%			
35.				$\frac{4}{5}$
36.				$\frac{5}{8}$
37.			0.20	
38.			1.35	
39.		55		
40.		12.5		

Solving Percent Equations In Exercises 41–46, solve the percent equation.

41. What number is 125% of 16?
42. What number is 0.8% of 3250?
43. 150 is $37\frac{1}{2}$% of what number?
44. 323 is 95% of what number?
45. 150 is what percent of 250?
46. 130.6 is what percent of 3265?

47. *Selling Price* An electronics store uses a markup rate of 78% of all items. The cost of a CD player is $48. What is the selling price of the CD player?

48. *Sale Price* A sporting goods store advertises 30% off the original price of all golf equipment. A set of golf clubs has an original price of $229.99. What is the sale price?

3.4

Comparing Measurements In Exercises 49–52, find a ratio that compares the relative sizes of the quantities. (Use the same units of measurement for both quantities.)

49. Eighteen inches to 4 yards

50. One pint to 2 gallons

51. Two hours to 90 minutes

52. Four meters to 150 centimeters

Comparing Unit Prices In Exercises 53 and 54, determine which product has the lower unit price.

53. (a) An 18-ounce container of cooking oil for $1.79

(b) A 24-ounce container of cooking oil for $1.99

54. (a) A 17.4-ounce box of pasta noodles for $1.32

(b) A 32-ounce box of pasta noodles for $2.62

Solving a Proportion In Exercises 55–60, solve the proportion.

55. $\dfrac{7}{16} = \dfrac{z}{8}$

56. $\dfrac{x}{12} = \dfrac{5}{4}$

57. $\dfrac{x+2}{4} = -\dfrac{1}{3}$

58. $\dfrac{x-4}{1} = \dfrac{9}{4}$

59. $\dfrac{x-3}{2} = \dfrac{x+6}{5}$

60. $\dfrac{x+1}{3} = \dfrac{x+2}{4}$

61. *Entertainment* A band charges $200 to play for three hours. How much would they charge to play for two hours?

62. *Resizing a Picture* You have a 4-by-6-inch photo of the student council that must be reduced to a size of 3.2 inches by 4.8 inches for the school yearbook. What percent does the photo need to be reduced to in order for it to fit in the allotted space?

3.5

Using the Distance Formula In Exercises 63–68, find the missing distance, rate, or time.

	Distance, d	Rate, r	Time, t
63.		65 mi/hr	8 hr
64.		45 mi/hr	2 hr
65.	855 m	5 m/min	
66.	205 mi	60 mi/hr	
67.	3000 mi		50 hr
68.	1000 km		25 hr

69. *Geometry* The width of a rectangular swimming pool is 4 feet less than its length. The perimeter of the pool is 112 feet. Find the dimensions of the pool.

70. *Geometry* The perimeter of an isosceles triangle is 65 centimeters. Find the length of the two equal sides if each is 10 centimeters longer than the third side. (An isosceles triangle has two sides of equal length.)

Simple Interest In Exercises 71 and 72, use the simple interest formula.

71. Find the total interest you will earn on a $1000 corporate bond that matures in 5 years and has an annual interest rate of 9.5%.

72. Find the annual interest rate on a certificate of deposit that pays $60 per year in interest on a principal of $750.

73. *Numbers of Coins* You have 30 coins in dimes and quarters with a combined value of $5.55. Determine the number of coins of each type.

74. *Bird Seed Mixture* A pet store owner mixes two types of bird seed that cost $1.25 per pound and $2.20 per pound to make 20 pounds of a mixture that costs $1.82 per pound. How many pounds of each kind of bird seed are in the mixture?

75. *Work Rate* One person can complete a task in 5 hours, and another can complete the same task in 6 hours. How long will it take both people working together to complete the task?

76. *Work Rate* The person in Exercise 75 who can complete the task in 5 hours begins working 1 hour before the second person. How long will they work together to complete the task?

3.6

Graphing an Inequality In Exercises 77–80, sketch the graph of the inequality.

77. $-3 \leq x < 1$

78. $-2.5 \leq x < 4$

79. $-7 < x$

80. $x \geq -2$

Solving an Inequality In Exercises 81–98, solve the inequality and sketch the solution on the real number line.

81. $x - 5 \leq -1$

82. $x + 8 > 5$

83. $-6x < -24$

84. $-16x \geq -48$

85. $5x + 3 > 18$

86. $3x - 11 \leq 7$

87. $8x + 1 \geq 10x - 11$

88. $12 - 3x < 4x - 2$

89. $\frac{1}{3} - \frac{1}{2}y < 12$

90. $\frac{x}{4} - 2 < \frac{3x}{8} + 5$

91. $-4(3 - 2x) \leq 3(2x - 6)$

92. $3(2 - y) \geq 2(1 + y)$

93. $-6 \leq 2x + 8 < 4$

94. $-13 \leq 3 - 4x < 13$

95. $5 > \frac{x + 1}{-3} > 0$

96. $12 \geq \frac{x - 3}{2} > 1$

97. $5x - 4 < 6$ and $3x + 1 > -8$

98. $6 - 2x \leq 1$ or $10 - 4x > -6$

99. *International Calling Card Rate* The cost of an international telephone call is $0.10 per minute. Your prepaid calling card has $12.50 left to pay for a call. How many minutes can you talk?

100. *Earnings* A country club waiter earns $6 per hour plus tips of at least 15% of the restaurant tab from each table served. What total amount of restaurant tabs assures the waiter of making at least $150 in a five-hour shift?

3.7

Solving an Equation In Exercises 101–108, solve the equation. (Some equations have no solution.)

101. $|x| = 6$

102. $|x| = -4$

103. $|4 - 3x| = 8$

104. $|2x + 3| = 7$

105. $|5x + 4| - 10 = -6$

106. $|x - 2| - 2 = 4$

107. $|3x - 4| = |x + 2|$

108. $|5x + 6| = |2x - 1|$

Solving an Inequality In Exercises 109–116, solve the inequality. (Some inequalities have no solution.)

109. $|x - 4| > 3$

110. $|t + 3| > 2$

111. $|3x| < 12$

112. $\left|\frac{t}{3}\right| < 1$

113. $|2x - 7| < 15$

114. $|4x - 1| > 7$

115. $|b + 2| - 6 > 1$

116. $|2y - 1| + 4 < -1$

Writing an Inequality In Exercises 117 and 118, write an absolute value inequality that represents the interval.

117.

118.

119. *Temperature* The storage temperature of a computer must satisfy the inequality $|t - 78.3| \leq 38.3$, where t is the temperature in degrees Fahrenheit. Sketch the graph of the solution of the inequality. What are the maximum and minimum temperatures?

120. *Temperature* The operating temperature of a computer must satisfy the inequality $|t - 77| \leq 27$, where t is the temperature in degrees Fahrenheit. Sketch the graph of the solution of the inequality. What are the maximum and minimum temperatures?

Chapter Test

Solutions in English & Spanish and tutorial videos at CollegePrepAlgebra.com

Take this test as you would take a test in class. After you are done, check your work against the answers in the back of the book.

In Exercises 1−8, solve the equation and check your solution.

1. $8x + 104 = 0$

2. $4x - 3 = 18$

3. $5 - 3x = -2x - 2$

4. $4 - (x - 3) = 5x + 1$

5. $\frac{2}{3}x = \frac{1}{9} + x$

6. $\frac{t + 2}{3} = \frac{2t}{9}$

7. $|2x + 6| = 16$

8. $|3x - 5| = |6x - 1|$

9. Solve $4.08(x + 10) = 9.50(x - 2)$. Round your answer to two decimal places.

10. The bill (including parts and labor) for the repair of an oven was $142. The cost of parts was $62 and the cost of labor was $32 per hour. How many hours were spent repairing the oven?

11. Write the fraction $\frac{5}{16}$ as a percent and as a decimal.

12. 324 is 27% of what number?

13. 90 is what percent of 250?

14. What number is 25% of 24?

15. Write the ratio of 40 inches to 2 yards as a fraction in simplest form. Use the same units for both quantities, and explain how you made this conversion.

16. Solve the proportion $\frac{2x}{3} = \frac{x + 4}{5}$.

17. Find the length x of the side of the larger triangle shown in the figure at the left. (Assume that the two triangles are similar, and use the fact that corresponding sides of similar triangles are proportional.)

18. You traveled 264 miles in 4 hours. What was your average speed?

19. You can paint a building in 9 hours. Your friend can paint the same building in 12 hours. How long will it take the two of you working together to paint the building?

20. How much must you deposit in an account to earn $500 per year at 8% simple interest?

21. Rewrite the statement "t is at least 8" using inequality notation.

22. A utility company has a fleet of vans. The annual operating cost per van is $C = 0.37m + 2700$, where m is the number of miles traveled by a van in a year. What is the maximum number of miles that will yield an annual operating cost that is less than or equal to $11,950?

In Exercises 23−28, solve and graph the inequality.

23. $21 - 3x \le 6$

24. $-(3 + x) < 2(3x - 5)$

25. $0 \le \frac{1 - x}{4} < 2$

26. $-7 < 4(2 - 3x) \le 20$

27. $|x - 3| \le 2$

28. $|5x - 3| > 12$

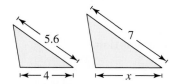

Figure for 17

Cumulative Test: Chapters 1−3

Solutions in English & Spanish and tutorial videos at CollegePrepAlgebra.com

Take this test as you would take a test in class. After you are done, check your work against the answers in the back of the book.

1. Place the correct inequality symbol ($<$ or $>$) between the real numbers.

$$-\frac{3}{4} \quad \boxed{} \quad \left|-\frac{7}{8}\right|$$

In Exercises 2−7, evaluate the expression.

2. $(-200)(2)(-3)$ 3. $\frac{3}{8} - \frac{5}{6}$ 4. $-\frac{2}{9} \div \frac{8}{75}$

5. $-(-2)^3$ 6. $3 + 2(6) - 1$ 7. $24 + 12 \div 3$

In Exercises 8 and 9, evaluate the expression when $x = -2$ and $y = 3$.

8. $-3x - (2y)^2$ 9. $\frac{5}{6}y + x^3$

10. Use exponential form to write the product $3 \cdot (x + y) \cdot (x + y) \cdot 3 \cdot 3$.

11. Use the Distributive Property to expand $-2x(x - 3)$.

12. Identify the property of real numbers illustrated by

$$2 + (3 + x) = (2 + 3) + x.$$

In Exercises 13−15, simplify the expression.

13. $(3x^3)(5x^4)$

14. $2x^2 - 3x + 5x^2 - (2 + 3x)$

15. $4(x^2 + x) + 7(2x - x^2)$

In Exercises 16−18, solve the equation and check your solution.

16. $12x - 3 = 7x + 27$ 17. $2x - \frac{5x}{4} = 13$

18. $5(x + 8) = -2x - 9$

19. Solve and graph the inequality.

$$-8(x + 5) \le 16$$

20. The sticker on a new car gives the fuel efficiency as 28.3 miles per gallon. In your own words, explain how to estimate the annual fuel cost for the buyer when the car will be driven approximately 15,000 miles per year and the fuel cost is $3.599 per gallon.

21. Write the ratio "24 ounces to 2 pounds" as a fraction in simplest form.

22. The original price of a digital camcorder is $1150. The camcorder is on sale for "20% off" the original price. Find the sale price.

23. The figure at the left shows two pieces of property. The assessed values of the properties are proportional to their areas. The value of the larger piece is $95,000. What is the value of the smaller piece?

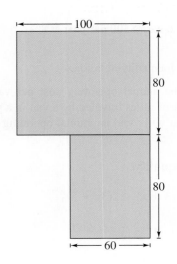

4
Graphs and Functions

4.1   Ordered Pairs and Graphs

4.2   Graphs of Equations in Two Variables

4.3   Relations, Functions, and Graphs

4.4   Slope and Graphs of Linear Equations

4.5   Equations of Lines

4.6   Graphs of Linear Inequalities

MASTERY IS WITHIN REACH!

"I used to be really afraid of math, so reading the textbook was difficult. I have learned that it just takes different strategies to read the textbook. It's my resource book when I do homework. I'm not afraid of math anymore because I know how to study it—finally."

James

See page 193 for suggestions about reading your textbook like a manual.

4.1 Ordered Pairs and Graphs

▶ Plot points on a rectangular coordinate system.

▶ Determine whether ordered pairs are solutions of equations.

▶ Use the verbal problem-solving method to plot points on a rectangular coordinate system.

The Rectangular Coordinate System

A **rectangular coordinate system** is formed by two real number lines intersecting at right angles. The horizontal number line is the **x-axis** and the vertical number line is the **y-axis**. (The plural of axis is *axes*.)

The point of intersection of the two axes is called the **origin**, and the axes separate the plane into four regions called **quadrants**. Each point in the plane corresponds to an **ordered pair** (x, y) of real numbers x and y called the **coordinates** of the point.

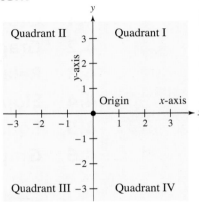

EXAMPLE 1 **Plotting Points on a Rectangular Coordinate System**

Plot the points $(-1, 2)$, $(3, 0)$, $(2, -1)$, $(3, 4)$, $(0, 0)$, and $(-2, -3)$ on a rectangular coordinate system.

SOLUTION

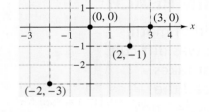

The point $(-1, 2)$ is one unit to the *left* of the vertical axis and two units *above* the horizontal axis.

One unit to the left of the vertical axis   Two units above the horizontal axis

$(-1, 2)$

Similarly, the point $(3, 0)$ is three units to the *right* of the vertical axis and *on* the horizontal axis. (It is on the horizontal axis because the y-coordinate is zero.) The other four points can be plotted in a similar way.

Exercises Within Reach ®

Solutions in English & Spanish and tutorial videos at CollegePrepAlgebra.com

Plotting Points In Exercises 1−8, **plot the points on a rectangular coordinate system.**

1. $(3, 2), (-4, 2), (2, -4)$

2. $(-1, 6), (-1, -6), (4, 6)$

3. $(-10, -4), (4, -4), (0, 0)$

4. $(-6, 4), (0, 0), (3, -2)$

5. $(-3, 4), (0, -1), (2, -2), (5, 0)$

6. $(-1, 3), (0, 2), (-4, -4), (-1, 0)$

7. $\left(\frac{3}{2}, -1\right), \left(-3, \frac{3}{4}\right), \left(\frac{1}{2}, -\frac{1}{2}\right)$

8. $\left(-\frac{2}{3}, 4\right), \left(\frac{1}{2}, -\frac{5}{2}\right), \left(-4, -\frac{5}{4}\right)$

Determining the Quadrant In Exercises 9−14, **determine the quadrant in which the point is located without plotting it.**

9. $(-3, 1)$

10. $(4, -3)$

11. $\left(-\frac{1}{8}, -\frac{2}{7}\right)$

12. $\left(\frac{3}{11}, \frac{7}{8}\right)$

13. $(-100, -365.6)$

14. $(-157.4, 305.6)$

Application   EXAMPLE 2   **Graphing Super Bowl Scores**

The scores of the Super Bowl games from 1992 through 2012 are shown in the table. Plot these points on a rectangular coordinate system. (*Source:* National Football League)

SOLUTION

The *x*-coordinates of the points represent the year, and the *y*-coordinates represent the winning and losing scores. The winning scores are shown as black dots, and the losing scores are shown as blue dots. Note that the break in the *x*-axis indicates that the numbers between 0 and 1992 have been omitted.

DATA
Year	Winning Score	Losing Score
1992	37	24
1993	52	17
1994	30	13
1995	49	26
1996	27	17
1997	35	21
1998	31	24
1999	34	19
2000	23	16
2001	34	7
2002	20	17
2003	48	21
2004	32	29
2005	24	21
2006	21	10
2007	29	17
2008	17	14
2009	27	23
2010	31	17
2011	31	25
2012	21	17

Spreadsheet at CollegePrepAlgebra.com

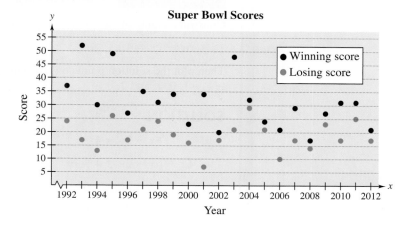

Exercises Within Reach®

Solutions in English & Spanish and tutorial videos at CollegePrepAlgebra.com

15. *Organizing Data* The table shows the normal average temperatures *y* (in degrees Fahrenheit) in Anchorage, Alaska, for each month *x* of the year, with *x* = 1 corresponding to January. (*Source:* National Climatic Data Center)

DATA
x	1	2	3	4	5	6
y	16	19	26	36	47	55

x	7	8	9	10	11	12
y	58	56	48	34	22	18

(a) Plot the data in the table. Did you use the same scale on both axes? Explain.

(b) Using the graph, find the month for which the normal temperature changed the least from the previous month.

16. *Organizing Data* The table shows the annual net profits *y* (in billions of dollars) for Hewlett-Packard for the years 2002 through 2011, where *x* represents the year. (*Source:* Hewlett-Packard Company)

DATA
x	2002	2003	2004	2005	2006
y	2.4	3.6	4.1	4.7	5.8

x	2007	2008	2009	2010	2011
y	7.3	8.3	7.7	8.8	7.1

(a) Plot the data in the table.

(b) Use the graph to determine the two consecutive years between which the greatest decrease occurred in the annual profits for Hewlett-Packard.

Ordered Pairs as Solutions of Equations

> ### Three Approaches to Problem Solving
>
> 1. **Algebraic Approach** Use algebra to find several solutions.
> 2. **Numerical Approach** Construct a table that shows several solutions.
> 3. **Graphical Approach** Draw a graph that shows several solutions.

EXAMPLE 3 **Constructing a Table of Values**

Construct a table of values showing five solution points for the equation

$$6x - 2y = 4.$$

Then plot the solution points on a rectangular coordinate system. Choose x-values of $-2, -1, 0, 1,$ and 2.

SOLUTION

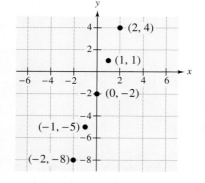

$6x - 2y = 4$	Write original equation.
$6x - 6x - 2y = 4 - 6x$	Subtract $6x$ from each side.
$-2y = -6x + 4$	Combine like terms.
$\dfrac{-2y}{-2} = \dfrac{-6x + 4}{-2}$	Divide each side by -2.
$y = 3x - 2$	Simplify.

Now, using the equation $y = 3x - 2$, you can construct a table of values.

x	-2	-1	0	1	2
$y = 3x - 2$	-8	-5	-2	1	4
Solution point	$(-2, -8)$	$(-1, -5)$	$(0, -2)$	$(1, 1)$	$(2, 4)$

From the table, you can plot the solution points as shown at the left.

Exercises Within Reach®

Solutions in English & Spanish and tutorial videos at CollegePrepAlgebra.com

Constructing a Table of Values **In Exercises 17–22, complete the table of values. Then plot the solution points on a rectangular coordinate system.**

17.

x	-2	0	2	4	6
$y = 3x - 4$					

18.

x	-2	0	2	4	6
$y = 2x + 1$					

19.

x	-2	0	4	6	8
$y = -\frac{3}{2}x + 5$					

20.

x	-4	-2	0	2	4
$y = -\frac{1}{2}x + 3$					

21.

x	-2	-1	0	1	2
$y = -4x - 5$					

22.

x	-2	0	$\frac{1}{2}$	2	4
$y = -\frac{7}{2}x + 3$					

> ### Guidelines for Verifying Solutions
>
> To verify that an ordered pair (x, y) is a solution of an equation with variables x and y, use the following steps.
>
> 1. Substitute the values of x and y into the equation.
>
> 2. Simplify each side of the equation.
>
> 3. If each side simplifies to the same number, then the ordered pair is a solution. If the two sides yield different numbers, then the ordered pair is not a solution.

EXAMPLE 4   **Verifying Solutions of an Equation**

Determine whether each ordered pair is a solution of $x + 3y = 6$.

a. $(1, 2)$ **b.** $(0, 2)$

SOLUTION

a.   For the ordered pair $(1, 2)$, substitute $x = 1$ and $y = 2$ into the original equation.

$$x + 3y = 6 \qquad \text{Write original equation.}$$
$$1 + 3(2) \overset{?}{=} 6 \qquad \text{Substitute 1 for } x \text{ and 2 for } y.$$
$$7 \neq 6 \qquad \text{Not a solution } \textbf{✗}$$

b.   For the ordered pair $(0, 2)$, substitute $x = 0$ and $y = 2$ into the original equation.

$$x + 3y = 6 \qquad \text{Write original equation.}$$
$$0 + 3(2) \overset{?}{=} 6 \qquad \text{Substitute 0 for } x \text{ and 2 for } y.$$
$$6 = 6 \qquad \text{Solution } \checkmark$$

Exercises Within Reach ®

 Solutions in English & Spanish and tutorial videos at CollegePrepAlgebra.com

Verifying Solutions of an Equation In Exercises 23−30, determine whether each ordered pair is a solution of the equation.

23. $y = 2x + 4$ (a) $(3, 10)$ (b) $(-1, 3)$ **24.** $y = 5x - 2$ (a) $(2, 0)$ (b) $(-2, -12)$
                           (c) $(0, 0)$ (d) $(-2, 0)$ (c) $(6, 28)$ (d) $(1, 1)$

25. $2y - 3x + 1 = 0$ (a) $(1, 1)$ (b) $(5, 7)$ **26.** $x - 8y + 10 = 0$ (a) $(-2, 1)$ (b) $(6, 2)$
                            (c) $(-3, -1)$ (d) $(-3, -5)$ (c) $(0, -1)$ (d) $(2, -2)$

27. $y = \frac{2}{3}x$ (a) $(6, 6)$ (b) $(-9, -6)$ **28.** $y = -\frac{7}{8}x$ (a) $(-5, -2)$ (b) $(0, 0)$
                            (c) $(0, 0)$ (d) $\left(-1, \frac{2}{3}\right)$ (c) $(8, 8)$ (d) $\left(\frac{3}{5}, 1\right)$

29. $y = 3 - 4x$ (a) $\left(-\frac{1}{2}, 5\right)$ (b) $(1, 7)$ **30.** $y = \frac{3}{2}x + 1$ (a) $\left(0, \frac{3}{2}\right)$ (b) $(4, 7)$
                            (c) $(0, 0)$ (d) $\left(-\frac{3}{4}, 0\right)$ (c) $\left(\frac{2}{3}, 2\right)$ (d) $(-2, -2)$

Determining Solutions of an Equation In Exercises 31 and 32, complete each ordered pair so that it satisfies the equation.

31. $y = 3x + 4$ (a) $\left(\boxed{}, 0\right)$ (b) $\left(4, \boxed{}\right)$ **32.** $y = -2x - 7$ (a) $\left(0, \boxed{}\right)$ (b) $\left(-1, \boxed{}\right)$

                            (c) $\left(\boxed{}, -2\right)$ (c) $\left(\boxed{}, 1\right)$

Applications

Application | **EXAMPLE 5** | **Finding the Total Cost**

You set up a small business to assemble computer keyboards. Your initial cost is $120,000, and your unit cost of assembling each keyboard is $40. Write an equation that relates your total cost to the number of keyboards produced. Then plot the total costs of producing 1000, 2000, 3000, 4000, and 5000 keyboards.

SOLUTION

The total cost equation must represent both the unit cost and the initial cost. A verbal model for this problem is as follows.

Verbal Model: $\boxed{\text{Total cost}} = \boxed{\text{Unit cost}} \cdot \boxed{\text{Number of keyboards}} + \boxed{\text{Initial cost}}$

Labels:
Total cost = C (dollars)
Unit cost = 40 (dollars per keyboard)
Number of keyboards = x (keyboards)
Initial cost = 120,000 (dollars)

Expression: $C = 40x + 120{,}000$

Using this equation, you can construct the following table of values.

x	1000	2000	3000	4000	5000
C = 40x + 120,000	160,000	200,000	240,0000	280,000	320,000

From the table, you can plot the ordered pairs.

Computer Keyboards

Total cost (in dollars) vs Number of keyboards

Exercises Within Reach ®

Solutions in English & Spanish and tutorial videos at CollegePrepAlgebra.com

33. *Spring Compression* The distance y (in centimeters) a spring is compressed by a force x (in kilograms) is given by $y = 0.066x$. Complete a table of values for $x = 20, 40, 60, 80,$ and 100 to determine the distance the spring is compressed for each of the specified forces. Plot the results on a rectangular coordinate system.

34. *Copier Value* A company buys a new copier for $9500. Its value y after x years is given by $y = -800x + 9500$. Complete a table of values for $x = 0, 2, 4, 6,$ and 8 to determine the value of the copier at each specified time. Plot the results on a rectangular coordinate system.

35. *Video Games* With an initial cost of $5000, a company will produce x units of a video game at $25 per unit. Write an equation that relates the total cost of producing x units to the number of units produced. Plot the cost of producing 100, 150, 200, 250, and 300 units.

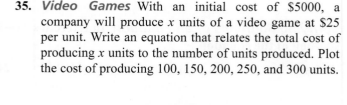

36. *Assembly Line* An employee earns $10 plus $0.50 for every x units produced per hour. Write an equation that relates the employee's total hourly wage to the number of units produced. Plot the hourly wages for producing 2, 5, 8, 10, and 20 units per hour.

Although graphs can help you visualize relationships between two variables, they can also be misleading, as shown in the following example.

Application   EXAMPLE 6   **Identifying Misleading Graphs**

The graphs shown below represent the yearly profits for a truck rental company. Which graph is misleading? Why?

a.

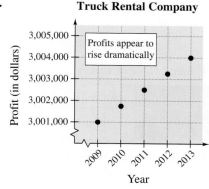

b.

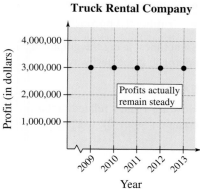

SOLUTION

a. This graph is misleading. The scale on the vertical axis makes it appear that the change in profits from 2009 to 2013 is dramatic, but the total change is only $3000, which is small in comparison with $3,000,000.

b. This graph is truthful. By showing the full scale on the *y*-axis, you can see that, relative to the overall size of the profit, there was almost no change from one year to the next.

Exercises Within Reach ®   Solutions in English & Spanish and tutorial videos at CollegePrepAlgebra.com

37. *Government Surplus* The graph below shows the estimated U.S. government surplus for the years 2012 through 2016. Is the graph misleading? Explain your reasoning. (*Source:* U.S. Office of Management and Budget)

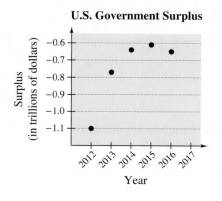

38. *Government Surplus* The graph below shows the estimated U.S. government surplus for the years 2012 through 2016. Is the graph misleading? Explain your reasoning. (*Source:* U.S. Office of Management and Budget)

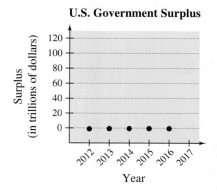

Concept Summary: Using a Graphical Approach to Solve Problems

What

Here are three ways to solve problems.

- **Algebraic Approach**
- **Numerical Approach**
- **Graphical Approach**

Sometimes, using the graphical approach is the best way.

EXAMPLE

How are x and y related in the equation $-x + 2y = 2$?

How

To use a graphical approach to solve a problem involving a linear equation:

1. Solve the equation for y.

$$-x + 2y = 2$$
$$2y = x + 2$$
$$y = \tfrac{1}{2}x + 1$$

2. Create a **table of values**.

x	−4	−2	0	2	4
$y = \tfrac{1}{2}x + 1$	−1	0	1	2	3

3. **Plot** the **ordered pairs** on a rectangular coordinate system.

Why

Using a graphical approach helps you see the relationship between the variables. For example, you can see from the graph that as x increases, y also increases.

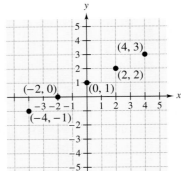

Exercises Within Reach ®

Worked-out solutions to odd-numbered exercises at CollegePrepAlgebra.com

Concept Summary Check

39. *Representing a Solution* Use the table of values above to write the coordinates of the point representing the solution of $-x + 2y = 2$ when $x = 2$.

40. *Creating a Table of Values* Explain how each y-value was determined in the table of values above.

41. *Identifying Points* Identify the point(s) in Quadrant III on the rectangular coordinate system above. Identify the point(s) on the y-axis.

42. *Describing a Graph* The phrase "x increases" describes a movement in what direction along the graph?

Extra Practice

Finding Coordinates In Exercises 43−46, determine the coordinates of the points.

43.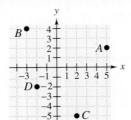

44.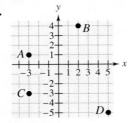

45.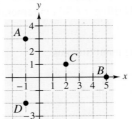

46.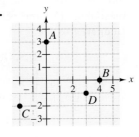

Plotting Points In Exercises 47−54, plot the points and connect them with line segments to form the figure.

47. Triangle: $(-1, 1), (2, -1), (3, 4)$

48. Triangle: $(0, 3), (-1, -2), (4, 8)$

49. Square: $(2, 4), (5, 1), (2, -2), (-1, 1)$

50. Rectangle: $(2, 1), (4, 2), (1, 8), (-1, 7)$

51. Parallelogram: $(5, 2), (7, 0), (1, -2), (-1, 0)$

52. Parallelogram: $(-1, 1), (0, 4), (5, 1), (4, -2)$

53. Rhombus: $(0, 0), (3, 2), (5, 5), (2, 3)$

54. Rhombus: $(0, 0), (1, 2), (3, 3), (2, 1)$

Solving for y In Exercises 55–58, **solve** the equation for *y*.

55. $7x + y = 8$

56. $2x + y = 1$

57. $10x - y = 2$

58. $12x - y = 7$

Determining Quadrants In Exercises 59 and 60, **determine** the quadrant(s) in which the point may be located without plotting it. Assume $x \neq 0$ and $y \neq 0$.

59. $(6, y)$, *y* is a real number.

60. $(x, -2)$, *x* is a real number.

61. *Graphical Interpretation* The table shows the numbers of hours *x* that a student studied for five different algebra exams, and the resulting scores *y*.

x	3.5	1	8	4.5	0.5
y	72	67	95	81	53

(a) Plot the data in the table.

(b) Use the graph to describe the relationship between the number of hours studied and the resulting exam score.

62. *Organizing Data* The table shows the speeds of a car *x* (in miles per hour) and the approximate stopping distance *y* (in feet).

x	20	30	40	50	60
y	63	109	164	229	303

(a) Plot the data in the table.

(b) The *x*-coordinates increase in equal increments of 10 miles per hour. Describe the pattern of the *y*-coordinates. What are the implications for the driver?

Explaining Concepts

63. (a) Plot the points $(3, 2)$, $(-5, 4)$, and $(6, -4)$ on a rectangular coordinate system.

(b) Change the sign of the *y*-coordinate of each point plotted in part (a). Plot the three new points on the same rectangular coordinate system used in part (a).

(c) What can you infer about the location of a point when the sign of its *y*-coordinate is changed?

64. (a) Plot the points $(3, 2)$, $(-5, 4)$, and $(6, -4)$ on a rectangular coordinate system.

(b) Change the sign of the *x*-coordinate of each point plotted in part (a). Plot the three new points on the same rectangular coordinate system used in part (a).

(c) What can you infer about the location of a point when the sign of its *x*-coordinate is changed?

65. *Think About It* The points $(6, -1)$, $(-2, -1)$, and $(-2, 4)$ are three vertices of a rectangle. Find the coordinates of the fourth vertex.

66. *Writing* Discuss the significance of the word "ordered" when referring to an ordered pair (x, y).

67. *Think About It* On a rectangular coordinate system, must the scales on the *x*-axis and *y*-axis be the same? If not, give an example in which the scales differ.

68. *Writing* Review the tables in Exercises 17–22 and observe that in some cases the *y*-coordinates of the solution points increase and in others the *y*-coordinates decrease. What factor in the equation causes this? Explain.

Cumulative Review

In Exercises 69–72, solve the equation.

69. $-y = 10$

70. $10 - t = 6$

71. $3x - 42 = 0$

72. $64 - 16x = 0$

In Exercises 73–76, solve the inequality.

73. $x + 3 > 2$

74. $y - 4 < -8$

75. $3x < 12$

76. $2(z - 4) > 10z$

4.2 Graphs of Equations in Two Variables

▶ Sketch graphs of equations using the point-plotting method.
▶ Find and use *x*- and *y*-intercepts as aids to sketching graphs.
▶ Use the verbal problem-solving method to write an equation and sketch its graph.

The Graph of an Equation in Two Variables

The solutions of an equation involving two variables can be represented by points on a rectangular coordinate system. The set of *all* such points is called the **graph** of the equation.

> ### The Point-Plotting Method of Sketching a Graph
>
> 1. If possible, rewrite the equation by isolating one of the variables.
> 2. Make a table of values showing several solution points.
> 3. Plot these points on a rectangular coordinate system.
> 4. Connect the points with a smooth curve or line.

EXAMPLE 1 Sketching the Graph of an Equation

Sketch the graph of $3x + y = 5$.

SOLUTION

Begin by solving the equation for *y*, so that *y* is isolated on the left.

$$3x + y = 5 \qquad \text{Write original equation.}$$
$$y = -3x + 5 \qquad \text{Subtract } 3x \text{ from each side.}$$

Next, create a table of values, as shown below.

x	−2	−1	0	1	2	3
y = −3x + 5	11	8	5	2	−1	−4
Solution point	(−2, 11)	(−1, 8)	(0, 5)	(1, 2)	(2, −1)	(3, −4)

Now, plot the solution points. It appears that all six points lie on a line, so complete the sketch by drawing a line through the points.

Exercises Within Reach ®   Solutions in English & Spanish and tutorial videos at CollegePrepAlgebra.com

Sketching the Graph of an Equation In Exercises 1 and 2, complete the table and use the results to sketch the graph of the equation.

1. $y = 9 - x$

x	−2	−1	0	1	2
y					

2. $y = x - 1$

x	−2	−1	0	1	2
y					

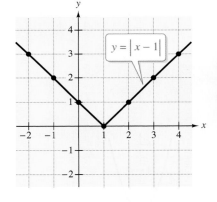

EXAMPLE 2   **Graphing a Nonlinear Equation**

Sketch the graph of $x^2 + y = 4$.

SOLUTION

Begin by solving the equation for y, so that y is isolated on the left.

$$y = -x^2 + 4$$

Next, create a table of values, as shown below.

x	−3	−2	−1	0	1	2	3
$y = -x^2 + 4$	−5	0	3	4	3	0	−5
Solution point	$(-3, -5)$	$(-2, 0)$	$(-1, 3)$	$(0, 4)$	$(1, 3)$	$(2, 0)$	$(3, -5)$

Now, plot the solution points. Finally, connect the points with a smooth curve.

EXAMPLE 3   **Graphing an Absolute Value Equation**

Sketch the graph of $y = |x - 1|$.

SOLUTION

x	−2	−1	0	1	2	3	4
$y = |x - 1|$	3	2	1	0	1	2	3
Solution point	$(-2, 3)$	$(-1, 2)$	$(0, 1)$	$(1, 0)$	$(2, 1)$	$(3, 2)$	$(4, 3)$

Now, plot the solution points. It appears that the points lie in a "V-shaped" pattern, with the point $(1, 0)$ lying at the bottom of the "V." Following this pattern, connect the points to form the graph shown at the left.

Exercises Within Reach ®

Solutions in English & Spanish and tutorial videos at CollegePrepAlgebra.com

Sketching the Graph of an Equation In Exercises 3−6, complete the table and use the results to sketch the graph of the equation.

3. $y = x^2 + 3$

x	−2	−1	0	1	2
y					

4. $x^2 + y = -1$

x	−2	−1	0	1	2
y					

5. $y = |x + 1|$

x	−3	−2	−1	0	1
y					

6. $y = |x| - 2$

x	−2	−1	0	1	2
y					

Sketching the Graph of an Equation In Exercises 7−12, sketch the graph of the equation and label the coordinates of at least three solution points.

7. $4x + y = 6$

8. $2x - y = 5$

9. $y = -x^2 + 9$

10. $y = x^2 - 1$

11. $y = 5 - |x|$

12. $y = |x| + 3$

Intercepts: Aids to Sketching Graphs

The point $(a, 0)$ is called an **x-intercept** of the graph of an equation if it is a solution point of the equation. To find the x-intercept(s), let $y = 0$ and solve the equation for x.

The point $(0, b)$ is called a **y-intercept** of the graph of an equation if it is a solution point of the equation. To find the y-intercept(s), let $x = 0$ and solve the equation for y.

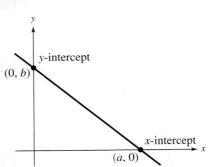

EXAMPLE 4 **Identifying the Intercepts of Graphs**

a.

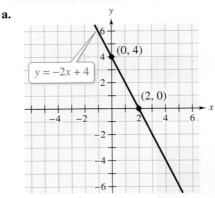

b.

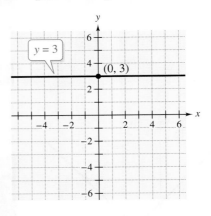

c.

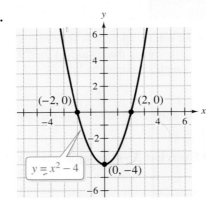

d.

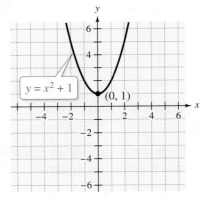

Exercises Within Reach ®

Solutions in English & Spanish and tutorial videos at CollegePrepAlgebra.com

Identifying the Intercepts of a Graph In Exercises 13−16, identify the *x*- and *y*-intercepts of the graph visually.

13.

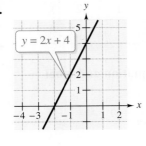

14.

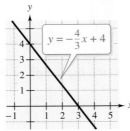

15.

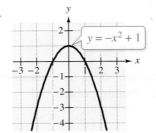

16.

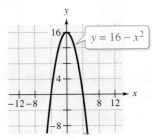

EXAMPLE 5   **Finding the Intercepts of a Graph**

Find the intercepts and sketch the graph of $y = 2x - 5$.

SOLUTION

To find any x-intercepts, let $y = 0$ and solve the resulting equation for x.

$y = 2x - 5$	Write original equation.
$0 = 2x - 5$	Let $y = 0$.
$\dfrac{5}{2} = x$	Solve equation for x.

To find any y-intercepts, let $x = 0$ and solve the resulting equation for y.

$y = 2x - 5$	Write original equation.
$y = 2(0) - 5$	Let $x = 0$.
$y = -5$	Solve equation for y.

So, the graph has one x-intercept, which occurs at the point $\left(\frac{5}{2}, 0\right)$, and one y-intercept, which occurs at the point $(0, -5)$. To sketch the graph of the equation, create a table of values. (Include the intercepts in the table.) Then plot the points and connect the points with a line, as shown at the left.

The graph shows the line $y = 2x - 5$ with the x-intercept: $\left(\frac{5}{2}, 0\right)$ and the y-intercept: $(0, -5)$.

x	-1	0	1	2	$\frac{5}{2}$	3	4
$y = 2x - 5$	-7	-5	-3	-1	0	1	3
Solution point	$(-1, -7)$	$(0, -5)$	$(1, -3)$	$(2, -1)$	$\left(\frac{5}{2}, 0\right)$	$(3, 1)$	$(4, 3)$

Exercises Within Reach ®

Solutions in English & Spanish and tutorial videos at CollegePrepAlgebra.com

Finding the Intercepts of a Graph In Exercises 17–20, find the x- and y-intercepts and sketch the graph of the equations.

17. $y = \frac{1}{2}x - 1$ **18.** $y = -\frac{1}{2}x + 3$ **19.** $2x + y = -2$ **20.** $3x - 2y = 1$

$y = \frac{3}{2}x - \frac{1}{2}$

$-2y = -3x + 1$

21. ***Car Rental*** A car rental costs \$50 per day plus an additional \$0.50 for each mile driven. The daily cost y is given by the equation

$y = 0.50x + 50$

where x is the number of miles driven. Find the y-intercept of the graph of the equation.

22. ***Hot-Air Balloon*** A hot-air balloon descends 4 feet per second from a height of 400 feet. The height y of the balloon after x seconds is represented by the equation

$y = 400 - 4x.$

Find the x-intercept of the graph of the equation.

Applications

Application | **EXAMPLE 6** **Depreciation**

The value of a $35,500 sport utility vehicle (SUV) depreciates over 10 years (the depreciation is the same each year). At the end of the 10 years, the salvage value is expected to be $5500.

a. Write an equation that relates the value of the SUV and its age in years.

b. Sketch the graph of the equation.

c. What is the y-intercept of the graph, and what does it represent in the context of the problem?

SOLUTION

a. The depreciation over the 10 years is $35,500 - 5500 = \$30,000$. Because the same amount is depreciated each year, it follows that the annual depreciation is $30,000/10 = \$3000$.

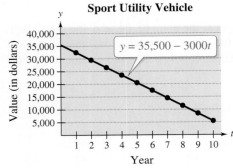

Sport Utility Vehicle

$y = 35,500 - 3000t$

| **Verbal Model:** | $\dfrac{\text{Value after } t \text{ years}}{} = \text{Original value} - \text{Annual depreciation} \cdot \text{Number of years}$ |

Labels:	Value after t years $= y$	(dollars)
	Original value $= 35,500$	(dollars)
	Annual depreciation $= 3000$	(dollars per year)
	Number of years $= t$	(years)

Equation: $y = 35,500 - 3000t$

b. A sketch of the graph of the depreciation equation is shown at the left.

c. To find the y-intercept of the graph, let $t = 0$ and solve the equation for y.

$$y = 35,500 - 3000t \qquad \text{Write original equation.}$$
$$y = 35,500 - 3000(0) \qquad \text{Substitute 0 for } t.$$
$$y = 35,500 \qquad \text{Simplify.}$$

So, the y-intercept is $(0, 35,500)$, and it corresponds to the original value of the SUV.

Exercises Within Reach ®

Solutions in English & Spanish and tutorial videos at CollegePrepAlgebra.com

23. Hot-Air Balloon A hot-air balloon at 1120 feet descends at a rate of 80 feet per minute. Let y represent the height of the balloon and let x represent the number of minutes the balloon descends.

(a) Write an equation that relates the height of the hot-air balloon and the number of minutes it descends.

(b) Sketch the graph of the equation.

(c) What is the y-intercept of the graph, and what does it represent in the context of the problem?

24. Fitness You run and walk on a trail that is 6 miles long. You run 4 miles per hour and walk 3 miles per hour. Let y be the number of hours you walk and let x be the number of hours you run.

(a) Write an equation that relates the number of hours you run and the number of hours you walk to the total length of the trail.

(b) Sketch the graph of the equation.

(c) What is the y-intercept of the graph, and what does it represent in the context of the problem?

Application EXAMPLE 7   **Life Expectancy**

The table shows the life expectancies y (in years) in the United States for a male child at birth for the years 2002 through 2007.

DATA Year	2002	2003	2004	2005	2006	2007
y	74.3	74.5	74.9	74.9	75.1	75.4

A model for this data is $y = 0.21t + 73.9$, where t is the year, with $t = 2$ corresponding to 2002. (*Source:* U.S. National Center for Health Statistics)

a. Plot the data and graph the model on the same set of coordinate axes.

b. Use the model to predict the life expectancy for a male child born in 2020.

SOLUTION

a. The points from the table are plotted with the graph of $y = 0.21t + 73.9$, as shown at the left.

b. To predict the life expectancy for a male child in 2020, let $t = 20$, and solve the equation for y.

$$y = 0.21t + 73.9 \qquad \text{Write model.}$$
$$= 0.21(20) + 73.9 \qquad \text{Substitute 20 for } t.$$
$$= 78.1 \qquad \text{Simplify.}$$

So, you can predict that the life expectancy in 2020 will be 78.1 years.

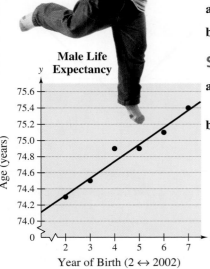

Male Life Expectancy

Age (years) vs. Year of Birth (2 ↔ 2002)

Exercises Within Reach ®

Solutions in English & Spanish and tutorial videos at CollegePrepAlgebra.com

25. *Life Expectancy* The table shows the life expectancies y (in years) in the United States for a female child at birth for the years 2002 through 2007.

DATA Year	y
2002	79.5
2003	79.6
2004	79.9
2005	79.9
2006	80.2
2007	80.4

Spreadsheet at CollegePrepAlgebra.com

A model for this data is $y = 0.18t + 79.1$, where t is the year, with $t = 2$ corresponding to 2002. (*Source:* U.S. National Center for Health Statistics)

(a) Plot the data and graph the model on the same set of coordinate axes.

(b) Use the model to predict the life expectancy for a female child born in 2020.

26. *Marital Status* The table shows the numbers y (in millions) of adults (over 18 years of age) never married in the United States for the years 2006 through 2011.

DATA Year	y
2006	55.3
2007	56.1
2008	58.3
2009	59.1
2010	61.5
2011	63.3

Spreadsheet at CollegePrepAlgebra.com

A model for this data is $y = 1.63t + 45.1$, where t is the year, with $t = 6$ corresponding to 2006. (*Source:* U.S. Census Bureau)

(a) Plot the data and graph the model on the same set of coordinate axes.

(b) Use the model to predict the number of adults over the age of 18 in 2020 who will never have married.

Concept Summary: *Graphing Equations in Two Variables*

What

You can use a **graph** to represent all the solutions of an equation in two variables.

EXAMPLE

Sketch the graph of

$y = |x| - 1$.

How

Use these steps to sketch the graph of such an equation.

1. Make a table of values. Include any x- and y-intercepts.

2. Plot the points.

3. Connect the points.

x	-2	-1	0	1	2
$y = |x| - 1$	1	0	-1	0	1

Why

You can use the graph of an equation in two variables to see the relationship between the variables. For example, the graph shows that y decreases as x increases to zero, and y increases as x increases above zero.

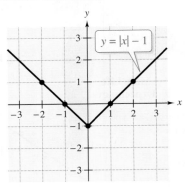

Exercises Within Reach ®

Worked-out solutions to odd-numbered exercises at CollegePrepAlgebra.com

Concept Summary Check

27. **Using a Table of Values** Use the table of values above to write the coordinates of the point representing the solution of $y = |x| - 1$ when $x = 2$.

28. **Using a Table of Values** Use the table of values above to determine the x- and y-intercepts of the graph of $y = |x| - 1$.

29. **Finding an Intercept** Explain how to find the y-intercept of the graph of an equation.

30. **Interpreting a Table of Values** Briefly describe the point-plotting method of sketching the graph of an equation.

Extra Practice

Sketching the Graph of an Equation In Exercises 31–38, sketch the graph of the equation and label the coordinates of at least three solution points.

31. $y = 3x$

32. $y = -2x$

33. $7x + 7y = 14$

34. $10x + 5y = 20$

35. $y = \frac{3}{8}x + 15$

36. $y = 14 - \frac{2}{3}x$

37. $y = |x - 5|$

38. $y = |x + 3|$

Finding Intercepts In Exercises 39–42, graphically estimate the x- and y-intercepts of the graph. Then check your results algebraically.

39. $x + 3y = 6$

40. $y = |x| - 3$

41. $y = 4 - |x|$

42. $y = x^2 - 4$

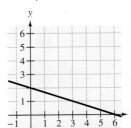
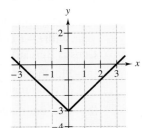
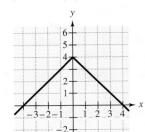
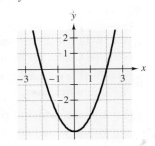

Comparing Graphs In Exercises 43–46, sketch the graphs of both equations on the same rectangular coordinate system. Are the graphs identical? If so, what property of real numbers is being illustrated?

43. $y_1 = \frac{1}{3}x - 1$

$y_2 = 1 - \frac{1}{3}x$

44. $y_1 = 3\left(\frac{1}{4}x\right)$

$y_2 = \left(3 \cdot \frac{1}{4}\right)x$

45. $y_1 = 2(x - 2)$

$y_2 = 2x - 4$

46. $y_1 = 2 + (x + 4)$

$y_2 = (2 + x) + 4$

47. *Creating a Model* Let y represent the distance traveled by a car that is moving at a constant speed of 35 miles per hour. Let t represent the number of hours the car has traveled. Write an equation that relates y and t, and sketch its graph.

48. *Creating a Model* The cost of printing a book is $500, plus $5 per book. Let C represent the total cost and let x represent the number of books. Write an equation that relates C and x, and sketch its graph.

Explaining Concepts

49. *Writing* Do all graphs of linear equations in two variables have a y-intercept? Explain.

50. *Reasoning* When the graph of a linear equation in two variables has a negative x-intercept and a positive y-intercept, does the line rise or fall from left to right? Through which quadrant(s) does the line pass? Use a graph to illustrate your answer.

51. *Writing* You walk toward a tree at a constant speed. Let x represent the time (in seconds) and let y represent the distance (in feet) between you and the tree. Sketch a possible graph of this situation. Explain how x and y are related. What does the x-intercept mean?

52. *Writing* How many solution points does a linear equation in two variables have? Explain.

Cumulative Review

In Exercises 53–56, evaluate the expression.

53. $-4 + (-7) - 3 + 1$

54. $-6 + 3 - (-1) + 11$

55. $-(-3) + 5 - 4 + 9$

56. $-18 - (-6) + 2 - 8$

In Exercises 57–62, evaluate the expression and write the result in simplest form.

57. $\frac{3}{4}\left(-\frac{2}{9}\right)$

58. $\left(-\frac{1}{6}\right)\left(-\frac{8}{15}\right)$

59. $\left(-\frac{7}{12}\right)\left(-\frac{18}{35}\right)$

60. $-\frac{6}{7} \div \frac{5}{21}$

61. $\frac{12}{5} \div \left(-\frac{1}{3}\right)$

62. $\left(-\frac{16}{25}\right) \div \left(-\frac{4}{5}\right)$

In Exercises 63–66, determine whether each ordered pair is a solution of the equation.

63. $y = 3x - 5$ (a) $(0, 5)$ (b) $(-1, -2)$
                       (c) $(3, 4)$ (d) $(-2, -11)$

64. $y = 2x + 1$ (a) $(-3, -5)$ (b) $(-1, 3)$
                       (c) $(5, 8)$ (d) $(2, 5)$

65. $3y - 4x = 7$ (a) $(1, 1)$ (b) $(-5, 9)$
                        (c) $(4, -3)$ (d) $(7, 7)$

66. $x - 2y = -2$ (a) $(-6, 2)$ (b) $(-2, 2)$
                        (c) $(4, 3)$ (d) $(2, 0)$

4.3 Relations, Functions, and Graphs

▶ Identify the domain and range of a relation.

▶ Determine whether relations are functions.

▶ Use function notation and evaluate functions.

▶ Identify the domain and range of a function.

Relations

> **Definition of Relation**
>
> A **relation** is any set of ordered pairs. The set of first components in the ordered pairs is the **domain** of the relation. The set of second components is the **range** of the relation.

In mathematics, relations are commonly described by ordered pairs of *numbers*. The set of *x*-coordinates is the domain, and the set of *y*-coordinates is the range. In the relation {(3, 5), (1, 2), (4, 4), (0, 3)}, the domain *D* and range *R* are the sets $D = \{3, 1, 4, 0\}$ and $R = \{5, 2, 4, 3\}$.

EXAMPLE 1   Analyzing a Relation

Find the domain and range of the relation {(0, 1), (1, 3), (2, 5), (3, 5), (0, 3)}. Then sketch a graphical representation of the relation.

SOLUTION

The domain is the set of all first components of the relation, and the range is the set of all second components.

$$D = \{0, 1, 2, 3\}$$

and

$$R = \{1, 3, 5\}$$

A graphical representation of the relation is shown at the left.

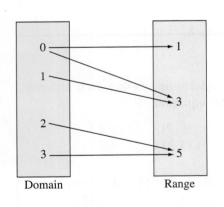

Domain   Range

You should note that it is not necessary to list repeated components of the domain and range of a relation.

Exercises Within Reach ®   Solutions in English & Spanish and tutorial videos at CollegePrepAlgebra.com

Analyzing a Relation In Exercises 1−6, **find the domain and range of the relation.**

1. {(−4, 3), (2, 5), (1, 2), (4, −3)}

2. {(−1, 5), (8, 3), (4, 6), (−5, −2)}

3. $\left\{(2, 16), (-9, -10), \left(\frac{1}{2}, 0\right)\right\}$

4. $\left\{\left(\frac{2}{3}, -4\right), \left(-6, \frac{1}{4}\right), (0, 0)\right\}$

5. {(−1, 3), (5, −7), (−1, 4), (8, −2), (1, −7)}

6. {(1, 1), (2, 4), (3, 9), (−2, 4), (−1, 1)}

Functions

Study Tip

The ordered pairs of a relation can be thought of in the form (input, output). For a *function*, a given input cannot yield two different outputs. For instance, if the input is a person's name and the output is that person's month of birth, then your name as the input can yield only your month of birth as the output.

> **Definition of Function**
>
> A **function** is a relation in which no two ordered pairs have the same first component and different second components.

This definition means that a given first component cannot be paired with two different second components. For instance, the pairs $(1, 3)$ and $(1, -1)$ could not be ordered pairs of a function.

EXAMPLE 2   **Testing Whether a Relation Is a Function**

Decide whether each relation represents a function.

a. Input: a, b, c
Output: $2, 3, 4$
$\{(a, 2), (b, 3), (c, 4)\}$

b.

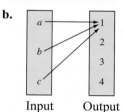

Input   Output

c.

Input x	Output y	(x, y)
3	1	(3, 1)
4	3	(4, 3)
5	4	(5, 4)
3	2	(3, 2)

SOLUTION

a. This set of ordered pairs *does* represent a function. No first component has two different second components.

b. This diagram *does* represent a function. No first component has two different second components.

c. This table *does not* represent a function. The first component 3 is paired with two different second components, 1 and 2.

Exercises Within Reach ®

Testing Whether a Relation Is a Function In Exercises 7–12, **determine whether the relation represents a function.**

7. *Domain*   *Range*
$-2 \longrightarrow 5$
$-1 \longrightarrow 6$
$0 \longrightarrow 7$
$1 \longrightarrow 8$
2

8. *Domain*   *Range*
$-2 \longrightarrow 7$
$-1 \longrightarrow 9$
0
1
2

9. Input: a, b, c; Output: $0, 4, 9$
$\{(a, 0), (b, 4), (c, 9)\}$

10. Input: $3, 5, 7$; Output: d, e, f
$\{(3, d), (5, e), (7, f), (7, d)\}$

11.

Input, x	Output, y	(x, y)
0	2	(0, 2)
1	4	(1, 4)
2	6	(2, 6)
3	8	(3, 8)
4	10	(4, 10)

12.

Input, x	Output, y	(x, y)
0	2	(0, 2)
1	4	(1, 4)
2	6	(2, 6)
1	8	(1, 8)
0	10	(0, 10)

In algebra, it is common to represent functions by equations in two variables rather than by ordered pairs. For instance, the equation $y = x^2$ represents the variable y as a function of x. The variable x is the **independent variable** (the input) and y is the **dependent variable** (the output). In this context, the domain of the function is the set of all *allowable* values of x, and the range is the *resulting* set of all values taken on by the dependent variable y.

Vertical Line Test

A set of points on a rectangular coordinate system is the graph of y as a function of x if and only if no vertical line intersects the graph at more than one point.

EXAMPLE 3 **Using the Vertical Line Test**

Use the Vertical Line Test to determine whether y is a function of x.

a. 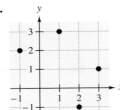   **b.** 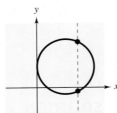   **c.**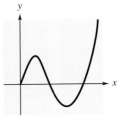

SOLUTION

a. From the graph, you can see that no vertical line intersects more than one point on the graph. So, the relation *does* represent y as a function of x.

b. From the graph, you can see that a vertical line intersects more than one point on the graph. So, the relation *does not* represent y as a function of x.

c. From the graph, you can see that no vertical line intersects more than one point on the graph. So, the relation *does* represent y as a function of x.

Exercises Within Reach ®

Solutions in English & Spanish and tutorial videos at CollegePrepAlgebra.com

Using the Vertical Line Test In Exercises 13–18, use the Vertical Line Test to determine whether y is a function of x.

13.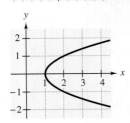

14.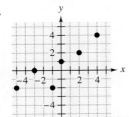

15.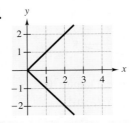

16.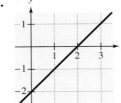

17.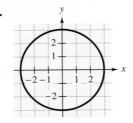

18.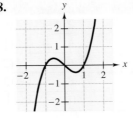

Function Notation

> ### Function Notation
>
> In the notation $f(x)$:
>
> > f is the **name** of the function.
> >
> > x is the **domain** (or input) value.
> >
> > $f(x)$ is a **range** (or output) value y for a given x.
>
> The symbol $f(x)$ is read as *the value of f at x* or simply *f of x*.

The process of finding the value of $f(x)$ for a given value of x is called **evaluating a function**. This is accomplished by substituting a given x-value (input) into the equation to obtain the value of $f(x)$ (output).

EXAMPLE 4   Evaluating a Function

Let $f(x) = x^2 + 1$. Find each value of the function.

a. $f(-2)$ b. $f(0)$

SOLUTION

a. $f(x) = x^2 + 1$ Write original function.

   $f(-2) = (-2)^2 + 1$ Substitute -2 for x.

   $= 4 + 1 = 5$ Simplify.

b. $f(x) = x^2 + 1$ Write original function.

   $f(0) = (0)^2 + 1$ Substitute 0 for x.

   $= 0 + 1 = 1$ Simplify.

EXAMPLE 5   Evaluating a Function

Let $g(x) = 3x - x^2$. Find each value of the function.

a. $g(2)$ b. $g(0)$

SOLUTION

a. Substituting 2 for x produces $g(2) = 3(2) - (2)^2 = 6 - 4 = 2$.

b. Substituting 0 for x produces $g(0) = 3(0) - (0)^2 = 0 - 0 = 0$.

Exercises Within Reach ®

Solutions in English & Spanish and tutorial videos at CollegePrepAlgebra.com

Evaluating a Function In Exercises 19–22, evaluate the function as indicated, and simplify.

19. $f(x) = 4x^2 + 2$ (a) $f(1)$ (b) $f(-1)$
    (c) $f(-4)$ (d) $f\left(-\frac{3}{2}\right)$

20. $g(t) = 5 - 2t^2$ (a) $g\left(\frac{5}{2}\right)$ (b) $g(-10)$
    (c) $g(0)$ (d) $g\left(\frac{3}{4}\right)$

21. $g(x) = 2x^2 - 3x + 1$ (a) $g(0)$ (b) $g(-2)$
    (c) $g(1)$ (d) $g\left(\frac{1}{2}\right)$

22. $h(x) = 1 - 4x - x^2$ (a) $h(0)$ (b) $h(-4)$
    (c) $h(10)$ (d) $h\left(\frac{3}{2}\right)$

The Domain and Range of a Function

The domain of a function may be explicitly described along with the function, or it may be *implied* by the context in which the function is used. For instance, if weekly pay is a function of hours worked (for a 40-hour work week), the implied domain is $0 \leq x \leq 40$. Certainly x cannot be negative in this context.

EXAMPLE 6   **Finding the Domain and Range of a Function**

Find the domain and range of each function.

a. $f: \{(-3, 0), (-1, 2), (0, 4), (2, 4), (4, -1)\}$

b.   Area of a square: $A = s^2$

SOLUTION

a.   The domain of f consists of all first components in the set of ordered pairs. So, the domain is

$\{-3, -1, 0, 2, 4\}$.   Domain

The range of f consists of all second components in the set of ordered pairs. So, the range is

$\{0, 2, 4, -1\}$.   Range

b.   For the area of a square, you must choose positive values for the side s. So, the domain is the set of all real numbers s such that

$s > 0$.   Domain

The area of a square must be a positive number. So, the range is the set of all real numbers A such that

$A > 0$.   Range

$A = s^2$

Exercises Within Reach ®

Solutions in English & Spanish and tutorial videos at CollegePrepAlgebra.com

Finding Domain and Range In Exercises 23–30, **find the domain and range of the function.**

23. $f: \{(0, 4), (1, 3), (2, 2), (3, 1), (4, 0)\}$

24. $f: \{(-2, -1), (-1, 0), (0, 1), (1, 2), (2, 3)\}$

25. $g: \{(-8, -1), (-6, 0), (2, 7), (5, 0), (12, 10)\}$

26. $g: \{(-4, 4), (3, 8), (4, 5), (9, -2), (10, -7)\}$

27. $h: \{(-5, 2), (-4, 2), (-3, 2), (-2, 2), (-1, 2)\}$

28. $h: \{(10, 100), (20, 200), (30, 300), (40, 400)\}$

29. Area of a circle: $A = \pi r^2$

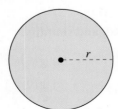

30. Perimeter of a square: $P = 4s$

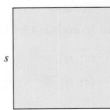

Application   EXAMPLE 7   **Estimating the Domain and Range of a Function**

The graph approximates the length of time L (in hours) between sunrise and sunset in Erie, Pennsylvania, for the year 2012. The variable t represents the day of the year.

Sunlight in Erie, PA

a. What is the domain of this function?

b. Estimate the range of this function.

SOLUTION

a. The year 2012 was a leap year with 366 days. So, the domain is

$$\{1, \quad 2, \quad 3, \quad 4, \ldots, 365, \quad 366\}. \quad \text{Domain}$$

Jan 1   Jan 2   Jan 3   Jan 4   Dec 30   Dec 31

b. The range is the set of different lengths of sunlight for the days of the year. These vary between about 9 hours and 15 hours. So, the range is

$$9 \le L \le 15. \qquad \text{Range}$$

Exercises Within Reach ®

Solutions in English & Spanish and tutorial videos at CollegePrepAlgebra.com

31. *Estimating Domain* The graph shows the SAT scores x and the grade-point averages y for 12 students.

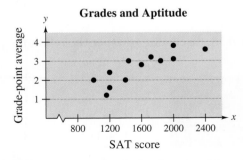

Grades and Aptitude

Estimate the greatest value in the domain of the relation.

32. *Estimating Range* Estimate the greatest value in the range of the relation in Exercise 31.

Concept Summary: Relations and Functions

What

A **relation** is any set of ordered pairs. A **function** is a special type of relation. How can you tell when a relation is a function?

EXAMPLE

Determine whether the relation represents a function.

$$\{(-2, -1), (0, 1), (3, 1), (4, 4)\}$$

How

One way to decide whether a relation represents a function is to use the Vertical Line Test. First, plot the ordered pairs on a rectangular coordinate system. Then decide whether any vertical line intersects more than one point on the graph.

EXAMPLE

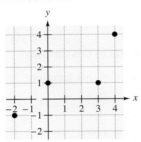

Because no vertical line intersects more than one point on the graph, the relation *does* represent a function.

Why

As you continue your study of mathematics, you will learn about many types of functions and how these functions model real-life situations.

Once you decide that a relation is a function, it is important that you (1) learn how to identify the **domain** and **range** of the function and (2) know how to **evaluate the function**.

Exercises Within Reach ®

Worked-out solutions to odd-numbered exercises at CollegePrepAlgebra.com

Concept Summary Check

33. *Vocabulary* Explain the difference between a relation and a function.

34. *Vocabulary* Explain the meanings of the terms *domain* and *range* in the context of a function.

35. *Writing* In your own words, explain how to use the Vertical Line Test.

36. *Reasoning* Using the function in the example above, explain how to find $f(3)$.

Extra Practice

Testing Whether a Relation Is a Function **In Exercises 37 and 38, determine whether the relation represents a function.**

37.

Input, x	1	3	5	3	1
Output, y	1	2	3	4	5
(x, y)	(1, 1)	(3, 2)	(5, 3)	(3, 4)	(1, 5)

38.

Input, x	2	4	6	8	10
Output, y	1	1	1	1	1
(x, y)	(2, 1)	(4, 1)	(6, 1)	(8, 1)	(10, 1)

Evaluating a Function **In Exercises 39–42, evaluate the function as indicated, and simplify.**

39. $g(u) = |u + 2|$ (a) $g(2)$ (b) $g(-2)$
          (c) $g(10)$ (d) $g\left(-\frac{5}{2}\right)$

40. $h(s) = |s| + 2$ (a) $h(4)$ (b) $h(-10)$
          (c) $h(-2)$ (d) $h\left(\frac{3}{2}\right)$

41. $h(x) = x^3 - 1$ (a) $h(0)$ (b) $h(1)$
          (c) $h(3)$ (d) $h\left(\frac{1}{2}\right)$

42. $f(x) = 16 - x^4$ (a) $f(-2)$ (b) $f(2)$
          (c) $f(1)$ (d) $f(3)$

43. *Demand* The demand for a product is a function of its price. Consider the demand function

$$f(p) = 20 - 0.5p$$

where p is the price in dollars.

   (a) Find $f(10)$ and $f(15)$.

   (b) Describe the effect a price increase has on demand.

44. *Maximum Load* The maximum safe load L (in pounds) for a wooden beam 2 inches wide and d inches high is $L(d) = 100d^2$.

   (a) Complete the table.

d	2	4	6	8
$L(d)$				

   (b) Describe the effect of an increase in height on the maximum safe load.

45. *Distance* The function $d(t) = 50t$ gives the distance (in miles) that a car will travel in t hours at an average speed of 50 miles per hour. Find the distance traveled for (a) $t = 2$, (b) $t = 4$, and (c) $t = 10$.

46. *Speed of Sound* The function $S(h) = 1116 - 4.04h$ approximates the speed of sound (in feet per second) at altitude h (in thousands of feet). Use the function to approximate the speed of sound for (a) $h = 0$, (b) $h = 10$, and (c) $h = 30$.

47. *Geometry* Write the formula for the perimeter P of a square with sides of length s. Is P a function of s? Explain.

48. *Geometry* Write the formula for the volume V of a cube with sides of length t. Is V a function of t? Explain.

Explaining Concepts

49. *Reasoning* Is it possible to find a relation that is not a function? If it is, find one.

50. *Reasoning* Is it possible to find a function that is not a relation? If it is, find one.

51. *Logic* Is it possible for the number of elements in the domain of a relation to be greater than the number of elements in the range of the relation? Explain.

52. *Writing* Determine whether the statement uses the word *function* in a way that is mathematically correct. Explain your reasoning.

   (a) The amount of money in your savings account is a function of your salary.

   (b) The speed at which a free-falling baseball strikes the ground is a function of the height from which it is dropped.

Cumulative Review

In Exercises 53–56, determine whether the inequalities are equivalent.

53. $4x - 3 < 9$, $4x < 6$

54. $6x + 5 > 11$, $6x > 6$

55. $8x - 5 \leq 5x + 2$, $3x \leq 7$

56. $13 - 4x \geq 6x + 1$, $2x \geq -12$

In Exercises 57–64, solve the equation.

57. $|x| = 8$

58. $|g| = -4$

59. $|4h| = 24$

60. $\left|\frac{1}{5}m\right| = 2$

61. $|x + 4| = 5$

62. $|2t - 3| = 11$

63. $|6b + 8| = 2b$

64. $|n - 2| = |2n + 9|$

Mid-Chapter Quiz: Sections 4.1–4.3

Solutions in English & Spanish and tutorial videos at CollegePrepAlgebra.com

Take this quiz as you would take a quiz in class. After you are done, check your work against the answers in the back of the book.

1. Plot the points $(4, -2)$ and $\left(-1, -\frac{5}{2}\right)$ on a rectangular coordinate system.

2. Determine the quadrant(s) in which the point $(3, y)$ may be located without plotting it. Assume $y \neq 0$. (y is a real number.)

3. Determine whether each ordered pair is a solution of the equation $y = 9 - |x|$.

    (a) $(2, 7)$ (b) $(-3, 12)$ (c) $(-9, 0)$ (d) $(0, -9)$

In Exercises 4 and 5, find the x- and y-intercepts of the graph of the equation.

4. $x - 3y = 12$

5. $y = -7x + 2$

In Exercises 6–11, sketch the graph of the equation.

6. $y = x - 1$ 7. $y = 5 - 2x$ 8. $y = 4 - x^2$

9. $y = (x + 2)^2$ 10. $y = |x + 3|$ 11. $y = 1 - |x|$

In Exercises 12 and 13, find the domain and range of the relation.

12. $\{(1, 4), (2, 6), (3, 10), (2, 14), (1, 0)\}$

13. $\{(-3, 6), (-2, 6), (-1, 6), (0, 6)\}$

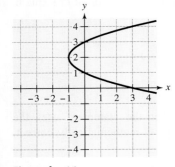

Figure for 14

14. Use the Vertical Line Test to determine whether y is a function of x in the relation shown in the figure.

In Exercises 15 and 16, evaluate the function as indicated, and simplify.

15. $f(x) = 3(x + 2) - 4$

    (a) $f(0)$ (b) $f(-3)$

16. $g(x) = 4 - x^2$

    (a) $g(-1)$ (b) $g(8)$

17. Find the domain of the function f: $\{(10, 1), (15, 3), (20, 9), (25, 27)\}$.

Applications

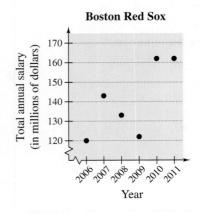

18. The scatter plot at the left shows the total annual salaries (in millions of dollars) of the Boston Red Sox in the years 2006 through 2011. Estimate the total salary each year. (*Source:* USA Today)

19. A new computer system sells for approximately $2000 and depreciates at the rate of $500 per year.

    (a) Write an equation that relates the value of the computer system and the number of years.

    (b) Sketch the graph of the equation.

    (c) What is the y-intercept of the graph, and what does it represent in the context of the problem?

Study Skills in Action

Reading Your Textbook Like a Manual

Many students avoid opening their textbooks for the same reason many people avoid going to the dentist—anxiety and frustration. The truth? Not opening your math textbook will cause more anxiety and frustration! Your textbook is a manual designed to help you master skills and understand and remember concepts. It contains many features and resources that can help you be successful in your math class.

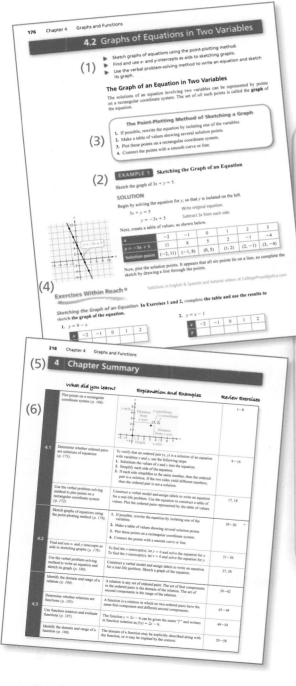

Smart Study Strategy

Use the Features of Your Textbook

To review what you learned in a previous class:

- Read the list of skills you should learn (1) at the beginning of this section. If you cannot remember how to perform a skill, review the appropriate example (2) in the section.

- Read and understand the contents of all tinted concept boxes (3)—these contain important definitions and rules.

To prepare for homework:

- Complete a few of the exercises (4) following each example. If you have difficulty with any of these, reread the example or seek help from a classmate or teacher.

To review for quizzes and tests:

- Make use of the Chapter Summary (5). Check off the concepts (6) you know, and review those you do not know.

- Complete the Review Exercises. Then take the Mid-Chapter Quiz, Chapter Test, or Cumulative Test, as appropriate.

4.4 Slope and Graphs of Linear Equations

▶ Determine the slope of a line through two points.

▶ Write linear equations in slope-intercept form and graph the equations.

▶ Use slopes to determine whether lines are parallel, perpendicular, or neither.

The Slope of a Line

The **slope** m of a nonvertical line that passes through the points (x_1, y_1) and (x_2, y_2) is

$$m = \frac{y_2 - y_1}{x_2 - x_1}$$

$$= \frac{\text{Change in } y}{\text{Change in } x}$$

$$= \frac{\text{Rise}}{\text{Run}}$$

where $x_1 \neq x_2$.

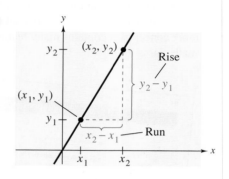

EXAMPLE 1   **Finding the Slope of a Line Through Two Points**

a. The slope of the line through $(x_1, y_1) = (-2, 0)$ and $(x_2, y_2) = (3, 1)$ is

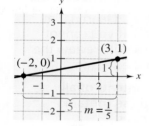

$$m = \frac{y_2 - y_1}{x_2 - x_1}$$

$$= \frac{1 - 0}{3 - (-2)} \quad \begin{array}{l} \text{Difference in } y\text{-values} \\ \text{Difference in } x\text{-values} \end{array}$$

$$= \frac{1}{5}. \quad \text{Simplify.}$$

The graph of the line is shown at the left.

b. The slope of the line through $(0, 0)$ and $(1, -1)$ is

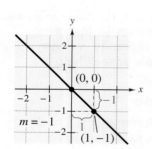

$$m = \frac{-1 - 0}{1 - 0} \quad \begin{array}{l} \text{Difference in } y\text{-values} \\ \text{Difference in } x\text{-values} \end{array}$$

$$= \frac{-1}{1} \quad \text{Simplify.}$$

$$= -1. \quad \text{Simplify.}$$

The graph of the line is shown at the left.

Exercises Within Reach ®

Finding the Slope of a Line In Exercises 1–6, **plot** the points and **find** the slope of the line that passes through the points.

1. $(0, 0), (4, 5)$

2. $(0, 0), (3, 6)$

3. $(0, 0), (8, -4)$

4. $(0, 0), (-1, 3)$

5. $(6, 0), (0, 4)$

6. $(0, -3), (5, 0)$

Slope of a Line

1. A line with positive slope ($m > 0$) *rises* from left to right.

2. A line with negative slope ($m < 0$) *falls* from left to right.

3. A line with zero slope ($m = 0$) is *horizontal*.

4. A line with undefined slope is *vertical*.

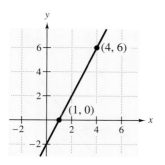

Line rises: positive slope

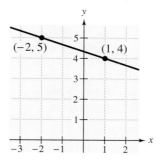

Line falls: negative slope

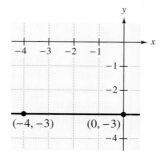

Horizontal line: zero slope

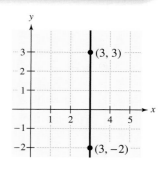

Vertical line: undefined slope

Application   **EXAMPLE 2**   **Finding the Slope of a Ladder**

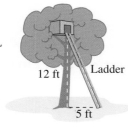

12 ft   Ladder

5 ft

Find the slope of the ladder leading up to the tree house.

SOLUTION

Consider the tree trunk as the *y*-axis and the level ground as the *x*-axis. The endpoints of the ladder are $(0, 12)$ and $(5, 0)$. So, the slope of the ladder is

$$m = \frac{y_2 - y_1}{x_2 - x_1} = \frac{0 - 12}{5 - 0} = -\frac{12}{5}.$$

Exercises Within Reach®

Solutions in English & Spanish and tutorial videos at CollegePrepAlgebra.com

Using Slope to Describe a Line In Exercises 7–14, **find the slope of the line that passes through the points. Use the slope to state whether the line rises, falls, is horizontal, or is vertical. Then sketch the line.**

7. $(-4, -1), (2, 6)$

8. $(5, 3), (-3, 1)$

9. $(-6, -1), (-6, 4)$

10. $(-4, -10), (-4, 0)$

11. $(3, -4), (8, -4)$

12. $(1, -2), (-2, -2)$

13. $\left(\frac{1}{4}, \frac{3}{2}\right), \left(\frac{9}{2}, -3\right)$

14. $\left(\frac{5}{4}, \frac{1}{4}\right), \left(\frac{7}{8}, 2\right)$

15. *Slide* The ladder of a straight slide in a playground is 8 feet high. The distance along the ground from the ladder to the foot of the slide is 12 feet. Approximate the slope of the slide.

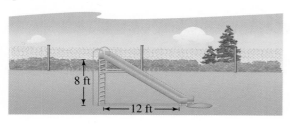

8 ft

12 ft

16. *Ladder* Find the slope of the ladder shown in the figure.

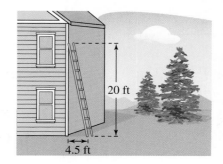

20 ft

4.5 ft

Slope as a Graphing Aid

You saw in Section 4.1 that before creating a table of values for an equation, it is helpful first to solve the equation for y. When you do this for a linear equation, you obtain some very useful information. Consider the following.

$3x - 2y = 4$	Original equation
$3x - 3x - 2y = -3x + 4$	Subtract $3x$ from each side.
$-2y = -3x + 4$	Combine like terms.
$\dfrac{-2y}{-2} = \dfrac{-3x + 4}{-2}$	Divide each side by -2.
$y = \dfrac{3}{2}x - 2$	Simplify.

Observe that the coefficient of x is the slope of the graph of this equation. Moreover, the constant term, -2, gives the y-intercept of the graph.

$$y = \frac{3}{2}x + -2$$

slope $\quad$ y-intercept $(0, -2)$

This form is called the **slope-intercept form** of the equation of the line.

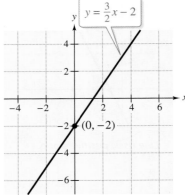

Slope-Intercept Form of the Equation of a Line

The graph of the equation

$$y = mx + b$$

is a line whose slope is m and whose y-intercept is $(0, b)$.

Exercises Within Reach ®

Solutions in English & Spanish and tutorial videos at CollegePrepAlgebra.com

Slope-Intercept Form In Exercises 17–22, write the equation in slope-intercept form. Use the equation to identify the slope and y-intercept.

17. $\frac{1}{2}x + y = 2$

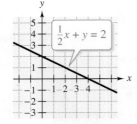

18. $\frac{3}{4}x - y = 3$

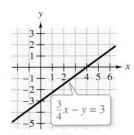

19. $2x + 3y = 6$

20. $4x + 2y = -8$

21. $3x - 4y + 2 = 0$

22. $2x - 3y + 1 = 0$

EXAMPLE 3   **Using Slope-Intercept Form**

Use the slope and y-intercept to sketch the graph of

$$x - 3y = -6.$$

SOLUTION

First, write the equation in slope-intercept form.

$x - 3y = -6$	Write original equation.
$-3y = -x - 6$	Subtract x from each side.
$y = \dfrac{-x - 6}{-3}$	Divide each side by -3.
$y = \dfrac{1}{3}x + 2$	Simplify to slope-intercept form.

So, the slope of the line is $m = \frac{1}{3}$ and the y-intercept is $(0, b) = (0, 2)$. Now you can sketch the graph of the equation. First, plot the y-intercept. Then, using a slope of $\frac{1}{3}$,

$$m = \frac{1}{3} = \frac{\text{Change in } y}{\text{Change in } x}$$

locate a second point on the line by moving three units to the right and one unit up (or one unit up and 3 units to the right). Finally, obtain the graph by drawing a line through the two points.

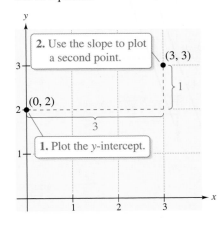 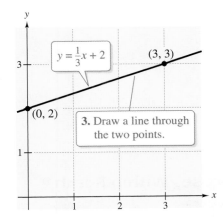

Solutions in English & Spanish and tutorial videos at CollegePrepAlgebra.com

Exercises Within Reach®

Using Slope-Intercept Form In Exercises 23−28, write the equation in slope-intercept form. Use the slope and y-intercept to sketch the graph of the line.

23. $2x - y = 3$

24. $x - y = -2$

25. $x - 2y = -2$

26. $x + 3y = 12$

27. $2x - 6y - 15 = 0$

28. $10x + 6y - 3 = 0$

Using Slope and a Point on the Line In Exercises 29−34, sketch the graph of the line through the point (3, 0) having the given slope.

29. $m = 0$

30. m is undefined.

31. $m = 2$

32. $m = -1$

33. $m = -\dfrac{2}{3}$

34. $m = \dfrac{3}{5}$

Parallel and Perpendicular Lines

You know from geometry that two lines in a plane are **parallel** if they do not intersect, and two lines in a plane are **perpendicular** if they intersect at right angles.

Study Tip

The phrase "if and only if" is used in mathematics as a way to write two statements in one. In the rule for parallel lines, the first statement says that if two distinct nonvertical lines have the same slope, they must be parallel. The second (or reverse) statement says that if two distinct nonvertical lines are parallel, they must have the same slope.

Parallel Lines and Perpendicular Lines

Parallel Lines: Two distinct nonvertical lines are parallel if and only if they have the same slope.

Perpendicular Lines: Two lines are perpendicular if and only if their slopes are negative reciprocals of each other. That is,

$$m_1 = -\frac{1}{m_2}.$$

EXAMPLE 4 **Parallel and Perpendicular Lines**

a. The lines $y = 3x$ and $y = 3x - 4$ each have a slope of 3. So, the lines are parallel.

b. The lines $y = 5x + 2$ and $y = -\frac{1}{5}x - 4$ have slopes that are negative reciprocals of each other. So, the lines are perpendicular.

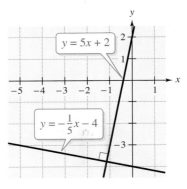

Exercises Within Reach ®

Solutions in English & Spanish and tutorial videos at CollegePrepAlgebra.com

Parallel and Perpendicular Lines In Exercises 35–38, determine whether the lines are parallel, perpendicular, or neither. Explain your reasoning.

35.

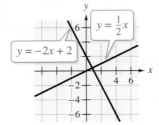

36.

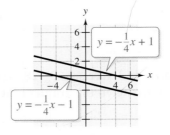

37.

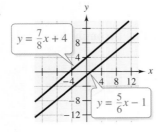

38.

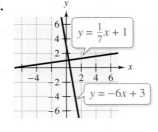

EXAMPLE 5   **Parallel or Perpendicular?**

Determine whether the pairs of lines are parallel, perpendicular, or neither.

a. $y = -3x - 2$, $y = \frac{1}{3}x + 1$ **b.** $y = \frac{1}{2}x + 1$, $y = \frac{1}{2}x - 1$

SOLUTION

a. This first line has a slope of $m_1 = -3$ and the second line has a slope of $m_2 = \frac{1}{3}$. Because these slopes are negative reciprocals of each other, the two lines must be perpendicular, as shown below on the left.

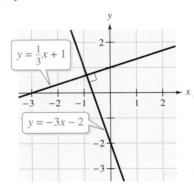

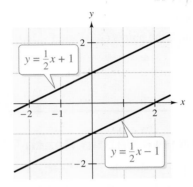

b. Both lines have a slope of $m = \frac{1}{2}$. So, the two lines must be parallel, as shown above on the right.

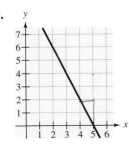

Parallel and perpendicular lines are common in architecture.

Exercises Within Reach®

Parallel or Perpendicular? **In Exercises 39−44, use the equations of the lines to determine whether the lines are parallel, perpendicular, or neither. Explain your reasoning.**

39. $y_1 = 2x - 3$
$y_2 = 2x + 1$

40. $y_1 = -\frac{1}{3}x - 3$
$y_2 = \frac{1}{3}x + 1$

41. $y_1 = 4x + 3$
$y_2 = 4x + 3$

42. $y_1 = 2x - 3$
$y_2 = -\frac{1}{2}x + 1$

43. $y_1 = -\frac{1}{3}x - 3$
$y_2 = 3x + 1$

44. $y_1 = \frac{3}{4}x + 5$
$y_2 = \frac{3}{4}x - 2$

Slopes of Parallel and Perpendicular Lines **In Exercises 45−48, determine the slope of a line that is (a) parallel and (b) perpendicular to the given line.**

45.
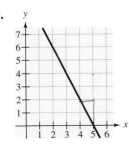

46.
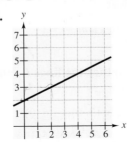

47.
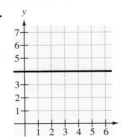

48.
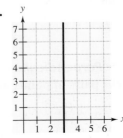

Concept Summary: *Finding the Slope of a Line*

What

The **slope** of a line is a measure of the "steepness" of the line. The slope also indicates whether the line *rises* or *falls* from left to right, is *horizontal*, or is *vertical*.

EXAMPLE

Find the slope of the line that passes through the points (2, 3) and (5, 1).

How

You can use a formula to find the slope of a nonvertical line that passes through any two points (x_1, y_1) and (x_2, y_2).

$$\text{Slope: } m = \frac{y_2 - y_1}{x_2 - x_1} = \frac{\text{Rise}}{\text{Run}}$$

EXAMPLE

Let $(x_1, y_1) = (2, 3)$ and let $(x_2, y_2) = (5, 1)$.

$$m = \frac{y_2 - y_1}{x_2 - x_1} = \frac{1 - 3}{5 - 2} = -\frac{2}{3}$$

Why

You can use slope to

- help sketch the graphs of linear equations.

- determine whether lines are **parallel**, **perpendicular**, or neither.

Exercises Within Reach ®

Worked-out solutions to odd-numbered exercises at CollegePrepAlgebra.com

Concept Summary Check

49. *Rise and Run* In the solution above, what is the rise from (2, 3) to (5, 1)? What is the run?

50. *Steepness* Which slope is steeper, $-\frac{2}{3}$ or $\frac{1}{3}$? Explain.

51. *The Sign of a Slope* Does a line with a negative slope rise or fall from left to right?

52. *Slope and Parallel Lines* How can you use slope to determine whether two lines are parallel?

Extra Practice

Identifying the Slope of a Line In Exercises 53 and 54, identify the line in the figure that has each slope.

53. (a) $m = \frac{3}{2}$

(b) $m = 0$

(c) $m = -\frac{2}{3}$

(d) $m = -2$

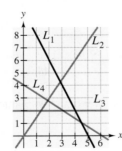

54. (a) $m = -\frac{3}{4}$

(b) $m = \frac{1}{2}$

(c) m is undefined.

(d) $m = 3$

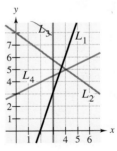

Using Slope and a Point on the Line In Exercises 55−58, plot the point and use the slope to find two additional points on the line. (There are many correct answers.)

55. $(-4, 0), m = \frac{2}{3}$ **56.** $(-1, -1), m = \frac{1}{4}$ **57.** $(0, 1), m = -2$ **58.** $(5, 6), m = -3$

Parallel or Perpendicular? In Exercises 59−62, determine whether the lines L_1 and L_2 that pass through the pairs of points are parallel, perpendicular, or neither.

59. L_1: $(0, -1), (5, 9)$

L_2: $(0, 3), (4, 1)$

60. L_1: $(-2, -1), (1, 5)$

L_2: $(1, 3), (5, 5)$

61. L_1: $(3, 6), (-6, 0)$

L_2: $(0, -1), \left(5, \frac{7}{3}\right)$

62. L_1: $(4, 8), (-4, 2)$

L_2: $(3, -5), \left(-1, \frac{1}{3}\right)$

Perp

Par

63. *Roof Pitch* Determine the slope, or pitch, of the roof of the house shown in the figure.

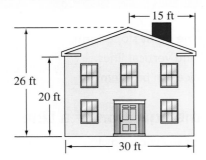

64. *Skateboarding Ramp* A wedge-shaped skateboarding ramp rises to a height of 12 inches over a 50-inch horizontal distance.

(a) Draw a diagram of the ramp and label the rise and run.

(b) Find the slope of the ramp.

65. *Skateboarding Ramp* Is a ramp that rises to a height of 12 inches over a 60-inch horizontal distance steeper than the ramp is Exercises 64? Explain.

66. *Gold Prices* The graph shows the average prices (in dollars) of a troy ounce of gold for the years 2006 through 2011.

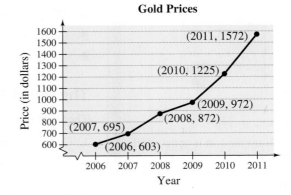

(a) Find the slopes of the five line segments.

(b) Find the slope of the line segment connecting the years 2006 and 2011. Interpret the meaning of this slope in the context of the problem.

Explaining Concepts

67. *Think About It* Can two perpendicular lines have positive slopes? Explain.

68. *Think About It* The slope of a line is $\frac{3}{2}$. How much will y change if x is increased by eight units? Explain.

69. *Reasoning* When a quantity y is increasing or decreasing at a constant rate over time t, the graph of y versus t is a line. What is another name for the rate of change?

70. *Reasoning* Explain how to use slopes to determine if the points $(-2, -3)$, $(1, 1)$, and $(3, 4)$ lie on the same line.

71. *Writing* When determining the slope of a line through two points, does the order of subtracting the coordinates of the points matter? Explain.

72. *Structure* The equations below give the heights h_1 and h_2 (in inches) of two bamboo plants over a period of 30 days, where t represents the time in days.

Plant 1: $h_1 = t + 30, \ 0 \le t \le 30$

Plant 2: $h_2 = 2t + 10, \ 0 \le t \le 30$

(a) Explain how you can use the equations to determine the heights of the plants on Day 0.

(b) Which plant grew at a faster rate? Explain.

(c) Were the two plants ever the same height? Explain.

Cumulative Review

In Exercises 73−80, simplify the expression.

73. $x^2 \cdot x^3$

74. $z^2 \cdot z^2$

75. $\left(-y^2\right)y$

76. $5x^2\left(x^5\right)$

77. $\left(25x^3\right)\left(2x^2\right)$

78. $\left(3yz\right)\left(6yz^3\right)$

79. $x^2 - 2x - x^2 + 3x + 2$

80. $x^2 - 5x - 2 + x$

In Exercises 81−84, find the *x*- and *y*-intercepts (if any) of the graph of the equation.

81. $y = 6x - 3$

82. $y = -\frac{4}{3}x - 8$

83. $2x + y = -3$

84. $3x - 5y = 15$

4.5 Equations of Lines

▶ Write equations of lines using the point-slope form.
▶ Write equations of horizontal and vertical lines.
▶ Use linear models to solve application problems.

The Point-Slope Form of the Equation of a Line

There are two basic types of problems in analytic geometry.

1. Given an equation, sketch its graph.

Algebra ⟹ Geometry

2. Given a graph, write its equation.

Geometry ⟹ Algebra

In Section 4.4, you worked primarily with the first type of problem. In this section, you will study the second type.

> ### Point-Slope Form of the Equation of a Line
>
> The **point-slope form** of the equation of a line with slope m that passes through the point (x_1, y_1) is $y - y_1 = m(x - x_1)$.

EXAMPLE 1 Using Point-Slope Form

Write an equation of the line that passes through the point $(1, -2)$ and has slope $m = 3$.

SOLUTION

$$y - y_1 = m(x - x_1) \qquad \text{Use point-slope form.}$$
$$y - (-2) = 3(x - 1) \qquad \text{Substitute } -2 \text{ for } y_1, 1 \text{ for } x_1, \text{ and } 3 \text{ for } m.$$
$$y + 2 = 3x - 3 \qquad \text{Simplify.}$$
$$y = 3x - 5 \qquad \text{Equation of line}$$

So, an equation of the line is $y = 3x - 5$.

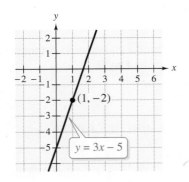

Exercises Within Reach ®

Solutions in English & Spanish and tutorial videos at CollegePrepAlgebra.com

Using Point-Slope Form In Exercises 1−4, use the point-slope form to write an equation of the line. Write the equation in slope-intercept form.

1.
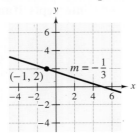

2.
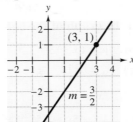

3.
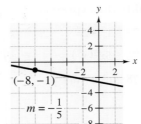

4.
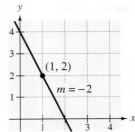

EXAMPLE 2   **An Equation of a Line Through Two Points**

Write an equation of the line that passes through the points $(3, 1)$ and $(-3, 4)$.

SOLUTION

Let $(x_1, y_1) = (3, 1)$ and $(x_2, y_2) = (-3, 4)$. The slope of the line passing through these points is

$$m = \frac{y_2 - y_1}{x_2 - x_1}$$ Formula for slope

$$= \frac{4 - 1}{-3 - 3}$$ Substitute for x_1, y_1, x_2, and y_2.

$$= \frac{3}{-6}$$ Subtract.

$$= -\frac{1}{2}.$$ Simplify.

Now, use the point-slope form to find an equation of the line.

$$y - y_1 = m(x - x_1)$$ Point-slope form

$$y - 1 = -\frac{1}{2}(x - 3)$$ Substitute 1 for y_1, 3 for x_1, and $-\frac{1}{2}$ for m.

$$y - 1 = -\frac{1}{2}x + \frac{3}{2}$$ Distributive Property

$$y = -\frac{1}{2}x + \frac{5}{2}$$ Equation of line

The graph of this line is shown at the left.

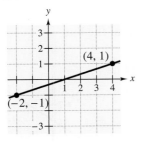

Exercises Within Reach®   Solutions in English & Spanish and tutorial videos at CollegePrepAlgebra.com

An Equation of a Line Through Two Points In Exercises 5–10, use the point-slope form to write an equation of the line. Write the equation in slope-intercept form.

5.

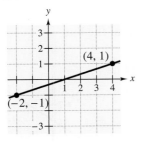

6.

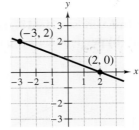

7.

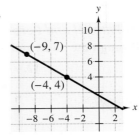

8.

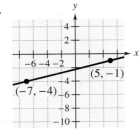

9.

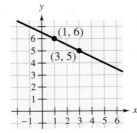

10.

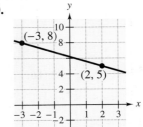

EXAMPLE 3 **Equations of Parallel and Perpendicular Lines**

Write an equation of the line that passes through the point $(2, -1)$ and is (a) parallel and (b) perpendicular to the line

$$y = \frac{2}{3}x - \frac{5}{3}.$$

SOLUTION

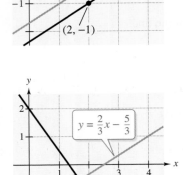

a. Because the line has a slope of $m = \frac{2}{3}$, it follows that any parallel line must have the same slope. So, an equation of the line through $(2, -1)$, parallel to the original line, is

$y - y_1 = m(x - x_1)$	Point-slope form
$y - (-1) = \frac{2}{3}(x - 2)$	Substitute -1 for y_1, 2 for x_1, and $\frac{2}{3}$ for m.
$y + 1 = \frac{2}{3}x - \frac{4}{3}$	Simplify.
$y = \frac{2}{3}x - \frac{7}{3}.$	Equation of parallel line

b. Because the line has a slope of $m = \frac{2}{3}$, it follows that any perpendicular line must have a slope of $-\frac{3}{2}$. So, an equation of the line through $(2, -1)$, perpendicular to the original line, is

$y - y_1 = m(x - x_1)$	Point-slope form
$y - (-1) = -\frac{3}{2}(x - 2)$	Substitute -1 for y_1, 2 for x_1, and $-\frac{3}{2}$ for m.
$y + 1 = -\frac{3}{2}x + 3$	Simplify.
$y = -\frac{3}{2}x + 2.$	Equation of perpendicular line

Exercises Within Reach ®

Solutions in English & Spanish and tutorial videos at CollegePrepAlgebra.com

Equations of Parallel and Perpendicular Lines In Exercises 11−18, **write an equation of the line that passes through the point and is (a) parallel and (b) perpendicular to the given line.**

11.
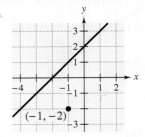

12.
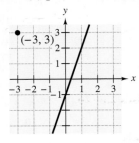

13.
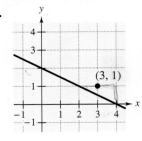

14.
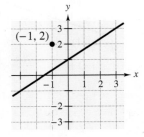

15. $(2, 1)$
$x - y = 3$

16. $(-3, 2)$
$x + y = 7$

17. $(-12, 4)$
$3x + 4y = 7$

18. $(15, -2)$
$5x - 3y = 0$

Equations of Horizontal and Vertical Lines

EXAMPLE 4 **Equations of Horizontal and Vertical Lines**

Write an equation for each line.

a. Vertical line through $(-3, 2)$
b. Line through $(-1, 2)$ and $(4, 2)$
c. Line through $(0, 2)$ and $(0, -2)$
d. Horizontal line through $(0, -4)$

SOLUTION

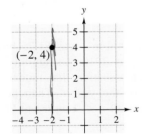

a. Because the line is vertical and passes through the point $(-3, 2)$, every point on the line has an x-coordinate of -3. So, an equation of the line is

$$x = -3. \qquad \text{Vertical line}$$

b. Because both points have the same y-coordinate, the line through $(-1, 2)$ and $(4, 2)$ is horizontal. So, its equation is

$$y = 2. \qquad \text{Horizontal line}$$

c. Because both points have the same x-coordinate, the line through $(0, 2)$ and $(0, -2)$ is vertical. So, its equation is

$$x = 0. \qquad \text{Vertical line (y-axis)}$$

d. Because the line is horizontal and passes through the point $(0, -4)$, every point on the line has a y-coordinate of -4. So, an equation of the line is

$$y = -4. \qquad \text{Horizontal line}$$

Exercises Within Reach ®

Solutions in English & Spanish and tutorial videos at CollegePrepAlgebra.com

An Equation of a Horizontal or Vertical Line **In Exercises 19−22, write an equation of the line of the given type that passes through the point shown.**

19. Vertical line

20. Horizontal line

21. Horizontal line

22. Vertical line

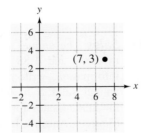

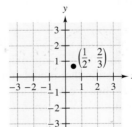

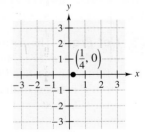

An Equation of a Horizontal or Vertical Line **In Exercises 23−26, write an equation of the line that passes through the points shown.**

23.

24.

25.

26.

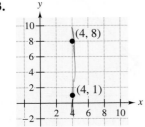

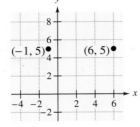

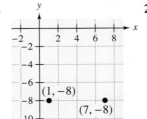

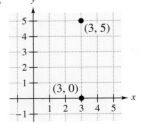

Applications

Application EXAMPLE 5 **Predicting Annual Sales**

The annual sales of AutoZone were $6.82 billion in 2009 and $7.36 billion in 2010. Using only this information, write a linear equation that models the annual sales in terms of the year. Then predict the sales for 2011. (*Source:* AutoZone, Inc.)

SOLUTION

Let $t = 9$ represent 2009. Then the two given values are represented by the data points $(9, 6.82)$ and $(10, 7.36)$. The slope of the line through these points is

$$m = \frac{7.36 - 6.82}{10 - 9}$$

$$= 0.54.$$

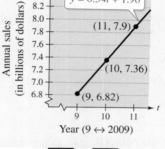

AutoZone

$y = 0.54t + 1.96$

$(11, 7.9)$

$(10, 7.36)$

$(9, 6.82)$

Year (9 ↔ 2009)

Using the point-slope form, you can find an equation that relates the annual sales y and the year t to be

$y - y_1 = m(t - t_1)$	Point-slope form
$y - 7.36 = 0.54(t - 10)$	Substitute 7.36 for y_1, 10 for t_1, and 0.54 for m.
$y - 7.36 = 0.54t - 5.4$	Distributive Property
$y = 0.54t + 1.96.$	Write in slope-intercept form.

Using this equation, a prediction of the annual sales in 2011 ($t = 11$) is

$$y = 0.54(11) + 1.96 = 7.9 \text{ billion}.$$

In this case, the prediction is fairly good—the actual annual sales in 2011 was $8.07 billion. The graph of this equation is shown at the left.

Exercises Within Reach ®

Solutions in English & Spanish and tutorial videos at CollegePrepAlgebra.com

27. *Net Profit* The net profit of Coach was $735 million in 2010 and $881 million in 2011. Using only this information, write a linear equation that models the net profit P in terms of the year t. Then predict the net profit for 2012. (Let $t = 0$ represent 2010.) (*Source:* Coach, Inc.)

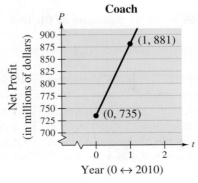

Coach

$(1, 881)$

$(0, 735)$

Year (0 ↔ 2010)

28. *Sales* The annual sales of Aaron's was $1.75 billion in 2009 and 1.88 billion in 2010. Using only this information, write a linear equation that models the annual sales S in terms of the year t. Then predict the annual sales for 2011. (Let $t = 9$ represent 2009.) (*Source:* Aaron's, Inc.)

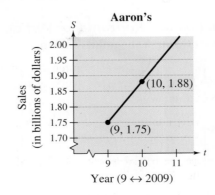

Aaron's

$(10, 1.88)$

$(9, 1.75)$

Year (9 ↔ 2009)

In the linear equation $y = mx + b$, you know that m represents the slope of the line. In applications, the slope of a line can often be interpreted as the *rate of change of y with respect to x*. Rates of change should always be described with appropriate units of measure.

Application   **EXAMPLE 6**   **Using Slope as a Rate of Change**

A rock climber is climbing up a 500-foot cliff. By 1 P.M., the rock climber has climbed 115 feet up the cliff. By 4 P.M., the climber has reached a height of 280 feet, as shown at the left. Find the average rate of change of the climber and use this rate of change to write a linear model that relates the height of the climber to the time.

SOLUTION

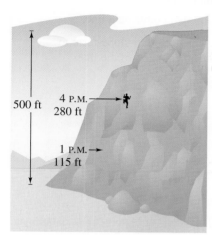

Let y represent the height of the climber and let t represent the time. Then the two points that represent the climber's two positions are $(t_1, y_1) = (1, 115)$ and $(t_2, y_2) = (4, 280)$. So, the average rate of change of the climber is

$$\text{Average rate of change} = \frac{y_2 - y_1}{t_2 - t_1} = \frac{280 - 115}{4 - 1} = 55 \text{ feet per hour.}$$

So, an equation that relates the height of the climber to the time is

$$y - y_1 = m(t - t_1) \qquad \text{Point-slope form}$$
$$y - 115 = 55(t - 1) \qquad \text{Substitute 115 for } y_1, 1 \text{ for } t_1, \text{ and 55 for } m.$$
$$y = 55t + 60. \qquad \text{Linear model}$$

Exercises Within Reach®

Solutions in English & Spanish and tutorial videos at CollegePrepAlgebra.com

29. *Depreciation* You purchase a boat for $25,000. After 1 year, its depreciated value is $22,700. The depreciation is linear.

   (a) Write a linear model that relates the value V of the boat to the time t in years.

   (b) Use the model to estimate the value of the boat after 3 years.

30. *Depreciation* A sub shop purchases a used pizza oven for $875. After 1 year, its depreciated value is $790. The depreciation is linear.

   (a) Write a linear model that relates the value V of the oven to the time t in years.

   (b) Use the model to estimate the value of the oven after 5 years.

31. *Bike Path* A city is paving a bike path. The same length of path is paved each day. After 4 days, 14 miles of the path remain to be paved. After 6 more days, 11 miles of the path remain to be paved. Find the average rate of change and use it to write a linear model that relates the distance remaining to be paved to the number of days.

32. *Swimming Pool* A swimming pool already contains a small amount of water when you start filling it at a constant rate. The pool contains 45 gallons of water after 5 minutes and 120 gallons after 30 minutes. Find the average rate of change and use it to write a linear model that relates the amount of water in the pool to the time.

Concept Summary: *Writing Equations of Lines*

What

You can use the **point-slope form** to write an equation of a line passing through any two points.

EXAMPLE

Write an equation of the line passing through the points (2, 3) and (3, 5).

How

Here are the steps to write an equation of the line.

1. Find the slope of the line.

$$m = \frac{y_2 - y_1}{x_2 - x_1} = \frac{5 - 3}{3 - 2} = 2$$

2. Use the point-slope form.

$$y - y_1 = m(x - x_1)$$
$$y - 3 = 2(x - 2)$$
$$y - 3 = 2x - 4$$
$$y = 2x - 1$$

Why

When you are given

*   the slope of a line and a point on the line, or

*   any two points on a line,

you can use point-slope form to write an equation of the line.

Exercises Within Reach ®

Worked-out solutions to odd-numbered exercises at CollegePrepAlgebra.com

Concept Summary Check

33. *Extra Steps* Why was finding the slope the first step of the solution above?

34. *Applying Point-Slope Form* What two pieces of information were used to write the equation in point-slope form in Step 2 above?

35. *The Form of an Equation* Describe the form of the final equation $y = 2x - 1$ in the solution above.

36. *Choosing a Form* When is it more convenient to apply the point-slope form than the slope-intercept form to write an equation of a line?

Extra Practice

Using Point-Slope Form **In Exercises 37−40, use the point-slope form to write an equation of the line that passes through the point and has the specified slope. Write the equation in slope-intercept form.**

37. $\left(0, \frac{3}{2}\right), m = \frac{2}{3}$

38. $\left(-\frac{5}{2}, 0\right), m = \frac{3}{4}$

39. $(2, 4), m = -0.8$

40. $(6, -3), m = 0.67$

Finding the Slope **In Exercises 41−44, determine the slope of the line.**

41. $y - 2 = 5(x + 3)$

42. $y + 3 = -2(x - 6)$

43. $3x - 2y + 10 = 0$

44. $5x + 4y - 8 = 0$

An Equation of a Line Through Two Points **In Exercises 45−48, write an equation of the line that passes through the points. Write the equation in general form.**

45. $\left(2, \frac{1}{2}\right), \left(\frac{1}{2}, \frac{5}{2}\right)$

46. $\left(\frac{1}{4}, 1\right), \left(-\frac{3}{4}, -\frac{2}{3}\right)$

47. $(1, 0.6), (2, -0.6)$

48. $(-8, 0.6), (2, -2.4)$

49. *Wages* A sales representative receives a monthly salary of $2000 plus a commission of 2% of the total monthly sales. Write a linear model that relates total monthly wages W to sales S.

50. *Reimbursed Expenses* A sales representative is reimbursed $250 per day for lodging and meals plus $0.30 per mile driven. Write a linear model that relates the daily cost C to the number of miles driven x.

51. *Graphical Interpretation* Match each of the situations labeled (a), (b), (c), and (d) with one of the graphs labeled (e), (f), (g), and (h). Then determine the slope of each line and interpret the slope in the context of the real-life situation.

(a) A friend is paying you $10 per week to repay a $100 loan.

(b) An employee is paid $12.50 per hour plus $1.50 for each unit produced per hour.

(c) A sales representative receives $40 per day for food plus $0.32 for each mile traveled.

(d) A television purchased for $600 depreciates $100 per year.

(e)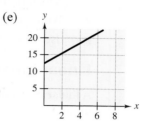

(f)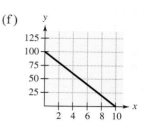

(g)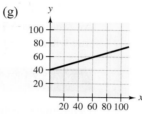

(h)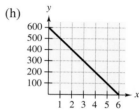

Explaining Concepts

52. *Writing* Can the equation of a vertical line be written in slope-intercept form? Explain.

53. *Writing* Explain how to find algebraically the x-intercept of the line given by $y = mx + b$.

54. *Think About It* Find the slope of the line given by $5x + 7y - 21 = 0$. Use the same process to find a formula for the slope of the line $ax + by + c = 0$, where $b \neq 0$.

55. *Think About It* What is implied about the graphs of the lines $a_1x + b_1y + c_1 = 0$ and $a_2x + b_2y + c_2 = 0$ if $\dfrac{a_1}{b_1} = \dfrac{a_2}{b_2}$?

56. *Research Project* Use a newspaper or weekly news magazine to find an example of data that are *increasing* linearly with time. Write a linear model that relates the data to time. Repeat the project for data that are *decreasing*.

Cumulative Review

In Exercises 57–60, simplify the expression.

57. $4(3 - 2x)$

58. $x^2(xy^3)$

59. $3x - 2(x - 5)$

60. $u - [3 + (u - 4)]$

In Exercises 61–64, solve for y in terms of x.

61. $3x + y = 4$

62. $4 - y + x = 0$

63. $4x - 5y = -2$

64. $3x + 4y - 5 = 0$

In Exercises 65–68, determine the slope of the line passing through the points.

65. $(3, 0)$ and $(4, 2)$

66. $(-1, 3)$ and $(2, -5)$

67. $(0, -2)$ and $(-7, -1)$

68. $(-2, 4)$ and $(-5, 10)$

4.6 Graphs of Linear Inequalities

▶ Determine whether an ordered pair is a solution of a linear inequality in two variables.

▶ Sketch graphs of linear inequalities in two variables.

▶ Use linear inequalities to model and solve real-life problems.

Linear Inequalities in Two Variables

A **linear inequality in two variables**, x and y, is an inequality that can be written in one of the forms below (where a and b are not both zero).

$$ax + by < c, \qquad ax + by > c, \qquad ax + by \le c, \qquad ax + by \ge c$$

An ordered pair (x_1, y_1) is a **solution** of a linear inequality in x and y if the inequality is true when x_1 and y_1 are substituted for x and y, respectively.

EXAMPLE 1   **Verifying Solutions of a Linear Inequality**

Determine whether each ordered pair is a solution of $3x - y \ge -1$.

a. $(0, 0)$ **b.** $(1, 4)$ **c.** $(-1, 2)$

SOLUTION

a.
$$3x - y \ge -1 \qquad \text{Write original inequality.}$$
$$3(0) - 0 \overset{?}{\ge} -1 \qquad \text{Substitute 0 for } x \text{ and 0 for } y.$$
$$0 \ge -1 \qquad \text{Inequality is satisfied.} \quad ✓$$

Because the inequality is satisfied, the point $(0, 0)$ *is* a solution.

b.
$$3x - y \ge -1 \qquad \text{Write original inequality.}$$
$$3(1) - 4 \overset{?}{\ge} -1 \qquad \text{Substitute 1 for } x \text{ and 4 for } y.$$
$$-1 \ge -1 \qquad \text{Inequality is satisfied.} \quad ✓$$

Because the inequality is satisfied, the point $(1, 4)$ *is* a solution.

c.
$$3x - y \ge -1 \qquad \text{Write original inequality.}$$
$$3(-1) - 2 \overset{?}{\ge} -1 \qquad \text{Substitute } -1 \text{ for } x \text{ and 2 for } y.$$
$$-5 \not\ge -1 \qquad \text{Inequality is not satisfied.} \quad ✗$$

Because the inequality is not satisfied, the point $(-1, 2)$ *is not* a solution.

Exercises Within Reach ®

Solutions in English & Spanish and tutorial videos at CollegePrepAlgebra.com

Verifying Solutions of a Linear Inequality In Exercises 1−4, determine whether each ordered pair is a solution of the inequality.

1. $x + 4y > 10$ (a) $(0, 0)$ (b) $(3, 2)$
      (c) $(1, 2)$ (d) $(-2, 4)$

2. $2x + 3y > 9$ (a) $(0, 0)$ (b) $(1, 1)$
      (c) $(2, 2)$ (d) $(-2, 5)$

3. $-3x + 5y \le 12$ (a) $(1, 2)$ (b) $(2, -3)$
      (c) $(1, 3)$ (d) $(2, 8)$

4. $5x + 3y < 100$ (a) $(25, 10)$ (b) $(6, 10)$
      (c) $(0, -12)$ (d) $(4, 5)$

The Graph of a Linear Inequality in Two Variables

> ### Sketching the Graph of a Linear Inequality in Two Variables
>
> 1. Replace the inequality sign by an equal sign and sketch the graph of the resulting equation. (Use a dashed line for < or > and a solid line for ≤ or ≥.)
> 2. Test one point in each of the half-planes formed by the graph in Step 1. If the point satisfies the inequality, then shade the entire half-plane to denote that every point in the region satisfies the inequality.

EXAMPLE 2   **Sketching the Graphs of Linear Inequalities**

Sketch the graphs of (a) $x > -2$ and (b) $y \leq 3$.

SOLUTION

a. The graph of the corresponding equation is a vertical line. The points that satisfy the inequality are those lying to the right of the line.

b. The graph of the corresponding equation is a horizontal line. The points that satisfy the inequality are those lying on or below the line.

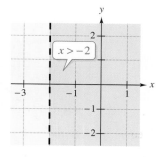

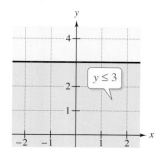

Exercises Within Reach ®

Solutions in English & Spanish and tutorial videos at CollegePrepAlgebra.com

Identifying the Type of Boundary In Exercises 5−8, state whether the boundary of the graph of the inequality should be dashed or solid.

5. $2x + 3y < 6$ **6.** $2x + 3y \leq 6$ **7.** $2x + 3y \geq 6$ **8.** $2x + 3y > 6$

Matching In Exercises 9−12, match the inequality with its graph.

(a) 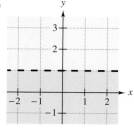   **(b)** 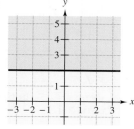   **(c)** 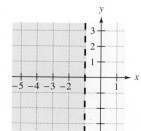   **(d)**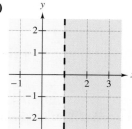

9. $x < -1$ **10.** $x > 1$ **11.** $y \geq 2$ **12.** $y < 1$

Sketching the Graph of a Linear Inequality In Exercises 13−16, sketch the graph of the linear inequality.

13. $y \geq 3$ **14.** $x > -4$ **15.** $x \leq 3$ **16.** $y > -5$

EXAMPLE 3 **Sketching the Graph of a Linear Inequality**

Sketch the graph of the linear inequality

$$x - y < 2.$$

SOLUTION

The graph of the corresponding equation

$$x - y = 2 \qquad \text{Write corresponding equation.}$$

is the line show below. Because the origin $(0, 0)$ does not lie on the line, use it as the test point.

$$x - y < 2 \qquad \text{Write original inequality}$$

$$0 - 0 \overset{?}{<} 2 \qquad \text{Substitute 0 for } x \text{ and 0 for } y.$$

$$0 < 2 \qquad \text{Inequality is satisfied.} \checkmark$$

Because $(0, 0)$ satisfies the inequality, the graph consists of the half-plane lying above the line. Try checking a point below the line. Regardless of which point you choose, you will see that it does not satisfy the inequality.

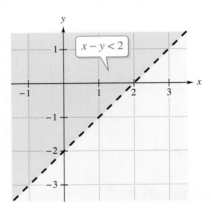

Exercises Within Reach®

Solutions in English & Spanish and tutorial videos at CollegePrepAlgebra.com

Matching In Exercises 17−20, match the inequality with its graph.

(a)
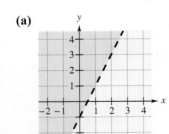

(b)
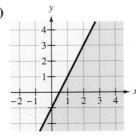

(c)
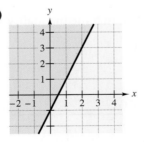

(d)
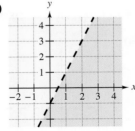

17. $2x - y \leq 1$ **18.** $2x - y < 1$ **19.** $2x - y \geq 1$ **20.** $2x - y > 1$

Sketching the Graph of a Linear Inequality In Exercises 21−28, sketch the graph of the linear inequality.

21. $y \leq 3x$ **22.** $y \geq 5x$ **23.** $y \leq 2x - 1$ **24.** $y \geq 2x - 1$

25. $x - y < 0$ **26.** $x + y > 0$ **27.** $y > -2x + 10$ **28.** $y < 3x + 1$

EXAMPLE 4   **Sketching the Graph of a Linear Inequality**

Use the slope-intercept form of a linear equation as an aid in sketching the graph of the inequality $5x + 4y \leq 12$.

SOLUTION

To begin, rewrite the inequality in slope-intercept form.

$5x + 4y \leq 12$ Write original inequality.

$4y \leq -5x + 12$ Subtract $5x$ from each side.

$y \leq -\dfrac{5}{4}x + 3$ Divide each side by 4.

From this form, you can conclude that the solution is the half-plane lying *on* or *below* the line $y = -\dfrac{5}{4}x + 3$. The graph is shown below. You can verify this by testing the solution point $(0, 0)$.

$5x + 4y \leq 12$ Write original inequality.

$5(0) + 4(0) \overset{?}{\leq} 12$ Substitute 0 for x and 0 for y.

$0 \leq 12$ Inequality is satisfied. ✓

$y \leq -\dfrac{5}{4}x + 3$

Exercises Within Reach ®

Solutions in English & Spanish and tutorial videos at CollegePrepAlgebra.com

Matching In Exercises 29−32, **match** the inequality with its graph.

(a)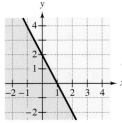

(b)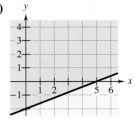

(c)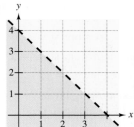

(d)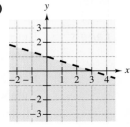

29. $y < -\dfrac{1}{3}x + 1$ **30.** $y \geq \dfrac{2}{5}x - 2$ **31.** $x + y < 4$ **32.** $y + 2x \leq 2$

Sketching the Graph of a Linear Inequality In Exercises 33−38, **sketch** the graph of the linear inequality.

33. $-3x + 2y < 6$ **34.** $x - 2y \leq -6$ **35.** $x \geq 3y - 5$

36. $x \leq -2y + 10$ **37.** $y - 3 > \dfrac{1}{2}(x - 4)$ **38.** $y + 1 < -2(x - 3)$

Applications

EXAMPLE 5 **Working to Meet a Budget**

Your budget requires you to earn *at least* $250 per week. You work two part-time jobs. One is at a fast-food restaurant, which pays $8 per hour, and the other is tutoring for $10 per hour. Let *x* represent the number of hours you work at the restaurant and let *y* represent the number of hours you work as a tutor. Write a linear inequality that represents the different numbers of hours you can work at each job in order to meet your budget requirements.

SOLUTION

To write the inequality, use the problem-solving method.

Verbal Model:	Hourly pay at fast-food restaurant	·	Number of hours at fast-food restaurant	+	Hourly pay tutoring	·	Number of hours tutoring	≥	Minimum weekly earnings

Labels: Hourly pay at fast-food restaurant = 8 (dollars per hour)
Number of hours at fast-food restaurant = *x* (hours)
Hourly pay for tutoring = 10 (dollars per hour)
Number of hours tutoring = *y* (hours)
Minimum weekly earnings = 250 (dollars)

Inequality: $8x + 10y \geq 250$

Exercises Within Reach ® Solutions in English & Spanish and tutorial videos at CollegePrepAlgebra.com

39. *Part-Time Jobs* Your budget requires you to earn at least $210 per week. You work two part-time jobs. One is at a grocery store, which pays $9 per hour, and the other is mowing lawns, which pays $12 per hour. Use the verbal model and labels below to write a linear inequality that represents the different numbers of hours you can work at each job in order to meet your budget requirements.

Verbal Model:	Hourly pay at grocery store	·	Number of hours at grocery store	+	Hourly pay mowing lawns	·	Number of hours mowing lawns	≥	Minimum weekly earnings

Labels: Hourly pay at grocery store = 9 (dollars per hour)
Number of hours at grocery store = *x* (hours)
Hourly pay mowing lawns = 12 (dollars per hour)
Number of hours mowing lawns = *y* (hours)
Minimum weekly earnings = 210 (dollars)

40. *Inventory* A store sells two models of central air conditioning units. The costs to the store of the two models are $2000 and $3000, and the owner of the store does not want more than $30,000 invested in the inventory for these two models. Write a linear inequality that represents the different numbers of each model that can be held in inventory.

Application   EXAMPLE 6   **Working to Meet a Budget**

Graph the inequality in Example 5 and find at least two ordered pairs (x, y) that identify the numbers of hours you can work at each job in order to meet your budget requirements.

SOLUTION

To sketch the graph, rewrite the inequality in slope-intercept form.

$$8x + 10y \geq 250 \qquad \text{Write original inequality.}$$

$$10y \geq -8x + 250 \qquad \text{Subtract } 8x \text{ from each side.}$$

$$y \geq -\frac{4}{5}x + 25 \qquad \text{Divide each side by 10.}$$

Graph the corresponding equation

$$y = -\frac{4}{5}x + 25$$

and shade the half-plane lying above the line, as shown at the left. From the graph, you can see that two solutions that will yield the desired weekly earnings of at least $250 are (10, 17) and (20, 15). In other words, you can work 10 hours at the restaurant and 17 hours as a tutor, or 20 hours at the restaurant and 15 hours as a tutor, to meet your budget requirements. There are many other solutions.

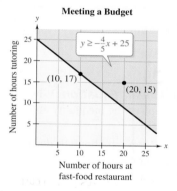

Meeting a Budget

$y \geq -\frac{4}{5}x + 25$

Number of hours tutoring

(10, 17)

(20, 15)

Number of hours at fast-food restaurant

Exercises Within Reach ®   Solutions in English & Spanish and tutorial videos at CollegePrepAlgebra.com

41. *Part Time Jobs* Graph the inequality in Exercise 39 and find three ordered pairs that are solutions of the inequality.

42. *Inventory* Graph the inequality in Exercise 40 and find three ordered pairs that are solutions of the inequality.

Concept Summary: Graphing Linear Inequalities

What

You can use a graph to represent all the **solutions** of a **linear inequality in two variables**.

EXAMPLE

Sketch the **graph** of the linear inequality $y < x + 1$.

How

Use these steps to sketch the graph of such an inequality.

1. Replace the inequality sign by an equal sign.
2. Sketch the graph of the resulting equation.
   - Use a dashed line for $<$ and $>$.
   - Use a solid line for $\leq$ and $\geq$.
3. Shade the **half-plane** that contains points that satisfy the inequality.

Why

You can use linear inequalities to model and solve many real-life problems. The graphs of inequalities help you see all the possible solutions. For instance, one solution of the inequality $y < x + 1$ is $(1, 1)$.

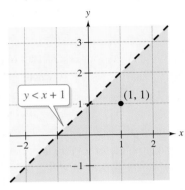

Exercises Within Reach ®

Worked-out solutions to odd-numbered exercises at CollegePrepAlgebra.com

Concept Summary Check

43. ***The Corresponding Linear Equation*** Why is a dashed line used in the graph of the inequality above?

44. ***Using the Graph*** Use the graph above to determine whether the point $(0, 2)$ represents a solution of the inequality $y < x + 1$. Explain.

45. ***Using the Graph*** Use the graph above to determine whether the point $(0, 1)$ represents a solution of the inequality $y < x + 1$. Explain.

46. ***Testing Half-Planes*** What point is often the most convenient test point to use when sketching the graph of a linear inequality?

Extra Practice

Writing a Linear Inequality In Exercises 47– 52, write the statement as a linear inequality. Then sketch the graph of the inequality.

47. y is more than six times x.

48. x is at most three times y.

49. The sum of x and y is at least 9.

50. The difference of x and y is less than 20.

51. y is no more than the sum of x and 3.

52. The sum of x and 7 is more than three times y.

Writing a Linear Inequality In Exercises 53 – 56, write an inequality that is represented by the graph.

53.

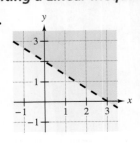

54.

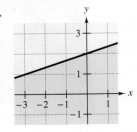

55.

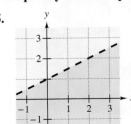

56.

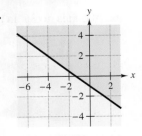

57. *Nutrition* A nutritionist recommends that the fat calories consumed per day should be at most 35% of the total calories consumed per day.

(a) Write a linear inequality that represents the different numbers of total calories and fat calories that are recommended for one day.

(b) Graph the inequality and find three ordered pairs that are solutions of the inequality.

58. *Money* A cash register must have at least $25 in change consisting of d dimes and q quarters.

(a) Write a linear inequality that represents the different numbers of dimes and quarters that can satisfy the requirement.

(b) Graph the inequality and find three ordered pairs that are solutions of the inequality.

59. *Manufacturing* Each table produced by a furniture company requires 1 hour in the assembly center. The matching chair requires $1\frac{1}{2}$ hours in the assembly center. A total of 12 hours per day is available in the assembly center.

(a) Write a linear inequality that represents the different numbers of hours that can be spent assembling tables and chairs.

(b) Graph the inequality and find three ordered pairs that are solutions of the inequality.

Explaining Concepts

60. *Think About It* Are there any points in the coordinate plane that are not solutions of either $y > x - 2$ or $y < x - 2$? Explain.

61. *Think About It* Write the inequality whose graph consists of all points above the x-axis.

62. *Think About It* Write an inequality whose graph has no points in the first quadrant.

63. *Writing* Does $2x < 2y$ have the same graph as $y > x$? Explain.

64. *Writing* Explain the difference between graphing the solution of the inequality $x \geq 1$ (a) on the real number line and (b) on a rectangular coordinate system.

Cumulative Review

In Exercises 65–70, find a ratio that compares the relative sizes of the quantities. (Use the same units of measurement for both quantities).

65. 36 feet to 8 feet

66. 8 dimes to 36 nickels

67. 4 hours to 45 minutes

68. 50 centimeters to 3 meters

69. 9 pounds to 8 ounces

70. 46 inches to 4 feet

In Exercises 71–76, write an equation of the line that passes through the points.

71. $(0, 0), (-3, 6)$

72. $(5, 8), (1, -4)$

73. $(5, 0), (0, -2)$

74. $(4, 3), (-4, 5)$

75. $(6, -1), (-3, -3)$

76. $\left(\frac{5}{6}, -2\right), \left(\frac{1}{6}, 4\right)$

4 Chapter Summary

What did you learn?	Explanation and Examples	Review Exercises
4.1 Plot points on a rectangular coordinate system *(p. 168)*.		1–8
Determine whether ordered pairs are solutions of equations *(p. 171)*.	To verify that an ordered pair (x, y) is a solution of an equation with variables x and y, use the following steps. **1.** Substitute the values of x and y into the equation. **2.** Simplify each side of the equation. **3.** If each side simplifies to the same number, then the ordered pair is a solution. If the two sides yield different numbers, then the ordered pair is not a solution.	9–16
Use the verbal problem-solving method to plot points on a rectangular coordinate system *(p. 172)*.	Construct a verbal model and assign labels to write an equation for a real-life problem. Use the equation to construct a table of values. Plot the ordered pairs represented by the table of values.	17, 18
4.2 Sketch graphs of equations using the point-plotting method *(p. 176)*.	**1.** If possible, rewrite the equation by isolating one of the variables. **2.** Make a table of values showing several solution points. **3.** Plot these points on a rectangular coordinate system. **4.** Connect the points with a smooth curve or line.	19–30
Find and use x- and y-intercepts as aids to sketching graphs *(p. 178)*.	To find the x-intercept(s), let $y = 0$ and solve the equation for x. To find the y-intercept(s), let $x = 0$ and solve the equation for y.	31–36
Use the verbal problem-solving method to write an equation and sketch its graph *(p. 180)*.	Construct a verbal model and assign labels to write an equation for a real-life problem. Sketch a graph of the equation.	37, 38
4.3 Identify the domain and range of a relation *(p. 184)*.	A relation is any set of ordered pairs. The set of first components in the ordered pairs is the domain of the relation. The set of second components is the range of the relation.	39–42
Determine whether relations are functions *(p. 185)*.	A function is a relation in which no two ordered pairs have the same first component and different second components.	43–48
Use function notation and evaluate functions *(p. 187)*.	The function $y = 2x - 6$ can be given the name "f" and written in function notation as $f(x) = 2x - 6$.	49–54
Identify the domain and range of a function *(p. 188)*.	The domain of a function may be explicitly described along with the function, or it may be implied by the context.	55–58

What did you learn?	Explanation and Examples	Review Exercises
4.4 Determine the slope of a line through two points *(p. 194)*.	The slope m of a nonvertical line that passes through the points (x_1, y_1) and (x_2, y_2) is $$m = \frac{y_2 - y_1}{x_2 - x_1} = \frac{\text{Change in } y}{\text{Change in } x} = \frac{\text{Rise}}{\text{Run}}$$ where $x_1 \neq x_2$. **1.** A line with positive slope ($m > 0$) *rises* from left to right. **2.** A line with negative slope ($m < 0$) *falls* from left to right. **3.** A line with zero slope ($m = 0$) is *horizontal*. **4.** A line with undefined slope is *vertical*.	59–70
Write linear equations in slope-intercept form and graph the equations *(p. 196)*.	The slope-intercept form of the equation of a line is $y = mx + b$. The graph of the equation $y = mx + b$ is a line whose slope is m and whose y-intercept is $(0, b)$.	71–76
Use slopes to determine whether lines are parallel, perpendicular, or neither *(p. 198)*.	Parallel Lines: Two distinct nonvertical lines are parallel if and only if they have the same slope. Perpendicular Lines: Two lines are perpendicular if and only if their slopes are negative reciprocals of each other. That is, $$m_1 = -\frac{1}{m_2}.$$	77–80
4.5 Write equations of lines using the point-slope form *(p. 202)*.	The point-slope form of the equation of a line with slope m that passes through the point (x_1, y_1) is $y - y_1 = m(x - x_1)$.	81–94
Write equations of horizontal and vertical lines *(p. 205)*.	A horizontal line has a slope of zero and an equation of the form $y = b$. A vertical line has an undefined slope and an equation of the form $x = a$.	95–98
Use linear models to solve application problems *(p. 206)*.	The slope of a line can often be interpreted as the *rate of change of y with respect to x*. Use linear models to make predictions about real-life situations.	99, 100
4.6 Determine whether an ordered pair is a solution of a linear inequality in two variables *(p. 210)*.	An ordered pair (x_1, y_1) is a solution of a linear inequality in x and y if the inequality is true when x_1 and y_1 are substituted for x and y, respectively.	101, 102
Sketch graphs of linear inequalities in two variables *(p. 211)*.	**1.** Replace the inequality sign by an equal sign and sketch the graph of the resulting equation. (Use a dashed line for $<$ or $>$ and a solid line for $\leq$ or $\geq$.) **2.** Test one point in each of the half-planes formed by the graph in Step 1. If the point satisfies the inequality, then shade the entire half-plane to denote that every point in the region satisfies the inequality.	103–112
Use linear inequalities to model and solve real-life problems *(p. 214)*.	Construct a verbal model and assign labels to write a linear inequality for a real-life problem. Graph the inequality to find the solutions for the problem.	113

Review Exercises

Worked-out solutions to odd-numbered exercises at CollegePrepAlgebra.com

4.1

Plotting Points In Exercises 1 and 2, plot the points on a rectangular coordinate system.

1. $(-1, 6), (4, -3), (-2, 2), (3, 5)$

2. $(0, -1), (-4, 2), (5, 1), (3, -4)$

Finding Coordinates In Exercises 3 and 4, determine the coordinates of the points.

3.

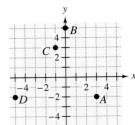

4.

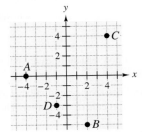

Determining the Quadrant In Exercises 5–8, determine the quadrant in which the point is located.

5. $(-5, 3)$

6. $(4, -6)$

7. $(-3, y), y < 0$

8. $(x, 5), x < 0$

Constructing a Table of Values In Exercises 9 and 10, complete the table of values. Then plot the solution points on a rectangular coordinate system.

9.

x	−1	0	1	2
y = 4x − 1				

10.

x	−1	0	1	2
y = $\frac{3}{2}$x + 5				

Solving for y In Exercises 11–14, solve the equation for y.

11. $3x + 4y = 12$

12. $2x + 3y = 6$

13. $9x - 3y = 12$

14. $-x - 3y = 9$

Verifying Solutions of an Equation In Exercises 15 and 16, determine whether each ordered pair is a solution of the equation.

15. $x - 3y = 4$
  (a) $(1, -1)$ (b) $(0, 0)$
  (c) $(2, 1)$ (d) $(5, -2)$

16. $y = \frac{1}{4}x + 2$
  (a) $(-4, 1)$ (b) $(-8, 0)$
  (c) $(12, 5)$ (d) $(0, 2)$

17. *Organizing Data* The data from a study measuring the relationship between the wattage x of a standard 120-volt light bulb and the energy rate y (in lumens) is shown in the table.

x	25	40	60	100	150	200
y	235	495	840	1675	2650	3675

  (a) Plot the data in the table.
  (b) Use the graph to describe the relationship between the wattage and the energy rate.

18. *Organizing Data* An employee earns $12 plus $0.25 for every x units produced per hour. Write an equation that relates the employee's total hourly wage to the number of units produced. Plot the hourly wages for producing 10, 20, 30, 40, and 50 units per hour.

4.2

Sketching the Graph of an Equation In Exercises 19–22, complete the table and use the results to sketch the graph of the equation.

19. $y = x - 5$

x	−2	−1	0	1	2
y					

20. $3x + y = -4$

x	−2	−1	0	1	2
y					

21. $y = x^2 - 1$

x	-2	-1	0	1	2
y					

22. $y = |x - 2|$

x	0	1	2	3	4
y					

Sketching the Graph of an Equation In Exercises 23–30, sketch the graph of the equation using the point-plotting method.

23. $y = 7$ **24.** $x = -2$

25. $y = 3x$ **26.** $y = -2x$

27. $y = 4 - \frac{1}{2}x$ **28.** $y = \frac{3}{2}x - 3$

29. $y - 2x - 4 = 0$ **30.** $3x + 2y + 6 = 0$

Finding the Intercepts of a Graph In Exercises 31–36, find the x- and y-intercepts and sketch the graph of the equation.

31. $y = 6x + 2$ **32.** $y = -3x + 5$

33. $y = \frac{2}{5}x - 2$ **34.** $y = \frac{1}{3}x + 1$

35. $2x - y = 4$ **36.** $4x + 2y = 8$

37. *Creating a Model* The cost of producing a DVD is $125, plus $3 per DVD. Let C represent the total cost and let x represent the number of DVDs. Write an equation that relates C and x, and sketch its graph.

38. *Creating a Model* Let y represent the distance traveled by a train that is moving at a constant speed of 80 miles per hour. Let t represent the number of hours the train has traveled. Write an equation that relates y and t, and sketch its graph.

4.3

Analyzing a Relation In Exercises 39–42, find the domain and range of the relation.

39. $\{(8, 3), (-2, 7), (5, 1), (3, 8)\}$

40. $\{(0, 1), (-1, 3), (4, 6), (-7, 5)\}$

41. $\{(2, -3), (-2, 3), (2, 4), (4, 0)\}$

42. $\{(1, 7), (-3, 4), (6, 4), (-2, 4)\}$

Testing Whether a Relation Is a Function In Exercises 43 and 44, determine whether the relation represents a function.

43. $\{(-1, 3), (3, 3), (0, 3), (7, 9), (10, 9)\}$

44. Input: a, b, c
Output: 4, 8, 9
$\{(a, 4), (b, 4), (b, 8), (c, 9)\}$

Using the Vertical Line Test In Exercises 45–48, use the Vertical Line Test to determine whether y is a function of x.

45. 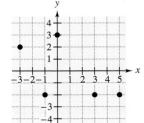   **46.**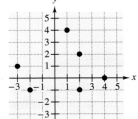

47. 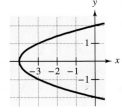   **48.**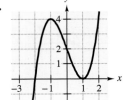

Evaluating a Function In Exercises 49–54, evaluate the function as indicated, and simplify.

49. $f(x) = \frac{3}{4}x$ (a) $f(-1)$ (b) $f(4)$
              (c) $f(10)$ (d) $f\left(-\frac{4}{3}\right)$

50. $f(x) = 2x - 7$ (a) $f(-1)$ (b) $f(3)$
              (c) $f\left(\frac{1}{2}\right)$ (d) $f(-4)$

51. $g(t) = -16t^2 + 64$ (a) $g(0)$ (b) $g\left(\frac{1}{4}\right)$
              (c) $g(1)$ (d) $g(2)$

52. $h(u) = u^3 + 2u^2 - 4$ (a) $h(0)$ (b) $h(3)$
              (c) $h(-1)$ (d) $h(-2)$

53. $f(x) = |2x + 3|$ (a) $f(0)$ (b) $f(5)$
              (c) $f(-4)$ (d) $f\left(-\frac{3}{2}\right)$

54. $f(x) = |x| - 4$ (a) $f(-1)$ (b) $f(1)$
              (c) $f(-4)$ (d) $f(2)$

Finding Domain and Range In Exercises 55–58, find the domain and range of the function.

55. $f: \{(1, 5), (2, 10), (3, 15), (4, -10), (5, -15)\}$

56. $g: \{(-3, 6), (-2, 4), (-1, 2), (0, 0), (1, -2)\}$

57. $f: \{(3, -1), (4, 6), (-2, -1), (0, -2), (7, 0)\}$

58. $g: \{(-8, 0), (3, -2), (10, 3), (-5, 1), (0, 0)\}$

4.4

Finding the Slope of a Line In Exercises 59 and 60, find the slope of the line that passes through the points.

59.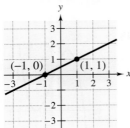

60.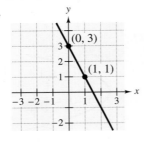

Finding the Slope of a Line In Exercises 61–68, plot the points and find the slope of the line that passes through the points. State whether the line rises, falls, is horizontal, or is vertical.

61. $(2, 1), (14, 6)$

62. $(1, 6), (4, 2)$

63. $(-1, 0), (6, -2)$

64. $(-1, -4), (-5, -10)$

65. $(4, 0), (4, 6)$

66. $(1, 3), (4, 3)$

67. $\left(0, \frac{5}{2}\right), \left(\frac{5}{6}, 0\right)$

68. $(0, 0), \left(3, \frac{4}{5}\right)$

69. *Loading Ramp* The bed of a truck is 4 feet above ground level. The end of the ramp used in loading the truck rests on the ground 6 feet behind the truck. Determine the slope of the ramp.

70. *Flight Path* An aircraft is on its approach to an airport. As it flies over a town, its altitude is 15,000 feet. The town is about 10 miles from the airport. Approximate the slope of the linear path followed by the aircraft during landing.

Using Slope-Intercept Form In Exercises 71–76, write the equation in slope-intercept form. Use the slope and the y-intercept to sketch the line.

71. $x + y = 6$

72. $x - y = -3$

73. $2x - y = -1$

74. $-4x + y = -2$

75. $3x + 6y = 12$

76. $7x + 21y = -14$

Parallel or Perpendicular? In Exercises 77–80, determine whether the lines L_1 and L_2 that pass through the pairs of points are parallel, perpendicular, or neither.

77. L_1: $(0, 3), (-2, 1)$
L_2: $(-8, -3), (4, 9)$

78. L_1: $(-3, -1), (2, 5)$
L_2: $(2, 11), (8, 6)$

79. L_1: $(3, 6), (-1, -5)$
L_2: $(-2, 3), (4, 7)$

80. L_1: $(-1, 2), (-1, 4)$
L_2: $(7, 3), (4, 7)$

4.5

Using Point-Slope Form In Exercises 81–86, use the point-slope form to write an equation of the line that passes through the point and has the specified slope. Write the equation in slope-intercept form.

81. $(4, -1), m = 2$

82. $(7, -3), m = -1$

83. $(-1, 3), m = -\frac{8}{3}$

84. $(4, -2), m = \frac{8}{5}$

85. $(3, 8), m$ is undefined.

86. $(-4, 6), m = 0$

An Equation of a Line Through Two Points In Exercises 87–92, write an equation of the line that passes through the points. Write the equation in general form.

87. $(-4, 0), (0, -2)$

88. $(-4, -2), (4, 6)$

89. $(0, 8), (6, 8)$

90. $(2, -6), (2, 5)$

91. $\left(0, \frac{4}{3}\right), (3, 0)$

92. $(-1.4, 0), (3.2, 9.2)$

Equations of Parallel and Perpendicular Lines In Exercises 93 and 94, write an equation of the line that passes through the point and is (a) parallel and (b) perpendicular to the given line.

93. $(-6, 3)$

$x - y = -2$

94. $\left(\frac{1}{5}, -\frac{4}{5}\right)$

$5x + y = 2$

An Equation of a Horizontal or Vertical Line In Exercises 95–98, write an equation of the line.

95. Horizontal line through $(-4, 5)$

96. Vertical line through $(-10, 4)$

97. 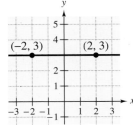   **98.**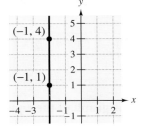

99. *Wages* A pharmaceutical salesperson receives a monthly salary of $5500 plus a commission of 7% of the total monthly sales. Write a linear model that relates the total monthly wages W to the sales S.

100. *Rental Demand* An apartment complex has 50 units. When the rent per unit is $425 per month, all 50 units are occupied. When the rent is $480 per month, the average number of occupied units drops to 47. Assume that the relationship between the monthly rent p and the demand x is linear.

    (a) Write a linear model that relates the monthly rent p to the demand x.

    (b) Use the model to estimate the number of units occupied when the rent is $475.

4.6

Verifying Solutions of a Linear Inequality In Exercises 101 and 102, determine whether each ordered pair is a solution of the inequality.

101. $x - y > 4$

    (a) $(-1, -5)$ (b) $(0, 0)$

    (c) $(3, -2)$ (d) $(8, 1)$

102. $-4y + 5x > 3$

    (a) $(1, 2)$ (b) $(-3, 6)$

    (c) $(-1, -3)$ (d) $(4, 4)$

Matching In Exercises 103–106, match the inequality with its graph.

103. $x \geq -1$ **104.** $y \leq x + 1$

105. $y > -\frac{1}{3}x$ **106.** $y < 2$

(a) 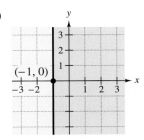   (b)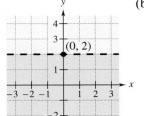

(c) 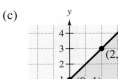   (d)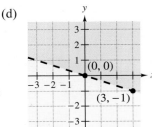

Sketching the Graph of a Linear Inequality In Exercises 107–112, sketch the graph of the linear inequality.

107. $x - 2 \geq 0$ **108.** $y + 3 < 0$

109. $2x + y < 1$ **110.** $3x - 4y > 2$

111. $x \leq 4y - 2$ **112.** $x \geq 3 - 2y$

113. *Manufacturing* Each DVD player produced by an electronics manufacturer requires 2 hours in the assembly center. Each camcorder produced by the same manufacturer requires 3 hours in the assembly center. A total of 120 hours per week is available in the assembly center.

    (a) Write a linear inequality that represents the different numbers of hours that can be spent assembling DVD players and camcorders.

    (b) Graph the inequality and find three ordered pairs that are solutions of the inequality.

Chapter Test

Solutions in English & Spanish and tutorial videos at CollegePrepAlgebra.com

Take this test as you would take a test in class. After you are done, check your work against the answers in the back of the book.

1. Plot the points $(-1, 2)$, $(1, 4)$, and $(2, -1)$ on a rectangular coordinate system. Connect the points with line segments to form a right triangle.

2. Determine whether each ordered pair is a solution of $y + 2x = 2$.
   (a) $(0, -2)$ (b) $(0, 2)$ (c) $(-4, 10)$ (d) $(-2, -2)$

3. What is the y-coordinate of any point on the x-axis?

4. Find the x- and y-intercepts of the graph of $8x - 2y = -16$.

5. Complete the table and use the results to sketch the graph of the equation.

   $3x + y = -4$

x	-2	-1	0	1	2
y					

In Exercises 6–8, sketch the graph of the equation.

6. $x + 2y = 6$ 7. $y = |x + 2|$ 8. $y = (x - 3)^2$

9. Does the table at the left represent y as a function of x? Explain.

10. Does the graph at the left represent y as a function of x? Explain.

11. Evaluate $f(x) = x^3 - 2x^2$ as indicated, and simplify.
    (a) $f(0)$ (b) $f(2)$ (c) $f(-2)$ (d) $f\left(\frac{1}{2}\right)$

12. Find the slope of the line that passes through the points $(-5, 0)$ and $\left(2, \frac{3}{2}\right)$. Then write an equation of the line in slope-intercept form.

13. A line with slope $m = -2$ passes through the point $(-3, 4)$. Plot the point and use the slope to find two additional points on the line. (There are many correct answers.)

14. Find the slope of a line *perpendicular* to the line $7x - 8y + 5 = 0$.

15. Write the equation in slope-intercept form of the line that passes through the point $(0, 6)$ with slope $m = -\frac{3}{8}$.

16. Write an equation of the vertical line that passes through the point $(3, -7)$.

17. Determine whether each ordered pair is a solution of $3x + 5y \leq 16$.
    (a) $(2, 2)$ (b) $(6, -1)$ (c) $(-2, 4)$ (d) $(7, -1)$

In Exercises 18–20, sketch the graph of the linear inequality.

18. $y \geq -2$ 19. $y < 5 - 2x$ 20. $-y + 4x > 3$

21. The unit sales y of a product are modeled by $y = 230x + 5000$, where x represents the time in years. Interpret the meaning of the slope in this model.

Input x	Output y	(x, y)
0	4	$(0, 4)$
1	5	$(1, 5)$
2	8	$(2, 8)$
1	-3	$(1, -3)$
0	-1	$(0, -1)$

Table for 9

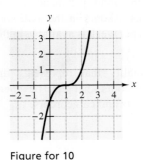

Figure for 10

5
Exponents and Polynomials

5.1 Integer Exponents and Scientific Notation

5.2 Adding and Subtracting Polynomials

5.3 Multiplying Polynomials: Special Products

5.4 Dividing Polynomials and Synthetic Division

MASTERY IS WITHIN REACH!

"I used to waste time reviewing for a test by just reading through my notes over and over again. Then I learned how to make note cards and keep working through them, saving the important ones for my mental cheat sheets. Now a couple of my friends and I go through our notes and make mental cheat sheets together. We are all doing well in our math class."

Stephanie

See page 243 for suggestions about managing test anxiety.

5.1 Integer Exponents and Scientific Notation

▶ Use the rules of exponents to simplify expressions.
▶ Rewrite exponential expressions involving negative and zero exponents.
▶ Write very large and very small numbers in scientific notation.

Rules of Exponents

Rules of Exponents

Let m and n be positive integers, and let a and b represent real numbers, variables, or algebraic expressions.

Rule	*Example*
1. Product: $a^m \cdot a^n = a^{m+n}$	$x^5(x^4) = x^{5+4} = x^9$
2. Product-to-Power: $(ab)^m = a^m \cdot b^m$	$(2x)^3 = 2^3(x^3) = 8x^3$
3. Power-to-Power: $(a^m)^n = a^{mn}$	$(x^2)^3 = x^{2 \cdot 3} = x^6$
4. Quotient: $\dfrac{a^m}{a^n} = a^{m-n}, m > n, a \neq 0$	$\dfrac{x^5}{x^3} = x^{5-3} = x^2, x \neq 0$
5. Quotient-to-Power: $\left(\dfrac{a}{b}\right)^m = \dfrac{a^m}{b^m}, b \neq 0$	$\left(\dfrac{x}{4}\right)^2 = \dfrac{x^2}{4^2} = \dfrac{x^2}{16}$

EXAMPLE 1 Using Rules of Exponents

a. $(x^2y^4)(3x) = 3(x^2 \cdot x)(y^4) = 3(x^{2+1})(y^4) = 3x^3y^4$

b. $-2(y^2)^3 = (-2)(y^{2 \cdot 3}) = -2y^6$

c. $(-2y^2)^3 = (-2)^3(y^2)^3 = -8(y^{2 \cdot 3}) = -8y^6$

d. $(3x^2)(-5x)^3 = 3(-5)^3(x^2 \cdot x^3) = 3(-125)(x^{2+3}) = -375x^5$

e. $\dfrac{14a^5b^3}{7a^2b^2} = 2(a^{5-2})(b^{3-2}) = 2a^3b$

f. $\left(\dfrac{x^2}{2y}\right)^3 = \dfrac{(x^2)^3}{(2y)^3} = \dfrac{x^{2 \cdot 3}}{2^3y^3} = \dfrac{x^6}{8y^3}$

g. $\dfrac{x^ny^{3n}}{x^2y^4} = x^{n-2}y^{3n-4}$

Exercises Within Reach ®

Solutions in English & Spanish and tutorial videos at CollegePrepAlgebra.com

Using Rules of Exponents In Exercises 1–16, use the rules of exponents to simplify the expression.

1. $(u^3v)(2v^2)$
2. $(x^5y^3)(2y^3)$
3. $-3(x^3)^2$
4. $-5(y^4)^3$
5. $(-5z^2)^3$
6. $(-5z^4)^2$
7. $(2u)^4(4u)$
8. $(3y)^3(2y^2)$
9. $\dfrac{27m^5n^6}{9mn^3}$
10. $\dfrac{-18m^3n^6}{-6mn^3}$
11. $-\left(\dfrac{2a}{3y}\right)^2$
12. $\left(\dfrac{5u}{3v}\right)^3$
13. $\dfrac{x^ny^{2n}}{x^3y^2}$
14. $\dfrac{x^{2n}y^n}{x^ny}$
15. $\dfrac{(-2x^2y)^3}{9x^2y^2}$
16. $\dfrac{(-2xy^3)^2}{6y^2}$

Integer Exponents

Study Tip

Notice that by definition, $a^0 = 1$ for all real *nonzero* values of *a*. Zero cannot have a zero exponent, because the expression 0^0 is undefined.

Definitions of Zero Exponents and Negative Exponents

Let a and b be real numbers such that $a \neq 0$ and $b \neq 0$, and let m be an integer.

1. $a^0 = 1$ **2.** $a^{-m} = \dfrac{1}{a^m}$ **3.** $\left(\dfrac{a}{b}\right)^{-m} = \left(\dfrac{b}{a}\right)^m$

EXAMPLE 2   **Using Rules of Exponents**

a. $3^0 = 1$ Definition of zero exponents

b. $3^{-2} = \dfrac{1}{3^2} = \dfrac{1}{9}$ Definition of negative exponents

c. $\left(\dfrac{3}{4}\right)^{-1} = \left(\dfrac{4}{3}\right)^1 = \dfrac{4}{3}$ Definition of negative exponents

---

Summary of Rules of Exponents

Let m and n be integers, and let a and b represent real numbers, variables, or algebraic expressions. (All denominators and bases are nonzero.)

Product and Quotient Rules	*Example*
1. $a^m \cdot a^n = a^{m+n}$	$x^4(x^3) = x^{4+3} = x^7$
2. $\dfrac{a^m}{a^n} = a^{m-n}$	$\dfrac{x^3}{x} = x^{3-1} = x^2$

Power Rules

3. $(ab)^m = a^m \cdot b^m$	$(3x)^2 = 3^2(x^2) = 9x^2$
4. $(a^m)^n = a^{mn}$	$(x^3)^3 = x^{3 \cdot 3} = x^9$
5. $\left(\dfrac{a}{b}\right)^m = \dfrac{a^m}{b^m}$	$\left(\dfrac{x}{3}\right)^2 = \dfrac{x^2}{3^2} = \dfrac{x^2}{9}$

Zero and Negative Exponent Rules

6. $a^0 = 1$	$(x^2 + 1)^0 = 1$
7. $a^{-m} = \dfrac{1}{a^m}$	$x^{-2} = \dfrac{1}{x^2}$
8. $\left(\dfrac{a}{b}\right)^{-m} = \left(\dfrac{b}{a}\right)^m$	$\left(\dfrac{x}{3}\right)^{-2} = \left(\dfrac{3}{x}\right)^2 = \dfrac{3^2}{x^2} = \dfrac{9}{x^2}$

Exercises Within Reach ®

Solutions in English & Spanish and tutorial videos at CollegePrepAlgebra.com

Using Rules of Exponents **In Exercises 17–24, use the rules of exponents to evaluate the expression.**

17. $(-3)^0$ **18.** 25^0

19. 5^{-2} **20.** 2^{-4}

21. $\left(\dfrac{2}{3}\right)^{-1}$ **22.** $\left(\dfrac{4}{5}\right)^{-3}$

23. -10^{-3} **24.** -20^{-2}

Study Tip

As you become accustomed to working with negative exponents, you will probably not write as many steps as shown in Example 4. For instance, to rewrite a fraction involving exponents, you might use the following simplified rule. *To move a factor from the numerator to the denominator or vice versa, change the sign of its exponent.* You can apply this rule to the expression in Example 4(a) by "moving" the factor x^{-2} to the numerator and changing the exponent to 2. That is,

$$\frac{3}{x^{-2}} = 3x^2.$$

Remember, you can move only *factors* in this manner, not terms.

EXAMPLE 3   Using Rules of Exponents

a. $2x^{-1} = 2(x^{-1}) = 2\left(\frac{1}{x}\right) = \frac{2}{x}$ Use negative exponent rule and simplify.

b. $(2x)^{-1} = \frac{1}{(2x)^1} = \frac{1}{2x}$ Use negative exponent rule and simplify.

EXAMPLE 4   Using Rules of Exponents

Rewrite each expression using only positive exponents. (Assume that $x \neq 0$.)

a. $\dfrac{3}{x^{-2}} = \dfrac{3}{\left(\dfrac{1}{x^2}\right)}$ Use negative exponent rule.

$= 3\left(\dfrac{x^2}{1}\right) = 3x^2$ Invert divisor and multiply.

b. $\dfrac{1}{(3x)^{-2}} = \dfrac{1}{\left[\dfrac{1}{(3x)^2}\right]}$ Use negative exponent rule.

$= \dfrac{1}{\left(\dfrac{1}{9x^2}\right)}$ Use product-to-power rule and simplify.

$= (1)\left(\dfrac{9x^2}{1}\right) = 9x^2$ Invert divisor and multiply.

Exercises Within Reach ®

Solutions in English & Spanish and tutorial videos at CollegePrepAlgebra.com

Using Rules of Exponents In Exercises 25–42, rewrite the expression using only positive exponents, and simplify. (Assume that any variables in the expression are nonzero.)

25. $7x^{-4}$

26. $3y^{-3}$

27. $(8x)^{-1}$

28. $(11x)^{-1}$

29. $(4x)^{-3}$

30. $(5u)^{-2}$

31. $y^4 \cdot y^{-2}$

32. $z^5 \cdot z^{-3}$

33. $\dfrac{1}{x^{-6}}$

34. $\dfrac{4}{y^{-1}}$

35. $\dfrac{(4t)^0}{t^{-2}}$

36. $\dfrac{(x^2)^0}{x^{-3}}$

37. $\dfrac{1}{(5y)^{-2}}$

38. $\dfrac{3}{(3a)^{-3}}$

39. $\dfrac{(5u)^{-4}}{(5u)^0}$

40. $\dfrac{(9n)^{-2}}{(9n^2)^0}$

41. $\left(\dfrac{x}{10}\right)^{-1}$

42. $\left(\dfrac{4}{z}\right)^{-2}$

EXAMPLE 5   **Using Rules of Exponents**

Rewrite each expression using only positive exponents. (Assume that $x \neq 0$ and $y \neq 0$.)

a. $(-5x^{-3})^2 = (-5)^2(x^{-3})^2$ Product-to-power rule

$\qquad\qquad = 25x^{-6}$ Power-to-power rule

$\qquad\qquad = \dfrac{25}{x^6}$ Negative exponent rule

b. $-\left(\dfrac{7x}{y^2}\right)^{-2} = -\left(\dfrac{y^2}{7x}\right)^2$ Negative exponent rule

$\qquad\qquad = -\dfrac{(y^2)^2}{(7x)^2}$ Quotient-to-power rule

$\qquad\qquad = -\dfrac{y^4}{49x^2}$ Power-to-power and product-to-power rules

c. $\dfrac{12x^2y^{-4}}{6x^{-1}y^2} = 2(x^{2-(-1)})(y^{-4-2})$ Quotient rule

$\qquad\qquad = 2x^3y^{-6}$ Simplify.

$\qquad\qquad = \dfrac{2x^3}{y^6}$ Negative exponent rule

EXAMPLE 6   **Using Rules of Exponents**

Rewrite each expression using only positive exponents. (Assume that $x \neq 0$ and $y \neq 0$.)

a. $\left(\dfrac{8x^{-1}y^4}{4x^3y^2}\right)^{-3} = \left(\dfrac{2y^2}{x^4}\right)^{-3}$ Simplify.

$\qquad\qquad = \left(\dfrac{x^4}{2y^2}\right)^3$ Negative exponent rule

$\qquad\qquad = \dfrac{x^{12}}{2^3y^6} = \dfrac{x^{12}}{8y^6}$ Quotient-to-power rule

b. $\dfrac{3xy^0}{x^2(5y)^0} = \dfrac{3x(1)}{x^2(1)} = \dfrac{3}{x}$ Zero exponent rule

Exercises Within Reach®

Using Rules of Exponents In Exercises 43–52, rewrite the expression using only positive exponents, and simplify. (Assume that any variables in the expression are nonzero.)

43. $(2x^2)^{-2}$

44. $(4a^{-2})^{-3}$

45. $-\left(\dfrac{5x}{y^3}\right)^{-3}$

46. $-\left(\dfrac{4n^2}{m^4}\right)^{-2}$

47. $\dfrac{6x^3y^{-3}}{12x^{-2}y}$

48. $\dfrac{2y^{-1}z^{-3}}{4yz^{-3}}$

49. $\left(\dfrac{3u^2v^{-1}}{3^3u^{-1}v^3}\right)^{-2}$

50. $\left(\dfrac{5^2x^3y^{-3}}{125xy}\right)^{-1}$

51. $\dfrac{5x^0y}{(6x)^0y^3}$

52. $\dfrac{(4x)^0y^4}{(3y)^2x^0}$

Scientific Notation

Exponents provide an efficient way of writing and computing with very large and very small numbers. For instance, a drop of water contains more than 33 billion billion molecules—that is, 33 followed by 18 zeros. It is convenient to write such numbers in **scientific notation**. This notation has the form $c \times 10^n$, where $1 \leq c < 10$ and n is an integer. So, the number of molecules in a drop of water can be written in scientific notation as follows.

$$33{,}000{,}000{,}000{,}000{,}000{,}000 = 3.3 \times 10^{19}$$

19 places

EXAMPLE 7 **Writing in Scientific Notation**

a. $0.0000684 = 6.84 \times 10^{-5}$ Small number $\Rightarrow$ negative exponent

Five places

b. $937{,}200{,}000.0 = 9.372 \times 10^8$ Large number $\Rightarrow$ positive exponent

Eight places

EXAMPLE 8 **Writing in Decimal Notation**

a. $2.486 \times 10^2 = 248.6$ Positive exponent $\Rightarrow$ large number

Two places

b. $1.81 \times 10^{-6} = 0.00000181$ Negative exponent $\Rightarrow$ small number

Six places

Exercises Within Reach ®

Solutions in English & Spanish and tutorial videos at CollegePrepAlgebra.com

Writing in Scientific Notation In Exercises 53−58, write the number in scientific notation.

53. 0.00031

54. 0.0000000000692

55. 3,600,000

56. 841,000,000,000

57. *Light Year:* 9,460,800,000,000 kilometers

58. *Thickness of a Soap Bubble:* 0.0000001 meter

Writing in Decimal Notation In Exercises 59−64, write the number in decimal notation.

59. 7.2×10^8

60. 7.413×10^{11}

61. 1.359×10^{-7}

62. 8.6×10^{-9}

63. *Interior Temperature of the Sun:* 1.5×10^7 degrees Celsius

64. *Width of an Air Molecule:* 9.0×10^{-9} meter

EXAMPLE 9   Using Scientific Notation

$$\frac{(2{,}400{,}000{,}000)(0.0000045)}{(0.00003)(1500)} = \frac{(2.4 \times 10^9)(4.5 \times 10^{-6})}{(3.0 \times 10^{-5})(1.5 \times 10^3)}$$

$$= \frac{(2.4)(4.5)(10^3)}{(4.5)(10^{-2})}$$

$$= (2.4)(10^5) = 2.4 \times 10^5$$

EXAMPLE 10   Using Scientific Notation with a Calculator

Use a calculator to evaluate each expression.

a. $65{,}000 \times 3{,}400{,}000{,}000$ **b.** $0.000000348 \div 870$

SOLUTION

a. 6.5 (EXP) 4 (×) 3.4 (EXP) 9 (=)   Scientific calculator

6.5 (EE) 4 (×) 3.4 (EE) 9 (ENTER)   Graphing calculator

The calculator display should read (2.21E 14), which implies that

$$(6.5 \times 10^4)(3.4 \times 10^9) = 2.21 \times 10^{14} = 221{,}000{,}000{,}000{,}000.$$

b. 3.48 (EXP) 7 (±) (÷) 8.7 (EXP) 2 (=)   Scientific calculator

3.48 (EE) ((−)) 7 (÷) 8.7 (EE) 2 (ENTER)   Graphing calculator

The calculator display should read (4E −10), which implies that

$$\frac{3.48 \times 10^{-7}}{8.7 \times 10^2} = 4.0 \times 10^{-10} = 0.0000000004.$$

Exercises Within Reach ®

Using Scientific Notation In Exercises 65−68, **evaluate** the expression by first writing each number in scientific notation.

65. $(4{,}500{,}000)(2{,}000{,}000{,}000)$

66. $(62{,}000{,}000)(0.0002)$

67. $\dfrac{64{,}000{,}000}{0.00004}$

68. $\dfrac{72{,}000{,}000{,}000}{0.00012}$

Using Scientific Notation with a Calculator In Exercises 69−72, use a calculator to **evaluate** the expression. Write the answer in scientific notation, $c \times 10^n$, with c rounded to two decimal places.

69. $7900 \times 5{,}700{,}000{,}000$

70. $0.0000000452 \div 767$

71. $\dfrac{(0.0000565)(2{,}850{,}000{,}000{,}000)}{0.00465}$

72. $\dfrac{(3{,}450{,}000{,}000)(0.000125)}{(52{,}000{,}000)(0.000003)}$

73. ***Masses of Earth and the Sun*** The masses of Earth and the Sun are approximately 5.98×10^{24} kilograms and 1.99×10^{30} kilograms, respectively. The mass of the Sun is approximately how many times that of Earth?

74. ***Light Year*** One light year (the distance light can travel in 1 year) is approximately 9.46×10^{15} meters. Approximate the time to the nearest minute for light to travel from the Sun to Earth if that distance is approximately 1.50×10^{11} meters.

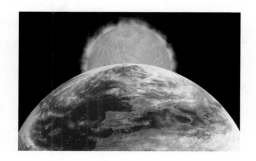

Concept Summary: *Using Scientific Notation*

What

You can use exponents and **scientific notation** to write very large or very small numbers.

EXAMPLE

Write each number in scientific notation.

a. 240,000,000,000

b. 0.000000012

How

A number written in scientific notation has the form $c \times 10^n$, where $1 \leq c < 10$ and n is an integer.

A positive exponent indicates a large number.

a. $240{,}000{,}000{,}000 = 2.4 \times 10^{11}$

        11 places

A negative exponent indicates a small number.

b. $0.000000012 = 1.2 \times 10^{-8}$

        8 places

Why

When multiplying or dividing very large or very small numbers, you can use scientific notation and the rules of exponents to find the products or quotients more efficiently.

Exercises Within Reach ®

Worked-out solutions to odd-numbered exercises at CollegePrepAlgebra.com

Concept Summary Check

75. *Scientific Notation* In the solution above, 240,000,000,000 is rewritten in the form $c \times 10^n$. What is the value of c?

76. *Scientific Notation* In the solution above, 0.000000012 is rewritten in the form $c \times 10^n$. What is the value of n?

77. *Writing* Explain how to write a small number in scientific notation.

78. *Writing* Explain how to write a large number in scientific notation.

Extra Practice

Using Rules of Exponents In Exercises 79–82, use the rules of exponents to simplify the expression.

79. $\left[\dfrac{(-5u^3v)^2}{10u^2v}\right]^2$

80. $\left[\dfrac{-5(u^3v)^2}{10u^2v}\right]^2$

81. $\dfrac{x^{2n+4}y^{4n}}{x^5y^{2n+1}}$

82. $\dfrac{x^{6n}y^{n-7}}{x^{4n+2}y^5}$

Using Rules of Exponents In Exercises 83–90, rewrite the expression using only positive exponents, and simplify. (Assume that any variables in the expression are nonzero.)

83. $(2x^3y^{-1})^{-3}(4xy^{-6})$

84. $(ab)^{-2}(a^2b^2)^{-1}$

85. $u^4(6u^{-3}v^0)(7v)^0$

86. $x^5(3x^0y^4)(7y)^0$

87. $[(x^{-4}y^{-6})^{-1}]^2$

88. $[(2x^{-3}y^{-2})^2]^{-2}$

89. $\dfrac{(2a^{-2}b^4)^3b}{(10a^3b)^2}$

90. $\dfrac{(5x^2y^{-5})^{-1}}{2x^{-5}y^4}$

Using Scientific Notation In Exercises 91–96, evaluate the expression without using a calculator.

91. $(2 \times 10^9)(3.4 \times 10^{-4})$

92. $(6.5 \times 10^6)(2 \times 10^4)$

93. $(5 \times 10^4)^2$

94. $(4 \times 10^6)^3$

95. $\dfrac{3.6 \times 10^{12}}{6 \times 10^5}$

96. $\dfrac{2.5 \times 10^{-3}}{5 \times 10^2}$

97. *Stars* A study by Australian astronomers estimated the number of stars within range of modern telescopes to be 70,000,000,000,000,000,000,000. Write this number in scientific notation. (*Source:* The Australian National University)

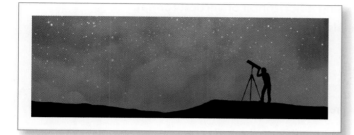

98. *Fishery Products* In 2010, the total supply of edible fishery products in the United States weighed 1.2389×10^{10} pounds. Write this weight in decimal notation. (*Source:* NOAA Fisheries)

99. *Federal Debt* In 2010, the resident population of the United States was about 309 million people, and it would have cost each resident about $43,783 to pay off the federal debt. Use these two numbers to approximate the federal debt in 2010. (*Source:* U.S. Census Bureau and U.S. Office of Management and Budget)

100. *Metal Expansion* When the temperature of a 200-foot-long iron steam pipe is increased by 75°C, the length of the pipe will increase by $75(200)(1.1 \times 10^{-5})$ foot. Find this amount and write the answer in decimal notation.

Explaining Concepts

101. *Think About It* Discuss whether you feel that using scientific notation to multiply or divide very large or very small numbers makes the process *easier* or *more difficult*. Support your position with an example.

102. *Think About It* You multiply an expression by a^5. The product is a^{12}. What was the original expression? Explain how you found your answer.

True or False? **In Exercises 103 and 104, determine whether the statement is true or false. Justify your answer.**

103. The value of $\dfrac{1}{3^{-3}}$ is less than 1.

104. The expression 0.142×10^{10} is in scientific notation.

Using Rules of Exponents **In Exercises 105–108, use the rules of exponents to explain why the statement is *false.***

105. $a^m \cdot b^n = ab^{m+n}$ ✗

106. $(ab)^m = a^m + b^m$ ✗

107. $(a^m)^n = a^{m+n}$ ✗

108. $\dfrac{a^m}{a^n} = a^m - a^n$ ✗

Cumulative Review

In Exercises 109–112, simplify the expression by combining like terms.

109. $3x + 4x - x$

110. $y - 3x + 4y - 2$

111. $a^2 + 2ab - b^2 + ab + 4b^2$

112. $x^2 + 5x^2y - 3x^2y + 4x^2$

In Exercises 113–118, sketch the graph of the linear inequalitiy.

113. $y < 7$

114. $x > -2$

115. $y \geq 8x$

116. $y > 3 + x$

117. $y \leq -2x + 4$

118. $x + 3y < -1$

5.2 Adding and Subtracting Polynomials

▶ Identify the degrees and leading coefficients of polynomials.

▶ Add polynomials using a horizontal or vertical format.

▶ Subtract polynomials using a horizontal or vertical format.

Basic Definitions

Study Tip

A polynomial with only one term is called a **monomial**. A polynomial with two unlike terms is called a **binomial**, and a polynomial with three unlike terms is called a **trinomial**. For example, $3x^2$ is a monomial, $-3x + 1$ is a binomial, and $4x^3 - 5x + 6$ is a trinomial.

Let $a_n, a_{n-1}, \ldots, a_2, a_1, a_0$ be real numbers and let n be a nonnegative integer. A **polynomial in x** is an expression of the form

$$a_n x^n + a_{n-1} x^{n-1} + \cdots + a_2 x^2 + a_1 x + a_0$$

where $a_n \neq 0$. The polynomial is of **degree n**, and the number a_n is called the **leading coefficient**. The number a_0 is called the **constant term**.

EXAMPLE 1   Identifying Polynomials

a. $3x^4 - 8x + x^{-1}$ is *not* a polynomial because the third term, x^{-1}, has a negative exponent.

b. $x^2 - 3x + 1$ *is* a polynomial of degree 2 with integer coefficients.

c. $x^3 + 3x^{1/2}$ is *not* a polynomial because the exponent in the second term, $3x^{1/2}$, is not an integer.

d. $-\dfrac{1}{3}x + \dfrac{x^3}{4}$ *is* a polynomial of degree 3 with rational coefficients.

EXAMPLE 2   Determining Degrees and Leading Coefficients

	Polynomial	Standard Form	Degree	Leading Coefficient
a.	$4x^2 - 5x^7 - 2 + 3x$	$-5x^7 + 4x^2 + 3x - 2$	7	-5
b.	$4 - 9x^2$	$-9x^2 + 4$	2	-9
c.	8	8	0	8
d.	$2 + x^3 - 5x^2$	$x^3 - 5x^2 + 2$	3	1

In part (c), note that a polynomial with only a constant term has a degree of zero.

Exercises Within Reach ®

Solutions in English & Spanish and tutorial videos at CollegePrepAlgebra.com

Identifying a Polynomial In Exercises 1−6, determine **whether the expression is a polynomial. If it is not, explain why.**

1. $9 - z$

2. $t^2 - 4$

3. $p^{3/4} - 16$

4. $9 - z^{1/2}$

5. $6x^{-1}$

6. $4 - 9x^4$

Determining the Degree and Leading Coefficient In Exercises 7−10, write **the polynomial in standard form. Then identify its degree and leading coefficient.**

7. $7x - 5x^2 + 10$

8. $1 - 4z + 12z^3$

9. $6m - 3m^5 - m^2 + 12$

10. $5x^3 - 3x^2 + 10$

Adding Polynomials

EXAMPLE 3　**Adding Polynomials Horizontally**

a.　$(2x^2 + 4x - 1) + (x^2 - 3)$　　　　　　　　　Original polynomials

　　$= (2x^2 + x^2) + (4x) + (-1 - 3)$　　　　　Group like terms.

　　$= 3x^2 + 4x - 4$　　　　　　　　　　　　Combine like terms.

b.　$(x^3 + 2x^2 + 4) + (3x^2 - x + 5)$　　　　　Original polynomials

　　$= (x^3) + (2x^2 + 3x^2) + (-x) + (4 + 5)$　Group like terms.

　　$= x^3 + 5x^2 - x + 9$　　　　　　　　　Combine like terms.

c.　$(2x^2 - x + 3) + (4x^2 - 7x + 2) + (-x^2 + x - 2)$　　Original polynomials

　　$= (2x^2 + 4x^2 - x^2) + (-x - 7x + x) + (3 + 2 - 2)$　Group like terms.

　　$= 5x^2 - 7x + 3$　　　　　　　　　　　　　　　　Combine like terms.

Study Tip

When you use a vertical format to add polynomials, be sure that you line up the like terms.

EXAMPLE 4　**Adding Polynomials Vertically**

a.　　$-4x^3 - 2x^2 + \ x - 5$
　　　$\underline{\ \ 2x^3 \qquad\quad + 3x + 4}$
　　　$-2x^3 - 2x^2 + 4x - 1$

b.　　$5x^3 + 2x^2 - \ x + 7$
　　　$\qquad\quad 3x^2 - 4x + 7$
　　　$\underline{-x^3 + 4x^2 - 2x - 8}$
　　　$4x^3 + 9x^2 - 7x + 6$

Exercises Within Reach®

Solutions in English & Spanish and tutorial videos at CollegePrepAlgebra.com

Adding Polynomials　In Exercises 11−16, use a horizontal format to find the sum.

11.　$(4w + 5) + (16w - 9)$

12.　$(-2x + 4) + (x - 6)$

13.　$(3z^2 - z + 2) + (z^2 - 4)$

14.　$(6x^4 + 8x) + (4x - 6)$

15.　$b^2 + (b^3 - 2b^2 + 3) + (b^3 - 3)$

16.　$(3x^2 - x) + 5x^3 + (-4x^3 + x^2 - 8)$

Adding Polynomials　In Exercises 17−22, use a vertical format to find the sum.

17.　$2x + 5$
　　　$\underline{3x + 8}$

18.　$11x + 5$
　　　$\underline{7x - 6}$

19.　$-2x + 10$
　　　$\underline{\ \ \ x - 38}$

20.　$4x^2 + 13$
　　　$\underline{3x^2 - 11}$

21.　$(x^2 - 2x + 2) + (x^2 + 4x) + 2x^2$

22.　$(5y + 10) + (y^2 - 3y - 2) + (2y^2 + 4y - 3)$

Geometry　In Exercises 23 and 24, find an expression for the perimeter of the figure.

23.

24.

Application   EXAMPLE 5   Modeling School Enrollments

The projected enrollments (in millions) at public colleges and private colleges for the years 2015 through 2020 can be modeled by the following, where t represents the year, with $t = 15$ corresponding to 2015. (*Source:* U.S. National Center for Education Statistics)

$$P = 0.20t + 12.7, \quad 15 \le t \le 20 \qquad \text{Public college enrollment}$$

$$R = -0.004t^2 + 0.21t + 3.7, \quad 15 \le t \le 20 \qquad \text{Private college enrollment}$$

a. Add the polynomials to find a model for the projected total enrollment at public *and* private colleges.

b. Make a bar graph for all three models.

SOLUTION

a. $T = 0.20t + 12.7 + (-0.004t^2 + 0.21t + 3.7)$

$\quad = -0.t^2 + 0.41t + 16.4$ Total school enrollment

b. A spreadsheet is useful for making a bar graph.

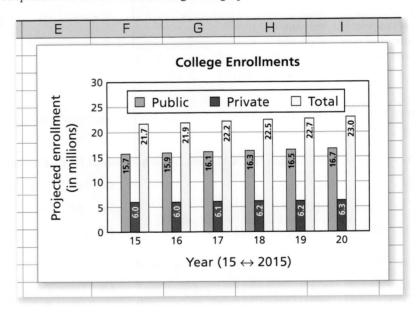

Exercises Within Reach®   Solutions in English & Spanish and tutorial videos at CollegePrepAlgebra.com

25. *The Civilian Labor Force* The numbers of men M and women W in the civilian labor force that were 65 years old or older in the years 2006 through 2010 can be modeled by the following, where t represents the year, with $t = 6$ corresponding to 2006. (*Source:* U.S. Bureau of Labor Statistics)

$M = 0.16t + 2.1, \quad 6 \le t \le 10$ Men
$W = -0.026t^2 + 0.57t - 0.1, \quad 6 \le t \le 10$ Women

  (a) Add the polynomials to find a model for the total number of people.

  (b) Make a bar graph for all three models.

26. *The Civilian Labor Force* The numbers of men M and women W in the civilian labor force that were 35 to 44 years old in the years 2006 through 2010 can be modeled by the following, where t represents the year, with $t = 6$ corresponding to 2006. (*Source:* U.S. Bureau of Labor Statistics)

$M = -0.051t^2 + 0.47t + 18.4, \quad 6 \le t \le 10$ Men
$W = -0.054t^2 + 0.57t + 14.9, \quad 6 \le t \le 10$ Women

  (a) Add the polynomials to find a model for the total number of people.

  (b) Make a bar graph for all three models.

Subtracting Polynomials

EXAMPLE 6 **Subtracting Polynomials Horizontally**

a. $(2x^2 + 3) - (3x^2 - 4) = 2x^2 + 3 - 3x^2 + 4$ Distributive Property

$\qquad\qquad\qquad\qquad\quad = (2x^2 - 3x^2) + (3 + 4)$ Group like terms.

$\qquad\qquad\qquad\qquad\quad = -x^2 + 7$ Combine like terms.

b. $(3x^3 - 4x^2 + 3) - (x^3 + 3x^2 - x - 4)$ Original polynomials

$\qquad\quad = 3x^3 - 4x^2 + 3 - x^3 - 3x^2 + x + 4$ Distributive Property

$\qquad\quad = (3x^3 - x^3) + (-4x^2 - 3x^2) + (x) + (3 + 4)$ Group like terms.

$\qquad\quad = 2x^3 - 7x^2 + x + 7$ Combine like terms.

EXAMPLE 7 **Subtracting Polynomials Vertically**

Study Tip

When using a vertical format, write the polynomial being subtracted underneath the one from which it is being subtracted. Be sure to line up like terms in vertical columns.

a.
$$
\begin{array}{r}
(3x^2 + 7x - 6) \\
-(3x^2 + 7x\;\;\;\;\;)
\end{array}
\Longrightarrow
\begin{array}{r}
3x^2 + 7x - 6 \\
-3x^2 - 7x \\
\hline
-6
\end{array}
$$
Change signs and add.

b.
$$
\begin{array}{r}
(5x^3 - 2x^2 + \;x\;\;\;\;\;\;) \\
-(\;\;\;\;\;\;\;4x^2 - 3x + 2)
\end{array}
\Longrightarrow
\begin{array}{r}
5x^3 - 2x^2 + \;x \\
-4x^2 + 3x - 2 \\
\hline
5x^3 - 6x^2 + 4x - 2
\end{array}
$$
Change signs and add.

c.
$$
\begin{array}{r}
(4x^4 - 2x^3 + 5x^2 - \;x + 8) \\
-(3x^4 - 2x^3 \;\;\;\;\;\;\;\;\;+ 3x - 4)
\end{array}
\Longrightarrow
\begin{array}{r}
4x^4 - 2x^3 + 5x^2 - \;x + \;8 \\
-3x^4 + 2x^3 \;\;\;\;\;\;\;\;\; - 3x + \;4 \\
\hline
x^4 \;\;\;\;\;\;\;\;\;+ 5x^2 - 4x + 12
\end{array}
$$

Exercises Within Reach ®

Solutions in English & Spanish and tutorial videos at CollegePrepAlgebra.com

Subtracting Polynomials **In Exercises 27−32, use a horizontal format to find the difference.**

27. $(11x - 8) - (2x + 3)$

28. $(5x + 1) - (18x - 7)$

29. $(x^2 - x) - (x - 2)$

30. $(x^2 - 4) - (x^2 - 4x)$

31. $(4z^3 - 6) - (-z^3 + z - 2)$

32. $(4t^3 - 3t + 5) - (3t^2 - 3t - 10)$

Subtracting Polynomials **In Exercises 33−40, use a vertical format to find the difference.**

33.
$$
\begin{array}{r}
2x - 2 \\
-(x - 1)
\end{array}
$$

34.
$$
\begin{array}{r}
9x + 7 \\
-(3x + 9)
\end{array}
$$

35.
$$
\begin{array}{r}
2x^2 - x + 2 \\
-(3x^2 + x - 1)
\end{array}
$$

36.
$$
\begin{array}{r}
y^4 - 2x - 3 \\
-(y^4 \;\;\;\;\;\; + 2)
\end{array}
$$

37. $(7x^2 - x) - (x^3 - 2x^2 + 10)$

38. $(y^2 - 3y + 8) - (y^3 + y^2 - 3)$

39. $(-3x^3 - 4x^2 + 2x - 5) - (2x^4 + 2x^3 - 4x + 5)$

40. $(12x^3 + 25x^2 - 15) - (-2x^3 + 18x^2 - 3x)$

EXAMPLE 8 **Combining Polynomials Horizontally**

a. $(x^2 - 2x + 1) - [(x^2 + x - 3) + (-2x^2 - 4x)]$ 　　Original polynomials

$= (x^2 - 2x + 1) - [(x^2 - 2x^2) + (x - 4x) + (-3)]$ 　　Group like terms.

$= (x^2 - 2x + 1) - [-x^2 - 3x - 3]$ 　　Combine like terms.

$= x^2 - 2x + 1 + x^2 + 3x + 3$ 　　Distributive Property

$= (x^2 + x^2) + (-2x + 3x) + (1 + 3)$ 　　Group like terms.

$= 2x^2 + x + 4$ 　　Combine like terms.

b. $(3x^2 - 7x + 2) - (4x^2 + 6x - 1) + (-x^2 + 4x + 5)$

$= 3x^2 - 7x + 2 - 4x^2 - 6x + 1 - x^2 + 4x + 5$

$= (3x^2 - 4x^2 - x^2) + (-7x - 6x + 4x) + (2 + 1 + 5)$

$= -2x^2 - 9x + 8$

c. $(-2x^2 + 4x - 3) - [(4x^2 - 5x + 8) - 2(-x^2 + x + 3)]$

$= (-2x^2 + 4x - 3) - [4x^2 - 5x + 8 + 2x^2 - 2x - 6]$

$= (-2x^2 + 4x - 3) - [(4x^2 + 2x^2) + (-5x - 2x) + (8 - 6)]$

$= (-2x^2 + 4x - 3) - [6x^2 - 7x + 2]$

$= -2x^2 + 4x - 3 - 6x^2 + 7x - 2$

$= (-2x^2 - 6x^2) + (4x + 7x) + (-3 - 2)$

$= -8x^2 + 11x - 5$

Exercises Within Reach ®

Solutions in English & Spanish and tutorial videos at CollegePrepAlgebra.com

Combining Polynomials In Exercises 41−54, **perform the indicated operations and simplify.**

41. $(6x - 5) - (8x + 15)$

42. $(2x^2 + 1) + (x^2 - 2x + 1)$

43. $-(x^3 - 2) + (4x^3 - 2x)$

44. $-(5x^2 - 1) - (-3x^2 + 5)$

45. $2(x^4 + 2x) + (5x + 2)$

46. $(z^4 - 2z^2) + 3(z^4 + 4)$

47. $5z - [3z - (10z + 8)]$

48. $9w^2 - [2w - (w^2 + 3w)]$

49. $(y^3 + 1) - [(y^2 + 1) + (3y - 7)]$

50. $(a^2 - a) - [(2a^2 + 3a) - (5a^2 - 12)]$

51. $2(t^2 + 5) - 3(t^2 + 5) + 5(t^2 + 5)$

52. $-10(u + 1) + 8(u - 1) - 3(u + 6)$

53. $8v - 6(3v - v^2) + 10(10v + 3)$

54. $3(x^2 - 2x + 3) - 4(4x + 1) - (3x^2 - 2x)$

Application   EXAMPLE 9   **Geometry: Area of a Region**

Find an expression for the area of the shaded region.

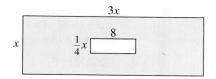

SOLUTION

To find a polynomial that represents the area of the shaded region, subtract the area of the inner rectangle from the area of the outer rectangle, as follows.

Area of shaded region	=	Area of outer rectangle	−	Area of inner rectangle

$$= 3x(x) - 8\left(\frac{1}{4}x\right)$$

$$= 3x^2 - 2x$$

Exercises Within Reach ®

Geometry **In Exercises 55−60, find an expression for the area of the shaded region of the figure.**

55.

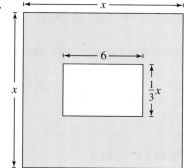

56.

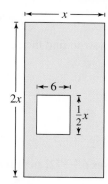

57.

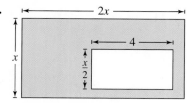

58.

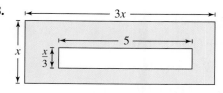

59.

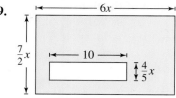

60.

Concept Summary: Adding and Subtracting Polynomials

What

The key to adding or subtracting **polynomials** is to recognize like terms. Like terms have the same degree.

EXAMPLE

Find the sum

$(3x^2 + 2x) + (x^2 - 4x)$.

How

To add two polynomials, use a horizontal or a vertical format to combine like terms.

EXAMPLE

Horizontal format:

$(3x^2 + 2x) + (x^2 - 4x)$

$\quad = (3x^2 + x^2) + (2x - 4x)$

$\quad = 4x^2 - 2x$

Vertical format: $3x^2 + 2x$

$\qquad\qquad\qquad\;\; \underline{x^2 - 4x}$

$\qquad\qquad\qquad\; 4x^2 - 2x$

Why

When you know how to add any two polynomials, you can subtract any two polynomials by *adding the opposite*. Use the Distributive Property to change all the signs in the polynomial that is being subtracted.

Exercises Within Reach ®

Worked-out solutions to odd-numbered exercises at CollegePrepAlgebra.com

Concept Summary Check

61. *Identifying Like Terms* What are the like terms in the expression $(3x^2 + 2x) + (x^2 - 4x)$?

62. *Combining Like Terms* Explain how to find the sum $3x^2 + x^2$.

63. *Adding Polynomials* What step in adding polynomials vertically corresponds to grouping like terms when adding polynomials horizontally?

64. *Writing* What does *adding the opposite* mean?

Extra Practice

Combining Polynomials In Exercises 65−72, perform the indicated operations and simplify.

65. $(x^5 - 3x^4 + x^3 - 5x + 1) - (4x^5 - x^3 + x - 5)$

66. $(t^4 + 5t^3 - t^2 + 8t - 10) - (t^4 + t^3 + 2t^2 + 4t - 7)$

67. $\left(\frac{2}{3}y^2 - \frac{3}{4}\right) + \left(\frac{5}{6}y^2 + 2\right)$

68. $\left(\frac{3}{4}x^3 - \frac{1}{2}\right) + \left(\frac{1}{8}x^3 + 3\right)$

69. $(0.1t^3 - 3.4t^2) + (1.5t^3 - 7.3)$

70. $(0.7x^2 - 0.2x + 2.5) + (7.4x - 3.9)$

71. $(2ab - 3) + (a^2 - 2ab) + (4b^2 - a^2)$

72. $(uv - 3) + (4uv - v^2) + (u^2 - 8uv)$

Geometry In Exercises 73 and 74, **find** an expression for the area of the shaded region of the figure.

73.
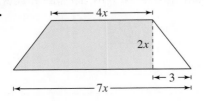

74.
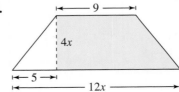

75. *Cost, Revenue, and Profit* The cost C (in dollars) of producing x dome tents is $C = 200 + 45x$. The revenue R (in dollars) for selling x dome tents is $R = 120x - x^2$, where $0 \le x \le 60$. The profit P is the difference between revenue and cost.

(a) Perform the subtraction required to find the polynomial representing profit P.

(b) Determine the profit when 40 tents are produced and sold.

(c) Find the change in profit when the number of tents produced and sold increases from 40 to 50.

76. *Cost, Revenue, and Profit* The cost C (in dollars) of producing x multimedia projectors is $C = 1000 + 150x$. The revenue R (in dollars) for selling x multimedia projectors is $R = 400x - \frac{1}{2}x^2$, where $0 \le x \le 600$. The profit P is the difference between revenue and cost.

(a) Perform the subtraction required to find the polynomial representing profit P.

(b) Determine the profit when 200 projectors are produced and sold.

(c) Find the change in profit when the number of projectors produced and sold decreases from 200 to 100.

Explaining Concepts

77. Is the sum of two binomials always a binomial? Explain.

78. Determine which of the two statements is always true. Is the other statement always false? Explain.

(a) A polynomial is a trinomial.

(b) A trinomial is a polynomial.

79. *Writing* In your own words, define "like terms." What are the only factors of like terms that can differ?

80. *Precision* Describe how to combine like terms. What operations are used?

81. *Structure* Is a polynomial an algebraic expression? Explain.

82. *Writing* Write a paragraph that explains how the adage "You can't add apples and oranges" might relate to adding two polynomials. Include several examples to illustrate the applicability of this statement.

Cumulative Review

In Exercises 83−86, solve the equation and check your solution.

83. $\dfrac{4x}{27} = \dfrac{8}{9}$

84. $\dfrac{x}{6} - \dfrac{x}{18} = 3$

85. $\dfrac{x + 3}{6} = \dfrac{2}{5}$

86. $\dfrac{x - 5}{2} = \dfrac{4x}{3}$

In Exercises 87 and 88, graph the equation.

87. $y = 2 - \frac{3}{2}x$

88. $y = |x - 1|$

In Exercises 89 and 90, determine the exponent that makes the statement true.

89. $2^{\rule{1em}{0.6em}} = \dfrac{1}{32}$

90. $(x^{\rule{1em}{0.6em}}y^2)^{-3} = \dfrac{1}{x^{12}y^6}$

In Exercises 91 and 92, use a calculator to evaluate the expression. Write your answer in scientific notation. Round your answer to four decimal places.

91. $(4.15 \times 10^3)^{-4}$

92. $\dfrac{1.5 \times 10^8}{2.3 \times 10^5}$

Mid-Chapter Quiz: Sections 5.1–5.2

Solutions in English & Spanish and tutorial videos at CollegePrepAlgebra.com

Take this quiz as you would take a quiz in class. After you are done, check your work against the answers in the back of the book.

In Exercises 1–4, simplify the expression. (Assume that no denominator is zero.)

1. $(9m^3)^2$

2. $(-3xy)(2x^2)^3$

3. $\dfrac{-12x^3y}{9x^5y^2}$

4. $\dfrac{3t^3}{(-6t)^2}$

In Exercises 5 and 6, rewrite the expression using only positive exponents.

5. $5x^{-2}y^{-3}$

6. $\dfrac{3x^{-2}y}{5z^{-1}}$

In Exercises 7 and 8, use the rules of exponents to simplify the expression using only positive exponents. (Assume that no variable is zero.)

7. $(3a^{-3}b^2)^{-2}$

8. $(4t^{-3})^0$

9. Write the number 8,168,000,000,000 in scientific notation.

10. Write the number 5.021×10^{-3} in decimal notation.

11. Explain why $x^2 + 2x - 3x^{-1}$ is not a polynomial.

12. Determine the degree and the leading coefficient of the polynomial $10 + x^2 - 4x^3$.

13. Write the polynomial $5x - 3 + x^2$ in standard form.

In Exercises 14–17, perform the indicated operations and simplify.

14. $(y^2 + 3y - 1) + (4 + 3y)$

15. $(3v^2 - 5) - (v^3 + 2v^2 - 6v)$

16. $9s - [6 - (s - 5)]$

17. $-3(4 - x) + 4(x^2 + 2) - (x^2 - 2x)$

In Exercises 18 and 19, use a vertical format to find the sum.

18. $\quad\;\; 2x^2 +\;\; x - 3$
   $\quad\underline{3x^3 - 2x^2 - 3x + 5}$

19. $\;x^3 - 3x^2 \qquad\;\; - 15$
   $\quad\underline{\qquad\; 2x^2 + 5x\; -\; 4}$

In Exercises 20 and 21, use a vertical format to find the difference.

20. $\quad\;\; x^2 - x + 2$
   $\quad\underline{-\quad\;\;(x - 4)}$

21. $\;\; 6x^4 + 3x^3 \qquad\;\; + 8$
   $\quad\underline{-(x^4 \qquad + 4x^2 + 2)}$

22. Find an expression for the perimeter of the figure at the left.

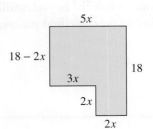

Study Skills in Action

Managing Test Anxiety

Test anxiety is different from the typical nervousness that usually occurs during tests. It interferes with the thinking process. After leaving the classroom, have you ever suddenly been able to recall what you could not remember during the test? It is likely that this was a result of test anxiety. Test anxiety is a learned reaction or response—no one is born with it. The good news is that most students can learn to manage test anxiety.

It is important to get as much information as you can into your long-term memory and to practice retrieving the information before you take a test. The more you practice retrieving information, the easier it will be during the test.

Smart Study Strategy

Make Mental Cheat Sheets

No, we are not asking you to cheat! Just prepare as if you were going to and then memorize the information you've gathered.

1 ▶ Write down important information on note cards. This can include:

- formulas
- examples of problems you find difficult
- concepts that always trip you up

2 ▶ Memorize the information on the note cards. Flash through the cards, placing the ones containing information you know in one stack and the ones containing information you do not know in another stack. Keep working on the information you do not know.

3 ▶ As soon as you receive your test, turn it over and write down all the information you remember, starting with things you have the greatest difficulty remembering. Having this information available should boost your confidence and free up mental energy for focusing on the test.

Do not wait until the night before the test to make note cards. Make them after you study each section. Then review them two or three times a week.

The FOIL Method

To multiply two binomials, you can combine the products of the **F**irst, **O**uter, **I**nner, and **L**ast terms.

$(2x + 1)(x - 5)$

$\qquad$ F $\qquad$ O $\qquad$ I $\qquad$ L

$= 2x(x) + 2x(-5) + 1(x) + 1(-5)$

$= 2x^2 - 10x + x - 5$

$= 2x^2 - 9x - 5$

Special Products:

$(a + b)(a - b) = a^2 - b^2$

$(a + b)^2 = a^2 + 2ab + b^2$

$(a - b)^2 = a^2 - 2ab + b^2$

5.3 Multiplying Polynomials: Special Products

▶ Find products with monomial multipliers.
▶ Multiply binomials using the Distributive Property and the FOIL Method.
▶ Multiply polynomials using a horizontal or vertical format.
▶ Identify and use special binomial products.

Monomial Multipliers

To multiply polynomials, you use many of the rules for simplifying algebraic expressions. You may want to review these rules in Section 2.2 and Section 5.1.

1. The Distributive Property
2. Combining like terms
3. Removing symbols of grouping
4. Rules of exponents

EXAMPLE 1 **Finding Products with Monomial Multipliers**

Find each product.

a. $(3x - 7)(-2x)$ **b.** $3x^2(5x - x^3 + 2)$ **c.** $(-x)(2x^2 - 3x)$

SOLUTION

a. $(3x - 7)(-2x) = 3x(-2x) - 7(-2x)$ Distributive Property

$= -6x^2 + 14x$ Write in standard form.

b. $3x^2(5x - x^3 + 2)$

$= (3x^2)(5x) - (3x^2)(x^3) + (3x^2)(2)$ Distributive Property

$= 15x^3 - 3x^5 + 6x^2$ Rules of exponents

$= -3x^5 + 15x^3 + 6x^2$ Write in standard form.

c. $(-x)(2x^2 - 3x) = (-x)(2x^2) - (-x)(3x)$ Distributive Property

$= -2x^3 + 3x^2$ Write in standard form.

Exercises Within Reach ®

Solutions in English & Spanish and tutorial videos at CollegePrepAlgebra.com

Finding a Product with a Monomial Multiplier **In Exercises 1–12, find the product.**

1. $-x(x^2 - 4)$

2. $-t(10 - 9t^2)$

3. $-3x(2x^2 + 5)$

4. $-5u(u^2 + 4)$

5. $-4x(3 + 3x^2 - 6x^3)$

6. $-5v(5 - 4v + 5v^2)$

7. $2x(x^2 - 2x + 8)$

8. $7y(y^2 - y + 5)$

9. $x^2(4x^2 - 3x + 1)$

10. $y^2(2y^2 + y - 5)$

11. $4t^3(t - 3)$

12. $-2t^4(t + 6)$

Multiplying Binomials

Study Tip
You can write the product of two binomials in just one step. This is called the **FOIL Method**. Note that the words first, outer, inner, and last refer to the positions of the terms in the original product.

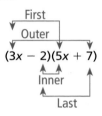

EXAMPLE 2   **Multiplying Binomials with the Distributive Property**

a. $(x - 1)(x + 5) = x(x + 5) - 1(x + 5)$ Distributive Property

$= x^2 + 5x - x - 5$ Distributive Property

$= x^2 + (5x - x) - 5$ Group like terms.

$= x^2 + 4x - 5$ Combine like terms.

b. $(2x + 3)(x - 2) = 2x(x - 2) + 3(x - 2)$ Distributive Property

$= 2x^2 - 4x + 3x - 6$ Distributive Property

$= 2x^2 + (-4x + 3x) - 6$ Group like terms.

$= 2x^2 - x - 6$ Combine like terms.

EXAMPLE 3   **Multiplying Binomials using the FOIL Method**

$$\text{F} \quad \text{O} \quad \text{I} \quad \text{L}$$

a. $(x + 4)(x - 4) = x^2 - 4x + 4x - 16$

$= x^2 - 16$ Combine like terms.

$$\text{F} \quad \text{O} \quad \text{I} \quad \text{L}$$

b. $(3x + 5)(2x + 1) = 6x^2 + 3x + 10x + 5$

$= 6x^2 + 13x + 5$ Combine like terms.

Exercises Within Reach®

Solutions in English & Spanish and tutorial videos at CollegePrepAlgebra.com

Multiplying with the Distributive Property In Exercises 13–16, use the Distributive Property to find the product.

13. $(x + 3)(x + 4)$ **14.** $(x - 5)(x + 10)$ **15.** $(3x - 5)(x + 1)$ **16.** $(7x - 2)(x - 3)$

Using the FOIL Method In Exercises 17 and 18, use the FOIL Method to complete the expression.

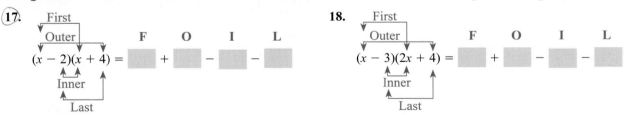

Multiplying Binomials using the FOIL Method In Exercises 19–26, use the FOIL Method to find the product.

19. $(2x - y)(x - 2y)$ **20.** $(x + y)(x + 2y)$ **21.** $(5x + 6)(3x + 1)$ **22.** $(4x + 3)(2x - 1)$

23. $(6 - 2x)(4x + 3)$ **24.** $(8x - 6)(5 - 4x)$ **25.** $(3x - 2y)(x - y)$ **26.** $(7x + 5y)(x + y)$

EXAMPLE 4 **A Geometric Model of a Polynomial Product**

Use the geometric model to show that

$$x^2 + 3x + 2 = (x + 1)(x + 2).$$

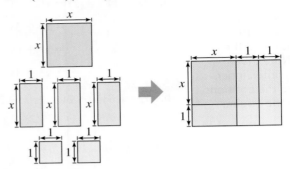

SOLUTION

The left part of the model shows that the sum of the areas of the six rectangles is

$$x^2 + (x + x + x) + (1 + 1) = x^2 + 3x + 2.$$

The right part of the model shows that the area of the rectangle is

$$(x + 1)(x + 2) = x^2 + 2x + x + 2$$
$$= x^2 + 3x + 2.$$

So, $x^2 + 3x + 2 = (x + 1)(x + 2)$.

Exercises Within Reach®

Solutions in English & Spanish and tutorial videos at CollegePrepAlgebra.com

A Geometric Model of a Polynomial Product In Exercises 27−30, write the polynomial product represented by the geometric model.

27.

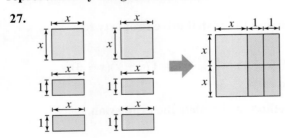

28.

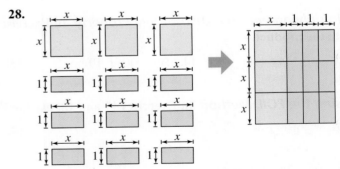

29.

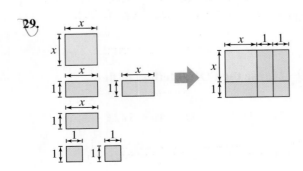

30.

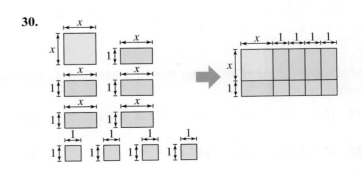

EXAMPLE 5   **Simplifying a Polynomial Expression**

Simplify the expression and write the result in standard form.

$(4x + 5)^2$

SOLUTION

$$\begin{aligned}
(4x + 5)^2 &= (4x + 5)(4x + 5) & \text{Repeated multiplication} \\
&= 16x^2 + 20x + 20x + 25 & \text{Use FOIL Method.} \\
&= 16x^2 + 40x + 25 & \text{Combine like terms.}
\end{aligned}$$

EXAMPLE 6   **Simplifying a Polynomial Expression**

Simplify the expression and write the result in standard form.

$(3x^2 - 2)(4x + 7) - (4x)^2$

SOLUTION

$$\begin{aligned}
(3x^2 - 2)(4x + 7) - (4x)^2 & \\
&= 12x^3 + 21x^2 - 8x - 14 - (4x)^2 & \text{Use FOIL Method.} \\
&= 12x^3 + 21x^2 - 8x - 14 - 16x^2 & \text{Square monomial.} \\
&= 12x^3 + 5x^2 - 8x - 14 & \text{Combine like terms.}
\end{aligned}$$

Exercises Within Reach ®

Solutions in English & Spanish and tutorial videos at CollegePrepAlgebra.com

Simplifying a Polynomial Expression In Exercises 31–40, simplify the expression and write the result in standard form.

31. $(2x + 4)^2$

32. $(7x - 3)^2$

33. $(8x + 2)^2$

34. $(5x - 1)^2$

35. $(3x^2 - 4)(x + 2)$

36. $(5x^2 - 2)(x - 1)$

37. $(3s + 1)(3s + 4) - (3s)^2$

38. $(2t + 5)(4t - 2) - (2t)^2$

39. $(4x^2 - 1)(2x + 8) + (-x^2)^3$

40. $(3 - 3x^2)(4 - 5x^2) - (-x^4)^2$

41. *Geometry* Add the areas of the four rectangular regions shown in the figure. What product does the geometric model represent?

42. *Geometry* Use the geometric model to write an equation that demonstrates the FOIL Method.

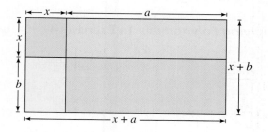

Multiplying Polynomials

EXAMPLE 7 **Multiplying Polynomials (Horizontal Format)**

a. $(x - 4)(x^2 - 4x + 2)$

$\qquad = x(x^2 - 4x + 2) - 4(x^2 - 4x + 2)$ Distributive Property

$\qquad = x^3 - 4x^2 + 2x - 4x^2 + 16x - 8$ Distributive Property

$\qquad = x^3 - 8x^2 + 18x - 8$ Combine like terms.

b. $(2x^2 - 7x + 1)(4x + 3)$

$\qquad = (2x^2 - 7x + 1)(4x) + (2x^2 - 7x + 1)(3)$ Distributive Property

$\qquad = 8x^3 - 28x^2 + 4x + 6x^2 - 21x + 3$ Distributive Property

$\qquad = 8x^3 - 22x^2 - 17x + 3$ Combine like terms.

EXAMPLE 8 **Multiplying Polynomials (Vertical Format)**

a.

$$
\begin{array}{r}
3x^2 + x - 5 \\
\times \qquad 2x - 1 \\
\hline
-3x^2 - x + 5 \\
6x^3 + 2x^2 - 10x \\
\hline
6x^3 - x^2 - 11x + 5
\end{array}
$$

Place polynomial with most terms on top.
Line up like terms.
$-1(3x^2 + x - 5)$
$2x(3x^2 + x - 5)$
Combine like terms in columns.

b.

$$
\begin{array}{r}
4x^3 + 8x - 1 \\
\times \qquad 2x^2 + 3 \\
\hline
12x^3 + 24x - 3 \\
8x^5 + 16x^3 - 2x^2 \\
\hline
8x^5 + 28x^3 - 2x^2 + 24x - 3
\end{array}
$$

Place polynomial with most terms on top.
Line up like terms.
$3(4x^3 + 8x - 1)$
$2x^2(4x^3 + 8x - 1)$
Combine like terms in columns.

Exercises Within Reach ®

Solutions in English & Spanish and tutorial videos at CollegePrepAlgebra.com

Multiplying Polynomials In Exercises 43−48, use a horizontal format to find the product.

43. $(x + 1)(x^2 + 2x - 1)$ **44.** $(x - 3)(x^2 - 3x + 4)$

45. $(x^2 - 2x + 1)(x - 5)$ **46.** $(x^2 - x + 1)(x + 4)$

47. $(x - 2)(5x^2 + 2x + 4)$ **48.** $(x + 9)(2x^2 - x - 4)$

Multiplying Polynomials In Exercises 49−54, use a vertical format to find the product.

49. $\begin{array}{r} x + 3 \\ \times\, x - 2 \end{array}$ **50.** $\begin{array}{r} 2x - 1 \\ \times\, 5x + 1 \end{array}$ **51.** $\begin{array}{r} 4x^2 - 6x + 9 \\ \times \qquad 2x + 3 \end{array}$ **52.** $\begin{array}{r} x^2 - 3x + 9 \\ \times \qquad x + 3 \end{array}$

53. $(3x^3 + x + 7)(x^2 + 1)$ **54.** $(5x^4 - 3x + 2)(2x^2 - 4)$

Study Tip

When multiplying two polynomials, it is best to write each in standard form before using either the horizontal or the vertical format.

EXAMPLE 9   **Multiplying Polynomials (Vertical Format)**

Write the polynomials in standard form and use a vertical format to find the product of $(x + 3x^2 - 4)$ and $(5 + 3x - x^2)$.

SOLUTION

$$
\begin{array}{r}
3x^2 + x - 4 \\
\times \quad -x^2 + 3x + 5 \\
\hline
15x^2 + 5x - 20 \\
9x^3 + 3x^2 - 12x \\
-3x^4 - x^3 + 4x^2 \\
\hline
-3x^4 + 8x^3 + 22x^2 - 7x - 20
\end{array}
$$

Write in standard form.
Write in standard form.
$5(3x^2 + x - 4)$
$3x(3x^2 + x - 4)$
$-x^2(3x^2 + x - 4)$
Combine like terms.

EXAMPLE 10   **Raising a Polynomial to a Power**

Use two steps to expand $(x - 3)^3$.

SOLUTION

Step 1: $(x - 3)^2 = (x - 3)(x - 3)$ Repeated multiplication

$\qquad\qquad\quad = x^2 - 3x - 3x + 9$ Use FOIL Method.

$\qquad\qquad\quad = x^2 - 6x + 9$ Combine like terms.

Step 2: $(x^2 - 6x + 9)(x - 3) = (x^2 - 6x + 9)(x) - (x^2 - 6x + 9)(3)$

$\qquad\qquad\qquad\qquad\qquad\quad = x^3 - 6x^2 + 9x - 3x^2 + 18x - 27$

$\qquad\qquad\qquad\qquad\qquad\quad = x^3 - 9x^2 + 27x - 27$

So, $(x - 3)^3 = x^3 - 9x^2 + 27x - 27$.

Exercises Within Reach ®   Solutions in English & Spanish and tutorial videos at CollegePrepAlgebra.com

Multiplying Polynomials In Exercises 55−58, use a vertical format to find the product.

55. $(x^2 - x + 2)(x^2 + x - 2)$

56. $(x^2 + 2x + 5)(2x^2 - x - 1)$

57. $(x + 3 - 2x^2)(5x + x^2 - 4)$

58. $(1 - x - x^2)(x + 1 - x^3)$

Raising a Polynomial to a Power In Exercises 59 and 60, expand the power.

59. $(x - 2)^3$

60. $(x + 3)^3$

Special Products

> ### Special Products
>
> Let a and b be real numbers, variables, or algebraic expressions.
>
> | *Special Product* | *Example* |
> |---|---|
> | **Sum and Difference of Two Terms:** | |
> | $(a + b)(a - b) = a^2 - b^2$ | $(2x - 5)(2x + 5) = 4x^2 - 25$ |
> | **Square of a Binomial:** | |
> | $(a + b)^2 = a^2 + 2ab + b^2$ | $(3x + 4)^2 = 9x^2 + 2(3x)(4) + 16$ |
> | | $= 9x^2 + 24x + 16$ |
> | $(a - b)^2 = a^2 - 2ab + b^2$ | $(x - 7)^2 = x^2 - 2(x)(7) + 49$ |
> | | $= x^2 - 14x + 49$ |

EXAMPLE 11 **Finding Special Products**

Difference   Sum   (1st term)^2 (2nd term)^2

a. $(5x - 6)(5x + 6) = (5x)^2 - (6)^2 = 25x^2 - 36$

2nd term   Twice the product of the terms
1st term   (1st term)^2 (2nd term)^2

b. $(3x + 7)^2 = (3x)^2 + 2(3x)(7) + (7)^2 = 9x^2 + 42x + 49$

2nd term   Twice the product of the terms
1st term   (1st term)^2 (2nd term)^2

c. $(4x - 9)^2 = (4x)^2 - 2(4x)(9) + (9)^2 = 16x^2 - 72x + 81$

2nd term   Twice the product of the terms
1st term   (1st term)^2 (2nd term)^2

d. $(6 - 5x^2)^2 = (6)^2 - 2(6)(5x^2) + (5x^2)^2$

$$= 36 - 60x^2 + (5)^2(x^2)^2 = 36 - 60x^2 + 25x^4$$

Exercises Within Reach ®

Solutions in English & Spanish and tutorial videos at CollegePrepAlgebra.com

Finding a Special Product In Exercises 61–76, use a special product pattern to find the product.

61. $(x + 3)(x - 3)$

62. $(x - 5)(x + 5)$

63. $(4t - 6)(4t + 6)$

64. $(3u + 7)(3u - 7)$

65. $(4x + y)(4x - y)$

66. $(5u + 12v)(5u - 12v)$

67. $(x + 6)^2$

68. $(a - 2)^2$

69. $(t - 3)^2$

70. $(x + 10)^2$

71. $(8 - 3z)^2$

72. $(1 - 5t)^2$

73. $(4 + 7s^2)^2$

74. $(3 + 8v^2)^2$

75. $(2x - 5y)^2$

76. $(4s + 3t)^2$

Application   EXAMPLE 12   **Finding the Dimensions of a Golf Tee**

A landscaper wants to reshape a square tee area for the ninth hole of a golf course. The new tee area will have one side 2 feet longer and the adjacent side 6 feet longer than the original tee. The area of the new tee will be 204 square feet greater than the area of the original tee. What are the dimensions of the original tee?

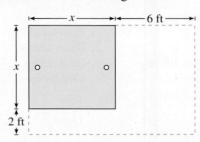

SOLUTION

Verbal Model: New area = Original area + 204

Labels: Original length = original width = x (feet)
Original area = x^2 (square feet)
New length = $x + 6$ (feet)
New width = $x + 2$ (feet)

Equation:

$(x + 6)(x + 2) = x^2 + 204$	Write equation.
$x^2 + 8x + 12 = x^2 + 204$	Multiply factors.
$8x + 12 = 204$	Subtract x^2 from each side.
$8x = 192$	Subtract 12 from each side.
$x = 24$	Divide each side by 8.

The original tee measured 24 feet by 24 feet.

Exercises Within Reach ®

Solutions in English & Spanish and tutorial videos at CollegePrepAlgebra.com

77. **Geometry** A park recreation manager wants to reshape a square sandbox. The new sandbox will have one side 2 feet longer and the adjacent side 3 feet longer than the original sandbox. The area of the new sandbox will be 26 square feet greater than the area of the original sandbox. What are the dimensions of the original sandbox?

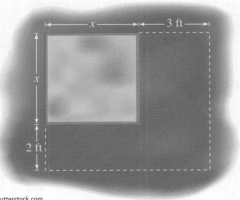

78. **Geometry** A carpenter wants to expand a square room. The new room will have one side 4 feet longer and the adjacent side 6 feet longer than the original room. The area of the new room will be 144 square feet greater than the area of the original room. What are the dimensions of the original room?

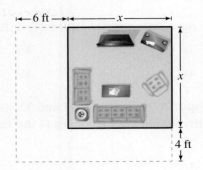

Concept Summary: *Multiplying Two Binomials*

What

Here are two methods you can use to multiply binomials.

1. The Distributive Property.

2. The **FOIL Method.**

Each method applies many of the rules for simplifying algebraic expressions.

EXAMPLE

Find the product $(x - y)(x + 2)$.

How

To use the FOIL Method, add the products of the following pairs of terms.

1. **First** terms
2. **Outer** terms
3. **Inner** terms
4. **Last** terms

$$(x - y)(x + 2) = x^2 + 2x - xy - 2y$$

First, Outer, Inner, Last; F O I L

Why

When multiplying two binomials, you can use the FOIL Method. It is a useful tool to help you remember all the products.

To multiply any two polynomials, use the Distributive Property.

Exercises Within Reach ®

Worked-out solutions to odd-numbered exercises at CollegePrepAlgebra.com

Concept Summary Check

79. *Identifying Binomials* Identify the binomials in the example above.

80. *Signs of Terms* Identify the positive terms and the negative terms in the two binomials in the example above.

81. *Identifying Terms* Identify the first terms, outer terms, inner terms, and last terms in the example above.

82. *Alternate Methods* Describe a method that you can use to multiply any two polynomials.

Extra Practice

Geometry **In Exercises 83 and 84, find a polynomial product that represents the area of the region. Then simplify the product.**

83.
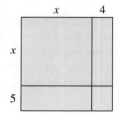

84.
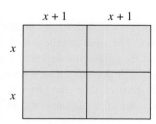

Simplifying a Polynomial Expression **In Exercises 85–90, perform the multiplication and simplify.**

85. $(x + 2)^2 - (x - 2)^2$ **86.** $(u + 5)^2 + (u - 5)^2$ **87.** $(x + 2)^2(x - 4)$

88. $(x - 4)^2(x - 1)$ **89.** $[(x + 1) + y]^2$ **90.** $[(x - 3) - y]^2$

Think About It **In Exercises 91 and 92, decide whether the equation is an identity. Explain. (An identity is true for any values of the variables.)**

91. $(x + y)^3 = x^3 + 3x^2y + 3xy^2 + y^3$ **92.** $(x - y)^3 = x^3 - 3x^2y + 3xy^2 - y^3$

Using a Result **In Exercises 93 and 94, use the result of Exercise 91 to find the product.**

93. $(x + 2)^3$ **94.** $(x + 1)^3$

95. *Geometry* The base of a triangular sail is $2x$ feet and its height is $(x + 10)$ feet (see figure). Find an expression for the area of the sail.

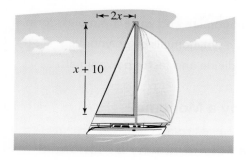

96. *Compound Interest* After 2 years, an investment of $500 compounded annually at interest rate r (in decimal form) will yield an amount $500(1 + r)^2$. Find this product.

Explaining Concepts

True or False? **In Exercises 97 and 98, determine whether the statement is true or false. Justify your answer.**

97. The expressions $(3x)^2$ and $3x^2$ represent the same quantity.

98. Because the product of two monomials is a monomial, it follows that the product of two binomials is a binomial.

99. *Writing* Explain why an understanding of the Distributive Property is essential in multiplying polynomials. Illustrate your explanation with an example.

100. *Reasoning* What is the degree of the product of two polynomials of degrees m and n? Explain.

101. *Reasoning* A polynomial with m terms is multiplied by a polynomial with n terms. How many monomial-by-monomial products must be found? Explain.

102. *Repeated Reasoning* Perform each multiplication.

(a) $(x - 1)(x + 1)$

(b) $(x - 1)(x^2 + x + 1)$

(c) $(x - 1)(x^3 + x^2 + x + 1)$

(d) From the pattern formed in the first three products, can you predict the product of

$$(x - 1)(x^4 + x^3 + x^2 + x + 1)?$$

Verify your prediction by multiplying.

Cumulative Review

In Exercises 103–106, perform the indicated operations and simplify.

103. $(12x - 3) + (3x - 4)$

104. $(9x - 5) - (x + 7)$

105. $(-8x + 11) - (-4x - 6)$

106. $-(5x - 10) + (-13x + 40)$

In Exercises 107–110, solve the percent equation.

107. What number is 25% of 45?

108. 78 is 10% of what number?

109. 20 is what percent of 60?

110. What number is 55% of 62?

In Exercises 111–114, solve the proportion.

111. $\dfrac{2}{5} = \dfrac{x}{10}$

112. $\dfrac{3}{2} = \dfrac{15}{y}$

113. $\dfrac{z}{6} = \dfrac{5}{8}$

114. $\dfrac{9}{w} = \dfrac{6}{7}$

5.4 Dividing Polynomials and Synthetic Division

▶ Divide polynomials by monomials and write in simplest form.

▶ Use long division to divide polynomials by polynomials.

▶ Use synthetic division to divide and factor polynomials.

Dividing a Polynomial by a Monomial

> **Dividing a Polynomial by a Monomial**
>
> Let u, v, and w represent real numbers, variables, or algebraic expressions such that $w \neq 0$.
>
> **1.** $\dfrac{u + v}{w} = \dfrac{u}{w} + \dfrac{v}{w}$ **2.** $\dfrac{u - v}{w} = \dfrac{u}{w} - \dfrac{v}{w}$

EXAMPLE 1 Dividing a Polynomial by a Monomial

Perform the division and simplify.

$$\frac{12x^2 - 20x + 8}{4x}$$

SOLUTION

$$\frac{12x^2 - 20x + 8}{4x} = \frac{12x^2}{4x} - \frac{20x}{4x} + \frac{8}{4x}$$ Divide each term in the numerator by $4x$.

$$= \frac{3(4x)(x)}{4x} - \frac{5(4x)}{4x} + \frac{2(4)}{4x}$$ Factor numerators.

$$= \frac{3(4x)(x)}{4x} - \frac{5(4x)}{4x} + \frac{2(4)}{4x}$$ Divide out common factors.

$$= 3x - 5 + \frac{2}{x}$$ Simplified form

Exercises Within Reach ®

Solutions in English & Spanish and tutorial videos at CollegePrepAlgebra.com

Dividing a Polynomial by a Monomial In Exercises 1−12, **perform** the division.

1. $(7x^3 - 2x^2) \div x$

2. $(3w^2 - 6w) \div w$

3. $(m^4 + 2m^2 - 7) \div m$

4. $(x^3 + x - 2) \div x$

5. $(4x^2 - 2x) \div (-x)$

6. $(5y^3 + 6y^2 - 3y) \div (-y)$

7. $\dfrac{50z^3 + 30z}{-5z}$

8. $\dfrac{18c^4 - 24c^2}{-6c}$

9. $\dfrac{4v^4 + 10v^3 - 8v^2}{4v^2}$

10. $\dfrac{6x^4 + 8x^3 - 18x^2}{3x^2}$

11. $(5x^2y - 8xy + 7xy^2) \div 2xy$

12. $(-14s^4t^2 + 7s^2t^2 - 18t) \div 2s^2t$

Long Division

EXAMPLE 2   **Long Division Algorithm for Positive Integers**

Use the long division algorithm to divide 6584 by 28.

SOLUTION

Think $\frac{65}{28} \approx 2$.

Think $\frac{98}{28} \approx 3$.

Think $\frac{144}{28} \approx 5$.

$$
\begin{array}{r}
235 \\
28\overline{)6584} \\
\end{array}
$$

56	Multiply 2 by 28.
98	Subtract and bring down 8.
84	Multiply 3 by 28.
144	Subtract and bring down 4.
140	Multiply 5 by 28.
4	Remainder

So, you have

$$6584 \div 28 = 235 + \frac{4}{28}$$

$$= 235 + \frac{1}{7}.$$

In Example 2, 6584 is the **dividend**, 28 is the **divisor**, 235 is the **quotient**, and 4 is the **remainder**.

Long Division of Polynomials

1. Write the dividend and divisor in descending powers of the variable.
2. Insert placeholders with zero coefficients for missing powers of the variable. (See Example 5.)
3. Perform the long division of the polynomials as you would with integers.
4. Continue the process until the degree of the remainder is less than that of the divisor.

Exercises Within Reach ®

Solutions in English & Spanish and tutorial videos at CollegePrepAlgebra.com

Using Long Division In Exercises 13−18, use the long division algorithm to perform the division.

13. Divide 1013 by 9.

14. Divide 3713 by 22.

15. $3235 \div 15$

16. $6344 \div 28$

17. $\dfrac{6055}{25}$

18. $\dfrac{4160}{12}$

Study Tip

Note that in Example 3, the division process requires $3x - 3$ to be subtracted from $3x + 4$. The difference

$$3x + 4$$
$$\underline{-(3x - 3)}$$

is implied and written simply as

$$3x + 4$$
$$\underline{3x - 3}$$
$$7.$$

EXAMPLE 3 **Long Division Algorithm for Polynomials**

Think $x^2/x = x$.
Think $3x/x = 3$.

$$
\begin{array}{r}
x + 3 \\
x - 1 \overline{\smash{)}\ x^2 + 2x + 4} \\
\underline{x^2 - x} \qquad\qquad \\
3x + 4 \\
\underline{3x - 3} \\
7
\end{array}
$$

Multiply x by $(x - 1)$.
Subtract and bring down 4.
Multiply 3 by $(x - 1)$.
Subtract.

The remainder is a fractional part of the divisor, so you can write

Dividend   Quotient   Remainder

$$\frac{x^2 + 2x + 4}{x - 1} = x + 3 + \frac{7}{x - 1}.$$

Divisor   Divisor

EXAMPLE 4 **Writing in Standard Form Before Dividing**

Divide $-13x^3 + 10x^4 + 8x - 7x^2 + 4$ by $3 - 2x$.

SOLUTION

First write the divisor and dividend in standard polynomial form.

$$
\begin{array}{r}
-5x^3 - x^2 + 2x - 1 \\
-2x + 3 \overline{\smash{)}\ 10x^4 - 13x^3 - 7x^2 + 8x + 4} \\
\underline{10x^4 - 15x^3} \qquad\qquad\qquad\qquad\qquad \\
2x^3 - 7x^2 \\
\underline{2x^3 - 3x^2} \\
-4x^2 + 8x \\
\underline{-4x^2 + 6x} \\
2x + 4 \\
\underline{2x - 3} \\
7
\end{array}
$$

Multiply $-5x^3$ by $(-2x + 3)$.
Subtract and bring down $-7x^2$.
Multiply $-x^2$ by $(-2x + 3)$.
Subtract and bring down 8x.
Multiply 2x by $(-2x + 3)$.
Subtract and bring down 4.
Multiply -1 by $(-2x + 3)$.
Subtract.

This shows that

Dividend   Quotient   Remainder

$$\frac{10x^4 - 13x^3 - 7x^2 + 8x + 4}{-2x + 3} = -5x^3 - x^2 + 2x - 1 + \frac{7}{-2x + 3}.$$

Divisor   Divisor

Exercises Within Reach ®   Solutions in English & Spanish and tutorial videos at CollegePrepAlgebra.com

Using Long Division In Exercises 19−26, use the long division algorithm to perform the division.

19. $\dfrac{x^2 - 8x + 15}{x - 3}$

20. $\dfrac{t^2 - 18t + 72}{t - 6}$

21. Divide $21 - 4x - x^2$ by $3 - x$.

22. Divide $5 + 4x - x^2$ by $1 + x$.

23. $(12 - 17t + 6t^2) \div (2t - 3)$

24. $(15 - 14u - 8u^2) \div (5 + 2u)$

25. $\dfrac{9x^3 - 3x^2 - 3x + 4}{3x + 2}$

26. $\dfrac{4y^3 + 12y^2 + 7y - 3}{2y + 3}$

EXAMPLE 5   **Accounting for Missing Powers of x**

Divide $x^3 - 2$ by $x - 1$.

SOLUTION

To account for the missing x^2- and x-terms, insert $0x^2$ and $0x$.

$$
\begin{array}{r}
x^2 + x + 1 \\
x - 1 \overline{\smash{)}\ x^3 + 0x^2 + 0x - 2} \\
\underline{x^3 - x^2} \\
x^2 + 0x \\
\underline{x^2 - x} \\
x - 2 \\
\underline{x - 1} \\
-1
\end{array}
$$

Insert $0x^2$ and $0x$.
Multiply x^2 by $(x - 1)$.
Subtract and bring down $0x$.
Multiply x by $(x - 1)$.
Subtract and bring down -2.
Multiply 1 by $(x - 1)$.
Subtract.

So, you have

$$\frac{x^3 - 2}{x - 1} = x^2 + x + 1 - \frac{1}{x - 1}.$$

EXAMPLE 6   **A Second-Degree Divisor**

Divide $x^4 + 6x^3 + 6x^2 - 10x - 3$ by $x^2 + 2x - 3$.

SOLUTION

Study Tip

If the remainder of a division problem is zero, the divisor is said to *divide evenly* into the dividend.

$$
\begin{array}{r}
x^2 + 4x + 1 \\
x^2 + 2x - 3 \overline{\smash{)}\ x^4 + 6x^3 + 6x^2 - 10x - 3} \\
\underline{x^4 + 2x^3 - 3x^2} \\
4x^3 + 9x^2 - 10x \\
\underline{4x^3 + 8x^2 - 12x} \\
x^2 + 2x - 3 \\
\underline{x^2 + 2x - 3} \\
0
\end{array}
$$

Multiply x^2 by $(x^2 + 2x - 3)$.
Subtract and bring down $-10x$.
Multiply $4x$ by $(x^2 + 2x - 3)$.
Subtract and bring down -3.
Multiply 1 by $(x^2 + 2x - 3)$.
Subtract.

So, $x^2 + 2x - 3$ divides evenly into $x^4 + 6x^3 + 6x^2 - 10x - 3$. That is,

$$\frac{x^4 + 6x^3 + 6x^2 - 10x - 3}{x^2 + 2x - 3} = x^2 + 4x + 1, \ x \neq -3, x \neq 1.$$

Exercises Within Reach®

Solutions in English & Spanish and tutorial videos at CollegePrepAlgebra.com

Using Long Division   **In Exercises 27−34, use the long division algorithm to perform the division.**

27. $\dfrac{x^2 + 16}{x + 4}$

28. $\dfrac{y^2 + 8}{y + 2}$

29. $\dfrac{x^3 + 125}{x + 5}$

30. $\dfrac{x^3 - 27}{x - 3}$

31. $(x^3 + 4x^2 + 7x + 7) \div (x^2 + 2x + 3)$

32. $(2x^3 + 2x^2 - 2x - 15) \div (2x^2 + 4x + 5)$

33. $(4x^4 - 3x^2 + x - 5) \div (x^2 - 3x + 2)$

34. $(8x^5 + 6x^4 - x^3 + 1) \div (2x^3 - x^2 - 3)$

Synthetic Division

Synthetic Division of a Third-Degree Polynomial

Use synthetic division to divide $ax^3 + bx^2 + cx + d$ by $x - k$, as follows.

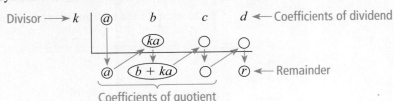

Vertical Pattern: Add terms.
Diagonal Pattern: Multiply by k.

EXAMPLE 7 **Using Synthetic Division**

Use synthetic division to divide $x^3 + 3x^2 - 4x - 10$ by $x - 2$.

SOLUTION

The coefficients of the dividend form the top of the synthetic division array. Because you are dividing by $x - 2$, write 2 at the top left of the array. To begin the algorithm, bring down the first coefficient. Then multiply this coefficient by 2, write the result in the second row, and add the two numbers in the second column. By continuing this pattern, you obtain the following.

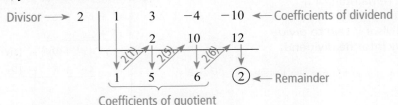

The bottom row shows the coefficients of the quotient. So, the quotient is

$$1x^2 + 5x + 6$$

and the remainder is 2. So, the result of the division problem is

$$\frac{x^3 + 3x^2 - 4x - 10}{x - 2} = x^2 + 5x + 6 + \frac{2}{x - 2}.$$

Exercises Within Reach ®

Solutions in English & Spanish and tutorial videos at CollegePrepAlgebra.com

---

Using Synthetic Division In Exercises 35−42, use synthetic division to divide.

35. $(x^2 + x - 6) \div (x - 2)$

36. $(x^2 + 5x - 6) \div (x + 6)$

37. $\dfrac{x^3 + 3x^2 - 1}{x + 4}$

38. $\dfrac{x^3 - 4x + 7}{x - 1}$

39. $\dfrac{5x^3 - 6x^2 + 8}{x - 4}$

40. $\dfrac{5x^3 + 6x + 8}{x + 2}$

41. $\dfrac{0.1x^2 + 0.8x + 1}{x - 0.2}$

42. $\dfrac{x^3 - 0.8x + 2.4}{x + 0.1}$

Synthetic division (or long division) can be used to factor polynomials. If the remainder in a synthetic division problem is zero, you know that the divisor divides *evenly* into the dividend.

> **EXAMPLE 8**   **Factoring a Polynomial**

Completely factor the polynomial $x^3 - 7x + 6$ given that one of its factors is $x - 1$.

SOLUTION

The polynomial $x^3 - 7x + 6$ can be factored completely using synthetic division. Because $x - 1$ is a factor of the polynomial, you can divide as follows.

$$
\begin{array}{c|cccc}
1 & 1 & 0 & -7 & 6 \\
  &   & 1 & 1 & -6 \\
\hline
  & 1 & 1 & -6 & \boxed{0} \leftarrow \text{Remainder}
\end{array}
$$

Because the remainder is zero, the divisor divides evenly into the dividend:

$$\frac{x^3 - 7x + 6}{x - 1} = x^2 + x - 6.$$

From this result, you can factor the original polynomial as follows.

$$x^3 - 7x + 6 = (x - 1)(x^2 + x - 6)$$
$$= (x - 1)(x + 3)(x - 2)$$

Study Tip

In Example 8, synthetic division is used to divide the polynomial by the factor $x - 1$. Long division could be used also.

Exercises Within Reach ®

Solutions in English & Spanish and tutorial videos at CollegePrepAlgebra.com

Factoring Completely In Exercises 43−50, completely factor the polynomial given one of its factors.

	Polynomial	*Factor*
43.	$x^3 - x^2 - 14x + 24$	$x - 3$
44.	$x^3 + x^2 - 32x - 60$	$x + 5$
45.	$4x^3 - 3x - 1$	$x - 1$
46.	$9x^3 + 51x^2 + 88x + 48$	$x + 3$
47.	$x^4 + 7x^3 + 3x^2 - 63x - 108$	$x + 4$
48.	$x^4 - 6x^3 - 8x^2 + 96x - 128$	$x - 4$
49.	$15x^2 - 2x - 8$	$x - \frac{4}{5}$
50.	$18x^2 - 9x - 20$	$x + \frac{5}{6}$

Finding a Constant In Exercises 51 and 52, find the constant c such that the denominator divides evenly into the numerator.

51. $\dfrac{x^3 + 2x^2 - 4x + c}{x - 2}$

52. $\dfrac{x^4 - 3x^2 + c}{x + 6}$

Concept Summary: Dividing Polynomials

What

You can use long division to divide polynomials just as you do for integers.

EXAMPLE

Divide $(-4 + x^2)$ by $(2 + x)$.

How

1. Write the **dividend** and **divisor** in standard form.
2. Insert placeholders with zero coefficients for missing powers of the variable.
3. Use the long division algorithm.

EXAMPLE

$$
\begin{array}{r}
x - 2 \\
x + 2 \overline{) x^2 + 0x - 4} \\
\underline{x^2 + 2x} \\
-2x - 4 \\
\underline{-2x - 4} \\
0
\end{array}
$$

Why

You can use long division of polynomials to check the product of two polynomials.

Exercises Within Reach ®

Worked-out solutions to odd-numbered exercises at CollegePrepAlgebra.com

Concept Summary Check

53. **Vocabulary** Identify the dividend, divisor, quotient, and remainder of the equation $1253 \div 12 = 104 + \frac{5}{12}$.

54. **Inserting Placeholders** What is the missing power of x in the dividend in the example above? Explain.

55. **Reasoning** Explain what it means for a divisor to divide *evenly* into a dividend.

56. **Reasoning** Explain how you can check polynomial division.

Extra Practice

Using Long Division In Exercises 57 and 58, use the long division algorithm to **perform** the division.

57. $\dfrac{x^3 - 9x}{x - 2}$

58. $\dfrac{2x^5 - 3x^3 + x}{x - 3}$

Simplifying an Expression In Exercises 59−62, simplify the expression.

59. $\dfrac{8u^2v}{2u} + \dfrac{3(uv)^2}{uv}$

60. $\dfrac{15x^3y}{10x^2} + \dfrac{3xy^2}{2y}$

61. $\dfrac{x^2 + 3x + 2}{x + 2} + (2x + 3)$

62. $\dfrac{x^2 + 2x - 3}{x - 1} - (3x - 4)$

Dividing Polynomials In Exercises 63 and 64, **perform** the division assuming that n is a positive integer.

63. $\dfrac{x^{3n} + 3x^{2n} + 6x^n + 8}{x^n + 2}$

64. $\dfrac{x^{3n} - x^{2n} + 5x^n - 5}{x^n - 1}$

Think About It In Exercises 65 and 66, the divisor, quotient, and remainder are given. Find the dividend.

	Divisor	Quotient	Remainder		Divisor	Quotient	Remainder
65.	$x - 6$	$x^2 + x + 1$	-4	66.	$x + 3$	$x^2 - 2x - 5$	8

67. *Geometry* The height of a cube is $x + 1$. The volume of the cube is $x^3 + 3x^2 + 3x + 1$. Use division to find the area of the base.

68. *Geometry* A rectangular house has a volume of $(x^3 + 55x^2 + 650x + 2000)$ cubic feet (the space in the attic is not included). The height of the house is $(x + 5)$ feet (see figure). Find the number of square feet of floor space on the first floor of the house.

Geometry **In Exercises 69 and 70, you are given the expression for the volume of the solid shown. Find the expression for the missing dimension.**

69. $V = x^3 + 18x^2 + 80x + 96$

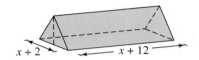

70. $V = 2h^3 + 3h^2 + h$

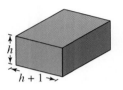

Explaining Concepts

71. *Error Analysis* Describe and correct the error.

$$\dfrac{6x + 5y}{x} = \dfrac{6x + 5y}{x} = 6 + 5y$$

72. *Error Analysis* Describe and correct the error.

$$\dfrac{x^2}{x + 1} = \dfrac{x^2}{x} + \dfrac{x^2}{1} = x + x^2$$

73. *Precision* Create a polynomial division problem and identify the dividend, divisor, quotient, and remainder.

74. *True or False?* If the divisor divides evenly into the dividend, then the divisor and quotient are factors of the dividend. Justify your answer.

Cumulative Review

In Exercises 75–80, solve the inequality.

75. $7 - 3x > 4 - x$

76. $2(x + 6) - 20 < 2$

77. $|x - 3| < 2$

78. $|x - 5| > 3$

79. $\left|\frac{1}{4}x - 1\right| \geq 3$

80. $\left|2 - \frac{1}{3}x\right| \leq 10$

In Exercises 81 and 82, determine the quadrants in which the point must be located.

81. $(-3, y)$, y is a real number.

82. $(x, 7)$, x is a real number.

83. Describe the location of the set of points whose x-coordinates are 0.

84. Find the coordinates of the point five units to the right of the y-axis and seven units below the x-axis.

5 Chapter Summary

What did you learn?	Explanation and Examples	Review Exercises
5.1 Use the rules of exponents to simplify expressions *(p. 226)*.	Let m and n be integers, and let a and b represent real numbers, variables, or algebraic expressions. *Rule* *Example* **1.** $a^m \cdot a^n = a^{m+n}$ $x^5(x^4) = x^{5+4} = x^9$ **2.** $(ab)^m = a^m \cdot b^m$ $(2x)^3 = 2^3(x^3) = 8x^3$ **3.** $(a^m)^n = a^{mn}$ $(x^2)^3 = x^{2 \cdot 3} = x^6$ **4.** $\dfrac{a^m}{a^n} = a^{m-n}, m > n, a \neq 0$ $\dfrac{x^5}{x^3} = x^{5-3} = x^2, x \neq 0$ **5.** $\left(\dfrac{a}{b}\right)^m = \dfrac{a^m}{b^m}, b \neq 0$ $\left(\dfrac{x}{4}\right)^2 = \dfrac{x^2}{4^2} = \dfrac{x^2}{16}$	1–14
Rewrite exponential expressions involving negative and zero exponents *(p. 227)*.	Let a and b be real numbers such that $a \neq 0$ and $b \neq 0$, and let m be an integer. $a^0 = 1$ $a^{-m} = \dfrac{1}{a^m}$ $\left(\dfrac{a}{b}\right)^{-m} = \left(\dfrac{b}{a}\right)^m$	15–30
Write very large and very small numbers in scientific notation *(p. 230)*.	$1{,}230{,}000 = 1.23 \times 10^6$ $0.000123 = 1.23 \times 10^{-4}$	31–42
5.2 Identify the degrees and leading coefficients of polynomials *(p. 234)*.	Let $a_n, a_{n-1}, \ldots, a_2, a_1, a_0$ be real numbers and let n be a nonnegative integer. A polynomial in x is an expression of the form $$a_n x^n + a_{n-1} x^{n-1} + \cdots + a_2 x^2 + a_1 x + a_0$$ where $a_n \neq 0$. The polynomial is of degree n, and the number a_n is called the leading coefficient. The number a_0 is called the constant term.	43–48
Add polynomials using a horizontal or vertical format *(p. 235)*.	The key to adding two polynomials is to recognize *like* terms—those having the same *degree*. To use a horizontal format, group like terms and add. To use a vertical format, line up like terms and add. *Horizontal format:* *Vertical format:* $(2x^2 + 3x) + (x^2 - 2x)$ $2x^2 + 3x$ $\quad = (2x^2 + x^2) + (3x - 2x)$ $\underline{x^2 - 2x}$ $\quad = 3x^2 + x$ $3x^2 + x$	49–60
Subtract polynomials using a horizontal or vertical format *(p. 237)*.	To subtract one polynomial from another, *add the opposite* by changing the sign of each term being subtracted. Then add like terms. *Horizontal format:* *Vertical format:* $(x^2 + 3) - (x^2 - 2)$ $(x^2 + 3) \Rightarrow x^2 + 3$ $\quad = x^2 + 3 - x^2 + 2$ $-(x^2 - 2) \Rightarrow \underline{-x^2 + 2}$ $\quad = (x^2 - x^2) + (3 + 2)$ $\phantom{-(x^2 - 2) \Rightarrow -x^2 + {}} 5$ $\quad = 5$	61–70

What did you learn?	*Explanation and Examples*	*Review Exercises*
5.3 Find products with monomial multipliers *(p. 244)*.	To multiply a polynomial by a monomial, use the Distributive Property. $$(3x - 7)(-2x) = 3x(-2x) - 7(-2x)$$ $$= -6x^2 + 14x$$	71–74
Multiply binomials using the Distributive Property and the FOIL Method *(p. 245)*.	*Distributive Property:* $$(3x - 2)(5x + 7) = 3x(5x + 7) - 2(5x + 7)$$ $$= 15x^2 + 11x - 14$$ *FOIL Method:* Product of **First terms** Product of **Outer terms** Product of **Inner terms** Product of **Last terms** $$(3x - 2)(5x + 7) = 15x^2 + 21x - 10x - 14$$ $$= 15x^2 + 11x - 14$$	75–86
Multiply polynomials using a horizontal or vertical format *(p. 248)*.	When multiplying two polynomials, it is best to write each in standard form before using either the horizontal or vertical format. To multiply using a horizontal format, use the Distributive Property. To multiply using a vertical format, line up the like terms to help with combining like terms after multiplying.	87–98
Identify and use special binomial products *(p. 250)*.	Let a and b be real numbers, variables, or algebraic expressions. *Sum and Difference of Two Terms:* $$(a + b)(a - b) = a^2 - b^2$$ *Square of a Binomial:* $$(a + b)^2 = a^2 + 2ab + b^2 \qquad (a - b)^2 = a^2 - 2ab + b^2$$	99–108
5.4 Divide polynomials by monomials and write in simplest form *(p. 254)*.	Divide each term of the polynomial by the monomial, then simplify each fraction.	109–112
Use long division to divide polynomials by polynomials *(p. 255)*.	**1.** Write the dividend and divisor in descending powers of the variable. **2.** Insert placeholders with zero coefficients for missing powers of the variable. **3.** Perform the long division of the polynomials as you would with integers. **4.** Continue the process until the degree of the remainder is less than the degree of the divisor.	113–118
Use synthetic division to divide and factor polynomials *(p. 258)*.	Divide $ax^3 + bx^2 + cx + d$ by $x - k$, as follows. Divisor ⟶ k $\boxed{a}$ b c d ⟵ Coefficients of dividend $\widehat{ka}$ ◯ ◯ $\widehat{a}$ $\widehat{b + ka}$ ◯ $\widehat{r}$ ⟵ Remainder ⏟ Coefficients of quotient ⏟ *Vertical Pattern:* Add terms. *Diagonal Pattern:* Multiply by k.	119–126

Review Exercises

Worked-out solutions to odd-numbered exercises at CollegePrepAlgebra.com

5.1

Using Rules of Exponents In Exercises 1−14, use the rules of exponents to **simplify** the expression.

1. $x^4 \cdot x^5$

2. $-3y^2 \cdot y^4$

3. $(u^2)^3$

4. $(v^4)^2$

5. $(-2z)^3$

6. $(-3y)^2(2)$

7. $-(u^2v)^2(-4u^3v)$

8. $(12x^2y)(3x^2y^4)^2$

9. $\dfrac{12z^5}{6z^2}$

10. $\dfrac{15m^3}{25m}$

11. $\dfrac{25g^4d^2}{80g^2d^2}$

12. $\dfrac{-48u^8v^6}{(-2u^2v)^3}$

13. $\left(\dfrac{72x^4}{6x^2}\right)^2$

14. $\left(-\dfrac{y^2}{2}\right)^3$

Using Rules of Exponents In Exercises 15−18, use the rules of exponents to **evaluate** the expression.

15. $(2^3 \cdot 3^2)^{-1}$

16. $(2^{-2} \cdot 5^2)^{-2}$

17. $\left(\dfrac{3}{4}\right)^{-3}$

18. $\left(\dfrac{1}{3^{-2}}\right)^2$

Using Rules of Exponents In Exercises 19−30, **rewrite** the expression using only positive exponents, and **simplify**. (Assume that any variables in the expression are nonzero.)

19. $(6y^4)(2y^{-3})$

20. $4(-3x)^{-3}$

21. $\dfrac{4x^{-2}}{2x}$

22. $\dfrac{15t^5}{24t^{-3}}$

23. $(x^3y^{-4})^0$

24. $(5x^{-2}y^4)^{-2}$

25. $\dfrac{7a^6b^{-2}}{14a^{-1}b^4}$

26. $\dfrac{2u^0v^{-2}}{10u^{-1}v^{-3}}$

27. $\left(\dfrac{3x^{-1}y^2}{12x^5y^{-3}}\right)^{-1}$

28. $\left(\dfrac{4x^{-3}z^{-1}}{8x^4z}\right)^{-2}$

29. $u^3(5u^0v^{-1})(9u)^2$

30. $a^4(16a^{-2}b^4)(2b)^{-3}$

Writing in Scientific Notation In Exercises 31−34, **write** the number in scientific notation.

31. 0.0000319

32. 0.0000008924

33. 17,350,000

34. *Circumference of the Sun:* 4,370,000,000 meters

Writing in Decimal Notation In Exercises 35−38, **write** the number in decimal notation.

35. 1.95×10^6

36. 7.025×10^4

37. 2.05×10^{-5}

38. 6.118×10^{-8}

Using Scientific Notation In Exercises 39−42, **evaluate** the expression without using a calculator.

39. $(6 \times 10^3)^2$

40. $(3 \times 10^{-3})(8 \times 10^7)$

41. $\dfrac{3.5 \times 10^7}{7 \times 10^4}$

42. $\dfrac{1}{(6 \times 10^{-3})^2}$

5.2

Determining the Degree and Leading Coefficient In Exercises 43–48, write the polynomial in standard form. Then identify its degree and leading coefficient.

43. $10x - 4 - 5x^3$

44. $2x^2 + 9$

45. $4x^3 - 2x + 5x^4 - 7x^2$

46. $6 - 3x + 6x^2 - x^3$

47. $7x^4 - 1 + 11x^2$

48. $12x^2 + 2x - 8x^5 + 1$

Adding Polynomials In Exercises 49–54, use a horizontal format to find the sum.

49. $(8x + 4) + (x - 4)$

50. $\left(\frac{1}{2}x + \frac{2}{3}\right) + \left(4x + \frac{1}{3}\right)$

51. $(3y^3 + 5y^2 - 9y) + (2y^3 - 3y + 10)$

52. $(6 - x + x^2) + (3x^2 + x)$

53. $(3u + 4u^2) + 5(u + 1) + 3u^2$

54. $6(u^2 + 2) + 12u + (u^2 - 5u + 2)$

Adding Polynomials In Exercises 55–58, use a vertical format to find the sum.

55. $x^3 + 2x - 3$
$4x + 5$

56. $-x^3 + 3$
$3x^3 + 2x^2 + 5$

57. $-x^4 - 2x^2 + 3$
$3x^4 - 5x^2 + 0$

58. $5z^3 - 4z - 7$
$ -z^2 - 4z$

59. **Geometry** The length of a rectangular wall is x units, and its height is $(x - 3)$ units (see figure). Find an expression for the perimeter of the wall.

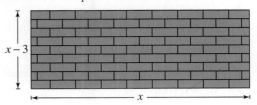

60. **Geometry** A rectangular garden has length $(t + 5)$ feet and width $2t$ feet (see figure). Find an expression for the perimeter of the garden.

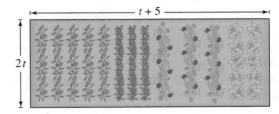

Subtracting Polynomials In Exercises 61–66, use a horizontal format to find the difference.

61. $(3t - 5) - (3t - 9)$

62. $\left(2x - \frac{1}{5}\right) - \left(\frac{1}{4}x + \frac{1}{4}\right)$

63. $(6x^2 - 9x - 5) - 3(4x^2 - 6x + 1)$

64. $(5t^2 + 2) - 2(4t^2 + 1)$

65. $4y^2 - [y - 3(y^2 + 2)]$

66. $(6a^3 + 3a) - 2[a - (a^3 + 2)]$

Subtracting Polynomials In Exercises 67 and 68, use a vertical format to find the difference.

67. $5x^2 + 2x - 27$
$-(2x^2 - 2x - 13)$

68. $12y^4 - 15y^2 + 7$
$-(18y^4 - 9)$

69. **Geometry** Find an expression for the area of the shaded region of the figure.

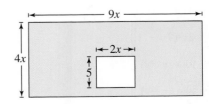

70. *Cost, Revenue, and Profit* The cost C (in dollars) of producing x units of a product is

$$C = 15 + 26x.$$

The revenue R (in dollars) for selling x units is

$R = 40x - \frac{1}{2}x^2$, where $0 \le x \le 20$.

The profit P is the difference between revenue and cost.

(a) Perform the subtraction required to find the polynomial representing profit P.

(b) Determine the profit when 14 units are produced and sold.

5.3

Finding A Product with a Monomial Multiplier **In Exercises 71–74, find the product.**

71. $2x(x + 4)$

72. $3y(y - 1)$

73. $(4x + 2)(-3x^2)$

74. $(5 - 7y)(-6y^2)$

Multiplying with the Distributive Property **In Exercises 75–78, use the Distributive Property to find the product.**

75. $(x + 10)(x + 2)$

76. $(x - 1)(x + 3)$

77. $(2x - 5)(7x + 2)$

78. $(3x - 2)(2x - 3)$

Multiplying Binomials using the FOIL Method **In Exercises 79–86, use the FOIL Method to find the product.**

79. $(x + 4)(x + 6)$

80. $(u + 5)(u - 2)$

81. $(4x - 3)(3x + 4)$

82. $(6x - 7)(2x + 5)$

83. $(3 - 4x)(7x - 6)$

84. $(6 - 2x)(7x - 10)$

85. $(x + y)(x + 3y)$

86. $(x - y)(x + 4y)$

Multiplying Polynomials **In Exercises 87–90, use a horizontal format to find the product.**

87. $(x^2 + 5x + 2)(x - 6)$

88. $(s^2 + 4s - 3)(s - 3)$

89. $(2t - 1)(t^2 - 3t + 3)$

90. $(4x + 2)(x^2 + 6x - 5)$

Multiplying Polynomials **In Exercises 91–94, use a vertical format to find the product.**

91.
$$\begin{array}{r} 3x^2 + x - 2 \\ \times \quad\quad 4x - 5 \\ \hline \end{array}$$

92.
$$\begin{array}{r} 5y^2 - 2y + 9 \\ \times \quad\quad 3y + 4 \\ \hline \end{array}$$

93.
$$\begin{array}{r} y^2 - 4y + 5 \\ \times \quad y^2 + 2y - 3 \\ \hline \end{array}$$

94.
$$\begin{array}{r} x^2 + 8x - 12 \\ \times \quad x^2 - 9x + 2 \\ \hline \end{array}$$

Raising a Polynomial to a Power **In Exercises 95 and 96, expand the power.**

95. $(2x + 1)^3$

96. $(3y - 2)^3$

97. *Geometry* The width of a rectangular window is $(2x + 6)$ inches, and its height is $(3x + 10)$ inches (see figure). Find an expression for the area of the window.

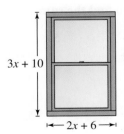

$3x + 10$

$\longmapsto 2x + 6 \longrightarrow$

98. *Geometry* The width of a rectangular parking lot is $(x + 25)$ meters, and its length is $(x + 30)$ meters (see figure). Find an expression for the area of the parking lot.

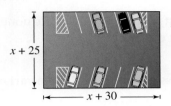

Finding a Special Product In Exercises 99−108, use a special product pattern to find the product.

99. $(x + 3)^2$

100. $(x - 5)^2$

101. $\left(\frac{1}{2}x - 4\right)^2$

102. $(4 + 3b)^2$

103. $(u - 6)(u + 6)$

104. $(r + 7)(r - 7)$

105. $(2r - 5t)^2$

106. $(3a + b)^2$

107. $(2x - 4y)(2x + 4y)$

108. $(4u + 5v)(4u - 5v)$

5.4

Dividing a Polynomial by a Monomial In Exercises 109−112, perform the division.

109. $(4x^3 - x) \div (2x)$

110. $(10x + 15) \div (5x)$

111. $\dfrac{3x^3y^2 - x^2y^2 + x^2y}{x^2y}$

112. $\dfrac{6a^3b^3 + 2a^2b - 4ab^2}{2ab}$

Using Long Division In Exercises 113−118, use the long division algorithm to perform the division.

113. $\dfrac{5x^2 + 2x + 3}{x + 2}$

114. $\dfrac{4x^4 - x^3 - 7x^2 + 18x}{x - 2}$

115. $\dfrac{x^4 - 3x^2 + 2}{x^2 - 1}$

116. $\dfrac{x^4 - 4x^3 + 3x}{x^2 - 1}$

117. $\dfrac{x^5 - 3x^4 + x^2 + 6}{x^3 - 2x^2 + x - 1}$

118. $\dfrac{x^6 + 4x^5 - 3x^2 + 5x}{x^3 + x^2 - 4x + 3}$

Using Synthetic Division In Exercises 119−124, use synthetic division to divide.

119. $\dfrac{x^2 + 3x + 5}{x + 1}$

120. $\dfrac{2x^2 + x - 10}{x - 2}$

121. $\dfrac{x^3 + 7x^2 + 3x - 14}{x + 2}$

122. $\dfrac{x^4 - 2x^3 - 15x^2 - 2x + 10}{x - 5}$

123. $(x^4 - 3x^2 - 25) \div (x - 3)$

124. $(2x^3 + 5x - 2) \div \left(x + \frac{1}{2}\right)$

Factoring Completely In Exercises 125 and 126, completely factor the polynomial given one of its factors.

	Polynomial	*Factor*
125.	$x^3 + 2x^2 - 5x - 6$	$x - 2$
126.	$2x^3 + x^2 - 2x - 1$	$x + 1$

Chapter Test

Solutions in English & Spanish and tutorial videos at CollegePrepAlgebra.com

Take this test as you would take a test in class. After you are done, check your work against the answers in the back of the book.

In Exercises 1–6, simplify the expression. (Assume that no variable is zero.)

1. $x^2 \cdot x^7$

2. $(5y^7)^2$

3. $\dfrac{-6a^2b}{-9ab}$

4. $(3x^{-2}y^3)^2$

5. $\left(-\dfrac{2u^2}{v^{-1}}\right)^3 \left(\dfrac{3v^2}{u^{-3}}\right)$

6. $\dfrac{(-3x^2y^{-1})^4}{6x^2y^0}$

7. (a) Write 690,000,000 in scientific notation.
   (b) Write 4.72×10^{-5} in decimal notation.

8. Determine the degree and the leading coefficient of $-3x^4 - 5x^2 + 2x - 10$.

In Exercises 9–20, simplify the expression. (Assume that no variable or denominator is zero.)

9. $(3z^2 - 3z + 7) + (8 - z^2)$

10. $(8u^3 + 3u^2 - 2u - 1) - (u^3 + 3u^2 - 2u)$

11. $6y + [2y - (3 - y^2)]$

12. $-5(x^2 - 1) + 3(4x + 7) - (x^2 + 26)$

13. $(x - 7)^2$

14. $(2x - 3)(2x + 3)$

15. $(z + 2)(2z^2 - 3z + 5)$

16. $(y + 3)(y^2 - 4)$

17. $\dfrac{4z^3 + z}{2z}$

18. $\dfrac{16x^2 - 12}{-8}$

19. $\dfrac{2x^4 - 15x^2 - 7}{x - 3}$

20. $\dfrac{t^4 + t^2 - 6t}{t^2 - 2}$

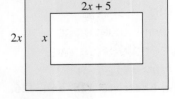

$3x + 2$

$2x + 5$

$2x$ x

Figure for 21

$x + 6$

$4x - 2$

Figure for 22

21. Find an expression for the area of the shaded region shown in the figure.

22. Find an expression for the area of the triangle shown in the figure.

23. After 2 years, an investment of $1500 compounded annually at interest rate r (in decimal form) will yield an amount of $1500(1 + r)^2$. Find this product.

24. The area of a rectangle is $x^2 - 2x - 3$, and its length is $x + 1$. Find the width of the rectangle.

6

Factoring and Solving Equations

6.1   Factoring Polynomials with Common Factors

6.2   Factoring Trinomials

6.3   More About Factoring Trinomials

6.4   Factoring Polynomials with Special Forms

6.5   Solving Polynomial Equations by Factoring

MASTERY IS WITHIN REACH!

"I always get to know the teachers because then they are more willing to help me. I usually talk to the teachers before class starts, just to visit. I used to avoid asking my teachers any questions, but now I realize that they are used to helping students, and it really is no big deal. I'm able to ask more questions about what is confusing me. I feel more connected in class too."

Manuel

See page 295 for suggestions about being confident.

6.1 Factoring Polynomials with Common Factors

▶ Find the greatest common factor of two or more expressions.
▶ Factor out the greatest common monomial factor from polynomials.
▶ Factor polynomials by grouping.

Greatest Common Factor

In Chapter 5, you used the Distributive Property to multiply polynomials. In this chapter, you will study the *reverse* process, which is **factoring**.

Multiplying Polynomials   *Factoring Polynomials*

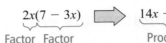   

EXAMPLE 1   **Finding the Greatest Common Factor**

To find the greatest common factor of $5x^2y^2$ and $30x^3y$, first factor each term.

$$5x^2y^2 = 5 \cdot x \cdot x \cdot y \cdot y = (5x^2y)(y)$$

$$30x^3y = 2 \cdot 3 \cdot 5 \cdot x \cdot x \cdot x \cdot y = (5x^2y)(6x)$$

So, you can conclude that the greatest common factor is $5x^2y$.

EXAMPLE 2   **Finding the Greatest Common Factor**

To find the greatest common factor of $8x^5$, $20x^3$, and $16x^4$, first factor each term.

$$8x^5 = 2 \cdot 2 \cdot 2 \cdot x \cdot x \cdot x \cdot x \cdot x = (4x^3)(2x^2)$$

$$20x^3 = 2 \cdot 2 \cdot 5 \cdot x \cdot x \cdot x = (4x^3)(5)$$

$$16x^4 = 2 \cdot 2 \cdot 2 \cdot 2 \cdot x \cdot x \cdot x \cdot x = (4x^3)(4x)$$

So, you can conclude that the greatest common factor is $4x^3$.

Exercises Within Reach®

Solutions in English & Spanish and tutorial videos at CollegePrepAlgebra.com

Finding the Greatest Common Factor In Exercises 1–12, **find the greatest common factor of the expressions.**

1. $z^2, -z^6$

2. t^4, t^7

3. $2x^2, 12x$

4. $36x^4, 18x^3$

5. u^2v, u^3v^2

6. $r^6s^4, -rs$

7. $9y^8z^4, -12y^5z^4$

8. $-15x^6y^3, 45xy^3$

9. $14x^2, 1, 7x^4$

10. $5y^4, 10x^2y^2, 1$

11. $28a^4b^2, 14a^3, 42a^2b^5$

12. $16x^2y, 12xy^2, 36x^2$

Common Monomial Factors

EXAMPLE 3   **Greatest Common Monomial Factor**

Factor out the greatest common monomial factor from $6x - 18$.

SOLUTION

The greatest common integer factor of $6x$ and 18 is 6. There is no common variable factor.

$$6x - 18 = 6(x) - 6(3) \qquad \text{Greatest common monomial factor is 6.}$$
$$= 6(x - 3) \qquad \text{Factor 6 out of each term.}$$

EXAMPLE 4   **Greatest Common Monomial Factor**

Factor out the greatest common monomial factor from $10y^3 - 25y^2$.

SOLUTION

For the terms $10y^2$ and $25y^2$, 5 is the greatest common integer factor and y^2 is the highest-power common variable factor.

$$10y^3 - 25y^2 = 5y^2(2y) - 5y^2(5) \qquad \text{Greatest common factor is } 5y^2.$$
$$= 5y^2(2y - 5) \qquad \text{Factor } 5y^2 \text{ out of each term.}$$

EXAMPLE 5   **Greatest Common Monomial Factor**

Factor out the greatest common monomial factor from $45x^3 - 15x^2 - 15$.

SOLUTION

The greatest common integer factor of $45x^3$, $15x^2$, and 15 is 15. There is no common variable factor.

$$45x^3 - 15x^2 - 15 = 15(3x^3) - 15(x^2) - 15(1)$$
$$= 15(3x^3 - x^2 - 1)$$

Study Tip

To find the greatest common monomial factor of a polynomial, answer these two questions.

1. What is the greatest integer factor common to each coefficient of the polynomial?
2. What is the highest-power variable factor common to each term of the polynomial?

Exercises Within Reach ®

Solutions in English & Spanish and tutorial videos at CollegePrepAlgebra.com

Greatest Common Monomial Factor **In Exercises 13−30, factor out the greatest common monomial factor from the polynomial. (*Note:* Some of the polynomials have no common monomial factor.)**

13. $3x + 3$

14. $5y - 5$

15. $8t - 16$

16. $4u + 12$

17. $24y^2 - 18$

18. $8z^3 + 12$

19. $x^2 + x$

20. $s^3 - s$

21. $25u^2 - 14u$

22. $36t^4 + 24t^2$

23. $2x^4 + 6x^3$

24. $9z^6 + 27z^4$

25. $7s^2 + 9t^2$

26. $12x^2 - 5y^3$

27. $12x^2 + 16x - 8$

28. $9 - 3y + 15y^2$

29. $100 + 75z + 50z^2$

30. $42t^3 - 21t^2 + 7$

EXAMPLE 6 **Greatest Common Monomial Factor**

Factor out the greatest common monomial factor from $3xy^2 - 15x^2y + 12xy$.

SOLUTION

$$3xy^2 - 15x^2y + 12xy = 3xy(y) - 3xy(5x) + 3xy(4) \qquad \text{Greatest common factor is } 3xy.$$
$$= 3xy(y - 5x + 4) \qquad \text{Factor } 3xy \text{ out of each term.}$$

EXAMPLE 7 **Greatest Common Monomial Factor**

Factor out the greatest common monomial factor from $35y^3 - 7y^2 - 14y$.

SOLUTION

$$35y^3 - 7y^2 - 14y = 7y(5y^2) - 7y(y) - 7y(2) \qquad \text{Greatest common factor is } 7y.$$
$$= 7y(5y^2 - y - 2) \qquad \text{Factor } 7y \text{ out of each term.}$$

EXAMPLE 8 **A Negative Common Monomial Factor**

Factor the polynomial $-2x^2 + 8x - 12$ in two ways.

a. Factor out a common monomial factor of 2.

b. Factor out a common monomial factor of -2.

SOLUTION

a. To factor out the common monomial factor of 2, write the following.

$$-2x^2 + 8x - 12 = 2(-x^2) + 2(4x) + 2(-6) \qquad \text{Factor each term.}$$
$$= 2(-x^2 + 4x - 6) \qquad \text{Factored form}$$

b. To factor -2 out the polynomial, write the following.

$$-2x^2 + 8x - 12 = -2(x^2) + (-2)(-4x) + (-2)(6) \qquad \text{Factor each term.}$$
$$= -2(x^2 - 4x + 6) \qquad \text{Factored form}$$

Check this result by multiplying $(x^2 - 4x + 6)$ by -2. When you do, you will obtain the original polynomial.

> **Study Tip**
>
> The greatest common monomial factor of the terms of a polynomial is usually considered to have a positive coefficient. However, sometimes it is convenient to factor a negative number out of a polynomial, as shown in Example 8.

Exercises Within Reach ®

Solutions in English & Spanish and tutorial videos at CollegePrepAlgebra.com

Greatest Common Monomial Factor In Exercises 31−38, factor out the greatest common monomial factor from the polynomial. (*Note:* Some of the polynomials have no common monomial factor.)

31. $9x^4 + 6x^3 + 18x^2$

32. $32a^5 - 2a^3 + 6a$

33. $5u^2 + 5u^2 + 5u$

34. $11y^3 - 22y^2 + 11y^2$

35. $10ab + 10a^2b$

36. $21x^2z^5 + 35x^6z$

37. $4xy - 8x^2y + 24x^4y^5$

38. $15m^4n^3 - 25m^7n + 30m^4n^8$

A Negative Common Monomial Factor In Exercises 39−44, factor out a negative real number from the polynomial and then write the polynomial factor in standard form.

39. $5 - 10x$

40. $3 - 6x$

41. $-15x^2 + 5x + 10$

42. $-4x^2 - 8x + 20$

43. $4 + 12x - 2x^2$

44. $8 - 4x - 12x^2$

Factoring by Grouping

EXAMPLE 9   **Common Binomial Factors**

Factor each expression.

a. $5x^2(7x - 1) - 3(7x - 1)$

b. $2x(3x - 4) + (3x - 4)$

c. $3y^2(y - 3) + 4(3 - y)$

SOLUTION

a. Each of the terms of this expression has a binomial factor of $(7x - 1)$.

$$5x^2(7x - 1) - 3(7x - 1) = (7x - 1)(5x^2 - 3)$$

b. Each of the terms of this expression has a binomial factor of $(3x - 4)$.

$$2x(3x - 4) + (3x - 4) = (3x - 4)(2x + 1)$$

Be sure you see that when $(3x - 4)$ is factored out of itself, you are left with the factor 1. This follows from the fact that $(3x - 4)(1) = (3x - 4)$.

c. $3y^2(y - 3) + 4(3 - y) = 3y^2(y - 3) - 4(y - 3)$ Write $4(3 - y)$ as $-4(y - 3)$.

$$= (y - 3)(3y^2 - 4)$$ Common factor is $(y - 3)$.

In Example 9, the polynomials were already grouped so that it was easy to determine the common binomial factors. In practice, you will have to do the grouping as well as the factoring. To see how this works, consider the expression

$$x^3 + 2x^2 + 3x + 6$$

and try to factor it. Note first that there is no common monomial factor to take out of all four terms. But suppose you *group* the first two terms together and the last two terms together.

$$x^3 + 2x^2 + 3x + 6 = (x^3 + 2x^2) + (3x + 6)$$ Group terms.

$$= x^2(x + 2) + 3(x + 2)$$ Factor out common monomial factor in each group.

$$= (x + 2)(x^2 + 3)$$ Factored form

When factoring by grouping, be sure to group terms that have a common monomial factor. For example, in the polynomial above, you should not group the first term x^3 with the fourth term 6.

Exercises Within Reach ®

Solutions in English & Spanish and tutorial videos at CollegePrepAlgebra.com

A Common Binomial Factor In Exercises 45−50, factor the polynomial by grouping.

45. $x(x - 3) + 5(x - 3)$

46. $x(x + 6) + 3(x + 6)$

47. $y(q - 5) - (q - 5)$

48. $a^2(b + 2) - (b + 2)$

49. $x^3(x - 4) + 2(4 - x)$

50. $x^3(x - 2) + 6(2 - x)$

EXAMPLE 10 **Factoring by Grouping**

Factor $x^3 + 2x^2 + x + 2$.

SOLUTION

$$x^3 + 2x^2 + x + 2 = (x^3 + 2x^2) + (x + 2) \qquad \text{Group terms.}$$

$$= x^2(x + 2) + (x + 2) \qquad \text{Factor out common monomial factor in each group.}$$

$$= (x + 2)(x^2 + 1) \qquad \text{Factored form}$$

Note that in Example 10 the polynomial is factored by grouping the first and second terms and the third and fourth terms. You could just as easily have grouped the first and third terms and the second and fourth terms, as follows.

$$x^3 + 2x^2 + x + 2 = (x^3 + x) + (2x^2 + 2) \qquad \text{Group terms.}$$

$$= x(x^2 + 1) + 2(x^2 + 1) \qquad \text{Factor out common monomial factor in each group.}$$

$$= (x^2 + 1)(x + 2) \qquad \text{Factored form}$$

EXAMPLE 11 **Factoring by Grouping**

Factor $3x^2 - 12x - 5x + 20$.

SOLUTION

$$3x^2 - 12x - 5x + 20 = (3x^2 - 12x) + (-5x + 20) \qquad \text{Group terms.}$$

$$= 3x(x - 4) - 5(x - 4) \qquad \text{Factor out common monomial factor in each group.}$$

$$= (x - 4)(3x - 5) \qquad \text{Factored form}$$

Note how a -5 is factored out so that the common binomial factor $x - 4$ appears.

Exercises Within Reach ®

Solutions in English & Spanish and tutorial videos at CollegePrepAlgebra.com

Factoring by Grouping In Exercises 51−62, factor the polynomial by grouping.

51. $x^2 + 10x + x + 10$

52. $x^2 - 5x + x - 5$

53. $x^2 + 3x + 4x + 12$

54. $x^2 - 6x + 5x - 30$

55. $x^2 + 3x - 5x - 15$

56. $x^2 + 4x + 2x + 8$

57. $4x^2 - 14x + 14x - 49$

58. $4x^2 - 6x + 6x - 9$

59. $6x^2 + 3x - 2x - 1$

60. $4x^2 + 20x - x - 5$

61. $8x^2 + 32x + x + 4$

62. $8x^2 - 4x - 2x + 1$

Application   EXAMPLE 12   Geometry: **Area of a Rectangle**

The area of a rectangle of width $(2x - 1)$ feet is $(2x^3 + 4x - x^2 - 2)$ square feet, as shown below. Factor this expression to determine the length of the rectangle.

← Length →

$2x - 1$ Area $= 2x^3 + 4x - x^2 - 2$

Study Tip

Notice in Example 12 that the polynomial is not written in standard form. You could have rewritten the polynomial before factoring and still obtained the same result.

$2x^3 + 4x - x^2 - 2$
$= 2x^3 - x^2 + 4x - 2$
$= (2x^3 - x^2) + (4x - 2)$
$= x^2(2x - 1) + 2(2x - 1)$
$= (2x - 1)(x^2 + 2)$

SOLUTION

Verbal Model: Area $=$ Length $\times$ Width

Labels: Area $= 2x^3 + 4x - x^2 - 2$ (square feet)
Width $= 2x - 1$ (feet)

Expression: $2x^3 + 4x - x^2 - 2 = $ (Length)$(2x - 1)$

$2x^3 + 4x - x^2 - 2 = (2x^3 + 4x) + (-x^2 - 2)$ Group terms.

$= 2x(x^2 + 2) - (x^2 + 2)$ Factor out common monomial factor in each group.

$= (x^2 + 2)(2x - 1)$ Factored form

You can see that the length of the rectangle is $(x^2 + 2)$ feet.

Exercises Within Reach®

Solutions in English & Spanish and tutorial videos at CollegePrepAlgebra.com

Geometry In Exercises 63–66, factor the polynomial to **find** an expression for the length of the rectangle.

63. Area $= 2x^2 + 2x$

$2x$

64. Area $= 12x^3 - 9x^2$

$3x^2$

65. Area $= 3x^2(4x - 1) - 2(4x - 1)$

$4x - 1$

66. Area $= x^2 + 2x + 10x + 20$

$x + 2$

Concept Summary: Factoring Polynomials by Grouping

What

You can use **factoring by grouping** to **factor** some polynomials.

EXAMPLE

Factor $x^3 + x^2 + 2x + 2$.

How

To factor polynomials by grouping, do the following.

1. Group terms that have a common monomial factor.
2. **Factor out** the **greatest common monomial factor** from each group.
3. Use the Distributive Property to write the factored form.

EXAMPLE

$x^3 + x^2 + 2x + 2$
$= (x^3 + x^2) + (2x + 2)$
$= x^2(x + 1) + 2(x + 1)$
$= (x + 1)(x^2 + 2)$

Why

You can use factoring by grouping to find an expression for the length of a rectangle given polynomial expressions for its area and its width.

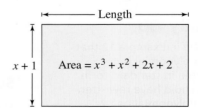

Exercises Within Reach ®

Worked-out solutions to odd-numbered exercises at CollegePrepAlgebra.com

Concept Summary Check

67. *Grouping Terms* Explain why the polynomial in the example above was rewritten as $(x^3 + x^2) + (2x + 2)$.

68. *Finding the GCF* Explain how to find the greatest common monomial factor of a binomial.

69. *The Common Binomial Factor* In the solution above, what is the common binomial factor of the two groups of terms?

70. *Checking the Result* Explain how you can check the factored form in the solution above.

Extra Practice

Factoring Out a Fraction In Exercises 71–76, determine the missing factor.

71. $\frac{1}{4}x + \frac{3}{4} = \frac{1}{4}(\boxed{})$

72. $\frac{5}{6}x - \frac{1}{6} = \frac{1}{6}(\boxed{})$

73. $2y - \frac{1}{5} = \frac{1}{5}(\boxed{})$

74. $3z + \frac{5}{4} = \frac{1}{4}(\boxed{})$

75. $\frac{7}{8}x + \frac{5}{16}y = \frac{1}{16}(\boxed{})$

76. $\frac{5}{12}u - \frac{5}{8}v = \frac{1}{24}(\boxed{})$

Factoring by Grouping In Exercises 77–80, factor the polynomial by grouping.

77. $(a + b)(a - b) + a(a + b)$

78. $(x + y)(x - y) - x(x - y)$

79. $ky^2 - 4ky + 2y - 8$

80. $ay^2 + 3ay + 3y + 9$

Geometry In Exercises 81 and 82, write an expression for the area of the shaded region and factor the expression if possible.

81.

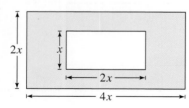

82.

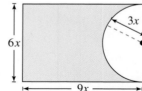

83. **Geometry** The surface area of a right circular cylinder is given by

    $$2\pi r^2 + 2\pi rh$$

    where r is the radius of the base of the cylinder and h is the height of the cylinder. Factor this expression.

84. **Simple Interest** A principal of P dollars earns simple interest at a rate r (in decimal form). The amount after t years is given by

    $$P + Prt.$$

    Factor this expression.

85. **Chemical Reaction** The rate of change in a chemical reaction is

    $$kQx - kx^2$$

    where Q is the original amount, x is the new amount, and k is a constant of proportionality. Factor this expression.

86. **Unit Price** The revenue R from selling x units of a product at a price of p dollars per unit is given by $R = xp$. For a pool table, the revenue is

    $$R = 900x - 0.1x^2.$$

    Factor the revenue model and determine an expression that represents the price p in terms of x.

Explaining Concepts

87. **Reasoning** Explain why $x^2(2x + 1)$ is in factored form.

88. **Reasoning** Explain why $3x$ is the greatest common monomial factor of $3x^3 + 3x^2 + 3x$.

89. **Creating an Example** Give an example of the use of the Distributive Property to factor out the greatest common monomial factor from a polynomial.

90. **Creating an Example** Give an example of the use of the Distributive Property to factor by grouping.

91. **Structure** Give an example of a trinomial with no common monomial factors.

92. **Structure** Give an example of a polynomial with four terms that can be factored by grouping.

Cumulative Review

In Exercises 93–96, determine whether each value of x is a solution of the equation.

93. $x + 2 = 3$ (a) $x = 5$

                         (b) $x = 1$

94. $x - 5 = 10$ (a) $x = 0$

                         (b) $x = 15$

95. $2x - 4 = 0$ (a) $x = 2$

                         (b) $x = -2$

96. $-7x - 8 = 6$ (a) $x = 1$

                         (b) $x = -2$

In Exercises 97–100, simplify the expression.

97. $\left(\dfrac{3y}{2x^3}\right)^2$

98. $\left(\dfrac{b^{-4}}{4a^{-5}}\right)^2$

99. $z^2 \cdot z^{-6}$

100. $(x^2y)^5(x^7y^6)^2$

In Exercises 101–104, perform the division and simplify. (Assume that no denominator is zero.)

101. $\dfrac{3m^2n}{m}$

102. $\dfrac{12x^3y^5}{3x^2y^5}$

103. $\dfrac{x^2 + 9x + 8}{x + 8}$

104. $\dfrac{2x^2 + 5x - 12}{2x - 3}$

6.2 Factoring Trinomials

▶ Factor trinomials of the form $x^2 + bx + c$.

▶ Factor trinomials in two variables.

▶ Factor trinomials completely.

Factoring Trinomials of the Form $x^2 + bx + c$

Guidelines for Factoring $x^2 + bx + c$

To factor $x^2 + bx + c$, you need to find two numbers m and n whose product is c and whose sum is b.

$$x^2 + bx + c = (x + m)(x + n)$$

1. If c is positive, then m and n have like signs that match the sign of b.
2. If c is negative, then m and n have unlike signs.
3. If $|b|$ is small relative to $|c|$, first try those factors of c that are closest to each other in absolute value.

EXAMPLE 1 Factoring a Trinomial

Factor the trinomial $x^2 + 5x - 6$.

SOLUTION

You need to find two numbers whose product is -6 and whose sum is 5.

The product of -1 and 6 is -6.

$$x^2 + 5x - 6 = (x - 1)(x + 6)$$

The sum of -1 and 6 is 5.

EXAMPLE 2 Factoring a Trinomial

Factor the trinomial $x^2 - x - 6$.

SOLUTION

The product of -3 and 2 is -6.

$$x^2 - x - 6 = (x - 3)(x + 2)$$

The sum of -3 and 2 is -1.

Study Tip

Use a list to help you find the two numbers with the required product and sum. For Example 2:

Factors of -6	Sum
1, -6	-5
-1, 6	5
2, -3	-1
-2, 3	1

Because -1 is the required sum, the correct factorization is $x^2 - x - 6 = (x - 3)(x + 2)$.

Exercises Within Reach ®

Solutions in English & Spanish and tutorial videos at CollegePrepAlgebra.com

Finding a Missing Factor In Exercises 1 and 2, determine the missing factor. Then check your answer by multiplying the factors.

1. $x^2 + 8x + 7 = (x + 7)()$

2. $x^2 + 2x - 3 = (x + 3)()$

Factoring a Trinomial In Exercises 3–6, factor the trinomial.

3. $x^2 + 6x + 8$

4. $x^2 + 12x + 35$

5. $x^2 + 2x - 15$

6. $x^2 + 4x - 21$

Study Tip

Not all trinomials are factorable using integer factors. For instance, $x^2 - 2x - 6$ is not factorable using integer factors because there is no pair of factors of -6 whose sum is -2. Such nonfactorable trinomials are called **prime polynomials**.

EXAMPLE 3 **Factoring Trinomials**

Factor the trinomial.

a. $x^2 - 5x + 6$

b. $14 + 5x - x^2$

c. $x^2 - 8x - 48$

d. $x^2 + 7x - 30$

SOLUTION

a. You need to find two numbers whose product is 6 and whose sum is -5.

The product of -2 and -3 is 6.

$$x^2 - 5x + 6 = (x - 2)(x - 3)$$

The sum of -2 and -3 is -5.

b. First, factor out -1 and write the polynomial factor in standard form. Then find two numbers whose product is -14 and whose sum is -5.

The product of -7 and 2 is -14.

$$14 + 5x - x^2 = -(x^2 - 5x - 14) = -(x - 7)(x + 2).$$

The sum of -7 and 2 is -5.

c. You need to find two numbers whose product is -48 and whose sum is -8.

The product of -12 and 4 is -48.

$$x^2 - 8x - 48 = (x - 12)(x + 4)$$

The sum of -12 and 4 is -8.

d. You need to find two numbers whose product is -30 and whose sum is 7.

The product of -3 and 10 is -30.

$$x^2 + 7x - 30 = (x - 3)(x + 10)$$

The sum of -3 and 10 is 7.

Exercises Within Reach ®

Solutions in English & Spanish and tutorial videos at CollegePrepAlgebra.com

Finding the Possible Binomial Factors In Exercises 7−10, find all possible products of the form $(x + m)(x + n)$ where $m \cdot n$ is the specified product. (Assume that m and n are integers.)

7. $m \cdot n = 11$

8. $m \cdot n = 5$

9. $m \cdot n = 14$

10. $m \cdot n = -10$

Factoring a Trinomial In Exercises 11−18, factor the trinomial.

11. $x^2 - 9x - 22$

12. $x^2 - 9x - 10$

13. $x^2 - 9x + 14$

14. $x^2 + 10x + 24$

15. $2x + 15 - x^2$

16. $3x + 18 - x^2$

17. $x^2 + 3x - 70$

18. $x^2 - 13x + 40$

Application ▎EXAMPLE 4▎ **Geometry: Area of a Rectangle**

The area of a rectangle shown in the figure is $(x^2 + 30x + 200)$ square feet. What is the area of the shaded region?

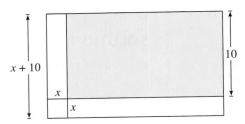

SOLUTION

Verbal Model: ▎Area = Length × Width▎

| *Labels:* | Area = $x^2 + 30x + 200$ | (square feet) |
| | Width = $x + 10$ | (feet) |

Equation: $x^2 + 30x + 200 = (\text{Length})(x + 10)$

$= (x + 20)(x + 10)$

The length of the rectangle is $(x + 20)$ feet. So, the length of the shaded region is 20 feet, which means that the area of the shaded region is 200 square feet.

Exercises Within Reach ®

Solutions in English & Spanish and tutorial videos at CollegePrepAlgebra.com

19. *Geometry* The area of the rectangle shown in the figure is $x^2 + 10x + 21$. What is the area of the shaded region?

20. *Geometry* The area of the rectangle shown in the figure is $x^2 + 17x + 70$. What is the area of the shaded region?

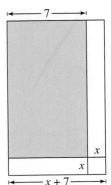

21. *Geometry* The area of the rectangle shown in the figure is $x^2 + 13x + 36$. What is the length of the shaded region?

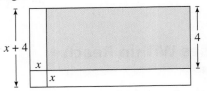

22. *Geometry* The area of the rectangle shown in the figure is $x^2 + 24x + 135$. What is the length of the shaded region?

Factoring Trinomials in Two Variables

Study Tip

With *any* factoring problem, remember that you can check your result by multiplying. For instance, in Example 5, you can check the result by multiplying $(x - 4y)$ by $(x + 3y)$ to obtain $x^2 - xy - 12y^2$.

EXAMPLE 5   Factoring a Trinomial in Two Variables

Factor the trinomial $x^2 - xy - 12y^2$.

SOLUTION

You need to find two numbers whose product is -12 and whose sum is -1.

The product of -4 and 3 is -12.

$$x^2 - xy - 12y^2 = (x - 4y)(x + 3y)$$

The sum of -4 and 3 is -1.

EXAMPLE 6   Factoring a Trinomial in Two Variables

Factor the trinomial $x^2 + 11xy + 10y^2$.

SOLUTION

You need to find two numbers whose product is 10 and whose sum is 11.

The product of 1 and 10 is 10.

$$x^2 + 11xy + 10y^2 = (x + y)(x + 10y)$$

The sum of 1 and 10 is 11.

EXAMPLE 7   Factoring a Trinomial in Two Variables

Factor the trinomial $y^2 - 6xy + 8x^2$.

SOLUTION

You need to find two numbers whose product is 8 and whose sum is -6.

The product of -2 and -4 is 8.

$$y^2 - 6xy + 8x^2 = (y - 2x)(y - 4x)$$

The sum of -2 and -4 is -6.

Exercises Within Reach ®

Solutions in English & Spanish and tutorial videos at CollegePrepAlgebra.com

Factoring a Trinomial In Exercises 23−30, factor the trinomial.

23. $x^2 - 7xz - 18z^2$

24. $u^2 - 4uv - 5v^2$

25. $x^2 - 5xy + 6y^2$

26. $x^2 + xy - 2y^2$

27. $x^2 + 8xy + 15y^2$

28. $x^2 + 15xy + 50y^2$

29. $a^2 + 2ab - 15b^2$

30. $y^2 + 4yz - 60z^2$

Factoring Completely

Some trinomials have a common monomial factor. In such cases, you should first factor out the common monomial factor. Then you can try to factor the resulting trinomial by the methods of this section. This "multiple-stage factoring process" is called **factoring completely**. The trinomial below is completely factored.

$$2x^2 - 4x - 6 = 2(x^2 - 2x - 3)$$ Factor out common monomial factor 2.

$$= 2(x - 3)(x + 1)$$ Factor trinomial.

EXAMPLE 8 **Factoring Completely**

Factor the trinomial $2x^2 - 12x + 10$ completely.

SOLUTION

$$2x^2 - 12x + 10 = 2(x^2 - 6x + 5)$$ Factor out common monomial factor 2.

$$= 2(x - 5)(x - 1)$$ Factor trinomial.

EXAMPLE 9 **Factoring Completely**

Factor the trinomial $3x^3 - 27x^2 + 54x$ completely.

SOLUTION

$$3x^3 - 27x^2 + 54x = 3x(x^2 - 9x + 18)$$ Factor out common monomial factor 3x.

$$= 3x(x - 3)(x - 6)$$ Factor trinomial.

EXAMPLE 10 **Factoring Completely**

Factor the trinomial $4y^4 + 32y^3 + 28y^2$ completely.

SOLUTION

$$4y^4 + 32y^3 + 28y^2 = 4y^2(y^2 + 8y + 7)$$ Factor out common monomial factor $4y^2$.

$$= 4y^2(y + 1)(y + 7)$$ Factor trinomial.

Exercises Within Reach ®

Solutions in English & Spanish and tutorial videos at CollegePrepAlgebra.com

Factoring Completely In Exercises 31−42, factor the trinomial completely.

31. $4x^2 - 32x + 60$

32. $4y^2 - 8y - 12$

33. $9x^2 + 18x - 18$

34. $6x^2 - 24x - 6$

35. $x^3 - 13x^2 + 30x$

36. $x^3 + x^2 - 2x$

37. $3x^3 + 18x^2 + 24x$

38. $4x^3 + 8x^2 - 12x$

39. $x^4 - 5x^3 + 6x^2$

40. $x^4 + 3x^3 - 10x^2$

41. $2x^4 - 20x^3 + 42x^2$

42. $5x^4 - 10x^3 - 240x^2$

Application   EXAMPLE 11   Geometry: Volume of an Open Box

An open box is to be made from a four-foot-by-six-foot sheet of metal by cutting equal squares from the corners and turning up the sides. The volume of the box can be modeled by $V = 4x^3 - 20x^2 + 24x$, $0 < x < 2$.

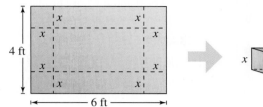

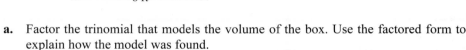

a. Factor the trinomial that models the volume of the box. Use the factored form to explain how the model was found.

b. Use a spreadsheet to approximate the size of the squares to be cut from the corners so that the box has the maximum volume.

SOLUTION

a. $4x^3 - 20x^2 + 24x = 4x(x^2 - 5x + 6)$ Factor out common monomial factor $4x$.

$= 4x(x - 3)(x - 2)$ Factored form

Because $4 = (-2)(-2)$, you can rewrite the factored form as

$4x(x - 3)(x - 2) = x[(-2)(x - 3)][(-2)(x - 2)]$

$= x(6 - 2x)(4 - 2x)$

$= (6 - 2x)(4 - 2x)(x)$.

The model was found by multiplying the length, width, and height of the box.

Length   Width   Height

$\text{Volume} = (6 - 2x)(4 - 2x)(x)$

b. From the spreadsheet at the left, you can see that the maximum volume of the box is about 8.45 cubic feet. This occurs when the value of x is about 0.8 foot.

	A	B
1	**x**	**Volume**
2	0.0	0.00
3	0.1	2.20
4	0.2	4.03
5	0.3	5.51
6	0.4	6.66
7	0.5	7.50
8	0.6	8.06
9	0.7	8.37
10	0.8	8.45
11	0.9	8.32
12	1.0	8.00
13	1.1	7.52
14	1.2	6.91
15	1.3	6.19
16	1.4	5.38
17	1.5	4.50
18	1.6	3.58
19	1.7	2.65
20	1.8	1.73
21	1.9	0.84
22	2.0	0.00

Exercises Within Reach®

43. *Geometry* The box in Example 11 is to be made from a six-foot-by-eight-foot sheet of metal. The volume of the box is modeled by

$V = 4x^3 - 28x^2 + 48x$, $0 < x < 3$.

(a) Factor the trinomial that models the volume of the box. Use the factored form to explain how the model was found.

(b) Use a spreadsheet to approximate the size of the squares to be cut from the corners so that the box has the maximum volume.

44. *Geometry* The box in Example 11 is to be made from an eight-foot-by-ten-foot sheet of metal. The volume of the box is modeled by

$V = 4x^3 - 36x^2 + 80x$, $0 < x < 4$.

(a) Factor the trinomial that models the volume of the box. Use the factored form to explain how the model was found.

(b) Use a spreadsheet to approximate the size of the squares to be cut from the corners so that the box has the maximum volume.

Concept Summary: Factoring Trinomials

What

The most common technique for factoring trinomials is guess, check, and revise. But the guidelines for factoring $x^2 + bx + c$ can help make the process more efficient.

EXAMPLE

Factor the trinomial $x^2 + 2x - 35$.

How

To factor $x^2 + bx + c$, find two numbers m and n whose product is c and whose sum is b.

EXAMPLE

The product of 7 and -5 is -35.

$$x^2 + 2x - 35 = (x + 7)(x - 5)$$

The sum of 7 and -5 is 2.

Why

Knowing how to factor trinomials can help you factor trinomials completely. Also, the techniques for factoring trinomials will help you when you learn to solve quadratic equations.

Exercises Within Reach ®

Worked-out solutions to odd-numbered exercises at CollegePrepAlgebra.com

Concept Summary Check

45. *Unlike Signs* For the trinomial $x^2 + 2x - 35$, how do you know that m and n will have unlike signs in the factored form $(x + m)(x + n)$?

46. *Possible Factors* When factoring $x^2 + 2x - 35$, why should you first try factors of 35 that are close to each other in absolute value?

47. *Factoring a Trinomial* To factor $x^2 + 2x - 35$ into the form $(x + m)(x + n)$, you need to find two numbers m and n with what product and what sum?

48. *Determining Signs* What are the signs of m and n for $x^2 + bx + c = (x + m)(x + n)$ when b is negative and c is positive?

Extra Practice

Factoring Completely In Exercises 49–56, factor the trinomial completely. (*Note:* some of the trinomials may be prime.)

49. $y^2 + 5y + 11$

50. $x^2 - x - 36$

51. $x^3 + 5x^2y + 6xy^2$

52. $x^2y - 6xy^2 + y^3$

53. $3z^2 + 5z + 6$

54. $7x^2 + 5x + 10$

55. $2x^3y + 4x^2y^2 - 6xy^3$

56. $2x^3y - 10x^2y^2 + 6xy^3$

Finding Coefficients In Exercises 57 and 58, find all integers b such that the trinomial can be factored.

57. $x^2 + bx + 18$

58. $x^2 + bx + 10$

Finding Constant Terms In Exercises 59 and 60, find two integers c such that the trinomial can be factored. (There are many correct answers.)

59. $x^2 + 3x + c$

60. $x^2 + 5x + c$

Geometric Model of Factoring In Exercises 61 and 62, factor the trinomial and draw a geometric model of the result. [The sample shows a geometric model for factoring $x^2 + 3x + 2 = (x + 1)(x + 2)$.]

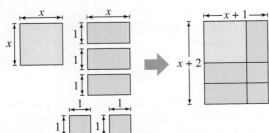

61. $x^2 + 4x + 3$

62. $x^2 + 5x + 6$

63. *Geometry* An open box is to be made from a sheet of metal that is y feet long and z feet wide by cutting equal squares from the corners and turning up the sides. The volume of the box can be modeled by

$$V = 4x^3 - 24x^2 + 32x, \quad 0 < x < 2.$$

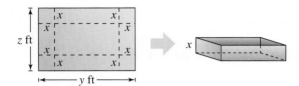

(a) Use x, y, and z to write an expression for the volume V of the box in the form

$V = $ Length $\times$ Width $\times$ Height. (Do not simplify.)

(b) Factor the trinomial that models the volume of the box.

(c) Use the expressions for the volume of the box in parts (a) and (b) to find the dimensions y and z.

64. *Geometry* An open box is to be made from a sheet of metal that is y feet long and z feet wide by cutting equal squares from the corners and turning up the sides. The volume of the box can be modeled by

$$V = 4x^3 - 44x^2 + 120x, \quad 0 < x < 5.$$

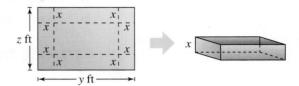

(a) Use x, y, and z to write an expression for the volume V of the box in the form

$V = $ Length $\times$ Width $\times$ Height. (Do not simplify.)

(b) Factor the trinomial that models the volume of the box.

(c) Use the expressions for the volume of the box in parts (a) and (b) to find the dimensions y and z.

Explaining Concepts

65. *Precision* State which of the following are factorizations of $2x^2 + 6x - 20$. For each correct factorization, state whether or not it is completely factored.

(a) $(2x - 4)(x + 5)$

(b) $(2x - 4)(2x + 10)$

(c) $(x - 2)(x + 5)$

(d) $2(x - 2)(x + 5)$

66. *Vocabulary* What is a prime trinomial?

67. *Structure* In factoring $x^2 - 4x + 3$, why is it unnecessary to test $(x - 1)(x + 3)$ and $(x + 1)(x - 3)$?

68. *Writing* In factoring the trinomial $x^2 + bx + c$, is the process easier if c is a prime number such as 5 or if c is a composite number such as 120? Explain.

Cumulative Review

In Exercises 69–76, solve the equation and check your solution.

69. $5x - 9 = 26$

70. $3x - 5 = 16$

71. $5 - 3x = 6$

72. $7 - 2x = 9$

73. $7x - 12 = 3x$

74. $10x + 24 = 2x$

75. $5x - 16 = 7x - 9$

76. $3x - 8 = 9x + 4$

In Exercises 77–84, find the greatest common factor of the expressions.

77. $6x^3, 52x^6$

78. $35t^2, 7t^8$

79. a^5b^4, a^3b^7

80. xy^3, x^2y^2

81. $18r^3s^3, -54r^5s^3$

82. $12xy^3, 28x^3y^4$

83. $16u^2v^5, 8u^4v^3, 2u^3v^7$

84. $21xy^4, 42x^2y^2, 9x^4y$

6.3 More About Factoring Trinomials

▶ Factor trinomials of the form $ax^2 + bx + c$.
▶ Factor trinomials completely.
▶ Factor trinomials by grouping.

Factoring Trinomials of the Form $ax^2 + bx + c$

To see how to factor a trinomial whose leading coefficient is not 1, consider the following.

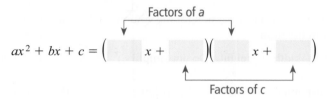

The goal is to find a combination of factors of a and c such that the outer and inner products add up to the middle term bx.

EXAMPLE 1　Factoring a Trinomial

Factor the trinomial $4x^2 - 4x - 3$.

SOLUTION

First, observe that $4x^2 - 4x - 3$ has no common monomial factor. For this trinomial, $a = 4$ and $c = -3$. You need to find a combination of the factors of 4 and -3 such that the outer and inner products add up to $-4x$. The possible combinations are as follows.

Factors	$O + I$	
Inner product $= 4x$		
$(x + 1)(4x - 3)$	$-3x + 4x = x$	x does not equal $-4x$.
Outer product $= -3x$		
$(x - 1)(4x + 3)$	$3x - 4x = -x$	$-x$ does not equal $-4x$.
$(x + 3)(4x - 1)$	$-x + 12x = 11x$	$11x$ does not equal $-4x$.
$(x - 3)(4x + 1)$	$x - 12x = -11x$	$-11x$ does not equal $-4x$.
$(2x + 1)(2x - 3)$	$-6x + 2x = -4x$	$-4x$ equals $-4x$.
$(2x - 1)(2x + 3)$	$6x - 2x = 4x$	$4x$ does not equal $-4x$.

So, the correct factorization is $4x^2 - 4x - 3 = (2x + 1)(2x - 3)$.

Exercises Within Reach ®

Solutions in English & Spanish and tutorial videos at CollegePrepAlgebra.com

Factoring a Trinomial In Exercises 1−4, **determine the missing factor.**

1. $2x^2 + 7x - 4 = (2x - 1)()$

2. $3x^2 + x - 4 = (3x + 4)()$

3. $3t^2 + 4t - 15 = (3t - 5)()$

4. $5t^2 + t - 18 = (5t - 9)()$

Factoring a Trinomial In Exercises 5−8, **factor the trinomial.**

5. $2x^2 + 5x + 3$

6. $3x^2 + 7x + 2$

7. $4y^2 + 5y + 1$

8. $3x^2 + 5x - 2$

Guidelines for Factoring $ax^2 + bx + c$ ($a > 0$)

1. When a trinomial has a common monomial factor, you should factor out the common factor before trying to find the binomial factors.
2. Because the resulting trinomial has no common monomial factors, you do not have to test any binomial factors that have a common monomial factor.
3. Switch the signs of the factors of c when the middle term $(O + I)$ is correct except in sign.

EXAMPLE 2   Factoring a Trinomial

Factor the trinomial $2x^2 + x - 15$.

SOLUTION

First, observe that $2x^2 + x - 15$ has no common monomial factor. For this trinomial, $a = 2$, which factors as $(1)(2)$, and $c = -15$, which factors as $(1)(-15)$, $(-1)(15)$, $(3)(-5)$, and $(-3)(5)$.

$$(2x + 1)(x - 15) = 2x^2 - 29x - 15$$

$$(2x + 15)(x - 1) = 2x^2 + 13x - 15$$

$$(2x + 3)(x - 5) = 2x^2 - 7x - 15$$

$$(2x + 5)(x - 3) = 2x^2 - x - 15 \qquad \text{Middle term has opposite sign.}$$

$$(2x - 5)(x + 3) = 2x^2 + x - 15 \qquad \Longleftarrow \quad \text{Correct factorization}$$

So, the correct factorization is $2x^2 + x - 15 = (2x - 5)(x + 3)$.

EXAMPLE 3   Factoring a Trinomial

Factor the trinomial $6x^2 + 5x - 4$.

SOLUTION

$$(x + 4)(6x - 1) = 6x^2 + 23x - 4 \qquad \text{23x does not equal 5x.}$$

$$(2x + 1)(3x - 4) = 6x^2 - 5x - 4 \qquad \text{Opposite sign}$$

$$(2x - 1)(3x + 4) = 6x^2 + 5x - 4 \qquad \text{Correct factorization}$$

So, the correct factorization is $6x^2 + 5x - 4 = (2x - 1)(3x + 4)$.

Exercises Within Reach ®

Solutions in English & Spanish and tutorial videos at CollegePrepAlgebra.com

Factoring a Trinomial In Exercises 9−16, factor the trinomial. (*Note:* Some of the trinomials may be prime.)

9. $6y^2 - 7y + 1$

10. $7a^2 - 9a + 2$

11. $5x^2 - 2x + 1$

12. $4z^2 - 8z + 1$

13. $4x^2 + 13x - 12$

14. $16y^2 + 24y - 27$

15. $9x^2 - 18x + 8$

16. $25a^2 - 40a + 12$

Factoring Completely

> **EXAMPLE 4** **Factoring Completely**

Factor $4x^3 - 30x^2 + 14x$ completely.

SOLUTION

Begin by factoring out the common monomial factor.

$$4x^3 - 30x^2 + 14x = 2x(2x^2 - 15x + 7)$$

Now, for the new trinomial $2x^2 - 15x + 7$, $a = 2$, and $c = 7$. The possible factorizations of the trinomial are as follows.

$$(2x - 7)(x - 1) = 2x^2 - 9x + 7$$

$$(2x - 1)(x - 7) = 2x^2 - 15x + 7 \quad \Leftarrow \quad \text{Correct factorization}$$

So, the complete factorization of the original trinomial is

$$4x^3 - 30x^2 + 14x = 2x(2x^2 - 15x + 7)$$

$$= 2x(2x - 1)(x - 7).$$

> **EXAMPLE 5** **A Negative Leading Coefficient**

Factor the trinomial $-5x^2 + 7x + 6$.

SOLUTION

This trinomial has a negative leading coefficient, so you should begin by factoring -1 out of the trinomial.

$$-5x^2 + 7x + 6 = (-1)(5x^2 - 7x - 6)$$

Now, for the new trinomial $5x^2 - 7x - 6$, $a = 5$, and $c = -6$. After testing the possible factorizations, you can conclude that

$$(x - 2)(5x + 3) = 5x^2 - 7x - 6. \quad \Leftarrow \quad \text{Correct factorization}$$

So, the correct factorization is

$$-5x^2 + 7x + 6 = (-1)(x - 2)(5x + 3)$$

$$= (-x + 2)(5x + 3). \qquad \text{Distributive Property}$$

Another correct factorization is $(x - 2)(-5x - 3)$.

Exercises Within Reach ®

Solutions in English & Spanish and tutorial videos at CollegePrepAlgebra.com

Factoring Completely In Exercises 17−22, factor the trinomial completely.

17. $2v^2 + 8v - 42$

18. $4z^2 - 12z - 40$

19. $9z^2 - 24z + 15$

20. $6x^2 + 8x - 8$

21. $16s^3 - 28s^2 + 6s$

22. $18v^3 + 3v^2 - 6v$

A Negative Leading Coefficient In Exercises 23−26, factor the trinomial.

23. $-2x^2 + 7x + 9$

24. $-5x^2 + x + 4$

25. $-6x^2 + 7x + 10$

26. $3 + 2x - 8x^2$

Application   EXAMPLE 6   **Geometry: The Dimensions of a Sandbox**

The sandbox shown in the figure below has a height of x feet and a width of $(x + 2)$ feet. The volume of the sandbox is $(2x^3 + 7x^2 + 6x)$ cubic feet.

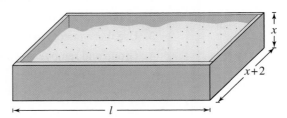

a. Find the length of the sandbox.

b. How many cubic yards of sand are needed to fill the sandbox when $x = 2$?

SOLUTION

a. *Verbal Model:*   Volume = Length × Width × Height

Labels: Volume = $2x^3 + 7x^2 + 6x$ (cubic feet)
Width = $x + 2$ (feet)
Height = x (feet)

Equation: $2x^3 + 7x^2 + 6x = x(2x^2 + 7x + 6)$
$= x(x + 2)(2x + 3)$

So, the length of the sandbox is $(2x + 3)$ feet.

b. When $x = 2$, the volume of the sandbox is

Volume $= 2(2^3) + 7(2^2) + 6(2)$

$= 2(8) + 7(4) + 12$

$= 56 \text{ ft}^3$.

One cubic yard contains 27 cubic feet. So, you need 56/27 or about 2 cubic yards of sand to fill the sandbox.

Exercises Within Reach ®   Solutions in English & Spanish and tutorial videos at CollegePrepAlgebra.com

27. *Geometry* The shower stall has a width of x feet and a depth of $(x + 2)$ feet. The volume of the shower stall is $(2x^3 + 3x^2 - 2x)$ cubic feet.

(a) Find the height of the shower stall.

(b) What is the volume of the shower stall when $x = 4$?

28. *Geometry* The box has a width of x centimeters and a height of $(5x - 1)$ centimeters. The volume of the box is $(15x^3 + 7x^2 - 2x)$ cubic centimeters.

(a) Find the length of the box.

(b) What is the volume of the box when $x = 5$?

Factoring Trinomials by Grouping

Guidelines for Factoring $ax^2 + bx + c$ by Grouping

1. If necessary, write the trinomial in standard form.
2. Choose factors of the product ac that add up to b.
3. Use these factors to rewrite the middle term as a sum or difference.
4. Group and remove any common monomial factors from the first two terms and the last two terms.
5. If possible, factor out the common binomial factor.

EXAMPLE 7 **Factoring a Trinomial by Grouping**

Use factoring by grouping to factor the trinomial $2x^2 + 5x - 3$.

SOLUTION

$2x^2 + 5x - 3 = 2x^2 + 6x - x - 3$	Rewrite middle term.
$= (2x^2 + 6x) + (-x - 3)$	Group terms.
$= 2x(x + 3) - (x + 3)$	Factor out common monomial factor in each group.
$= (x + 3)(2x - 1)$	Factor out common binomial factor.

So, the trinomial factors as $2x^2 + 5x - 3 = (x + 3)(2x - 1)$.

EXAMPLE 8 **Factoring a Trinomial by Grouping**

Use factoring by grouping to factor the trinomial $6x^2 - 11x - 10$.

SOLUTION

$6x^2 - 11x - 10 = 6x^2 - 15x + 4x - 10$	Rewrite middle term.
$= (6x^2 - 15x) + (4x - 10)$	Group terms.
$= 3x(2x - 5) + 2(2x - 5)$	Factor out common monomial factor in each group.
$= (2x - 5)(3x + 2)$	Factor out common binomial factor.

So, the trinomial factors as $6x^2 - 11x - 10 = (2x - 5)(3x + 2)$.

Exercises Within Reach ®

Solutions in English & Spanish and tutorial videos at CollegePrepAlgebra.com

Factoring a Trinomial by Grouping **In Exercises 29−36, factor the trinomial by grouping.**

29. $3x^2 + 4x + 1$

30. $2x^2 + 5x + 2$

31. $7x^2 + 20x - 3$

32. $5x^2 - 14x - 3$

33. $6x^2 + 5x - 4$

34. $12y^2 + 11y + 2$

35. $15x^2 - 11x + 2$

36. $12x^2 - 13x + 1$

Application   EXAMPLE 9   Geometry: **The Dimensions of a Swimming Pool**

The swimming pool shown in the figure has a depth of d feet and a length of $(5d + 2)$ feet. The volume of the swimming pool is $(15d^3 - 14d^2 - 8d)$ cubic feet.

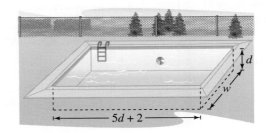

a. Find the width of the swimming pool.

b. How many cubic feet of water are needed to fill the swimming pool when $d = 6$?

SOLUTION

a. *Verbal Model:*   Volume = Length × Width × Depth

   Labels: Volume = $15d^3 - 14d^2 - 8d$ (cubic feet)
                Width = $5d + 2$ (feet)
                Depth = d (feet)

   Equation: $15d^3 - 14d^2 - 8d = d(15d^2 - 14d - 8)$

$$= d(15d^2 + 6d - 20d - 8)$$

$$= d[3d(5d + 2) - 4(5d + 2)]$$

$$= d(5d + 2)(3d - 4)$$

So, the width of the swimming pool is $(3d - 4)$ feet.

b. When $d = 6$, the dimensions of the swimming pool are 6 feet by 32 feet by 14 feet. So, the volume is $6(32)(14) = 2688$ cubic feet. There are about 7.5 gallons in 1 cubic foot. So, the swimming pool holds about 20,160 gallons of water.

Exercises Within Reach ®

Solutions in English & Spanish and tutorial videos at CollegePrepAlgebra.com

37. *Geometry* The fire box of a wood stove is x inches deep and $(2x - 7)$ inches wide. The volume of the fire box is $(2x^3 - 3x^2 - 14x)$ cubic inches.

(a) Find the height of the fire box.

(b) What is the volume of the fire box when $x = 15$?

38. *Geometry* The block of ice has a width of x inches and a length of $(2x + 5)$ inches. The volume of the block is $(6x^3 + 17x^2 + 5x)$ cubic inches.

(a) Find the height of the block of ice.

(b) What is the volume of the block of ice when $x = 10$?

Concept Summary: *Factoring Trinomials of the Form* $ax^2 + bx + c$

What

Using some simple guidelines can help you factor trinomials of the form $ax^2 + bx + c$, where $a > 0$ and $a \neq 1$.

EXAMPLE

Factor the trinomial

$3x^2 + 13x - 16$.

How

Guidelines for factoring $ax^2 + bx + c$

1. If necessary, write the trinomial in standard form.

2. Factor out any common monomial factors.

3. Choose factors of a and c such that the outer and inner products add up to the middle term bx.

Factors of a

$$ax^2 + bx + c = (\quad x + \quad)(\quad x + \quad)$$

Factors of c

EXAMPLE

$(3x - 1)(x + 16) = 3x^2 + 47x - 16$
$(3x - 16)(x + 1) = 3x^2 - 13x - 16$
$(3x + 16)(x - 1) = 3x^2 + 13x - 16$
So, $3x^2 + 13x - 16 = (3x + 16)(x - 1)$

Why

Trinomials are used in many geometric applications. Knowing how to factor trinomials will help you solve these applications.

Exercises Within Reach ®

Worked-out solutions to odd-numbered exercises at CollegePrepAlgebra.com

Concept Summary Check

39. **Common Monomial Factor** Does the trinomial $3x^2 + 13x - 16$ have a common monomial factor?

40. **Testing Factors** How do you know that $(3x - 1)(x + 16)$ is not the correct factorization of $3x^2 + 13x - 16$?

41. **Reasoning** Use the first step in the solution above to explain why the factorization $(3x + 1)(x - 16)$ was not tested.

42. **Reasoning** Use the second step in the solution above to explain why the factorization $(3x + 16)(x - 1)$ was tested.

Extra Practice

Factoring Completely In Exercises 43 – 48, factor the trinomial completely.

43. $-15x^4 - 2x^3 + 8x^2$

44. $15y^2 - 7y^3 - 2y^4$

45. $6x^3 + 24x^2 - 192x$

46. $35x + 28x^2 - 7x^3$

47. $18u^4 + 18u^3 - 27u^2$

48. $12x^5 - 16x^4 + 8x^3$

Finding Coefficients In Exercises 49 – 54, find all integers b such that the trinomial can be factored.

49. $3x^2 + bx + 10$

50. $4x^2 + bx + 3$

51. $2x^2 + bx - 6$

52. $5x^2 + bx - 6$

53. $6x^2 + bx + 20$

54. $8x^2 + bx - 18$

Finding Constant Terms In Exercises 55 – 60, find two integers c such that the trinomial can be factored. (There are many correct answers.)

55. $4x^2 + 3x + c$

56. $2x^2 + 5x + c$

57. $3x^2 - 10x + c$

58. $8x^2 - 3x + c$

59. $6x^2 - 5x + c$

60. $4x^2 - 9x + c$

61. Geometry The area of the rectangle shown in the figure is $2x^2 + 9x + 10$. What is the area of the shaded region?

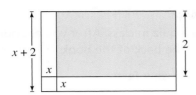

62. Geometry The area of the rectangle shown in the figure is $3x^2 + 10x + 3$. What is the area of the shaded region?

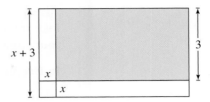

Explaining Concepts

63. Writing What is the first step in factoring $ax^2 + bx + c$ by grouping?

64. Error Analysis Describe and correct the error.

$$9x^2 - 9x - 54 = (3x + 6)(3x - 9)$$
$$= 3(x + 2)(x - 3)$$

65. Reasoning Without multiplying the factors, explain why $(2x + 3)(x + 5)$ is not a factorization of $2x^2 + 7x - 15$.

66. Structure Give an example of a prime trinomial of the form $ax^2 + bx + c$.

67. Structure Give an example of a trinomial of the form $ax^3 + bx^2 + cx$ that has a common monomial factor of $2x$.

68. Reasoning Can a trinomial with a leading coefficient not equal to 1 have two identical factors? If so, give an example.

69. Reasoning How many possible factorizations are there to consider for a trinomial of the form $ax^2 + bx + c$, when a and c are prime? Explain your reasoning.

70. Explanation Many people think the technique of factoring a trinomial by grouping is more efficient than the *guess, check, and revise* strategy, especially when the coefficients a and c have many factors. Try factoring $6x^2 - 13x + 6$, $2x^2 + 5x - 12$, and $3x^2 + 11x - 4$ using both methods. Which method do you prefer? Explain the advantages and disadvantages of each method.

Cumulative Review

In Exercises 71−74, write the prime factorization of the number.

71. 500

72. 315

73. 792

74. 2275

In Exercises 75−78, perform the multiplication and simplify.

75. $(2x - 5)(x + 7)$

76. $(3x - 2)^2$

77. $(7y + 2)(7y - 2)$

78. $(3y + 8)(9y + 3)$

In Exercises 79−82, factor the trinomial.

79. $x^2 - 4x - 45$

80. $y^2 + 2y - 15$

81. $z^2 + 22z + 40$

82. $x^2 - 12x + 20$

Mid-Chapter Quiz: Sections 6.1–6.3

Solutions in English & Spanish and tutorial videos at CollegePrepAlgebra.com

Take this quiz as you would take a quiz in class. After you are done, check your work against the answers in the back of the book.

In Exercises 1–4, determine the missing factor.

1. $\frac{2}{3}x - \frac{1}{3} = \frac{1}{3}\big($ ⬚ $\big)$

2. $x^2y - xy^2 = xy\big($ ⬚ $\big)$

3. $y^2 + y - 42 = (y + 7)\big($ ⬚ $\big)$

4. $3y^2 - y - 30 = (3y - 10)\big($ ⬚ $\big)$

In Exercises 5–16, factor the polynomial completely.

5. $9x^2 + 21$

6. $5a^3 - 25a^2$

7. $x(x + 7) - 6(x + 7)$

8. $t^3 - 3t^2 + t - 3$

9. $y^2 + 11y + 30$

10. $u^2 + u - 56$

11. $x^3 - x^2 - 30x$

12. $2x^2y + 8xy - 64y$

13. $2y^2 - 3y - 27$

14. $6 - 13z - 5z^2$

15. $12x^2 - 5x - 2$

16. $10s^4 - 14s^3 + 2s^2$

17. Find all integers b such that the trinomial

   $x^2 + bx + 12$

   can be factored. Describe the method you used.

18. Find two integers c such that the trinomial

   $x^2 - 10x + c$

   can be factored. Describe the method you used. (There are many correct answers.)

19. Find all possible products of the form

   $(3x + m)(x + n)$

   such that $m \cdot n = 6$.

Application

20. The area of the rectangle shown in the figure is $3x^2 + 38x + 80$. What is the area of the shaded region?

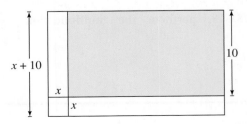

Study Skills in Action

Being Confident

How does someone "get" confidence? Confidence is linked to another attribute called self-efficacy. Self-efficacy is the belief that one has the ability to accomplish a specific task. It is possible for a student to have high self-efficacy when it comes to writing a personal essay, but to have low self-efficacy when it comes to learning math.

A good way to foster self-efficacy is by building a support system. A support system should include teachers and family members who can encourage and guide you. It can also include other students or other adults who can help you study and stay focused.

Friends who share the same desire to do well can be the best type of support.

Smart Study Strategy

Build a Support System

1 ▶ **Surround yourself with positive friends.** Find another student in class with whom to study. Make sure this person is not anxious about math because you do not want another student's anxiety to increase your own. Arrange to meet and compare notes, homework, and so on at least two times per week. Friends can encourage each other.

2 ▶ **Find a place to study where other students are also studying.** Libraries, learning centers, and tutoring centers are great places to study. While studying in such places, you will be able to ask for assistance when you have questions. You do not want to study alone if you typically get down on yourself with lots of negative self-talk.

3 ▶ **Establish a relationship with a learning assistant.** Get to know someone who can help you find assistance for any type of academic issue. Learning assistants, tutors, family members, and teachers are excellent resources.

4 ▶ **Seek out assistance before you are overwhelmed.** Ask your teacher when you need help. Teachers are more than willing to help their students. Go with a friend if you are nervous about meeting with your teacher.

5 ▶ **Be your own support.** Listen to what you tell yourself when frustrated with studying math. Replace any negative self-talk dialog with more positive statements. Here are some examples of positive statements:

> "I may not have done well in the past, but I'm learning how to study math, and will get better."

> "It does not matter what others believe—I know that I can get through this class."

> "Wow, I messed up on a quiz. I need to talk to someone and figure out what I need to do differently."

6.4 Factoring Polynomials with Special Forms

▶ Factor the difference of two squares.

▶ Factor a polynomial completely.

▶ Identify and factor perfect square trinomials.

▶ Factor the sum or difference of two cubes.

Difference of Two Squares

> ### Difference of Two Squares
>
> Let a and b be real numbers, variables, or algebraic expressions.
>
> $$a^2 - b^2 = (a + b)(a - b)$$
>
> ↑   ↑   ↑
>
> Difference   Opposite signs

To recognize perfect square terms, look for coefficients that are squares of integers and for variables raised to *even* powers. Here are some examples.

Original Polynomials		*Difference of Squares*		*Factored Form*
$x^2 - 1$	⟹	$(x)^2 - (1)^2$	⟹	$(x + 1)(x - 1)$
$4x^2 - 9$	⟹	$(2x)^2 - (3)^2$	⟹	$(2x + 3)(2x - 3)$

EXAMPLE 1   **Factoring the Difference of Two Squares**

a. $x^2 - 36 = x^2 - 6^2$ Write as difference of two squares.

$\qquad\qquad = (x + 6)(x - 6)$ Factored form

b. $x^2 - \frac{4}{25} = x^2 - \left(\frac{2}{5}\right)^2$ Write as difference of two squares.

$\qquad\qquad = \left(x + \frac{2}{5}\right)\left(x - \frac{2}{5}\right)$ Factored form

c. $81x^2 - 49 = (9x)^2 - 7^2$ Write as difference of two squares.

$\qquad\qquad = (9x + 7)(9x - 7)$ Factored form

d. $(x + 1)^2 - 4 = (x + 1)^2 - 2^2$ Write as difference of two squares.

$\qquad\qquad = [(x + 1) + 2][(x + 1) - 2]$ Factored form

$\qquad\qquad = (x + 3)(x - 1)$ SImplify.

Exercises Within Reach ®

Solutions in English & Spanish and tutorial videos at CollegePrepAlgebra.com

Factoring the Difference to Two Squares In Exercises 1−8, factor the difference of two squares.

1. $x^2 - 9$ **2.** $y^2 - 49$ **3.** $u^2 - \frac{1}{4}$ **4.** $v^2 - \frac{4}{9}$

5. $16y^2 - 9$ **6.** $36z^2 - 121$ **7.** $(x - 1)^2 - 4$ **8.** $(t + 2)^2 - 9$

Application   EXAMPLE 2   **Removing a Common Monomial Factor First**

A hammer is dropped from the roof of a building. The height of the hammer is given by the expression $-16t^2 + 64$, where t is the time in seconds.

	A	B
1	**t**	**Height**
2	0	64
3	0.1	63.84
4	0.2	63.36
5	0.3	62.56
6	0.4	61.44
7	0.5	60
8	0.6	58.24
9	0.7	56.16
10	0.8	53.76
11	0.9	51.04
12	1	48
13	1.1	44.64
14	1.2	40.96
15	1.3	36.96
16	1.4	32.64

Spreadsheet at CollegePrepAlgebra.com

a. Factor the expression.

b. How many seconds does it take the hammer to fall to a height of 41 feet?

SOLUTION

a. $-16t^2 + 64 = -16(t^2 - 4)$ Factor out common monomial factor.

$= -16(t^2 - 2^2)$ Write a difference of two squares.

$= -16(t + 2)(t - 2)$ Factored form

b. Use a spreadsheet to find the heights of the hammer at 0.1-second intervals of the time t. From the spreadsheet at the left, you can see that the hammer falls to a height of 41 feet in about 1.2 seconds.

Exercises Within Reach ®   Solutions in English & Spanish and tutorial videos at CollegePrepAlgebra.com

9. **Free-Falling Object** The height of an object that is dropped from the top of the U.S. Steel Tower in Pittsburgh is given by the expression $-16t^2 + 841$, where t is the time in seconds.

(a) Factor this expression.

(b) Use a spreadsheet to determine how many seconds it takes the object to fall to a height of 805 feet. (Use 0.5-second intervals for t.

10. **Geometry** An *annulus* is the region between two concentric circles. The area of the annulus shown in the figure is $\pi R^2 - \pi r^2$.

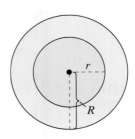

(a) Factor this expression.

(b) Use a spreadsheet to determine the value of R for which the annulus has an area of about 8 square feet when $r = 3$ feet. (Use 3.14 for π and increments of 0.1 for R.)

Factoring Completely

EXAMPLE 3 **Removing a Common Monomial Factor First**

Factor the polynomial $20x^3 - 5x$.

SOLUTION

$$20x^3 - 5x = 5x(4x^2 - 1)$$ Factor out common monomial factor $5x$.

$$= 5x[(2x)^2 - 1^2]$$ Write as difference of two squares.

$$= 5x(2x + 1)(2x - 1)$$ Factored form

EXAMPLE 4 **Factoring Completely**

Factor the polynomial completely.

a. $x^4 - 16$ **b.** $48x^4 - 3$

SOLUTION

a. Recognizing $x^4 - 16$ as a difference of two squares, you can write

$$x^4 - 16 = (x^2)^2 - 4^2$$ Write as difference of two squares.

$$= (x^2 + 4)(x^2 - 4).$$ Factored form

Note that the second factor, $(x^2 - 4)$, is itself a difference of two squares, and so

$$x^4 - 16 = (x^2 + 4)(x^2 - 4)$$ Factor as difference of two squares.

$$= (x^2 + 4)(x + 2)(x - 2).$$ Factor completely.

b. Start by removing the common monomial factor.

$$48x^4 - 3 = 3(16x^4 - 1)$$ Remove common monomial factor 3.

Recognizing $16x^4 - 1$ as the difference of two squares, you can write

$$48x^4 - 3 = 3(16x^4 - 1)$$ Factor out common monomial.

$$= 3[(4x^2)^2 - 1^2]$$ Write as difference of two squares.

$$= 3(4x^2 + 1)(4x^2 - 1)$$ Recognize $4x^2 - 1$ as a difference of two squares.

$$= 3(4x^2 + 1)[(2x)^2 - 1^2]$$ Write as difference of two squares.

$$= 3(4x^2 + 1)(2x + 1)(2x - 1).$$ Factor completely.

Study Tip

Note in Example 4 that no attempt is made to factor the *sum of two squares*. A second-degree polynomial that is the sum of two squares cannot be factored as the product of binomials (using integer coefficients). In general, *the sum of two squares is not factorable*.

Exercises Within Reach ®

Solutions in English & Spanish and tutorial videos at CollegePrepAlgebra.com

Removing a Common Monomial Factor First In Exercises 11−16, factor the polynomial.

11. $2x^2 - 72$

12. $3x^2 - 27$

13. $4x - 25x^3$

14. $a^3 - 16a$

15. $8y^3 - 50y$

16. $20x^3 - 180x$

Factoring Completely In Exercises 17−22, factor the polynomial completely.

17. $y^4 - 81$

18. $z^4 - 625$

19. $1 - x^4$

20. $256 - u^4$

21. $2x^4 - 162$

22. $5x^4 - 80$

Perfect Square Trinomials

A **perfect square trinomial** is the square of a binomial. For instance,

$$x^2 + 4x + 4 = (x + 2)(x + 2)$$
$$= (x + 2)^2$$

is the square of the binomial $(x + 2)$.

Perfect Square Trinomials

Let a and b be real numbers, variables, or algebraic expressions.

1. $a^2 + 2ab + b^2 = (a + b)^2$ **2.** $a^2 - 2ab + b^2 = (a - b)^2$

  Same sign   Same sign

Study Tip

To recognize a perfect square trinomial, remember that the first and last terms must be perfect squares and positive, and the middle term must be twice the product of a and b. (The middle term can be positive or negative.) Watch for squares of fractions.

$$4x^2 - \frac{4}{3}x + \frac{1}{9}$$

$$(2x)^2 \qquad \left(\frac{1}{3}\right)^2$$

$$2(2x)\left(\frac{1}{3}\right)$$

EXAMPLE 5   Identifying Perfect Square Trinomials

Which of the following are prefect square trinomials?

a. $m^2 - 4m + 4$ **b.** $4x^2 - 2x + 1$

c. $y^2 + 6y - 9$ **d.** $x^2 + x + \frac{1}{4}$

SOLUTION

a. This polynomial *is* a perfect square trinomial. It factors as $(m - 2)^2$.

b. This polynomial *is not* a perfect square trinomial because the middle term is not twice the product of $2x$ and 1.

c. This polynomial *is not* a perfect square trinomial because the last term, -9, is not positive.

d. This polynomial *is* a perfect square trinomial. It factors as $\left(x + \frac{1}{2}\right)^2$.

Exercises Within Reach ®

Solutions in English & Spanish and tutorial videos at CollegePrepAlgebra.com

Identifying a Perfect Square Trinomial In Exercises 23−28, **determine whether the polynomial is a perfect square trinomial.**

23. $9b^2 + 24b + 16$

24. $y^2 - 2y + 6$

25. $m^2 - 2m - 1$

26. $16n^2 + 2n + 1$

27. $4k^2 - 20k + 25$

28. $x^2 + 20x + 100$

Geometric Model of Factoring In Exercises 29 and 30, **write the factoring problem represented by the geometric factoring model.**

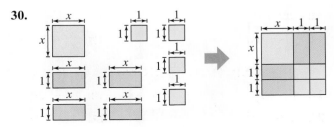

29.

30.

EXAMPLE 6   **Factoring a Perfect Square Trinomial**

Factor the trinomial $y^2 - 6y + 9$.

SOLUTION

$$y^2 - 6y + 9 = y^2 - 2(3y) + 3^2$$ Recognize the pattern.

$$= (y - 3)^2$$ Write in factored form.

EXAMPLE 7   **Factoring a Perfect Square Trinomial**

Factor the trinomial $16x^2 + 40x + 25$.

SOLUTION

$$16x^2 + 40x + 25 = (4x)^2 + 2(4x)(5) + 5^2$$ Recognize the pattern.

$$= (4x + 5)^2$$ Write in factored form.

EXAMPLE 8   **Factoring a Perfect Square Trinomial**

Factor the trinomial $9x^2 - 24xy + 16y^2$.

SOLUTION

$$9x^2 - 24xy + 16y^2 = (3x)^2 - 2(3x)(4y) + (4y)^2$$ Recognize the pattern.

$$= (3x - 4y)^2$$ Write in factored form.

Exercises Within Reach ®

Solutions in English & Spanish and tutorial videos at CollegePrepAlgebra.com

Factoring a Perfect Square Trinomial   **In Exercises 31−42, factor the perfect square trinomial.**

31. $x^2 - 8x + 16$ **32.** $x^2 + 10x + 25$ **33.** $x^2 + 14x + 49$

34. $a^2 - 12a + 36$ **35.** $4t^2 + 4t + 1$ **36.** $9x^2 - 12x + 4$

37. $25y^2 - 10y + 1$ **38.** $16z^2 + 24z + 9$ **39.** $x^2 - 6xy + 9y^2$

40. $16x^2 - 8xy + y^2$ **41.** $4y^2 + 20yz + 25z^2$ **42.** $36u^2 + 84uv + 49v^2$

43. *Geometry* The vaccine cooler shown in the figure has a square base and a height of x. The volume of the cooler is $(x^3 + 16x^2 + 64x)$ cubic centimeters. Find the dimensions of the cooler.

44. *Geometry* The building shown in the figure has a square base and a height of x. The volume of the building is $(x^3 - 80x^2 + 1600x)$ cubic feet. Find the dimensions of the building.

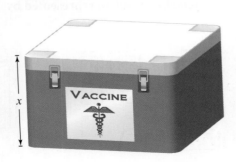

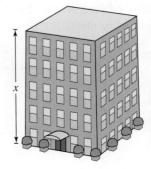

Sum or Difference of Two Cubes

Study Tip

When using either of the factoring patterns at the right, pay special attention to the signs. Remembering the "like" and "unlike" patterns for the signs is helpful.

Sum or Difference of Two Cubes

Let a and b be real numbers, variables, or algebraic expressions.

Like signs

1. $a^3 + b^3 = (a + b)(a^2 - ab + b^2)$

Unlike signs

Like signs

2. $a^3 - b^3 = (a - b)(a^2 + ab + b^2)$

Unlike signs

EXAMPLE 9   **Factoring the Sum or Difference of Two Cubes**

a. $y^3 + 27 = y^3 + 3^3$ Write as sum of two cubes.

$ = (y + 3)[y^2 - (y)(3) + 3^2]$ Factored form

$ = (y + 3)(y^2 - 3y + 9)$ Simplify.

b. $64 - x^3 = 4^3 - x^3$ Write as difference of two cubes.

$ = (4 - x)[4^2 + (4)(x) + x^2]$ Factored form

$ = (4 - x)(16 + 4x + x^2)$ Simplify.

c. $2x^3 - 16 = 2(x^3 - 8)$ Factor out common monomial factor 2.

$ = 2(x^3 - 2^3)$ Write as difference of two cubes.

$ = 2(x - 2)[x^2 + (x)(2) + 2^2]$ Factored form

$ = 2(x - 2)(x^2 + 2x + 4)$ Simplify.

Exercises Within Reach ®

Solutions in English & Spanish and tutorial videos at CollegePrepAlgebra.com

Factoring the Sum or Difference of Two Cubes In Exercises 45–54, factor the sum or difference of two cubes.

45. $x^3 - 8$

46. $x^3 - 27$

47. $y^3 + 64$

48. $z^3 + 125$

49. $1 + 8t^3$

50. $1 + 27s^3$

51. $27u^3 - 8$

52. $64v^3 - 125$

53. $27x^3 + 64y^3$

54. $27y^3 + 125z^3$

Concept Summary: Factoring Polynomials with Special Forms

What

When you are asked to factor a polynomial, you should identify whether the polynomial involves a special form.

EXAMPLE

Factor the polynomial.

a. $3x^2 - 12$

b. $16 - 2t^3$

c. $x^3 + 4x^2 + 4x$

How

Factor out any common monomial factor and then compare the remaining polynomial to the special polynomial forms.

EXAMPLE

Difference of two squares

a. $3x^2 - 12 = 3\overbrace{(x^2 - 4)}$

Difference of two cubes

b. $16 - 2t^3 = 2\overbrace{(8 - t^3)}$

Perfect square trinomial

c. $x^3 + 4x^2 + 4x = x\overbrace{(x^2 + 4x + 4)}$

Why

Recognizing polynomials involving special forms can help you factor such polynomials *completely*.

EXAMPLE

a. $3x^2 - 12 = 3(x + 2)(x - 2)$

b. $16 - 2t^3 = 2(2 - t)(4 + 2t + t^2)$

c. $x^3 + 4x^2 + 4x = x(x + 2)^2$

Exercises Within Reach ®

Worked-out solutions to odd-numbered exercises at CollegePrepAlgebra.com

Concept Summary Check

55. *Special Polynomial Forms* Are any of the polynomials $3x^2 - 12$, $16 - 2t^3$, and $x^3 + 4x^2 + 4x$ in one of the special polynomial forms?

56. *Finding Special Polynomial Forms* What was done to each polynomial in the example above to reveal a special polynomial form?

57. *Analyzing a Perfect Square Trinomial* The polynomial $x^2 + 4x + 4$ is a perfect square trinomial of the form $a^2 + 2ab + b^2$. Identify a and b.

58. *Factoring a Polynomial* What is the first thing you should do when you factor a polynomial?

Extra Practice

Finding Coefficients In Exercises 59−62, find two real numbers b such that the expression is a perfect square trinomial.

59. $x^2 + bx + 1$ **60.** $x^2 + bx + 100$ **61.** $4x^2 + bx + 81$ **62.** $4x^2 + bx + 9$

Finding a Constant Term In Exercises 63−66, find a real number c such that the expression is a perfect square trinomial.

63. $x^2 + 6x + c$ **64.** $x^2 + 10x + c$ **65.** $y^2 - 4y + c$ **66.** $z^2 - 14z + c$

Factoring a Polynomial In Exercises 67−78, factor the polynomial completely. (*Note:* Some of the polynomials may be prime.)

67. $y^4 - 25y^2$ **68.** $x^6 - 49x^4$ **69.** $x^2 - 2x + 1$

70. $81 + 18x + x^2$ **71.** $(t - 1)^2 - 121$ **72.** $(x - 3)^2 - 100$

73. $x^2 + 81$ **74.** $x^2 + 16$ **75.** $x^4 - 81$

76. $2x^4 - 32$ **77.** $x^3 + 4x^2 - x + 4$ **78.** $y^3 + 3y^2 - 4y - 12$

Mental Math In Exercises 79−82, evaluate the quantity mentally using the two samples as models.

$$29^2 = (30 - 1)^2 = 30^2 - 2 \cdot 30 \cdot 1 + 1^2$$
$$= 900 - 60 + 1 = 841$$

$$48 \cdot 52 = (50 - 2)(50 + 2)$$
$$= 50^2 - 2^2 = 2496$$

79. 21^2 **80.** 49^2 **81.** $59 \cdot 61$ **82.** $28 \cdot 32$

Rewriting a Trinomial **In Exercise 83 and 84, write the polynomial as the difference of two squares. Use the result to factor the polynomial completely.**

83. $x^2 + 6x + 8 = (x^2 + 6x + 9) - 1$

$$= \boxed{}^2 - \boxed{}^2$$

84. $x^2 + 8x + 12 = (x^2 + 8x + 16) - 4$

$$= \boxed{}^2 - \boxed{}^2$$

85. *Modeling* The figure below shows two cubes: a large cube whose volume is a^3 and a smaller cube whose volume is b^3. If the smaller cube is removed from the larger, the remaining solid has a volume of $a^3 - b^3$ and is composed of three rectangular boxes, labeled Box 1, Box 2, and Box 3. Find the volume of each box and describe how these results are related to the following special product pattern.

$$a^3 - b^3 = (a - b)(a^2 + ab + b^2)$$
$$= (a - b)a^2 + (a - b)ab + (a - b)b^2$$

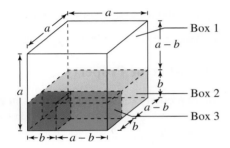

86. *Geometry* From the eight vertices of a cube of dimension x, cubes of dimension y are removed (see figure).

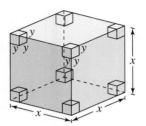

(a) Write an expression for the volume of the solid that remains after the eight cubes at the vertices are removed.

(b) Factor the expression for the volume in part (a).

(c) In the context of this problem, y must be less than what multiple of x? Explain your answer geometrically and from the result of part (b).

Explaining Concepts

True or False? **In Exercise 87–90, determine whether the statement is true or false. Justify your answer.**

87. The expression $x(x + 2) - 2(x + 2)$ is completely factored.

88. $x^2 + 4 = (x + 2)^2$

89. $x^3 - 27 = (x - 3)^3$

90. Because the sum of two squares cannot be factored, it follows that the sum of two cubes cannot be factored.

91. *Writing* Explain how to identify and factor the difference of two squares.

92. *Writing* Explain how to identify and factor a perfect square trinomial.

Cumulative Review

In Exercises 93–96, solve the equation and check your solution.

93. $7 + 5x = 7x - 1$

94. $2 - 5(x - 1) = 2[x + 10(x - 1)]$

95. $2(x + 1) = 0$

96. $\frac{3}{4}(12x - 8) = 10$

In Exercises 97–100, factor the trinomial.

97. $2x^2 + 7x + 3$

98. $3y^2 - 5y - 12$

99. $6m^2 + 7m - 20$

100. $15x^2 - 28x + 12$

6.5 Solving Polynomial Equations by Factoring

▶ Use the Zero-Factor Property to solve equations.

▶ Solve quadratic equations by factoring.

▶ Solve higher-degree polynomial equations by factoring.

▶ Solve application problems by factoring.

The Zero-Factor Property

Zero-Factor Property

Let a and b be real numbers, variables, or algebraic expressions. If a and b are factors such that

$$ab = 0$$

then $a = 0$ or $b = 0$. This property also applies to three or more factors.

EXAMPLE 1   **Using the Zero-Factor Property**

The Zero-Factor Property is the primary property for solving equations in algebra. For instance, to solve the equation

$$(x - 1)(x + 2) = 0 \qquad \text{Original equation}$$

you can use the Zero-Factor Property to conclude that either $(x - 1)$ or $(x + 2)$ equals 0. Setting the first factor equal to 0 implies that $x = 1$ is a solution.

$$x - 1 = 0 \quad \Longrightarrow \quad x = 1 \qquad \text{First solution}$$

Similarly, setting the second factor equal to 0 implies that $x = -2$ is a solution.

$$x + 2 = 0 \quad \Longrightarrow \quad x = -2 \qquad \text{Second solution}$$

So, the equation $(x - 1)(x + 2) = 0$ has exactly two solutions: $x = 1$ and $x = -2$. Check these solutions by substituting them into the original equation.

Exercises Within Reach ®

Solutions in English & Spanish and tutorial videos at CollegePrepAlgebra.com

Understanding the Zero-Factor Property In Exercises 1−6, determine whether the equation is written in the correct form to apply the Zero-Factor Property.

1. $x(x - 1) = 2$

2. $(x - 1)(x + 1) = 1$

3. $(x + 2) + (x - 1) = 0$

4. $x(x - 3) = 0$

5. $(x - 1)(x + 2) = 0$

6. $3(x^2 + x) = 0$

Using the Zero-Factor Property In Exercises 7−16, use the Zero-Factor Property to solve the equation.

7. $x(x - 4) = 0$

8. $z(z + 6) = 0$

9. $(y - 3)(y + 10) = 0$

10. $(s - 7)(s + 4) = 0$

11. $25(a + 4)(a - 2) = 0$

12. $17(t - 3)(t + 8) = 0$

13. $(2t + 5)(3t + 1) = 0$

14. $(5x - 3)(2x - 8) = 0$

15. $(x - 3)(2x + 1)(x + 4) = 0$

16. $(y - 39)(2y + 7)(y + 12) = 0$

Solving Quadratic Equations by Factoring

Definition of Quadratic Equation

A **quadratic equation** is an equation that can be written in the general form

$$ax^2 + bx + c = 0 \qquad \text{Quadratic equation}$$

where a, b, and c are real number with $a \neq 0$.

Study Tip

In Section 3.1, you learned that the basic idea in solving a linear equation is to *isolate the variable*. Notice in Example 2 that the basic idea in solving a quadratic equation is to factor the left side so that the equation can be converted into two linear equations.

EXAMPLE 2   **Solving a Quadratic Equation by Factoring**

Solve $x^2 - x - 6 = 0$.

SOLUTION

First, make sure that the right side of the equation is zero. Next, factor the left side of the equation. Finally, apply the Zero-Factor Property to find the solutions.

$x^2 - x - 6 = 0$	Write original equation.
$(x + 2)(x - 3) = 0$	Factor left side of equation.
$x + 2 = 0 \quad\Longrightarrow\quad x = -2$	Set 1st factor equal to 0 and solve for x.
$x - 3 = 0 \quad\Longrightarrow\quad x = 3$	Set 2nd factor equal to 0 and solve for x.

The solutions are $x = -2$ and $x = 3$.

Guidelines for Solving Quadratic Equations

1. Write the quadratic equation in general form.

2. Factor the left side of the equation.

3. Set each factor with a variable equal to zero.

4. Solve each linear equation.

5. Check each solution in the original equation.

Exercises Within Reach ®

Solutions in English & Spanish and tutorial videos at CollegePrepAlgebra.com

Solving a Quadratic Equation by Factoring In Exercises 17–26, **solve** the equation by factoring.

17. $x^2 + 2x = 0$

18. $x^2 - 5x = 0$

19. $x^2 - 8x = 0$

20. $x^2 + 16x = 0$

21. $x^2 - 25 = 0$

22. $x^2 - 121 = 0$

23. $x^2 - 3x - 10 = 0$

24. $x^2 - x - 12 = 0$

25. $x^2 - 10x + 24 = 0$

26. $x^2 - 13x + 42 = 0$

Study Tip

Be sure you see that one side of an equation must be zero to apply the Zero-Factor Property. For instance, in Example 3, you cannot simply factor the left side to obtain $x(2x + 5) = 12$ and assume that $x = 12$ and $2x + 5 = 12$ yield correct solutions. In fact, neither of the resulting solutions satisfies the original equation.

EXAMPLE 3   Solving a Quadratic Equation by Factoring

Solve $2x^2 + 5x = 12$.

SOLUTION

$2x^2 + 5x = 12$	Write original equation.
$2x^2 + 5x - 12 = 0$	Write in general form.
$(2x - 3)(x + 4) = 0$	Factor left side of equation.
$2x - 3 = 0 \implies x = \frac{3}{2}$	Set 1st factor equal to 0 and solve for x.
$x + 4 = 0 \implies x = -4$	Set 2nd factor equal to 0 and solve for x.

The solutions are $x = \frac{3}{2}$ and $x = -4$.

EXAMPLE 4   A Quadratic Equation with a Repeated Solution

Solve $x^2 - 2x + 16 = 6x$.

SOLUTION

$x^2 - 2x + 16 = 6x$	Write original equation.
$x^2 - 8x + 16 = 0$	Write in general form.
$(x - 4)^2 = 0$	Factor.
$x - 4 = 0 \quad \text{or} \quad x - 4 = 0$	Set factors equal to 0.
$x = 4$	Solve for x.

Note that even though the left side of this equation has two factors, the factors are the same. So, the only solution of the equation is

$$x = 4.$$

This solution is called a **repeated solution**.

Exercises Within Reach ®

Solutions in English & Spanish and tutorial videos at CollegePrepAlgebra.com

Solving a Quadratic Equation by Factoring In Exercises 27–40, solve the equation by factoring.

27. $4x^2 + 15x = 25$

28. $14x^2 + 9x = -1$

29. $7 + 13x - 2x^2 = 0$

30. $11 + 32y - 3y^2 = 0$

31. $3y^2 - 2 = -y$

32. $-2x - 15 = -x^2$

33. $-13x + 36 = -x^2$

34. $x^2 - 15 = -2x$

35. $m^2 - 8m + 18 = 2$

36. $a^2 + 4a + 10 = 6$

37. $x^2 + 16x + 57 = -7$

38. $x^2 - 12x + 21 = -15$

39. $4z^2 - 12z + 15 = 6$

40. $16t^2 + 48t + 40 = 4$

Solving Higher-Degree Equations by Factoring

> **EXAMPLE 5**   **Solving a Polynomial Equation with Three Factors**

$3x^3 = 15x^2 + 18x$	Original equation
$3x^3 - 15x^2 - 18x = 0$	Write in general form.
$3x(x^2 - 5x - 6) = 0$	Factor out common factor.
$3x(x - 6)(x + 1) = 0$	Factor.
$3x = 0 \implies x = 0$	Set 1st factor equal to 0.
$x - 6 = 0 \implies x = 6$	Set 2nd factor equal to 0.
$x + 1 = 0 \implies x = -1$	Set 3rd factor equal to 0.

The solutions are $x = 0$, $x = 6$, and $x = -1$. Check these three solutions.

> **EXAMPLE 6**   **Solving a Polynomial Equation with Four Factors**

$x^4 + x^3 - 4x^2 - 4x = 0$	Original equation
$x(x^3 + x^2 - 4x - 4) = 0$	Factor out common factor.
$x[(x^3 + x^2) + (-4x - 4)] = 0$	Group terms.
$x[x^2(x + 1) - 4(x + 1)] = 0$	Factor grouped terms.
$x[(x + 1)(x^2 - 4)] = 0$	Distributive Property
$x(x + 1)(x + 2)(x - 2) = 0$	Difference of two squares
$x = 0 \implies x = 0$	
$x + 1 = 0 \implies x = -1$	
$x + 2 = 0 \implies x = -2$	
$x - 2 = 0 \implies x = 2$	

The solutions are $x = 0$, $x = -1$, $x = -2$, and $x = 2$. Check these four solutions.

Exercises Within Reach ®

Solutions in English & Spanish and tutorial videos at CollegePrepAlgebra.com

Solving a Polynomial Equation by Factoring In Exercises 41−54, solve the equation by factoring.

41. $x^3 - 19x^2 + 84x = 0$

42. $x^3 + 18x^2 + 45x = 0$

43. $6t^3 = t^2 + t$

44. $3u^3 = 5u^2 + 2u$

45. $z^2(z + 2) - 4(z + 2) = 0$

46. $16(3 - u) - u^2(3 - u) = 0$

47. $a^3 + 2a^2 - 9a - 18 = 0$

48. $x^3 - 2x^2 - 4x + 8 = 0$

49. $c^3 - 3c^2 - 9c + 27 = 0$

50. $v^3 + 4v^2 - 4v - 16 = 0$

51. $x^4 - 3x^3 - x^2 + 3x = 0$

52. $x^4 + 2x^3 - 9x^2 - 18x = 0$

53. $8x^4 + 12x^3 - 32x^2 - 48x = 0$

54. $9x^4 - 15x^3 - 9x^2 + 15x = 0$

Applications

Application **EXAMPLE 7** Geometry: **Dimensions of a Room**

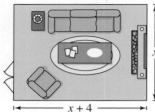

A rectangular room has an area of 192 square feet. The length of the room is 4 feet more than its width, as shown at the left. Find the dimensions of the room.

SOLUTION

Verbal Model: ☐ Length • Width ☐ = ☐ Area ☐

Labels: Length = $x + 4$ (feet)
Width = x (feet)
Area = 192 (square feet)

Equation:
$$(x + 4)x = 192$$
$$x^2 + 4x - 192 = 0$$
$$(x + 16)(x - 12) = 0$$
$$x = -16 \quad \text{or} \quad x = 12$$

Because the negative solution does not make sense, choose the positive solution $x = 12$. When the width of the room is 12 feet, the length of the room is

Length = $x + 4 = 12 + 4 = 16$ feet.

So, the dimensions of the room are 12 feet by 16 feet.

Exercises Within Reach ®

Solutions in English & Spanish and tutorial videos at CollegePrepAlgebra.com

55. *Geometry* The rectangular floor of a storage shed has an area of 540 square feet. The length of the floor is 7 feet more than its width (see figure). Find the dimensions of the floor.

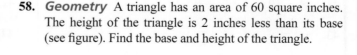

56. *Geometry* The outside dimensions of a picture frame are 28 centimeters and 20 centimeters (see figure). The picture alone has an area of 468 square centimeters. Find the width w of the frame.

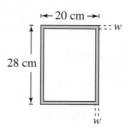

57. *Geometry* A triangle has an area of 27 square inches. The height of the triangle is $1\frac{1}{2}$ times its base (see figure). Find the base and height of the triangle.

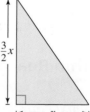

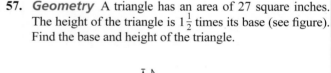

58. *Geometry* A triangle has an area of 60 square inches. The height of the triangle is 2 inches less than its base (see figure). Find the base and height of the triangle.

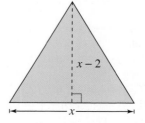

Application   EXAMPLE 8   Free-Falling Object

A rock is dropped into a well from a height of 64 feet above the water (see figure). The height h (in feet) of the rock relative to the surface of the water is modeled by the position function $h(t) = -16t^2 + 64$, where t is the time (in seconds) since the rock was dropped. How long does it take the rock to reach the water?

SOLUTION

The surface of the water corresponds to a height of 0 feet. So, substitute 0 for $h(t)$ in the equation, and solve for t.

$0 = -16t^2 + 64$	Substitute 0 for $h(t)$.
$16t^2 - 64 = 0$	Write in general form.
$16(t^2 - 4) = 0$	Factor out common factor.
$16(t + 2)(t - 2) = 0$	Difference of two squares
$t = -2 \quad \text{or} \quad t = 2$	Solutions using Zero-Factor Property

Because a time of -2 seconds does not make sense, choose the positive solution $t = 2$, and conclude that the rock reaches the water 2 seconds after it is dropped.

64 ft

Exercises Within Reach ®

Solutions in English & Spanish and tutorial videos at CollegePrepAlgebra.com

59. *Free-Falling Object* A tool is dropped from a construction project 400 feet above the ground. The height h (in feet) of the tool is modeled by the position function

$$h(t) = -16t^2 + 400$$

where t is the time in seconds. How long does it take for the tool to reach the ground?

60. *Free-Falling Object* A penny is dropped from the roof of a building 256 feet above the ground. The height h (in feet) of the penny is modeled by the position function

$$h(t) = -16t^2 + 256$$

where t is the time in seconds. How long does it take for the penny to reach the ground?

61. *Free-Falling Object* You throw a baseball upward with an initial velocity of 30 feet per second. The height h (in feet) of the baseball relative to your glove is modeled by the position function $h(t) = -16t^2 + 30t$, where t is the time in seconds. How long does it take for the ball to reach your glove?

62. *Free-Falling Object* An object is thrown upward from the Royal Gorge Bridge in Colorado, 1053 feet above the Arkansas River, with an initial velocity of 48 feet per second. The height h (in feet) of the object is modeled by the position function $h(t) = -16t^2 + 48t + 1053$, where t is the time in seconds. How long does it take for the object to reach the river?

Concept Summary: *Solving Quadratic Equations by Factoring*

What

To solve a **quadratic equation** by factoring, use the guidelines for solving quadratic equations.

EXAMPLE

Solve $2x^2 + 18x = 0$.

How

- Write the quadratic equation in **general form**.
- Factor the left side of the equation.
- Set each factor with a variable equal to zero.
- Solve each linear equation.

EXAMPLE

$$2x^2 + 18x = 0$$
$$2x(x + 9) = 0$$
$$2x = 0 \implies x = 0$$
$$x + 9 = 0 \implies x = -9$$

Why

You can solve quadratic equations to answer many real-life problems, such as finding the length of time it will take a falling object to reach the ground. You can use the same guidelines to solve polynomial equations of higher degrees.

Exercises Within Reach ®

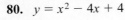

Worked-out solutions to odd-numbered exercises at CollegePrepAlgebra.com

Concept Summary Check

63. **General Form** Is the equation $2x^2 + 18x = 0$ in general form? Explain.

64. **Factoring the Left Side** What are the factors of the left side of the equation $2x^2 + 18x = 0$?

65. **Using a Property** How is the Zero-Factor Property used in the example above?

66. **Verifying Solutions** How can you verify the solutions in the example above?

Extra Practice

Solving a Quadratic Equation by Factoring In Exercises 67−78, solve the equation by factoring.

67. $8x^2 = 5x$

68. $5x^2 = 7x$

69. $x(x - 5) = 36$

70. $s(s + 4) = 96$

71. $x(x + 2) - 10(x + 2) = 0$

72. $x(x - 15) + 3(x - 15) = 0$

73. $(x - 4)(x + 5) = 10$

74. $(u - 6)(u + 4) = -21$

75. $81 - (x + 4)^2 = 0$

76. $(s + 5)^2 - 49 = 0$

77. $(t - 2)^2 = 16$

78. $(s + 4)^2 = 49$

Graphical Reasoning In Exercises 79−82, determine the x-intercepts of the graph and **explain** how the x-intercepts correspond to the solutions of the polynomial equation when $y = 0$.

79. $y = x^2 - 9$

80. $y = x^2 - 4x + 4$

81. $y = x^3 - 6x^2 + 9x$

82. $y = x^3 - 3x^2 - x + 3$

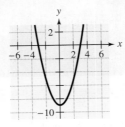
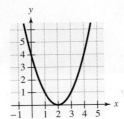
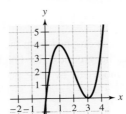
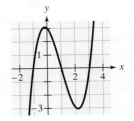

Think About It In Exercises 83 and 84, find a quadratic equation with the given solutions.

83. $x = -2, \quad x = 6$

84. $x = -2, \quad x = 4$

85. ***Number Problem*** The sum of a positive number and its square is 240. Find the number.

86. ***Number Problem*** Find two consecutive positive integers whose product is 132.

87. ***Free-Falling Object*** An object falls from the roof of a building 80 feet above the ground toward a balcony 16 feet above the ground. The object's height h (in feet) relative to the ground after t seconds is modeled by the equation $h = -16t^2 + 80$. How long does it take for the object to reach the balcony?

88. ***Free-Falling Object*** Your friend stands 96 feet above you on a cliff. You throw an object upward with an initial velocity of 80 feet per second. The object's height h (in feet) after t seconds is modeled by the equation $h = -16t^2 + 80t$. How long does it take for the object to reach your friend on the way up? On the way down?

89. ***Break-Even Analysis*** The revenue R from the sale of x home theater systems is given by $R = 140x - x^2$. The cost of producing x systems is given by $C = 2000 + 50x$. How many home theater systems can be produced and sold in order to break even?

90. ***Break-Even Analysis*** The revenue R from the sale of x digital cameras is given by $R = 120x - x^2$. The cost of producing x digital cameras is given by $C = 1200 + 40x$. How many cameras can be produced and sold in order to break even?

Explaining Concepts

91. ***Structure*** What is the maximum number of solutions of an nth-degree polynomial equation? Give an example of a third-degree equation that has only one real number solution.

92. ***Structure*** What is the maximum number of first-degree factors that an nth-degree polynomial equation can have? Explain.

93. ***Think About It*** A quadratic equation has a repeated solution. Describe the x-intercept(s) of the graph of the equation formed by replacing 0 with y in the general form of the equation.

94. ***Reasoning*** A third-degree polynomial equation has two solutions. What must be special about one of the solutions? Explain.

95. ***Reasoning*** There are some polynomial equations that have real number solutions but cannot be solved by factoring. Explain how this can be.

96. ***Using a Graphing Calculator*** The polynomial equation $x^3 - x - 3 = 0$ *cannot* be solved algebraically using any of the techniques described in this book. It does, however, have one solution that is a real number.

(a) Use a graphing calculator to graph the related equation $y = x^3 - x - 3$. Use the graph to estimate the solution of the original equation.

(b) Use the *table* feature of a graphing calculator to create a table and estimate the solution.

Cumulative Review

In Exercises 97–100, find the unit price (in dollars per ounce) of the product.

97. A 12-ounce soda for $0.75

98. A 12-ounce package of brown-and-serve rolls for $1.89

99. A 30-ounce can of pumpkin pie filling for $2.13

100. Turkey meat priced at $0.94 per pound

In Exercises 101–104, find the domain of the function.

101. $f(x) = \dfrac{x + 3}{x + 1}$

102. $f(x) = \dfrac{12}{x - 2}$

103. $g(x) = \sqrt{3 - x}$

104. $h(x) = \sqrt{x^2 - 4}$

6 Chapter Summary

What did you learn?	Explanation and Examples	Review Exercises
6.1 Find the greatest common factor of two or more expressions *(p. 270)*.	$60x^4 = 2 \cdot 2 \cdot 3 \cdot 5 \cdot x \cdot x \cdot x \cdot x = (20x^3)(3x)$ $40x^3 = 2 \cdot 2 \cdot 2 \cdot 5 \cdot x \cdot x \cdot x = (20x^3)(2)$ The greatest common factor of $60x^4$ and $40x^3$ is $20x^3$.	1−8
Factor out the greatest common monomial factor from polynomials *(p. 271)*.	Use the Distributive Property to factor out the greatest common monomial factor from each term of a polynomial.	9−22
Factor polynomials by grouping *(p. 273)*.	For polynomials with four terms, group terms that have a common monomial factor. Factor the two groupings and then look for a common binomial factor.	23−32
6.2 Factor trinomials of the form $x^2 + bx + c$ *(p. 278)*.	To factor $x^2 + bx + c$, you need to find two numbers m and n whose product is c and whose sum is b. $$x^2 + bx + c = (x + m)(x + n)$$ **1.** If c is positive, then m and n have like signs that match the sign of b. **2.** If c is negative, then m and n have unlike signs. **3.** If $|b|$ is small relative to $|c|$, first try those factors of c that are closest to each other in absolute value.	33−48
Factor trinomials in two variables *(p. 281)*.	To factor $x^2 + bxy + cy^2$, find two factors of c whose sum is b. The product of -4 and 3 is -12. $$x^2 - xy - 12y^2 = (x - 4y)(x + 3y)$$ The sum of -4 and 3 is -1.	49−54
Factor trinomials completely *(p. 282)*.	When a trinomial has a common monomial factor, you should factor out the common factor first. Then factor the resulting trinomial. Be sure to include the common monomial factor in the final factored form.	55−62
6.3 Factor trinomials of the form $ax^2 + bx + c$ *(p. 286)*.	**1.** When a trinomial has a common monomial factor, you should factor out the common factor before trying to find the binomial factors. **2.** Because the resulting trinomial has no common monomial factors, you do not have to test any binomial factors that have a common monomial factor. **3.** Switch the signs of the factors of c when the middle term (O + I) is correct except in sign.	63−80
Factor trinomials completely *(p. 288)*.	Remember that when a trinomial has a common monomial factor, you should factor out the common factor first. The complete factorization will show all monomial and binomial factors.	81−92

What did you learn?	Explanation and Examples	Review Exercises
6.3 Factor trinomials by grouping *(p. 290)*.	1. If necessary, write the trinomial in standard form. 2. Choose factors of the product ac that add up to b. 3. Use these factors to rewrite the middle term as a sum or difference. 4. Group and remove any common monomial factors from the first two terms and the last two terms. 5. If possible, factor out the common binomial factor.	93–100
Factor the difference of two squares *(p. 296)*.	$a^2 - b^2 = (a + b)(a - b)$ Difference Opposite signs	101−110
Factor a polynomial completely *(p. 298)*.	1. Factor out any common factors. 2. Factor according to one of the special polynomial forms: difference of two squares, sum or difference of two cubes, or perfect square trinomials. 3. Factor trinomials, $ax^2 + bx + c$, with $a = 1$ or $a \neq 1$. 4. Factor by grouping—for polynomials with four terms. 5. Check to see whether the factors themselves can be factored.	111−118
6.4 Identify and factor perfect square trinomials *(p. 299)*.	$a^2 + 2ab + b^2 = (a + b)^2$ Same sign $a^2 - 2ab + b^2 = (a - b)^2$ Same sign	119 − 130
Factor the sum or difference of two cubes *(p. 301)*.	Like signs $a^3 + b^3 = (a + b)(a^2 - ab + b^2)$ Unlike signs Like signs $a^3 - b^3 = (a - b)(a^2 + ab + b^2)$ Unlike signs	131 − 136
6.5 Use the Zero-Factor Property to solve equations *(p. 304)*.	Let a and b be real numbers, variables, or algebraic expressions. If a and b are factors such that $ab = 0$, then $a = 0$ or $b = 0$. This property also applies to three or more factors.	137−142
Solve quadratic equations by factoring *(p. 305)*.	1. Write the quadratic equation in general form. 2. Factor the left side of the equation. 3. Set each factor with a variable equal to zero. 4. Solve each linear equation. 5. Check each solution in the original equation.	143−150
Solve higher-degree polynomial equations by factoring *(p. 307)*.	Use the same steps as for solving quadratic equations.	151−158
Solve application problems by factoring *(p. 308)*.	When you solve application problems by factoring, check your answers. Eliminate answers that are not appropriate in the context of the problem.	159−163

Review Exercises

Worked-out solutions to odd-numbered exercises at CollegePrepAlgebra.com

6.1

Finding the Greatest Common Factor In Exercises 1–8, find the greatest common factor of the expressions.

1. t^2, t^5
2. $-y^3, y^8$
3. $3x^4, 21x^2$
4. $14z^2, 21z$
5. $14x^2y^3, -21x^3y^5$
6. $-15y^2z^2, 5y^2z$
7. $8x^2y, 24xy^2, 4xy$
8. $27ab^5, 9ab^6, 18a^2b^3$

Greatest Common Monomial Factor In Exercises 9–20, factor out the greatest common monomial factor from the polynomial.

9. $3x - 6$
10. $7 + 21x$
11. $3t - t^2$
12. $u^2 - 6u$
13. $5x^2 + 10x^3$
14. $7y - 21y^4$
15. $8a^2 - 12a^3$
16. $14x - 26x^4$
17. $5x^3 + 5x^2 - 5x$
18. $6u - 9u^2 + 15u^3$
19. $8y^2 + 4y + 12$
20. $3z^4 - 21z^3 + 10z$

Geometry In Exercises 21 and 22, write an expression for the area of the shaded region and factor the expression.

21.
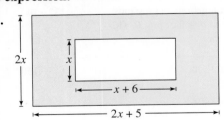

22.
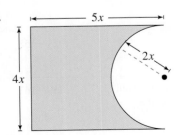

Factoring by Grouping In Exercises 23–32, factor the polynomial by grouping.

23. $x(x + 1) - 3(x + 1)$
24. $5(y - 3) - y(y - 3)$
25. $2u(u - 2) + 5(u - 2)$
26. $7(x + 8) + 3x(x + 8)$
27. $y^3 + 3y^2 + 2y + 6$
28. $z^3 - 5z^2 + z - 5$
29. $x^3 + 2x^2 + x + 2$
30. $x^3 - 5x^2 + 5x - 25$
31. $x^2 - 4x + 3x - 12$
32. $2x^2 + 6x - 5x - 15$

6.2

Factoring a Trinomial In Exercises 33–44, factor the trinomial.

33. $x^2 - 3x - 28$
34. $x^2 - 3x - 40$
35. $u^2 + 5u - 36$
36. $y^2 + 15y + 56$
37. $x^2 - 2x - 24$
38. $x^2 + 8x + 15$
39. $y^2 + 10y + 21$
40. $a^2 - 7a + 12$
41. $b^2 + 13b - 30$
42. $z^2 - 9z + 18$
43. $w^2 + 3w - 40$
44. $x^2 - 7x - 8$

Finding Coefficients In Exercises 45–48, find all integers b such that the trinomial can be factored.

45. $x^2 + bx + 9$
46. $y^2 + by + 25$
47. $z^2 + bz + 11$
48. $x^2 + bx + 14$

Factoring a Trinomial In Exercises 49–54, factor the trinomial.

49. $x^2 + 9xy - 10y^2$
50. $u^2 + 3uv - 4v^2$
51. $y^2 - 6xy - 27x^2$
52. $v^2 + 18uv + 32u^2$
53. $x^2 - 2xy - 8y^2$
54. $a^2 - ab - 30b^2$

Factoring Completely In Exercises 55–62, factor the trinomial completely.

55. $4x^2 - 24x + 32$

56. $3u^2 - 6u - 72$

57. $x^3 + 9x^2 + 18x$

58. $y^3 - 8y^2 + 15y$

59. $3x^2 + 18x - 81$

60. $8x^2 - 48x + 64$

61. $4x^3 + 36x^2 + 56x$

62. $2y^3 - 4y^2 - 30y$

6.3

Factoring a Trinomial In Exercises 63–76, factor the trinomial.

63. $3x^2 + 2x - 5$

64. $8x^2 - 18x + 9$

65. $2x^2 - 5x - 7$

66. $2x^2 - 3x - 35$

67. $6x^2 + 7x + 2$

68. $16x^2 + 13x - 3$

69. $4y^2 - 3y - 1$

70. $5x^2 - 12x + 7$

71. $3x^2 + 7x - 6$

72. $45y^2 - 8y - 4$

73. $3x^2 + 5x - 2$

74. $7x^2 - 4x - 3$

75. $2x^2 - 3x + 1$

76. $3x^2 + 8x + 4$

Finding Coefficients In Exercises 77 and 78, find all integers b such that the trinomial can be factored.

77. $x^2 + bx - 24$

78. $2x^2 + bx - 16$

Finding Constant Terms In Exercises 79 and 80, find two integers c such that the trinomial can be factored. (There are many correct answers.)

79. $2x^2 - 4x + c$

80. $5x^2 + 6x + c$

Factoring Completely In Exercises 81–90, factor the trinomial completely.

81. $3x^2 + 33x + 90$

82. $4x^2 + 12x - 16$

83. $6y^2 + 39y - 21$

84. $10b^2 - 38b + 24$

85. $6u^3 + 3u^2 - 30u$

86. $8x^3 - 8x^2 - 30x$

87. $8y^3 - 20y^2 + 12y$

88. $14x^3 + 26x^2 - 4x$

89. $6x^3 + 14x^2 - 12x$

90. $12y^3 + 36y^2 + 15y$

91. **Geometry** The pastry box shown in the figure has a height of x inches and a width of $(x + 1)$ inches. The volume of the box is $(3x^3 + 4x^2 + x)$. find the length of the box.

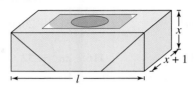

92. **Geometry** The area of the rectangle shown in the figure is $2x^2 + 5x + 3$. What is the area of the shaded region?

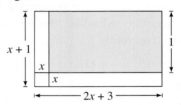

Factoring a Trinomial by Grouping In Exercises 93–100, factor the trinomial by grouping.

93. $2x^2 - 13x + 21$

94. $3a^2 - 13a - 10$

95. $4y^2 + y - 3$

96. $6z^2 - 43z + 7$

97. $6x^2 + 11x - 10$

98. $21x^2 - 25x - 4$

99. $14x^2 + 17x + 5$

100. $5t^2 + 27t - 18$

6.4

Factoring the Difference of Two Squares In Exercises 101–110, factor the difference of two squares.

101. $a^2 - 100$

102. $36 - b^2$

103. $25 - 4y^2$

104. $16b^2 - 1$

105. $12x^2 - 27$

106. $100x^2 - 64$

107. $(u + 1)^2 - 4$

108. $(y - 2)^2 - 9$

109. $16 - (z - 5)^2$

110. $81 - (x + 9)^2$

Factoring Completely In Exercises 111−118, factor the polynomial completely.

111. $3y^3 - 75y$

112. $16b^3 - 36b$

113. $s^3t - st^3$

114. $5x^3 - 20xy^2$

115. $x^4 - 81$

116. $2a^4 - 32$

117. $x^3 - 2x^2 + 4x - 8$

118. $b^3 - 3b^2 + 9b - 27$

Identifying a Perfect Square Trinomial In Exercises 119−122, determine whether the polynomial is a perfect square trinomial.

119. $x^2 + 25x + 10$

120. $x^2 + 12x + 36$

121. $4y^2 - 8y + 4$

122. $9b^2 - 18b - 9$

Factoring a Perfect Square Trinomial In Exercises 123−130, factor the perfect square trinomial.

123. $x^2 - 8x + 16$

124. $y^2 + 24y + 144$

125. $9s^2 + 12s + 4$

126. $16x^2 - 40x + 25$

127. $y^2 + 4yz + 4z^2$

128. $u^2 - 2uv + v^2$

129. $x^2 + \frac{2}{3}x + \frac{1}{9}$

130. $y^2 - \frac{4}{3}y + \frac{4}{9}$

Factoring the Sum or Difference of Two Cubes In Exercises 131−136, factor the sum or difference of two cubes.

131. $a^3 + 1$

132. $z^3 + 8$

133. $27 - 8t^3$

134. $z^3 - 125$

135. $8x^3 + y^3$

136. $125a^3 - 27b^3$

6.5

Using the Zero-Factor Property In Exercises 137−142, use the Zero-Factor Property to solve the equation.

137. $4x(x - 2) = 0$

138. $-3x(2x + 6) = 0$

139. $(2x + 1)(x - 3) = 0$

140. $(x - 7)(3x - 8) = 0$

141. $(x + 10)(4x - 1)(5x + 9) = 0$

142. $3x(x + 8)(2x - 7) = 0$

Solving a Quadratic Equation by Factoring In Exercises 143−150, solve the quadratic equation by factoring.

143. $3s^2 - 2s - 8 = 0$

144. $5v^2 - 12v - 9 = 0$

145. $m(2m - 1) + 3(2m - 1) = 0$

146. $4w(2w + 8) - 7(2w + 8) = 0$

147. $z(5 - z) + 36 = 0$

148. $(x + 3)^2 - 25 = 0$

149. $v^2 - 100 = 0$

150. $x^2 - 121 = 0$

Solving a Polynomial by Factoring In Exercises 151−158, solve the equation by factoring.

151. $2y^4 + 2y^3 - 24y^2 = 0$

152. $9x^4 - 15x^3 - 6x^2 = 0$

153. $x^3 - 11x^2 + 18x = 0$

154. $x^3 + 20x^2 + 36x = 0$

155. $b^3 - 6b^2 - b + 6 = 0$

156. $q^3 + 3q^2 - 4q - 12 = 0$

157. $x^4 - 5x^3 - 9x^2 + 45x = 0$

158. $2x^4 + 6x^3 - 50x^2 - 150x = 0$

159. *Number Problem* Find two consecutive positive odd integers whose product is 99.

160. *Number Problem* Find two consecutive positive even integers whose product is 168.

161. *Geometry* A rectangle has an area of 900 square inches. The length of the rectangle is $2\frac{1}{4}$ times its width. Find the dimensions of the rectangle.

162. *Geometry* A closed box with a square base stands 12 inches tall. The total surface area of the outside of the box is 512 square inches. What are the dimensions of the base? (*Hint:* The surface area is given by $S = 2x^2 + 4xh$.)

163. *Free-Falling Object* An object is thrown upward from the Trump Tower in New York City, which is 664 feet tall, with an initial velocity of 45 feet per second. The height h (in feet) of the object is modeled by the position equation $h = -16t^2 + 45t + 664$, where t is the time (in seconds). How long does it take the object to reach the ground?

Chapter Test

Solutions in English & Spanish and tutorial videos at CollegePrepAlgebra.com

Take this test as you would take a test in class. After you are done, check your work against the answers in the back of the book.

In Exercises 1–10, factor the polynomial completely.

1. $9x^2 - 63x^5$

2. $z(z + 17) - 10(z + 17)$

3. $t^2 - 2t - 80$

4. $6x^2 - 11x + 4$

5. $3y^3 + 72y^2 - 75y$

6. $4 - 25v^2$

7. $x^3 + 8$

8. $100 - (z + 11)^2$

9. $x^3 + 2x^2 - 9x - 18$

10. $16 - z^4$

11. Determine the missing factor: $\frac{2}{5}x - \frac{3}{5} = \frac{1}{5}\left(\right)$.

12. Find all integers b such that $x^2 + bx + 5$ can be factored.

13. Find a real number c such that $x^2 + 12x + c$ is a perfect square trinomial.

14. Explain why $(x + 1)(3x - 6)$ is not a complete factorization of $3x^2 - 3x - 6$.

In Exercises 15–18, solve the equation.

15. $(x + 4)(2x - 3) = 0$

16. $3x^2 + 7x - 6 = 0$

17. $y(2y - 1) = 6$

18. $2x^3 + 10x^2 + 8x = 0$

19. The suitcase shown below has a height of x inches and a width of $(x + 2)$ inches. The volume of the suitcase is $(x^3 + 6x^2 + 8x)$ cubic inches. Find the length of the suitcase.

20. The width of a rectangle is 5 inches less than its length. The area of the rectangle is 84 square inches. Find the dimensions of the rectangle.

21. An object is thrown upward from the top of the Aon Center in Chicago at a height of 1136 feet, with an initial velocity of 14 feet per second. The height h (in feet) of the object is modeled by the position equation

$$h = -16t^2 + 14t + 1136$$

where t is the time measured in seconds. How long will it take for the object to reach the ground?

22. Find two consecutive positive even integers whose product is 624.

Cumulative Test: Chapters 4–6

Solutions in English & Spanish and tutorial videos at CollegePrepAlgebra.com

Take this test as you would take a test in class. After you are done, check your work against the answers in the back of the book.

1. Determine the quadrants in which the point $(-2, y)$ may be located. Assume $y \neq 0$. (y is a real number.)

2. Determine whether each ordered pair is a solution of the equation $9x - 4y + 36 = 0$.

    (a) $(-1, -1)$　　(b) $(8, 27)$　　(c) $(-4, 0)$　　(d) $(3, -2)$

In Exercises 3 and 4, sketch the graph of the equation and determine any intercepts of the graph.

3. $y = 3 + |x|$

4. $x + 2y = 6$

5. Determine whether the relation at the left represents a function.

6. The slope of a line is $-\frac{1}{4}$ and a point on the line is $(2, 1)$. Find the coordinates of a second point on the line. Explain why there are many correct answers.

7. Write the slope-intercept form of the equation of the line that passes through the point $\left(0, -\frac{3}{2}\right)$ and has slope $m = \frac{2}{5}$.

In Exercises 8 and 9, sketch the lines and determine whether they are parallel, perpendicular, or neither.

8. $y_1 = \frac{2}{3}x - 3, y_2 = -\frac{3}{2}x + 1$

9. $y_1 = 2 - 0.4x, y_2 = -\frac{2}{5}x$

10. Subtract: $(x^3 - 3x^2) - (x^3 + 2x^2 - 5)$.

11. Multiply: $(6z)(-7z)(z^2)$.

12. Multiply: $(3x + 5)(x - 4)$.

13. Multiply: $(5x - 3)(5x + 3)$.

14. Simplify: $(5x + 6)^2$.

15. Divide: $(6x^2 + 72x) \div 6x$.

16. Divide: $\dfrac{x^2 - 3x - 2}{x - 4}$.

17. Simplify: $\dfrac{(3xy^2)^{-2}}{6x^{-3}}$.

18. Factor: $2u^2 - 6u$.

19. Factor and simplify: $(x - 4)^2 - 36$.

20. Factor completely: $x^3 + 8x^2 + 16x$.

21. Factor completely: $x^3 + 2x^2 - 4x - 8$.

22. Solve: $u(u - 12) = 0$.

23. Solve: $5x^2 - 12x - 9 = 0$.

24. Rewrite the expression $\left(\dfrac{x}{2}\right)^{-2}$ using only positive exponents, and simplify.

25. A sales representative is reimbursed $150 per day for lodging and meals, plus $0.45 per mile driven. Write a linear equation giving the daily cost C to the company in terms of x, the number of miles driven. Find the cost for a day when the representative drives 70 miles.

26. You must perform two songs that are no longer than 10 minutes combined for a jazz band audition. Write a linear inequality that represents the numbers of minutes that can be spent performing each song. Then find the possible lengths of the second song when the first song is 6 minutes long.

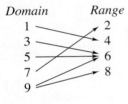

Domain　　Range

Figure for 5

7

Rational Expressions, Equations, and Functions

7.1 Rational Expressions and Functions

7.2 Multiplying and Dividing Rational Expressions

7.3 Adding and Subtracting Rational Expressions

7.4 Complex Fractions

7.5 Solving Rational Equations

7.6 Applications and Variation

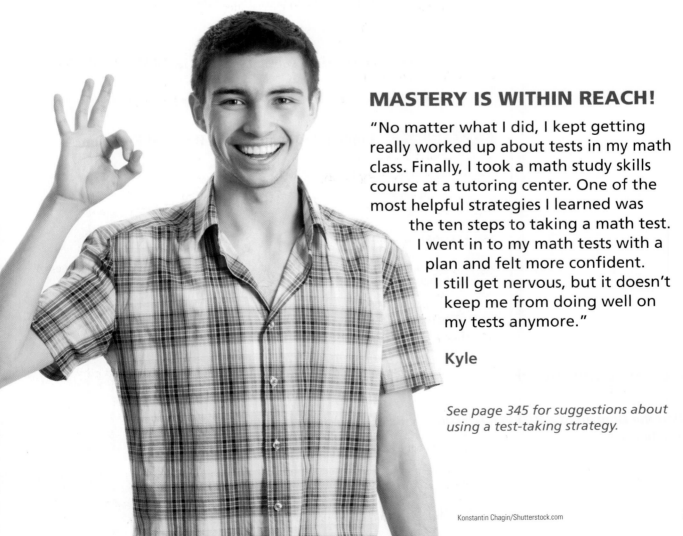

MASTERY IS WITHIN REACH!

"No matter what I did, I kept getting really worked up about tests in my math class. Finally, I took a math study skills course at a tutoring center. One of the most helpful strategies I learned was the ten steps to taking a math test. I went in to my math tests with a plan and felt more confident. I still get nervous, but it doesn't keep me from doing well on my tests anymore."

Kyle

See page 345 for suggestions about using a test-taking strategy.

7.1 Rational Expressions and Functions

▶ Find the domain of a rational function.

▶ Simplify rational expressions.

▶ Use rational expressions to model and solve real-life problems.

The Domain of a Rational Function

Study Tip

Every polynomial is also a rational expression because you can consider the denominator to be 1. The domain of every polynomial is the set of all real numbers.

Definition of a Rational Expression

Let u and v be polynomials. The algebraic expression

$$\frac{u}{v}$$

is a **rational expression**. The **domain** of this rational expression is the set of all real numbers for which $v \neq 0$.

Definition of a Rational Function

Let $u(x)$ and $v(x)$ be polynomial functions. The function

$$f(x) = \frac{u(x)}{v(x)}$$

is a **rational function**. The **domain** of f is the set of all real numbers for which $v(x) \neq 0$.

EXAMPLE 1   **Finding the Domain of a Rational Function**

Find the domain of each rational function.

a. $f(x) = \dfrac{4}{x-2}$ **b.** $g(x) = \dfrac{5x}{x^2-16}$

SOLUTION

a. The denominator is 0 when $x - 2 = 0$ or $x = 2$. So, the domain is all real values of x such that $x \neq 2$.

b. The denominator is 0 when $x^2 - 16 = 0$. Solving this equation by factoring, you find that the denominator is 0 when $x = -4$ or $x = 4$. So, the domain is all real values of x such that $x \neq -4$ and $x \not\equiv 4$.

Exercises Within Reach ®   Solutions in English & Spanish and tutorial videos at CollegePrepAlgebra.com

Finding the Domain of a Rational Function In Exercises 1−6, find the domain of the rational function.

1. $f(x) = \dfrac{4}{x-3}$

2. $g(x) = \dfrac{-2}{x-7}$

3. $f(z) = \dfrac{z+2}{z(z-4)}$

4. $f(x) = \dfrac{x^2}{x(x-1)}$

5. $f(t) = \dfrac{5t}{t^2-16}$

6. $f(x) = \dfrac{x}{x^2-4}$

$(x-2)(x+2)$

Application   **EXAMPLE 2**   **An Application Involving a Restricted Domain**

You have started a small business that manufactures lamps. The initial investment for the business is \$120,000. The cost of manufacturing each lamp is \$15. So, your total cost of producing x lamps is

$$C = 15x + 120{,}000. \qquad \text{Cost function}$$

Your average cost per lamp depends on the number of lamps produced. For instance, the average cost per lamp $\overline{C}$ of producing 100 lamps is

$$\overline{C} = \frac{15(100) + 120{,}000}{100} \qquad \text{Substitute 100 for } x.$$

$$= \frac{121{,}500}{100} \qquad \text{Simplify.}$$

$$= \$1215. \qquad \text{Average cost per lamp for 100 lamps}$$

The average cost per lamp decreases as the number of lamps increases. For instance, the average cost per lamp $\overline{C}$ of producing 1000 lamps is

$$\overline{C} = \frac{15(1000) + 120{,}000}{1000} \qquad \text{Substitute 1000 for } x.$$

$$= \frac{135{,}000}{1000} \qquad \text{Simplify.}$$

$$= \$135. \qquad \text{Average cost per lamp for 1000 lamps}$$

In general, the average cost of producing x lamps is

$$\overline{C} = \frac{15x + 120{,}000}{x}. \qquad \text{Average cost per lamp for } x \text{ lamps}$$

What is the domain of this rational function?

SOLUTION

If you were considering this function from only a mathematical point of view, you would say that the domain is all real values of x such that $x \neq 0$. However, because this function is a mathematical model representing a real-life situation, you must decide which values of x make sense in real life. For this model, the variable x represents the number of lamps that you produce. Assuming that you cannot produce a fractional number of lamps, you can conclude that the domain is the set of positive integers—that is,

$$\text{Domain} = \{1, 2, 3, 4, \ldots\}.$$

Exercises Within Reach ®   Solutions in English & Spanish and tutorial videos at CollegePrepAlgebra.com

Describing the Domain of a Rational Function **In Exercises 7–10, describe the domain.**

7. A rectangle of length x inches has an area of 500 square inches. The perimeter P of the rectangle is given by

$$P = 2\left(x + \frac{500}{x}\right).$$

8. The cost C in millions of dollars for the government to seize $p\%$ of an illegal drug as it enters the country is given by

$$C = \frac{528p}{100 - p}.$$

9. The inventory cost I when x units of a product are ordered from a supplier is given by

$$I = \frac{0.25x + 2000}{x}.$$

10. The average cost $\overline{C}$ for a manufacturer to produce x units of a product is given by

$$\overline{C} = \frac{1.35x + 4570}{x}.$$

Simplifying Rational Expressions

> ### Simplifying Rational Expressions
>
> Let u, v, and w represent real numbers, variables, or algebraic expressions such that $v \neq 0$ and $w \neq 0$. Then the following is valid.
>
> $$\frac{uw}{vw} = \frac{u\cancel{w}}{v\cancel{w}} = \frac{u}{v}$$

EXAMPLE 3   Simplifying a Rational Expression

Simplify the rational expression $\dfrac{2x^3 - 6x}{6x^2}$.

SOLUTION

First note that the domain of the rational expression is all real values of x such that $x \neq 0$. Then, completely factor both the numerator and denominator.

$$\frac{2x^3 - 6x}{6x^2} = \frac{2x(x^2 - 3)}{2x(3x)} \qquad \text{Factor numerator and denominator.}$$

$$= \frac{\cancel{2x}(x^2 - 3)}{\cancel{2x}(3x)} \qquad \text{Divide out common factor } 2x.$$

$$= \frac{x^2 - 3}{3x} \qquad \text{Simplified form}$$

In simplified form, the domain of the rational expression is the same as that of the original expression—all real values of x such that $x \neq 0$.

EXAMPLE 4   Simplifying a Rational Expression

Simplify the rational expression $\dfrac{x^2 + 2x - 15}{3x - 9}$.

SOLUTION

The domain of the rational expression is all real values of x such that $x \neq 3$.

$$\frac{x^2 + 2x - 15}{3x - 9} = \frac{(x + 5)(x - 3)}{3(x - 3)} \qquad \text{Factor numerator and denominator.}$$

$$= \frac{(x + 5)\cancel{(x - 3)}}{3\cancel{(x - 3)}} \qquad \text{Divide out common factor } (x - 3).$$

$$= \frac{x + 5}{3}, \quad x \neq 3 \qquad \text{Simplified form}$$

> **Study Tip**
>
> Dividing out common factors can change the implied domain. In Example 4, the domain restriction $x \neq 3$ must be listed because it is no longer implied by the simplified expression.

Exercises Within Reach ®

Solutions in English & Spanish and tutorial videos at CollegePrepAlgebra.com

Simplifying a Rational Expression   **In Exercises 11−16, simplify the rational expression.**

11. $\dfrac{3x^2 - 9x}{12x^2}$

12. $\dfrac{8x^3 + 4x^2}{20x}$

13. $\dfrac{x^2(x - 8)}{x(x - 8)}$

14. $\dfrac{a^2 b(b - 3)}{b^3(b - 3)^2}$

15. $\dfrac{u^2 - 12u + 36}{u - 6}$

16. $\dfrac{z^2 + 22z + 121}{3z + 33}$

EXAMPLE 5   **Simplifying a Rational Expression**

Simplify the rational expression $\dfrac{x^3 - 16x}{x^2 - 2x - 8}$.

SOLUTION

The domain of the rational expression is all real values of x such that $x \neq -2$ and $x \neq 4$.

$$\frac{x^3 - 16x}{x^2 - 2x - 8} = \frac{x(x^2 - 16)}{(x + 2)(x - 4)} \qquad \text{Partially factor.}$$

$$= \frac{x(x + 4)(x - 4)}{(x + 2)(x - 4)} \qquad \text{Factor completely.}$$

$$= \frac{x(x + 4)\cancel{(x - 4)}}{(x + 2)\cancel{(x - 4)}} \qquad \text{Divide out common factor } (x - 4).$$

$$= \frac{x(x + 4)}{x + 2}, \ x \neq 4 \qquad \text{Simplified form}$$

When you simplify a rational expression, keep in mind that you must list any domain restrictions that are no longer implied in the simplified expression. For instance, in Example 5 the restriction $x \neq 4$ is listed so that the domains agree for the original and simplified expressions. The example does not list $x \neq -2$ because this restriction is apparent by looking at either expression.

EXAMPLE 6   **Simplification Involving a Change in Sign**

Simplify the rational expression $\dfrac{2x^2 - 9x + 4}{12 + x - x^2}$.

SOLUTION

The domain of the rational expression is all real values of x such that $x \neq -3$ and $x \neq 4$.

$$\frac{2x^2 - 9x + 4}{12 + x - x^2} = \frac{(2x - 1)(x - 4)}{(4 - x)(3 + x)} \qquad \text{Factor numerator and denominator.}$$

$$= \frac{(2x - 1)(x - 4)}{-(x - 4)(3 + x)} \qquad (4 - x) = -(x - 4)$$

$$= \frac{(2x - 1)\cancel{(x - 4)}}{-\cancel{(x - 4)}(3 + x)} \qquad \text{Divide out common factor } (x - 4).$$

$$= -\frac{2x - 1}{x + 3}, \ x \neq 4 \qquad \text{Simplified form}$$

Study Tip

Be sure to *factor completely* the numerator and denominator of a rational expression in order to find any common factors. You may need to use a change in signs. Remember that the Distributive Property allows you to write $(b - a)$ as $-(a - b)$.

Exercises Within Reach ®   Solutions in English & Spanish and tutorial videos at CollegePrepAlgebra.com

Simplifying a Rational Expression In Exercises 17–22, simplify the rational expression.

17. $\dfrac{y^3 - 4y}{y^2 + 4y - 12}$

18. $\dfrac{x^3 - 4x}{x^2 - 5x + 6}$

19. $\dfrac{3x^2 - 7x - 20}{12 + x - x^2}$

20. $\dfrac{2x^2 + 3x - 5}{7 - 6x - x^2}$

21. $\dfrac{2x^2 + 19x + 24}{2x^2 - 3x - 9}$

22. $\dfrac{2y^2 + 13y + 20}{2y^2 + 17y + 30}$

Study Tip

As you study the examples and work the exercises in this section and the next four sections, keep in mind that you are *rewriting expressions in simpler forms*. You are not solving equations. Equal signs are used in the steps of the simplification process only to indicate that the new form of the expression is *equivalent* to the original form.

EXAMPLE 7 **Rational Expressions Involving Two Variables**

a.
$$\frac{3xy + y^2}{2y} = \frac{y(3x + y)}{2y}$$ Factor numerator.

$$= \frac{\cancel{y}(3x + y)}{2\cancel{y}}$$ Divide out common factor y.

$$= \frac{3x + y}{2}, \ y \neq 0$$ Simplified form

b.
$$\frac{4x^2y - y^3}{2x^2y - xy^2} = \frac{(4x^2 - y^2)y}{(2x - y)xy}$$ Partially factor.

$$= \frac{(2x - y)(2x + y)y}{(2x - y)xy}$$ Factor completely.

$$= \frac{\cancel{(2x - y)}(2x + y)\cancel{y}}{\cancel{(2x - y)}x\cancel{y}}$$ Divide out common factors $(2x - y)$ and y.

$$= \frac{2x + y}{x}, \ y \neq 0, \ y \neq 2x$$ Simplified form

The domain of the original rational expression is all real values of x and y such that $x \neq 0$, $y \neq 0$, and $y \neq 2x$.

c.
$$\frac{2x^2 + 2xy - 4y^2}{5x^3 - 5xy^2} = \frac{2(x^2 + xy - 2y^2)}{5x(x^2 - y^2)}$$ Partially factor.

$$= \frac{2(x - y)(x + 2y)}{5x(x - y)(x + y)}$$ Factor completely.

$$= \frac{2\cancel{(x - y)}(x + 2y)}{5x\cancel{(x - y)}(x + y)}$$ Divide out common factor $(x - y)$.

$$= \frac{2(x + 2y)}{5x(x + y)}, \ x \neq y$$ Simplified form

The domain of the original rational expression is all real values of x and y such that $x \neq 0$ and $x \neq \pm y$.

Exercises Within Reach ®

Solutions in English & Spanish and tutorial videos at CollegePrepAlgebra.com

Simplifying a Rational Expression In Exercises 23−32, simplify the rational expression.

23. $\dfrac{3xy^2}{xy^2 + x}$

24. $\dfrac{x + 3x^2y}{3xy + 1}$

25. $\dfrac{y^2 - 64x^2}{5(3y + 24x)}$

26. $\dfrac{x^2 - 25z^2}{2(3x + 15z)}$

27. $\dfrac{5xy + 3x^2y^2}{xy^3}$

28. $\dfrac{4u^2v - 12uv^2}{18uv}$

29. $\dfrac{u^2 - 4v^2}{u^2 + uv - 2v^2}$

30. $\dfrac{x^2 + 4xy}{x^2 - 16y^2}$

31. $\dfrac{3m^2 - 12n^2}{m^2 + 4mn + 4n^2}$

32. $\dfrac{x^2 + xy - 2y^2}{x^2 + 3xy + 2y^2}$

Application

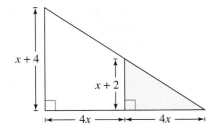

Application   **EXAMPLE 8**   Geometry: **Finding a Ratio**

Find the ratio of the area of the shaded portion of the triangle to the total area of the triangle.

SOLUTION

The area of the shaded portion of the triangle is given by

$$\text{Area} = \frac{1}{2}(4x)(x + 2)$$

$$= \frac{1}{2}(4x^2 + 8x)$$

$$= 2x^2 + 4x.$$

The total area of the triangle is given by

$$\text{Area} = \frac{1}{2}(4x + 4x)(x + 4)$$

$$= \frac{1}{2}(8x)(x + 4)$$

$$= \frac{1}{2}(8x^2 + 32x)$$

$$= 4x^2 + 16x.$$

So, the ratio of the area of the shaded portion of the triangle to the total area of the triangle is

$$\frac{2x^2 + 4x}{4x^2 + 16x} = \frac{2x(x + 2)}{4x(x + 4)}$$

$$= \frac{x + 2}{2(x + 4)}, \; x > 0.$$

Exercises Within Reach ®   Solutions in English & Spanish and tutorial videos at CollegePrepAlgebra.com

Geometry **In Exercises 33–36, find the ratio of the area of the shaded portion to the total area of the figure.**

33.

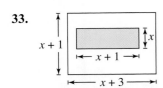

34.

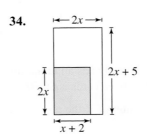

35.

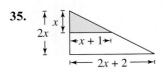

36.

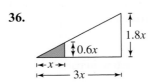

Concept Summary: *Simplifying Rational Expressions*

What

A **rational expression** is in **simplified form** when its numerator and denominator have no common factors (other than ± 1).

EXAMPLE

Simplify $\dfrac{3x}{3x + 12}$.

How

Be sure to completely factor the numerator and denominator of a rational expression before concluding that there are no common factors.

EXAMPLE

$$\dfrac{3x}{3x + 12} \quad \text{Original expression}$$

$$= \dfrac{3 \cdot x}{3(x + 4)} \quad \text{Factor completely.}$$

$$= \dfrac{\cancel{3} \cdot x}{\cancel{3}(x + 4)} \quad \text{Divide out factors.}$$

$$= \dfrac{x}{x + 4} \quad \text{Simplified form}$$

Why

Simplifying a rational expression into a more usable form is a skill frequently used in algebra.

Exercises Within Reach ®

Worked-out solutions to odd-numbered exercises at CollegePrepAlgebra.com

Concept Summary Check

37. *Vocabulary* How do you determine whether a rational expression is in simplified form?

38. *Reasoning* Can you divide out common terms from the numerator and denominator of a rational expression? Explain.

39. *Precision* After factoring completely, what is one additional step that is sometimes needed to find common factors in the numerator and denominator of a rational expression?

40. *Reasoning* Is the following expression in simplified form? Explain your reasoning.

$$\dfrac{5 + x}{5 + (x + 2)}$$

Extra Practice

Finding a Missing Factor In Exercises 41−44, determine the missing factor.

41. $\dfrac{(x + 5)()}{3x^2(x - 2)} = \dfrac{x + 5}{3x}, \; x \neq 2$

42. $\dfrac{(3y - 7)()}{y^2 - 4} = \dfrac{3y - 7}{y + 2}, \; y \neq 2$

43. $\dfrac{(8x)()}{x^2 - 2x - 15} = \dfrac{8x}{x - 5}, \; x \neq -3$

44. $\dfrac{(3 - z)()}{z^3 + 2z^2} = \dfrac{3 - z}{z^2}, \; z \neq -2$

45. *Average Cost* A greeting card company has an initial investment of $60,000. The cost of producing one dozen cards is $6.50.

(a) Write the total cost C as a function of x, the number of dozens of cards produced.

(b) Write the average cost per dozen $\overline{C} = C/x$ as a function of x, the number of dozens of cards produced.

(c) Determine the domain of the function in part (b).

(d) Find the value of $\overline{C}$ (11,000).

46. *Distance Traveled* A car starts on a trip and travels at an average speed of 55 miles per hour. Two hours later, a second car starts on the same trip and travels at an average speed of 65 miles per hour.

   (a) Find the distance each vehicle has traveled when the second car has been on the road for t hours.

   (b) Use the result of part (a) to write the distance between the first car and the second car as a function of t.

   (c) Write the ratio of the distance the second car has traveled to the distance the first car has traveled as a function of t.

47. *Geometry* One swimming pool is circular and another is rectangular. The rectangular pool's width is three times its depth. Its length is 6 feet more than its width. The circular pool has a diameter that is twice the width of the rectangular pool, and it is 2 feet deeper. Find the ratio of the circular pool's volume to the rectangular pool's volume.

48. *Geometry* A circular pool has a radius five times its depth. A rectangular pool has the same depth as the circular pool. Its width is 4 feet more than three times its depth and its length is 2 feet less than six times its depth. Find the ratio of the rectangular pool's volume to the circular pool's volume.

Explaining Concepts

49. *Writing* Describe the process for finding the implied domain restrictions of a rational function.

50. *Writing* Describe a situation in which you would need to indicate a domain restriction to the right of a rational function.

51. *Precision* Give an example of a rational function whose domain is the set of all real numbers and whose denominator is a second-degree polynomial function.

52. *Error Analysis* Describe the error.

$$\frac{2x^2}{x^2+4} = \frac{2x^2}{x^2+4} = \frac{2}{1+4} = \frac{2}{5}$$

53. *Logic* A student writes the following incorrect solution for simplifying a rational expression. Discuss the student's errors and misconceptions, and construct a correct solution.

$$\frac{x^2+7x}{x+7} = \frac{x^2}{x} + \frac{7x}{7}$$
$$= x + x$$
$$= 2x$$

54. *Reasoning* Is the following statement true? Explain.

$$\frac{6x-5}{5-6x} = -1$$

55. *Writing* Explain how you can use a given polynomial function $f(x)$ to write a rational function $g(x)$ that is equivalent to $f(x)$, $x \neq 2$.

56. *Think About It* Is it possible for a rational function $f(x)$ (without added domain restrictions) to be undefined on an interval $[a, b]$, where a and b are real numbers such that $a < b$? Explain.

Cumulative Review

In Exercises 57−60, find the product.

57. $\frac{1}{4}\left(\frac{3}{4}\right)$

58. $\frac{2}{3}\left(-\frac{5}{6}\right)$

59. $\frac{1}{3}\left(\frac{3}{5}\right)(5)$

60. $\left(-\frac{3}{7}\right)\left(\frac{2}{5}\right)\left(-\frac{1}{6}\right)$

In Exercises 61−64, perform the indicated multiplication.

61. $(-2a^3)(-2a)$

62. $6x^2(-3x)$

63. $(-3b)(b^2 - 3b + 5)$

64. $ab^2(3a - 4ab + 6a^2b^2)$

7.2 Multiplying and Dividing Rational Expressions

▶ Multiply rational expressions and simplify.
▶ Divide rational expressions and simplify.

Multiplying Rational Expressions

> **Multiplying Rational Expressions**
>
> Let u, v, w, and z represent real numbers, variables, or algebraic expressions such that $v \neq 0$ and $z \neq 0$. Then the product of u/v and w/z is
>
> $$\frac{u}{v} \cdot \frac{w}{z} = \frac{uw}{vz}.$$

EXAMPLE 1 **Multiplying Rational Expressions**

Multiply the rational expressions.

$$\frac{4x^3y}{3xy^4} \cdot \frac{-6x^2y^2}{10x^4}$$

SOLUTION

$$\frac{4x^3y}{3xy^4} \cdot \frac{-6x^2y^2}{10x^4} = \frac{(4x^3y) \cdot (-6x^2y^2)}{(3xy^4) \cdot (10x^4)}$$ Multiply numerators and denominators.

$$= \frac{-24x^5y^3}{30x^5y^4}$$ Simplify.

$$= \frac{-4(6)(x^5)(y^3)}{5(6)(x^5)(y^3)(y)}$$ Factor and divide out common factors.

$$= -\frac{4}{5y}, \ x \neq 0$$ Simplified form

Exercises Within Reach ®

Solutions in English & Spanish and tutorial videos at CollegePrepAlgebra.com

Finding a Missing Factor In Exercises 1–4, determine the missing factor.

1. $\dfrac{7x^2}{3y(\boxed{})} = \dfrac{7}{3y}, \ x \neq 0$

2. $\dfrac{14x(x-3)^2}{(x-3)(\boxed{})} = \dfrac{2x}{x-3}$

3. $\dfrac{3x(x+2)^2}{(x-4)(\boxed{})} = \dfrac{3x}{x-4}, \ x \neq -2$

4. $\dfrac{(x+1)^3}{x(\boxed{})} = \dfrac{x+1}{x}, \ x \neq -1$

Multiplying Rational Expressions In Exercises 5–12, multiply and simplify.

5. $4x \cdot \dfrac{7}{12x}$

6. $\dfrac{8}{7y} \cdot (42y)$

7. $\dfrac{8s^3}{9s} \cdot \dfrac{6s^2}{32s}$

8. $\dfrac{3x^4}{7x} \cdot \dfrac{8x^2}{9}$

9. $16u^4 \cdot \dfrac{12}{8u^2}$

10. $18x^4 \cdot \dfrac{4}{15x}$

11. $\dfrac{8}{3+4x} \cdot (9+12x)$

12. $(6-4x) \cdot \dfrac{10}{3-2x}$

EXAMPLE 2   **Multiplying Rational Expressions**

Multiply the rational expressions.

a. $\dfrac{x}{5x^2 - 20x} \cdot \dfrac{x - 4}{2x^2 + x - 3}$

b. $\dfrac{4x^2 - 4x}{x^2 + 2x - 3} \cdot \dfrac{x^2 + x - 6}{4x}$

SOLUTION

a. $\dfrac{x}{5x^2 - 20x} \cdot \dfrac{x - 4}{2x^2 + x - 3}$

$= \dfrac{x \cdot (x - 4)}{(5x^2 - 20x) \cdot (2x^2 + x - 3)}$ Multiply numerators and denominators.

$= \dfrac{x(x - 4)}{5x(x - 4)(x - 1)(2x + 3)}$ Factor.

$= \dfrac{x(x - 4)}{5x(x - 4)(x - 1)(2x + 3)}$ Divide out common factors.

$= \dfrac{1}{5(x - 1)(2x + 3)},\ x \neq 0,\ x \neq 4$ Simplified form

b. $\dfrac{4x^2 - 4x}{x^2 + 2x - 3} \cdot \dfrac{x^2 + x - 6}{4x}$

$= \dfrac{4x(x - 1)(x + 3)(x - 2)}{(x - 1)(x + 3)(4x)}$ Multiply and factor.

$= \dfrac{4x(x - 1)(x + 3)(x - 2)}{(x - 1)(x + 3)(4x)}$ Divide out common factors.

$= x - 2,\ x \neq 0,\ x \neq 1,\ x \neq -3$ Simplified form

Exercises Within Reach ®

Multiplying Rational Expressions **In Exercises 13−22, multiply and simplify.**

13. $\dfrac{8u^2v}{3u + v} \cdot \dfrac{u + v}{12u}$

14. $\dfrac{1 - 3xy}{4x^2y} \cdot \dfrac{46x^4y^2}{15 - 45xy}$

15. $\dfrac{12 - r}{3} \cdot \dfrac{3}{r - 12}$

16. $\dfrac{8 - z}{8 + z} \cdot \dfrac{z + 8}{z - 8}$

17. $\dfrac{(2x - 3)(x + 8)}{x^3} \cdot \dfrac{x}{3 - 2x}$

18. $\dfrac{x + 14}{x^3(10 - x)} \cdot \dfrac{x(x - 10)}{5}$

19. $\dfrac{4r - 12}{r - 2} \cdot \dfrac{r^2 - 4}{r - 3}$

20. $\dfrac{5y - 20}{5y + 15} \cdot \dfrac{2y + 6}{y - 4}$

21. $\dfrac{2t^2 - t - 15}{t + 2} \cdot \dfrac{t^2 - t - 6}{t^2 - 6t + 9}$

22. $\dfrac{y^2 - 16}{y^2 + 8y + 16} \cdot \dfrac{3y^2 - 5y - 2}{y^2 - 6y + 8}$

EXAMPLE 3 **Multiplying Rational Expressions**

Multiply the rational expressions.

a. $\dfrac{x - y}{y^2 - x^2} \cdot \dfrac{x^2 - xy - 2y^2}{3x - 6y}$

b. $\dfrac{x^2 - 3x + 2}{x + 2} \cdot \dfrac{3x}{x - 2} \cdot \dfrac{2x + 4}{x^2 - 5x}$

SOLUTION

a. $\dfrac{x - y}{y^2 - x^2} \cdot \dfrac{x^2 - xy - 2y^2}{3x - 6y}$

$= \dfrac{(x - y)(x - 2y)(x + y)}{(y + x)(y - x)(3)(x - 2y)}$ Multiply and factor.

$= \dfrac{(x - y)(x - 2y)(x + y)}{(y + x)(-1)(x - y)(3)(x - 2y)}$ $(y - x) = -1(x - y)$

$= \dfrac{\cancel{(x - y)}\cancel{(x - 2y)}(x + y)}{(x + y)(-1)\cancel{(x - y)}(3)\cancel{(x - 2y)}}$ Divide out common factors.

$= -\dfrac{1}{3}, \quad x \neq y, \ x \neq -y, \ x \neq 2y$ Simplified form

b. $\dfrac{x^2 - 3x + 2}{x + 2} \cdot \dfrac{3x}{x - 2} \cdot \dfrac{2x + 4}{x^2 - 5x}$

$= \dfrac{(x - 1)(x - 2)(3)(x)(2)(x + 2)}{(x + 2)(x - 2)(x)(x - 5)}$ Multiply and factor.

$= \dfrac{(x - 1)\cancel{(x - 2)}(3)\cancel{(x)}(2)\cancel{(x + 2)}}{\cancel{(x + 2)}\cancel{(x - 2)}\cancel{(x)}(x - 5)}$ Divide out common factors.

$= \dfrac{6(x - 1)}{x - 5}, \quad x \neq 0, \ x \neq 2, \ x \neq -2$ Simplified form

Exercises Within Reach ®

Multiplying Rational Expressions In Exercises 23−30, **multiply and simplify.**

23. $(4y^2 - x^2) \cdot \dfrac{xy}{(x - 2y)^2}$

24. $(u - 2v)^2 \cdot \dfrac{u + 2v}{2v - u}$

25. $\dfrac{x^2 + 2xy - 3y^2}{(x + y)^2} \cdot \dfrac{x^2 - y^2}{x + 3y}$

26. $\dfrac{(x - 2y)^2}{x + 2y} \cdot \dfrac{x^2 + 7xy + 10y^2}{x^2 - 4y^2}$

27. $\dfrac{x + 5}{x - 5} \cdot \dfrac{2x^2 - 9x - 5}{3x^2 + x - 2} \cdot \dfrac{x^2 - 1}{x^2 + 7x + 10}$

28. $\dfrac{t^2 + 4t + 3}{2t^2 - t - 10} \cdot \dfrac{t}{t^2 + 3t + 2} \cdot \dfrac{2t^2 + 4t^3}{t^2 + 3t}$

29. $\dfrac{9 - x^2}{2x + 3} \cdot \dfrac{4x^2 + 8x - 5}{4x^2 - 8x + 3} \cdot \dfrac{6x^4 - 2x^3}{8x^2 + 4x}$

30. $\dfrac{16x^2 - 1}{4x^2 + 9x + 5} \cdot \dfrac{5x^2 - 9x - 18}{x^2 - 12x + 36} \cdot \dfrac{12 + 4x - x^2}{4x^2 - 13x + 3}$

Dividing Rational Expressions

> ### Dividing Rational Expressions
>
> Let u, v, w, and z represent real numbers, variables, or algebraic expressions such that $v \neq 0$, $w \neq 0$, and $z \neq 0$. Then the quotient of u/v and w/z is
>
> $$\frac{u}{v} \div \frac{w}{z} = \frac{u}{v} \cdot \frac{z}{w} = \frac{uz}{vw}.$$

Study Tip

Don't forget to add domain restrictions as needed in division problems. In Example 4(a), an implied domain restriction in the original expression is $x \neq 1$. Because this restriction is not implied by the final expression, it must be added as a written restriction.

EXAMPLE 4   **Dividing Rational Expressions**

Divide the rational expressions.

a. $\dfrac{x}{x+3} \div \dfrac{4}{x-1}$

b. $\dfrac{2x}{3x-12} \div \dfrac{x^2-2x}{x^2-6x+8}$

SOLUTION

a.
$$\frac{x}{x+3} \div \frac{4}{x-1} = \frac{x}{x+3} \cdot \frac{x-1}{4} \qquad \text{Invert divisor and multiply.}$$

$$= \frac{x(x-1)}{(x+3)(4)} \qquad \text{Multiply numerators and denominators.}$$

$$= \frac{x(x-1)}{4(x+3)}, \quad x \neq 1 \qquad \text{Simplified form}$$

b.
$$\frac{2x}{3x-12} \div \frac{x^2-2x}{x^2-6x+8}$$

$$= \frac{2x}{3x-12} \cdot \frac{x^2-6x+8}{x^2-2x} \qquad \text{Invert divisor and multiply.}$$

$$= \frac{(2)(x)(x-2)(x-4)}{(3)(x-4)(x)(x-2)} \qquad \text{Factor.}$$

$$= \frac{(2)(x)\cancel{(x-2)}\cancel{(x-4)}}{(3)\cancel{(x-4)}(x)\cancel{(x-2)}} \qquad \text{Divide out common factors.}$$

$$= \frac{2}{3}, \quad x \neq 0, \ x \neq 2, \ x \neq 4 \qquad \text{Simplified form}$$

Remember that the original expression is equivalent to $\frac{2}{3}$ except for $x = 0$, $x = 2$, and $x = 4$.

Exercises Within Reach ®

Solutions in English & Spanish and tutorial videos at CollegePrepAlgebra.com

Dividing Rational Expressions **In Exercises 31−38, divide and simplify.**

31. $\dfrac{x}{x+2} \div \dfrac{3}{x+1}$

32. $\dfrac{x+3}{4} \div \dfrac{x-2}{x}$

33. $x^2 \div \dfrac{3x}{4}$

34. $\dfrac{u}{10} \div u^2$

35. $\dfrac{2x}{5} \div \dfrac{x^2}{15}$

36. $\dfrac{3y^2}{20} \div \dfrac{y}{15}$

37. $\dfrac{4x}{3x-3} \div \dfrac{x^2+2x}{x^2+x-2}$

38. $\dfrac{5x+5}{2x} \div \dfrac{x^2-3x}{x^2-2x-3}$

EXAMPLE 5 **Dividing Rational Expressions**

a. $\dfrac{x^2 - y^2}{2x + 2y} \div \dfrac{2x^2 - 3xy + y^2}{6x + 2y}$

$= \dfrac{x^2 - y^2}{2x + 2y} \cdot \dfrac{6x + 2y}{2x^2 - 3xy + y^2}$ Invert divisor and multiply.

$= \dfrac{(x + y)(x - y)(2)(3x + y)}{(2)(x + y)(2x - y)(x - y)}$ Factor.

$= \dfrac{\cancel{(x + y)}\cancel{(x - y)}(2)(3x + y)}{\cancel{(2)}\cancel{(x + y)}(2x - y)\cancel{(x - y)}}$ Divide out common factors.

$= \dfrac{3x + y}{2x - y},\ x \neq y,\ x \neq -y,\ y \neq -3x$ Simplified form

b. $\dfrac{x^2 - 14x + 49}{x^2 - 49} \div \dfrac{3x - 21}{x^2 + 2x - 35}$

$= \dfrac{x^2 - 14x + 49}{x^2 - 49} \cdot \dfrac{x^2 + 2x - 35}{3x - 21}$ Invert divisor and multiply.

$= \dfrac{(x - 7)(x - 7)(x + 7)(x - 5)}{(x - 7)(x + 7)(3)(x - 7)}$ Factor.

$= \dfrac{\cancel{(x - 7)}\cancel{(x - 7)}\cancel{(x + 7)}(x - 5)}{\cancel{(x - 7)}\cancel{(x + 7)}(3)\cancel{(x - 7)}}$ Divide out common factors.

$= \dfrac{x - 5}{3},\ x \neq 5,\ x \neq 7,\ x \neq -7$ Simplified form

Exercises Within Reach ®

Solutions in English & Spanish and tutorial videos at CollegePrepAlgebra.com

Dividing Rational Expressions **In Exercises 39−48, divide and simplify.**

39. $\dfrac{7xy^2}{10x^2y} \div \dfrac{21x^3}{45xy}$

40. $\dfrac{25x^2y}{60x^3y^2} \div \dfrac{5x^4y^3}{16x^2y}$

41. $\dfrac{3(a + b)}{4} \div \dfrac{(a + b)^2}{2}$

42. $\dfrac{x^2 + 9}{5(x + 2)} \div \dfrac{x + 3}{5(x^2 - 4)}$

43. $\dfrac{2x + 2y}{3} \div \dfrac{x^2 - y^2}{x - y}$

44. $\dfrac{x^2 + 2x - xy - 2y}{x^2 - y^2} \div \dfrac{2x + 4}{x + y}$

45. $\dfrac{(x^3y)^2}{(x + 2y)^2} \div \dfrac{x^2y}{(x + 2y)^3}$

46. $\dfrac{x^2 - y^2}{2x^2 - 8x} \div \dfrac{(x - y)^2}{2xy}$

47. $\dfrac{x^2 + 2x - 15}{x^2 + 11x + 30} \div \dfrac{x^2 - 8x + 15}{x^2 + 2x - 24}$

48. $\dfrac{y^2 + 5y - 14}{y^2 + 10y + 21} \div \dfrac{y^2 + 5y + 6}{y^2 + 7y + 12}$

Application   EXAMPLE 6   **Amount Spent on Meals and Beverages**

The annual amount A (in millions of dollars) Americans spent on meals and beverages purchased for the home, and the population P (in millions) of the United States, for the years 2003 through 2009 can be modeled by

$$A = \frac{32{,}684.374t + 396{,}404.73}{0.014t + 1}, \quad 3 \le t \le 9$$

and

$$P = 2.798t + 282.31, \quad 3 \le t \le 9$$

where t represents the year, with $t = 3$ corresponding to 2003. Find a model T for the amount Americans spent *per person* on meals and beverages for the home. (*Source:* U.S. Department of Agriculture and U.S. Census Bureau)

SOLUTION

To find a model T for the amount Americans spent per person on meals and beverages for the home, divide the total amount by the population.

$$T = \frac{32{,}684.374t + 396{,}404.73}{0.014t + 1} \div (2.798t + 282.31) \qquad \text{Divide amount spent by population.}$$

$$= \frac{32{,}684.374t + 396{,}404.73}{0.014t + 1} \cdot \frac{1}{2.798t + 282.31} \qquad \text{Invert divisor and multiply.}$$

$$= \frac{32{,}684.374t + 396{,}404.73}{(0.014t + 1)(2.798t + 282.31)}, \quad 3 \le t \le 9 \qquad \text{Model}$$

Exercises Within Reach ®

Solutions in English & Spanish and tutorial videos at CollegePrepAlgebra.com

49. *Per Capita Income* The total annual amount I (in millions of dollars) of personal income earned in Alabama, and its population P (in millions), for the years 2004 through 2009 can be modeled by

$$I = 6591.43t + 102{,}139.0, \quad 4 \le t \le 9 \text{ and}$$

$$P = \frac{0.184t + 4.3}{0.029t + 1}, \quad 4 \le t \le 9$$

where t represents the year, with $t = 4$ corresponding to 2004. Find a model Y for the annual per capita income for these years. (*Source:* U.S. Bureau of Economic Analysis and U.S. Census Bureau)

50. *Per Capita Income* The total annual amount I (in millions of dollars) of personal income earned in Montana, and its population P (in millions), for the years 2004 through 2009 can be modeled by

$$I = \frac{4651.460t + 14{,}528.41}{0.070t + 1}, \quad 4 \le t \le 9 \text{ and}$$

$$P = \frac{0.028t + 0.88}{0.018t + 1}, \quad 4 \le t \le 9$$

where t represents the year, with $t = 4$ corresponding to 2004. Find a model Y for the annual per capita income for these years. (*Source:* U.S. Bureau of Economic Analysis and U.S. Census Bureau)

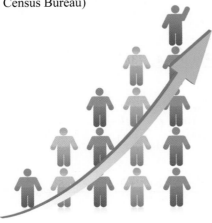

Concept Summary: *Multiplying and Dividing Rational Expressions*

What

The rules for multiplying and dividing rational expressions are the same as the rules for multiplying and dividing numerical fractions.

EXAMPLE

Divide:

$$\frac{x^2 - 2x - 24}{x - 6} \div \frac{x^2 + 6x + 8}{x + 3}.$$

How

To divide two rational expressions, multiply the first expression by the reciprocal of the second.

EXAMPLE

$$\frac{x^2 - 2x - 24}{x - 6} \div \frac{x^2 + 6x + 8}{x + 3} \quad \text{Original}$$

$$= \frac{x^2 - 2x - 24}{x - 6} \cdot \frac{x + 3}{x^2 + 6x + 8} \quad \begin{array}{l}\text{Invert and}\\\text{multiply.}\end{array}$$

$$= \frac{(x - 6)(x + 4)(x + 3)}{(x - 6)(x + 2)(x + 4)} \quad \text{Factor.}$$

$$= \frac{(\cancel{x - 6})(\cancel{x + 4})(x + 3)}{(\cancel{x - 6})(x + 2)(\cancel{x + 4})} \quad \begin{array}{l}\text{Divide out}\\\text{common}\\\text{factors.}\end{array}$$

$$= \frac{x + 3}{x + 2}, x \neq 6, x \neq -3, x \neq -4 \quad \begin{array}{l}\text{Simplified}\\\text{form}\end{array}$$

Why

Knowing how to multiply and divide rational expressions will help you solve rational equations.

Exercises Within Reach ®

Worked-out solutions to odd-numbered exercises at CollegePrepAlgebra.com

Concept Summary Check

51. *Writing* Explain how to multiply two rational expressions.

52. *Reasoning* Why is factoring used in multiplying rational expressions?

53. *Writing* In your own words, explain how to divide rational expressions.

54. *Logic* In dividing rational expressions, explain how you can lose implied domain restrictions when you invert the divisor.

Extra Practice

Using Order of Operations In Exercises 55–62, **perform** the operations and simplify.
(In Exercises 61 and 62, *n* is a positive integer.)

55. $\left[\dfrac{x^2}{9} \cdot \dfrac{3(x + 4)}{x^2 + 2x} \right] \div \dfrac{x}{x + 2}$

56. $\left(\dfrac{x^2 + 6x + 9}{x^2} \cdot \dfrac{2x + 1}{x^2 - 9} \right) \div \dfrac{4x^2 + 4x + 1}{x^2 - 3x}$

57. $\left[\dfrac{xy + y}{4x} \div (3x + 3) \right] \div \dfrac{y}{3x}$

58. $\dfrac{3u^2 - u - 4}{u^2} \div \dfrac{3u^2 + 12u + 4}{u^4 - 3u^3}$

59. $\dfrac{2x^2 + 5x - 25}{3x^2 + 5x + 2} \cdot \dfrac{3x^2 + 2x}{x + 5} \div \left(\dfrac{x}{x + 1} \right)^2$

60. $\dfrac{t^2 - 100}{4t^2} \cdot \dfrac{t^3 - 5t^2 - 50t}{t^4 + 10t^3} \div \dfrac{(t - 10)^2}{5t}$

61. $x^3 \cdot \dfrac{x^{2n} - 9}{x^{2n} + 4x^n + 3} \div \dfrac{x^{2n} - 2x^n - 3}{x}$

62. $\dfrac{x^{n + 1} - 8x}{x^{2n} + 2x^n + 1} \cdot \dfrac{x^{2n} - 4x^n - 5}{x} \div x^n$

Probability In Exercises 63–66, consider an experiment in which a marble is tossed into a rectangular box with dimensions $2x$ centimeters by $(4x + 2)$ centimeters. The probability that the marble will come to rest in the unshaded portion of the box is equal to the ratio of the unshaded area to the total area of the figure. **Find** the probability in simplified form.

63.

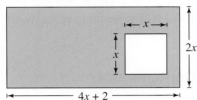

64.

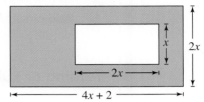

65.

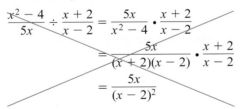

66.

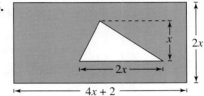

Explaining Concepts

67. *Writing* Describe how the operation of division is used in the process of simplifying a product of rational expressions.

68. *Writing* In a quotient of two rational expressions, the denominator of the divisor is x. Describe a set of circumstances in which you will *not* need to list $x \neq 0$ as a domain restriction after dividing.

69. *Logic* Explain what is missing in the following statement.

$$\frac{x - a}{x - b} \div \frac{x - a}{x - b} = 1$$

70. *Logic* When two rational expressions are multiplied, the resulting expression is a polynomial. Explain how the total number of factors in the numerators of the expressions you multiplied compares to the total number of factors in the denominators.

71. *Error Analysis* Describe and correct the errors.

$$\frac{x^2 - 4}{5x} \div \frac{x + 2}{x - 2} = \frac{5x}{x^2 - 4} \cdot \frac{x + 2}{x - 2}$$
$$= \frac{5x}{(x + 2)(x - 2)} \cdot \frac{x + 2}{x - 2}$$
$$= \frac{5x}{(x - 2)^2}$$

72. *Repeated Reasoning* Complete the table for the given values of x. Round your answers to five decimal places.

x	60	100	1000
$\dfrac{x - 10}{x + 10}$			
$\dfrac{x + 50}{x - 50}$			
$\dfrac{x - 10}{x + 10} \cdot \dfrac{x + 50}{x - 50}$			

x	10,000	100,000	1,000,000
$\dfrac{x - 10}{x + 10}$			
$\dfrac{x + 50}{x - 50}$			
$\dfrac{x - 10}{x + 10} \cdot \dfrac{x + 50}{x - 50}$			

What kind of pattern do you see? Try to explain what is going on. Can you see why?

Cumulative Review

In Exercises 73–76, evaluate the expression.

73. $\frac{1}{8} + \frac{3}{8} + \frac{5}{8}$

74. $\frac{3}{7} - \frac{2}{7}$

75. $\frac{3}{5} + \frac{4}{15}$

76. $\frac{7}{6} - \frac{9}{7}$

In Exercises 77 and 78, solve the equation by factoring.

77. $x^2 + 3x = 0$

78. $x^2 + 3x - 10 = 0$

7.3 Adding and Subtracting Rational Expressions

▶ Add or subtract rational expressions with like denominators, and simplify.

▶ Add or subtract rational expressions with unlike denominators, and simplify.

Adding or Subtracting with Like Denominators

Adding or Subtracting with Like Denominators

If u, v, and w are real numbers, variables, or algebraic expressions, and $w \neq 0$, the following rules are valid.

1. $\dfrac{u}{w} + \dfrac{v}{w} = \dfrac{u+v}{w}$ Add fractions with like denominators.

2. $\dfrac{u}{w} - \dfrac{v}{w} = \dfrac{u-v}{w}$ Subtract fractions with like denominators.

EXAMPLE 1 Adding and Subtracting Rational Expressions

a. $\dfrac{x}{4} + \dfrac{5-x}{4} = \dfrac{x+(5-x)}{4} = \dfrac{5}{4}$ Add numerators.

b. $\dfrac{7}{2x-3} - \dfrac{3x}{2x-3} = \dfrac{7-3x}{2x-3}$ Subtract numerators.

c. $\dfrac{x}{x^2-2x-3} - \dfrac{3}{x^2-2x-3} = \dfrac{x-3}{x^2-2x-3}$ Subtract numerators.

$\qquad\qquad\qquad\qquad\quad = \dfrac{(1)(x-3)}{(x-3)(x+1)}$ Factor.

$\qquad\qquad\qquad\qquad\quad = \dfrac{1}{x+1}, \; x \neq 3$ Simplified form

Exercises Within Reach ®

Solutions in English & Spanish and tutorial videos at CollegePrepAlgebra.com

Adding and Subtracting Rational Expressions In Exercises 1–10, combine and simplify.

1. $\dfrac{5x}{6} + \dfrac{4x}{6}$

2. $\dfrac{7y}{12} + \dfrac{9y}{12}$

3. $\dfrac{2}{3a} - \dfrac{11}{3a}$

4. $\dfrac{6}{19x} - \dfrac{7}{19x}$

5. $\dfrac{x}{9} - \dfrac{x+2}{9}$

6. $\dfrac{4-y}{4} + \dfrac{3y}{4}$

7. $\dfrac{2x-1}{x(x-3)} + \dfrac{1-x}{x(x-3)}$

8. $\dfrac{3-2n}{n(n+2)} - \dfrac{1-3n}{n(n+2)}$

9. $\dfrac{c}{c^2+3c-4} - \dfrac{1}{c^2+3c-4}$

10. $\dfrac{2v}{2v^2-5v-12} + \dfrac{3}{2v^2-5v-12}$

Adding or Subtracting with Unlike Denominators

The **least common multiple (LCM)** of two (or more) polynomials can be helpful when adding or subtracting rational expressions with *unlike* denominators. The least common multiple of two (or more) polynomials is the simplest polynomial that is a multiple of each of the original polynomials. This means that the LCM must contain all the *different* factors in each polynomial, with each factor raised to the greatest power of its occurrence in any one of the polynomials.

EXAMPLE 2   **Finding Least Common Multiples**

Find the least common multiple of the expressions.

a. $6x, 2x^2, 9x^3$ **b.** $x^2 - x, 2x - 2$ **c.** $3x^2 + 6x, x^2 + 4x + 4$

SOLUTION

a. The least common multiple of
$$6x = 2 \cdot 3 \cdot x, \quad 2x^2 = 2 \cdot x^2, \quad \text{and} \quad 9x^3 = 3^2 \cdot x^3$$
is $2 \cdot 3^2 \cdot x^3 = 18x^3$.

b. The least common multiple of
$$x^2 - x = x(x - 1) \quad \text{and} \quad 2x - 2 = 2(x - 1)$$
is $2x(x - 1)$.

c. The least common multiple of
$$3x^2 + 6x = 3x(x + 2) \quad \text{and} \quad x^2 + 4x + 4 = (x + 2)^2$$
is $3x(x + 2)^2$.

To add or subtract rational expressions with *unlike* denominators, you must first rewrite the rational expressions so that they have a common denominator. You can always find a common denominator of two (or more) rational expressions by multiplying their denominators. However, if you use the **least common denominator (LCD)**, which is the least common multiple of the denominators, you may have less simplifying to do. After the rational expressions have been written with a common denominator, you can simply add or subtract using the rules given at the beginning of this section.

Exercises Within Reach ®

Solutions in English & Spanish and tutorial videos at CollegePrepAlgebra.com

Finding the Least Common Multiple In Exercises 11−22, find the least common multiple of the expressions.

11. $5x^2, 20x^3$

12. $14t^2, 42t^5$

13. $9y^3, 12y$

14. $18m^2, 45m$

15. $15x^2, 3(x + 5)$

16. $6x^2, 15x(x - 1)$

17. $63z^2(z + 1), 14(z + 1)^4$

18. $18y^3, 27y(y - 3)^2$

19. $8t(t + 2), 14(t^2 - 4)$

20. $6(x^2 - 4), 2x(x + 2)$

21. $2y^2 + y - 1, 4y^2 - 2y$

22. $t^3 + 3t^2 + 9t, 2t^2(t^2 - 9)$

EXAMPLE 3 **Adding with Unlike Denominators**

Add the rational expressions: $\dfrac{7}{6x} + \dfrac{5}{8x}$.

SOLUTION

By factoring the denominators, $6x = 2 \cdot 3 \cdot x$ and $8x = 2^3 \cdot x$, you can conclude that the least common denominator is $2^3 \cdot 3 \cdot x = 24x$.

$$\dfrac{7}{6x} + \dfrac{5}{8x} = \dfrac{7(4)}{6x(4)} + \dfrac{5(3)}{8x(3)}$$ Rewrite expressions using LCD of $24x$.

$$= \dfrac{28}{24x} + \dfrac{15}{24x}$$ Like denominators

$$= \dfrac{28 + 15}{24x} = \dfrac{43}{24x}$$ Add fractions and simplify.

EXAMPLE 4 **Subtracting with Unlike Denominators**

Subtract the rational expressions: $\dfrac{3}{x - 3} - \dfrac{5}{x + 2}$.

SOLUTION

The only factors of the denominators are $x - 3$ and $x + 2$. So, the least common denominator is $(x - 3)(x + 2)$.

$$\dfrac{3}{x - 3} - \dfrac{5}{x + 2}$$ Write original expressions.

$$= \dfrac{3(x + 2)}{(x - 3)(x + 2)} - \dfrac{5(x - 3)}{(x - 3)(x + 2)}$$ Rewrite expressions using LCD of $(x - 3)(x + 2)$.

$$= \dfrac{3x + 6}{(x - 3)(x + 2)} - \dfrac{5x - 15}{(x - 3)(x + 2)}$$ Distributive Property

$$= \dfrac{3x + 6 - 5x + 15}{(x - 3)(x + 2)}$$ Subtract fractions and use the Distributive Property.

$$= \dfrac{-2x + 21}{(x - 3)(x + 2)}$$ Simplified form

Exercises Within Reach ®

Solutions in English & Spanish and tutorial videos at CollegePrepAlgebra.com

Adding and Subtracting Rational Expressions **In Exercises 23−32, combine and simplify.**

23. $\dfrac{5}{4x} - \dfrac{3}{5}$

24. $\dfrac{10}{b} + \dfrac{1}{10b}$

25. $\dfrac{7}{a} + \dfrac{14}{a^2}$

26. $\dfrac{1}{6u^2} - \dfrac{2}{9u}$

27. $25 + \dfrac{10}{x + 4}$

28. $\dfrac{30}{x - 6} - 4$

29. $\dfrac{x}{x + 3} - \dfrac{5}{x - 2}$

30. $\dfrac{1}{x + 4} - \dfrac{1}{x + 2}$

31. $\dfrac{12}{x^2 - 9} - \dfrac{2}{x - 3}$

32. $\dfrac{12}{x^2 - 4} - \dfrac{3}{x + 2}$

EXAMPLE 5   **Adding with Unlike Denominators**

Find the sum of the rational expressions.

$$\frac{6x}{x^2 - 4} + \frac{3}{2 - x}$$

SOLUTION

$$\frac{6x}{x^2 - 4} + \frac{3}{2 - x}$$ Original expressions

$$= \frac{6x}{(x + 2)(x - 2)} + \frac{3}{(-1)(x - 2)}$$ Factor denominators.

$$= \frac{6x}{(x + 2)(x - 2)} - \frac{3(x + 2)}{(x + 2)(x - 2)}$$ Rewrite expressions using
LCD of $(x + 2)(x - 2)$.

$$= \frac{6x}{(x + 2)(x - 2)} - \frac{3x + 6}{(x + 2)(x - 2)}$$ Distributive Property

$$= \frac{6x - (3x + 6)}{(x + 2)(x - 2)}$$ Subtract.

$$= \frac{6x - 3x - 6}{(x + 2)(x - 2)}$$ Distributive Property

$$= \frac{3x - 6}{(x + 2)(x - 2)}$$ Simplify.

$$= \frac{3(x - 2)}{(x + 2)(x - 2)}$$ Factor numerator.

$$= \frac{3(x - 2)}{(x + 2)(x - 2)}$$ Divide out common factor.

$$= \frac{3}{x + 2}, \ x \neq 2$$ Simplified form

Exercises Within Reach ®

Solutions in English & Spanish and tutorial videos at CollegePrepAlgebra.com

Adding and Subtracting Rational Expressions In Exercises 33−42, combine and simplify.

33. $\dfrac{20}{x - 4} + \dfrac{20}{4 - x}$

34. $\dfrac{15}{2 - t} - \dfrac{7}{t - 2}$

35. $\dfrac{3x}{x - 8} - \dfrac{6}{8 - x}$

36. $\dfrac{1}{y - 6} + \dfrac{y}{6 - y}$

37. $\dfrac{3x}{3x - 2} + \dfrac{2}{2 - 3x}$

38. $\dfrac{y}{5y - 3} - \dfrac{3}{3 - 5y}$

39. $\dfrac{3}{x - 5} + \dfrac{2}{x + 5}$

40. $\dfrac{7}{2x - 3} + \dfrac{3}{2x + 3}$

41. $\dfrac{9}{5x} + \dfrac{3}{x - 1}$

42. $\dfrac{3}{x - 1} + \dfrac{5}{4x}$

EXAMPLE 6 **Subtracting with Unlike Denominators**

$$\frac{x}{x^2 - 5x + 6} - \frac{1}{x^2 - x - 2} \qquad \text{Original expressions}$$

$$= \frac{x}{(x - 3)(x - 2)} - \frac{1}{(x - 2)(x + 1)} \qquad \text{Factor denominators.}$$

$$= \frac{x(x + 1)}{(x - 3)(x - 2)(x + 1)} - \frac{1(x - 3)}{(x - 3)(x - 2)(x + 1)} \qquad \begin{array}{l}\text{Rewrite expressions} \\ \text{using LCD of } (x - 3) \\ (x - 2)(x + 1).\end{array}$$

$$= \frac{x^2 + x}{(x - 3)(x - 2)(x + 1)} - \frac{x - 3}{(x - 3)(x - 2)(x + 1)} \qquad \text{Distributive Property}$$

$$= \frac{(x^2 + x) - (x - 3)}{(x - 3)(x - 2)(x + 1)} \qquad \text{Subtract fractions.}$$

$$= \frac{x^2 + x - x + 3}{(x - 3)(x - 2)(x + 1)} \qquad \text{Distributive Property}$$

$$= \frac{x^2 + 3}{(x - 3)(x - 2)(x + 1)} \qquad \text{Simplified form}$$

EXAMPLE 7 **Combining Rational Expressions**

$$\frac{4x}{x^2 - 16} + \frac{x}{x + 4} - \frac{2}{x} = \frac{4x}{(x + 4)(x - 4)} + \frac{x}{x + 4} - \frac{2}{x}$$

$$= \frac{4x(x)}{x(x + 4)(x - 4)} + \frac{x(x)(x - 4)}{x(x + 4)(x - 4)} - \frac{2(x + 4)(x - 4)}{x(x + 4)(x - 4)}$$

$$= \frac{4x^2 + x^2(x - 4) - 2(x^2 - 16)}{x(x + 4)(x - 4)}$$

$$= \frac{4x^2 + x^3 - 4x^2 - 2x^2 + 32}{x(x + 4)(x - 4)}$$

$$= \frac{x^3 - 2x^2 + 32}{x(x + 4)(x - 4)}$$

Exercises Within Reach ®

Solutions in English & Spanish and tutorial videos at CollegePrepAlgebra.com

Combining Rational Expressions In Exercises 43−52, combine and simplify.

43. $\dfrac{x}{x^2 - x - 30} - \dfrac{1}{x + 5}$

44. $\dfrac{x}{x^2 - 9} + \dfrac{3}{x^2 - 5x + 6}$

45. $\dfrac{4}{x - 4} + \dfrac{16}{(x - 4)^2}$

46. $\dfrac{3}{x - 2} - \dfrac{1}{(x - 2)^2}$

47. $\dfrac{y}{x^2 + xy} - \dfrac{x}{xy + y^2}$

48. $\dfrac{5}{x + y} + \dfrac{5}{x^2 - y^2}$

49. $\dfrac{4}{x} - \dfrac{2}{x^2} + \dfrac{4}{x + 3}$

50. $\dfrac{5}{2} - \dfrac{1}{2x} - \dfrac{3}{x + 1}$

51. $\dfrac{3u}{u^2 - 2uv + v^2} + \dfrac{2}{u - v} - \dfrac{u}{u - v}$

52. $\dfrac{1}{x - y} - \dfrac{3}{x + y} + \dfrac{3x - y}{x^2 - y^2}$

Application   EXAMPLE 8   **Marital Status**

The four types of marital status are never married, married, widowed, and divorced. For the years 2005 through 2010, the total number of people in the United States 18 years old or older P (in millions) and the total number of these people who were never married N (in millions) can be modeled by

$$P = \frac{10.443t + 201.35}{0.033t + 1} \quad \text{and} \quad N = \frac{-1.471t + 49.75}{-0.043t + 1}, \quad 5 \le t \le 10$$

where t represents the year, with $t = 5$ corresponding to 2005. Find a rational model T for the total number of people whose marital status was married, widowed, or divorced during this time period. (*Source:* U.S. Census Bureau)

SOLUTION

To find a model for T, find the difference of P and N.

$$T = \frac{10.443t + 201.35}{0.033t + 1} - \frac{-1.471t + 49.75}{-0.043t + 1} \qquad \text{Subtract } N \text{ from } P.$$

$$= \frac{(10.443t + 201.35)(-0.043t + 1) - (0.033t + 1)(-1.471t + 49.75)}{(0.033t + 1)(-0.043t + 1)} \qquad \text{Basic definition}$$

$$= \frac{-0.400506t^2 + 1.6142t + 151.6}{(0.033t + 1)(-0.043t + 1)} \qquad \text{Simplify.}$$

Exercises Within Reach ®

Solutions in English & Spanish and tutorial videos at CollegePrepAlgebra.com

Undergraduate Students **In Exercises 53 and 54, use the following models, which give the numbers (in millions) of males M and females F enrolled as undergraduate students from 2005 through 2010.**

$$M = \frac{-0.150t + 5.53}{-0.049t + 1}, \ 5 \le t \le 10 \qquad\qquad F = \frac{-0.313t + 7.71}{-0.055t + 1}, \ 5 \le t \le 10$$

In these models, t represents the year, with $t = 5$ corresponding to 2005. (*Source:* U.S. Department of Education)

53. Find a rational model T for the total number of undergraduate students (in millions) from 2005 through 2010.

54. Use the model you found in Exercise 53 to complete the table showing the total number of undergraduate students (rounded to the nearest million) each year from 2005 through 2010.

Year		
Undergraduates (in millions)		

Year		
Undergraduates (in millions)		

Concept Summary: *Adding and Subtracting Rational Expressions*

What

The rules for adding and subtracting rational expressions are the same as the rules for adding and subtracting numerical fractions.

EXAMPLE

Add: $\dfrac{7}{4x^2} + \dfrac{5}{2x}$.

How

To add (or subtract) rational expressions with unlike denominators do the following.

1. Use the **least common denominator** (LCD) to rewrite the fractions so they have like denominators.
2. Add (or subtract) the numerators.
3. Simplify the result.

EXAMPLE

$\dfrac{7}{4x^2} + \dfrac{5}{2x}$

$= \dfrac{7}{4x^2} + \dfrac{5(2x)}{2x(2x)}$ Rewrite.

$= \dfrac{7}{4x^2} + \dfrac{10x}{4x^2}$ Simplify.

$= \dfrac{10x + 7}{4x^2}$ Add.

Why

Knowing how to add and subtract rational expressions will help you solve rational equations.

Exercises Within Reach®

Worked-out solutions to odd-numbered exercises at CollegePrepAlgebra.com

Concept Summary Check

55. *True or False?* Two rational expressions with *like* denominators have a common denominator.

56. *Logic* When adding or subtracting rational expressions, how do you rewrite each rational expression as an equivalent expression whose denominator is the LCD?

57. *Writing* In your own words, describe how to add or subtract rational expressions with *like* denominators.

58. *Writing* In your own words, describe how to add or subtract rational expressions with *unlike* denominators.

Extra Practice

Adding and Subtracting Rational Expressions **In Exercises 59−62, combine and simplify.**

59. $\dfrac{4}{x^2} - \dfrac{4}{x^2 + 1}$

60. $\dfrac{3}{y^2 - 3} + \dfrac{2}{3y^2}$

61. $\dfrac{x + 2}{x - 1} - \dfrac{2}{x + 6} - \dfrac{14}{x^2 + 5x - 6}$

62. $\dfrac{-2x - 10}{x^2 + 8x + 15} + \dfrac{2}{x + 3} + \dfrac{x}{x + 5}$

63. *Work Rate* After working together for *t* hours on a common task, two workers have completed fractional parts of the job equal to *t*/4 and *t*/6. What fractional part of the task has been completed?

64. *Work Rate* After working together for *t* hours on a common task, two workers have completed fractional parts of the job equal to *t*/3 and *t*/5. What fractional part of the task has been completed?

65. Rewriting a Fraction The fraction $4/(x^3 - x)$ can be rewritten as a sum of three fractions, as follows.

$$\frac{4}{x^3 - x} = \frac{A}{x} + \frac{B}{x + 1} + \frac{C}{x - 1}$$

The numbers A, B, and C are the solutions of the system

$$\begin{cases} A + B + C = 0 \\ \quad\; -B + C = 0. \\ -A \qquad\quad = 4 \end{cases}$$

Solve the system and verify that the sum of the three resulting fractions is the original fraction.

66. Rewriting a Fraction The fraction

$$\frac{x + 1}{x^3 - x^2}$$

can be rewritten as a sum of three fractions, as follows.

$$\frac{x + 1}{x^3 - x^2} = \frac{A}{x} + \frac{B}{x^2} + \frac{C}{x - 1}$$

The numbers A, B, and C are the solutions of the system

$$\begin{cases} A \qquad\quad + C = 0 \\ -A + B \qquad = 1. \\ \quad\;\; -B \qquad = 1 \end{cases}$$

Solve the system and verify that the sum of the three resulting fractions is the original fraction.

Explaining Concepts

67. Error Analysis Describe the error.

$$\frac{x - 1}{x + 4} - \frac{4x - 11}{x + 4} = \frac{x - 1 - 4x - 11}{x + 4}$$

$$= \frac{-3x - 12}{x + 4}$$

$$= \frac{-3(x + 4)}{x + 4}$$

$$= -3, \; x \neq -4$$

68. Error Analysis Describe the error.

$$\frac{2}{x} - \frac{3}{x + 1} + \frac{x + 1}{x^2}$$

$$= \frac{2x(x + 1) - 3x^2 + (x + 1)^2}{x^2(x + 1)}$$

$$= \frac{2x^2 + x - 3x^2 + x^2 + 1}{x^2(x + 1)}$$

$$= \frac{x + 1}{x^2(x + 1)} = \frac{1}{x^2}, \; x \neq -1$$

69. Reasoning Is it possible for the least common denominator of two fractions to be the same as one of the fraction's denominators? If so, give an example.

70. Precision Evaluate each expression at the given value of the variable in two different ways: (1) combine and simplify the rational expressions first and then evaluate the simplified expression at the given value of the variable, and (2) substitute the given value of the variable first and then simplify the resulting expression. Do you get the same result with each method? Discuss which method you prefer and why. List the advantages and/or disadvantages of each method.

(a) $\dfrac{1}{m - 4} - \dfrac{1}{m + 4} + \dfrac{3m}{m^2 + 16}$, $m = 2$

(b) $\dfrac{x - 2}{x^2 - 9} + \dfrac{3x + 2}{x^2 - 5x + 6}$, $x = 4$

(c) $\dfrac{3y^2 + 16y - 8}{y^2 + 2y - 8} - \dfrac{y - 1}{y - 2} + \dfrac{y}{y + 4}$, $y = 3$

Cumulative Review

In Exercises 71–74, find the sum or difference.

71. $5v + (4 - 3v)$

72. $(2v + 7) + (9v + 8)$

73. $(x^2 - 4x + 3) - (6 - 2x)$

74. $(5y + 2) - (2y^2 + 8y - 5)$

In Exercises 75–78, factor the trinomial, if possible.

75. $x^2 - 7x + 12$

76. $c^2 + 6c + 10$

77. $2a^2 - 9a - 18$

78. $6w^2 + 14w - 12$

Mid-Chapter Quiz: Sections 7.1–7.3

Solutions in English & Spanish and tutorial videos at CollegePrepAlgebra.com

Take this quiz as you would take a quiz in class. After you are done, check your work against the answers in the back of the book.

In Exercises 1 and 2, find the domain of the rational function.

1. $f(x) = \dfrac{x}{x^2 + x}$

2. $f(x) = \dfrac{x^2 + 4x}{x^2 - 4}$

In Exercises 3–8, simplify the rational expression.

3. $\dfrac{9y^2}{6y}$

4. $\dfrac{6u^4v^3}{15uv^3}$

5. $\dfrac{4x^2 - 1}{x - 2x^2}$

6. $\dfrac{(z + 3)^2}{2z^2 + 5z - 3}$

7. $\dfrac{5a^2b + 3ab^3}{a^2b^2}$

8. $\dfrac{2mn^2 - n^3}{2m^2 + mn - n^2}$

In Exercises 9–20, perform the indicated operations and simplify.

9. $\dfrac{11t^2}{6} \cdot \dfrac{9}{33t}$

10. $(x^2 + 2x) \cdot \dfrac{5}{x^2 - 4}$

11. $\dfrac{4}{3(x - 1)} \cdot \dfrac{12x}{6(x^2 + 2x - 3)}$

12. $\dfrac{32z^4}{5x^5y^5} \div \dfrac{80z^5}{25x^8y^6}$

13. $\dfrac{a - b}{9a + 9b} \div \dfrac{a^2 - b^2}{a^2 + 2a + 1}$

14. $\dfrac{5u}{3(u + v)} \cdot \dfrac{2(u^2 - v^2)}{3v} \div \dfrac{25u^2}{18(u - v)}$

15. $\dfrac{5x - 6}{x - 2} + \dfrac{2x - 5}{x - 2}$

16. $\dfrac{x}{x^2 - 9} - \dfrac{4(x - 3)}{x + 3}$

17. $\dfrac{x^2 + 2}{x^2 - x - 2} + \dfrac{1}{x + 1} - \dfrac{x}{x - 2}$

18. $\dfrac{9t^2}{3 - t} \div \dfrac{6t}{t - 3}$

19. $\dfrac{10}{x^2 + 2x} \div \dfrac{15}{x^2 + 3x + 2}$

20. $\dfrac{3x^{-1} - y^{-1}}{(x - y)^{-1}}$

Application

21. You open a floral shop with a setup cost of $25,000. The cost of creating one dozen floral arrangements is $144.

    (a) Write the total cost C as a function of x, the number of dozens of floral arrangements created.

    (b) Write the average cost per dozen $\overline{C} = C/x$ as a function of x, the number of dozens of floral arrangements created.

    (c) Find the value of $\overline{C}(500)$.

Study Skills in Action

Using a Test-Taking Strategy

What do runners do before a race? They design a strategy for running their best. They make sure they get enough rest, eat sensibly, and get to the track early to warm up. In the same way, it is important for students to get a good night's sleep, eat a healthy meal, and get to school early to allow time to focus before a test.

The biggest difference between a runner's race and a math test is that a math student does not have to reach the finish line first! In fact, many students would increase their scores if they used all the test time instead of worrying about being the last student left in the class. This is why it is important to have a strategy for taking the test.

These runners are focusing on their techniques, not on whether other runners are ahead of or behind them.

Smart Study Strategy

Use Ten Steps for Test Taking

1 ▶ **Do a memory data dump.** As soon as you get the test, turn it over and write down anything that you still have trouble remembering sometimes (formulas, calculations, rules).

2 ▶ **Preview the test.** Look over the test and mark the questions you know how to do easily. These are the problems you should do first.

3 ▶ **Do a second memory data dump.** As you previewed the test, you may have remembered other information. Write this information on the back of the test.

4 ▶ **Develop a test progress schedule.** Based on how many points each question is worth, decide on a progress schedule. You should always have more than half the test done before half the time has elapsed.

5 ▶ **Answer the easiest problems first.** Solve the problems you marked while previewing the test.

6 ▶ **Skip difficult problems.** Skip the problems that you suspect will give you trouble.

7 ▶ **Review the skipped problems.** After solving all the problems that you know how to do easily, go back and reread the problems you skipped.

8 ▶ **Try your best at the remaining problems that confuse you.** Even if you cannot completely solve a problem, you may be able to get partial credit for a few correct steps.

9 ▶ **Review the test.** Look for any careless errors you may have made.

10 ▶ **Use all the allowed test time.** The test is not a race against the other students.

7.4 Complex Fractions

▶ Simplify complex fractions using rules for dividing rational expressions.

▶ Simplify complex fractions that have a sum or difference in the numerator and/or denominator.

Complex Fractions

Problems involving the division of two rational expressions are sometimes written as complex fractions. A **complex fraction** is a fraction that has a fraction in its numerator or denominator, or both.

$$\dfrac{\left.\dfrac{x+2}{3}\right\}}{\left.\dfrac{x-2}{x}\right\}}$$

→ Numerator fraction
→ Main fraction line
→ Denominator fraction

EXAMPLE 1 **Simplifying a Complex Fraction**

Simplify the complex fraction.

$$\frac{\left(\dfrac{5}{14}\right)}{\left(\dfrac{25}{8}\right)}$$

SOLUTION

$$\frac{\left(\dfrac{5}{14}\right)}{\left(\dfrac{25}{8}\right)} = \frac{5}{14} \cdot \frac{8}{25}$$ Invert divisor and multiply.

$$= \frac{5 \cdot 2 \cdot 2 \cdot 2}{2 \cdot 7 \cdot 5 \cdot 5}$$ Multiply and factor.

$$= \frac{\cancel{5} \cdot \cancel{2} \cdot 2 \cdot 2}{\cancel{2} \cdot 7 \cdot \cancel{5} \cdot 5}$$ Divide out common factors.

$$= \frac{4}{35}$$ Simplified form

Exercises Within Reach ®   Solutions in English & Spanish and tutorial videos at CollegePrepAlgebra.com

Simplifying a Complex Fraction In Exercises 1−4, simplify the complex fraction.

1. $\dfrac{\left(\dfrac{3}{16}\right)}{\left(\dfrac{9}{12}\right)}$

2. $\dfrac{\left(\dfrac{20}{21}\right)}{\left(\dfrac{8}{7}\right)}$

3. $\dfrac{\left(\dfrac{8x^2y}{3z^2}\right)}{\left(\dfrac{4xy}{9z^5}\right)}$

4. $\dfrac{\left(\dfrac{36x^4}{5y^4z^5}\right)}{\left(\dfrac{9xy^2}{20z^5}\right)}$

EXAMPLE 2   **Simplifying Complex Fractions**

Simplify each complex fraction.

a. $\dfrac{\left(\dfrac{4y^3}{(5x)^2}\right)}{\left(\dfrac{(2y)^2}{10x^3}\right)}$

b. $\dfrac{\left(\dfrac{x+1}{x+2}\right)}{\left(\dfrac{x+1}{x+5}\right)}$

SOLUTION

a. $\dfrac{\left(\dfrac{4y^3}{(5x)^2}\right)}{\left(\dfrac{(2y)^2}{10x^3}\right)} = \dfrac{4y^3}{25x^2} \cdot \dfrac{10x^3}{4y^2}$ Invert divisor and multiply.

$= \dfrac{4y^2 \cdot y \cdot 2 \cdot 5x^2 \cdot x}{5 \cdot 5x^2 \cdot 4y^2}$ Multiply and factor.

$= \dfrac{4y^2 \cdot y \cdot 2 \cdot 5x^2 \cdot x}{5 \cdot 5x^2 \cdot 4y^2}$ Divide out common factors.

$= \dfrac{2xy}{5}, \ x \neq 0, \ y \neq 0$ Simplified form

b. $\dfrac{\left(\dfrac{x+1}{x+2}\right)}{\left(\dfrac{x+1}{x+5}\right)} = \dfrac{x+1}{x+2} \cdot \dfrac{x+5}{x+1}$ Invert divisor and multiply.

$= \dfrac{(x+1)(x+5)}{(x+2)(x+1)}$ Multiply numerators and denominators.

$= \dfrac{(x+1)(x+5)}{(x+2)(x+1)}$ Divide out common factor.

$= \dfrac{x+5}{x+2}, \ x \neq -1, \ x \neq -5$ Simplified form

Study Tip

Domain restrictions result from the values that make any denominator zero in a complex fraction. In Example 2(a), note that the original expression has three denominators: $(5x)^2$, $10x^3$, and $(2y)^2/10x^3$. The domain restrictions that result from these denominators are $x \neq 0$ and $y \neq 0$.

Exercises Within Reach ®   Solutions in English & Spanish and tutorial videos at CollegePrepAlgebra.com

Simplifying a Complex Fraction In Exercises 5–10, simplify the complex fraction.

5. $\dfrac{\left(\dfrac{6x^3}{(5y)^2}\right)}{\left(\dfrac{(3x)^2}{15y^4}\right)}$

6. $\dfrac{\left(\dfrac{(3r)^3}{10t^4}\right)}{\left(\dfrac{9r}{(2t)^2}\right)}$

7. $\dfrac{\left(\dfrac{y}{3-y}\right)}{\left(\dfrac{y^2}{y-3}\right)}$

8. $\dfrac{\left(\dfrac{x}{x-4}\right)}{\left(\dfrac{x}{4-x}\right)}$

9. $\dfrac{\left(\dfrac{25x^2}{x-5}\right)}{\left(\dfrac{10x}{5+4x-x^2}\right)}$

10. $\dfrac{\left(\dfrac{5x}{x+7}\right)}{\left(\dfrac{10}{x^2+8x+7}\right)}$

EXAMPLE 3   **Simplifying a Complex Fraction**

Simplify the complex fraction.

$$\frac{\left(\dfrac{x^2 + 4x + 3}{x - 2}\right)}{2x + 6}$$

SOLUTION

$$\frac{\left(\dfrac{x^2 + 4x + 3}{x - 2}\right)}{2x + 6} = \frac{\left(\dfrac{x^2 + 4x + 3}{x - 2}\right)}{\left(\dfrac{2x + 6}{1}\right)}$$ Rewrite denominator.

$$= \frac{x^2 + 4x + 3}{x - 2} \cdot \frac{1}{2x + 6}$$ Invert divisor and multiply.

$$= \frac{(x + 1)(x + 3)}{(x - 2)(2)(x + 3)}$$ Multiply and factor.

$$= \frac{(x + 1)\cancel{(x + 3)}}{(x - 2)(2)\cancel{(x + 3)}}$$ Divide out common factor.

$$= \frac{x + 1}{2(x - 2)}, \quad x \ne -3$$ Simplified form

Exercises Within Reach ®

Solutions in English & Spanish and tutorial videos at CollegePrepAlgebra.com

Simplifying a Complex Fraction In Exercises 11−22, simplify the complex fraction.

11. $\dfrac{\left(\dfrac{x^2 + 3x - 10}{x + 4}\right)}{3x - 6}$

12. $\dfrac{\left(\dfrac{x^2 - 2x - 8}{x - 1}\right)}{5x - 20}$

13. $\dfrac{2x - 14}{\left(\dfrac{x^2 - 9x + 14}{x + 3}\right)}$

14. $\dfrac{4x + 16}{\left(\dfrac{x^2 + 9x + 20}{x - 1}\right)}$

15. $\dfrac{\left(\dfrac{6x^2 - 17x + 5}{3x^2 + 3x}\right)}{\left(\dfrac{3x - 1}{3x + 1}\right)}$

16. $\dfrac{\left(\dfrac{6x^2 - 13x - 5}{5x^2 + 5x}\right)}{\left(\dfrac{2x - 5}{5x + 1}\right)}$

17. $\dfrac{\left(\dfrac{16x^2 + 8x + 1}{3x^2 + 8x - 3}\right)}{\left(\dfrac{4x^2 - 3x - 1}{x^2 + 6x + 9}\right)}$

18. $\dfrac{\left(\dfrac{9x^2 - 24x + 16}{x^2 + 10x + 25}\right)}{\left(\dfrac{6x^2 - 5x - 4}{2x^2 + 3x - 35}\right)}$

19. $\dfrac{x^2 + x - 6}{x^2 - 4} \div \dfrac{x + 3}{x^2 + 4x + 4}$

20. $\dfrac{t^3 + t^2 - 9t - 9}{t^2 - 5t + 6} \div \dfrac{t^2 + 6t + 9}{t - 2}$

21. $\dfrac{\left(\dfrac{x^2 - 3x - 10}{x^2 - 4x + 4}\right)}{\left(\dfrac{21 + 4x - x^2}{x^2 - 5x - 14}\right)}$

22. $\dfrac{\left(\dfrac{x^2 + 5x + 6}{4x^2 - 20x + 25}\right)}{\left(\dfrac{x^2 - 5x - 24}{4x^2 - 25}\right)}$

Complex Fractions with Sums or Differences

Complex fractions can have numerators and/or denominators that are sums or differences of fractions. One way to simplify such a complex fraction is to combine the terms so that the numerator and denominator each consist of a single fraction. Then divide by inverting the denominator and multiplying.

EXAMPLE 4   **Simplifying a Complex Fraction**

Simplify the complex fraction.

$$\frac{\left(\dfrac{x}{3} + \dfrac{2}{3}\right)}{\left(1 - \dfrac{2}{x}\right)}$$

SOLUTION

$$\frac{\left(\dfrac{x}{3} + \dfrac{2}{3}\right)}{\left(1 - \dfrac{2}{x}\right)} = \frac{\left(\dfrac{x}{3} + \dfrac{2}{3}\right)}{\left(\dfrac{x}{x} - \dfrac{2}{x}\right)} \qquad \text{Rewrite with least common denominator.}$$

$$= \frac{\left(\dfrac{x + 2}{3}\right)}{\left(\dfrac{x - 2}{x}\right)} \qquad \text{Add fractions.}$$

$$= \frac{x + 2}{3} \cdot \frac{x}{x - 2} \qquad \text{Invert divisor and multiply.}$$

$$= \frac{x(x + 2)}{3(x - 2)}, \ x \neq 0 \qquad \text{Simplified form}$$

Study Tip

Another way of simplifying the complex fraction in Example 4 is to multiply the numerator and denominator by $3x$, the least common denominator of all the fractions in the numerator and denominator. This produces the same result, as shown below.

$$\frac{\left(\dfrac{x}{3} + \dfrac{2}{3}\right)}{\left(1 - \dfrac{2}{x}\right)} = \frac{\left(\dfrac{x}{3} + \dfrac{2}{3}\right)}{\left(1 - \dfrac{2}{x}\right)} \cdot \frac{3x}{3x}$$

$$= \frac{\dfrac{x}{3}(3x) + \dfrac{2}{3}(3x)}{(1)(3x) - \dfrac{2}{x}(3x)}$$

$$= \frac{x^2 + 2x}{3x - 6}$$

$$= \frac{x(x + 2)}{3(x - 2)}, \ x \neq 0$$

Exercises Within Reach ®

Solutions in English & Spanish and tutorial videos at CollegePrepAlgebra.com

Simplifying a Complex Fraction In Exercises 23−30, simplify the complex fraction.

23. $\dfrac{\left(1 + \dfrac{4}{y}\right)}{y}$

24. $\dfrac{x}{\left(\dfrac{3}{x} + 2\right)}$

25. $\dfrac{\left(\dfrac{4}{x} + 3\right)}{\left(\dfrac{4}{x} - 3\right)}$

26. $\dfrac{\left(\dfrac{1}{t} - 1\right)}{\left(\dfrac{1}{t} + 1\right)}$

27. $\dfrac{\left(\dfrac{x}{2}\right)}{\left(2 + \dfrac{3}{x}\right)}$

28. $\dfrac{\left(1 - \dfrac{2}{x}\right)}{\left(\dfrac{x}{2}\right)}$

29. $\dfrac{\left(3 + \dfrac{9}{x - 3}\right)}{\left(4 + \dfrac{12}{x - 3}\right)}$

30. $\dfrac{\left(4 + \dfrac{16}{x - 4}\right)}{\left(5 + \dfrac{20}{x - 4}\right)}$

EXAMPLE 5 Simplifying Complex Fractions

a.
$$\frac{\left(\frac{2}{x+2}\right)}{\left(\frac{3}{x+2} + \frac{2}{x}\right)} = \frac{\left(\frac{2}{x+2}\right)(x)(x+2)}{\left(\frac{3}{x+2}\right)(x)(x+2) + \left(\frac{2}{x}\right)(x)(x+2)}$$

$x(x+2)$ is the least common denominator.

$$= \frac{2x}{3x + 2(x+2)}$$

Multiply and simplify.

$$= \frac{2x}{3x + 2x + 4}$$

Distributive Property

$$= \frac{2x}{5x + 4}, \quad x \neq -2, x \neq 0$$

Simplify.

b.
$$\frac{5 + x^{-2}}{8x^{-1} + x} = \frac{\left(5 + \frac{1}{x^2}\right)}{\left(\frac{8}{x} + x\right)}$$

Rewrite with positive exponents.

$$= \frac{\left(\frac{5x^2}{x^2} + \frac{1}{x^2}\right)}{\left(\frac{8}{x} + \frac{x^2}{x}\right)}$$

Rewrite with least common denominators.

$$= \frac{\left(\frac{5x^2 + 1}{x^2}\right)}{\left(\frac{x^2 + 8}{x}\right)}$$

Add fractions.

$$= \frac{5x^2 + 1}{x^2} \cdot \frac{x}{x^2 + 8}$$

Invert divisor and multiply.

$$= \frac{x(5x^2 + 1)}{x^2(x^2 + 8)}$$

Multiply.

$$= \frac{x(5x^2 + 1)}{x(x)(x^2 + 8)}$$

Divide out common factor.

$$= \frac{5x^2 + 1}{x(x^2 + 8)}$$

Simplified form

Exercises Within Reach ®

Simplifying a Complex Fraction In Exercises 31 and 32, simplify the complex fraction.

31. $\dfrac{\left(\dfrac{1}{x} - \dfrac{1}{x+1}\right)}{\left(\dfrac{1}{x+1}\right)}$

32. $\dfrac{\left(\dfrac{5}{y} - \dfrac{6}{2y+1}\right)}{\left(\dfrac{5}{2y+1}\right)}$

Simplifying an Expression In Exercises 33−36, simplify the expression.

33. $\dfrac{2y - y^{-1}}{10 - y^{-2}}$

34. $\dfrac{9x - x^{-1}}{3 + x^{-1}}$

35. $\dfrac{7x^2 + 2x^{-1}}{5x^{-3} + x}$

36. $\dfrac{3x^{-2} - x}{4x^{-1} + 6x}$

Application   EXAMPLE 6   **Monthly Payment**

The approximate annual percent interest rate r (in decimal form) of a monthly installment loan is

$$r = \frac{\left[\dfrac{24(MN - P)}{N}\right]}{\left(P + \dfrac{MN}{12}\right)}$$

where N is the total number of payments, M is the monthly payment, and P is the amount financed.

a. Simplify the expression.

b. Approximate the annual percent interest rate for a four-year home-improvement loan of $15,000 with monthly payments of $350.

SOLUTION

a. $r = \dfrac{\left[\dfrac{24(MN - P)}{N}\right]}{\left(P + \dfrac{MN}{12}\right)} = \dfrac{24(MN - P)}{N} \div \dfrac{12P + MN}{12}$

$$= \frac{24(MN - P)}{N} \cdot \frac{12}{12P + MN}$$

$$= \frac{288(MN - P)}{N(12P + MN)}$$

b. $r = \dfrac{288(MN - P)}{N(12P + MN)} = \dfrac{288[350(48) - 15{,}000]}{48[12(15{,}000) + 350(48)]}$

$$= \frac{288(1800)}{48(196{,}800)}$$

$$\approx 0.0549$$

The rate is about 5.5%.

Exercises Within Reach ®   Solutions in English & Spanish and tutorial videos at CollegePrepAlgebra.com

37. *Electronics* When two resistors of resistance R_1 and R_2 (all in ohms) are connected in parallel, the total resistance (in ohms) is modeled by

$$\frac{1}{\left(\dfrac{1}{R_1} + \dfrac{1}{R_2}\right)}.$$

Simplify this complex fraction.

38. *Using Results* Use the simplified fraction in Exercise 37 to find the total resistance when $R_1 = 10$ ohms and $R_2 = 20$ ohms.

Concept Summary: Simplifying Complex Fractions

What

When simplifying a **complex fraction**, the rules for dividing rational expressions still apply.

EXAMPLE

Simplify $\dfrac{\left(\dfrac{x+2}{3}\right)}{\left(\dfrac{x+2}{x}\right)}$.

How

To simplify a complex fraction, invert the denominator fraction and multiply.

EXAMPLE

$$\frac{\left(\dfrac{x+2}{3}\right)}{\left(\dfrac{x+2}{x}\right)} = \frac{x+2}{3} \cdot \frac{x}{x+2}$$

$$= \frac{x\cancel{(x+2)}}{3\cancel{(x+2)}}$$

$$= \frac{x}{3}, \; x \neq 0, \; x \neq -2$$

Why

Knowing how to simplify "messy" expressions, such as complex fractions, will help you as you continue your study of algebra.

A useful strategy in algebra is to *rewrite complicated problems into simpler forms.*

Exercises Within Reach ®

Worked-out solutions to odd-numbered exercises at CollegePrepAlgebra.com

Concept Summary Check

39. *Vocabulary* Define the term *complex fraction*. Give an example.

40. *Precision* Describe how to rewrite a complex fraction as a product.

41. *Writing* Describe the method for simplifying complex fractions that involves the use of a least common denominator.

42. *Reasoning* Explain how you can find the implied domain restrictions for a complex fraction.

Extra Practice

Simplifying a Complex Fraction In Exercises 43−48, simplify the complex fraction.

43. $\dfrac{\left(\dfrac{y}{x} - \dfrac{x}{y}\right)}{\left(\dfrac{x+y}{xy}\right)}$

44. $\dfrac{\left(\dfrac{x}{y} - \dfrac{y}{x}\right)}{\left(\dfrac{x-y}{xy}\right)}$

45. $\dfrac{\left(\dfrac{x}{x-3} - \dfrac{2}{3}\right)}{\left(\dfrac{10}{3x} + \dfrac{x^2}{x-3}\right)}$

46. $\dfrac{\left(\dfrac{1}{2x} - \dfrac{6}{x+5}\right)}{\left(\dfrac{x}{x-5} + \dfrac{1}{x}\right)}$

47. $\dfrac{\left(\dfrac{10}{x+1}\right)}{\left(\dfrac{1}{2x+2} + \dfrac{3}{x+1}\right)}$

48. $\dfrac{\left(\dfrac{2}{x+5}\right)}{\left(\dfrac{2}{x+5} + \dfrac{1}{4x+20}\right)}$

Simplifying an Expression In Exercises 49−52, simplify the expression.

49. $\dfrac{x^{-1} + y^{-1}}{x^{-1} - y^{-1}}$

50. $\dfrac{x^{-1} - y^{-1}}{x^{-2} - y^{-2}}$

51. $\dfrac{x^{-2} - y^{-2}}{(x+y)^2}$

52. $\dfrac{x-y}{x^{-2} - y^{-2}}$

Simplifying an Expression In Exercises 53 and 54, use the function to find and simplify the expression for $\dfrac{f(2+h) - f(2)}{h}$.

53. $f(x) = \dfrac{1}{x}$

54. $f(x) = \dfrac{x}{x-1}$

55. *Average of Two Numbers* Determine the average of two real numbers $x/5$ and $x/6$.

56. *Average of Two Numbers* Determine the average of two real numbers $2x/3$ and $3x/5$.

57. *Average of Two Numbers* Determine the average of two real numbers $2x/3$ and $x/4$.

58. *Average of Two Numbers* Determine the average of two real numbers $4/a^2$ and $2/a$.

59. *Average of Two Numbers* Determine the average of two real numbers $(b + 5)/4$ and $2/b$.

60. *Average of Two Numbers* Determine the average of two real numbers $5/(2s)$ and $(s + 1)/5$.

61. *Number Problem* Find three real numbers that divide the real number line between $x/9$ and $x/6$ into four equal parts (see figure).

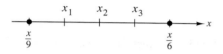

62. *Number Problem* Find two real numbers that divide the real number line between $x/3$ and $5x/4$ into three equal parts (see figure).

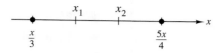

63. *Electronics* When three resistors of resistance R_1, R_2, and R_3 (all in ohms) are connected in parallel, the total resistance (in ohms) is modeled by

$$\frac{1}{\left(\dfrac{1}{R_1} + \dfrac{1}{R_2} + \dfrac{1}{R_3}\right)}.$$

Simplify this complex fraction.

64. *Using Results* Use the simplified fraction found in Exercise 63 to determine the total resistance when $R_1 = 10$ ohms, $R_2 = 15$ ohms, and $R_3 = 20$ ohms.

Explaining Concepts

65. *Reasoning* Is the simplified form of a complex fraction a complex fraction? Explain.

66. *Writing* Describe the effect of multiplying two rational expressions by their least common denominator.

Error Analysis In Exercises 67 and 68, describe and correct the error.

67. $\dfrac{\left(\dfrac{a}{b}\right)}{b} = \dfrac{a}{\left(\dfrac{b}{b}\right)} = \dfrac{a}{1} = a,\ b \neq 0$

68. $\dfrac{\left(\dfrac{a}{b}\right)}{\left(\dfrac{b}{c}\right)} = \dfrac{a}{b} \cdot \dfrac{b}{c} = \dfrac{a}{c},\ b \neq 0$

Cumulative Review

In Exercises 69 and 70, use the rules of exponents to simplify the expression.

69. $(2y)^3(3y)^2$

70. $\dfrac{27x^4y^2}{9x^3y}$

In Exercises 71 and 72, factor the trinomial.

71. $3x^2 + 5x - 2$

72. $x^2 + xy - 2y^2$

In Exercises 73–76, divide and simplify.

73. $\dfrac{x^2}{2} \div 4x$

74. $\dfrac{4x^3}{3} \div \dfrac{2x^2}{9}$

75. $\dfrac{(x + 1)^2}{x + 2} \div \dfrac{x + 1}{(x + 2)^3}$

76. $\dfrac{x^2 - 4x + 4}{x - 3} \div \dfrac{x^2 - 3x + 2}{x^2 - 6x + 9}$

7.5 Solving Rational Equations

▶ Solve rational equations containing constant denominators.

▶ Solve rational equations containing variable denominators.

▶ Use rational equations to model and solve real-life problems.

Equations Containing Constant Denominators

EXAMPLE 1 Solving a Rational Equation

Solve $\dfrac{3}{5} = \dfrac{x}{2} + 1$.

SOLUTION

The least common denominator of the fractions is 10, so begin by multiplying each side of the equation by 10.

$$10\left(\frac{3}{5}\right) = 10\left(\frac{x}{2} + 1\right) \qquad \text{Multiply each side by LCD of 10.}$$

$$6 = 5x + 10 \qquad \text{Distribute and simplify.}$$

$$-4 = 5x \quad \Longrightarrow \quad -\frac{4}{5} = x \qquad \begin{array}{l}\text{Subtract 10 from each side,}\\ \text{then divide each side by 5.}\end{array}$$

The solution is $x = -\frac{4}{5}$. You can check this in the original equation as follows.

> **Check**
>
> $$\frac{3}{5} \stackrel{?}{=} \frac{-4/5}{2} + 1 \qquad \text{Substitute } -\frac{4}{5} \text{ for } x \text{ in the original equation.}$$
>
> $$\frac{3}{5} \stackrel{?}{=} -\frac{4}{5} \cdot \frac{1}{2} + 1 \qquad \text{Invert divisor and multiply.}$$
>
> $$\frac{3}{5} = -\frac{2}{5} + 1 \qquad \text{Solution checks.} ✔$$

> **Study Tip**
>
> A *rational equation* is an equation containing one or more rational expressions.

Exercises Within Reach ®

Solutions in English & Spanish and tutorial videos at CollegePrepAlgebra.com

Checking Solutions In Exercises 1 and 2, determine whether each value of x is a solution of the equation.

Equation		*Values*		*Equation*		*Values*	
1. $\dfrac{x}{3} - \dfrac{x}{5} = \dfrac{4}{3}$		(a) $x = 0$	(b) $x = -2$	**2.** $\dfrac{x}{4} + \dfrac{3}{4x} = 1$		(a) $x = -1$	(b) $x = 1$
		(c) $x = \frac{1}{8}$	(d) $x = 10$			(c) $x = 3$	(d) $x = \frac{1}{2}$

Solving a Rational Equation In Exercises 3−8, solve the equation.

3. $\dfrac{x}{6} - 1 = \dfrac{2}{3}$

4. $\dfrac{y}{8} + 7 = -\dfrac{1}{2}$

5. $\dfrac{1}{4} = \dfrac{z + 1}{8}$

6. $\dfrac{a}{2} = \dfrac{a + 2}{3}$

7. $\dfrac{x}{4} + \dfrac{x}{2} = \dfrac{2x}{3}$

8. $\dfrac{x}{4} - \dfrac{x}{6} = \dfrac{1}{4}$

EXAMPLE 2   **Solving Rational Equations**

Solve (a) $\dfrac{x-3}{6} = 7 - \dfrac{x}{12}$ and (b) $\dfrac{x^2}{3} + \dfrac{x}{2} = \dfrac{5}{6}$.

SOLUTION

a. The least common denominator of the fractions is 12, so begin by multiplying each side of the equation by 12.

$$\frac{x-3}{6} = 7 - \frac{x}{12}$$ Write original equation.

$$12\left(\frac{x-3}{6}\right) = 12\left(7 - \frac{x}{12}\right)$$ Multiply each side by LCD of 12.

$$2x - 6 = 84 - x$$ Distribute and simplify.

$$3x - 6 = 84$$ Add x to each side.

$$3x = 90 \implies x = 30$$ Add 6 to each side, then divide each side by 3.

The solution is $x = 30$. Check this in the original equation.

b. The least common denominator of the fractions is 6, so begin by multiplying each side of the equation by 6.

$$\frac{x^2}{3} + \frac{x}{2} = \frac{5}{6}$$ Write original equation.

$$6\left(\frac{x^2}{3} + \frac{x}{2}\right) = 6\left(\frac{5}{6}\right)$$ Multiply each side by LCD of 6.

$$\frac{6x^2}{3} + \frac{6x}{2} = \frac{30}{6}$$ Distributive Property

$$2x^2 + 3x = 5$$ Simplify.

$$2x^2 + 3x - 5 = 0$$ Subtract 5 from each side.

$$(2x + 5)(x - 1) = 0$$ Factor.

$$2x + 5 = 0 \implies x = -\tfrac{5}{2}$$ Set 1st factor equal to 0.

$$x - 1 = 0 \implies x = 1$$ Set 2nd factor equal to 0.

The solutions are $x = -\tfrac{5}{2}$ and $x = 1$. Check these in the original equation.

Exercises Within Reach ®

Solutions in English & Spanish and tutorial videos at CollegePrepAlgebra.com

Solving a Rational Equation In Exercises 9−16, solve the equation.

9. $\dfrac{z+2}{3} = 4 - \dfrac{z}{12}$

10. $\dfrac{2y-9}{6} = 3y - \dfrac{3}{4}$

11. $\dfrac{x-5}{5} + 3 = -\dfrac{x}{4}$

12. $\dfrac{4x-2}{7} - \dfrac{5}{14} = 2x$

13. $\dfrac{x^2}{2} - \dfrac{3x}{5} = -\dfrac{1}{10}$

14. $\dfrac{x^2}{3} - \dfrac{x}{6} = \dfrac{1}{6}$

15. $\dfrac{t}{2} = 12 - \dfrac{3t^2}{2}$

16. $\dfrac{x}{12} = \dfrac{1}{10} - \dfrac{x^2}{15}$

Equations Containing Variable Denominators

| EXAMPLE 3 | **An Equation Containing Variable Denominators** |

Solve $\dfrac{8}{3} = \dfrac{7}{x} - \dfrac{1}{3x}$.

SOLUTION

The least common denominator of the fractions is $3x$, so begin by multiplying each side of the equation by $3x$.

$$\dfrac{8}{3} = \dfrac{7}{x} - \dfrac{1}{3x} \qquad \text{Write original equation.}$$

$$3x\left(\dfrac{8}{3}\right) = 3x\left(\dfrac{7}{x} - \dfrac{1}{3x}\right) \qquad \text{Multiply each side by LCD of } 3x.$$

$$\dfrac{24x}{3} = \dfrac{21x}{x} - \dfrac{3x}{3x} \qquad \text{Distributive Property}$$

$$8x = 21 - 1 \qquad \text{Simplify.}$$

$$x = \dfrac{20}{8} = \dfrac{5}{2} \qquad \text{Subtract, divide each side by 8, and simplify.}$$

The solution is $x = \frac{5}{2}$. Check this in the original equation.

> **Check**
>
> $$\dfrac{8}{3} \overset{?}{=} \dfrac{7}{(5/2)} - \dfrac{1}{3(5/2)} \qquad \text{Substitute } \tfrac{5}{2} \text{ for } x.$$
>
> $$15\left(\dfrac{8}{3}\right) \overset{?}{=} 15\left(\dfrac{14}{5} - \dfrac{2}{15}\right) \qquad \begin{array}{l}\text{Invert divisors and multiply,}\\ \text{and then multiply by LCD of 15.}\end{array}$$
>
> $$40 = 42 - 2 \qquad \text{Solution checks. } \checkmark$$

Exercises Within Reach®

Solutions in English & Spanish and tutorial videos at CollegePrepAlgebra.com

Solving a Rational Equation In Exercises 17−30, solve the equation.

17. $\dfrac{9}{25 - y} = -\dfrac{1}{4}$

18. $-\dfrac{6}{u + 3} = \dfrac{2}{3}$

19. $5 - \dfrac{12}{a} = \dfrac{5}{3}$

20. $\dfrac{5}{b} - 18 = 21$

21. $\dfrac{4}{x} - \dfrac{7}{5x} = -\dfrac{1}{2}$

22. $\dfrac{5}{3} = \dfrac{6}{7x} + \dfrac{2}{x}$

23. $\dfrac{12}{y + 5} + \dfrac{1}{2} = 2$

24. $\dfrac{7}{8} - \dfrac{16}{t - 2} = \dfrac{3}{4}$

25. $\dfrac{5}{x} = \dfrac{25}{3(x + 2)}$

26. $\dfrac{10}{x + 4} = \dfrac{15}{4(x + 1)}$

27. $\dfrac{8}{3x + 5} = \dfrac{1}{x + 2}$

28. $\dfrac{500}{3x + 5} = \dfrac{50}{x - 3}$

29. $\dfrac{3}{x + 2} - \dfrac{1}{x} = \dfrac{1}{5x}$

30. $\dfrac{12}{x + 5} + \dfrac{5}{x} = \dfrac{20}{x}$

EXAMPLE 4   **An Equation with No Solution**

Solve $\dfrac{5x}{x-2} = 7 + \dfrac{10}{x-2}$.

SOLUTION

The least common denominator of the fractions is $x - 2$, so begin by multiplying each side of the equation by $x - 2$.

$$\dfrac{5x}{x-2} = 7 + \dfrac{10}{x-2} \qquad \text{Write original equation.}$$

$$(x-2)\!\left(\dfrac{5x}{x-2}\right) = (x-2)\!\left(7 + \dfrac{10}{x-2}\right) \qquad \text{Multiply each side by } x-2.$$

$$5x = 7(x-2) + 10 \qquad \text{Distribute and simplify.}$$

$$5x = 7x - 14 + 10 \qquad \text{Distributive Property}$$

$$5x = 7x - 4 \qquad \text{Combine like terms.}$$

$$-2x = -4 \qquad \text{Subtract } 7x \text{ from each side.}$$

$$x = 2 \qquad \text{Divide each side by } -2.$$

At this point, the solution appears to be $x = 2$. However, by performing a check, you can see that this "trial solution" is extraneous.

Check

$$\dfrac{5x}{x-2} = 7 + \dfrac{10}{x-2} \qquad \text{Write original equation.}$$

$$\dfrac{5(2)}{2-2} \stackrel{?}{=} 7 + \dfrac{10}{2-2} \qquad \text{Substitute 2 for } x.$$

$$\dfrac{10}{0} \stackrel{?}{=} 7 + \dfrac{10}{0} \qquad \text{Solution does not check. } \boldsymbol{\times}$$

Because the check results in *division by zero*, you can conclude that 2 is extraneous. So, the original equation has no solution.

Exercises Within Reach ®

Solutions in English & Spanish and tutorial videos at CollegePrepAlgebra.com

Solving a Rational Equation   **In Exercises 31−38, solve the equation.**

31. $\dfrac{1}{x-4} + 2 = \dfrac{2x}{x-4}$

32. $\dfrac{-2x}{x-1} + 2 = \dfrac{10}{x-1}$

33. $\dfrac{4}{x(x-1)} + \dfrac{3}{x} = \dfrac{4}{x-1}$

34. $\dfrac{10}{x(x-2)} + \dfrac{4}{x} = \dfrac{5}{x-2}$

35. $\dfrac{2}{x-10} - \dfrac{3}{x-2} = \dfrac{6}{x^2 - 12x + 20}$

36. $\dfrac{5}{x+2} + \dfrac{2}{x^2 - 6x - 16} = -\dfrac{4}{x-8}$

37. $1 - \dfrac{6}{4-x} = \dfrac{x+2}{x^2 - 16}$

38. $\dfrac{4}{2x+3} + \dfrac{17}{5x-3} = 3$

Study Tip

Although cross-multiplication can be a little quicker than multiplying by the least common denominator, remember that *it can be used only with equations that have a single fraction on each side* of the equation.

EXAMPLE 5 **Cross-Multiplying**

$$\frac{2x}{x+4} = \frac{3}{x-1}$$ Original equation

$$2x(x-1) = 3(x+4)$$ Cross-multiply.

$$2x^2 - 2x = 3x + 12$$ Distributive Property

$$2x^2 - 5x - 12 = 0$$ Subtract $3x$ and 12 from each side.

$$(2x + 3)(x - 4) = 0$$ Factor.

$$2x + 3 = 0 \quad \Longrightarrow \quad x = -\frac{3}{2}$$ Set 1st factor equal to 0.

$$x - 4 = 0 \quad \Longrightarrow \quad x = 4$$ Set 2nd factor equal to 0.

The solutions are $x = -\frac{3}{2}$ and $x = 4$. Check these in the original equation.

EXAMPLE 6 **An Equation That Has Two Solutions**

$$\frac{3x}{x+1} = \frac{12}{x^2-1} + 2$$ Original equation

$$(x^2 - 1)\left(\frac{3x}{x+1}\right) = (x^2 - 1)\left(\frac{12}{x^2-1} + 2\right)$$ Multiply each side by LCD of $x^2 - 1$.

$$(x - 1)(3x) = 12 + 2(x^2 - 1)$$ Distribute and simplify.

$$3x^2 - 3x = 12 + 2x^2 - 2$$ Distributive Property

$$x^2 - 3x - 10 = 0$$ Subtract $2x^2$ and 10 from each side.

$$(x + 2)(x - 5) = 0$$ Factor.

$$x + 2 = 0 \quad \Longrightarrow \quad x = -2$$ Set 1st factor equal to 0.

$$x - 5 = 0 \quad \Longrightarrow \quad x = 5$$ Set 2nd factor equal to 0.

The solutions are $x = -2$ and $x = 5$. Check these in the original equation.

Exercises Within Reach ®

Solutions in English & Spanish and tutorial videos at CollegePrepAlgebra.com

Solving a Rational Equation In Exercises 39−46, solve the equation.

39. $\dfrac{2x}{5} = \dfrac{x^2 - 5x}{5x}$

40. $\dfrac{3x}{4} = \dfrac{x^2 + 3x}{8x}$

41. $\dfrac{y+1}{y+10} = \dfrac{y-2}{y+4}$

42. $\dfrac{x-3}{x+1} = \dfrac{x-6}{x+5}$

43. $\dfrac{x}{x-2} + \dfrac{3x}{x-4} = -\dfrac{2(x-6)}{x^2 - 6x + 8}$

44. $\dfrac{2(x+1)}{x^2 - 4x + 3} + \dfrac{6x}{x-3} = \dfrac{3x}{x-1}$

45. $\dfrac{x}{3} = \dfrac{1 + \dfrac{4}{x}}{1 + \dfrac{2}{x}}$

46. $\dfrac{2x}{3} = \dfrac{1 + \dfrac{2}{x}}{1 + \dfrac{1}{x}}$

Application

Application   EXAMPLE 7   **Finding a Save Percentage in Hockey**

A hockey goalie faces 799 shots and saves 707 of them. How many additional consecutive saves does the goalie need to obtain a save percentage (in decimal form) of .900?

SOLUTION

Verbal Model: $\dfrac{\text{Desired save}}{\text{percentage}} = \dfrac{707 + (\text{Additional saves})}{799 + (\text{Additional saves})}$

Labels: Desired save percentage = 0.900   (percentage in decimal form)
Additional saves = x (saves)

Equation:

$0.900 = \dfrac{707 + x}{799 + x}$ Write equation.

$0.900(799 + x) = 707 + x$ Cross-multiply.

$719.1 + 0.9x = 707 + x$ Distributive Property

$12.1 = 0.1x$ Subtract 707 and 0.9x from each side.

$121 = x$ Divide each side by 0.1.

The goalie needs 121 additional consecutive saves to attain a save percentage of 0.900.

Exercises Within Reach ®

Solutions in English & Spanish and tutorial videos at CollegePrepAlgebra.com

47. *Batting Average* A softball player bats 47 times and hits the ball safely 8 times. How many additional consecutive times must the player hit the ball safely to obtain a batting average of .250?

48. *Batting Average* A softball player bats 45 times and hits the ball safely 9 times. How many additional consecutive times must the player hit the ball safely to obtain a batting average of .280?

49. *Speed* One person runs 1.5 miles per hour faster than a second person. The first person runs 4 miles in the same time the second person runs 3 miles. Find the speed of each person.

50. *Speed* One person runs 2 miles per hour faster than a second person. The first person runs 5 miles in the same time the second person runs 4 miles. Find the speed of each person.

Concept Summary: Solving Rational Equations

What

The goal when solving rational equations is the same as when solving linear equations: You want to isolate the variable.

EXAMPLE

Solve $\dfrac{2}{x} - \dfrac{2}{5} = \dfrac{4}{x}$.

How

Use the following steps to solve rational equations.

1. Find the LCD of all fractions in the equation.
2. Multiply each side of the equation by the LCD.
3. Simplify each term.
4. Use the properties of equality to solve the resulting linear equation.

EXAMPLE

$$\frac{2}{x} - \frac{2}{5} = \frac{4}{x} \qquad \text{Original equation}$$

$$5x\left(\frac{2}{x} - \frac{2}{5}\right) = 5x\left(\frac{4}{x}\right) \qquad \text{Multiply by LCD.}$$

$$\frac{10x}{x} - \frac{10x}{5} = \frac{20x}{x} \qquad \text{Distributive Property}$$

$$10 - 2x = 20 \qquad \text{Simplify.}$$

$$-2x = 10 \qquad \text{Subtract 10.}$$

$$x = -5 \qquad \text{Divide by } -2.$$

Why

You can use rational equations to model and solve many real-life problems. For example, knowing how to solve a rational equation can help you determine how many additional consecutive saves a goalie needs to obtain a specific save percentage.

Exercises Within Reach ®

Worked-out solutions to odd-numbered exercises at CollegePrepAlgebra.com

Concept Summary Check

51. *Writing* What is a rational equation?

52. *Writing* Describe how to solve a rational equation.

53. *Domain Restrictions* Explain the domain restrictions that may exist for a rational equation.

54. *Cross-Multiplication* When can you use cross-multiplication to solve a rational equation? Explain.

Extra Practice

Think About It In Exercises 55−62, if the exercise is an equation, solve it; if it is an expression, simplify it.

55. $\dfrac{1}{2} = \dfrac{18}{x^2}$

56. $\dfrac{1}{4} = \dfrac{16}{z^2}$

57. $\dfrac{x+5}{4} - \dfrac{3x-8}{3} = \dfrac{4-x}{12}$

58. $\dfrac{2x-7}{10} - \dfrac{3x+1}{5} = \dfrac{6-x}{5}$

59. $\dfrac{16}{x^2-16} + \dfrac{x}{2x-8} = \dfrac{1}{2}$

60. $\dfrac{16}{x^2-16} + \dfrac{x}{2x-8} + \dfrac{1}{2}$

61. $\dfrac{5}{x+3} + \dfrac{5}{3} + 3$

62. $\dfrac{5}{x+3} + \dfrac{5}{3} = 3$

Using a Graph In Exercises 63–66, (a) use the graph to determine any *x*-intercepts of the graph and (b) set *y* = 0 and solve the resulting rational equation to confirm the result of part (a).

63. $y = \dfrac{x + 2}{x - 2}$

64. $y = \dfrac{2x}{x + 4}$

65. $y = x - \dfrac{1}{x}$

66. $y = x - \dfrac{2}{x} - 1$

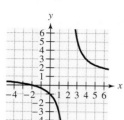
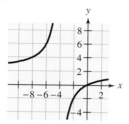
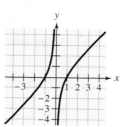
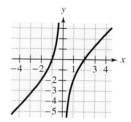

67. **Number Problem** Find a number such that the sum of two times the number and three times its reciprocal is $\dfrac{203}{10}$.

68. **Painting** A painter can paint a room in 4 hours, while his partner can paint the room in 6 hours. How long would it take to paint the room if both worked together?

69. **Roofing** A roofer requires 15 hours to shingle a roof, while an apprentice requires 21 hours. How long would it take to shingle the roof if both worked together?

70. **Wind Speed** A plane has a speed of 300 miles per hour in still air. The plane travels a distance of 680 miles with a tail wind in the same time it takes to travel 520 miles into a head wind. Find the speed of the wind.

Explaining Concepts

71. **Writing** Define the term *extraneous solution*. How do you identify an extraneous solution?

72. **Using a Graph** Explain how you can use the graph of a rational equation to estimate the solution of the equation.

73. **Precision** Explain why the equation $\dfrac{n}{x} + n = \dfrac{n}{x}$ has no solution if *n* is any real nonzero number.

74. **Reasoning** Does multiplying a rational equation by its LCD produce an equivalent equation? Explain.

Cumulative Review

In Exercises 75–78, factor the expression.

75. $x^2 - 81$

76. $x^2 - 121$

77. $4x^2 - \dfrac{1}{4}$

78. $49 - (x - 2)^2$

In Exercises 79 and 80, find the domain of the rational function.

79. $f(x) = \dfrac{2x^2}{5}$

80. $f(x) = \dfrac{4}{x - 6}$

7.6 Applications and Variation

▶ Solve application problems involving direct variation.

▶ Solve application problems involving inverse variation.

▶ Solve application problems involving joint variation.

Direct Variation

> ### Direct Variation
>
> The following statements are equivalent.
> 1. y varies directly as x.
> 2. y is directly proportional to x.
> 3. $y = kx$ for some constant k.
>
> The number k is called the **constant of proportionality**.

Application   EXAMPLE 1   **Direct Variation**

The total revenue R (in dollars) obtained from selling x ice show tickets is directly proportional to the number of tickets sold x. When 10,000 tickets are sold, the total revenue is \$142,500. Find a mathematical model that relates the total revenue R to the number of tickets sold x.

SOLUTION

Ice Show

Revenue (in dollars)

$R = 14.25x$

Tickets sold

Because the total revenue is directly proportional to the number of tickets sold, the linear model is

$R = kx$.

To find the value of the constant k, use the fact that $R = 142,500$ when $x = 10,000$. Substituting these values into the model produces

$142,500 = k(10,000)$ Substitute 142,500 for R and 10,000 for x.

which implies that

$$k = \frac{142,500}{10,000} = 14.25.$$

So, the equation relating the total revenue to the total number of tickets sold is

$R = 14.25x$.   Direct variation model

The graph of this equation is shown at the left.

Exercises Within Reach ®

Solutions in English & Spanish and tutorial videos at CollegePrepAlgebra.com

1. **Revenue** The total revenue R (in dollars) is directly proportional to the number of units sold x. When 500 units are sold, the total revenue is \$4825. Find a mathematical model that relates the total revenue R to the number of units sold x.

2. **Revenue** The total revenue R (in dollars) is directly proportional to the number of units sold x. When 25 units are sold, the total revenue is \$300. Find a mathematical model that relates the total revenue R to the number of units sold x.

Application   EXAMPLE 2   **Direct Variation**

Hooke's Law for springs states that the distance a spring is stretched (or compressed) is directly proportional to the force on the spring. A force of 20 pounds stretches a spring 5 inches.

a. Find a mathematical model that relates the distance the spring is stretched to the force applied to the spring.

b. How far will a force of 30 pounds stretch the spring?

SOLUTION

a. For this problem, let d represent the distance (in inches) that the spring is stretched and let F represent the force (in pounds) that is applied to the spring. Because the distance d is directly proportional to the force F, the model is

$$d = kF.$$

To find the value of the constant k, use the fact that $d = 5$ when $F = 20$. Substituting these values into the model produces

$$5 = k(20) \qquad \text{Substitute 5 for } d \text{ and 20 for } F.$$

$$\frac{5}{20} = k \qquad \text{Divide each side by 20.}$$

$$\frac{1}{4} = k. \qquad \text{Simplify.}$$

So, the equation relating distance to force is

$$d = \frac{1}{4}F. \qquad \text{Direct variation model}$$

b. When $F = 30$, the distance is

$$d = \frac{1}{4}(30) = 7.5 \text{ inches.}$$

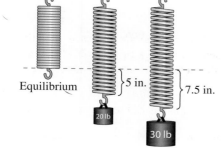

Equilibrium } 5 in. } 7.5 in.
20 lb   30 lb

Exercises Within Reach ®

Solutions in English & Spanish and tutorial videos at CollegePrepAlgebra.com

3. *Hooke's Law* A force of 50 pounds stretches a spring 5 inches.

(a) How far will a force of 20 pounds stretch the spring?

(b) What force is required to stretch the spring 1.5 inches?

4. *Hooke's Law* A force of 50 pounds stretches a spring 3 inches.

(a) How far will a force of 20 pounds stretch the spring?

(b) What force is required to stretch the spring 1.5 inches?

5. *Hooke's Law* A baby weighing $10\frac{1}{2}$ pounds compresses the spring of a baby scale 7 millimeters. Determine the weight of a baby that compresses the spring 12 millimeters.

6. *Hooke's Law* An apple weighing 14 ounces compresses the spring of a produce scale 3 millimeters. Determine the weight of a grapefruit that compresses the spring 5 millimeters.

> ## Direct Variation as nth Power
>
> The following statements are equivalent.
> 1. y varies directly as the nth power x.
> 2. y is directly proportional to the nth power of x.
> 3. $y = kx^n$ for some constant k.

Application **EXAMPLE 3** **Direct Variation as a Power**

The distance a ball rolls down an inclined plane is directly proportional to the square of the time it rolls. During the first second, a ball rolls down a plane a distance of 6 feet.

   a. Find a mathematical model that relates the distance traveled to the time.

   b. How far will the ball roll during the first 2 seconds?

SOLUTION

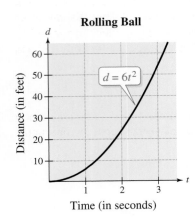

Rolling Ball

$d = 6t^2$

Distance (in feet)

Time (in seconds)

   a. Letting d be the distance (in feet) that the ball rolls and letting t be the time (in seconds), you obtain the model

$$d = kt^2.$$

Because $d = 6$ when $t = 1$, you obtain

$$d = kt^2 \qquad \text{Write original equation.}$$

$$6 = k(1)^2 \quad\Longrightarrow\quad 6 = k \qquad \text{Substitute 6 for } d \text{ and 1 for } t.$$

So, the equation relating distance to time is

$$d = 6t^2. \qquad \text{Direct variation as 2nd power model}$$

The graph of this equation is shown at the left.

   b. When $t = 2$, the distance traveled is

$$d = 6(2)^2 = 6(4) = 24 \text{ feet.}$$

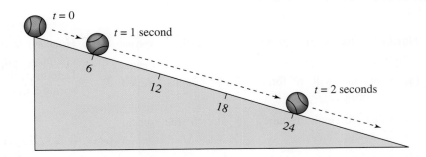

Exercises Within Reach ®

Solutions in English & Spanish and tutorial videos at CollegePrepAlgebra.com

  7. *Stopping Distance* The stopping distance d of an automobile is directly proportional to the square of its speed s. On one road, a car requires 75 feet to stop from a speed of 30 miles per hour. How many feet does the car require to stop from a speed of 48 miles per hour on the same road?

  8. *Frictional Force* The frictional force F (between the tires of a car and the road) that is required to keep a car on a curved section of a highway is directly proportional to the square of the speed s of the car. By what factor does the force F change when the speed of the car is doubled on the same curve?

Inverse Variation

> ### Inverse Variation
>
> **1.** The following three statements are equivalent.
>
>   **a.** y varies inversely as x.
>
>   **b.** y is inversely proportional to x.
>
>   **c.** $y = \dfrac{k}{x}$ for some constant k.
>
> **2.** If $y = \dfrac{k}{x^n}$, then y is inversely proportional to the nth power of x.

Application   EXAMPLE 4   **Inverse Variation**

The marketing department of a large company has found that the demand for one of its hand tools varies inversely as the price of the product. (When the price is low, more people are willing to buy the product than when the price is high.) When the price of the tool is $7.50, the monthly demand is 50,000 tools. Approximate the monthly demand when the price is reduced to $6.

SOLUTION

Let x represent the number of tools that are sold each month (the demand), and let p represent the price per tool (in dollars). Because the demand is inversely proportional to the price, the model is

$$x = \frac{k}{p}.$$

Hand Tools

By substituting $x = 50{,}000$ when $p = 7.50$, you obtain

$$50{,}000 = \frac{k}{7.50} \qquad \text{Substitute 50,000 for } x \text{ and 7.50 for } p.$$

$$375{,}000 = k. \qquad \text{Multiply each side by 7.50.}$$

So, the inverse variation model is $x = \dfrac{375{,}000}{p}$.

The graph of this equation is shown at the left. To find the demand that corresponds to a price of $6, substitute 6 for p in the equation and obtain

$$x = \frac{375{,}000}{6} = 62{,}500 \text{ tools.}$$

So, when the price is lowered from $7.50 per tool to $6.00 per tool, you can expect the monthly demand to increase from 50,000 tools to 62,500 tools.

Exercises Within Reach®

Solutions in English & Spanish and tutorial videos at CollegePrepAlgebra.com

9. *Demand* A company has found that the daily demand x for its boxes of chocolates is inversely proportional to the price p. When the price is $5, the demand is 800 boxes. Approximate the demand when the price is increased to $6.

10. *Demand* A company has found that the daily demand for its boxes of greeting cards is inversely proportional to the price. When the price is $6, the demand is 600 boxes. Approximate the demand when the price is increased to $7.

Application   EXAMPLE 5   **Direct Variation and Inverse Variation**

A computer hardware manufacturer determines that the demand for its USB flash drive is directly proportional to the amount spent on advertising and inversely proportional to the price of the flash drive. When $40,000 is spent on advertising and the price per unit is $20, the monthly demand is 10,000 flash drives.

a. When the amount of advertising is increased to $50,000, how much can the price increase and still maintain a monthly demand of 10,000 flash drives?

b. If you were in charge of the advertising department, would you recommend this increased expense in advertising?

SOLUTION

a. Let x represent the number of flash drives that are sold each month (the demand), let a represent the amount spent on advertising (in dollars), and let p represent the price per unit (in dollars). Because the demand is directly proportional to the advertising expense and inversely proportional to the price, the model is

$$x = \frac{ka}{p}.$$

By substituting 10,000 for x when $a = 40,000$ and $p = 20$, you obtain

$10,000 = \dfrac{k(40,000)}{20}$	Substitute 10,000 for x, 40,000 for a, and 20 for p.
$200,000 = 40,000k$	Multiply each side by 20.
$5 = k.$	Divide each side by 40,000.

So, the model is

$$x = \frac{5a}{p}. \qquad \text{Direct and inverse variation model}$$

To find the price that corresponds to a demand of 10,000 and an advertising expense of $50,000, substitute 10,000 for x and 50,000 for a in the model and solve for p.

$$10,000 = \frac{5(50,000)}{p} \quad \Longrightarrow \quad p = \frac{5(50,000)}{10,000} = \$25$$

So, the price can increase $25 − $20 = $5.

b. The total revenue for selling 10,000 units at $20 each is $200,000, and the revenue for selling 10,000 units at $25 each is $250,000. Because increasing the advertising expense from $40,000 to $50,000 increases the revenue by $50,000, you should recommend the increase in advertising expenses.

Exercises Within Reach ®

Solutions in English & Spanish and tutorial videos at CollegePrepAlgebra.com

11. *Revenue* The weekly demand for a company's frozen pizzas varies directly as the amount spent on advertising and inversely as the price per pizza. At $5 per pizza, when $500 is spent each week on ads, the demand is 2000 pizzas. When advertising is increased to $600, what price yields a demand of 2000 pizzas? Is this increase worthwhile in terms of revenue?

12. *Revenue* The monthly demand for a company's sports caps varies directly as the amount spent on advertising and inversely as the square of the price per cap. At $15 per cap, when $2500 is spent each week on ads, the demand is 300 caps. When advertising is increased to $3000, what price yields a demand of 300 caps? Is this increase worthwhile in terms of revenue?

Joint Variation

> ### Joint Variation
>
> 1. The following three statements are equivalent.
>   **a.** z varies jointly as x and y.
>   **b.** z is jointly proportional to x and y.
>   **c.** $z = kxy$ for some constant k.
> 2. If $z = kx^ny^m$, then z is jointly proportional to the nth power of x and the mth power of y.

Application   **EXAMPLE 6**   **Joint Variation**

The simple interest earned by a savings account is jointly proportional to the time and the principal. After one quarter (3 months), the interest for a principal of $6000 is $120. How much interest would a principal of $7500 earn in 5 months?

SOLUTION

Let I represent the interest earned (in dollars), let P represent the principal (in dollars), and let t represent the time (in years). Because the interest is jointly proportional to the time and the principal, the model is

$$I = ktP.$$

Because $I = 120$ when $P = 6000$ and $t = \frac{1}{4}$, you have

$$120 = k\left(\frac{1}{4}\right)(6000) \qquad \text{Substitute 120 for } I, \tfrac{1}{4} \text{ for } t, \text{ and 6000 for } P.$$

$$120 = 1500k \qquad \text{Simplify.}$$

$$0.08 = k. \qquad \text{Divide each side by 1500.}$$

So, the model that relates interest to time and principal is

$$I = 0.08tP. \qquad \text{Joint variation model}$$

To find the interest earned on a principal of $7500 over a five-month period of time, substitute $P = 7500$ and $t = \frac{5}{12}$ into the model to obtain an interest of

$$I = 0.08\left(\frac{5}{12}\right)(7500)$$

$$= \$250.$$

Exercises Within Reach ®

13. *Simple Interest* The simple interest earned by an account varies jointly as the time and the principal. A principal of $600 earns $10 interest in 4 months. How much would $900 earn in 6 months?

14. *Simple Interest* The simple interest earned by an account varies jointly as the time and the principal. In 2 years, a principal of $5000 earns $650 interest. How much would $1000 earn in 1 year?

Concept Summary: *Variation*

What

You can write mathematical models for three types of variation: **direct variation**, **inverse variation**, and **joint variation**.

EXAMPLE

Write a model for the statement.
1. y is directly proportional to x.
2. y is inversely proportional to x.
3. z is jointly proportional to x and y.

How

Use the definitions of direct, inverse, and joint variation to write the models. Let k be the **constant of proportionality**.

EXAMPLE

1. $y = kx$
2. $y = \dfrac{k}{x}$
3. $z = kxy$

Why

You can use variation models to solve many real-life problems. For instance, knowing how to use direct and inverse variation models can help you determine whether a company should increase the amount it spends on advertising.

Exercises Within Reach ®

Worked-out solutions to odd-numbered exercises at CollegePrepAlgebra.com

Concept Summary Check

15. *Direct Variation* In a problem, y varies directly as x and the constant of proportionality is positive. If one of the variables increases, how does the other change? Explain.

16. *Inverse Variation* In a problem, y varies inversely as x and the constant of proportionality is positive. If one of the variables increases, how does the other change? Explain.

17. *Reasoning* Are the following statements equivalent? Explain.

  (a) y varies directly as x.

  (b) y is directly proportional to the square of x.

18. *Complete the Sentence* The direct variation model $y = kx$ can be described as "y varies directly as x," or "y is _____ _____ to x."

Extra Practice

Writing a Model **In Exercises 19–24, write a model for the statement.**

19. I varies directly as V.

20. V is directly proportional to t.

21. p varies inversely as d.

22. P is inversely proportional to the square root of $1 + r$.

23. A varies jointly as l and w.

24. V varies jointly as h and the square of r.

Writing an Equation **In Exercises 25–30, find the constant of proportionality and write an equation that relates the variables.**

25. h is directly proportional to r, and $h = 28$ when $r = 12$.

26. n varies inversely as m, and $n = 32$ when $m = 1.5$.

27. F varies jointly as x and y, and $F = 500$ when $x = 15$ and $y = 8$.

28. V varies jointly as h and the square of b, and $V = 288$ when $h = 6$ and $b = 12$.

29. d varies directly as the square of x and inversely with r, and $d = 3000$ when $x = 10$ and $r = 4$.

30. z is directly proportional to x and inversely proportional to the square root of y, and $z = 720$ when $x = 48$ and $y = 81$.

Determining the Type of Variation **In Exercises 31 and 32, determine whether the variation model is of the form $y = kx$ or $y = k/x$, and find k.**

31.

x	10	20	30	40	50
y	$\frac{2}{5}$	$\frac{1}{5}$	$\frac{2}{15}$	$\frac{1}{10}$	$\frac{2}{25}$

32.

x	10	20	30	40	50
y	-3	-6	-9	-12	-15

33. **Power Generation** The power P generated by a wind turbine varies directly as the cube of the wind speed w. The turbine generates 400 watts of power in a 20-mile-per-hour wind. Find the power it generates in a 30-mile-per-hour wind.

34. **Environment** The graph shows the percent p of oil that remained in Chedabucto Bay, Nova Scotia, after an oil spill. The cleaning of the spill was left primarily to natural actions. After about a year, the percent that remained varied inversely as time. Find a model that relates p to t, where t is the number of years since the spill. Then use it to find the percent of oil that remained $6\frac{1}{2}$ years after the spill, and compare the result with the graph.

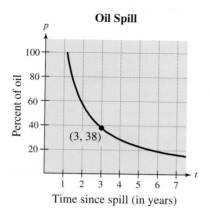

Oil Spill

(3, 38)

Percent of oil

Time since spill (in years)

35. **Engineering** The load P that can be safely supported by a horizontal beam varies jointly as the product of the width W of the beam and the square of the depth D, and inversely as the length L (see figure).

   (a) Write a model for the statement.

   (b) How does P change when the width and length of the beam are both doubled?

   (c) How does P change when the width and depth of the beam are doubled?

   (d) How does P change when all three of the dimensions are doubled?

   (e) How does P change when the depth of the beam is cut in half?

   (f) A beam with width 3 inches, depth 8 inches, and length 120 inches can safely support 2000 pounds. Determine the safe load of a beam made from the same material if its depth is increased to 10 inches.

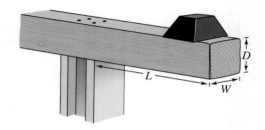

Explaining Concepts

True or False? In Exercises 36 and 37, determine whether the statement is true or false. Justify your answer.

36. In a situation involving both direct and inverse variation, y can vary directly as x and inversely as x at the same time.

37. In a joint variation problem where z varies jointly as x and y, if x increases, then z and y must both increase.

38. **Precision** If y varies directly as the square of x and x is doubled, how does y change? Use the rules of exponents to explain your answer.

39. **Precision** If y varies inversely as the square of x and x is doubled, how does y change? Use the rules of exponents to explain your answer.

40. **Think About It** Describe a real-life problem for each type of variation (direct, inverse, and joint).

Cumulative Review

In Exercises 41−44, write the expression using exponential notation.

41. $(6)(6)(6)(6)$

42. $(-4)(-4)(-4)$

43. $\left(\frac{1}{5}\right)\left(\frac{1}{5}\right)\left(\frac{1}{5}\right)\left(\frac{1}{5}\right)\left(\frac{1}{5}\right)$

44. $-\left(-\frac{3}{4}\right)\left(-\frac{3}{4}\right)\left(-\frac{3}{4}\right)$

In Exercises 45−48, use synthetic division to divide.

45. $(x^2 - 5x - 14) \div (x + 2)$

46. $(3x^2 - 5x + 2) \div (x + 1)$

47. $\dfrac{4x^5 - 14x^4 + 6x^3}{x - 3}$

48. $\dfrac{x^5 - 3x^2 - 5x + 1}{x - 2}$

7 Chapter Summary

What did you learn?	Explanation and Examples	Review Exercises
7.1 Find the domain of a rational function *(p. 320).*	Let $u(x)$ and $v(x)$ be polynomial functions. The function $$f(x) = \frac{u(x)}{v(x)}$$ is a rational function. The domain of f is the set of all real numbers for which $v(x) \neq 0$.	1–8
Simplify rational expressions *(p. 322).*	To simplify a rational expression, factor completely the numerator and denominator, then divide out common factors. You may need to use a change in signs. $$\frac{uw}{vw} = \frac{u\cancel{w}}{v\cancel{w}} = \frac{u}{v}, \ w \neq 0$$ Domain restrictions of the original expression that are not implied by the simplified form must be listed.	9–16
Use rational expressions to model and solve real-life problems *(p. 325).*	You can use rational expressions to model many real-life situations including geometry and business applications.	17, 18
7.2 Multiply rational expressions and simplify *(p. 328).*	1. Multiply the numerators and the denominators. $$\frac{u}{v} \cdot \frac{w}{z} = \frac{uw}{vz}$$ 2. Factor the numerator and the denominator. 3. Simplify by dividing out the common factors. 4. List any domain restrictions not implied by the simplified form.	19–26
Divide rational expressions and simplify *(p. 331).*	Invert the divisor and multiply using the steps for multiplying rational expressions. $$\frac{u}{v} \div \frac{w}{z} = \frac{u}{v} \cdot \frac{z}{w} = \frac{uz}{vw}$$	27–32
7.3 Add or subtract rational expressions with like denominators, and simplify *(p. 336).*	Combine the numerators using the rules for adding or subtracting with like denominators. 1. $\dfrac{u}{w} + \dfrac{v}{w} = \dfrac{u+v}{w}$ 2. $\dfrac{u}{w} - \dfrac{v}{w} = \dfrac{u-v}{w}$ Simplify the resulting rational expression.	33–42

	What did you learn?	Explanation and Examples	Review Exercises
7.3	Add or subtract rational expressions with unlike denominators, and simplify *(p. 338).*	**1.** Find the least common denominator (LCD) of the rational expressions. **2.** Rewrite each rational expression so that it has the LCD in its denominator. **3.** Combine these rational expressions using the rules for adding or subtracting with like denominators.	43–52
7.4	Simplify complex fractions using rules for dividing rational expressions *(p. 346).*	When the numerator and denominator of the complex fraction each consist of a single fraction, use the rules for dividing rational expressions.	53–58
	Simplify complex fractions that have a sum or difference in the numerator and/or denominator *(p. 349).*	One way to simplify a complex fraction when a sum or difference is present in the numerator and/or denominator is to combine the terms so that the numerator and denominator each consist of a single fraction.	59–64
7.5	Solve rational equations containing constant denominators *(p. 354).*	**1.** Multiply each side of the equation by the LCD of all the fractions in the equation. This will clear the equation of fractions. **2.** Solve the resulting equation.	65–70
	Solve rational equations containing variable denominators *(p. 356).*	**1.** Determine the domain restrictions of the equation. **2.** Multiply each side of the equation by the LCD of all the fractions in the equation. *Alternatively,* cross-multiplication can be used in the special case of an equation with a single fraction on each side. $$\frac{2x}{x+4} = \frac{3}{x-1} \implies 2x(x-1) = 3(x+4)$$ **3.** Solve the resulting equation. **4.** Check the solution(s).	71–86
	Use rational equations to model and solve real-life problems *(p. 359).*	Many application problems require the use of rational equations. (See Example 7.)	87–92
7.6	Solve application problems involving direct variation *(p. 362).*	Direct variation: $y = kx$ Direct variation as *n*th power: $y = kx^n$	93, 94
	Solve application problems involving inverse variation *(p. 365).*	Inverse variation: $y = \dfrac{k}{x}$ Inverse variation as *n*th power: $y = \dfrac{k}{x^n}$	95–98
	Solve application problems involving joint variation *(p. 367).*	Joint variation: $z = kxy$ Joint variation as *n*th and *m*th powers: $z = kx^n y^m$	99, 100

Review Exercises

Worked-out solutions to odd-numbered exercises at CollegePrepAlgebra.com

7.1

Finding the Domain of a Rational Function In Exercises 1−6, find the domain of the rational function.

1. $f(y) = \dfrac{3y}{y-8}$

2. $g(t) = \dfrac{t+4}{t+12}$

3. $f(x) = \dfrac{2x}{x^2+1}$

4. $g(t) = \dfrac{t+2}{t^2+4}$

5. $g(u) = \dfrac{u}{u^2-7u+6}$

6. $f(x) = \dfrac{x-12}{x(x^2-16)}$

7. **Geometry** A rectangle of width w inches has an area of 36 square inches. The perimeter P of the rectangle is given by

$$P = 2\left(w + \frac{36}{w}\right).$$

Describe the domain of the function.

8. **Average Cost** The average cost $\overline{C}$ for a manufacturer to produce x units of a product is given by

$$\overline{C} = \frac{15{,}000 + 0.75x}{x}.$$

Describe the domain of the function.

Simplifying a Rational Expression In Exercises 9−16, simplify the rational expression.

9. $\dfrac{6x^4y^2}{15xy^2}$

10. $\dfrac{2(y^3z)^2}{28(yz^2)^2}$

11. $\dfrac{5b-15}{30b-120}$

12. $\dfrac{4a}{10a^2+26a}$

13. $\dfrac{9x-9y}{y-x}$

14. $\dfrac{x+3}{x^2-x-12}$

15. $\dfrac{x^2-5x}{2x^2-50}$

16. $\dfrac{x^2+3x+9}{x^3-27}$

Geometry In Exercises 17 and 18, find the ratio of the area of the shaded portion to the total area of the figure.

17.

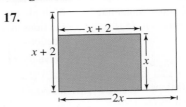

18.

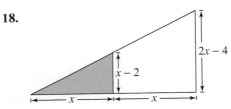

7.2

Multiplying Rational Expressions In Exercises 19−26, multiply and simplify.

19. $\dfrac{4}{x} \cdot \dfrac{x^2}{12}$

20. $\dfrac{3}{y^3} \cdot 5y^3$

21. $\dfrac{7}{8} \cdot \dfrac{2x}{y} \cdot \dfrac{y^2}{14x^2}$

22. $\dfrac{15(x^2y)^3}{3y^3} \cdot \dfrac{12y}{x}$

23. $\dfrac{60z}{z+6} \cdot \dfrac{z^2-36}{5}$

24. $\dfrac{x^2-16}{6} \cdot \dfrac{3}{x^2-8x+16}$

25. $\dfrac{u}{u-3} \cdot \dfrac{3u-u^2}{4u^2}$

26. $x^2 \cdot \dfrac{x+1}{x^2-x} \cdot \dfrac{(5x-5)^2}{x^2+6x+5}$

Dividing Rational Expressions In Exercises 27–32, divide **and simplify.**

27. $24x^4 \div \dfrac{6x}{5}$

28. $\dfrac{8u^2}{3} \div \dfrac{u}{9}$

29. $25y^2 \div \dfrac{xy}{5}$

30. $\dfrac{6}{z^2} \div 4z^2$

31. $\dfrac{x^2 + 3x + 2}{3x^2 + x - 2} \div (x + 2)$

32. $\dfrac{x^2 - 14x + 48}{x^2 - 6x} \div (3x - 24)$

7.3

Adding and Subtracting Rational Expressions In Exercises 33–42, combine **and simplify.**

33. $\dfrac{4x}{5} + \dfrac{11x}{5}$

34. $\dfrac{7y}{12} - \dfrac{4y}{12}$

35. $\dfrac{15}{3x} - \dfrac{3}{3x}$

36. $\dfrac{4}{5x} + \dfrac{1}{5x}$

37. $\dfrac{8 - x}{4x} + \dfrac{5}{4x}$

38. $\dfrac{3}{5x} - \dfrac{x - 1}{5x}$

39. $\dfrac{2(3y + 4)}{2y + 1} + \dfrac{3 - y}{2y + 1}$

40. $\dfrac{4x - 2}{3x + 1} - \dfrac{x + 1}{3x + 1}$

41. $\dfrac{4x}{x + 2} + \dfrac{3x - 7}{x + 2} - \dfrac{9}{x + 2}$

42. $\dfrac{3}{2y - 3} - \dfrac{y - 10}{2y - 3} + \dfrac{5y}{2y - 3}$

Adding and Subtracting Rational Expressions In Exercises 43–52, combine **and simplify.**

43. $\dfrac{3}{5x^2} + \dfrac{4}{10x}$

44. $\dfrac{3}{z} - \dfrac{5}{2z^2}$

45. $\dfrac{1}{x + 5} + \dfrac{3}{x - 12}$

46. $\dfrac{2}{x - 10} + \dfrac{3}{4 - x}$

47. $5x + \dfrac{2}{x - 3} - \dfrac{3}{x + 2}$

48. $4 - \dfrac{4x}{x + 6} + \dfrac{7}{x - 5}$

49. $\dfrac{6}{x - 5} - \dfrac{4x + 7}{x^2 - x - 20}$

50. $\dfrac{5}{x + 2} + \dfrac{25 - x}{x^2 - 3x - 10}$

51. $\dfrac{5}{x + 3} - \dfrac{4x}{(x + 3)^2} - \dfrac{1}{x - 3}$

52. $\dfrac{8}{y} - \dfrac{3}{y + 5} + \dfrac{4}{y - 2}$

7.4

Simplifying a Complex Fraction In Exercises 53–58, simplify **the complex fraction.**

53. $\dfrac{\left(\dfrac{6}{x}\right)}{\left(\dfrac{2}{x^3}\right)}$

54. $\dfrac{xy}{\left(\dfrac{5x^2}{2y}\right)}$

55. $\dfrac{\left(\dfrac{x}{x - 2}\right)}{\left(\dfrac{2x}{2 - x}\right)}$

56. $\dfrac{\left(\dfrac{y^2}{5 - y}\right)}{\left(\dfrac{y}{y - 5}\right)}$

57. $\dfrac{\left(\dfrac{6x^2}{x^2 + 2x - 35}\right)}{\left(\dfrac{x^3}{x^2 - 25}\right)}$

58. $\dfrac{\left[\dfrac{24 - 18x}{(2 - x)^2}\right]}{\left(\dfrac{60 - 45x}{x^2 - 4x + 4}\right)}$

Simplifying a Complex Fraction In Exercises 59−64, simplify the complex fraction.

59. $\dfrac{3t}{\left(5 - \dfrac{2}{t}\right)}$

60. $\dfrac{\left(\dfrac{1}{x} - \dfrac{1}{2}\right)}{2x}$

61. $\dfrac{\left(x - 3 + \dfrac{2}{x}\right)}{\left(1 - \dfrac{2}{x}\right)}$

62. $\dfrac{3x - 1}{\left(\dfrac{2}{x^2} + \dfrac{5}{x}\right)}$

63. $\dfrac{\left(\dfrac{1}{a^2 - 16} - \dfrac{1}{a}\right)}{\left(\dfrac{1}{a^2 + 4a} + 4\right)}$

64. $\dfrac{\left(\dfrac{1}{x^2} - \dfrac{1}{y^2}\right)}{\left(\dfrac{1}{x} + \dfrac{1}{y}\right)}$

7.5

Solving a Rational Equation In Exercises 65−70, solve the equation.

65. $\dfrac{x}{15} + \dfrac{3}{5} = 1$

66. $\dfrac{x}{6} + \dfrac{5}{3} = 3$

67. $\dfrac{3x}{8} = -15 + \dfrac{x}{4}$

68. $\dfrac{t + 1}{6} = \dfrac{1}{2} - 2t$

69. $\dfrac{x^2}{6} - \dfrac{x}{12} = \dfrac{1}{2}$

70. $\dfrac{x^2}{4} = -\dfrac{x}{12} + \dfrac{1}{6}$

Solving a Rational Equation In Exercises 71−86, solve the equation.

71. $8 - \dfrac{12}{t} = \dfrac{1}{3}$

72. $5 + \dfrac{2}{x} = \dfrac{1}{4}$

73. $\dfrac{2}{y} - \dfrac{1}{3y} = \dfrac{1}{3}$

74. $\dfrac{7}{4x} - \dfrac{6}{8x} = 1$

75. $r = 2 + \dfrac{24}{r}$

76. $\dfrac{2}{x} - \dfrac{x}{6} = \dfrac{2}{3}$

77. $\dfrac{t}{4} = \dfrac{4}{t}$

78. $\dfrac{20}{u} = \dfrac{u}{5}$

79. $\dfrac{3}{y + 1} - \dfrac{8}{y} = 1$

80. $\dfrac{4x}{x - 5} + \dfrac{2}{x} = -\dfrac{4}{x - 5}$

81. $\dfrac{2x}{x - 3} - \dfrac{3}{x} = 0$

82. $\dfrac{6x}{x - 3} = 9 + \dfrac{18}{x - 3}$

83. $\dfrac{12}{x^2 + x - 12} - \dfrac{1}{x - 3} = -1$

84. $\dfrac{3}{x - 1} + \dfrac{6}{x^2 - 3x + 2} = 2$

85. $\dfrac{5}{x^2 - 4} - \dfrac{6}{x - 2} = -5$

86. $\dfrac{3}{x^2 - 9} + \dfrac{4}{x + 3} = 1$

87. *Average Speeds* You and a friend ride bikes for the same amount of time. You ride 24 miles and your friend rides 15 miles. Your friend's average speed is 6 miles per hour slower than yours. What are the average speeds of you and your friend?

88. *Average Speed* You drive 220 miles to see a friend. The return trip takes 20 minutes less than the original trip, and your average speed is 5 miles per hour faster. What is your average speed on the return trip?

89. *Population Growth* The Parks and Wildlife Commission introduces 80,000 fish into a large lake. The population P (in thousands) of the fish is approximated by the model

$$P = \frac{20(4 + 3t)}{1 + 0.05t}$$

where t is the time in years. Find the time required for the population to increase to 400,000 fish.

90. *Average Cost* The average cost $\overline{C}$ of producing x units of a product is given by

$$\overline{C} = 1.5 + \frac{4200}{x}.$$

Determine the number of units that must be produced to obtain an average cost of $2.90 per unit.

91. *Partnership Costs* A group of people starting a business agree to share equally in the cost of a $60,000 piece of machinery. If they could find two more people to join the group, each person's share of the cost would decrease by $5000. How many people are presently in the group?

92. *Work Rate* One painter works $1\frac{1}{2}$ times as fast as another painter. It takes them 4 hours working together to paint a room. Find the time it takes each painter to paint the room working alone.

7.6

93. *Hooke's Law* A force of 100 pounds stretches a spring 4 inches. Find the force required to stretch the spring 6 inches.

94. *Stopping Distance* The stopping distance d of an automobile is directly proportional to the square of its speed s. How does the stopping distance change when the speed of the car is doubled?

95. *Travel Time* The travel time between two cities is inversely proportional to the average speed. A train travels between the cities in 3 hours at an average speed of 65 miles per hour. How long would it take to travel between the cities at an average speed of 80 miles per hour?

96. *Demand* A company has found that the daily demand x for its cordless telephones is inversely proportional to the price p. When the price is $25, the demand is 1000 telephones. Approximate the demand when the price is increased to $28.

97. *Revenue* The monthly demand for brand X athletic shoes varies directly as the amount spent on advertising and inversely as the square of the price per pair. When $20,000 is spent on monthly advertising and the price per pair of shoes is $55, the demand is 900 pairs. When advertising is increased to $25,000, what price yields a demand of 900 pairs? Is this increase worthwhile in terms of revenue?

98. *Revenue* The seasonal demand for Ace brand sunglasses varies directly as the amount spent on advertising and inversely as the square of the price per pair. When $125,000 is spent on advertising and the price per pair is $35, the demand is 5000 pairs. If advertising is increased to $135,000, what price will yield a demand of 5000 pairs? Is this increase worthwhile in terms of revenue?

99. *Cost* The cost of constructing a wooden box with a square base varies jointly as the height of the box and the square of the width of the box. A box of height 16 inches and of width 6 inches costs $28.80. How much would a box of height 14 inches and of width 8 inches cost?

100. *Simple Interest* The simple interest earned by a savings account is jointly proportional to the time and the principal. After three quarters (9 months), the interest for a principal of $12,000 is $675. How much interest would a principal of $8200 earn in 18 months?

Chapter Test

Solutions in English & Spanish and tutorial videos at CollegePrepAlgebra.com

Take this test as you would take a test in class. After you are done, check your work against the answers in the back of the book.

1. Find the domain of $f(x) = \dfrac{x+1}{x^2 - 6x + 5}$.

In Exercises 2 and 3, simplify the rational expression.

2. $\dfrac{4 - 2x}{x - 2}$

3. $\dfrac{2a^2 - 5a - 12}{5a - 20}$

4. Find the least common multiple of x^2, $3x^3$, and $(x + 4)^2$.

In Exercises 5–16, perform the operation and simplify.

5. $\dfrac{4z^3}{5} \cdot \dfrac{25}{12z^2}$

6. $\dfrac{y^2 + 8y + 16}{2(y - 2)} \cdot \dfrac{8y - 16}{(y + 4)^3}$

7. $\dfrac{(2xy^2)^3}{15} \div \dfrac{12x^3}{21}$

8. $(4x^2 - 9) \div \dfrac{2x + 3}{2x^2 - x - 3}$

9. $\dfrac{3}{x - 3} + \dfrac{x - 2}{x - 3}$

10. $2x + \dfrac{1 - 4x^2}{x + 1}$

11. $\dfrac{5x}{x + 2} - \dfrac{2}{x^2 - x - 6}$

12. $\dfrac{3}{x} - \dfrac{5}{x^2} + \dfrac{2x}{x^2 + 2x + 1}$

13. $\dfrac{\left(\dfrac{3x}{x + 2}\right)}{\left(\dfrac{12}{x^3 + 2x^2}\right)}$

14. $\dfrac{\left(9x - \dfrac{1}{x}\right)}{\left(\dfrac{1}{x} - 3\right)}$

15. $\dfrac{\left(\dfrac{3}{x^2} + \dfrac{1}{y}\right)}{\left(\dfrac{1}{x + y}\right)}$

16. $\dfrac{6x^2 - 4x + 8}{2x}$

In Exercises 17–19, solve the equation.

17. $\dfrac{3}{h + 2} = \dfrac{1}{6}$

18. $\dfrac{2}{x + 5} - \dfrac{3}{x + 3} = \dfrac{1}{x}$

19. $\dfrac{1}{x + 1} + \dfrac{1}{x - 1} = \dfrac{2}{x^2 - 1}$

20. Find a mathematical model that relates u to v when v varies directly as the square root of u, and $v = \frac{3}{2}$ when $u = 36$.

21. When the temperature of a gas is not allowed to change, the absolute pressure P of the gas is inversely proportional to its volume V, according to Boyle's Law. A large balloon is filled with 180 cubic meters of helium at atmospheric pressure (1 atm) at sea level. What is the volume of the helium when the balloon rises to an altitude at which the atmospheric pressure is 0.75 atm? (Assume that the temperature does not change.)

Systems of Equations and Inequalities

8.1   Solving Systems of Equations by Graphing and Substitution

8.2   Solving Systems of Equations by Elimination

8.3   Linear Systems in Three Variables

8.4   Matrices and Linear Systems

8.5   Determinants and Linear Systems

8.6   Systems of Linear Inequalities

MASTERY IS WITHIN REACH!

"I have learned how note cards can be used to help review and memorize math concepts. I started using them in my math class, and it really helps me, especially when I only have short amounts of time to study. I don't have to pull out all of my books to study. I just keep my note cards in my backpack."

Sidney

See page 405 for suggestions about using note cards.

8.1 Solving Systems of Equations by Graphing and Substitution

▶ Determine whether ordered pairs are solutions of systems of equations.
▶ Solve systems of equations graphically.
▶ Use the method of substitution to solve systems of equations algebraically.
▶ Use systems of equations to model and solve real-life problems.

Systems of Equations

Many problems in business and science involve **systems of equations**. These systems consist of two or more equations involving two or more variables.

$$\begin{cases} ax + by = c \\ dx + ey = f \end{cases} \qquad \begin{array}{l} \text{Equation 1} \\ \text{Equation 2} \end{array}$$

A **solution** of such a system is an ordered pair (x, y) of real numbers that satisfies *each* equation in the system. When you find the set of all solutions of the system of equations, you are **solving the system of equations**.

EXAMPLE 1   **Checking Solutions of a System of Equations**

Determine whether each ordered pair is a solution of the system of equations.

$$\begin{cases} x + y = 6 \\ 2x - 5y = -2 \end{cases} \qquad \begin{array}{l} \text{Equation 1} \\ \text{Equation 2} \end{array}$$

a. $(3, 3)$ **b.** $(4, 2)$

SOLUTION

a. To determine whether the ordered pair $(3, 3)$ is a solution of the system of equations, substitute 3 for x and 3 for y in *each* of the equations.

$3 + 3 = 6$ ✓   Substitute 3 for x and 3 for y in Equation 1.

$2(3) - 5(3) \neq -2$ ✗   Substitute 3 for x and 3 for y in Equation 2.

Because the check fails in Equation 2, you can conclude that the ordered pair $(3, 3)$ *is not* a solution of the original system of equations.

b. By substituting 4 for x and 2 for y in each of the original equations, you can determine that the ordered pair $(4, 2)$ is a solution of *both* equations.

$4 + 2 = 6$ ✓   Substitute 4 for x and 2 for y in Equation 1.

$2(4) - 5(2) = -2$ ✓   Substitute 4 for x and 2 for y in Equation 2.

So, $(4, 2)$ *is* a solution of the original system of equations.

Exercises Within Reach ®   Solutions in English & Spanish and tutorial videos at CollegePrepAlgebra.com

Checking Solutions In Exercises 1−4, determine whether each ordered pair is a solution of the system of equations.

1. $\begin{cases} x + 2y = 9 \\ -2x + 3y = 10 \end{cases}$ (a) $(1, 4)$
(b) $(3, -1)$

2. $\begin{cases} 5x - 4y = 34 \\ x - 2y = 8 \end{cases}$ (a) $(0, 3)$
(b) $(6, -1)$

3. $\begin{cases} -2x + 7y = 46 \\ 3x + y = 0 \end{cases}$ (a) $(-3, 2)$
(b) $(-2, 6)$

4. $\begin{cases} -5x - 2y = 23 \\ x + 4y = -19 \end{cases}$ (a) $(-3, -4)$
(b) $(3, 7)$

Solving a System of Linear Equations by Graphing

EXAMPLE 2   **Solving a System of Linear Equations**

Solve the system of linear equations by graphing each equation and locating the point of intersection.

$$\begin{cases} x + y = -2 & \text{Equation 1} \\ 2x - 3y = -9 & \text{Equation 2} \end{cases}$$

SOLUTION

One way to begin is to write each equation in slope-intercept form.

Equation 1	Equation 2
$x + y = -2$	$2x - 3y = -9$
$y = -x - 2$	$-3y = -2x - 9$
	$y = \frac{2}{3}x + 3$

Then use a numerical approach by creating two tables of values.

Table of Values for Equation 1

x	−4	−3	−2	−1	0	1	2
y = −x − 2	2	1	0	−1	−2	−3	−4

Table of Values for Equation 2

x	−4	−3	−2	−1	0	1	2
$y = \frac{2}{3}x + 3$	$\frac{1}{3}$	1	$-\frac{5}{3}$	$\frac{7}{3}$	3	$\frac{11}{3}$	$\frac{13}{3}$

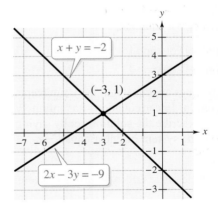

The tables show that the point $(-3, 1)$ is a common solution point. Another way to solve the system is to sketch the graphs of both equations, as shown at the left. From the graph, it appears that the lines intersect at the point $(-3, 1)$. To verify this, substitute the coordinates of the point into each of the two original equations.

Substitute into Equation 1	Substitute into Equation 2
$x + y = -2$	$2x - 3y = -9$
$-3 + 1 \overset{?}{=} -2$	$2(-3) - 3(1) \overset{?}{=} -9$
$-2 = -2$ ✔	$-9 = -9$ ✔

Because both equations are satisfied, the point $(-3, 1)$ is the solution of the system.

Exercises Within Reach ®

Solutions in English & Spanish and tutorial videos at CollegePrepAlgebra.com

Solving a System of Linear Equations In Exercises 5−10, solve the system by graphing.

5. $\begin{cases} y = -x + 3 \\ y = x + 1 \end{cases}$

6. $\begin{cases} y = 2x - 1 \\ y = x + 1 \end{cases}$

7. $\begin{cases} y = 2x - 4 \\ y = -\frac{1}{2}x + 1 \end{cases}$

8. $\begin{cases} y = \frac{1}{2}x + 2 \\ y = -x + 8 \end{cases}$

9. $\begin{cases} x - y = 3 \\ x + y = 3 \end{cases}$

10. $\begin{cases} x - y = 0 \\ x + y = 4 \end{cases}$

A system of linear equations can have exactly one solution, infinitely many solutions, or no solution.

Graphs			
Graphical interpretation	The two lines intersect.	The two lines coincide (are identical).	The two lines are parallel.
Intersection	Single point of intersection	Infinitely many points of intersection	No point of intersection
Slopes of lines	Slopes are not equal.	Slopes are equal.	Slopes are equal.
Number of solutions	Exactly one solution	Infinitely many solutions	No solution
Type of system	**Consistent system**	**Dependent (consistent) system**	**Inconsistent system**

EXAMPLE 3   A System with Infinitely Many Solutions

Solve the system of linear equations.

$$\begin{cases} x - y = 2 & \text{Equation 1} \\ -3x + 3y = -6 & \text{Equation 2} \end{cases}$$

SOLUTION

Begin by writing each equation in slope-intercept form.

$$\begin{cases} y = x - 2 & \text{Slope-intercept form of Equation 1} \\ y = x - 2 & \text{Slope-intercept form of Equation 2} \end{cases}$$

From these forms, you can see that the slopes of the lines are equal and the y-intercepts are the same (see the graph). So, the original system of linear equations has infinitely many solutions and is a dependent system. You can describe the solution set by saying that each point on the line $y = x - 2$ is a solution of the system of linear equations.

Exercises Within Reach ®

Solutions in English & Spanish and tutorial videos at CollegePrepAlgebra.com

Determining the Number of Solutions In Exercises 11–14, use the graphs of the equations to determine the number of solutions of the system of linear equations.

11.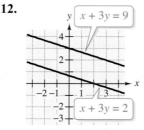

12.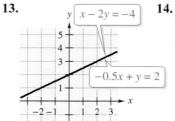

13.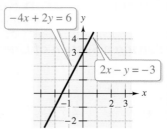

14.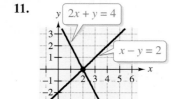

The Method of Substitution

Solving a system of equations graphically is limited by the ability to sketch an accurate graph. An accurate solution is difficult to obtain if one or both coordinates of a solution point are fractional or irrational. One analytic way to determine an exact solution of a system of two equations in two variables is to convert the system to *one* equation in *one* variable by an appropriate substitution.

Study Tip

The term **back-substitute** implies that you work backwards. After solving for one of the variables, substitute that value back into one of the equations in the original (or revised) system to find the value of the other variable.

The Method of Substitution

1. Solve one of the equations for one variable in terms of the other.
2. Substitute the expression obtained in Step 1 into the other equation to obtain an equation in one variable.
3. Solve the equation obtained in Step 2.
4. Back-substitute the solution from Step 3 into the expression obtained in Step 1 to find the value of the other variable.
5. Check the solution to see that it satisfies *both* of the original equations.

EXAMPLE 4   **The Method of Substitution**

Solve the system of linear equations.

$$\begin{cases} -x + y = 1 & \text{Equation 1} \\ 2x + y = -2 & \text{Equation 2} \end{cases}$$

SOLUTION

Begin by solving for y in Equation 1.

$$y = x + 1 \qquad \text{Revised Equation 1}$$

Next, substitute this expression for y in Equation 2 and solve for x.

$$
\begin{aligned}
2x + y &= -2 & &\text{Equation 2} \\
2x + (x + 1) &= -2 & &\text{Substitute } x + 1 \text{ for } y. \\
3x + 1 &= -2 & &\text{Combine like terms.} \\
x &= -1 & &\text{Solve for } x.
\end{aligned}
$$

Finally, *back-substitute* this x-value into the revised Equation 1.

$$
\begin{aligned}
y &= x + 1 & &\text{Revised Equation 1} \\
y &= -1 + 1 = 0 & &\text{Substitute } -1 \text{ for } x.
\end{aligned}
$$

So, the solution is $(-1, 0)$. Check this solution by substituting $x = -1$ and $y = 0$ into both of the original equations.

Exercises Within Reach ®

Solutions in English & Spanish and tutorial videos at CollegePrepAlgebra.com

Solving a System of Linear Equations In Exercises 15–18, solve the system by the method of substitution.

15. $\begin{cases} y = 2x - 1 \\ y = -x + 5 \end{cases}$ **16.** $\begin{cases} y = -2x + 9 \\ y = 3x - 1 \end{cases}$ **17.** $\begin{cases} x - y = 0 \\ 2x + y = 9 \end{cases}$ **18.** $\begin{cases} x - y = 0 \\ 5x - 3y = 10 \end{cases}$

EXAMPLE 5 **The Method of Substitution**

Solve the system of linear equations.

$$\begin{cases} 5x + 3y = 18 & \text{Equation 1} \\ 2x - 7y = -1 & \text{Equation 2} \end{cases}$$

SOLUTION

Because neither variable has a coefficient of 1, you can choose to solve for either variable first. For instance, you can begin by solving for x in Equation 1.

$5x + 3y = 18$	Original Equation 1
$5x = -3y + 18$	Subtract $3y$ from each side.
$x = -\dfrac{3}{5}y + \dfrac{18}{5}$	Revised Equation 1

Next, substitute this expression for x in Equation 2 and solve for y.

$2x - 7y = -1$	Equation 2
$2\left(-\dfrac{3}{5}y + \dfrac{18}{5}\right) - 7y = -1$	Substitute $-\dfrac{3}{5}y + \dfrac{18}{5}$ for x.
$-\dfrac{6}{5}y + \dfrac{36}{5} - 7y = -1$	Distributive Property
$-6y + 36 - 35y = -5$	Multiply each side by 5.
$36 - 41y = -5$	Combine like terms.
$-41y = -41$	Subtract 36 from each side.
$y = 1$	Divide each side by -41.

Finally, back-substitute this y-value into the revised Equation 1.

$x = -\dfrac{3}{5}y + \dfrac{18}{5}$	Revised Equation 1
$x = -\dfrac{3}{5}(1) + \dfrac{18}{5}$	Substitute 1 for y.
$x = 3$	Simplify.

So, the solution is $(3, 1)$. Check this in the original system.

Exercises Within Reach ®

Solutions in English & Spanish and tutorial videos at CollegePrepAlgebra.com

Solving a System of Linear Equations In Exercises 19−24, solve the system by the method of substitution.

19. $\begin{cases} 8x + 4y = -2 \\ -12x + 5y = -8 \end{cases}$

20. $\begin{cases} 5x - 2y = 10 \\ 3x + 2y = 6 \end{cases}$

21. $\begin{cases} 3x + 7y = 2 \\ 5x - 3y = -26 \end{cases}$

22. $\begin{cases} 4x + 3y = 15 \\ 2x - 5y = 1 \end{cases}$

23. $\begin{cases} 2x - 3y = 16 \\ 3x + 4y = 7 \end{cases}$

24. $\begin{cases} -2x + 7y = 9 \\ 3x + 2y = -1 \end{cases}$

EXAMPLE 6   **The Method of Substitution: No-Solution Case**

Solve the system of linear equations.

$$\begin{cases} x - 3y = 2 & \text{Equation 1} \\ -2x + 6y = 2 & \text{Equation 2} \end{cases}$$

SOLUTION

Begin by solving for x in Equation 1 to obtain $x = 3y + 2$. Then substitute this expression for x in Equation 2.

$$\begin{aligned} -2x + 6y &= 2 & \text{Equation 2} \\ -2(3y + 2) + 6y &= 2 & \text{Substitute } 3y + 2 \text{ for } x. \\ -6y - 4 + 6y &= 2 & \text{Distributive Property} \\ -4 &= 2 & \text{False statement} \end{aligned}$$

Because $-4 = 2$ is a false statement, you can conclude that the original system is inconsistent and has no solution. The graph confirms this result.

EXAMPLE 7   **The Method of Substitution: Many-Solutions Case**

Solve the system of linear equations.

$$\begin{cases} 9x + 3y = 15 & \text{Equation 1} \\ 3x + y = 5 & \text{Equation 2} \end{cases}$$

SOLUTION

Begin by solving for y in Equation 2 to obtain $y = -3x + 5$. Then substitute this expression for y in Equation 1.

$$\begin{aligned} 9x + 3y &= 15 & \text{Equation 1} \\ 9x + 3(-3x + 5) &= 15 & \text{Substitute } -3x + 5 \text{ for } y. \\ 9x - 9x + 15 &= 15 & \text{Distributive Property} \\ 15 &= 15 & \text{Simplify.} \end{aligned}$$

Study Tip

By writing both equations in Example 7 in slope-intercept form, you will get identical equations. This means that the lines coincide and the system has infinitely many solutions.

The equation $15 = 15$ is true for any value of x. This implies that any solution of Equation 2 is also a solution of Equation 1. In other words, the original system of linear equations is *dependent* and has infinitely many solutions. The solutions consist of all ordered pairs (x, y) lying on the line $3x + y = 5$, such as $(-1, 8)$, $(0, 5)$, and $(1, 2)$.

Exercises Within Reach ®

Solutions in English & Spanish and tutorial videos at CollegePrepAlgebra.com

Solving a System of Linear Equations In Exercises 25–30, solve the system by the method of substitution.

25. $\begin{cases} 4x - y = 2 \\ 2x - \frac{1}{2}y = 1 \end{cases}$

26. $\begin{cases} x - 5y = 5 \\ 3x - 15y = 15 \end{cases}$

27. $\begin{cases} -5x + 4y = 14 \\ 5x - 4y = 4 \end{cases}$

28. $\begin{cases} 3x - 2y = 3 \\ -6x + 4y = -6 \end{cases}$

29. $\begin{cases} -6x + 1.5y = 6 \\ 8x - 2y = -8 \end{cases}$

30. $\begin{cases} 0.3x - 0.3y = 0 \\ x - y = 4 \end{cases}$

Applications

Application **EXAMPLE 8** **A Mixture Problem**

A roofing contractor buys 30 bundles of shingles and 4 rolls of roofing paper for $732. In a second purchase (at the same prices), the contractor pays $194 for 8 bundles of shingles and 1 roll of roofing paper. Find the price per bundle of shingles and the price per roll of roofing paper.

SOLUTION

Verbal Model: $30 \left(\begin{array}{c}\text{Price of}\\\text{a bundle}\end{array}\right) + 4 \left(\begin{array}{c}\text{Price of}\\\text{a roll}\end{array}\right) = 732$

$8 \left(\begin{array}{c}\text{Price of}\\\text{a bundle}\end{array}\right) + 1 \left(\begin{array}{c}\text{Price of}\\\text{a roll}\end{array}\right) = 194$

Labels: Price of a bundle of shingles = x (dollars)

Price of a roll of roofing paper = y (dollars)

System: $\begin{cases} 30x + 4y = 732 & \text{Equation 1} \\ 8x + y = 194 & \text{Equation 2} \end{cases}$

Solving the second equation for y produces $y = 194 - 8x$, and substituting this expression for y in the first equation produces the following.

$30x + 4(194 - 8x) = 732$	Substitute $194 - 8x$ for y.
$30x + 776 - 32x = 732$	Distributive Property
$-2x = -44$	Simplify.
$x = 22$	Divide each side by -2.

Back-substituting 22 for x in revised Equation 2 produces

$$y = 194 - 8(22) = 18.$$

So, you can conclude that the contractor pays $22 per bundle of shingles and $18 per roll of roofing paper. Check this in the original statement of the problem.

Exercises Within Reach ®

Solutions in English & Spanish and tutorial videos at CollegePrepAlgebra.com

31. **Hay Mixture** A farmer wants to mix two types of hay. The first type sells for $125 per ton and the second type sells for $75 per ton. The farmer wants a total of 100 tons of hay at a cost of $90 per ton. How many tons of each type of hay should be used in the mixture?

32. **Seed Mixture** Ten pounds of mixed birdseed sells for $6.97 per pound. The mixture is obtained from two kinds of birdseed, with one variety priced at $5.65 per pound and the other at $8.95 per pound. How many pounds of each variety of birdseed are used in the mixture?

Application   EXAMPLE 9   **Break-Even Analysis**

A small business invests $14,000 to produce a new energy bar. Each bar costs $0.80 to produce and sells for $1.50. How many bars must be sold before the business breaks even?

SOLUTION

Verbal Model:

| Total cost | = | Cost per bar | · | Number of bars | + | Initial cost |

| Total revenue | = | Price per bar | · | Number of bars |

Labels:

Total cost = C	(dollars)
Cost per bar = 0.80	(dollars per bar)
Number of bars = x	(bars)
Initial cost = 14,000	(dollars)
Total revenue = R	(dollars)
Price per bar = 1.50	(dollars per bar)

System:

$$\begin{cases} C = 0.80x + 14{,}000 & \text{Equation 1} \\ R = 1.50x & \text{Equation 2} \end{cases}$$

Energy Bars

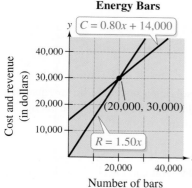

y | $C = 0.80x + 14{,}000$

(20,000, 30,000)

$R = 1.50x$

x

Cost and revenue (in dollars)

Number of bars

Because the break-even point occurs when $R = C$, you have

$1.50x = 0.80x + 14{,}000$ $R = C$

$0.7x = 14{,}000$ Subtract $0.80x$ from each side.

$x = 20{,}000.$ Divide each side by 0.7.

So, it follows that the business must sell 20,000 bars before it breaks even. Profit P (or loss) for the business can be determined by the equation $P = R - C$. Note in the graph at the left, that sales less than the break-even point correspond to a loss for the business, whereas sales greater than the break-even point correspond to a profit for the business. The following table helps confirm this conclusion.

Units, x	0	5000	10,000	15,000	20,000	25,000
Revenue, R	$0	$7500	$15,000	$22,500	$30,000	$37,500
Cost, C	$14,000	$18,000	$22,000	$26,000	$30,000	$34,000
Profit, P	−$14,000	−$10,500	−$7000	−$3500	$0	$3500

Exercises Within Reach ®

Solutions in English & Spanish and tutorial videos at CollegePrepAlgebra.com

33. *Break-Even Analysis* A small business invests $8000 to produce a new candy bar. Each bar costs $1.20 to produce and sells for $2.00. How many candy bars must be sold before the business breaks even?

34. *Break-Even Analysis* A business spends $10,000 to produce a new board game. Each game costs $1.50 to produce and is sold for $9.99. How many games must be sold before the business breaks even?

Concept Summary: *Solving Systems of Linear Equations by Substitution*

What

You can **solve a system of equations** using an algebraic method called the method of substitution.

EXAMPLE

Solve the system of linear equations.

$$\begin{cases} y = x + 2 & \text{Equation 1} \\ x + y = 6 & \text{Equation 2} \end{cases}$$

How

The goal is to reduce the system of two linear equations in two variables to a single equation in one variable. Then solve the equation.

EXAMPLE

Substitute $x + 2$ for y in Equation 2. Then solve for x.

$$x + (x + 2) = 6$$
$$x = 2$$

Now substitute 2 for x in Equation 1 to find the value of y.

$$y = 2 + 2$$
$$y = 4$$

Why

Solving a system of equations graphically is limited by your ability to sketch an accurate graph, especially when the coordinates of a solution point are fractional or irrational.

One way to obtain an exact solution is to solve the system by substitution.

Exercises Within Reach ®

Worked-out solutions to odd-numbered exercises at CollegePrepAlgebra.com

Concept Summary Check

35. *Precision* In your own words, explain the basic steps in solving a system of linear equations by the method of substitution.

36. *Reasoning* When solving a system of linear equations by the method of substitution, how do you recognize that it has no solution?

37. *Reasoning* When solving a system of linear equations by the method of substitution, how do you recognize that it has infinitely many solutions?

38. *Writing* Explain how you can check the solution of a system of linear equations algebraically.

Extra Practice

Solving a System of Linear Equations In Exercises 39−44, **solve the system by graphing.**

39. $\begin{cases} x + y = 4 \\ 4x + 4y = 2 \end{cases}$

40. $\begin{cases} 3x + 3y = 1 \\ x + y = 3 \end{cases}$

41. $\begin{cases} -2x + y = -1 \\ x - 2y = -1 \end{cases}$

42. $\begin{cases} -x + 10y = 30 \\ x + 10y = 10 \end{cases}$

43. $\begin{cases} x - 2y = 4 \\ 2x - 4y = 8 \end{cases}$

44. $\begin{cases} 2x + 3y = 6 \\ 4x + 6y = 12 \end{cases}$

Solving a System of Linear Equations In Exercises 45−48, **solve the system by the method of substitution.**

45. $\begin{cases} \dfrac{x}{4} + \dfrac{y}{2} = 1 \\ \dfrac{x}{2} - \dfrac{y}{3} = 1 \end{cases}$

46. $\begin{cases} -\dfrac{x}{6} + \dfrac{y}{12} = 1 \\ \dfrac{x}{2} + \dfrac{y}{8} = 1 \end{cases}$

47. $\begin{cases} 2(x - 5) = y + 2 \\ 3x = 4(y + 2) \end{cases}$

48. $\begin{cases} 3(x - 2) + 5 = 4(y + 3) - 2 \\ 2x + 7 = 2y + 8 \end{cases}$

Think About It In Exercises 49–52, find the value of *a* or *b* such that the system of linear equations is inconsistent.

49. $\begin{cases} x + by = 1 \\ x + 2y = 2 \end{cases}$

50. $\begin{cases} ax + 3y = 6 \\ 5x - 5y = 2 \end{cases}$

51. $\begin{cases} -6x + y = 4 \\ 2x + by = 3 \end{cases}$

52. $\begin{cases} 6x - 3y = 4 \\ ax - y = -2 \end{cases}$

Think About It In Exercises 53 and 54, find a system of linear equations that has the given solution. (There are many correct answers.)

53. $(2, 1)$

54. $(-4, -3)$

Number Problem In Exercises 55 and 56, find two positive integers that satisfy the given requirements.

55. The sum of the greater number and twice the lesser number is 61, and their difference is 7.

56. The sum of the two numbers is 52, and the greater number is 8 less than twice the lesser number.

57. *Investment* A total of $12,000 is invested in two bonds that pay 8.5% and 10% simple interest. The annual interest is $1140. How much is invested in each bond?

58. *Investment* A total of $25,000 is invested in two funds paying 8% and 8.5% simple interest. The annual interest is $2060. How much is invested in each fund?

Explaining Concepts

Think About It In Exercises 59 and 60, the graphs of the two equations appear to be parallel. **Explain** whether the system has a solution. If so, find the solution.

59. $\begin{cases} x - 200y = -200 \\ x - 199y = 198 \end{cases}$

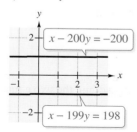

60. $\begin{cases} 25x - 24y = 0 \\ 13x - 12y = 24 \end{cases}$

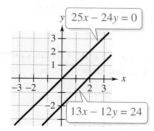

61. *Writing* Describe any advantages of the method of substitution over the graphical method of solving a system of linear equations.

62. *True or False?* It is possible for a consistent system of linear equations to have exactly two solutions. Justify your answer.

Cumulative Review

In Exercises 63–66, evaluate the expression.

63. $\frac{2}{3} + \frac{1}{3}$

64. $\frac{5}{8} + \frac{3}{2}$

65. $\frac{2}{7} - \frac{1}{3}$

66. $\frac{3}{5} - \frac{7}{6}$

In Exercises 67–70, solve the equation and check your solution.

67. $x - 6 = 5x$

68. $2 - 3x = 14 + x$

69. $y - 3(4y - 2) = 1$

70. $y + 6(3 - 2y) = 4$

In Exercises 71–74, solve the rational equation.

71. $\frac{x}{5} + \frac{2x}{5} = 3$

72. $\frac{3x}{5} + \frac{4x}{8} = \frac{11}{10}$

73. $\frac{x - 3}{x + 1} = \frac{4}{3}$

74. $\frac{3}{x} = \frac{9}{2(x + 2)}$

8.2 Solving Systems of Equations by Elimination

▶ Solve systems of linear equations algebraically using the method of elimination.

▶ Solve systems with no solution or infinitely many solutions.

▶ Use the method of elimination to solve application problems.

The Method of Elimination

The Method of Elimination

1. Obtain opposite coefficients of x (or y) by multiplying all terms of one or both equations by suitable constants.

2. Add the equations to eliminate one variable and solve the resulting equation.

3. Back-substitute the value obtained in Step 2 into either of the original equations and solve for the other variable.

4. Check your solution in *both* of the original equations.

EXAMPLE 1 **The Method of Elimination**

Solve the system of linear equations.

$$\begin{cases} 4x + 3y = 1 & \text{Equation 1} \\ 2x - 3y = 5 & \text{Equation 2} \end{cases}$$

SOLUTION

Begin by noting that the coefficients of y are opposites. So, by adding the two equations, you can eliminate y.

$$\begin{cases} 4x + 3y = 1 & \text{Equation 1} \\ \underline{2x - 3y = 5} & \text{Equation 2} \end{cases}$$
$$6x = 6 \qquad \text{Add equations.}$$

So, $x = 1$. Back-substitute this x-value into Equation 1 and solve for y.

$$4(1) + 3y = 1 \qquad \text{Substitute 1 for } x \text{ in Equation 1.}$$
$$3y = -3 \qquad \text{Subtract 4 from each side.}$$
$$y = -1 \qquad \text{Divide each side by 3.}$$

The solution is $(1, -1)$. Check this in both of the original equations.

Study Tip

Try solving the system in Example 1 by substitution. Notice that the method of elimination is more efficient for this system.

Exercises Within Reach ®

Solutions in English & Spanish and tutorial videos at CollegePrepAlgebra.com

Solving a System of Linear Equations In Exercises 1–6, solve the system by the method of elimination.

1. $\begin{cases} x - y = 4 \\ x + y = 12 \end{cases}$

2. $\begin{cases} x + y = 7 \\ x - y = 3 \end{cases}$

3. $\begin{cases} -x + 2y = 12 \\ x + 6y = 20 \end{cases}$

4. $\begin{cases} x + 2y = 14 \\ x - 2y = 10 \end{cases}$

5. $\begin{cases} 3x - 5y = 1 \\ 2x + 5y = 9 \end{cases}$

6. $\begin{cases} -2x + 3y = -4 \\ 2x - 4y = 6 \end{cases}$

To obtain opposite coefficients for one of the variables, you often need to multiply one or both of the equations by a suitable constant. This is demonstrated in the following example.

EXAMPLE 2   The Method of Elimination

Solve the system of linear equations.

$$\begin{cases} 2x - 3y = -7 \\ 3x + y = -5 \end{cases}$$

Equation 1
Equation 2

SOLUTION

For this system, you can obtain opposite coefficients of y by multiplying Equation 2 by 3.

$$\begin{cases} 2x - 3y = -7 \\ 3x + y = -5 \end{cases} \Longrightarrow \begin{array}{l} 2x - 3y = -7 \\ 9x + 3y = -15 \\ \hline 11x = -22 \end{array}$$

Equation 1
Multiply Equation 2 by 3.
Add equations.

So, $x = -2$. Back-substitute this x-value into Equation 2 and solve for y.

$$3x + y = -5$$ Equation 2

$$3(-2) + y = -5$$ Substitute -2 for x.

$$-6 + y = -5$$ Simplify.

$$y = 1$$ Add 6 to each side.

The solution is $(-2, 1)$. Check this in both of the original equations, as follows.

Substitute into Equation 1   *Substitute into Equation 2*

$$2x - 3y = -7 \qquad\qquad 3x + y = -5$$

$$2(-2) - 3(1) \overset{?}{=} -7 \qquad 3(-2) + 1 \overset{?}{=} -5$$

$$-4 - 3 \overset{?}{=} -7 \qquad\qquad -6 + 1 \overset{?}{=} -5$$

$$-7 = -7 \checkmark \qquad\qquad -5 = -5 \checkmark$$

Exercises Within Reach ®

Solutions in English & Spanish and tutorial videos at CollegePrepAlgebra.com

Solving a System of Linear Equations In Exercises 7–12, solve the system by the method of elimination.

7. $\begin{cases} 2a + 5b = 3 \\ 2a + b = 9 \end{cases}$

8. $\begin{cases} 4a + 5b = 9 \\ 2a + 5b = 7 \end{cases}$

9. $\begin{cases} -x + 2y = 6 \\ 2x + 5y = 6 \end{cases}$

10. $\begin{cases} -4x + 8y = 0 \\ 3x - 2y = 2 \end{cases}$

11. $\begin{cases} 2x - 5y = -1 \\ 2x - y = 1 \end{cases}$

12. $\begin{cases} 7x + 8y = 6 \\ 3x - 4y = 10 \end{cases}$

EXAMPLE 3 The Method of Elimination

Solve the system of linear equations.

$$\begin{cases} 5x + 3y = 6 & \text{Equation 1} \\ 2x - 4y = 5 & \text{Equation 2} \end{cases}$$

SOLUTION

You can obtain opposite coefficients of y by multiplying Equation 1 by 4 and Equation 2 by 3.

$$\begin{cases} 5x + 3y = 6 \\ 2x - 4y = 5 \end{cases} \implies \begin{array}{r} 20x + 12y = 24 \\ 6x - 12y = 15 \\ \hline 26x \qquad = 39 \end{array}$$

Multiply Equation 1 by 4.

Multiply Equation 2 by 3.

Add equations.

From this equation, you can see that $x = \frac{3}{2}$. Back-substitute this x-value into Equation 2 and solve for y.

$$2x - 4y = 5 \qquad \text{Equation 2}$$

$$2\left(\frac{3}{2}\right) - 4y = 5 \qquad \text{Substitute } \frac{3}{2} \text{ for } x.$$

$$3 - 4y = 5 \qquad \text{Simplify.}$$

$$-4y = 2 \qquad \text{Subtract 3 from each side.}$$

$$y = -\frac{1}{2} \qquad \text{Divide each side by } -4.$$

The solution is $\left(\frac{3}{2}, -\frac{1}{2}\right)$. You can check this as follows.

Substitute into Equation 1

$$5x + 3y = 6$$

$$5\left(\frac{3}{2}\right) + 3\left(-\frac{1}{2}\right) \overset{?}{=} 6$$

$$\frac{15}{2} - \frac{3}{2} = 6 \ \checkmark$$

Substitute into Equation 2

$$2x - 4y = 5$$

$$2\left(\frac{3}{2}\right) - 4\left(-\frac{1}{2}\right) \overset{?}{=} 5$$

$$3 + 2 = 5 \ \checkmark$$

The graph of this system is shown at the left. From the graph, it appears that the solution $\left(\frac{3}{2}, -\frac{1}{2}\right)$ is reasonable.

Exercises Within Reach ®

Solutions in English & Spanish and tutorial videos at CollegePrepAlgebra.com

Solving a System of Linear Equations In Exercises 13–18, solve the system by the method of elimination.

13. $\begin{cases} 4x + 5y = 7 \\ 6x - 2y = -18 \end{cases}$

14. $\begin{cases} 5x + 3y = 18 \\ 2x - 7y = -1 \end{cases}$

15. $\begin{cases} 3x + 2y = 10 \\ 2x + 5y = 3 \end{cases}$

16. $\begin{cases} 5u + 6v = 14 \\ 3u + 5v = 7 \end{cases}$

17. $\begin{cases} 2x + 3y = 16 \\ 5x - 10y = 30 \end{cases}$

18. $\begin{cases} 3x - 4y = 1 \\ 4x + 3y = 1 \end{cases}$

The next example shows how the method of elimination works with a system of linear equations with decimal coefficients.

EXAMPLE 4   **The Method of Elimination**

Solve the system of linear equations.

$$\begin{cases} 0.02x - 0.05y = -0.38 & \text{Equation 1} \\ 0.03x + 0.04y = 1.04 & \text{Equation 2} \end{cases}$$

SOLUTION

Because the coefficients in this system have two decimal places, begin by multiplying each equation by 100. This produces a system in which the coefficients are all integers.

$$\begin{cases} 2x - 5y = -38 & \text{Revised Equation 1} \\ 3x + 4y = 104 & \text{Revised Equation 2} \end{cases}$$

Now you can obtain opposite coefficients of x by multiplying Equation 1 by 3 and Equation 2 by -2.

$$\begin{cases} 2x - 5y = -38 \\ 3x + 4y = 104 \end{cases}$$

$$\begin{array}{ll} 6x - 15y = -114 & \text{Multiply Equation 1 by 3.} \\ -6x - 8y = -208 & \text{Multiply Equation 2 by } -2. \\ \hline {-23y} = -322 & \text{Add equations.} \end{array}$$

> **Study Tip**
>
> When multiplying an equation by a negative number, be sure to distribute the negative sign to each term of the equation.

So, the y-coordinate of the solution is

$$y = \frac{-322}{-23} = 14.$$

Back-substitute this y-value into revised Equation 2 and solve for x.

$$\begin{array}{ll} 3x + 4(14) = 104 & \text{Substitute 14 for } y \text{ in revised Equation 2.} \\ 3x + 56 = 104 & \text{Simplify.} \\ 3x = 48 & \text{Subtract 56 from each side.} \\ x = 16 & \text{Divide each side by 3.} \end{array}$$

So, the solution is $(16, 14)$. Check this in both of the original equations.

Exercises Within Reach ®

Solutions in English & Spanish and tutorial videos at CollegePrepAlgebra.com

Solving a System of Linear Equations In Exercises 19–24, solve the system by the method of elimination.

19. $\begin{cases} 0.02x - 0.05y = -0.19 \\ 0.03x + 0.04y = 0.52 \end{cases}$

20. $\begin{cases} 0.05x - 0.03y = 0.21 \\ 0.01x + 0.01y = 0.09 \end{cases}$

21. $\begin{cases} 0.1x - 0.1y = 0 \\ 0.8x + 0.3y = 1.5 \end{cases}$

22. $\begin{cases} x - 2y = 0 \\ 0.2x + 0.8y = 2.4 \end{cases}$

23. $\begin{cases} 6r + 5s = 3 \\ \frac{3}{2}r - \frac{5}{4}s = \frac{3}{4} \end{cases}$

24. $\begin{cases} \frac{1}{4}x - y = \frac{1}{2} \\ 4x + 4y = 3 \end{cases}$

The No-Solution and Many-Solutions Cases

> **EXAMPLE 5**   The Method of Elimination: No-Solution Case

Solve the system of linear equations.

$$\begin{cases} 2x - 6y = 5 & \text{Equation 1} \\ 3x - 9y = 2 & \text{Equation 2} \end{cases}$$

SOLUTION

You can obtain opposite coefficients by multiplying Equation 1 by 3 and Equation 2 by -2.

$$\begin{cases} 2x - 6y = 5 \\ 3x - 9y = 2 \end{cases} \implies \begin{array}{r} 6x - 18y = 15 \\ -6x + 18y = -4 \\ \hline 0 = 11 \end{array} \quad \begin{array}{l} \text{Multiply Equation 1 by 3.} \\ \text{Multiply Equation 2 by } -2. \\ \text{Add equations.} \end{array}$$

Because $0 = 11$ is a false statement, you can conclude that the system is inconsistent and has no solution. The lines corresponding to the two equations of this system are shown at the left. Note that the two lines are parallel and have no point of intersection.

> **EXAMPLE 6**   The Method of Elimination: Many-Solutions Case

Solve the system of linear equations.

$$\begin{cases} 2x - 6y = -5 & \text{Equation 1} \\ -4x + 12y = 10 & \text{Equation 2} \end{cases}$$

SOLUTION

You can obtain opposite coefficients of x by multiplying Equation 1 by 2.

$$\begin{cases} 2x - 6y = -5 \\ -4x + 12y = 10 \end{cases} \implies \begin{array}{r} 4x - 12y = -10 \\ -4x + 12y = 10 \\ \hline 0 = 0 \end{array} \quad \begin{array}{l} \text{Multiply Equation 1 by 2.} \\ \text{Equation 2} \\ \text{Add equations.} \end{array}$$

Because $0 = 0$ is a true statement, you can conclude that the system is dependent and has infinitely many solutions. The solutions consist of all ordered pairs (x, y) lying on the line $2x - 6y = -5$.

Study Tip

By writing both equations in Example 6 in slope-intercept form, you will obtain identical equations. This shows that the system has infinitely many solutions.

Exercises Within Reach ®

Solutions in English & Spanish and tutorial videos at CollegePrepAlgebra.com

Solving a System of Linear Equations In Exercises 25−28, solve the system by the method of elimination.

25. $\begin{cases} -3x - 12y = 3 \\ 5x + 20y = -5 \end{cases}$

26. $\begin{cases} 7x + 10y = 0 \\ 21x + 30y = 0 \end{cases}$

27. $\begin{cases} 0.4a + 0.7b = 3 \\ 0.8a + 1.4b = 7 \end{cases}$

28. $\begin{cases} 0.2u - 0.1v = 1 \\ -0.8u + 0.4v = 3 \end{cases}$

Application

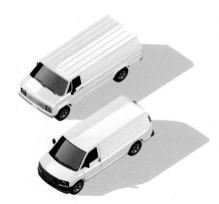

Application   EXAMPLE 7   **Solving a Mixture Problem**

A company with two stores buys six large delivery vans and five small delivery vans. The first store receives 4 of the large vans and 2 of the small vans for a total cost of $200,000. The second store receives 2 of the large vans and 3 of the small vans for a total cost of $160,000. What is the cost of each type of van?

SOLUTION

The two unknowns in this problem are the costs of the two types of vans.

Verbal Model: $4\left(\dfrac{\text{Cost of}}{\text{large van}}\right) + 2\left(\dfrac{\text{Cost of}}{\text{small van}}\right) = \$200{,}000$

$2\left(\dfrac{\text{Cost of}}{\text{large van}}\right) + 3\left(\dfrac{\text{Cost of}}{\text{small van}}\right) = \$160{,}000$

Labels: Cost of large van $= x$ (dollars)
Cost of small van $= y$ (dollars)

System: $\begin{cases} 4x + 2y = 200{,}000 & \text{Equation 1} \\ 2x + 3y = 160{,}000 & \text{Equation 2} \end{cases}$

To solve this system of linear equations, use the method of elimination. To obtain coefficients of x that are opposites, multiply Equation 2 by -2.

$\begin{cases} 4x + 2y = 200{,}000 \\ 2x + 3y = 160{,}000 \end{cases}$ $\Longrightarrow$

$\begin{array}{ll} 4x + 2y = 200{,}000 & \text{Equation 1} \\ -4x - 6y = -320{,}000 & \text{Multiply Equation 2 by } -2. \\ \hline -4y = -120{,}000 & \text{Add equations.} \\ y = 30{,}000 & \text{Divide each side by } -4. \end{array}$

So, the cost of each small van is $y = \$30{,}000$. Back-substitute this value into Equation 1 to find the cost of each large van.

$\begin{array}{ll} 4x + 2y = 200{,}000 & \text{Equation 1} \\ 4x + 2(30{,}000) = 200{,}000 & \text{Substitute 30,000 for } y. \\ 4x = 140{,}000 & \text{Simplify.} \\ x = 35{,}000 & \text{Divide each side by 4.} \end{array}$

The cost of each large van is $x = \$35{,}000$. Check this solution in the original statement of the problem.

Exercises Within Reach ®

Solutions in English & Spanish and tutorial videos at CollegePrepAlgebra.com

29. *Comparing Costs* A band charges $500 to play for 4 hours plus $50 for each additional hour. A DJ charges $300 to play for 4 hours plus $75 for each additional hour. After how many hours will the cost of the DJ exceed the cost of the band?

30. *Comparing Costs* An SUV costs $26,445 and costs an average of $0.18 per mile to maintain. A hybrid model of the SUV costs $31,910 and costs an average of $0.13 per mile to maintain. After how many miles will the cost of the gas-only SUV exceed the cost of the hybrid?

Concept Summary: *Solving Systems of Linear Equations by Elimination*

What

You can solve a system of linear equations using an algebraic method called the **method of elimination**.

EXAMPLE

Solve the system of linear equations.

$$\begin{cases} 3x + 5y = 7 & \text{Equation 1} \\ -3x - 2y = -1 & \text{Equation 2} \end{cases}$$

How

The key step in this method is to obtain opposite coefficients for one of the variables so that adding the two equations eliminates this variable.

EXAMPLE

$$\begin{array}{ll} 3x + 5y = 7 & \text{Equation 1} \\ \underline{-3x - 2y = -1} & \text{Equation 2} \\ 3y = 6 & \text{Add equations.} \end{array}$$

After eliminating the variable, solve for the other variable. Then use back-substitution to find the value of the eliminated variable.

Why

When solving a system of linear equations, choose the method that is most efficient.

1. Use graphing to approximate the solution.

2. To find exact solutions, use substitution or elimination.
   - When one of the variables has a coefficient of 1, use substitution.
   - When the coefficients of one of the variables are opposites, use elimination.

3. When you are not sure, use elimination. It is usually more efficient.

Exercises Within Reach ®

Worked-out solutions to odd-numbered exercises at CollegePrepAlgebra.com

Concept Summary Check

31. *Solving a System* What is the solution of the system of linear equations in the example above?

32. *Precision* Explain how to solve a system of linear equations by elimination.

33. *Reasoning* When solving a system by the method of elimination, how do you recognize that it has no solution?

34. *Reasoning* When solving a system by the method of elimination, how do you recognize that it has infinitely many solutions?

Extra Practice

Solving a System of Linear Equations In Exercises 35−38, solve the system by the method of elimination.

35. $\begin{cases} -\dfrac{x}{4} + y = 1 \\ \dfrac{x}{4} + \dfrac{y}{2} = 1 \end{cases}$

36. $\begin{cases} \dfrac{x}{3} - \dfrac{y}{5} = 1 \\ \dfrac{x}{12} + \dfrac{y}{40} = 1 \end{cases}$

37. $\begin{cases} 3(x + 5) - 7 = 2(3 - 2y) \\ 2x + 1 = 4(y + 2) \end{cases}$

38. $\begin{cases} \frac{1}{2}(x - 4) + 9 = y - 10 \\ -5(x + 3) = 8 - 2(y - 3) \end{cases}$

Describing a System In Exercises 39−42, determine whether the system is consistent or inconsistent.

39. $\begin{cases} 4x - 5y = 3 \\ -8x + 10y = -6 \end{cases}$

40. $\begin{cases} -10x + 15y = 25 \\ 2x - 3y = -24 \end{cases}$

41. $\begin{cases} -2x + 5y = 3 \\ 5x + 2y = 8 \end{cases}$

42. $\begin{cases} x + 10y = 12 \\ -2x + 5y = 2 \end{cases}$

Number Problem In Exercises 43 and 44, find two integers that satisfy the given requirements.

43. The sum of two numbers x and y is 82 and the difference of the numbers is 14.

44. The sum of two numbers x and y is 154 and the difference of the numbers is 38.

45. *Geometry* Find an equation of the line of slope $m = \frac{1}{3}$ passing through the intersection of the lines

$$3x + 4y = 7 \quad \text{and} \quad 5x - 4y = 1.$$

46. *Geometry* Find an equation of the line of slope $m = -2$ passing through the intersection of the lines

$$2x + 5y = 11 \quad \text{and} \quad 4x - y = 11.$$

47. *Gasoline Mixture* Twelve gallons of regular unleaded gasoline plus 8 gallons of premium unleaded gasoline cost $76.48. Premium unleaded gasoline costs $0.11 more per gallon than regular unleaded. Find the price per gallon for each grade of gasoline.

48. *Alcohol Mixture* How many liters of a 40% alcohol solution must be mixed with a 65% solution to obtain 20 liters of a 50% alcohol solution?

49. *Acid Mixture* Thirty liters of a 46% acid solution is obtained by mixing a 40% solution with a 70% solution. How many liters of each solution must be used to obtain the desired mixture?

50. *Nut Mixture* Ten pounds of mixed nuts sells for $6.87 per pound. The mixture is obtained from two kinds of nuts, peanuts priced at $5.70 per pound and cashews at $8.70 per pound. How many pounds of each variety of nut are used in the mixture?

Explaining Concepts

51. *Creating an Example* Explain how to "clear" a system of decimals. Give an example to justify your answer. (There are many correct answers.)

52. *Creating a System* Write a system of linear equations that is more efficiently solved by the method of elimination than by the method of substitution. (There are many correct answers.)

53. *Creating a System* Write a system of linear equations that is more efficiently solved by the method of substitution than by the method of elimination. (There are many correct answers.)

54. *Reasoning* Consider the system of linear equations.

$$\begin{cases} x + y = 8 \\ 2x + 2y = k \end{cases}$$

(a) Find the value(s) of k for which the system has an infinite number of solutions.

(b) Find one value of k for which the system has no solution. (There are many correct answers.)

(c) Can the system have a single solution for some value of k? Why or why not?

Cumulative Review

In Exercises 55–60, plot the points and find the slope (if possible) of the line that passes through the points. If not possible, state why.

55. $(-6, 4), (-3, -4)$

56. $(4, 6), (8, -2)$

57. $\left(\frac{7}{2}, \frac{9}{2}\right), \left(\frac{4}{3}, -3\right)$

58. $\left(-\frac{3}{4}, -\frac{7}{4}\right), \left(-1, \frac{5}{2}\right)$

59. $(-3, 6), (-3, 2)$

60. $(6, 2), (10, 2)$

In Exercises 61–64, solve and graph the inequality.

61. $x \le 3$

62. $x > -4$

63. $x + 5 < 6$

64. $3x - 7 \ge 2x + 9$

In Exercises 65–68, solve the system by the method of substitution.

65. $\begin{cases} y = x \\ x + 3y = 20 \end{cases}$

66. $\begin{cases} x + y = 9 \\ 2x + 2y = 18 \end{cases}$

67. $\begin{cases} 2x + y = 5 \\ 5x + 3y = 12 \end{cases}$

68. $\begin{cases} 5x + 6y = 21 \\ 25x + 30y = 10 \end{cases}$

8.3 Linear Systems in Three Variables

▶ Solve systems of linear equations in row-echelon form using back-substitution.

▶ Solve systems of linear equations using the method of Gaussian elimination.

▶ Solve application problems using the method of Gaussian elimination.

Row-Echelon Form

When the method of elimination is used to solve a system of linear equations, the goal is to rewrite the system in a form to which back-substitution can be applied. This method can be applied to a system of linear equations in more than two variables, as shown below. The system on the right is in **row-echelon form**, which means that it has a "stair-step" pattern with leading coefficients of 1.

$$\begin{cases} x - 2y + 2z = 9 \\ -x + 3y = -4 \\ 2x - 5y + z = 10 \end{cases} \implies \begin{cases} x - 2y + 2z = 9 \\ y + 2z = 5 \\ z = 3 \end{cases}$$

EXAMPLE 1 Using Back-Substitution

In the following system of linear equations, you know the value of z from Equation 3.

$$\begin{cases} x - 2y + 2z = 9 & \text{Equation 1} \\ y + 2z = 5 & \text{Equation 2} \\ z = 3 & \text{Equation 3} \end{cases}$$

To solve for y, substitute $z = 3$ into Equation 2 to obtain

$$y + 2(3) = 5 \implies y = -1 \qquad \text{Substitute 3 for } z.$$

Finally, substitute $y = -1$ and $z = 3$ into Equation 1 to obtain

$$x - 2(-1) + 2(3) = 9 \implies x = 1. \qquad \text{Substitute } -1 \text{ for } y \text{ and 3 for } z.$$

The solution is $x = 1$, $y = -1$, and $z = 3$, which can also be written as the **ordered triple** $(1, -1, 3)$. Check this in the original system of equations.

Study Tip

When checking a solution, remember that the solution must satisfy each equation in the original system.

Exercises Within Reach ®

Solutions in English & Spanish and tutorial videos at CollegePrepAlgebra.com

Using Back-Substitution In Exercises 1–4, use back-substitution to solve the system of linear equations.

1. $\begin{cases} x - 2y + 4z = 4 \\ 3y - z = 2 \\ z = -5 \end{cases}$

2. $\begin{cases} 5x + 4y - z = 0 \\ 10y - 3z = 11 \\ z = 3 \end{cases}$

3. $\begin{cases} x - 2y + 4z = 4 \\ y = 3 \\ y + z = 2 \end{cases}$

4. $\begin{cases} x = 10 \\ 3x + 2y = 2 \\ x + y + 2z = 0 \end{cases}$

The Method of Gaussian Elimination

Study Tip

Two systems of equations are **equivalent systems** when they have the same solution set. Rewriting a system of linear equations in row-echelon form usually involves a chain of equivalent systems, each of which is obtained by using one of the three basic row operations. This process is called **Gaussian elimination**.

> ### Operations That Produce Equivalent Systems
>
> Each of the following **row operations** on a system of linear equations produces an *equivalent* system of linear equations.
>
> **1.** Interchange two equations.
> **2.** Multiply one of the equations by a nonzero constant.
> **3.** Add a multiple of one of the equations to another equation to replace the latter equation.

EXAMPLE 2   **Using Gaussian Elimination to Solve a System**

Solve the system of linear equations.

$$\begin{cases} x - 2y + 2z = 9 & \text{Equation 1} \\ -x + 3y = -4 & \text{Equation 2} \\ 2x - 5y + z = 10 & \text{Equation 3} \end{cases}$$

SOLUTION

Equation 1 has a leading coefficient of 1, so leave it alone to keep the x in the upper left position. Begin by eliminating the other x terms from the first column, as follows.

$$\begin{cases} x - 2y + 2z = 9 \\ y + 2z = 5 \\ 2x - 5y + z = 10 \end{cases}$$

> Adding the first equation to the second equation produces a new second equation.

$$\begin{cases} x - 2y + 2z = 9 \\ y + 2z = 5 \\ -y - 3z = -8 \end{cases}$$

> Adding -2 times the first equation to the third equation produces a new third equation.

Now work on the second column. (You need to eliminate y from the third equation.)

$$\begin{cases} x - 2y + 2z = 9 \\ y + 2z = 5 \\ -z = -3 \end{cases}$$

> Adding the second equation to the third equation produces a new third equation.

Finally, you need a coefficient of 1 for z in the third equation.

$$\begin{cases} x - 2y + 2z = 9 \\ y + 2z = 5 \\ z = 3 \end{cases}$$

> Multiplying the third equation by -1 produces a new third equation.

This is the same system that was solved in Example 1, and, as in that example, you can conclude by back-substitution that the solution is

$$x = 1, \quad y = -1, \quad \text{and} \quad z = 3. \qquad \text{The solution is } (1, -1, 3).$$

Exercises Within Reach ®

Solutions in English & Spanish and tutorial videos at CollegePrepAlgebra.com

Using Gaussian Elimination In Exercises 5−8, solve the system of linear equations.

5. $\begin{cases} x + z = 4 \\ y = 2 \\ 4x + z = 7 \end{cases}$

6. $\begin{cases} x + y = 6 \\ 3x - y = 2 \\ z = 3 \end{cases}$

7. $\begin{cases} x + y + z = 6 \\ 2x - y + z = 3 \\ 3x - z = 0 \end{cases}$

8. $\begin{cases} x + y + z = 2 \\ -x + 3y + 2z = 8 \\ 4x + y = 4 \end{cases}$

EXAMPLE 3 **Using Gaussian Elimination to Solve a System**

Solve the system of linear equations.
$\begin{cases} 4x + y - 3z = 11 & \text{Equation 1} \\ 2x - 3y + 2z = 9 & \text{Equation 2} \\ x + y + z = -3 & \text{Equation 3} \end{cases}$

SOLUTION

$\begin{cases} x + y + z = -3 \\ 2x - 3y + 2z = 9 \\ 4x + y - 3z = 11 \end{cases}$ — Interchange the first and third equations.

$\begin{cases} x + y + z = -3 \\ -5y = 15 \\ 4x + y - 3z = 11 \end{cases}$ — Adding -2 times the first equation to the second equation produces a new second equation.

$\begin{cases} x + y + z = -3 \\ -5y = 15 \\ -3y - 7z = 23 \end{cases}$ — Adding -4 times the first equation to the third equation produces a new third equation.

$\begin{cases} x + y + z = -3 \\ y = -3 \\ -3y - 7z = 23 \end{cases}$ — Multiplying the second equation by $-\frac{1}{5}$ produces a new second equation.

$\begin{cases} x + y + z = -3 \\ y = -3 \\ -7z = 14 \end{cases}$ — Adding 3 times the second equation to the third equation produces a new third equation.

$\begin{cases} x + y + z = -3 \\ y = -3 \\ z = -2 \end{cases}$ — Multiplying the third equation by $-\frac{1}{7}$ produces a new third equation.

Now you can back-substitute $z = -2$ and $y = -3$ into Equation 1 to find that $x = 2$. So,

$x = 2$, $y = -3$, and $z = -2$. The solution is $(2, -3, -2)$.

Exercises Within Reach ®

Solutions in English & Spanish and tutorial videos at CollegePrepAlgebra.com

Using Gaussian Elimination **In Exercises 9–16, solve the system of linear equations.**

9. $\begin{cases} x + y + z = -3 \\ 4x + y - 3z = 11 \\ 2x - 3y + 2z = 9 \end{cases}$

10. $\begin{cases} x - y + 2z = -4 \\ 3x + y - 4z = -6 \\ 2x + 3y - 4z = 4 \end{cases}$

11. $\begin{cases} x + 6y + 2z = 9 \\ 3x - 2y + 3z = -1 \\ 5x - 5y + 2z = 7 \end{cases}$

12. $\begin{cases} 2x + 2z = 2 \\ 5x + 3y = 4 \\ 3y - 4z = 4 \end{cases}$

13. $\begin{cases} 6y + 4z = -12 \\ 3x + 3y = 9 \\ 2x - 3z = 10 \end{cases}$

14. $\begin{cases} 2x - 4y + z = 0 \\ 3x + 2z = -1 \\ -6x + 3y + 2z = -10 \end{cases}$

15. $\begin{cases} 2x + y + 3z = 1 \\ 2x + 6y + 8z = 3 \\ 6x + 8y + 18z = 5 \end{cases}$

16. $\begin{cases} 3x - y - 2z = 5 \\ 2x + y + 3z = 6 \\ 6x - y - 4z = 9 \end{cases}$

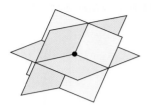

Solution: one point

Solution: one line

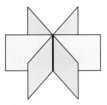

Solution: one plane

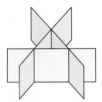

Solution: none

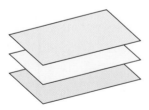

Solution: none

EXAMPLE 4   **An Inconsistent System**

Solve the system of linear equations.

$$\begin{cases} x - 3y + z = 1 & \text{Equation 1} \\ 2x - y - 2z = 2 & \text{Equation 2} \\ x + 2y - 3z = -1 & \text{Equation 3} \end{cases}$$

SOLUTION

$$\begin{cases} x - 3y + z = 1 \\ 5y - 4z = 0 \\ x + 2y - 3z = -1 \end{cases}$$

Adding -2 times the first equation to the second equation produces a new second equation.

$$\begin{cases} x - 3y + z = 1 \\ 5y - 4z = 0 \\ 5y - 4z = -2 \end{cases}$$

Adding -1 times the first equation to the third equation produces a new third equation.

$$\begin{cases} x - 3y + z = 1 \\ 5y - 4z = 0 \\ 0 = -2 \end{cases}$$

Adding -1 times the second equation to the third equation produces a new third equation.

Because the third "equation" is a false statement, you can conclude that this system is inconsistent and therefore has no solution. Moreover, because this system is equivalent to the original system, you can conclude that the original system also has no solution.

The Number of Solutions of a Linear System

For a system of linear equations, exactly one of the following is true.

1. There is exactly one solution.

2. There are infinitely many solutions.

3. There is no solution.

The graph of a system of three linear equations in three variables consists of *three planes*. When these planes intersect in a single point, the system has exactly one solution (see figure). When the three planes intersect in a line or a plane, the system has infinitely many solutions (see figure). When the three planes have no point in common, the system has no solution (see figure).

Exercises Within Reach ®

Solutions in English & Spanish and tutorial videos at CollegePrepAlgebra.com

Using Gaussian Elimination **In Exercises 17–20, solve the system of linear equations.**

17. $\begin{cases} x + 2y + 6z = 5 \\ -x + y - 2z = 3 \\ x - 4y - 2z = 1 \end{cases}$

18. $\begin{cases} x + y + 8z = 3 \\ 2x + y + 11z = 4 \\ x + 3z = 0 \end{cases}$

19. $\begin{cases} y + z = 5 \\ 2x + 4z = 4 \\ 2x - 3y = -14 \end{cases}$

20. $\begin{cases} 5x + 2y = -8 \\ z = 5 \\ 3x - y + z = 9 \end{cases}$

EXAMPLE 5 **A System with Infinitely Many Solutions**

Solve the system of linear equations.

$$\begin{cases} x + y - 3z = -1 & \text{Equation 1} \\ y - z = 0 & \text{Equation 2} \\ -x + 2y = 1 & \text{Equation 3} \end{cases}$$

SOLUTION

Begin by rewriting the system in row-echelon form.

$$\begin{cases} x + y - 3z = -1 \\ y - z = 0 \\ 3y - 3z = 0 \end{cases}$$

Adding the first equation to the third equation produces a new third equation.

$$\begin{cases} x + y - 3z = -1 \\ y - z = 0 \\ 0 = 0 \end{cases}$$

Adding -3 times the second equation to the third equation produces a new third equation.

This means that Equation 3 depends on Equations 1 and 2 in the sense that it gives no additional information about the variables. So, the original system is equivalent to the system

$$\begin{cases} x + y - 3z = -1 \\ y - z = 0 \end{cases}.$$

In the last equation, solve for y in terms of z to obtain $y = z$. Back-substituting for y in the previous equation produces

$$x = 2z - 1.$$

Finally, letting $z = a$, where a is any real number, you can see that there are an infinite number of solutions to the original system, all of the form

$$x = 2a - 1, y = a, \text{ and } z = a.$$

So, every ordered triple of the form

$$(2a - 1, a, a), \qquad a \text{ is a real number.}$$

is a solution of the system.

Exercises Within Reach ®

Solutions in English & Spanish and tutorial videos at CollegePrepAlgebra.com

Using Gaussian Elimination In Exercises 21−26, solve the system of linear equations.

21. $\begin{cases} 2x + z = 1 \\ 5y - 3z = 2 \\ 6x + 20y - 9z = 11 \end{cases}$

22. $\begin{cases} 3x + y + z = 2 \\ 4x + 2z = 1 \\ 5x - y + 3z = 0 \end{cases}$

23. $\begin{cases} x + 4y - 2z = 2 \\ -3x + y + z = -2 \\ 5x + 7y - 5z = 6 \end{cases}$

24. $\begin{cases} x - 2y - z = 3 \\ 2x + y - 3z = 1 \\ x + 8y - 3z = -7 \end{cases}$

25. $\begin{cases} x + 2y - 7z = -4 \\ 2x + y + z = 13 \\ 3x + 9y - 36z = -33 \end{cases}$

26. $\begin{cases} 2x + y - 3z = 4 \\ 4x + 2z = 10 \\ -2x + 3y - 13z = -8 \end{cases}$

Application

Application

EXAMPLE 6   **Vertical Motion**

The height at time t of an object that is moving in a (vertical) line with constant acceleration a is given by the **position equation**

$$s = \frac{1}{2}at^2 + v_0t + s_0.$$

The height s is measured in feet, the acceleration a is measured in feet per second squared, the time t is measured in seconds, v_0 is the initial velocity (at time $t = 0$), and s_0 is the initial height. Find the values of a, v_0, and s_0 for a projected object given that $s = 164$ feet at 1 second, $s = 180$ feet at 2 seconds, and $s = 164$ feet at 3 seconds.

SOLUTION

By substituting the three values of t and s into the position equation, you obtain three linear equations in a, v_0, and s_0.

When $t = 1$, $s = 164$: $\frac{1}{2}a(1)^2 + v_0(1) + s_0 = 164$

When $t = 2$, $s = 180$: $\frac{1}{2}a(2)^2 + v_0(2) + s_0 = 180$

When $t = 3$, $s = 164$: $\frac{1}{2}a(3)^2 + v_0(3) + s_0 = 164$

By multiplying the first and third equations by 2, this system can be rewritten as

$$\begin{cases} a + 2v_0 + 2s_0 = 328 & \text{Equation 1} \\ 2a + 2v_0 + s_0 = 180 & \text{Equation 2} \\ 9a + 6v_0 + 2s_0 = 328 & \text{Equation 3} \end{cases}$$

and you can apply Gaussian elimination to obtain

$$\begin{cases} a + 2v_0 + 2s_0 = 328 & \text{Equation 1} \\ -2v_0 - 3s_0 = -476. & \text{Equation 2} \\ 2s_0 = 232 & \text{Equation 3} \end{cases}$$

From the third equation, $s_0 = 116$, so back-substitution into Equation 2 yields

$-2v_0 - 3(116) = -476$ Substitute 116 for s_0.

$-2v_0 = -128$ Simplify.

$v_0 = 64.$ Divide each side by -2.

Finally, back-substituting $v_0 = 64$ and $s_0 = 116$ into Equation 1 yields

$a + 2(64) + 2(116) = 328$ Substitute 64 for v_0 and 116 for s_0.

$a = -32.$ Simplify.

So, the position equation for this object is $s = -16t^2 + 64t + 116$.

Exercises Within Reach ®

Solutions in English & Spanish and tutorial videos at CollegePrepAlgebra.com

Vertical Motion In Exercises 27 and 28, find the position equation $s = \frac{1}{2}at^2 + v_0t + s_0$ for an object that has the indicated heights at the specified times.

27. $s = 128$ feet at $t = 1$ second

$s = 80$ feet at $t = 2$ seconds

$s = 0$ feet at $t = 3$ seconds

28. $s = 48$ feet at $t = 1$ second

$s = 64$ feet at $t = 2$ seconds

$s = 48$ feet at $t = 3$ seconds

Concept Summary: *Solving Systems of Linear Equations by Gaussian Elimination*

What

You can solve a system of linear equations in three variables by rewriting the system in **row-echelon form**. This usually involves a chain of **equivalent systems**, each of which is obtained by using one of three basic **row operations**. This process is called **Gaussian elimination**.

How

Each of the following row operations on a system of linear equations produces an equivalent system of linear equations.

1. Interchange two equations.
2. Multiply one of the equations by a nonzero constant.
3. Add a multiple of one of the equations to another equation to replace the latter equation.

Why

When solving an equivalent system, you can conclude that the solution is also the solution of the original system.

Exercises Within Reach ®

Worked-out solutions to odd-numbered exercises at CollegePrepAlgebra.com

Concept Summary Check

29. *Gaussian Elimination* How can the process of Gaussian elimination help you to solve a system of equations? In general, after applying Gaussian elimination in a system of equations, what are the next steps you take to find the solution of the system?

30. *Logic* Describe the three row operations that you can use to produce an equivalent system of equations while applying Gaussian elimination.

31. *Vocabulary* Give an example of a system of three linear equations in three variables that is in row-echelon form.

32. *Writing* Show how to use back-substitution to solve the system of equations you wrote in Exercise 31.

Extra Practice

Finding an Equivalent System **In Exercises 33 and 34, determine whether the two systems of linear equations are equivalent. Justify your answer.**

33.
$$\begin{cases} x + 3y - z = 6 \\ 2x - y + 2z = 1 \\ 3x + 2y - z = 2 \end{cases} \quad \begin{cases} x + 3y - z = 6 \\ -7y + 4z = -11 \\ -7y + 2z = -16 \end{cases}$$

34.
$$\begin{cases} x - 2y + 3z = 9 \\ -x + 3y = -4 \\ 2x - 5y + 5z = 17 \end{cases} \quad \begin{cases} x - 2y + 3z = 9 \\ y + 3z = 5 \\ -y - z = -1 \end{cases}$$

Finding a System in Three Variables **In Exercises 35 and 36, find a system of linear equations in three variables with integer coefficients that has the given point as a solution. (There are many correct answers.)**

35. $(4, -3, 2)$

36. $(5, 7, -10)$

37. *Geometry* The sum of the measures of two angles of a triangle is twice the measure of the third angle. The measure of the second angle is 28° less than the measure of the third angle. Find the measures of the three angles.

38. *Geometry* The measure of one angle of a triangle is two-thirds the measure of a second angle, and the measure of the second angle is 12° greater than the measure of the third angle. Find the measures of the three angles.

39. *Coffee* A coffee manufacturer sells a 10-pound package that consists of three flavors of coffee. Vanilla coffee costs $6 per pound, hazelnut coffee costs $6.50 per pound, and French roast coffee costs $7 per pound. The package contains the same amount of hazelnut coffee as French roast coffee. The cost of the 10-pound package is $66. How many pounds of each type of coffee are in the package?

40. *Hot Dogs* A vendor sells three sizes of hot dogs at prices of $1.50, $2.50, and $3.25. On a day when the vendor had a total revenue of $289.25 from sales of 143 hot dogs, 4 times as many $1.50 hot dogs were sold as $3.25 hot dogs. How many hot dogs were sold at each price?

41. *School Orchestra* The table shows the percents of each section of the North High School orchestra that were chosen to participate in the city orchestra, the county orchestra, and the state orchestra. Thirty members of the city orchestra, 17 members of the county orchestra, and 10 members of the state orchestra are from North High. How many members are in each section of North High's orchestra?

Orchestra	String	Wind	Percussion
City orchestra	40%	30%	50%
County orchestra	20%	25%	25%
State orchestra	10%	15%	25%

42. *Sports* The table shows the percents of each unit of the North High School football team that were chosen for academic honors, as city all-stars, and as county all-stars. Of all the players on the football team, 5 were awarded with academic honors, 13 were named city all-stars, and 4 were named county all-stars. How many members of each unit are there on the football team?

	Defense	Offense	Special teams
Academic honors	0%	10%	20%
City all-stars	10%	20%	50%
County all-stars	10%	0%	20%

Explaining Concepts

43. *Logic* You apply Gaussian elimination to a system of three equations in the variables x, y, and z. From the row-echelon form, the solution $(1, -3, 4)$ is apparent *without* applying back-substitution or any other calculations. Explain why.

44. *Reasoning* A system of three linear equations in three variables has an infinite number of solutions. Is it possible that the graphs of two of the three equations are parallel planes? Explain.

45. *Precision* Two ways that a system of three linear equations in three variables can have no solution are shown on page 399. Describe the graph for a third type of situation that results in no solution.

46. *Think About It* Describe the graphs and numbers of solutions possible for a system of three linear equations in three variables in which at least two of the equations are dependent.

47. *Think About It* Describe the graphs and numbers of solutions possible for a system of three linear equations in three variables when each pair of equations is consistent and *not* dependent.

48. *Writing* Write a system of four linear equations in four unknowns, and use Gaussian elimination with back-substitution to solve it.

Cumulative Review

In Exercises 49−52, identify the terms and coefficients of the algebraic expression.

49. $3x + 2$

50. $4x^2 + 5x - 4$

51. $14t^5 - t + 25$

52. $5s^2 + 3st + 2t^2$

In Exercises 53−56, solve the system of linear equations by the method of elimination.

53. $\begin{cases} 2x + 3y = 17 \\ \quad\quad 4y = 12 \end{cases}$

54. $\begin{cases} x - 2y = 11 \\ 3x + 3y = 6 \end{cases}$

55. $\begin{cases} 3x - 4y = -30 \\ 5x + 4y = 14 \end{cases}$

56. $\begin{cases} 3x + 5y = 1 \\ 4x + 15y = 5 \end{cases}$

Mid-Chapter Quiz: Sections 8.1–8.3

Solutions in English & Spanish and tutorial videos at CollegePrepAlgebra.com

Take this quiz as you would take a quiz in class. After you are done, check your work against the answers in the back of the book.

1. Is $(4, 2)$ a solution of $3x + 4y = 4$ *and* $5x - 3y = 14$? Explain.

2. Is $(2, -1)$ a solution of $2x - 3y = 7$ *and* $3x + 5y = 1$? Explain.

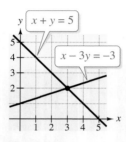

Figure for 3

In Exercises 3–5, use the given graphs to solve the system of linear equations.

3. $\begin{cases} x + y = 5 \\ x - 3y = -3 \end{cases}$

4. $\begin{cases} x + 2y = 6 \\ 3x - 4y = 8 \end{cases}$

5. $\begin{cases} x + 2y = 2 \\ x - 2y = 6 \end{cases}$

In Exercises 6–8, solve the system by graphing.

6. $\begin{cases} x = -3 \\ x + y = 8 \end{cases}$

7. $\begin{cases} y = \frac{3}{2}x - 1 \\ y = -x + 4 \end{cases}$

8. $\begin{cases} 4x + y = 0 \\ -x + y = 5 \end{cases}$

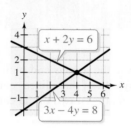

Figure for 4

In Exercises 9–11, solve the system by the method of substitution.

9. $\begin{cases} x - y = 4 \\ y = 2 \end{cases}$

10. $\begin{cases} y = -\frac{2}{3}x + 5 \\ y = 2x - 3 \end{cases}$

11. $\begin{cases} 2x - y = -7 \\ 4x + 3y = 16 \end{cases}$

In Exercises 12–15, use elimination or Gaussian elimination to solve the linear system.

12. $\begin{cases} x + 10y = 18 \\ 5x + 2y = 42 \end{cases}$

13. $\begin{cases} x - 3y = 6 \\ 3x + y = 8 \end{cases}$

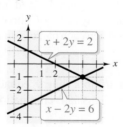

Figure for 5

14. $\begin{cases} a + b + c = 1 \\ 4a + 2b + c = 2 \\ 9a + 3b + c = 4 \end{cases}$

15. $\begin{cases} x + 4z = 17 \\ -3x + 2y - z = -20 \\ x - 5y + 3z = 19 \end{cases}$

In Exercises 16 and 17, write a system of linear equations having the given solution. (There are many correct answers.)

16. $(10, -12)$

17. $(2, -5, 10)$

Applications

18. Twenty gallons of a 30% brine solution is obtained by mixing a 20% solution with a 50% solution. How many gallons of each solution are required?

19. In a triangle, the measure of one angle is 14° less than twice the measure of a second angle. The measure of the third angle is 30° greater than the measure of the second angle. Find the measures of the three angles.

Study Skills in Action

Viewing Math as a Foreign Language

Learning math requires more than just completing homework problems. For instance, learning the material in a chapter may require using approaches similar to those used for learning a foreign language in that you must:

- understand and memorize vocabulary words;

- understand and memorize mathematical rules (as you would memorize grammatical rules); and

- apply rules to mathematical expressions or equations (like creating sentences using correct grammar rules).

You should understand the vocabulary words and rules in a chapter as well as memorize and say them out loud. Strive to speak the mathematical language with fluency, just as a student learning a foreign language must strive to do.

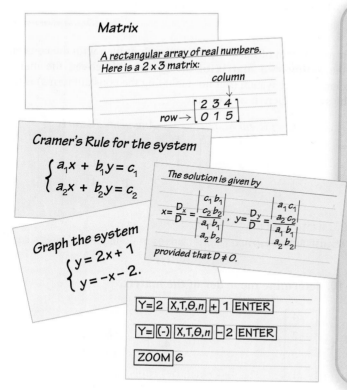

Smart Study Strategy

Make Note Cards

Invest in three different colors of 4 × 6 note cards. Use one color for each of the following: vocabulary words; rules; and calculator keystrokes.

- Write vocabulary words on note cards, one word per card. Write the definition and an example on the other side. If possible, put definitions in your own words.

- Write rules on note cards, one per card. Include an example and an explanation on the other side.

- Write each kind of calculation on a separate note card. Include the keystrokes required to perform the calculation on the other side.

Use the note cards as references while doing your homework. Quiz yourself once a day.

8.4 Matrices and Linear Systems

▶ Form augmented matrices and form linear systems from augmented matrices.

▶ Perform elementary row operations to solve systems of linear equations.

▶ Use matrices and Gaussian elimination to solve systems of linear equations.

Augmented and Coefficient Matrices

A **matrix** is a rectangular array of numbers. The plural of matrix is *matrices*.

EXAMPLE 1   Determining the Order of a Matrix

Determine the order of each matrix.

a. $\begin{bmatrix} 1 & -2 & 4 \\ 0 & 1 & -2 \end{bmatrix}$ **b.** $\begin{bmatrix} 0 & 0 \\ 0 & 0 \end{bmatrix}$ **c.** $\begin{bmatrix} 1 & -3 \\ -2 & 0 \\ 4 & -2 \end{bmatrix}$

SOLUTION

a. This matrix has two rows and three columns, so the order is 2×3.

b. This matrix has two rows and two columns, so the order is 2×2.

c. This matrix has three rows and two columns, so the order is 3×2.

> **Study Tip**
>
> The **order** of a matrix is always given as *row by column*. A matrix with the same number of rows as columns is called a **square matrix**. For instance, the 2×2 matrix in Example 1(b) is square.

A matrix derived from a system of linear equations (each written with the constant term on the right) is the **augmented matrix** of the system. Moreover, the matrix derived from the coefficients of the system (but not including the constant terms) is the **coefficient matrix** of the system. Here is an example.

System	Coefficient Matrix	Augmented Matrix
$\begin{cases} x - 4y + 3z = 5 \\ -x + 3y - z = -3 \\ 2x - 4z = 6 \end{cases}$	$\begin{bmatrix} 1 & -4 & 3 \\ -1 & 3 & -1 \\ 2 & 0 & -4 \end{bmatrix}$	$\begin{bmatrix} 1 & -4 & 3 & \vdots & 5 \\ -1 & 3 & -1 & \vdots & -3 \\ 2 & 0 & -4 & \vdots & 6 \end{bmatrix}$

Exercises Within Reach ®

Solutions in English & Spanish and tutorial videos at CollegePrepAlgebra.com

Determining the Order of a Matrix In Exercises 1−10, determine the order of the matrix.

1. $\begin{bmatrix} 3 & -2 \\ -4 & 0 \\ 2 & -7 \\ -1 & -3 \end{bmatrix}$ **2.** $\begin{bmatrix} 3 & 4 \\ 2 & -1 \\ 8 & 10 \\ -6 & -6 \\ 12 & 50 \end{bmatrix}$ **3.** $\begin{bmatrix} 4 \\ -2 \\ 0 \\ 1 \end{bmatrix}$ **4.** $\begin{bmatrix} 5 & -8 & 32 \\ 7 & 15 & 28 \end{bmatrix}$

5. $\begin{bmatrix} -2 & 5 \\ 0 & -1 \end{bmatrix}$ **6.** $\begin{bmatrix} 4 & 0 & -5 \\ -1 & 8 & 9 \\ 0 & -3 & 4 \end{bmatrix}$ **7.** $[5]$ **8.** $\begin{bmatrix} 6 \\ -13 \\ 22 \end{bmatrix}$

9. $[13 \quad 12 \quad -9 \quad 0]$ **10.** $[1 \quad -1 \quad 2 \quad 3]$

EXAMPLE 2   **Forming Coefficient and Augmented Matrices**

Form the coefficient matrix and the augmented matrix for each system.

a. $\begin{cases} -x + 5y = 2 \\ 7x - 2y = -6 \end{cases}$

b. $\begin{cases} 3x + 2y - z = 1 \\ x + 2z = -3 \\ -2x - y = 4 \end{cases}$

Study Tip

In Example 2(b), note the use of 0 for the missing *y*-variable in the second equation and also for the missing *z*-variable in the third equation.

SOLUTION

	System	*Coefficient Matrix*	*Augmented Matrix*
a.	$\begin{cases} -x + 5y = 2 \\ 7x - 2y = -6 \end{cases}$	$\begin{bmatrix} -1 & 5 \\ 7 & -2 \end{bmatrix}$	$\begin{bmatrix} -1 & 5 & \vdots & 2 \\ 7 & -2 & \vdots & -6 \end{bmatrix}$
b.	$\begin{cases} 3x + 2y - z = 1 \\ x + 2z = -3 \\ -2x - y = 4 \end{cases}$	$\begin{bmatrix} 3 & 2 & -1 \\ 1 & 0 & 2 \\ -2 & -1 & 0 \end{bmatrix}$	$\begin{bmatrix} 3 & 2 & -1 & \vdots & 1 \\ 1 & 0 & 2 & \vdots & -3 \\ -2 & -1 & 0 & \vdots & 4 \end{bmatrix}$

EXAMPLE 3   **Forming Linear Systems from Their Matrices**

Write the system of linear equations represented by each augmented matrix.

a. $\begin{bmatrix} 3 & -5 & \vdots & 4 \\ -1 & 2 & \vdots & 0 \end{bmatrix}$

b. $\begin{bmatrix} 1 & 3 & \vdots & 2 \\ 0 & 1 & \vdots & -3 \end{bmatrix}$

c. $\begin{bmatrix} 2 & 0 & -8 & \vdots & 1 \\ -1 & 1 & 1 & \vdots & 2 \\ 5 & -1 & 7 & \vdots & 3 \end{bmatrix}$

SOLUTION

a. $\begin{cases} 3x - 5y = 4 \\ -x + 2y = 0 \end{cases}$

b. $\begin{cases} x + 3y = 2 \\ y = -3 \end{cases}$

c. $\begin{cases} 2x - 8z = 1 \\ -x + y + z = 2 \\ 5x - y + 7z = 3 \end{cases}$

Exercises Within Reach ®

Solutions in English & Spanish and tutorial videos at CollegePrepAlgebra.com

Forming Coefficient and Augmented Matrices In Exercises 11−14, form (a) the coefficient matrix and (b) the augmented matrix for the system of linear equations.

11. $\begin{cases} 4x - 5y = -2 \\ -x + 8y = 10 \end{cases}$

12. $\begin{cases} 8x + 3y = 25 \\ 3x - 9y = 12 \end{cases}$

13. $\begin{cases} x + y = 0 \\ 5x - 2y - 2z = 12 \\ 2x + 4y + z = 5 \end{cases}$

14. $\begin{cases} 9x - 3y + z = 13 \\ 12x - 8z = 5 \\ 3x + 4y - z = 6 \end{cases}$

Forming a Linear System In Exercises 15−18, write the system of linear equations represented by the augmented matrix. (Use variables *x*, *y*, and *z*.)

15. $\begin{bmatrix} 4 & 3 & \vdots & 8 \\ 1 & -2 & \vdots & 3 \end{bmatrix}$

16. $\begin{bmatrix} 9 & -4 & \vdots & 0 \\ 6 & 1 & \vdots & -4 \end{bmatrix}$

17. $\begin{bmatrix} 1 & 0 & 2 & \vdots & -10 \\ 0 & 3 & -1 & \vdots & 5 \\ 4 & 2 & 0 & \vdots & 3 \end{bmatrix}$

18. $\begin{bmatrix} 4 & -1 & 3 & \vdots & 5 \\ 2 & 0 & -2 & \vdots & -1 \\ -1 & 6 & 0 & \vdots & 3 \end{bmatrix}$

Elementary Row Operations

Study Tip

Although elementary row operations are simple to perform, they involve a lot of arithmetic. So that you can check your work, you should get in the habit of noting the elementary row operations performed in each step. People use different schemes to do this. The scheme that is used in this text is to write an abbreviated version of the row operation at the left of the row that has been changed, as shown in Example 4.

Elementary Row Operations

Any of the following **elementary row operations** performed on an augmented matrix will produce a matrix that is row-equivalent to the original matrix. Two matrices are **row-equivalent** if one can be obtained from the other by a sequence of elementary row operations.

1. Interchange two rows.
2. Multiply a row by a nonzero constant.
3. Add a multiple of a row to another row.

EXAMPLE 4 Performing Elementary Row Operations

a. Interchange the first and second rows.

Original Matrix

$$\begin{bmatrix} 0 & 1 & 3 & 4 \\ -1 & 2 & 0 & 3 \\ 2 & -3 & 4 & 1 \end{bmatrix}$$

New Row-Equivalent Matrix

$$\begin{matrix} R_2 \\ R_1 \end{matrix} \begin{bmatrix} -1 & 2 & 0 & 3 \\ 0 & 1 & 3 & 4 \\ 2 & -3 & 4 & 1 \end{bmatrix}$$

b. Multiply the first row by $\frac{1}{2}$.

Original Matrix

$$\begin{bmatrix} 2 & -4 & 6 & -2 \\ 1 & 3 & -3 & 0 \\ 5 & -2 & 1 & 2 \end{bmatrix}$$

New Row-Equivalent Matrix

$$\frac{1}{2}R_1 \rightarrow \begin{bmatrix} 1 & -2 & 3 & -1 \\ 1 & 3 & -3 & 0 \\ 5 & -2 & 1 & 2 \end{bmatrix}$$

c. Add -2 times the first row to the third row.

Original Matrix

$$\begin{bmatrix} 1 & 2 & -4 & 3 \\ 0 & 3 & -2 & -1 \\ 2 & 1 & 5 & -2 \end{bmatrix}$$

New Row-Equivalent Matrix

$$\begin{bmatrix} 1 & 2 & -4 & 3 \\ 0 & 3 & -2 & -1 \\ 0 & -3 & 13 & -8 \end{bmatrix}$$
$$-2R_1 + R_3 \rightarrow$$

d. Add 6 times the first row to the second row.

Original Matrix

$$\begin{bmatrix} 1 & 2 & 2 & -4 \\ -6 & -11 & 3 & 18 \\ 0 & 0 & 4 & 7 \end{bmatrix}$$

New Row-Equivalent Matrix

$$6R_1 + R_2 \rightarrow \begin{bmatrix} 1 & 2 & 2 & -4 \\ 0 & 1 & 15 & -6 \\ 0 & 0 & 4 & 7 \end{bmatrix}$$

Exercises Within Reach ®

Solutions in English & Spanish and tutorial videos at CollegePrepAlgebra.com

Performing an Elementary Row Operation In Exercises 19−24, fill in the entries of the row-equivalent matrix formed by performing the indicated elementary row operation.

19. $\begin{bmatrix} 1 & 1 & -4 & 2 \\ 0 & 0 & 8 & 3 \\ 0 & 4 & 5 & 5 \end{bmatrix} \begin{matrix} R_3 \\ R_2 \end{matrix} \begin{bmatrix} \square & \square & \square & \square \\ \square & \square & \square & \square \\ \square & \square & \square & \square \end{bmatrix}$

20. $\begin{bmatrix} 0 & 0 & -5 & 2 \\ 0 & -7 & -3 & 3 \\ 1 & 4 & 5 & 4 \end{bmatrix} \begin{matrix} R_3 \\ R_1 \end{matrix} \begin{bmatrix} \square & \square & \square & \square \\ \square & \square & \square & \square \\ \square & \square & \square & \square \end{bmatrix}$

21. $\begin{bmatrix} 9 & -18 & 27 \\ 3 & 4 & 5 \end{bmatrix} \frac{1}{9}R_1 \rightarrow \begin{bmatrix} \square & \square & \square \\ \square & \square & \square \end{bmatrix}$

22. $\begin{bmatrix} 1 & 21 & 7 \\ 0 & -7 & 14 \end{bmatrix} -\frac{1}{7}R_2 \rightarrow \begin{bmatrix} \square & \square & \square \\ \square & \square & \square \end{bmatrix}$

23. $\begin{bmatrix} 1 & 4 & 3 \\ 2 & 8 & 6 \end{bmatrix} -2R_1 + R_2 \rightarrow \begin{bmatrix} \square & \square & \square \\ \square & \square & \square \end{bmatrix}$

24. $\begin{bmatrix} 1 & 4 & 5 \\ 4 & -7 & 3 \end{bmatrix} -4R_1 + R_2 \rightarrow \begin{bmatrix} \square & \square & \square \\ \square & \square & \square \end{bmatrix}$

EXAMPLE 5   **Solving a System of Linear Equations**

Linear System

$$\begin{cases} x - 2y + 2z = 9 \\ -x + 3y = -4 \\ 2x - 5y + z = 10 \end{cases}$$

Associated Augmented Matrix

$$\begin{bmatrix} 1 & -2 & 2 & \vdots & 9 \\ -1 & 3 & 0 & \vdots & -4 \\ 2 & -5 & 1 & \vdots & 10 \end{bmatrix}$$

Add the first equation to the second equation.

Add the first row to the second row.

$$\begin{cases} x - 2y + 2z = 9 \\ y + 2z = 5 \\ 2x - 5y + z = 10 \end{cases}$$

$$R_1 + R_2 \rightarrow \begin{bmatrix} 1 & -2 & 2 & \vdots & 9 \\ 0 & 1 & 2 & \vdots & 5 \\ 2 & -5 & 1 & \vdots & 10 \end{bmatrix}$$

Add -2 times the first equation to the third equation.

Add -2 times the first row to the third row.

$$\begin{cases} x - 2y + 2z = 9 \\ y + 2z = 5 \\ -y - 3z = -8 \end{cases}$$

$$-2R_1 + R_3 \rightarrow \begin{bmatrix} 1 & -2 & 2 & \vdots & 9 \\ 0 & 1 & 2 & \vdots & 5 \\ 0 & -1 & -3 & \vdots & -8 \end{bmatrix}$$

Add the second equation to the third equation.

Add the second row to the third row.

$$\begin{cases} x - 2y + 2z = 9 \\ y + 2z = 5 \\ -z = -3 \end{cases}$$

$$R_2 + R_3 \rightarrow \begin{bmatrix} 1 & -2 & 2 & \vdots & 9 \\ 0 & 1 & 2 & \vdots & 5 \\ 0 & 0 & -1 & \vdots & -3 \end{bmatrix}$$

Multiply the third equation by -1.

Multiply the third row by -1.

$$\begin{cases} x - 2y + 2z = 9 \\ y + 2z = 5 \\ z = 3 \end{cases}$$

$$-R_3 \rightarrow \begin{bmatrix} 1 & -2 & 2 & \vdots & 9 \\ 0 & 1 & 2 & \vdots & 5 \\ 0 & 0 & 1 & \vdots & 3 \end{bmatrix}$$

At this point, you can use back-substitution to find that the solution is $x = 1$, $y = -1$, and $z = 3$. The solution can be written as the ordered triple $(1, -1, 3)$.

Study Tip

The last matrix in Example 5 is in row-echelon form. The term *echelon* refers to the stair-step pattern formed by the nonzero elements of the matrix.

Definition of Row-Echelon Form of a Matrix

A matrix in **row-echelon form** has the following properties.

1. All rows consisting entirely of zeros occur at the bottom of the matrix.
2. For each row that does not consist entirely of zeros, the first nonzero entry is 1 (called a **leading 1**).
3. For two successive (nonzero) rows, the leading 1 in the higher row is farther to the left than the leading 1 in the lower row.

Exercises Within Reach ®

Solutions in English & Spanish and tutorial videos at CollegePrepAlgebra.com

Solving a System of Linear Equations **In Exercises 25 and 26, use matrices to solve the system of linear equations.**

25. $\begin{cases} x - 2y - z = 6 \\ y + 4z = 5 \\ 4x + 2y + 3z = 8 \end{cases}$

26. $\begin{cases} x - 3z = -2 \\ 3x + y - 2z = 5 \\ 2x + 2y + z = 4 \end{cases}$

Solving a System of Equations

> ### Gaussian Elimination with Back-Substitution
>
> To use matrices and Gaussian elimination to solve a system of linear equations, use the following steps.
>
> 1. Write the augmented matrix of the system of linear equations.
> 2. Use elementary row operations to rewrite the augmented matrix in row-echelon form.
> 3. Write the system of linear equations corresponding to the matrix in row-echelon form, and use back-substitution to find the solution.

EXAMPLE 6 Gaussian Elimination with Back-Substitution

Solve the system of linear equations.

$$\begin{cases} 2x - 3y = -2 \\ x + 2y = 13 \end{cases}$$

SOLUTION

$$\begin{bmatrix} 2 & -3 & \vdots & -2 \\ 1 & 2 & \vdots & 13 \end{bmatrix}$$ Augmented matrix for system of linear equations

$$\begin{matrix} R_2 \\ R_1 \end{matrix} \begin{bmatrix} 1 & 2 & \vdots & 13 \\ 2 & -3 & \vdots & -2 \end{bmatrix}$$ First column has leading 1 in upper left corner.

$$-2R_1 + R_2 \rightarrow \begin{bmatrix} 1 & 2 & \vdots & 13 \\ 0 & -7 & \vdots & -28 \end{bmatrix}$$ First column has a zero under its leading 1.

$$-\tfrac{1}{7}R_2 \rightarrow \begin{bmatrix} 1 & 2 & \vdots & 13 \\ 0 & 1 & \vdots & 4 \end{bmatrix}$$ Second column has leading 1 in second row.

The system of linear equations that corresponds to the (row-echelon) matrix is

$$\begin{cases} x + 2y = 13 \\ y = 4 \end{cases}.$$

Using back-substitution, you can find that the solution of the system is $x = 5$ and $y = 4$, which can be written as the ordered pair $(5, 4)$. Check this solution in the original system, as follows.

> **Check**
>
> Equation 1: $2(5) - 3(4) = -2$ ✔
>
> Equation 2: $5 + 2(4) = 13$ ✔

Exercises Within Reach ®

Solutions in English & Spanish and tutorial videos at CollegePrepAlgebra.com

Solving a System of Linear Equations In Exercises 27−30, use matrices to solve the system of linear equations.

27. $\begin{cases} 6x - 4y = 2 \\ 5x + 2y = 7 \end{cases}$ 28. $\begin{cases} 2x + 6y = 16 \\ 2x + 3y = 7 \end{cases}$ 29. $\begin{cases} 12x + 10y = -14 \\ 4x - 3y = -11 \end{cases}$ 30. $\begin{cases} -x - 5y = -10 \\ 2x - 3y = 7 \end{cases}$

EXAMPLE 7   **Gaussian Elimination with Back-Substitution**

Solve the system of linear equations.

$$\begin{cases} 3x + 3y & = & 9 \\ 2x & - 3z = & 10 \\ & 6y + 4z = -12 \end{cases}$$

SOLUTION

$$\begin{bmatrix} 3 & 3 & 0 & \vdots & 9 \\ 2 & 0 & -3 & \vdots & 10 \\ 0 & 6 & 4 & \vdots & -12 \end{bmatrix}$$

Augmented matrix for system of linear equations

$$\tfrac{1}{3}R_1 \rightarrow \begin{bmatrix} 1 & 1 & 0 & \vdots & 3 \\ 2 & 0 & -3 & \vdots & 10 \\ 0 & 6 & 4 & \vdots & -12 \end{bmatrix}$$

First column has leading 1 in upper left corner.

$$-2R_1 + R_2 \rightarrow \begin{bmatrix} 1 & 1 & 0 & \vdots & 3 \\ 0 & -2 & -3 & \vdots & 4 \\ 0 & 6 & 4 & \vdots & -12 \end{bmatrix}$$

First column has zeros under its leading 1.

$$-\tfrac{1}{2}R_2 \rightarrow \begin{bmatrix} 1 & 1 & 0 & \vdots & 3 \\ 0 & 1 & \tfrac{3}{2} & \vdots & -2 \\ 0 & 6 & 4 & \vdots & -12 \end{bmatrix}$$

Second column has leading 1 in second row.

$$-6R_2 + R_3 \rightarrow \begin{bmatrix} 1 & 1 & 0 & \vdots & 3 \\ 0 & 1 & \tfrac{3}{2} & \vdots & -2 \\ 0 & 0 & -5 & \vdots & 0 \end{bmatrix}$$

Second column has zero under its leading 1.

$$-\tfrac{1}{5}R_3 \rightarrow \begin{bmatrix} 1 & 1 & 0 & \vdots & 3 \\ 0 & 1 & \tfrac{3}{2} & \vdots & -2 \\ 0 & 0 & 1 & \vdots & 0 \end{bmatrix}$$

Third column has leading 1 in third row.

The system of linear equations that corresponds to the (row-echelon) matrix is

$$\begin{cases} x + y & = & 3 \\ y + \tfrac{3}{2}z & = -2. \\ z & = & 0 \end{cases}$$

Using back-substitution, you can find that the solution is

$$x = 5 \text{ and } y = -2, \text{ and } z = 0$$

which can be written as the ordered triple $(5, -2, 0)$.

Exercises Within Reach ®

Solutions in English & Spanish and tutorial videos at CollegePrepAlgebra.com

Solving a System of Linear Equations In Exercises 31–34, use matrices to solve the system of linear equations.

31. $\begin{cases} 2x + 4y & = & 10 \\ 2x + 2y + 3z = & 3 \\ -3x + y + 2z = -3 \end{cases}$

32. $\begin{cases} 2x - y + 3z = 24 \\ 2y - z = 14 \\ 7x - 5y = 6 \end{cases}$

33. $\begin{cases} -2x - 2y - 15z = & 0 \\ x + 2y + 2z = 18 \\ 3x + 3y + 22z = & 2 \end{cases}$

34. $\begin{cases} 2x + 4y + 5z = 5 \\ x + 3y + 3z = 2 \\ 2x + 4y + 4z = 2 \end{cases}$

EXAMPLE 8 **A System with No Solution**

Linear System

$$\begin{cases} 6x - 10y = -4 \\ 9x - 15y = 5 \end{cases}$$

Associated Augmented Matrix

$$\begin{bmatrix} 6 & -10 & \vdots & -4 \\ 9 & -15 & \vdots & 5 \end{bmatrix}$$

$$\frac{1}{6}R_1 \rightarrow \begin{bmatrix} 1 & -\frac{5}{3} & \vdots & -\frac{2}{3} \\ 9 & -15 & \vdots & 5 \end{bmatrix}$$

$$-9R_1 + R_2 \rightarrow \begin{bmatrix} 1 & -\frac{5}{3} & \vdots & -\frac{2}{3} \\ 0 & 0 & \vdots & 11 \end{bmatrix}$$

The "equation" that corresponds to the second row of this matrix is $0 = 11$. Because this is a false statement, the system of equations has no solution.

EXAMPLE 9 **A System with Infinitely Many Solutions**

Linear System

$$\begin{cases} 12x - 6y = -3 \\ -8x + 4y = 2 \end{cases}$$

Associated Augmented Matrix

$$\begin{bmatrix} 12 & -6 & \vdots & -3 \\ -8 & 4 & \vdots & 2 \end{bmatrix}$$

$$\frac{1}{12}R_1 \rightarrow \begin{bmatrix} 1 & -\frac{1}{2} & \vdots & -\frac{1}{4} \\ -8 & 4 & \vdots & 2 \end{bmatrix}$$

$$8R_1 + R_2 \rightarrow \begin{bmatrix} 1 & -\frac{1}{2} & \vdots & -\frac{1}{4} \\ 0 & 0 & \vdots & 0 \end{bmatrix}$$

Because the second row of the matrix is all zeros, the system of equations has an infinite number of solutions, represented by all points (x, y) on the line

$$x - \frac{1}{2}y = -\frac{1}{4}.$$

Because this line can be written as

$$x = \frac{1}{2}y - \frac{1}{4}$$

you can write the solution set as

$\left(\frac{1}{2}a - \frac{1}{4}, a \right)$, where a is any real number.

Exercises Within Reach ®

Solutions in English & Spanish and tutorial videos at CollegePrepAlgebra.com

Solving a System of Linear Equations In Exercises 35−38, use matrices to solve the system of linear equations.

35. $\begin{cases} x + y - 5z = 3 \\ x - 2z = 1 \\ 2x - y - z = 0 \end{cases}$

36. $\begin{cases} 2x + 3z = 3 \\ 4x - 3y + 7z = 5 \\ 8x - 9y + 15z = 9 \end{cases}$

37. $\begin{cases} 2x + 4z = 1 \\ x + y + 3z = 0 \\ x + 3y + 5z = 0 \end{cases}$

38. $\begin{cases} 3x + y - 2z = 2 \\ 6x + 2y - 4z = 1 \\ -3x - y + 2z = 1 \end{cases}$

Application   EXAMPLE 10   **Investment Portfolio**

You have $219,000 to invest in municipal bonds, blue-chip stocks, and growth stocks. The municipal bonds pay 6% annually. Over the investment period, you expect blue-chip stocks to return 10% annually and growth stocks to return 15% annually. You want a combined annual return of 8%, and you also want to have only one-fourth of the portfolio invested in stocks. How much should be allocated to each type of investment?

SOLUTION

Let M, B, and G represent the amounts invested in municipal bonds, blue-chip stocks, and growth stocks, respectively. This situation is represented by the following system.

$$\begin{cases} M + B + G = 219{,}000 & \text{Equation 1: Total investment is \$219,000.} \\ 0.06M + 0.10B + 0.15G = 17{,}520 & \text{Equation 2: Combined annual return is 8\%.} \\ B + G = 54{,}750 & \text{Equation 3: } \tfrac{1}{4} \text{ of investment is in stocks.} \end{cases}$$

$$\begin{bmatrix} 1 & 1 & 1 & \vdots & 219{,}000 \\ 0.06 & 0.10 & 0.15 & \vdots & 17{,}520 \\ 0 & 1 & 1 & \vdots & 54{,}750 \end{bmatrix}$$

Augmented matrix for system of linear equations

$$-0.06R_1 + R_2 \rightarrow \begin{bmatrix} 1 & 1 & 1 & \vdots & 219{,}000 \\ 0 & 0.04 & 0.09 & \vdots & 4{,}380 \\ 0 & 1 & 1 & \vdots & 54{,}750 \end{bmatrix}$$

First column has zeros under its leading 1.

$$25R_2 \rightarrow \begin{bmatrix} 1 & 1 & 1 & \vdots & 219{,}000 \\ 0 & 1 & 2.25 & \vdots & 109{,}500 \\ 0 & 1 & 1 & \vdots & 54{,}750 \end{bmatrix}$$

Second column has leading 1 in second row.

$$-R_2 + R_3 \rightarrow \begin{bmatrix} 1 & 1 & 1 & \vdots & 219{,}000 \\ 0 & 1 & 2.25 & \vdots & 109{,}500 \\ 0 & 0 & -1.25 & \vdots & -54{,}750 \end{bmatrix}$$

Second column has zero under its leading 1.

$$-0.8R_3 \rightarrow \begin{bmatrix} 1 & 1 & 1 & \vdots & 219{,}000 \\ 0 & 1 & 2.25 & \vdots & 109{,}500 \\ 0 & 0 & 1 & \vdots & 43{,}800 \end{bmatrix}$$

Third column has leading 1 in third row and matrix is in row-echelon form.

From the row-echelon form, you can see that $G = 43{,}800$. By back-substituting G into the revised second equation, you can determine the value of B.

$$B + 2.25(43{,}800) = 109{,}500 \implies B = 10{,}950$$

By back-substituting B and G into Equation 1, you can solve for M.

$$M + 10{,}950 + 43{,}800 = 219{,}000 \implies M = 164{,}250$$

So, you should invest $164,250 in municipal bonds, $10,950 in blue-chip stocks, and $43,800 in growth or speculative stocks.

Exercises Within Reach ®

Solutions in English & Spanish and tutorial videos at CollegePrepAlgebra.com

39. *Investment* A corporation borrows $1,500,000 to expand its line of clothing. Some of the money is borrowed at 8%, some at 9%, and the remainder at 12%. The annual interest payment to the lenders is $133,000. The amount borrowed at 8% is four times the amount borrowed at 12%. How much is borrowed at each rate?

40. *Nut Mixture* A grocer wants to mix three kinds of nuts to obtain 50 pounds of a mixture priced at $4.10 per pound. Peanuts cost $3.00 per pound, pecans cost $4.00 per pound, and cashews cost $6.00 per pound. Three-quarters of the mixture is composed of peanuts and pecans. How many pounds of each variety should the grocer use?

Concept Summary: *Using Matrices to Solve Systems of Linear Equations*

What

You can use **matrices** to solve a system of linear equations.

EXAMPLE

Use matrices to solve the system of linear equations.

$$\begin{cases} x - 2y + 3z = 9 \\ -x + 3y = -4 \\ 2x - 5y + 5z = 17 \end{cases}$$

How

Form an **augmented matrix** by using the coefficients and constants of the system.

Augmented Matrix

$$\begin{bmatrix} 1 & -2 & 3 & \vdots & 9 \\ -1 & 3 & 0 & \vdots & -4 \\ 2 & -5 & 5 & \vdots & 17 \end{bmatrix}$$

Then use Gaussian elimination and **elementary row operations** to write the matrix in **row-echelon form**.

Once the matrix is written in row-echelon form, use the matrix to write a new system of equations and then use back-substitution to find the solution.

Why

Using matrices makes solving a system less complex because you do not need to keep writing the variables.

Exercises Within Reach ®

Worked-out solutions to odd-numbered exercises at CollegePrepAlgebra.com

Concept Summary Check

41. *Vocabulary* A matrix contains exactly four entries. What are the possible orders of the matrix? State the numbers of rows and columns in each possible order.

42. *Logic* For a given system of equations, which has more entries, the coefficient matrix or the augmented matrix? Explain.

43. *Vocabulary* What is the primary difference between performing row operations on a system of equations and performing elementary row operations?

44. *Writing* After using matrices to perform Gaussian elimination, what steps are generally needed to find the solution of the original system of equations?

Extra Practice

Solving a System of Linear Equations In Exercises 45–48, use matrices to solve the system of linear equations.

45. $\begin{cases} 4x + 3y = 10 \\ 2x - y = 10 \\ -2x + z = -9 \end{cases}$

46. $\begin{cases} 4x - y + z = 4 \\ -6x + 3y - 2z = -5 \\ 2x + 5y - z = 7 \end{cases}$

47. $\begin{cases} 2x + y - 2z = 4 \\ 3x - 2y + 4z = 6 \\ -4x + y + 6z = 12 \end{cases}$

48. $\begin{cases} 3x + 3y + z = 4 \\ 2x + 6y + z = 5 \\ -x - 3y + 2z = -5 \end{cases}$

49. *Ticket Sales* A theater owner wants to sell 1500 total tickets at his three theaters for a total revenue of $10,050. Tickets cost $1.50 at theater A, $7.50 at theater B, and $8.50 at theater C. Theaters B and C each have twice as many seats as theater A. How many tickets must be sold at each theater to reach the owner's goal?

50. *Investment* An inheritance of $25,000 is divided among three investments yielding a total of $1890 in simple interest per year. The interest rates for the three investments are 5%, 7%, and 10%. The 5% and 7% investments are $2000 and $3000 less than the 10% investment, respectively. Find the amount placed in each investment.

51. *Number Problem* The sum of three positive numbers is 33. The second number is 3 greater than the first, and the third is four times the first. Find the three numbers.

52. *Number Problem* The sum of three positive numbers is 24. The second number is 4 greater than the first, and the third is three times the first. Find the three numbers.

53. *Production* A company produces computer chips, resistors, and transistors. Each computer chip requires 2 units of copper, 2 units of zinc, and 1 unit of glass. Each resistor requires 1 unit of copper, 3 units of zinc, and 2 units of glass. Each transistor requires 3 units of copper, 2 units of zinc, and 2 units of glass. There are 70 units of copper, 80 units of zinc, and 55 units of glass available for use. Find the numbers of computer chips, resistors, and transistors the company can produce.

54. *Production* A gourmet baked goods company specializes in chocolate muffins, chocolate cookies, and chocolate brownies. Each muffin requires 2 units of chocolate, 3 units of flour, and 2 units of sugar. Each cookie requires 1 unit of chocolate, 1 unit of flour, and 1 unit of sugar. Each brownie requires 2 units of chocolate, 1 unit of flour, and 1.5 units of sugar. There are 550 units of chocolate, 525 units of flour, and 500 units of sugar available for use. Find the numbers of chocolate muffins, chocolate cookies, and chocolate brownies the company can produce.

Explaining Concepts

55. *Reasoning* The entries in a matrix consist of the whole numbers from 1 to 15. The matrix has more than one row and there are more columns than rows. What is the order of the matrix? Explain.

56. *Vocabulary* Give an example of a matrix in *row-echelon form*. (There are many correct answers.)

57. *Writing* Describe the row-echelon form of an augmented matrix that corresponds to a system of linear equations that is inconsistent.

58. *Writing* Describe the row-echelon form of an augmented matrix that corresponds to a system of linear equations that has an infinite number of solutions.

59. *Logic* An augmented matrix in row-echelon form represents a system of three variables in three equations that has exactly one solution. The matrix has six nonzero entries, and three of them are in the last column. Discuss the possible entries in the first three columns of this matrix.

60. *Precision* An augmented matrix in row-echelon form represents a system of three variables in three equations with exactly one solution. What is the smallest number of nonzero entries that this matrix can have? Explain.

Cumulative Review

In Exercises 61−64, evaluate the expression.

61. $6(-7)$

62. $45 \div (-5)$

63. $5(4) - 3(-2)$

64. $\dfrac{(-45) - (-20)}{-5}$

In Exercises 65 and 66, solve the system of linear equations.

65. $\begin{cases} x = 4 \\ 3y + 2z = -4 \\ x + y + z = 3 \end{cases}$

66. $\begin{cases} x - 2y - 3z = 4 \\ 2x + 2y + z = -4 \\ -2x + z = 0 \end{cases}$

8.5 Determinants and Linear Systems

▶ Find determinants of 2 × 2 matrices and 3 × 3 matrices.

▶ Use determinants and Cramer's Rule to solve systems of linear equations.

▶ Use determinants to find areas of regions, to test for collinear points, and to find equations of lines.

The Determinant of a Matrix

Study Tip

Note that det(A) and $|A|$ are used interchangeably to represent the determinant of A. Although vertical bars are also used to denote the absolute value of a real number, the context will show which use is intended.

> ### Definition of the Determinant of a 2 × 2 Matrix
>
> $$\det(A) = |A| = \begin{vmatrix} a_1 & b_1 \\ a_2 & b_2 \end{vmatrix} = a_1 b_2 - a_2 b_1$$

A convenient method for remembering the formula for the determinant of a 2 × 2 matrix is shown in the diagram below.

$$\det(A) = |A| = \begin{vmatrix} a_1 & b_1 \\ a_2 & b_2 \end{vmatrix} = a_1 b_2 - a_2 b_1$$

Note that the determinant is given by the difference of the products of the two diagonals of the matrix.

EXAMPLE 1 **Finding the Determinant of a 2 × 2 Matrix**

Find the determinant of each matrix.

a. $A = \begin{bmatrix} 2 & -3 \\ 1 & 4 \end{bmatrix}$ **b.** $B = \begin{bmatrix} -1 & 2 \\ 2 & -4 \end{bmatrix}$ **c.** $C = \begin{bmatrix} 1 & 3 \\ 2 & 5 \end{bmatrix}$

SOLUTION

a. $\det(A) = \begin{vmatrix} 2 & -3 \\ 1 & 4 \end{vmatrix} = 2(4) - 1(-3) = 8 + 3 = 11$

b. $\det(B) = \begin{vmatrix} -1 & 2 \\ 2 & -4 \end{vmatrix} = (-1)(-4) - 2(2) = 4 - 4 = 0$

c. $\det(C) = \begin{vmatrix} 1 & 3 \\ 2 & 5 \end{vmatrix} = 1(5) - 2(3) = 5 - 6 = -1$

Exercises Within Reach ®

Solutions in English & Spanish and tutorial videos at CollegePrepAlgebra.com

Finding the Determinant In Exercises 1−12, find the determinant of the matrix.

1. $\begin{bmatrix} 2 & 1 \\ 3 & 4 \end{bmatrix}$ **2.** $\begin{bmatrix} -3 & 1 \\ 5 & 2 \end{bmatrix}$ **3.** $\begin{bmatrix} 5 & 2 \\ -6 & 3 \end{bmatrix}$ **4.** $\begin{bmatrix} 2 & -2 \\ 4 & 3 \end{bmatrix}$

5. $\begin{bmatrix} -4 & 0 \\ 9 & 0 \end{bmatrix}$ **6.** $\begin{bmatrix} 4 & -3 \\ 0 & 0 \end{bmatrix}$ **7.** $\begin{bmatrix} 3 & -3 \\ -6 & 6 \end{bmatrix}$ **8.** $\begin{bmatrix} -2 & 3 \\ 6 & -9 \end{bmatrix}$

9. $\begin{bmatrix} -7 & 6 \\ \frac{1}{2} & 3 \end{bmatrix}$ **10.** $\begin{bmatrix} \frac{2}{3} & \frac{5}{6} \\ 14 & -2 \end{bmatrix}$ **11.** $\begin{bmatrix} 0.4 & 0.7 \\ 0.7 & 0.4 \end{bmatrix}$ **12.** $\begin{bmatrix} -1.2 & 4.5 \\ 0.4 & -0.9 \end{bmatrix}$

Study Tip

The *signs* of the terms used in expanding by minors follow the alternating pattern shown below.

$$\begin{bmatrix} + & - & + \\ - & + & - \\ + & - & + \end{bmatrix}$$

Expanding by Minors

$$\det(A) = \begin{vmatrix} a_1 & b_1 & c_1 \\ a_2 & b_2 & c_2 \\ a_3 & b_3 & c_3 \end{vmatrix}$$

$$= a_1(\text{minor of } a_1) - b_1(\text{minor of } b_1) + c_1(\text{minor of } c_1)$$

$$= a_1 \begin{vmatrix} b_2 & c_2 \\ b_3 & c_3 \end{vmatrix} - b_1 \begin{vmatrix} a_2 & c_2 \\ a_3 & c_3 \end{vmatrix} + c_1 \begin{vmatrix} a_2 & b_2 \\ a_3 & b_3 \end{vmatrix}$$

This pattern is called **expanding by minors** along the first row. A similar pattern can be used to expand by minors along any row or column.

EXAMPLE 2   **Finding the Determinant of a 3 × 3 Matrix**

Find the determinant of **a.** $A = \begin{bmatrix} -1 & 1 & 2 \\ 0 & 2 & 3 \\ 3 & 4 & 2 \end{bmatrix}$ and **b.** $B = \begin{bmatrix} 1 & 2 & 1 \\ 3 & 0 & 2 \\ 4 & 0 & -1 \end{bmatrix}$.

SOLUTION

a. By expanding by minors along the *first column*, you obtain

$$\det(A) = \begin{vmatrix} -1 & 1 & 2 \\ 0 & 2 & 3 \\ 3 & 4 & 2 \end{vmatrix} = (-1)\begin{vmatrix} 2 & 3 \\ 4 & 2 \end{vmatrix} - (0)\begin{vmatrix} 1 & 2 \\ 4 & 2 \end{vmatrix} + (3)\begin{vmatrix} 1 & 2 \\ 2 & 3 \end{vmatrix}$$

$$= (-1)(4 - 12) - (0)(2 - 8) + (3)(3 - 4)$$

$$= 8 - 0 - 3$$

$$= 5.$$

b. By expanding by minors along the *second column*, you obtain

$$\det(B) = \begin{vmatrix} 1 & 2 & 1 \\ 3 & 0 & 2 \\ 4 & 0 & -1 \end{vmatrix} = -(2)\begin{vmatrix} 3 & 2 \\ 4 & -1 \end{vmatrix} + (0)\begin{vmatrix} 1 & 1 \\ 4 & -1 \end{vmatrix} - (0)\begin{vmatrix} 1 & 1 \\ 3 & 2 \end{vmatrix}$$

$$= -(2)(-3 - 8) + 0 - 0$$

$$= 22.$$

Exercises Within Reach ®   Solutions in English & Spanish and tutorial videos at CollegePrepAlgebra.com

Finding the Determinant In Exercises 13−20, **find the determinant of the matrix.**
Expand by minors along the row or column that appears to make the computation easiest.

13. $\begin{bmatrix} 2 & 3 & -1 \\ 6 & 0 & 0 \\ 4 & 1 & 1 \end{bmatrix}$

14. $\begin{bmatrix} 10 & 2 & -4 \\ 8 & 0 & -2 \\ 4 & 0 & 2 \end{bmatrix}$

15. $\begin{bmatrix} 1 & 1 & 2 \\ 3 & 1 & 0 \\ -2 & 0 & 3 \end{bmatrix}$

16. $\begin{bmatrix} 2 & 1 & 3 \\ 1 & 4 & 4 \\ 1 & 0 & 2 \end{bmatrix}$

17. $\begin{bmatrix} 2 & 4 & 6 \\ 0 & 3 & 1 \\ 0 & 0 & -5 \end{bmatrix}$

18. $\begin{bmatrix} 2 & 3 & 1 \\ 0 & 5 & -2 \\ 0 & 0 & -2 \end{bmatrix}$

19. $\begin{bmatrix} -2 & 2 & 3 \\ 1 & -1 & 0 \\ 0 & 1 & 4 \end{bmatrix}$

20. $\begin{bmatrix} -2 & 3 & 0 \\ 3 & 1 & -4 \\ 0 & 4 & 2 \end{bmatrix}$

Cramer's Rule

Cramer's Rule

1. For the system of linear equations
$$\begin{cases} a_1x + b_1y = c_1 \\ a_2x + b_2y = c_2 \end{cases}$$
the solution is given by $x = \dfrac{D_x}{D} = \dfrac{\begin{vmatrix} c_1 & b_1 \\ c_2 & b_2 \end{vmatrix}}{\begin{vmatrix} a_1 & b_1 \\ a_2 & b_2 \end{vmatrix}}$, $\quad y = \dfrac{D_y}{D} = \dfrac{\begin{vmatrix} a_1 & c_1 \\ a_2 & c_2 \end{vmatrix}}{\begin{vmatrix} a_1 & b_1 \\ a_2 & b_2 \end{vmatrix}}$, $\quad D \neq 0.$

2. For the system of linear equations
$$\begin{cases} a_1x + b_1y + c_1z = d_1 \\ a_2x + b_2y + c_2z = d_2 \\ a_3x + b_3y + c_3z = d_3 \end{cases}$$
the solution is given by

$$x = \frac{D_x}{D} = \frac{\begin{vmatrix} d_1 & b_1 & c_1 \\ d_2 & b_2 & c_2 \\ d_3 & b_3 & c_3 \end{vmatrix}}{\begin{vmatrix} a_1 & b_1 & c_1 \\ a_2 & b_2 & c_2 \\ a_3 & b_3 & c_3 \end{vmatrix}}, \quad y = \frac{D_y}{D} = \frac{\begin{vmatrix} a_1 & d_1 & c_1 \\ a_2 & d_2 & c_2 \\ a_3 & d_3 & c_3 \end{vmatrix}}{\begin{vmatrix} a_1 & b_1 & c_1 \\ a_2 & b_2 & c_2 \\ a_3 & b_3 & c_3 \end{vmatrix}}, \quad z = \frac{D_z}{D} = \frac{\begin{vmatrix} a_1 & b_1 & d_1 \\ a_2 & b_2 & d_2 \\ a_3 & b_3 & d_3 \end{vmatrix}}{\begin{vmatrix} a_1 & b_1 & c_1 \\ a_2 & b_2 & c_2 \\ a_3 & b_3 & c_3 \end{vmatrix}}, \quad D \neq 0.$$

EXAMPLE 3 **Using Cramer's Rule for a 2 × 2 System**

Use Cramer's Rule to solve the system of linear equations.
$$\begin{cases} 4x - 2y = 10 \\ 3x - 5y = 11 \end{cases}$$

SOLUTION

The determinant of the coefficient matrix is
$$D = \begin{vmatrix} 4 & -2 \\ 3 & -5 \end{vmatrix} = -20 - (-6) = -14$$

$$x = \frac{D_x}{D} = \frac{\begin{vmatrix} 10 & -2 \\ 11 & -5 \end{vmatrix}}{-14} = \frac{-50 - (-22)}{-14} = \frac{-28}{-14} = 2$$

$$y = \frac{D_y}{D} = \frac{\begin{vmatrix} 4 & 10 \\ 3 & 11 \end{vmatrix}}{-14} = \frac{44 - 30}{-14} = \frac{14}{-14} = -1$$

The solution is $(2, -1)$. Check this in the original system of equations.

Exercises Within Reach ®

Solutions in English & Spanish and tutorial videos at CollegePrepAlgebra.com

Using Cramer's Rule In Exercises 21−26, use Cramer's Rule to solve the system of linear equations. (If not possible, state the reason.)

21. $\begin{cases} x + 2y = 5 \\ -x + y = 1 \end{cases}$

22. $\begin{cases} 2x - y = -10 \\ 3x + 2y = -1 \end{cases}$

23. $\begin{cases} 3x + 4y = -2 \\ 5x + 3y = 4 \end{cases}$

24. $\begin{cases} 3x + 2y = -3 \\ 4x + 5y = -11 \end{cases}$

25. $\begin{cases} 13x - 6y = 17 \\ 26x - 12y = 8 \end{cases}$

26. $\begin{cases} -0.4x + 0.8y = 1.6 \\ 2x - 4y = 5 \end{cases}$

Study Tip

When using Cramer's Rule, remember that the method *does not* apply when the determinant of the coefficient matrix is zero.

EXAMPLE 4   **Using Cramer's Rule for a 3 × 3 System**

Use Cramer's Rule to solve the system of linear equations.

$$\begin{cases} -x + 2y - 3z = 1 \\ 2x \quad\;\; + \;\; z = 0 \\ 3x - 4y + 4z = 2 \end{cases}$$

SOLUTION

The determinant of the coefficient matrix is $D = 10$.

$$x = \frac{D_x}{D} = \frac{\begin{vmatrix} 1 & 2 & -3 \\ 0 & 0 & 1 \\ 2 & -4 & 4 \end{vmatrix}}{10} = \frac{8}{10} = \frac{4}{5}$$

$$y = \frac{D_y}{D} = \frac{\begin{vmatrix} -1 & 1 & -3 \\ 2 & 0 & 1 \\ 3 & 2 & 4 \end{vmatrix}}{10} = \frac{-15}{10} = -\frac{3}{2}$$

$$z = \frac{D_z}{D} = \frac{\begin{vmatrix} -1 & 2 & 1 \\ 2 & 0 & 0 \\ 3 & -4 & 2 \end{vmatrix}}{10} = \frac{-16}{10} = -\frac{8}{5}$$

The solution is $\left(\frac{4}{5}, -\frac{3}{2}, -\frac{8}{5}\right)$. Check this in the original system of equations.

Exercises Within Reach ®

Solutions in English & Spanish and tutorial videos at CollegePrepAlgebra.com

Using Cramer's Rule **In Exercises 27−34, use Cramer's Rule to solve the system of linear equations. (If not possible, state the reason.)**

27. $\begin{cases} 4x - y + z = -5 \\ 2x + 2y + 3z = 10 \\ 5x - 2y + 6z = 1 \end{cases}$

28. $\begin{cases} 4x - 2y + 3z = -2 \\ 2x + 2y + 5z = 16 \\ 8x - 5y - 2z = 4 \end{cases}$

29. $\begin{cases} 4x + 3y + 4z = 1 \\ 4x - 6y + 8z = 8 \\ -x + 9y - 2z = -7 \end{cases}$

30. $\begin{cases} 5x + 4y - 6z = -10 \\ -4x + 2y + 3z = -1 \\ 8x + 4y + 12z = 2 \end{cases}$

31. $\begin{cases} 2x + 3y + 5z = 4 \\ 3x + 5y + 9z = 7 \\ 5x + 9y + 17z = 13 \end{cases}$

32. $\begin{cases} 5x - 3y + 2z = 2 \\ 2x + 2y - 3z = 3 \\ x - 7y + 8z = -4 \end{cases}$

33. $\begin{cases} 3x - 2y + 3z = 8 \\ x + 3y + 6z = -3 \\ x + 2y + 9z = -5 \end{cases}$

34. $\begin{cases} 6x + 4y - 8z = -22 \\ -2x + 2y + 3z = 13 \\ -2x + 2y - z = 5 \end{cases}$

Applications

In addition to Cramer's Rule, determinants have many other practical applications. For instance, you can use a determinant to find the area of a triangle whose vertices are given by three points on a rectangular coordinate system.

Area of a Triangle

The area of a triangle with vertices (x_1, y_1), (x_2, y_2), and (x_3, y_3) is

$$\text{Area} = \pm \frac{1}{2} \begin{vmatrix} x_1 & y_1 & 1 \\ x_2 & y_2 & 1 \\ x_3 & y_3 & 1 \end{vmatrix}$$

where the symbol ($\pm$) indicates that the appropriate sign should be chosen to yield a positive area.

Application **EXAMPLE 5** **Geometry: Finding the Area of a Triangle**

Find the area of the triangle whose vertices are (2, 0), (1, 3), and (3, 2), as shown at the left.

SOLUTION

Let $(x_1, y_1) = (2, 0)$, $(x_2, y_2) = (1, 3)$, and $(x_3, y_3) = (3, 2)$. To find the area of the triangle, evaluate the determinant by expanding by minors along the first row.

$$\begin{vmatrix} x_1 & y_1 & 1 \\ x_2 & y_2 & 1 \\ x_3 & y_3 & 1 \end{vmatrix} = \begin{vmatrix} 2 & 0 & 1 \\ 1 & 3 & 1 \\ 3 & 2 & 1 \end{vmatrix}$$

$$= 2 \begin{vmatrix} 3 & 1 \\ 2 & 1 \end{vmatrix} - 0 \begin{vmatrix} 1 & 1 \\ 3 & 1 \end{vmatrix} + 1 \begin{vmatrix} 1 & 3 \\ 3 & 2 \end{vmatrix}$$

$$= 2(1) - 0 + 1(-7)$$

$$= -5$$

Using this value, you can conclude that the area of the triangle is

$$\text{Area} = -\frac{1}{2} \begin{vmatrix} 2 & 0 & 1 \\ 1 & 3 & 1 \\ 3 & 2 & 1 \end{vmatrix}$$

$$= -\frac{1}{2}(-5) = \frac{5}{2}.$$

Exercises Within Reach ®

Solutions in English & Spanish and tutorial videos at CollegePrepAlgebra.com

Finding the Area of a Triangle In Exercises 35–42, use a determinant to find the area of the triangle with the given vertices.

35. (0, 3), (4, 0), (8, 5)

36. (2, 0), (0, 5), (6, 3)

37. (−3, 4), (1, −2), (6, 1)

38. (−2, −3), (2, −3), (0, 4)

39. (−2, 1), (3, −1), (1, 6)

40. (−1, 4), (−4, 0), (1, 3)

41. $\left(0, \frac{1}{2}\right), \left(\frac{5}{2}, 0\right)$ (4, 3)

42. $\left(\frac{1}{4}, 0\right), \left(0, \frac{3}{4}\right)$, (8, −2)

Application   EXAMPLE 6   **Finding the Area of a Region**

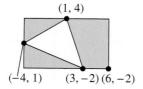
(1, 2)   (6, 2)

(−3, −1)

(2, −2)

Find the area of the shaded region of the figure.

SOLUTION

Let $(x_1, y_1) = (-3, -1)$, $(x_2, y_2) = (2, -2)$, and $(x_3, y_3) = (1, 2)$.

To find the area of the triangle, evaluate the determinant by expanding by minors along the first column.

$$\begin{vmatrix} x_1 & y_1 & 1 \\ x_2 & y_2 & 1 \\ x_3 & y_3 & 1 \end{vmatrix} = \begin{vmatrix} -3 & -1 & 1 \\ 2 & -2 & 1 \\ 1 & 2 & 1 \end{vmatrix}$$

$$= -3 \begin{vmatrix} -2 & 1 \\ 2 & 1 \end{vmatrix} - 2 \begin{vmatrix} -1 & 1 \\ 2 & 1 \end{vmatrix} + 1 \begin{vmatrix} -1 & 1 \\ -2 & 1 \end{vmatrix}$$

$$= -3(-4) - 2(-3) + 1(1)$$

$$= 19$$

Using this value, you can conclude that the area of the triangle is

$$\text{Area} = \frac{1}{2}(19)$$

$$= 9.5 \text{ square units.}$$

Now find the area of the shaded region.

Verbal Model: Length of rectangle · Height of rectangle − Area of triangle

Labels: Length of rectangle $= 6 - (-3) = 9$ (units)
Height of rectangle $= 2 - (-2) = 4$ (units)
Area of triangle $= 9.5$ (square units)

Expression: $A = (9)(4) - 9.5 = 36 - 9.5 = 26.5$

The area of the shaded region is 26.5 square units.

Exercises Within Reach ®

Solutions in English & Spanish and tutorial videos at CollegePrepAlgebra.com

Finding the Area of a Region In Exercises 43−46, **find the area of the shaded region of the figure.**

43.

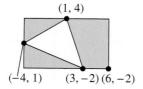
(1, 4)

(−4, 1)   (3, −2) (6, −2)

44.

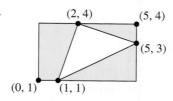
(2, 4)   (5, 4)

(5, 3)

(0, 1)   (1, 1)

45.

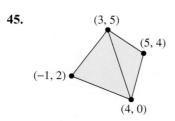
(3, 5)

(5, 4)

(−1, 2)

(4, 0)

46. (−1, 2)   (5, 2)

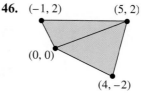
(0, 0)

(4, −2)

Test for Collinear Points

Three points (x_1, y_1), (x_2, y_2), and (x_3, y_3) are collinear (lie on the same line) if and only if

$$\begin{vmatrix} x_1 & y_1 & 1 \\ x_2 & y_2 & 1 \\ x_3 & y_3 & 1 \end{vmatrix} = 0.$$

Application   **EXAMPLE 7**   **Testing for Collinear Points**

Determine whether the points $(-2, -2)$, $(1, 1)$, and $(7, 5)$ are collinear.

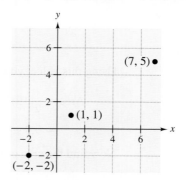

SOLUTION

Letting $(x_1, y_1) = (-2, -2)$, $(x_2, y_2) = (1, 1)$, and $(x_3, y_3) = (7, 5)$, you have

$$\begin{vmatrix} x_1 & y_1 & 1 \\ x_2 & y_2 & 1 \\ x_3 & y_3 & 1 \end{vmatrix} = \begin{vmatrix} -2 & -2 & 1 \\ 1 & 1 & 1 \\ 7 & 5 & 1 \end{vmatrix}$$

$$= -2\begin{vmatrix} 1 & 1 \\ 5 & 1 \end{vmatrix} - (-2)\begin{vmatrix} 1 & 1 \\ 7 & 1 \end{vmatrix} + 1\begin{vmatrix} 1 & 1 \\ 7 & 5 \end{vmatrix}$$

$$= -2(-4) - (-2)(-6) + 1(-2)$$

$$= -6.$$

Because the value of this determinant *is not* zero, you can conclude that the three points *are not* collinear.

Exercises Within Reach®   Solutions in English & Spanish and tutorial videos at CollegePrepAlgebra.com

Testing for Collinear Points **In Exercises 47−52, determine whether the points are collinear.**

47. $(-1, 11), (0, 8), (2, 2)$

48. $(-1, -1), (1, 9), (2, 13)$

49. $(2, -4), (5, 2), (10, 10)$

50. $(1, 8), (3, 2), (6, -7)$

51. $\left(-2, \frac{1}{3}\right), (2, 1), \left(3, \frac{1}{5}\right)$

52. $\left(0, \frac{1}{2}\right), \left(1, \frac{7}{6}\right), \left(9, \frac{13}{2}\right)$

> ### Two-Point Form of the Equation of a Line
>
> An equation of the line passing through the distinct points (x_1, y_1) and (x_2, y_2) is given by
>
> $$\begin{vmatrix} x & y & 1 \\ x_1 & y_1 & 1 \\ x_2 & y_2 & 1 \end{vmatrix} = 0.$$

Application   EXAMPLE 8   **Finding an Equation of a Line**

Find an equation of the line passing through $(-2, 1)$, and $(3, -2)$.

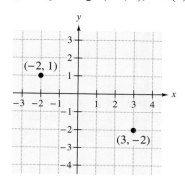

SOLUTION

Applying the determinant formula for the equation of a line produces

$$\begin{vmatrix} x & y & 1 \\ -2 & 1 & 1 \\ 3 & -2 & 1 \end{vmatrix} = 0.$$

To evaluate this determinant, you can expand by minors along the first row to obtain the following.

$$x \begin{vmatrix} 1 & 1 \\ -2 & 1 \end{vmatrix} - y \begin{vmatrix} -2 & 1 \\ 3 & 1 \end{vmatrix} + 1 \begin{vmatrix} -2 & 1 \\ 3 & -2 \end{vmatrix} = 0$$

$$3x + 5y + 1 = 0$$

So, an equation of the line is $3x + 5y + 1 = 0$.

Exercises Within Reach ®

Solutions in English & Spanish and tutorial videos at CollegePrepAlgebra.com

Finding an Equation of a Line In Exercises 53−60, use a determinant to find an equation of the line passing through the points.

53. $(-2, -1), (4, 2)$

54. $(-1, 3), (2, -6)$

55. $(10, 7), (-2, -7)$

56. $(-8, 3), (4, 6)$

57. $\left(-2, \frac{3}{2}\right), (3, -3)$

58. $\left(-\frac{1}{2}, 3\right), \left(\frac{5}{2}, 1\right)$

59. $(2, 3.6), (8, 10)$

60. $(3, 1.6), (5, -2.2)$

Concept Summary: Finding the Determinant of a Matrix

What

Associated with each square matrix is a real number called a **determinant**.

EXAMPLE

Find the determinant of the matrix.

$$A = \begin{vmatrix} 2 & 4 \\ 1 & 6 \end{vmatrix}$$

How

To find the determinant of a square matrix, find the difference of the products of the two diagonals of the matrix.

$$\det(A) = \begin{vmatrix} 2 & 4 \\ 1 & 6 \end{vmatrix} = 2(6) - 1(4) = 12 - 4 = 8$$

Why

You can use determinants to do the following.

1. Solve systems of linear equations.
2. Find areas of triangles.
3. Test for collinear points.
4. Find equations of lines.

Exercises Within Reach ®

Worked-out solutions to odd-numbered exercises at CollegePrepAlgebra.com

Concept Summary Check

61. *Writing* Explain how to find the determinant of a 2×2 matrix.

62. *Vocabulary* The determinant of a matrix can be represented by vertical bars, similar to the vertical bars used for absolute value. Does this mean that every determinant is nonnegative?

63. *Logic* Is it possible to find the determinant of a 2×3 matrix? Explain.

64. *Reasoning* When one column of a 3×3 matrix is all zeros, what is the determinant of the matrix? Explain.

Extra Practice

Finding the Determinant **In Exercises 65–70, find the determinant of the matrix. Expand by minors along the row or column that appears to make the computation easiest.**

65. $\begin{bmatrix} 5 & -3 & 2 \\ 7 & 5 & -7 \\ 0 & 6 & -1 \end{bmatrix}$

66. $\begin{bmatrix} 3 & -1 & 2 \\ 1 & -1 & 2 \\ -2 & 3 & 10 \end{bmatrix}$

67. $\begin{bmatrix} -\frac{1}{2} & -1 & 6 \\ 8 & -\frac{1}{4} & -4 \\ 1 & 2 & 1 \end{bmatrix}$

68. $\begin{bmatrix} \frac{1}{2} & \frac{3}{2} & \frac{1}{2} \\ 4 & 8 & 10 \\ -2 & -6 & 12 \end{bmatrix}$

69. $\begin{bmatrix} 0.6 & 0.4 & -0.6 \\ 0.1 & 0.5 & -0.3 \\ 8 & -2 & 12 \end{bmatrix}$

70. $\begin{bmatrix} 0.4 & 0.3 & 0.3 \\ -0.2 & 0.6 & 0.6 \\ 3 & 1 & 1 \end{bmatrix}$

71. *Area of a Region* A large region of forest has been infested with gypsy moths. The region is roughly triangular, as show in the figure. Find the area of this region. (*Note:* The measurements in the figure are in miles.)

72. *Area of a Region* You have purchased a triangular tract of land, as shown in the figure. What is the area of this tract of land? (*Note:* The measurements in the figure are in feet.)

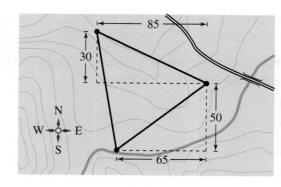

73. *Electrical Networks* When Kirchhoff's Laws are applied to the electrical network shown in the figure, the currents I_1, I_2, and I_3 are the solution of the system

$$\begin{cases} I_1 + I_2 - I_3 = 0 \\ I_1 \qquad + 2I_3 = 12 \\ I_1 - 2I_2 \qquad = -4 \end{cases}$$

Find the currents.

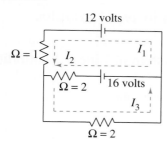

74. *Electrical Networks* When Kirchhoff's Laws are applied to the electrical network shown in the figure, the currents I_1, I_2, and I_3 are the solution of the system

$$\begin{cases} I_1 - I_2 + I_3 = 0 \\ I_2 + 4I_3 = 8 \\ 4I_1 + I_2 \qquad = 16 \end{cases}$$

Find the currents.

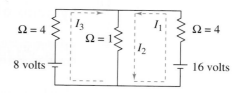

75. (a) Use Cramer's Rule to solve the system of linear equations.

$$\begin{cases} kx + 3ky = 2 \\ (2 + k)x + ky = 5 \end{cases}$$

(b) State the values of k for which Cramer's Rule does not apply.

76. (a) Use Cramer's Rule to solve the system of linear equations.

$$\begin{cases} kx + (1 - k)y = 1 \\ (1 - k)x + ky = 3 \end{cases}$$

(b) For what value(s) of k will the system be inconsistent?

Explaining Concepts

77. *Vocabulary* Explain the difference between a square matrix and its determinant.

78. *Writing* What is meant by the minor of an entry of a square matrix?

79. *Writing* When two rows of a 3×3 matrix have identical entries, what is the value of the determinant? Explain.

80. *Precision* What conditions must be met in order to use Cramer's Rule to solve a system of linear equations?

Cumulative Review

In Exercises 81–84, sketch the graph of the linear inequality.

81. $4x - 2y < 0$

82. $2x + 8y \geq 0$

83. $-x + 3y > 12$

84. $-3x - y \leq 2$

In Exercises 85–88, write an equation of the line that passes through the points. Write the equation in general form.

85. $(0, 0), (4, 2)$

86. $(1, 2), (6, 3)$

87. $(-1, 2), (5, 2)$

88. $(-3, 3), (8, -6)$

In Exercises 89–92, determine whether the set of ordered pairs represents a function.

89. $\{(0, 0), (2, 1), (4, 2), (6, 3)\}$

90. $\{(0, 2), (1, 4), (4, 1), (0, 4)\}$

91. $\{(-4, 5), (-1, 0), (3, -2), (3, -4)\}$

92. $\{(-3, 1), (-1, 3), (1, 3), (3, 1)\}$

8.6 Systems of Linear Inequalities

▶ Solve systems of linear inequalities in two variables.

▶ Use systems of linear inequalities to model and solve real-life problems.

Systems of Linear Inequalities in Two Variables

EXAMPLE 1 Graphing a System of Linear Inequalities

Sketch the graph of the system of linear inequalities.

$$\begin{cases} 2x - y \le 5 \\ x + 2y \ge 2 \end{cases}$$

SOLUTION

Begin by rewriting each inequality in slope-intercept form.

$$\begin{cases} 2x - y \le 5 \\ x + 2y \ge 2 \end{cases} \implies \begin{aligned} y &\ge 2x - 5 \\ y &\ge -\tfrac{1}{2}x + 1 \end{aligned}$$

Then sketch the line for the corresponding equation of each inequality.

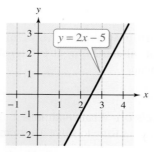

Graph of $2x - y \le 5$ is all points on and above $y = 2x - 5$.

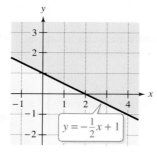

Graph of $x + 2y \ge 2$ is all points on and above $y = -\tfrac{1}{2}x + 1$.

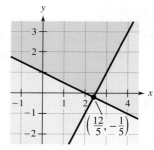

Graph of system is the purple wedge-shaped region.

Exercises Within Reach ®

Solutions in English & Spanish and tutorial videos at CollegePrepAlgebra.com

Graphing a System of Linear Inequalities In Exercises 1–12, sketch the graph of the system of linear inequalities.

1. $\begin{cases} x + y \le 3 \\ x - y \le 1 \end{cases}$

2. $\begin{cases} x + y \ge 2 \\ x - y \le 2 \end{cases}$

3. $\begin{cases} 2x - 4y \le 6 \\ x + y \ge 2 \end{cases}$

4. $\begin{cases} 4x + 10y \le 5 \\ x - y \le 4 \end{cases}$

5. $\begin{cases} x + 2y \le 6 \\ x - 2y \le 0 \end{cases}$

6. $\begin{cases} 2x + y \le 0 \\ x - y \le 8 \end{cases}$

7. $\begin{cases} x - 2y > 4 \\ 2x + y > 6 \end{cases}$

8. $\begin{cases} 3x + y < 6 \\ x + 2y > 2 \end{cases}$

9. $\begin{cases} x + y > -1 \\ x + y < 3 \end{cases}$

10. $\begin{cases} x - y > 2 \\ x - y < -4 \end{cases}$

11. $\begin{cases} y \ge \tfrac{4}{3}x + 1 \\ y \le 5x - 2 \end{cases}$

12. $\begin{cases} y \ge \tfrac{1}{2}x + \tfrac{1}{2} \\ y \le 4x - \tfrac{1}{2} \end{cases}$

Graphing a System of Linear Inequalities

1. Sketch the line that corresponds to each inequality. (Use dashed lines for inequalities with $<$ or $>$ and solid lines for inequalities with $\leq$ or $\geq$.)

2. Lightly shade the half-plane that is the graph of each linear inequality. (Colored pencils may help distinguish different half-planes.)

3. The graph of the system is the intersection of the half-planes. (If you use colored pencils, it is the region that is shaded with *every* color.)

EXAMPLE 2   **Graphing a System of Linear Inequalities**

Sketch the graph of the system of linear inequalities: $\begin{cases} y < 4 \\ y > 1 \end{cases}$.

SOLUTION

The graph of the first inequality is the half-plane below the horizontal line

$y = 4$.   Upper boundary

The graph of the second inequality is the half-plane above the horizontal line

$y = 1$.   Lower boundary

The graph of the system is the horizontal band that lies *between* the two horizontal lines (where $y < 4$ *and* $y > 1$), as shown below.

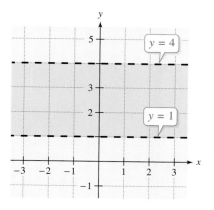

Exercises Within Reach ®

Solutions in English & Spanish and tutorial videos at CollegePrepAlgebra.com

Graphing a System of Linear Inequalities In Exercises 13–20, sketch the graph of the system of linear inequalities.

13. $\begin{cases} x > -4 \\ x \leq 2 \end{cases}$

14. $\begin{cases} y \leq 4 \\ y > -2 \end{cases}$

15. $\begin{cases} x < 3 \\ x > -2 \end{cases}$

16. $\begin{cases} y > -1 \\ y \leq 2 \end{cases}$

17. $\begin{cases} x \leq 5 \\ x > -6 \end{cases}$

18. $\begin{cases} y > -7 \\ y < 6 \end{cases}$

19. $\begin{cases} x < 3 \\ x > -3 \end{cases}$

20. $\begin{cases} y \leq 5 \\ y \geq -5 \end{cases}$

EXAMPLE 3 **Graphing a System of Linear Inequalities**

Sketch the graph of the system of linear inequalities, and label the vertices.

$$\begin{cases} x - y < & 2 \\ x & > -2 \\ & y \le & 3 \end{cases}$$

SOLUTION

Begin by sketching the half-planes represented by the three linear inequalities. The graph of

$$x - y < 2$$

is the half-plane lying above the line $y = x - 2$, the graph of

$$x > -2$$

is the half-plane lying to the right of the line $x = -2$, and the graph of

$$y \le 3$$

is the half-plane lying on and below the line $y = 3$. As shown below, the region that is common to all three of these half-planes is a triangle. The vertices of the triangle are found as follows.

Vertex A: $(-2, -4)$

Solution of the system

$$\begin{cases} x - y = & 2 \\ x & = -2 \end{cases}$$

Vertex B: $(5, 3)$

Solution of the system

$$\begin{cases} x - y = 2 \\ y = 3 \end{cases}$$

Vertex C: $(-2, 3)$

Solution of the system

$$\begin{cases} x = -2 \\ y = & 3 \end{cases}$$

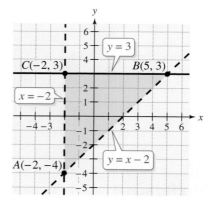

Exercises Within Reach ®

Solutions in English & Spanish and tutorial videos at CollegePrepAlgebra.com

Graphing a System of Linear Inequalities In Exercises 21−26, sketch the graph of the system of linear inequalities, and label the vertices.

21. $\begin{cases} x + y \le 4 \\ x \ge 0 \\ y \ge 0 \end{cases}$

22. $\begin{cases} 2x + y \le 6 \\ x \ge 0 \\ y \ge 0 \end{cases}$

23. $\begin{cases} 4x - 2y > 8 \\ x \ge 0 \\ y \le 0 \end{cases}$

24. $\begin{cases} 2x - 6y > 6 \\ x \le 0 \\ y \le 0 \end{cases}$

25. $\begin{cases} y > -5 \\ x \le 2 \\ y \le x + 2 \end{cases}$

26. $\begin{cases} y \ge -1 \\ x < 3 \\ y \ge x - 1 \end{cases}$

EXAMPLE 4   **Graphing a System of Linear Inequalities**

Sketch the graph of the system of linear inequalities, and label the vertices.

$$\begin{cases} x + y \le 5 \\ 3x + 2y \le 12 \\ x \ge 0 \\ y \ge 0 \end{cases}$$

SOLUTION

Begin by sketching the half-planes represented by the four linear inequalities. The graph of $x + y \le 5$ is the half-plane lying on and below the line $y = -x + 5$. The graph of $3x + 2y \le 12$ is the half-plane lying on and below the line $y = -\frac{3}{2}x + 6$. The graph of $x \ge 0$ is the half-plane lying on and to the right of the y-axis, and the graph of $y \ge 0$ is the half-plane lying on and above the x-axis. As shown below, the region that is common to all four of these half-planes is a four-sided polygon. The vertices of the region are found as follows.

Vertex A: $(0, 5)$

Solution of the system

$$\begin{cases} x + y = 5 \\ x = 0 \end{cases}$$

Vertex B: $(2, 3)$

Solution of the system

$$\begin{cases} x + y = 5 \\ 3x + 2y = 12 \end{cases}$$

Vertex C: $(4, 0)$

Solution of the system

$$\begin{cases} 3x + 2y = 12 \\ y = 0 \end{cases}$$

Vertex D: $(0, 0)$

Solution of the system

$$\begin{cases} x = 0 \\ y = 0 \end{cases}$$

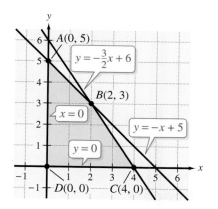

Exercises Within Reach ®   Solutions in English & Spanish and tutorial videos at CollegePrepAlgebra.com

Graphing a System of Linear Inequalities In Exercises 27–30, sketch the graph of the system of linear inequalities, and label the vertices.

27. $\begin{cases} x \ge 1 \\ x - 2y \le 3 \\ 3x + 2y \ge 9 \\ x + y \le 6 \end{cases}$

28. $\begin{cases} x + y \le 4 \\ x + y \ge -1 \\ x - y \ge -2 \\ x - y \le 2 \end{cases}$

29. $\begin{cases} x - y \le 8 \\ 2x + 5y \le 25 \\ x \ge 0 \\ y \ge 0 \end{cases}$

30. $\begin{cases} 4x - y \le 13 \\ -x + 2y \le 22 \\ x \ge 0 \\ y \ge 0 \end{cases}$

EXAMPLE 5 **Finding the Boundaries of a Region**

Write a system of inequalities that describes the region shown below.

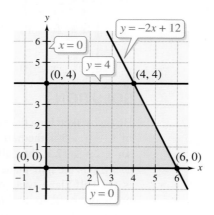

SOLUTION

Three of the boundaries of the region are horizontal or vertical—they are easy to find. To find the diagonal boundary line, you can use the techniques of Section 4.5 to find the equation of the line passing through the points $(4, 4)$ and $(6, 0)$. Use the formula for slope to find $m = -2$, and then use the point-slope form with point $(6, 0)$ and $m = -2$ to obtain

$$y - 0 = -2(x - 6).$$

So, the equation is $y = -2x + 12$. The system of linear inequalities that describes the region is as follows.

$$\begin{cases} y \leq & 4 \\ y \geq & 0 \\ x \geq & 0 \\ y \leq -2x + 12 \end{cases}$$

Region lies on and below line $y = 4$.
Region lies on and above x-axis.
Region lies on and to the right of y-axis.
Region lies on and below line $y = -2x + 12$.

Exercises Within Reach ®

Solutions in English & Spanish and tutorial videos at CollegePrepAlgebra.com

Finding the Boundaries of a Region In Exercises 31−34, write a system of linear inequalities that describes the shaded region.

31.

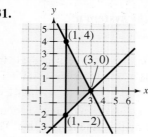

32.

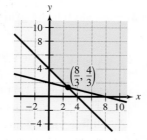

33.

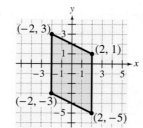

34.

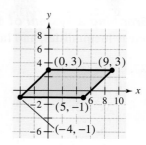

Application

Application   EXAMPLE 6   **Nutrition**

The minimum daily requirements for the liquid portion of a diet are 300 calories, 36 units of vitamin A, and 90 units of vitamin C. A cup of dietary drink X provides 60 calories, 12 units of vitamin A, and 10 units of vitamin C. A cup of dietary drink Y provides 60 calories, 6 units of vitamin A, and 30 units of vitamin C. Write a system of linear inequalities that describes how many cups of each drink should be consumed each day to meet the minimum daily requirements for calories and vitamins.

SOLUTION

Begin by letting x and y represent the following.

$$x = \text{number of cups of dietary drink X}$$

$$y = \text{number of cups of dietary drink Y}$$

To meet the minimum daily requirements, the following inequalities must be satisfied.

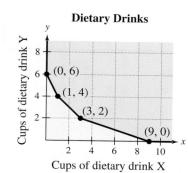

Dietary Drinks

$$\begin{cases} 60x + 60y \geq 300 & \text{Calories} \\ 12x + 6y \geq 36 & \text{Vitamin A} \\ 10x + 30y \geq 90 & \text{Vitamin C} \\ x \geq 0 \\ y \geq 0 \end{cases}$$

The last two inequalities are included because x and y cannot be negative. The graph of this system of inequalities is shown at the left.

Exercises Within Reach ®

Solutions in English & Spanish and tutorial videos at CollegePrepAlgebra.com

35. *Production* A furniture company can sell all the tables and chairs it produces. Each table requires 1 hour in the assembly center and $1\frac{1}{3}$ hours in the finishing center. Each chair requires $1\frac{1}{2}$ hours in the assembly center and $\frac{3}{4}$ hour in the finishing center. The company's assembly center is available 12 hours per day, and its finishing center is available 16 hours per day. Write a system of linear inequalities that describes the different production levels. Graph the system.

36. *Production* An electronics company can sell all the HD TVs and DVD players it produces. Each HD TV requires 3 hours on the assembly line and $1\frac{1}{4}$ hours on the testing line. Each DVD player requires $2\frac{1}{2}$ hours on the assembly line and 1 hour on the testing line. The company's assembly line is available 20 hours per day, and its testing line is available 16 hours per day. Write a system of linear inequalities that describes the different production levels. Graph the system.

Concept Summary: Graphing Systems of Linear Inequalities

What

The graph of a **system of linear inequalities** shows *all* of the **solutions** of the system.

EXAMPLE

Sketch the graph of the system of linear inequalities.

$$\begin{cases} y < 2x + 1 \\ y \geq -3x - 2 \end{cases}$$

How

Sketch the line that corresponds to each inequality. Be sure to use dashed lines and solid lines appropriately.

Then shade the half-plane that is the graph of each linear inequality.

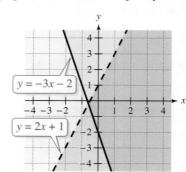

Why

Many practical problems in business, science, and engineering involve multiple constraints. These problems can be solved with systems of inequalities.

Exercises Within Reach ®

Worked-out solutions to odd-numbered exercises at CollegePrepAlgebra.com

Concept Summary Check

37. Vocabulary What is a system of linear inequalities in two variables?

38. Reasoning Explain when you should use dashed lines and when you should use solid lines in sketching a system of linear inequalities.

39. Precision Does the point of intersection of each pair of boundary lines correspond to a vertex? Explain.

40. Logic Is it possible for a system of linear inequalities to have no solution? Explain.

Extra Practice

Graphing a System of Linear Inequalities In Exercises 41−46, sketch the graph of the system of linear inequalities.

41. $\begin{cases} x + y \leq 1 \\ -x + y \leq 1 \\ y \geq 0 \end{cases}$

42. $\begin{cases} 3x + 2y < 6 \\ x - 3y \geq 1 \\ y \geq 0 \end{cases}$

43. $\begin{cases} x + y \leq 5 \\ x - 2y \geq 2 \\ y \geq 3 \end{cases}$

44. $\begin{cases} 2x + y \geq 2 \\ x - 3y \leq 2 \\ y \leq 1 \end{cases}$

45. $\begin{cases} -3x + 2y < 6 \\ x - 4y > -2 \\ 2x + y < 3 \end{cases}$

46. $\begin{cases} x + 2y > 14 \\ -2x + 3y > 15 \\ x + 3y < 3 \end{cases}$

47. Investment A person plans to invest up to $25,000 in two different interest-bearing accounts, account X and account Y. Account Y is to contain at least $4000. Moreover, account X should have at least three times the amount in account Y. Write a system of linear inequalities that describes the various amounts that can be deposited in each account. Graph the system.

48. Nutrition A veterinarian is asked to design a special canine dietary supplement using two different dog foods. Each ounce of food X contains 12 units of calcium, 8 units of iron, and 6 units of protein. Each ounce of food Y contains 10 units of calcium, 10 units of iron, and 8 units of protein. The minimum daily requirements of the diet are 200 units of calcium, 100 units of iron, and 120 units of protein. Write a system of linear inequalities that describes the different amounts of dog food X and dog food Y that can be used.

49. *Ticket Sales* For a concert event, there are $30 reserved seat tickets and $20 general admission tickets. There are 2000 reserved seats available, and fire regulations limit the number of paid ticket holders to 3000. The promoter must take in at least $75,000 in ticket sales. Write a system of linear inequalities that describes the different numbers of tickets that can be sold.

51. *Geometry* The figure shows the chorus platform on a stage. Write a system of linear inequalities that describes the part of the audience that can see the full chorus. (Each unit in the coordinate system represents 1 meter.)

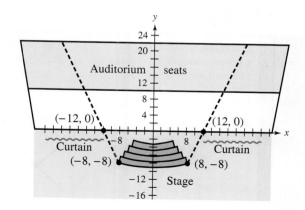

50. *Geometry* The figure shows a cross section of a roped-off swimming area at a beach. Write a system of linear inequalities that describes the cross section. (Each unit in the coordinate system represents 1 foot.)

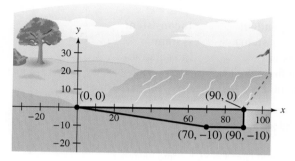

Explaining Concepts

52. *Writing* Explain the meaning of the term *half-plane.* Give an example of an inequality whose graph is a half-plane.

53. *Precision* Explain how you can check any single point (x_1, y_1) to determine whether the point is a solution of a system of linear inequalities.

54. *Reasoning* Explain how to determine the vertices of the solution region for a system of linear inequalities.

55. *Think About It* Describe the difference between the solution set of a system of linear equations and the solution set of a system of linear inequalities.

Cumulative Review

In Exercises 56–61, find the *x*- and *y*-intercepts (if any) of the graph of the equation.

56. $y = 4x + 2$

57. $y = 8 - 3x$

58. $-x + 3y = -3$

59. $3x - 6y = 12$

60. $y = |x + 2|$

61. $y = |x - 1| - 2$

In Exercises 62–65, evaluate the function as indicated, and simplify.

62. $f(x) = 3x - 7$

  (a) $f(-1)$

  (b) $f\left(\frac{2}{3}\right)$

63. $f(x) = x^2 + x$

  (a) $f(3)$

  (b) $f(-2)$

64. $f(x) = 3x - x^2$

  (a) $f(0)$

  (b) $f(2m)$

65. $f(x) = \dfrac{x + 3}{x - 1}$

  (a) $f(8)$

  (b) $f(k - 2)$

8 Chapter Summary

What did you learn?	Explanation and Examples	Review Exercises
8.1 Determine whether ordered pairs are solutions of systems of equations *(p. 378).*	Substitute the values of x and y into each of the equations and simplify. If the check does *not* fail, the ordered pair is a solution.	1–4
Solve systems of equations graphically *(p. 379).*	A system of equations can have one solution, infinitely many solutions, or no solution. **Consistent system** (one solution) **Dependent (consistent) system** (infinitely many solutions) **Inconsistent system** (no solution)	5–14
Use the method of substitution to solve systems of equations algebraically *(p. 381).*	1. Solve one of the equations for one variable in terms of the other. 2. Substitute the expression obtained in Step 1 into the other equation to obtain an equation in one variable. 3. Solve the equation obtained in Step 2. 4. Back-substitute the solution from Step 3 into the expression found in Step 1 to find the value of the other variable. 5. Check the solution to see that it satisfies both of the original equations.	15–28
Use systems of equations to model and solve real-life problems *(p. 384).*	You can use verbal models to write a system of linear equations that represents a real-life problem.	29, 30
8.2 Solve systems of linear equations algebraically using the method of elimination *(p. 388).*	1. Obtain opposite coefficients of x (or y) by multiplying all terms of one or both equations by suitable constants. 2. Add the equations to eliminate one variable. Solve the resulting equation. 3. Back-substitute the value from Step 2 into either of the original equations and solve for the other variable. 4. Check the solution in both of the original equations.	31–40
Solve systems with no solution or infinitely many solutions *(p. 392).*	As with the method of substitution, the key is to recognize the occurrence of a false or true statement.	41–44
Use the method of elimination to solve application problems *(p. 393).*	Using the algebraic method of elimination is an accurate way to solve applications involving more than one unknown.	45–48

	What did you learn?	Explanation and Examples	Review Exercises
8.3	Solve systems of linear equations in row-echelon form using back-substitution *(p. 396)*.	A system of equations in *row-echelon form* has a stair-step pattern with leading coefficients of 1. You can use back-substitution to solve a system in row-echelon form.	49 − 52
	Solve systems of linear equations using the method of Gaussian elimination *(p. 397)*.	*Gaussian elimination* is the process of forming a chain of equivalent systems by performing one row operation at a time to obtain an equivalent system in row-echelon form.	53 − 56
	Solve application problems using the method of Gaussian elimination *(p. 401)*.	You can use Gaussian elimination to find the solution of application problems involving vertical motion and the position equation.	57, 58
8.4	Form augmented matrices and form linear systems from augmented matrices *(p. 406)*.	A matrix derived from a system of linear equations (each written in standard form with the constant term on the right) is the augmented matrix of the system.	59 − 62
	Perform elementary row operations to solve systems of linear equations *(p. 408)*.	1. Interchange two rows. 2. Multiply a row by a nonzero constant. 3. Add a multiple of a row to another row.	63 − 70
	Use matrices and Gaussian elimination to solve systems of linear equations *(p. 410)*.	1. Write the augmented matrix of the system of equations. 2. Use elementary row operations to rewrite the augmented matrix in row-echelon form. 3. Write the system of equations corresponding to the matrix in row-echelon form. Then use back-substitution to find the solution.	63 − 70
8.5	Find determinants of 2×2 matrices and 3×3 matrices *(p. 416)*.	$\begin{vmatrix} a_1 & b_1 \\ a_2 & b_2 \end{vmatrix} = a_1 b_2 - a_2 b_1$ $\begin{vmatrix} a_1 & b_1 & c_1 \\ a_2 & b_2 & c_2 \\ a_3 & b_3 & c_3 \end{vmatrix} = a_1 \begin{vmatrix} b_2 & c_2 \\ b_3 & c_3 \end{vmatrix} - b_1 \begin{vmatrix} a_2 & c_2 \\ a_3 & c_3 \end{vmatrix} + c_1 \begin{vmatrix} a_2 & b_2 \\ a_3 & b_3 \end{vmatrix}$	71 − 76
	Use determinants and Cramer's Rule to solve systems of linear equations *(p. 418)*.	Cramer's Rule is not as general as the elimination method because Cramer's Rule requires that the coefficient matrix of the system be square *and* that the system have exactly one solution.	77 − 80
	Use determinants to find areas of regions, to test for collinear points, and to find equations of lines *(p. 420)*.	These are examples of practical applications of determinants.	81 − 88
8.6	Solve systems of linear inequalities in two variables *(p. 426)*.	1. Sketch a dashed or solid line corresponding to each inequality. 2. Shade the half-plane for each inequality. 3. The intersection of all half-planes represents the system.	89 − 92
	Use systems of linear inequalities to model and solve real-life problems *(p. 431)*.	A system of linear inequalities can be used to find a number of acceptable solutions to real-life problems.	93, 94

Review Exercises

Worked-out solutions to odd-numbered exercises at CollegePrepAlgebra.com

8.1

Checking Solutions In Exercises 1−4, determine whether each ordered pair is a solution of the system of equations.

System	Ordered Pairs

1. $\begin{cases} 3x - 5y = \ 11 \\ -x + 2y = -4 \end{cases}$ (a) $(2, -1)$ (b) $(3, -2)$

2. $\begin{cases} 10x + 8y = -2 \\ 2x - 5y = \ 26 \end{cases}$ (a) $(4, -4)$ (b) $(3, -4)$

3. $\begin{cases} 0.2x + 0.4y = \ 5 \\ x + \ 3y = 30 \end{cases}$ (a) $(0.5, -0.7)$ (b) $(15, 5)$

4. $\begin{cases} -\frac{1}{2}x - \frac{2}{3}y = \frac{1}{2} \\ x + \ y = 1 \end{cases}$ (a) $(-5, 6)$ (b) $(7, -3)$

Solving a System of Linear Equations In Exercises 5−14, solve the system by graphing.

5. $\begin{cases} y = \ x - 4 \\ y = 2x - 9 \end{cases}$

6. $\begin{cases} y = -\frac{5}{3}x + \ 6 \\ y = \ x - 10 \end{cases}$

7. $\begin{cases} x + y = 2 \\ x - y = 0 \end{cases}$

8. $\begin{cases} x - y = 9 \\ -x + y = 1 \end{cases}$

9. $\begin{cases} 2x + 3 = 3y \\ y = \frac{2}{3}x \end{cases}$

10. $\begin{cases} x + \ y = -1 \\ 3x + 2y = \ 0 \end{cases}$

11. $\begin{cases} -3x - 3y = -6 \\ x + \ y = \ 2 \end{cases}$

12. $\begin{cases} 2x - 3y = 11 \\ x + \ y = \ 3 \end{cases}$

13. $\begin{cases} x + 7y = 6 \\ -3x + \ y = 4 \end{cases}$

14. $\begin{cases} 5x - 4y = 12 \\ -x + 3y = \ 2 \end{cases}$

Solving a System of Linear Equations In Exercises 15−28, solve the system by the method of substitution.

15. $\begin{cases} y = 2x \\ y = \ x + 4 \end{cases}$

16. $\begin{cases} x = -2y + 13 \\ x = \ \frac{y}{2} + \ 3 \end{cases}$

17. $\begin{cases} x = 3y - 2 \\ x = 6 - y \end{cases}$

18. $\begin{cases} y = -4x + 1 \\ y = \ x - 4 \end{cases}$

19. $\begin{cases} x - 2y = \ 6 \\ 3x + 2y = 10 \end{cases}$

20. $\begin{cases} 5x + \ y = \ 20 \\ 7x - 5y = -4 \end{cases}$

21. $\begin{cases} 2x - \ y = \ 2 \\ 6x + 8y = 39 \end{cases}$

22. $\begin{cases} 3x + 4y = \ 1 \\ x - 7y = -3 \end{cases}$

23. $\begin{cases} \frac{3}{5}x - \ y = \ 8 \\ 2x - 3y = 25 \end{cases}$

24. $\begin{cases} -x + 8y = -115 \\ 2x + \frac{2}{7}y = \ 2 \end{cases}$

25. $\begin{cases} x = y + 3 \\ x = y + 1 \end{cases}$

26. $\begin{cases} y = 3x + \ 4 \\ 9x = 3y - 12 \end{cases}$

27. $\begin{cases} -6x + y = -3 \\ 12x - 2y = 6 \end{cases}$

28. $\begin{cases} 3x + 4y = 7 \\ 6x + 8y = 10 \end{cases}$

29. **Break-Even Analysis** A small business invests $25,000 to produce a one-time-use camera. Each camera costs $4.45 to produce and sells for $8.95. How many one-time-use cameras must be sold before the business breaks even?

30. **Seed Mixture** Fifteen pounds of mixed birdseed sells for $8.85 per pound. The mixture is obtained from two kinds of birdseed, with one variety priced at $7.05 per pound and the other at $9.30 per pound. How many pounds of each variety of birdseed are used in the mixture?

8.2

Solving a System of Linear Equations In Exercises 31–44, solve the system by the method of elimination.

31. $\begin{cases} 2x + 4y = 2 \\ -2x - 7y = 4 \end{cases}$

32. $\begin{cases} 3x - y = 5 \\ 2x + y = 5 \end{cases}$

33. $\begin{cases} 3x - 2y = 9 \\ x + y = 3 \end{cases}$

34. $\begin{cases} 5x + 4y = 2 \\ -x + y = -22 \end{cases}$

35. $\begin{cases} x + 2y = 2 \\ x - 4y = 20 \end{cases}$

36. $\begin{cases} 2x + 6y = 16 \\ 2x + 3y = 7 \end{cases}$

37. $\begin{cases} \frac{1}{2}x - \frac{3}{5}y = \frac{1}{6} \\ -3x + 6y = 1 \end{cases}$

38. $\begin{cases} \frac{2}{3}x + \frac{1}{12}y = \frac{3}{4} \\ 3x - 4y = 2 \end{cases}$

39. $\begin{cases} 0.2x - 0.1y = 0.07 \\ 0.4x - 0.5y = -0.01 \end{cases}$

40. $\begin{cases} 0.2x + 0.1y = 0.03 \\ 0.3x - 0.1y = -0.13 \end{cases}$

41. $\begin{cases} 2x - 5y = 2 \\ 6x - 15y = 4 \end{cases}$

42. $\begin{cases} 8x - 6y = 4 \\ -4x + 3y = -2 \end{cases}$

43. $\begin{cases} 6x - 3y = 27 \\ -2x + y = -9 \end{cases}$

44. $\begin{cases} -\frac{1}{4}x + \frac{2}{3}y = 1 \\ 3x - 8y = 1 \end{cases}$

45. **Ticket Sales** Five hundred tickets were sold for a fundraising dinner. The receipts totaled $3400.00. Adult tickets were $7.50 each and children's tickets were $4.00 dollars each. How many tickets of each type were sold?

46. **Ticket Sales** A fundraising dinner, was held on two consecutive nights. On the first night, 100 adult tickets and 175 children's tickets were sold, for a total of $937.50. On the second night, 200 adult tickets and 316 children's tickets were sold, for a total of $1790.00. Find the price of each type of ticket.

47. **Alcohol Mixture** Fifty gallons of a 90% alcohol solution is obtained by mixing a 100% solution with a 75% solution. How many gallons of each solution must be used to obtain the desired mixture?

48. **Acid Mixture** Forty gallons of a 60% acid solution is obtained by mixing a 75% solution with a 50% solution. How many gallons of each solution must be used to obtain the desired mixture?

8.3

Using Back-Substitution **In Exercises 49–52, use back-substitution to solve the system of linear equations.**

49. $\begin{cases} x = 3 \\ x + 2y = 7 \\ -3x - y + 4z = 9 \end{cases}$ **50.** $\begin{cases} 2x + 3y = 9 \\ 4x - 6z = 12 \\ y = 5 \end{cases}$

51. $\begin{cases} x + 2y = 6 \\ 3y = 9 \\ x + 2z = 12 \end{cases}$ **52.** $\begin{cases} 3x - 2y + 5z = -10 \\ 3y = 18 \\ 6x - 4y = -6 \end{cases}$

Using Gaussian Elimination **In Exercises 53–56, solve the system of linear equations.**

53. $\begin{cases} -x + y + 2z = 1 \\ 2x + 3y + z = -2 \\ 5x + 4y + 2z = 4 \end{cases}$

54. $\begin{cases} 2x + 3y + z = 10 \\ 2x - 3y - 3z = 22 \\ 4x - 2y + 3z = -2 \end{cases}$

55. $\begin{cases} x - y - z = 1 \\ -2x + y + 3z = -5 \\ 3x + 4y - z = 6 \end{cases}$

56. $\begin{cases} -3x + y + 2z = -13 \\ -x - y + z = 0 \\ 2x + 2y - 3z = -1 \end{cases}$

57. **Investment** An inheritance of $20,000 is divided among three investments yielding a total of $1780 in interest per year. The interest rates for the three investments are 7%, 9%, and 11%. The amounts invested at 9% and 11% are $3000 and $1000 less than the amount invested at 7%, respectively. Find the amount invested at each rate.

58. **Vertical Motion** Find the position equation

$$s = \frac{1}{2}at^2 + v_0t + s_0$$

for an object that has the indicated heights at the specified times.

$s = 192$ feet at $t = 1$ second
$s = 152$ feet at $t = 2$ seconds
$s = 80$ feet at $t = 3$ seconds

8.4

Forming Coefficient and Augmented Matrices **In Exercises 59 and 60, form (a) the coefficient matrix and (b) the augmented matrix for the system of linear equations.**

59. $\begin{cases} 7x - 5y = 11 \\ x - y = -5 \end{cases}$

60. $\begin{cases} x + 2y + z = 4 \\ 3x - z = 2 \\ -x + 5y - 2z = -6 \end{cases}$

Forming a Linear System **In Exercises 61 and 62, write the system of linear equations represented by the matrix. (Use variables x, y, and z.)**

61. $\begin{bmatrix} 4 & -1 & 0 & \vdots & 2 \\ 6 & 3 & 2 & \vdots & 1 \\ 0 & 1 & 4 & \vdots & 0 \end{bmatrix}$

62. $\begin{bmatrix} 7 & 8 & \vdots & -26 \\ 4 & -9 & \vdots & -12 \end{bmatrix}$

Solving a System of Linear Equations **In Exercises 63–70, use matrices to solve the system.**

63. $\begin{cases} 5x + 4y = 2 \\ -x + y = -22 \end{cases}$ **64.** $\begin{cases} 2x - 5y = 2 \\ 3x - 7y = 1 \end{cases}$

65. $\begin{cases} 0.2x - 0.1y = 0.07 \\ 0.4x - 0.5y = -0.01 \end{cases}$

66. $\begin{cases} 2x + y = 0.3 \\ 3x - y = -1.3 \end{cases}$

67. $\begin{cases} x + 4y + 4z = 7 \\ -3x + 2y + 3z = 0 \\ 4x - 2z = -2 \end{cases}$

68. $\begin{cases} -x + 3y - z = -4 \\ 2x + 6z = 14 \\ -3x - y + z = 10 \end{cases}$

69. $\begin{cases} 2x + 3y + 3z = 3 \\ 6x + 6y + 12z = 13 \\ 12x + 9y - z = 2 \end{cases}$

70. $\begin{cases} -x + 2y + 3z = 4 \\ 2x - 4y - z = -13 \\ 3x + 2y - 4z = -1 \end{cases}$

8.5

Finding the Determinant In Exercises 71–76, find the determinant of the matrix.

71. $\begin{bmatrix} 9 & 8 \\ 10 & 10 \end{bmatrix}$

72. $\begin{bmatrix} -3.4 & 1.2 \\ -5 & 2.5 \end{bmatrix}$

73. $\begin{bmatrix} 8 & 6 & 3 \\ 6 & 3 & 0 \\ 3 & 0 & 2 \end{bmatrix}$

74. $\begin{bmatrix} 7 & -1 & 10 \\ -3 & 0 & -2 \\ 12 & 1 & 1 \end{bmatrix}$

75. $\begin{bmatrix} 8 & 3 & 2 \\ 1 & -2 & 4 \\ 6 & 0 & 5 \end{bmatrix}$

76. $\begin{bmatrix} 4 & 0 & 10 \\ 0 & 10 & 0 \\ 10 & 0 & 34 \end{bmatrix}$

Using Cramer's Rule In Exercises 77–80, use Cramer's Rule to solve the system of linear equations. (If not possible, state the reason.)

77. $\begin{cases} 7x + 12y = 63 \\ 2x + 3y = 15 \end{cases}$

78. $\begin{cases} 12x + 42y = -17 \\ 30x - 18y = 19 \end{cases}$

79. $\begin{cases} -x + y + 2z = 1 \\ 2x + 3y + z = -2 \\ 5x + 4y + 2z = 4 \end{cases}$

80. $\begin{cases} 2x + y + 2z = 4 \\ 2x + 2y = 5 \\ 2x - y + 6z = 2 \end{cases}$

Finding the Area of a Triangle In Exercises 81–84, use a determinant to find the area of the triangle with the given vertices.

81. $(1, 0), (5, 0), (5, 8)$

82. $(-6, 0), (6, 0), (0, 5)$

83. $(1, 2), (4, -5), (3, 2)$

84. $\left(\frac{3}{2}, 1\right), \left(4, -\frac{1}{2}\right), (4, 2)$

Testing for Collinear Points In Exercises 85 and 86, determine whether the points are collinear.

85. $(1, 2), (5, 0), (10, -2)$

86. $(-4, 3), (1, 1), (6, -1)$

Finding an Equation of a Line In Exercises 87 and 88, use a determinant to find an equation of the line passing through the points.

87. $(-4, 0), (4, 4)$

88. $\left(-\frac{5}{2}, 3\right), \left(\frac{7}{2}, 1\right)$

8.6

Graphing a System of Linear Inequalities In Exercises 89–92, sketch the graph of the system of linear inequalities.

89. $\begin{cases} x + y < 5 \\ x > 2 \\ y \ge 0 \end{cases}$

90. $\begin{cases} \frac{1}{2}x + y > 4 \\ x < 6 \\ y < 3 \end{cases}$

91. $\begin{cases} x + 2y \le 160 \\ 3x + y \le 180 \\ x \ge 0 \\ y \ge 0 \end{cases}$

92. $\begin{cases} 2x + 3y \le 24 \\ 2x + y \le 16 \\ x \ge 0 \\ y \ge 0 \end{cases}$

93. Soup Distribution A charitable organization can purchase up to 500 cartons of soup to be divided between a soup kitchen and a homeless shelter. These two organizations need at least 150 cartons and 220 cartons, respectively. Write a system of linear inequalities that describes the various numbers of cartons that can go to each organization. Graph the system.

94. Inventory Costs A warehouse operator has up to 24,000 square feet of floor space in which to store two products. Each unit of product X requires 20 square feet of floor space and costs $12 per day to store. Each unit of product Y requires 30 square feet of floor space and costs $8 per day to store. The total storage cost per day cannot exceed $12,400. Write a system of linear inequalities that describes the various ways the two products can be stored. Graph the system.

Chapter Test

Solutions in English & Spanish and tutorial videos at CollegePrepAlgebra.com

Take this test as you would take a test in class. After you are done, check your work against the answers in the back of the book.

$$\begin{cases} 2x - 2y = 2 \\ -x + 2y = 0 \end{cases}$$

1. Determine whether each ordered pair is a solution of the system at the left.
   (a) $(2, 1)$ (b) $(4, 3)$

In Exercises 2–10, use the indicated method to solve the system.

2. *Graphical:* $\begin{cases} x - 2y = -1 \\ 2x + 3y = 12 \end{cases}$

3. *Substitution:* $\begin{cases} 4x - y = 1 \\ 4x - 3y = -5 \end{cases}$

4. *Substitution:* $\begin{cases} 2x - 2y = -2 \\ 3x + y = 9 \end{cases}$

5. *Elimination:* $\begin{cases} 3x - 4y = -14 \\ -3x + y = 8 \end{cases}$

6. *Elimination:* $\begin{cases} x + 2y - 4z = 0 \\ 3x + y - 2z = 5 \\ 3x - y + 2z = 7 \end{cases}$

7. *Matrices:* $\begin{cases} x \quad\quad - 3z = -10 \\ \quad -2y + 2z = 0 \\ x - 2y \quad\quad = -7 \end{cases}$

8. *Matrices:* $\begin{cases} x - 3y + z = -3 \\ 3x + 2y - 5z = 18 \\ y + z = -1 \end{cases}$

9. *Cramer's Rule:* $\begin{cases} 2x - 7y = 7 \\ 3x + 7y = 13 \end{cases}$

10. *Any Method:* $\begin{cases} 3x - 2y + z = 12 \\ x - 3y \quad\quad = 2 \\ -3x \quad\quad - 9z = -6 \end{cases}$

11. Find the determinant of the matrix shown at the left.

12. Use a determinant to find the area of the triangle with vertices $(0, 0)$, $(5, 4)$, and $(6, 0)$.

13. Graph the system of linear inequalities.
$$\begin{cases} x - 2y > -3 \\ 2x + 3y \le 22 \\ y \ge 0 \end{cases}$$

14. A midsize car costs $24,000 and costs an average of $0.28 per mile to maintain. A minivan costs $26,000 and costs an average of $0.24 per mile to maintain. Determine after how many miles the total costs of the two vehicles will be the same. (Each model is driven the same number of miles).

15. An inheritance of $25,000 is divided among three investments yielding a total of $1275 in interest per year. The interest rates for the three investments are 4.5%, 5%, and 8%. The amounts invested at 5% and 8% are $4000 and $10,000 less than the amount invested at 4.5%, respectively. Find the amount invested at each rate.

16. Two types of tickets are sold for a concert. Reserved seat tickets cost $30 per ticket and floor seat tickets cost $40 per ticket. The promoter of the concert can sell at most 9000 reserved seat tickets and 4000 floor seat tickets. Gross receipts must total at least $300,000 in order for the concert to be held. Write a system of linear inequalities that describes the different numbers of tickets that can be sold. Graph the system.

Radicals and Complex Numbers

9.1   Radicals and Rational Exponents

9.2   Simplifying Radical Expressions

9.3   Adding and Subtracting Radical Expressions

9.4   Multiplying and Dividing Radical Expressions

9.5   Radical Equations and Applications

9.6   Complex Numbers

MASTERY IS WITHIN REACH!

"I learned about study groups by accident last year. I started studying in the learning center. When I saw someone else from my class come into the center, I mentioned something to the tutor, and she suggested asking her over to study with us. We did, and it actually turned into a group of three or four students every session. I learned a lot more and enjoyed it more too."

Jennifer

See page 467 for suggestions about studying in a group.

9.1 Radicals and Rational Exponents

▶ Determine the nth roots of numbers and evaluate radical expressions.

▶ Use the rules of exponents to evaluate or simplify expressions with rational exponents.

▶ Evaluate radical functions and find the domains of radical functions.

Roots and Radicals

Definition of nth Root of a Number

Let a and b be real numbers and let n be an integer such that $n \geq 2$. If

$$a = b^n$$

then b is an **nth root of a**. When $n = 2$, the root is a **square root**. When $n = 3$, the root is a **cube root**.

Study Tip

In the definition at the right, "the nth root that has the same sign as a" means that the principal nth root of a is positive if a is positive and negative if a is negative. For instance, $\sqrt{4} = 2$ and $\sqrt[3]{-8} = -2$. Furthermore, to denote the negative square root of a number, you must use a negative sign in front of the radical. For instance, $-\sqrt{4} = -2$.

Principal nth Root of a Number

Let a be a real number that has at least one (real number) nth root. The **principal nth root of a** is the nth root that has the same sign as a, and it is denoted by the **radical**

$$\sqrt[n]{a}. \qquad \text{Principal } n\text{th root}$$

The positive integer n is the **index** of the radical, and the number a is the **radicand**. When $n = 2$, omit the index and write $\sqrt{a}$ rather than $\sqrt[2]{a}$.

EXAMPLE 1   **Finding Roots of Numbers**

a. $\sqrt{36} = 6$ because $6 \cdot 6 = 6^2 = 36$.

b. $-\sqrt{36} = -6$ because $6 \cdot 6 = 6^2 = 36$. So, $(-1)(\sqrt{36}) = (-1)(6) = -6$.

c. $\sqrt{-4}$ is not real because there is no real number that when multiplied by itself yields -4.

d. $\sqrt[3]{8} = 2$ because $2 \cdot 2 \cdot 2 = 2^3 = 8$.

e. $\sqrt[3]{-8} = -2$ because $(-2)(-2)(-2) = (-2)^3 = -8$.

Exercises Within Reach ®

Solutions in English & Spanish and tutorial videos at CollegePrepAlgebra.com

Finding a Root of a Number In Exercises 1−16, find the root if it exists.

1. $\sqrt{64}$

2. $\sqrt{25}$

3. $-\sqrt{100}$

4. $-\sqrt{49}$

5. $\sqrt{-25}$

6. $\sqrt{-1}$

7. $-\sqrt{-1}$

8. $-\sqrt{-4}$

9. $\sqrt[3]{27}$

10. $\sqrt[3]{64}$

11. $\sqrt[3]{-27}$

12. $\sqrt[3]{-64}$

13. $-\sqrt[3]{1}$

14. $-\sqrt[3]{8}$

15. $-\sqrt[3]{-27}$

16. $-\sqrt[3]{-64}$

Study Tip

The square roots of perfect squares are rational numbers, so $\sqrt{25}$, $\sqrt{49}$, and $\sqrt{\frac{4}{9}}$ are rational numbers. However, square roots such as $\sqrt{5}$, $\sqrt{6}$, and $\sqrt{\frac{2}{5}}$ are irrational numbers. Similarly, $\sqrt[3]{27}$ and $\sqrt[4]{16}$ are rational numbers, whereas $\sqrt[3]{6}$ and $\sqrt[4]{21}$ are irrational numbers.

Properties of *n*th Roots

Property	*Example*
1. If a is a positive real number and n is even, then a has exactly two (real) nth roots, which are denoted by $\sqrt[n]{a}$ and $-\sqrt[n]{a}$.	The two real square roots of 81 are $\sqrt{81} = 9$ and $-\sqrt{81} = -9$.
2. If a is any real number and n is odd, then a has only one (real) nth root, which is denoted by $\sqrt[n]{a}$.	$\sqrt[3]{27} = 3$ $\sqrt[3]{-64} = -4$
3. If a is a negative real number and n is even, then a has no (real) nth root.	$\sqrt{-64}$ is not a real number.

EXAMPLE 2   Classifying Perfect *n*th Powers

State whether each number is a perfect square, a perfect cube, both, or neither.

a. 81   **b.** -125 **c.** 64   **d.** 32

SOLUTION

a. 81 is a perfect square because $9^2 = 81$. It is not a perfect cube.

b. -125 is a perfect cube because $(-5)^3 = -125$. It is not a perfect square.

c. 64 is a perfect square because $8^2 = 64$, and it is also a perfect cube because $4^3 = 64$.

d. 32 is not a perfect square or a perfect cube. (It is, however, a perfect fifth power, because $2^5 = 32$.)

Exercises Within Reach ®

Solutions in English & Spanish and tutorial videos at CollegePrepAlgebra.com

Finding a Perfect Square In Exercises 17–20, **find the perfect square that represents the area of the square.**

17.    **18.**   **19.**   **20.**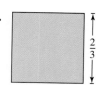

Classifying a Perfect nth Power In Exercises 21–26, **state whether the number is a perfect square, a perfect cube, or neither.**

21. 49

22. -27

23. 1728

24. 964

25. 96

26. 225

Finding the Side Length In Exercises 27–30, **use the area *A* of the square or the volume *V* of the cube to find the side length *l*.**

27.   **28.** 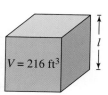   **29.** 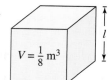   **30.**

Inverse Properties of *n*th Powers and *n*th Roots

Let a be a real number, and let n be an integer such that $n \geq 2$.

Property	Example
1. If a has a principal nth root, then $(\sqrt[n]{a})^n = a.$	$(\sqrt{5})^2 = 5$
2. If n is odd, then $\sqrt[n]{a^n} = a.$	$\sqrt[3]{5^3} = 5$
If n is even, then $\sqrt[n]{a^n} = |a|.$	$\sqrt{(-5)^2} = |-5| = 5$

EXAMPLE 3 Inverse Operations

a. $(\sqrt{4})^2 = (2)^2 = 4$ and $\sqrt{4} = \sqrt{2^2} = 2$

b. $(\sqrt[3]{27})^3 = (3)^3 = 27$ and $\sqrt[3]{27} = \sqrt[3]{3^3} = 3$

c. $(\sqrt[4]{16})^4 = (2)^4 = 16$ and $\sqrt[4]{16} = \sqrt[4]{2^4} = 2$

d. $(\sqrt[5]{-243})^5 = (-3)^5 = -243$ and $\sqrt[5]{-243} = \sqrt[5]{(-3)^5} = -3$

EXAMPLE 4 Evaluating Radical Expressions

Evaluate each radical expression.

a. $\sqrt[3]{4^3}$ b. $\sqrt[3]{(-2)^3}$ c. $(\sqrt{7})^2$ d. $\sqrt{(-3)^2}$

SOLUTION

a. Because the index of the radical is odd, you can write
$$\sqrt[3]{4^3} = 4.$$

b. Because the index of the radical is odd, you can write
$$\sqrt[3]{(-2)^3} = -2.$$

c. Because the radicand is positive, $\sqrt{7}$ is real and you can write
$$(\sqrt{7})^2 = 7.$$

d. Because the index of the radical is even, you must include absolute value signs, and write
$$\sqrt{(-3)^2} = |-3| = 3.$$

Exercises Within Reach ®

Solutions in English & Spanish and tutorial videos at CollegePrepAlgebra.com

Evaluating an Expression In Exercises 31–46, evaluate the radical expression. If not possible, state the reason.

31. $\sqrt{8^2}$

32. $\sqrt{12^2}$

33. $\sqrt[3]{5^3}$

34. $\sqrt[3]{10^3}$

35. $\sqrt[3]{(-7)^3}$

36. $\sqrt[3]{\left(-\frac{1}{4}\right)^3}$

37. $\sqrt{(-10)^2}$

38. $\sqrt{(-12)^2}$

39. $\sqrt{-\left(\frac{3}{10}\right)^2}$

40. $\sqrt{-9^2}$

41. $-\sqrt[3]{\left(\frac{1}{5}\right)^3}$

42. $-\sqrt[3]{9^3}$

43. $-\sqrt[4]{2^4}$

44. $\sqrt[4]{3^4}$

45. $-\sqrt[5]{7^5}$

46. $\sqrt[5]{(-2)^5}$

Rational Exponents

Study Tip

The numerator of a rational exponent denotes the *power* to which the base is raised, and the denominator denotes the *root* to be taken.

> ### Definition of Rational Exponents
>
> Let a be a real number, and let n be an integer such that $n \geq 2$. If the principal nth root of a exists, then $a^{1/n}$ is defined as
>
> $$a^{1/n} = \sqrt[n]{a}.$$
>
> If m is a positive integer that has no common factor with n, then
>
> $$a^{m/n} = (a^{1/n})^m = (\sqrt[n]{a})^m \quad \text{and} \quad a^{m/n} = (a^m)^{1/n} = \sqrt[n]{a^m}.$$

> ### Summary of Rules of Exponents
>
> Let r and s be rational numbers, and let a and b be real numbers, variables, or algebraic expressions. (All denominators and bases are nonzero.)
>
> | *Product and Quotient Rules* | *Example* |
> |---|---|
> | 1. $a^r \cdot a^s = a^{r+s}$ | $4^{1/2}(4^{1/3}) = 4^{5/6}$ |
> | 2. $\dfrac{a^r}{a^s} = a^{r-s}$ | $\dfrac{x^2}{x^{1/2}} = x^{2-(1/2)} = x^{3/2}$ |
>
> | *Power Rules* | |
> |---|---|
> | 3. $(ab)^r = a^r \cdot b^r$ | $(2x)^{1/2} = 2^{1/2}(x^{1/2})$ |
> | 4. $(a^r)^s = a^{rs}$ | $(x^3)^{1/2} = x^{3/2}$ |
> | 5. $\left(\dfrac{a}{b}\right)^r = \dfrac{a^r}{b^r}$ | $\left(\dfrac{x}{3}\right)^{2/3} = \dfrac{x^{2/3}}{3^{2/3}}$ |
>
> | *Zero and Negative Exponent Rules* | |
> |---|---|
> | 6. $a^0 = 1$ | $(3x)^0 = 1$ |
> | 7. $a^{-r} = \dfrac{1}{a^r}$ | $4^{-3/2} = \dfrac{1}{4^{3/2}} = \dfrac{1}{(2)^3} = \dfrac{1}{8}$ |
> | 8. $\left(\dfrac{a}{b}\right)^{-r} = \left(\dfrac{b}{a}\right)^r$ | $\left(\dfrac{x}{4}\right)^{-1/2} = \left(\dfrac{4}{x}\right)^{1/2} = \dfrac{2}{x^{1/2}}$ |

Exercises Within Reach ®

Solutions in English & Spanish and tutorial videos at CollegePrepAlgebra.com

Using a Definition **In Exercises 47−50, determine the missing description.**

Radical Form	*Rational Exponent Form*
47. $\sqrt{36} = 6$	
48. $\sqrt[3]{27^2} = 9$	
49.	$256^{3/4} = 64$
50.	$125^{1/3} = 5$

Writing in Radical Form **In Exercises 51−54, rewrite the expression in radical form.**

51. $x^{1/3}$ **52.** $n^{1/4}$ **53.** $y^{2/5}$ **54.** $u^{5/4}$

| EXAMPLE 5 | **Evaluating Expressions with Rational Exponents** |

a. $8^{4/3} = (8^{1/3})^4 = \left(\sqrt[3]{8}\right)^4 = 2^4 = 16$ Root is 3. Power is 4.

b. $(4^2)^{3/2} = 4^{2 \cdot (3/2)} = 4^{6/2} = 4^3 = 64$ Power-to-power rule

c. $25^{-3/2} = \dfrac{1}{25^{3/2}} = \dfrac{1}{(\sqrt{25})^3} = \dfrac{1}{5^3} = \dfrac{1}{125}$ Root is 2. Power is 3.

d. $\left(\dfrac{64}{125}\right)^{2/3} = \dfrac{64^{2/3}}{125^{2/3}} = \dfrac{\left(\sqrt[3]{64}\right)^2}{\left(\sqrt[3]{125}\right)^2} = \dfrac{4^2}{5^2} = \dfrac{16}{25}$ Root is 3. Power is 2.

e. $-16^{1/2} = -\sqrt{16} = -(4) = -4$ Root is 2. Power is 1.

f. $(-16)^{1/2} = \sqrt{-16}$ is not a real number.   Root is 2. Power is 1.

| EXAMPLE 6 | **Using Rules of Exponents** |

Use rational exponents to rewrite and simplify each expression.

a. $x\sqrt[4]{x^3} = x(x^{3/4}) = x^{1 + (3/4)} = x^{7/4}$

b. $\dfrac{\sqrt[3]{x^2}}{\sqrt{x^3}} = \dfrac{x^{2/3}}{x^{3/2}} = x^{(2/3) - (3/2)} = x^{-5/6} = \dfrac{1}{x^{5/6}}$

c. $\sqrt[3]{x^2 y} = (x^2 y)^{1/3} = (x^2)^{1/3} y^{1/3} = x^{2/3} y^{1/3}$

d. $\sqrt{\sqrt[3]{x}} = \sqrt{x^{1/3}} = (x^{1/3})^{1/2} = x^{(1/3)(1/2)} = x^{1/6}$

e. $\dfrac{(2x - 1)^{4/3}}{\sqrt[3]{2x - 1}} = \dfrac{(2x - 1)^{4/3}}{(2x - 1)^{1/3}} = (2x - 1)^{(4/3) - (1/3)} = (2x - 1)^{3/3} = 2x - 1$

Exercises Within Reach ®

Solutions in English & Spanish and tutorial videos at CollegePrepAlgebra.com

Evaluating an Expression with a Rational Exponent In Exercises 55−66, evaluate the expression.

55. $27^{2/3}$ **56.** $27^{4/3}$ **57.** $(3^3)^{2/3}$

58. $(8^2)^{3/2}$ **59.** $32^{-2/5}$ **60.** $81^{-3/4}$

61. $\left(\dfrac{8}{27}\right)^{2/3}$ **62.** $\left(\dfrac{256}{625}\right)^{1/4}$ **63.** $-36^{1/2}$

64. $-121^{1/2}$ **65.** $(-27)^{-2/3}$ **66.** $(-243)^{-3/5}$

Using Rules of Exponents In Exercises 67−76, use rational exponents to rewrite and simplify the expression.

67. $x\sqrt[3]{x^6}$ **68.** $t\sqrt[5]{t^2}$

69. $\dfrac{\sqrt[4]{t}}{\sqrt{t^5}}$ **70.** $\dfrac{\sqrt[3]{x^4}}{\sqrt{x^3}}$

71. $\sqrt[4]{x^3 y}$ **72.** $\sqrt[3]{u^4 v^2}$

73. $\sqrt{\sqrt[4]{y}}$ **74.** $\sqrt[3]{\sqrt{2x}}$

75. $\dfrac{(x + y)^{3/4}}{\sqrt[4]{x + y}}$ **76.** $\dfrac{(a - b)^{1/3}}{\sqrt[3]{a - b}}$

Radical Functions

| EXAMPLE 7 | **Evaluating a Radical Function** |

Evaluate each radical function when $x = 4$.

a. $f(x) = \sqrt[3]{x - 31}$ b. $g(x) = \sqrt{16 - 3x}$

SOLUTION

a. $f(4) = \sqrt[3]{4 - 31} = \sqrt[3]{-27} = -3$

b. $g(4) = \sqrt{16 - 3(4)} = \sqrt{16 - 12} = \sqrt{4} = 2$

Domain of a Radical Function

Let n be an integer that is greater than or equal to 2.

1. If n is odd, the domain of $f(x) = \sqrt[n]{x}$ is the set of all real numbers.

2. If n is even, the domain of $f(x) = \sqrt[n]{x}$ is the set of all nonnegative real numbers.

| EXAMPLE 8 | **Finding the Domain of a Radical Function** |

a. The domain of $f(x) = \sqrt[3]{x}$ is the set of all real numbers because for any real number x, the expression $\sqrt[3]{x}$ is a real number.

b. The domain of $f(x) = \sqrt{x^3}$ is the set of all nonnegative real numbers. For instance, 1 is in the domain but -1 is not because $\sqrt{(-1)^3} = \sqrt{-1}$ is not a real number.

| EXAMPLE 9 | **Finding the Domain of a Radical Function** |

Find the domain of $f(x) = \sqrt{2x - 1}$.

SOLUTION

The domain of f consists of all x such that $2x - 1 \geq 0$. Using the methods described in Section 3.6, you can solve this inequality as follows.

$2x - 1 \geq 0$ Write original inequality.

$2x \geq 1$ Add 1 to each side.

$x \geq \frac{1}{2}$ Divide each side by 2.

So, the domain is the set of all real numbers x such that $x \geq \frac{1}{2}$.

Study Tip

In general, when the index n of a radical function is even, the domain of the function includes all real values for which the expression under the radical is greater than or equal to zero.

Exercises Within Reach ®

Solutions in English & Spanish and tutorial videos at CollegePrepAlgebra.com

Evaluating a Radical Function **In Exercises 77 and 78, evaluate the function for each indicated** x**-value, if possible, and simplify.**

77. $f(x) = \sqrt{2x + 9}$

     (a) $f(0)$ (b) $f(8)$ (c) $f(-6)$ (d) $f(36)$

78. $f(x) = \sqrt[3]{2x - 1}$

     (a) $f(0)$ (b) $f(-62)$ (c) $f(-13)$ (d) $f(63)$

Finding the Domain of a Radical Function **In Exercises 79−84, describe the domain of the function.**

79. $f(x) = \sqrt[5]{x}$

80. $f(x) = \sqrt[3]{2x}$

81. $f(x) = 3\sqrt{x}$

82. $h(x) = \sqrt[4]{x}$

83. $h(x) = \sqrt{2x + 9}$

84. $f(x) = \sqrt{3x - 5}$

Concept Summary: *Using Rational Exponents*

What

You can use **rational exponents** and the rules of exponents to simplify some **radical** expressions.

EXAMPLE

Simplify the radical expression $\sqrt[3]{x} \cdot \sqrt[4]{x^3}$.

How

Use the definition of rational exponents to write the radicals in exponent form.

- $a^{1/n} = \sqrt[n]{a}$
- $a^{m/n} = \left(\sqrt[n]{a}\right)^m = \sqrt[n]{a^m}$

Then use the rules of exponents to simplify the expression.

EXAMPLE

$$\sqrt[3]{x} \cdot \sqrt[4]{x^3} = x^{1/3} \cdot x^{3/4}$$
$$= x^{(1/3) + (3/4)}$$
$$= x^{(4/12) + (9/12)}$$
$$= x^{13/12}$$

Why

As you continue your study of mathematics, you will use and build upon the concepts that you learned in this section.

Exercises Within Reach ®

Worked-out solutions to odd-numbered exercises at CollegePrepAlgebra.com

Concept Summary Check

85. *Structure* Write an expression that represents the principal cube root of x in radical form and in rational exponent form.

86. *Structure* Write an expression that represents the principal fourth root of the cube of x in radical form and in rational exponent form.

87. *Identifying a Rule* What rule of exponents is used to write $x^{1/3} \cdot x^{3/4}$ as $x^{1/3 \,+\, 3/4}$?

88. *Writing the Radical Form* Write the simplified form of the expression $\sqrt[3]{x} \cdot \sqrt[4]{x^3}$ in radical form in two ways.

Extra Practice

Finding Square Roots In Exercises 89−92, find **all of the square roots of the perfect square.**

89. $\frac{9}{16}$

90. $\frac{25}{36}$

91. 0.16

92. 0.25

Finding Cube Roots In Exercises 93−96, find **all of the cube roots of the perfect cube.**

93. $\frac{1}{1000}$

94. $-\frac{8}{125}$

95. 0.001

96. -0.008

Using Rules of Exponents In Exercises 97−108, use rational exponents to rewrite and simplify the expression.

97. $u^2 \sqrt[3]{u}$

98. $y \sqrt[4]{y^2}$

99. $\dfrac{\sqrt{x}}{\sqrt{x^3}}$

100. $\dfrac{\sqrt[3]{x^2}}{\sqrt[3]{x^4}}$

101. $\sqrt[4]{y^3} \cdot \sqrt[3]{y}$

102. $\sqrt[6]{x^5} \cdot \sqrt[3]{x^4}$

103. $z^2 \sqrt[3]{y^5 z^4}$

104. $x^2 \sqrt[3]{xy^4}$

105. $\sqrt[4]{\sqrt{x^3}}$

106. $\sqrt[5]{\sqrt[3]{y^4}}$

107. $\dfrac{(3u - 2v)^{2/3}}{\sqrt{(3u - 2v)^3}}$

108. $\dfrac{\sqrt[4]{2x + y}}{(2x + y)^{3/2}}$

Mathematical Modeling **In Exercises 109 and 110, use the formula for the *declining balances method***

$$r = 1 - \left(\frac{S}{C}\right)^{1/n}$$

to find the depreciation rate *r* (in decimal form). In the formula, *n* is the useful life of the item (in years), *S* is the salvage value (in dollars), and *C* is the original cost (in dollars).

109. A $75,000 truck depreciates over an eight-year period, as shown in the graph. Find *r*. (Round your answer to three decimal places.)

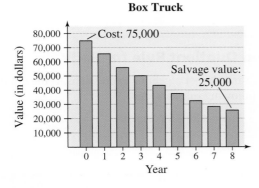

110. A $125,000 stretch limousine depreciates over a 10-year period, as shown in the graph. Find *r*. (Round your answer to three decimal places.)

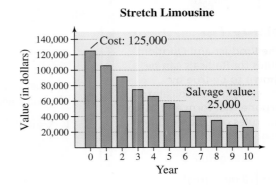

111. *Geometry* Find the dimensions of a piece of carpet for a square classroom with 529 square feet of floor space.

112. *Geometry* Find the dimensions of a square mirror with an area of 1024 square inches.

Explaining Concepts

113. *Vocabulary* What is the *n*th root of a number?

114. *Structure* Explain how you can determine the domain of a function that has the form of a fraction with radical expressions in both the numerator and the denominator.

115. *Precision* Is it true that $\sqrt{2} = 1.414$? Explain.

116. *Number Sense* Given that *x* represents a real number, state the conditions on *n* for each of the following.

(a) $\sqrt[n]{x^n} = x$ (b) $\sqrt[n]{x^n} = |x|$

117. *Investigation* Find all possible "last digits" of perfect squares. (For instance, the last digit of 81 is 1 and the last digit of 64 is 4.) Is it possible that 4,322,788,986 is a perfect square?

118. *Number Sense* Use what you know about the domains of radical functions to write a set of rules for the domain of a rational exponent function of the form $f(x) = x^{1/n}$. Can the same rules be used for a function of the form $f(x) = x^{m/n}$? Explain.

Cumulative Review

In Exercises 119−122, solve the equation.

119. $\dfrac{a}{5} = \dfrac{a-3}{2}$ **120.** $\dfrac{x}{3} - \dfrac{3x}{4} = \dfrac{5x}{12}$

121. $\dfrac{2}{u+4} = \dfrac{5}{8}$

122. $\dfrac{6}{b} + 22 = 24$

In Exercises 123−126, write a model for the statement.

123. *s* is directly proportional to the square of *t*.

124. *r* varies inversely as the fourth power of *x*.

125. *a* varies jointly as *b* and *c*.

126. *x* is directly proportional to *y* and inversely proportional to *z*.

9.2 Simplifying Radical Expressions

▶ Use the Product and Quotient Rules for Radicals to simplify radical expressions.

▶ Use rationalization techniques to simplify radical expressions.

▶ Use the Pythagorean Theorem in application problems.

Simplifying Radicals

Study Tip

The Product and Quotient Rules for Radicals can be shown to be true by converting the radicals to exponential form and using the rules of exponents on page 445.

Using Rule 3

$$\sqrt[n]{uv} = (uv)^{1/n}$$
$$= u^{1/n}v^{1/n}$$
$$= \sqrt[n]{u}\,\sqrt[n]{v}$$

Using Rule 5

$$\sqrt[n]{\frac{u}{v}} = \left(\frac{u}{v}\right)^{1/n}$$
$$= \frac{u^{1/n}}{v^{1/n}} = \frac{\sqrt[n]{u}}{\sqrt[n]{v}}$$

Product and Quotient Rule for Radicals

Let u and v be real numbers, variables, or algebraic expressions. If the nth roots of u and v are real, then the following rules are true.

1. $\sqrt[n]{uv} = \sqrt[n]{u}\,\sqrt[n]{v}$ Product Rule for Radicals

2. $\sqrt[n]{\dfrac{u}{v}} = \dfrac{\sqrt[n]{u}}{\sqrt[n]{v}},\ v \neq 0$ Quotient Rule for Radicals

---

EXAMPLE 1 **Removing Constant Factors from Radicals**

Simplify each radical expression by removing as many factors as possible.

a. $\sqrt{75}$ **b.** $\sqrt{72}$ **c.** $\sqrt{162}$

SOLUTION

a. $\sqrt{75} = \sqrt{25 \cdot 3} = \sqrt{25}\sqrt{3} = 5\sqrt{3}$ 25 is a perfect square factor of 75.

b. $\sqrt{72} = \sqrt{36 \cdot 2} = \sqrt{36}\sqrt{2} = 6\sqrt{2}$ 36 is a perfect square factor of 72.

c. $\sqrt{162} = \sqrt{81 \cdot 2} = \sqrt{81}\sqrt{2} = 9\sqrt{2}$ 81 is a perfect square factor of 162.

Exercises Within Reach ®

Solutions in English & Spanish and tutorial videos at CollegePrepAlgebra.com

Removing a Constant Factor from a Radical In Exercises 1 and 2, simplify the radical expression by removing as many factors as possible. Does the figure support your answer? Explain.

1. $\sqrt{28}$

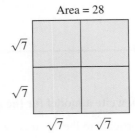

2. $\sqrt{54}$

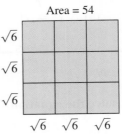

Removing a Constant Factor from a Radical In Exercises 3−14, simplify the radical expression.

3. $\sqrt{8}$ **4.** $\sqrt{12}$ **5.** $\sqrt{18}$ **6.** $\sqrt{27}$

7. $\sqrt{45}$ **8.** $\sqrt{125}$ **9.** $\sqrt{96}$ **10.** $\sqrt{84}$

11. $\sqrt{153}$ **12.** $\sqrt{147}$ **13.** $\sqrt{1183}$ **14.** $\sqrt{1176}$

EXAMPLE 2   **Removing Variable Factors from Radicals**

Simplify each radical expression.

a. $\sqrt{25x^2}$ b. $\sqrt{12x^3}$

c. $\sqrt{144x^4}$ d. $\sqrt{72x^3y^2}$

SOLUTION

a. $\sqrt{25x^2} = \sqrt{5^2x^2} = \sqrt{5^2}\sqrt{x^2}$ Product Rule for Radicals

$\qquad = 5|x|$ $\sqrt{x^2} = |x|$

b. $\sqrt{12x^3} = \sqrt{2^2x^2(3x)} = \sqrt{2^2}\sqrt{x^2}\sqrt{3x}$ Product Rule for Radicals

$\qquad = 2x\sqrt{3x}$ $\sqrt{2^2}\sqrt{x^2} = 2x,\ x \geq 0$

c. $\sqrt{144x^4} = \sqrt{12^2(x^2)^2} = \sqrt{12^2}\sqrt{(x^2)^2}$ Product Rule for Radicals

$\qquad = 12x^2$ $\sqrt{12^2}\sqrt{(x^2)^2} = 12|x^2| = 12x^2$

d. $\sqrt{72x^3y^2} = \sqrt{6^2x^2y^2} \cdot \sqrt{2x}$ Product Rule for Radicals

$\qquad = \sqrt{6^2}\sqrt{x^2}\sqrt{y^2} \cdot \sqrt{2x}$ Product Rule for Radicals

$\qquad = 6x|y|\sqrt{2x}$ $\sqrt{6^2}\sqrt{x^2}\sqrt{y^2} = 6x|y|,\ x \geq 0$

EXAMPLE 3   **Removing Factors from Radicals**

Simplify each radical expression.

a. $\sqrt[3]{40}$

b. $\sqrt[4]{x^5}$

SOLUTION

a. $\sqrt[3]{40} = \sqrt[3]{8(5)} = \sqrt[3]{2^3} \cdot \sqrt[3]{5}$ Product Rule for Radicals

$\qquad = 2\sqrt[3]{5}$ $\sqrt[3]{2^3} = 2$

b. $\sqrt[4]{x^5} = \sqrt[4]{x^4(x)} = \sqrt[4]{x^4}\sqrt[4]{x}$ Product Rule for Radicals

$\qquad = x\sqrt[4]{x}$ $\sqrt[4]{x^4} = x,\ x \geq 0$

Exercises Within Reach ®

Solutions in English & Spanish and tutorial videos at CollegePrepAlgebra.com

Removing Factors from a Radical In Exercises 15–34, simplify the radical expression.

15. $\sqrt{4y^2}$ 16. $\sqrt{100x^2}$ 17. $\sqrt{9x^5}$ 18. $\sqrt{64x^3}$

19. $\sqrt{48y^4}$ 20. $\sqrt{32x}$ 21. $\sqrt{117y^6}$ 22. $\sqrt{160x^8}$

23. $\sqrt{120x^2y^3}$ 24. $\sqrt{125u^4v^6}$ 25. $\sqrt{192a^5b^7}$ 26. $\sqrt{363x^{10}y^9}$

27. $\sqrt[3]{48}$ 28. $\sqrt[3]{54}$ 29. $\sqrt[3]{112}$ 30. $\sqrt[4]{112}$

31. $\sqrt[4]{x^7}$ 32. $\sqrt[4]{x^9}$ 33. $\sqrt[4]{x^6}$ 34. $\sqrt[4]{x^{14}}$

Study Tip

When you write a number as a product of prime numbers, you are writing its *prime factorization*. To find the perfect nth root factor of 486 in Example 4(a), you can write the prime factorization of 486.

$$486 = 2 \cdot 3 \cdot 3 \cdot 3 \cdot 3$$
$$= 2 \cdot 3^5$$

From its prime factorization, you can see that 3^5 is a fifth root factor of 486.

$$\sqrt[5]{486} = \sqrt[5]{2 \cdot 3^5}$$
$$= \sqrt[5]{3^5}\sqrt[5]{2}$$
$$= 3\sqrt[5]{2}$$

EXAMPLE 4 **Removing Factors from Radicals**

a. $\sqrt[5]{486x^7} = \sqrt[5]{243x^5(2x^2)}$ Factor out perfect 5th powers.

$\quad = \sqrt[5]{3^5x^5} \cdot \sqrt[5]{2x^2}$ Product Rule for Radicals

$\quad = 3x\sqrt[5]{2x^2}$ $\sqrt[5]{3^5}\sqrt[5]{x^5} = 3x$

b. $\sqrt[3]{128x^3y^5} = \sqrt[3]{64x^3y^3(2y^2)}$ Factor out perfect 3rd powers.

$\quad = \sqrt[3]{4^3x^3y^3} \cdot \sqrt[3]{2y^2}$ Product Rule for Radicals

$\quad = 4xy\sqrt[3]{2y^2}$ $\sqrt[3]{4^3}\sqrt[3]{x^3}\sqrt[3]{y^3} = 4xy$

EXAMPLE 5 **Removing Factors from Radicals**

a. $\sqrt{\dfrac{81}{25}} = \dfrac{\sqrt{81}}{\sqrt{25}} = \dfrac{9}{5}$ Quotient Rule for Radicals

b. $\dfrac{\sqrt{56x^2}}{\sqrt{8}} = \sqrt{\dfrac{56x^2}{8}}$ Quotient Rule for Radicals

$\quad = \sqrt{7x^2}$ Simplify.

$\quad = \sqrt{7} \cdot \sqrt{x^2}$ Product Rule for Radicals

$\quad = \sqrt{7}\,|x|$ $\sqrt{x^2} = |x|$

EXAMPLE 6 **Removing Factors from Radicals**

$$-\sqrt[3]{\dfrac{y^5}{27x^3}} = -\dfrac{\sqrt[3]{y^3y^2}}{\sqrt[3]{27x^3}}$$ Quotient Rule for Radicals

$$= -\dfrac{\sqrt[3]{y^3} \cdot \sqrt[3]{y^2}}{\sqrt[3]{27} \cdot \sqrt[3]{x^3}}$$ Product Rule for Radicals

$$= -\dfrac{y\sqrt[3]{y^2}}{3x}$$ Simplify.

Exercises Within Reach®   Solutions in English & Spanish and tutorial videos at CollegePrepAlgebra.com

Removing Factors from a Radical **In Exercises 35−54, simplify the radical expression.**

35. $\sqrt[3]{40x^5}$ 36. $\sqrt[3]{81a^7}$ 37. $\sqrt[4]{324y^6}$ 38. $\sqrt[4]{160x^6}$

39. $\sqrt[3]{x^4y^3}$ 40. $\sqrt[3]{a^5b^6}$ 41. $\sqrt[4]{4x^4y^6}$ 42. $\sqrt[4]{128u^4v^7}$

43. $\sqrt[5]{32x^5y^6}$ 44. $\sqrt[3]{16x^4y^5}$ 45. $\sqrt{\dfrac{16}{9}}$ 46. $\sqrt{\dfrac{36}{49}}$

47. $\sqrt[3]{\dfrac{35}{64}}$ 48. $\sqrt[4]{\dfrac{5}{16}}$ 49. $\dfrac{\sqrt{39y^2}}{\sqrt{3}}$ 50. $\dfrac{\sqrt{56w^3}}{\sqrt{2}}$

51. $\sqrt{\dfrac{32a^4}{b^2}}$ 52. $\sqrt{\dfrac{18x^2}{z^6}}$ 53. $\sqrt[5]{\dfrac{32x^2}{y^5}}$ 54. $\sqrt[3]{\dfrac{16z^3}{y^6}}$

Rationalization Techniques

Study Tip

When rationalizing a denominator, remember that for square roots, you want a perfect square in the denominator, for cube roots, you want a perfect cube, and so on. For instance, to find the rationalizing factor needed to create a perfect square in the denominator of Example 7(c), you can write the prime factorization of 18.

$$18 = 2 \cdot 3 \cdot 3$$
$$= 2 \cdot 3^2$$

From its prime factorization, you can see that 3^2 is a square root factor of 18. You need one more factor of 2 to create a perfect square in the denominator.

$$2 \cdot (2 \cdot 3^2) = 2 \cdot 2 \cdot 3^2$$
$$= 2^2 \cdot 3^2$$
$$= 4 \cdot 9 = 36$$

Simplifying Radical Expressions

A radical expression is said to be in *simplest form* when all three of the statements below are true.

1. All possible *n*th-powered factors have been removed from each radical.
2. No radical contains a fraction.
3. No denominator of a fraction contains a radical.

EXAMPLE 7   **Rationalizing Denominators**

a. $\sqrt{\dfrac{3}{5}} = \dfrac{\sqrt{3}}{\sqrt{5}} = \dfrac{\sqrt{3}}{\sqrt{5}} \cdot \dfrac{\sqrt{5}}{\sqrt{5}} = \dfrac{\sqrt{15}}{\sqrt{5^2}} = \dfrac{\sqrt{15}}{5}$ Multiply by $\sqrt{5}/\sqrt{5}$ to create a perfect square in the denominator.

b. $\dfrac{4}{\sqrt[3]{9}} = \dfrac{4}{\sqrt[3]{9}} \cdot \dfrac{\sqrt[3]{3}}{\sqrt[3]{3}} = \dfrac{4\sqrt[3]{3}}{\sqrt[3]{27}} = \dfrac{4\sqrt[3]{3}}{3}$ Multiply by $\sqrt[3]{3}/\sqrt[3]{3}$ to create a perfect cube in the denominator.

c. $\dfrac{8}{3\sqrt{18}} = \dfrac{8}{3\sqrt{18}} \cdot \dfrac{\sqrt{2}}{\sqrt{2}} = \dfrac{8\sqrt{2}}{3\sqrt{36}} = \dfrac{8\sqrt{2}}{3\sqrt{6^2}}$ Multiply by $\sqrt{2}/\sqrt{2}$ to create a perfect square in the denominator.

$$= \dfrac{8\sqrt{2}}{3(6)} = \dfrac{4\sqrt{2}}{9}$$

EXAMPLE 8   **Rationalizing Denominators**

a. $\sqrt{\dfrac{8x}{12y^5}} = \sqrt{\dfrac{(4)(2)x}{(4)(3)y^5}} = \sqrt{\dfrac{2x}{3y^5}} = \dfrac{\sqrt{2x}}{\sqrt{3y^5}} \cdot \dfrac{\sqrt{3y}}{\sqrt{3y}} = \dfrac{\sqrt{6xy}}{\sqrt{3^2y^6}} = \dfrac{\sqrt{6xy}}{3|y^3|}$

b. $\sqrt[3]{\dfrac{54x^6y^3}{5z^2}} = \dfrac{\sqrt[3]{(3^3)(2)(x^6)(y^3)}}{\sqrt[3]{5z^2}} \cdot \dfrac{\sqrt[3]{25z}}{\sqrt[3]{25z}} = \dfrac{3x^2y\sqrt[3]{50z}}{\sqrt[3]{5^3z^3}} = \dfrac{3x^2y\sqrt[3]{50z}}{5z}$

Exercises Within Reach ®

Solutions in English & Spanish and tutorial videos at CollegePrepAlgebra.com

Rationalizing the Denominator **In Exercises 55−74, rationalize the denominator and simplify further, if possible.**

55. $\sqrt{\dfrac{1}{3}}$

56. $\sqrt{\dfrac{1}{5}}$

57. $\dfrac{1}{\sqrt{7}}$

58. $\dfrac{12}{\sqrt{3}}$

59. $\sqrt[4]{\dfrac{5}{4}}$

60. $\sqrt[3]{\dfrac{9}{25}}$

61. $\dfrac{6}{\sqrt[3]{32}}$

62. $\dfrac{10}{\sqrt[5]{16}}$

63. $\dfrac{1}{\sqrt{y}}$

64. $\dfrac{2}{\sqrt{3c}}$

65. $\sqrt{\dfrac{4}{x}}$

66. $\sqrt{\dfrac{4}{x^3}}$

67. $\dfrac{1}{x\sqrt{2}}$

68. $\dfrac{1}{3x\sqrt{x}}$

69. $\dfrac{6}{\sqrt{3b^3}}$

70. $\dfrac{1}{\sqrt{xy}}$

71. $\sqrt[3]{\dfrac{2x}{3y}}$

72. $\sqrt[3]{\dfrac{20x^2}{9y^2}}$

73. $\sqrt[3]{\dfrac{24x^3y^4}{25z}}$

74. $\sqrt[3]{\dfrac{3y^4z^6}{16x^5}}$

Applications of Radicals

Radicals commonly occur in applications involving right triangles. Recall that a right triangle is one that contains a right (or 90°) angle, as shown below.

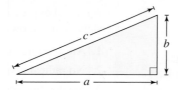

The relationship among the three sides of a right triangle is described by the **Pythagorean Theorem**, which states that if a and b are the lengths of the legs and c is the length of the hypotenuse, then

$$c = \sqrt{a^2 + b^2} \quad \text{and} \quad a = \sqrt{c^2 - b^2}. \qquad \text{Pythagorean Theorem: } a^2 + b^2 = c^2$$

Application **EXAMPLE 9** **Geometry: The Pythagorean Theorem**

Find the length of the hypotenuse of the right triangle shown below.

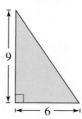

SOLUTION

Letting $a = 6$ and $b = 9$, use the Pythagorean Theorem to find c, as follows.

$$c = \sqrt{a^2 + b^2} \qquad \text{Pythagorean Theorem}$$
$$= \sqrt{6^2 + 9^2} \qquad \text{Substitute 6 for } a \text{ and 9 for } b.$$
$$= \sqrt{117} \qquad \text{Simplify.}$$
$$= \sqrt{9}\sqrt{13} \qquad \text{Product Rule for Radicals}$$
$$= 3\sqrt{13} \qquad \text{Simplify.}$$

Exercises Within Reach ®

Solutions in English & Spanish and tutorial videos at CollegePrepAlgebra.com

Geometry In Exercises 75−78, find the length of the hypotenuse of the right triangle.

75.

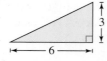

76.

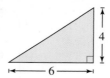

77.

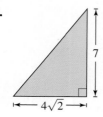

78.

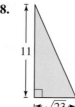

Application EXAMPLE 10 **Dimensions of a Softball Diamond**

A softball diamond has the shape of a square with 60-foot sides, as shown below. The catcher is 5 feet behind home plate. How far does the catcher have to throw to reach second base?

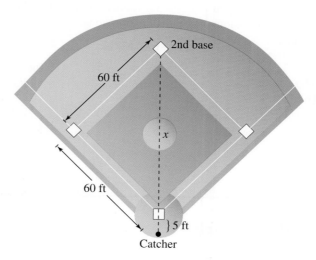

Let x be the hypotenuse of a right triangle with 60-foot sides. So, by the Pythagorean Theorem, you have the following.

$x = \sqrt{60^2 + 60^2}$	Pythagorean Theorem
$= \sqrt{7200}$	Simplify.
$= \sqrt{3600}\,\sqrt{2}$	Product Rule for Radicals
$= 60\sqrt{2}$	Simplify.
≈ 84.9 feet	Use a calculator.

So, the distance from home plate to second base is approximately 84.9 feet. Because the catcher is 5 feet behind home plate, the catcher must make a throw of

$$x + 5 \approx 84.9 + 5$$
$$= 89.9 \text{ feet.}$$

Exercises Within Reach ®   Solutions in English & Spanish and tutorial videos at CollegePrepAlgebra.com

79. **Geometry** A ladder is to reach a window that is 26 feet high. The ladder is placed 10 feet from the base of the wall (see figure). How long must the ladder be?

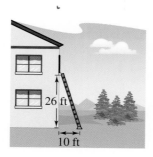

80. **Geometry** A string is attached to opposite corners of a piece of wood that is 6 inches wide and 14 inches long (see figure). How long must the string be?

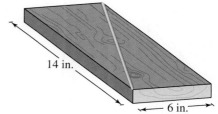

Concept Summary: *Simplifying Radical Expressions*

What

You already knew how to simplify algebraic expressions. Now you can simplify radical expressions involving constants.

EXAMPLE

Simplify $\sqrt{\dfrac{54}{16}}$.

How

Use the Product and Quotient Rules for Radicals to simplify these types of expressions.

EXAMPLE

$$\sqrt{\frac{54}{16}} = \frac{\sqrt{54}}{\sqrt{16}} \qquad \text{Quotient Rule for Radicals}$$

$$= \frac{\sqrt{9} \cdot \sqrt{6}}{\sqrt{16}} \qquad \text{Product Rule for Radicals}$$

$$= \frac{3\sqrt{6}}{4} \qquad \text{Simplify.}$$

Why

You can also use these rules to simplify radical expressions involving variables. It is trickier than simplifying radicals involving only constants. Just remember, $\sqrt{x^2} = |x|$.

Exercises Within Reach ®

Worked-out solutions to odd-numbered exercises at CollegePrepAlgebra.com

Concept Summary Check

81. *Stating a Rule* State the Product Rule for Radicals in words and give an example.

82. *Stating a Rule* State the Quotient Rule for Radicals in words and give an example.

83. *Writing* In your own words, describe the three conditions that must be true for a radical expression to be in simplest form.

84. *Precision* Explain why the Product Rule for Radicals cannot be applied to the expression $\sqrt{-8} \cdot \sqrt{-2}$.

Extra Practice

Simplifying a Radical Expression In Exercises 85–98, simplify the radical expression.

85. $\sqrt{0.04}$

86. $\sqrt{0.25}$

87. $\sqrt{0.0072}$

88. $\sqrt{0.0027}$

89. $\sqrt{\dfrac{60}{3}}$

90. $\sqrt{\dfrac{208}{4}}$

91. $\sqrt{\dfrac{13}{25}}$

92. $\sqrt{\dfrac{15}{36}}$

93. $\sqrt[3]{\dfrac{54a^4}{b^9}}$

94. $\sqrt[4]{\dfrac{3u^2}{16v^8}}$

95. $\sqrt{4 \times 10^{-4}}$

96. $\sqrt{8.5 \times 10^3}$

97. $\sqrt[3]{2.4 \times 10^6}$

98. $\sqrt[4]{3.2 \times 10^5}$

99. *Frequency* The frequency f (in cycles per second) of a vibrating string is given by

$$f = \frac{1}{100} \sqrt{\frac{400 \times 10^6}{5}}.$$

Use a calculator to approximate this number. (Round the result to two decimal places.)

100. *Period of a Pendulum* The time t (in seconds) for a pendulum of length L (in feet) to go through one complete cycle (its period) is given by

$$t = 2\pi \sqrt{\frac{L}{32}}.$$

Find the period of a pendulum whose length is 4 feet. (Round your answer to two decimal places.)

101. *Geometry* The foundation of a house is 40 feet long and 30 feet wide. The roof rises 8 feet vertically and forms two congruent right triangles with the side of the house, as shown. (Assume there is no overhang.)

(a) Use the Pythagorean Theorem to find the length of the hypotenuse of each right triangle formed by the roof line.

(b) Use the result of part (a) to determine the total area of the roof.

Explaining Concepts

102. *Number Sense* When is $\sqrt{x^2} \neq x$? Explain.

103. *Writing* Explain why $\sqrt{8}$ is not in simplest form.

104. *Precision* Describe how you would simplify $\dfrac{1}{\sqrt{3}}$.

105. *Number Sense* Enter any positive real number into your calculator and find its square root. Then repeatedly take the square root of the result.

$$\sqrt{x}, \ \sqrt{\sqrt{x}}, \ \sqrt{\sqrt{\sqrt{x}}}, \dots$$

What real number do the results appear to be approaching?

106. *Think About It* Square the real number $\dfrac{5}{\sqrt{3}}$ and note that the radical is eliminated from the denominator. Is this equivalent to rationalizing the denominator? Why or why not?

107. *Writing* Let u be a positive real number. Explain why $\sqrt[3]{u} \cdot \sqrt[4]{u} \neq \sqrt[12]{u}$.

108. *Writing* Explain how to find a perfect nth root factor in the radicand of an nth root radical.

Cumulative Review

In Exercises 109 and 110, solve the system of equations by graphing.

109. $\begin{cases} 2x + 3y = 12 \\ 4x - y = 10 \end{cases}$

110. $\begin{cases} 3x + 2y = -4 \\ y = 3x + 7 \end{cases}$

In Exercises 111 and 112, solve the system of equations by the method of substitution.

111. $\begin{cases} y = x + 2 \\ y - x = 8 \end{cases}$

112. $\begin{cases} x - 3y = -2 \\ 7y - 4x = 6 \end{cases}$

In Exercises 113 and 114, solve the system of equations by the method of elimination.

113. $\begin{cases} x + 4y + 3z = 2 \\ 2x + y + z = 10 \\ -x + y + 2z = 8 \end{cases}$

114. $\begin{cases} 1.5x - 3 = -2y \\ 3x + 4y = 6 \end{cases}$

9.3 Adding and Subtracting Radical Expressions

▶ Use the Distributive Property to add and subtract like radicals.

▶ Use radical expressions in application problems.

Adding and Subtracting Radical Expressions

Two or more radical expressions are called **like radicals** when they have the same index and the same radicand. For instance, the expressions $\sqrt{2}$ and $3\sqrt{2}$ are like radicals, whereas the expressions $\sqrt{3}$ and $\sqrt[3]{3}$ are not. Two radical expressions that are like radicals can be added or subtracted by adding or subtracting their coefficients.

EXAMPLE 1   Combining Radical Expressions

Simplify each expression by combining like radicals.

a. $\sqrt{7} + 5\sqrt{7} - 2\sqrt{7}$

b. $6\sqrt{x} - \sqrt[3]{4} - 5\sqrt{x} + 2\sqrt[3]{4}$

c. $3\sqrt[3]{x} + 2\sqrt[3]{x} + \sqrt{x} - 8\sqrt{x}$

SOLUTION

a. $\sqrt{7} + 5\sqrt{7} - 2\sqrt{7} = (1 + 5 - 2)\sqrt{7}$ Distributive Property

$\qquad\qquad\qquad\qquad = 4\sqrt{7}$ Simplify.

b. $6\sqrt{x} - \sqrt[3]{4} - 5\sqrt{x} + 2\sqrt[3]{4}$

$\qquad = (6\sqrt{x} - 5\sqrt{x}) + \left(-\sqrt[3]{4} + 2\sqrt[3]{4}\right)$ Group like radicals.

$\qquad = (6 - 5)\sqrt{x} + (-1 + 2)\sqrt[3]{4}$ Distributive Property

$\qquad = \sqrt{x} + \sqrt[3]{4}$ Simplify.

c. $3\sqrt[3]{x} + 2\sqrt[3]{x} + \sqrt{x} - 8\sqrt{x}$

$\qquad = (3 + 2)\sqrt[3]{x} + (1 - 8)\sqrt{x}$ Distributive Property

$\qquad = 5\sqrt[3]{x} - 7\sqrt{7}$ Simplify.

Exercises Within Reach ®

Solutions in English & Spanish and tutorial videos at CollegePrepAlgebra.com

Combining Radical Expressions In Exercises 1−18, simplify the expression by combining like radicals.

1. $3\sqrt{2} - \sqrt{2}$

2. $6\sqrt{5} - 2\sqrt{5}$

3. $4\sqrt[3]{y} + 9\sqrt[3]{y}$

4. $3\sqrt[3]{3} + 6\sqrt[3]{3}$

5. $\sqrt{7} + 3\sqrt{7} - 2\sqrt{7}$

6. $\sqrt{15} + 4\sqrt{15} - 2\sqrt{15}$

7. $8\sqrt{2} + 6\sqrt{2} - 5\sqrt{2}$

8. $2\sqrt{6} + 8\sqrt{6} - 3\sqrt{6}$

9. $2\sqrt{2} + 5\sqrt{2} - \sqrt{2} + 3\sqrt{2}$

10. $4\sqrt{5} - 3\sqrt{5} - 2\sqrt{5} + 12\sqrt{5}$

11. $\sqrt[4]{5} - 6\sqrt[4]{13} + 3\sqrt[4]{5} - \sqrt[4]{13}$

12. $9\sqrt[3]{17} + 7\sqrt[3]{2} - 4\sqrt[3]{17} + \sqrt[3]{2}$

13. $9\sqrt[3]{7} - \sqrt{3} + 4\sqrt[3]{7} + 2\sqrt{3}$

14. $5\sqrt{7} - 8\sqrt[4]{11} + \sqrt{7} + 9\sqrt[4]{11}$

15. $7\sqrt{x} + 5\sqrt[3]{9} - 3\sqrt{x} + 3\sqrt[3]{9}$

16. $4\sqrt{5} - 3\sqrt[4]{x} - 7\sqrt{5} - 16\sqrt[4]{x}$

17. $3\sqrt[4]{x} + 5\sqrt[4]{x} - 3\sqrt{x} - 11\sqrt[4]{x}$

18. $8\sqrt[3]{x} - 7\sqrt{x} - 6\sqrt[3]{x} + \sqrt[3]{x}$

Study Tip

It is important to realize that the expression $\sqrt{a} + \sqrt{b}$ is not equal to $\sqrt{a + b}$. For instance, you may be tempted to add $\sqrt{6} + \sqrt{3}$ and get $\sqrt{9} = 3$. But remember, you cannot add unlike radicals. So, $\sqrt{6} + \sqrt{3}$ cannot be simplified further.

$\boxed{\text{EXAMPLE 2}}$ **Simplifying Before Combining Radical Expressions**

Simplify each expression by combining like radicals.

a. $\sqrt{45x} + 3\sqrt{20x}$

b. $5\sqrt{x^3} - x\sqrt{4x}$

c. $6\sqrt{\dfrac{24}{x^4}} - 3\sqrt{\dfrac{54}{x^4}}$

d. $\sqrt{50y^5} - \sqrt{32y^5} + 3y^2\sqrt{2y}$

SOLUTION

a. $\sqrt{45x} + 3\sqrt{20x} = 3\sqrt{5x} + 6\sqrt{5x}$ Simplify radicals.

      $= 9\sqrt{5x}$ Combine like radicals.

b. $5\sqrt{x^3} - x\sqrt{4x} = 5x\sqrt{x} - 2x\sqrt{x}$ Simplify radicals.

      $= 3x\sqrt{x}$ Combine like radicals.

c. $6\sqrt{\dfrac{24}{x^4}} - 3\sqrt{\dfrac{54}{x^4}} = 6\dfrac{\sqrt{4 \cdot 6}}{x^2} - 3\dfrac{\sqrt{9 \cdot 6}}{x^2}$ Inverse Property of Radicals

      $= 6\dfrac{2\sqrt{6}}{x^2} - 3\dfrac{3\sqrt{6}}{x^2}$ Inverse Property of Radicals

      $= \dfrac{12\sqrt{6}}{x^2} - \dfrac{9\sqrt{6}}{x^2}$ Multiply.

      $= \dfrac{12\sqrt{6} - 9\sqrt{6}}{x^2}$ Add fractions.

      $= \dfrac{3\sqrt{6}}{x^2}$ Combine like terms.

d. $\sqrt{50y^5} - \sqrt{32y^5} + 3y^2\sqrt{2y}$

      $= \sqrt{5^2y^4(2y)} - \sqrt{4^2y^4(2y)} + 3y^2\sqrt{2y}$ Factor radicands.

      $= 5y^2\sqrt{2y} - 4y^2\sqrt{2y} + 3y^2\sqrt{2y}$ Inverse Property of Radicals

      $= 4y^2\sqrt{2y}$ Combine like radicals.

Exercises Within Reach ®   Solutions in English & Spanish and tutorial videos at CollegePrepAlgebra.com

Combining Radical Expressions **In Exercises 19−34, simplify the expression by combining like radicals.**

19. $8\sqrt{27} - 3\sqrt{3}$

20. $9\sqrt{50} - 4\sqrt{2}$

21. $3\sqrt{45} + 7\sqrt{20}$

22. $5\sqrt{12} + 16\sqrt{27}$

23. $\sqrt{16x} - \sqrt{9x}$

24. $\sqrt{25y} + \sqrt{64y}$

25. $5\sqrt{9x} - 3\sqrt{x}$

26. $4\sqrt{y} + 2\sqrt{16y}$

27. $\sqrt{18y} + 4\sqrt{72y}$

28. $4\sqrt{12x} + \sqrt{75x}$

29. $6\sqrt{x^3} - x\sqrt{9x}$

30. $8\sqrt{x^3} + x\sqrt{81x}$

31. $9\sqrt{\dfrac{40}{x^4}} - 2\sqrt{\dfrac{90}{x^4}}$

32. $11\sqrt{\dfrac{28}{x^8}} - 2\sqrt{\dfrac{112}{x^8}}$

33. $\sqrt{48x^5} + 2x\sqrt{27x^3} - 3x^2\sqrt{3x}$

34. $\sqrt{20x^5} - 4x\sqrt{80x^3} + 9x^2\sqrt{5x}$

EXAMPLE 3 **Simplifying Before Combining Radical Expressions**

Simplify each expression by combining like radicals.

a. $\sqrt[3]{54y^5} + 4\sqrt[3]{2y^2}$ **b.** $\sqrt[3]{6x^4} + \sqrt[3]{48x} - \sqrt[3]{162x^4}$ **c.** $\sqrt[3]{\dfrac{6x}{16}} + 2\sqrt[3]{3x}$

SOLUTION

a. $\sqrt[3]{54y^5} + 4\sqrt[3]{2y^2} = 3y\sqrt[3]{2y^2} + 4\sqrt[3]{2y^2}$ Simplify radical.

$\qquad\qquad\qquad\qquad = (3y + 4)\sqrt[3]{2y^2}$ Distributive Property

b. $\sqrt[3]{6x^4} + \sqrt[3]{48x} - \sqrt[3]{162x^4}$ Write original expression.

$\qquad = x\sqrt[3]{6x} + 2\sqrt[3]{6x} - 3x\sqrt[3]{6x}$ Simplify radicals.

$\qquad = (x + 2 - 3x)\sqrt[3]{6x}$ Distributive Property

$\qquad = (2 - 2x)\sqrt[3]{6x}$ Combine like terms.

c. $\sqrt[3]{\dfrac{6x}{16}} + 2\sqrt[3]{3x} = \sqrt[3]{\dfrac{2 \cdot 3x}{2 \cdot 8}} + 2\sqrt[3]{3x}$ Factor numerator and denominator.

$\qquad\qquad\qquad = \sqrt[3]{\dfrac{\cancel{2} \cdot 3x}{\cancel{2} \cdot 8}} + 2\sqrt[3]{3x}$ Simplify fraction.

$\qquad\qquad\qquad = \sqrt[3]{\dfrac{3x}{2^3}} + 2\sqrt[3]{3x}$ Write 8 as 2^3.

$\qquad\qquad\qquad = \dfrac{1}{2}\sqrt[3]{3x} + 2\sqrt[3]{3x}$ Inverse Property of Radicals

$\qquad\qquad\qquad = \dfrac{5}{2}\sqrt[3]{3x}$ Combine like radicals.

Exercises Within Reach ®

Solutions in English & Spanish and tutorial videos at CollegePrepAlgebra.com

Combining Radical Expressions In Exercises 35−44, simplify the expression by combining like radicals.

35. $2\sqrt[3]{54} + 12\sqrt[3]{16}$

36. $4\sqrt[4]{48} - \sqrt[4]{243}$

37. $\sqrt[3]{6x^4} + \sqrt[3]{48x}$

38. $\sqrt[3]{54x} - \sqrt[3]{2x^4}$

39. $5\sqrt[3]{24u^2} + 2\sqrt[3]{81u^5}$

40. $3\sqrt[3]{16z^2} + 4\sqrt[3]{54z^5}$

41. $\sqrt[3]{3x^4} + \sqrt[3]{81x} + \sqrt[3]{24x^4}$

42. $\sqrt[3]{2x^4} + \sqrt[3]{128x^4} - \sqrt[3]{54x}$

43. $\sqrt[3]{\dfrac{6x}{24}} + 4\sqrt[3]{2x}$

44. $\sqrt[3]{\dfrac{21x^2}{81}} + 3\sqrt[3]{7x^2}$

Geometry In Exercises 45 and 46, the figure shows a stack of cube-shaped boxes. Use the volume of each box to find an expression for the height *h* of the stack.

45.

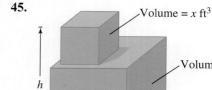

Volume = x ft^3

Volume = $8x$ ft^3

h

46.

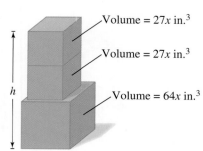

Volume = $27x$ in.^{3}

Volume = $27x$ in.^{3}

Volume = $64x$ in.^{3}

h

EXAMPLE 4   **Rationalizing Denominators Before Simplifying**

Simplify each expression by combining like radicals.

a. $\sqrt{7} - \dfrac{5}{\sqrt{7}}$

b. $\sqrt{12y} - \dfrac{y}{\sqrt{3y}}$

SOLUTION

a. $\sqrt{7} - \dfrac{5}{\sqrt{7}} = \sqrt{7} - \left(\dfrac{5}{\sqrt{7}} \cdot \dfrac{\sqrt{7}}{\sqrt{7}} \right)$ Multiply by $\sqrt{7}/\sqrt{7}$ to remove the radical from the denominator.

$= \sqrt{7} - \dfrac{5\sqrt{7}}{7}$ Simplify.

$= \left(1 - \dfrac{5}{7} \right)\sqrt{7}$ Distributive Property

$= \dfrac{2}{7}\sqrt{7}$ Simplify.

b. $\sqrt{12y} - \dfrac{y}{\sqrt{3y}} = \sqrt{2^2(3y)} - \left(\dfrac{y}{\sqrt{3y}} \cdot \dfrac{\sqrt{3y}}{\sqrt{3y}} \right)$ Factor and multiply.

$= 2\sqrt{3y} - \dfrac{y\sqrt{3y}}{3y}$ Inverse Property of Radicals

$= 2\sqrt{3y} - \dfrac{\cancel{y}\sqrt{3y}}{3\cancel{y}}$ Divide out common factor.

$= 2\sqrt{3y} - \dfrac{\sqrt{3y}}{3}$ Simplify.

$= \dfrac{6\sqrt{3y}}{3} - \dfrac{\sqrt{3y}}{3}$ Rewrite with common denominator.

$= \dfrac{5\sqrt{3y}}{3}, \; y > 0$ Combine like terms.

Exercises Within Reach ®

Solutions in English & Spanish and tutorial videos at CollegePrepAlgebra.com

Rationalizing a Denominator Before Simplifying **In Exercises 47–56, perform the addition or subtraction and simplify your answer.**

47. $\sqrt{5} - \dfrac{3}{\sqrt{5}}$

48. $\sqrt{10} + \dfrac{5}{\sqrt{10}}$

49. $\sqrt{32} + \sqrt{\dfrac{1}{2}}$

50. $\sqrt{\dfrac{1}{5}} - \sqrt{45}$

51. $\sqrt{18y} - \dfrac{y}{\sqrt{2y}}$

52. $\dfrac{x}{\sqrt{3x}} + \sqrt{27x}$

53. $\dfrac{2}{\sqrt{3x}} + \sqrt{3x}$

54. $2\sqrt{7x} - \dfrac{4}{\sqrt{7x}}$

55. $\sqrt{7y^3} - \sqrt{\dfrac{9}{7y^3}}$

56. $\sqrt{\dfrac{4}{3x^3}} + \sqrt{3x^3}$

Applications

Application EXAMPLE 5 Geometry: **Finding the Perimeter of a Triangle**

Write and simplify an expression for the perimeter of the triangle shown below.

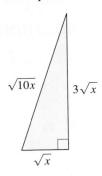

SOLUTION

$$P = a + b + c$$ Formula for perimeter of a triangle

$$= \sqrt{x} + 3\sqrt{x} + \sqrt{10x}$$ Substitute.

$$= (1 + 3)\sqrt{x} + \sqrt{10x}$$ Distributive Property

$$= 4\sqrt{x} + \sqrt{10x}$$ Simplify.

Exercises Within Reach ®

Solutions in English & Spanish and tutorial videos at CollegePrepAlgebra.com

Finding the Perimeter of a Figure In Exercises 57−64, write and simplify an expression for the perimeter of the figure.

57.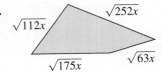

58.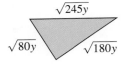

59.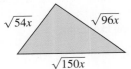

60.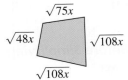

61.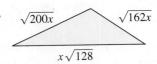

62.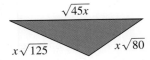

63.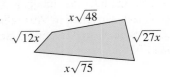

64.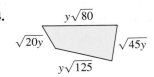

Application   **EXAMPLE 6**   **Out-of-Pocket Expense**

Paying for College

School Year
(5 ↔ 2005-2006)

The annual cost C of tuition and fees per student at a college and the average annual amount A of financial aid received per student at the college from 2005 through 2013 can be modeled by the functions

$$C = -186 - 180.2t + 3131.6\sqrt{t}, \quad 5 \le t \le 13 \qquad \text{Tuition and fees}$$

$$A = 4722 + 486.6t - 1695.9\sqrt{t}, \quad 5 \le t \le 13 \qquad \text{Aid amount}$$

where t represents the school year, with $t = 5$ corresponding to 2005–2006.

a. Find a function that models the average annual out-of-pocket expense E for tuition and fees at the college from 2005 through 2013.

b. Estimate a student's out-of-pocket expense for tuition and fees for the 2011–2012 school year.

SOLUTION

a. The difference of the cost and the aid gives the out-of-pocket expense.

$$C - A = (-186 - 180.2t + 3131.6\sqrt{t}) - (4722 + 486.6t - 1695.9\sqrt{t})$$

$$= -186 - 180.2t + 3131.6\sqrt{t} - 4722 - 486.6t + 1695.9\sqrt{t}$$

$$= -(186 + 4722) - (180.2t + 486.6t) + (3131.6\sqrt{t} + 1695.9\sqrt{t})$$

$$= -4908 - 666.8t + 4827.5\sqrt{t}$$

So, the average out-of-pocket expense E for tuition and fees is given by

$$E = -4908 - 666.8t + 4827.5\sqrt{t}.$$

b. Substitute 11 for t in the model for E.

$$E = -4908 - 666.8(11) + 4827.5\sqrt{11} \approx 3768.21$$

The average out-of-pocket expense for the 2011–2012 school year was $3768.21.

Exercises Within Reach ®   Solutions in English & Spanish and tutorial videos at CollegePrepAlgebra.com

65. *Immigration* The number of immigrants from Guatemala G and El Salvador S to become permanent residents of the United States from 2006 through 2011 can be modeled by the functions

$$G = -3856 + 102{,}701\sqrt{t} - 58{,}994t + 8854\sqrt{t^3},$$
$$6 \le t \le 11$$

$$S = 2{,}249{,}527 - 2{,}230{,}479\sqrt{t} + 742{,}197t - 82{,}167\sqrt{t^3},$$
$$6 \le t \le 11$$

where t represents the year, with $t = 6$ corresponding to 2006. (*Source:* U.S. Department of Homeland Security)

(a) Find a function that models the total number T of immigrants from Guatemala and El Salvador to become permanent residents of the United States from 2006 through 2011.

(b) Estimate T in 2009.

66. *Immigration* The number of immigrants from Oceania O and Australia A to become permanent residents of the United States from 2006 through 2011 can be modeled by the functions

$$O = 339{,}061 - 334{,}718\sqrt{t} + 111{,}948t - 12{,}482\sqrt{t^3},$$
$$6 \le t \le 11$$

$$A = 224{,}864 - 228{,}070\sqrt{t} + 77{,}745t - 8807\sqrt{t^3},$$
$$6 \le t \le 11$$

where t represents the year, with $t = 6$ corresponding to 2006. (*Source:* U.S. Department of Homeland Security)

(a) Find a function that models the number X of immigrants from Oceania excluding Australia to become permanent residents of the United States from 2006 through 2011.

(b) Estimate X in 2011.

Concept Summary: *Adding and Subtracting Radical Expressions*

What

When two or more radical expressions have the same index and the same radicand, they are called **like radicals**.

You can add and subtract radical expressions that contain like radicals.

EXAMPLE

Simplify the expression

$$2\sqrt{54x} + \sqrt{6x} + \sqrt{24x^2} - 3\sqrt{6}.$$

How

Use these steps to add or subtract radical expressions.

1. Write each radical in simplest form.
2. Combine like radicals.
   a. Use the Distributive Property to factor out any like radicals.
   b. Combine any like terms inside the parentheses.

EXAMPLE

$$2\sqrt{54x} + \sqrt{6x} + \sqrt{24x^2} - 3\sqrt{6}$$
$$= 6\sqrt{6x} + \sqrt{6x} + 2x\sqrt{6} - 3\sqrt{6}$$
$$= (6 + 1)\sqrt{6x} + (2x - 3)\sqrt{6}$$
$$= 7\sqrt{6x} + (2x - 3)\sqrt{6}$$

Why

You can use radical expressions to model real-life problems. Knowing how to add and subtract radical expressions will help you solve such problems. For example, you can find a radical function for the average annual out-of-pocket expenses for the students at a college.

Exercises Within Reach ®

Worked-out solutions to odd-numbered exercises at CollegePrepAlgebra.com

Concept Summary Check

67. *Writing Radicals in Simplest Form* Which expression in the solution above is the result of writing the radicals in simplest form?

68. *Writing Radicals in Simplest Form* What process is used to write the radicals in simplest form in the solution above?

69. *Identifying Like Radicals* Identify the like radicals in the expression $6\sqrt{6x} + \sqrt{6x} + 2x\sqrt{6} - 3\sqrt{6}$.

70. *Simplifying an Expression* Explain why $(2x - 3)\sqrt{6}$ cannot be simplified further.

Extra Practice

Combining Radical Expressions In Exercises 71−86, simplify the expression by combining like radicals.

71. $3\sqrt{x + 1} + 10\sqrt{x + 1}$

72. $7\sqrt{2a - 3} - 4\sqrt{2a - 3}$

73. $\sqrt[3]{16t^4} - \sqrt[3]{54t^4}$

74. $10\sqrt[3]{z} - \sqrt[3]{z^4}$

75. $\sqrt{5a} + 2\sqrt{45a^3}$

76. $4\sqrt{3x^3} - \sqrt{12x}$

77. $\sqrt{9x - 9} + \sqrt{x - 1}$

78. $\sqrt{4y + 12} + \sqrt{y + 3}$

79. $\sqrt{x^3 - x^2} + \sqrt{4x - 4}$

80. $\sqrt{9x - 9} - \sqrt{x^3 - x^2}$

81. $2\sqrt[3]{a^4b^2} + 3a\sqrt[3]{ab^2}$

82. $3y\sqrt[4]{2x^5y^3} - x\sqrt[4]{162xy^7}$

83. $\sqrt{4r^7s^5} + 3r^2\sqrt{r^3s^5} - 2rs\sqrt{r^5s^3}$

84. $x\sqrt[3]{27x^5y^2} - x^2\sqrt[3]{x^2y^2} + z\sqrt[3]{x^8y^2}$

85. $\sqrt[3]{128x^9y^{10}} - 2x^2y\sqrt[3]{16x^3y^7}$

86. $5\sqrt[3]{320x^5y^8} + 2x\sqrt[3]{135x^2y^8}$

Comparing Radical Expressions In Exercises 87−90, place the correct symbol (<, >, or =) between the expressions.

87. $\sqrt{7} + \sqrt{18}$ ▨ $\sqrt{7 + 18}$

88. $\sqrt{10} - \sqrt{6}$ ▨ $\sqrt{10 - 6}$

89. 5 ▨ $\sqrt{9^2 - 4^2}$

90. 5 ▨ $\sqrt{3^2 + 4^2}$

91. *Geometry* The foundation of a house is 40 feet long and 30 feet wide. The roof rises 5 feet vertically and forms two congruent right triangles with the side of the house, as shown. (Assume there is no overhang.)

(a) Use the Pythagorean Theorem to find the length of the hypotenuse of each right triangle formed by the roof line.

(b) You are replacing the drip edge around the entire perimeter of the roof. Find the total length of drip edge you need.

92. *Geometry* The four corners are cut from a four-foot-by-eight-foot sheet of plywood, as shown in the figure. Find the perimeter of the remaining piece of plywood.

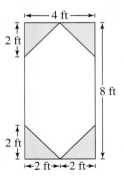

Explaining Concepts

93. *Reasoning* Will the sum of two radicals always be a radical? Give an example to support your answer.

94. *Reasoning* Will the difference of two radicals always be a radical? Give an example to support your answer.

95. *Number Sense* Is $\sqrt{2x} + \sqrt{2x}$ equal to $\sqrt{8x}$? Explain.

96. *Structure* Explain how adding two monomials compares to adding two radicals.

97. *Error Analysis* Find and correct the error(s) in each solution.

(a) $7\sqrt{3} + 4\sqrt{2} = 11\sqrt{5}$

(b) $3\sqrt[3]{k} - 6\sqrt{k} = 3\sqrt[3]{k}$

98. *Structure* Is $\sqrt{2} - \dfrac{1}{\sqrt{2}}$ in simplest form? Explain.

Cumulative Review

In Exercises 99−104, combine the rational expressions and simplify.

99. $\dfrac{7z - 2}{2z} - \dfrac{4z + 1}{2z}$

100. $\dfrac{2x + 1}{3x} + \dfrac{3 - 4x}{3x}$

101. $\dfrac{2x + 3}{x - 3} + \dfrac{6 - 5x}{x - 3}$

102. $\dfrac{4m + 6}{m + 2} - \dfrac{3m + 4}{m + 2}$

103. $\dfrac{2v}{v - 5} - \dfrac{3}{5 - v}$

104. $\dfrac{4}{x - 4} + \dfrac{2x}{x + 1}$

In Exercises 105−108, simplify the complex fraction.

105. $\dfrac{\left(\dfrac{2}{3}\right)}{\left(\dfrac{4}{15}\right)}$

106. $\dfrac{\left(\dfrac{27a^3}{4b^2c}\right)}{\left(\dfrac{9ac^2}{10b^2}\right)}$

107. $\dfrac{3w - 9}{\left(\dfrac{w^2 - 10w + 21}{w + 1}\right)}$

108. $\dfrac{\left(\dfrac{x^2 + 2x - 8}{x - 8}\right)}{2x + 8}$

Mid-Chapter Quiz: Sections 9.1–9.3

Solutions in English & Spanish and tutorial videos at CollegePrepAlgebra.com

Take this quiz as you would take a quiz in class. After you are done, check your work against the answers in the back of the book.

In Exercises 1–4, evaluate the expression.

1. $\sqrt{255}$

2. $\sqrt[4]{\frac{81}{16}}$

3. $49^{1/2}$

4. $(-27)^{2/3}$

In Exercises 5 and 6, evaluate the function for each indicated x-value, if possible, and simplify.

5. $f(x) = \sqrt{3x - 5}$

     (a) $f(0)$ (b) $f(2)$ (c) $f(10)$

6. $g(x) = \sqrt{9 - x}$

     (a) $g(-7)$ (b) $g(5)$ (c) $g(9)$

In Exercises 7 and 8, describe the domain of the function.

7. $g(x) = \dfrac{12}{\sqrt[3]{x}}$

8. $h(x) = \sqrt{3x + 10}$

In Exercises 9–14, simplify the radical expression.

9. $\sqrt{27x^2}$

10. $\sqrt[4]{32x^8}$

11. $\sqrt{\dfrac{4u^3}{9}}$

12. $\sqrt[3]{\dfrac{16}{u^6}}$

13. $\sqrt{125x^3y^2z^4}$

14. $2a\sqrt[3]{16a^3b^5}$

In Exercises 15 and 16, rationalize the denominator and simplify further, if possible.

15. $\dfrac{24}{\sqrt{12}}$

16. $\dfrac{21x^2}{\sqrt{7x}}$

In Exercises 17–22, simplify the expression by combining like radicals.

17. $2\sqrt{3} - 4\sqrt{7} + \sqrt{3}$

18. $\sqrt{200y} - 3\sqrt{8y}$

19. $5\sqrt{12} + 2\sqrt{3} - \sqrt{75}$

20. $\sqrt{25x + 50} - \sqrt{x + 2}$

21. $6x\sqrt[3]{5x^2} + 2\sqrt[3]{40x^4}$

22. $3\sqrt{x^3y^4z^5} + 2xy^2\sqrt{xz^5} - xz^2\sqrt{xy^4z}$

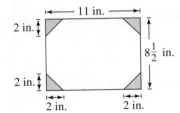

Application

23. The four corners are cut from an $8\frac{1}{2}$-inch-by-11-inch sheet of paper, as shown in the figure at the left. Find the perimeter of the remaining piece of paper.

Study Skills in Action

Studying in a Group

Many students endure unnecessary frustration because they study by themselves. Studying in a group or with a partner has many benefits. First, the combined memory and comprehension of the members minimizes the likelihood of any member getting "stuck" on a particular problem. Second, discussing math often helps clarify unclear areas. Third, regular study groups keep many students from procrastinating. Finally, study groups often build a camaraderie that helps students stick with the class when it gets tough.

These students are keeping each other motivated.

Smart Study Strategy

Form a Weekly Study Group

1 ▶ **Set up the group.**

- Select students who are just as dedicated to doing well in the math class as you are.

- Find a regular meeting place that has minimal distractions. Try to find a place that has a white board.

- Compare schedules, and select at least one time a week to meet, allowing at least 1.5 hours for study time.

2 ▶ **Organize the study time.** If you are unsure about how to structure your time during the first few study sessions, try using the guidelines at the right.

- Review and compare notes - 20 minutes
- Identify and review the key rules, definitions, etc. - 20 minutes
- Demonstrate at least one homework problem for each key concept - 40 minutes
- Make small talk (saving this until the end improves your chances of getting through all the math) - 10 minutes

3 ▶ **Set up rules for the group.** Consider using the following rules.

- Members must attend regularly, be on time, and participate.

- The sessions will focus on the key math concepts, not on the needs of one student.

- Students who keep the group from being productive will be asked to leave the group.

4 ▶ **Inform the teacher.** Let the teacher know about your study group. Ask for advice about maintaining a productive group.

9.4 Multiplying and Dividing Radical Expressions

▶ Use the Distributive Property or the FOIL Method to multiply radical expressions.

▶ Determine the products of conjugates.

▶ Simplify quotients involving radicals by rationalizing the denominators.

Multiplying Rational Expressions

EXAMPLE 1 **Multiplying Radical Expressions**

Find each product and simplify.

a. $\sqrt{6} \cdot \sqrt{3}$

b. $\sqrt[3]{5} \cdot \sqrt[3]{16}$

SOLUTION

a. $\sqrt{6} \cdot \sqrt{3} = \sqrt{6 \cdot 3} = \sqrt{18} = \sqrt{9 \cdot 2} = 3\sqrt{2}$

b. $\sqrt[3]{5} \cdot \sqrt[3]{16} = \sqrt[3]{5 \cdot 16} = \sqrt[3]{80} = \sqrt[3]{8 \cdot 10} = 2\sqrt[3]{10}$

EXAMPLE 2 **Multiplying Radical Expressions**

Find each product and simplify.

a. $\sqrt{3}(2 + \sqrt{5})$ **b.** $\sqrt{2}(4 - \sqrt{8})$ **c.** $\sqrt{6}(\sqrt{12} - \sqrt{3})$

SOLUTION

a. $\sqrt{3}(2 + \sqrt{5}) = 2\sqrt{3} + \sqrt{3}\sqrt{5}$ Distributive Property

$= 2\sqrt{3} + \sqrt{15}$ Product Rule for Radicals

b. $\sqrt{2}(4 - \sqrt{8}) = 4\sqrt{2} - \sqrt{2}\sqrt{8}$ Distributive Property

$= 4\sqrt{2} - \sqrt{16} = 4\sqrt{2} - 4$ Product Rule for Radicals

c. $\sqrt{6}(\sqrt{12} - \sqrt{3}) = \sqrt{6}\sqrt{12} - \sqrt{6}\sqrt{3}$ Distributive Property

$= \sqrt{72} - \sqrt{18}$ Product Rule for Radicals

$= 6\sqrt{2} - 3\sqrt{2} = 3\sqrt{2}$ Find perfect square factors.

Exercises Within Reach ®

Solutions in English & Spanish and tutorial videos at CollegePrepAlgebra.com

Multiplying Radical Expression In Exercises 1−20, multiply and simplify.

1. $\sqrt{2} \cdot \sqrt{8}$

2. $\sqrt{6} \cdot \sqrt{18}$

3. $\sqrt{3} \cdot \sqrt{15}$

4. $\sqrt{5} \cdot \sqrt{10}$

5. $\sqrt[3]{12} \cdot \sqrt[3]{6}$

6. $\sqrt[3]{9} \cdot \sqrt[3]{3}$

7. $\sqrt[4]{8} \cdot \sqrt[4]{2}$

8. $\sqrt[4]{54} \cdot \sqrt[4]{3}$

9. $\sqrt{7}(3 - \sqrt{7})$

10. $\sqrt{3}(4 + \sqrt{3})$

11. $\sqrt{2}(\sqrt{20} + 8)$

12. $\sqrt{7}(\sqrt{14} + 3)$

13. $\sqrt{6}(\sqrt{12} - \sqrt{3})$

14. $\sqrt{10}(\sqrt{5} + \sqrt{6})$

15. $4\sqrt{3}(\sqrt{3} - \sqrt{5})$

16. $3\sqrt{5}(\sqrt{5} - \sqrt{2})$

17. $\sqrt{y}(\sqrt{y} + 4)$

18. $\sqrt{x}(5 - \sqrt{x})$

19. $\sqrt{a}(4 - \sqrt{a})$

20. $\sqrt{z}(\sqrt{z} + 5)$

In Section 5.3, you learned the FOIL Method for multiplying two binomials. This method can be used when the binomials involve radicals.

EXAMPLE 3   **Using the FOIL Method to Multiply**

Find each product and simplify.

a. $\left(2\sqrt{7} - 4\right)\left(\sqrt{7} + 1\right)$ **b.** $(3 - \sqrt{x})(1 + \sqrt{x})$

SOLUTION

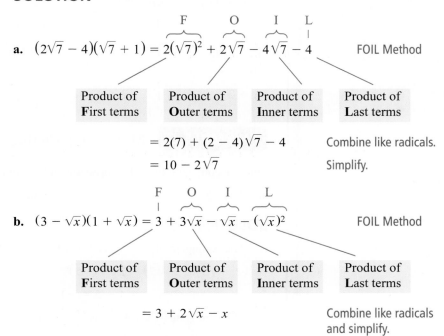

$$\textbf{a.} \quad \left(2\sqrt{7} - 4\right)\left(\sqrt{7} + 1\right) = \overset{F}{2(\sqrt{7})^2} + \overset{O}{2\sqrt{7}} - \overset{I}{4\sqrt{7}} - \overset{L}{4} \qquad \text{FOIL Method}$$

Product of First terms	Product of Outer terms	Product of Inner terms	Product of Last terms

$$= 2(7) + (2 - 4)\sqrt{7} - 4 \qquad \text{Combine like radicals.}$$
$$= 10 - 2\sqrt{7} \qquad \text{Simplify.}$$

$$\textbf{b.} \quad (3 - \sqrt{x})(1 + \sqrt{x}) = \overset{F}{3} + \overset{O}{3\sqrt{x}} - \overset{I}{\sqrt{x}} - \overset{L}{(\sqrt{x})^2} \qquad \text{FOIL Method}$$

Product of First terms	Product of Outer terms	Product of Inner terms	Product of Last terms

$$= 3 + 2\sqrt{x} - x \qquad \text{Combine like radicals and simplify.}$$

Exercises Within Reach ®

Multiplying Radical Expressions In Exercises 21–30, multiply and simplify.

21. $\left(\sqrt{5} + 3\right)\left(\sqrt{3} - 5\right)$

22. $\left(\sqrt{7} + 6\right)\left(\sqrt{2} + 6\right)$

23. $\left(\sqrt{20} + 2\right)^2$

24. $\left(4 - \sqrt{20}\right)^2$

25. $\left(\sqrt{5} - \sqrt{3}\right)\left(\sqrt{5} - \sqrt{3}\right)$

26. $\left(\sqrt{2} + \sqrt{7}\right)\left(\sqrt{2} + \sqrt{7}\right)$

27. $\left(10 + \sqrt{2x}\right)^2$

28. $\left(5 - \sqrt{3v}\right)^2$

29. $\left(9\sqrt{x} + 2\right)\left(5\sqrt{x} - 3\right)$

30. $\left(16\sqrt{u} - 3\right)\left(\sqrt{u} - 1\right)$

Geometry In Exercises 31–34, write and simplify an expression for the area of the rectangle.

31.

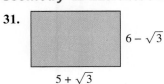

$6 - \sqrt{3}$

$5 + \sqrt{3}$

32.

$8 - \sqrt{2}$

$6 + \sqrt{2}$

33.

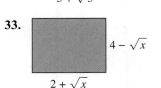

$4 - \sqrt{x}$

$2 + \sqrt{x}$

34.

$8 - \sqrt{x}$

$5 + \sqrt{x}$

Conjugates

The expressions $3 + \sqrt{6}$ and $3 - \sqrt{6}$ are called **conjugates** of each other. Notice that they differ only in the sign between the terms. The product of two conjugates is the difference of two squares, which is given by the special product formula $(a + b)(a - b) = a^2 - b^2$. Here are some other examples.

EXAMPLE 4   Multiplying Conjugates

Expression	Conjugate	Product
$1 - \sqrt{3}$	$1 + \sqrt{3}$	$(1)^2 - (\sqrt{3})^2 = 1 - 3 = -2$
$\sqrt{5} + \sqrt{2}$	$\sqrt{5} - \sqrt{2}$	$(\sqrt{5})^2 - (\sqrt{2})^2 = 5 - 2 = 3$
$\sqrt{10} - 3$	$\sqrt{10} + 3$	$(\sqrt{10})^2 - (3)^2 = 10 - 9 = 1$
$\sqrt{x} + 2$	$\sqrt{x} - 2$	$(\sqrt{x})^2 - (2)^2 = x - 4,\ x \geq 0$

EXAMPLE 5   Multiplying Conjugates

Find the conjugate of each expression and multiply each expression by its conjugate.

a. $2 - \sqrt{5}$

b. $\sqrt{3} + \sqrt{x}$

SOLUTION

a. The conjugate of $2 - \sqrt{5}$ is $2 + \sqrt{5}$.

$(2 - \sqrt{5})(2 + \sqrt{5}) = 2^2 - (\sqrt{5})^2$ Special product formula

$= 4 - 5 = -1$ Simplify.

b. The conjugate of $\sqrt{3} + \sqrt{x}$ is $\sqrt{3} - \sqrt{x}$.

$(\sqrt{3} + \sqrt{x})(\sqrt{3} - \sqrt{x}) = (\sqrt{3})^2 - (\sqrt{x})^2$ Special product formula

$= 3 - x,\ x \geq 0$ Simplify.

Exercises Within Reach ®

Solutions in English & Spanish and tutorial videos at CollegePrepAlgebra.com

Multiplying Conjugates In Exercises 35−48, find the conjugate of the expression. Then multiply the expression by its conjugate and simplify.

35. $2 + \sqrt{5}$

36. $\sqrt{2} - 9$

37. $\sqrt{11} - \sqrt{3}$

38. $\sqrt{10} + \sqrt{7}$

39. $\sqrt{15} + 3$

40. $\sqrt{14} - 3$

41. $\sqrt{x} - 3$

42. $\sqrt{t} + 7$

43. $\sqrt{2u} - \sqrt{3}$

44. $\sqrt{5a} + \sqrt{2}$

45. $2\sqrt{2} + \sqrt{4}$

46. $4\sqrt{3} + \sqrt{2}$

47. $\sqrt{x} + \sqrt{y}$

48. $3\sqrt{u} + \sqrt{3v}$

Geometry In Exercises 49 and 50, find the area of the rectangle.

49.

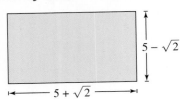

$5 - \sqrt{2}$

$5 + \sqrt{2}$

50.

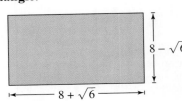

$8 - \sqrt{6}$

$8 + \sqrt{6}$

Dividing Radical Expressions

To simplify a quotient involving radicals, you rationalize the denominator. For single-term denominators, you can use the rationalization process described in Section 9.2. To rationalize a denominator involving two terms, multiply both the numerator and denominator by the *conjugate of the denominator*.

EXAMPLE 6   **Simplifying Quotients Involving Radicals**

Simplify (a) $\dfrac{\sqrt{3}}{1 - \sqrt{5}}$ and (b) $\dfrac{4}{2 - \sqrt{3}}$.

SOLUTION

a. $\dfrac{\sqrt{3}}{1 - \sqrt{5}} = \dfrac{\sqrt{3}}{1 - \sqrt{5}} \cdot \dfrac{1 + \sqrt{5}}{1 + \sqrt{5}}$ Multiply numerator and denominator by conjugate of denominator.

$= \dfrac{\sqrt{3}\left(1 + \sqrt{5}\right)}{1^2 - \left(\sqrt{5}\right)^2}$ Special product formula

$= \dfrac{\sqrt{3} + \sqrt{15}}{1 - 5}$ Simplify.

$= -\dfrac{\sqrt{3} + \sqrt{15}}{4}$ Simplify.

b. $\dfrac{4}{2 - \sqrt{3}} = \dfrac{4}{2 - \sqrt{3}} \cdot \dfrac{2 + \sqrt{3}}{2 + \sqrt{3}}$ Multiply numerator and denominator by conjugate of denominator.

$= \dfrac{4\left(2 + \sqrt{3}\right)}{2^2 - \left(\sqrt{3}\right)^2}$ Special product formula

$= \dfrac{8 + 4\sqrt{3}}{4 - 3}$ Simplify.

$= 8 + 4\sqrt{3}$ Simplify.

Exercises Within Reach ®

Solutions in English & Spanish and tutorial videos at CollegePrepAlgebra.com

Simplifying a Quotient Involving a Radical In Exercises 51−60, simplify the expression.

51. $\dfrac{1}{2 + \sqrt{5}}$

52. $\dfrac{1}{3 + \sqrt{10}}$

53. $\dfrac{\sqrt{2}}{2 - \sqrt{6}}$

54. $\dfrac{\sqrt{5}}{3 - \sqrt{7}}$

55. $\dfrac{5}{9 - \sqrt{6}}$

56. $\dfrac{8}{8 - \sqrt{8}}$

57. $\dfrac{6}{\sqrt{11} - 2}$

58. $\dfrac{8}{\sqrt{7} + 3}$

59. $\dfrac{7}{\sqrt{3} + 5}$

60. $\dfrac{9}{\sqrt{7} - 4}$

EXAMPLE 7 **Simplifying a Quotient Involving Radicals**

$$\frac{5\sqrt{2}}{\sqrt{7} + \sqrt{2}} = \frac{5\sqrt{2}}{\sqrt{7} + \sqrt{2}} \cdot \frac{\sqrt{7} - \sqrt{2}}{\sqrt{7} - \sqrt{2}}$$

Multiply numerator and denominator by conjugate of denominator.

$$= \frac{5\sqrt{2}(\sqrt{7} - \sqrt{2})}{(\sqrt{7})^2 - (\sqrt{2})^2}$$

Special product formula

$$= \frac{5(\sqrt{14} - \sqrt{4})}{7 - 2}$$

Simplify.

$$= \frac{5(\sqrt{14} - 2)}{5}$$

Simplify.

$$= \frac{\cancel{5}(\sqrt{14} - 2)}{\cancel{5}}$$

Divide out common factor.

$$= \sqrt{14} - 2$$

Simplest form

EXAMPLE 8 **Simplifying a Quotient Involving Radicals**

Perform the division and simplify.

$$1 \div \left(\sqrt{x} - \sqrt{x + 1}\right)$$

SOLUTION

$$\frac{1}{\sqrt{x} - \sqrt{x + 1}} = \frac{1}{\sqrt{x} - \sqrt{x + 1}} \cdot \frac{\sqrt{x} + \sqrt{x + 1}}{\sqrt{x} + \sqrt{x + 1}}$$

Multiply numerator and denominator by conjugate of denominator.

$$= \frac{\sqrt{x} + \sqrt{x + 1}}{(\sqrt{x})^2 - (\sqrt{x + 1})^2}$$

Special product formula

$$= \frac{\sqrt{x} + \sqrt{x + 1}}{x - (x + 1)}$$

Simplify.

$$= \frac{\sqrt{x} + \sqrt{x + 1}}{-1}$$

Combine like terms.

$$= -\sqrt{x} - \sqrt{x + 1}$$

Simplify.

Exercises Within Reach ®

Solutions in English & Spanish and tutorial videos at CollegePrepAlgebra.com

Simplifying a Quotient Involving Radicals **In Exercises 61−68, simplify the expression.**

61. $\dfrac{\sqrt{5}}{\sqrt{6} - \sqrt{5}}$

62. $\dfrac{\sqrt{2}}{\sqrt{2} - \sqrt{3}}$

63. $\dfrac{4\sqrt{3}}{\sqrt{5} + \sqrt{3}}$

64. $\dfrac{5\sqrt{7}}{\sqrt{6} + \sqrt{7}}$

65. $1 \div \left(\sqrt{x} - \sqrt{x + 3}\right)$

66. $1 \div \left(\sqrt{x} + \sqrt{x - 2}\right)$

67. $\dfrac{3x}{\sqrt{15} - \sqrt{3}}$

68. $\dfrac{5y}{\sqrt{12} + \sqrt{10}}$

EXAMPLE 9   **Simplifying Quotient Involving Radicals**

Perform each division and simplify.

a. $6 \div (\sqrt{x} - 2)$

b. $(2 - \sqrt{3}) \div (\sqrt{6} + \sqrt{2})$

SOLUTION

a. $\dfrac{6}{\sqrt{x} - 2} = \dfrac{6}{\sqrt{x} - 2} \cdot \dfrac{\sqrt{x} + 2}{\sqrt{x} + 2}$ Multiply numerator and denominator by conjugate of denominator.

$= \dfrac{6(\sqrt{x} + 2)}{(\sqrt{x})^2 - 2^2}$ Special product formula

$= \dfrac{6\sqrt{x} + 12}{x - 4}$ Simplify.

b. $\dfrac{2 - \sqrt{3}}{\sqrt{6} + \sqrt{2}} = \dfrac{2 - \sqrt{3}}{\sqrt{6} + \sqrt{2}} \cdot \dfrac{\sqrt{6} - \sqrt{2}}{\sqrt{6} - \sqrt{2}}$ Multiply numerator and denominator by conjugate of denominator.

$= \dfrac{2\sqrt{6} - 2\sqrt{2} - \sqrt{18} + \sqrt{6}}{(\sqrt{6})^2 - (\sqrt{2})^2}$ FOIL Method and special product formula

$= \dfrac{3\sqrt{6} - 2\sqrt{2} - 3\sqrt{2}}{6 - 2}$ Simplify.

$= \dfrac{3\sqrt{6} - 5\sqrt{2}}{4}$ Simplify.

Exercises Within Reach ®

Solutions in English & Spanish and tutorial videos at CollegePrepAlgebra.com

Simplifying a Quotient Involving Radicals In Exercises 69–78, simplify the expression.

69. $8 \div (\sqrt{x} - 3)$

70. $7 \div (\sqrt{x} - 5)$

71. $\dfrac{\sqrt{5t}}{\sqrt{5} - \sqrt{t}}$

72. $\dfrac{\sqrt{2x}}{\sqrt{x} - \sqrt{2}}$

73. $\dfrac{8a}{\sqrt{3a} + \sqrt{a}}$

74. $\dfrac{7z}{\sqrt{5z} - \sqrt{z}}$

75. $(\sqrt{7} + 2) \div (\sqrt{7} - 2)$

76. $(5 - \sqrt{3}) \div (3 + \sqrt{3})$

77. $(\sqrt{x} - 5) \div (2\sqrt{x} - 1)$

78. $(2\sqrt{t} + 1) \div (2\sqrt{t} - 1)$

Geometry In Exercises 79 and 80, find the exact width *w* of the rectangle.

79.

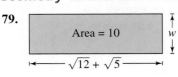

Area = 10 ⟋ *w*

$\sqrt{12} + \sqrt{5}$

80.

Area = 32 ⟋ *w*

$2\sqrt{5} + \sqrt{8}$

Concept Summary: Multiplying and Dividing Radical Expressions

What

You can use previously learned techniques to multiply and divide radical expressions.

EXAMPLE

Simplify each expression.

a. $(1 + \sqrt{2})(3 + \sqrt{2})$

b. $\dfrac{1}{\sqrt{2} - 1}$

How

- To multiply radical expressions, use the Distributive Property, FOIL Method, and special product formulas.

- To simplify quotients involving radicals, rationalize the denominators.

EXAMPLE

a. $(1 + \sqrt{2})(3 + \sqrt{2})$
$$= 3 + \sqrt{2} + 3\sqrt{2} + 2$$
$$= 5 + 4\sqrt{2}$$

b. $\dfrac{1}{\sqrt{2} - 1} \cdot \dfrac{\sqrt{2} + 1}{\sqrt{2} + 1} = \dfrac{\sqrt{2} + 1}{2 - 1}$
$$= \sqrt{2} + 1$$

Why

You can use radical expressions to model real-life problems. Knowing how to multiply and divide radical expressions will help you solve such problems. For example, you can multiply radical expressions to find the area of a cross section of the strongest beam that can be cut from a log.

Exercises Within Reach ®

Worked-out solutions to odd-numbered exercises at CollegePrepAlgebra.com

Concept Summary Check

81. **Multiplying Radical Expressions** In the solution above, what technique is used to multiply $(1 + \sqrt{2})$ and $(3 + \sqrt{2})$?

82. **Product Rule for Radicals** In the solution above, what term of the expression $3 + \sqrt{2} + 3\sqrt{2} + 2$ is obtained using the Product Rule for Radicals?

83. **Describing a Relationship** Describe the relationship between $\sqrt{2} + 1$ and $\sqrt{2} - 1$.

84. **Identifying a Process** What process is used to rewrite $\dfrac{1}{\sqrt{2} - 1}$ as $\sqrt{2} + 1$ in the example above?

Extra Practice

Multiplying Radical Expressions In Exercises 85–90, multiply and simplify.

85. $(2\sqrt{2x} - \sqrt{5})(2\sqrt{2x} + \sqrt{5})$

86. $(\sqrt{7} - 3\sqrt{3t})(\sqrt{7} + 3\sqrt{3t})$

87. $(\sqrt[3]{t} + 1)(\sqrt[3]{t^2} + 4\sqrt[3]{t} - 3)$

88. $(\sqrt[3]{x} - 2)(\sqrt[3]{x^2} - 2\sqrt[3]{x} + 1)$

89. $2\sqrt[3]{x^4y^5}\left(\sqrt[3]{8x^{12}y^4} + \sqrt[3]{16xy^9}\right)$

90. $\sqrt[4]{8x^3y^5}\left(\sqrt[4]{4x^5y^7} - \sqrt[4]{3x^7y^6}\right)$

Evaluating a Function In Exercises 91 and 92, evaluate the function as indicated.

91. $f(x) = x^2 - 6x + 1$

   (a) $f(2 - \sqrt{3})$ (b) $f(3 - 2\sqrt{2})$

92. $g(x) = x^2 + 8x + 11$

   (a) $g(-4 + \sqrt{5})$ (b) $g(-4\sqrt{2})$

Rationalizing the Numerator In the study of calculus, students sometimes rewrite an expression by rationalizing the numerator. In Exercises 93–96, rationalize the numerator. (*Note:* The results will not be in simplest radical form.)

93. $\dfrac{\sqrt{10}}{\sqrt{3x}}$

94. $\dfrac{\sqrt{5}}{\sqrt{7x}}$

95. $\dfrac{\sqrt{7} + \sqrt{3}}{5}$

96. $\dfrac{\sqrt{2} - \sqrt{5}}{4}$

97. *Geometry* The width w and height h of the strongest rectangular beam that can be cut from a log with a diameter of 24 inches (see figure) are given by

$$w = 8\sqrt{3} \quad \text{and} \quad h = \sqrt{24^2 - (8\sqrt{3})^2}.$$

Find the area of the rectangular cross section of the beam, and write the area in simplest form.

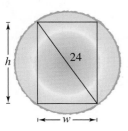
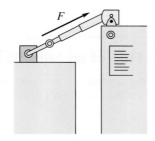

Figure for 97   Figure for 98

98. *Force* The force required to slide a steel block weighing 500 pounds across a milling machine is

$$\frac{500k}{\dfrac{1}{\sqrt{k^2 + 1}} + \dfrac{k^2}{\sqrt{k^2 + 1}}}$$

where k is the friction constant (see figure). Simplify this expression.

99. *Basketball* The area of the circular cross section of a basketball is 70 square inches. The area enclosed by a basketball hoop is about 254 square inches. Find the ratio of the diameter of the basketball to the diameter of the hoop.

100. *Geometry* The ratio of the width of the Temple of Hephaestus to its height (see figure) is approximately

$$\frac{w}{h} \approx \frac{2}{\sqrt{5} - 1}.$$

This number is called the **golden section**. Early Greeks believed that the most aesthetically pleasing rectangles were those whose sides had this ratio.

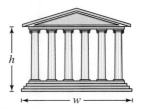

(a) Rationalize the denominator to simplify the expression. Then approximate the value of the expression. Round your answer to two decimal places.

(b) Use the Pythagorean Theorem, a straightedge, and a compass to construct a rectangle whose sides have the golden section as their ratio.

Explaining Concepts

101. *Number Sense* Let a and b be integers, but not perfect squares. Describe the circumstances (if any) for which each expression represents a rational number. Explain.

(a) $a\sqrt{b}$ (b) $\sqrt{a}\sqrt{b}$

102. *Number Sense* Given that a and b are positive integers, what type of number is the product of the expression $\sqrt{a} + \sqrt{b}$ and its conjugate? Explain.

103. *Exploration* Find the conjugate of $\sqrt{a} + \sqrt{b}$. Multiply the conjugates. Next, find the conjugate of $\sqrt{b} + \sqrt{a}$. Multiply the conjugates. Explain how changing the order of the terms affects the conjugate and the product of the conjugates.

104. *Exploration* Rationalize the denominators of $\dfrac{1}{\sqrt{a} + \sqrt{b}}$ and $\dfrac{1}{\sqrt{b} + \sqrt{a}}$. Explain how changing the order of the terms in the denominator affects the rationalized form of the quotient.

Cumulative Review

In Exercises 105–108, solve the equation. If there is exactly one solution, check your solution. If not, justify your answer.

105. $3x - 18 = 0$ **106.** $7t - 4 = 4t + 8$

107. $3x - 4 = 3x$ **108.** $3(2x + 5) = 6x + 15$

In Exercises 109–112, solve the equation by factoring.

109. $x^2 - 144 = 0$

110. $4x^2 - 25 = 0$

111. $x^2 + 2x - 15 = 0$

112. $6x^2 - x - 12 = 0$

In Exercises 113–116, simplify the radical expression.

113. $\sqrt{32x^2y^5}$ **114.** $\sqrt[3]{32x^2y^5}$

115. $\sqrt[4]{32x^2y^5}$ **116.** $\sqrt[5]{32x^2y^5}$

9.5 Radical Equations and Applications

▶ Solve a radical equation by raising each side to the *n*th power.

▶ Solve application problems involving radical equations.

Solving Radical Equations

Raising Each Side of an Equation to the *n*th Power

Let u and v be real numbers, variables, or algebraic expressions, and let n be a positive integer. If $u = v$, then it follows that

$$u^n = v^n.$$

This is called *raising each side of an equation to the nth power*.

EXAMPLE 1 **Solving an Equation Having One Radical**

$\sqrt{x} - 8 = 0$	Original equation
$\sqrt{x} = 8$	Isolate radical.
$(\sqrt{x})^2 = 8^2$	Square each side.
$x = 64$	Simplify.

Check

$\sqrt{64} - 8 \overset{?}{=} 0$	Substitute 64 for x in original equation.
$8 - 8 = 0$	Solution checks. ✔

So, the equation has one solution: $x = 64$.

Exercises Within Reach ®   Solutions in English & Spanish and tutorial videos at CollegePrepAlgebra.com

Checking Solutions of an Equation In Exercises 1–4, determine whether each value of x is a solution of the equation.

Equation	*Values of x*			*Equation*	*Values of x*	
1. $\sqrt{x} - 10 = 0$	(a) $x = -4$	(b) $x = -100$		**2.** $\sqrt{3x} - 6 = 0$	(a) $x = \frac{2}{3}$	(b) $x = 2$
	(c) $x = \sqrt{10}$	(d) $x = 100$			(c) $x = 12$	(d) $x = -\frac{1}{3}\sqrt{6}$
3. $\sqrt[3]{x-4} = 4$	(a) $x = -60$	(b) $x = 68$		**4.** $\sqrt[4]{2x} + 2 = 6$	(a) $x = 128$	(b) $x = 2$
	(c) $x = 20$	(d) $x = 0$			(c) $x = -2$	(d) $x = 0$

Solving an Equation In Exercises 5–14, solve the equation and check your solution.

5. $\sqrt{x} = 12$ **6.** $\sqrt{x} = 5$ **7.** $\sqrt{y} = 7$ **8.** $\sqrt{t} = 4$

9. $\sqrt[3]{z} = 3$ **10.** $\sqrt[4]{x} = 3$ **11.** $\sqrt{y} - 7 = 0$ **12.** $\sqrt{t} - 13 = 0$

13. $\sqrt{x} - 8 = 0$ **14.** $\sqrt{x} - 10 = 0$

Checking solutions of a radical equation is especially important because raising each side of an equation to the *n*th power to remove the radical(s) often introduces *extraneous* solutions.

EXAMPLE 2   Solving an Equation Having One Radical

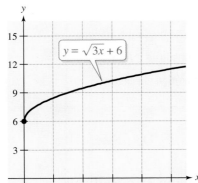

$$\sqrt{3x} + 6 = 0 \qquad \text{Original equation}$$

$$\sqrt{3x} = -6 \qquad \text{Isolate radical.}$$

$$(\sqrt{3x})^2 = (-6)^2 \qquad \text{Square each side.}$$

$$3x = 36 \qquad \text{Simplify.}$$

$$x = 12 \qquad \text{Divide each side by 3.}$$

Check

$$\sqrt{3(12)} + 6 \overset{?}{=} 0 \qquad \text{Substitute 12 for } x \text{ in original equation.}$$

$$6 + 6 \neq 0 \qquad \text{Solution does not check.} ✗$$

The solution $x = 12$ is an extraneous solution. So, the original equation has no solution. You can also check this graphically, as shown at the left. Notice that the graph does not cross the *x*-axis and so it has no *x*-intercept.

EXAMPLE 3   Solving an Equation Having One Radical

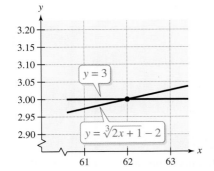

$$\sqrt[3]{2x + 1} - 2 = 3 \qquad \text{Original equation}$$

$$\sqrt[3]{2x + 1} = 5 \qquad \text{Isolate radical.}$$

$$(\sqrt[3]{2x + 1})^3 = 5^3 \qquad \text{Cube each side.}$$

$$2x + 1 = 125 \qquad \text{Simplify.}$$

$$2x = 124 \qquad \text{Subtract 1 from each side.}$$

$$x = 62 \qquad \text{Divide each side by 2.}$$

Substituting 62 for x in the original equation shows that $x = 62$ is the solution of the equation. You can also check the solution graphically by determining the point of intersection of the graphs of $y = \sqrt[3]{2x + 1} - 2$ (left side of equation) and $y = 3$ (right side of equation), as shown at the left.

Exercises Within Reach ®

Solving an Equation In Exercises 15–26, solve the equation and check your solution. (Some of the equations have no solution.)

15. $\sqrt{u} + 13 = 0$

16. $\sqrt{y} + 15 = 0$

17. $\sqrt{10x} = 30$

18. $\sqrt{8x} = 6$

19. $\sqrt{-3x} = 9$

20. $\sqrt{-4y} = 4$

21. $\sqrt{3y + 1} = 4$

22. $\sqrt{3 - 2x} = 2$

23. $\sqrt{9 - 2x} + 8 = -1$

24. $\sqrt{2t - 7} + 2 = -3$

25. $\sqrt[3]{y + 1} - 2 = 4$

26. $4\sqrt[3]{x + 4} - 6 = 1$

EXAMPLE 4 Solving an Equation Having Two Radicals

$\sqrt{5x + 3} = \sqrt{x + 11}$	Original equation
$(\sqrt{5x + 3})^2 = (\sqrt{x + 11})^2$	Square each side.
$5x + 3 = x + 11$	Simplify.
$4x + 3 = 11$	Subtract x from each side.
$4x = 8$	Subtract 3 from each side.
$x = 2$	Divide each side by 4.

Check

$\sqrt{5x + 3} = \sqrt{x + 11}$	Write original equation.
$\sqrt{5(2) + 3} \overset{?}{=} \sqrt{2 + 11}$	Substitute 2 for x.
$\sqrt{13} = \sqrt{13}$	Solution checks. ✔

So, the equation has one solution: $x = 2$.

EXAMPLE 5 Solving an Equation Having Two Radicals

$\sqrt[4]{3x} + \sqrt[4]{2x - 5} = 0$	Original equation
$\sqrt[4]{3x} = -\sqrt[4]{2x - 5}$	Isolate radicals.
$(\sqrt[4]{3x})^4 = (-\sqrt[4]{2x - 5})^4$	Raise each side to fourth power.
$3x = 2x - 5$	Simplify.
$x = -5$	Subtract $2x$ from each side.

Check

$\sqrt[4]{3x} + \sqrt[4]{2x - 5} = 0$	Write original equation.
$\sqrt[4]{3(-5)} + \sqrt[4]{2(-5) - 5} \overset{?}{=} 0$	Substitute -5 for x.
$\sqrt[4]{-15} + \sqrt[4]{-15} \neq 0$	Solution does not check. ✗

The solution does not check because it yields fourth roots of negative radicands. So, this equation has no solution. Try checking this graphically. If you graph both sides of the equation, you will discover that the graphs do not intersect.

Exercises Within Reach ®

Solutions in English & Spanish and tutorial videos at CollegePrepAlgebra.com

Solving an Equation In Exercises 27–36, solve the equation and check your solution. (Some of the equations have no solution.

27. $\sqrt{x + 3} = \sqrt{2x - 1}$ **28.** $\sqrt{3t + 1} = \sqrt{t + 15}$ **29.** $\sqrt{3x + 4} = \sqrt{4x + 3}$ **30.** $\sqrt{2x - 7} = \sqrt{3x - 12}$

31. $\sqrt{3y - 5} - 3\sqrt{y} = 0$ **32.** $\sqrt{2u + 10} - 2\sqrt{u} = 0$ **33.** $\sqrt[3]{3x - 4} = \sqrt[3]{x + 10}$ **34.** $2\sqrt[3]{10 - 3x} = \sqrt[3]{2 - x}$

35. $\sqrt[3]{2x + 15} - \sqrt[3]{x} = 0$ **36.** $\sqrt[4]{2x} + \sqrt[4]{x + 3} = 0$

EXAMPLE 6   **An Equation that Converts to a Quadratic Equation**

$\sqrt{x} + 2 = x$	Original equation
$\sqrt{x} = x - 2$	Isolate radical.
$(\sqrt{x})^2 = (x - 2)^2$	Square each side.
$x = x^2 - 4x + 4$	Simplify.
$-x^2 + 5x - 4 = 0$	Write in general form.
$(-1)(x - 4)(x - 1) = 0$	Factor.
$x - 4 = 0 \implies x = 4$	Set 1st factor equal to 0.
$x - 1 = 0 \implies x = 1$	Set 2nd factor equal to 0.

Check

First Solution

$\sqrt{4} + 2 \stackrel{?}{=} 4$

$2 + 2 = 4$ ✓

Second Solution

$\sqrt{1} + 2 \stackrel{?}{=} 1$

$1 + 2 \neq 1$ ✗

From the check you can see that $x = 1$ is an extraneous solution. So, the only solution is $x = 4$.

EXAMPLE 7   **Repeatedly Squaring Each Side of an Equation**

$\sqrt{3t + 1} = 2 - \sqrt{3t}$	Original equation
$(\sqrt{3t + 1})^2 = (2 - \sqrt{3t})^2$	Square each side (1st time).
$3t + 1 = 4 - 4\sqrt{3t} + 3t$	Simplify.
$-3 = -4\sqrt{3t}$	Isolate radical.
$(-3)^2 = (-4\sqrt{3t})^2$	Square each side (2nd time).
$9 = 16(3t)$	Simplify.
$\dfrac{3}{16} = t$	Divide each side by 48 and simplify.

The solution is $t = \frac{3}{16}$. Check this in the original equation.

Exercises Within Reach ®

Solutions in English & Spanish and tutorial videos at CollegePrepAlgebra.com

An Equation that Converts to a Quadratic Equation In Exercises 37−42, solve the equation and check your solution(s).

37. $\sqrt{x^2 - 2} = x + 4$

38. $\sqrt{x^2 - 4} = x - 2$

39. $\sqrt{2x} = x - 4$

40. $\sqrt{x} = 6 - x$

41. $\sqrt{8x + 1} = x + 2$

42. $\sqrt{3x + 7} = x + 3$

Repeatedly Squaring Each Side of an Equation In Exercises 43−50, solve the equation and check your solution.

43. $\sqrt{z + 2} = 1 + \sqrt{z}$

44. $\sqrt{2x + 5} = 7 - \sqrt{2x}$

45. $\sqrt{2t + 3} = 3 - \sqrt{2t}$

46. $\sqrt{x} + \sqrt{x + 2} = 2$

47. $\sqrt{x + 5} - \sqrt{x} = 1$

48. $\sqrt{x + 1} = 2 - \sqrt{x}$

49. $\sqrt{x - 6} + 3 = \sqrt{x + 9}$

50. $\sqrt{x + 3} - \sqrt{x - 1} = 1$

Applications

Application   EXAMPLE 8   **Electricity**

The amount of power consumed by an electrical appliance is given by $I = \sqrt{P/R}$, where I is the current measured in amps, R is the resistance measured in ohms, and P is the power measured in watts. Find the power used by an electric heater for which $I = 10$ amps and $R = 16$ ohms.

SOLUTION

$$10 = \sqrt{\frac{P}{16}}$$ Substitute 10 for I and 16 for R in original equation.

$$10^2 = \left(\sqrt{\frac{P}{16}}\right)^2$$ Square each side.

$$100 = \frac{P}{16} \quad\Longrightarrow\quad 1600 = P$$ Simplify and multiply each side by 16.

So, the solution is $P = 1600$ watts. Check this in the original equation.

EXAMPLE 9   **Velocity of a Falling Object**

The velocity of a free-falling object can be determined from the equation $v = \sqrt{2gh}$, where v is the velocity measured in feet per second, $g = 32$ feet per second per second, and h is the distance (in feet) the object has fallen. Find the height from which a rock was dropped when it strikes the ground with a velocity of 50 feet per second.

SOLUTION

$$v = \sqrt{2gh}$$ Write original equation.

$$50 = \sqrt{2(32)h}$$ Substitute 50 for v and 32 for g.

$$50^2 = \left(\sqrt{64h}\right)^2$$ Square each side.

$$2500 = 64h$$ Simplify.

$$39 \approx h$$ Divide each side by 64.

So, the rock was dropped from a height of about 39 feet. Check this in the original equation.

Study Tip

An alternative way to solve the problem in Example 8 would be first to solve the equation for P.

$$I = \sqrt{\frac{P}{R}}$$

$$I^2 = \left(\sqrt{\frac{P}{R}}\right)^2$$

$$I^2 = \frac{P}{R}$$

$$I^2 R = P$$

At this stage, you can substitute the known values of I and R to obtain

$$P = (10)^2 16 = 1600.$$

Exercises Within Reach ®   Solutions in English & Spanish and tutorial videos at CollegePrepAlgebra.com

Height In Exercises 51 and 52, use the formula $t = \sqrt{d/16}$, which gives the time t (in seconds) for a free-falling object to fall d feet.

51. A construction worker drops a nail from a building and observes it strike a water puddle after approximately 2 seconds. Estimate the height from which the nail was dropped.

52. A farmer drops a stone down a well and hears it strike the water after approximately 4.5 seconds. Estimate the depth of the well.

Free-Falling Object In Exercises 53 and 54, use the equation for the velocity of a free-falling object, $v = \sqrt{2gh}$, as described in Example 9.

53. A cliff diver strikes the water with a velocity of $32\sqrt{5}$ feet per second. Find the height from which the diver dove.

54. A stone strikes the water with a velocity of 130 feet per second. Estimate to two decimal places the height from which the stone was dropped.

Application   **EXAMPLE 10**   **Geometry: The Pythagorean Theorem**

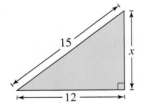

The distance between a house on shore and a playground on shore is 40 meters. The distance between the playground and a house on an island is 50 meters, as shown at the left. What is the distance between the two houses?

SOLUTION

You can see that the distances form a right triangle. So, you can use the Pythagorean Theorem to find the distance between the two houses.

$c = \sqrt{a^2 + b^2}$	Pythagorean Theorem
$50 = \sqrt{40^2 + b^2}$	Substitute 40 for a and 50 for c.
$50 = \sqrt{1600 + b^2}$	Simplify.
$50^2 = \left(\sqrt{1600 + b^2}\right)^2$	Square each side.
$2500 = 1600 + b^2$	Simplify.
$0 = b^2 - 900$	Write in general form.
$0 = (b + 30)(b - 30)$	Factor.
$b + 30 = 0 \implies b = -30$	Set 1st factor equal to 0.
$b + 30 = 0 \implies b = 30$	Set 2nd factor equal to 0.

Choose the positive solution to obtain a distance of 30 meters. Check this solution in the original equation.

Exercises Within Reach ®

Solutions in English & Spanish and tutorial videos at CollegePrepAlgebra.com

Geometry In Exercises 55−58, **find the length x of the unknown side of the right triangle.**
(Round your answer to two decimal places, if necessary.)

55.
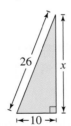

56.

57.
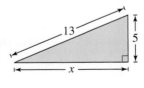

58.
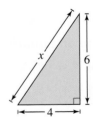

59. ***Length of a Ramp*** A ramp is 20 feet long and rests on a porch that is 4 feet high (see figure). Find the distance x between the porch and the base of the ramp.

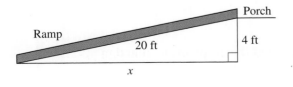

60. ***Ladder*** A ladder is 17 feet long, and the bottom of the ladder is 8 feet from the side of a house. How far does the ladder reach up the side of the house?

Concept Summary: *Solving Radical Equations*

What

Solving radical equations is somewhat like solving equations that contain fractions.

EXAMPLE

Solve $\sqrt{2x} + 1 = 5$.

How

First, try to get rid of the radicals and obtain a linear or polynomial equation. Then, solve the equation using the standard procedures.

EXAMPLE

$\sqrt{2x} + 1 = 5$	Original equation
$\sqrt{2x} = 4$	Isolate radical.
$(\sqrt{2x})^2 = (4)^2$	Square each side.
$2x = 16$	Simplify.
$x = 8$	Divide by 2.

Why

Knowing how to solve radical equations will help as you work through and solve problems involving the Pythagorean Theorem.

Exercises Within Reach ®

Worked-out solutions to odd-numbered exercises at CollegePrepAlgebra.com

Concept Summary Check

61. *Solving an Equation* In the example above, why is each side of the equation squared?

62. *Solving an Equation* Explain how to solve the equation $\sqrt[3]{2x} + 1 = 5$.

63. *Checking a Solution* One reason to check a solution in the original equation is the discover errors made in solving the equation. Describe another reason.

64. *Checking the Solution Graphically* Explain how to check the solution in the example above graphically.

Extra Practice

Solving an Equation In Exercises 65−68, solve the equation and check your solution(s).

65. $3y^{1/3} = 18$

66. $2x^{3/4} = 54$

67. $(x + 4)^{2/3} = 4$

68. $(u - 2)^{4/3} = 81$

Think About It In Exercises 69−72, use the given function to find the indicated value(s) of x.

69. For $f(x) = \sqrt{x} - \sqrt{x - 9}$, find x such that $f(x) = 1$.

70. For $g(x) = \sqrt{x} + \sqrt{x - 5}$, find x such that $g(x) = 5$.

71. For $h(x) = \sqrt{x - 2} - \sqrt{4x + 1}$, find x such that $h(x) = -3$.

72. For $f(x) = \sqrt{2x + 7} - \sqrt{x + 15}$, find x such that $f(x) = -1$.

Finding x-Intercepts In Exercises 73−76, find the x-intercept(s) of the graph of the function without graphing the function.

73. $f(x) = \sqrt{x + 5} - 3 + \sqrt{x}$

74. $f(x) = \sqrt{6x + 7} - 2 - \sqrt{2x + 3}$

75. $f(x) = \sqrt{3x - 2} - 1 - \sqrt{2x - 3}$

76. $f(x) = \sqrt{5x + 6} - 1 - \sqrt{3x + 3}$

77. *Airline Passengers* An airline offers daily flights between Chicago and Denver. The total monthly cost C (in millions of dollars) of these flights is given by

$$C = \sqrt{0.2x + 1}, \; x \geq 0$$

where x is measured in thousands of passengers (see figure). The total cost of the flights for June is $2.5 million. Approximately how many passengers flew in June?

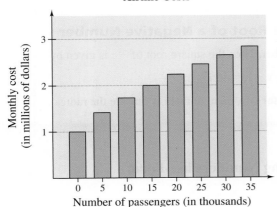

Airline Costs

Monthly cost (in millions of dollars) vs. Number of passengers (in thousands)

78. *Killer Whales* The weight w (in pounds) of a killer whale can be modeled by

$$w = 280 + 325\sqrt{t}, \quad 0 \leq t \leq 144$$

where t represents the age (in months) of the killer whale. At what age did the killer whale weigh about 3400 pounds?

79. *Geometry* The lateral surface area of a cone (see figure) is given by $S = \pi r \sqrt{r^2 + h^2}$. Solve the equation for h. Then find the height of a cone with a lateral surface area of $364\pi\sqrt{2}$ square centimeters and a radius of 14 centimeters.

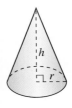

80. *Geometry* Determine the length and width of a rectangle with a perimeter of 92 inches and a diagonal of 34 inches.

Explaining Concepts

81. *Error Analysis* Describe the error.

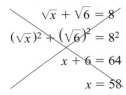

$$\sqrt{x} + \sqrt{6} = 8$$
$$(\sqrt{x})^2 + (\sqrt{6})^2 = 8^2$$
$$x + 6 = 64$$
$$x = 58$$

82. *Precision* Does raising each side of an equation to the nth power always yield an equivalent equation? Explain.

83. *Think About It* Explain how to find the values of a and b for which the equation $x + \sqrt{x - a} = b$ has a solution of $x = 20$. (There are many correct values for a and b).

84. *Number Sense* Explain how you can tell that $\sqrt{x - 9} = -4$ has no solution without solving the equation.

Cumulative Review

In Exercises 85–88, determine whether the two lines are parallel, perpendicular, or neither.

85. L_1: $y = 4x + 2$

  L_2: $y = 4x - 1$

86. L_1: $y = 3x - 8$

  L_2: $y = -3x - 8$

87. L_1: $y = -x + 5$

  L_2: $y = x - 3$

88. L_1: $y = 2x$

  L_2: $y = \frac{1}{2}x + 4$

In Exercises 89 and 90, use matrices to solve the system of linear equations.

89. $\begin{cases} 4x - y = 10 \\ -7x - 2y = -25 \end{cases}$

90. $\begin{cases} 3x - 2y = 5 \\ 6x - 5y = 14 \end{cases}$

In Exercises 91–94, simplify the expression.

91. $a^{3/5} \cdot a^{1/5}$

92. $\dfrac{m^2}{m^{2/3}}$

93. $\left(\dfrac{x^{1/2}}{x^{1/8}}\right)^4$

94. $\dfrac{(a + b)^{3/4}}{\sqrt[4]{a + b}}$

9.6 Complex Numbers

▶ Perform operations on numbers in *i*-form.

▶ Add, subtract, and multiply complex numbers.

▶ Use complex conjugates to find the quotient of two complex numbers.

The Imaginary Unit *i*

The Square Root of a Negative Number

Let c be a positive real number. Then the square root of $-c$ is given by

$$\sqrt{-c} = \sqrt{c(-1)} = \sqrt{c}\sqrt{-1} = \sqrt{c}\,i$$

When writing $\sqrt{-c}$ in the **i-form**, $\sqrt{c}\,i$, note that i is outside the radical.

EXAMPLE 1 **Writing Numbers in *i*-Form**

Write each number in *i*-form.

a. $\sqrt{-36}$ **b.** $\sqrt{-\dfrac{16}{25}}$ **c.** $\sqrt{-54}$ **d.** $\dfrac{\sqrt{-48}}{\sqrt{-3}}$

SOLUTION

a. $\sqrt{-36} = \sqrt{36(-1)} = \sqrt{36}\sqrt{-1} = 6i$

b. $\sqrt{-\dfrac{16}{25}} = \sqrt{\dfrac{16}{25}(-1)} = \sqrt{\dfrac{16}{25}}\sqrt{-1} = \dfrac{4}{5}i$

c. $\sqrt{-54} = \sqrt{54(-1)} = \sqrt{54}\sqrt{-1} = 3\sqrt{6}\,i$

d. $\dfrac{\sqrt{-48}}{\sqrt{-3}} = \dfrac{\sqrt{48}\sqrt{-1}}{\sqrt{3}\sqrt{-1}} = \dfrac{\sqrt{48}\,i}{\sqrt{3}\,i} = \sqrt{\dfrac{48}{3}} = \sqrt{16} = 4$

Exercises Within Reach ®

Solutions in English & Spanish and tutorial videos at CollegePrepAlgebra.com

Writing a Number in i-Form In Exercises 1−18, write the number in *i*-form.

1. $\sqrt{-4}$

2. $\sqrt{-9}$

3. $-\sqrt{-144}$

4. $-\sqrt{-49}$

5. $\sqrt{-\dfrac{4}{25}}$

6. $\sqrt{-\dfrac{9}{64}}$

7. $-\sqrt{-\dfrac{36}{121}}$

8. $-\sqrt{-\dfrac{9}{25}}$

9. $\sqrt{-8}$

10. $\sqrt{-75}$

11. $\sqrt{-7}$

12. $\sqrt{-15}$

13. $\dfrac{\sqrt{-12}}{\sqrt{-3}}$

14. $\dfrac{\sqrt{-45}}{\sqrt{-5}}$

15. $\sqrt{-\dfrac{18}{25}}$

16. $\sqrt{-\dfrac{20}{49}}$

17. $\sqrt{-0.09}$

18. $\sqrt{-0.0004}$

To perform operations with square roots of negative numbers, you must *first* write the numbers in *i*-form. You can then add, subtract, and multiply as follows.

$ai + bi = (a + b)i$	Addition
$ai - bi = (a - b)i$	Subtraction
$(ai)(bi) = ab(i^2) = ab(-1) = -ab$	Multiplication

Study Tip

When performing operations with numbers in *i*-form, you sometimes need to be able to evaluate powers of the imaginary unit *i*. The first several powers of *i* are as follows.

$i^1 = i$
$i^2 = -1$
$i^3 = i(i^2) = i(-1) = -i$
$i^4 = (i^2)(i^2) = (-1)(-1) = 1$
$i^5 = i(i^4) = i(1) = i$
$i^6 = (i^2)(i^4) = (-1)(1) = -1$
$i^7 = (i^3)(i^4) = (-i)(1) = -i$
$i^8 = (i^4)(i^4) = (1)(1) = 1$

Note how the pattern of values i, -1, $-i$, and 1 repeats itself for powers greater than 4.

EXAMPLE 2   **Operations with Square Roots of Negative Numbers**

Perform each operation.

a. $\sqrt{-9} + \sqrt{-49}$

b. $\sqrt{-32} - 2\sqrt{-2}$

SOLUTION

a. $\sqrt{-9} + \sqrt{-49} = \sqrt{9}\sqrt{-1} + \sqrt{49}\sqrt{-1}$ Product Rule for Radicals

$\qquad\qquad\qquad = 3i + 7i$ Write in *i*-form.

$\qquad\qquad\qquad = 10i$ Simplify.

b. $\sqrt{-32} - 2\sqrt{-2} = \sqrt{32}\sqrt{-1} - 2\sqrt{2}\sqrt{-1}$ Product Rule for Radicals

$\qquad\qquad\qquad = 4\sqrt{2}i - 2\sqrt{2}i$ Write in *i*-form.

$\qquad\qquad\qquad = 2\sqrt{2}i$ Simplify.

EXAMPLE 3   **Multiplying Square Roots of Negative Numbers**

Find each product.

a. $\sqrt{-15}\sqrt{-15}$

b. $\sqrt{-5}(\sqrt{-45} - \sqrt{-4})$

SOLUTION

a. $\sqrt{-15}\sqrt{-15} = (\sqrt{15}i)(\sqrt{15}i)$ Write in *i*-form.

$\qquad\qquad\quad = (\sqrt{15})^2 i^2$ Multiply.

$\qquad\qquad\quad = 15(-1)$ $i^2 = -1$

$\qquad\qquad\quad = -15$ Simplify.

b. $\sqrt{-5}(\sqrt{-45} - \sqrt{-4}) = \sqrt{5}i(3\sqrt{5}i - 2i)$ Write in *i*-form.

$\qquad\qquad\qquad = (\sqrt{5}i)(3\sqrt{5}i) - (\sqrt{5}i)(2i)$ Distributive Property

$\qquad\qquad\qquad = 3(5)(-1) - 2\sqrt{5}(-1)$ Multiply.

$\qquad\qquad\qquad = -15 + 2\sqrt{5}$ Simplify.

Exercises Within Reach ®

Solutions in English & Spanish and tutorial videos at CollegePrepAlgebra.com

Operations with Square Roots of Negative Numbers In Exercises 19−34, perform the operation(s).

19. $\sqrt{-16} + \sqrt{-36}$

20. $\sqrt{-25} - \sqrt{-9}$

21. $\sqrt{-50} - \sqrt{-8}$

22. $\sqrt{-500} + \sqrt{-45}$

23. $\sqrt{-12}\sqrt{-2}$

24. $\sqrt{-25}\sqrt{-6}$

25. $\sqrt{-18}\sqrt{-3}$

26. $\sqrt{-7}\sqrt{-7}$

27. $\sqrt{-0.16}\sqrt{-1.21}$

28. $\sqrt{-0.49}\sqrt{-1.44}$

29. $\sqrt{-3}(\sqrt{-3} + \sqrt{-4})$

30. $\sqrt{-12}(\sqrt{-3} - \sqrt{-12})$

31. $\sqrt{-2}(3 - \sqrt{-8})$

32. $\sqrt{-9}(1 + \sqrt{-16})$

33. $(\sqrt{-16})^2$

34. $(\sqrt{-2})^2$

Complex Numbers

A number of the form $a + bi$, where a and b are real numbers, is called a **complex number**. The real number a is called the **real part** of the complex number $a + bi$, and the number bi is called the **imaginary part**.

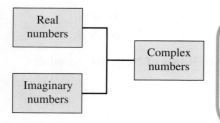

The real numbers and the imaginary numbers make up the complex numbers.

Definition of a Complex Number

If a and b are real numbers, the number $a + bi$ is a **complex number**, and it is said to be written in **standard form**. If $b = 0$, the number $a + bi = a$ is a real number. If $b \neq 0$, the number $a + bi$ is called an **imaginary number**. A number of the form bi, where $b \neq 0$, is called a **pure imaginary number**.

EXAMPLE 4   Equality of Two Complex Numbers

To determine whether the complex numbers $\sqrt{9} + \sqrt{-48}$ and $3 - 4\sqrt{3}\,i$ are equal, begin by writing the first number in standard form.

$$\sqrt{9} + \sqrt{-48} = \sqrt{3^2} + \sqrt{4^2(3)(-1)} = 3 + 4\sqrt{3}\,i$$

The two numbers are not equal because their imaginary parts differ in sign.

EXAMPLE 5   Equality of Two Complex Numbers

To find the values of x and y that satisfy the equation $3x - \sqrt{-25} = -6 + 3yi$, begin by writing the left side of the equation in standard form.

$$3x - 5i = -6 + 3yi \qquad \text{Each side is in standard form.}$$

For these two numbers to be equal, their real parts must be equal to each other and their imaginary parts must be equal to each other.

Real Parts	Imaginary Parts
$3x = -6$	$3yi = -5i$
$x = -2$	$3y = -5$
	$y = -\frac{5}{3}$

So, $x = -2$ and $y = -\frac{5}{3}$.

Exercises Within Reach ®

Solutions in English & Spanish and tutorial videos at CollegePrepAlgebra.com

Equality of Two Complex Numbers In Exercises 35−38, determine whether the complex numbers are equal.

35. $\sqrt{1} + \sqrt{-25}$ and $1 + 5i$

36. $\sqrt{16} + \sqrt{-9}$ and $4 - 3i$

37. $\sqrt{27} - \sqrt{-8}$ and $3\sqrt{3} + 2\sqrt{2}\,i$

38. $\sqrt{18} - \sqrt{-12}$ and $3\sqrt{2} - 2\sqrt{3}\,i$

Equality of Two Complex Numbers In Exercises 39−46, determine the values of a and b that satisfy the equation.

39. $3 - 4i = a + bi$

40. $-8 + 6i = a + bi$

41. $5 - 4i = (a + 3) + (b - 1)i$

42. $-10 + 12i = 2a + (5b - 3)i$

43. $-4 - \sqrt{-8} = a + bi$

44. $\sqrt{-36} - 3 = a + bi$

45. $\sqrt{a} + \sqrt{-49} = 8 + bi$

46. $\sqrt{100} + \sqrt{b} = a + 2\sqrt{3}\,i$

To add or subtract two complex numbers, you add (or subtract) the real and imaginary parts separately. This is similar to combining like terms of a polynomial.

$$(a + bi) + (c + di) = (a + c) + (b + d)i$$ Addition of complex numbers

$$(a + bi) - (c + di) = (a - c) + (b - d)i$$ Subtraction of complex numbers

Study Tip
Note in part (b) of Example 6 that the sum of two complex numbers can be a real number.

EXAMPLE 6   **Adding and Subtracting Complex Numbers**

a. $(3 - i) + (-2 + 4i) = (3 - 2) + (-1 + 4)i = 1 + 3i$

b. $3i + (5 - 3i) = 5 + (3 - 3)i = 5$

c. $4 - (-1 + 5i) + (7 + 2i) = [4 - (-1) + 7] + (-5 + 2)i = 12 - 3i$

d. $(6 + 3i) + (2 - \sqrt{-8}) - \sqrt{-4} = (6 + 3i) + (2 - 2\sqrt{2}i) - 2i$

$$= (6 + 2) + (3 - 2\sqrt{2} - 2)i$$

$$= 8 + (1 - 2\sqrt{2})i$$

EXAMPLE 7   **Multiplying Complex Numbers**

a. $(7i)(-3i) = -21i^2$ Multiply.

$$= -21(-1) = 21$$ $i^2 = -1$

b. $(1 - i)(\sqrt{-9}) = (1 - i)(3i)$ Write in i-form.

$$= 3i - 3i^2$$ Distributive Property

$$= 3i - 3(-1) = 3 + 3i$$ $i^2 = -1$

c. $(2 - i)(4 + 3i) = 8 + 6i - 4i - 3i^2$ FOIL Method

$$= 8 + 6i - 4i - 3(-1)$$ $i^2 = -1$

$$= 11 + 2i$$ Combine like terms.

d. $(3 + 2i)(3 - 2i) = 3^2 - (2i)^2$ Special product formula

$$= 9 - 4i^2$$ Simplify.

$$= 9 - 4(-1) = 13$$ $i^2 = -1$

Exercises Within Reach ®

Solutions in English & Spanish and tutorial videos at CollegePrepAlgebra.com

Adding and Subtracting Complex Numbers In Exercises 47−54, **perform** the operation(s) and write the result in standard form.

47. $(-4 - 7i) + (-10 - 33i)$

48. $(15 + 10i) - (2 + 10i)$

49. $13i - (14 - 7i)$

50. $17i + (9 - 14i)$

51. $6 - (3 - 4i) + 2i$

52. $22 + (-5 + 8i) + 10i$

53. $15i - (3 - 25i) + \sqrt{-81}$

54. $(-1 + i) - \sqrt{2} - \sqrt{-2}$

Multiplying Complex Numbers In Exercises 55−62, **perform** the operation and write the result in standard form.

55. $(3i)(-8i)$

56. $(-5i)(4i)$

57. $(9 - 2i)(\sqrt{-4})$

58. $(11 + 3i)(\sqrt{-25})$

59. $(4 + 3i)(-7 + 4i)$

60. $(3 + 5i)(2 - 15i)$

61. $(6 + 3i)(6 - 3i)$

62. $(5 - 2i)(5 + 2i)$

Complex Conjugates

Complex numbers of the form $a + bi$ and $a - bi$ are called **complex conjugates**. In general, the product of complex conjugates has the following form.

$$(a + bi)(a - bi) = a^2 - (bi)^2 = a^2 - b^2 i^2 = a^2 - b^2(-1) = a^2 + b^2$$

EXAMPLE 8 Complex Conjugates

Complex Number	Complex Conjugate	Product
$4 - 5i$	$4 + 5i$	$4^2 + 5^2 = 41$
$3 + 2i$	$3 - 2i$	$3^2 + 2^2 = 13$
$-2 = -2 + 0i$	$-2 = -2 - 0i$	$(-2)^2 + 0^2 = 4$
$i = 0 + i$	$-i = 0 - i$	$0^2 + 1^2 = 1$

EXAMPLE 9 Writing Quotients in Standard Form

a.

$$\frac{2 - i}{4i} = \frac{2 - i}{4i} \cdot \frac{(-4i)}{(-4i)}$$ Multiply numerator and denominator by complex conjugate of denominator.

$$= \frac{-8i + 4i^2}{-16i^2}$$ Multiply fractions.

$$= \frac{-8i + 4(-1)}{-16(-1)}$$ $i^2 = -1$

$$= \frac{-8i - 4}{16}$$ Simplify.

$$= -\frac{1}{4} - \frac{1}{2}i$$ Write in standard form.

b.

$$\frac{5}{3 - 2i} = \frac{5}{3 - 2i} \cdot \frac{3 + 2i}{3 + 2i}$$ Multiply numerator and denominator by complex conjugate of denominator.

$$= \frac{5(3 + 2i)}{(3 - 2i)(3 + 2i)}$$ Multiply fractions.

$$= \frac{5(3 + 2i)}{3^2 + 2^2}$$ Product of complex conjugates

$$= \frac{15 + 10i}{13}$$ Simplify.

$$= \frac{15}{13} + \frac{10}{13}i$$ Write in standard form.

Exercises Within Reach ®

Solutions in English & Spanish and tutorial videos at CollegePrepAlgebra.com

Complex Conjugates In Exercises 63–70, multiply the number by its complex conjugate and simplify.

63. $-2 - 8i$ **64.** $10 - 3i$ **65.** $2 + i$ **66.** $3 + 2i$

67. $10i$ **68.** 20 **69.** -12 **70.** $-12i$

Writing a Quotient in Standard Form In Exercises 71–74, write the quotient in standard form.

71. $\dfrac{2 + i}{-5i}$ **72.** $\dfrac{1 + i}{3i}$ **73.** $\dfrac{-12}{2 + 7i}$ **74.** $\dfrac{15}{2(1 - i)}$

EXAMPLE 10   **Writing a Quotient in Standard Form**

Write $\dfrac{8-i}{8+i}$ in standard form.

SOLUTION

$\dfrac{8-i}{8+i} = \dfrac{8-i}{8+i} \cdot \dfrac{8-i}{8-i}$ Multiply numerator and denominator by complex conjugate of denominator.

$= \dfrac{(8-i)(8-i)}{(8+i)(8-i)}$ Multiply fractions.

$= \dfrac{64 - 16i + i^2}{8^2 + 1^2}$ FOIL Method and product of complex conjugates

$= \dfrac{64 - 16i + (-1)}{8^2 + 1^2}$ $i^2 = -1$

$= \dfrac{63 - 16i}{65}$ Simplify.

$= \dfrac{63}{65} - \dfrac{16}{65}i$ Write in standard form.

EXAMPLE 11   **Writing a Quotient in Standard Form**

Write $\dfrac{2+3i}{4-2i}$ in standard form.

SOLUTION

$\dfrac{2+3i}{4-2i} = \dfrac{2+3i}{4-2i} \cdot \dfrac{4+2i}{4+2i}$ Multiply numerator and denominator by complex conjugate of denominator.

$= \dfrac{(2+3i)(4+2i)}{(4-2i)(4+2i)}$ Multiply fractions.

$= \dfrac{8 + 16i + 6i^2}{4^2 + 2^2}$ FOIL Method and product of complex conjugates

$= \dfrac{8 + 16i + 6(-1)}{4^2 + 2^2}$ $i^2 = -1$

$= \dfrac{2 + 16i}{20}$ Simplify.

$= \dfrac{1}{10} + \dfrac{4}{5}i$ Write in standard form.

Exercises Within Reach ®   Solutions in English & Spanish and tutorial videos at CollegePrepAlgebra.com

Writing a Quotient in Standard Form **In Exercises 75–80, write the quotient in standard form.**

75. $\dfrac{5-i}{5+i}$

76. $\dfrac{9+i}{9-i}$

77. $\dfrac{4-i}{3+i}$

78. $\dfrac{2+i}{5-i}$

79. $\dfrac{4+5i}{3-7i}$

80. $\dfrac{5+3i}{7-4i}$

Concept Summary: Complex Numbers

What

The **imaginary unit** i is used to define the square root of a negative number.

- $i = \sqrt{-1}$
- $\sqrt{-c} = \sqrt{c}\,i$, where $c > 0$.

A number of the form $a + bi$, where a and b are real numbers, is called a **complex number**.

Performing operations with complex numbers is similar to performing operations with polynomials.

EXAMPLE

Perform each operation.

a. $(2 - i) + (4 + 2i)$

b. $(1 + i)(3 - 2i)$

c. $\dfrac{3}{i}$

How

To perform operations with complex numbers:

- Combine like terms to add and subtract.
- Use the Commutative, Associative, and Distributive Properties, along with the FOIL Method to multiply.
- Use **complex conjugates** to simplify quotients.

EXAMPLE

a. $(2 - i) + (4 + 2i)$

$= (2 + 4) + (-1 + 2)i = 6 + i$

b. $(1 + i)(3 - 2i)$

$= 3 - 2i + 3i - 2i^2$

$= 3 - 2i + 3i - 2(-1) = 5 + i$

c. $\dfrac{3}{i} = \dfrac{3}{i} \cdot \dfrac{-i}{-i} = \dfrac{-3i}{-(-1)} = -3i$

Why

The solutions of some quadratic equations are complex numbers. For instance, when a quadratic equation has no real solution, it has exactly two complex solutions. Knowing how to perform operations with complex numbers will help you identify such solutions.

Exercises Within Reach ®

Worked-out solutions to odd-numbered exercises at CollegePrepAlgebra.com

Concept Summary Check

81. *Writing a Number in i-Form* Write $\sqrt{-2}$ in i-form.

82. *Identifying Like Terms* Identify any like terms in the expression $(2 - i) + (4 + 2i)$.

83. *Multiplying Complex Numbers* What method is used to multiply $(1 + i)$ and $(3 - 2i)$ in the solution above?

84. *Complex Conjugate* What is the complex conjugate of i? What is the product of i and its complex conjugate?

Extra Practice

Operations with Complex Numbers In Exercises 85–96, perform the operation(s) and write the result in standard form.

85. $\sqrt{-48} + \sqrt{-12} - \sqrt{-27}$

86. $\sqrt{-32} - \sqrt{-18} + \sqrt{-50}$

87. $(-5i)(-i)(\sqrt{-49})$

88. $(10i)(\sqrt{-36})(-5i)$

89. $(-3i)^3$

90. $(2i)^4$

91. $(-2 + \sqrt{-5})(-2 - \sqrt{-5})$

92. $(-3 - \sqrt{-12})(4 - \sqrt{-12})$

93. $(2 + 5i)^2$

94. $(8 - 3i)^2$

95. $(3 + i)^3$

96. $(2 - 2i)^3$

Evaluating Powers of i In Exercises 97–104, simplify the expression.

97. i^9

98. i^{11}

99. i^{42}

100. i^{24}

101. i^{35}

102. i^{64}

103. $(-i)^6$

104. $(-i)^4$

Operations with Complex Numbers **In Exercises 105−108, perform the operation by first writing each quotient in standard form.**

105. $\dfrac{5}{3+i} + \dfrac{1}{3-i}$

106. $\dfrac{1}{1-2i} + \dfrac{4}{1+2i}$

107. $\dfrac{3i}{1+i} + \dfrac{2}{2+3i}$

108. $\dfrac{i}{4-3i} - \dfrac{5}{2+i}$

Operations with Complex Conjugates **In Exercises 109−112, perform the operations.**

109. $(a+bi) + (a-bi)$

110. $(a+bi)(a-bi)$

111. $(a+bi) - (a-bi)$

112. $(a+bi)^2 + (a-bi)^2$

113. *Cube Roots* The principal cube root of 125, $\sqrt[3]{125}$, is 5. Evaluate the expression x^3 for each value of x.

(a) $x = \dfrac{-5 + 5\sqrt{3}\,i}{2}$

(b) $x = \dfrac{-5 - 5\sqrt{3}\,i}{2}$

114. *Cube Roots* The principal cube root of 27, $\sqrt[3]{27}$, is 3. Evaluate the expression x^3 for each value of x.

(a) $x = \dfrac{-3 + 3\sqrt{3}\,i}{2}$

(b) $\dfrac{-3 - 3\sqrt{3}\,i}{2}$

115. *Pattern Recognition* Compare the results of Exercises 113 and 114. Use the result to list the cube roots in simplest form of each number.

(a) 1

(b) 8

(c) 64

116. *Algebraic Properties* Consider the complex number $1 + 5i$.

(a) Find the additive inverse of the number.

(b) Find the multiplicative inverse of the number.

Explaining Concepts

117. *Writing Rules* Look back at Exercises 109–112. Based on your results, write a general rule for each exercise about operations on complex conjugates of the form $a + bi$ and $a - bi$.

118. *True or False?* Some numbers are both real and imaginary. Justify your answer.

119. *Error Analysis* Describe and correct the error.

$$\sqrt{-3}\sqrt{-3} = \sqrt{(-3)(-3)} = \sqrt{9} = 3$$

120. *Precision* Explain why the Product Rule for Radicals cannot be used to produce the second expression in Exercise 119.

121. *Number Sense* The denominator of a quotient is a pure imaginary number of the form bi. How can you use the complex conjugate of bi to write the quotient in standard form? Can you use the number i instead of the conjugate of bi? Explain.

122. *Number Sense* The polynomial $x^2 + 1$ is prime *with respect to the integers*. It is not, however, prime *with respect to the complex numbers*. Show how $x^2 + 1$ can be factored using complex numbers.

Cumulative Review

In Exercises 123−126, use the Zero-Factor Property to solve the equation.

123. $(x-5)(x+7) = 0$

124. $z(z-2) = 0$

125. $3y(y-3)(y+4) = 0$

126. $(3x-2)(4x+1)(x+9) = 0$

In Exercises 127−130, solve the equation and check your solution.

127. $\sqrt{x} = 9$

128. $\sqrt[3]{t} = 8$

129. $\sqrt{x} - 5 = 0$

130. $\sqrt{2x+3} - 7 = 0$

9 Chapter Summary

	What did you learn?	Explanation and Examples	Review Exercises
9.1	Determine the *n*th roots of numbers and evaluate radical expressions *(p. 442)*.	**1.** If *a* is a positive real number and *n* is even, then *a* has exactly two (real) *n*th roots, which are denoted by $\sqrt[n]{a}$ and $-\sqrt[n]{a}$. **2.** If *a* is any real number and *n* is odd, then *a* has only one (real) *n*th root, which is denoted by $\sqrt[n]{a}$. **3.** If *a* is a negative real number and *n* is even, then *a* has no (real) *n*th root.	1—12
	Use the rules of exponents to evaluate or simplify expressions with rational exponents *(p. 445)*.	**1.** $a^{1/n} = \sqrt[n]{a}$ **2.** $a^{m/n} = (a^{1/n})^m = (\sqrt[n]{a})^m$ **3.** $a^{m/n} = (a^m)^{1/n} = \sqrt[n]{a^m}$ See page 445 for the rules of exponents as they apply to rational exponents.	13—34
	Evaluate radical functions and find the domains of radical functions *(p. 447)*.	Let *n* be an integer that is greater than or equal to 2. **1.** If *n* is odd, the domain of $f(x) = \sqrt[n]{x}$ is the set of all real numbers. **2.** If *n* is even, the domain of $f(x) = \sqrt[n]{x}$ is the set of all nonnegative real numbers.	35—44
9.2	Use the Product and Quotient Rules for Radicals to simplify radical expressions *(p. 450)*.	Let *u* and *v* be real numbers, variables, or algebraic expressions. If the *n*th roots of *u* and *v* are real, then the following rules are true. **1.** $\sqrt[n]{uv} = \sqrt[n]{u}\sqrt[n]{v}$ **2.** $\sqrt[n]{\dfrac{u}{v}} = \dfrac{\sqrt[n]{u}}{\sqrt[n]{v}}, v \neq 0$	45—54
	Use rationalization techniques to simplify radical expressions *(p. 453)*.	A radical expression is said to be in *simplest form* when all three of the statements below are true. **1.** All possible *n*th powered factors have been removed from each radical. **2.** No radical contains a fraction. **3.** No denominator of a fraction contains a radical.	55—58
	Use the Pythagorean Theorem in application problems *(p. 454)*.	In a right triangle, if *a* and *b* are the lengths of the legs and *c* is the length of the hypotenuse, then $c = \sqrt{a^2 + b^2}$ and $a = \sqrt{c^2 - b^2}$.	59—62
9.3	Use the Distributive Property to add and subtract like radicals *(p. 458)*.	**1.** Write each radical in simplest form. **2.** Combine like radicals: Use the Distributive Property to factor out the like radical, and then combine the terms inside the parentheses. $\sqrt{4x} + \sqrt{9x}$ $= 2\sqrt{x} + 3\sqrt{x}$ $= (2 + 3)\sqrt{x}$ $= 5\sqrt{x}$	63—70

	What did you learn?	*Explanation and Examples*	*Review Exercises*
9.3	Use radical expressions in application problems *(p. 462)*.	Radical expressions can be used to model real-life quantities, such as the average cost of tuition (See Example 6).	71, 72
9.4	Use the Distributive Property or the FOIL Method to multiply radical expressions *(p. 468)*.	**Distributive Property:** $\sqrt{2}(2 + \sqrt{2}) = 2\sqrt{2} + (\sqrt{2})^2$ $= 2\sqrt{2} + 2$ **FOIL Method:** $$\begin{array}{cccc} \mathbf{F} & \mathbf{O} & \mathbf{I} & \mathbf{L} \end{array}$$ $(\sqrt{2} + 1)(\sqrt{2} - 3) = (\sqrt{2})^2 - 3\sqrt{2} + \sqrt{2} - 3$ $= 2 - 2\sqrt{2} - 3 = -1 - 2\sqrt{2}$	73−78
	Determine the products of conjugates *(p. 470)*.	The product of two conjugates $(a + b)$ and $(a - b)$ is the difference of two squares. $(a + b)(a - b) = a^2 - b^2$	79−82
	Simplify quotients involving radicals by rationalizing the denominators *(p. 471)*.	To simplify a quotient involving radicals, rationalize the denominator by multiplying both the numerator and denominator by the conjugate of the denominator.	83−86
9.5	Solve a radical equation by raising each side to the nth power *(p. 476)*.	Let u and v be real numbers, variables, or algebraic expressions, and let n be a positive integer. If $u = v$, then it follows that $u^n = v^n$. $\sqrt{x} = 9$ Let $u = \sqrt{x}, v = 9$. $(\sqrt{x})^2 = 9^2$ $u^2 = v^2$ $x = 81$ Simplify.	87−96
	Solve application problems involving radical equations *(p. 480)*.	Real-life problems involving electricity, the velocity of a falling object, and distances can result in radical equations (See Examples 8−10).	97−102
9.6	Perform operations on numbers in i-form *(p. 484)*.	To perform operations with square roots of negative numbers, first write the numbers in i-form. Let c be a positive real number. Then the square root of $-c$ is given by $\sqrt{-c} = \sqrt{c(-1)} = \sqrt{c}\sqrt{-1} = \sqrt{c}\,i$. When writing $\sqrt{-c}$ in the i-form, $\sqrt{c}\,i$, note that i is outside the radical.	103−116
	Add, subtract, and multiply complex numbers *(p. 486)*.	If a and b are real numbers, then $a + bi$ is a complex number in standard form. To add (or subtract) complex numbers, add (or subtract) the real and imaginary parts separately. To multiply complex numbers, use the Commutative, Associative, and Distributive Properties, along with the FOIL Method.	117−126
	Use complex conjugates to find the quotient of two complex numbers *(p. 488)*.	Complex numbers of the form $(a + bi)$ and $(a - bi)$ are called complex conjugates. To write the quotient of two complex numbers in standard form, multiply the numerator and denominator by the complex conjugate of the denominator, and simplify.	127−132

Review Exercises

Worked-out solutions to odd-numbered exercises at CollegePrepAlgebra.com

9.1

Evaluating an Expression In Exercises 1−10, evaluate the radical expression. If not possible, state the reason.

1. $-\sqrt{81}$

2. $\sqrt{-16}$

3. $-\sqrt[3]{64}$

4. $\sqrt[3]{-125}$

5. $-\sqrt{\left(\frac{3}{4}\right)^2}$

6. $\sqrt{\left(-\frac{9}{13}\right)^2}$

7. $\sqrt[3]{-\left(\frac{1}{5}\right)^3}$

8. $-\sqrt[3]{\left(-\frac{27}{64}\right)^3}$

9. $\sqrt{-2^2}$

10. $-\sqrt{-3^2}$

Finding the Side Length In Exercises 11 and 12, use the area A of the square or the volume V of the cube to find the side length l.

11.

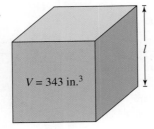

$A = \frac{16}{81}$ cm² l

12.

$V = 343$ in.³ l

Using a Definition In Exercises 13−16, determine the missing description.

Radical Form	Rational Exponent Form
13. $\sqrt[3]{27} = 3$	
14. $\sqrt[3]{0.125} = 0.5$	
15.	$216^{1/3} = 6$
16.	$16^{1/4} = 2$

Evaluating an Expression with a Rational Exponent In Exercises 17−22, evaluate the expression.

17. $27^{4/3}$

18. $16^{3/4}$

19. $(-25)^{3/2}$

20. $-(4^3)^{2/3}$

21. $8^{-4/3}$

22. $243^{-2/5}$

Using Rules of Exponents In Exercises 23−34, use rational exponents to rewrite and simplify the expression.

23. $x^{3/4} \cdot x^{-1/6}$

24. $a^{2/3} \cdot a^{3/5}$

25. $z\sqrt[3]{z^2}$

26. $x^2\sqrt[4]{x^3}$

27. $\dfrac{\sqrt[4]{x^3}}{\sqrt{x^4}}$

28. $\dfrac{\sqrt{x^3}}{\sqrt[3]{x^2}}$

29. $\sqrt[3]{a^3b^2}$

30. $\sqrt[4]{m^3n^8}$

31. $\sqrt[4]{\sqrt{x}}$

32. $\sqrt{\sqrt[3]{x^4}}$

33. $\dfrac{(3x + 2)^{2/3}}{\sqrt[3]{3x + 2}}$

34. $\dfrac{\sqrt[5]{3x + 6}}{(3x + 6)^{4/5}}$

Evaluating a Radical Function In Exercises 35−38, evaluate the function for each indicated x-value, if possible, and simplify.

35. $f(x) = \sqrt{x - 2}$

   (a) $f(-7)$

   (b) $f(51)$

36. $f(x) = \sqrt{6x - 5}$

   (a) $f(5)$

   (b) $f(-1)$

37. $g(x) = \sqrt[3]{2x - 1}$

   (a) $g(0)$

   (b) $g(14)$

38. $g(x) = \sqrt[4]{x + 5}$

   (a) $g(-4)$

   (b) $g(76)$

Finding the Domain of a Radical Function In Exercises 39−44, describe the domain of the function.

39. $f(x) = \sqrt[7]{x}$

40. $g(x) = \sqrt[3]{4x}$

41. $g(x) = \sqrt{6x}$

42. $f(x) = \sqrt[4]{2x}$

43. $f(x) = \sqrt{9 - 2x}$

44. $g(x) = \sqrt[3]{x + 2}$

9.2

Removing a Constant Factor from a Radical In Exercises 45−48, simplify the radical expression.

45. $\sqrt{63}$

46. $\sqrt{28}$

47. $\sqrt{242}$

48. $\sqrt{245}$

Removing Factors from a Radical In Exercises 49−54, simplify the radical expression.

49. $\sqrt{36u^5v^2}$

50. $\sqrt{24x^3y^4}$

51. $\sqrt{0.25x^4y}$

52. $\sqrt{0.16s^6t^3}$

53. $\sqrt[3]{48a^3b^4}$

54. $\sqrt[4]{48u^4v^6}$

Rationalizing the Denominator In Exercises 55−58, rationalize the denominator and simplify further, if possible.

55. $\sqrt{\dfrac{5}{6}}$

56. $\dfrac{4y}{\sqrt{10z}}$

57. $\dfrac{2}{\sqrt[3]{2x}}$

58. $\sqrt[3]{\dfrac{16t}{s^2}}$

Geometry In Exercises 59−62, find the length of the hypotenuse of the right triangle.

59.

60.

61.

62.

9.3

Combining Radical Expressions In Exercises 63−70, simplify the expression by combining like radicals.

63. $2\sqrt{24} + 7\sqrt{6} - \sqrt{54}$

64. $9\sqrt{50} - 5\sqrt{8} + \sqrt{48}$

65. $5\sqrt{x} - \sqrt[3]{x} + 9\sqrt{x} - 8\sqrt[3]{x}$

66. $\sqrt{3x} - \sqrt[4]{6x^2} + 2\sqrt[4]{6x^2} - 4\sqrt{3x}$

67. $10\sqrt[4]{y+3} - 3\sqrt[4]{y+3}$

68. $5\sqrt[3]{x-3} + 4\sqrt[3]{x-3}$

69. $2x\sqrt[3]{24x^2y} - \sqrt[3]{3x^5y}$

70. $4xy^2\sqrt[4]{243x} + 2y^2\sqrt[4]{48x^5}$

Dining Hall In Exercises 71 and 72, a campus dining hall is undergoing renovations. The four corners of the hall are to be walled off and used as storage units (see figure.)

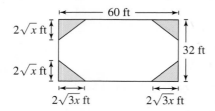

71. Find the perimeter of one of the storage units.

72. Find the perimeter of the newly designed dining hall.

9.4

Multiplying Radical Expressions In Exercises 73–78, multiply and simplify.

73. $\sqrt{15} \cdot \sqrt{20}$

74. $\sqrt{36} \cdot \sqrt{60}$

75. $\sqrt{10}\left(\sqrt{2} + \sqrt{5}\right)$

76. $\sqrt{12}\left(\sqrt{6} - \sqrt{8}\right)$

77. $\left(\sqrt{3} - \sqrt{x}\right)\left(\sqrt{3} + \sqrt{x}\right)$

78. $\left(4 - 3\sqrt{2}\right)^2$

Multiplying Conjugates In Exercises 79–82, find the conjugate of the expression. Then multiply the expression by its conjugate and simplify.

79. $3 - \sqrt{7}$

80. $\sqrt{6} + 9$

81. $\sqrt{x} + 20$

82. $9 - \sqrt{2y}$

Simplifying a Quotient Involving a Radical In Exercises 83–86, simplify the expression.

83. $\dfrac{\sqrt{2} - 1}{\sqrt{3} - 4}$

84. $\dfrac{2 + \sqrt{20}}{3 + \sqrt{5}}$

85. $\left(\sqrt{x} + 10\right) \div \left(\sqrt{x} - 10\right)$

86. $\left(3\sqrt{s} + 4\right) \div \left(\sqrt{s} + 2\right)$

9.5

Solving an Equation In Exercises 87–96, solve the equation and check your solution(s). (Some of the equations have no solution.)

87. $\sqrt{2x} - 8 = 0$

88. $\sqrt{4x} + 6 = 9$

89. $\sqrt[4]{3x-1} + 6 = 3$

90. $\sqrt[3]{5x-7} - 3 = -1$

91. $\sqrt[3]{5x+2} - \sqrt[3]{7x-8} = 0$

92. $\sqrt[4]{9x-2} - \sqrt[4]{8x} = 0$

93. $\sqrt{2(x+5)} = x + 5$

94. $y - 2 = \sqrt{y+4}$

95. $\sqrt{1+6x} = 2 - \sqrt{6x}$

96. $\sqrt{2+9b} + 1 = 3\sqrt{b}$

97. ***Plasma TV*** The screen of a plasma television has a diagonal of 42 inches and a width of 36.8 inches. Find the height of the screen.

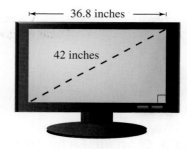

98. ***Geometry*** Determine the length and width of a rectangle with a perimeter of 82 inches and a diagonal of 29 inches.

99. ***Period of a Pendulum*** The time t (in seconds) for a pendulum of length L (in feet) to go through one complete cycle (its period) is given by

$$t = 2\pi\sqrt{\dfrac{L}{32}}.$$

How long is the pendulum of a grandfather clock with a period of 1.9 seconds?

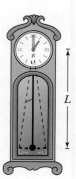

100. *Height* The time t (in seconds) for a free-falling object to fall d feet is given by

$$t = \sqrt{\frac{d}{16}}.$$

A child drops a pebble from a bridge and observes it strike the water after approximately 4 seconds. Estimate the height from which the pebble was dropped.

Free-Falling Object In Exercises 101 and 102, the velocity of a free-falling object can be determined from the equation

$$v = \sqrt{2gh}$$

where v is the velocity (in feet per second), $g = 32$ feet per second per second, and h is the distance (in feet) the object has fallen.

101. Find the height from which a brick was dropped when it strikes the ground with a velocity of 64 feet per second.

102. Find the height from which a wrench was dropped when it strikes the ground with a velocity of 112 feet per second.

9.6

Writing a Number in i-Form In Exercises 103−110, write the number in *i*-form.

103. $\sqrt{-48}$

104. $\sqrt{-0.16}$

105. $10 - 3\sqrt{-27}$

106. $3 + 2\sqrt{-500}$

107. $\frac{3}{4} - 5\sqrt{-\frac{3}{25}}$

108. $\frac{2}{3} + 4\sqrt{-\frac{5}{16}}$

109. $8.4 + 20\sqrt{-0.81}$

110. $-0.5 + 3\sqrt{-1.21}$

Operations with Square Roots of Negative Numbers In Exercises 111−116, perform the operation(s) and write the result in standard form.

111. $\sqrt{-9} - \sqrt{-1}$

112. $\sqrt{-16} + \sqrt{-64}$

113. $\sqrt{-81} + \sqrt{-36}$

114. $\sqrt{-121} - \sqrt{-84}$

115. $\sqrt{-10}(\sqrt{-4} - \sqrt{-7})$

116. $\sqrt{-5}(\sqrt{-10} + \sqrt{-15})$

Equality of Two Complex Numbers In Exercises 117−120, determine the values of a and b that satisfy the equation.

117. $12 - 5i = (a + 2) + (b - 1)i$

118. $-48 + 9i = (a - 5) + (b + 10)i$

119. $\sqrt{-49} + 4 = a + bi$

120. $-3 - \sqrt{-4} = a + bi$

Operations with Complex Numbers In Exercises 121−126, perform the operation and write the result in standard form.

121. $(-4 + 5i) - (-12 + 8i)$

122. $(-6 + 3i) + (-1 + i)$

123. $(4 - 3i)(4 + 3i)$

124. $(12 - 5i)(2 + 7i)$

125. $(6 - 5i)^2$

126. $(2 - 9i)^2$

Writing a Quotient in Standard Form In Exercises 127−132, write the quotient in standard form.

127. $\frac{7}{3i}$

128. $\frac{4}{5i}$

129. $\frac{-3i}{4 - 6i}$

130. $\frac{5i}{2 + 9i}$

131. $\frac{3 - 5i}{6 + i}$

132. $\frac{2 + i}{1 - 9i}$

Chapter Test

Solutions in English & Spanish and tutorial videos at CollegePrepAlgebra.com

Take this test as you would take a test in class. After you are done, check your work against the answers in the back of the book.

In Exercises 1 and 2, evaluate each expression.

1. (a) $16^{3/2}$

   (b) $\sqrt{5}\sqrt{20}$

2. (a) $125^{-2/3}$

   (b) $\sqrt{3}\sqrt{12}$

3. For $f(x) = \sqrt{9 - 5x}$, find $f(-8)$ and $f(0)$.

4. Describe the domain of $g(x) = \sqrt{7x - 3}$.

In Exercises 5–7, simplify the expression.

5. (a) $\left(\dfrac{x^{1/2}}{x^{1/3}}\right)^2$

   (b) $5^{1/4} \cdot 5^{7/4}$

6. (a) $\sqrt{\dfrac{32}{9}}$

   (b) $\sqrt[3]{24}$

7. (a) $\sqrt{24x^3}$

   (b) $\sqrt[4]{16x^5y^8}$

In Exercises 8 and 9, rationalize the denominator of the expression and simplify.

8. $\dfrac{2}{\sqrt[3]{9y}}$

9. $\dfrac{10}{\sqrt{6} - \sqrt{2}}$

10. Subtract: $6\sqrt{18x} - 3\sqrt{32x}$

11. Multiply and simplify: $\sqrt{5}(\sqrt{15x} + 3)$

12. Multiply and simplify: $(4 - \sqrt{2x})^2$

13. Factor: $7\sqrt{27} + 14y\sqrt{12} = 7\sqrt{3}\,(\,\rule{2cm}{0.4pt}\,)$

In Exercises 14–16, solve the equation.

14. $\sqrt{6z} + 5 = 17$

15. $\sqrt{x^2 - 1} = x - 2$

16. $\sqrt{x} - x + 6 = 0$

In Exercises 17–20, perform the operation and write the result in standard form.

17. $(2 + 3i) - \sqrt{-25}$

18. $(3 - 5i)^2$

19. $\sqrt{-16}(1 + \sqrt{-4})$

20. $(3 - 2i)(1 + 5i)$

21. Write $\dfrac{5 - 2i}{3 + i}$ in standard form.

22. The velocity v (in feet per second) of an object is given by $v = \sqrt{2gh}$, where $g = 32$ feet per second per second and h is the distance (in feet) the object has fallen. Find the height from which a rock was dropped when it strikes the ground with a velocity of 96 feet per second.

Cumulative Test: Chapters 7–9

Take this test as you would take a test in class. After you are done, check your work against the answers in the back of the book.

In Exercises 1–6, perform the operation(s) and simplify.

1. $\dfrac{x^2 + 8x + 16}{18x^2} \cdot \dfrac{2x^4 + 4x^3}{x^2 - 16}$

2. $\dfrac{x^2 + 4x}{2x^2 - 7x + 3} \div \dfrac{x^2 - 16}{x - 3}$

3. $\dfrac{5x}{x + 2} - \dfrac{2}{x^2 - x - 6}$

4. $\dfrac{2}{x} - \dfrac{x}{x^3 + 3x^2} + \dfrac{1}{x + 3}$

5. $\dfrac{\left(\dfrac{3x}{x + 2}\right)}{\left(\dfrac{12}{x^3 + 2x^2}\right)}$

6. $\dfrac{\left(\dfrac{x}{y} - \dfrac{y}{x}\right)}{\left(\dfrac{x - y}{xy}\right)}$

[handwritten: $2x^2 - 6x - 1x + 3$; $2x(x-3) - 1(x-3)$]

In Exercises 7 and 8, solve the equation.

7. $\dfrac{1}{x} + \dfrac{4}{10 - x} = 1$

8. $\dfrac{x - 3}{x} + 1 = \dfrac{x - 4}{x - 6}$

9. Find the domain of $f(x) = \dfrac{3(x - 1)}{8x - 3}$.

10. Determine whether each ordered pair is a solution of the system of linear equations.

$$\begin{cases} 2x - y = 2 \\ -x + 3y = 4 \end{cases}$$ (a) $(2, 2)$ (b) $(0, 4)$

In Exercises 11–16, use the indicated method to solve the system.

11. *Graphical:* $\begin{cases} x - y = 1 \\ 2x + y = 5 \end{cases}$

12. *Substitution:* $\begin{cases} 4x + 2y = 8 \\ x - 5y = 13 \end{cases}$

13. *Elimination:* $\begin{cases} 4x - 3y = 8 \\ -2x + y = -6 \end{cases}$

14. *Elimination:* $\begin{cases} x + y + z = -1 \\ x \quad\quad = 0 \\ 2x + y \quad = 1 \end{cases}$

15. *Matrices:* $\begin{cases} x + y + z = 1 \\ 5x + 4y + 3z = 0 \\ 6x + 3y + 2z = 1 \end{cases}$

16. *Cramer's Rule:* $\begin{cases} 2x - y = 4 \\ 3x + y = -5 \end{cases}$

17. Graph the solution of the system of inequalities.

$$\begin{cases} x - 2y < 0 \\ -2x + y > 2 \\ y > 0 \end{cases}$$

In Exercises 18–21, simplify the expression.

18. $\sqrt{24x^2y^3}$

19. $\sqrt[3]{80a^{15}b^8}$

20. $(12a^{-4}b^6)^{1/2}$

21. $\left(\dfrac{t^{1/2}}{t^{1/4}}\right)^2$

22. Add: $10\sqrt{20x} + 3\sqrt{125x}$

23. Multiply and simplify: $\left(\sqrt{2x} - 3\right)^2$

24. Simplify: $\dfrac{3}{\sqrt{10} - \sqrt{x}}$

In Exercises 25–28, solve the equation.

25. $\sqrt{x - 5} - 6 = 0$

26. $\sqrt{3 - x} + 10 = 11$

27. $\sqrt{x + 5} - \sqrt{x - 7} = 2$

28. $\sqrt{x - 4} = \sqrt{x + 7} - 1$

In Exercises 29–31, perform the operation and write the result in standard form.

29. $\sqrt{-2}\left(\sqrt{-8} + 3\right)$

30. $(-4 + 11i) - (3 - 5i)$

31. $(5 + 2i)^2$

32. Write $\dfrac{2 - 3i}{6 - 2i}$ in standard form.

33. After working together for t hours on a common task, two workers have completed fractional parts of the job equal to $t/2$ and $2t/7$. What fractional part of the task has been completed?

34. A force of 50 pounds stretches a spring 7 centimeters.

    (a) How far will a force of 20 pounds stretch the spring?

    (b) What force is required to stretch the spring 4.2 centimeters?

35. A total of $50,000 is invested in two funds paying 8% and 8.5% simple interest. The combined yearly interest is $4150. How much is invested at each rate?

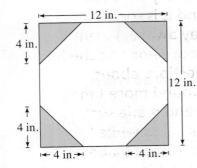

12 in.
4 in.
4 in.
12 in.
4 in.
4 in.

Figure for 36

36. The four corners are cut from a 12-inch-by-12-inch piece of glass, as shown in the figure. Find the perimeter of the remaining piece of glass.

37. A guy wire on a 180-foot cell phone tower is attached to the top of the tower and to an anchor 90 feet from the base of the tower. Find the length of the guy wire.

38. The time t (in seconds) for a free-falling object to fall d feet is given by

$$t = \sqrt{\dfrac{d}{16}}.$$

A construction worker drops a nail from a building and observes it strike a water puddle after approximately 5 seconds. Estimate the height from which the nail was dropped.

10

Quadratic Equations, Functions, and Inequalities

10.1 Solving Quadratic Equations

10.2 Completing the Square

10.3 The Quadratic Formula

10.4 Graphs of Quadratic Functions

10.5 Applications of Quadratic Equations

10.6 Quadratic and Rational Inequalities

MASTERY IS WITHIN REACH!

"I am on the swim team and spend a lot of time practicing and traveling. The only time I sit still is in class and it is sometimes hard to stay awake. I went to talk to a counselor and she had lots of suggestions about organization. I found more time to study my math and made sure that I was really learning while I did my homework. Things made more sense in class and it was easier to understand and remember everything."

Tamika

See page 527 for suggestions about improving your memory.

Darrin Henry/Shutterstock.com

10.1 Solving Quadratic Equations

▶ Solve quadratic equations by factoring.

▶ Solve quadratic equations by the Square Root Property.

▶ Use substitution to solve equations of quadratic form.

Solving Quadratic Equations by Factoring

From Section 6.5, remember that the first step in solving a quadratic equation by factoring is to write the equation in general form. Next, factor the left side. Finally, set each factor equal to zero and solve for x. Be sure to check each solution in the original equation.

EXAMPLE 1 **Solving Quadratic Equations by Factoring**

a.

$x^2 + 5x = 24$	Original equation
$x^2 + 5x - 24 = 0$	Write in general form.
$(x + 8)(x - 3) = 0$	Factor.
$x + 8 = 0 \implies x = -8$	Set 1st factor equal to 0.
$x - 3 = 0 \implies x = 3$	Set 2nd factor equal to 0.

b.

$3x^2 = 4 - 11x$	Original equation
$3x^2 + 11x - 4 = 0$	Write in general form.
$(3x - 1)(x + 4) = 0$	Factor.
$3x - 1 = 0 \implies x = \dfrac{1}{3}$	Set 1st factor equal to 0.
$x + 4 = 0 \implies x = -4$	Set 2nd factor equal to 0.

c.

$9x^2 + 12 = 3 + 12x + 5x^2$	Original equation
$4x^2 - 12x + 9 = 0$	Write in general form.
$(2x - 3)(2x - 3) = 0$	Factor.
$2x - 3 = 0 \implies x = \dfrac{3}{2}$	Set factor equal to 0.

Check each solution in its original equation.

> **Study Tip**
>
> In Example 1(c), the quadratic equation produces two identical solutions. This is called a **double** or **repeated solution**.

Exercises Within Reach ®

Solutions in English & Spanish and tutorial videos at CollegePrepAlgebra.com

Solving a Quadratic Equation by Factoring In Exercises 1–12, solve the equation by factoring.

1. $x^2 - 15x + 54 = 0$

2. $x^2 + 15x + 44 = 0$

3. $x^2 - x - 30 = 0$

4. $x^2 - 2x - 48 = 0$

5. $x^2 + 4x = 45$

6. $x^2 - 7x = 18$

7. $x^2 - 16x + 64 = 0$

8. $x^2 + 60x + 900 = 0$

9. $9x^2 - 10x - 16 = 0$

10. $8x^2 - 10x + 3 = 0$

11. $u(u - 9) - 12(u - 9) = 0$

12. $16x(x - 8) - 12(x - 8) = 0$

The Square Root Property

> ### Square Root Property
>
> The equation $u^2 = d$, where $d > 0$, has exactly two solutions:
>
> $$u = \sqrt{d} \quad \text{and} \quad u = -\sqrt{d}.$$
>
> These solutions can also be written as $u = \pm\sqrt{d}$. This solution process is also called **extracting square roots**.

EXAMPLE 2   **Using the Square Root Property**

a.

$3x^2 = 15$	Original equation
$x^2 = 5$	Divide each side by 3.
$x = \pm\sqrt{5}$	Square Root Property

The solutions are $x = \sqrt{5}$ and $x = -\sqrt{5}$. Check these in the original equation.

b.

$(x - 2)^2 = 10$	Original equation
$x - 2 = \pm\sqrt{10}$	Square Root Property
$x = 2 \pm \sqrt{10}$	Add 2 to each side.

The solutions are $x = 2 + \sqrt{10} \approx 5.16$ and $x = 2 - \sqrt{10} \approx -1.16$. Check these in the original equation.

c.

$(3x - 6)^2 - 8 = 0$	Original equation
$(3x - 6)^2 = 8$	Add 8 to each side.
$3x - 6 = \pm 2\sqrt{2}$	Square Root Property and rewrite $\sqrt{8}$ as $2\sqrt{2}$.
$3x = 6 \pm 2\sqrt{2}$	Add 6 to each side.
$x = 2 \pm \dfrac{2\sqrt{2}}{3}$	Divide each side by 3.

The solutions are $x = 2 + \dfrac{2\sqrt{2}}{3} \approx 2.94$ and $x = 2 - \dfrac{2\sqrt{2}}{3} \approx 1.06$. Check these in the original equation.

Exercises Within Reach ®

Solutions in English & Spanish and tutorial videos at CollegePrepAlgebra.com

Using the Square Root Property In Exercises 13−28, solve the equation by using the Square Root Property.

13. $6x^2 = 54$

14. $5t^2 = 5$

15. $25x^2 = 16$

16. $9z^2 = 121$

17. $\dfrac{w^2}{4} = 49$

18. $\dfrac{x^2}{6} = 24$

19. $4x^2 - 25 = 0$

20. $16y^2 - 121 = 0$

21. $(x + 4)^2 = 64$

22. $(m - 12)^2 = 400$

23. $(x - 3)^2 = 0.25$

24. $(x + 2)^2 = 0.81$

25. $(x - 2)^2 = 7$

26. $(y + 4)^2 = 27$

27. $(2x + 1)^2 - 50 = 0$

28. $(3x - 5)^2 - 48 = 0$

> ### Square Root Property (Complex Square Root)
>
> The equation $u^2 = d$, where $d < 0$, has exactly two solutions:
>
> $$u = \sqrt{|d|}\,i \quad \text{and} \quad u = -\sqrt{|d|}\,i.$$
>
> These solutions can also be written as $u = \pm\sqrt{|d|}\,i.$

EXAMPLE 3 **Square Root Property (Complex Square Root)**

a.

$x^2 + 8 = 0$	Original equation
$x^2 = -8$	Subtract 8 from each side.
$x = \pm\sqrt{8}\,i = \pm2\sqrt{2}\,i$	Square Root Property

The solutions are $x = 2\sqrt{2}\,i$ and $x = -2\sqrt{2}\,i$. Check these in the original equation.

b.

$(x - 4)^2 = -3$	Original equation
$x - 4 = \pm\sqrt{3}\,i$	Square Root Property
$x = 4 \pm \sqrt{3}\,i$	Add 4 to each side.

The solutions are $x = 4 + \sqrt{3}\,i$ and $x = 4 - \sqrt{3}\,i$. Check these in the original equation.

c.

$2(3x - 5)^2 + 32 = 0$	Original equation
$2(3x - 5)^2 = -32$	Subtract 32 from each side.
$(3x - 5)^2 = -16$	Divide each side by 2.
$3x - 5 = \pm4i$	Square Root Property
$3x = 5 \pm 4i$	Add 5 to each side.
$x = \dfrac{5}{3} \pm \dfrac{4}{3}i$	Divide each side by 3.

The solutions are $x = \dfrac{5}{3} + \dfrac{4}{3}i$ and $x = \dfrac{5}{3} - \dfrac{4}{3}i$. Check these in the original equation.

Exercises Within Reach ®

Solutions in English & Spanish and tutorial videos at CollegePrepAlgebra.com

Using the Square Root Property In Exercises 29−42, solve the equation by using the Square Root Property.

29. $z^2 = -36$

30. $x^2 = -16$

31. $x^2 + 4 = 0$

32. $p^2 + 9 = 0$

33. $9u^2 + 17 = 0$

34. $25x^2 + 4 = 0$

35. $(t - 3)^2 = -25$

36. $(x + 5)^2 = -81$

37. $(3z + 4)^2 + 144 = 0$

38. $(2y - 3)^2 + 25 = 0$

39. $(4m + 1)^2 = -80$

40. $(6y - 5)^2 = -8$

41. $36(t + 3)^2 = -100$

42. $4(x - 4)^2 = -169$

Application   EXAMPLE 4   **Diameter of a Softball**

The surface area of a sphere of radius r is given by

$$S = 4\pi r^2.$$

The surface area of a softball is

$$\frac{144}{\pi} \text{ square inches.}$$

Find the diameter d of the softball.

SOLUTION

$S = 4\pi r^2$	Write original equation.
$\dfrac{144}{\pi} = 4\pi r^2$	Substitute $\dfrac{144}{\pi}$ for S.
$\dfrac{144}{\pi(4\pi)} = \dfrac{4\pi r^2}{4\pi}$	Divide each side by 4π.
$\dfrac{36}{\pi^2} = r^2$	Simplify.
$\pm\sqrt{\dfrac{36}{\pi^2}} = r$	Square Root Property
$\pm\dfrac{6}{\pi} = r$	Simplify.

Choose the positive root to obtain $r = \dfrac{6}{\pi}$. Then,

$d = 2r$	Diameter is twice radius.
$\quad = 2\left(\dfrac{6}{\pi}\right)$	Substitute $\dfrac{6}{\pi}$ for r.
$\quad = \dfrac{12}{\pi}$	Simplify.
$\quad \approx 3.82 \text{ inches.}$	Use calculator to approximate.

So, the diameter of the softball is about 3.82 inches.

Exercises Within Reach ®

Solutions in English & Spanish and tutorial videos at CollegePrepAlgebra.com

43. **Geometry** The surface area S of a basketball is

    $\dfrac{900}{\pi}$ square inches.

    Find the radius r of the basketball.

44. **Unisphere** The Unisphere is the world's largest man-made globe. It was built as the symbol of the 1964−1965 New York World's Fair. A sphere with the same diameter as the Unisphere globe would have a surface area of 45,239 square feet. What is the diameter of the Unisphere? (*Source:* The World's Fair and Exposition Information and Reference Guide)

Equations of Quadratic Form

Both the factoring method and the Square Root Property can be applied to nonquadratic equations that are of *quadratic form*. An equation is said to be of **quadratic form** if it has the form

$$au^2 + bu + c = 0$$

where u is an algebraic expression. Here are some examples.

Equation	Written in Quadratic Form
$x^4 + 5x^2 + 4 = 0$	$(x^2)^2 + 5(x^2) + 4 = 0$
$x - 5\sqrt{x} + 6 = 0$	$(\sqrt{x})^2 - 5(\sqrt{x}) + 6 = 0$
$2x^{2/3} + 5x^{1/3} - 3 = 0$	$2(x^{1/3})^2 + 5(x^{1/3}) - 3 = 0$
$18 + 2x^2 + (x^2 + 9)^2 = 8$	$(x^2 + 9)^2 + 2(x^2 + 9) - 8 = 0$

To solve an equation of quadratic form, it helps to make a substitution and rewrite the equation in terms of u, as demonstrated in Examples 5 and 6.

EXAMPLE 5 Solving an Equation of Quadratic Form

Solve $x^4 - 13x^2 + 36 = 0$.

SOLUTION

Begin by writing the original equation in quadratic form, as follows.

$x^4 - 13x^2 + 36 = 0$	Write original equation.
$(x^2)^2 - 13(x^2) + 36 = 0$	Write in quadratic form.

Next, let $u = x^2$ and substitute u into the equation written in quadratic form. Then, factor and solve the equation.

$u^2 - 13u + 36 = 0$	Substitute u for x^2.
$(u - 4)(u - 9) = 0$	Factor.
$u - 4 = 0 \implies u = 4$	Set 1st factor equal to 0.
$u - 9 = 0 \implies u = 9$	Set 2nd factor equal to 0.

At this point you have found the "u-solutions." To find the "x-solutions," replace u with x^2 and solve for x.

$$u = 4 \implies x^2 = 4 \implies x = \pm 2$$
$$u = 9 \implies x^2 = 9 \implies x = \pm 3$$

The solutions are $x = 2$, $x = -2$, $x = 3$, and $x = -3$. Check these in the original equation.

Exercises Within Reach ®   Solutions in English & Spanish and tutorial videos at CollegePrepAlgebra.com

Solving an Equation of Quadratic Form In Exercises 45–50, solve the equation of quadratic form.

45. $x^4 - 5x^2 + 4 = 0$

46. $x^4 - 10x^2 + 25 = 0$

47. $x^4 - 5x^2 + 6 = 0$

48. $x^4 - 10x^2 + 21 = 0$

49. $(x^2 - 4)^2 + 2(x^2 - 4) - 3 = 0$

50. $(x^2 - 1)^2 + (x^2 - 1) - 6 = 0$

Study Tip

When solving equations involving square roots, be sure to check for extraneous solutions.

EXAMPLE 6   **Solving Equations of Quadratic Form**

a. $x - 5\sqrt{x} + 6 = 0$ Original equation

This equation is of quadratic form with $u = \sqrt{x}$.

$(\sqrt{x})^2 - 5(\sqrt{x}) + 6 = 0$ Write in quadratic form.

$u^2 - 5u + 6 = 0$ Substitute u for $\sqrt{x}$.

$(u - 2)(u - 3) = 0$ Factor.

$u - 2 = 0$ ⟹ $u = 2$ Set 1st factor equal to 0.

$u - 3 = 0$ ⟹ $u = 3$ Set 2nd factor equal to 0.

Now, using the u-solutions of 2 and 3, you obtain the x-solutions as follows.

$u = 2$ ⟹ $\sqrt{x} = 2$ ⟹ $x = 4$

$u = 3$ ⟹ $\sqrt{x} = 3$ ⟹ $x = 9$

b. $x^{2/3} - x^{1/3} - 6 = 0$ Original equation

This equation is of quadratic form with $u = x^{1/3}$.

$(x^{1/3})^2 - (x^{1/3}) - 6 = 0$ Write in quadratic form.

$u^2 - u - 6 = 0$ Substitute u for $x^{1/3}$.

$(u + 2)(u - 3) = 0$ Factor.

$u + 2 = 0$ ⟹ $u = -2$ Set 1st factor equal to 0.

$u - 3 = 0$ ⟹ $u = 3$ Set 2nd factor equal to 0.

Now, using the u-solutions of -2 and 3, you obtain the x-solutions as follows.

$u = -2$ ⟹ $x^{1/3} = -2$ ⟹ $x = -8$

$u = 3$ ⟹ $x^{1/3} = 3$ ⟹ $x = 27$

Exercises Within Reach ®

Solutions in English & Spanish and tutorial videos at CollegePrepAlgebra.com

Solving an Equation of Quadratic Form In Exercises 51−64, solve the equation of quadratic form.

51. $x - 3\sqrt{x} - 4 = 0$

52. $x - \sqrt{x} - 6 = 0$

53. $x - 7\sqrt{x} + 10 = 0$

54. $x - 11\sqrt{x} + 24 = 0$

55. $x^{2/3} - x^{1/3} - 6 = 0$

56. $x^{2/3} + 3x^{1/3} - 10 = 0$

57. $2x^{2/3} - 7x^{1/3} + 5 = 0$

58. $5x^{2/3} - 13x^{1/3} + 6 = 0$

59. $x^{2/5} - 3x^{1/5} + 2 = 0$

60. $x^{2/5} + 5x^{1/5} + 6 = 0$

61. $2x^{2/5} - 7x^{1/5} + 3 = 0$

62. $2x^{2/5} + 3x^{1/5} + 1 = 0$

63. $x^{1/3} - x^{1/6} - 6 = 0$

64. $x^{1/3} + 2x^{1/6} - 3 = 0$

Concept Summary: Solving Quadratic Equations by Using the Square Root Property

What

Two ways to solve quadratic equations are (1) by factoring and (2) by using the **Square Root Property**.

EXAMPLE

Solve $x^2 = -144$ by using the Square Root Property.

How

To use the Square Root Property, take the square root of each side of the equation.

EXAMPLE

$x^2 = -144$ Original equation

$x = \pm 12i$ Square Root Property

The solutions are $x = -12i$ and $x = 12i$.

Notice that the solutions of the quadratic equation are complex numbers.

Why

Many real-life situations can be modeled by quadratic equations. Knowing how to solve these types of equations will help you answer many real-life problems.

Exercises Within Reach ®

Worked-out solutions to odd-numbered exercises at CollegePrepAlgebra.com

Concept Summary Check

65. *Writing* Describe how to solve a quadratic equation by using the Square Root Property.

66. *True or False?* Determine whether the following statement is true or false. Justify your answer.

The only solution of the equation $x^2 = 25$ is $x = 5$.

67. *Reasoning* Does the equation $4x^2 + 9 = 0$ have two real solutions or two complex solutions? Explain your reasoning.

68. *Writing* Describe two ways to solve the equation $2x^2 - 18 = 0$.

Extra Practice

Solving a Quadratic Equation In Exercises 69−78, solve the equation.

69. $x^2 + 900 = 0$

70. $z^2 + 256 = 0$

71. $\frac{2}{3}x^2 = 6$

72. $\frac{1}{3}x^2 = 4$

73. $(p - 2)^2 - 108 = 0$

74. $(y + 12)^2 - 400 = 0$

75. $(p - 2)^2 + 108 = 0$

76. $(y + 12)^2 + 400 = 0$

77. $(x + 2)^2 + 18 = 0$

78. $(x + 2)^2 - 18 = 0$

Solving an Equation of Quadratic Form In Exercises 79−88, solve the equation of quadratic form.

79. $x^{1/2} - 3x^{1/4} + 2 = 0$

80. $x^{1/2} - 5x^{1/4} + 6 = 0$

81. $\frac{1}{x^2} - \frac{3}{x} + 2 = 0$

82. $\frac{1}{x^2} - \frac{1}{x} - 6 = 0$

83. $4x^{-2} - x^{-1} - 5 = 0$

84. $2x^{-2} - x^{-1} - 1 = 0$

85. $(x^2 - 3x)^2 - 2(x^2 - 3x) - 8 = 0$

86. $(x^2 - 6x)^2 - 2(x^2 - 6x) - 35 = 0$

87. $16\left(\frac{x - 1}{x - 8}\right)^2 + 8\left(\frac{x - 1}{x - 8}\right) + 1 = 0$

88. $9\left(\frac{x + 2}{x + 3}\right)^2 - 6\left(\frac{x + 2}{x + 3}\right) + 1 = 0$

Free-Falling Object **In Exercises 89−92, the height *h* (in feet) of a falling object at any time *t* (in seconds) is modeled by $h = -16t^2 + s_0$, where s_0 is the initial height (in feet). Use the model to find the time it takes for an object to fall to the ground given s_0.**

89. $s_0 = 256$ **90.** $s_0 = 48$ **91.** $s_0 = 128$ **92.** $s_0 = 500$

93. *Free-Falling Object* The height *h* (in feet) of an object thrown vertically upward from the top of a tower that is 144 feet tall is given by

$$h = 144 + 128t - 16t^2$$

where *t* measures the time (in seconds) from when the object is released. How long does it take for the object to reach the ground?

94. *Profit* The monthly profit *P* (in dollars) a company makes depends on the amount *x* (in dollars) the company spends on advertising according to the model

$$P = 800 + 120x - \frac{1}{2}x^2.$$

Find the amount the company must spend on advertising to make a monthly profit of $8000.

Compound Interest **The amount *A* in an account after 2 years, when a principal of *P* dollars is invested at annual interest rate *r* (in decimal form) compounded annually, is given by $A = P(1 + r)^2$. In Exercises 95 and 96, find *r*.**

95. $P = \$1500, A = \$1685.40$ **96.** $P = \$5000, A = \$5724.50$

Explaining Concepts

97. *Reasoning* For a quadratic equation $ax^2 + bx + c = 0$, where *a*, *b*, and *c* are real numbers with $a \neq 0$, explain why *b* and *c* can equal 0, but *a* cannot.

98. *Vocabulary* Is the equation $x^6 - 6x^3 + 9 = 0$ of quadratic form? Explain your reasoning.

99. *Logic* Is it possible for a quadratic equation of the form $x^2 = m$ to have one real solution and one complex solution? Explain your reasoning.

100. *Precision* Describe a procedure for solving an equation of quadratic form. Give an example.

Cumulative Review

In Exercises 101−104, solve the inequality and sketch the solution on the real number line.

101. $3x - 8 > 4$

102. $4 - 5x \geq 12$

103. $2x - 6 \leq 9 - x$

104. $x - 4 < 6$ or $x + 3 > 8$

In Exercises 105 and 106, solve the system of linear equations.

105. $x + y - z = 4$
$2x + y + 2z = 10$
$x - 3y - 4z = -7$

106. $2x - y + z = -6$
$x + 5y - z = 7$
$-x - 2y - 3z = 8$

In Exercises 107−112, combine the radical expressions, if possible, and simplify.

107. $5\sqrt{3} - 2\sqrt{3}$

108. $8\sqrt{27} + 4\sqrt{27}$

109. $16\sqrt[3]{y} - 9\sqrt[3]{x}$

110. $12\sqrt{x - 1} + 6\sqrt{x - 1}$

111. $\sqrt{16m^4n^3} + m\sqrt{m^2n}$

112. $x^2y\sqrt[4]{32x^2} + x\sqrt[4]{2x^6y^4} - y\sqrt[4]{162x^{10}}$

10.2 Completing the Square

▶ Rewrite quadratic expressions in completed square form.
▶ Solve quadratic equations by completing the square.

Constructing Perfect Square Trinomials

In this section, you will study a technique for rewriting an equation in a completed square form. This technique is called *completing the square*. Note that prior to completing the square, the coefficient of the second-degree term must be 1.

> ### Completing the Square
>
> To **complete the square** for the expression $x^2 + bx$, add $(b/2)^2$, which is the square of half the coefficient of x. Consequently,
>
> $$x^2 + bx + \left(\frac{b}{2}\right)^2 = \left(x + \frac{b}{2}\right)^2.$$
>
> $\underbrace{\qquad}_{\text{(half)}^2}$

EXAMPLE 1 **Constructing a Perfect Square Trinomial**

What term should be added to $x^2 - 8x$ to make it a perfect square trinomial?

SOLUTION

For this expression, the coefficient of the x-term is -8. Add the square of half of this coefficient to the expression to make it a perfect square trinomial.

$$x^2 - 8x + \left(-\frac{8}{2}\right)^2 = x^2 - 8x + 16 \qquad \text{Add } \left(-\frac{8}{2}\right)^2 = 16 \text{ to the expression.}$$

You can then rewrite the expression as the square of a binomial, $(x - 4)^2$.

Exercises Within Reach ®   Solutions in English & Spanish and tutorial videos at CollegePrepAlgebra.com

Constructing a Perfect Square Trinomial In Exercises 1–16, **add a term to the expression to make it a perfect square trinomial.**

1. $x^2 + 8x +$ ▭

2. $x^2 + 12x +$ ▭

3. $y^2 - 20y +$ ▭

4. $y^2 - 2y +$ ▭

5. $x^2 + 14x +$ ▭

6. $x^2 - 24x +$ ▭

7. $t^2 + 5t +$ ▭

8. $u^2 + 7u +$ ▭

9. $x^2 - 9x +$ ▭

10. $y^2 - 11y +$ ▭

11. $a^2 - \frac{1}{3}a +$ ▭

12. $y^2 + \frac{4}{3}y +$ ▭

13. $y^2 + \frac{8}{5}y +$ ▭

14. $x^2 - \frac{9}{5}x +$ ▭

15. $r^2 - 0.4r +$ ▭

16. $s^2 + 4.6s +$ ▭

Solving Equations by Completing the Square

Completing the square can be used to solve quadratic equations. When using this procedure, remember to *preserve the equality* by adding the same constant to each side of the equation.

Study Tip

In Example 2, completing the square is used for the sake of illustration. This particular equation would be easier to solve by factoring. Try reworking the problem by factoring to see that you obtain the same two solutions.

EXAMPLE 2   **Completing the Square: Leading Coefficient Is 1**

Solve $x^2 + 12x = 0$ by completing the square.

SOLUTION

$$x^2 + 12x = 0$$ Write original equation.

$$x^2 + 12x + 6^2 = 36$$ Add $6^2 = 36$ to each side.

$$\left(\tfrac{12}{2}\right)^2$$

$$(x + 6)^2 = 36$$ Completed square form

$$x + 6 = \pm\sqrt{36}$$ Square Root Property

$$x = -6 \pm 6$$ Subtract 6 from each side.

$$x = -6 + 6 \text{ or } x = -6 - 6$$ Separate solutions.

$$x = 0 \qquad\qquad x = -12$$ Simplify.

The solutions are $x = 0$ and $x = -12$. Check these in the original equation.

Exercises Within Reach ®

Solutions in English & Spanish and tutorial videos at CollegePrepAlgebra.com

Solving an Equation In Exercises 17–32, solve the equation first by completing the square and then by factoring.

17. $x^2 - 20x = 0$

18. $x^2 + 32x = 0$

19. $x^2 + 6x = 0$

20. $t^2 - 10t = 0$

21. $y^2 - 5y = 0$

22. $t^2 - 9t = 0$

23. $t^2 - 8t + 7 = 0$

24. $y^2 - 4y + 4 = 0$

25. $x^2 + 7x + 12 = 0$

26. $z^2 + 3z - 10 = 0$

27. $x^2 - 3x - 18 = 0$

28. $a^2 + 12a + 32 = 0$

29. $x^2 + 8x + 7 = 0$

30. $x^2 - 10x + 9 = 0$

31. $x^2 - 10x + 21 = 0$

32. $x^2 - 10x + 24 = 0$

EXAMPLE 3 **Completing the Square: Leading Coefficient Is 1**

$$x^2 - 6x + 7 = 0 \qquad \text{Original equation}$$

$$x^2 - 6x = -7 \qquad \text{Subtract 7 from each side.}$$

$$x^2 - 6x + (-3)^2 = -7 + 9 \qquad \text{Add } (-3)^2 = 9 \text{ to each side.}$$

$$\left(-\frac{6}{2}\right)^2$$

$$(x - 3)^2 = 2 \qquad \text{Completed square form}$$

$$x - 3 = \pm\sqrt{2} \qquad \text{Square Root Property}$$

$$x = 3 \pm \sqrt{2} \qquad \text{Add 3 to each side.}$$

$$x = 3 + \sqrt{2} \text{ or } x = 3 - \sqrt{2} \qquad \text{Separate solutions.}$$

The solutions are $x = 3 + \sqrt{2} \approx 4.41$ and $x = 3 - \sqrt{2} \approx 1.59$. Check these in the original equation.

EXAMPLE 4 **Completing the Square: Leading Coefficient Is Not 1**

$$2x^2 - x - 2 = 0 \qquad \text{Original equation}$$

$$2x^2 - x = 2 \qquad \text{Add 2 to each side.}$$

$$x^2 - \frac{1}{2}x = 1 \qquad \text{Divide each side by 2.}$$

$$x^2 - \frac{1}{2}x + \left(-\frac{1}{4}\right)^2 = 1 + \frac{1}{16} \qquad \text{Add } \left(-\frac{1}{4}\right)^2 = \frac{1}{16} \text{ to each side.}$$

$$\left(x - \frac{1}{4}\right)^2 = \frac{17}{16} \qquad \text{Completed square form}$$

$$x - \frac{1}{4} = \pm\frac{\sqrt{17}}{4} \qquad \text{Square Root Property}$$

$$x = \frac{1}{4} \pm \frac{\sqrt{17}}{4} \qquad \text{Add } \frac{1}{4} \text{ to each side.}$$

The solutions are $x = \frac{1}{4} + \frac{\sqrt{17}}{4} \approx 1.28$ and $x = \frac{1}{4} - \frac{\sqrt{17}}{4} \approx -0.78$. Check these in the original equation.

Exercises Within Reach ®

Solutions in English & Spanish and tutorial videos at CollegePrepAlgebra.com

Completing the Square In Exercises 33−38, **solve the equation by completing the square. Give the solutions in exact form and in decimal form rounded to two decimal places.**

33. $x^2 - 4x - 3 = 0$

34. $x^2 - 6x + 7 = 0$

35. $x^2 + 4x - 3 = 0$

36. $x^2 + 6x + 7 = 0$

37. $3x^2 + 9x + 5 = 0$

38. $5x^2 - 15x + 7 = 0$

EXAMPLE 5   **Completing the Square: Leading Coefficient Is Not 1**

Solve $3x^2 - 6x + 1 = 0$ by completing the square.

SOLUTION

$$3x^2 - 6x + 1 = 0 \qquad \text{Write original equation.}$$

$$3x^2 - 6x = -1 \qquad \text{Subtract 1 from each side.}$$

$$x^2 - 2x = -\frac{1}{3} \qquad \text{Divide each side by 3.}$$

$$x^2 - 2x + (-1)^2 = -\frac{1}{3} + 1 \qquad \text{Add } (-1)^2 = 1 \text{ to each side.}$$

$$(x - 1)^2 = \frac{2}{3} \qquad \text{Completed square form}$$

$$x - 1 = \pm\sqrt{\frac{2}{3}} \qquad \text{Square Root Property}$$

$$x - 1 = \pm\frac{\sqrt{6}}{3} \qquad \text{Rationalize the denominator.}$$

$$x = 1 \pm \frac{\sqrt{6}}{3} \qquad \text{Add 1 to each side.}$$

The solutions are $x = 1 + \dfrac{\sqrt{6}}{3} \approx 1.82$ and $x = 1 - \dfrac{\sqrt{6}}{3} \approx 0.18$. Check these in the original equation.

Exercises Within Reach ®

Solutions in English & Spanish and tutorial videos at CollegePrepAlgebra.com

Completing the Square In Exercises 39–46, solve the equation by completing the square. Give the solutions in exact form and in decimal form rounded to two decimal places.

39. $2x^2 + 8x + 3 = 0$

40. $3x^2 - 24x - 5 = 0$

41. $4y^2 + 4y - 9 = 0$

42. $7u^2 - 8u - 3 = 0$

43. $x\left(x - \dfrac{2}{3}\right) = 14$

44. $2x\left(x + \dfrac{4}{3}\right) = 5$

45. $0.1x^2 + 0.5x = -0.2$

46. $0.2x^2 + 0.1x = 0.5$

EXAMPLE 6 **A Quadratic Equation with Complex Solutions**

Solve $x^2 - 4x + 8 = 0$ by completing the square.

SOLUTION

$x^2 - 4x + 8 = 0$	Write original equation.
$x^2 - 4x = -8$	Subtract 8 from each side.
$x^2 - 4x + (-2)^2 = -8 + 4$	Add $(-2)^2 = 4$ to each side.
$(x - 2)^2 = -4$	Completed square form
$x - 2 = \pm 2i$	Square Root Property
$x = 2 \pm 2i$	Add 2 to each side.

The solutions are $x = 2 + 2i$ and $x = 2 - 2i$. Check these in the original equation. Note that when a quadratic equation has no real solutions, the graph of the corresponding equation has no x-intercepts.

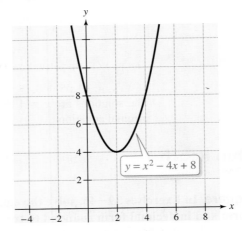

$y = x^2 - 4x + 8$

Exercises Within Reach®

Solutions in English & Spanish and tutorial videos at CollegePrepAlgebra.com

Completing the Square In Exercises 47–54, solve the equation by completing the square. Give the solutions in exact form and in decimal form rounded to two decimal places.

47. $z^2 + 4z + 13 = 0$

48. $z^2 - 6z + 18 = 0$

49. $x^2 + 9 = 4x$

50. $x^2 + 10 = 6x$

51. $-x^2 + x - 1 = 0$

52. $-x^2 - x - 1 = 0$

53. $4z^2 - 3z + 2 = 0$

54. $5x^2 - 3x + 10 = 0$

Application   **EXAMPLE 7**   **Geometry: Dimensions of an iPhone**

The first generation of the iPhone has an approximate volume of 4.968 cubic inches. Its width is 0.46 inch and its face has the dimensions x inches by $(x + 2.1)$ inches. Find the dimensions of the face in inches. (*Source:* Apple, Inc.)

SOLUTION

$lwh = V$	Formula for volume of a rectangular solid
$(x)(0.46)(x + 2.1) = 4.968$	Substitute 4.968 for V, x for l, 0.46 for w, and $x + 2.1$ for h.
$0.46x^2 + 0.966x = 4.968$	Multiply factors.
$x^2 + 2.1x = 10.8$	Divide each side by 0.46.
$x^2 + 2.1x + \left(\dfrac{2.1}{2}\right)^2 = 10.8 + 1.1025$	Add $\left(\dfrac{2.1}{2}\right)^2 = 1.1025$ to each side.
$(x + 1.05)^2 = 11.9025$	Completed square form
$x + 1.05 = \pm\sqrt{11.9025}$	Square Root Property
$x = -1.05 \pm \sqrt{11.9025}$	Subtract 1.05 from each side.

Choosing the positive root, you obtain

$x = -1.05 + 3.45 = 2.4$ inches	Length of face

and

$x + 2.1 = 2.4 + 2.1 = 4.5$ inches.	Height of face

Exercises Within Reach ®   Solutions in English & Spanish and tutorial videos at CollegePrepAlgebra.com

55. *Geometric Modeling* Use the figure shown below.

   (a) Find the area of the two adjoining rectangles and large square in the figure.

   (b) Find the area of the small square in the lower right-hand corner of the figure and add it to the area found in part (a).

   (c) Find the dimensions and the area of the entire figure after adjoining the small square in the lower right-hand corner of the figure. Note that you have shown geometrically the technique of completing the square.

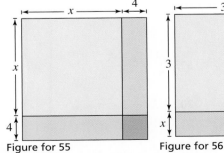

Figure for 55

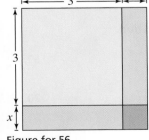

Figure for 56

56. *Geometric Modeling* Repeat Exercise 55 for the figure shown above.

57. *Geometry* An open box with a rectangular base of x inches by $(x + 4)$ inches has a height of 6 inches (see figure). The volume of the box is 840 cubic inches. Find the dimensions of the box.

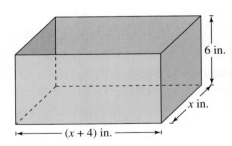

58. *Geometry* An open box with a rectangular base of $2x$ inches by $(6x - 2)$ inches has a height of 9 inches (see figure). The volume of the box is 1584 cubic inches. Find the dimensions of the box.

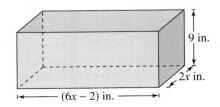

Concept Summary: Solving Quadratic Equations by Completing the Square

What

Another way to solve a quadratic equation is by **completing the square**.

EXAMPLE

Solve $2x^2 + 12x + 10 = 0$ by completing the square.

How

1. Write the equation in the form $x^2 + bx = c$.
2. Add $(b/2)^2$ to each side.
3. Write the equation in completed square form and use the Square Root Property to solve.

EXAMPLE

$$2x^2 + 12x + 10 = 0$$

$$x^2 + 6x = -5 \qquad \text{Step 1}$$

$$x^2 + 6x + 3^2 = -5 + 9 \qquad \text{Step 2}$$

$$\left.\begin{array}{r} (x + 3)^2 = 4 \\ x + 3 = \pm 2 \\ x = -3 \pm 2 \\ x = -1 \quad \text{and} \quad x = -5 \end{array}\right\} \text{Step 3}$$

Why

You can solve *any* quadratic equation by completing the square.

You can also use this method to identify quadratic equations that have no real solutions.

Exercises Within Reach ®

Worked-out solutions to odd-numbered exercises at CollegePrepAlgebra.com

Concept Summary Check

59. *Writing* What is a perfect square trinomial?

60. *Completing the Square* What term must be added to $x^2 + 5x$ to complete the square? Explain how you found the term.

61. *Reasoning* When using the method of completing the square to solve $2x^2 - 7x = 6$, what is the first step? Is the resulting equation equivalent to the original equation? Explain.

62. *Think About It* Is it possible for a quadratic equation to have no real number solution? If so, give an example.

Extra Practice

Completing the Square In Exercises 63 and 64, solve the equation by completing the square. Give the solutions in exact form and in decimal form rounded to two decimal places.

63. $0.2x^2 + 0.1x = -0.5$

64. $0.75x^2 + 1.25x + 1.5 = 0$

Solving an Equation In Exercises 65−70, solve the equation.

65. $\dfrac{x}{2} - \dfrac{1}{x} = 1$

66. $\dfrac{x}{2} + \dfrac{5}{x} = 4$

67. $\dfrac{x^2}{8} = \dfrac{x + 3}{2}$

68. $\dfrac{x^2 + 2}{24} = \dfrac{x - 1}{3}$

69. $\sqrt{2x + 1} = x - 3$

70. $\sqrt{3x - 2} = x - 2$

71. *Geometry* You have 200 meters of fencing to enclose two adjacent rectangular corrals (see figure). The total area of the enclosed region is 1400 square meters. What are the dimensions of each corral? (The corrals are the same size.)

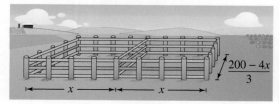

72. *Geometry* A kennel is adding a rectangular outdoor enclosure along one side of the kennel wall (see figure). The other three sides of the enclosure will be formed by a fence. The kennel has 111 feet of fencing and plans to use 1215 square feet of land for the enclosure. What are the dimensions of the enclosure?

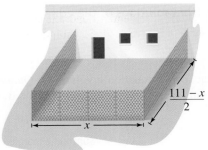

73. *Revenue* The revenue R (in dollars) from selling x pairs of running shoes is given by

$$R = x\left(80 - \frac{1}{2}x\right).$$

Find the number of pairs of running shoes that must be sold to produce a revenue of $2750.

74. *Revenue* The revenue R (in dollars) from selling x golf clubs is given by

$$R = x\left(150 - \frac{1}{10}x\right).$$

Find the number of golf clubs that must be sold to produce a revenue of $15,033.60.

Explaining Concepts

75. *Writing* Explain the use of the Square Root Property when solving a quadratic equation by the method of completing the square.

76. *True or False?* If you solve a quadratic equation by completing the square and obtain solutions that are rational numbers, then you could have solved the equation by factoring. Justify your answer.

77. *Think About It* Consider the quadratic equation $(x - 1)^2 = d$.

(a) What value(s) of d will produce a quadratic equation that has exactly one (repeated) solution?

(b) Describe the value(s) of d that will produce two different solutions, both of which are *rational* numbers.

(c) Describe the value(s) of d that will produce two different solutions, both of which are *irrational* numbers.

(d) Describe the value(s) of d that will produce two different solutions, both of which are *complex* numbers.

78. *Error Analysis* You teach an algebra class and one of your students hands in the following solution. Describe and correct the error(s).

Solve $x^2 + 6x - 13 = 0$ by completing the square.

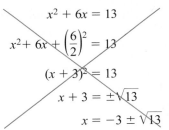

$$x^2 + 6x = 13$$
$$x^2 + 6x + \left(\frac{6}{2}\right)^2 = 13$$
$$(x + 3)^2 = 13$$
$$x + 3 = \pm\sqrt{13}$$
$$x = -3 \pm \sqrt{13}$$

Cumulative Review

In Exercises 79−86, perform the operation and simplify the expression.

79. $3\sqrt{5}\sqrt{500}$

80. $2\sqrt{2x^2}\sqrt{27x}$

81. $(3 + \sqrt{2})(3 - \sqrt{2})$

82. $(\sqrt[3]{6} - 2)(\sqrt[3]{4} + 1)$

83. $(3 + \sqrt{2})^2$

84. $(2\sqrt{x} - 5)^2$

85. $\dfrac{8}{\sqrt{10}}$

86. $\dfrac{5}{\sqrt{12} - 2}$

In Exercises 87 and 88, rewrite the expression using the specified rule, where a and b are nonnegative real numbers.

87. Product Rule: $\sqrt{ab} = $ [   ].

88. Quotient Rule: $\sqrt{\dfrac{a}{b}} = $ [   ].

10.3 The Quadratic Formula

▶ Use the Quadratic Formula to solve quadratic equations.

▶ Determine the types of solutions of quadratic equations using the discriminant.

▶ Write quadratic equations from solutions of the equations.

The Quadratic Formula

A fourth technique for solving a quadratic equation involves the *Quadratic Formula*. This formula is derived by completing the square for a general quadratic equation.

$$ax^2 + bx + c = 0 \qquad \text{General form, } a \neq 0$$

$$ax^2 + bx = -c \qquad \text{Subtract } c \text{ from each side.}$$

$$x^2 + \frac{b}{a}x = -\frac{c}{a} \qquad \text{Divide each side by } a.$$

$$x^2 + \frac{b}{a}x + \left(\frac{b}{2a}\right)^2 = -\frac{c}{a} + \left(\frac{b}{2a}\right)^2 \qquad \text{Add } \left(\frac{b}{2a}\right)^2 \text{ to each side.}$$

$$\left(x + \frac{b}{2a}\right)^2 = \frac{b^2 - 4ac}{4a^2} \qquad \text{Simplify.}$$

$$x + \frac{b}{2a} = \pm\sqrt{\frac{b^2 - 4ac}{4a^2}} \qquad \text{Square Root Property}$$

$$x = -\frac{b}{2a} \pm \frac{\sqrt{b^2 - 4ac}}{2|a|} \qquad \text{Subtract } \frac{b}{2a} \text{ from each side.}$$

$$x = \frac{-b \pm \sqrt{b^2 - 4ac}}{2a} \qquad \text{Simplify.}$$

Notice in the derivation of the Quadratic Formula that, because $\pm 2|a|$ represents the same numbers as $\pm 2a$, you can omit the absolute value bars.

Study Tip

The Quadratic Formula is one of the most important formulas in algebra, and you should memorize it. It helps to try to memorize a verbal statement of the rule. For instance, you might try to remember the following verbal statement of the Quadratic Formula: "The opposite of *b*, plus or minus the square root of *b* squared minus 4*ac*, all divided by 2*a*."

> ### The Quadratic Formula
>
> The solutions of $ax^2 + bx + c = 0$, $a \neq 0$, are given by the **Quadratic Formula**
>
> $$x = \frac{-b \pm \sqrt{b^2 - 4ac}}{2a}.$$

Exercises Within Reach ®

Solutions in English & Spanish and tutorial videos at CollegePrepAlgebra.com

Writing a Quadratic Equation in General Form In Exercises 1−4, write the quadratic equation in general form.

1. $2x^2 = 7 - 2x$

2. $7x^2 + 15x = 5$

3. $x(10 - x) = 5$

4. $x(2x + 9) = 12$

EXAMPLE 1   **The Quadratic Formula: Two Distinct Solutions**

Study Tip

In Example 1, the solutions are rational numbers, which means that the equation could have been solved by factoring. Try solving the equation by factoring.

$x^2 + 6x = 16$	Original equation
$x^2 + 6x - 16 = 0$	Write in general form.
$x = \dfrac{-b \pm \sqrt{b^2 - 4ac}}{2a}$	Quadratic Formula
$x = \dfrac{-6 \pm \sqrt{6^2 - 4(1)(-16)}}{2(1)}$	Substitute 1 for a, 6 for b, and -16 for c.
$x = \dfrac{-6 \pm \sqrt{100}}{2}$	Simplify.
$x = \dfrac{-6 \pm 10}{2}$	Simplify.
$x = 2$ or $x = -8$	Solutions

The solutions are $x = 2$ and $x = -8$. Check these in the original equation.

EXAMPLE 2   **The Quadratic Formula: Two Distinct Solutions**

Study Tip

If the leading coefficient of a quadratic equation is negative, you should begin by multiplying each side of the equation by -1, as shown in Example 2. This will produce a positive leading coefficient, which is easier to work with.

$-x^2 - 4x + 8 = 0$	Leading coefficient is negative.
$x^2 + 4x - 8 = 0$	Multiply each side by -1.
$x = \dfrac{-b \pm \sqrt{b^2 - 4ac}}{2a}$	Quadratic Formula
$x = \dfrac{-4 \pm \sqrt{4^2 - 4(1)(-8)}}{2(1)}$	Substitute 1 for a, 4 for b, and -8 for c.
$x = \dfrac{-4 \pm \sqrt{48}}{2}$	Simplify.
$x = \dfrac{-4 \pm 4\sqrt{3}}{2}$	Simplify.
$x = -2 \pm 2\sqrt{3}$	Solutions

The solutions are $x = -2 + 2\sqrt{3}$ and $x = -2 - 2\sqrt{3}$. Check these in the original equation.

Exercises Within Reach ®

Solutions in English & Spanish and tutorial videos at CollegePrepAlgebra.com

Solving a Quadratic Equation In Exercises 5−8, solve the equation first by using the Quadratic Formula and then by factoring.

5. $x^2 - 11x + 28 = 0$

6. $x^2 - 12x + 27 = 0$

7. $x^2 + 6x + 8 = 0$

8. $x^2 + 9x + 14 = 0$

Solving a Quadratic Equation In Exercises 9−14, solve the equation by using the Quadratic Formula.

9. $x^2 - 2x - 4 = 0$

10. $x^2 - 2x - 6 = 0$

11. $t^2 + 4t + 1 = 0$

12. $y^2 + 6y - 8 = 0$

13. $-x^2 + 10x - 23 = 0$

14. $-u^2 + 12u - 29 = 0$

Study Tip

Example 3 could have been solved as follows, without dividing each side by 2 in the first step.

$x =$

$$\frac{-(-24) \pm \sqrt{(-24)^2 - 4(18)(8)}}{2(18)}$$

$$x = \frac{24 \pm \sqrt{576 - 576}}{36}$$

$$x = \frac{24 \pm 0}{36}$$

$$x = \frac{2}{3}$$

While the result is the same, dividing each side by 2 simplifies the equation before the Quadratic Formula is applied. This allows you to work with smaller numbers.

EXAMPLE 3 **The Quadratic Formula: One Repeated Solution**

$18x^2 - 24x + 8 = 0$	Original equation
$9x^2 - 12x + 4 = 0$	Divide each side by 2.
$x = \dfrac{-b \pm \sqrt{b^2 - 4ac}}{2a}$	Quadratic Formula
$x = \dfrac{-(-12) \pm \sqrt{(-12)^2 - 4(9)(4)}}{2(9)}$	Substitute 9 for a, -12 for b, and 4 for c.
$x = \dfrac{12 \pm \sqrt{144 - 144}}{18}$	Simplify.
$x = \dfrac{12 \pm \sqrt{0}}{18}$	Simplify.
$x = \dfrac{2}{3}$	Solution

The only solution is $x = \frac{2}{3}$. Check this in the original equation.

EXAMPLE 4 **The Quadratic Formula: Complex Solutions**

$2x^2 - 4x + 5 = 0$	Original equation
$x = \dfrac{-b \pm \sqrt{b^2 - 4ac}}{2a}$	Quadratic Formula
$x = \dfrac{-(-4) \pm \sqrt{(-4)^2 - 4(2)(5)}}{2(2)}$	Substitute 2 for a, -4 for b, and 5 for c.
$x = \dfrac{4 \pm \sqrt{-24}}{4}$	Simplify.
$x = \dfrac{4 \pm 2\sqrt{6}i}{4}$	Write in i-form.
$x = \dfrac{2(2 \pm \sqrt{6}i)}{2 \cdot 2}$	Factor numerator and denominator.
$x = \dfrac{\cancel{2}(2 \pm \sqrt{6}i)}{\cancel{2} \cdot 2}$	Divide out common factor.
$x = 1 \pm \dfrac{\sqrt{6}}{2}i$	Solutions

The solutions are $x = 1 + \dfrac{\sqrt{6}}{2}i$ and $x = 1 - \dfrac{\sqrt{6}}{2}i$. Check these in the original equation.

Exercises Within Reach ®

Solutions in English & Spanish and tutorial videos at CollegePrepAlgebra.com

Solving a Quadratic Equation In Exercises 15 and 16, solve the equation first by using the Quadratic Formula and then by factoring.

15. $16x^2 + 8x + 1 = 0$

16. $9x^2 + 12x + 4 = 0$

Solving a Quadratic Equation In Exercises 17 and 18, solve the equation by using the Quadratic Formula.

17. $2x^2 + 3x + 3 = 0$

18. $2x^2 - 2x + 3 = 0$

The Discriminant

> ### Using the Discriminant
>
> Let a, b, and c be rational numbers such that $a \neq 0$. The **discriminant** of the quadratic equation $ax^2 + bx + c = 0$ is given by $b^2 - 4ac$, and can be used to classify the solutions of the equation as follows.
>
> | *Discriminant* | *Solution Type* |
> |---|---|
> | **1.** Perfect square | Two distinct rational solutions (Example 1) |
> | **2.** Positive nonperfect square | Two distinct irrational solutions (Example 2) |
> | **3.** Zero | One repeated rational solution (Example 3) |
> | **4.** Negative number | Two distinct complex solutions (Example 4) |

Study Tip

By reexamining Examples 1 through 4, you can see that the equations with rational or repeated solutions could have been solved by *factoring*. In general, quadratic equations (with integer coefficients) for which the discriminant is either zero or a perfect square are factorable using integer coefficients. Consequently, a quick test of the discriminant will help you decide which solution method to use to solve a quadratic equation.

EXAMPLE 5   **Using the Discriminant**

Determine the type of solution(s) for each quadratic equation.

a. $x^2 - x + 2 = 0$ **b.** $2x^2 - 3x - 2 = 0$

c. $x^2 - 2x + 1 = 0$ **d.** $x^2 - 2x - 1 = 9$

SOLUTION

Equation	*Discriminant*	*Solution Type*
a. $x^2 - x + 2 = 0$	$b^2 - 4ac = (-1)^2 - 4(1)(2)$ $= 1 - 8 = -7$	Two distinct complex solutions
b. $2x^2 - 3x - 2 = 0$	$b^2 - 4ac = (-3)^2 - 4(2)(-2)$ $= 9 + 16 = 25$	Two distinct rational solutions
c. $x^2 - 2x + 1 = 0$	$b^2 - 4ac = (-2)^2 - 4(1)(1)$ $= 4 - 4 = 0$	One repeated rational solution
d. $x^2 - 2x - 1 = 9$	$b^2 - 4ac = (-2)^2 - 4(1)(-10)$ $= 4 + 40 = 44$	Two distinct irrational solutions

Exercises Within Reach ®   Solutions in English & Spanish and tutorial videos at CollegePrepAlgebra.com

Using the Discriminant **In Exercises 19–26, use the discriminant to determine the type of solution(s) of the quadratic equation.**

19. $x^2 + x + 1 = 0$ **20.** $x^2 + x - 1 = 0$

21. $3x^2 - 2x - 5 = 0$ **22.** $5x^2 + 7x + 3 = 0$

23. $9x^2 - 24x + 16 = 0$ **24.** $2x^2 + 10x + 6 = 0$

25. $3x^2 - x = -2$ **26.** $4x^2 - 16x = -16$

Summary of Methods for Solving Quadratic Equations

Method   *Example*

1. Factoring
$$3x^2 + x = 0$$
$$x(3x + 1) = 0 \implies x = 0 \quad \text{and} \quad x = -\frac{1}{3}$$

2. Square Root Property
$$(x + 2)^2 = 7$$
$$x + 2 = \pm\sqrt{7} \implies x = -2 + \sqrt{7} \quad \text{and} \quad x = -2 - \sqrt{7}$$

3. Completing the square
$$x^2 + 6x = 2$$
$$x^2 + 6x + 3^2 = 2 + 9$$
$$(x + 3)^2 = 11 \implies x = -3 + \sqrt{11} \quad \text{and} \quad x = -3 - \sqrt{11}$$

4. Quadratic Formula
$$3x^2 - 2x + 2 = 0 \implies x = \frac{-(-2) \pm \sqrt{(-2)^2 - 4(3)(2)}}{2(3)} = \frac{1}{3} \pm \frac{\sqrt{5}}{3}i$$

Exercises Within Reach ®

Solutions in English & Spanish and tutorial videos at CollegePrepAlgebra.com

Choosing a Method In Exercises 27–42, solve the quadratic equation by using the most convenient method.

27. $z^2 - 169 = 0$

28. $t^2 = 144$

29. $5y^2 + 15y = 0$

30. $12u^2 + 30u = 0$

31. $25(x - 3)^2 - 36 = 0$

32. $9(x + 4)^2 + 16 = 0$

33. $2y(y - 18) + 3(y - 18) = 0$

34. $4y(y + 7) - 5(y + 7) = 0$

35. $x^2 + 8x + 25 = 0$

36. $y^2 + 21y + 108 = 0$

37. $3x^2 - 13x + 169 = 0$

38. $2x^2 - 15x + 225 = 0$

39. $25x^2 + 80x + 61 = 0$

40. $14x^2 + 11x - 40 = 0$

41. $7x(x + 2) + 5 = 3x(x + 1)$

42. $5x(x - 1) - 7 = 4x(x - 2)$

Using Technology and Algebra In Exercises 43–46, use a graphing calculator to graph the function. Use the graph to approximate any *x*-intercepts of the graph. Set $y = 0$ and solve the resulting equation. Compare the result with the *x*-intercepts of the graph.

43. $y = x^2 - 4x + 3$

44. $y = 5x^2 - 18x + 6$

45. $y = -0.03x^2 + 2x - 0.4$

46. $y = 3.7x^2 - 10.2x + 3.2$

Writing Quadratic Equations from Solutions

Using the Zero-Factor Property, you know that the equation $(x + 5)(x - 2) = 0$ has two solutions, $x = -5$ and $x = 2$. You can use the Zero-Factor Property in reverse to find a quadratic equation given its solutions. This process is demonstrated in Example 6.

> ### Reverse of Zero-Factor Property
>
> Let a and b be real numbers, variables, or algebraic expressions. If $a = 0$ or $b = 0$, then a and b are factors such that $ab = 0$.

EXAMPLE 6   **Writing a Quadratic Equation from Its Solutions**

Write a quadratic equation that has the solutions $x = 4$ and $x = -7$.

SOLUTION

Using the solutions $x = 4$ and $x = -7$, you can write the following.

$x = 4$ and $x = -7$	Solutions
$x - 4 = 0$ $x + 7 = 0$	Obtain zero on one side of each equation.
$(x - 4)(x + 7) = 0$	Reverse of Zero-Factor Property
$x^2 + 3x - 28 = 0$	FOIL Method

So, a quadratic equation that has the solutions $x = 4$ and $x = -7$ is

$$x^2 + 3x - 28 = 0.$$

This is not the only quadratic equation with the solutions $x = 4$ and $x = -7$. You can obtain other quadratic equations with these solutions by multiplying $x^2 + 3x - 28 = 0$ by any nonzero real number.

Exercises Within Reach ®

Solutions in English & Spanish and tutorial videos at CollegePrepAlgebra.com

Writing a Quadratic Equation In Exercises 47−58, write a quadratic equation that has the given solutions.

47. $x = 5, x = -2$

48. $x = -2, x = 3$

49. $x = 1, x = 7$

50. $x = 2, x = 8$

51. $x = 1 + \sqrt{2}, x = 1 - \sqrt{2}$

52. $x = -3 + \sqrt{5}, x = -3 - \sqrt{5}$

53. $x = 5i, x = -5i$

54. $x = 2i, x = -2i$

55. $x = 12$

56. $x = -4$

57. $x = \dfrac{1}{2}$

58. $x = -\dfrac{3}{4}$

Concept Summary: Solving Quadratic Equations by Using the Quadratic Formula

What

You can use the **Quadratic Formula** to solve quadratic equations.

EXAMPLE

Use the Quadratic Formula to solve $x^2 + 3x + 2 = 0$.

How

1. Identify the values of a, b, and c from the general form of the equation.
2. Substitute these values into the Quadratic Formula.
3. Simplify to obtain the solution.

EXAMPLE

$$x = \frac{-b \pm \sqrt{b^2 - 4ac}}{2a} \quad \text{Quadratic Formula}$$

$$x = \frac{-3 \pm \sqrt{3^2 - 4(1)(2)}}{2(1)} \quad \text{Substitute.}$$

$$x = \frac{-3 \pm \sqrt{1}}{2} = \frac{-3 \pm 1}{2} \quad \text{Simplify.}$$

$$x = -1 \quad \text{and} \quad x = -2$$

Why

You now know four methods for solving quadratic equations.

1. Factoring
2. Square Root Property
3. Completing the square
4. The Quadratic Formula

Remember that you can use the Quadratic Formula or completing the square to solve *any* quadratic equation.

Exercises Within Reach ®

Worked-out solutions to odd-numbered exercises at CollegePrepAlgebra.com

Concept Summary Check

59. *Vocabulary* State the Quadratic Formula in words.

60. *Four Methods* State the four methods used to solve quadratic equations.

61. *Reasoning* To solve the quadratic equation $3x^2 = 3 - x$ using the Quadratic Formula, what are the values of a, b, and c?

62. *Think About It* The discriminant of a quadratic equation is -25. What type of solution(s) does the equation have?

Extra Practice

Solving an Equation In Exercises 63−66, solve the equation.

63. $\dfrac{x^2}{4} - \dfrac{2x}{3} = 1$

64. $\dfrac{x^2 - 9x}{6} = \dfrac{x - 1}{2}$

65. $\sqrt{x + 3} = x - 1$

66. $\sqrt{2x - 3} = x - 2$

67. *Geometry* A rectangle has a width of x inches, a length of $(x + 6.3)$ inches, and an area of 58.14 square inches. Find its dimensions.

68. *Geometry* A rectangle has a length of $(x + 1.5)$ inches, a width of x inches, and an area of 18.36 square inches. Find its dimensions.

69. *Depth of a River* The depth d (in feet) of a river is given by

$$d = -0.25t^2 + 1.7t + 3.5, \quad 0 \le t \le 7$$

where t is the time (in hours) after a heavy rain begins. When is the river 6 feet deep?

70. *Free-Falling Object* A stone is thrown vertically upward at a velocity of 20 feet per second from a bridge that is 40 feet above the level of the water. The height h (in feet) of the stone at time t (in seconds) after it is thrown is given by

$$h - 16t^2 + 20t + 40.$$

(a) Find the time when the stone is again 40 feet above the water.

(b) Find the time when the stone strikes the water.

(c) Does the stone reach a height of 50 feet? Use the discriminant to justify your answer.

71. *Fuel Economy* The fuel economy y (in miles per gallon) of a car is given by

$$y = -0.013x^2 + 1.25x + 5.6, \quad 5 \le x \le 75$$

where x is the speed (in miles per hour) of the car.

(a) Use a graphing calculator to graph the model.

(b) Use the graph in part (a) to find the speeds at which you can travel and have a fuel economy of 32 miles per gallon. Verify your results algebraically.

72. (a) Determine the two solutions, x_1 and x_2, of each quadratic equation. Use the values of x_1 and x_2 to fill in the boxes.

Equation	x_1, x_2	$x_1 + x_2$	$x_1 x_2$
(i) $x^2 - x - 6 = 0$			
(ii) $2x^2 + 5x - 3 = 0$			
(iii) $4x^2 - 9 = 0$			
(iv) $x^2 - 10x + 34 = 0$			

(b) Consider a general quadratic equation

$$ax^2 + bx + c = 0$$

whose solutions are x_1 and x_2. Use the results of part (a) to determine how the coefficients a, b, and c are related to both the sum $(x_1 + x_2)$ and the product $(x_1 x_2)$ of the solutions.

Explaining Concepts

Choosing a Method **In Exercises 73–76, determine the method of solving the quadratic equation that would be most convenient. Explain your reasoning.**

73. $(x - 3)^2 = 25$

74. $x^2 + 8x - 12 = 0$

75. $2x^2 - 9x + 12 = 0$

76. $8x^2 - 40x = 0$

77. *Precision* Explain how the discriminant of $ax^2 + bx + c = 0$ is related to the number of x-intercepts of the graph of $y = ax^2 + bx + c$.

78. *Error Analysis* Describe and correct the student's error in writing a quadratic equation that has solutions $x = 2$ and $x = 4$.

$$(x + 2)(x + 4) = 0$$
$$x^2 + 6x + 8 = 0$$

Cumulative Review

In Exercises 79–82, find the distance between the points.

79. $(-1, 11), (2, 2)$

80. $(-2, 4), (3, -3)$

81. $(-6, -2), (-3, -4)$

82. $(-4, 7), (0, 4)$

In Exercises 83–86, sketch the graph of the function.

83. $f(x) = (x - 1)^2$

84. $f(x) = \frac{1}{2}x^2$

85. $f(x) = (x - 2)^2 + 4$

86. $f(x) = (x + 3)^2 - 1$

Mid-Chapter Quiz: Sections 10.1–10.3

Solutions in English & Spanish and tutorial videos at CollegePrepAlgebra.com

Take this quiz as you would take a quiz in class. After you are done, check your work against the answers in the back of the book.

In Exercises 1–8, solve the quadratic equation by the specified method.

1. Factoring:
   $2x^2 - 72 = 0$

2. Factoring:
   $2x^2 + 3x - 20 = 0$

3. Square Root Property:
   $3x^2 = 36$

4. Square Root Property
   $(u - 3)^2 - 16 = 0$

5. Completing the square:
   $m^2 + 7m + 2 = 0$

6. Completing the square:
   $2y^2 + 6y - 5 = 0$

7. Quadratic Formula:
   $x^2 + 4x - 6 = 0$

8. Quadratic Formula:
   $6v^2 - 3v - 4 = 0$

In Exercises 9–16, solve the equation by using the most convenient method.

9. $x^2 + 5x + 7 = 0$

10. $36 - (t - 4)^2 = 0$

11. $x(x - 10) + 3(x - 10) = 0$

12. $x(x - 3) = 10$

13. $4b^2 - 12b + 9 = 0$

14. $3m^2 + 10m + 5 = 0$

15. $x - 4\sqrt{x} - 21 = 0$

16. $x^4 + 7x^2 + 12 = 0$

In Exercises 17 and 18, solve the equation of quadratic form.

17. $x - 4\sqrt{x} + 3 = 0$

18. $x^4 - 14x^2 + 24 = 0$

Applications

19. The revenue R (in dollars) from selling x handheld video games is given by

    $R = x(180 - 1.5x)$.

    Find the number of handheld video games that must be sold to produce a revenue of $5400.

20. A rectangle has a length of x meters, a width of $(100 - x)$ meters, and an area of 2275 square meters. Find its dimensions.

21. The path of a baseball after it is hit is given by

    $h = -0.003x^2 + 1.19x + 5.2$

    where h is the height (in feet) of the baseball and x is the horizontal distance (in feet) of the ball from home plate. The ball hits the top of the outfield fence that is 10 feet high. How far is the outfield fence from home plate.

Study Skills in Action

Improving Your Memory

Have you ever driven on a highway for ten minutes when all of a sudden you kind of woke up and wondered where the last ten miles had gone? It was like the car was on autopilot. The same thing happens to many students as they sit through classes. The longer students sit through classes on "autopilot," the more likely they will "crash" when it comes to studying outside of class on their own.

While on autopilot, you do not process and retain new information effectively. You can improve your memory by learning how to focus during class and while studying on your own.

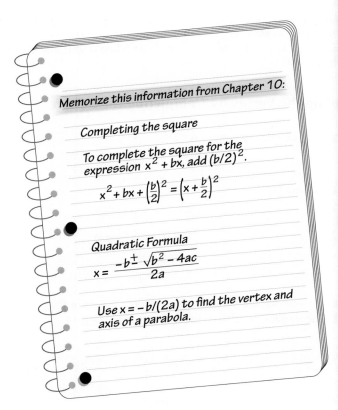

Memorize this information from Chapter 10:

Completing the square

To complete the square for the expression $x^2 + bx$, add $(b/2)^2$.

$$x^2 + bx + \left(\frac{b}{2}\right)^2 = \left(x + \frac{b}{2}\right)^2$$

Quadratic Formula

$$x = \frac{-b \pm \sqrt{b^2 - 4ac}}{2a}$$

Use $x = -b/(2a)$ to find the vertex and axis of a parabola.

Smart Study Strategy

Keep Your Mind Focused

During class
- When you sit down at your desk, get all other issues out of your mind by reviewing your notes from the last class and focusing just on math.
- Repeat in your mind what you are writing in your notes.
- When the math is particularly difficult, ask your teacher for help.

While completing homework
- Before doing homework, review the concept boxes and examples. Talk through the examples out loud.
- Complete homework as though you were also preparing for a quiz. Memorize the different types of problems, formulas, rules, and so on.

Between classes
- Review the concept boxes, the Concept Summaries, and the Concept Summary Check exercises.

Preparing for a test
- Review all your notes that pertain to the upcoming test. Review examples of each type of problem that could appear on the test.

10.4 Graphs of Quadratic Functions

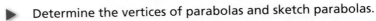

▶ Determine the vertices of parabolas and sketch parabolas.
▶ Write an equation of a parabola given the vertex and a point on the graph.
▶ Use parabolas to solve application problems.

Graphs of Quadratic Functions

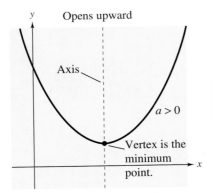

Graphs of Quadratic Functions

The graph of $f(x) = ax^2 + bx + c$, $a \neq 0$, is a **parabola**. The completed square form

$$f(x) = a(x - h)^2 + k \qquad \text{Standard form}$$

is the **standard form** of the function. The **vertex** of the parabola occurs at the point (h, k), and the vertical line passing through the vertex is the **axis** of the parabola.

Every parabola is *symmetric* about its axis, which means that if it were folded along its axis, the two parts would match.

If a is positive, the graph of $f(x) = ax^2 + bx + c$ opens upward, and if a is negative, the graph opens downward, as shown at the left. Observe in the graphs that the y-coordinate of the vertex identifies the minimum function value when $a > 0$ and the maximum function value when $a < 0$.

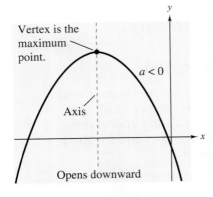

EXAMPLE 1 **Finding the Vertex by Completing the Square**

Find the vertex of the parabola given by $f(x) = x^2 - 6x + 5$.

SOLUTION

Begin by writing the function in standard form.

$f(x) = x^2 - 6x + 5$	Write original function.
$f(x) = x^2 - 6x + (-3)^2 - (-3)^2 + 5$	Complete the square.
$f(x) = (x^2 - 6x + 9) - 9 + 5$	Group terms.
$f(x) = (x - 3)^2 - 4$	Standard form

From the standard form, you can see that the vertex of the parabola occurs at the point $(3, -4)$, as shown at the left. The minimum value of the function is $f(3) = -4$.

Exercises Within Reach® Solutions in English & Spanish and tutorial videos at CollegePrepAlgebra.com

Finding the Vertex of a Parabola In Exercises 1–10, write the equation of the parabola in standard form, and find the vertex of its graph.

1. $y = x^2 - 2x$

2. $y = x^2 + 2x$

3. $y = x^2 - 4x + 7$

4. $y = x^2 + 6x - 5$

5. $y = x^2 + 6x + 5$

6. $y = x^2 - 4x + 5$

7. $y = -x^2 + 6x - 10$

8. $y = -x^2 + 4x - 8$

9. $y = 2x^2 + 6x + 2$

10. $y = 3x^2 - 3x - 9$

In Example 1, the vertex of the graph was found by *completing the square.* Another approach to finding the vertex is to complete the square once for a general function and then use the resulting formula to find the vertex.

$$f(x) = ax^2 + bx + c \qquad \text{Quadratic function}$$

$$= a\left(x^2 + \frac{b}{a}x\right) + c \qquad \text{Factor } a \text{ out of first two terms.}$$

$$= a\left[x^2 + \frac{b}{a}x + \left(\frac{b}{2a}\right)^2\right] + c - \frac{b^2}{4a} \qquad \text{Complete the square.}$$

$$= a\left(x + \frac{b}{2a}\right)^2 + c - \frac{b^2}{4a} \qquad \text{Standard form}$$

From this form you can see that the vertex occurs when $x = -\dfrac{b}{2a}$.

EXAMPLE 2   Finding the Vertex Using a Formula

Find the vertex of the parabola given by $f(x) = 3x^2 - 9x$.

SOLUTION

From the original function, it follows that $a = 3$ and $b = -9$. So, the x-coordinate of the vertex is

$$x = \frac{-b}{2a}$$

$$= \frac{-(-9)}{2(3)}$$

$$= \frac{3}{2}.$$

Substitute $\frac{3}{2}$ for x in the original equation to find the y-coordinate.

$$f\left(-\frac{b}{2a}\right) = f\left(\frac{3}{2}\right)$$

$$= 3\left(\frac{3}{2}\right)^2 - 9\left(\frac{3}{2}\right)$$

$$= -\frac{27}{4}$$

So, the vertex of the parabola is $\left(\frac{3}{2}, -\frac{27}{4}\right)$, the minimum value of the function is $f\left(\frac{3}{2}\right) = -\frac{27}{4}$, and the parabola opens upward, as shown in the figure.

Exercises Within Reach ®

Solutions in English & Spanish and tutorial videos at CollegePrepAlgebra.com

Finding the Vertex of a Parabola In Exercises 11−16, **find** the vertex of the graph of the function by using the formula $x = -\dfrac{b}{2a}$.

11. $f(x) = x^2 - 8x + 15$

12. $f(x) = x^2 + 4x + 1$

13. $g(x) = -x^2 - 2x + 1$

14. $h(x) = -x^2 + 14x - 14$

15. $y = 4x^2 + 4x + 4$

16. $y = 9x^2 - 12x$

Sketching a Parabola

1. Determine the vertex and axis of the parabola by completing the square or by using the formula $x = -\dfrac{b}{2a}$.

2. Plot the vertex, axis, *x*- and *y*-intercepts, and a few additional points on the parabola. (Using the symmetry about the axis can reduce the number of points you need to plot.)

3. Use the fact that the parabola opens *upward* when $a > 0$ and opens *downward* when $a < 0$ to complete the sketch.

EXAMPLE 3   **Sketching a Parabola**

To sketch the parabola given by

$$y = x^2 + 6x + 8$$

begin by writing the equation in standard form.

$y = x^2 + 6x + 8$	Write original equation.
$y = x^2 + 6x + 3^2 - 3^2 + 8$	Complete the square.
$y = (x^2 + 6x + 9) - 9 + 8$	Group terms.
$y = (x + 3)^2 - 1$	Standard form

The vertex occurs at the point $(-3, -1)$ and the axis is the line $x = -3$. After plotting this information, calculate a few additional points on the parabola, as shown in the table. Note that the *y*-intercept is $(0, 8)$ and the *x*-intercepts are solutions of the equation

$$x^2 + 6x + 8 = (x + 4)(x + 2) = 0.$$

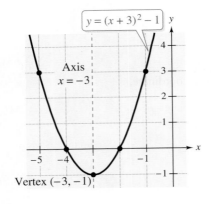

x	−5	−4	−3	−2	−1
$y = (x + 3)^2 - 1$	3	0	−1	0	3
Solution point	$(-5, 3)$	$(-4, 0)$	$(-3, -1)$	$(-2, 0)$	$(-1, 3)$

The graph of the parabola is shown at the left. Note that the parabola opens upward because the leading coefficient (in general form) is positive.

Exercises Within Reach ®

Sketching a Parabola In Exercises 17−24, sketch the parabola. Identify the vertex and any *x*-intercepts.

17. $g(x) = x^2 - 4$

18. $h(x) = x^2 - 9$

19. $f(x) = -x^2 + 4$

20. $f(x) = -x^2 + 9$

21. $y = (x - 4)^2$

22. $y = -(x + 4)^2$

23. $y = x^2 - 9x - 18$

24. $y = x^2 + 4x + 2$

Writing an Equation of a Parabola

To write an equation of a parabola with a vertical axis, use the fact that its standard equation has the form $y = a(x - h)^2 + k$, where (h, k) is the vertex.

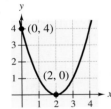

Vertex $(-2, 1)$

Axis
$x = -2$

$(0, -3)$

EXAMPLE 4   **Writing an Equation of a Parabola**

Write an equation of the parabola with vertex $(-2, 1)$ and y-intercept $(0, -3)$, as shown at the left.

SOLUTION

Because the vertex occurs at $(h, k) = (-2, 1)$, the equation has the form

$y = a(x - h)^2 + k$	Standard form
$y = a[x - (-2)]^2 + 1$	Substitute -2 for h and 1 for k.
$y = a(x + 2)^2 + 1$	Simplify.

To find the value of a, use the fact that y-intercept is $(0, -3)$.

$y = a(x + 2)^2 + 1$	Write standard form.
$-3 = a(0 + 2)^2 + 1$	Substitute 0 for x and -3 for y.
$-1 = a$	Simplify.

So, the standard form of the equation of the parabola is $y = -(x + 2)^2 + 1$.

Exercises Within Reach ®

Writing an Equation of a Parabola In Exercises 25−30, **write an equation of the parabola.**

25.

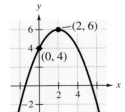

$(0, 4)$

$(2, 0)$

26.

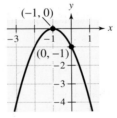

$(-1, 0)$

$(0, -1)$

27.

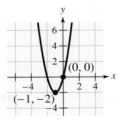

$(2, 6)$

$(0, 4)$

28.

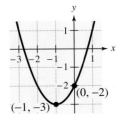

$(0, 0)$

$(-1, -2)$

29.

$(0, -2)$

$(-1, -3)$

30.

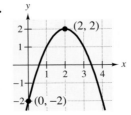

$(2, 2)$

$(0, -2)$

EXAMPLE 5 **Writing an Equation of a Parabola**

Write an equation of the parabola with vertex $(3, -4)$ that passes through the point $(5, -2)$, as shown below.

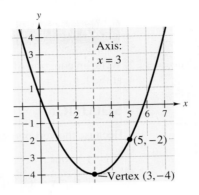

Axis: $x = 3$

$(5, -2)$

Vertex $(3, -4)$

SOLUTION

Because the vertex occurs at $(h, k) = (3, -4)$, the equation has the form

$y = a(x - h)^2 + k$ Standard form

$y = a(x - 3)]^2 + (-4)$ Substitute 3 for h and -4 for k.

$y = a(x - 3)^2 - 4.$ Simplify.

To find the value of a, use the fact that the parabola passes through the point $(5, -2)$.

$y = a(x - 3)^2 - 4$ Write standard form.

$-2 = a(5 - 3)^2 - 4$ Substitute 5 for x and -2 for y.

$\dfrac{1}{2} = a$ Simplify.

So, the standard form of the equation of the parabola is $y = \dfrac{1}{2}(x - 3)^2 - 4$.

Exercises Within Reach®   Solutions in English & Spanish and tutorial videos at CollegePrepAlgebra.com

Writing an Equation of a Parabola In Exercises 31–40, write an equation of the parabola that satisfies the conditions.

31. Vertex: $(2, 1)$; $a = 1$

32. Vertex: $(-3, -3)$; $a = 1$

33. Vertex: $(2, -4)$; Point on the graph: $(0, 0)$

34. Vertex: $(-2, -4)$; Point on the graph: $(0, 0)$

35. Vertex: $(-2, -1)$; Point on the graph: $(1, 8)$

36. Vertex: $(4, 2)$; Point on the graph: $(2, -4)$

37. Vertex: $(-1, 1)$; Point on the graph: $(-4, 7)$

38. Vertex: $(5, 2)$; Point on the graph: $(10, 3)$

39. Vertex: $(2, -2)$; Point on the graph: $(7, 8)$

40. Vertex: $(-4, 4)$; Point on the graph: $(-8, 12)$

Application

Application   EXAMPLE 6   **Bridge**

Each cable of a bridge is suspended (in the shape of a parabola) between two towers that are 1280 meters apart. The top of each tower is 152 meters above the roadway. The cables touch the roadway at the midpoint between the towers. Write an equation that models the cables of the bridge.

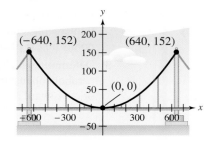

SOLUTION

From the figure, you can see that the vertex of the parabola occurs at (0, 0). So, the equation has the form

$$y = a(x - h)^2 + k \qquad \text{Standard form}$$

$$y = a(x - 0)^2 + 0 \qquad \text{Substitute 0 for } h \text{ and 0 for } k.$$

$$y = ax^2. \qquad \text{Simplify.}$$

To find the value of a, use the fact that the parabola passes through the point (640, 152).

$$y = ax^2 \qquad \text{Write standard form.}$$

$$152 = a(640)^2 \qquad \text{Substitute 640 for } x \text{ and 152 for } y.$$

$$\frac{19}{51,200} = a \qquad \text{Simplify.}$$

So, an equation that models the cables of the bridge is $y = \dfrac{19}{51,200}x^2$.

Exercises Within Reach ®

Solutions in English & Spanish and tutorial videos at CollegePrepAlgebra.com

41. **Roller Coaster Design** A structural engineer must design a parabolic arc for the bottom of a roller coaster track. The vertex of the parabola is placed at the origin, and the parabola must pass through the points $(-30, 15)$ and $(30, 15)$ (see figure). Write an equation of the parabolic arc.

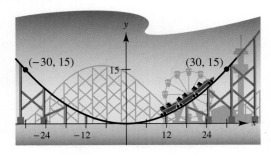

42. **Highway Design** A highway department engineer must design a parabolic arc to create a turn in a freeway around a park. The vertex of the parabola is placed at the origin, and the parabola must connect with roads represented by the equations $y = -0.4x - 100$ when $x < -500$ and $y = 0.4x - 100$ when $x > 500$ (see figure). Write an equation of the parabolic arc.

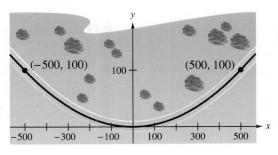

Concept Summary: Graphing Quadratic Functions

What

The graph of the quadratic function

$$y = ax^2 + bx + c, a \neq 0$$

is called a **parabola**.

EXAMPLE

Sketch the parabola given by
$y = x^2 + 2x + 2$.

How

Use the following guidelines to sketch a parabola.

1. Determine the **vertex** and **axis** of the parabola.

2. Plot the vertex, axis, x- and y-intercepts, and a few additional points on the parabola.

3. Determine whether the parabola opens upward ($a > 0$) or downward ($a < 0$).

Why

You can use parabolas to model many real-life situations. For example, you can use a parabola to model the cables of a suspension bridge.

EXAMPLE

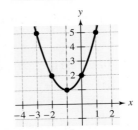

Exercises Within Reach ®   Worked-out solutions to odd-numbered exercises at CollegePrepAlgebra.com

Concept Summary Check

43. *Describing a Graph* In your own words, describe the graph of the quadratic function $f(x) = ax^2 + bx + c$.

44. *Writing* Explain how to find the vertex of the graph of a quadratic function.

45. *Writing* Explain how to find any x- or y-intercepts of the graph of a quadratic function.

46. *Reasoning* Explain how to determine whether the graph of a quadratic function opens upward or downward.

Extra Practice

Identifying a Transformation In Exercises 47–54, identify the transformation of the graph of $f(x) = x^2$, and sketch the graph of h.

47. $h(x) = x^2 - 1$

48. $h(x) = x^2 + 3$

49. $h(x) = (x + 2)^2$

50. $h(x) = (x - 4)^2$

51. $h(x) = -(x + 5)^2$

52. $h(x) = -x^2 - 6$

53. $h(x) = -(x - 2)^2 - 3$

54. $h(x) = -(x + 1)^2 + 5$

55. *Path of a Ball* The height y (in feet) of a ball thrown by a child is given by

$$y = -\frac{1}{12}x^2 + 2x + 4$$

where x is the horizontal distance (in feet) from where the ball is thrown.

(a) How high is the ball when it leaves the child's hand?

(b) How high is the ball at its maximum height?

(c) How far from the child does the ball strike the ground?

56. *Path of a Toy Rocket* A child launches a toy rocket from a table. The height y (in feet) of the rocket is given by

$$y = -\frac{1}{5}x^2 + 6x + 3$$

where x is the horizontal distance (in feet) from where the rocket is launched.

(a) How high is the rocket when it is launched?

(b) How high is the rocket at its maximum height?

(c) How far from where it is launched does the rocket land?

57. *Path of a Golf Ball* The height y (in yards) of a golf ball hit by a professional golfer is given by

$$y = -\frac{1}{480}x^2 + \frac{1}{2}x$$

where x is the horizontal distance (in yards) from where you hit the ball.

(a) How high is the ball when it is hit?

(b) How high is that ball at its maximum height?

(c) How far from where the ball is hit does it strike the ground?

58. *Path of a Softball* The height y (in feet) of a softball that you hit is given by

$$y = -\frac{1}{70}x^2 + 2x + 2$$

where x is the horizontal distance (in feet) from where you hit the ball.

(a) How high is the ball when you hit it?

(b) How high is that ball at its maximum height?

(c) How far from where you hit the ball does it strike the ground?

59. *Path of a Diver* The path of a diver is given by

$$y = -\frac{4}{9}x^2 + \frac{24}{9}x + 10$$

where y is the height in feet and x is the horizontal distance from the end of the diving board in feet. What is the maximum height of the diver?

60. *Path of a Diver* Repeat Exercise 59 when the path of the diver is modeled by

$$y = -\frac{4}{3}x^2 + \frac{10}{3}x + 10.$$

Explaining Concepts

61. *Writing* How is the discriminant related to the graph of a quadratic function?

62. *Think About It* Is it possible for the graph of a quadratic function to have two y-intercepts? Explain.

63. *Logic* Explain how to determine the maximum (or minimum) value of a quadratic function.

64. *Reasoning* The domain of a quadratic function is the set of real numbers. Explain how to find the range.

Cumulative Review

In Exercises 65–72, find the slope-intercept form of the equation of the line that passes through the two points.

65. $(0, 0), (4, -2)$ **66.** $(0, 0), (100, 75)$

67. $(-1, -2), (3, 6)$ **68.** $(1, 5), (6, 0)$

69. $\left(\frac{3}{2}, 8\right), \left(\frac{11}{2}, \frac{5}{2}\right)$ **70.** $(0, 2), (7.3, 15.4)$

71. $(0, 8), (5, 8)$ **72.** $(-3, 2), (-3, 5)$

In Exercises 73–76, write the number in *i*-form.

73. $\sqrt{-64}$

74. $\sqrt{-32}$

75. $\sqrt{-0.0081}$

76. $\sqrt{-\frac{20}{16}}$

10.5 Applications of Quadratic Equations

▶ Use quadratic equations to solve application problems.

Applications

Application EXAMPLE 1 **An Investment Problem**

A car dealer buys a fleet of cars from a car rental agency for a total of $120,000. The dealer regains this $120,000 investment by selling all but 4 of the cars at an average profit of $2500 each. How many cars has the dealer sold, and what is the average price per car?

SOLUTION

Although this problem is stated in terms of average price and average profit per car, you can use a model that assumes that each car has sold for the same price.

Verbal Model: | Selling price per car | = | Cost per car | + | Profit per car |

Labels:
Number of cars sold $= x$ (cars)
Number of cars bought $= x + 4$ (cars)
Selling price per car $= 120,000/x$ (dollars per car)
Cost per car $= 120,000/(x + 4)$ (dollars per car)
Profit per car $= 2500$ (dollars per car)

Equation:

$$\frac{120,000}{x} = \frac{120,000}{x + 4} + 2500$$

$$120,000(x + 4) = 120,000x + 2500x(x + 4), \quad x \neq 0, \ x \neq -4$$

$$120,000x + 480,000 = 120,000x + 2500x^2 + 10,000x$$

$$0 = 2500x^2 + 10,000x - 480,000$$

$$0 = x^2 + 4x - 192$$

$$0 = (x - 12)(x + 16)$$

$$x - 12 = 0 \implies x = 12$$

$$x + 16 = 0 \implies x = -16$$

By choosing the positive value, it follows that the dealer sold 12 cars at an average price of $120,000/12 = \$10,000$ per car.

Exercises Within Reach ®

Solutions in English & Spanish and tutorial videos at CollegePrepAlgebra.com

1. *Selling Price* A store owner buys a case of eggs for $21.60. The owner regains this investment by selling all but 6 dozen of the eggs at a profit of $0.30 per dozen. How many dozen eggs has the owner sold and what is the selling price per dozen?

2. *Selling Price* A computer store manager buys several computers of the same model for $12,600. The store can regain this investment by selling all but 4 of the computers at a profit of $360 per computer. To do this, how many computers must be sold, and at what price?

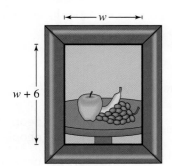

Application   EXAMPLE 2   Geometry: **The Dimensions of a Picture**

A picture is 6 inches taller than it is wide and has an area of 216 square inches, as shown at the left. What are the dimensions of the picture?

SOLUTION

Verbal Model: Area of picture = Width • Height

Labels: Picture width = w (inches)
              Picture height = $w + 6$ (inches)
              Area = 216   (square inches)

Equation: $216 = w(w + 6)$

                $0 = w^2 + 6w - 216$

                $0 = (w + 18)(w - 12)$

                $w + 18 = 0$ ➡ $w = -18$

                $w - 12 = 0$ ➡ $w = 12$

By choosing the positive value of w, you can conclude that the width of the picture is 12 inches and the height of the picture is $12 + 6 = 18$ inches.

Application   EXAMPLE 3   **An Interest Problem**

The amount A after 2 years in an account earning r percent (in decimal form) compounded annually is given by $A = P(1 + r)^2$, where P is the original investment. Find the interest rate when an investment of \$6000 increases to \$6933.75 over a two-year period.

SOLUTION

$A = P(1 + r)^2$	Write given formula.
$6933.75 = 6000(1 + r)^2$	Substitute 6933.75 for A and 6000 for P.
$1.155625 = (1 + r)^2$	Divide each side by 6000.
$\pm 1.075 = 1 + r$	Square Root Property
$0.075 = r$	Choose positive solution.

The annual interest rate is $r = 0.075 = 7.5\%$.

Exercises Within Reach ®   Solutions in English & Spanish and tutorial videos at CollegePrepAlgebra.com

3. **Geometry** A picture frame is 4 inches taller than it is wide and has an area of 192 square inches. What are the dimensions of the picture frame?

4. **Geometry** The height of triangle is 8 inches less than its base. The area of the triangle is 192 square inches. Find the dimensions of the triangle.

Compound Interest The amount A after 2 years in an account earning r percent (in decimal form) compounded annually is given by $A = P(1 + r)^2$, where P is the original investment. In Exercises 5–8, **find the interest rate r.**

5. $P = \$10,000$
   $A = \$11,990.25$

6. $P = \$3000$
   $A = \$3499.20$

7. $P = \$500$
   $A = \$572.45$

8. $P = \$250$
   $A = \$280.90$

Application EXAMPLE 4 **Reduced Rates**

A ski club charters a bus for a ski trip at a cost of $720. When four nonmembers accept invitations from the club to go on the trip, the bus fare per skier decreases by $6. How many club members are going on the trip?

SOLUTION

Verbal Model: Fare per skier • Number of skiers = 720

Labels: Number of ski club members = x (people)
Number of skiers = $x + 4$ (people)

Original fare per skier = $\dfrac{720}{x}$ (dollars per person)

New fare per skier = $\dfrac{720}{x} - 6$ (dollars per person)

Equation: $\left(\dfrac{720}{x} - 6\right)(x + 4) = 720$ Original equation.

$\left(\dfrac{720 - 6x}{x}\right)(x + 4) = 720$ Rewrite 1st factor.

$(720 - 6x)(x + 4) = 720x, \ x \neq 0$ Multiply each side by x.

$720x + 2880 - 6x^2 - 24x = 720x$ Multiply factors.

$-6x^2 - 24x + 2880 = 0$ Subtract 720x from each side.

$x^2 + 4x - 480 = 0$ Divide each side by -6.

$(x + 24)(x - 20) = 0$ Factor left side of equation.

$x + 24 = 0 \quad \Longrightarrow \quad x = -24$ Set 1st factor equal to 0.

$x - 20 = 0 \quad \Longrightarrow \quad x = 20$ Set 2nd factor equal to 0.

By choosing the positive value of x, you can conclude that 20 ski club members are going on the trip.

Exercises Within Reach ®   Solutions in English & Spanish and tutorial videos at CollegePrepAlgebra.com

9. **Reduced Rates** A service organization pays $210 for a block of tickets to a baseball game. The block contains three more tickets than the organization needs for its member. By inviting 3 more people to attend (and share in the cost), the organization lowers the price per person by $3.50. How many people are going to the game?

10. **Reduced Fares** A science club charters a bus to attend a science fair at a cost of $480. To lower the bus fare per person, the club invites nonmembers to go along. When 3 nonmembers join the trip, the fare per person is decreased by $1. How many people are going to the science fair?

Application EXAMPLE 5 **Work-Rate Problem**

An office has two copy machines. Machine B is known to take 12 minutes longer than machine A to copy the company's monthly report. Using both machines together, it takes 8 minutes to copy the report. How long would it take each machine alone to copy the report?

SOLUTION

Verbal Model: $\boxed{\begin{array}{c}\text{Work done by}\\\text{machine A}\end{array}} + \boxed{\begin{array}{c}\text{Work done by}\\\text{machine B}\end{array}} = \boxed{\begin{array}{c}1\text{ complete}\\\text{job}\end{array}}$

$\boxed{\begin{array}{c}\text{Rate}\\\text{for A}\end{array}} \cdot \boxed{\begin{array}{c}\text{Time}\\\text{for both}\end{array}} + \boxed{\begin{array}{c}\text{Rate}\\\text{for B}\end{array}} \cdot \boxed{\begin{array}{c}\text{Time}\\\text{for both}\end{array}} = 1$

Labels:
Time for machine A $= t$	(minutes)
Rate for machine A $= 1/t$	(job per minute)
Time for machine B $= t + 12$	(minutes)
Rate for machine B $= 1/(t + 12)$	(job per minute)
Time for both machines $= 8$	(minutes)
Rate for both machines $= 1/8$	(job per minute)

Equation:

$\dfrac{1}{t}(8) + \dfrac{1}{t + 12}(8) = 1$ Original equation

$8\left(\dfrac{1}{t} + \dfrac{1}{t + 12}\right) = 1$ Distributive Property

$8\left[\dfrac{t + 12 + t}{t(t + 12)}\right] = 1$ Rewrite with common denominator.

$8t(t + 12)\left[\dfrac{2t + 12}{t(t + 12)}\right] = t(t + 12)$ Multiply each side by $t(t + 12)$.

$8(2t + 12) = t^2 + 12t$ Simplify.

$16t + 96 = t^2 + 12t$ Distributive Property

$0 = t^2 - 4t - 96$ Subtract $16t + 96$ from each side.

$0 = (t - 12)(t + 8)$ Factor right side of equation.

$t - 12 = 0 \implies t = 12$ Set 1st factor equal to 0.

$t + 8 = 0 \implies t = -8$ Set 2nd factor equal to 0.

By choosing the positive value of t, you can conclude that machine A would take 12 minutes and machine B would take $12 + 12 = 24$ minutes.

Exercises Within Reach ®

11. Work Rate An office has two printers. Machine B is known to take 3 minutes longer than machine A to produce the company's monthly financial report. Using both machines together, it takes 6 minutes to produce the report. How long would it take each machine alone to produce the report?

12. Work Rate A builder works with two plumbing companies. Company A is known to take 3 days longer than Company B to install the plumbing in a particular style of house. Using both companies together, it takes 4 days to install the plumbing. How long would it take each company alone to install the plumbing?

Application EXAMPLE 6 Geometry: **The Pythagorean Theorem**

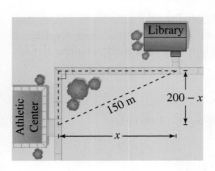

An L-shaped sidewalk from the athletic center to the library on a college campus is 200 meters long, as shown at the left. By cutting diagonally across the grass, students shorten the walking distance to 150 meters. What are the lengths of the two legs of the sidewalk?

SOLUTION

Common Formula:	$a^2 + b^2 = c^2$	Pythagorean Theorem

Labels:	Length of one leg $= x$	(meters)
	Length of other leg $= 200 - x$	(meters)
	Length of diagonal $= 150$	(meters)

Equation:
$$x^2 + (200 - x)^2 = 150^2$$
$$x^2 + 40,000 - 400x + x^2 = 22,500$$
$$2x^2 - 400x + 40,000 = 22,500$$
$$2x^2 - 400x + 17,500 = 0$$
$$x^2 - 200x + 8750 = 0$$

Using the Quadratic Formula, you can find the solutions as follows.

$$x = \frac{-(-200) \pm \sqrt{(-200)^2 - 4(1)(8750)}}{2(1)}$$ Substitute 1 for a, -200 for b, and 8750 for c.

$$= \frac{200 \pm \sqrt{5000}}{2}$$

$$= \frac{200 \pm 50\sqrt{2}}{2}$$

$$= 100 \pm 25\sqrt{2}$$

Both solutions are positive, so it does not matter which you choose. When you let $x = 100 + 25\sqrt{2} \approx 135.4$ meters, the length of the other leg is $200 - x \approx 200 - 135.4 = 64.6$ meters.

Exercises Within Reach®

Solutions in English & Spanish and tutorial videos at CollegePrepAlgebra.com

13. *Geometry* An L-shaped sidewalk from the library to the gym on a high school campus is 100 yards long, as shown in the figure. By cutting diagonally across the grass, students shorten the walking distance to 80 yards. What are the lengths of the two legs of the sidewalk?

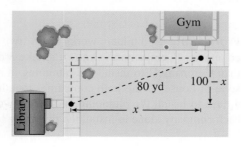

14. *Delivery Route* You deliver pizzas to an insurance office and an apartment complex (see figure). Your total mileage in driving to the insurance office and then the apartment complex is 12 miles. By using a direct route, you are able to drive just 9 miles to return to the pizza shop. Find the possible distances from the pizza shop to the insurance office.

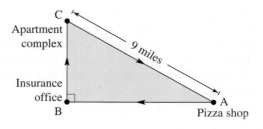

Application   EXAMPLE 7   **The Height of a Model Rocket**

A model rocket is projected straight upward from ground level according to the height equation

$$h = -16t^2 + 192t, \ t \geq 0$$

where h is the height in feet and t is the time in seconds.

a.   After how many seconds is the height 432 feet?

b.   After how many seconds does the rocket hit the ground?

SOLUTION

a.

$h = -16t^2 + 192t$	Write original equation.
$432 = -16t^2 + 192t$	Substitute 432 for h.
$16t^2 - 192t + 432 = 0$	Write in general form.
$t^2 - 12t + 27 = 0$	Divide each side by 16.
$(t - 3)(t - 9) = 0$	Factor left side of equation.
$t - 3 = 0 \implies t = 3$	Set 1st factor equal to 0.
$t - 9 = 0 \implies t = 9$	Set 2nd factor equal to 0.

432 ft

The rocket attains a height of 432 feet at two different times—once (going up) after 3 seconds, and again (coming down) after 9 seconds.

b.   To find the time it takes for the rocket to hit the ground, let the height be 0.

$0 = -16t^2 + 192t$	Substitute 0 for h in original equation.
$0 = t^2 - 12t$	Divide each side by -16.
$0 = t(t - 12)$	Factor right side of equation.
$t = 0 \quad \text{or} \quad t = 12$	Solutions

The rocket hits the ground after 12 seconds. (Note that the time of $t = 0$ seconds corresponds to the time of lift-off.)

Exercises Within Reach ®

Solutions in English & Spanish and tutorial videos at CollegePrepAlgebra.com

15. *Height* You are hitting baseballs. When you toss the ball into the air, your hand is 5 feet above the ground (see figure). You hit the ball when it falls back to a height of 3.5 feet. You toss the ball with an initial velocity of 18 feet per second. The height h of the ball t seconds after leaving your hand is given by $h = 5 + 18t - 16t^2$. About how much time passes before you hit the ball?

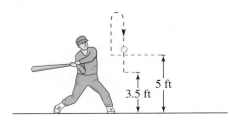

5 ft

3.5 ft

16. *Height* A model rocket is projected straight upward from ground level according to the height equation

$$h = -16t^2 + 160t$$

where h is the height of the rocket in feet and t is the time in seconds.

(a)   After how many seconds is the height 336 feet?

(b)   After how many seconds does the rocket hit the ground?

(c)   What is the maximum height of the rocket?

Concept Summary: *Using Quadratic Equations to Solve Problems*

What

You can use quadratic equations to model many real-life problems.

EXAMPLE

A postcard is 2 inches wider than it is tall and has an area of 35 square inches. What are the dimensions of the postcard?

How

EXAMPLE

Here is one way to model this problem.

Create a Verbal Model:

$$\boxed{\text{Area of postcard}} = \boxed{\text{Height}} \cdot \boxed{\text{Width}}$$

Assign labels:

Postcard height $= t$

Postcard width $= t + 2$

Area $= 35$

Write an equation:

$35 = t(t + 2)$

Why

Notice that the equation that models the problem is a quadratic equation. You can now solve the problem by solving the quadratic equation.

EXAMPLE

$35 = t(t + 2)$

$0 = t^2 + 2t - 35$

$0 = (t + 7)(t - 5)$

So, $t = -7$ and $t = 5$.

Because the height of the postcard cannot be negative, the height is 5 inches and the width is 7 inches.

Exercises Within Reach ®

Worked-out solutions to odd-numbered exercises at CollegePrepAlgebra.com

Concept Summary Check

Problem Solving In Exercises 17−20, a problem situation is given. **Describe two quantities that can be set equal to each other to write an equation that can be used to solve the problem.**

17. You know the length of the hypotenuse and the sum of the lengths of the legs of a right triangle. You want to find the lengths of the legs.

18. You know the area of a rectangle and you know how many units longer the length is than the width. You want to find the length and width.

19. You know the amount invested in an unknown number of product units. You know the number of units remaining when the investment is regained, and the profit per unit sold. You want to find the number of units sold and the price per unit.

20. You know the time in minutes for two machines to complete a task together and you know how many more minutes it takes one machine than the other to complete the task alone. You want to find the time to complete the task alone for each machine.

Extra Practice

Geometry In Exercises 21−28, **find the perimeter or area of the rectangle, as indicated.**

	Width	Length	Perimeter	Area		Width	Length	Perimeter	Area
21.	$1.4l$	l	54 in.		22.	w	$3.5w$	60 m	
23.	w	$2.5w$		250 ft^2	24.	w	$1.5w$		216 cm^2
25.	w	$w + 3$	54 km		26.	$l - 6$	l	108 ft	
27.	$l - 20$	l		12,000 m^2	28.	w	$w + 5$		500 ft^2

29. **Selling Price** A flea market vendor buys a box of DVD movies for $50. The vendor regains this investment by selling all but 15 of the DVDs at a profit of $3 each. How many DVDs has the vendor sold, and at what price?

30. **Selling Price** A running club buys a case of sweatshirts for $750 to sell at a fundraiser. The club needs to sell all but 20 of the sweatshirts at a profit of $10 per sweatshirt to regain the $750 investment. How many sweatshirts must be sold, and at what price, to do this?

Number Problem **In Exercises 31 and 32, find two positive integers that satisfy the requirement.**

31. The product of two consecutive integers is 182.

32. The product of two consecutive odd integers is 323.

33. ***Open Conduit*** An open-topped rectangular conduit for carrying water in a manufacturing process is made by folding up the edges of a sheet of aluminum that is 48 inches wide (see figure). A cross section of the conduit must have an area of 288 square inches. Find the width and height of the conduit.

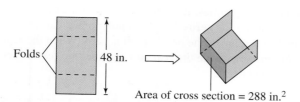

Folds

48 in.

Area of cross section = 288 in.²

34. ***Speed*** A company uses a pickup truck for deliveries. The cost per hour for fuel is $C = v^2/300$, where v is the speed in miles per hour. The driver is paid $15 per hour. The cost of wages and fuel for an 80-mile trip at constant speed is $36. Find the possible speeds.

35. ***Average Speed*** A truck traveled the first 100 miles of a trip at one speed and the last 135 miles at an average speed of 5 miles per hour less. The entire trip took 5 hours. What was the average speed for the first part of the trip?

36. ***Fenced Area*** A family builds a fence around three sides of their property (see figure). In total, they use 550 feet of fencing. By their calculations, the lot is 1 acre (43,560 square feet). Is this correct? Explain your reasoning.

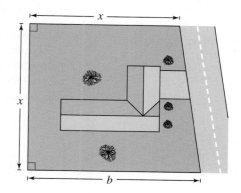

37. ***Distance*** Find any points on the line

$$y = 9$$

that are 10 units from the point (2, 3).

38. ***Distance*** Find any points on the line

$$y = 14$$

that are 13 units from the point (1, 2).

Explaining Concepts

39. ***Reasoning*** To solve some of the problems in this section, you wrote rational equations. Explain why these types of problems are included as applications of quadratic equations.

40. ***Writing*** In Exercises 21–28 you are asked to find the perimeter or area of a rectangle. To do this, you must write an equation that can be solved for the length or width of the rectangle. Explain how you can tell when the equation will be a *quadratic equation* or a *linear equation*.

41. ***Think About It*** In a *reduced rates* problem such as Example 4, does the cost per person decrease by the same amount for each additional person? Explain.

42. ***Think About It*** In a *height of an object* problem such as Example 7, suppose you try solving the height equation using a height greater than the maximum height reached by the object. What type of result will you get for t? Explain.

Cumulative Review

In Exercises 43 and 44, solve the inequality and sketch the solution on the real number line.

43. $5 - 3x > 17$

44. $-3 < 2x + 3 < 5$

In Exercises 45 and 46, solve the equation by completing the square.

45. $x^2 - 8x = 0$

46. $x^2 - 2x - 2 = 0$

10.6 Quadratic and Rational Inequalities

▶ Use test intervals to solve quadratic inequalities.
▶ Use test intervals to solve rational inequalities.
▶ Use inequalities to solve application problems.

Test Intervals and Quadratic Inequalities

When working with polynomial inequalities, it is important to realize that the value of a polynomial can change signs only at its **zeros**. That is, a polynomial can change signs only at the x-values for which the value of the polynomial is zero. When the real zeros of a polynomial are put in order, they divide the real number line into **test intervals** in which the polynomial has no sign changes.

Finding Test Intervals for a Polynomial

1. Find all real zeros of the polynomial, and arrange the zeros in increasing order. The zeros of a polynomial are called its **critical numbers**.
2. Use the critical numbers of the polynomial to determine its test intervals.
3. Choose a representative x-value in each test interval and evaluate the polynomial at that value. When the value of the polynomial is negative, the polynomial has negative values for *all* x-values in the interval. When the value of the polynomial is positive, the polynomial has positive values for *all* x-values in the interval.

EXAMPLE 1 **Test Intervals for a Linear Polynomial**

The first-degree polynomial $x + 2$ has a zero at $x = -2$, and it changes signs at that zero. You can picture this result on the real number line, as shown below.

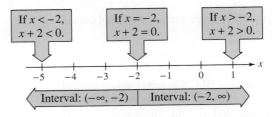

Note in the figure that the zero of the polynomial divides the real number line into two test intervals. The value of the polynomial is negative for every x-value in the first test interval $(-\infty, -2)$, and positive for every x-value in the second test interval $(-2, \infty)$.

Exercises Within Reach ®   Solutions in English & Spanish and tutorial videos at CollegePrepAlgebra.com

Finding Test Intervals In Exercises 1−8, determine the intervals for which the polynomial is entirely negative and entirely positive.

1. $x - 4$

2. $3 - x$

3. $3 - \frac{1}{2}x$

4. $\frac{2}{3}x - 8$

5. $4x(x - 5)$

6. $7x(3 - x)$

7. $4 - x^2$

8. $x^2 - 36$

EXAMPLE 2   **Solving a Quadratic Inequality**

Solve the inequality $x^2 - 5x < 0$.

SOLUTION

First find the critical numbers of $x^2 - 5x < 0$ by finding the solutions of the equation $x^2 - 5x = 0$.

$x^2 - 5x = 0$	Write corresponding equation.
$x(x - 5) = 0$	Factor.
$x = 0, \ x = 5$	Critical numbers

This implies that the test intervals are $(-\infty, 0)$, $(0, 5)$, and $(5, \infty)$. To test an interval, choose a convenient value in the interval and determine whether the value satisfies the inequality.

Study Tip

In Example 2, note that the original inequality contains a "less than" symbol. This means that the solution set does not contain the endpoints of the test interval (0, 5).

Test interval	Representative x-value	Is inequality satisfied?
$(-\infty, 0)$	$x = -1$	$(-1)^2 - 5(-1) \overset{?}{<} 0$ $6 \not< 0$
$(0, 5)$	$x = 1$	$1^2 - 5(1) \overset{?}{<} 0$ $-4 < 0$
$(5, \infty)$	$x = 6$	$6^2 - 5(6) \overset{?}{<} 0$ $6 \not< 0$

Of the three x-values tested above, only the value $x = 1$ satisfies the inequality $x^2 - 5x < 0$. So, you can conclude that the solution set of the inequality is $0 < x < 5$, as shown below.

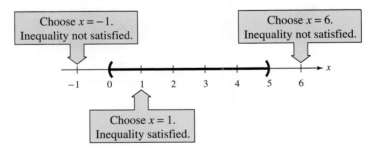

Choose $x = -1$.
Inequality not satisfied.

Choose $x = 6$.
Inequality not satisfied.

Choose $x = 1$.
Inequality satisfied.

Exercises Within Reach ®   Solutions in English & Spanish and tutorial videos at CollegePrepAlgebra.com

Solving a Quadratic Inequality In Exercises 9–16, solve the inequality and graph the solution on the real number line.

9. $3x(x - 2) < 0$

10. $5x(x - 8) > 0$

11. $3x(2 - x) \geq 0$

12. $5x(8 - x) > 0$

13. $x^2 + 4x > 0$

14. $x^2 - 5x \geq 0$

15. $x^2 - 3x - 10 \geq 0$

16. $x^2 + 8x + 7 < 0$

EXAMPLE 3   **Solving a Quadratic Inequality**

Solve the inequality $2x^2 + 5x \geq 12$.

SOLUTION

Begin by writing the inequality in the general form $2x^2 + 5x - 12 \geq 0$. Next, find the critical numbers by finding the solutions of the equation $2x^2 + 5x - 12 = 0$.

$$2x^2 + 5x - 12 = 0 \qquad \text{Write corresponding equation.}$$

$$(x + 4)(2x - 3) = 0 \qquad \text{Factor.}$$

$$x = -4, \; x = \frac{3}{2} \qquad \text{Critical numbers}$$

This implies that the test intervals are $(-\infty, -4)$, $\left(-4, \frac{3}{2}\right)$, and $\left(\frac{3}{2}, \infty\right)$. To test an interval, choose a convenient value in the interval and determine whether the value satisfies the inequality.

Test interval	Representative x-value	Is inequality satisfied?
$(-\infty, -4)$	$x = -5$	$2(-5)^2 + 5(-5) \overset{?}{\geq} 12$ $25 \geq 12$
$\left(-4, \frac{3}{2}\right)$	$x = 0$	$2(0)^2 + 5(0) \overset{?}{\geq} 12$ $0 \ngeq 12$
$\left(\frac{3}{2}, \infty\right)$	$x = 2$	$2(2)^2 + 5(2) \overset{?}{\geq} 12$ $18 \geq 12$

From this table, you can see that the solution set of the inequality $2x^2 + 5x \geq 12$ is $x \leq -4$ or $x \geq \frac{3}{2}$, as shown below.

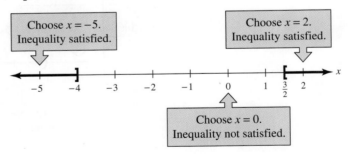

Choose $x = -5$. Inequality satisfied.

Choose $x = 2$. Inequality satisfied.

Choose $x = 0$. Inequality not satisfied.

Exercises Within Reach ®   Solutions in English & Spanish and tutorial videos at CollegePrepAlgebra.com

Solving a Quadratic Inequality In Exercises 17–22, **solve** the inequality and **graph** the solution on the real number line.

17. $x^2 > 4$

18. $z^2 \leq 9$

19. $x^2 + 5x \leq 36$

20. $t^2 - 4t > 12$

21. $u^2 + 2u - 2 > 1$

22. $t^2 - 15t < -50$

EXAMPLE 4   **Unusual Solution Sets**

a. The solution set of the quadratic inequality

$$x^2 + 2x + 4 > 0$$

consists of the entire set of real numbers, $-\infty < x < \infty$. This is true because the value of the quadratic $x^2 + 2x + 4$ is positive for every real value of x. You can see in the graph at the left that the entire parabola lies above the x-axis.

b. The solution set of the quadratic inequality

$$x^2 + 2x + 1 \le 0$$

consists of the single number -1. This is true because $x^2 + 2x + 1 = (x + 1)^2$ has just one critical number, $x = -1$, and it is the only value that satisfies the inequality. You can see in the graph at the left that the parabola meets the x-axis only when $x = -1$.

c. The solution set of the quadratic inequality

$$x^2 + 3x + 5 < 0$$

is empty. This is true because the value of the quadratic $x^2 + 3x + 5$ is not less than zero for any value of x. No point on the parabola lies below the x-axis, as shown in the graph below on the left.

d. The solution set of the quadratic inequality

$$x^2 - 4x + 4 > 0$$

consists of all real numbers *except* the number 2. So, the solution set of the quadratic inequality is $x < 2$ or $x > 2$. You can see in the graph below on the right that the parabola lies above the x-axis *except* at $x = 2$, where it meets the x-axis.

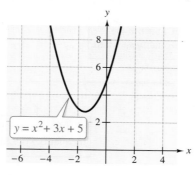

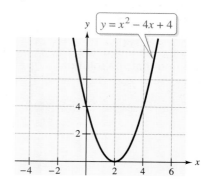

Exercises Within Reach ®

Solutions in English & Spanish and tutorial videos at CollegePrepAlgebra.com

Unusual Solution Sets **In Exercises 23–28, solve the inequality and graph the solution on the real number line. (Some of the inequalities have no solutions.)**

23. $x^2 + 4x + 5 < 0$

24. $x^2 + 14x + 49 < 0$

25. $x^2 + 6x + 10 > 0$

26. $x^2 + 2x + 1 \ge 0$

27. $y^2 + 16y + 64 \le 0$

28. $4x^2 + 28x + 49 \le 0$

Rational Inequalities

The concepts of critical numbers and test intervals can be extended to inequalities involving rational expressions. To do this, use the fact that the value of a rational expression can change sign only at its *zeros* (the x-values for which its numerator is zero) and its *undefined values* (the x-values for which its denominator is zero). These two types of numbers make up the **critical numbers** of a rational inequality.

EXAMPLE 5 Solving a Rational Inequality

To solve the inequality $\dfrac{x}{x-2} > 0$, first find the critical numbers. The numerator is zero when $x = 0$, and the denominator is zero when $x = 2$. So, the two critical numbers are 0 and 2, which implies that the test intervals are $(-\infty, 0)$, $(0, 2)$, and $(2, \infty)$. To test an interval, choose a convenient value in the interval and determine whether the value satisfies the inequality.

Test interval	Representative x-value	Is inequality satisfied?	
$(-\infty, 0)$	$x = -1$	$\dfrac{-1}{-1-2} \overset{?}{>} 0$	$\dfrac{1}{3} > 0$
$(0, 2)$	$x = 1$	$\dfrac{1}{1-2} \overset{?}{>} 0$	$-1 \not> 0$
$(2, \infty)$	$x = 3$	$\dfrac{3}{3-2} \overset{?}{>} 0$	$3 > 0$

From this table, you can see that the solution set of the inequality is $x < 0$ or $x > 2$, as shown below.

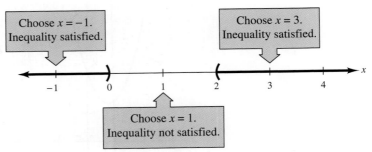

Choose $x = -1$.
Inequality satisfied.

Choose $x = 3$.
Inequality satisfied.

Choose $x = 1$.
Inequality not satisfied.

Study Tip

When solving a rational inequality, you should begin by writing the inequality in general form, with the rational expression (as a single fraction) on the left and zero on the right. For instance, the first step in solving

$$\frac{2x}{x+3} < 4$$

is to write it as

$$\frac{2x}{x+3} - 4 < 0$$

$$\frac{2x - 4(x+3)}{x+3} < 0$$

$$\frac{-2x - 12}{x+3} < 0.$$

Try solving this inequality. You should find that the solution set is $x < -6$ or $x > -3$.

Exercises Within Reach ®

Solutions in English & Spanish and tutorial videos at CollegePrepAlgebra.com

Solving a Rational Inequality In Exercises 29–36, solve the inequality and graph the solution on the real number line.

29. $\dfrac{5}{x-3} > 0$

30. $\dfrac{3}{4-x} > 0$

31. $\dfrac{-5}{x-3} > 0$

32. $\dfrac{-3}{4-x} > 0$

33. $\dfrac{x+4}{x-2} > 0$

34. $\dfrac{x-5}{x+2} < 0$

35. $\dfrac{3}{y-1} \leq -1$

36. $\dfrac{2}{x-3} \geq -1$

Application

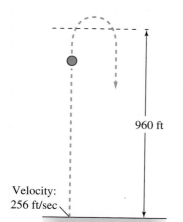

960 ft

Velocity:
256 ft/sec

| Application | EXAMPLE 6 | **The Height of a Projectile** |

A projectile is fired straight upward from ground level with an initial velocity of 256 feet per second, as shown at the left. Its height h at any time t is given by

$$h = -16t^2 + 256t$$

where h is measured in feet and t is measured in seconds. During what interval of time will the height of the projectile exceed 960 feet?

SOLUTION

To solve this problem, begin by writing the inequality in general form.

$-16t^2 + 256t > 960$	Write original inequality.
$-16t^2 + 256t - 960 > 0$	Write in general form.

Next, find the critical numbers for $-16t^2 + 256t - 960 > 0$ by finding the solutions of the equation $-16t^2 + 256t - 960 = 0$.

$-16t^2 + 256t - 960 = 0$	Write corresponding equation.
$t^2 - 16t + 60 = 0$	Divide each side by -16.
$(t - 6)(t - 10) = 0$	Factor.
$t = 6, \ t = 10$	Critical numbers

This implies that the test intervals are

$(-\infty, 6), (6, 10),$ and $(10, \infty).$ Test intervals

To test an interval, choose a convenient value in the interval and determine whether the value satisfies the inequality.

Test interval	Representative t-value	Is inequality satisfied?
$(-\infty, 6)$	$t = 0$	$-16(0)^2 + 256(0) \not> 960$
$(6, 10)$	$t = 7$	$-16(7)^2 + 256(7) > 960$
$(10, \infty)$	$t = 11$	$-16(11)^2 + 256(11) \not> 960$

So, the height will exceed 960 feet for values of t such that $6 < t < 10$.

Exercises Within Reach ®

Solutions in English & Spanish and tutorial videos at CollegePrepAlgebra.com

37. *Height* A projectile is fired straight upward from ground level with an initial velocity of 128 feet per second. Its height h at any time t is given by

$$h = -16t^2 + 128t$$

where h is measured in feet and t is measured in seconds. During what interval of time will the height of the projectile exceed 240 feet?

38. *Height* A projectile is fired straight upward from ground level with an initial velocity of 88 feet per second. Its height h at any time t is given by

$$h = -16t^2 + 88t$$

where h is measured in feet and t is measured in seconds. During what interval of time will the height of the projectile exceed 50 feet?

Concept Summary: Solving Quadratic and Rational Inequalities

What

You can use critical numbers and test intervals to solve quadratic and rational inequalities.

EXAMPLE

Solve the inequality

$x(x - 4) < 0.$

How

1. Find the critical numbers of the polynomial or rational inequality.

2. Use the critical numbers to determine the test intervals.

3. Choose a value in each interval and determine if the value satisfies the inequality.

EXAMPLE

$x(x - 4) = 0$

$x = 0, \ x = 4$

This implies that the test intervals are $(-\infty, 0)$, $(0, 4)$, and $(4, \infty)$.

By testing $x = -1, x = 1$, and $x = 5$, you can conclude that the solution set is $0 < x < 4$.

Why

By using the critical numbers to determine the test intervals, you need to test only a few values to find the solution set of the inequality.

Exercises Within Reach ®

Worked-out solutions to odd-numbered exercises at CollegePrepAlgebra.com

Concept Summary Check

39. *Vocabulary* The test intervals of a polynomial are $(-\infty, -1)$, $(-1, 3)$, $(3, \infty)$. What are the critical numbers of the polynomial?

40. *Writing* In your own words, describe a procedure for solving quadratic inequalities.

41. *Writing* Is $x = 4$ a solution of the inequality $x(x - 4) < 0$? Explain.

42. *Logic* How is the procedure for finding the critical numbers of a quadratic inequality different from the procedure for finding the critical numbers of a rational inequality?

Extra Practice

Finding Critical Numbers In Exercises 43−46, find the critical numbers.

43. $x(2x - 5)$ **44.** $5x(x - 3)$ **45.** $4x^2 - 81$ **46.** $9y^2 - 16$

Solving an Inequality In Exercises 47−58, solve the inequality and graph the solution on the real number line.

47. $6 - (x - 2)^2 < 0$

48. $(y + 3)^2 - 6 \geq 0$

49. $16 \leq (u + 5)^2$

50. $25 \geq (x - 3)^2$

51. $\dfrac{u - 6}{3u - 5} \leq 0$

52. $\dfrac{3(u - 3)}{u + 1} < 0$

53. $\dfrac{2(4 - t)}{4 + t} > 0$

54. $\dfrac{2}{x - 5} \geq 3$

55. $\dfrac{1}{x + 2} > -3$

56. $\dfrac{4x}{x + 2} < -1$

57. $\dfrac{6x}{x - 4} < 5$

58. $\dfrac{x - 3}{x - 6} \leq 4$

59. *Compound Interest* You are investing $1000 in a certificate of deposit for 2 years, and you want the interest for that time period to exceed $150. The interest is compounded annually. What interest rate should you have? [*Hint:* Solve the inequality $1000(1 + r)^2 > 1150$.]

60. *Compound Interest* You are investing $500 in a certificate of deposit for 2 years, and you want the interest for that time period to exceed $50. The interest is compounded annually. What interest rate should you have? [*Hint:* Solve the inequality $500(1 + r)^2 > 550$.]

61. *Geometry* You have 64 feet of fencing to enclose a rectangular region. Determine the interval for the length such that the area will exceed 240 square feet.

62. *Cost, Revenue, and Profit* The revenue and cost equations for a computer desk are given by

$$R = x(50 - 0.0002x) \text{ and } C = 12x + 150,000$$

where R and C are measured in dollars and x represents the number of desks sold. How many desks must be sold to obtain a profit of at least $1,650,000?

63. *Cost, Revenue, and Profit* The revenue and cost equations for a digital camera are given by

$$R = x(125 - 0.0005x)$$

and

$$C = 3.5x + 185,000$$

where R and C are measured in dollars and x represents the number of cameras sold. How many cameras must be sold to obtain a profit of at least $6,000,000?

64. *Antibiotics* The concentration C (in milligrams per liter) of an antibiotic t minutes after it is administered is given by

$$C(t) = \frac{21.9 - 0.043t}{1 + 0.005t}, \quad 30 \le t \le 500.$$

(a) Use a graphing calculator to graph the concentration function.

(b) How long does it take for the concentration of the antibiotic to fall below 5 milligrams per liter?

Explaining Concepts

65. *Reasoning* Explain why the critical numbers of a polynomial are not included in its test intervals.

66. *Precision* Explain the difference in the solution sets of $x^2 - 4 < 0$ and $x^2 - 4 \le 0$.

67. *Reasoning* The graph of a quadratic function g lies completely above the x-axis. What is the solution set of the inequality $g(x) < 0$? Explain your reasoning.

68. *Using a Graph* Explain how you can use the graph of $f(x) = x^2 - x - 6$ to check the solution of $x^2 - x - 6 > 0$.

Cumulative Review

In Exercises 69−74, perform the operation and simplify.

69. $\dfrac{4xy^3}{x^2y} \cdot \dfrac{y}{8x}$

70. $\dfrac{2x^2 - 2}{x^2 - 6x - 7} \cdot (x^2 - 10x + 21)$

71. $\dfrac{x^2 - x - 6}{4x^3} \cdot \dfrac{x + 1}{x^2 + 5x + 6}$

72. $\dfrac{32x^3y}{y^9} \div \dfrac{8x^4}{y^6}$

73. $\dfrac{x^2 + 8x + 16}{x^2 - 6x} \div (3x - 24)$

74. $\dfrac{x^2 + 6x - 16}{3x^2} \div \dfrac{x + 8}{6x}$

In Exercises 75−78, evaluate the expression for the specified value. Round your result to the nearest hundredth, if necessary.

75. $x^2; x = -\dfrac{1}{3}$

76. $1000 - 20x^3; x = 4.02$

77. $\dfrac{100}{x^4}; x = 1.06$

78. $\dfrac{50}{1 - \sqrt{x}}; x = 0.1024$

10 Chapter Summary

What did you learn?	Explanation and Examples	Review Exercises
10.1 Solve quadratic equations by factoring *(p. 502)*.	1. Write the equation in general form. 2. Factor the left side. 3. Set each factor equal to zero and solve for x.	1–10
Solve quadratic equations by the Square Root Property *(p. 503)*.	1. The equation $u^2 = d$, where $d > 0$, has exactly two solutions: $u = \pm\sqrt{d}$. 2. The equation $u^2 = d$, where $d < 0$, has exactly two solutions: $u = \pm\sqrt{|d|}\,i$.	11–22
Use substitution to solve equations of quadratic form *(p. 506)*.	An equation is said to be of quadratic form if it has the form $au^2 + bu + c = 0$, where u is an algebraic expression. To solve an equation of quadratic form, it helps to make a substitution and rewrite the equation in terms of u.	23–30
10.2 Rewrite quadratic expressions in completed square form *(p. 510)*.	To complete the square for the expression $x^2 + bx$, add $(b/2)^2$, which is the square of half the coefficient of x. Consequently, $$x^2 + bx + \left(\frac{b}{2}\right)^2 = \left(x + \frac{b}{2}\right)^2.$$	31–36
Solve quadratic equations by completing the square *(p. 511)*.	1. Prior to completing the square, the coefficient of the second-degree term must be 1. 2. Preserve the equality by adding the same constant to each side of the equation. 3. Use the Square Root Property to solve the quadratic equation.	37–42
10.3 Use the Quadratic Formula to solve quadratic equations *(p. 518)*.	The solutions of $ax^2 + bx + c = 0$, $a \neq 0$, are given by the Quadratic Formula $$x = \frac{-b \pm \sqrt{b^2 - 4ac}}{2a}.$$	43–48
Determine the types of solutions of quadratic equations using the discriminant *(p. 521)*.	Let a, b, and c be rational numbers such that $a \neq 0$. The discriminant of the quadratic equation $ax^2 + bx + c = 0$ is given by $b^2 - 4ac$, and can be used to classify the solutions of the equation as follows. *Discriminant* — *Solution Type* 1. Perfect square — Two distinct rational solutions 2. Positive nonperfect square — Two distinct irrational solutions 3. Zero — One repeated rational solution 4. Negative number — Two distinct complex solutions	49–56

	What did you learn?	Explanation and Examples	Review Exercises
10.3	Write quadratic equations from solutions of the equations *(p. 523)*.	You can use the Zero-Factor Property in reverse to find a quadratic equation given its solutions.	57–62
10.4	Determine the vertices of parabolas and sketch parabolas *(p. 528)*.	**1.** Determine the vertex and axis of the parabola by completing the square or by using the formula $x = -b/(2a)$. **2.** Plot the vertex, axis, x- and y-intercepts, and a few additional points on the parabola. (Using the symmetry about the axis can reduce the number of points you need to plot.) **3.** Use the fact that the parabola opens upward when $a > 0$ and opens downward when $a < 0$ to complete the sketch.	63–70
	Write an equation of a parabola given the vertex and a point on the graph *(p. 531)*.	To write an equation of a parabola with a vertical axis, use the fact that its standard equation has the form $y = a(x - h)^2 + k$, where (h, k) is the vertex.	71–74
	Use parabolas to solve application problems *(p. 533)*.	You can solve applications involving the path of an object using parabolas.	75, 76
10.5	Use quadratic equations to solve application problems *(p. 536)*.	The following are samples of applications of quadratic equations. **1.** Investment **2.** Interest **3.** Height of a projectile **4.** Geometric dimensions **5.** Work rate **6.** Structural design **7.** Falling object **8.** Reduced rate	77–86
10.6	Use test intervals to solve quadratic inequalities *(p. 545)*.	**1.** For a polynomial expression, find all the real zeros. **2.** Arrange the zeros in increasing order. These numbers are called *critical numbers*. **3.** Use the critical numbers to determine the test intervals. **4.** Choose a representative x-value in each test interval and evaluate the expression at that value. When the value of the expression is negative, the expression has negative values for all x-values in the interval. When the value of the expression is positive, the expression has positive values for all x-values in the interval.	87–96
	Use test intervals to solve rational inequalities *(p. 548)*.	For a rational expression, find all the real zeros and those x-values for which the function is undefined. Then use Steps 2–4 as described above.	97–100
	Use inequalities to solve application problems *(p. 549)*.	Not only can you solve real-life applications using quadratic and rational equations, but you can also solve these types of problems using quadratic and rational inequalities.	101, 102

Review Exercises

Worked-out solutions to odd-numbered exercises at CollegePrepAlgebra.com

10.1

Solving a Quadratic Equation by Factoring In Exercises 1−10, solve the equation by factoring.

1. $x^2 + 12x = 0$
2. $u^2 - 18u = 0$
3. $3y^2 - 27 = 0$
4. $2z^2 - 72 = 0$
5. $4y^2 + 20y + 25 = 0$
6. $x^2 + \frac{8}{3}x + \frac{16}{9} = 0$
7. $2x^2 - 2x - 180 = 0$
8. $9x^2 + 18x - 135 = 0$
9. $2x^2 - 9x - 18 = 0$
10. $3x^2 - 19x + 20 = 0$

Using the Square Root Property In Exercises 11−22, solve the equation by using the Square Root Property.

11. $z^2 = 144$
12. $2x^2 = 98$
13. $y^2 - 12 = 0$
14. $y^2 - 45 = 0$
15. $(x - 16)^2 = 400$
16. $(x + 3)^2 = 900$
17. $z^2 = -121$
18. $u^2 = -225$
19. $y^2 + 50 = 0$
20. $x^2 + 48 = 0$
21. $(y + 4)^2 + 18 = 0$
22. $(x - 2)^2 + 24 = 0$

Solving an Equation of Quadratic Form In Exercises 23−30, solve the equation of quadratic form.

23. $x^4 - 4x^2 - 5 = 0$
24. $x^4 - 10x^2 + 9 = 0$
25. $x - 4\sqrt{x} + 3 = 0$
26. $x - 4\sqrt{x} + 13 = 0$
27. $(x^2 - 2x)^2 - 4(x^2 - 2x) - 5 = 0$
28. $(\sqrt{x} - 2)^2 + 2(\sqrt{x} - 2) - 3 = 0$
29. $x^{2/3} + 3x^{1/3} - 28 = 0$
30. $x^{2/5} + 4x^{1/5} + 3 = 0$

10.2

Constructing a Perfect Square Trinomial In Exercises 31−36, add a term to the expression to make it a perfect square trinomial.

31. $z^2 + 18z + \rule{2em}{0.8em}$
32. $y^2 - 80y + \rule{2em}{0.8em}$
33. $x^2 - 15x + \rule{2em}{0.8em}$
34. $x^2 + 21x + \rule{2em}{0.8em}$
35. $y^2 + \frac{2}{5}y + \rule{2em}{0.8em}$
36. $x^2 - \frac{3}{4}x + \rule{2em}{0.8em}$

Completing the Square In Exercises 37−42, solve the equation by completing the square. Give the solutions in exact form and in decimal form rounded to two decimal places.

37. $x^2 - 6x - 3 = 0$
38. $x^2 + 12x + 6 = 0$
39. $v^2 + 5v + 4 = 0$
40. $u^2 - 5u + 6 = 0$
41. $y^2 - \frac{2}{3}y + 2 = 0$
42. $t^2 + \frac{1}{2}t - 1 = 0$

10.3

Solving a Quadratic Equation In Exercises 43−48, solve the equation by using the Quadratic Formula.

43. $v^2 + v - 42 = 0$
44. $4x^2 + 12x + 9 = 0$
45. $5x^2 - 16x + 2 = 0$
46. $3x^2 + 12x + 4 = 0$
47. $8x^2 - 6x + 2 = 0$
48. $x^2 - 4x + 8 = 0$

Using the Discriminant In Exercises 49−56, use the discriminant to determine the type of solution(s) of the quadratic equation.

49. $x^2 + 4x + 4 = 0$

50. $y^2 - 26y + 169 = 0$

51. $s^2 - s - 20 = 0$

52. $r^2 - 5r - 45 = 0$

53. $4t^2 + 16t + 10 = 0$

54. $8x^2 + 85x - 33 = 0$

55. $v^2 - 6v + 21 = 0$

56. $9y^2 + 1 = 0$

Writing a Quadratic Equation In Exercises 57−62, write a quadratic equation that has the given solutions.

57. $x = 3, x = -7$

58. $x = -2, x = 8$

59. $x = 5 + \sqrt{7}, x = 5 - \sqrt{7}$

60. $x = 2 + \sqrt{2}, x = 2 - \sqrt{2}$

61. $x = 6 + 2i, x = 6 - 2i$

62. $x = 3 + 4i, x = 3 - 4i$

10.4

Finding the Vertex of a Parabola In Exercises 63−66, write the equation of the parabola in standard form, and find the vertex of its graph.

63. $y = x^2 - 8x + 3$

64. $y = 8 - 8x - x^2$

65. $y = 2x^2 - x + 3$

66. $y = 3x^2 + 2x - 6$

Sketching a Parabola In Exercises 67−70, sketch the parabola. Identify the vertex and any x-intercepts.

67. $y = x^2 + 8x$

68. $y = -x^2 + 3x$

69. $f(x) = -x^2 - 2x + 4$

70. $f(x) = x^2 + 3x - 10$

Writing an Equation of a Parabola In Exercises 71−74, write an equation of the parabola that satisfies the conditions.

71. Vertex: $(2, -5)$
    Point on the graph: $(0, 3)$

72. Vertex: $(-4, 0)$
    Point on the graph: $(0, -6)$

73. Vertex: $(5, 0)$
    Point on the graph: $(1, 1)$

74. Vertex: $(-2, 5)$
    Point on the graph: $(-4, 11)$

75. *Path of a Ball* The height y (in feet) of a ball thrown by a child is given by $y = -\frac{1}{10}x^2 + 3x + 6$, where x is the horizontal distance (in feet) from where the ball is thrown.

    (a) How high is the ball when it leaves the child's hand?

    (b) How high is the ball at its maximum height?

    (c) How far from the child does the ball strike the ground?

76. *Path of an Object* You use a fishing rod to cast a lure into the water. The height y (in feet) of the lure is given by

$$y = -\frac{1}{90}x^2 + \frac{1}{5}x + 9,$$

where x is the horizontal distance (in feet) from where the lure is released.

(a) How high is the lure when it is released?

(b) How high is the lure at its maximum height?

(c) How far from its release point does the lure land?

10.5

77. *Selling Price* A car dealer buys a fleet of used cars for a total of $80,000. The dealer regains this investment by selling all but 4 of these cars at a profit of $1000 each. How many cars has the dealer sold, and at what price?

78. *Selling Price* A manager of a computer store buys several computers of the same model for $27,000. The store can regain this investment by selling all but 5 of the computers at a profit of $900 per computer. To do this, how many computers must be sold, and at what price?

79. *Geometry* The length of a rectangle is 12 inches greater than its width. The area of the rectangle is 85 square inches. Find the dimensions of the rectangle.

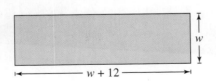

80. *Geometry* The height of a triangle is 3 inches greater than its base. The area of the triangle is 44 square inches. Find the base and height of the triangle.

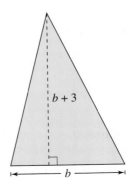

81. *Reduced Rates* A Little League baseball team pays $96 for a block of tickets to a ball game. The block contains three more tickets than the team needs. By inviting 3 more people to attend (and share in the cost), the team lowers the price per ticket by $1.60. How many people are going to the game?

82. *Compound Interest* You want to invest $35,000 for 2 years at an annual interest rate of r (in decimal form). Interest on the account is compounded annually. Find the interest rate when a deposit of $35,000 increases to $40,221.44 over a two-year period.

83. *Geometry* You leave campus to pick up two student council officers for a meeting. You drive a total of 20 miles to pick up the secretary and then the treasurer (see figure). By using a direct route, you are able to drive just 16 miles to return to campus. Find the possible distances from campus to the secretary's location.

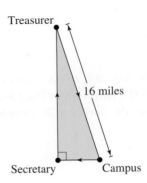

84. *Geometry* A corner lot has an L-shaped sidewalk along its sides. The total length of the sidewalk is 69 feet. By cutting diagonally across the lot, the walking distance is shortened to 51 feet. What are the lengths of the two legs of the sidewalk?

85. *Work-Rate Problem* Working together, two people can complete a task in 10 hours. Working alone, one person takes 2 hours longer than the other. How long would it take each person to do the task alone?

86. *Height* A model rocket is projected straight upward at an initial velocity of 64 feet per second from a height of 192 feet. The height h at any time t is given by

$$h = -16t^2 + 64t + 192$$

where t is the time in seconds.

(a) After how many seconds is the height 256 feet?

(b) After how many seconds does the object hit the ground?

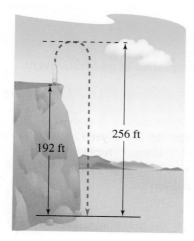

256 ft

192 ft

10.6

Finding Critical Numbers In Exercises 87–90, find the critical numbers.

87. $2x(x + 7)$

88. $x(x - 2) + 4(x - 2)$

89. $x^2 - 6x - 27$ **90.** $2x^2 + 11x + 5$

Solving a Quadratic Inequality In Exercises 91–96, solve the inequality and graph the solution on the real number line.

91. $5x(7 - x) > 0$

92. $-2x(x - 10) \le 0$

93. $16 - (x - 2)^2 \le 0$

94. $(x - 5)^2 - 36 > 0$

95. $2x^2 + 3x - 20 < 0$

96. $-3x^2 + 10x + 8 \ge 0$

landysh/Shutterstock.com

Solving a Rational Inequality In Exercises 97–100, solve the inequality and graph the solution on the real number line.

97. $\dfrac{x + 3}{2x - 7} \ge 0$

98. $\dfrac{3x + 2}{x - 3} > 0$

99. $\dfrac{x + 4}{x - 1} < 0$

100. $\dfrac{2x - 9}{x - 1} \le 0$

101. *Height* A projectile is fired straight upward from ground level with an initial velocity of 312 feet per second. Its height h at any time t is given by

$$h = -16t^2 + 312t$$

where h is measured in feet and t is measured in seconds. During what interval of time will the height of the projectile exceed 1200 feet?

102. *Average Cost* The cost C of producing x notebooks is

$$C = 100{,}000 + 0.9x, \; x > 0.$$

Write the average cost

$$\overline{C} = \frac{C}{x}$$

as a function of x. Then determine how many notebooks must be produced for the average cost per unit to be less than \$2.

Chapter Test

Solutions in English & Spanish and tutorial videos at CollegePrepAlgebra.com

Take this test as you would take a test in class. After you are done, check your work against the answers in the back of the book.

In Exercises 1–6, solve the equation by the specified method.

1. Factoring:

   $x(x - 3) - 10(x - 3) = 0$

2. Factoring:

   $6x^2 - 34x - 12 = 0$

3. Square Root Property:

   $(x - 2)^2 = 0.09$

4. Square Root Property

   $(x + 4)^2 + 100 = 0$

5. Completing the square:

   $2x^2 - 6x + 3 = 0$

6. Completing the square:

   $2y(y - 2) = 7$

In Exercises 7 and 8, solve the equation of quadratic form.

7. $\dfrac{1}{x^2} - \dfrac{6}{x} + 4 = 0$

8. $x^{2/3} - 9x^{1/3} + 8 = 0$

9. Find the discriminant and explain what it means in terms of the type of solutions of the quadratic equation $5x^2 - 12x + 10 = 0$.

10. Write a quadratic equation that has the solutions -7 and -3.

In Exercises 11 and 12, sketch the parabola. Identify the vertex and any x-intercepts.

11. $y = -x^2 + 2x - 4$

12. $y = x^2 - 2x - 15$

In Exercises 13–15, solve the inequality and graph the solution on the real number line.

13. $16 \le (x - 2)^2$

14. $2x(x - 3) < 0$

15. $\dfrac{x + 1}{x - 5} \le 0$

16. The width of a rectangle is 22 feet less than its length. The area of the rectangle is 240 square feet. Find the dimensions of the rectangle.

17. An English club charters a bus trip to a Shakespearean festival. The cost of the bus is $1250. To lower the bus fare per person, the club invites nonmembers to go along. When 10 nonmembers join the trip, the fare per person is decreased by $6.25. How many club members are going on the trip?

18. An object is dropped from a height of 75 feet. Its height h (in feet) at any time t is given by $h = -16t^2 + 75$, where t is measured in seconds. Find the time required for the object to fall to a height of 35 feet.

19. Two buildings are connected by an L-shaped walkway. The total length of the walkway is 155 feet. By cutting diagonally across the grass, the walking distance is shortened to 125 feet. What are the lengths of the two legs of the walkway?

11
Exponential and Logarithmic Functions

11.1 Exponential Functions

11.2 Composite and Inverse Functions

11.3 Logarithmic Functions

11.4 Properties of Logarithms

11.5 Solving Exponential and Logarithmic Equations

11.6 Applications

MASTERY IS WITHIN REACH!

"I failed my first test because, for some reason, I thought just showing up and listening would be enough. I was wrong. Now, I always review my notes, sit close to the front of the class, and ask questions. I try to learn and remember as much as possible in class because I am so busy juggling band and school."

Kayla

See page 587 for suggestions about making the most of class time.

11.1 Exponential Functions

▶ Evaluate and graph exponential functions.
▶ Evaluate the natural base e and graph natural exponential functions.
▶ Use exponential functions to solve application problems.

Exponential Functions

Whereas polynomial and rational functions have terms with variable bases and constant exponents, **exponential functions** have terms with constant bases and variable exponents.

Polynomial or Rational Function

Constant Exponents
$$f(x) = x^2, \quad f(x) = x^{-3}$$
Variable Bases

Exponential Function

Variable Exponents
$$f(x) = 2^x, \quad f(x) = 3^{-x}$$
Constant Bases

Definition of Exponential Function

The **exponential function f with base a** is denoted by $f(x) = a^x$, where $a > 0$, $a \neq 1$, and x is any real number.

Rules of Exponential Functions

Let a be a positive real number, and let x and y be real numbers, variables, or algebraic expressions.

1. $a^x \cdot a^y = a^{x+y}$ Product rule   **2.** $\dfrac{a^x}{a^y} = a^{x-y}$ Quotient rule

3. $(a^x)^y = a^{xy}$ Power-to-power rule   **4.** $a^{-x} = \dfrac{1}{a^x} = \left(\dfrac{1}{a}\right)^x$ Negative exponent rule

EXAMPLE 1   **Using a Calculator**

To evaluate exponential functions with a calculator, you can use the exponential key $\boxed{y^x}$ or $\boxed{\wedge}$. For example, to evaluate $3^{-1.3}$, you can use the following keystrokes.

Keystrokes	Display	
3 $\boxed{y^x}$ 1.3 $\boxed{+/-}$ $\boxed{=}$	0.239741	Scientific
3 $\boxed{\wedge}$ $\boxed{(}$ $\boxed{(-)}$ 1.3 $\boxed{)}$ $\boxed{\text{ENTER}}$	0.239741	Graphing

Exercises Within Reach ®

Using a Calculator In Exercises 1–6, use a calculator to evaluate the expression. (Round your answer to three decimal places.)

1. $2^{-2.3}$ **2.** $5^{1.4}$ **3.** $5^{\sqrt{2}}$

4. $4^{-\pi}$ **5.** $6^{1/3}$ **6.** $6^{-1/3}$

EXAMPLE 2   **Evaluating Exponential Functions**

Evaluate each function. Use a calculator only if it is necessary or more efficient.

Function		*Values*
a.	$f(x) = 2^x$	$x = 3, x = -4, x = \pi$
b.	$g(x) = 12^x$	$x = 3, x = -0.1, x = \frac{5}{7}$
c.	$h(x) = (1.04)^{2x}$	$x = 0, x = -2, x = \sqrt{2}$

SOLUTION

Evaluation	*Comment*
a. $f(3) = 2^3 = 8$	Calculator is not necessary.
$f(-4) = 2^{-4} = \frac{1}{2^4} = \frac{1}{16}$	Calculator is not necessary.
$f(\pi) = 2^\pi \approx 8.825$	Calculator is necessary.
b. $g(3) = 12^3 \approx 1728$	Calculator is more efficient
$g(-0.1) = 12^{-0.1} \approx 0.780$	Calculator is necessary.
$g\left(\frac{5}{7}\right) = 12^{5/7} \approx 5.900$	Calculator is necessary.
c. $h(0) = (1.04)^{2 \cdot 0} = (1.04)^0 = 1$	Calculator is not necessary.
$h(-2) = (1.04)^{2(-2)} \approx 0.855$	Calculator is more efficient.
$h(\sqrt{2}) = (1.04)^{2\sqrt{2}} \approx 1.117$	Calculator is necessary.

Exercises Within Reach ®   Solutions in English & Spanish and tutorial videos at CollegePrepAlgebra.com

Evaluating an Exponential Function **In Exercises 7–16, evaluate the function as indicated. Use a calculator only if it is necessary or more efficient. (Round your answers to three decimal places.)**

7. $f(x) = 3^x$
(a) $x = -2$
(b) $x = 0$
(c) $x = 1$

8. $F(x) = 3^{-x}$
(a) $x = -2$
(b) $x = 0$
(c) $x = 1$

9. $g(x) = 2.2^{-x}$
(a) $x = 1$
(b) $x = 3$
(c) $x = \sqrt{6}$

10. $G(x) = 4.2^x$
(a) $x = -1$
(b) $x = -2$
(c) $x = \sqrt{2}$

11. $f(t) = 500\left(\frac{1}{2}\right)^t$
(a) $t = 0$
(b) $t = 1$
(c) $t = \pi$

12. $g(s) = 1200\left(\frac{2}{3}\right)^s$
(a) $s = 0$
(b) $s = 2$
(c) $s = \sqrt{2}$

13. $f(x) = 1000(1.05)^{2x}$
(a) $x = 0$
(b) $x = 5$
(c) $x = 10$

14. $g(t) = 10,000(1.03)^{4t}$
(a) $t = 1$
(b) $t = 3$
(c) $t = 5.5$

15. $h(x) = \dfrac{5000}{(1.06)^{8x}}$
(a) $x = 5$
(b) $x = 10$
(c) $x = 20$

16. $P(t) = \dfrac{10,000}{(1.01)^{12t}}$
(a) $t = 2$
(b) $t = 10$
(c) $t = 20$

EXAMPLE 3 **Graphing Exponential Functions**

In the same coordinate plane, sketch the graph of each function. Determine the domain and range of each function.

a. $f(x) = 2^x$ **b.** $g(x) = 4^x$

SOLUTION

The table lists some values of each function, and the figure at the left shows the graph of each function. From the graphs, you can see that the domain of each function is the set of all real numbers and that the range of each function is the set of all positive real numbers.

x	-2	-1	0	1	2	3
2^x	$\frac{1}{4}$	$\frac{1}{2}$	1	2	4	8
4^x	$\frac{1}{16}$	$\frac{1}{4}$	1	4	16	64

EXAMPLE 4 **Graphing Exponential Functions**

In the same coordinate plane, sketch the graph of each function.

a. $f(x) = 2^{-x}$ **b.** $g(x) = 4^{-x}$

SOLUTION

The table lists some values of each function, and the figure at the left shows the graph of each function.

x	-3	-2	-1	0	1	2
2^{-x}	8	4	2	1	$\frac{1}{2}$	$\frac{1}{4}$
4^{-x}	64	16	4	1	$\frac{1}{4}$	$\frac{1}{16}$

Exercises Within Reach ®

Matching In Exercises 17–20, match the function with its graph.

17. $f(x) = 2^x$

18. $g(x) = 6^x$

19. $h(x) = 2^{-x}$

20. $k(x) = 6^{-x}$

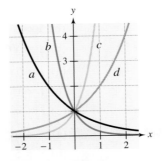

Graphing an Exponential Function In Exercises 21–24, sketch the graph of the function. Determine the domain and range.

21. $f(x) = 3^x$

22. $h(x) = \frac{1}{2}(3^x)$

23. $f(x) = 3^{-x} = \left(\frac{1}{3}\right)^x$

24. $h(x) = \frac{1}{2}(3^{-x})$

Study Tip

An **asymptote** of a graph is a line to which the graph becomes arbitrarily close as $|x|$ or $|y|$ increases without bound. In other words, when a graph has an asymptote, it is possible to move far enough out on the graph so that there is almost no difference between the graph and the asymptote.

Graph of $y = a^x$

- Domain: $(-\infty, \infty)$
- Range: $(0, \infty)$
- Intercept: $(0, 1)$
- Increasing
  (moves up to the right)
- Asymptote: x-axis

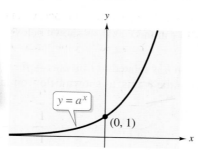

Graph of $y = a^{-x} = \left(\dfrac{1}{a}\right)^x$

- Domain: $(-\infty, \infty)$
- Range: $(0, \infty)$
- Intercept: $(0, 1)$
- Decreasing
  (moves down to the right)
- Asymptote: x-axis

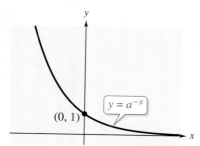

EXAMPLE 5   **Transformations of Graphs**

Use transformations to analyze and sketch the graph of each function.

a. $g(x) = 3^{x + 1}$

b. $h(x) = 3^x - 2$

SOLUTION

Consider the function $f(x) = 3^x$.

a. The function g is related to f by $g(x) = f(x + 1)$. To sketch the graph of g, shift the graph of f one unit to the left, as shown below on the left. Note that the y-intercept of g is $(0, 3)$.

b. The function h is related to f by $h(x) = f(x) - 2$. To sketch the graph of g, shift the graph of f two units downward, as shown below on the right. Note that the y-intercept of h is $(0, -1)$ and the horizontal asymptote is $y = -2$.

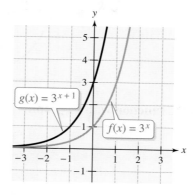

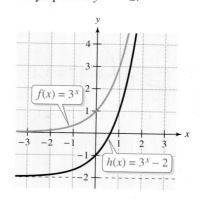

Exercises Within Reach ®

Solutions in English & Spanish and tutorial videos at CollegePrepAlgebra.com

Transformations of Graphs **In Exercises 25−32, use transformations to analyze and sketch the graph of the function.**

25. $g(x) = 3^x - 1$

26. $g(x) = 3^x + 1$

27. $g(x) = 5^{x - 1}$

28. $g(x) = 5^{x + 3}$

29. $g(x) = 2^x + 3$

30. $g(x) = 2^{x + 3}$

31. $g(x) = 2^{x - 4}$

32. $g(x) = 2^x - 4$

EXAMPLE 6 **Reflections of Graphs**

Use transformations to analyze and sketch the graph of each function.

a. $g(x) = -3^x$ **b.** $h(x) = 3^{-x}$

SOLUTION

Consider the function $f(x) = 3^x$.

a. The function g is related to f by $g(x) = -f(x)$. To sketch the graph of g, reflect the graph of f in the x-axis, as shown below on the left. Note that the y-intercept of g is $(0, -1)$.

b. The function h is related to f by $h(x) = f(-x)$. To sketch the graph of h, reflect the graph of f in the y-axis, as shown below on the right.

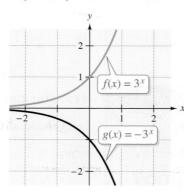
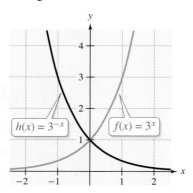

Exercises Within Reach ®

Solutions in English & Spanish and tutorial videos at CollegePrepAlgebra.com

Matching In Exercises 33−36, **match the function with its graph.**

(a)

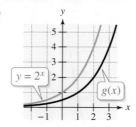

(b)

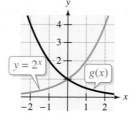

(c)

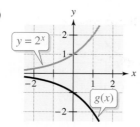

(d)

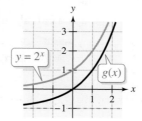

33. $g(x) = 2^{-x}$ **34.** $g(x) = 2^x - 1$ **35.** $g(x) = 2^{x-1}$ **36.** $g(x) = -2^x$

Reflections of Graphs In Exercises 37−40, use transformations to analyze and sketch the graph of the function.

37. $g(x) = -4^x$ **38.** $g(x) = 4^{-x}$ **39.** $g(x) = 5^{-x}$ **40.** $g(x) = -5^x$

The Natural Exponential Function

So far, integers or rational numbers have been used as bases of exponential functions. In many applications of exponential functions, the convenient choice for a base is the following irrational number, denoted by the letter "e."

$$e \approx 2.71828\ldots \qquad \text{Natural base}$$

This number is called the **natural base**. The function

$$f(x) = e^x \qquad \text{Natural exponential function}$$

is called the **natural exponential function**. To evaluate the natural exponential function, you need a calculator, preferably one with a natural exponential key $\boxed{e^x}$. Here are some examples.

Value	Keystrokes	Display	
e^2	2 $\boxed{e^x}$	7.3890561	Scientific
e^2	$\boxed{e^x}$ 2 $\boxed{)}$ $\boxed{\text{ENTER}}$	7.3890561	Graphing
e^{-3}	3 $\boxed{+/-}$ $\boxed{e^x}$	0.0497871	Scientific
e^{-3}	$\boxed{e^x}$ $\boxed{(-)}$ 3 $\boxed{)}$ $\boxed{\text{ENTER}}$	0.0497871	Graphing

EXAMPLE 7   Graphing the Natural Exponential Function

When evaluating the natural exponential function, remember that e is the constant number 2.71828 . . . and x is a variable. After evaluating this function at several values, you can sketch its graph, as shown at the left.

x	-2	-1.5	-1	-0.5	0	0.5	1	1.5
$f(x) = e^x$	0.135	0.223	0.368	0.607	1.000	1.649	2.718	4.482

From the graph, notice the following characteristics of the natural exponential function.

- Domain: $(-\infty, \infty)$
- Range: $(0, \infty)$
- Intercept: $(0, 1)$
- Increasing (moves up to the right)
- Asymptote: x-axis

Exercises Within Reach®

Solutions in English & Spanish and tutorial videos at CollegePrepAlgebra.com

Using a Calculator In Exercises 41−44, use a calculator to evaluate the expression. (**Round your answer to three decimal places.**)

41. $e^{1/3}$

42. $e^{-1/3}$

43. $3(2e^{1/2})^3$

44. $(9e^2)^{3/2}$

Evaluating an Exponential Function In Exercises 45 and 46, use a calculator to evaluate the function as indicated. (**Round your answers to three decimal places.**)

45. $g(x) = 10e^{-0.5x}$
    (a) $x = -4$
    (b) $x = 4$
    (c) $x = 8$

46. $A(t) = 200e^{0.1t}$
    (a) $t = 10$
    (b) $t = 20$
    (c) $t = 40$

Transformations of Graphs In Exercises 47−50, use transformations to analyze and sketch the graph of the function.

47. $g(x) = -e^x$

48. $g(x) = e^{-x}$

49. $g(x) = e^x + 1$

50. $g(x) = e^{x+1}$

Applications

Application **EXAMPLE 8** **Radioactive Decay**

A particular radioactive element has a half-life of 25 years. For an initial mass of 10 grams, the mass y (in grams) that remains after t years is given by

$$y = 10\left(\frac{1}{2}\right)^{t/25}, \quad t \geq 0$$

How much of the initial mass remains after 120 years?

SOLUTION

When t = 120, the mass is given by

$$y = 10\left(\frac{1}{2}\right)^{120/25} \qquad \text{Substitute 120 for } t.$$

$$= 10\left(\frac{1}{2}\right)^{4.8} \qquad \text{Simplify.}$$

$$\approx 0.359. \qquad \text{Use a calculator.}$$

So, after 120 years, the mass has decayed from an initial amount of 10 grams to only 0.359 gram. Note that the graph of the function shows the 25-year half-life. That is, after 25 years the mass is 5 grams (half of the original), after another 25 years the mass is 2.5 grams, and so on.

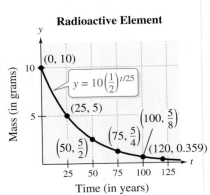

Radioactive Element

$y = 10\left(\frac{1}{2}\right)^{t/25}$

Points: (0, 10), (25, 5), $\left(50, \frac{5}{2}\right)$, $\left(75, \frac{5}{4}\right)$, $\left(100, \frac{5}{8}\right)$, (120, 0.359)

Mass (in grams) / Time (in years)

Exercises Within Reach ®

Solutions in English & Spanish and tutorial videos at CollegePrepAlgebra.com

51. *Radioactive Decay* Cesium-137 is a man-made radioactive element with a half-life of 30 years. A large amount of cesium-137 was introduced into the atmosphere by nuclear weapons testing in the 1950s and 1960s. By now, much of that substance has decayed. Most of the cesium-137 currently in the atmosphere is from nuclear accidents like those at Chernobyl in the Ukraine and Fukushima in Japan. Cesium-137 is one of the most dangerous substances released during those events.

After t years, 16 grams of cesium-137 decays to a mass y (in grams) given by

$$y = 16\left(\frac{1}{2}\right)^{t/30}, \quad t \geq 0.$$

How much of the initial mass remains after 80 years?

52. *Radioactive Substance* In July of 1999, an individual bought several leaded containers from a metals recycler and found two of them labeled "radioactive." An investigation showed that the containers, originally obtained from Ohio State University, apparently had been used to store iodine-131 starting in January of 1999. Because iodine-131 has a half-life of only 8 days, no elevated radiation levels were detected. (*Source:* United States Nuclear Regulatory Commission)

Suppose 6 grams of iodine-131 is stored in January. The mass y (in grams) that remains after t days is given by

$$y = 6\left(\frac{1}{2}\right)^{t/8}, t \geq 0.$$

How much of the substance is left in July, after 180 days have passed?

Formulas for Compound Interest

After t years, the balance A in an account with principal P and annual interest rate r (in decimal form) is given by one of the following formulas.

1. For n compoundings per year: $A = P\left(1 + \dfrac{r}{n}\right)^{nt}$

2. For continuous compounding: $A = Pe^{rt}$

Application   **EXAMPLE 9**   **Comparing Three Types of Compounding**

A total of $15,000 is invested at an annual interest rate of 8%. Find the balance after 6 years for each type of compounding.

a. Quarterly   **b.** Monthly   **c.** Continuous

SOLUTION

a. Letting $P = 15,000$, $r = 0.08$, $n = 4$, and $t = 6$, the balance after 6 years when compounded quarterly is

$$A = 15,000\left(1 + \frac{0.08}{4}\right)^{4(6)}$$

$$\approx \$24,126.56.$$

b. Letting $P = 15,000$, $r = 0.08$, $n = 12$, and $t = 6$, the balance after 6 years when compounded monthly is

$$A = 15,000\left(1 + \frac{0.08}{12}\right)^{12(6)}$$

$$\approx \$24,202.53.$$

c. Letting $P = 15,000$, $r = 0.08$, and $t = 6$, the balance after 6 years when compounded continuously is

$$A = 15,000e^{0.08(6)}$$

$$\approx \$24,241.12.$$

Note that the balance is greater with continuous compounding than with quarterly or monthly compounding.

Exercises Within Reach ®

Solutions in English & Spanish and tutorial videos at CollegePrepAlgebra.com

Compound Interest In Exercises 53−56, complete the table to determine the balance A for P dollars invested at rate r for t years, compounded n times per year.

n	1	4	12	365	Continuous compounding
A					

	Principal	Rate	Time		Principal	Rate	Time
53.	$P = \$100$	$r = 7\%$	$t = 15$ years	**54.**	$P = \$600$	$r = 4\%$	$t = 5$ years
55.	$P = \$2000$	$r = 9.5\%$	$t = 10$ years	**56.**	$P = \$1500$	$r = 6.5\%$	$t = 20$ years

---

Concept Summary: *Graphing Exponential Functions*

What

The **exponential function** f with base a is

$$f(x) = a^x$$

where $a > 0$, $a \neq 1$, and x is any real number.

You can use the graph of $y = a^x$ to sketch the graphs of functions of the form $f(x) = b \pm a^{x+c}$.

EXAMPLE

Sketch the graph of $h(x) = -2^x + 1$.

How

One way to sketch the graph of h is to relate it to the graph of $f(x) = 2^x$.

EXAMPLE

The function h is related to f by $h(x) = -f(x) + 1$. To sketch the graph of h, reflect the graph of f in the x-axis and shift it one unit upward.

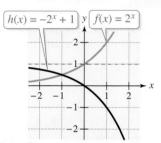

Why

Many real-life situations are modeled by exponential functions. Two such situations are radioactive decay and compound interest.

Exercises Within Reach ®

Worked-out solutions to odd-numbered exercises at CollegePrepAlgebra.com

Concept Summary Check

57. *Describing a Transformation* Describe how to transform the graph of $f(x) = 2^x$ to obtain the graph of $g(x) = -2^x$.

58. *Describing a Transformation* Describe how to transform the graph of $g(x) = -2^x$ to obtain the graph of $h(x) = -2^x + 1$.

59. *Analyzing Behavior* How does a reflection in the x-axis affect the behavior of an exponential function?

60. *Analyzing the Asymptote* How does a shift upward affect the asymptote of an exponential function?

Extra Practice

Using Rules of Exponential Functions In Exercises 61−68, simplify the expression.

61. $3^x \cdot 3^{x+2}$

62. $e^{3x} \cdot e^{-x}$

63. $3(e^x)^{-2}$

64. $4(e^{2x})^{-1}$

65. $\dfrac{e^{x+2}}{e^x}$

66. $\dfrac{3^{2x+3}}{3^{x+1}}$

67. $\sqrt[3]{-8e^{3x}}$

68. $\sqrt{4e^{6x}}$

Compound Interest In Exercises 69 and 70, complete the table to determine the principal P that will yield a balance of A dollars when invested at rate r for t years, compounded n times per year.

	Balance	Rate	Time
69.	$A = \$5000$	$r = 7\%$	$t = 10$ years
70.	$A = \$100,000$	$r = 9\%$	$t = 20$ years

n	1	4	12	365	Continuous compounding
P					

71. *Demand* The daily demand x and the price p for a collectible are related by $p = 25 - 0.4e^{0.02x}$. Find the prices for demands of (a) $x = 100$ units and (b) $x = 125$ units.

72. *Compound Interest* A sum of \$5000 is invested at an annual interest rate of 6%, compounded monthly. Find the balance in the account after 5 years.

73. *Property Value* The value of a piece of property doubles every 15 years. You buy the property for \$64,000. Its value t years after the date of purchase should be $V(t) = 64{,}000(2)^{t/15}$. Use the model to approximate the values of the property (a) 5 years and (b) 20 years after its purchase.

74. *Depreciation* The value of a car that originally cost $16,000 depreciates so that each year it is worth $\frac{3}{4}$ of its value from the previous year. Find a model for $V(t)$, the value of the car after t years. Sketch a graph of the model, and determine the value of the car 2 years and 4 years after its purchase.

75. *Savings Plan* You decide to start saving pennies according to the following pattern. You save 1 penny the first day, 2 pennies the second day, 4 the third day, 8 the fourth day, and so on. Each day you save twice the number of pennies you saved on the previous day. Write an exponential function that models this problem. How many pennies do you save on the thirtieth day?

76. *Exploration* Consider the function
$$f(x) = \left(1 + \frac{1}{x}\right)^x.$$

(a) Use a calculator to complete the table.

x	1	10	100	1000	10,000
$f(x)$					

(b) Use the table to sketch the graph of f. Does the graph appear to approach a horizontal asymptote?

(c) From parts (a) and (b), what conclusions can you make about the value of f as x gets larger and larger?

Explaining Concepts

77. *Reasoning* Explain why $y = 1^x$ is not an exponential function.

78. *Structure* Compare the graphs of $f(x) = 3^x$ and $g(x) = \left(\frac{1}{3}\right)^x$.

79. *Number Sense* Does e equal $\dfrac{271,801}{99,990}$? Explain.

80. *Analyzing Graphs* Consider the graphs of the functions $f(x) = 3^x$, $g(x) = 4^x$, and $h(x) = 4^{-x}$.

(a) What point do the graphs of f, g, and h have in common?

(b) Describe the asymptote for each graph.

(c) State whether each graph increases or decreases as x increases.

(d) Compare the graphs of f and g.

81. *Reasoning* Use the characteristics of the exponential function with base 2 to explain why $2^{\sqrt{2}}$ is greater than 2 but less than 4.

82. *Reasoning* Identify the graphs of $y_1 = e^{0.2x}$, $y_2 = e^{0.5x}$, and $y_3 = e^x$ in the figure. Describe the effect on the graph of $y = e^{kx}$ when $k > 0$ is changed.

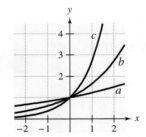

83. *Reasoning* Consider functions of the form $f(x) = k^x$, where k is positive. Describe the real values of k for which the values of f will *increase*, *decrease*, and *remain constant* as x increases.

84. *Repeated Reasoning* Look back at your answers to the compound interest problems in Exercises 53−56. In terms of the interest earned, would you say that the difference between quarterly compounding and annual compounding is greater than the difference between *hourly* compounding and daily compounding? Explain.

Cumulative Review

In Exercises 85 and 86, find the domain of the function.

85. $g(s) = \sqrt{s - 4}$

86. $h(t) = \dfrac{\sqrt{t^2 - 1}}{t - 2}$

In Exercises 87 and 88, sketch the graph of the equation. Use the Vertical Line Test to determine whether y is a function of x.

87. $y^2 = x - 1$

88. $x = y^4 + 1$

11.2 Composite and Inverse Functions

▶ Form composite functions and find the domains of composite functions.

▶ Find inverse functions algebraically.

▶ Compare the graph of a function with the graph of its inverse.

Composite Functions

Definition of Composition of Two Functions

The **composition** of the functions f and g is given by $(f \circ g)(x) = f(g(x))$. The domain of the **composite function** $(f \circ g)$ is the set of all x in the domain of g such that $g(x)$ is in the domain of f.

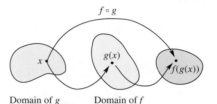

Domain of g Domain of f

Study Tip

A composite function can be viewed as a function within a function, where the composition

$(f \circ g)(x) = f(g(x))$

has f as the "outer" function and g as the "inner" function. This is reversed in the composition

$(g \circ f)(x) = g(f(x))$.

EXAMPLE 1   **Finding the Composition of Two Functions**

Given $f(x) = 2x + 4$ and $g(x) = 3x - 1$, find the composition of f with g. Then evaluate the composite function when $x = 1$ and when $x = -3$.

SOLUTION

$$(f \circ g)(x) = f(g(x)) \qquad \text{Definition of } f \circ g$$

$$= f(3x - 1) \qquad g(x) = 3x - 1 \text{ is the inner function.}$$

$$= 2(3x - 1) + 4 \qquad \text{Input } 3x - 1 \text{ into the outer function } f.$$

$$= 6x - 2 + 4 \qquad \text{Distributive Property}$$

$$= 6x + 2 \qquad \text{Simplify.}$$

When $x = 1$, the value of this composite function is

$$(f \circ g)(1) = 6(1) + 2 = 8.$$

When $x = -3$, the value of this composite function is

$$(f \circ g)(-3) = 6(-3) + 2 = -16.$$

Exercises Within Reach ®   Solutions in English & Spanish and tutorial videos at CollegePrepAlgebra.com

Finding the Composition of Two Functions In Exercises 1 and 2, find the compositions.

1. $f(x) = 2x + 3$, $g(x) = x - 6$

    (a) $(f \circ g)(x)$ (b) $(g \circ f)(x)$

    (c) $(f \circ g)(4)$ (d) $(g \circ f)(7)$

2. $f(x) = x - 5$, $g(x) = 3x + 2$

    (a) $(f \circ g)(x)$ (b) $(g \circ f)(x)$

    (c) $(f \circ g)(3)$ (d) $(g \circ f)(3)$

EXAMPLE 2   **Comparing the Compositions of Functions**

Given $f(x) = 2x - 3$ and $g(x) = x^2 + 1$, find each composition.

a. $(f \circ g)(x)$ **b.** $(g \circ f)(x)$

SOLUTION

a. $(f \circ g)(x) = f(g(x))$ Definition of $f \circ g$

$\qquad = f(x^2 + 1)$ $g(x) = x^2 + 1$ is the inner function.

$\qquad = 2(x^2 + 1) - 3$ Input $x^2 + 1$ into the outer function f.

$\qquad = 2x^2 + 2 - 3$ Distributive Property

$\qquad = 2x^2 - 1$ Simplify.

b. $(g \circ f)(x) = g(f(x))$ Definition of $g \circ f$

$\qquad = g(2x - 3)$ $f(x) = 2x - 3$ is the inner function.

$\qquad = (2x - 3)^2 + 1$ Input $2x - 3$ into the outer function g.

$\qquad = 4x^2 - 12x + 9 + 1$ Expand.

$\qquad = 4x^2 - 12x + 10$ Simplify.

Note that $(f \circ g)(x) \neq (g \circ f)(x)$

EXAMPLE 3   **Finding the Domain of a Composite Function**

Find the domain of the composition of f with g when $f(x) = x^2$ and $g(x) = \sqrt{x}$.

SOLUTION

$(f \circ g)(x) = f(g(x))$ Definition of $f \circ g$

$\qquad = f(\sqrt{x})$ $g(x) = \sqrt{x}$ is the inner function.

$\qquad = (\sqrt{x})^2$ Input $\sqrt{x}$ into the outer function f.

$\qquad = x, \; x \geq 0$ Domain of $f \circ g$ is all $x \geq 0$.

The domain of the inner function $g(x) = \sqrt{x}$ is the set of all nonnegative real numbers. The simplified form of $f \circ g$ has no restriction on this set of numbers. So, the restriction $x \geq 0$ must be added to the composition of this function. The domain of $f \circ g$ is the set of all nonnegative real numbers.

Study Tip

To determine the domain of a composite function, first write the composite function in simplest form. Then use the fact that its domain either is equal to or is a restriction of the domain of the "inner" function.

Exercises Within Reach ®   Solutions in English & Spanish and tutorial videos at CollegePrepAlgebra.com

Finding the Composition of Two Functions In Exercises 3 and 4, find the compositions.

3. $f(x) = x^2 + 3, \; g(x) = x + 2$

   (a) $(f \circ g)(x)$ (b) $(g \circ f)(x)$

   (c) $(f \circ g)(2)$ (d) $(g \circ f)(-3)$

4. $f(x) = 2x + 1, \; g(x) = x^2 - 5$

   (a) $(f \circ g)(x)$ (b) $(g \circ f)(x)$

   (c) $(f \circ g)(-1)$ (d) $(g \circ f)(3)$

Finding the Domains of Composite Functions In Exercises 5 and 6, find the compositions
(a) $f \circ g$ and (b) $g \circ f$. Then find the domain of each composition.

5. $f(x) = \sqrt{x + 2}$

   $g(x) = x - 4$

6. $f(x) = \sqrt{x - 5}$

   $g(x) = x + 3$

Inverse Functions

Let f be a function from the set $A = \{1, 2, 3, 4\}$ to the set $B = \{3, 4, 5, 6\}$ such that

$$f(x) = x + 2: \quad \{(1, 3), (2, 4), (3, 5), (4, 6)\}.$$

By interchanging the first and second coordinates of each of these ordered pairs, you can form another function that is called the **inverse function** of f, denoted by f^{-1}. It is a function from the set B to the set A, and can be written as follows.

$$f^{-1}(x) = x - 2: \quad \{(3, 1), (4, 2), (5, 3), (6, 4)\}.$$

Interchanging the ordered pairs of a function f produces a function only when f is one-to-one. A function f is **one-to-one** when each value of the dependent variable corresponds to exactly one value of the independent variable.

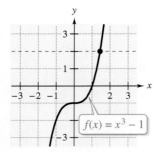

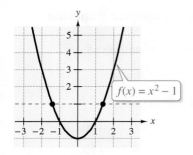

> ### Horizontal Line Test for Inverse Functions
>
> A function f has an inverse function f^{-1} if and only if f is one-to-one. Graphically, a function f has an inverse function f^{-1} if and only if no *horizontal* line intersects the graph of f at more than one point.

EXAMPLE 4 **Applying the Horizontal Line Test**

Use the Horizontal Line Test to determine whether the function is one-to-one and so has an inverse function.

a. The graph of the function $f(x) = x^3 - 1$ is shown at the top left. Because no horizontal line intersects the graph of f at more than one point, you can conclude that f is a one-to-one function and *does* have an inverse function.

b. The graph of the function $f(x) = x^2 - 1$ is shown at the bottom left. Because it is possible to find a horizontal line that intersects the graph of f at more than one point, you can conclude that f *is not* a one-to-one function and *does not* have an inverse function.

> ### Definition of Inverse Function
>
> Let f and g be two functions such that
>
> $$f(g(x)) = x \quad \text{for every } x \text{ in the domain of } g$$
>
> and
>
> $$g(f(x)) = x \quad \text{for every } x \text{ in the domain of } f.$$
>
> The function g is called the **inverse function** of the function f, and is denoted by f^{-1} (read "f-inverse"). So, $f(f^{-1}(x)) = x$ and $f^{-1}(f(x)) = x$. The domain of f must be equal to the range of f^{-1}, and vice versa.

Exercises Within Reach ®   Solutions in English & Spanish and tutorial videos at CollegePrepAlgebra.com

Applying the Horizontal Line Test In Exercises 7–12, sketch the graph of the function. Then use the Horizontal Line Test to determine whether the function is one-to-one and so has an inverse function.

7. $f(x) = x^2 - 2$

8. $f(x) = \frac{1}{5}x$

9. $f(x) = x^2, \ x \geq 0$

10. $f(x) = \sqrt{-x}$

11. $g(x) = \sqrt{25 - x^2}$

12. $g(x) = |x - 4|$

Study Tip

You can graph a function and use the Horizontal Line Test to see whether the function is one-to-one before trying to find its inverse function.

Finding an Inverse Function Algebraically

1. In the equation for $f(x)$, replace $f(x)$ with y.

2. Interchange x and y.

3. Solve the new equation for y. (If the new equation does not represent y as a function of x, the function f does not have an inverse function.)

4. Replace y with $f^{-1}(x)$.

5. Verify that f and f^{-1} are inverse functions of each other by showing that $f(f^{-1}(x)) = x = f^{-1}(f(x))$.

EXAMPLE 5   **Finding Inverse Functions**

a.

$f(x) = 2x + 3$	Original function
$y = 2x + 3$	Replace $f(x)$ with y.
$x = 2y + 3$	Interchange x and y.
$y = \dfrac{x - 3}{2}$	Solve for y.
$f^{-1}(x) = \dfrac{x - 3}{2}$	Replace y with $f^{-1}(x)$.

You can verify that $f(f^{-1}(x)) = x = f^{-1}(f(x))$, as follows.

$$f(f^{-1}(x)) = f\left(\frac{x - 3}{2}\right) = 2\left(\frac{x - 3}{2}\right) + 3 = (x - 3) + 3 = x$$

$$f^{-1}(f(x)) = f^{-1}(2x + 3) = \frac{(2x + 3) - 3}{2} = \frac{2x}{2} = x$$

b.

$f(x) = x^3 + 3$	Original function
$y = x^3 + 3$	Replace $f(x)$ with y.
$x = y^3 + 3$	Interchange x and y.
$y = \sqrt[3]{x - 3}$	Solve for y.
$f^{-1}(x) = \sqrt[3]{x - 3}$	Replace y with $f^{-1}(x)$.

You can verify that $f(f^{-1}(x)) = x = f^{-1}(f(x))$, as follows.

$$f(f^{-1}(x)) = f\left(\sqrt[3]{x - 3}\right) = \left(\sqrt[3]{x - 3}\right)^3 + 3 = (x - 3) + 3 = x$$

$$f^{-1}(f(x)) = f^{-1}(x^3 + 3) = \sqrt[3]{(x^3 + 3) - 3} = \sqrt[3]{x^3} = x$$

Exercises Within Reach ®

Solutions in English & Spanish and tutorial videos at CollegePrepAlgebra.com

Finding an Inverse Function In Exercises 13−16, **find the inverse function of f.**

13. $f(x) = 3 - 4x$ **14.** $f(t) = 6t + 1$ **15.** $f(t) = t^3 - 1$ **16.** $f(t) = t^5 + 8$

Graphs of Inverse Functions

The graphs of f and f^{-1} are related to each other in the following way. If the point (a, b) lies on the graph of f, then the point (b, a) must lie on the graph of f^{-1}, and vice versa. This means that the graph of f^{-1} is a reflection of the graph of f in the line $y = x$, as shown in the figure below.

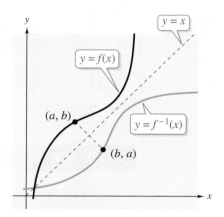

EXAMPLE 6   The Graphs of f and f^{-1}

Sketch the graphs of the inverse functions $f(x) = 2x - 3$ and $f^{-1}(x) = \frac{1}{2}(x + 3)$ on the same rectangular coordinate system, and show that the graphs are reflections of each other in the line $y = x$.

SOLUTION

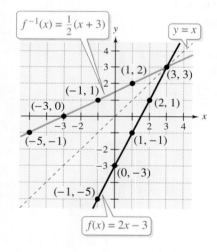

The graphs of f and f^{-1} are shown at the left. Visually, it appears that the graphs are reflections of each other. You can verify this reflective property by testing a few points on each graph. Note in the following list that if the point (a, b) is on the graph of f, then the point (b, a) is on the graph of f^{-1}.

$f(x) = 2x - 3$	$f^{-1}(x) = \frac{1}{2}(x + 3)$
$(-1, -5)$	$(-5, -1)$
$(0, -3)$	$(-3, 0)$
$(1, -1)$	$(-1, 1)$
$(2, 1)$	$(1, 2)$
$(3, 3)$	$(3, 3)$

Exercises Within Reach ®   Solutions in English & Spanish and tutorial videos at CollegePrepAlgebra.com

The Graphs of f and f^{-1} In Exercises 17–22, sketch the graphs of f and f^{-1} on the same rectangular coordinate system. Show that the graphs are reflections of each other in the line $y = x$.

17. $f(x) = x + 4$

$\quad f^{-1}(x) = x - 4$

18. $f(x) = x - 7$

$\quad f^{-1}(x) = x + 7$

19. $f(x) = 3x - 1$

$\quad f^{-1}(x) = \frac{1}{3}(x + 1)$

20. $f(x) = 5 - 4x$

$\quad f^{-1}(x) = -\frac{1}{4}(x - 5)$

21. $f(x) = x^2 - 1, \ x \geq 0$

$\quad f^{-1}(x) = \sqrt{x + 1}$

22. $f(x) = (x + 2)^2, \ x \geq -2$

$\quad f^{-1}(x) = \sqrt{x} - 2$

EXAMPLE 7   **Verifying Inverse Functions Graphically**

Graphically verify that f and g are inverse functions of each other.

$$f(x) = x^2, \ x \geq 0 \quad \text{and} \quad g(x) = \sqrt{x}$$

SOLUTION

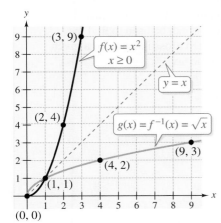

You can graphically verify that f and g are inverse functions of each other by graphing the functions on the same rectangular coordinate system, as shown at the left. Visually, it appears that the graphs are reflections of each other in the line $y = x$. You can verify this reflective property by testing a few points on each graph. Note in the following list that if the point (a, b) is on the graph of f, then the point (b, a) is on the graph of g.

$f(x) = x^2, \ x \geq 0$	$g(x) = f^{-1}(x) = \sqrt{x}$
$(0, 0)$	$(0, 0)$
$(1, 1)$	$(1, 1)$
$(2, 4)$	$(4, 2)$
$(3, 9)$	$(9, 3)$

So, f and g are inverse functions of each other.

Exercises Within Reach ®

Solutions in English & Spanish and tutorial videos at CollegePrepAlgebra.com

Matching **In Exercises 23−26, match the graph with the graph of its inverse function.**

(a) 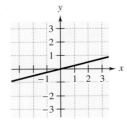 (b) 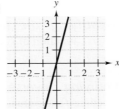 (c) 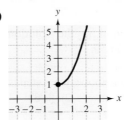 (d)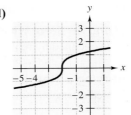

23. 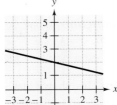 24. 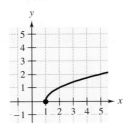 25. 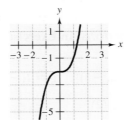 26.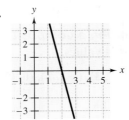

Verifying Inverse Functions Graphically **In Exercises 27−30, sketch the graphs of f and g on the same rectangular coordinate system. Plot the indicated points on the graph of f. Then plot the reflections of these points in the line $y = x$ to verify that f and g are inverse functions.**

27. $f(x) = \frac{1}{3}x$

    $g(x) = 3x$

28. $f(x) = \frac{1}{5}x - 1$

    $g(x) = 5x + 5$

29. $f(x) = \sqrt{x - 4}$

    $g(x) = x^2 + 4, \ x \geq 0$

30. $f(x) = \sqrt{4 - x}$

    $g(x) = 4 - x^2, \ x \geq 0$

Concept Summary: *Finding Inverse Functions*

What

You can form functions using one or more other functions by creating **composite functions** and **inverse functions**.

You can think of a composite function as a "function within a function." You can think of inverse functions as functions that "undo" each other.

EXAMPLE

Find the inverse function of

$f(x) = 2x + 6.$

How

Given an equation for $f(x)$, here is how to find the inverse function f^{-1}.

1. Replace $f(x)$ with y.
2. Interchange x and y.
3. Solve for y.
4. Replace y with $f^{-1}(x)$.
5. Verify that f and f^{-1} are inverse functions.

EXAMPLE

$$f(x) = 2x + 6$$
$$y = 2x + 6$$
$$x = 2y + 6$$
$$y = \frac{x - 6}{2}$$
$$f^{-1}(x) = \frac{x - 6}{2}$$

Why

Inverse functions are used in many real-life problems. For instance, the inverse of a function used to find income tax based on adjusted gross income can be used to find the adjusted gross income for a given income tax.

Exercises Within Reach ®

Worked-out solutions to odd-numbered exercises at CollegePrepAlgebra.com

Concept Summary Check

31. *Applying a Step* Which equation in the solution above is obtained by replacing $f(x)$ with y?

32. *Identifying a Step* In the solution above, what step is performed after obtaining the equation $x = 2y + 6$?

33. *Identifying the Inverse Function* What is the inverse function of $f(x) = 2x + 6$?

34. *Verifying Inverse Functions* To show that two functions f and g are inverse functions, you must show that both $f(g(x))$ and $g(f(x))$ are equal to what?

Extra Practice

Evaluating a Composition of Functions In Exercises 35−38, use the functions f and g to find the indicated values.

$f = \{(-2, 3), (-1, 1), (0, 0), (1, -1), (2, -3)\}, \quad g = \{(-3, 1), (-1, -2), (0, 2), (2, 2), (3, 1)\}$

35. $(f \circ g)(-3)$ **36.** $(f \circ g)(2)$ **37.** $(g \circ f)(-2)$ **38.** $(g \circ f)(2)$

Verifying Inverse Functions In Exercises 39 and 40, verify algebraically that the functions f and g are inverse functions of each other.

39. $f(x) = 1 - 2x, \ g(x) = \frac{1}{2}(1 - x)$ **40.** $f(x) = 2x - 1, \ g(x) = \frac{1}{2}(x + 1)$

Finding an Inverse Function In Exercises 41−44, find the inverse function (if it exists).

41. $g(x) = x^2 + 4$ **42.** $h(x) = (4 - x)^2$ **43.** $h(x) = \sqrt{x}$ **44.** $h(x) = \sqrt{x + 5}$

Restricting the Domain In Exercises 45 and 46, restrict the domain of f so that f is a one-to-one function. Then find the inverse function f^{-1} and state its domain. (Note: There is more than one correct answer.)

45. $f(x) = (x - 2)^2$ **46.** $f(x) = 9 - x^2$

47. *Ripples* You are standing on a bridge over a calm pond and drop a pebble, causing ripples of concentric circles in the water. The radius (in feet) of the outermost ripple is given by $r(t) = 0.6t$, where t is time in seconds after the pebble hits the water. The area of the circle is given by the function $A(r) = \pi r^2$. Find an equation for the composition $A(r(t))$. Describe the input and output of this composite function. What is the area of the circle after 3 seconds?

48. *Daily Production Cost* The daily cost of producing x units in a manufacturing process is

$$C(x) = 8.5x + 300.$$

The number of units produced in t hours during a day is given by

$$x(t) = 12t, \ 0 \le t \le 8.$$

Find, simplify, and interpret $(C \circ x)(t)$.

49. *Hourly Wage* Your wage is \$9.00 per hour plus \$0.65 for each unit produced per hour. So, your hourly wage y in terms of the number of units produced x is $y = 9 + 0.65x$.

(a) Find the inverse function.

(b) Determine the number of units produced when your hourly wage averages \$14.20.

50. *Federal Income Tax* In 2012, the function $T = 0.15(x - 8700) + 870$ represented the federal income tax owed by a single person whose adjusted gross income x was between \$8700 and \$35,350. (*Source: Internal Revenue Service*)

(a) Find the inverse function.

(b) What does each variable represent in the inverse function?

(c) Use the context of the problem to determine the domain of the inverse function.

(d) Determine the adjusted gross income for a single person who owed \$3315 in federal income taxes in 2012.

Explaining Concepts

True or False? In Exercises 51–54, decide whether the statement is true or false. If true, explain your reasoning. If false, give an example.

51. If the inverse function of f exists, then the y-intercept of f is an x-intercept of f^{-1}. Explain.

52. There exists no function f such that $f = f^{-1}$.

53. If the inverse function of f exists, then the domains of f and f^{-1} are the same.

54. If the inverse function of f exists and the graph of f passes through the point $(2, 2)$, then the graph of f^{-1} also passes through the point $(2, 2)$.

55. *Writing* Describe how to find the inverse of a function given by a set of ordered pairs. Give an example.

56. *Reasoning* Why must a function be one-to-one in order for its inverse to be a function?

Cumulative Review

In Exercises 57–60, identify the transformation of the graph of $f(x) = x^2$.

57. $h(x) = -x^2$

58. $g(x) = (x - 4)^2$

59. $k(x) = (x + 3)^2 - 5$

60. $v(x) = x^2 + 1$

In Exercises 61–64, factor the expression completely.

61. $16 - (y + 2)^2$

62. $2x^3 - 6x$

63. $5 - u + 5u^2 - u^3$

64. $t^2 + 10t + 25$

In Exercises 65–68, graph the equation.

65. $3x - 4y = 6$ **66.** $y = 3 - \frac{1}{2}x$

67. $y = -(x - 2)^2 + 1$

68. $y = x^2 - 6x + 5$

11.3 Logarithmic Functions

▶ Evaluate and graph logarithmic functions.
▶ Evaluate and graph natural logarithmic functions.
▶ Use the change-of-base formula to evaluate logarithms.

Logarithmic Functions

Definition of Logarithmic Function

Let a and x be positive real numbers such that $a \neq 1$. The **logarithm of x with base a** is denoted by $\log_a x$ and is defined as follows.

$$y = \log_a x \quad \text{if and only if} \quad x = a^y$$

The function $f(x) = \log_a x$ is the **logarithmic function with base a.** Note that the inverse function of $f(x) = a^x$ is $f^{-1}(x) = \log_a x$.

EXAMPLE 1   Evaluating Logarithms

Evaluate each logarithm.

a. $\log_3 9$ **b.** $\log_4 2$ **c.** $\log_5 1$ **d.** $\log_{10} \dfrac{1}{10}$ **e.** $\log_3(-1)$

SOLUTION

In each case you should answer the question, "To what power must the base be raised to obtain the given number?"

a. The power to which 3 must be raised to obtain 9 is 2. That is,

$$3^2 = 9 \quad \Longrightarrow \quad \log_3 9 = 2.$$

b. The power to which 4 must be raised to obtain 2 is $\frac{1}{2}$. That is,

$$4^{1/2} = 2 \quad \Longrightarrow \quad \log_4 2 = \frac{1}{2}.$$

c. The power to which 5 must be raised to obtain 1 is 0. That is,

$$5^0 = 1 \quad \Longrightarrow \quad \log_5 1 = 0.$$

d. The power to which 10 must be raised to obtain $\frac{1}{10}$ is -1. That is,

$$10^{-1} = \frac{1}{10} \quad \Longrightarrow \quad \log_{10} \frac{1}{10} = -1.$$

e. There is no power to which 3 can be raised to obtain -1. The reason for this is that for any value of x, 3^x is a positive number. So, $\log_3(-1)$ is undefined.

Study Tip

Study the results of parts (c), (d), and (e) carefully. Each of the logarithms illustrates an important special property of logarithms that you should know.

Exercises Within Reach ®

Solutions in English & Spanish and tutorial videos at CollegePrepAlgebra.com

Evaluating a Logarithm In Exercises 1−12, **evaluate the logarithm. (If not possible, state the reason.)**

1. $\log_2 8$ 2. $\log_3 27$ 3. $\log_9 3$ 4. $\log_{125} 5$

5. $\log_{16} 8$ 6. $\log_{81} 9$ 7. $\log_4 1$ 8. $\log_3 1$

9. $\log_2 \frac{1}{16}$ 10. $\log_3 \frac{1}{9}$ 11. $\log_2(-3)$ 12. $\log_4(-4)$

Properties of Logarithms

Let a and x be positive real numbers such that $a \neq 1$. Then the following properties are true.

1. $\log_a 1 = 0$ because $a^0 = 1$.

2. $\log_a a = 1$ because $a^1 = a$.

3. $\log_a a^x = x$ because $a^x = a^x$.

The logarithmic function with base 10 is called the **common logarithmic function**. On most calculators, this function can be evaluated with the common logarithmic key $\boxed{\text{LOG}}$, as illustrated in the next example.

EXAMPLE 2   **Evaluating Common Logarithms**

Evaluate each logarithm. Use a calculator only if necessary.

a. $\log_{10} 100$ **b.** $\log_{10} 0.01$ **c.** $\log_{10} 5$ **d.** $\log_{10} 2.5$

SOLUTION

a. The power to which 10 must be raised to obtain 100 is 2. That is,

$$10^2 = 100 \implies \log_{10} 100 = 2.$$

b. The power to which 10 must be raised to obtain 0.01 or $\frac{1}{100}$ is -2. That is,

$$10^{-2} = \frac{1}{100} \implies \log_{10} 0.01 = -2.$$

c. There is no simple power to which 10 can be raised to obtain 5, so you should use a calculator to evaluate $\log_{10} 5$.

Keystrokes	Display	
5 $\boxed{\text{LOG}}$	0.69897	Scientific
$\boxed{\text{LOG}}$ 5 $\boxed{)}$ $\boxed{\text{ENTER}}$	0.69897	Graphing

So, rounded to three decimal places, $\log_{10} 5 \approx 0.699$.

d. There is no simple power to which 10 can be raised to obtain 2.5, so you should use a calculator to evaluate $\log_{10} 2.5$.

Keystrokes	Display	
2.5 $\boxed{\text{LOG}}$	0.39794	Scientific
$\boxed{\text{LOG}}$ 2.5 $\boxed{)}$ $\boxed{\text{ENTER}}$	0.39794	Graphing

So, rounded to three decimal places, $\log_{10} 2.5 \approx 0.398$.

Study Tip

Be sure you see that the value of a logarithm can be zero or negative, as in Example 2(b), but you *cannot* take the logarithm of zero or a negative number. This means that the logarithms $\log_{10}(-10)$ and $\log_5 0$ are undefined.

Exercises Within Reach ®

Solutions in English & Spanish and tutorial videos at CollegePrepAlgebra.com

Evaluating a Common Logarithm In Exercises 13−24, evaluate the logarithm. Use a calculator only if necessary.

13. $\log_{10} 1000$

14. $\log_{10} 10,000$

15. $\log_{10} 0.1$

16. $\log_{10} 0.00001$

17. $\log_{10} \frac{1}{10,000}$

18. $\log_{10} \frac{1}{100}$

19. $\log_{10} 42$

20. $\log_{10} 7561$

21. $\log_{10} 0.023$

22. $\log_{10} 0.149$

23. $\log_{10}\left(\sqrt{5} + 3\right)$

24. $\log_{10} \frac{\sqrt{3}}{2}$

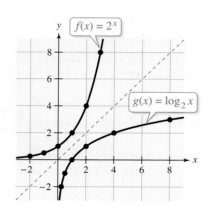

EXAMPLE 3 **Graphing a Logarithmic Function**

On the same rectangular coordinate system, sketch the graph of each function.

a. $f(x) = 2^x$

b. $g(x) = \log_2 x$

SOLUTION

a. Begin by making a table of values for $f(x) = 2^x$.

x	−2	−1	0	1	2	3
$f(x) = 2^x$	$\frac{1}{4}$	$\frac{1}{2}$	1	2	4	8

By plotting these points and connecting them with a smooth curve, you obtain the graph shown at the left.

b. Because $g(x) = \log_2 x$ is the inverse function of $f(x) = 2^x$, the graph of g is obtained by reflecting the graph of f in the line $y = x$, as shown at the left.

Notice from the graph of $g(x) = \log_2 x$, shown above, that the domain of the function is the set of positive numbers and the range is the set of all real numbers. The basic characteristics of the graph of a logarithmic function are summarized in the figure below. Note that the graph has one x-intercept at $(1, 0)$. Also note that $x = 0$ (y-axis) is a vertical asymptote of the graph.

Study Tip

In Example 3, the inverse property of logarithmic functions is used to sketch the graph of $g(x) = \log_2 x$. You could also use a standard point-plotting approach or a graphing calculator.

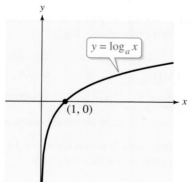

Graph of $y = \log_a x$, $a > 1$

- Domain: $(0, \infty)$
- Range: $(-\infty, \infty)$
- Intercept: $(1, 0)$
- Increasing (moves up to the right)
- Asymptote: y-axis

Exercises Within Reach ®

Solutions in English & Spanish and tutorial videos at CollegePrepAlgebra.com

Graphing a Logarithmic Function In Exercises 25−30, sketch the graph of f. Then use the graph of f to sketch the graph of g.

25. $f(x) = 3^x$

$g(x) = \log_3 x$

27. $f(x) = 6^x$

$g(x) = \log_6 x$

29. $f(x) = \left(\frac{1}{2}\right)^x$

$g(x) = \log_{1/2} x$

26. $f(x) = 4^x$

$g(x) = \log_4 x$

28. $f(x) = 5^x$

$g(x) = \log_5 x$

30. $f(x) = \left(\frac{1}{3}\right)^x$

$g(x) = \log_{1/3} x$

EXAMPLE 4   Sketching the Graphs of Logarithmic Functions

The graph of each function is similar to the graph of $f(x) = \log_{10} x$, as shown in the figures. From the graph, you can determine the domain of the function.

a. Because $g(x) = \log_{10}(x - 1) = f(x - 1)$, the graph of g can be obtained by shifting the graph of f one unit to the right. The vertical asymptote of the graph of g is $x = 1$. The domain of g is $(1, \infty)$.

b. Because $h(x) = 2 + \log_{10} x = 2 + f(x)$, the graph of h can be obtained by shifting the graph of f two units upward. The vertical asymptote of the graph of h is $x = 0$. The domain of h is $(0, \infty)$.

c. Because $k(x) = -\log_{10} x = -f(x)$, the graph of k can be obtained by reflecting the graph of f in the x-axis. The vertical asymptote of the graph of k is $x = 0$. The domain of k is $(0, \infty)$.

d. Because $j(x) = \log_{10}(-x) = f(-x)$, the graph of j can be obtained by reflecting the graph of f in the y-axis. The vertical asymptote of the graph of j is $x = 0$. The domain of j is $(-\infty, 0)$.

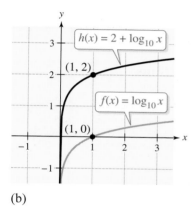

(a)

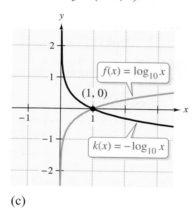

(b)

(c)

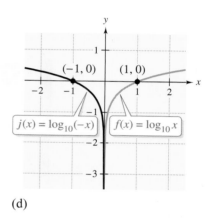

(d)

Exercises Within Reach ®

Matching In Exercises 31−34, match the function with its graph.

(a)

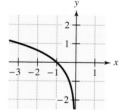

(b)

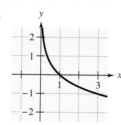

(c)

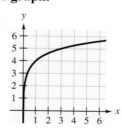

(d)

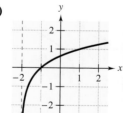

31. $f(x) = 4 + \log_3 x$ **32.** $f(x) = -\log_3 x$ **33.** $f(x) = \log_3(-x)$ **34.** $f(x) = \log_3(x + 2)$

Identifying a Transformation In Exercises 35−40, identify the transformation of the graph of $f(x) = \log_2 x$. Then sketch the graph of h. Determine the vertical asymptote and the domain of the graph of h.

35. $h(x) = 3 + \log_2 x$ **36.** $h(x) = -5 + \log_2 x$

37. $h(x) = \log_2(x - 2)$ **38.** $h(x) = \log_2(x + 5)$

39. $h(x) = \log_2(-x)$ **40.** $h(x) = -\log_2 x$

The Natural Logarithmic Function

The Natural Logarithmic Function

The function defined by

$$f(x) = \log_e x = \ln x$$

where $x > 0$, is called the **natural logarithmic function**.

Properties of Natural Logarithms

Let x be a positive real number. Then the following properties are true.

1. $\ln 1 = 0$ because $e^0 = 1$.
2. $\ln e = 1$ because $e^1 = e$.
3. $\ln e^x = x$ because $e^x = e^x$.

EXAMPLE 5 **Evaluating Natural Logarithms**

Evaluate each expression.

a. $\ln e^2$

b. $\ln \dfrac{1}{e}$

SOLUTION

Using the property that $\ln e^x = x$, you obtain the following.

a. $\ln e^2 = 2$

b. $\ln \dfrac{1}{e} = \ln e^{-1} = -1$

EXAMPLE 6 **Graphing the Natural Logarithmic Function**

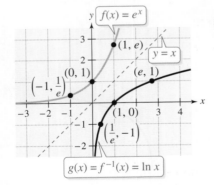

Because the functions $f(x) = e^x$ and $g(x) = \ln x$ are inverse functions of each other, their graphs are reflections of each other in the line $y = x$, as shown in the figure.

From the graph, notice the following characteristics of the natural logarithmic function.

- Domain: $(0, \infty)$
- Range: $(-\infty, \infty)$
- Intercept: $(1, 0)$
- Increasing (moves up to the right)
- Asymptote: y-axis

Exercises Within Reach ®

Solutions in English & Spanish and tutorial videos at CollegePrepAlgebra.com

Evaluating a Natural Logarithm In Exercises 41−44, evaluate the expression.

41. $\ln e^3$

42. $\ln e^6$

43. $\ln \dfrac{1}{e^2}$

44. $\ln \dfrac{1}{e^4}$

Graphing a Natural Logarithmic Function In Exercises 45−48, sketch the graph of the function.

45. $f(x) = 3 + \ln x$

46. $h(x) = 2 + \ln x$

47. $f(x) = -\ln x$

48. $f(x) = \ln(-x)$

Change of Base

> ### Change-of-Base Formula
>
> Let a, b, and x be positive real numbers such that $a \neq 1$ and $b \neq 1$. Then $\log_a x$ is given as follows.
>
> $$\log_a x = \frac{\log_b x}{\log_b a} \quad \text{or} \quad \log_a x = \frac{\ln x}{\ln a}$$

EXAMPLE 7   **Changing Bases to Evaluate Logarithms**

a.   Use *common* logarithms to evaluate $\log_3 5$.

b.   Use *natural* logarithms to evaluate $\log_6 2$.

SOLUTION

Using the change-of-base formula, you can convert to common and natural logarithms by writing

$$\log_3 5 = \frac{\log_{10} 5}{\log_{10} 3} \quad \text{and} \quad \log_6 2 = \frac{\ln 2}{\ln 6}.$$

Now, use the following keystrokes.

a.
Keystrokes	Display	
5 (LOG) (÷) 3 (LOG) (=)	1.4649735	Scientific
(LOG) 5 ()) (÷) (LOG) 3 ()) (ENTER)	1.4649735	Graphing

So, $\log_3 5 \approx 1.465$.

b.
Keystrokes	Display	
2 (LN) (÷) 6 (LN) (=)	0.3868528	Scientific
(LN) 2 ()) (÷) (LN) 6 ()) (ENTER)	0.3868528	Graphing

So, $\log_6 2 \approx 0.387$.

Study Tip

In Example 7(a), $\log_3 5$ could have been evaluated using natural logarithms in the change-of-base formula.

$$\log_3 5 = \frac{\ln 5}{\ln 3} \approx 1.465$$

Notice that you get the same answer whether you use natural logarithms or common logarithms in the change-of-base formula.

Exercises Within Reach ®

Solutions in English & Spanish and tutorial videos at CollegePrepAlgebra.com

Using the Change-of-Base Formula **In Exercises 49−62, use (a) common logarithms and (b) natural logarithms to evaluate the expression. (Round your answer to four decimal places.)**

49. $\log_9 36$

50. $\log_7 411$

51. $\log_5 14$

52. $\log_6 9$

53. $\log_2 0.72$

54. $\log_{12} 0.6$

55. $\log_{15} 1250$

56. $\log_{20} 125$

57. $\log_{1/4} 16$

58. $\log_{1/3} 18$

59. $\log_4 \sqrt{42}$

60. $\log_5 \sqrt{21}$

61. $\log_2(1 + e)$

62. $\log_4(2 + e^3)$

Concept Summary: Evaluating Logarithms

What

The function $f(x) = \log_a x$, where a and x are positive real numbers such that $a \neq 1$, is the **logarithmic function with base a**.

The exponential function with base a and the logarithmic function with base a are inverse functions.

$$f(x) = a^x \qquad f^{-1}(x) = \log_a x$$

EXAMPLE

Evaluate $\log_3 81$.

How

To evaluate a logarithm, you should answer the question, " To what power must the base be raised to obtain the given number?"

EXAMPLE

The power to which 3 must be raised to obtain 81 is 4. That is.

$$3^4 = 81 \implies \log_3 81 = 4.$$

Why

Logarithmic functions are used in many scientific and business applications. For instance, you can use a logarithmic function to determine the length of a home mortgage given any monthly payment.

Exercises Within Reach ®

Worked-out solutions to odd-numbered exercises at CollegePrepAlgebra.com

Concept Summary Check

63. *Inverse Functions* What is the inverse of the function $y = a^x$?

64. *The Base of a Logarithm* What is the base of the logarithm $\log_3 81$?

65. *The Base of a Natural Logarithm* What is the base of the logarithm $\ln 5$?

66. *The Change-of-Base Formula* Write an expression involving natural logarithms that is equivalent to $\log_3 81$.

Extra Practice

Writing Exponential Form **In Exercises 67−74, write the logarithmic equation in exponential form.**

67. $\log_7 49 = 2$

68. $\log_{11} 121 = 2$

69. $\log_2 \frac{1}{32} = -5$

70. $\log_3 \frac{1}{27} = -3$

71. $\log_{36} 6 = \frac{1}{2}$

72. $\log_{64} 4 = \frac{1}{3}$

73. $\log_8 4 = \frac{2}{3}$

74. $\log_{16} 8 = \frac{3}{4}$

Writing Logarithmic Form **In Exercises 75−78, write the exponential equation in logarithmic form.**

75. $6^2 = 36$

76. $3^5 = 243$

77. $8^{2/3} = 4$

78. $81^{3/4} = 27$

Graphing a Logarithmic Function **In Exercises 79−86, sketch the graph of the function. Identify the vertical asymptote.**

79. $h(s) = -2 \log_3 s$

80. $g(x) = -\frac{1}{2} \log_3 x$

81. $f(x) = \log_{10}(10x)$

82. $g(x) = \log_4(4x)$

83. $y = \log_4(x - 1) - 2$

84. $y = \log_4(x - 2) + 3$

85. $y = -\log_3 x + 2$

86. $f(x) = -\log_6(x + 2)$

87. *American Elk* The antler spread a (in inches) and shoulder height h (in inches) of an adult male American elk are related by the model

$$h = 116 \log_{10}(a + 40) - 176.$$

Approximate to one decimal place the shoulder height of a male American elk with an antler spread of 55 inches.

88. *Meteorology* Most tornadoes last less than 1 hour and travel about 20 miles. The speed of the wind S (in miles per hour) near the center of a tornado and the distance d (in miles) the tornado travels are related by the model $S = 93 \log_{10} d + 65$. On March 18, 1925, a large tornado struck portions of Missouri, Illinois, and Indiana, covering a distance of 220 miles. Approximate to one decimal place the speed of the wind near the center of this tornado.

89. *Tractrix* A person walking along a dock (the y-axis) drags a boat by a 10-foot rope (see figure). The boat travels along a path known as a tractrix. The equation of the path is

$$y = 10 \ln\left(\frac{10 + \sqrt{100 - x^2}}{x}\right) - \sqrt{100 - x^2}.$$

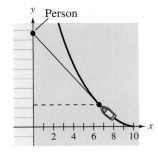

(a) Use a graphing calculator to graph the function. What is the domain of the function?

(b) Identify any asymptotes.

(c) Determine the position of the person when the x-coordinate of the position of the boat is $x = 2$.

90. *Home Mortgage* The model

$$t = 10.042 \ln\left(\frac{x}{x - 1250}\right), \quad x > 1250$$

approximates the length t (in years) of a home mortgage of \$150,000 at 10% interest in terms of the monthly payment x.

(a) Use a graphing calculator to graph the model. Describe the change in the length of the mortgage as the monthly payment increases.

(b) Use the graph in part (a) to approximate the length of the mortgage when the monthly payment is \$1316.35.

(c) Use the result of part (b) to find the total amount paid over the term of the mortgage. What amount of the total is interest costs?

Think About It In Exercises 91–96, answer the question for the function $f(x) = \log_{10} x$. (Do not use a calculator.)

91. What is the domain of f?

92. Find the inverse function of f.

93. Describe the values of $f(x)$ for $1000 \le x \le 10{,}000$.

94. Describe the values of x, given that $f(x)$ is negative.

95. By what amount will x increase, given that $f(x)$ is increased by 1 unit?

96. Find the ratio of a to b when $f(a) = 3 + f(b)$.

Explaining Concepts

True or False? In Exercises 97 and 98, determine whether the statement is true or false. Justify your answer.

97. The statement $8 = 2^3$ is equivalent to $2 = \log_8 3$.

98. The graph of $f(x) = \ln x$ is the reflection of the graph of $f(x) = e^x$ in the x-axis.

99. *Vocabulary* Explain the difference between common logarithms and natural logarithms.

100. *Structure* Explain the relationship between the domain of the graph of $f(x) = \log_5 x$ and the range of the graph of $g(x) = 5^x$.

101. *Think About It* Discuss how shifting or reflecting the graph of a logarithmic function affects the domain and the range.

102. *Number Sense* Explain why $\log_a x$ is defined only when $0 < a < 1$ and $a > 1$.

Cumulative Review

In Exercises 103–106, use the rules of exponents to simplify the expression.

103. $(-m^6 n)(m^4 n^3)$ **104.** $(m^2 n^4)^3 (mn^2)$

105. $\dfrac{36 x^4 y}{8 x y^3}$ **106.** $-\left(\dfrac{3x}{5y}\right)^5$

In Exercises 107–110, perform the indicated operation(s) and simplify. (Assume all variables are positive.)

107. $25\sqrt{3x} - 3\sqrt{12x}$ **108.** $(\sqrt{x} + 3)(\sqrt{x} - 3)$

109. $\sqrt{u}(\sqrt{20} - \sqrt{5})$ **110.** $(2\sqrt{t} + 3)^2$

Mid-Chapter Quiz: Sections 11.1–11.3

Solutions in English & Spanish and tutorial videos at CollegePrepAlgebra.com

Take this quiz as you would take a quiz in class. After you are done, check your work against the answers in the back of the book.

1. Given $f(x) = \left(\frac{4}{3}\right)^x$, find (a) $f(2)$, (b) $f(0)$, (c) $f(-1)$, and (d) $f(1.5)$.

2. Identify the horizontal asymptote of the graph of $g(x) = 3^{x-1}$.

In Exercises 3–6, match the function with its graph.

3. $f(x) = 2^x$

4. $g(x) = -2^x$

5. $h(x) = 2^{-x}$

6. $s(x) = 2^x - 2$

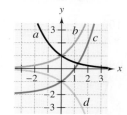

7. Given $f(x) = 2x - 3$ and $g(x) = x^3$, find each composition.

   (a) $(f \circ g)(x)$ (b) $(g \circ f)(x)$ (c) $(f \circ g)(-2)$ (d) $(g \circ f)(4)$

8. Verify algebraically and graphically that $f(x) = 5 - 2x$ and $g(x) = \frac{1}{2}(5 - x)$ are inverse functions of each other.

In Exercises 9 and 10, find the inverse function of f.

9. $f(x) = 10x + 3$ 10. $f(t) = \frac{1}{2}t^3 + 2$

11. Write the logarithmic equation $\log_9 \frac{1}{81} = -2$ in exponential form.

12. Write the exponential equation $2^6 = 64$ in logarithmic form.

13. Evaluate $\log_5 125$ without a calculator.

In Exercises 14 and 15, sketch the graph of the function. Identify the vertical asymptote.

14. $f(t) = \ln(t + 3)$ 15. $h(x) = 1 + \ln x$

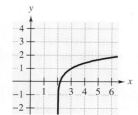

16. Use the graph of f shown at the left to determine h and k in $f(x) = \log_5(x - h) + k$.

17. Use a calculator and the change-of-base formula to evaluate $\log_3 782$.

Applications

18. You deposit $1200 in an account at an annual interest rate of $6\frac{1}{4}\%$. Complete the table to determine the balance A in the account after 15 years when the account is compounded n times per year.

n	1	4	12	365	Continuous compounding
A					

19. After t years, 14 grams of a radioactive element with a half-life of 40 years decays to a mass y (in grams) given by $y = 14\left(\frac{1}{2}\right)^{t/40}$, $t \geq 0$. How much of the initial mass remains after 125 years?

Study Skills in Action

Making the Most of Class Time

Have you ever slumped at your desk while in class and thought, "I'll just get the notes down and study later—I'm too tired"? Learning math is a team effort, between teacher and student. The more you understand in class, the more you will be able to learn while studying outside of class.

Approach math class with the intensity of a navy pilot during a mission briefing. The pilot has strategic plans to learn during the briefing. He or she listens intensely, takes notes, and memorizes important information. The goal is for the pilot to leave the briefing with a clear picture of the mission. It is the same with a student in a math class.

These students are sitting in the front row, where they are more likely to pay attention.

Smart Study Strategy

Take Control of Your Class Time

1 ▶ **Sit where you can easily see and hear the teacher, and the teacher can see you.** The teacher may be able to tell when you are confused just by the look on your face, and may adjust the lesson accordingly. In addition, sitting in this strategic place will keep your mind from wandering.

2 ▶ **Pay attention to what the teacher says about the math, not just what is written on the board.** Write problems on the left side of your notes and what the teacher says about the problems on the right side.

3 ▶ **If the teacher is moving through the material too fast, ask a question.** Questions help to slow the pace for a few minutes and also to clarify what is confusing to you.

4 ▶ **Try to memorize new information while learning it.** Repeat in your head what you are writing in your notes. That way you are reviewing the information twice.

5 ▶ **Ask for clarification.** If you don't understand something at all and don't even know how to phrase a question, just ask for clarification. You might say something like, "Could you please explain the steps in this problem one more time?"

6 ▶ **Think as intensely as if you were going to take a quiz on the material at the end of class.** This kind of mindset will help you to process new information.

7 ▶ **If the teacher asks for someone to go up to the board, volunteer.** The student at the board often receives additional attention and instruction to complete the problem.

8 ▶ **At the end of class, identify concepts or problems on which you still need clarification.** Make sure you see the teacher or a tutor as soon as possible.

11.4 Properties of Logarithms

▶ Use the properties of logarithms to evaluate logarithms.

▶ Rewrite, expand, or condense logarithmic expressions.

▶ Use the properties of logarithms to solve application problems.

Properties of Logarithms

Properties of Logarithms

Let a be a positive real number such that $a \neq 1$, and let n be a real number. If u and v are real numbers, variables, or algebraic expressions such that $u > 0$ and $v > 0$, then the following properties are true.

	Logarithm with Base a	Natural Logarithm
1. Product Property:	$\log_a(uv) = \log_a u + \log_a v$	$\ln(uv) = \ln u + \ln v$
2. Quotient Property:	$\log_a \dfrac{u}{v} = \log_a u - \log_a v$	$\ln \dfrac{u}{v} = \ln u - \ln v$
3. Power Property:	$\log_a u^n = n \log_a u$	$\ln u^n = n \ln u$

EXAMPLE 1 Using Properties of Logarithms

Use $\ln 2 \approx 0.693$, $\ln 3 \approx 1.099$, and $\ln 5 \approx 1.609$ to approximate each expression.

a. $\ln \dfrac{2}{3}$ **b.** $\ln 10$ **c.** $\ln 30$

SOLUTION

a. $\ln \dfrac{2}{3} = \ln 2 - \ln 3$ Quotient Property

     $\approx 0.693 - 1.099 = -0.406$ Substitute for ln 2 and ln 3.

b. $\ln 10 = \ln(2 \cdot 5)$ Factor.

     $= \ln 2 + \ln 5$ Product Property

     $\approx 0.693 + 1.609$ Substitute for ln 2 and ln 5.

     $= 2.302$ Simplify.

c. $\ln 30 = \ln(2 \cdot 3 \cdot 5)$ Factor.

     $= \ln 2 + \ln 3 + \ln 5$ Product Property

     $\approx 0.693 + 1.099 + 1.609$ Substitute for ln 2, ln 3, and ln 5.

     $= 3.401$ Simplify.

Exercises Within Reach ®   Solutions in English & Spanish and tutorial videos at CollegePrepAlgebra.com

Using Properties of Logarithms In Exercises 1−8, use **ln 3 ≈ 1.0986** and **ln 5 ≈ 1.6094** to **approximate** the expression. Use a calculator to verify your result.

1. $\ln \frac{5}{3}$ **2.** $\ln \frac{3}{5}$ **3.** $\ln 9$ **4.** $\ln 15$

5. $\ln 75$ **6.** $\ln 45$ **7.** $\ln \sqrt{45}$ **8.** $\ln \sqrt[3]{25}$

When using the properties of logarithms, it helps to state the properties *verbally*. For instance, the verbal form of the Product Property

$$\ln(uv) = \ln u + \ln v$$

is: *The log of a product is the sum of the logs of the factors.* Similarly, the verbal form of the Quotient Property

$$\ln \frac{u}{v} = \ln u - \ln v$$

is: *The log of a quotient is the difference of the logs of the numerator and denominator.*

Study Tip

Remember that you can verify results, such as those given in Examples 1 and 2, with a calculator.

EXAMPLE 2   **Using Properties of Logarithms**

Use the properties of logarithms to verify that $-\ln 2 = \ln \frac{1}{2}$.

SOLUTION

Using the Power Property, you can write the following.

$-\ln 2 = (-1)\ln 2$	Rewrite coefficient as -1.
$\quad = \ln 2^{-1}$	Power Property
$\quad = \ln \frac{1}{2}$	Rewrite 2^{-1} as $\frac{1}{2}$.

Exercises Within Reach ®   Solutions in English & Spanish and tutorial videos at CollegePrepAlgebra.com

Using Properties of Logarithms **In Exercises 9−14, use the properties of logarithms to verify the statement.**

9. $-3 \log_4 2 = \log_4 \frac{1}{8}$

10. $-2 \ln \frac{1}{3} = \ln 9$

11. $-3 \log_{10} 3 + \log_{10} \frac{3}{2} = \log_{10} \frac{1}{18}$

12. $-2 \ln 2 + \ln 24 = \ln 6$

13. $-\ln \frac{1}{7} = \ln 56 - \ln 8$

14. $-\log_5 10 = \log_5 10 - \log_5 100$

Using Properties of Logarithms **In Exercises 15−34, use the properties of logarithms to evaluate the expression.**

15. $\log_{12} 12^3$

16. $\log_3 81$

17. $\log_4 \left(\frac{1}{16}\right)^2$

18. $\log_7 \left(\frac{1}{49}\right)^3$

19. $\log_5 \sqrt[3]{5}$

20. $\ln \sqrt{e}$

21. $\ln 14^0$

22. $\ln\left(\frac{7.14}{7.14}\right)$

23. $\ln e^{-9}$

24. $\ln e^7$

25. $\log_8 4 + \log_8 16$

26. $\log_{10} 5 + \log_{10} 20$

27. $\log_3 54 - \log_3 2$

28. $\log_3 324 - \log_3 4$

29. $\log_2 5 - \log_2 40$

30. $\log_4\left(\frac{3}{16}\right) + \log_4\left(\frac{1}{3}\right)$

31. $\ln e^8 + \ln e^4$

32. $\ln e^5 - \ln e^2$

33. $\ln \frac{e^3}{e^2}$

34. $\ln(e^2 \cdot e^4)$

Rewriting Logarithmic Expressions

In Examples 1 and 2, the properties of logarithms were used to rewrite logarithmic expressions involving the log of a *constant*. A more common use of these properties is to rewrite the log of a *variable expression*.

EXAMPLE 3 **Expanding Logarithmic Expressions**

Use the properties of logarithms to expand each expression.

a. $\log_{10} 7x^3$ **b.** $\log_6 \dfrac{8x^3}{y}$ **c.** $\ln \dfrac{\sqrt{3x-5}}{7}$

SOLUTION

a. $\log_{10} 7x^3 = \log_{10} 7 + \log_{10} x^3$ Product Property

$= \log_{10} 7 + 3 \log_{10} x$ Power Property

b. $\log_6 \dfrac{8x^3}{y} = \log_6 8x^3 - \log_6 y$ Quotient Property

$= \log_6 8 + \log_6 x^3 - \log_6 y$ Product Property

$= \log_6 8 + 3 \log_6 x - \log_6 y$ Power Property

c. $\ln \dfrac{\sqrt{3x-5}}{7} = \ln\left[\dfrac{(3x-5)^{1/2}}{7}\right]$ Rewrite using rational exponent.

$= \ln(3x-5)^{1/2} - \ln 7$ Quotient Property

$= \dfrac{1}{2} \ln(3x-5) - \ln 7$ Power Property

Exercises Within Reach ®

Solutions in English & Spanish and tutorial videos at CollegePrepAlgebra.com

Expanding a Logarithmic Expression In Exercises 35–58, use the properties of logarithms to expand the expression. (Assume all variables are positive.)

35. $\log_3 11x$

36. $\log_2 3x$

37. $\ln 3y$

38. $\ln 5x$

39. $\log_7 x^2$

40. $\log_3 x^3$

41. $\log_4 x^{-3}$

42. $\log_2 s^{-4}$

43. $\log_4 \sqrt{3x}$

44. $\log_3 \sqrt[3]{5y}$

45. $\log_2 \dfrac{z}{17}$

46. $\log_{10} \dfrac{7}{y}$

47. $\log_9 \dfrac{\sqrt{x}}{12}$

48. $\ln \dfrac{\sqrt{x}}{x+9}$

49. $\ln x^2(y+2)$

50. $\ln y(y+1)^2$

51. $\log_4[x^6(x+7)^2]$

52. $\log_8[(x-y)^3 z^6]$

53. $\log_3 \sqrt[3]{x+1}$

54. $\log_5 \sqrt{xy}$

55. $\ln \sqrt{x(x+2)}$

56. $\ln \sqrt[3]{x(x+5)}$

57. $\ln \dfrac{xy^2}{z^3}$

58. $\log_5 \dfrac{x^2 y^5}{z^7}$

When you rewrite a logarithmic expression as in Example 3, you are *expanding* the expression. The reverse procedure is demonstrated in Example 4, and is called *condensing* a logarithmic expression.

EXAMPLE 4   **Condensing Logarithmic Expressions**

Use the properties of logarithms to condense each expression.

a. $\ln x - \ln 3$ **b.** $\dfrac{1}{2} \log_3 x + \log_3 5$ **c.** $3(\ln 4 + \ln x)$

SOLUTION

a. $\ln x - \ln 3 = \ln \dfrac{x}{3}$ Quotient Property

b. $\dfrac{1}{2} \log_3 x + \log_3 5 = \log_3 x^{1/2} + \log_3 5$ Power Property

$\qquad\qquad\qquad\quad = \log_3 5\sqrt{x}$ Product Property

c. $3(\ln 4 + \ln x) = 3(\ln 4x)$ Product Property

$\qquad\qquad\quad = \ln(4x)^3$ Power Property

$\qquad\qquad\quad = \ln 64x^3$ Simplify.

When you expand or condense a logarithmic expression, it is possible to change the domain of the expression. For instance, the domain of the function

$\qquad f(x) = 2 \ln x$ Domain is the set of positive real numbers.

is the set of positive real numbers, whereas the domain of

$\qquad g(x) = \ln x^2$ Domain is the set of nonzero real numbers.

is the set of nonzero real numbers. So, when you expand or condense a logarithmic expression, you should check to see whether the rewriting has changed the domain of the expression. In such cases, you should restrict the domain appropriately. For instance, you can write

$\qquad f(x) = 2 \ln x$

$\qquad\qquad = \ln x^2, \ x > 0.$

Exercises Within Reach ®

Solutions in English & Spanish and tutorial videos at CollegePrepAlgebra.com

Condensing a Logarithmic Expression **In Exercises 59–68, use the properties of logarithms to condense the expression.**

59. $\log_{12} x - \log_{12} 3$

60. $\log_6 12 - \log_6 y$

61. $\log_3 5 + \log_3 x$

62. $\log_5 2x + \log_5 3y$

63. $7 \log_2 x + 3 \log_2 z$

64. $2 \log_{10} x + \dfrac{1}{2} \log_{10} y$

65. $4(\ln x + \ln y)$

66. $\dfrac{1}{3}(\ln 10 + \ln 4x)$

67. $\log_4(x + 8) - 3 \log_4 x$

68. $5 \log_3 x + \log_3(x - 6)$

Applications

EXAMPLE 5 **Human Memory Model**

In an experiment, students attended several lectures on a subject. Every month for a year after that, the students were tested to see how much of the material they remembered. The average scores for the group are given by the human memory model

$$f(t) = 80 - \ln(t + 1)^9, \ 0 \le t \le 12$$

where t is the time in months. Find the average score for the group after 8 months.

SOLUTION

To make the calculations easier, rewrite the model using the Power Property, as follows.

$$f(t) = 80 - 9\ln(t + 1), \ 0 \le t \le 12$$

After 8 months, the average score was

$$
\begin{aligned}
f(8) &= 80 - 9\ln(8 + 1) && \text{Substitute 8 for } t. \\
&= 80 - 9\ln 9 && \text{Add.} \\
&\approx 80 - 19.8 && \text{Simplify.} \\
&= 60.2. && \text{Average score after 8 months}
\end{aligned}
$$

The graph of the function is shown at the left.

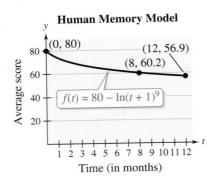

Human Memory Model

(0, 80) (12, 56.9) (8, 60.2) $f(t) = 80 - \ln(t + 1)^9$

Average score / Time (in months)

Exercises Within Reach ® Solutions in English & Spanish and tutorial videos at CollegePrepAlgebra.com

69. *Human Memory Model* Students participating in an experiment attended several lectures on a subject. Every month for a year after that, the students were tested to see how much of the material they remembered. The average scores for the group are given by the human memory model

$$f(t) = 80 - \log_{10}(t + 1)^{12}, \ 0 \le t \le 12$$

where t is the time in months. (See figure.) Use the Power Property to rewrite the model. Then find the average scores for the group after 2 months and 8 months.

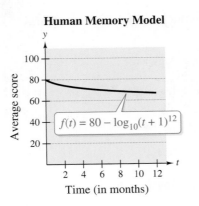

Human Memory Model

$f(t) = 80 - \log_{10}(t + 1)^{12}$

Average score / Time (in months)

70. *Human Memory Model* Students participating in an experiment attended several lectures on a subject. Every month for a year after that, the students were tested to see how much of the material they remembered. The average scores for the group are given by the human memory model

$$f(t) = 75 - \ln(t + 1)^6, \ 0 \le t \le 12$$

where t is the time in months. (See figure.) Use the Power Property to rewrite the model. Then find the average scores for the group after 1 month and 11 months.

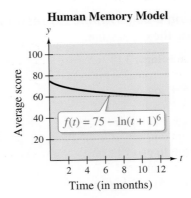

Human Memory Model

$f(t) = 75 - \ln(t + 1)^6$

Average score / Time (in months)

Application   EXAMPLE 6   **Sound Intensity**

The relationship between the number B of decibels and the intensity I of a sound (in watts per square centimeter) is given by

$$B = 10 \log_{10}\left(\frac{I}{10^{-16}}\right).$$

Use properties of logarithms to write the formula in simpler form, and determine the number of decibels of a thunderclap with an intensity of 10^{-1} watt per square centimeter.

SOLUTION

$B = 10 \log_{10}\left(\dfrac{I}{10^{-16}}\right)$	Write formula for decibels.
$= 10(\log_{10} I - \log_{10} 10^{-16})$	Quotient Property
$= 10 \log_{10} I - 10(-16 \log_{10} 10)$	Power Property
$= 10 \log_{10} I - 10(-16)$	$\log_{10} 10 = 1$
$= 10 \log_{10} I + 160$	Simplify.

When the intensity is $I = 10^{-1}$ watt per square centimeter, the number of decibels is

$B = 10 \log_{10} I + 160$	Write formula for decibels.
$= 10 \log_{10}(10^{-1}) + 160$	Substitute 10^{-1} for I.
$= 10(-1) + 160$	$\log_{10}(10^{-1}) = -1$
$= -10 + 160$	Multiply.
$= 150$ decibels.	Add.

Exercises Within Reach ®   Solutions in English & Spanish and tutorial videos at CollegePrepAlgebra.com

71. **Sound Intensity** The relationship between the number B of decibels and the intensity I of a sound (in watts per square meter) is given by

$$B = 10 \log_{10}\left(\frac{I}{10^{-12}}\right).$$

Use properties of logarithms to write the formula in simpler form, and determine the number of decibels in the front row of the concert when the intensity is 10^{-1} watt per square meter.

72. **Sound Intensity** The relationship between the number B of decibels and the intensity I of a sound (in watts per square centimeter) is given by

$$B = 10 \log_{10}\left(\frac{I}{10^{-16}}\right).$$

Find the number of decibels in the middle of an arena during a concert when the intensity is 10^{-6} watt per square centimeter.

Concept Summary: Using Properties of Logarithms

What

Here are three properties of logarithms.

1. Product Property
2. Quotient Property
3. Power Property

You can use these properties to evaluate, rewrite, expand, or condense logarithmic expressions.

EXAMPLE

Show that $2(\log_2 3 + \log_2 x - \log_2 y)$ and $\log_2\left(\dfrac{3x}{y}\right)^2$ are equivalent.

How

Use the Product Property, the Quotient Property, and the Power Property.

EXAMPLE

$2(\log_2 3 + \log_2 x - \log_2 y)$

$= 2(\log_2 3x - \log_2 y)$ Product Property

$= 2\left(\log_2 \dfrac{3x}{y}\right)$ Quotient Property

$= \log_2\left(\dfrac{3x}{y}\right)^2$ Power Property

Why

Knowing how to use the properties of logarithms correctly can help you rewrite logarithmic expressions into forms that are easier to evaluate. Just be careful that when doing so, you identify whether you change the domain of the expression.

Exercises Within Reach®

Worked-out solutions to odd-numbered exercises at CollegePrepAlgebra.com

Concept Summary Check

73. *Rewriting an Expression* In the solution above, is the final expression obtained by expanding or by condensing the original expression?

74. *Using the Product Property* Use the Product Property of Logarithms to condense the expression $\log_2 3 + \log_2 x$.

75. *Condensing an Expression* What property can you use to condense the expression $2\left(\log_2 \dfrac{3x}{y}\right)$?

76. *Using the Quotient Property* Use the Quotient Property of Logarithms to expand the expression $\log_2\left(\dfrac{3x}{y}\right)$.

Extra Practice

Condensing a Logarithmic Expression In Exercises 77−86, use the properties of logarithms to condense the expression.

77. $3 \ln x + \ln y - 2 \ln z$

78. $4 \ln 2 + 2 \ln x - \frac{1}{2} \ln y$

79. $2[\ln x - \ln(x + 1)]$

80. $5\left[\ln x - \frac{1}{2} \ln(x + 4)\right]$

81. $\frac{1}{3} \log_5(x + 3) - \log_5(x - 6)$

82. $\frac{1}{4} \log_6(x + 1) - 5 \log_6(x - 4)$

83. $5 \log_6(c + d) - \frac{1}{2} \log_6(m - n)$

84. $2 \log_5(x + y) + 3 \log_5 w$

85. $\frac{1}{5}(3 \log_2 x - 4 \log_2 y)$

86. $\frac{1}{3}[\ln(x - 6) - 4 \ln y - 2 \ln z]$

Simplifying a Logarithmic Expression In Exercises 87−92, simplify the expression.

87. $\ln 3e^2$

88. $\log_3(3^2 \cdot 4)$

89. $\log_5 \sqrt{50}$

90. $\log_2 \sqrt{22}$

91. $\log_8 \dfrac{8}{x^3}$

92. $\ln \dfrac{6}{e^5}$

Molecular Transport **In Exercises 93 and 94, use the following information. The energy E (in kilocalories per gram molecule) required to transport a substance from the outside to the inside of a living cell is given by**

$$E = 1.4(\log_{10} C_2 - \log_{10} C_1)$$

where C_1 and C_2 are the concentrations of the substance outside and inside the cell, respectively.

93. Condense the expression.

94. The concentration of a substance inside a cell is twice the concentration outside the cell. How much energy is required to transport the substance from outside to inside the cell?

Explaining Concepts

True or False? **In Exercises 95–100, use properties of logarithms to determine whether the equation is true or false. Justify your answer.**

95. $\log_2 8x = 3 + \log_2 x$

96. $\log_3(u + v) = \log_3 u + \log_3 v$

97. $\log_3(u + v) = \log_3 u \cdot \log_3 v$

98. $\dfrac{\log_6 10}{\log_6 3} = \log_6 10 - \log_6 3$

99. If $f(x) = \log_a x$, then $f(ax) = 1 + f(x)$.

100. If $f(x) = \log_a x$, then $f(a^n) = n$.

True or False? **In Exercises 101–104, determine whether the statement is true or false given that $f(x) = \ln x$. Justify your answer.**

101. $f(0) = 0$

102. $f(2x) = \ln 2 + \ln x$

103. $f(x - 3) = \ln x - \ln 3, \ x > 3$

104. $\sqrt{f(x)} = \frac{1}{2} \ln x$

105. *Think About It* Explain how you can show that
$$\dfrac{\ln x}{\ln y} \neq \ln \dfrac{x}{y}.$$

106. *Think About It* Without a calculator, approximate the natural logarithms of as many integers as possible between 1 and 20 using $\ln 2 \approx 0.6931$, $\ln 3 \approx 1.0986$, $\ln 5 \approx 1.6094$, and $\ln 7 \approx 1.9459$. Explain the method you used. Then verify your results with a calculator and explain any differences in the results.

Cumulative Review

In Exercises 107–112, solve the equation.

107. $\dfrac{2}{3}x + \dfrac{2}{3} = 4x - 6$

108. $x^2 - 10x + 17 = 0$

109. $\dfrac{5}{2x} - \dfrac{4}{x} = 3$ 110. $\dfrac{1}{x} + \dfrac{2}{x - 5} = 0$

111. $|x - 4| = 3$ 112. $\sqrt{x + 2} = 7$

In Exercises 113–116, sketch the parabola. Identify the vertex and any x-intercepts.

113. $g(x) = -(x + 2)^2$ 114. $f(x) = x^2 - 16$

115. $g(x) = -2x^2 + 4x - 7$

116. $h(x) = x^2 + 6x + 14$

In Exercises 117–120, find the compositions (a) $f \circ g$ and (b) $g \circ f$. Then find the domain of each composition.

117. $f(x) = 4x + 9$
    $g(x) = x - 5$

118. $f(x) = \sqrt{x}$
    $g(x) = x - 3$

119. $f(x) = \dfrac{1}{x}$
    $g(x) = x + 2$

120. $f(x) = \dfrac{5}{x^2 - 4}$
    $g(x) = x + 1$

11.5 Solving Exponential and Logarithmic Equations

▶ Use one-to-one properties to solve exponential and logarithmic equations.
▶ Use inverse properties to solve exponential and logarithmic equations.
▶ Use exponential or logarithmic equations to solve application problems.

Using One-to-One Properties

One-to-One Properties of Exponential and Logarithmic Equations

Let a be a positive real number such that $a \neq 1$, and let x and y be real numbers. Then the following properties are true.

1. $a^x = a^y$ if and only if $x = y$.
2. $\log_a x = \log_a y$ if and only if $x = y$ $(x > 0,\ y > 0)$.

EXAMPLE 1   **Solving Exponential and Logarithmic Equations**

Solve each equation.

a. $4^{x+2} = 64$ **b.** $\ln(2x - 3) = \ln 11$

SOLUTION

a.

$4^{x+2} = 64$	Write original equation.
$4^{x+2} = 4^3$	Rewrite with like bases.
$x + 2 = 3$	One-to-one property
$x = 1$	Subtract 2 from each side.

The solution is $x = 1$. Check this in the original equation.

b.

$\ln(2x - 3) = \ln 11$	Write original equation.
$2x - 3 = 11$	One-to-one property
$2x = 14$	Add 3 to each side.
$x = 7$	Divide each side by 2.

The solution is $x = 7$. Check this in the original equation.

Exercises Within Reach ®   Solutions in English & Spanish and tutorial videos at CollegePrepAlgebra.com

Using a One-to-One Property In Exercises 1–14, solve the equation.

1. $7^x = 7^3$

2. $4^x = 4^6$

3. $e^{1-x} = e^4$

4. $e^{x+3} = e^8$

5. $3^{2-x} = 81$

6. $4^{2x-1} = 64$

7. $6^{2x} = 36$

8. $5^{3x} = 25$

9. $\ln 5x = \ln 22$

10. $\ln 4x = \ln 30$

11. $\ln(3 - x) = \ln 10$

12. $\ln(2x - 3) = \ln 17$

13. $\log_3(4 - 3x) = \log_3(2x + 9)$

14. $\log_4(2x - 6) = \log_4(5x + 6)$

Using Inverse Properties

Solving Exponential Equations

To solve an exponential equation, first isolate the exponential expression. Then *take the logarithm* of each side of the equation and solve for the variable.

Inverse Properties of Exponents and Logarithms

Base a	*Natural Base e*
1. $\log_a(a^x) = x$	$\ln(e^x) = x$
2. $a^{(\log_a x)} = x$	$e^{(\ln x)} = x$

EXAMPLE 2   **Solving Exponential Equations**

Solve each exponential equation.

a. $2^x = 7$ **b.** $4^{x-3} = 9$ **c.** $2e^x = 10$

SOLUTION

a.

$2^x = 7$	Write original equation.
$\log_2(2^x) = \log_2 7$	Take logarithm of each side.
$x = \log_2 7$	Inverse property

The solution is $x = \log_2 7 \approx 2.807$. Check this in the original equation.

b.

$4^{x-3} = 9$	Write original equation.
$\log_4(4^{x-3}) = \log_4 9$	Take logarithm of each side.
$x - 3 = \log_4 9$	Inverse property
$x = \log_4 9 + 3$	Add 3 to each side.

The solution is $x = \log_4 9 + 3 \approx 4.585$. Check this in the original equation.

c.

$2e^x = 10$	Write original equation.
$e^x = 5$	Divide each side by 2.
$\ln(e^x) = \ln 5$	Take logarithm of each side.
$x = \ln 5$	Inverse property

The solution is $x = \ln 5 \approx 1.609$. Check this in the original equation.

Study Tip

Remember that to evaluate a logarithm such as $\log_2 7$, you need to use the change-of-base formula.

$$\log_2 7 = \frac{\ln 7}{\ln 2} \approx 2.807$$

Similarly,

$$\log_4 9 + 3 = \frac{\ln 9}{\ln 4} + 3$$

$$\approx 1.585 + 3$$

$$= 4.585.$$

Exercises Within Reach ®

Solutions in English & Spanish and tutorial videos at CollegePrepAlgebra.com

Solving an Exponential Equation In Exercises 15–26, solve the exponential equation. (Round your answer to two decimal places, if necessary.)

15. $3^x = 91$

16. $4^x = 40$

17. $5^x = 8.2$

18. $2^x = 3.6$

19. $3^{2-x} = 8$

20. $5^{3-x} = 15$

21. $10^{x+6} = 250$

22. $12^{x-1} = 324$

23. $\frac{1}{4}e^x = 5$

24. $\frac{2}{3}e^x = 1$

25. $4e^{-x} = 24$

26. $6e^{-x} = 3$

EXAMPLE 3   **Solving Exponential Equations**

Solve each exponential equation.

a. $5 + e^{x+1} = 20$ **b.** $23 - 5e^{x+1} = 3$

SOLUTION

a.

$5 + e^{x+1} = 20$	Write original equation.
$e^{x+1} = 15$	Subtract 5 from each side.
$\ln e^{x+1} = \ln 15$	Take the logarithm of each side.
$x + 1 = \ln 15$	Inverse Property
$x = -1 + \ln 15$	Subtract 1 from each side.

The solution is $x = -1 + \ln 15 \approx 1.708$. You can check this as follows.

Check

$5 + e^{x+1} = 20$	Write original equation.
$5 + e^{-1 + \ln 15 + 1} \overset{?}{=} 20$	Substitute $-1 + \ln 15$ for x.
$5 + e^{\ln 15} \overset{?}{=} 20$	Simplify.
$5 + 15 = 20$	Solution checks.

b.

$23 - 5e^{x+1} = 3$	Write original equation.
$-5e^{x+1} = -20$	Subtract 23 from each side.
$e^{x+1} = 4$	Divide each side by -5.
$\ln e^{x+1} = \ln 4$	Take the logarithm of each side.
$x + 1 = \ln 4$	Inverse property
$x = \ln 4 - 1$	Subtract 1 from each side.

The solution is $x = \ln 4 - 1 \approx 0.386$. You can check this as follows.

Check

$23 - 5e^{x+1} = 3$	Write original equation.
$23 - 5e^{(\ln 4 - 1) + 1} \overset{?}{=} 3$	Substitute $\ln 4 - 1$ for x.
$23 - 5e^{\ln 4} \overset{?}{=} 3$	Simplify.
$23 - 5(4) = 3$	Solution checks.

Exercises Within Reach®

Solutions in English & Spanish and tutorial videos at CollegePrepAlgebra.com

Solving an Exponential Equation In Exercises 27−36, solve the exponential equation. (Round your answer to two decimal places.)

27. $7 + e^{2-x} = 28$ **28.** $5^{x+6} - 4 = 12$ **29.** $4 + e^{2x} = 10$ **30.** $10 + e^{4x} = 18$

31. $17 - e^{x/4} = 14$ **32.** $50 - e^{x/2} = 35$ **33.** $8 - 12e^{-x} = 7$ **34.** $6 - 3e^{-x} = -15$

35. $4(1 + e^{x/3}) = 84$ **36.** $50(3 - e^{2x}) = 125$

> ## Solving Logarithmic Equations
>
> To solve a logarithmic equation, first isolate the logarithmic expression. Then *exponentiate* each side of the equation and solve for the variable.

EXAMPLE 4   **Solving Logarithmic Equations**

Solve each logarithmic equation.

a. $2 \log_4 x = 5$

b. $\frac{1}{4} \log_2 x = \frac{1}{2}$

SOLUTION

a.

$2 \log_4 x = 5$	Write original equation.
$\log_4 x = \dfrac{5}{2}$	Divide each side by 2.
$4^{\log_4 x} = 4^{5/2}$	Exponentiate each side.
$x = 4^{5/2}$	Inverse Property
$x = 32$	Simplify.

The solution is $x = 32$. Check this in the original equation, as follows.

> ## Check
>
> | | |
> |---|---|
> | $2 \log_4 x = 5$ | Write original equation. |
> | $2 \log_4(32) \overset{?}{=} 5$ | Substitute 32 for x. |
> | $2(2.5) \overset{?}{=} 5$ | Use a calculator. |
> | $5 = 5$ | Solution checks. ✔ |

b.

$\dfrac{1}{4} \log_2 x = \dfrac{1}{2}$	Write original equation.
$\log_2 x = 2$	Multiply each side by 4.
$2^{\log_2 x} = 2^2$	Exponentiate each side.
$x = 4$	Inverse property

The solution is $x = 4$. Check this in the original equation.

Exercises Within Reach ®

Solutions in English & Spanish and tutorial videos at CollegePrepAlgebra.com

Solving a Logarithmic Equation In Exercises 37−46, solve the logarithmic equation. (Round your answer to two decimal places, if necessary.)

37. $\log_{10} x = -1$

38. $\log_{10} x = 3$

39. $4 \log_3 x = 28$

40. $6 \log_2 x = 18$

41. $\frac{1}{6} \log_3 x = \frac{1}{3}$

42. $\frac{1}{8} \log_5 x = \frac{1}{2}$

43. $\log_{10} 4x = 2$

44. $\log_3 6x = 4$

45. $2 \log_4(x + 5) = 3$

46. $5 \log_{10}(x + 2) = 15$

EXAMPLE 5 **Solving Logarithmic Equations**

a. $3 \log_{10} x = 6$ Original equation

$\log_{10} x = 2$ Divide each side by 3.

$x = 10^2$ Exponential form

$x = 100$ Simplify.

The solution is $x = 100$. Check this in the original equation.

b. $20 \ln 0.2x = 30$ Original equation

$\ln 0.2x = 1.5$ Divide each side by 20.

$e^{\ln 0.2x} = e^{1.5}$ Exponentiate each side.

$0.2x = e^{1.5}$ Inverse property

$x = 5e^{1.5}$ Divide each side by 0.2.

The solution is $x = 5e^{1.5} \approx 22.408$. Check this in the original equation.

Study Tip

When checking approximate solutions to exponential and logarithmic equations, be aware that the check will not be exact because the solutions are approximate.

EXAMPLE 6 **Checking for Extraneous Solutions**

$\log_6 x + \log_6(x - 5) = 2$ Original equation

$\log_6[x(x - 5)] = 2$ Condense the left side.

$x(x - 5) = 6^2$ Exponential form

$x^2 - 5x - 36 = 0$ Write in general form.

$(x - 9)(x + 4) = 0$ Factor.

$x - 9 = 0 \implies x = 9$ Set 1st factor equal to 0.

$x + 4 = 0 \implies x = -4$ Set 2nd factor equal to 0.

Check the possible solutions $x = 9$ and $x = -4$ in the original equation.

First Solution

$\log_6(9) + \log_6(9 - 5) \stackrel{?}{=} 2$

$\log_6(9 \cdot 4) \stackrel{?}{=} 2$

$\log_6 36 = 2$ ✓

Second Solution

$\log_6(-4) + \log_6(-4 - 5) \stackrel{?}{=} 2$

$\log_6(-4) + \log_6(-9) \neq 2$ ✗

Of the two possible solutions, only $x = 9$ checks. So, $x = -4$ is extraneous.

Exercises Within Reach ®   Solutions in English & Spanish and tutorial videos at CollegePrepAlgebra.com

Solving a Logarithmic Equation In Exercises 47−52, solve the logarithmic equation. (Round your answer to two decimal places, if necessary.)

47. $2 \log_{10} x = 10$

48. $7 \log_2 x = 35$

49. $3 \ln 0.1x = 4$

50. $16 \ln 0.25x = 48$

51. $\log_{10} x + \log_{10}(x - 3) = 1$

52. $\log_{10} x + \log_{10}(x + 1) = 0$

Application

Application   **EXAMPLE 7**   **Compound Interest**

A deposit of $5000 is placed in a savings account for 2 years. The interest is compounded continuously. At the end of 2 years, the balance in the account is $5416.44. What is the annual interest rate for this account?

SOLUTION

Formula: $A = Pe^{rt}$

Labels: Principal = P = 5000   (dollars)
Amount = A = 5416.44   (dollars)
Time = t = 2   (years)
Annual interest rate = r (percent in decimal form)

Equation: $5416.44 = 5000e^{2r}$ Substitute for A, P, and t.
$1.083288 = e^{2r}$ Divide each side by 5000.
$\ln 1.083288 = \ln(e^{2r})$ Take logarithm of each side.
$0.08 \approx 2r$ ⟹ $0.04 \approx r$ Inverse property

The interest rate is about 4%. Check this solution.

Exercises Within Reach ®

Solutions in English & Spanish and tutorial videos at CollegePrepAlgebra.com

Compound Interest In Exercises 53 and 54, use the formula for continuous compounding $A = Pe^{rt}$, where A is the account balance after t years for the principal P and annual interest rate r (in decimal form).

53. A deposit of $10,000 is placed in a savings account for 2 years. The interest is compounded continuously. At the end of 2 years, the balance in the account is $11,051.71. What is the annual interest rate for this account?

54. A deposit of $2500 is placed in a savings account for 2 years. The interest is compounded continuously. At the end of 2 years, the balance in the account is $2847.07. What is the annual interest rate for this account?

55. *Friction* In order to restrain an untrained horse, a trainer partially wraps a rope around a cylindrical post in a corral (see figure). The horse is pulling on the rope with a force of 200 pounds. The force F (in pounds) needed to hold back the horse is $F = 200e^{-0.5\pi\theta/180}$, where θ is the angle of wrap (in degrees). Find the smallest value of θ for which a force of 80 pounds will hold the horse by solving for θ when $F = 80$.

56. *Online Retail* The projected online retail sales S (in billions of dollars) in the United States for the years 2009 through 2014 are modeled by the equation $S = 67.2e^{0.0944t}$, for $9 \le t \le 14$, where t is the time in years, with $t = 9$ corresponding to 2009. Find the year when S is about $230 billion by solving for t when $S = 230$. (*Source:* Forrester Research, Inc.)

Concept Summary: Solving Exponential and Logarithmic Equations

What

You can use the one-to-one properties and the inverse properties to solve exponential and logarithmic equations.

EXAMPLE

Solve each equation.

a. $3^{x-2} = 3^3$

b. $\log_6 3x = 2$

How

One-to-One Properties Let a, x, and y be real numbers ($a > 0$, $a \neq 1$).

1. $a^x = a^y$ if and only if $x = y$.
2. $\log_a x = \log_a y$ if and only if $x = y$ ($x > 0$, $y > 0$).

Inverse Properties

1. $\log_a(a^x) = x$ $\qquad \ln(e^x) = x$
2. $a^{(\log_a x)} = x$ $\qquad e^{(\ln x)} = x$

EXAMPLE

a. $3^{x-2} = 3^3$

$\qquad x - 2 = 3$ $\qquad$ One-to-one property

$\qquad\quad\ x = 5$ $\qquad$ Add 2 to each side.

b. $\log_6 3x = 2$

$\qquad 6^{\log_6 3x} = 6^2$ $\qquad$ Exponentiate each side.

$\qquad\qquad 3x = 6^2$ $\qquad$ Inverse property

$\qquad\qquad\ x = 12$ $\qquad$ Divide each side by 3.

Why

You will use exponential and logarithmic equations often in your study of mathematics and its applications. For instance, you will solve a logarithmic equation to determine the intensity of a sound.

Exercises Within Reach®

Worked-out solutions to odd-numbered exercises at CollegePrepAlgebra.com

Concept Summary Check

57. *Applying a One-to-One Property* Explain why you can apply a one-to-one property to solve the equation $3^{x-2} = 3^3$.

58. *Applying a One-to-One Property* Explain how to apply a one-to-one property to the equation $3^{x-2} = 3^3$.

59. *Applying an Inverse Property* Explain how to apply an inverse property to the equation $\log_6 3x = 2$.

60. *Applying an Inverse Property* Explain how to apply an inverse property to solve an exponential equation.

Extra Practice

Solving an Exponential or a Logarithmic Equation In Exercises 61−80, solve the equation, if possible. (Round your answer to two decimal places, if necessary.)

61. $5^x = \frac{1}{125}$

62. $3^x = \frac{1}{243}$

63. $2^{x+2} = \frac{1}{16}$

64. $3^{x+2} = \frac{1}{27}$

65. $\log_6 3x = \log_6 18$

66. $\log_5 2x = \log_5 36$

67. $\log_4(x-8) = \log_4(-4)$

68. $\log_5(2x-3) = \log_5(4x-5)$

69. $\frac{1}{5}(4^{x+2}) = 300$

70. $3(2^{t+4}) = 350$

71. $6 + 2^{x-1} = 1$

72. $24 + e^{4-x} = 22$

73. $\log_5(x+3) - \log_5 x = 1$

74. $\log_3(x-2) + \log_3 5 = 3$

75. $\log_2(x-1) + \log_2(x+3) = 3$

76. $\log_6(x-5) + \log_6 x = 2$

77. $\log_{10} 4x - \log_{10}(x-2) = 1$

78. $\log_2 3x - \log_2(x+4) = 3$

79. $\log_2 x + \log_2(x+2) - \log_2 3 = 4$

80. $\log_3 2x + \log_3(x-1) - \log_3 4 = 1$

81. *Doubling Time* Solve the exponential equation

$$5000 = 2500e^{0.09t}$$

for t to determine the number of years for an investment of $2500 to double in value when compounded continuously at the rate of 9%.

82. *Doubling Rate* Solve the exponential equation

$$10,000 = 5000e^{10r}$$

for r to determine the interest rate required for an investment of $5000 to double in value when compounded continuously for 10 years.

83. *Sound Intensity* The relationship between the number B of decibels and the intensity I of a sound in watts per square centimeter is given by

$$B = 10 \log_{10}\left(\frac{I}{10^{-16}}\right).$$

Determine the intensity of a sound that registers 80 decibels on a decibel meter.

84. *Sound Intensity* The relationship between the number B of decibels and the intensity I of a sound in watts per square centimeter is given by

$$B = 10 \log_{10}\left(\frac{I}{10^{-16}}\right).$$

Determine the intensity of a sound that registers 110 decibels on a decibel meter.

Newton's Law of Cooling In Exercises 85 and 86, use Newton's Law of Cooling

$$kt = \ln \frac{T - S}{T_0 - S}$$

where T is the temperature of a body (in °F), t is the number of hours elapsed, S is the temperature of the environment, and T_0 is the initial temperature of the body.

85. A corpse is discovered in a motel room at 10 P.M., and its temperature is 85°F. Three hours later, the temperature of the corpse is 78°F. The temperature of the motel room is a constant 65°F.

    (a) What is the constant k?

    (b) Find the time of death assuming the body temperature is 98.6°F at the time of death.

    (c) What is the temperature of the corpse two hours after death?

86. A corpse is discovered in the bedroom of a home at 7 A.M., and its temperature is 92°F. Two hours later, the temperature of the corpse is 88°F. The temperature of the bedroom is a constant 68°F.

    (a) What is the constant k?

    (b) Find the time of death assuming the body temperature is 98.6°F at the time of death.

    (c) What is the temperature of the corpse three hours after death?

Explaining Concepts

87. *Think About It* Which equation can be solved without logarithms, $2^{x-1} = 32$ or $2^{x-1} = 30$? Explain.

88. *Writing* Explain how to solve $10^{2x-1} = 5316$.

89. *Writing* In your own words, state the guidelines for solving exponential and logarithmic equations.

90. *Think About It* Why is it possible for a logarithmic equation to have an extraneous solution?

Cumulative Review

In Exercises 91–94, solve the equation by using the Square Root Property.

91. $x^2 = -25$

92. $x^2 - 49 = 0$

93. $9n^2 - 16 = 0$

94. $(2a + 3)^2 = 18$

In Exercises 95 and 96, solve the equation of quadratic form.

95. $t^4 - 13t^2 + 36 = 0$

96. $u + 2\sqrt{u} - 15 = 0$

In Exercises 97–100, find the perimeter or area of the rectangle, as indicated.

	Width	Length	Perimeter	Area
97.	2.5x	x	42 in.	
98.	w	1.6w	78 ft	
99.	w	w + 4		192 km²
100.	x − 3	x		270 cm²

11.6 Applications

▶ Use exponential equations to solve compound interest problems.
▶ Use exponential equations to solve growth and decay problems.
▶ Use logarithmic equations to solve intensity problems.

Compound Interest

In Section 11.1, you were introduced to two formulas for compound interest. Recall that in these formulas, A is the balance, P is the principal, r is the annual interest rate (in decimal form), and t is the time in years.

n Compoundings per Year

$$A = P\left(1 + \frac{r}{n}\right)^{nt}$$

Continuous Compounding

$$A = Pe^{rt}$$

Application | **EXAMPLE 1** **Finding the Annual Interest Rate**

An investment of $50,000 is made in an account that compounds interest quarterly. After 4 years, the balance in the account is $71,381.07. What is the annual interest rate for this account?

SOLUTION

Formula: $\quad A = P\left(1 + \frac{r}{n}\right)^{nt}$

Labels: Principal $= P = 50{,}000$ (dollars)
Amount $= A = 71{,}381.07$ (dollars)
Time $= t = 4$ (years)
Number of compoundings per year $= n = 4$
Annual interest rate $= r$ (percent in decimal form)

Equation:

$$71{,}381.07 = 50{,}000\left(1 + \frac{r}{4}\right)^{(4)(4)} \quad \text{Substitute for } A, P, n, \text{ and } t.$$

$$1.42762 \approx \left(1 + \frac{r}{4}\right)^{16} \quad \text{Divide each side by 50,000.}$$

$$(1.42762)^{1/16} \approx 1 + \frac{r}{4} \quad \text{Raise each side to } \tfrac{1}{16} \text{ power.}$$

$$1.0225 \approx 1 + \frac{r}{4} \quad \text{Simplify.}$$

$$0.09 \approx r \quad \text{Subtract 1 from each side and then multiply each side by 4.}$$

The annual interest rate is about 9%. Check this in the original problem.

Study Tip

To remove an exponent from one side of an equation, you can often raise each side of the equation to the *reciprocal* power. For instance, in Example 1, the exponent 16 is eliminated from the right side by raising each side to the reciprocal power $\frac{1}{16}$.

Exercises Within Reach ®

Solutions in English & Spanish and tutorial videos at CollegePrepAlgebra.com

Finding an Annual Interest Rate In Exercises 1 and 2, **find the annual interest rate.**

	Principal	Balance	Time	Compounding
1.	$500	$1004.83	10 years	Monthly
2.	$3000	$21,628.70	20 years	Quarterly

Application   EXAMPLE 2   **Doubling Time for Continuous Compounding**

An investment is made in a trust fund at an annual interest rate of 8.75%, compounded continuously. How long will it take for the investment to double?

SOLUTION

Study Tip

In "doubling time" problems, you do not need to know the value of the principal P to find the doubling time. As shown in Example 2, the factor P divides out the equation and so does not affect the doubling time.

$A = Pe^{rt}$	Formula for continuous compounding
$2P = Pe^{0.0875t}$	Substitute known values.
$2 = e^{0.0875t}$	Divide each side by P.
$\ln 2 = 0.0875t$	Inverse property
$\dfrac{\ln 2}{0.0875} = t$	Divide each side by 0.0875.
$7.92 \approx t$	Use a calculator.

It will take about 7.92 years for the investment to double.

Application   EXAMPLE 3   **Finding the Type of Compounding**

You deposit $1000 in an account. At the end of 1 year, your balance is $1077.63. The bank tells you that the annual interest rate for the account is 7.5%. How was the interest compounded?

SOLUTION

If the interest had been compounded continuously at 7.5%, then the balance would have been $A = 1000e^{(0.075)(1)} = \$1077.88$. Because the actual balance is slightly less than this, you should use the formula for interest that is compounded n times per year.

$$1077.63 = 1000\left(1 + \frac{0.075}{n}\right)^n$$

At this point, it is not clear what you should do to solve the equation for n. However, by completing a table like the one shown below, you can see that $n = 12$. So, the interest was compounded monthly.

n	1	4	12	365
$1000\left(1 + \dfrac{0.075}{n}\right)^n$	1075	1077.14	1077.63	1077.88

Exercises Within Reach ®

Solutions in English & Spanish and tutorial videos at CollegePrepAlgebra.com

Doubling Time In Exercises 3−6, find the time for an investment to double at the given annual interest rate, compounded continuously.

3. 8%   **4.** 5%   **5.** 6.75%   **6.** 9.75%

Compound Interest In Exercises 7 and 8, determine the type of compounding. Solve the problem by trying the more common types of compounding.

Principal	Balance	Time	Rate		Principal	Balance	Time	Rate
7. $5000	$8954.24	10 years	6%		**8.** $5000	$9096.98	10 years	6%

In Example 3, notice that an investment of $1000 compounded monthly produced a balance of $1077.63 at the end of 1 year. Because $77.63 of this amount is interest, the **effective yield** for the investment is

$$\text{Effective yield} = \frac{\text{Year's interest}}{\text{Amount invested}} = \frac{77.63}{1000}$$

$$= 0.07763$$

$$= 7.763\%.$$

In other words, the effective yield for an investment collecting compound interest is the *simple interest rate* that would yield the same balance at the end of 1 year.

Application **EXAMPLE 4** **Finding the Effective Yield**

An investment is made in an account that pays 6.75% interest, compounded continuously. What is the effective yield for this investment?

SOLUTION

Notice that you do not have to know the principal or the time that the money will be left in the account. Instead, you can choose an arbitrary principal, such as $1000. Then, because effective yield is based on the balance at the end of 1 year, you can use the following formula.

$$A = Pe^{rt}$$

$$= 1000e^{0.0675(1)}$$

$$\approx 1069.83$$

Now, because the account would earn $69.83 in interest after 1 year for a principal of $1000, you can conclude that the effective yield is

$$\text{Effective yield} = \frac{69.83}{1000} \qquad \text{Divide interest by principal.}$$

$$= 0.06983 \qquad \text{Simplify.}$$

$$= 6.983\%. \qquad \text{Write in percent form.}$$

Exercises Within Reach ®

Solutions in English & Spanish and tutorial videos at CollegePrepAlgebra.com

Finding an Effective Yield In Exercises 9–16, find the effective yield.

	Rate	Compounding		Rate	Compounding
9.	8%	Continuously	**10.**	9.5%	Continuously
11.	7%	Monthly	**12.**	8%	Yearly
13.	6%	Quarterly	**14.**	9%	Quarterly
15.	$5\frac{1}{4}\%$	Daily	**16.**	8%	Monthly

17. *Finding Effective Yield* Is it necessary to know the principal P to find the effective yield in Exercises 9–16? Explain.

18. *Compounding and Effective Yield* Consider the results of Exercises 9, 12, and 16. When the interest is compounded more frequently, what inference can you make about the difference between the effective yield and the stated annual percentage rate?

Growth and Decay

> ### Exponential Growth and Decay
>
> The mathematical model for exponential growth or decay is given by
>
> $$y = Ce^{kt}.$$
>
> For this model, C is the original amount, and y is the amount after time t. The number k is a constant that is determined by the rate of growth (or decay). When $k > 0$, the model represents **exponential growth**, and when $k < 0$, it represents **exponential decay**.

Application   **EXAMPLE 5** **Website Growth**

A college created an algebra tutoring website in 2010. The number of hits per year at the website has grown exponentially. The website had 4080 hits in 2010 and 8568 hits in 2013. Predict the number of hits in 2018.

SOLUTION

t (year)	Ce^{kt} (hits)
0	$Ce^{k(0)} = 4080$
3	$Ce^{k(3)} = 8568$
8	$Ce^{k(8)} = ?$

In the exponential growth model $y = Ce^{kt}$, let $t = 0$ represent 2010. Next, use the information given in the problem to set up the table shown. Because $Ce^{k(0)} = Ce^0 = 4080$, you can conclude that $C = 4080$. Then, using this value of C, you can solve for k as follows.

$$Ce^{k(3)} = 8568 \qquad \text{From table}$$

$$4080e^{3k} = 8568 \qquad \text{Substitute 4080 for } C.$$

$$e^{3k} = 2.1 \qquad \text{Divide each side by 4080.}$$

$$3k = \ln 2.1 \qquad \text{Inverse property}$$

$$k = \frac{1}{3}\ln 2.1 \approx 0.2473 \qquad \text{Divide each side by 3 and simplify.}$$

Finally, you can use this value of k in the model from the table to predict the number of hits in 2018 to be $4080e^{0.2473(8)} \approx 29{,}503$.

Exercises Within Reach ®

19. **Computer Virus** In 2005, a computer worm called "Samy" interrupted the operations of a social networking website by inserting the payload message "but most of all, Samy is my hero" in the personal profile pages of the website's users. It is said that the "Samy" worm's message spread from 73 users to 1 million users within 20 hours.

    (a) Find the constants C and k to obtain an exponential growth model $y = Ce^{kt}$ for the "Samy" worm.

    (b) Use your model from part (a) to estimate how long it took the "Samy" worm to drop its payload message in 5300 personal profile pages.

20. **Album Downloads** In 2006, about 27.6 million albums were purchased through downloading in the United States. In 2011, the number had increased to about 104.8 million. Use an exponential growth model to predict the number of albums that will be purchased through downloading in 2015. (*Source:* Recording Industry Association of America)

Application EXAMPLE 6 **Radioactive Decay**

Radioactive iodine-125 is a by-product of some types of nuclear reactors. Its half-life is 60 days. That is, after 60 days, a given amount of radioactive iodine-125 will have decayed to half the original amount. A nuclear accident occurs and releases 20 grams of radioactive iodine-125. How long will it take for the radioactive iodine to decay to 1 gram?

SOLUTION

Use the model for exponential decay, $y = Ce^{kt}$, and the information given in the problem to set up the table shown. Because

t (days)	Ce^{kt} (grams)
0	$Ce^{k(0)} = 20$
60	$Ce^{k(60)} = 10$
?	$Ce^{k(t)} = 1$

$$Ce^{k(0)} = Ce^0 = 20$$

you can conclude that

$$C = 20.$$

Then, using this value of C, you can solve for k as follows.

$Ce^{k(60)} = 10$	From table
$20e^{60k} = 10$	Substitute 20 for C.
$e^{60k} = \dfrac{1}{2}$	Divide each side by 20.
$60k = \ln \dfrac{1}{2}$	Inverse property
$k = \dfrac{1}{60} \ln \dfrac{1}{2}$	Divide each side by 60.
$k \approx -0.01155$	Simplify.

Finally, you can use this value of k in the model from the table to find the time when the amount is 1 gram, as follows.

$Ce^{kt} = 1$	From table
$20e^{-0.01155t} = 1$	Substitute for C and k.
$e^{-0.01155t} = \dfrac{1}{20}$	Divide each side by 20.
$-0.01155t = \ln \dfrac{1}{20}$	Inverse property
$t = \dfrac{1}{-0.01155} \ln \dfrac{1}{20}$	Divide each side by -0.01155.
$t \approx 259.4$ days	Simplify.

So, 20 grams of radioactive iodine-125 will decay to 1 gram after about 259.4 days. This solution is shown graphically at the left.

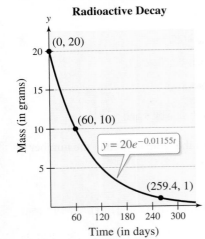

Radioactive Decay

(0, 20)

(60, 10)

$y = 20e^{-0.01155t}$

(259.4, 1)

Mass (in grams)

Time (in days)

Exercises Within Reach ®   Solutions in English & Spanish and tutorial videos at CollegePrepAlgebra.com

21. ***Radioactive Decay*** Radioactive radium (^{226}Ra) has a half-life of 1620 years. Starting with 5 grams of this substance, how much will remain after 1000 years?

22. ***Radioactive Decay*** Carbon 14 (^{14}C) has a half-life of 5730 years. Starting with 5 grams of this substance, how much will remain after 1000 years?

Intensity Models

On the Richter scale, the magnitude R of an earthquake is given by the intensity model

$$R = \log_{10} I$$

where I is the intensity of the shock wave.

Application   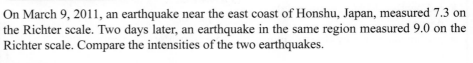 EXAMPLE 7   **Earthquake Intensity**

On March 9, 2011, an earthquake near the east coast of Honshu, Japan, measured 7.3 on the Richter scale. Two days later, an earthquake in the same region measured 9.0 on the Richter scale. Compare the intensities of the two earthquakes.

SOLUTION

The intensity of the earthquake on March 9 is given as follows.

$R = \log_{10} I$	Intensity model
$7.3 = \log_{10} I$	Substitute 7.3 for R.
$10^{7.3} = I$	Inverse property

On March 11, 2011, an earthquake of magnitude 9.0 in Honshu, Japan, had devastating results. It left more than 330,000 buildings destroyed or damaged, 130,900 people displaced, and 15,700 people dead.

The intensity of the earthquake on March 11 can be found in a similar way.

$R = \log_{10} I$	Intensity model
$9.0 = \log_{10} I$	Substitute 9.0 for R.
$10^{9.0} = I$	Inverse property

The ratio of these two intensities is

$$\frac{I \text{ for March 11}}{I \text{ for March 9}} = \frac{10^{9.0}}{10^{7.3}}$$

$$= 10^{9.0 - 7.3}$$

$$= 10^{1.7}$$

$$\approx 50.$$

So, the earthquake on March 11 had an intensity that was about 50 times greater than the intensity of the earthquake on March 9.

Exercises Within Reach ®

Solutions in English & Spanish and tutorial videos at CollegePrepAlgebra.com

Earthquake Intensity In Exercises 23−26, compare the intensities of the two earthquakes.

	Location	Date	Magnitude		Location	Date	Magnitude
23.	Chile	5/22/1960	9.5	**24.**	Alaska	3/28/1964	9.2
	Chile	2/11/2011	6.8		Alaska	9/2/2011	6.8
25.	New Zealand	6/13/2011	6.0	**26.**	India	9/18/2011	6.9
	New Zealand	7/6/2011	7.6		India	10/29/2011	3.5

Concept Summary: Using Exponential and Logarithmic Equations to Solve Problems

What

Many real-life situations involve exponential and logarithmic equations.

- investing (compound interest)
- population growth (**exponential growth**)
- radioactive decay (**exponential decay**)
- earthquake intensity (logarithmic equation)

EXAMPLE

How long will it take an investment to double at an annual interest rate of 6%, compounded continuously?

How

You can use a formula containing an exponential equation to solve problems involving continuous compounding.

EXAMPLE

$$A = Pe^{rt}$$

$$2P = Pe^{0.06t}$$

$$2 = e^{0.06t}$$

$$\ln 2 = 0.06t$$

$$11.55 \approx t$$

Why

Knowing the correct formula to use is the first step in solving many real-life problems involving exponential and logarithmic equations. Here are a few other formulas you should know.

Compound interest (n compoundings per year):

$$A = P\left(1 + \frac{r}{n}\right)^{nt}$$

Exponential growth or decay:

$$y = Ce^{kt}$$

Earthquake magnitude: $R = \log_{10} I$

Exercises Within Reach®

Worked-out solutions to odd-numbered exercises at CollegePrepAlgebra.com

Concept Summary Check

27. Choosing a Formula Why is the formula $A = Pe^{rt}$ used in the solution above?

28. Using a Formula Why is $2P$ substituted for A in the solution above?

29. Using a Formula Why is 0.06 substituted for r in the solution above?

30. Interpreting an Equation Does the equation $120 = 16e^{0.2t}$ represent a situation involving exponential growth or exponential decay?

Extra Practice

Doubling Time In Exercises 31–34, find the time for the investment to double.

	Principal	Rate	Compounding
31.	$2500	7.5%	Monthly
33.	$900	$5\frac{3}{4}$%	Quarterly

	Principal	Rate	Compounding
32.	$250	6.5%	Yearly
34.	$1500	$7\frac{1}{4}$%	Monthly

Exponential Growth and Decay In Exercises 35–38, find the constant k such that the graph of $y = Ce^{kt}$ passes through the points.

35.

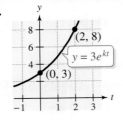

36.

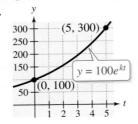

37.

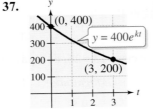

38.

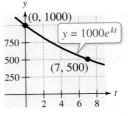

Acidity In Exercises 39–42, use the acidity model $pH = -\log_{10}[H^+]$, where acidity (pH) is a measure of the hydrogen ion concentration $[H^+]$ (measured in moles of hydrogen per liter) of a solution.

39. Find the pH of a solution that has a hydrogen ion concentration of 9.2×10^{-8}.

40. Compute the hydrogen ion concentration of a solution that has a pH of 4.7.

41. A blueberry has a pH of 2.5 and a liquid antacid has a pH of 9.5. The hydrogen ion concentration of the blueberry is how many times the concentration of the antacid?

42. When pH of a solution decreases by 1 unit, the hydrogen ion concentration increases by what factor?

43. *World Population* Projections of the world population P (in billions) every 5 years from 2010 through 2050 can be modeled by

$$P = \frac{10.9}{1 + 0.80e^{-0.031t}}, \quad 10 \le t \le 50$$

where t represents the year, with $t = 10$ corresponding to 2010 (see figure). Use the model to predict the world population in 2022. (*Source:* U.S. Census Bureau)

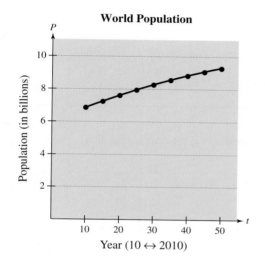

World Population

44. *World Population* Use the model P given in Exercise 43 to predict the world population in 2038.

45. *Population Growth* The population p of a species of wild rabbit t years after it is introduced into a new habitat is given by

$$p(t) = \frac{5000}{1 + 4e^{-t/6}}.$$

(a) Determine the size of the population of rabbits that was introduced into the habitat.

(b) Determine the size of the population of rabbits after 9 years.

(c) After how many years will the size of the population of rabbits be 2000?

46. *Advertising Effect* The sales S (in thousands of units) of a brand of jeans after the company spends x hundred dollars in advertising are given by

$$S = 10(1 - e^{kx}).$$

(a) Write S as a function of x given that 2500 pairs of jeans are sold when $500 is spent on advertising.

(b) How many pairs of jeans are sold when $700 is spent on advertising?

Explaining Concepts

47. *Structure* Explain how to determine whether an exponential model of the form $y = Ce^{kt}$ models growth or decay.

48. *Structure* The formulas for periodic and continuous compounding have the four variables A, P, r, and t in common. Explain what each variable measures.

49. *Think About It* For what types of compounding is the effective yield on an investment greater than the annual interest rate? Explain.

50. *Number Sense* If the reading on the Richter scale is increased by 1, the intensity of the earthquake is increased by what factor? Explain.

Cumulative Review

In Exercises 51–54, solve the equation by using the Quadratic Formula.

51. $x^2 - 7x - 5 = 0$ **52.** $x^2 + 5x - 3 = 0$

53. $3x^2 + 9x + 4 = 0$ **54.** $3x^2 + 4x = -2x + 5$

In Exercises 55–58, solve the inequality and graph the solution on the real number line.

55. $\dfrac{4}{x - 4} > 0$ **56.** $\dfrac{x - 1}{x + 2} < 0$

57. $\dfrac{2x}{x - 3} > 1$ **58.** $\dfrac{x - 5}{x + 2} \le -1$

11 Chapter Summary

	What did you learn?	*Explanation and Examples*	*Review Exercises*
11.1	Evaluate and graph exponential functions *(p. 560)*.	*Graph of $y = a^x$* Domain: $(-\infty, \infty)$ Range: $(0, \infty)$ Intercept: $(0, 1)$ Increases from left to right. The x-axis is a horizontal asymptote.	1–10
	Evaluate the natural base e and graph natural exponential functions *(p. 565)*.	The natural exponential function is simply an exponential function with a special base, the natural base $e \approx 2.71828$.	11–16
	Use exponential functions to solve application problems *(p. 566)*.	*Formulas for compound interest:* **1.** For n compoundings per year: $A = P\left(1 + \dfrac{r}{n}\right)^{nt}$ **2.** For continuous compounding: $A = Pe^{rt}$	17–20
11.2	Form composite functions and find the domains of composite functions *(p. 570)*.	$(f \circ g)(x) = f(g(x))$ The domain of $(f \circ g)$ is the set of all x in the domain of g such that $g(x)$ is in the domain of f.	21–26
	Find inverse functions algebraically *(p. 573)*.	**1.** In the equation for $f(x)$, replace $f(x)$ with y. **2.** Interchange x and y. **3.** Solve for y. (If y is not a function of x, the original equation does not have an inverse.) **4.** Replace y with $f^{-1}(x)$. **5.** Verify that $f(f^{-1}(x)) = x = f^{-1}(f(x))$.	27–34
	Compare the graph of a function with the graph of its inverse *(p. 574)*.		35, 36
11.3	Evaluate and graph logarithmic functions *(p. 578)*.	Let a and x be positive real numbers such that $a \neq 1$. $y = \log_a x$ if and only if $x = a^y$ *Graph of $y = \log_a x$* Domain: $(0, \infty)$ Range: $(-\infty, \infty)$ Intercept: $(1, 0)$ Increases from left to right. The y-axis is a vertical asymptote.	37–50

	What did you learn?	Explanation and Examples	Review Exercises
11.3	Evaluate and graph natural logarithmic functions *(p. 582)*.	The natural logarithmic function $\ln x$ is defined by $f(x) = \log_e x = \ln x$.	51–56
	Use the change-of-base formula to evaluate logarithms *(p. 583)*.	$\log_a x = \dfrac{\log_b x}{\log_b a}$ or $\log_a x = \dfrac{\ln x}{\ln a}$	57–60
11.4	Use the properties of logarithms to evaluate logarithms *(p. 588)*.	Let a be a positive real number such that $a \neq 1$, and let n be a real number. If u and v are real numbers, variables, or algebraic expressions such that $u > 0$ and $v > 0$, then the following properties are true. *Logarithm with base a* *Natural logarithm* **1.** $\log_a(uv) = \log_a u + \log_a v$ $\ln(uv) = \ln u + \ln v$ **2.** $\log_a \dfrac{u}{v} = \log_a u - \log_a v$ $\ln \dfrac{u}{v} = \ln u - \ln v$ **3.** $\log_a u^n = n \log_a u$ $\ln u^n = n \ln u$	61–66
	Rewrite, expand, or condense logarithmic expressions *(p. 590)*.	In the properties of logarithms stated above, the left side of each equation gives the condensed form and the right side gives the expanded form.	67–86
	Use the properties of logarithms to solve application problems *(p. 592)*.	The relationship between the number B of decibels and the intensity I of a sound (in watts per square centimeter) is given by $B = 10 \log_{10}\left(\dfrac{I}{10^{-16}}\right)$.	87, 88
11.5	Use one-to-one properties to solve exponential and logarithmic equations *(p. 596)*.	Let a be a positive real number such that $a \neq 1$, and let x and y be real numbers. Then the following properties are true. **1.** $a^x = a^y$ if and only if $x = y$. **2.** $\log_a x = \log_a y$ if and only if $x = y$ $(x > 0, y > 0)$.	89–94
	Use inverse properties to solve exponential and logarithmic equations *(p. 597)*.	*Base a* *Natural base e* **1.** $\log_a(a^x) = x$ $\ln(e^x) = x$ **2.** $a^{(\log_a x)} = x$ $e^{(\ln x)} = x$	95–110
	Use exponential or logarithmic equations to solve application problems *(p. 601)*.	Exponential or logarithmic equations can be used to solve many real-life problems involving compound interest and the physical sciences.	111, 112
11.6	Use exponential equations to solve compound interest problems *(p. 604)*.	In the following compound interest formulas, A is the balance, P is the principal, r is the annual interest rate (in decimal form), and t is the time in years. *n Compoundings per Year:* $A = P\left(1 + \dfrac{r}{n}\right)^{nt}$ *Continuous Compounding:* $A = Pe^{rt}$	113–124
	Use exponential equations to solve growth and decay problems *(p. 607)*.	The mathematical model for exponential growth or decay is $y = Ce^{kt}$ where C is the original amount, y is the amount after time t, and k is a constant determined by the rate of growth (or decay). The model represents growth when $k > 0$ and decay when $k < 0$.	125, 126
	Use logarithmic problems to solve intensity problems *(p. 609)*.	On the Richter scale, the magnitude R of an earthquake is given by the intensity model $R = \log_{10} I$, where I is the intensity of the shock wave.	127, 128

Review Exercises

Worked-out solutions to odd-numbered exercises at CollegePrepAlgebra.com

11.1

Evaluating an Exponential Function In Exercises 1−4, evaluate the function as indicated. Use a calculator only if it is necessary or more efficient. (Round your answers to three decimal places.)

1. $f(x) = 4^x$

    (a) $x = -3$

    (b) $x = 1$

    (c) $x = 2$

2. $g(x) = 4^{-x}$

    (a) $x = -2$

    (b) $x = 0$

    (c) $x = 2$

3. $g(t) = 5^{-t/3}$

    (a) $t = -3$

    (b) $t = \pi$

    (c) $t = 6$

4. $h(s) = 1 - 3^{0.2s}$

    (a) $s = 0$

    (b) $s = 2$

    (c) $s = \sqrt{10}$

Graphing an Exponential Function In Exercises 5−10, sketch the graph of the function.

5. $f(x) = 3^x$

6. $f(x) = 3^{-x}$

7. $f(x) = 3^x - 3$

8. $f(x) = 3^x + 5$

9. $f(x) = 3^{x+1}$

10. $f(x) = 3^{x-1}$

Evaluating an Exponential Function In Exercises 11 and 12, use a calculator to evaluate the exponential function as indicated. (Round your answers to three decimal places, if necessary.)

11. $f(x) = 3e^{-2x}$

    (a) $x = 3$

    (b) $x = 0$

    (c) $x = -19$

12. $g(x) = e^{x/5} + 11$

    (a) $x = 12$

    (b) $x = -8$

    (c) $x = 18.4$

Transformations of Graphs In Exercises 13−16, use transformations to analyze and sketch the graph of the function.

13. $y = e^{-x} + 1$

14. $y = -e^x + 1$

15. $g(x) = e^{x+2}$

16. $h(t) = e^{x-2}$

Compound Interest In Exercises 17 and 18, complete the table to determine the balance A for P dollars invested at rate r for t years, compounded n times per year.

n	1	4	12	365	Continuous compounding
A					

	Principal	Rate	Time
17.	$P = \$5000$	$r = 10\%$	$t = 40$ years
18.	$P = \$10{,}000$	$r = 9.5\%$	$t = 30$ years

19. **Radioactive Decay** After t years, 21 grams of a radioactive element with a half-life of 25 years decays to a mass y (in grams) given by $y = 21\left(\frac{1}{2}\right)^{t/25}$, $t \geq 0$. How much of the initial mass remains after 58 years?

20. **Depreciation** The value of a truck that originally cost $38,000 depreciates so that each year it is worth $\frac{2}{3}$ of its value from the previous year. Find a model for $V(t)$, the value of the truck after t years. Sketch a graph of the model, and determine the value of the truck 6 years after its purchase.

11.2

Finding the Composition of Two Functions In Exercises 21−24, find the compositions.

21. $f(x) = x + 2, g(x) = x^2$

    (a) $(f \circ g)(2)$ (b) $(g \circ f)(-1)$

22. $f(x) = \sqrt[3]{x}$, $g(x) = x + 2$

    (a) $(f \circ g)(6)$ (b) $(g \circ f)(64)$

23. $f(x) = \sqrt{x + 1}$, $g(x) = x^2 - 1$

    (a) $(f \circ g)(5)$ (b) $(g \circ f)(-1)$

24. $f(x) = \dfrac{1}{x - 4}$, $g(x) = \dfrac{x + 1}{2x}$

    (a) $(f \circ g)(1)$ (b) $(g \circ f)\left(\dfrac{1}{5}\right)$

Finding the Domains of Composite Functions In Exercises 25 and 26, find the compositions (a) $f \circ g$ and (b) $g \circ f$. Then find the domain of each composition.

25. $f(x) = \sqrt{x + 6}$, $g(x) = 2x$

26. $f(x) = \dfrac{2}{x - 4}$, $g(x) = x^2$

Applying the Horizontal Line Test In Exercises 27 and 28, use the Horizontal Line Test to determine whether the function is one-to-one and so has an inverse function.

27. $f(x) = x^2 - 25$ **28.** $f(x) = \frac{1}{4}x^3$

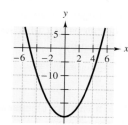
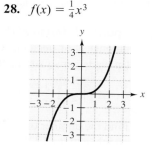

Finding an Inverse Function In Exercises 29−34, find the inverse function.

29. $f(x) = 3x + 4$

30. $f(x) = 2x - 3$

31. $h(x) = \sqrt{5x}$

32. $g(x) = x^2 + 2$, $x \geq 0$

33. $f(t) = t^3 + 4$

34. $h(t) = \sqrt[3]{t - 1}$

Verifying Inverse Functions Graphically In Exercises 35 and 36, sketch the graphs of f and g on the same rectangular coordinate system. Plot the indicated points on the graph of f. Then plot the reflections of these points in the line $y = x$ to verify that f and g are inverse functions.

35. $f(x) = 3x + 4$

    $g(x) = \frac{1}{3}(x - 4)$

36. $f(x) = \frac{1}{3}\sqrt[3]{x}$

    $g(x) = 27x^3$

11.3

Evaluating a Logarithm In Exercises 37−44, evaluate the logarithm.

37. $\log_{10} 1000$

38. $\log_{27} 3$

39. $\log_3 \frac{1}{9}$

40. $\log_4 \frac{1}{16}$

41. $\log_2 64$

42. $\log_{10} 0.01$

43. $\log_3 1$

44. $\log_2 \sqrt{4}$

Graphing a Logarithmic Function In Exercises 45−50, sketch the graph of the function. Identify the vertical asymptote.

45. $f(x) = \log_3 x$

46. $f(x) = -\log_3 x$

47. $f(x) = -1 + \log_3 x$

48. $f(x) = 1 + \log_3 x$

49. $f(x) = \log_2(x - 4)$

50. $f(x) = \log_4(x + 1)$

Evaluating a Natural Logarithm In Exercises 51 and 52, evaluate the expression.

51. $\ln e^7$

52. $\ln \dfrac{1}{e^3}$

Graphing a Natural Logarithmic Function In Exercises 53–56, sketch the graph of the function.

53. $y = \ln(x - 3)$

54. $y = -\ln(x + 2)$

55. $y = 5 - \ln x$

56. $y = 3 + \ln x$

Using the Change-of-Base Formula In Exercises 57–60, use (a) common logarithms and (b) natural logarithms to evaluate the expression. (Round your answer to four decimal places.)

57. $\log_4 9$

58. $\log_{1/2} 5$

59. $\log_8 160$

60. $\log_3 0.28$

11.4

Using Properties of Logarithms In Exercises 61–66, use $\log_5 2 \approx 0.4307$ and $\log_5 3 \approx 0.6826$ to approximate the expression. (Round your answer to four decimal places.)

61. $\log_5 18$

62. $\log_5 \sqrt{6}$

63. $\log_5 \dfrac{1}{2}$

64. $\log_5 \dfrac{2}{3}$

65. $\log_5 (12)^{2/3}$

66. $\log_5 (5^2 \cdot 6)$

Expanding a Logarithmic Expression In Exercises 67–72, use the properties of logarithms to expand the expression. (Assume all variables are positive.)

67. $\log_4 6x^4$

68. $\log_{12} 2x^{-5}$

69. $\log_5 \sqrt{x + 2}$

70. $\ln \sqrt[3]{\dfrac{x}{5}}$

71. $\ln \dfrac{x + 2}{x + 3}$

72. $\ln x(x + 4)^2$

Condensing a Logarithmic Expression In Exercises 73–80, use the properties of logarithms to condense the expression.

73. $5 \log_2 y$

74. $-\dfrac{2}{3} \ln 3y$

75. $\log_8 16x + \log_8 2x^2$

76. $\log_4 6x - \log_4 10$

77. $-2(\ln 2x - \ln 3)$

78. $5(1 + \ln x + \ln 2)$

79. $4[\log_2 k - \log_2(k - t)]$

80. $\dfrac{1}{3}(\log_8 a + 2 \log_8 b)$

True or False? In Exercises 81–86, use the properties of logarithms to determine whether the equation is true or false. If it is false, state why or give an example to show that it is false.

81. $\log_2 4x = 2 \log_2 x$

82. $\dfrac{\ln 5x}{\ln 10x} = \ln \dfrac{1}{2}$

83. $\log_{10} 10^{2x} = 2x$

84. $e^{\ln t} = t, \ t > 0$

85. $\log_4 \dfrac{16}{x} = 2 - \log_4 x$

86. $6 \ln x + 6 \ln y = \ln(xy)^6, \ x > 0, y > 0$

87. **Sound Intensity** The relationship between the number B of decibels and the intensity I of a sound (in watts per square meter) is given by

$$B = 10 \log_{10}\left(\frac{I}{10^{-12}}\right).$$

Find the number of decibels of a vacuum cleaner with an intensity of 10^{-4} watt per square meter.

88. **Human Memory Model** A psychologist finds that the percent p of retention in a group of subjects can be modeled by

$$p = \frac{\log_{10}(10^{68})}{\log_{10}(t + 1)^{20}}$$

where t is the time in months after the subjects' initial testing. Use properties of logarithms to write the formula in simpler form, and determine the percent of retention after 5 months.

11.5

Using a One-to-one Property In Exercises 89−94, solve the equation.

89. $2^x = 64$

90. $6^x = 216$

91. $4^{x-3} = \frac{1}{16}$

92. $3^{x-2} = 81$

93. $\log_7(x + 6) = \log_7 12$

94. $\ln(8 - x) = \ln 3$

Solving an Exponential Equation In Exercises 95−100, solve the exponential equation, if possible. (Round your answer to two decimal places.)

95. $3^x = 500$

96. $8^x = 1000$

97. $2e^{0.5x} = 45$

98. $125e^{-0.4x} = 40$

99. $12(1 - 4^x) = 18$

100. $25(1 - e^t) = 12$

Solving a Logarithmic Equation In Exercises 101−110, solve the logarithmic equation. (Round your answer to two decimal places.)

101. $\ln x = 7.25$

102. $\ln x = -0.5$

103. $\log_{10} 4x = 2.1$

104. $\log_2 2x = -0.65$

105. $\log_3(2x + 1) = 2$

106. $\log_5(x - 10) = 2$

107. $\frac{1}{3} \log_2 x + 5 = 7$

108. $4 \log_5(x + 1) = 4.8$

109. $\log_3 x + \log_3 7 = 4$

110. $2 \log_4 x - \log_4(x - 1) = 1$

111. Compound Interest A deposit of $5000 is placed in a savings account for 2 years. The interest is compounded continuously. At the end of 2 years, the balance in the account is $5751.37. What is the annual interest rate for this account?

112. Sound Intensity The relationship between the number B of decibels and the intensity I of a sound (in watts per square centimeter) is given by

$$B = 10 \log_{10}\left(\frac{I}{10^{-16}}\right).$$

Determine the intensity of a firework display I that registers 130 decibels on a decibel meter.

11.6

Finding an Annual Interest Rate In Exercises 113−118, find the annual interest rate.

	Principal	Balance	Time	Compounding
113.	$250	$410.90	10 years	Quarterly
114.	$1000	$1348.85	5 years	Monthly
115.	$5000	$15,399.30	15 years	Daily
116.	$10,000	$35,236.45	20 years	Yearly
117.	$1800	$46,422.61	50 years	Continuous
118.	$7500	$15,877.50	15 years	Continuous

Finding an Effective Yield In Exercises 119−124, find the effective yield.

	Rate	Compounding
119.	5.5%	Daily
120.	6%	Monthly
121.	7.5%	Quarterly
122.	8%	Yearly
123.	7.5%	Continuously
124.	3.75%	Continuously

Radioactive Decay In Exercises 125 and 126, determine the amount of the radioactive isotope that will remain after 1000 years.

	Isotope	Half-Life (Years)	Initial Quantity	Amount After 1000 Years
125.	^{226}Ra	1620	3.5 g	▬ g
126.	^{14}C	5730	10 g	▬ g

Earthquake Intensity In Exercises 127 and 128, compare the intensities of the two earthquakes.

	Location	Date	Magnitude
127.	San Francisco, California	4/18/1906	8.3
	San Francisco, California	3/5/2012	4.0
128.	Virginia	8/23/2011	5.8
	Colorado	8/23/2011	5.3

Chapter Test

Solutions in English & Spanish and tutorial videos at CollegePrepAlgebra.com

Take this test as you would take a test in class. After you are done, check your work against the answers in the back of the book.

1. Evaluate $f(t) = 54\left(\frac{2}{3}\right)^t$ when $t = -1, 0, \frac{1}{2}$, and 2.

2. Sketch a graph of the function $f(x) = 2^{x-5}$ and identify the horizontal asymptote.

3. Find the compositions (a) $(f \circ g)$ and (b) $(g \circ f)$. Then find the domain of each composition.
$$f(x) = 2x^2 + x \qquad g(x) = 5 - 3x$$

4. Find the inverse function of $f(x) = 9x - 4$.

5. Verify algebraically that the functions f and g are inverse functions of each other.
$$f(x) = -\tfrac{1}{2}x + 3 \qquad g(x) = -2x + 6$$

6. Evaluate $\log_4 \frac{1}{256}$ without a calculator.

7. Describe the relationship between the graphs of $f(x) = \log_5 x$ and $g(x) = 5^x$.

8. Use the properties of logarithms to expand $\log_8\left(4\sqrt{x}/y^4\right)$.

9. Use the properties of logarithms to condense $\ln x - 4 \ln y$.

In Exercises 10–17, solve the equation. Round your answer to two decimal places, if necessary.

10. $\log_2 x = 5$

11. $9^{2x} = 182$

12. $400e^{0.08t} = 1200$

13. $3 \ln(2x - 3) = 10$

14. $12(7 - 2^x) = -300$

15. $\log_2 x + \log_2 4 = 5$

16. $\ln x - \ln 2 = 4$

17. $30(e^x + 9) = 300$

18. Determine the balance after 20 years when $2000 is invested at 7% compounded (a) quarterly and (b) continuously.

19. Determine the principal that will yield $100,000 when invested at 9% compounded quarterly for 25 years.

20. A principal of $500 yields a balance of $1006.88 in 10 years when the interest is compounded continuously. What is the annual interest rate?

21. A car that originally cost $20,000 has a depreciated value of $15,000 after 1 year. Find the value of the car when it is 5 years old by using the exponential model $y = Ce^{kt}$.

In Exercises 22–24, the population p of a species of fox t years after it is introduced into a new habitat is given by

$$p(t) = \frac{2400}{1 + 3e^{-t/4}}.$$

22. Determine the size of the population that was introduced into the habitat.

23. Determine the size of the population after 4 years.

24. After how many years will the size of the population be 1200?

12

Conics

12.1 Circles and Parabolas

12.2 Ellipses

12.3 Hyperbolas

12.4 Solving Nonlinear Systems of Equations

MASTERY IS WITHIN REACH!

"My teacher told me that if I just put a little more effort into studying and getting ready for the final, I could possibly get an A. So, I pulled out my old tests. That is when I noticed that most of my mistakes involved word problems and not reading directions carefully. I got help from a tutor on word problems and made sure I correctly read the instructions on the final. It worked."

Franklin

See page 637 for suggestions about avoiding test-taking errors.

maximino/Shutterstock.com

12.1 Circles and Parabolas

▶ Recognize the four basic conics: circles, parabolas, ellipses, and hyperbolas.
▶ Graph and write equations of circles.
▶ Graph and write equations of parabolas.

The Conics

In Section 10.4, you saw that the graph of a second-degree equation of the form $y = ax^2 + bx + c$ is a parabola. A parabola is one of four types of **conics** or **conic sections**. The other three types are circles, ellipses, and hyperbolas. All four types have equations of second degree. Each figure can be obtained by intersecting a plane with a double-napped cone, as shown below.

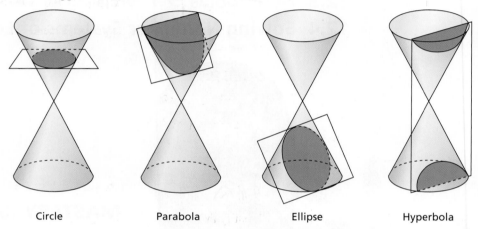

Circle   Parabola   Ellipse   Hyperbola

Conics occur in many practical applications. Reflective surfaces in satellite dishes, flashlights, and telescopes often have a parabolic shape. The orbits of planets are elliptical, and the orbits of comets are usually elliptical or hyperbolic. Ellipses and parabolas are also used in building archways and bridges.

Exercises Within Reach ®

Identifying a Conic In Exercises 1–4, Identify the type of conic shown in the graph.

1.

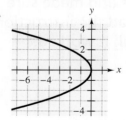

2.

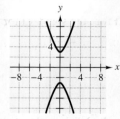

3.

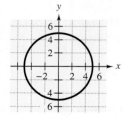

4.

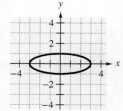

Circles

A **circle** in the rectangular coordinate system consists of all points (x, y) that are a given positive distance r from a fixed point, called the **center** of the circle. The distance r is called the **radius** of the circle.

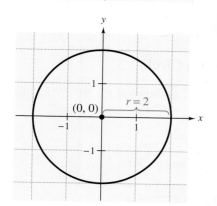

Standard Equation of a Circle (Center at Origin)

The **standard form of the equation of a circle centered at the origin** is

$$x^2 + y^2 = r^2. \qquad \text{Circle with center at } (0, 0)$$

The positive number r is called the **radius** of the circle.

EXAMPLE 1   **Writing an Equation of a Circle**

Write an equation of the circle that is centered at the origin and has a radius of 2, as shown at the left.

SOLUTION

$$x^2 + y^2 = r^2 \qquad \text{Standard form with center at } (0, 0)$$
$$x^2 + y^2 = 2^2 \qquad \text{Substitute 2 for } r.$$
$$x^2 + y^2 = 4 \qquad \text{Equation of circle}$$

EXAMPLE 2   **Sketching a Circle**

Identify the radius of the circle given by $4x^2 + 4y^2 - 25 = 0$. Then sketch the circle.

SOLUTION

Begin by writing the equation in standard form.

$$4x^2 + 4y^2 - 25 = 0 \qquad \text{Write original equation.}$$
$$4x^2 + 4y^2 = 25 \qquad \text{Add 25 to each side.}$$
$$x^2 + y^2 = \frac{25}{4} \qquad \text{Divide each side by 4.}$$
$$x^2 + y^2 = \left(\frac{5}{2}\right)^2 \qquad \text{Standard form}$$

From the standard form of the equation of this circle centered at the origin, you can see that the radius is $\frac{5}{2}$. The graph of the circle is shown at the left.

Exercises Within Reach ®

Solutions in English & Spanish and tutorial videos at CollegePrepAlgebra.com

Writing an Equation of a Circle **In Exercises 5–8, write the standard form of the equation of the circle centered at the origin.**

5. Radius: 5   **6.** Radius: 9   **7.** Radius: $\frac{2}{3}$ **8.** Radius: $\frac{5}{2}$

Sketching a Circle **In Exercises 9–12, identify the center and radius of the circle, and sketch its graph.**

9. $x^2 + y^2 = 16$ **10.** $x^2 + y^2 = 1$ **11.** $x^2 + y^2 = 36$ **12.** $x^2 + y^2 = 15$

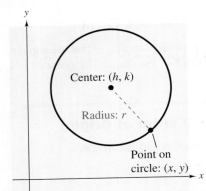

Consider a circle whose radius is r and whose center is the point (h, k), as shown at the left. Let (x, y) be any point on the circle. To find an equation for this circle, you can use a variation of the Distance Formula and write

$$\text{Radius} = r = \sqrt{(x - h)^2 + (y - k)^2}.$$ Distance Formula

By squaring each side of this equation, you obtain the equation shown below, which is called the *standard form of the equation of a circle centered at (h, k)*.

Standard Equation of a Circle [Center at (h, k)]

The **standard form of the equation of a circle centered at (h, k)** is

$$(x - h)^2 + (y - k)^2 = r^2.$$

EXAMPLE 3 **Writing an Equation of a Circle**

The point $(2, 5)$ lies on a circle whose center is $(5, 1)$, as shown at the left. Write the standard form of the equation of this circle.

SOLUTION

The radius r of the circle is the distance between $(2, 5)$ and $(5, 1)$.

$$
\begin{aligned}
r &= \sqrt{(2 - 5)^2 + (5 - 1)^2} && \text{Distance Formula}\\
&= \sqrt{(-3)^2 + 4^2} && \text{Simplify.}\\
&= \sqrt{9 + 16} && \text{Simplify.}\\
&= \sqrt{25} && \text{Simplify.}\\
&= 5 && \text{Radius}
\end{aligned}
$$

Using $(h, k) = (5, 1)$ and $r = 5$, the equation of the circle is

$$
\begin{aligned}
(x - h)^2 + (y - k)^2 &= r^2 && \text{Standard form}\\
(x - 5)^2 + (y - 1)^2 &= 5^2 && \text{Substitute for } h, k, \text{ and } r.\\
(x - 5)^2 + (y - 1)^2 &= 25. && \text{Equation of circle}
\end{aligned}
$$

From the graph, you can see that the center of the circle is shifted five units to the right and one unit upward from the origin.

Exercises Within Reach ®

Solutions in English & Spanish and tutorial videos at CollegePrepAlgebra.com

Writing an Equation of a Circle In Exercises 13−18, write **the standard form of the equation of the circle centered at (h, k).**

13. Center: $(4, 3)$
  Radius: 10

14. Center: $(-4, 8)$
  Radius: 7

15. Center: $(6, -5)$
  Radius: 3

16. Center: $(-5, -2)$
  Radius: $\frac{5}{2}$

17. Center: $(-2, 1)$
  Passes through the point $(0, 1)$

18. Center: $(8, 2)$
  Passes through the point $(8, 0)$

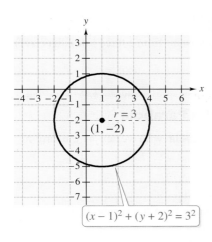

$(x-1)^2 + (y+2)^2 = 3^2$

EXAMPLE 4   **Writing an Equation in Standard Form**

Write the equation $x^2 + y^2 - 2x + 4y - 4 = 0$ in standard form. Then sketch the circle represented by the equation.

SOLUTION

$$x^2 + y^2 - 2x + 4y - 4 = 0 \qquad \text{Write original equation.}$$

$$\left(x^2 - 2x + \quad\right) + \left(y^2 + 4y + \quad\right) = 4 \qquad \text{Group terms.}$$

$$[x^2 - 2x + (-1)^2] + (y^2 + 4y + 2^2) = 4 + 1 + 4 \qquad \text{Complete the squares.}$$

$$(x - 1)^2 + (y + 2)^2 = 3^2 \qquad \text{Standard form}$$

The circle is centered at $(1, -2)$ with a radius of 3, as shown at the left.

Application   EXAMPLE 5   **Mechanical Drawing**

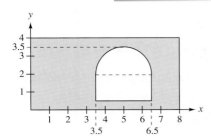

In a mechanical drawing class, you have to program a computer to model the metal piece shown at the left. Find an equation that represents the semicircular portion of the hole in the metal piece.

SOLUTION

The center of the circle is $(h, k) = (5, 2)$ and the radius of the circle is $r = 1.5$. This implies that the equation of the entire circle is

$$(x - h)^2 + (y - k)^2 = r^2 \qquad \text{Standard form}$$

$$(x - 5)^2 + (y - 2)^2 = 1.5^2 \qquad \text{Substitute for } h, k, \text{ and } r.$$

$$(x - 5)^2 + (y - 2)^2 = 2.25. \qquad \text{Equation of circle}$$

To find the equation of the upper portion of the circle, solve this standard equation for y.

$$(x - 5)^2 + (y - 2)^2 = 2.25$$

$$(y - 2)^2 = 2.25 - (x - 5)^2$$

$$y - 2 = \pm\sqrt{2.25 - (x - 5)^2}$$

$$y = 2 \pm \sqrt{2.25 - (x - 5)^2}$$

Finally, take the positive square root to obtain the equation of the upper portion of the circle.

$$y = 2 + \sqrt{2.25 - (x - 5)^2}$$

Exercises Within Reach ®

Solutions in English & Spanish and tutorial videos at CollegePrepAlgebra.com

Sketching a Circle **In Exercises 19 and 20, identify the center and radius of the circle, and sketch its graph.**

19. $x^2 + y^2 + 2x + 6y + 6 = 0$

20. $x^2 + y^2 + 6x - 4y - 3 = 0$

Dog Leash **A dog is leashed to a side of a house. The boundary has a diameter of 80 feet.**

21. Write an equation that represents the semicircle.

22. The dog stands on the semicircle, 10 feet from the fence, as shown at the right. Use the equation you found in Exercise 21 to find how far the dog is from the house.

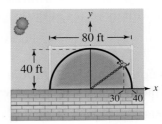

Parabolas

A **parabola** is the set of all points (x, y) that are equidistant from a fixed line (**directrix**) and a fixed point (**focus**) not on the line.

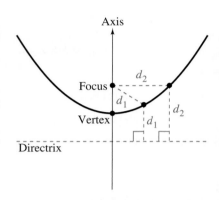

<table>
<tr><td colspan="2">

Standard Equation of a Parabola

The **standard form of the equation of a parabola** with vertex at the origin $(0, 0)$ is

$x^2 = 4py$, $p \neq 0$	Vertical axis
$y^2 = 4px$, $p \neq 0$.	Horizontal axis

The focus lies on the axis p units (*directed distance*) from the vertex. If the vertex is at (h, k), then the standard form of the equation is

$(x - h)^2 = 4p(y - k)$, $p \neq 0$	Vertical axis; directrix: $y = k - p$
$(y - k)^2 = 4p(x - h)$, $p \neq 0$.	Horizontal axis; directrix: $x = h - p$

</td></tr>
</table>

EXAMPLE 6   Writing an Equation of a Parabola

Write the standard form of the equation of the parabola with the given vertex and focus.

a. Vertex: $(0, 0)$
    Focus: $(0, -2)$

b. Vertex: $(3, -2)$
    Focus: $(4, -2)$

SOLUTION

a. $x^2 = 4py$

  $x^2 = 4(-2)y$

  $x^2 = -8y$

b. $(y - k)^2 = 4p(x - h)$

  $[y - (-2)]^2 = 4(1)(x - 3)$

  $(y + 2)^2 = 4(x - 3)$

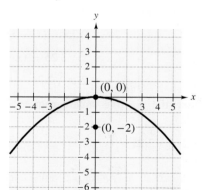

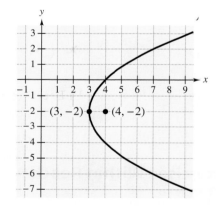

Study Tip

If the focus of a parabola is above or to the right of the vertex, p is positive. If the focus is below or to the left of the vertex, p is negative.

Exercises Within Reach ®

Solutions in English & Spanish and tutorial videos at CollegePrepAlgebra.com

Writing an Equation of a Parabola In Exercises 23−26, write the standard form of the equation of the parabola with its vertex at the origin.

23. Focus: $\left(0, -\frac{3}{2}\right)$ **24.** Focus: $\left(\frac{5}{4}, 0\right)$ **25.** Focus: $(-2, 0)$ **26.** Focus: $(0, -2)$

Writing an Equation of a Parabola In Exercises 27 and 28, write the standard form of the equation of the parabola with its vertex at (h, k).

27. Vertex: $(3, 2)$, Focus: $(1, 2)$

28. Vertex: $(-1, 2)$, Focus: $(-1, 0)$

EXAMPLE 7   **Sketching a Parabola**

Sketch the parabola given by $y = \frac{1}{8}x^2$, and identify its vertex and focus.

SOLUTION

Because the equation can be written in the standard form $x^2 = 4py$, it is a parabola whose vertex is at the origin. You can identify the focus of the parabola by writing its equation in standard form.

$y = \frac{1}{8}x^2$	Write original equation.
$\frac{1}{8}x^2 = y$	Interchange sides of the equation.
$x^2 = 8y$	Multiply each side by 8.
$x^2 = 4(2)y$	Rewrite 8 in the form $4p$.

From this standard form, you can see that $p = 2$. Because the parabola opens upward, as shown at the left, you can conclude that the focus lies $p = 2$ units above the vertex. So, the focus is $(0, 2)$.

Parabolas occur in a wide variety of applications. For instance, a parabolic reflector can be formed by revolving a parabola around its axis. The light rays emanating from the focus of a parabolic reflector used in a flashlight are all parallel to one another, as shown at the right.

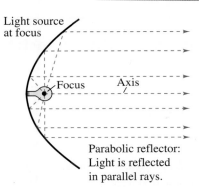

Parabolic reflector: Light is reflected in parallel rays.

Exercises Within Reach ®

Solutions in English & Spanish and tutorial videos at CollegePrepAlgebra.com

Sketching a Parabola In Exercises 29–42, sketch the parabola, and identify its vertex and focus.

29. $y = \frac{1}{2}x^2$

30. $y = 2x^2$

31. $y^2 = -10x$

32. $y^2 = 3x$

33. $x^2 + 8y = 0$

34. $x + y^2 = 0$

35. $(x - 1)^2 + 8(y + 2) = 0$

36. $(x + 3) + (y - 2)^2 = 0$

37. $\left(y + \frac{1}{2}\right)^2 = 2(x - 5)$

38. $\left(x + \frac{1}{2}\right)^2 = 4(y - 3)$

39. $y = \frac{1}{3}(x^2 - 2x + 10)$

40. $4x - y^2 - 2y - 33 = 0$

41. $y^2 + 6y + 8x + 25 = 0$

42. $y^2 - 4y - 4x = 0$

Concept Summary: *Writing and Graphing Equations of Circles*

What

The standard form of the equation of a **circle** with **center** (h, k) and **radius** r is

$$(x - h)^2 + (y - k)^2 = r^2.$$

EXAMPLE

Identify the center and radius of the circle given by

$$x^2 + y^2 - 4x - 2y + 1 = 0.$$

Then sketch the graph.

How

To identify the center and radius of the circle, write the equation in standard form.

EXAMPLE

$x^2 + y^2 - 4x - 2y + 1 = 0$ Original

$(x - 2)^2 + (y - 1)^2 = 4$ Standard form

From the equation, you can identify the center and radius.

Center: $(2, 1)$

Radius: $\sqrt{4} = 2$

Why

Knowing how to write equations of circles in standard form allows you to easily sketch the graphs of circles. For example, here is the graph of the circle $(x - 2)^2 + (y - 1)^2 = 4$.

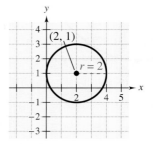

You can use a similar process to sketch parabolas.

Exercises Within Reach ®

Worked-out solutions to odd-numbered exercises at CollegePrepAlgebra.com

Concept Summary Check

43. *Reasoning* Explain how to identify the center and radius of the circle given by the equation $x^2 + y^2 - 36 = 0$.

44. *Translation* Is the center of the circle given by the equation $(x + 2)^2 + (y + 4)^2 = 20$ shifted two units to the right and four units upward from the origin? Explain your reasoning.

45. *Standard Form* Which standard form of the equation of a parabola should you use to write an equation for a parabola with vertex $(2, -3)$ and focus $(2, 1)$? Explain your reasoning.

46. *Think About It* Given the equation of a parabola, explain how to determine whether the parabola opens upward, downward, to the right, or to the left.

Extra Practice

Sketching a Circle In Exercises 47–50, **identify the center and radius of the circle, and sketch its graph.**

47. $\left(x + \frac{9}{4}\right)^2 + (y - 4)^2 = 16$

48. $(x - 5)^2 + \left(y + \frac{3}{4}\right)^2 = 1$

49. $x^2 + y^2 + 10x - 4y - 7 = 0$

50. $x^2 + y^2 - 14x + 8y + 56 = 0$

51. *Observation Wheel* Write an equation that represents the circular wheel of the Singapore Flyer in Singapore, which has a diameter of 150 meters. Place the origin of the rectangular coordinate system at the center of the wheel.

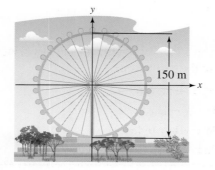

52. *Mirror* Write an equation that represents the circular mirror, with a diameter of 3 feet, shown in the figure. The wall hangers of the mirror are shown as two points on the circle. Use the equation to determine the height of the left wall hanger.

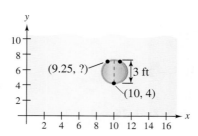

53. *Suspension Bridge* Each cable of a suspension bridge is suspended (in the shape of a parabola) between two towers that are 120 meters apart, and the top of each tower is 20 meters above the roadway. The cables touch the roadway at the midpoint between the two towers (see figure).

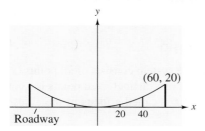

(a) Write an equation that represents the parabolic shape of each cable.

(b) Complete the table by finding the height of the suspension cables y over the roadway at a distance of x meters from the center of the bridge.

x	0	20	40	60
y				

54. *Beam Deflection* A simply supported beam is 16 meters long and has a load at the center (see figure). The deflection of the beam at its center is 3 centimeters. Assume that the shape of the deflected beam is parabolic.

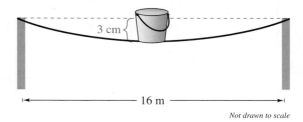

Not drawn to scale

(a) Write an equation of the parabola. (Assume that the origin is at the center of the deflected beam.)

(b) How far from the center of the beam is the deflection equal to 1 centimeter?

Explaining Concepts

55. *Precision* The point $(-4, 3)$ lies on a circle with center $(-1, 1)$. does the point $(3, 2)$ lie on the same circle? Explain your reasoning.

56. *Logic* A student claims that

$$x^2 + y^2 - 6y = -5$$

does not represent a circle. Is the student correct? Explain your reasoning.

57. *Reasoning* Is y a function of x in the equation $y^2 = 6x$? Explain.

58. *Reasoning* Is it possible for a parabola to intersect its directrix? Explain.

59. *Think About It* If the vertex and focus of a parabola are on a horizontal line, is the directrix of the parabola vertical? Explain.

Cumulative Review

In Exercises 60−65, solve the equation by completing the square.

60. $x^2 + 4x = 6$

61. $x^2 + 6x = -4$

62. $x^2 - 2x - 3 = 0$

63. $4x^2 - 12x - 10 = 0$

64. $2x^2 + 5x - 8 = 0$

65. $9x^2 - 12x = 14$

In Exercises 66−69, use the properties of logarithms to expand the expression. (Assume all variables are positive.)

66. $\log_8 x^{10}$

67. $\log_{10} \sqrt{xy^3}$

68. $\ln 5x^2y$

69. $\ln \dfrac{x}{y^4}$

In Exercises 70−73, use the properties of logarithms to condense the expression.

70. $\log_{10} x + \log_{10} 6$

71. $2 \log_3 x - \log_3 y$

72. $3 \ln x + \ln y - \ln 9$

73. $4(\ln x + \ln y) - \ln(x^4 + y^4)$

12.2 Ellipses

▶ Graph and write equations of ellipses centered at the origin.

▶ Graph and write equations of ellipses centered at (h, k).

Ellipses Centered at the Origin

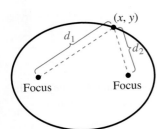

$d_1 + d_2$ is constant.

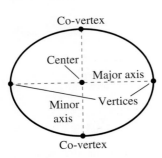

An **ellipse** in the rectangular coordinate system consists of all points (x, y) such that the sum of the distances between (x, y) and two distinct fixed points is a constant, as shown at the left. Each of the two fixed points is called a **focus** of the ellipse. (The plural of focus is *foci*.)

Standard Equation of an Ellipse (Center at Origin)

The **standard form of the equation of an ellipse centered at the origin** with major and minor axes of lengths $2a$ and $2b$ is

$$\frac{x^2}{a^2} + \frac{y^2}{b^2} = 1 \quad \text{or} \quad \frac{x^2}{b^2} + \frac{y^2}{a^2} = 1, \quad 0 < b < a.$$

The vertices lie on the major axis, a units from the center, and the co-vertices lie on the minor axis, b units from the center, as shown below.

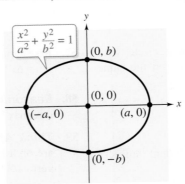

Major axis is horizontal.
Minor axis is vertical.

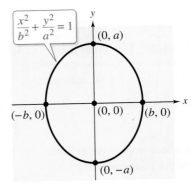

Major axis is vertical.
Minor axis is horizontal.

Exercises Within Reach ®

Solutions in English & Spanish and tutorial videos at CollegePrepAlgebra.com

Identifying the Standard Form In Exercises 1–4, determine whether the standard form of the equation of the ellipse is $\dfrac{x^2}{a^2} + \dfrac{y^2}{b^2} = 1$ or $\dfrac{x^2}{b^2} + \dfrac{y^2}{a^2} = 1$.

1.

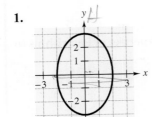

2.

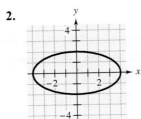

3.

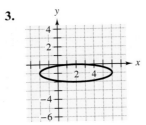

4.

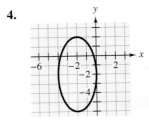

EXAMPLE 1   **Writing the Standard Equation of an Ellipse**

Write an equation of the ellipse that is centered at the origin, with vertices $(-3, 0)$ and $(3, 0)$ and co-vertices $(0, -2)$ and $(0, 2)$.

SOLUTION

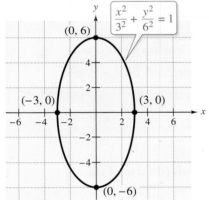

Begin by plotting the vertices and co-vertices, as shown at the left. The center of the ellipse is $(0, 0)$ and the major axis is horizontal. So, the standard form of the equation is

$$\frac{x^2}{a^2} + \frac{y^2}{b^2} = 1. \qquad \text{Major axis is horizontal.}$$

The distance from the center to either vertex is $a = 3$, and the distance from the center to either co-vertex is $b = 2$. So, the standard form of the equation of the ellipse is

$$\frac{x^2}{3^2} + \frac{y^2}{2^2} = 1. \qquad \text{Standard form}$$

EXAMPLE 2   **Sketching an Equation**

Sketch the ellipse given by $4x^2 + y^2 = 36$. Identify the vertices and co-vertices.

SOLUTION

To sketch an ellipse, it helps first to write its equation in standard form.

$$4x^2 + y^2 = 36 \qquad \text{Write original equation.}$$

$$\frac{x^2}{9} + \frac{y^2}{36} = 1 \qquad \text{Divide each side by 36 and simplify.}$$

$$\frac{x^2}{3^2} + \frac{y^2}{6^2} = 1 \qquad \text{Standard form}$$

Because the denominator of the y^2-term is larger than the denominator of the x^2-term, you can conclude that the major axis is vertical. Moreover, because $a = 6$, the vertices are $(0, -6)$ and $(0, 6)$. Finally, because $b = 3$, the co-vertices are $(-3, 0)$ and $(3, 0)$, as shown at the left.

Exercises Within Reach ®

Solutions in English & Spanish and tutorial videos at CollegePrepAlgebra.com

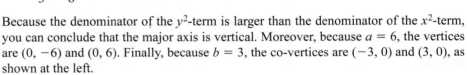

Writing the Standard Equation of an Ellipse In Exercises 5−8, write the standard form of the equation of the ellipse centered at the origin.

	Vertices	Co-vertices			Vertices	Co-vertices
5.	$(-4, 0), (4, 0)$	$(0, -3), (0, 3)$		**6.**	$(-2, 0), (2, 0)$	$(0, -1), (0, 1)$
7.	$(0, -6), (0, 6)$	$(-3, 0), (3, 0)$		**8.**	$(0, -5), (0, 5)$	$(-1, 0), (1, 0)$

Sketching an Ellipse In Exercises 9−12, sketch the ellipse. Identify the vertices and co-vertices.

9. $\dfrac{x^2}{16} + \dfrac{y^2}{4} = 1$

10. $\dfrac{x^2}{9} + \dfrac{y^2}{25} = 1$

11. $4x^2 + y^2 = 4$

12. $4x^2 + 9y^2 = 36$

Ellipses Centered at (*h, k*)

Standard Equation of an Ellipse [Center at (*h, k*)]

The **standard form of the equation of an ellipse centered at (*h, k*)** with major and minor axes of lengths $2a$ and $2b$, where $0 < b < a$, is

$$\frac{(x - h)^2}{a^2} + \frac{(y - k)^2}{b^2} = 1 \qquad \text{Major axis is horizontal.}$$

or

$$\frac{(x - h)^2}{b^2} + \frac{(y - k)^2}{a^2} = 1. \qquad \text{Major axis is vertical.}$$

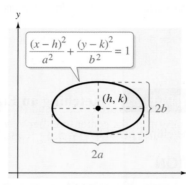

Major axis is horizontal.
Minor axis is vertical.

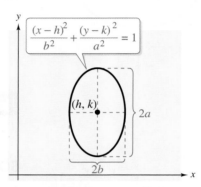

Major axis is vertical.
Minor axis is horizontal.

The foci lie on the major axis, c units form the center, with $c^2 = a^2 - b^2$.

Exercises Within Reach ®

Solutions in English & Spanish and tutorial videos at CollegePrepAlgebra.com

Identifying the Standard Form In Exercises 13−16, determine whether the standard form of the equation of the ellipse is $\dfrac{(x - h)^2}{a^2} + \dfrac{(y - k)^2}{b^2} = 1$ or $\dfrac{(x - h)^2}{b^2} + \dfrac{(y - k)^2}{a^2} = 1$.

13.

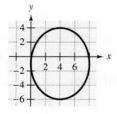

14.

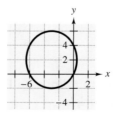

15.

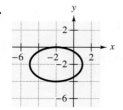

16.

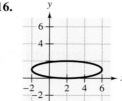

EXAMPLE 3   **Writing the Standard Equation of an Ellipse**

Write the standard form of the equation of the ellipse with vertices $(-2, 2)$ and $(4, 2)$ and co-vertices $(1, 3)$ and $(1, 1)$, as shown below.

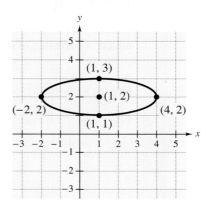

SOLUTION

Because the vertices are $(-2, 2)$ and $(4, 2)$, the center of the ellipse is $(h, k) = (1, 2)$. The distance from the center to either vertex is $a = 3$, and the distance from the center to either co-vertex is $b = 1$. Because the major axis is horizontal, the standard form of the equation is

$$\frac{(x - h)^2}{a^2} + \frac{(y - k)^2}{b^2} = 1. \qquad \text{Major axis is horizontal.}$$

Substitute the values of h, k, a, and b to obtain

$$\frac{(x - 1)^2}{3^2} + \frac{(y - 2)^2}{1^2} = 1. \qquad \text{Standard form}$$

From the graph, you can see that the center of the ellipse is shifted one unit to the right and two units upward from the origin.

Exercises Within Reach ®   Solutions in English & Spanish and tutorial videos at CollegePrepAlgebra.com

Writing the Standard Equation of an Ellipse **In Exercises 17−20, write the standard form of the equation of the ellipse.**

17.
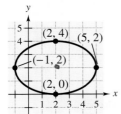

18.
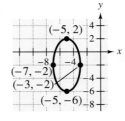

19.
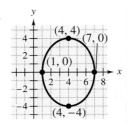

20.
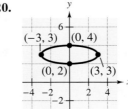

EXAMPLE 4   **Sketching an Ellipse**

Sketch the ellipse given by $4x^2 + y^2 - 8x + 6y + 9 = 0$.

SOLUTION

Begin by writing the equation in standard form. In the fourth step, note that 9 and 4 are added to *each* side of the equation.

$4x^2 + y^2 - 8x + 6y + 9 = 0$	Write original equation.
$\left(4x^2 - 8x + \rule{1.5em}{0.8em}\right) + \left(y^2 + 6y + \rule{1.5em}{0.8em}\right) = -9$	Group terms.
$4\left(x^2 - 2x + \rule{1.5em}{0.8em}\right) + \left(y^2 + 6y + \rule{1.5em}{0.8em}\right) = -9$	Factor 4 out of x-terms.
$4(x^2 - 2x + 1) + (y^2 + 6y + 9) = -9 + 4(1) + 9$	Complete the squares.
$4(x - 1)^2 + (y + 3)^2 = 4$	Simplify.
$\dfrac{(x - 1)^2}{1} + \dfrac{(y + 3)^2}{4} = 1$	Divide each side by 4.
$\dfrac{(x - 1)^2}{1^2} + \dfrac{(y + 3)^2}{2^2} = 1$	Standard form

You can see that the center of the ellipse is $(h, k) = (1, -3)$. Because the denominator of the y^2-term is larger than the denominator of the x^2-term, you can conclude that the major axis is vertical. Because the denominator of the x^2-term is $b^2 = 1^2$, you can locate the endpoints of the minor axis one unit to the right of the center and one unit to the left of the center. Because the denominator of the y^2-term is $a^2 = 2^2$, you can locate the endpoints of the major axis two units up from the center and two units down from the center, as shown below on the left. To complete the graph, sketch an oval shape that is determined by the vertices and co-vertices, as shown below on the right.

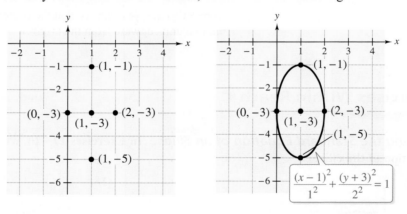

Sketching an Ellipse In Exercises 21−26, **find the center and vertices of the ellipse, and sketch its graph.**

21. $4(x - 2)^2 + 9(y + 2)^2 = 36$

22. $2(x + 5)^2 + 8(y - 2)^2 = 72$

23. $9x^2 + 4y^2 + 36x - 24y + 36 = 0$

24. $9x^2 + 4y^2 - 36x + 8y + 31 = 0$

25. $25x^2 + 9y^2 - 200x + 54y + 256 = 0$

26. $25x^2 + 16y^2 - 150x - 128y + 81 = 0$

Application   EXAMPLE 5   Semielliptical Archway

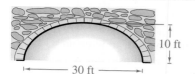

You are responsible for designing a semielliptical archway, as shown at the left. The height of the archway is 10 feet, and its width is 30 feet. Write an equation that represents the archway and use the equation to sketch an accurate diagram of the archway.

SOLUTION

To make the equation simple, place the origin at the center of the ellipse. This means that the standard form of the equation is

$$\frac{x^2}{a^2} + \frac{y^2}{b^2} = 1.$$ Major axis is horizontal.

Because the major axis is horizontal, it follows that $a = 15$ and $b = 10$, which implies that the equation of the ellipse is

$$\frac{x^2}{15^2} + \frac{y^2}{10^2} = 1.$$ Standard form

To write an equation that represents the archway, solve this equation for y.

$$\frac{x^2}{225} + \frac{y^2}{100} = 1$$ Simplify denominators.

$$\frac{y^2}{100} = 1 - \frac{x^2}{225}$$ Subtract $\frac{x^2}{225}$ from each side.

$$y^2 = 100\left(1 - \frac{x^2}{225}\right)$$ Multiply each side by 100.

$$y = 10\sqrt{1 - \frac{x^2}{225}}$$ Take the positive square root of each side.

To make an accurate sketch of the archway, calculate several y-values, as shown in the table. Then use the values in the table to sketch the archway, as shown at the left.

x	y
±15	0
±12.5	5.53
±10	7.45
±7.5	8.66
±5	9.43
±2.5	9.86
0	10

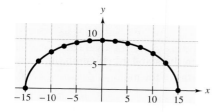

Exercises Within Reach ®

Solutions in English & Spanish and tutorial videos at CollegePrepAlgebra.com

27. *Motorsports* Most sprint car dirt tracks are elliptical in shape. Write an equation of an elliptical race track with a major axis that is 1230 feet long and a minor axis that is 580 feet long.

28. *Bicycle Chainwheel* The pedals of a bicycle drive a chainwheel, which drives a smaller sprocket wheel on the rear axle (see figure). Many chainwheels are circular. Some, however, are slightly elliptical, which tends to make pedaling easier. Write an equation of an elliptical chainwheel with a major axis that is 8 inches long and a minor axis that is $7\frac{1}{2}$ inches long.

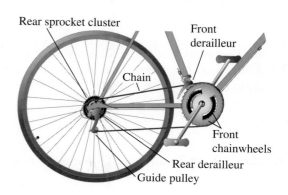

Rear sprocket cluster

Front derailleur

Chain

Front chainwheels

Rear derailleur

Guide pulley

Concept Summary: Writing Equations of Ellipses

What

The standard form of the equation of an **ellipse** centered at the origin is

$\frac{x^2}{a^2} + \frac{y^2}{b^2} = 1$ or $\frac{x^2}{b^2} + \frac{y^2}{a^2} = 1$,

where $0 < b < a$.

The **vertices** lie on the **major axis**, a units from the **center**. The **co-vertices** lie on the **minor axis**, b units from the center.

EXAMPLE

Write an equation of the ellipse centered at the origin, with vertices $(0, -6)$ and $(0, 6)$ and co-vertices $(-3, 0)$ and $(3, 0)$.

How

To write the standard form of the equation of an ellipse centered at $(0, 0)$, follow these steps.

1. Plot the vertices and co-vertices.
2. Decide whether the major axis is horizontal or vertical.
3. Identify the values of a and b.

EXAMPLE

From the graph of the ellipse (see figure), you can see the vertices and co-vertices. You can also see that the major axis is vertical, $a = 6$, and $b = 3$. So an equation of the ellipse is $\frac{x^2}{3^2} + \frac{y^2}{6^2} = 1$.

Why

Understanding how to write an equation of an ellipse centered at the origin will help you when you learn how to write an equation of an ellipse *not* centered at the origin.

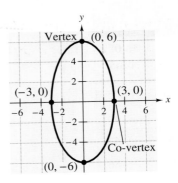

Exercises Within Reach ®

Worked-out solutions to odd-numbered exercises at CollegePrepAlgebra.com

Concept Summary Check

29. *Vocabulary* Define an ellipse and write the standard form of the equation of an ellipse centered at the origin.

30. *Writing an Equation* What points do you need to know in order to write the equation of an ellipse?

31. *Think About It* From the standard equation, how can you determine the lengths of the major and minor axes of an ellipse?

32. *Writing* From the standard equation, how can you determine the orientation of the major and minor axes of an ellipse?

Extra Practice

Writing the Standard Equation of an Ellipse **In Exercises 33 and 34, write the standard form of the equation of the ellipse centered at the origin.**

33. Major axis (vertical) 10 units, minor axis 6 units

34. Major axis (horizontal) 24 units, minor axis 10 units

Sketching an Ellipse **In Exercises 35 and 36, sketch the ellipse. Identify the vertices and co-vertices.**

35. $16x^2 + 25y^2 - 9 = 0$

36. $64x^2 + 36y^2 - 49 = 0$

37. *Wading Pool* You are building a wading pool that is in the shape of an ellipse. Your plans give the following equation for the elliptical shape of the pool, measured in feet.

$\frac{x^2}{324} + \frac{y^2}{196} = 1$

Find the longest distance and shortest distance across the pool.

38. *Oval Office* In the White House, the Oval Office is in the shape of an ellipse. The perimeter (in meters) of the floor can be modeled by the equation

$\frac{x^2}{19.36} + \frac{y^2}{30.25} = 1.$

Find the longest distance and shortest distance across the office.

Airplane In Exercises 39 and 40, an airplane with enough fuel to fly 800 miles safely will take off from airport A and land at airport B. Answer the following questions given the situation in each exercise.

(a) Explain why the region in which the airplane can fly is bounded by an ellipse (see figure).

(b) Let (0, 0) represent the center of the ellipse. Find the coordinates of each airport.

(c) Suppose the plane flies from airport A straight past airport B to a vertex of the ellipse, and then straight back to airport B. How far does the plane fly? Use your answer to find the coordinates of the vertices.

(d) Write an equation of the ellipse. (*Hint:* $c^2 = a^2 - b^2$)

(e) The area of an ellipse is given by $A = \pi ab$. Find the area of the ellipse.

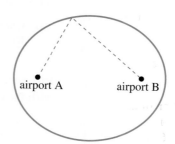

39. Airport A is 500 miles from airport B.

40. Airport A is 650 miles form airport B.

Explaining Concepts

41. ***Vocabulary*** Describe the relationship between circles and ellipses. How are they similar? How do they differ?

42. ***Area*** The area of an ellipse is given by $A = \pi ab$. Explain how this area is related to the area of a circle.

43. ***Vocabulary*** Explain the significance of the foci in an ellipse.

44. ***Reasoning*** Explain how to write an equation of an ellipse when you know the coordinates of the vertices and co-vertices.

45. ***Logic*** From the standard form of the equation, explain how you can determine if the graph of an ellipse intersects the x- or y-axis.

Cumulative Review

In Exercises 46–53, evaluate the function as indicated and sketch the graph of the function.

46. $f(x) = 4^x$

   (a) $x = 3$

   (b) $x = -1$

47. $f(x) = 3^{-x}$

   (a) $x = -2$

   (b) $x = 2$

48. $g(x) = 5^{x-1}$

   (a) $x = 4$

   (b) $x = 0$

49. $g(x) = 6e^{0.5x}$

   (a) $x = -1$

   (b) $x = 2$

50. $h(x) = \log_{10} 2x$

   (a) $x = 5$

   (b) $x = 500$

51. $h(x) = \log_{16} 4x$

   (a) $x = 4$

   (b) $x = 64$

52. $f(x) = \ln(-x)$

   (a) $x = -6$

   (b) $x = 3$

53. $f(x) = \log_4(x - 3)$

   (a) $x = 3$

   (b) $x = 35$

Mid-Chapter Quiz: Sections 12.1–12.2

Solutions in English & Spanish and tutorial videos at CollegePrepAlgebra.com

Take this quiz as you would take a quiz in class. After you are done, check your work against the answers in the back of the book.

In Exercises 1 and 2, write the standard form of the equation of the circle centered at (h, k).

1. Center: $(0, 0)$
   Radius: 5

2. Center: $(3, -5)$
   Passes through the point $(0, -1)$

In Exercises 3 and 4, write the standard form of the equation of the parabola with its vertex at (h, k).

3. Vertex: $(-2, 1)$
   Focus: $(0, 1)$

4. Vertex: $(2, 3)$
   Focus: $(2, 1)$

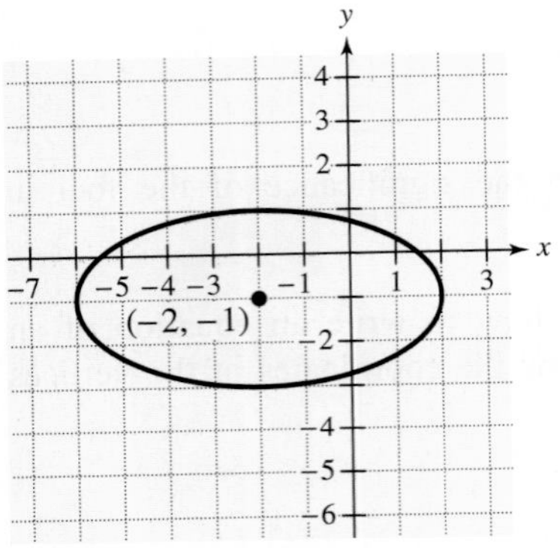

5. Write the standard form of the equation of the ellipse shown in the figure.

6. Write the standard form of the equation of the ellipse with vertices $(0, -10)$ and $(0, 10)$ and co-vertices $(-6, 0)$ and $(6, 0)$.

In Exercises 7 and 8, write the equation of the circle in standard form. Then find the center and the radius of the circle.

7. $x^2 + y^2 + 6y - 7 = 0$

8. $x^2 + y^2 + 2x - 4y + 4 = 0$

In Exercises 9 and 10, write the equation of the parabola in standard form. Then find the vertex and the focus of the parabola.

9. $x = y^2 - 6y - 7$

10. $x^2 - 8x + y + 12 = 0$

In Exercises 11 and 12, write the equation of the ellipse in standard form. Then find the center and the vertices of the ellipse.

11. $4x^2 + y^2 - 16x - 20 = 0$

12. $4x^2 + 9y^2 - 48x + 36y + 144 = 0$

In Exercises 13–18, sketch the graph of the equation.

13. $(x + 5)^2 + (y - 1)^2 = 9$

14. $9x^2 + y^2 = 81$

15. $x = -y^2 - 4y$

16. $x^2 + (y + 4)^2 = 1$

17. $y = x^2 - 2x + 1$

18. $4(x + 3)^2 + (y - 2)^2 = 16$

Study Skills in Action

Avoiding Test-Taking Errors

For some students, the day they get their math tests back is just as nerve-racking as the day they take the test. Do you look at your grade, sigh hopelessly, and stuff the test in your book bag? This kind of response is not going to help you to do better on the next test. When professional football players lose a game, the coach does not let them just forget about it. They review all their mistakes and discuss how to correct them. That is what you need to do with every math test.

There are six types of test errors, as listed below. Look at your test and see what types of errors you make. Then decide what you can do to avoid making them again. Many students need to do this with a tutor or teacher the first time through.

Smart Study Strategy

Analyze Your Errors

Type of error	Corrective action
1 ▶ **Misreading Directions:** You do not correctly read or understand directions.	Read the instructions in the textbook exercises at least twice and make sure you understand what they mean. Make this a habit in time for the next test.
2 ▶ **Careless Errors:** You understand how to do a problem but make careless errors, such as not carrying a sign, miscopying numbers, and so on.	Pace yourself during a test to avoid hurrying. Also, make sure you write down every step of the solution neatly. Use a finger to move from one step to the next, looking for errors.
3 ▶ **Concept Errors:** You do not understand how to apply the properties and rules needed to solve a problem.	Find a tutor who will work with you on the next chapter. Visit the teacher to make sure you understand the math.
4 ▶ **Application Errors:** You can do numerical problems that are similar to your homework problems but struggle with problems that vary, such as application problems.	Do not just mimic the steps of solving an application problem. Explain out loud why you are doing each step. Ask the teacher or tutor for different types of problems.
5 ▶ **Test-Taking Errors:** You hurry too much, do not use all of the allowed time, spend too much time on one problem, and so on.	Refer to the *Ten Steps for Test-Taking* on page 345.
6 ▶ **Study Errors:** You do not study the right material or do not learn it well enough to remember it on a test without resources such as notes.	Take a practice test. Work with a study group. Confer with your teacher. Do not try to learn a whole chapter's worth of material in one night—cramming does not work in math!

12.3 Hyperbolas

▶ Analyze hyperbolas centered at the origin.

▶ Analyze hyperbolas centered at (h, k).

Hyperbolas Centered at the Origin

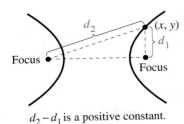

$d_2 - d_1$ is a positive constant.

A **hyperbola** in the rectangular coordinate system consists of all points (x, y) such that the difference of the distances between (x, y) and two fixed points is a positive constant, as shown at the left. The two fixed points are called the **foci** of the hyperbola. The line on which the foci lie is called the **transverse axis** of the hyperbola.

Standard Equation of a Hyperbola (Center at Origin)

The **standard form of the equation of a hyperbola centered at the origin** is

$$\frac{x^2}{a^2} - \frac{y^2}{b^2} = 1 \qquad \text{or} \qquad \frac{y^2}{a^2} - \frac{x^2}{b^2} = 1$$

Transverse axis is horizontal.   Transverse axis is vertical.

where a and b are positive real numbers. The **vertices** of the hyperbola lie on the transverse axis, a units from the center, as shown below.

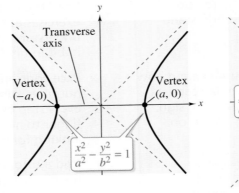

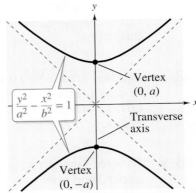

Identifying the Standard Form In Exercises 1−4, determine whether the standard form of the equation of the hyperbola is $\frac{x^2}{a^2} - \frac{y^2}{b^2} = 1$ or $\frac{y^2}{a^2} - \frac{x^2}{b^2} = 1$.

1.

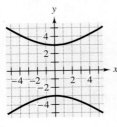

2.

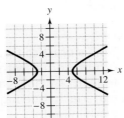

3.

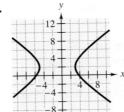

4.

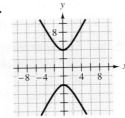

A hyperbola has two disconnected parts, each of which is called a **branch** of the hyperbola. The two branches approach a pair of intersecting lines called the **asymptotes** of the hyperbola. The two asymptotes intersect at the center of the hyperbola.

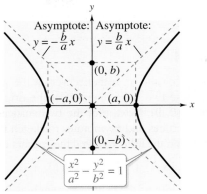

Transverse axis is horizontal.

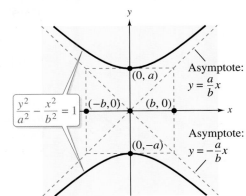

Transverse axis is vertical.

Study Tip

To sketch a hyperbola, form a **central rectangle** that is centered at the origin and has side lengths of 2a and 2b. The asymptotes pass through the corners of the central rectangle, and the vertices of the hyperbola lie at the centers of opposite sides of the central rectangle.

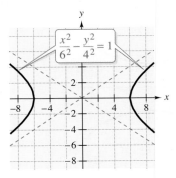

EXAMPLE 1   **Sketching a Hyperbola**

Identify the vertices of the hyperbola given by the equation, and sketch the hyperbola.

$$\frac{x^2}{36} - \frac{y^2}{16} = 1$$

SOLUTION

From the standard form of the equation

$$\frac{x^2}{6^2} - \frac{y^2}{4^2} = 1$$

you can see that the center of the hyperbola is the origin and the transverse axis is horizontal. So, the vertices lie six units to the left and right of the center at the points

$$(-6, 0) \text{ and } (6, 0).$$

Because $a = 6$ and $b = 4$, you can sketch the hyperbola by first drawing a central rectangle with a width of $2a = 12$ and a height of $2b = 8$, as shown in the top left figure. Next, draw the asymptotes of the hyperbola through the corners of the central rectangle, and plot the vertices. Finally, draw the hyperbola, as shown in the bottom left figure.

Exercises Within Reach ®

Sketching a Hyperbola In Exercises 5−10, identify the vertices of the hyperbola, and sketch its graph.

5. $\dfrac{x^2}{9} - \dfrac{y^2}{25} = 1$

6. $\dfrac{x^2}{4} - \dfrac{y^2}{9} = 1$

7. $\dfrac{y^2}{9} - \dfrac{x^2}{25} = 1$

8. $\dfrac{y^2}{4} - \dfrac{x^2}{9} = 1$

9. $y^2 - x^2 = 9$

10. $y^2 - x^2 = 1$

EXAMPLE 2 **Writing an Equation of a Hyperbola**

Write the standard form of the equation of the hyperbola with a vertical transverse axis and vertices $(0, 3)$ and $(0, -3)$. The equations of the asymptotes of the hyperbola are $y = \frac{3}{5}x$ and $y = -\frac{3}{5}x$.

SOLUTION

To begin, sketch the lines that represent the asymptotes, as shown below on the left. Note that these two lines intersect at the origin, which implies that the center of the hyperbola is $(0, 0)$. Next, plot the two vertices at the points $(0, 3)$ and $(0, -3)$. You can use the vertices and asymptotes to sketch the central rectangle of the hyperbola, as shown below on the left. Note that the corners of the central rectangle occur at the points

$$(-5, 3), (5, 3), (-5, -3), \text{ and } (5, -3).$$

Because the width of the central rectangle is $2b = 10$, it follows that $b = 5$. Similarly, because the height of the central rectangle is $2a = 6$, it follows that $a = 3$. Now that you know the values of a and b, you can use the standard form of the equation of a hyperbola to write an equation.

$$\frac{y^2}{a^2} - \frac{x^2}{b^2} = 1 \qquad \text{Transverse axis is vertical.}$$

$$\frac{y^2}{3^2} - \frac{x^2}{5^2} = 1 \qquad \text{Substitute 3 for } a \text{ and 5 for } b.$$

$$\frac{y^2}{9} - \frac{x^2}{25} = 1 \qquad \text{Simplify.}$$

The graph is shown below on the right.

Study Tip

For a hyperbola, note that a and b are not determined in the same way as for an ellipse, where a is always greater than b. In the standard form of the equation of a hyperbola, a^2 is always the denominator of the positive term.

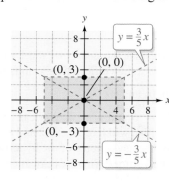

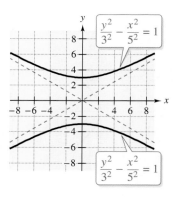

Exercises Within Reach ®   Solutions in English & Spanish and tutorial videos at CollegePrepAlgebra.com

Writing an Equation of a Hyperbola In Exercises 11−16, write the standard form of the equation of the hyperbola centered at the origin.

	Vertices	Asymptotes			Vertices	Asymptotes	
11.	$(-4, 0), (4, 0)$	$y = 2x$	$y = -2x$	**12.**	$(-2, 0), (2, 0)$	$y = \frac{1}{3}x$	$y = -\frac{1}{3}x$
13.	$(0, -4), (0, 4)$	$y = \frac{1}{2}x$	$y = -\frac{1}{2}x$	**14.**	$(0, -2), (0, 2)$	$y = 3x$	$y = -3x$
15.	$(-9, 0), (9, 0)$	$y = \frac{2}{3}x$	$y = -\frac{2}{3}x$	**16.**	$(-1, 0), (1, 0)$	$y = \frac{1}{2}x$	$y = -\frac{1}{2}x$

Hyperbolas Centered at (*h, k*)

Standard Equation of a Hyperbola [Center at (*h, k*)]

The **standard form of the equation of a hyperbola centered at (*h, k*)** is

$$\frac{(x - h)^2}{a^2} - \frac{(y - k)^2}{b^2} = 1$$ Transverse axis is horizontal.

or

$$\frac{(y - k)^2}{a^2} - \frac{(x - h)^2}{b^2} = 1$$ Transverse axis is vertical.

where *a* and *b* are positive real numbers. The vertices lie on the transverse axis, *a* units from the center, as shown below.

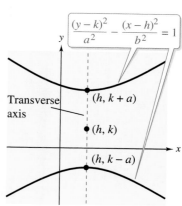

EXAMPLE 3   Sketching a Hyperbola

Sketch the hyperbola given by $\dfrac{(y - 1)^2}{9} - \dfrac{(x + 2)^2}{4} = 1$.

SOLUTION

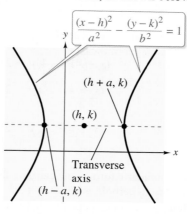

From the form of the equation, you can see that the transverse axis is vertical, and the center of the hyperbola is $(h, k) = (-2, 1)$. Because $a = 3$ and $b = 2$, you can begin by sketching a central rectangle that is six units high and four units wide, centered at $(-2, 1)$. Then, sketch the asymptotes by drawing lines through the corners of the central rectangle. Sketch the hyperbola, as shown at the left. From the graph, you can see that the center of the hyperbola is shifted two units to the left and one unit upward from the origin.

Exercises Within Reach ®

Solutions in English & Spanish and tutorial videos at CollegePrepAlgebra.com

Sketching a Hyperbola In Exercises 17−20, **find the center and vertices of the hyperbola, and sketch its graph.**

17. $\dfrac{(x - 1)^2}{4} - \dfrac{(y + 2)^2}{1} = 1$

18. $\dfrac{(x - 2)^2}{4} - \dfrac{(y - 3)^2}{9} = 1$

19. $(y + 4)^2 - (x - 3)^2 = 25$

20. $(y + 6)^2 - (x - 2)^2 = 1$

EXAMPLE 4   **Sketching a Hyperbola**

Sketch the hyperbola given by

$$x^2 - 4y^2 + 8x + 16y - 4 = 0.$$

SOLUTION

Complete the square to write the equation in standard form.

$x^2 - 4y^2 + 8x + 16y - 4 = 0$	Write original equation.
$\left(x^2 + 8x + \boxed{}\right) - \left(4y^2 - 16y + \boxed{}\right) = 4$	Group terms.
$\left(x^2 + 8x + \boxed{}\right) - 4\left(y^2 - 4y + \boxed{}\right) = 4$	Factor 4 out of y-terms.
$(x^2 + 8x + 16) - 4(y^2 - 4y + 4) = 4 + 16 - 4(4)$	Complete the squares.
$(x + 4)^2 - 4(y - 2)^2 = 4$	Simplify.
$\dfrac{(x + 4)^2}{4} - \dfrac{(y - 2)^2}{1} = 1$	Divide each side by 4.
$\dfrac{(x + 4)^2}{2^2} - \dfrac{(y - 2)^2}{1^2} = 1$	Standard form

From the standard form, you can see that the transverse axis is horizontal, and the center of the hyperbola is $(h, k) = (-4, 2)$. Because $a = 2$ and $b = 1$, you can begin by sketching a central rectangle that is four units wide and two units high, centered at $(-4, 2)$. Then, sketch the asymptotes by drawing lines through the corners of the central rectangle. Sketch the hyperbola, as shown below. From the graph, you can see that the center of the hyperbola is shifted four units to the left and two units upward from the origin.

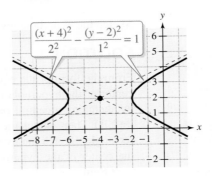

$$\frac{(x + 4)^2}{2^2} - \frac{(y - 2)^2}{1^2} = 1$$

Exercises Within Reach®

Solutions in English & Spanish and tutorial videos at CollegePrepAlgebra.com

Sketching a Hyperbola In Exercises 21−24, find the center and vertices of the hyperbola, and sketch its graph.

21. $9x^2 - y^2 - 36x - 6y + 18 = 0$

22. $x^2 - 9y^2 + 36y - 72 = 0$

23. $4x^2 - y^2 + 24x + 4y + 28 = 0$

24. $25x^2 - 4y^2 + 100x + 8y + 196 = 0$

Application   EXAMPLE 5   Navigation

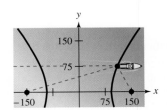

Long-distance radio navigation for aircraft and ships uses synchronized pulses transmitted by widely separated transmitting stations. The locations of two transmitting stations that are 300 miles apart are represented by the points $(-150, 0)$ and $(150, 0)$ (see figure). A ship's location is given by $(x, 75)$. The difference in the arrival times of pulses transmitted simultaneously to the ship from the two stations is constant at any point on the hyperbola given by

$$\frac{x^2}{8649} - \frac{y^2}{13{,}851} = 1$$

which passes through the ship's location and has the two stations as foci. Use the equation to find the x-coordinate of the ship's location.

SOLUTION

To find the x-coordinate of the ship's location, substitute $y = 75$ into the equation for the hyperbola and solve for x.

$$\frac{x^2}{8649} - \frac{y^2}{13{,}851} = 1 \qquad \text{Write equation for hyperbola.}$$

$$\frac{x^2}{8649} - \frac{75^2}{13{,}851} = 1 \qquad \text{Substitute 75 for } y.$$

$$x^2 = 8649\left[1 + \frac{75^2}{13{,}851}\right] \qquad \text{Isolate the variable.}$$

$$x^2 \approx 12{,}161.4 \qquad \text{Use a calculator.}$$

$$x \approx 110.3 \qquad \text{Take positive square root of each side.}$$

The ship is about 110.3 miles to the right of the y-axis.

Exercises Within Reach ®   Solutions in English & Spanish and tutorial videos at CollegePrepAlgebra.com

25. *Optics* Hyperbolic mirrors are used in some telescopes. The figure shows a cross section of a hyperbolic mirror as the right branch of a hyperbola. A property of the mirror is that a light ray directed at the focus $(48, 0)$ is reflected to the other focus $(0, 0)$. Use the equation of the hyperbola

$$89x^2 - 55y^2 - 4272x + 31{,}684 = 0$$

to find the coordinates of the mirror's vertex.

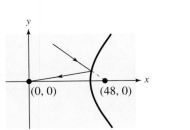

26. *Art* A sculpture has a hyperbolic cross section (see figure). Write an equation that models the curved sides of the sculpture.

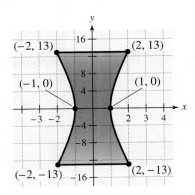

Concept Summary: *Sketching Hyperbolas*

What

You can use the standard form of the equation of a **hyperbola** to sketch its graph.

EXAMPLE

Sketch the hyperbola.

$$\frac{x^2}{9} - \frac{y^2}{4} = 1$$

How

1. Write the equation in standard form.
2. Determine the center and the values of a and b.
3. Use the values of $2a$ and $2b$ to sketch the **central rectangle**.
4. Draw the **asymptotes** through the corners of the central rectangle.
5. Plot the **vertices** and finally sketch the hyperbola.

EXAMPLE

$$\frac{x^2}{3^2} - \frac{y^2}{2^2} = 1 \qquad \text{Standard form}$$

The center of the hyperbola is the origin. Because $a = 3$ and $b = 2$, the central rectangle has a width of $2a = 2(3) = 6$ units and a height of $2b = 2(2) = 4$ units. The graph is shown at the right.

Why

Understanding how to sketch the graph of a hyperbola centered at the origin will help you when you learn how to sketch the graph of a hyperbola *not* centered at the origin.

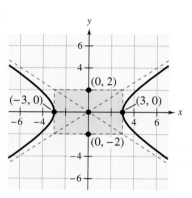

Exercises Within Reach ®

Worked-out solutions to odd-numbered exercises at CollegePrepAlgebra.com

Concept Summary Check

27. *Vocabulary* You are given the equation of a hyperbola in the standard form

$$\frac{x^2}{a^2} - \frac{y^2}{b^2} = 1.$$

Explain how you can sketch the central rectangle for the hyperbola. Explain how you can use the central rectangle to sketch the asymptotes of the hyperbola.

28. *Reasoning* You are given the vertices and the equations of the asymptotes of a hyperbola. Explain how you can determine the values of a and b in the standard form of the equation of the hyperbola.

29. *Logic* What are the dimensions of the central rectangle and the coordinates of the center of the hyperbola whose equation in standard form is

$$\frac{(y - k)^2}{a^2} - \frac{(x - h)^2}{b^2} = 1?$$

30. *Reasoning* Given the equation of a hyperbola in the general polynomial form

$$ax^2 - by^2 + cx + dy + e = 0$$

what process can you use to find the center of the hyperbola?

Writing an Equation of a Hyperbola In Exercises 31−34, write the standard form of the equation of the hyperbolas.

31.

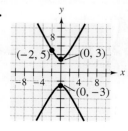

32.

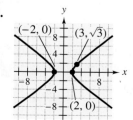

33.

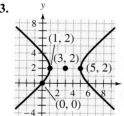

34.

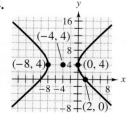

Sketching a Hyperbola In Exercises 35–38, identify the vertices and asymptotes of the hyperbola, and sketch its graph.

35. $\dfrac{x^2}{1} - \dfrac{y^2}{9/4} = 1$

36. $\dfrac{y^2}{1/4} - \dfrac{x^2}{25/4} = 1$

37. $4y^2 - x^2 + 16 = 0$

38. $4y^2 - 9x^2 - 36 = 0$

Identifying a Conic In Exercises 39–44, determine whether the graph represented by the equation is a circle, a parabola, an ellipse, or a hyperbola.

39. $\dfrac{(x-3)^2}{4^2} + \dfrac{(y-4)^2}{6^2} = 1$

40. $\dfrac{(x+2)^2}{25} + \dfrac{(y-2)^2}{25} = 1$

41. $x^2 - y^2 = 1$

42. $2x + y^2 = 0$

43. $y^2 - x^2 - 2y + 8x - 19 = 0$

44. $9x^2 + y^2 - 18x - 8y + 16 = 0$

45. *Aeronautics* When an airplane travels faster than the speed of sound, the sound waves form a cone behind the airplane. When the airplane is flying parallel to the ground, the sound waves intersect the ground in a hyperbola with the airplane directly above its center (see figure). A sonic boom can be heard along the hyperbola. You hear a sonic boom that is audible along a hyperbola with the equation

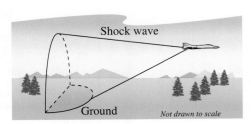

Shock wave

Ground   *Not drawn to scale*

$$\dfrac{x^2}{100} - \dfrac{y^2}{4} = 1$$

where x and y are measured in miles. What is the shortest horizontal distance you could be from the airplane?

Explaining Concepts

46. *Think About It* Describe the part of the hyperbola

$$\dfrac{(x-3)^2}{4} - \dfrac{(y-1)^2}{9} = 1$$

given by each equation.

(a) $x = 3 - \frac{2}{3}\sqrt{9 + (y-1)^2}$

(b) $y = 1 + \frac{3}{2}\sqrt{(x-3)^2 - 4}$

47. *Reasoning* Consider the definition of a hyperbola. How many hyperbolas have a given pair of points as foci? Explain your reasoning.

48. *Precision* How many hyperbolas pass through a given point and have a given pair of points as foci? Explain your reasoning.

49. *Project* Cut cone-shaped pieces of styrofoam to demonstrate how to obtain each type of conic section: circle, parabola, ellipse, and hyperbola. Discuss how you could write directions for someone else to form each conic section. Compile a list of real-life situations and/or everyday objects in which conic sections may be seen.

Cumulative Review

In Exercises 50 and 51, solve the system of equations by graphing.

50. $\begin{cases} -x + 3y = 8 \\ 4x - 12y = -32 \end{cases}$

51. $\begin{cases} x - 3y = 5 \\ 2x - 6y = -5 \end{cases}$

In Exercises 52 and 53, solve the system of linear equations by the method of elimination.

52. $\begin{cases} x + y = 3 \\ x - y = 2 \end{cases}$

53. $\begin{cases} 4x + 3y = 3 \\ x - 2y = 9 \end{cases}$

12.4 Solving Nonlinear Systems of Equations

▶ Solve nonlinear systems of equations graphically.

▶ Solve nonlinear systems of equations by substitution.

▶ Solve nonlinear systems of equations by elimination.

Solving Nonlinear Systems of Equations by Graphing

A **nonlinear system of equations** is a system that contains at least one nonlinear equation. Nonlinear systems of equations can have no solution, one solution, or two or more solutions.

Solving a Nonlinear System Graphically

1. Sketch the graph of each equation in the system.

2. Locate the point(s) of intersection of the graphs (if any) and graphically approximate the coordinates of the point(s).

3. Check the coordinates by substituting them into each equation in the original system. If the coordinates do not check, you may have to use an algebraic approach, as discussed later in this section.

EXAMPLE 1   **Solving a Nonlinear System Graphically**

Find all solutions of the nonlinear system of equations.

$$\begin{cases} x = (y - 3)^2 & \text{Equation 1} \\ x + y = 5 & \text{Equation 2} \end{cases}$$

SOLUTION

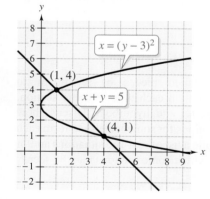

Begin by sketching the graph of each equation. Solve the first equation for y.

$x = (y - 3)^2$	Write original equation.
$\pm\sqrt{x} = y - 3$	Take the square root of each side.
$3 \pm \sqrt{x} = y$	Add 3 to each side.

The graph of $y = 3 \pm \sqrt{x}$ is a parabola with its vertex at $(0, 3)$. The graph of the second equation is a line with a slope of -1 and a y-intercept of $(0, 5)$. The system appears to have two solutions: $(4, 1)$ and $(1, 4)$, as shown at the left. Check these solutions in the original system.

Exercises Within Reach ®

Solutions in English & Spanish and tutorial videos at CollegePrepAlgebra.com

Solving a System Graphically In Exercises 1−4, **graph** the equations to determine whether the system has any solutions. **Find** any solutions that exist.

1. $\begin{cases} x + y = 2 \\ x^2 - y = 0 \end{cases}$

2. $\begin{cases} y = 4 \\ x^2 - y = 0 \end{cases}$

3. $\begin{cases} x^2 + y = 9 \\ x - y = -3 \end{cases}$

4. $\begin{cases} x - y^2 = 0 \\ x - y = 2 \end{cases}$

EXAMPLE 2 **Solving a Nonlinear System Graphically**

Find all solutions of the nonlinear system of equations.

$$\begin{cases} x^2 + y^2 = 25 & \text{Equation 1} \\ x - y = 1 & \text{Equation 2} \end{cases}$$

SOLUTION

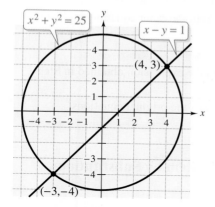

Begin by sketching the graph of each equation. The graph of the first equation is a circle centered at the origin with a radius of 5. The graph of the second equation is a line with a slope of 1 and a y-intercept of $(0, -1)$. The system appears to have two solutions: $(-3, -4)$ and $(4, 3)$, as shown at the left.

Check

To check $(-3, -4)$, substitute -3 for x and -4 for y in each equation.

$$(-3)^2 + (-4)^2 \stackrel{?}{=} 25 \qquad \text{Substitute } -3 \text{ for } x \text{ and } -4 \text{ for } y \text{ in Equation 1.}$$

$$9 + 16 = 25 \qquad \text{Solution checks in Equation 1.} \checkmark$$

$$(-3) - (-4) \stackrel{?}{=} 1 \qquad \text{Substitute } -3 \text{ for } x \text{ and } -4 \text{ for } y \text{ in Equation 2.}$$

$$-3 + 4 = 1 \qquad \text{Solution checks in Equation 2.} \checkmark$$

To check $(4, 3)$, substitute 4 for x and 3 for y in each equation.

$$4^2 + 3^2 \stackrel{?}{=} 25 \qquad \text{Substitute 4 for } x \text{ and 3 for } y \text{ in Equation 1}$$

$$16 + 9 = 25 \qquad \text{Solution checks in Equation 1.} \checkmark$$

$$4 - 3 \stackrel{?}{=} 1 \qquad \text{Substitute 4 for } x \text{ and 3 for } y \text{ in Equation 2}$$

$$1 = 1 \qquad \text{Solution checks in Equation 2.} \checkmark$$

Exercises Within Reach ®

Solutions in English & Spanish and tutorial videos at CollegePrepAlgebra.com

Solving a System Graphically In Exercises 5–12, **graph the equations to determine whether the system has any solutions. Find any solutions that exist.**

5. $\begin{cases} x^2 + y^2 = 100 \\ x + y = 2 \end{cases}$

6. $\begin{cases} x^2 + y^2 = 169 \\ x + y = 7 \end{cases}$

7. $\begin{cases} x^2 + y^2 = 25 \\ 2x - y = -5 \end{cases}$

8. $\begin{cases} x = 0 \\ x^2 + y^2 = 9 \end{cases}$

9. $\begin{cases} x - 2y = 4 \\ x^2 - y = 0 \end{cases}$

10. $\begin{cases} y = \sqrt{x - 2} \\ x - 2y = 1 \end{cases}$

11. $\begin{cases} x^2 - y^2 = 16 \\ 3x - y = 12 \end{cases}$

12. $\begin{cases} 9x^2 - 4y^2 = 36 \\ 5x - 2y = 0 \end{cases}$

Solving Nonlinear Systems of Equations by Substitution

Method of Substitution

To solve a system of two equations in two variables, use the steps below.

1. Solve one of the equations for one variable in terms of the other.
2. Substitute the expression obtained in Step 1 into the other equation to obtain an equation in one variable.
3. Solve the equation obtained in Step 2.
4. Back-substitute the solution from Step 3 into the expression obtained in Step 1 to find the value of the other variable.
5. Check the solution to see that it satisfies *both* of the original equations.

EXAMPLE 3   **Solving a Nonlinear System by Substitution**

Solve the nonlinear system of equations.

$$\begin{cases} 4x^2 + y^2 = 4 & \text{Equation 1} \\ -2x + y = 2 & \text{Equation 2} \end{cases}$$

SOLUTION

Begin by solving for y in Equation 2 to obtain $y = 2x + 2$. Next, substitute this expression for y into Equation 1.

$4x^2 + y^2 = 4$	Write Equation 1.
$4x^2 + (2x + 2)^2 = 4$	Substitute $2x + 2$ for y.
$4x^2 + 4x^2 + 8x + 4 = 4$	Expand.
$8x^2 + 8x = 0$	Simplify.
$8x(x + 1) = 0$	Factor.
$8x = 0 \implies x = 0$	Set 1st factor equal to 0.
$x + 1 = 0 \implies x = -1$	Set 2nd factor equal to 0.

Finally, back-substitute these values of x into the revised Equation 2 to solve for y.

For $x = 0$: $y = 2(0) + 2 = 2$
For $x = -1$: $y = 2(-1) + 2 = 0$

So, the system of equations has two solutions: $(0, 2)$ and $(-1, 0)$. The graph of the system is shown in the figure. Check these solutions in the original system.

Exercises Within Reach ®

Solutions in English & Spanish and tutorial videos at CollegePrepAlgebra.com

Solving a System by Substitution **In Exercises 13−16, solve the system by the method of substitution.**

13. $\begin{cases} y = 2x^2 \\ y = 6x - 4 \end{cases}$

14. $\begin{cases} y = 5x^2 \\ y = -5x + 10 \end{cases}$

15. $\begin{cases} x^2 + y^2 = 4 \\ x + y = 2 \end{cases}$

16. $\begin{cases} 2x^2 - y^2 = -8 \\ x - y = 6 \end{cases}$

EXAMPLE 4   **Solving a Nonlinear System: The No Solution Case**

Solve the nonlinear system of equations.

$$\begin{cases} x^2 - y = 0 & \text{Equation 1} \\ x - y = 1 & \text{Equation 2} \end{cases}$$

SOLUTION

Begin by solving for y in Equation 2 to obtain $y = x - 1$. Next, substitute this expression for y into Equation 1.

$$x^2 - y = 0 \qquad\qquad \text{Write Equation 1.}$$

$$x^2 - (x - 1) = 0 \qquad\qquad \text{Substitute } x - 1 \text{ for } y.$$

$$x^2 - x + 1 = 0 \qquad\qquad \text{Distributive Property}$$

Use the Quadratic Formula, because this equation cannot be factored.

$$x = \frac{-(-1) \pm \sqrt{(-1)^2 - 4(1)(1)}}{2(1)} \qquad \text{Use Quadratic Formula.}$$

$$= \frac{1 \pm \sqrt{1 - 4}}{2} = \frac{1 \pm \sqrt{-3}}{2} \qquad \text{Simplify.}$$

Now, because the Quadratic Formula yields a negative number inside the radical, you can conclude that the equation $x^2 - x + 1 = 0$ has no real solution. So, the system has no real solution. The graph of the system is shown in the figure. From the graph, you can see that the parabola and the line have no point of intersection.

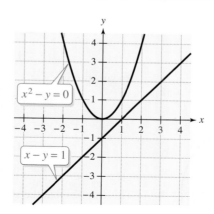

Exercises Within Reach ®

Solutions in English & Spanish and tutorial videos at CollegePrepAlgebra.com

Solving a System by Substitution In Exercises 17–24, solve the system by the method of substitution.

17. $\begin{cases} x^2 - 4y^2 = 16 \\ x^2 + y^2 = 1 \end{cases}$

18. $\begin{cases} 2x^2 - y^2 = 12 \\ 3x^2 - y^2 = -4 \end{cases}$

19. $\begin{cases} y = x^2 - 3 \\ x^2 + y^2 = 9 \end{cases}$

20. $\begin{cases} x^2 + y^2 = 25 \\ x - 3y = -5 \end{cases}$

21. $\begin{cases} 16x^2 + 9y^2 = 144 \\ 4x + 3y = 12 \end{cases}$

22. $\begin{cases} 4x^2 + 16y^2 = 64 \\ x + 2y = 4 \end{cases}$

23. $\begin{cases} x^2 - y^2 = 9 \\ x^2 + y^2 = 1 \end{cases}$

24. $\begin{cases} x^2 + y^2 = 1 \\ x + y = 7 \end{cases}$

Solving Nonlinear Systems of Equations by Elimination

EXAMPLE 5 Solving a Nonlinear System by Elimination

Solve the nonlinear system of equations.

$$\begin{cases} 4x^2 + y^2 = 64 & \text{Equation 1} \\ x^2 + y^2 = 52 & \text{Equation 2} \end{cases}$$

SOLUTION

Both equations have y^2 as a term (and no other terms containing y). To eliminate y, multiply Equation 2 by -1 and then add.

$$\begin{array}{rcl} 4x^2 + y^2 &=& 64 \\ -x^2 - y^2 &=& -52 \\ \hline 3x^2 &=& 12 \end{array}$$ Add equations.

After eliminating y, solve the remaining equation for x.

$3x^2 = 12$	Write resulting equation.
$x^2 = 4$	Divide each side by 3.
$x = \pm 2$	Take square root of each side.

By substituting $x = 2$ into Equation 2, you obtain

$x^2 + y^2 = 52$	Write Equation 2.
$(2)^2 + y^2 = 52$	Substitute 2 for x.
$y^2 = 48$	Subtract 4 from each side.
$y = \pm 4\sqrt{3}.$	Take square root of each side and simplify.

By substituting $x = -2$, you obtain the same values of y, as follows.

$x^2 + y^2 = 52$	Write Equation 2.
$(-2)^2 + y^2 = 52$	Substitute -2 for x.
$y^2 = 48$	Subtract 4 from each side.
$y = \pm 4\sqrt{3}.$	Take square root of each side and simplify.

This implies that the system has four solutions:

$$\left(2, 4\sqrt{3}\right), \quad \left(2, -4\sqrt{3}\right), \quad \left(-2, 4\sqrt{3}\right), \quad \left(-2, -4\sqrt{3}\right).$$

Check these solutions in the original system. The graph of the system is shown in the figure. Notice that the graph of Equation 1 is an ellipse and the graph of Equation 2 is a circle.

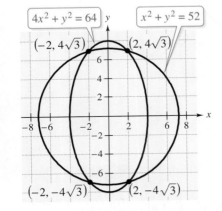

Exercises Within Reach ®

Solutions in English & Spanish and tutorial videos at CollegePrepAlgebra.com

Solving a System by Elimination In Exercises 25−28, solve the system by the method of elimination.

25. $\begin{cases} x^2 + 2y = 1 \\ x^2 + y^2 = 4 \end{cases}$

26. $\begin{cases} x + y^2 = 5 \\ 2x^2 + y^2 = 6 \end{cases}$

27. $\begin{cases} -x + y^2 = 10 \\ x^2 - y^2 = -8 \end{cases}$

28. $\begin{cases} x^2 + y = 9 \\ x^2 - y^2 = 7 \end{cases}$

Application   EXAMPLE 6   Avalanche Rescue System

RECCO® is an avalanche rescue system utilized by rescue organizations worldwide. RECCO technology enables quick directional pinpointing of a victim's exact location using harmonic radar. The two-part system consists of a detector used by rescuers and reflectors that are integrated into apparel, helmets, protection gear, or boots. The range of the detector through snow is 30 meters. Two rescuers are 30 meters apart on the surface. What is the maximum depth of a reflection that is in range of both rescuers?

SOLUTION

Let the first rescuer be located at the origin and let the second rescuer be located 30 meters (units) to the right. The range of each detector is circular and can be modeled by the following equations.

$$x^2 + y^2 = 30^2 \qquad \text{Range of first rescuer}$$
$$(x - 30)^2 + y^2 = 30^2 \qquad \text{Range of second rescuer}$$

Using methods demonstrated earlier in this section, you will find that these two equations intersect when

$$x = 15 \text{ and } y \approx \pm 25.98.$$

You are concerned only about the lower portions of the circles. So, the maximum depth in range of both rescuers is point R, as shown below, which is about 26 meters beneath the surface.

RECCO technology is often used in conjunction with other rescue methods such as avalanche dogs, transceivers, and probe lines.

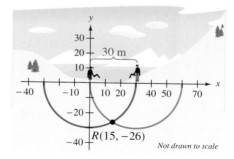

Exercises Within Reach ®

Solutions in English & Spanish and tutorial videos at CollegePrepAlgebra.com

29. *Busing Boundary* To be eligible to ride the school bus to East High School, a student must live at least 1 mile from the school (see figure). Describe the portion of Clarke Street for which the residents are *not* eligible to ride the school bus. Use a coordinate system in which the school is at (0, 0) and each unit represents 1 mile.

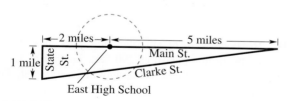

30. *Search Team* A search team of three members splits to search an area of the woods. Each member carries a family service radio with a circular range of 3 miles. The team members agree to communicate from their bases every hour. The second member sets up base 3 miles north of the first member. Where should the third member set up base to be as far east as possible but within direct communication range of each of the other two searchers? Use a coordinate system in which the first member is at (0, 0) and each unit represents 1 mile.

Concept Summary: Solving Nonlinear Systems of Equations

What

One way to solve a **nonlinear system of equations** is by graphing.

EXAMPLE

Find all solutions of the nonlinear system of equations.

$$\begin{cases} y = 2x - 1 & \text{Equation 1} \\ y = 2x^2 - 1 & \text{Equation 2} \end{cases}$$

How

Sketch the graph of each equation. Then approximate and check the point(s) of intersection (if any).

EXAMPLE

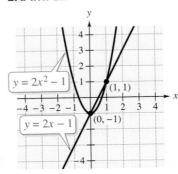

The system has two solutions: $(0, -1)$, $(1, 1)$.

Why

Not all systems of equations are linear. There are many examples of nonlinear systems of equations in real-life applications. You can use graphing, substitution, and elimination to solve these systems just as you did with linear systems of equations.

Exercises Within Reach ®

Worked-out solutions to odd-numbered exercises at CollegePrepAlgebra.com

Concept Summary Check

31. *Vocabulary* How is a system of nonlinear equations different from a system of linear equations?

32. *Solving a System* Identify the different methods you can use to solve a system of nonlinear equations.

33. *Choosing a Method* If one of the equations in a system is linear, which algebraic method usually works best for solving the system?

34. *Choosing a Method* If both of the equations in a system are conics, which algebraic method usually works best for solving the system?

Extra Practice

Solving a System by Substitution **In Exercises 35–38, solve the system by the method of substitution.**

35. $\begin{cases} y = x^2 - 5 \\ 3x + 2y = 10 \end{cases}$

36. $\begin{cases} x + y = 4 \\ x^2 - y^2 = 4 \end{cases}$

37. $\begin{cases} y = \sqrt{4 - x} \\ x + 3y = 6 \end{cases}$

38. $\begin{cases} y = \sqrt{25 - x^2} \\ x + y = 7 \end{cases}$

Solving a System by Elimination **In Exercises 39–42, solve the system by the method of elimination.**

39. $\begin{cases} \dfrac{x^2}{4} + y^2 = 1 \\ x^2 + \dfrac{y^2}{4} = 1 \end{cases}$

40. $\begin{cases} x^2 - y^2 = 1 \\ \dfrac{x^2}{2} + y^2 = 1 \end{cases}$

41. $\begin{cases} y^2 - x^2 = 10 \\ x^2 + y^2 = 16 \end{cases}$

42. $\begin{cases} x^2 + y^2 = 25 \\ x^2 + 2y^2 = 36 \end{cases}$

43. *Sailboat* A sail for a sailboat is shaped like a right triangle that has a perimeter of 36 meters and a hypotenuse of 15 meters. Find the dimensions of the sail.

44. *Dog Park* A rectangular dog park has a diagonal sidewalk that measures 290 feet. The perimeter of each triangle formed by the diagonal is 700 feet. Find the dimensions of the dog park.

45. *Hyperbolic Mirror* In a hyperbolic mirror, light rays directed to one focus are reflected to the other focus. The mirror in the figure has the equation

$$\frac{x^2}{9} - \frac{y^2}{16} = 1.$$

At which point on the mirror will light from the point $(0, 10)$ reflect to the focus?

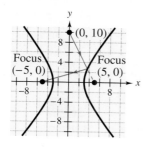

46. *Miniature Golf* You are playing miniature golf and your golf ball is at $(-15, 25)$ (see figure). A wall at the end of the enclosed area is part of a hyperbola whose equation is

$$\frac{x^2}{19} - \frac{y^2}{81} = 1.$$

Using the reflective property of hyperbolas given in Exercise 45, at which point on the wall must your ball hit for it to go into the hole? (The ball bounces off the wall only once.)

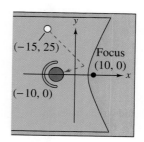

Explaining Concepts

47. *Writing* Explain how to solve a nonlinear system of equations using the method of substitution.

48. *Writing* Explain how to solve a nonlinear system of equations using the method of elimination.

49. *Precision* What is the maximum number of points of intersection of a line and a hyperbola? Explain.

50. *Precision* A circle and a parabola can have 0, 1, 2, 3, or 4 points of intersection. Sketch the circle given by $x^2 + y^2 = 4$. Discuss how this circle could intersect a parabola with an equation of the form $y = x^2 + C$. Then find the values of C for each of the five cases described below.

(a) No points of intersection

(b) One point of intersection

(c) Two points of intersection

(d) Three points of intersection

(e) Four points of intersection

Cumulative Review

In Exercises 51−62, solve the equation and check your solution(s).

51. $\sqrt{6 - 2x} = 4$

52. $\sqrt{x + 3} = -9$

53. $\sqrt{x} = x - 6$

54. $\sqrt{x + 14} = \sqrt{x} + 3$

55. $3^x = 243$

56. $4^x = 256$

57. $5^{x-1} = 310$

58. $e^{0.5x} = 8$

59. $\log_{10} x = 0.01$

60. $\log_4 8x = 3$

61. $2 \ln(x + 1) = -2$

62. $\ln(x + 3) - \ln x = \ln 1$

12 Chapter Summary

What did you learn?	Explanation and Examples	Review Exercises
Recognize the four basic conics: circles, parabolas, ellipses, and hyperbolas *(p. 620)*.	Circle Parabola Ellipse Hyperbola	1–10
12.1 Graph and write equations of circles *(p. 621)*.	*Standard equation of a circle with radius r and center (0, 0):* $x^2 + y^2 = r^2$ *Standard equation of a circle with radius r and center (h, k):* $(x - h)^2 + (y - k)^2 = r^2$	11–18
Graph and write equations of parabolas *(p. 624)*.	*Standard equation of a parabola with vertex (0, 0):* $x^2 = 4py$ Vertical axis $y^2 = 4px$ Horizontal axis *Standard equation of a parabola with vertex at (h, k):* $(x - h)^2 = 4p(y - k)$ Vertical axis $(y - k)^2 = 4p(x - h)$ Horizontal axis The focus of a parabola lies on the axis, a directed distance of p units from the vertex.	19–24
Graph and write equations of ellipses centered at the origin *(p. 628)*.	*Standard equation of an ellipse centered at the origin:* $\dfrac{x^2}{a^2} + \dfrac{y^2}{b^2} = 1$ Major axis is horizontal. $\dfrac{x^2}{b^2} + \dfrac{y^2}{a^2} = 1$ Major axis is vertical.	25–32
12.2 Graph and write equations of ellipses centered at (h, k) *(p. 630)*.	*Standard equation of an ellipse centered at (h, k):* $\dfrac{(x - h)^2}{a^2} + \dfrac{(y - k)^2}{b^2} = 1$ Major axis is horizontal. $\dfrac{(x - h)^2}{b^2} + \dfrac{(y - k)^2}{a^2} = 1$ Major axis is vertical. In all of the standard equations for ellipses, $0 < b < a$. The vertices of the ellipse lie on the major axis, a units from the center (the major axis has length $2a$). The co-vertices lie on the minor axis, b units from the center (the minor axis has length $2b$).	33–40

What did you learn?   **Explanation and Examples**   **Review Exercises**

	What did you learn?	Explanation and Examples	Review Exercises
12.3	Analyze hyperbolas centered at the origin *(p. 638)*.	*Standard form of a hyperbola centered at the origin:* $\dfrac{x^2}{a^2} - \dfrac{y^2}{b^2} = 1$ Transverse axis is horizontal. $\dfrac{y^2}{a^2} - \dfrac{x^2}{b^2} = 1$ Transverse axis is vertical.	41–48
	Analyze hyperbolas centered at (h, k) *(p. 641)*.	*Standard equation of a hyperbola centered at (h, k):* $\dfrac{(x - h)^2}{a^2} - \dfrac{(y - k)^2}{b^2} = 1$ Transverse axis is horizontal. $\dfrac{(y - k)^2}{a^2} - \dfrac{(x - h)^2}{b^2} = 1$ Transverse axis is vertical. A hyperbola's vertices lie on the transverse axis, a units from the center. A hyperbola's central rectangle has side lengths of $2a$ and $2b$. A hyperbola's asymptotes pass through opposite corners of its central rectangle.	49–54
12.4	Solve nonlinear systems of equations graphically *(p. 646)*.	1. Sketch the graph of each equation in the system. 2. Graphically approximate the coordinates of any points of intersection of the graphs. 3. Check the coordinates by substituting them into each equation in the original system. If the coordinates do not check, you may have to use an algebraic approach such as substitution or elimination.	55–58
	Solve nonlinear systems of equations by substitution *(p. 648)*.	1. Solve one equation for one variable in terms of the other. 2. Substitute the expression obtained in Step 1 into the other equation to obtain an equation in one variable. 3. Solve the equation obtained in Step 2. 4. Back-substitute the solution from Step 3 into the expression obtained in Step 1 to find the value of the other variable. 5. Check that the solution satisfies *both* of the original equations.	59–66, 77–82
	Solve nonlinear systems of equations by elimination *(p. 650)*.	1. Obtain coefficients for x (or y) that are opposites by multiplying one or both equations by suitable constants. 2. Add the equations to eliminate one variable. Solve the resulting equation. 3. Back-substitute the value from Step 2 into either of the original equations and solve for the other variable. 4. Check the solution in both of the original equations.	67–76

Review Exercises

Worked-out solutions to odd-numbered exercises at CollegePrepAlgebra.com

12.1

Identifying a Conic In Exercises 1−10, identify the type of conic shown in the graph.

1.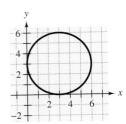

2.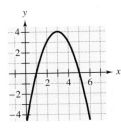

3.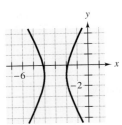

4.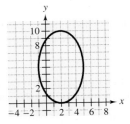

5.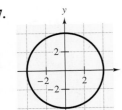

6.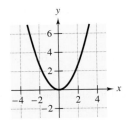

7.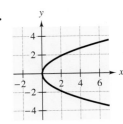

8.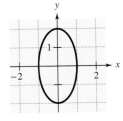

9.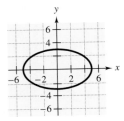

10.

Writing an Equation of a Circle In Exercises 11 and 12, write the standard form of the equation of the circle centered at the origin.

11. Radius: 6

12. Passes through the point $(-1, 3)$

Sketching a Circle In Exercises 13 and 14, identify the center and radius of the circle, and sketch its graph.

13. $x^2 + y^2 = 64$

14. $9x^2 + 9y^2 - 49 = 0$

Writing an Equation of a Circle In Exercises 15 and 16, write the standard form of the equation of the circle centered at (h, k).

15. Center: $(2, 6)$, Radius: 3

16. Center: $(-2, 3)$, Passes through the point $(1, 1)$

Sketching a Circle In Exercises 17 and 18, identify the center and radius of the circle, and sketch its graph.

17. $x^2 + y^2 + 6x + 8y + 21 = 0$

18. $x^2 + y^2 - 8x + 16y + 75 = 0$

Writing an Equation of a Parabola In Exercises 19−22, write the standard form of the equation of the parabola. Then sketch the parabola.

19. Vertex: $(0, 0)$, Focus: $(6, 0)$

20. Vertex: $(0, 0)$, Focus: $(0, 1)$

21. Vertex: $(0, 5)$, Focus: $(2, 5)$

22. Vertex: $(-3, 2)$, Focus: $(-3, 0)$

Sketching a Parabola **In Exercises 23 and 24, sketch the parabola, and identify its vertex and focus.**

23. $y = \frac{1}{2}x^2 - 8x + 7$

24. $x = y^2 + 10y - 4$

12.2

Writing the Standard Equation of an Ellipse **In Exercises 25–28, write the standard form of the equation of the ellipse centered at the origin.**

25. Vertices: $(0, -5), (0, 5)$

Co-vertices: $(-2, 0), (2, 0)$

26. Vertices: $(-8, 0), (8, 0)$

Co-vertices: $(0, -3), (0, 3)$

27. Major axis (vertical) 6 units, minor axis 4 units

28. Major axis (horizontal) 12 units, minor axis 2 units

Sketching an Ellipse **In Exercises 29–32, sketch the ellipse. Identify the vertices and co-vertices.**

29. $\dfrac{x^2}{64} + \dfrac{y^2}{16} = 1$

30. $\dfrac{x^2}{9} + y^2 = 1$

31. $36x^2 + 9y^2 - 36 = 0$

32. $100x^2 + 4y^2 - 4 = 0$

Writing the Standard Equation of an Ellipse **In Exercises 33–36, write the standard form of the equation of the ellipse.**

33. Vertices: $(-2, 4), (8, 4)$

Co-vertices: $(3, 0), (3, 8)$

34. Vertices: $(0, 5), (12, 5)$

Co-vertices: $(6, 2), (6, 8)$

35. Vertices: $(0, 0), (0, 8)$

Co-vertices: $(-3, 4), (3, 4)$

36. Vertices: $(5, -3), (5, 13)$

Co-vertices: $(3, 5), (7, 5)$

Sketching an Ellipse **In Exercises 37–40, find the center and vertices of the ellipse, and sketch its graph.**

37. $9(x + 1)^2 + 4(y - 2)^2 = 144$

38. $x^2 + 25y^2 - 4x - 21 = 0$

39. $16x^2 + y^2 + 6y - 7 = 0$

40. $x^2 + 4y^2 + 10x - 24y + 57 = 0$

12.3

Sketching a Hyperbola **In Exercises 41–44, identify the vertices and asymptotes of the hyperbola, and sketch its graph.**

41. $x^2 - y^2 = 25$

42. $y^2 - x^2 = 16$

43. $\dfrac{y^2}{25} - \dfrac{x^2}{4} = 1$

44. $\dfrac{x^2}{16} - \dfrac{y^2}{25} = 1$

Writing an Equation of a Hyperbola **In Exercises 45–48, write the standard form of the equation of the hyperbola centered at the origin.**

	Vertices	Asymptotes	
45.	$(-2, 0), (2, 0)$	$y = \frac{3}{2}x$	$y = -\frac{3}{2}x$
46.	$(0, -6), (0, 6)$	$y = 3x$	$y = -3x$
47.	$(0, -8), (0, 8)$	$y = \frac{4}{5}x$	$y = -\frac{4}{5}x$
48.	$(-3, 0), (3, 0)$	$y = \frac{4}{3}x$	$y = -\frac{4}{3}x$

Sketching a Hyperbola **In Exercises 49–52, find the center and vertices of the hyperbola, and sketch its graph.**

49. $\dfrac{(x - 3)^2}{9} - \dfrac{(y + 1)^2}{4} = 1$

50. $\dfrac{(x + 4)^2}{25} - \dfrac{(y - 7)^2}{64} = 1$

51. $8y^2 - 2x^2 + 48y + 16x + 8 = 0$

52. $25x^2 - 4y^2 - 200x - 40y = 0$

Writing an Equation of a Hyperbola **In Exercises 53 and 54, write the standard form of the equation of the hyperbola.**

53.

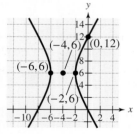

54.

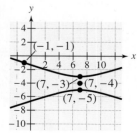

12.4

Solving a System Graphically **In Exercises 55–58, graph the equations to determine whether the system has any solutions. Find any solutions that exist.**

55. $\begin{cases} y = x^2 \\ y = 3x \end{cases}$

56. $\begin{cases} y = 2 + x^2 \\ y = 8 - x \end{cases}$

57. $\begin{cases} x^2 + y^2 = 16 \\ -x + y = 4 \end{cases}$

58. $\begin{cases} 2x^2 - y^2 = -8 \\ y = x + 6 \end{cases}$

Solving a System by Substitution **In Exercises 59–66, solve the system by the method of substitution.**

59. $\begin{cases} y = 5x^2 \\ y = -15x - 10 \end{cases}$

60. $\begin{cases} y^2 = 16x \\ 4x - y = -24 \end{cases}$

61. $\begin{cases} x^2 + y^2 = 1 \\ x + y = -1 \end{cases}$

62. $\begin{cases} y^2 - x^2 = 9 \\ x + y = 1 \end{cases}$

63. $\begin{cases} 4x + y^2 = 2 \\ 2x - y = -11 \end{cases}$

64. $\begin{cases} x^2 + y^2 = 10 \\ 2x - y = 5 \end{cases}$

65. $\begin{cases} x^2 + y^2 = 9 \\ x + 2y = 3 \end{cases}$

66. $\begin{cases} x^2 + y^2 = 4 \\ x - 2y = 4 \end{cases}$

Solving a System by Elimination **In Exercises 67−76, solve the system by the method of elimination.**

67. $\begin{cases} 6x^2 - y^2 = 15 \\ x^2 + y^2 = 13 \end{cases}$

68. $\begin{cases} x^2 + y^2 = 16 \\ -x^2 + \dfrac{y^2}{16} = 1 \end{cases}$

69. $\begin{cases} x^2 + y^2 = 7 \\ x^2 - y^2 = 1 \end{cases}$

70. $\begin{cases} x^2 + y^2 = 25 \\ y^2 - x^2 = 7 \end{cases}$

71. $\begin{cases} x^2 - y^2 = 4 \\ x^2 + y^2 = 4 \end{cases}$

72. $\begin{cases} x^2 + y^2 = 25 \\ x^2 - y^2 = -36 \end{cases}$

73. $\begin{cases} x^2 + y^3 = 13 \\ 2x^2 + 3y^2 = 30 \end{cases}$

74. $\begin{cases} 3x^2 - y^2 = 4 \\ x^2 + 4y^2 = 10 \end{cases}$

75. $\begin{cases} 4x^2 + 9y^2 = 36 \\ 2x^2 - 9y^2 = 18 \end{cases}$

76. $\begin{cases} 5x^2 - 2y^2 = -13 \\ 3x^2 + 4y^2 = 39 \end{cases}$

77. *Geometry* A circuit board has a perimeter of 28 centimeters and a diagonal of 10 centimeters. Find the dimensions of the circuit board.

78. *Geometry* A ceramic tile has a perimeter of 6 inches and a diagonal of $\sqrt{5}$ inches. Find the dimensions of the tile.

79. *Ice Rink* A rectangular ice rink has an area of 3000 square feet. The diagonal across the rink is 85 feet. Find the dimensions of the rink.

80. *Cell Phone* A cell phone has a rectangular external display that contains 19,200 pixels with a diagonal of 200 pixels. Find the resolution (dimensions in pixels) of the external display.

81. *Geometry* A piece of wire 100 inches long is cut into two pieces. Each of the two pieces is then bent into a square. The area of one square is 144 square inches greater than the area of the other square. Find the length of each piece of wire.

82. *Geometry* You have 250 feet of fencing to enclose two corrals of equal size (see figure). The combined area of the corrals is 2400 square feet. Find the dimensions of each corral.

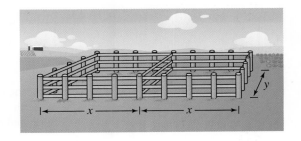

Chapter Test

Solutions in English & Spanish and tutorial videos at CollegePrepAlgebra.com

Take this test as you would take a test in class. After you are done, check your work against the answers in the back of the book.

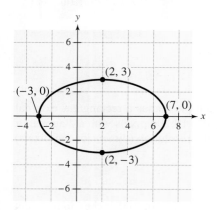

1. Write the standard form of the equation of the circle shown at the left.

In Exercises 2 and 3, write the standard form of the equation of the circle, and sketch its graph.

2. $x^2 + y^2 - 2x - 6y + 1 = 0$

3. $x^2 + y^2 + 4x - 6y + 4 = 0$

4. Identify the vertex and the focus of the parabola $x = -3y^2 + 12y - 8$, and sketch its graph.

5. Write the standard form of the equation of the parabola with vertex $(7, -2)$ and focus $(7, 0)$.

6. Write the standard form of the equation of the ellipse shown at the left.

In Exercises 7 and 8, find the center and vertices of the ellipse, and sketch its graph.

7. $16x^2 + 4y^2 = 64$

8. $25x^2 + 4y^2 - 50x - 24y - 39 = 0$

In Exercises 9 and 10, write the standard form of the equation of the hyperbola.

9. Vertices: $(-3, 0), (3, 0)$; Asymptotes: $y = \pm\frac{2}{3}x$

10. Vertices: $(0, -5), (0, 5)$; Asymptotes: $y = \pm\frac{5}{2}x$

In Exercises 11 and 12, find the center and vertices of the hyperbola, and sketch its graph.

11. $4x^2 - 2y^2 - 24x + 20 = 0$

12. $16y^2 - 25x^2 + 64y + 200x - 736 = 0$

In Exercises 13-15, solve the nonlinear system of equations.

13. $\begin{cases} \dfrac{x^2}{16} + \dfrac{y^2}{9} = 1 \\ 3x + 4y = 12 \end{cases}$ 14. $\begin{cases} x^2 + y^2 = 16 \\ \dfrac{x^2}{16} - \dfrac{y^2}{9} = 1 \end{cases}$ 15. $\begin{cases} x^2 + y^2 = 10 \\ x^2 \quad\ = y^2 + 2 \end{cases}$

16. Write the equation of the circular orbit of a satellite 1000 miles above the surface of Earth. Place the origin of the rectangular coordinate system at the center of Earth and assume the radius of Earth is 4000 miles.

17. A rectangle has a perimeter of 56 inches and a diagonal of 20 inches. Find the dimensions of the rectangle.

Cumulative Test: Chapters 10–12

Solutions in English & Spanish and tutorial videos at CollegePrepAlgebra.com

Take this test as you would take a test in class. After you are done, check your work against the answers in the back of the book.

In Exercises 1–4, solve the equation by the specified method.

1. Factoring:

$4x^2 - 9x - 9 = 0$

2. Square Root Property:

$(x - 5)^2 - 64 = 0$

3. Completing the square:

$x^2 - 10x - 25 = 0$

4. Quadratic Formula:

$3x^2 + 6x + 2 = 0$

5. Solve the equation of quadratic form: $x^4 - 8x^2 + 15 = 0$

In Exercises 6 and 7, solve the inequality and graph the solution on the real number line.

6. $3x^2 + 8x \le 3$

7. $\dfrac{3x + 4}{2x - 1} < 0$

8. Find a quadratic equation having the solutions -2 and 6.

9. Find the compositions (a) $f \circ g$ and (b) $g \circ f$ for $f(x) = 2x^2 - 3$ and $g(x) = 5x - 1$. Then find the domain of each composition.

10. Find the inverse function of $f(x) = \dfrac{5 - 3x}{4}$.

11. Evaluate $f(x) = 7 + 2^{-x}$ when $x = 1, 0.5$, and 3.

12. Sketch the graph of $f(x) = 4^{x - 1}$ and identify the horizontal asymptote.

13. Describe the relationship between the graphs of $f(x) = e^x$ and $g(x) = \ln x$.

14. Sketch the graph of $\log_3(x - 1)$ and identify the vertical asymptote.

15. Evaluate $\log_4 \frac{1}{16}$.

16. Use the properties of logarithms to condense $3(\log_2 x + \log_2 y) - \log_2 z$.

17. Use the properties of logarithms to expand $\log_{10} \dfrac{\sqrt{x + 1}}{x^4}$.

In Exercises 18–21, solve the equation.

18. $\log_x\left(\frac{1}{9}\right) = -2$

19. $4 \ln x = 10$

20. $500(1.08)^t = 2000$

21. $3(1 + e^{2x}) = 20$

22. If the inflation rate averages 2.8% over the next 5 years, the approximate cost C of goods and services t years from now is given by

$$C(t) = P(1.028)^t, \ 0 \le t \le 5$$

where P is the percent cost. The price of an oil change is presently $29.95. Estimate the price 5 years from now.

23. Determine the effective yield of an 8% interest rate compounded continuously.

24. Determine the length of time for an investment of $1500 to quadruple in value when the investment earns 7% compounded continuously.

25. Write the standard form of the equation of the circle, and sketch its graph.

$$x^2 + y^2 - 6x + 14y - 6 = 0$$

26. Identify the vertex and focus of the parabola, and sketch its graph.

$$y = 2x^2 - 20x + 5$$

27. Write the standard form of the equation of the ellipse shown at the left.

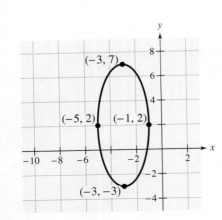

28. Find the center and vertices of the ellipse, and sketch its graph.

$$4x^2 + y^2 = 4$$

29. Write the standard form of the equation of the hyperbola with vertices $(0, -3)$ and $(0, 3)$ and asymptotes $y = \pm 3x$.

30. Find the center and vertices of the hyperbola, and sketch its graph.

$$x^2 - 9y^2 + 18y = 153$$

In Exercises 31 and 32, solve the nonlinear system of equations.

31. $\begin{cases} y = x^2 - x - 1 \\ 3x - y = 4 \end{cases}$

32. $\begin{cases} x^2 + 5y^2 = 21 \\ -x + y^2 = 5 \end{cases}$

33. A rectangle has an area of 32 square feet and a perimeter of 24 feet. Find the dimensions of the rectangle.

34. A rectangle has an area of 21 square feet and a perimeter of 20 feet. Find the dimensions of the rectangle.

13

Sequences, Series, and the Binomial Theorem

13.1 Sequences and Series

13.2 Arithmetic Sequences

13.3 Geometric Sequences

13.4 The Binomial Theorem

MASTERY IS WITHIN REACH!

"When I was in my math classes, I had to get an early start on getting ready for my last test. I had to get a good grade in math because I was competing with other students to get into a good university. I studied with a friend from class a couple of weeks before the test. I stuck to studying for the test more since we did it together. We both did great!"

Mindy

See page 681 for suggestions about preparing for the final exam.

AlexandreNunes/Shutterstock.com

13.1 Sequences and Series

▶ Write the terms of sequences.

▶ Find the apparent nth term of a sequence.

▶ Sum the terms of a sequence to obtain a series.

Sequences

A mathematical **sequence** is simply an ordered list of numbers. Each number in the list is a **term** of the sequence. A sequence can have a finite number of terms or an infinite number of terms.

> ### Sequences
>
> An **infinite sequence** $a_1, a_2, a_3, \ldots, a_n, \ldots$ is a function whose domain is the set of positive integers.
>
> A **finite sequence** $a_1, a_2, a_3, \ldots, a_n$ is a function whose domain is the finite set $\{1, 2, 3, \ldots, n\}$.

On occasion it is convenient to begin subscripting a sequence with 0 instead of 1 so that the terms of the sequence become $a_0, a_1, a_2, a_3, \ldots$. When this is the case, the domain includes 0.

EXAMPLE 1   Writing the Terms of a Sequence

Write the first six terms of the sequence with the given nth term.

a. $a_n = n^2 - 1$ Begin sequence with $n = 1$.

b. $a_n = 3(2^n)$ Begin sequence with $n = 0$.

SOLUTION

a. $a_1 = (1)^2 - 1 = 0$ $a_2 = (2)^2 - 1 = 3$ $a_3 = (3)^2 - 1 = 8$

$a_4 = (4)^2 - 1 = 15$ $a_5 = (5)^2 - 1 = 24$ $a_6 = (6)^2 - 1 = 35$

The sequence can be written as $0, 3, 8, 15, 24, 35, \ldots, n^2 - 1, \ldots$.

b. $a_0 = 3(2^0) = 3 \cdot 1 = 3$ $a_1 = 3(2^1) = 3 \cdot 2 = 6$

$a_2 = 3(2^2) = 3 \cdot 4 = 12$ $a_3 = 3(2^3) = 3 \cdot 8 = 24$

$a_4 = 3(2^4) = 3 \cdot 16 = 48$ $a_5 = 3(2^5) = 3 \cdot 32 = 96$

The sequence can be written as $3, 6, 12, 24, 48, 96, \ldots, 3(2^n), \ldots$.

Exercises Within Reach ®

Solutions in English & Spanish and tutorial videos at CollegePrepAlgebra.com

Writing the Terms of a Sequence In Exercises 1–10, write the first five terms of the sequence with the given nth term. (Assume that n begins with 1.)

1. $a_n = 2n$ **2.** $a_n = 3n$ **3.** $a_n = \left(\frac{1}{4}\right)^n$ **4.** $a_n = \left(\frac{1}{3}\right)^n$

5. $a_n = 5n - 2$ **6.** $a_n = 2n + 3$ **7.** $a_n = \dfrac{4}{n + 3}$

8. $a_n = \dfrac{9}{5 + n}$ **9.** $a_n = \dfrac{3n}{5n - 1}$ **10.** $a_n = \dfrac{2n}{6n - 3}$

EXAMPLE 2   **A Sequence Whose Terms Alternate in Sign**

Write the first six terms of the sequence whose nth term is

$$a_n = \frac{(-1)^n}{2n - 1}.$$ Begin sequence with $n = 1$.

SOLUTION

$$a_1 = \frac{(-1)^1}{2(1) - 1} = -\frac{1}{1} \qquad a_2 = \frac{(-1)^2}{2(2) - 1} = \frac{1}{3} \qquad a_3 = \frac{(-1)^3}{2(3) - 1} = -\frac{1}{5}$$

$$a_4 = \frac{(-1)^4}{2(4) - 1} = \frac{1}{7} \qquad a_5 = \frac{(-1)^5}{2(5) - 1} = -\frac{1}{9} \qquad a_6 = \frac{(-1)^6}{2(6) - 1} = \frac{1}{11}$$

The sequence can be written as $-1, \frac{1}{3}, -\frac{1}{5}, \frac{1}{7}, -\frac{1}{9}, \frac{1}{11}, \ldots, \frac{(-1)^n}{2n - 1}, \ldots$.

Some very important sequences in mathematics involve terms that are defined with special types of products call *factorials*.

Definition of Factorial

When n is a positive integer, n **factorial** is defined as

$$n! = 1 \cdot 2 \cdot 3 \cdot 4 \cdots (n - 1) \cdot n.$$

As a special case, zero factorial is defined as $0! = 1$.

EXAMPLE 3   **Writing Factorials**

The first several factorial values are as follows.

$0! = 1$ $1! = 1$

$2! = 1 \cdot 2 = 2$ $3! = 1 \cdot 2 \cdot 3 = 6$

$4! = 1 \cdot 2 \cdot 3 \cdot 4 = 24$ $5! = 1 \cdot 2 \cdot 3 \cdot 4 \cdot 5 = 120$

Exercises Within Reach ®

Solutions in English & Spanish and tutorial videos at CollegePrepAlgebra.com

Writing the Terms of a Sequence In Exercises 11–16, write the first five terms of the sequence with the given nth term. (Assume that n begins with 1.)

11. $a_n = (-1)^n 2n$

12. $a_n = (-1)^{n + 1} 3n$

13. $a_n = \left(-\frac{1}{2}\right)^{n + 1}$

14. $a_n = \left(-\frac{2}{3}\right)^{n - 1}$

15. $a_n = \frac{(-1)^n}{n^2}$

16. $a_n = \frac{(-1)^{n + 1}}{n}$

Writing a Factorial In Exercises 17–20, write the product represented by the factorial. Then evaluate the product.

17. $6!$

18. $7!$

19. $9!$

20. $10!$

Many calculators have a factorial key, denoted by $\boxed{n!}$. If your calculator has such a key, try using it to evaluate $n!$ for several values of n. You will see that as n increases, the value of $n!$ becomes very large. For instance, $10! = 3,628,800$.

EXAMPLE 4   **Writing Sequences Involving Factorials**

Write the first six terms of the sequence with the given nth term.

a. $a_n = \dfrac{1}{n!}$ Begin sequence with $n = 0$.

b. $a_n = \dfrac{2^n}{n!}$ Begin sequence with $n = 0$.

SOLUTION

a. $a_0 = \dfrac{1}{0!} = \dfrac{1}{1} = 1$ 　　　　　$a_1 = \dfrac{1}{1!} = \dfrac{1}{1} = 1$

$a_2 = \dfrac{1}{2!} = \dfrac{1}{1 \cdot 2} = \dfrac{1}{2}$ 　　　$a_3 = \dfrac{1}{3!} = \dfrac{1}{1 \cdot 2 \cdot 3} = \dfrac{1}{6}$

$a_4 = \dfrac{1}{4!} = \dfrac{1}{1 \cdot 2 \cdot 3 \cdot 4} = \dfrac{1}{24}$ 　$a_5 = \dfrac{1}{5!} = \dfrac{1}{1 \cdot 2 \cdot 3 \cdot 4 \cdot 5} = \dfrac{1}{120}$

b. $a_0 = \dfrac{2^0}{0!} = \dfrac{1}{1} = 1$ 　　　　$a_1 = \dfrac{2^1}{1!} = \dfrac{2}{1} = 2$

$a_2 = \dfrac{2^2}{2!} = \dfrac{2 \cdot 2}{1 \cdot 2} = \dfrac{4}{2} = 2$ 　$a_3 = \dfrac{2^3}{3!} = \dfrac{2 \cdot 2 \cdot 2}{1 \cdot 2 \cdot 3} = \dfrac{8}{6} = \dfrac{4}{3}$

$a_4 = \dfrac{2^4}{4!} = \dfrac{2 \cdot 2 \cdot 2 \cdot 2}{1 \cdot 2 \cdot 3 \cdot 4} = \dfrac{2}{3}$ 　$a_5 = \dfrac{2^5}{5!} = \dfrac{2 \cdot 2 \cdot 2 \cdot 2 \cdot 2}{1 \cdot 2 \cdot 3 \cdot 4 \cdot 5} = \dfrac{4}{15}$

EXAMPLE 5   **Simplifying Expressions Involving Factorials**

a. $\dfrac{7!}{5!} = \dfrac{1 \cdot 2 \cdot 3 \cdot 4 \cdot 5 \cdot 6 \cdot 7}{1 \cdot 2 \cdot 3 \cdot 4 \cdot 5} = 6 \cdot 7 = 42$

b. $\dfrac{n!}{(n-1)!} = \dfrac{1 \cdot 2 \cdot 3 \cdot \ldots \cdot (n-1) \cdot n}{1 \cdot 2 \cdot 3 \cdot \ldots \cdot (n-1)} = n$

Exercises Within Reach®   Solutions in English & Spanish and tutorial videos at CollegePrepAlgebra.com

Writing the Terms of a Sequence In Exercises 21–24, write the first five terms of the sequence with the given nth term. (Assume that n begins with 1.)

21. $a_n = \dfrac{n!}{n}$ 　　　　　　　　**22.** $a_n = \dfrac{n!}{n+1}$

23. $a_n = \dfrac{(n+1)!}{n!}$ 　　　　　　**24.** $a_n = \dfrac{n!}{(n-1)!}$

Simplifying an Expression Involving a Factorial In Exercises 25–32, simplify the expression.

25. $\dfrac{5!}{4!}$ 　　**26.** $\dfrac{6!}{8!}$ 　　**27.** $\dfrac{25!}{20!5!}$ 　　**28.** $\dfrac{20!}{15!5!}$

29. $\dfrac{n!}{(n+1)!}$ 　　**30.** $\dfrac{(n+2)!}{n!}$ 　　**31.** $\dfrac{(n+1)!}{(n-1)!}$ 　　**32.** $\dfrac{(2n)!}{(2n-1)!}$

Finding the *n*th Term of a Sequence

Sometimes you have the first several terms of a sequence and need to find a formula (the *n*th term) to generate those terms. Pattern recognition is crucial in finding a form for the *n*th term.

EXAMPLE 6   **Finding the *n*th Term of a Sequence**

Write an expression for the *n*th term of each sequence.

a. $\dfrac{1}{2}, \dfrac{1}{4}, \dfrac{1}{8}, \dfrac{1}{16}, \dfrac{1}{32}, \ldots$

b. $1, -4, 9, -16, 25, \ldots$

SOLUTION

a.

n:	1	2	3	4	5	...	*n*
Terms:	$\dfrac{1}{2}$	$\dfrac{1}{4}$	$\dfrac{1}{8}$	$\dfrac{1}{16}$	$\dfrac{1}{32}$	...	a_n

Pattern: The numerators are 1 and the denominators are increasing powers of 2.

So, an expression for the *n*th term is $\dfrac{1}{2^n}$.

b.

n:	1	2	3	4	5	...	*n*
Terms:	1	-4	9	-16	25	...	a_n

Pattern: The terms have alternating signs, with those in the even positions being negative. The absolute value of each term is the square of *n*.

So, an expression for the *n*th term is $(-1)^{n+1}n^2$.

Exercises Within Reach ®

Solutions in English & Spanish and tutorial videos at CollegePrepAlgebra.com

Matching In Exercises 33−36, **match the sequence with the graph of its first 10 terms.**

(a)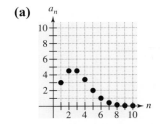

(b)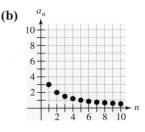

(c)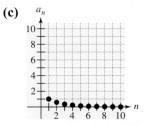

(d)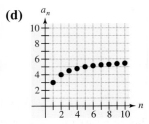

33. $\dfrac{6}{n+1}$

34. $a_n = \dfrac{6n}{n+1}$

35. $a_n = (0.6)^{n-1}$

36. $a_n = \dfrac{3^n}{n!}$

Finding the nth Term In Exercises 37−42, **write an expression for the *n*th term of the sequence.**
(Assume that *n* begins with 1.)

37. $1, 3, 5, 7, 9, \ldots$

38. $2, -4, 6, -8, 10, \ldots$

39. $0, 3, 8, 15, 24, \ldots$

40. $1, 8, 27, 64, 125, \ldots$

41. $-\dfrac{1}{5}, \dfrac{1}{25}, -\dfrac{1}{125}, \dfrac{1}{625}, -\dfrac{1}{3125}, \ldots$

42. $1, \dfrac{1}{4}, \dfrac{1}{9}, \dfrac{1}{16}, \dfrac{1}{25}, \ldots$

Series

> ### Definition of Series
>
> For an infinite sequence $a_1, a_2, a_3, \ldots, a_n, \ldots$
>
> **1.** the sum of the first n terms
>
> $$S_n = a_1 + a_2 + a_3 + \cdots + a_n$$
>
> is called a **partial sum**, and
>
> **2.** the sum of all the terms
>
> $$a_1 + a_2 + a_3 + \cdots + a_n + \cdots$$
>
> is called an **infinite series**, or simply a **series**.

EXAMPLE 7   **Finding Partial Sums**

Find the indicated partial sums for each sequence.

a. Find S_1, S_2, and S_5 for $a_n = 3n - 1$.

b. Find S_2, S_3, and S_4 for $a_n = \dfrac{(-1)^n}{n + 1}$.

SOLUTION

a. The first five terms of the sequence $a_n = 3n - 1$ are

$$a_1 = 2, a_2 = 5, a_3 = 8, a_4 = 11, \text{ and } a_5 = 14$$

So, the partial sums are

$$S_1 = 2, S_2 = 2 + 5 = 7, \text{ and } S_5 = 2 + 5 + 8 + 11 + 14 = 40.$$

b. The first four terms of the sequence $a_n = \dfrac{(-1)^n}{n + 1}$ are

$$a_1 = -\frac{1}{2}, a_2 = \frac{1}{3}, a_3 = -\frac{1}{4}, \text{ and } a_4 = \frac{1}{5}.$$

So, the partial sums are

$$S_2 = -\frac{1}{2} + \frac{1}{3} = -\frac{1}{6}, \text{ and } S_3 = -\frac{1}{2} + \frac{1}{3} - \frac{1}{4} = -\frac{5}{12}, \text{ and}$$

$$S_4 = -\frac{1}{2} + \frac{1}{3} - \frac{1}{4} + \frac{1}{5} = -\frac{13}{60}.$$

Exercises Within Reach ®   Solutions in English & Spanish and tutorial videos at CollegePrepAlgebra.com

Finding Partial Sums In Exercises 43−46, **find the indicated partial sums for the sequence.**

43. Find S_1, S_2, and S_6 for $a_n = 2n + 5$.

44. Find S_3, S_4, and S_{10} for $a_n = n^3 - 1$.

45. Find S_2, S_3, and S_9 for $a_n = \dfrac{1}{n}$.

46. Find S_1, S_2, and S_5 for $a_n = \dfrac{(-1)^{n + 1}}{n + 1}$.

> ## Definition of Sigma Notation
>
> The sum of the first n terms of the sequence whose nth term is a_n is
>
> $$\sum_{i=1}^{n} a_i = a_1 + a_2 + a_3 + a_4 + \ldots + a_n$$
>
> where i is the **index of summation**, n is the **upper limit of summation**, and 1 is the **lower limit of summation**.

Study Tip

In Example 8(a), the index of summation is i and the summation begins with $i = 1$. Any letter can be used as the index of summation, and the summation can begin with any integer. For instance, in Example 8(b), the index of summation is k and the summation begins with $k = 0$.

EXAMPLE 8   Finding Sums in Sigma Notation

a. $\displaystyle\sum_{i=1}^{6} 2i = 2(1) + 2(2) + 2(3) + 2(4) + 2(5) + 2(6)$

$\qquad\qquad = 2 + 4 + 6 + 8 + 10 + 12$

$\qquad\qquad = 42$

b. $\displaystyle\sum_{k=0}^{8} \frac{1}{k!} = \frac{1}{0!} + \frac{1}{1!} + \frac{1}{2!} + \frac{1}{3!} + \frac{1}{4!} + \frac{1}{5!} + \frac{1}{6!} + \frac{1}{7!} + \frac{1}{8!}$

$\qquad\qquad = 1 + 1 + \frac{1}{2} + \frac{1}{6} + \frac{1}{24} + \frac{1}{120} + \frac{1}{720} + \frac{1}{5040} + \frac{1}{40{,}320}$

$\qquad\qquad \approx 2.71828$

Note that this sum is approximately $e = 2.71828\ldots.$.

EXAMPLE 9   Writing a Sum in Sigma Notation

Write the sum in sigma notation: $\dfrac{2}{2} + \dfrac{2}{3} + \dfrac{2}{4} + \dfrac{2}{5} + \dfrac{2}{6}.$

SOLUTION

To write this sum in sigma notation, you must find a pattern for the terms. You can see that the terms have numerators of 2 and denominators that range over the integers from 2 to 6. So, one possible sigma notation is

$$\sum_{i=1}^{5} \frac{2}{i+1} = \frac{2}{2} + \frac{2}{3} + \frac{2}{4} + \frac{2}{5} + \frac{2}{6}.$$

Exercises Within Reach ®

Solutions in English & Spanish and tutorial videos at CollegePrepAlgebra.com

Finding a Partial Sum In Exercises 47–54, find the partial sum.

47. $\displaystyle\sum_{k=1}^{5} 6$

48. $\displaystyle\sum_{k=1}^{4} 5k$

49. $\displaystyle\sum_{i=0}^{6} (2i + 5)$

50. $\displaystyle\sum_{i=0}^{4} (2i + 3)$

51. $\displaystyle\sum_{j=0}^{3} \frac{1}{j^2 + 1}$

52. $\displaystyle\sum_{j=1}^{5} \frac{(-1)^{j+1}}{j^2}$

53. $\displaystyle\sum_{n=0}^{5} \left(-\frac{1}{3}\right)^n$

54. $\displaystyle\sum_{k=1}^{4} \left(\frac{5}{3}\right)^{k-1}$

Writing a Sum in Sigma Notation In Exercises 55–58, write the sum in sigma notation. (Begin with $k = 1$.)

55. $1 + 2 + 3 + 4 + 5$

56. $5 + 10 + 15 + 20 + 25 + 30$

57. $\frac{1}{2} + \frac{2}{3} + \frac{3}{4} + \frac{4}{5} + \frac{5}{6} + \cdots + \frac{11}{12}$

58. $\frac{2}{4} + \frac{4}{5} + \frac{6}{6} + \frac{8}{7} + \cdots + \frac{40}{23}$

Concept Summary: *Finding the nth Term of a Sequence*

What

A **sequence** is an ordered list of numbers. You can use the first several **terms** of a sequence to write an expression for the nth term a_n of the sequence.

EXAMPLE

Write an expression for the nth term of the sequence $\frac{2}{3}, \frac{2}{6}, \frac{2}{9}, \frac{2}{12}, \dots$.

How

Use pattern recognition to find a form for the nth term.

EXAMPLE

$n:$ 1   2   3   4   . . . n

$Terms:$ $\frac{2}{3}$ $\frac{2}{6}$ $\frac{2}{9}$ $\frac{2}{12}$ . . . a_n

$Pattern:$ Each numerator is 2 and each denominator is three times n.

So, an expression for the nth term is $a_n = \frac{2}{3n}$.

Why

You can use the expression for the nth term of a sequence to represent **partial sums** for the sequence in **sigma notation**.

For instance, for the sequence whose nth term is $a_n = \frac{2}{3n}$, the partial sum S_5 is

$$S_5 = \sum_{n=1}^{5} \frac{2}{3n}.$$

Exercises Within Reach®

Worked-out solutions to odd-numbered exercises at CollegePrepAlgebra.com

Concept Summary Check

59. *Identifying a Term* What is the third term of the sequence in the example above?

60. *Identifying a Domain Value* In the example above, what domain value of the sequence corresponds to the term $\frac{2}{12}$?

61. *Finding a Term of a Sequence* Explain how to find the 6th term of the sequence in the example above.

62. *Vocabulary* Explain the difference between a *sequence* and a *series*.

Extra Practice

63. *Compound Interest* A deposit of $500 is made in an account that earns 7% interest compounded yearly. The balance in the account after N years is given by

$$A_N = 500(1 + 0.07)^N, \quad N = 1, 2, 3, \dots.$$

(a) Compute the first eight terms of the sequence.

(b) Find the balance in this account after 40 years by computing A_{40}.

(c) Use a graphing calculator to graph the first 40 terms of the sequence.

(d) The graph shows how the terms of the sequence increase. Do the terms keep increasing by the same amount? Explain.

64. *Soccer Ball Design* The number of degrees a_n in each angle of a regular n-sided polygon is

$$a_n = \frac{180(n-2)}{n}, \quad n \geq 3.$$

The surface of a soccer ball is made of regular hexagons and pentagons. When a soccer ball is taken apart and flattened, as shown in the figure, the sides do not meet each other. Use the terms a_5 and a_6 to explain why there are gaps between adjacent hexagons.

65. *Stars* Stars are formed by placing n equally spaced points on a circle and connecting each point with a second point on the circle (see figure). The measure of degrees d_n of the angle at each tip of the star is given by

$$d_n = \frac{180(n-4)}{n}, \quad n \geq 5.$$

Write the first six terms of this sequence.

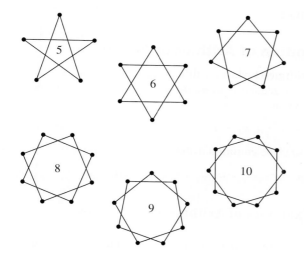

66. *Stars* The stars in Exercise 65 were formed by placing n equally spaced points on a circle and connecting each point with the second point from it on the circle. The stars in the figure for this exercise were formed in a similar way except that each point was connected with the third point from it. For these stars, the measure in degrees d_n of the angle at each point is given by

$$d_n = \frac{180(n-6)}{n}, \quad n \geq 7.$$

Write the first five terms of this sequence.

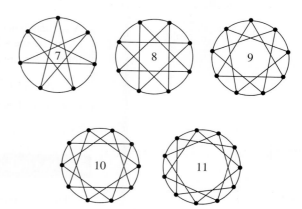

Explaining Concepts

67. *Structure* In your own words, explain why a sequence is a function.

68. *Number Sense* The nth term of a sequence is $a_n = (-1)^n n$. Which terms of the sequence are negative? Explain.

69. *Number Sense* Explain the difference between $a_n = 4n!$ and $a_n = (4n)!$.

True or False? In Exercises 70–72, determine whether the statement is true or false. Justify your answer.

70. $\displaystyle\sum_{i=1}^{4}(i^2 + 2i) = \sum_{i=1}^{4} i^2 + \sum_{i=1}^{4} 2i$

71. $\displaystyle\sum_{k=1}^{4} 3k = 3\sum_{k=1}^{4} k$

72. $\displaystyle\sum_{j=1}^{4} 2^j = \sum_{j=3}^{6} 2^{j-2}$

Cumulative Review

In Exercises 73–76, evaluate the expression for the specified value of the variable.

73. $-2n + 15; n = 3$ **74.** $-20n + 100; n = 4$

75. $25 - 3(n + 4); n = 8$

76. $-\frac{3}{2}(n - 1) + 6; n = 10$

In Exercises 77–80, identify the center and radius of the circle, and sketch its graph.

77. $x^2 + y^2 = 36$

78. $4x^2 + 4y^2 = 9$

79. $x^2 + y^2 + 4x - 12 = 0$

80. $x^2 + y^2 - 10x - 2y - 199 = 0$

In Exercises 81–84, identify the vertex and focus of the parabola, and sketch its graph.

81. $x^2 = 6y$

82. $y^2 = 9x$

83. $x^2 + 8y + 32 = 0$

84. $y^2 - 10x + 6y + 29 = 0$

13.2 Arithmetic Sequences

▶ Recognize, write, and find the nth terms of arithmetic sequences.
▶ Find the nth partial sum of an arithmetic sequence.
▶ Use arithmetic sequences to solve application problems.

Arithmetic Sequences

> ### Definition of Arithmetic Sequence
>
> A sequence is called **arithmetic** when the differences between consecutive terms are the same. So, the sequence $a_1, a_2, a_3, a_4, \ldots, a_n, \ldots$ is arithmetic when there is a number d such that
>
> $$a_2 - a_1 = d, a_3 - a_2 = d, a_4 - a_3 = d$$
>
> and so on. The number d is the **common difference** of the sequence.

EXAMPLE 1 **Examples of Arithmetic Sequences**

a. The sequence whose nth term is $3n + 2$ is arithmetic. For this sequence, the common difference between consecutive terms is 3.

$$5, 8, 11, 14, \ldots, 3n + 2, \ldots \qquad \text{Begin with } n = 1.$$
$$8 - 5 = 3$$

b. The sequence whose nth term is $7 - 5n$ is arithmetic. For this sequence, the common difference between consecutive terms is -5.

$$2, -3, -8, -13, \ldots, 7 - 5n, \ldots \qquad \text{Begin with } n = 1.$$
$$-3 - 2 = -5$$

c. The sequence whose nth term is $\frac{1}{4}(n + 3)$ is arithmetic. For this sequence, the common difference between consecutive terms is $\frac{1}{4}$.

$$1, \frac{5}{4}, \frac{3}{2}, \frac{7}{4}, \ldots, \frac{1}{4}(n + 3), \ldots \qquad \text{Begin with } n = 1.$$
$$\frac{5}{4} - 1 = \frac{1}{4}$$

Exercises Within Reach ®

Solutions in English & Spanish and tutorial videos at CollegePrepAlgebra.com

Finding the Common Difference In Exercises 1−6, **find the common difference of the** arithmetic sequence.

1. $2, 5, 8, 11, \ldots$

2. $-8, 0, 8, 16, \ldots$

3. $100, 94, 88, 82, \ldots$

4. $3200, 2800, 2400, 2000, \ldots$

5. $4, \frac{9}{2}, 5, \frac{11}{2}, 6, \ldots$

6. $1, \frac{5}{3}, \frac{7}{3}, 3, \ldots$

Finding the Common Difference In Exercises 7−12, **find the common difference of the** arithmetic sequence with the given nth term.

7. $a_n = 4n + 5$

8. $a_n = 7n + 6$

9. $a_n = 8 - 3n$

10. $a_n = 12 - 4n$

11. $a_n = \frac{1}{2}(n + 1)$

12. $a_n = \frac{1}{3}(n + 4)$

> ## The *n*th Term of an Arithmetic Sequence
>
> The *n*th term of an arithmetic sequence has the form
>
> $$a_n = a_1 + (n - 1)d$$
>
> where d is the common difference between the terms of the sequence, and a_1 is the first term.

Study Tip

The *n*th term of an arithmetic sequence can be derived from the following pattern.

$a_1 = a_1$	1st term
$a_2 = a_1 + d$	2nd term
$a_3 = a_1 + 2d$	3rd term
$a_4 = a_1 + 3d$	4th term
$a_5 = a_1 + 4d$	5th term

1 less

$\vdots$ $\qquad$ $\vdots$

$a_n = a_1 + (n - 1)d$ $\quad$ *n*th term

1 less

EXAMPLE 2   **Finding the *n*th Term of an Arithmetic Sequence**

a.   Find a formula for the *n*th term of the arithmetic sequence whose common difference is 2 and whose first term is 5.

b.   Find a formula for the *n*th term of the arithmetic sequence whose common difference is 3 and whose first term is 2.

SOLUTION

a.   You know that the formula for the *n*th term is of the form $a_n = a_1 + (n - 1)d$. Moreover, because the common difference is $d = 2$ and the first term is $a_1 = 5$, the formula must have the form

$$a_n = 5 + 2(n - 1). \qquad \text{Substitute 5 for } a_1 \text{ and 2 for } d.$$

So, the formula for the *n*th term is $a_n = 2n + 3$, and the sequence has the following form.

$$5, 7, 9, 11, 13, \ldots, 2n - 3, \ldots$$

b.   You know that the formula for the *n*th term is of the form $a_n = a_1 + (n - 1)d$. Moreover, because the common difference is $d = 3$ and the first term is $a_1 = 2$, the formula must have the form

$$a_n = 2 + 3(n - 1). \qquad \text{Substitute 2 for } a_1 \text{ and 3 for } d.$$

So, the formula for the *n*th term is $a_n = 3n - 1$, and the sequence has the following form.

$$2, 5, 8, 11, 14, \ldots, 3n - 1, \ldots$$

Exercises Within Reach ®

Solutions in English & Spanish and tutorial videos at CollegePrepAlgebra.com

Finding the nth Term   **In Exercises 13−22, find a formula for the *n*th term of the arithmetic sequence.**

13. $a_1 = 4$, $d = 3$

14. $a_1 = 7$, $d = 2$

15. $a_1 = \frac{1}{2}$, $d = \frac{3}{2}$

16. $a_1 = \frac{5}{3}$, $d = \frac{1}{3}$

17. $a_1 = 100$, $d = -5$

18. $a_1 = -6$, $d = -1$

19. $a_3 = 6$, $d = \frac{3}{2}$

20. $a_6 = 5$, $d = \frac{3}{2}$

21. $a_1 = 5$, $a_5 = 15$

22. $a_2 = 93$, $a_6 = 65$

EXAMPLE 3 Finding a Term

Find the ninth term of the arithmetic sequence that begins with 2 and 9.

SOLUTION

For this sequence, the common difference is $d = 9 - 2 = 7$. Because the common difference is $d = 7$ and the first term is $a_1 = 2$, the formula must have the form

$$a_n = 2 + 7(n - 1). \qquad \text{Substitute 2 for } a_1 \text{ and 7 for } d.$$

So, a formula for the nth term is

$$a_n = 7n - 5$$

which implies that the ninth term is

$$a_n = 7(9) - 5 = 58.$$

When you know the nth term and the common difference of an arithmetic sequence, you can find the $(n + 1)$th term by using the **recursion formula**

$$a_{n + 1} = a_n + d.$$

EXAMPLE 4 Using a Recursion Formula

The 12th term of an arithmetic sequence is 52 and the common difference is 3.

a. What is the 13th term of the sequence? **b.** What is the first term?

SOLUTION

a. You know that $a_{12} = 52$ and $d = 3$. So, using the recursion formula $a_{13} = a_{12} + d$, you can determine that the 13th term of sequence is

$$a_{13} = 52 + 3 = 55.$$

b. Using $n = 12$, $d = 3$, and $a_{12} = 52$ in the formula $a_n = a_1 + (n - 1)d$ yields

$$52 = a_1 + (12 - 1)(3)$$

$$19 = a_1.$$

Exercises Within Reach ®

Solutions in English & Spanish and tutorial videos at CollegePrepAlgebra.com

Finding a Term In Exercises 23 and 24, the first two terms of an arithmetic sequence are given. Find the missing term.

23. $a_1 = 5$, $a_2 = 11$, $a_{10} =$ ▢

24. $a_1 = 3$, $a_2 = 13$, $a_9 =$ ▢

Using a Recursion Formula In Exercises 25–28, write the first five terms of the arithmetic sequence defined recursively.

25. $a_1 = 14$
$a_{k + 1} = a_k + 6$

26. $a_1 = 3$
$a_{k + 1} = a_k - 2$

27. $a_1 = 23$
$a_{k + 1} = a_k - 5$

28. $a_1 = -16$
$a_{k + 1} = a_k + 5$

Using a Recursion Formula In Exercises 29 and 30, find (a) a_1 and (b) a_5.

29. $a_4 = 23$, $d = 6$

30. $a_4 = \frac{15}{4}$, $d = -\frac{3}{4}$

The Partial Sum of an Arithmetic Sequence

Study Tip

You can use the formula for the nth partial sum of an arithmetic sequence to find the sum of consecutive numbers. For instance, the sum of the integers from 1 to 100 is

$$\sum_{i=1}^{100} i = \frac{100}{2}(1 + 100)$$

$$= 50(101)$$

$$= 5050.$$

The nth Partial Sum of an Arithmetic Sequence

The nth partial sum of the arithmetic sequence whose nth term is a_n is

$$\sum_{i=1}^{n} a_i = a_1 + a_2 + a_3 + a_4 + \cdots + a_n$$

$$= \frac{n}{2}(a_1 + a_n).$$

Or, equivalently, you can find the sum of the first n terms of an arithmetic sequence by multiplying the average of the first and nth terms by n.

EXAMPLE 5   **Finding the nth Partial Sum**

Find the sum of the first 20 terms of the arithmetic sequence whose nth term is $4n + 1$.

SOLUTION

The first term of this sequence is $a_1 = 4(1) + 1 = 5$ and the 20th term is $a_{20} = 4(20) + 1 = 81$. So, the sum of the first 20 terms is given by

$$\sum_{i=1}^{n} a_i = \frac{n}{2}(a_1 + a_n) \qquad \text{nth partial sum formula}$$

$$\sum_{i=1}^{20} (4i + 1) = \frac{20}{2}(a_1 + a_{20}) \qquad \text{Substitute 20 for n.}$$

$$= 10(5 + 81) \qquad \text{Substitute 5 for a_1 and 81 for a_{20}.}$$

$$= 10(86) \qquad \text{Simplify.}$$

$$= 860. \qquad \text{nth partial sum}$$

Exercises Within Reach ®   Solutions in English & Spanish and tutorial videos at CollegePrepAlgebra.com

Finding the nth Partial Sum In Exercises 31−40, find the partial sum.

31. $\displaystyle\sum_{k=1}^{20} k$

32. $\displaystyle\sum_{k=1}^{30} 4k$

33. $\displaystyle\sum_{k=1}^{50} (k + 3)$

34. $\displaystyle\sum_{n=1}^{30} (n + 2)$

35. $\displaystyle\sum_{k=1}^{10} (5k - 2)$

36. $\displaystyle\sum_{k=1}^{100} (4k - 1)$

37. $\displaystyle\sum_{n=1}^{500} \frac{n}{2}$

38. $\displaystyle\sum_{n=1}^{300} \frac{n}{3}$

39. $\displaystyle\sum_{n=1}^{30} \left(\frac{1}{3}n - 4\right)$

40. $\displaystyle\sum_{n=1}^{75} (0.3n + 5)$

EXAMPLE 6   **Finding the nth Partial Sum**

Find the sum of the even integers from 2 to 100.

SOLUTION

Because the integers

$$2, 4, 6, 8, \ldots, 100$$

form an arithmetic sequence, you can find the sum as follows.

$$\sum_{i=1}^{n} a_i = \frac{n}{2}(a_1 + a_n) \qquad \text{nth partial sum formula}$$

$$\sum_{i=1}^{50} 2i = \frac{50}{2}(a_1 + a_{50}) \qquad \text{Substitute 50 for n.}$$

$$= 25(2 + 100) \qquad \text{Substitute 2 for a_1 and 100 for a_{50}.}$$

$$= 25(102) \qquad \text{Simplify.}$$

$$= 2550 \qquad \text{nth partial sum}$$

EXAMPLE 7   **Finding the nth Partial Sum**

Find the sum.

$$1 + 3 + 5 + 7 + 9 + 11 + 13 + 15 + 17 + 19$$

SOLUTION

To begin, notice that the sequence is arithmetic (with a common difference of 2). Moreover, the sequence has 10 terms. So, the sum of the sequence is

$$\sum_{i=1}^{n} a_i = \frac{n}{2}(a_1 + a_n) \qquad \text{nth partial sum formula}$$

$$\sum_{i=1}^{10} (2i - 1) = \frac{10}{2}(a_1 + a_{10}) \qquad \text{Substitute 10 for n.}$$

$$= 5(1 + 19) \qquad \text{Substitute 1 for a_1 and 19 for a_{10}.}$$

$$= 5(20) \qquad \text{Simplify.}$$

$$= 100. \qquad \text{nth partial sum}$$

Exercises Within Reach ®

Solutions in English & Spanish and tutorial videos at CollegePrepAlgebra.com

Finding the nth Partial Sum In Exercises 41−48, find the nth partial sum of the arithmetic sequence.

41. $5, 12, 19, 26, 33, \ldots,$ $n = 12$

42. $2, 12, 22, 32, 42, \ldots,$ $n = 20$

43. $-50, -38, -26, -14, -2, \ldots,$ $n = 50$

44. $-16, -8, 0, 8, 16, \ldots,$ $n = 30$

45. $1, 4.5, 8, 11.5, 15, \ldots,$ $n = 12$

46. $2.2, 2.8, 3.4, 4.0, 4.6, \ldots,$ $n = 12$

47. $a_1 = 0.5, a_4 = 1.7, \ldots,$ $n = 10$

48. $a_1 = 15, a_{100} = 307, \ldots,$ $n = 100$

49. **Number Problem** Find the sum of the first 75 positive integers.

50. **Number Problem** Find the sum of the first 50 positive odd integers.

Application

Application

EXAMPLE 8   **Total Sales**

Your business sells $100,000 worth of handmade furniture during its first year. You have a goal of increasing annual sales by $25,000 each year for 9 years. If you meet this goal, how much will you sell during your first 10 years of business?

SOLUTION

The annual sales during the first 10 years form the following arithmetic sequence.

$100,000, $125,000, $150,000, $175,000, $200,000, $225,000, $250,000, $275,000, $300,000, $325,000

Using the formula for the nth partial sum of an arithmetic sequence, you can find the total sales during the first 10 years as follows.

$$\text{Total sales} = \frac{n}{2}(a_1 + a_n)$$ nth partial sum formula

$$= \frac{10}{2}(100,000 + 325,000)$$ Substitute for n, a_1 and a_n.

$$= 5(425,000)$$ Simplify.

$$= \$2,125,000$$ Simplify.

Furniture Business

From the bar graph shown at the left, notice that the annual sales for your company follow a *linear growth* pattern. In other words, saying that a quantity increases arithmetically is the same as saying that it increases linearly.

Exercises Within Reach ®

Solutions in English & Spanish and tutorial videos at CollegePrepAlgebra.com

51. *Salary* In your new job as an actuary, your starting salary will be $54,000 with an increase of $3000 at the end of each of the first 5 years. How much will you be paid through the end of your first 6 years of employment with the company?

52. *Wages* You earn 5 dollars on the first day of the month, 10 dollars on the second day, 15 dollars on the third day, and so on. Determine the total amount that you will earn during a 30-day month.

53. *Baling Hay* In the first two trips baling hay around a large field (see figure), a farmer obtains 93 bales and 89 bales, respectively. The farmer estimates that the number will continue to decrease in a linear pattern. Estimate the total number of bales obtained after completing the six remaining trips around the field.

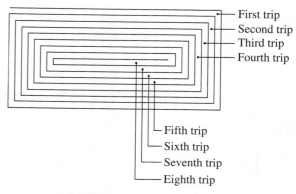

54. *Ticket Prices* There are 20 rows of seats on the main floor of an outdoor arena: 20 seats in the first row, 21 seats in the second row, 22 seats in the third row, and so on. How much should you charge per ticket to obtain $15,000 for the sale of all the seats on the main floor?

Concept Summary: Writing Formulas for Arithmetic Sequences

What

An **arithmetic sequence** is a sequence whose consecutive terms have a **common difference**. You can use a formula to represent an arithmetic sequence.

EXAMPLE

Find a formula for the nth term of the arithmetic sequence.

$8, 12, 16, 20, 24, \ldots$

How

The formula for the nth term of an arithmetic sequence is of the form $a_n = a_1 + (n-1)d$.

EXAMPLE

From the sequence, you can see that the common difference is $d = 4$ and the first term is $a_1 = 8$.

$$
\begin{aligned}
a_n &= a_1 + (n-1)d & \text{Write formula.} \\
&= 8 + 4(n-1) & \text{Substitute.} \\
&= 4n + 4 & \text{Simplify.}
\end{aligned}
$$

Why

Once you find a formula for the nth term of an arithmetic sequence, you can use this formula to find any term in the sequence.

Exercises Within Reach®

Worked-out solutions to odd-numbered exercises at CollegePrepAlgebra.com

Concept Summary Check

55. *Vocabulary* In an arithmetic sequence, the common difference between consecutive terms is 3. How can you use the value of one term to find the value of the next term in the sequence?

56. *Writing* Explain how you can use the first two terms of an arithmetic sequence to write a formula for the nth term of the sequence.

57. *Reasoning* Explain how you can use the average of the first term and nth term of an arithmetic sequence to find the nth partial sum of the sequence.

58. *Logic* In an arithmetic sequence, you know the common difference d between consecutive terms. What else do you need to know to find a_6?

Extra Practice

Writing the Terms of a Sequence In Exercises 59−62, write **the first five terms of the arithmetic sequence.**

59. $a_1 = 7, \ d = 5$

60. $a_1 = 8, \ d = 3$

61. $a_1 = 11, \ d = 4$

62. $a_1 = 18, \ d = 10$

Writing the Terms of a Sequence In Exercises 63−66, write **the first five terms of the arithmetic sequence. (Assume that n begins with 1.)**

63. $a_n = 3n + 4$

64. $a_n = 5n - 4$

65. $a_n = \frac{5}{2}n - 1$

66. $a_n = \frac{2}{3}n + 2$

Matching In Exercises 67−70, match the arithmetic sequence with its graph.

(a)
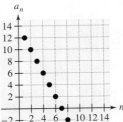

(b)
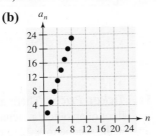

(c)

(d)

67. $a_n = -\frac{1}{2}n + 6$

68. $a_n = -2n + 10$

69. $a_1 = 12$
    $a_{n+1} = a_n - 2$

70. $a_1 = 2$
    $a_{n+1} = a_n + 3$

71. *Free-Falling Object* A free-falling object falls 16 feet during the first second, 48 feet during the second second, 80 feet during the third second, and so on. What total distance does the object fall in 8 seconds?

72. *Free-Falling Object* A free-falling object falls 4.9 meters during the first second, 14.7 meters during the second second, 24.5 meters during the third second, and so on. What total distance does the object fall in 5 seconds?

73. *Clock Chimes* A clock chimes once at 1:00, twice at 2:00, three times at 3:00, and so on. The clock also chimes once at 15-minute intervals that are not on the hour. How many times does the clock chime in a 12-hour period?

74. *Clock Chimes* A clock chimes once at 1:00, twice at 2:00, three times at 3:00, and so on. The clock also chimes once on the half-hour. How many times does the clock chime in a 12-hour period?

Explaining Concepts

75. *Vocabulary* Explain what a recursion formula does.

76. *Reasoning* Explain how to use the nth term a_n and the common difference d of an arithmetic sequence to write a recursion formula for the term a_{n+2} of the sequence.

77. *Writing* Is it possible to use the nth term a_n and the common difference d of an arithmetic sequence to write a recursion formula for the term a_{2n}? Explain.

78. *Reasoning* Each term of an arithmetic sequence is multiplied by a constant C. Is the resulting sequence arithmetic? If so, how does the common difference compare with the common difference of the original sequence?

79. *Pattern Recognition*

(a) Compute the sums of positive odd integers.

$1 + 3 = \boxed{}$

$1 + 3 + 5 = \boxed{}$

$1 + 3 + 5 + 7 = \boxed{}$

$1 + 3 + 5 + 7 + 9 = \boxed{}$

$1 + 3 + 5 + 7 + 9 + 11 = \boxed{}$

(b) Do the partial sums of the positive odd integers form an arithmetic sequence? Explain.

(c) Use the sums in part (a) to make a conjecture about the sums of positive odd integers. Check your conjecture for the sum below.

$1 + 3 + 5 + 7 + 9 + 11 + 13 = \boxed{}$.

Cumulative Review

In Exercises 80−83, find the center and vertices of the ellipse.

80. $\dfrac{(x-4)^2}{25} + \dfrac{(y+5)^2}{9} = 1$

81. $\dfrac{(x+2)^2}{4} + (y-8)^2 = 1$

82. $9x^2 + 4y^2 - 18x + 24y + 9 = 0$

83. $x^2 + 4y^2 - 8x + 12 = 0$

In Exercises 84−87, write the sum in sigma notation. (Begin with $k = 1$.)

84. $3 + 4 + 5 + 6 + 7 + 8 + 9$

85. $3 + 6 + 9 + 12 + 15$

86. $12 + 15 + 18 + 21 + 24$

87. $2 + 2^2 + 2^3 + 2^4 + 2^5$

Mid-Chapter Quiz: Sections 13.1–13.2

Solutions in English & Spanish and tutorial videos at CollegePrepAlgebra.com

Take this quiz as you would take a quiz in class. After you are done, check your work against the answers in the back of the book.

In Exercises 1–4, write the first five terms of the sequence. (Assume that n begins with 1.)

1. $a_n = 4n$

2. $a_n = 2n + 5$

3. $a_n = 32\left(\dfrac{1}{4}\right)^{n-1}$

4. $a_n = \dfrac{(-3)^n n}{n+4}$

In Exercises 5–10, find the partial sum.

5. $\displaystyle\sum_{k=1}^{4} 10k$

6. $\displaystyle\sum_{i=1}^{10} 4$

7. $\displaystyle\sum_{j=1}^{5} \dfrac{60}{j+1}$

8. $\displaystyle\sum_{n=1}^{4} \dfrac{12}{n}$

9. $\displaystyle\sum_{n=1}^{5} (3n-1)$

10. $\displaystyle\sum_{k=1}^{4} (k^2 - 1)$

In Exercises 11–14, write the sum in sigma notation. (Begin with $k = 1$.)

11. $\dfrac{2}{3(1)} + \dfrac{2}{3(2)} + \dfrac{2}{3(3)} + \cdots + \dfrac{2}{3(20)}$

12. $\dfrac{1}{1^3} - \dfrac{1}{2^3} + \dfrac{1}{3^3} - \cdots + \dfrac{1}{25^3}$

13. $0 + \dfrac{1}{2} + \dfrac{2}{3} + \dfrac{3}{4} + \cdots + \dfrac{19}{20}$

14. $\dfrac{1}{2} + \dfrac{4}{2} + \dfrac{9}{2} + \cdots + \dfrac{100}{2}$

In Exercises 15 and 16, find the common difference of the arithmetic sequence.

15. $1, \dfrac{3}{2}, 2, \dfrac{5}{2}, 3, \ldots$

16. $100, 94, 88, 82, 76, \ldots$

In Exercises 17 and 18, find a formula for the nth term of the arithmetic sequence.

17. $a_1 = 20, \quad a_4 = 11$

18. $a_1 = 32, \quad d = -4$

19. Find the sum of the first 200 positive even numbers.

20. You save $0.50 on one day, $1.00 the next day, $1.50 the next day, and so on. How much will you have accumulated at the end of one year (365 days)?

Study Skills in Action

Preparing for the Final Exam

At the end of the school year, most students are busy with projects, papers, and tests. Teachers may speed up the pace in class to get through all the material. If something unexpected is going to happen to a student, it often happens during this time.

Getting through the last couple of weeks of a math class can be challenging. This is why it is important to plan your review time for the final exam at least three weeks before the test.

These students are planning how they will study for the final exam.

Smart Study Strategy

Form a Final Exam Study Group

1 ▶ Form a study group of three or four students several weeks before the final exam. The intent of this group is to review what you have already learned while continuing to learn new material.

2 ▶ Find out what material you must know for the final, even if the teacher has not yet covered it. As a group, meet with the teacher outside of class. A group is likely to receive more attention and can ask more questions.

3 ▶ Ask for or create a practice final and have the teacher look at it. Make sure the problems are at an appropriate level of difficulty. Look for sample problems in old tests and in cumulative tests in the textbook. Review what the textbook and your notes say as you look for problems. This will refresh your memory.

4 ▶ Have each group member take the practice final exam. Then have each member identify what he or she needs to study. Make sure you can complete the problems with the speed and accuracy that are necessary to complete the real final exam.

5 ▶ Decide when the group is going to meet during the next couple of weeks and what you will cover during each session. A tutoring or learning center is an ideal setting in which to meet. Many libraries have small study rooms that study groups can reserve. Set up several study times for each week.

6 ▶ During the study group sessions, make sure you stay on track. Prepare for each study session by knowing what material in the textbook you are going to cover and having the class notes for that material. When you have questions, assign a group member to go to the teacher for answers. Then this member can relay the correct information to the other group members. Save socializing for after the final exam.

13.3 Geometric Sequences

▶ Recognize, write, and find the nth terms of geometric sequences.

▶ Find the sums of finite and infinite geometric sequences.

▶ Use geometric sequences to solve application problems.

Geometric Sequences

> ### Definition of Geometric Sequence
>
> A sequence is called **geometric** when the ratios of consecutive terms are the same. So, the sequence $a_1, a_2, a_3, a_4, \ldots, a_n, \ldots$ is geometric when there is a number r, with $r \neq 0$, such that
>
> $$\frac{a_2}{a_1} = r, \frac{a_3}{a_2} = r, \frac{a_4}{a_3} = r$$
>
> and so on. The number r is the **common ratio** of the sequence.

EXAMPLE 1   Examples of Geometric Sequences

a. The sequence whose nth term is 2^n is geometric. For this sequence, the common ratio between consecutive terms is 2.

$$2, 4, 8, 16, \ldots, 2^n, \ldots$$

$\frac{4}{2} = 2$

Begin with $n = 1$.

b. The sequence whose nth term is $4(3^n)$ is geometric. For this sequence, the common ratio between consecutive terms is 3.

$$12, 36, 108, 324, \ldots, 4(3^n), \ldots$$

$\frac{36}{12} = 3$

Begin with $n = 1$.

c. The sequence whose nth term is $\left(-\frac{1}{3}\right)^n$ is geometric. For this sequence, the common ratio between consecutive terms is $-\frac{1}{3}$.

$$-\frac{1}{3}, \frac{1}{9}, -\frac{1}{27}, \frac{1}{81}, \ldots, \left(-\frac{1}{3}\right)^n, \ldots$$

$\frac{1/9}{-1/3} = -\frac{1}{3}$

Begin with $n = 1$.

Exercises Within Reach ®

Solutions in English & Spanish and tutorial videos at CollegePrepAlgebra.com

Identifying a Geometric Sequence In Exercises 1−10, determine **whether the sequence is geometric. If so, find the common ratio.**

1. $3, 6, 12, 24, \ldots$

2. $2, 6, 18, 54, \ldots$

3. $1, \pi, \pi^2, \pi^3, \ldots$

4. $e, e^2, e^3, e^4, \ldots$

5. $10, 15, 20, 25, \ldots$

6. $1, 8, 27, 64, 125, \ldots$

7. $64, 32, 16, 8, \ldots$

8. $10, 20, 40, 80, \ldots$

9. $a_n = 4\left(-\frac{1}{2}\right)^n$

10. $a_n = -2\left(\frac{1}{3}\right)^n$

The *n*th Term of a Geometric Sequence

The *n*th term of a geometric sequence has the form

$$a_n = a_1 r^{n-1}$$

where r is the common ratio of consecutive terms of the sequence. So, every geometric sequence can be written in the following form.

$$a_1, \, a_1 r, \, a_1 r^2, \, a_1 r^3, \, a_1 r^4, \, \ldots, \, a_1 r^{n-1}, \, \ldots$$

Study Tip

When you know the *n*th term of a geometric sequence, you can multiply that term by r to find the $(n+1)$th term. That is, $a_{n+1} = r a_n$.

EXAMPLE 2   **Finding the *n*th Term of a Geometric Sequence**

Find a formula for the *n*th term of the geometric sequence whose common ratio is 3 and whose first term is 1. Then find the eighth term of the sequence.

SOLUTION

Because the common ratio is 3 and the first term is 1, the formula for the *n*th term must be

$$a_n = a_1 r^{n-1} = (1)(3)^{n-1} = 3^{n-1}.$$ Substitute 1 for a_1 and 3 for r.

The sequence has the form $1, 3, 9, 27, 81, \ldots, 3^{n-1}, \ldots$.

The eighth term of the sequence is $a_8 = 3^{8-1} = 3^7 = 2187$.

EXAMPLE 3   **Finding the *n*th Term of a Geometric Sequence**

Find a formula for the *n*th term of the geometric sequence whose first two terms are 4 and 2.

SOLUTION

Because the common ratio is

$$r = \frac{a_2}{a_1} = \frac{2}{4} = \frac{1}{2}$$

the formula for the *n*th term must be

$$a_n = a_1 r^{n-1}$$ Formula for geometric sequence

$$= 4\left(\frac{1}{2}\right)^{n-1}.$$ Substitute 4 for a_1 and $\frac{1}{2}$ for r.

The sequence has the form $4, 2, 1, \dfrac{1}{2}, \dfrac{1}{4}, \ldots, 4\left(\dfrac{1}{2}\right)^{n-1}, \ldots$.

Exercises Within Reach ®

Solutions in English & Spanish and tutorial videos at CollegePrepAlgebra.com

Finding the nth Term In Exercises 11−18, **find a formula for the *n*th term of the geometric sequence. Then find a_7.**

11. $a_1 = 1, \ r = 2$

12. $a_1 = 5, \ r = 4$

13. $a_1 = 9, \ r = \frac{2}{3}$

14. $a_1 = 10, \ r = -\frac{1}{5}$

15. $a_1 = 2, \ a_2 = 4$

16. $a_1 = 25, \ a_2 = 125$

17. $a_1 = 8, \ a_2 = 2$

18. $a_1 = 18, \ a_2 = 8$

The Sum of a Geometric Sequence

> ### The *n*th Partial Sum of a Geometric Sequence
>
> The *n*th partial sum of the geometric sequence whose *n*th term is $a_n = a_1 r^{n-1}$ is given by
>
> $$\sum_{i=1}^{n} a_1 r^{i-1} = a_1 + a_1 r + a_1 r^2 + a_1 r^3 + \cdots + a_1 r^{n-1} = a_1\left(\frac{r^n - 1}{r - 1}\right).$$

EXAMPLE 4 **Finding the *n*th Partial Sum**

Find the sum $1 + 2 + 4 + 8 + 16 + 32 + 64 + 128$.

SOLUTION

This is a geometric sequence whose common ratio is $r = 2$. Because the first term of the sequence is $a_1 = 1$, follows that the sum is

$$\sum_{i=1}^{8} 2^{i-1} = (1)\left(\frac{2^8 - 1}{2 - 1}\right) = \frac{256 - 1}{2 - 1} = 255.$$ Substitute 1 for a_1 and 2 for r.

EXAMPLE 5 **Finding the *n*th Partial Sum**

Find the sum of the first five terms of the geometric sequence whose *n*th term is $a_n = \left(\frac{2}{3}\right)^n$.

SOLUTION

$$\sum_{i=1}^{5}\left(\frac{2}{3}\right)^i = \frac{2}{3}\left[\frac{(2/3)^5 - 1}{(2/3) - 1}\right]$$ Substitute $\frac{2}{3}$ for a_1 and $\frac{2}{3}$ for r.

$$= \frac{2}{3}\left[\frac{(32/243) - 1}{-1/3}\right]$$ Simplify.

$$= \frac{422}{243} \approx 1.737$$ Use a calculator to simplify.

Exercises Within Reach ®

Solutions in English & Spanish and tutorial videos at CollegePrepAlgebra.com

Finding the nth Partial Sum In Exercises 19−26, find the *n*th partial sum of the geometric sequence. Round to the nearest hundredth, if necessary.

19. $4, 12, 36, 108, \ldots,$ $n = 8$

20. $5, 10, 20, 40, 80, \ldots,$ $n = 10$

21. $1, -3, 9, -27, 81, \ldots,$ $n = 10$

22. $3, -6, 12, -24, 48, \ldots,$ $n = 12$

23. $8, 4, 2, 1, \frac{1}{2}, \ldots,$ $n = 15$

24. $9, 6, 4, \frac{8}{3}, \frac{16}{9}, \ldots,$ $n = 10$

25. $a_n = \left(\frac{3}{4}\right)^n,$ $n = 6$

26. $a_n = \left(\frac{5}{6}\right)^n,$ $n = 4$

Using Sigma Notation In Exercises 27−30, find the partial sum. Round to the nearest hundredth, if necessary.

27. $\displaystyle\sum_{i=1}^{10} 2^{i-1}$

28. $\displaystyle\sum_{i=1}^{6} 3^{i-1}$

29. $\displaystyle\sum_{i=1}^{12} 3\left(\frac{3}{2}\right)^{i-1}$

30. $\displaystyle\sum_{i=1}^{20} 12\left(\frac{2}{3}\right)^{i-1}$

Sum of an Infinite Geometric Series

If $a_1, a_1r, a_1r^2, \ldots, a_1r^n, \ldots$ is an infinite geometric sequence and $|r| < 1$, then the sum of the terms of the corresponding infinite geometric series is

$$S = \sum_{i=0}^{\infty} a_1 r^i = \frac{a_1}{1-r}.$$

EXAMPLE 6   **Finding the Sum of an Infinite Geometric Series**

Find each sum.

a. $\displaystyle\sum_{i=1}^{\infty} 5\left(\frac{3}{4}\right)^{i-1}$ b. $\displaystyle\sum_{n=0}^{\infty} 4\left(\frac{3}{10}\right)^{n}$ c. $\displaystyle\sum_{i=0}^{\infty} \left(-\frac{3}{5}\right)^{i}$

SOLUTION

a.   The series is geometric, with $a_1 = 5\left(\frac{3}{4}\right)^{1-1} = 5$ and $r = \frac{3}{4}$. So,

$$\sum_{i=1}^{\infty} 5\left(\frac{3}{4}\right)^{i-1} = \frac{5}{1-(3/4)}$$

$$= \frac{5}{1/4} = 20.$$

b.   The series is geometric, with $a_1 = 4\left(\frac{3}{10}\right)^{0} = 4$ and $r = \frac{3}{10}$. So,

$$\sum_{n=0}^{\infty} 4\left(\frac{3}{10}\right)^{n} = \frac{4}{1-(3/10)} = \frac{4}{7/10} = \frac{40}{7}.$$

c.   The series is geometric, with $a_1 = \left(-\frac{3}{5}\right)^{0} = 1$ and $r = -\frac{3}{5}$. So,

$$\sum_{i=0}^{\infty} \left(-\frac{3}{5}\right)^{i} = \frac{1}{1-(-3/5)} = \frac{1}{1+(3/5)} = \frac{5}{8}.$$

Exercises Within Reach ®

Solutions in English & Spanish and tutorial videos at CollegePrepAlgebra.com

Finding the Sum of an Infinite Geometric Series In Exercises 31–42, find the sum.

31. $\displaystyle\sum_{n=1}^{\infty} 8\left(\frac{3}{4}\right)^{n-1}$

32. $\displaystyle\sum_{n=1}^{\infty} 10\left(\frac{5}{6}\right)^{n-1}$

33. $\displaystyle\sum_{n=0}^{\infty} 2\left(\frac{2}{3}\right)^{n}$

34. $\displaystyle\sum_{n=0}^{\infty} 4\left(\frac{1}{4}\right)^{n}$

35. $\displaystyle\sum_{n=0}^{\infty} \left(-\frac{3}{7}\right)^{n}$

36. $\displaystyle\sum_{n=0}^{\infty} \left(-\frac{5}{9}\right)^{n}$

37. $\displaystyle\sum_{n=1}^{\infty} \left(\frac{1}{2}\right)^{n-1}$

38. $\displaystyle\sum_{n=1}^{\infty} \left(\frac{1}{3}\right)^{n-1}$

39. $\displaystyle\sum_{n=1}^{\infty} \left(-\frac{1}{2}\right)^{n-1}$

40. $\displaystyle\sum_{n=1}^{\infty} \left(-\frac{1}{3}\right)^{n-1}$

41. $\displaystyle\sum_{n=0}^{\infty} \left(\frac{1}{10}\right)^{n}$

42. $\displaystyle\sum_{n=0}^{\infty} \left(\frac{1}{8}\right)^{n}$

Applications

Application EXAMPLE 7 **A Lifetime Salary**

You accept a job as a meteorologist that pays a salary of $45,000 the first year. During the next 39 years, you receive a 6% raise each year. What is your total salary over the 40-year period?

SOLUTION

Using a geometric sequence, your salary during the first year is $a_1 = 45,000$. Then, with a 6% raise each year, your salary for the next 2 years will be as follows.

$$a_2 = 45,000 + 45,000(0.06) = 45,000(1.06)^1$$

$$a_3 = 45,000(1.06) + 45,000(1.06)(0.06) = 45,000(1.06)^2$$

From this pattern, you can see that the common ratio of the geometric sequence is $r = 1.06$. Using the formula for the nth partial sum of a geometric sequence, you can find the total salary over the 40-year period.

$$\text{Total salary} = a_1\left(\frac{r^n - 1}{r - 1}\right)$$

$$= 45,000\left[\frac{(1.06)^{40} - 1}{1.06 - 1}\right]$$

$$= 45,000\left[\frac{(1.06)^{40} - 1}{0.06}\right]$$

$$\approx \$6,964,288$$

Exercises Within Reach®

Solutions in English & Spanish and tutorial videos at CollegePrepAlgebra.com

43. **Salary** You accept a job as an archaeologist that pays a salary of $30,000 the first year. During the next 39 years, you receive a 5% raise each year. What is your total salary over the 40-year period?

44. **Salary** You accept a job as marine biologist that pays a salary of $45,000 the first year. During the next 39 years, you receive a 5.5% raise each year. What is your total salary over the 40-year period?

45. **Wages** You work at a company that pays $0.01 for the first day, $0.02 for the second day, $0.04 for the third day, and so on. The daily wage keeps doubling. What is your total income for working (a) 29 days and (b) 30 days?

46. **Wages** You work at a company that pays $0.01 for the first day, $0.03 for the second day, $0.09 for the third day, and so on. The daily wage keeps tripling. What is your total income for working (a) 25 days and (b) 26 days?

Application   **EXAMPLE 8**   **Increasing Annuity**

You deposit $100 in an account each month for 2 years. The account pays an annual interest rate of 9%, compounded monthly. What is your balance at the end of 2 years? (This type of savings plan is called an **increasing annuity**.)

SOLUTION

The first deposit would earn interest for the full 24 months, the second deposit would earn interest for 23 months, the third deposit would earn interest for 22 months, and so on. Using the formula for compound interest, you can find the balance of the account at the end of 2 years.

$$\text{Total} = a_1 + a_2 + \cdots + a_{24}$$

$$= 100\left(1 + \frac{0.09}{12}\right)^1 + 100\left(1 + \frac{0.09}{12}\right)^2 + \cdots + 100\left(1 + \frac{0.09}{12}\right)^{24}$$

$$= 100(1.0075)^1 + 100(1.0075)^2 + \cdots + 100(1.0075)^{24}$$

$$= 100(100.75)\left(\frac{1.0075^{24} - 1}{1.0075 - 1}\right)$$

$$= \$2638.49$$

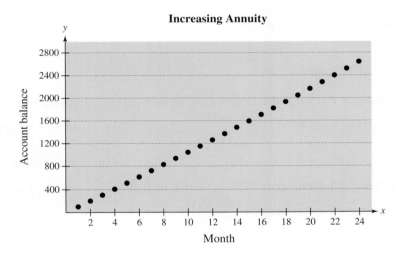

Increasing Annuity

Exercises Within Reach ®

Solutions in English & Spanish and tutorial videos at CollegePrepAlgebra.com

Increasing Annuity In Exercises 47−52, **find the balance A in an increasing annuity in which a principal of P dollars is invested each month for t years, compounded monthly at rate r.**

47. $P = \$50$ $t = 10$ years $r = 9\%$

48. $P = \$50$ $t = 5$ years $r = 7\%$

49. $P = \$30$ $t = 40$ years $r = 8\%$

50. $P = \$200$ $t = 30$ years $r = 10\%$

51. $P = \$100$ $t = 30$ years $r = 6\%$

52. $P = \$100$ $t = 25$ years $r = 8\%$

Concept Summary: Finding Sums of Geometric Sequences

What

A **geometric sequence** is a sequence whose consecutive terms have a **common ratio**.

You can use formulas to find partial sums of geometric sequences and sums of infinite geometric series.

EXAMPLE

Find the sum.

a. $\displaystyle\sum_{i=1}^{5} 3(2)^{i-1}$

b. $\displaystyle\sum_{n=0}^{\infty} 5\left(\frac{4}{5}\right)^{n}$

How

The nth partial sum of a geometric sequence:

$$\sum_{i=1}^{n} a_1 r^{i-1} = a_1\left(\frac{r^n - 1}{r - 1}\right)$$

Sum of an infinite geometric series:

$$\sum_{i=0}^{\infty} a_1 r^{i} = \frac{a_1}{1 - r}$$

EXAMPLE

a. $\displaystyle\sum_{i=1}^{5} 3(2)^{i-1} = 3\left(\frac{2^5 - 1}{2 - 1}\right)$
$$= 3(31) = 93$$

b. $\displaystyle\sum_{n=0}^{\infty} 5\left(\frac{4}{5}\right)^{n} = \frac{5}{1 - (4/5)} = \frac{5}{1/5} = 25$

Why

Many physics problems and compound interest problems involve sums of geometric sequences. You can use the formulas to find these sums efficiently.

Exercises Within Reach®

Worked-out solutions to odd-numbered exercises at CollegePrepAlgebra.com

Concept Summary Check

53. *Describing a Partial Sum* Describe the partial sum represented by $\displaystyle\sum_{i=1}^{6} 3(2)^{i-1}$.

54. *Identifying the First Term* What is the first term of the geometric series with the sum $\displaystyle\sum_{n=0}^{\infty} 5\left(\frac{4}{5}\right)^{n}$?

55. *Identifying the Common Ratio* What is the common ratio of the geometric series with the sum $\displaystyle\sum_{n=0}^{\infty} 5\left(\frac{4}{5}\right)^{n}$?

56. *An Infinite Geometric Series* Explain why the sum $\displaystyle\sum_{n=0}^{\infty} 5\left(\frac{4}{5}\right)^{n}$ is finite.

Extra Practice

Writing the Terms of a Sequence In Exercises 57−60, write the first five terms of the geometric sequence.

57. $a_1 = 5$, $r = -2$

58. $a_1 = -12$, $r = -1$

59. $a_1 = -4$, $r = -\frac{1}{2}$

60. $a_1 = 3$, $r = -\frac{3}{2}$

Finding the Sum of a Geometric Series In Exercises 61 and 62, find the sum.

61. $8 + 6 + \frac{9}{2} + \frac{27}{8} + \cdots$

62. $3 - 1 + \frac{1}{3} - \frac{1}{9} + \cdots$

63. *Geometry* An equilateral triangle has an area of 1 square unit. The triangle is divided into four smaller triangles and the center triangle is shaded (see figure). Each of the three unshaded triangles is then divided into four smaller triangles and each center triangle is shaded. This process is repeated one more time. What is the total area of the shaded regions?

64. *Geometry* A square has an area of 1 square unit. The square is divided into nine smaller squares and the center square is shaded (see figure). Each of the eight unshaded squares is then divided into nine smaller squares and each center square is shaded. This process is repeated one more time. What is the total area of the shaded regions?

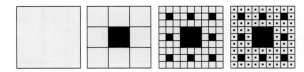

65. *Cooling* The temperature of water in an ice cube tray is 70°F when it is placed in a freezer. Its temperature n hours after being placed in the freezer is 20% less than 1 hour earlier. Find a formula for the nth term of the geometric sequence that gives the temperature of the water after n hours in the freezer. Then find the temperature after 6 hours in the freezer.

Explaining Concepts

67. *Structure* What is the general formula for the nth term of a geometric sequence?

68. *Writing* How can you determine whether a sequence is geometric?

69. *Writing* Explain the difference between an arithmetic sequence and a geometric sequence.

70. *Structure* Give an example of a geometric sequence whose terms alternate is sign.

71. *Think About It* Explain why the terms of a geometric sequence decrease when $a_1 > 0$ and $0 < r < 1$.

66. *Bungee Jumping* A bungee jumper drops 100 feet from a bridge and rebounds 75% of that distance for a total distance of 175 feet (see figure). Each successive drop and rebound covers 75% of the distance of the previous drop and rebound. Evaluate the expression below for the total distance traveled during 10 drops and rebounds.

$$175 + 175(0.75) + \cdots + 175(0.75)^9$$

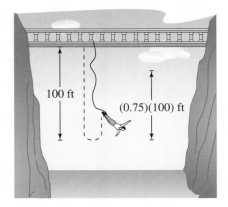

72. *Geometry* A unit square is divided into two equal rectangles. One of the resulting rectangles is then divided into two equal rectangles, as shown in the figure. This process is repeated indefinitely.

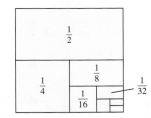

(a) Explain why the areas of the rectangles (from largest to smallest) form a geometric sequence.

(b) Find a formula for the nth term of the geometric sequence.

(c) Use the formula for the sum of an infinite geometric series to show that the combined area of the rectangles is 1.

Cumulative Review

In Exercises 73 and 74, solve the system of equations by the method of substitution.

73. $\begin{cases} y = 2x^2 \\ y = 2x + 4 \end{cases}$ **74.** $\begin{cases} x^2 + y^2 = 1 \\ x^2 + y = 1 \end{cases}$

In Exercises 75–78, find the annual interest rate.

	Principal	Balance	Time	Compounding
75.	$1000	$2219.64	10 years	Monthly
76.	$2000	$3220.65	8 years	Quarterly
77.	$2500	$10,619.63	20 years	Yearly
78.	$3500	$25,861.70	40 years	Continuously

In Exercises 79 and 80, sketch the hyperbola. Identify the vertices and asymptotes.

79. $\dfrac{x^2}{16} - \dfrac{y^2}{9} = 1$ **80.** $\dfrac{y^2}{1} - \dfrac{x^2}{4} = 1$

13.4 The Binomial Theorem

▶ Use the Binomial Theorem to calculate binomial coefficients.
▶ Use Pascal's Triangle to calculate binomial coefficients.
▶ Expand binomial expressions.

Binomial Coefficients

The Binomial Theorem

In the expansion of $(x + y)^n$

$$(x + y)^n = x^n + nx^{n-1}y + \cdots + {}_nC_r x^{n-r}y^r + \cdots + nxy^{n-1} + y^n$$

the coefficient of $x^{n-r}y^r$ is given by

$${}_nC_r = \frac{n!}{(n-r)!\,r!}.$$

EXAMPLE 1   Finding Binomial Coefficients

a. ${}_8C_2 = \dfrac{8!}{6! \cdot 2!} = \dfrac{(8 \cdot 7) \cdot \cancel{6!}}{\cancel{6!} \cdot 2!} = \dfrac{8 \cdot 7}{2 \cdot 1} = 28$

b. ${}_{10}C_3 = \dfrac{10!}{7! \cdot 3!} = \dfrac{(10 \cdot 9 \cdot 8) \cdot \cancel{7!}}{\cancel{7!} \cdot 3!} = \dfrac{10 \cdot 9 \cdot 8}{3 \cdot 2 \cdot 1} = 120$

c. ${}_7C_0 = \dfrac{7!}{7! \cdot 0!} = \dfrac{\cancel{7!}}{\cancel{7!} \cdot 1} = 1$

d. ${}_8C_8 = \dfrac{8!}{0! \cdot 8!} = \dfrac{\cancel{8!}}{1 \cdot \cancel{8!}} = 1$

e. ${}_7C_3 = \dfrac{7!}{4! \cdot 3!} = \dfrac{(7 \cdot 6 \cdot 5) \cdot \cancel{4!}}{\cancel{4!} \cdot 3!} = \dfrac{7 \cdot 6 \cdot 5}{3 \cdot 2 \cdot 1} = 35$

f. ${}_7C_4 = \dfrac{7!}{3! \cdot 4!} = \dfrac{(7 \cdot 6 \cdot 5) \cdot \cancel{4!}}{3! \cdot \cancel{4!}} = \dfrac{7 \cdot 6 \cdot 5}{3 \cdot 2 \cdot 1} = 35$

g. ${}_{12}C_1 = \dfrac{12!}{11! \cdot 1!} = \dfrac{(12) \cdot \cancel{11!}}{\cancel{11!} \cdot 1!} = \dfrac{12}{1} = 12$

h. ${}_{12}C_{11} = \dfrac{12!}{1! \cdot 11!} = \dfrac{(12) \cdot \cancel{11!}}{1! \cdot \cancel{11!}} = \dfrac{12}{1} = 12$

Study Tip

In Example 1, note that the answers to parts (e) and (f) are the same and that the answers to parts (g) and (h) are the same. In general $_nC_r = {}_nC_{n-r}$.

Exercises Within Reach ®

Solutions in English & Spanish and tutorial videos at CollegePrepAlgebra.com

Finding a Binomial Coefficient In Exercises 1−10, evaluate the binomial coefficient $_nC_r$.

1. ${}_6C_4$ 2. ${}_9C_3$

3. ${}_{10}C_5$ 4. ${}_{12}C_9$

5. ${}_{12}C_{12}$ 6. ${}_8C_1$

7. ${}_{20}C_6$ 8. ${}_{15}C_{10}$

9. ${}_{20}C_{14}$ 10. ${}_{15}C_5$

Pascal's Triangle

There is a convenient way to remember a pattern for binomial coefficients. By arranging the coefficients in a triangular pattern, you obtain the following array, which is called **Pascal's Triangle**. This triangle is named after the famous French mathematician Blaise Pascal (1623–1662).

Pascal's Triangle

```
                    1
                 1     1
              1     2     1          1 + 2 = 3
           1     3     3     1
        1     4     6     4     1
     1     5    10    10     5     1
  1     6    15    20    15     6     1     10 + 5 = 15
1     7    21    35    35    21     7     1
```

Study Tip

The top row in Pascal's Triangle is called the *zeroth row* because it corresponds to the binomial expansion

$$(x + y)^0 = 1.$$

Similarly, the next row is called the *first row* because it corresponds to the binomial expansion

$$(x + y)^1 = 1(x) + 1(y).$$

In general, the *nth row* in Pascal's Triangle gives the coefficients of $(x + y)^n$.

The first and last numbers in each row of Pascal's Triangle are 1. Every other number in each row is formed by adding the two numbers immediately above the number. Pascal noticed that numbers in this triangle are precisely the same numbers that are the coefficients of binomial expansions, as follows.

$(x + y)^0 = 1$	0th row
$(x + y)^1 = 1x + 1y$	1st row
$(x + y)^2 = 1x^2 + 2xy + 1y^2$	2nd row
$(x + y)^3 = 1x^3 + 3x^2y + 3xy^2 + 1y^3$	3rd row
$(x + y)^4 = 1x^4 + 4x^3y + 6x^2y^2 + 4xy^3 + 1y^4$	$\vdots$

$$(x + y)^5 = 1x^5 + 5x^4y + 10x^3y^2 + 10x^2y^3 + 5xy^4 + 1y^5$$

$$(x + y)^6 = 1x^6 + 6x^5y + 15x^4y^2 + 20x^3y^3 + 15x^2y^4 + 6xy^5 + 1y^6$$

$$(x + y)^7 = 1x^7 + 7x^6y + 21x^5y^2 + 35x^4y^3 + 35x^3y^4 + 21x^2y^5 + 7xy^6 + 1y^7$$

Exercises Within Reach ®

11. *Pascal's Triangle* Find the eighth row of Pascal's Triangle using the diagram below.

```
                    1
                 1     1
              1     2     1
           1     3     3     1
        1     4     6     4     1
     1     5    10    10     5     1
  1     6    15    20    15     6     1
1     7    21    35    35    21     7     1
```

12. *Pascal's Triangle* Use your answer to Exercise 11 to expand $(x + y)^8$.

EXAMPLE 2 Using Pascal's Triangle

Use the fifth row of Pascal's Triangle to evaluate $_5C_2$.

SOLUTION

1	5	10	10	5	1
$_5C_0$	$_5C_1$	$_5C_2$	$_5C_3$	$_5C_4$	$_5C_5$

So, $_5C_2 = 10$.

Check

$$_5C_2 = \frac{5!}{3! \cdot 2!} = \frac{(5 \cdot 4) \cdot 3!}{3! \cdot 2!} = \frac{5 \cdot 4}{2 \cdot 1} = 10 \checkmark$$

EXAMPLE 3 Using Pascal's Triangle

Use the sixth row of Pascal's Triangle to evaluate $_6C_4$.

SOLUTION

1	6	15	20	15	6	1
$_6C_0$	$_6C_1$	$_6C_2$	$_6C_3$	$_6C_4$	$_6C_5$	$_6C_6$

So, $_6C_4 = 15$.

Check

$$_6C_4 = \frac{6!}{2! \cdot 4!} = \frac{(6 \cdot 5) \cdot 4!}{2! \cdot 4!} = \frac{6 \cdot 5}{2 \cdot 1} = 15 \checkmark$$

Exercises Within Reach ®

Solutions in English & Spanish and tutorial videos at CollegePrepAlgebra.com

Using Pascal's Triangle In Exercises 13−22, use Pascal's Triangle to evaluate $_nC_r$.

13. $_6C_2$

14. $_9C_3$

15. $_7C_3$

16. $_9C_5$

17. $_8C_4$

18. $_{10}C_6$

19. $_5C_3$

20. $_8C_6$

21. $_7C_4$

22. $_{10}C_2$

Binomial Expansions

> **EXAMPLE 4**   **Expanding a Binomial**

Write the expansion of the expression $(x + 1)^5$.

SOLUTION

The binomial coefficients from the fifth row of Pascal's Triangle are

$1, 5, 10, 10, 5, 1.$

So, the expansion is as follows.

$$(x + 1)^5 = (1)x^5 + (5)x^4(1) + (10)x^3(1^2) + (10)x^2(1^3) + (5)x(1^4) + (1)(1^5)$$

$$= x^5 + 5x^4 + 10x^3 + 10x^2 + 5x + 1$$

To expand binomials representing *differences*, rather than sums, you alternate signs. For example, $(x - 1)^3 = x^3 - 3x^2 + 3x - 1$.

> **EXAMPLE 5**   **Expanding a Binomial**

Write the expansion of each expression.

a. $(x - 3)^4$ **b.** $(2x - 1)^3$

SOLUTION

a. The binomial coefficients from the fourth row of Pascal's Triangle are

$1, 4, 6, 4, 1.$

So, the expansion is as follows.

$$(x - 3)^4 = (1)x^4 - (4)x^3(3) + (6)x^2(3^2) - (4)x(3^3) + (1)(3^4)$$

$$= x^4 - 12x^3 + 54x^2 - 108x + 81$$

b. The binomial coefficients from the third row of Pascal's Triangle are

$1, 3, 3, 1.$

So, the expansion is as follows.

$$(2x - 1)^3 = (1)(2x)^3 - (3)(2x)^2(1) + (3)(2x)(1^2) - (1)(1^3)$$

$$= 8x^3 - 12x^2 + 6x - 1$$

Exercises Within Reach ®

Solutions in English & Spanish and tutorial videos at CollegePrepAlgebra.com

Using Pascal's Triangle In Exercises 23−26, use Pascal's Triangle to expand the expression.

23. $(t + 5)^3$ **24.** $(y + 2)^4$

25. $(m - n)^5$ **26.** $(r - s)^7$

Using the Binomial Theorem In Exercises 27−30, use the Binomial Theorem to expand the expression.

27. $(x + 3)^6$ **28.** $(m - 4)^4$

29. $(u - v)^3$ **30.** $(x - y)^4$

EXAMPLE 6 **Expanding a Binomial**

Write the expansion of the expression $(x - 2y)^4$.

SOLUTION

Use the fourth row of Pascal's Triangle, as follows.

$$(x - 2y)^4 = (1)x^4 - (4)x^3(2y) + (6)x^2(2y)^2 - (4)x(2y)^3 + (1)(2y)^4$$
$$= x^4 - 8x^3y + 24x^2y^2 - 32xy^3 + 16y^4$$

EXAMPLE 7 **Expanding a Binomial**

Write the expansion of the expression $(x^2 + 4)^3$.

SOLUTION

Use the third row of Pascal's Triangle, as follows.

$$(x^2 + 4)^3 = (1)(x^2)^3 + (3)(x^2)^2(4) + (3)x^2(4^2) + (1)(4^3)$$
$$= x^6 + 12x^4 + 48x^2 + 64$$

EXAMPLE 8 **Expanding a Binomial**

Write the expansion of the expression $(x^2 - 4)^3$.

SOLUTION

Use the third row of Pascal's Triangle, as follows.

$$(x^2 - 4)^3 = (1)(x^2)^3 - (3)(x^2)^2(4) + (3)x^2(4^2) - (1)(4^3)$$
$$= x^6 - 12x^4 + 48x^2 - 64$$

Exercises Within Reach ®

Solutions in English & Spanish and tutorial videos at CollegePrepAlgebra.com

Using the Binomial Theorem In Exercises 31−42, use the **Binomial Theorem** to expand the expression.

31. $(3a - 1)^5$

32. $(1 - 4b)^3$

33. $(2y + z)^6$

34. $(3c + d)^6$

35. $(x^2 + 2)^4$

36. $(5 + y^2)^5$

37. $(3a + 2b)^4$

38. $(4u - 3v)^3$

39. $\left(x + \dfrac{2}{y}\right)^4$

40. $\left(s + \dfrac{1}{t}\right)^5$

41. $(2x^2 - y)^5$

42. $(x - 4y^3)^4$

Sometimes you will need to find a specific term in a binomial expansion. Instead of writing out the entire expansion, you can use the fact that from the Binomial Theorem, the $(r + 1)$th term is

$$_nC_rx^{n-r}y^r.$$

EXAMPLE 9   Finding a Term in a Binomial Expansion

a. Find the sixth term in the expansion of $(a + 2b)^8$.

b. Find the coefficient of the term a^6b^5 in the expansion of $(3a - 2b)^{11}$.

SOLUTION

a. In this case, $6 = r + 1$ means that $r = 5$. Because $n = 8$, $x = a$, and $y = 2b$, the sixth term in the binomial expansion is

$$_nC_rx^{n-r}y^r = {_8C_5}a^{8-5}(2b)^5$$
$$= 56 \cdot a^3 \cdot (2b)^5$$
$$= 56(2^5)\, a^3b^5$$
$$= 1792a^3b^5.$$

b. In this case, $n = 11$, $r = 5$, $x = 3a$, and $y = -2b$. Substitute these values to obtain

$$_nC_rx^{n-r}y^r = {_{11}C_5}(3a)^6(-2b)^5$$
$$= 462(729a^6)(-32b^5)$$
$$= -10{,}777{,}536a^6b^5.$$

So, the coefficient is $-10{,}777{,}536$.

Exercises Within Reach ®

Solutions in English & Spanish and tutorial videos at CollegePrepAlgebra.com

Finding a Specific Term In Exercises 43–48, **find** the specified term in the expansion of the binomial.

43. $(x + y)^{10}$, 4th term

44. $(x - y)^6$, 7th term

45. $(a + 6b)^9$, 5th term

46. $(3a - b)^{12}$, 10th term

47. $(4x + 3y)^9$, 8th term

48. $(5a + 6b)^5$, 5th term

Finding the Coefficient of a Term In Exercises 49–52, **find** the coefficient of the given term in the expansion of the binomial.

	Expression	*Term*		*Expression*	*Term*
49.	$(x + 1)^{10}$	x^7	**50.**	$(x^2 - 3)^4$	x^4
51.	$(x + 3)^{12}$	x^9	**52.**	$(3 - y^3)^5$	x^9

Concept Summary: *Using Pascal's Triangle to Expand Binomials*

What

You can use **Pascal's Triangle** to **expand a binomial**.

EXAMPLE

Use Pascal's Triangle to expand $(x - 2)^3$.

$$
\begin{array}{c}
1 \\
1 \quad 1 \\
1 \quad 2 \quad 1 \\
1 \quad 3 \quad 3 \quad 1 \\
1 \quad 4 \quad 6 \quad 4 \quad 1 \\
1 \quad 5 \quad 10 \quad 10 \quad 5 \quad 1 \\
1 \quad 6 \quad 15 \quad 20 \quad 15 \quad 6 \quad 1
\end{array}
$$

How

Because the exponent is 3, use the third row of Pascal's Triangle.

EXAMPLE

The **binomial coefficients** from the third row of Pascal's Triangle are

$$1, 3, 3, 1.$$

So, the expansion is as follows.

$$(x - 2)^3 = (1)x^3 - (3)x^2(2) + (3)x(2)^2$$
$$- (1)(2)^3$$
$$= x^3 - 6x^2 + 12x - 8$$

Why

Knowing how to form Pascal's Triangle will allow you to expand complex binomials easily.

Using Pascal's Triangle to find

$$(x - 2)^3$$

is easier than finding the product

$$(x - 2)(x - 2)(x - 2).$$

Exercises Within Reach®

Concept Summary Check

53. *Writing* In your own words, explain how to form the rows in Pascal's Triangle.

54. *Reasoning* How many terms are in the expansion of $(x + y)^{10}$?

55. *Writing* How do the expansions of $(x + y)^n$ and $(x - y)^n$ differ?

56. *Reasoning* Which row of Pascal's Triangle would you use to evaluate $_{10}C_3$?

Extra Practice

Using the Binomial Theorem In Exercises 57–62, use the Binomial Theorem to expand the expression.

57. $(\sqrt{x} + 5)^3$

58. $(2\sqrt{t} - 1)^3$

59. $(x^{2/3} - y^{1/3})^3$

60. $(u^{3/5} + 2)^5$

61. $(3\sqrt{t} + \sqrt[4]{t})^4$

62. $(x^{3/4} - 2x^{5/4})^4$

Evaluating $_nC_r$ In Exercises 63–68, use a graphing calculator to evaluate $_nC_r$.

63. $_{30}C_6$

64. $_{40}C_8$

65. $_{52}C_5$

66. $_{100}C_4$

67. $_{800}C_{797}$

68. $_{1000}C_2$

Using the Binomial Theorem In Exercises 69 and 70, use the Binomial Theorem to expand the complex number.

69. $(1 + i)^4$

70. $(2 - i)^5$

Using the Binomial Theorem **In Exercises 71–74, use the Binomial Theorem to** approximate **the quantity rounded to three decimal places. For example:**

$$(1.02)^{10} = (1 + 0.02)^{10} \approx 1 + 10(0.02) + 45(0.02)^2.$$

71. $(1.02)^8$

72. $(2.005)^{10}$

73. $(2.99)^{12}$

74. $(1.98)^9$

Probability **In the study of probability, it is sometimes necessary to use the expansion** $(p + q)^n$, **where** $p + q = 1$. **In Exercises 75–78, use the Binomial Theorem to** expand **the expression.**

75. $\left(\frac{1}{2} + \frac{1}{2}\right)^5$

76. $\left(\frac{2}{3} + \frac{1}{3}\right)^4$

77. $\left(\frac{1}{4} + \frac{3}{4}\right)^4$

78. $\left(\frac{2}{5} + \frac{3}{5}\right)^3$

79. *Pascal's Triangle* Rows 0 through 6 of Pascal's Triangle are shown. Find the sum of the numbers in each row. Describe the pattern.

```
                1
              1   1
            1   2   1
          1   3   3   1
        1   4   6   4   1
      1   5   10   10   5   1
    1   6   15   20   15   6   1
```

80. *Pascal's Triangle* Use each encircled group of numbers to form a 2 × 2 matrix. Find the determinant of each matrix. Describe the pattern.

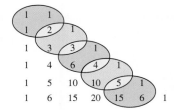

Explaining Concepts

81. *Precision* In the expansion of $(x + y)^{10}$, is $_6C_4$ the coefficient of the x^4y^6 term? Explain.

82. *Number Sense* Which of the following is equal to $_{11}C_5$? Explain.

(a) $\dfrac{11 \cdot 10 \cdot 9 \cdot 8 \cdot 7}{5 \cdot 4 \cdot 3 \cdot 2 \cdot 1}$

(b) $\dfrac{11 \cdot 10 \cdot 9 \cdot 8 \cdot 7}{6 \cdot 5 \cdot 4 \cdot 3 \cdot 2 \cdot 1}$

83. *Reasoning* When finding the seventh term of a binomial expansion by evaluating $_nC_r x^{n-r}y^r$, what value should you substitute for r? Explain.

84. *Reasoning* In the expansion of $(x + 2)^9$, are the coefficients of the x^3-term and the x^6-term identical? Explain.

Cumulative Review

In Exercises 85 and 86, find the partial sum of the arithmetic sequence.

85. $\displaystyle\sum_{i=1}^{15} (2 + 3i)$

86. $\displaystyle\sum_{k=1}^{25} (9k - 5)$

In Exercises 87 and 88, find the partial sum of the geometric sequence. Round to the nearest hundredth, if necessary.

87. $\displaystyle\sum_{k=1}^{8} 5^{k-1}$

88. $\displaystyle\sum_{i=1}^{20} 10\left(\frac{3}{4}\right)^{i-1}$

13 Chapter Summary

	What did you learn?	Explanation and Examples	Review Exercises
13.1	Write the terms of sequences *(p. 664)*.	An infinite sequence $a_1, a_2, a_3, \ldots, a_n, \ldots$ is a function whose domain is the set of positive integers. A finite sequence $a_1, a_2, a_3, \ldots, a_n$ is a function whose domain is the finite set $\{1, 2, 3, \ldots, n\}$. When n is a positive integer, n factorial is defined as $n! = 1 \cdot 2 \cdot 3 \cdot 4 \cdot \cdots \cdot (n-1) \cdot n$. As a special case, zero factorial is defined as $0! = 1$.	1–8
	Find the apparent nth term of a sequence *(p. 667)*.	Sometimes you have the first several terms of a sequence and need to find a formula (the nth term) to generate those terms. Pattern recognition is crucial in finding a form for the nth term.	9–16
	Sum the terms of a sequence to obtain a series *(p. 668)*.	For an infinite sequence $a_1, a_2, a_3, \ldots, a_n, \ldots$ **1.** the sum of the first n terms is called a partial sum, and **2.** the sum of all the terms is called an infinite series, or simply a series. The sum of the first n terms of the sequence whose nth term is a_n is $$\sum_{i=1}^{n} a_i = a_1 + a_2 + a_3 + a_4 + \ldots + a_n$$ where i is the index of summation, n is the upper limit of summation, and 1 is the lower limit of summation.	17–24
13.2	Recognize, write, and find the nth terms of arithmetic sequences *(p. 672)*.	A sequence is called arithmetic when the differences between consecutive terms are the same. The difference d between each pair of consecutive terms is called the common difference of the sequence. The nth term of an arithmetic sequence has the form $a_n = a_1 + (n-1)d$ where d is the common difference between the terms of the sequence, and a_1 is the first term.	25–36
	Find the nth partial sum of an arithmetic sequence *(p. 675)*.	The nth partial sum of the arithmetic sequence whose nth term is a_n is $$\sum_{i=1}^{n} a_i = a_1 + a_2 + a_3 + a_4 + \ldots + a_n$$ $$= \frac{n}{2}(a_1 + a_n).$$	37–42
	Use arithmetic sequences to solve application problems *(p. 677)*.	Saying that a quantity increases arithmetically is the same as saying that it increases linearly.	43–46

What did you learn?	Explanation and Examples	Review Exercises
13.3 Recognize, write, and find the nth terms of geometric sequences *(p. 682).*	The nth term of a geometric sequence has the form $$a_n = a_1 r^{n-1}$$ where r is the common ratio of consecutive terms of the sequence. So, every geometric sequence can be written in the following form. $$a_1, a_1 r, a_1 r^2, a_1 r^3, a_1 r^4, \ldots, a_1 r^{n-1}, \ldots$$	47–60
Find the sums of finite and infinite geometric sequences *(p. 684).*	The nth partial sum of the geometric sequence whose nth term is $a_n = a_1 r^{n-1}$ is given by $$\sum_{i=1}^{n} a_1 r^{i-1} = a_1 + a_1 r + a_1 r^2 + a_1 r^3 + \cdots + a_1 r^{n-1}$$ $$= a_1\left(\frac{r^n - 1}{r - 1}\right).$$ If $a_1, a_1 r, a_1 r^2, \ldots, a_1 r^n, \ldots$ is an infinite geometric sequence and $|r| < 1$, then the sum of the terms of the corresponding infinite geometric series is $$S = \sum_{i=0}^{\infty} a_1 r^i = \frac{a_1}{1 - r}.$$	61–72
Use geometric sequences to solve application problems *(p. 686).*	You can use a partial sum of a geometric sequence to find the value of an increasing annuity.	73–76
13.4 Use the Binomial Theorem to calculate binomial coefficients *(p. 690).*	In the expansion of $(x + y)^n$ $$(x + y)^n = x^n + nx^{n-1}y + \cdots + {}_nC_r x^{n-r}y^r +$$ $$\cdots + nxy^{n-1} + y^n$$ the coefficient of $x^{n-r}y^r$ is given by $${}_nC_r = \frac{n!}{(n-r)!\,r!}.$$	77–84
Use Pascal's Triangle to calculate binomial coefficients *(p. 691).*	The first and last numbers in each row of Pascal's Triangle are 1. Every other number in each row is formed by adding the two numbers immediately above the number. $$1$$ $$1 \quad 1$$ $$1 \quad 2 \quad 1$$ $$1 \quad 3 \quad 3 \quad 1$$ $$1 \quad 4 \quad 6 \quad 4 \quad 1$$ $$1 \quad 5 \quad 10 \quad 10 \quad 5 \quad 1$$ $$1 \quad 6 \quad 15 \quad 20 \quad 15 \quad 6 \quad 1$$	85–88
Expand binomial expressions *(p. 693).*	When you write out the coefficients of a binomial raised to a power, you are expanding a binomial. The formulas for binomial coefficients give you an easy way to expand binomials.	89–102

Review Exercises

Worked-out solutions to odd-numbered exercises at CollegePrepAlgebra.com

13.1

Writing the Terms of a Sequence In Exercises 1–8, write the first five terms of the sequence with the given nth term. (Assume that n begins with 1.)

1. $a_n = 3n + 5$

2. $a_n = \frac{1}{2}n - 4$

3. $a_n = \dfrac{n}{3n - 1}$

4. $a_n = 3^n + n$

5. $a_n = (n + 1)!$

6. $a_n = (-1)^n n!$

7. $a_n = \dfrac{n!}{2n}$

8. $a_n = \dfrac{(n + 1)!}{(2n)!}$

Finding the nth Term In Exercises 9–16, write an expression for the nth term of the sequence. (Assume that n begins with 1.)

9. $4, 7, 10, 13, 16, \ldots$

10. $3, -6, 9, -12, 15, \ldots$

11. $\dfrac{1}{2}, \dfrac{1}{5}, \dfrac{1}{10}, \dfrac{1}{17}, \dfrac{1}{26}, \ldots$

12. $\dfrac{0}{2}, \dfrac{1}{3}, \dfrac{2}{4}, \dfrac{3}{5}, \dfrac{4}{6}, \ldots$

13. $3, 1, -1, -3, -5, \ldots$

14. $3, 7, 11, 15, 19, \ldots$

15. $\dfrac{3}{2}, \dfrac{12}{5}, \dfrac{27}{10}, \dfrac{48}{17}, \dfrac{75}{26}, \ldots$

16. $-1, \dfrac{1}{2}, -\dfrac{1}{4}, \dfrac{1}{8}, -\dfrac{1}{16}, \ldots$

Finding a Partial Sum In Exercises 17–20, find the partial sum.

17. $\displaystyle\sum_{k=1}^{4} 7$

18. $\displaystyle\sum_{k=1}^{4} \dfrac{(-1)^k}{k}$

19. $\displaystyle\sum_{i=1}^{5} \dfrac{i - 2}{i + 1}$

20. $\displaystyle\sum_{n=1}^{4} \left(\dfrac{1}{n} - \dfrac{1}{n + 2} \right)$

Writing a Sum in Sigma Notation In Exercises 21–24, write the sum in sigma notation. (Begin with $k = 1$.)

21. $[5(1) - 3] + [5(2) - 3] + [5(3) - 3] + [5(4) - 3]$

22. $1(1 - 5) + 2(2 - 5) + 3(3 - 5)$
$\qquad + 4(4 - 5) + 5(5 - 5)$

23. $\dfrac{1}{3(1)} + \dfrac{1}{3(2)} + \dfrac{1}{3(3)} + \dfrac{1}{3(4)} + \dfrac{1}{3(5)} + \dfrac{1}{3(6)}$

24. $\left(-\dfrac{1}{3}\right)^0 + \left(-\dfrac{1}{3}\right)^1 + \left(-\dfrac{1}{3}\right)^2 + \left(-\dfrac{1}{3}\right)^3 + \left(-\dfrac{1}{3}\right)^4$

13.2

Finding the Common Difference In Exercises 25 and 26, find the common difference of the arithmetic sequence.

25. $50, 44.5, 39, 33.5, 28, \ldots$

26. $9, 12, 15, 18, 21, \ldots$

Writing the Terms of a Sequence In Exercises 27–30, write the first five terms of the arithmetic sequence with the given nth term. (Assume that n begins with 1.)

27. $a_n = 132 - 5n$

28. $a_n = 2n + 3$

29. $a_n = \frac{1}{3}n + \frac{5}{3}$

30. $a_n = -\frac{3}{5}n + 1$

Using a Recursion Formula In Exercises 31 and 32, write the first five terms of the arithmetic sequence defined recursively.

31. $a_1 = 80$
$\quad a_{k+1} = a_k - \frac{5}{2}$

32. $a_1 = 30$
$\quad a_{k+1} = a_k - 12$

Finding the nth Term In Exercises 33–36, find a formula for the nth term of the arithmetic sequence.

33. $a_1 = 10$, $d = 4$

34. $a_1 = 32$, $d = -2$

35. $a_1 = 1000$, $a_2 = 950$

36. $a_2 = 150$, $a_5 = 201$

Finding the nth Partial Sum In Exercises 37–40, find the partial sum.

37. $\sum_{k=1}^{12} (7k - 5)$

38. $\sum_{k=1}^{10} (100 - 10k)$

39. $\sum_{j=1}^{120} \left(\frac{1}{4}j + 1\right)$

40. $\sum_{j=1}^{50} \frac{3j}{2}$

Finding the nth Partial Sum In Exercises 41 and 42, find the nth partial sum of the arithmetic sequence.

41. $5.25, 6.5, 7.75, 9, \ldots, n = 60$

42. $\frac{5}{2}, 3, \frac{7}{2}, 4, \frac{9}{2}, \ldots, n = 150$

43. **Number Problem** Find the sum of the first 50 positive integers that are multiplies of 4.

44. **Number Problem** Find the sum of the integers from 225 to 300.

45. **Auditorium Seating** Each row in a small auditorium has three more seats than the preceding row. The front row seats 22 people and there are 12 rows of seats. Find the seating capacity of the auditorium.

46. **Wages** You earn $25 on the first day of the month and $100 on the last day of the month. Each day you are paid $2.50 more than the previous day. How much do you earn in a 31-day month?

13.3

Identifying a Geometric Sequence In Exercises 47 and 48, determine whether the sequence is geometric. If so, find the common ratio.

47. $8, 20, 50, 125, \frac{625}{2}, \ldots$

48. $27, -18, 12, -8, \frac{16}{3}, \ldots$

Writing the Terms of a Sequence In Exercises 49–54, write the first five terms of the geometric sequence.

49. $a_1 = 10$, $r = 3$

50. $a_1 = 2$, $r = -5$

51. $a_1 = 100$, $r = -\frac{1}{2}$

52. $a_1 = 20$, $r = \frac{1}{5}$

53. $a_1 = 4$, $r = \frac{3}{2}$

54. $a_1 = 32$, $r = -\frac{3}{4}$

Finding the nth Term In Exercises 55–60, find a formula for the nth term of the geometric sequence.

55. $a_1 = 1$, $r = -\frac{2}{3}$

56. $a_1 = 100$, $r = 1.07$

57. $a_1 = 24$, $a_2 = 72$

58. $a_1 = 16$, $a_2 = -4$

59. $a_1 = 12$, $a_4 = -\frac{3}{2}$

60. $a_2 = 1$, $a_3 = \frac{1}{3}$

Finding the nth Partial Sum In Exercises 61–64, find the nth partial sum of the geometric sequence. Round to the nearest hundredth, if necessary.

61. $200, 280, 392, 548.8, \ldots, \quad n = 12$

62. $25, 22.5, 20.25, 18.225, \ldots, \quad n = 20$

63. $27, -36, 48, -64, \ldots, \quad n = 14$

64. $-1024, 512, -256, 128, \ldots, \quad n = 9$

Using Sigma Notation In Exercises 65–68, find the partial sum. Round to the nearest thousandth, if necessary.

65. $\sum_{n=1}^{12} 2^n$

66. $\sum_{n=1}^{12} (-2)^n$

67. $\sum_{k=1}^{8} 5\left(-\frac{3}{4}\right)^k$

68. $\sum_{k=1}^{12} (-0.6)^{k-1}$

Finding the Sum of an Infinite Geometric Series In Exercises 69–72, find the sum

69. $\sum_{i=1}^{\infty} \left(\frac{7}{8}\right)^{i-1}$

70. $\sum_{i=1}^{\infty} \left(\frac{3}{5}\right)^{i-1}$

71. $\sum_{k=0}^{\infty} 4\left(\frac{2}{3}\right)^k$

72. $\sum_{k=0}^{\infty} 1.3\left(\frac{1}{10}\right)^k$

73. *Depreciation* A company pays \$120,000 for a machine. During the next 5 years, the machine depreciates at the rate of 30% per year. (That is, at the end of each year, the depreciated value is 70% of what it was at the beginning of the year.)

   (a) Find a formula for the nth term of the geometric sequence that gives the value of the machine n full years after it was purchased.

   (b) Find the depreciated value of the machine at the end of 5 full years.

74. *Population Increase* A city of 85,000 people is growing at the rate of 1.2% per year. (That is, at the end of each year, the population is 1.012 times what it was at the beginning of the year.)

   (a) Find a formula for the nth term of the geometric sequence that gives the population after n years.

   (b) Estimate the population after 50 years.

75. *Internet* On its first day, a website has 1000 visits. During the next 89 days, the number of visits increases by 12.5% each day. What is the total number of visits during the 90-day period?

76. *Increasing Annuity* You deposit \$200 in an account each month for 10 years. The account pays an annual interest rate of 8%, compounded monthly. What is your balance at the end of 10 years?

13.4

Finding a Binomial Coefficient In Exercises 77–84, evaluate the binomial coefficient $_nC_r$.

77. $_8C_3$

78. $_{12}C_2$

79. $_{15}C_4$

80. $_{100}C_1$

81. $_{40}C_4$

82. $_{32}C_8$

83. $_{25}C_6$

84. $_{48}C_5$

Using Pascal's Triangle In Exercises 85–88, use Pascal's Triangle to evaluate $_nC_r$.

85. $_4C_2$

86. $_9C_9$

87. $_{10}C_3$

88. $_6C_3$

Using Pascal's Triangle In Exercises 89–92, use Pascal's Triangle to expand the expression.

89. $(x - 5)^4$

90. $(x + y)^7$

91. $(5x + 2)^3$

92. $(x - 3y)^4$

Using the Binomial Theorem In Exercises 93–98, use the Binomial Theorem to expand the expression.

93. $(x + 1)^{10}$

94. $(y - 2)^6$

95. $(3x - 2y)^4$

96. $(4u + v)^5$

97. $(u^2 + v^3)^5$

98. $(x^4 + y^5)^4$

Finding a Specific Term In Exercises 99 and 100, find the specified term in the expansion of the binomial.

99. $(x + 2)^{10}$, 7th term

100. $(2x - 3y)^5$, 4th term

Finding the Coefficient of a Term In Exercises 101 and 102, find the coefficient of the given term in the expansion of the binomial.

Expression	Term
101. $(x - 3)^{10}$	x^5
102. $(3x + 4y)^6$	x^2y^4

Chapter Test

Solutions in English & Spanish and tutorial videos at CollegePrepAlgebra.com

Take this test as you would take a test in class. After you are done, check your work against the answers in the back of the book.

1. Write the first five terms of the sequence whose nth term is $a_n = \left(-\frac{3}{5}\right)^{n-1}$. (Assume that n begins with 1.)

2. Write the first five terms of the sequence whose nth term is $a_n = 3n^2 - n$. (Assume that n begins with 1.)

In Exercises 3–5, find the partial sum.

3. $\displaystyle\sum_{n=1}^{12} 5$

4. $\displaystyle\sum_{k=0}^{8} (2k-3)$

5. $\displaystyle\sum_{n=1}^{5} (3-4n)$

6. Write the sum in sigma notation: $\dfrac{2}{3(1)+1} + \dfrac{2}{3(2)+1} + \cdots + \dfrac{2}{3(12)+1}$.

7. Write the sum in sigma notation:

$$\left(\frac{1}{2}\right)^0 + \left(\frac{1}{2}\right)^2 + \left(\frac{1}{2}\right)^4 + \left(\frac{1}{2}\right)^6 + \left(\frac{1}{2}\right)^8 + \left(\frac{1}{2}\right)^{10}.$$

8. Write the first five terms of the arithmetic sequence whose first term is $a_1 = 12$ and whose common difference is $d = 4$.

9. Find a formula for the nth term of the arithmetic sequence whose first term is $a_1 = 5000$ and whose common difference is $d = -100$.

10. Find the sum of the first 50 positive integers that are multiples of 3.

11. Find the common ratio of the geometric sequence: $-4, 3, -\frac{9}{4}, \frac{27}{16}, \ldots\,.$

12. Find a formula for the nth term of the geometric sequence whose first term is $a_1 = 4$ and whose common ratio is $r = \frac{1}{2}$.

In Exercises 13 and 14, find the partial sum.

13. $\displaystyle\sum_{n=1}^{8} 2(2^n)$

14. $\displaystyle\sum_{n=1}^{10} 3\left(\frac{1}{2}\right)^n$

In Exercises 15 and 16, find the sum.

15. $\displaystyle\sum_{i=1}^{\infty} \left(\frac{1}{2}\right)^i$

16. $\displaystyle\sum_{i=1}^{\infty} 10(0.4)^{i-1}$

17. Evaluate the binomial coefficient $_{20}C_3$.

18. Use Pascal's Triangle to expand the expression $(x-2)^5$.

19. Find the coefficient of the term $x^3 y^5$ in the expansion of the expression $(x+y)^8$.

20. A free-falling object falls 4.9 meters during the first second, 14.7 more meters during the second second, 24.5 more meters during the third second, and so on. Assume the pattern continues. What is the total distance the object falls in 10 seconds?

21. You deposit $80 each month in an increasing annuity that pays 4.8% compounded monthly. What is the balance after 45 years?

Appendix A Review of Elementary Algebra Topics

A.1 The Real Number System

▶ Sets and Real Numbers
▶ Operations with Real Numbers
▶ Properties of Real Numbers

Sets and Real Numbers

Real numbers are used in everyday life to describe quantities such as age, miles per gallon, container size, and population. Real numbers are represented by symbols such as

$$-5, 9, 0, \tfrac{4}{3}, 0.666\ldots, 28.21, \sqrt{2}, \pi, \text{ and } \sqrt[3]{-32}.$$

Here are some important *subsets* of the set of real numbers.

$$\{1, 2, 3, 4, \ldots\} \qquad \text{Set of natural numbers}$$

$$\{0, 1, 2, 3, 4, \ldots\} \qquad \text{Set of whole numbers}$$

$$\{\ldots, -3, -2, -1, 0, 1, 2, 3, \ldots\} \qquad \text{Set of integers}$$

A real number is *rational* if it can be written as the ratio p/q of two integers, where $q \neq 0$. For instance, the numbers

$$\tfrac{1}{3} = 0.3333\ldots = 0.\overline{3}, \tfrac{1}{8} = 0.125, \text{ and } \tfrac{125}{111} = 1.126126\ldots = 1.\overline{126}$$

are rational. The decimal representation of a rational number either repeats or terminates.

$$\tfrac{173}{55} = 3.1\overline{45} \qquad\qquad\qquad \text{A rational number that repeats}$$

$$\tfrac{1}{2} = 0.5 \qquad\qquad\qquad\qquad \text{A rational number that terminates}$$

A real number that cannot be written as the ratio of two integers is called *irrational*. Irrational numbers have infinite nonrepeating decimal representations. For instance, the numbers

$$\sqrt{2} \approx 1.4142136 \quad \text{and} \quad \pi \approx 3.1415927$$

are irrational. (The symbol $\approx$ means "is approximately equal to.")

Real numbers are represented graphically by a *real number line*. The point 0 on the real number line is the *origin*. Numbers to the right of 0 are positive, and numbers to the left are negative, as shown in Figure A.1. The term *nonnegative* describes a number that is either positive or zero.

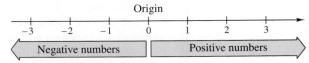

Figure A.1 The Real Number Line

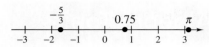

Every real number corresponds to exactly one point on the real number line.

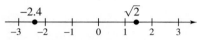

Every point on the real number line corresponds to exactly one real number.

Figure A.2

Figure A.3

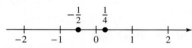

Figure A.4

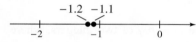

Figure A.5

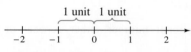

−1 is the opposite of 1.

Figure A.6

As illustrated in Figure A.2, there is a *one-to-one correspondence* between real numbers and points on the real number line.

The real number line provides you with a way of comparing any two real numbers. For any two (different) numbers on the real number line, one of the numbers must be to the left of the other number. A "less than" comparison is denoted by the inequality symbol $<$, a "greater than" comparison is denoted by $>$, a "less than or equal to" comparison is denoted by $\leq$, and a "greater than or equal to" comparison is denoted by $\geq$. When you are asked to order two numbers, you are simply being asked to say which of the two numbers is greater.

EXAMPLE 1   Ordering Real Numbers

Place the correct inequality symbol ($<$ or $>$) between each pair of numbers.

a.   2 ▨ −1   b. $-\frac{1}{2}$ ▨ $\frac{1}{4}$ c.   −1.1 ▨ −1.2

SOLUTION

a. $2 > -1$, because 2 lies to the *right* of −1.   See Figure A.3.

b. $-\frac{1}{2} < \frac{1}{4}$, because $-\frac{1}{2} = -\frac{2}{4}$ lies to the *left* of $\frac{1}{4}$.   See Figure A.4.

c. $-1.1 > -1.2$, because −1.1 lies to the *right* of −1.2.   See Figure A.5.

Two real numbers are *opposites* of each other when they lie the same distance from, but on opposite sides of, zero. For instance, −1 is the opposite of 1, as shown in Figure A.6.

The *absolute value* of a real number is its distance from zero on the real number line. A pair of vertical bars, $|\ \ |$, is used to denote absolute value. The absolute value of a real number is either positive or zero (never negative).

EXAMPLE 2   Evaluating Absolute Values

a. $|5| = 5$, because the distance between 5 and 0 is 5.

b. $|0| = 0$, because the distance between 0 and itself is 0.

c. $\left|-\frac{2}{3}\right| = \frac{2}{3}$, because the distance between $-\frac{2}{3}$ and 0 is $\frac{2}{3}$.

Operations with Real Numbers

There are four basic arithmetic operations with real numbers: addition, subtraction, multiplication, and division.

The result of adding two real numbers is called the *sum* of the two numbers. Subtraction of one real number from another can be described as adding the opposite of the second number to the first number. For instance,

$$7 - 5 = 7 + (-5) = 2 \text{ and } 10 - (-13) = 10 + 13 = 23.$$

The result of subtracting one real number from another is called the *difference* of the two numbers.

Study Tip

In the fraction

$$\frac{a}{b}$$

a is the *numerator* and *b* is the *denominator*.

EXAMPLE 3   **Adding and Subtracting Real Numbers**

a. $-25 + 12 = -13$

b. $5 + (-10) = -5$

c. $-13.8 - 7.02 = -13.8 + (-7.02) = -20.82$

d.   To add two fractions with unlike denominators, you must first rewrite one (or both) of the fractions so that they have a common denominator. To do this, find the least common multiple (LCM) of the denominators.

$$\frac{1}{3} + \frac{2}{9} = \frac{1(3)}{3(3)} + \frac{2}{9}$$ LCM of 3 and 9 is 9.

$$= \frac{3}{9} + \frac{2}{9} = \frac{5}{9}$$ Rewrite with like denominators and add numerators.

The result of multiplying two real numbers is called their *product*, and each of the numbers is called a *factor* of the product. The product of zero and any other number is zero. Multiplication is denoted in a variety of ways. For instance,

$$3 \times 2, \ 3 \cdot 2, \ 3(2), \text{ and } (3)(2)$$

all denote the product of "3 times 2," which you know is 6.

EXAMPLE 4   **Multiplying Real Numbers**

a. $(6)(-4) = -24$

b. $(-1.2)(-0.4) = 0.48$

c.   To find the product of more than two numbers, find the product of their absolute values. When there is an *even* number of negative factors, the product is positive. When there is an *odd* number of negative factors, the product is negative. For instance, in the product $6(2)(-5)(-8)$, there are two negative factors, so the product must be positive, and you can write $6(2)(-5)(-8) = 480$.

d.   To multiply two fractions, multiply their numerators and their denominators. For instance, the product of $\frac{2}{3}$ and $\frac{4}{5}$ is

$$\left(\frac{2}{3}\right)\left(\frac{4}{5}\right) = \frac{(2)(4)}{(3)(5)} = \frac{8}{15}.$$

The *reciprocal* of a nonzero real number *a* is defined as the number by which *a* must be multiplied to obtain 1. The reciprocal of the fraction a/b is b/a.

To divide one real number by a second (nonzero) real number, multiply the first number by the reciprocal of the second number. The result of dividing two real numbers is called the *quotient* of the two numbers. Division is denoted in a variety of ways. For instance,

$$12 \div 4, \ 12/4, \ \frac{12}{4}, \text{ and } 4\overline{)12}$$

all denote the quotient of "12 divided by 4," which you know is 3.

EXAMPLE 5 Dividing Real Numbers

a. $-30 \div 5 = -30\left(\dfrac{1}{5}\right) = -\dfrac{30}{5} = -6$

b. $-\dfrac{9}{14} \div -\dfrac{1}{3} = -\dfrac{9}{14}\left(-\dfrac{3}{1}\right) = \dfrac{27}{14}$

c. $\dfrac{5}{16} \div 2\dfrac{3}{4} = \dfrac{5}{16} \div \dfrac{11}{4} = \dfrac{5}{16}\left(\dfrac{4}{11}\right) = \dfrac{5(4)}{4(4)(11)} = \dfrac{5}{44}$

Let n be a positive integer and let a be a real number. Then the product of n factors of a is given by

$$a^n = \underbrace{a \cdot a \cdot a \cdot \cdots \cdot a}_{n \text{ factors}}.$$

In the exponential form a^n, a is called the *base* and n is called the *exponent*.

EXAMPLE 6 Evaluating Exponential Expressions

a. $(-2)^5 = (-2)(-2)(-2)(-2)(-2) = -32$

b. $\left(\dfrac{1}{5}\right)^3 = \left(\dfrac{1}{5}\right)\left(\dfrac{1}{5}\right)\left(\dfrac{1}{5}\right) = \dfrac{1}{125}$

c. $(-7)^2 = (-7)(-7) = 49$

One way to help avoid confusion when communicating algebraic ideas is to establish an order of operations. This order is summarized below.

Order of Operations

1. Perform operations inside *symbols of grouping*—( ) or [ ]—or *absolute value symbols*, starting with the innermost set of symbols.

2. Evaluate all *exponential* expressions.

3. Perform all *multiplications* and *divisions* from left to right.

4. Perform all *additions* and *subtractions* from left to right.

EXAMPLE 7 Order of Operations

a. $20 - 2 \cdot 3^2 = 20 - 2 \cdot 9 = 20 - 18 = 2$

b. $-4 + 2(-2 + 5)^2 = -4 + 2(3)^2 = -4 + 2(9) = -4 + 18 = 14$

c. $\dfrac{2 \cdot 5^2 - 10}{3^2 - 4} = (2 \cdot 5^2 - 10) \div (3^2 - 4)$ Rewrite using parentheses.

$= (50 - 10) \div (9 - 4)$ Evaluate exponential expressions and multiply within symbols of grouping.

$= 40 \div 5 = 8$ Simplify.

Properties of Real Numbers

Below is a review of the properties of real numbers. In this list, a verbal description of each property is given, as well as an example.

Properties of Real Numbers: Let *a*, *b*, and *c* be real numbers.

Property   *Example*

1. *Commutative Property of Addition:*
   Two real numbers can be added in either order.

   $a + b = b + a$ $3 + 5 = 5 + 3$

2. *Commutative Property of Multiplication:*
   Two real numbers can be multiplied in either order.

   $ab = ba$ $4 \cdot (-7) = -7 \cdot 4$

3. *Associative Property of Addition:*
   When three real numbers are added, it makes no difference which two are added first.

   $(a + b) + c = a + (b + c)$ $(2 + 6) + 5 = 2 + (6 + 5)$

4. *Associative Property of Multiplication:*
   When three real numbers are multiplied, it makes no difference which two are multiplied first.

   $(ab)c = a(bc)$ $(3 \cdot 5) \cdot 2 = 3 \cdot (5 \cdot 2)$

5. *Distributive Property:*
   Multiplication distributes over addition.

   $a(b + c) = ab + ac$ $3(8 + 5) = 3 \cdot 8 + 3 \cdot 5$
   $(a + b)c = ac + bc$ $(3 + 8)5 = 3 \cdot 5 + 8 \cdot 5$

6. *Additive Identity Property:*
   The sum of zero and a real number equals the number itself.

   $a + 0 = 0 + a = a$ $3 + 0 = 0 + 3 = 3$

7. *Multiplicative Identity Property:*
   The product of 1 and a real number equals the number itself.

   $a \cdot 1 = 1 \cdot a = a$ $4 \cdot 1 = 1 \cdot 4 = 4$

8. *Additive Inverse Property:*
   The sum of a real number and its opposite is zero.

   $a + (-a) = 0$ $3 + (-3) = 0$

9. *Multiplicative Inverse Property:*
   The product of a nonzero real number and its reciprocal is 1.

   $a \cdot \dfrac{1}{a} = 1, \; a \neq 0$ $8 \cdot \dfrac{1}{8} = 1$

A.2 Fundamentals of Algebra

▶ Algebraic Expressions
▶ Constructing Verbal Models
▶ Equations

Algebraic Expressions

One characteristic of algebra is the use of letters to represent numbers. The letters are *variables*, and combinations of letters and numbers are *algebraic expressions*. The *terms* of an algebraic expression are those parts separated by addition. For example, in the expression $-x^2 + 5x + 8$, $-x^2$ and $5x$ are the *variable terms* and 8 is the *constant term*. The numerical factor of a term is called the *coefficient*. The coefficient of the variable term $-x^2$ is -1 and the coefficient of $5x$ is 5.

To *evaluate* an algebraic expression, substitute numerical values for each of the variables in the expression.

EXAMPLE 1   **Evaluating Algebraic Expressions**

a.   Evaluate the expression $-3x + 5$ when $x = 3$.

$$-3(3) + 5 = -9 + 5 = -4$$

b.   Evaluate the expression $3x^2 + 2xy - y^2$ when $x = 3$ and $y = -1$.

$$3(3)^2 + 2(3)(-1) - (-1)^2 = 3(9) + (-6) - 1 = 20$$

The properties of real numbers listed on page A5 can be used to rewrite and simplify algebraic expressions. To simplify an algebraic expression generally means to remove symbols of grouping such as parentheses or brackets and to combine like terms. In an algebraic expression, two terms are said to be like terms when they are both constant terms or when they have the same variable factor(s). To combine like terms in an algebraic expression, add their respective coefficients and attach the common variable factor.

EXAMPLE 2   **Combining Like Terms**

a. $2x + 3y - 6x - y = (2x - 6x) + (3y - y)$ Group like terms.

$ = (2 - 6)x + (3 - 1)y$ Distributive Property

$ = -4x + 2y$ Simplest form

b. $4x^2 + 5x - x^2 - 8x = (4x^2 - x^2) + (5x - 8x)$ Group like terms.

$ = (4 - 1)x^2 + (5 - 8)x$ Distributive Property

$ = 3x^2 - 3x$ Simplest form

EXAMPLE 3   **Removing Symbols of Grouping**

a. $-2(a + 5) + 4(a - 8) = -2a - 10 + 4a - 32$ Distributive Property

$\qquad\qquad\qquad\qquad = (-2a + 4a) + (-10 - 32)$ Group like terms.

$\qquad\qquad\qquad\qquad = 2a - 42$ Combine like terms.

b. $3x^2 - [9x + 3x(2x - 1)] = 3x^2 - [9x + 6x^2 - 3x]$ Distributive Property

$\qquad\qquad\qquad\qquad\qquad = 3x^2 - [6x^2 + 6x]$ Combine like terms.

$\qquad\qquad\qquad\qquad\qquad = 3x^2 - 6x^2 - 6x$ Distributive Property

$\qquad\qquad\qquad\qquad\qquad = -3x^2 - 6x$ Combine like terms.

Constructing Verbal Models

When you translate a verbal sentence or phrase into an algebraic expression, watch for key words and phrases that indicate the four different operations of arithmetic.

EXAMPLE 4   **Translating Verbal Phrases**

a. *Verbal Description:* Seven more than 3 times x

$\qquad$ *Algebraic Expression:* $3x + 7$

b. *Verbal Description:* Four times the sum of y and 9

$\qquad$ *Algebraic Expression:* $4(y + 9)$

c. *Verbal Description:* Five decreased by the product of 2 and a number

$\qquad$ *Label:* The number $= x$ $\qquad$ *Algebraic Expression:* $5 - 2x$

d. *Verbal Description:* One more than the product of 8 and a number, all divided by 6

$\qquad$ *Label:* The number $= x$ $\qquad$ *Algebraic Expression:* $\dfrac{8x + 1}{6}$

Study Tip

When verbal phrases are translated into algebraic expressions, products are often overlooked, as demonstrated in Example 5.

EXAMPLE 5   **Constructing Verbal Models**

A cash register contains x quarters. Write an expression for this amount of money in dollars.

SOLUTION

Verbal Model:	Value of coin	$\cdot$	Number of coins

Labels:	Value of coin $= 0.25$	(dollars per quarter)
	Number of coins $= x$	(quarters)

Expression:	$0.25x$	(dollars)

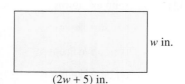

(2w + 5) in.

Figure A.7

EXAMPLE 6   **Constructing Verbal Models**

The width of a rectangle is w inches. The length of the rectangle is 5 inches more than twice its width. Write an expression for the perimeter of the rectangle.

SOLUTION

Draw a rectangle, as shown in Figure A.7. Next, use a verbal model to solve the problem. Use the formula (perimeter) = 2(length) + 2(width).

Verbal Model: 2 • Length + 2 • Width

Labels: Length = $2w + 5$ (inches)
               Width = w (inches)

Expression: $2(2w + 5) + 2w = 4w + 10 + 2w = 6w + 10$ (inches)

Equations

An *equation* is a statement that equates two algebraic expressions. Solving an equation involving x means finding all values of x for which the equation is true. Such values are *solutions* and are said to *satisfy* the equation. Example 7 shows how to check whether a given value is a solution of an equation.

EXAMPLE 7   **Checking a Solution of an Equation**

Determine whether $x = -3$ is a solution of $-3x - 5 = 4x + 16$.

$$-3(-3) - 5 \overset{?}{=} 4(-3) + 16 \qquad \text{Substitute } -3 \text{ for } x \text{ in original equation.}$$

$$9 - 5 \overset{?}{=} -12 + 16 \qquad \text{Simplify.}$$

$$4 = 4 \qquad \text{Solution checks.} \ ✔$$

EXAMPLE 8   **Using a Verbal Model to Construct an Equation**

You are given a speeding ticket for $80 for speeding on a road where the speed limit is 45 miles per hour. You are fined $10 for each mile per hour over the speed limit. How fast were you driving? Write an algebraic equation that models the situation.

SOLUTION

Verbal Model: Fine • Speed over limit = Amount of ticket

Labels: Fine = 10   (dollars per mile per hour)
               Your speed = x (mile per hour)
               Speed over limit = $x - 45$ (mile per hour)
               Amount of ticket = 80   (dollars)

Algebraic Model: $10(x - 45) = 80$

Study Tip

For more review on the fundamentals of algebra, refer to Chapter 2.

A.3 Equations, Inequalities, and Problem Solving

▶ Equations
▶ Inequalities
▶ Problem Solving

Equations

A *linear equation in one variable x* is an equation that can be written in the standard form

$$ax + b = 0$$

where a and b are real numbers with $a \neq 0$. To solve a linear equation, you want to isolate x on one side of the equation by a sequence of *equivalent equations*, each having the same solution(s) as the original equation. The operations that yield equivalent equations are as follows.

Operations That Yield Equivalent Equations

1. Remove symbols of grouping, combine like terms, or simplify fractions on one or both sides of the equation.
2. Add (or subtract) the same quantity to (from) each side of the equation.
3. Multiply (or divide) each side of the equation by the same nonzero quantity.
4. Interchange the two sides of the equation.

EXAMPLE 1   Solving a Linear Equation in Standard Form

Solve $3x - 6 = 0$. Then check the solution.

SOLUTION

$3x - 6 = 0$	Write original equation.
$3x - 6 + 6 = 0 + 6$	Add 6 to each side.
$3x = 6$	Combine like terms.
$\dfrac{3x}{3} = \dfrac{6}{3}$	Divide each side by 3.
$x = 2$	Simplify.

Check

$3x - 6 = 0$	Write original equation.
$3(2) - 6 \overset{?}{=} 0$	Substitute 2 for x.
$0 = 0$	Solution checks. ✔

So, the solution is $x = 2$.

EXAMPLE 2   **Solving a Linear Equation in Nonstandard Form**

Solve $5x + 4 = 3x - 8$.

SOLUTION

$$5x + 4 = 3x - 8 \qquad \text{Write original equation.}$$
$$5x - 3x + 4 = 3x - 3x - 8 \qquad \text{Subtract } 3x \text{ from each side.}$$
$$2x + 4 = -8 \qquad \text{Combine like terms.}$$
$$2x + 4 - 4 = -8 - 4 \qquad \text{Subtract 4 from each side.}$$
$$2x = -12 \qquad \text{Combine like terms.}$$
$$\frac{2x}{2} = \frac{-12}{2} \qquad \text{Divide each side by 2.}$$
$$x = -6 \qquad \text{Simplify.}$$

The solution is $x = -6$. Check this in the original equation.

Linear equations often contain parentheses or other symbols of grouping. In most cases, it helps to remove symbols of grouping as a first step in solving an equation. This is illustrated in Example 3.

EXAMPLE 3   **Solving a Linear Equation Involving Parentheses**

Solve $2(x + 4) = 5(x - 8)$.

SOLUTION

$$2(x + 4) = 5(x - 8) \qquad \text{Write original equation.}$$
$$2x + 8 = 5x - 40 \qquad \text{Distributive Property}$$
$$2x - 5x + 8 = 5x - 5x - 40 \qquad \text{Subtract } 5x \text{ from each side.}$$
$$-3x + 8 = -40 \qquad \text{Combine like terms.}$$
$$-3x + 8 - 8 = -40 - 8 \qquad \text{Subtract 8 from each side.}$$
$$-3x = -48 \qquad \text{Combine like terms.}$$
$$\frac{-3x}{-3} = \frac{-48}{-3} \qquad \text{Divide each side by } -3.$$
$$x = 16 \qquad \text{Simplify.}$$

The solution is $x = 16$. Check this in the original equation.

Study Tip

Recall that when finding the least common multiple of a set of numbers, you should first consider all multiples of each number. Then, you should choose the smallest of the common multiples of the numbers.

To solve an equation involving fractional expressions, find the least common multiple (LCM) of the denominators and multiply each side by the LCM.

EXAMPLE 4   **Solving a Linear Equation Involving Fractions**

Solve $\dfrac{x}{3} + \dfrac{3x}{4} = 2$.

SOLUTION

$$12\left(\dfrac{x}{3} + \dfrac{3x}{4}\right) = 12(2)$$ Multiply each side of original equation by LCM 12.

$$12 \cdot \dfrac{x}{3} + 12 \cdot \dfrac{3x}{4} = 24$$ Distributive Property

$$4x + 9x = 24$$ Clear fractions.

$$13x = 24$$ Combine like terms.

$$x = \dfrac{24}{13}$$ Divide each side by 13.

The solution is $x = \frac{24}{13}$. Check this in the original equation.

To solve an equation involving an absolute value, remember that the expression inside the absolute value signs can be positive or negative. This results in two separate equations, each of which must be solved.

EXAMPLE 5   **Solving an Equation Involving Absolute Value**

Solve $|4x - 3| = 13$.

SOLUTION

$$|4x - 3| = 13$$ Write original equation.

$$4x - 3 = -13 \quad \text{or} \quad 4x - 3 = 13$$ Equivalent equations

$$4x = -10 \qquad\qquad 4x = 16$$ Add 3 to each side.

$$x = -\dfrac{5}{2} \qquad\qquad x = 4$$ Divide each side by 4.

The solutions are $x = -\frac{5}{2}$ and $x = 4$. Check this in the original equation.

Inequalities

The simplest type of inequality is a *linear inequality* in one variable. For instance, $2x + 3 > 4$ is a linear inequality in x. The procedures for solving linear inequalities in one variable are much like those for solving linear equations, as described on page A9. The exception is that when each side of an inequality is multiplied or divided by a negative number, the direction of the inequality symbol *must be reversed*.

EXAMPLE 6   **Solving a Linear Inequality**

Solve and graph the inequality $-5x - 7 > 3x + 9$.

SOLUTION

$$-5x - 7 > 3x + 9 \qquad \text{Write original equation.}$$

$$-8x - 7 > 9 \qquad \text{Subtract } 3x \text{ from each side.}$$

$$-8x > 16 \qquad \text{Add 7 to each side.}$$

$$x < -2 \qquad \begin{array}{l}\text{Divide each side by } -8 \text{ and reverse} \\ \text{the direction of the inequality symbol.}\end{array}$$

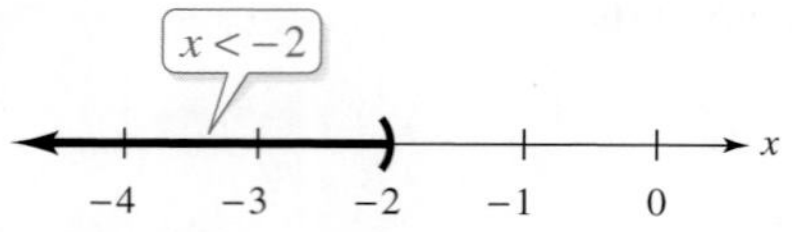

The solution set in interval notation is $(-\infty, -2)$ and in set notation is $\{x \mid x < -2\}$. The graph of the solution set is shown in Figure A.8.

$x < -2$

$-4 \quad -3 \quad -2 \quad -1 \quad 0$

Figure A.8

Two inequalities joined by the word *and* or the word *or* constitute a *compound inequality*. Sometimes it is possible to write a compound inequality as a *double inequality*. For instance, you can write $-3 < 6x - 1$ *and* $6x - 1 < 3$ more simply as $-3 < 6x - 1 < 3$. A compound inequality formed by the word *and* is called *conjunctive* and may be rewritten as a double inequality. A compound inequality joined by the word *or* is called *disjunctive* and cannot be rewritten as a double inequality.

EXAMPLE 7   **Solving a Conjunctive Inequality**

Solve and graph the inequality $2x + 3 \geq 4$ and $3x - 8 < -2$.

SOLUTION

$$2x + 3 \geq 4 \quad \text{and} \quad 3x - 8 < -2$$

$$2x \geq 1 \qquad\qquad 3x < 6$$

$$x \geq \tfrac{1}{2} \qquad\qquad x < 2$$

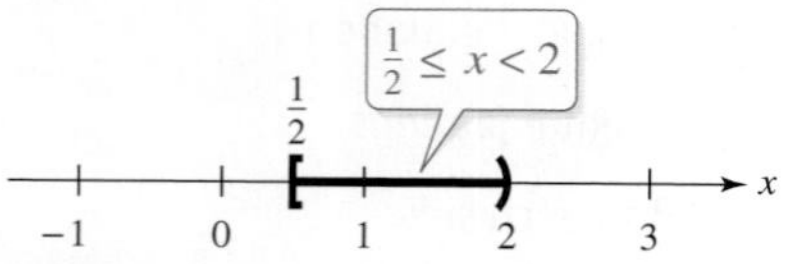

The solution set in interval notation is $\left[\tfrac{1}{2}, 2\right)$ and in set notation is $\left\{x \mid \tfrac{1}{2} \leq x < 2\right\}$. The graph of the solution set is shown in Figure A.9.

$\tfrac{1}{2} \leq x < 2$

$-1 \quad 0 \quad 1 \quad 2 \quad 3$

Figure A.9

Study Tip

The word *or* is represented by the symbol $\cup$, which is read as *union*.

EXAMPLE 8   **Solving a Disjunctive Inequality**

Solve and graph the inequality $x - 8 > -3$ or $-6x + 1 \geq -5$.

SOLUTION

$$x - 8 > -3 \quad \text{or} \quad -6x + 1 \geq -5$$

$$x > 5 \qquad\qquad -6x \geq -6$$

$$\qquad\qquad x \leq 1$$

The solution set in interval notation is $(-\infty, 1] \cup (5, \infty)$ and in set notation is $\{x \mid x > 5 \text{ or } x \leq 1\}$. The graph of the solution set is shown in Figure A.10.

$x \leq 1 \qquad\qquad x > 5$

$-1 \quad 0 \quad 1 \quad 2 \quad 3 \quad 4 \quad 5 \quad 6$

Figure A.10

To solve an absolute value inequality, use the following rules.

> ## Solving an Absolute Value Inequality
>
> Let x be a variable or an algebraic expression and let a be a real number such that $a > 0$.
>
> **1.** The solutions of $|x| < a$ are all values of x that lie between $-a$ and a.
>
> $$|x| < a \text{ if and only if } -a < x < a$$
>
> **2.** The solutions of $|x| > a$ are all values of x that are less than $-a$ or greater than a.
>
> $$|x| > a \text{ if and only if } x < -a \text{ or } x > a$$
>
> These rules are also valid when $<$ is replaced by $\leq$ and $>$ is replaced by $\geq$.

EXAMPLE 9   Solving Absolute Value Inequalities

Solve and graph the inequality.

a. $|4x + 3| > 9$ **b.** $|2x - 7| \leq 1$

SOLUTION

a.

$|4x + 3| > 9$	Write original inequality.
$4x + 3 < -9 \quad \text{or} \quad 4x + 3 > 9$	Equivalent inequalities
$4x < -12 \qquad\qquad 4x > 6$	Subtract 3 from each side.
$x < -3 \qquad\qquad x > \frac{3}{2}$	Divide each side by 4.

The solution set consists of all real numbers that are less than -3 or greater than $\frac{3}{2}$. The solution set in interval notation is $(-\infty, -3) \cup \left(\frac{3}{2}, \infty\right)$ and in set notation is $\left\{x \mid x < -3 \text{ or } x > \frac{3}{2}\right\}$. The graph is shown in Figure A.11.

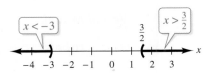

Figure A.11

b.

$|2x - 7| \leq 1$	Write original inequality.
$-1 \leq 2x - 7 \leq 1$	Equivalent double inequality
$6 \leq 2x \leq 8$	Add 7 to all three parts.
$3 \leq x \leq 4$	Divide all three parts by 2.

The solution set consists of all real numbers that are greater than or equal to 3 and less than or equal to 4. The solution set in interval notation is [3, 4] and in set notation is $\{x \mid 3 \leq x \leq 4\}$. The graph is shown in Figure A.12.

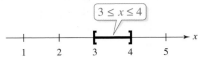

Figure A.12

Problem Solving

Algebra is used to solve word problems that relate to real-life situations. The following guidelines summarize the problem-solving strategy that you should use when solving word problems.

Guidelines for Solving Word Problems

1. Write a *verbal model* that describes the problem.

2. Assign *labels* to fixed quantities and variable quantities.

3. Rewrite the verbal model as an *algebraic equation* using the assigned labels.

4. *Solve* the resulting algebraic equation.

5. *Check* to see that your solution satisfies the original problem as stated.

EXAMPLE 10   Finding the Percent of Monthly Expenses

Your family has an annual income of $77,520 and the following monthly expenses: mortgage ($1500), car payment ($510), food ($400), utilities ($325), and credit cards ($300). The total expenses for one year represent what percent of your family's annual income?

SOLUTION

The total amount of your family's monthly expenses is

$$1500 + 510 + 400 + 325 + 300 = \$3035.$$

The total monthly expenses for one year are

$$3035 \cdot 12 = \$36,420.$$

Verbal Model:	Expenses = Percent · Income

Labels: Expenses = 36,420 (dollars)
Percent = p (in decimal form)
Income = 77,520 (dollars)

Equation: $36,420 = p \cdot 77,520$ Original equation

$\dfrac{36,420}{77,520} = p$ Divide each side by 77,520.

$0.470 \approx p$ Use a calculator.

Your family's total expenses for one year are about 0.470 or 47.0% of your family's annual income.

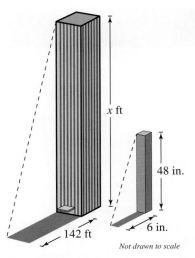

Figure A.13

x ft

48 in.

142 ft

6 in.

Not drawn to scale

EXAMPLE 11   Geometry: **Similar Triangles**

To determine the height of the Aon Center Building (in Chicago), you measure the shadow cast by the building and find it to be 142 feet long, as shown in Figure A.13. Then you measure the shadow cast by a four-foot post and find it to be 6 inches long. Estimate the building's height.

SOLUTION

To solve this problem, you use a property from geometry that states that the ratios of corresponding sides of similar triangles are equal.

Verbal Model: $\dfrac{\text{Height of building}}{\text{Length of building's shadow}} = \dfrac{\text{Height of post}}{\text{Length of post's shadow}}$

Labels:		
Height of building $= x$	(feet)
Length of building's shadow $= 142$	(feet)
Height of post $= 4$ feet $= 48$ inches	(inches)
Length of post's shadow $= 6$	(inches)

Equation:

$\dfrac{x}{142} = \dfrac{48}{6}$ Original proportion

$x \cdot 6 = 142 \cdot 48$ Cross-multiply.

$x = 1136$ Divide each side by 6.

So, you can estimate the Aon Center Building to be 1136 feet high.

EXAMPLE 12   Geometry: **Dimensions of a Room**

A rectangular kitchen is twice as long as it is wide, and its perimeter is 84 feet. Find the dimensions of the kitchen.

SOLUTION

For this problem, it helps to sketch a diagram, as shown in Figure A.14.

Verbal Model: $2 \cdot \text{Length} + 2 \cdot \text{Width} = \text{Perimeter}$

Labels:		
Length $= l = 2w$	(feet)
Width $= w$	(feet)
Perimeter $= 84$	(feet)

Equation:

$2(2w) + 2w = 84$ Original equation

$6w = 84$ Combine like terms.

$w = 14$ Divide each side by 6.

Because the length is twice the width, you have $l = 2w = 2(14) = 28$. So, the dimensions of the room are 14 feet by 28 feet.

w

l

Figure A.14

Study Tip

For more review on equations and inequalities, refer to Chapter 3.

A.4 Graphs and Functions

▶ The Rectangular Coordinate System
▶ Graphs of Equations
▶ Functions
▶ Slope and Linear Equations
▶ Graphs of Linear Inequalities

The Rectangular Coordinate System

You can represent ordered pairs of real numbers by points in a plane. This plane is called a *rectangular coordinate system*. A rectangular coordinate system is formed by two real lines, the *x-axis* (horizontal line) and the *y-axis* (vertical line), intersecting at right angles. The point of intersection of the axes is called the *origin*, and the axes divide the plane into four regions called *quadrants*.

Each point in the plane corresponds to an *ordered pair* (x, y) of real numbers x and y, called the *coordinates* of the point. The *x-coordinate* tells how far to the left or right the point is from the vertical axis, and the *y-coordinate* tells how far up or down the point is from the horizontal axis, as shown in Figure A.15.

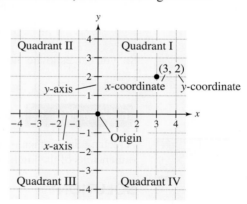

Figure A.15

EXAMPLE 1 **Finding Coordinates of Points**

Determine the coordinates of each of the points shown in Figure A.16, and then determine the quadrant in which each point is located.

SOLUTION

Point A lies two units to the *right* of the vertical axis and one unit *below* the horizontal axis. So, point A must be given by $(2, -1)$. The coordinates of the other four points can be determined in a similar way. The results are as follows.

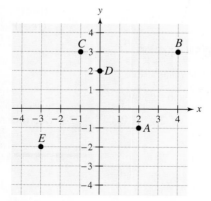

Figure A.16

Point	Coordinates	Quadrant
A	$(2, -1)$	IV
B	$(4, 3)$	I
C	$(-1, 3)$	II
D	$(0, 2)$	None
E	$(-3, -2)$	III

Graphs of Equations

The solutions of an equation involving two variables can be represented by points on a rectangular coordinate system. The *graph of an equation* is the set of all points that are solutions of the equation.

The simplest way to sketch the graph of an equation is the *point-plotting method.* With this method, you construct a table of values consisting of several solution points of the equation, plot these points, and then connect the points with a smooth curve or line.

> **EXAMPLE 2**   **Sketching the Graph of an Equation**

Sketch the graph of $y = x^2 - 2$.

SOLUTION

Begin by choosing several x-values and then calculating the corresponding y-values. For instance, if you choose $x = -2$, the corresponding y-value is

$y = x^2 - 2$	Original equation
$y = (-2)^2 - 2$	Substitute -2 for x.
$y = 4 - 2$	Simplify.
$y = 2.$	Simplify.

Then, create a table using these values, as shown below.

x	−2	−1	0	1	2	3
$y = x^2 - 2$	2	−1	−2	−1	2	7
Solution point	$(-2, 2)$	$(-1, -1)$	$(0, -2)$	$(1, -1)$	$(2, 2)$	$(3, 7)$

Next, plot the solution points, as shown in Figure A.17. Finally, connect the points with a smooth curve, as shown in Figure A.18.

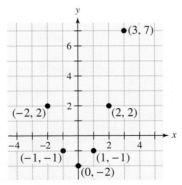

Figure A.17

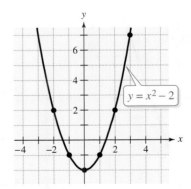

Figure A.18

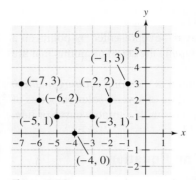

Figure A.19

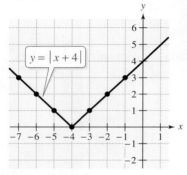

Figure A.20

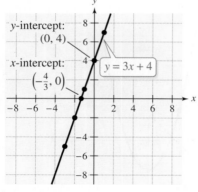

Figure A.21

EXAMPLE 3 **Sketching the Graph of an Equation**

Sketch the graph of $y = |x + 4|$.

SOLUTION

Begin by creating a table of values, as shown below. Plot the solution points, as shown in Figure A.19. It appears that the points lie in a "V-shaped" pattern, with the point $(-4, 0)$ lying at the bottom of the "V." Following this pattern, connect the points to form the graph shown in Figure A.20.

x	-7	-6	-5	-4	-3	-2	-1
$y = |x + 4|$	3	2	1	0	1	2	3
Solution point	$(-7, 3)$	$(-6, 2)$	$(-5, 1)$	$(-4, 0)$	$(-3, 1)$	$(-2, 2)$	$(-1, 3)$

Intercepts of a graph are the points at which the graph intersects the x- or y-axis. To find x-intercepts, let $y = 0$ and solve the equation for x. To find y-intercepts, let $x = 0$ and solve the equation for y.

EXAMPLE 4 **Finding the Intercepts of a Graph**

Find the intercepts and sketch the graph of $y = 3x + 4$.

SOLUTION

To find any x-intercepts, let $y = 0$ and solve the resulting equation for x.

$y = 3x + 4$	Write original equation.
$0 = 3x + 4$	Let $y = 0$.
$-\dfrac{4}{3} = x$	Solve equation for x.

To find any y-intercepts, let $x = 0$ and solve the resulting equation for y.

$y = 3x + 4$	Write original equation.
$y = 3(0) + 4$	Let $x = 0$.
$y = 4$	Solve equation for y.

So, the x-intercept is $\left(-\frac{4}{3}, 0\right)$ and the y-intercept is $(0, 4)$. To sketch the graph of the equation, create a table of values (including intercepts), as shown below. Then plot the points and connect them with a line, as shown in Figure A.21.

x	-3	-2	$-\frac{4}{3}$	-1	0	1
$y = 3x + 4$	-5	-2	0	1	4	7
Solution point	$(-3, -5)$	$(-2, -2)$	$\left(-\frac{4}{3}, 0\right)$	$(-1, 1)$	$(0, 4)$	$(1, 7)$

Functions

A *relation* is any set of ordered pairs, which can be thought of as (input, output). A *function* is a relation in which no two ordered pairs have the same first component and different second components.

EXAMPLE 5   **Testing Whether Relations Are Functions**

Decide whether the relation represents a function.

a.

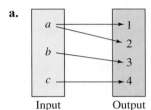

Input   Output

b. Input: 2, 5, 7
Output: 1, 2, 3
$\{(2, 1), (5, 2), (7, 3)\}$

SOLUTION

a. This diagram *does not* represent a function. The first component a is paired with two different second components, 1 and 2.

b. This set of ordered pairs *does* represent a function. No first component has two different second components.

The graph of an equation represents y as a function of x if and only if no vertical line intersects the graph more than once. This is called the *Vertical Line Test*.

EXAMPLE 6   **Using the Vertical Line Test for Functions**

Use the Vertical Line Test to determine whether y is a function of x.

a.

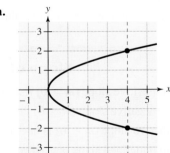

b.

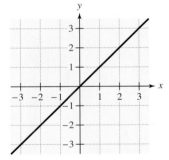

SOLUTION

a. From the graph, you can see that a vertical line intersects more than one point on the graph. So, the relation *does not* represent y as a function of x.

b. From the graph, you can see that no vertical line intersects more than one point on the graph. So, the relation *does* represent y as a function of x.

Slope and Linear Equations

The graph in Figure A.21 on page A18 is an example of a graph of a linear equation. The equation is written in *slope-intercept form*, $y = mx + b$, where m is the slope and $(0, b)$ is the y-intercept. Linear equations can be written in other forms, as shown below.

Forms of Linear Equations

1. General form: $ax + by + c = 0$

2. Slope-intercept form: $y = mx + b$

3. Point-slope form: $y - y_1 = m(x - x_1)$

The *slope* of a nonvertical line is the number of units the line rises or falls vertically for each unit of horizontal change from left to right. To find the slope m of the line through (x_1, y_1) and (x_2, y_2), use the following formula.

$$m = \frac{y_2 - y_1}{x_2 - x_1} = \frac{\text{Change in } y}{\text{Change in } x}$$

EXAMPLE 7   Finding the Slope of a Line Through Two Points

Find the slope of the line passing through $(3, 1)$ and $(-6, 0)$.

SOLUTION

Let $(x_1, y_1) = (3, 1)$ and $(x_2, y_2) = (-6, 0)$. The slope of the line through these points is

$$m = \frac{y_2 - y_1}{x_2 - x_1} = \frac{0 - 1}{-6 - 3} = \frac{-1}{-9} = \frac{1}{9}.$$

The graph of the line is shown in Figure A.22.

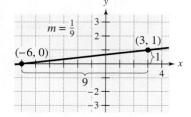

Figure A.22

You can make several generalizations about the slopes of lines.

Slope of a Line

1. A line with positive slope ($m > 0$) rises from left to right.

2. A line with negative slope ($m < 0$) falls from left to right.

3. A line with zero slope ($m = 0$) is horizontal.

4. A line with undefined slope is vertical.

5. Parallel lines have equal slopes: $m_1 = m_2$.

6. Perpendicular lines have negative reciprocal slopes: $m_1 = -\dfrac{1}{m_2}$.

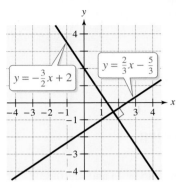

Figure A.23

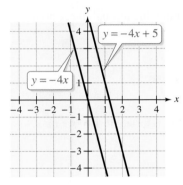

Figure A.24

EXAMPLE 8   Parallel or Perpendicular?

Determine whether the pairs of lines are parallel, perpendicular, or neither.

a. $y = \frac{2}{3}x - \frac{5}{3}$

  $y = -\frac{3}{2}x + 2$

b. $4x + y = 5$

  $-8x - 2y = 0$

SOLUTION

a. The first line has a slope of $m_1 = \frac{2}{3}$ and the second line has a slope of $m_2 = -\frac{3}{2}$. Because these slopes are negative reciprocals of each other, the two lines must be perpendicular, as shown in Figure A.23.

b. To begin, write each equation in slope-intercept form.

$4x + y = 5$	Write first equation.
$y = -4x + 5$	Slope-intercept form

So, the first line has a slope of $m_1 = -4$.

$-8x - 2y = 0$	Write second equation.
$-2y = 8x$	Add 8x to each side.
$y = -4x$	Slope-intercept form

So, the second line has a slope of $m_2 = -4$. Because both lines have the same slope, they must be parallel, as shown in Figure A.24.

You can use the *point-slope form* of the equation of a line to write the equation of a line when you are given its slope and a point on the line.

EXAMPLE 9   Writing an Equation of a Line

Write an equation of the line that passes through the point $(3, 4)$ and has slope $m = -2$.

SOLUTION

Use the point-slope form with $(x_1, y_1) = (3, 4)$ and $m = -2$.

$y - y_1 = m(x - x_1)$	Point-slope form
$y - 4 = -2(x - 3)$	Substitute 4 for y_1, 3 for x_1, and -2 for m.
$y - 4 = -2x + 6$	Simplify.
$y = -2x + 10$	Equation of line

So, an equation of the line in slope-intercept form is $y = -2x + 10$. The graph of this line is shown in Figure A.25.

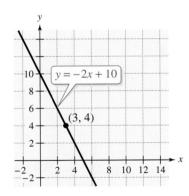

Figure A.25

The point-slope form can also be used to write the equation of a line passing through any two points. To use this form, substitute the formula for slope into the point-slope form, as follows.

$$y - y_1 = m(x - x_1) \qquad \text{Point-slope form}$$

$$y - y_1 = \frac{y_2 - y_1}{x_2 - x_1}(x - x_1) \qquad \text{Substitute formula for slope.}$$

EXAMPLE 10 An Equation of a Line Passing Through Two Points

Write an equation of the line that passes through the points $(5, -1)$ and $(2, 0)$.

SOLUTION

Let $(x_1, y_1) = (5, -1)$ and $(x_2, y_2) = (2, 0)$. The slope of the line through these points is

$$m = \frac{y_2 - y_1}{x_2 - x_1} = \frac{0 - (-1)}{2 - 5} = \frac{1}{-3} = -\frac{1}{3}.$$

Now, use the point-slope form to find an equation of the line.

$$y - y_1 = m(x - x_1) \qquad \text{Point-slope form}$$

$$y - (-1) = -\tfrac{1}{3}(x - 5) \qquad \text{Substitute } -1 \text{ for } y_1,\ 5 \text{ for } x_1,\ \text{and } -\tfrac{1}{3} \text{ for } m.$$

$$y + 1 = -\tfrac{1}{3}x + \tfrac{5}{3} \qquad \text{Simplify.}$$

$$y = -\tfrac{1}{3}x + \tfrac{2}{3} \qquad \text{Equation of line}$$

The graph of this line is shown in Figure A.26.

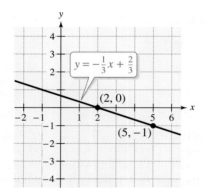

Figure A.26

The slope and y-intercept of a line can be used as an aid when you are sketching a line.

EXAMPLE 11 Using the Slope and y-Intercept to Sketch a Line

Use the slope and y-intercept to sketch the graph of $-x + 2y = -4$.

SOLUTION

First, write the equation in slope-intercept form.

$$-x + 2y = -4 \qquad \text{Write original equation.}$$

$$2y = x - 4 \qquad \text{Add } x \text{ to each side.}$$

$$y = \tfrac{1}{2}x - 2 \qquad \text{Slope-intercept form}$$

So, the slope of the line is $m = \tfrac{1}{2}$ and the y-intercept is $(0, b) = (0, -2)$. Now, plot the y-intercept and locate a second point by using the slope. Because the slope is $m = \tfrac{1}{2}$, move two units to the right and one unit upward from the y-intercept. Then draw a line through the two points, as shown in Figure A.27.

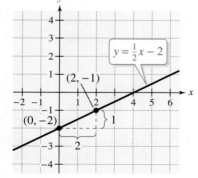

Figure A.27

You know that a horizontal line has a slope of $m = 0$. So, the equation of a horizontal line is $y = b$. A vertical line has an undefined slope, so it has an equation of the form $x = a$.

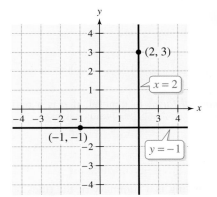

Figure A.28

EXAMPLE 12   Equations of Horizontal and Vertical Lines

a. Write an equation of the horizontal line passing through $(-1, -1)$.

b. Write an equation of the vertical line passing through $(2, 3)$.

SOLUTION

a. The line is horizontal and passes through the point $(-1, -1)$, so every point on the line has a y-coordinate of -1. The equation of the line is $y = -1$.

b. The line is vertical and passes through the point $(2, 3)$ so every point on the line has an x-coordinate of 2. The equation of the line is $x = 2$.

The graphs of these two lines are shown in Figure A.28.

Graphs of Linear Inequalities

The statements $3x - 2y < 6$ and $2x + 3y \geq 1$ are linear inequalities in two variables. An ordered pair (x_1, y_1) is a solution of a linear inequality in x and y if the inequality is true when x_1 and y_1 are substituted for x and y, respectively. The *graph of a linear inequality* is the collection of all solution points of the inequality. To sketch the graph of a linear inequality, begin by sketching the graph of the corresponding linear equation (use a dashed line for $<$ and $>$ and a solid line for $\leq$ and $\geq$). The graph of the equation separates the plane into two regions, called *half-planes*. In each half-plane, either *all* points in the half-plane are solutions of the inequality or *no* point in the half-plane is a solution of the inequality. To determine whether the points in an entire half-plane satisfy the inequality, simply test one point in the region. If the point satisfies the inequality, then shade the entire half-plane to denote that every point in the region satisfies the inequality.

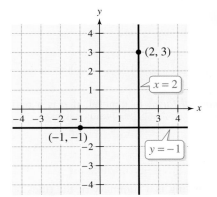

Figure A.29

Study Tip

For more review on graphs and functions, refer to Chapter 4.

EXAMPLE 13   Sketching the Graph of a Linear Inequality

Use the slope-intercept form of a linear equation to graph $-2x + 2y > 7$.

SOLUTION

To begin, rewrite the inequality in slope-intercept form.

$$2y > 2x + 7 \qquad \text{Add } 2x \text{ to each side.}$$

$$y > x + \tfrac{7}{2} \qquad \text{Write in slope-intercept form.}$$

From this form, you can conclude that the solution is the half-plane lying above the line $y = x + \tfrac{7}{2}$. The graph is shown in Figure A.29.

A.5 Exponents and Polynomials

▶ Exponents
▶ Polynomials
▶ Operations with Polynomials

Exponents

Repeated multiplication can be written in *exponential form*. In general, if a is a real number and n is a positive integer, then

$$a^n = \underbrace{a \cdot a \cdot a \cdot \cdots \cdot a}_{n \text{ factors}}$$

where n is the *exponent* and a is the *base*. The following is a summary of the rules of exponents. In Rule 6 below, be sure you see how to use a negative exponent.

Summary of Rules of Exponents

Let m and n be integers, and let a and b be real numbers, variables, or algebraic expressions, such that $a \neq 0$ and $b \neq 0$.

Rule	*Example*
1. *Product Rule:* $a^m a^n = a^{m+n}$	$y^2 \cdot y^4 = y^{2+4} = y^6$
2. *Quotient Rule:* $\dfrac{a^m}{a^n} = a^{m-n}$	$\dfrac{x^7}{x^4} = x^{7-4} = x^3$
3. *Product-to-Power Rule:* $(ab)^m = a^m b^m$	$(5x)^4 = 5^4 x^4 = 625x^4$
4. *Quotient-to-Power Rule:* $\left(\dfrac{a}{b}\right)^m = \dfrac{a^m}{b^m}$	$\left(\dfrac{2}{x}\right)^3 = \dfrac{2^3}{x^3} = \dfrac{8}{x^3}$
5. *Power-to-Power Rule:* $(a^m)^n = a^{mn}$	$(y^3)^{-4} = y^{3(-4)} = y^{-12}$
6. *Negative Exponent Rule:* $a^{-n} = \dfrac{1}{a^n}$	$y^{-4} = \dfrac{1}{y^4}$
7. *Zero Exponent Rule:* $a^0 = 1$	$(x^2 + 1)^0 = 1$

EXAMPLE 1 Using Rules of Exponents

Use the rules of exponents to simplify each expression.

a. $(a^2 b^4)(3ab^{-2})$ **b.** $(2xy^2)^3$ **c.** $3a(-4a^2)^0$ **d.** $\left(\dfrac{4x}{y^3}\right)^3$

SOLUTION

a. $(a^2 b^4)(3ab^{-2}) = (3)(a^2)(a)(b^4)(b^{-2})$ Regroup factors.

$= (3)(a^{2+1})(b^{4-2})$ Apply rules of exponents.

$= 3a^3 b^2$ Simplify.

b. $(2xy^2)^3 = (2)^3(x)^3(y^2)^3$ Apply rules of exponents.

$= 8x^3y^{2 \cdot 3}$ Apply rules of exponents.

$= 8x^3y^6$ Simplify.

c. $3a(-4a^2)^0 = (3a)(-4)^0(a^2)^0$ Apply rules of exponents.

$= 3a(1)(a^{2 \cdot 0})$ Apply rules of exponents.

$= 3a, a \neq 0$ Simplify.

d. $\left(\dfrac{4x}{y^3}\right)^3 = \dfrac{4^3x^3}{(y^3)^3}$ Apply rules of exponents.

$= \dfrac{64x^3}{y^{3 \cdot 3}}$ Apply rules of exponents.

$= \dfrac{64x^3}{y^9}$ Simplify.

EXAMPLE 2   **Rewriting with Positive Exponents**

Use rules of exponents to simplify each expression using only positive exponents. (Assume that no variable is equal to zero.)

a. x^{-1} **b.** $\dfrac{1}{3x^{-2}}$

c. $\dfrac{25a^3b^{-4}}{5a^{-2}b}$ **d.** $\left(\dfrac{2x^{-1}}{xy^0}\right)^{-2}$

SOLUTION

a. $x^{-1} = \dfrac{1}{x}$ Apply rules of exponents.

b. $\dfrac{1}{3x^{-2}} = \dfrac{x^2}{3}$ Apply rules of exponents.

c. $\dfrac{25a^3b^{-4}}{5a^{-2}b} = 5a^{3-(-2)}b^{-4-1}$ Apply rules of exponents.

$= 5a^5b^{-5}$ Simplify.

$= \dfrac{5a^5}{b^5}$ Apply rules of exponents.

d. $\left(\dfrac{2x^{-1}}{xy^0}\right)^{-2} = \dfrac{(2)^{-2}x^2}{x^{-2}y^0}$ Apply rules of exponents.

$= \dfrac{x^2 \cdot x^2}{2^2}$ Simplify.

$= \dfrac{x^4}{4}$ Apply rules of exponents.

It is convenient to write very large or very small numbers in *scientific notation*. This notation has the form $c \times 10^n$, where $1 \le c < 10$ and n is an integer. A positive exponent indicates that the number is large (10 or more) and a negative exponent indicates that the number is small (less than 1).

> **EXAMPLE 3** **Scientific Notation**

Write each number in scientific notation.

a. 0.0000782
b. 836,100,000

SOLUTION

a. $0.0000782 = 7.82 \times 10^{-5}$
b. $836,100,000 = 8.361 \times 10^8$

> **EXAMPLE 4** **Decimal Notation**

Write each number in decimal notation.

a. 9.36×10^{-6}
b. 1.345×10^2

SOLUTION

a. $9.36 \times 10^{-6} = 0.00000936$
b. $1.345 \times 10^2 = 134.5$

Polynomials

The most common type of algebraic expression is the *polynomial*. Some examples are

$$-x + 1, \quad 2x^2 - 5x + 4, \quad \text{and} \quad 3x^3.$$

A polynomial in x is an expression of the form

$$a_n x^n + a_{n-1} x^{n-1} + \cdots + a_2 x^2 + a_1 x + a_0$$

where $a_n, a_{n-1}, \ldots, a_2, a_1, a_0$ are real numbers, n is a nonnegative integer, and $a_n \ne 0$. The polynomial is of *degree n*, a_n is called the *leading coefficient*, and a_0 is called the *constant term*.

Polynomials with one, two, and three terms are called *monomials*, *binomials*, and *trinomials*, respectively. In *standard form*, a polynomial is written with descending powers of x.

> **EXAMPLE 5** **Determining Degrees and Leading Coefficients**

	Polynomial	Standard Form	Degree	Leading Coefficient
a.	$3x - x^2 + 4$	$-x^2 + 3x + 4$	2	-1
b.	-5	-5	0	-5
c.	$8 - 4x^3 + 7x + 2x^5$	$2x^5 - 4x^3 + 7x + 8$	5	2

Operations with Polynomials

You can add and subtract polynomials in much the same way that you add and subtract real numbers. Simply add or subtract the *like terms* (terms having the same variables to the same powers) by adding their coefficients. For instance, $-3xy^2$ and $5xy^2$ are like terms and their sum is

$$-3xy^2 + 5xy^2 = (-3 + 5)xy^2 = 2xy^2.$$

To subtract one polynomial from another, add the opposite by changing the sign of each term of the polynomial that is being subtracted and then adding the resulting like terms. You can add and subtract polynomials using either a horizontal or a vertical format.

EXAMPLE 6   **Adding and Subtracting Polynomials**

a. $3x^2 - 2x + 4$
$\underline{-x^2 + 7x - 9}$
$2x^2 + 5x - 5$

b. $(5x^3 - 7x^2 - 3) + (x^3 + 2x^2 - x + 8)$ Original polynomials

$= (5x^3 + x^3) + (-7x^2 + 2x^2) - x + (-3 + 8)$ Group like terms.

$= 6x^3 - 5x^2 - x + 5$ Combine like terms.

c. $(4x^3 + 3x - 6)$ ⟹ $4x^3 + 3x - 6$
$\underline{-(3x^3 + x + 10)}$ ⟹ $\underline{-3x^3 - x - 10}$
$x^3 + 2x - 16$

d. $(7x^4 - x^2 - x + 2) - (3x^4 - 4x^2 + 3x)$ Original polynomials

$= 7x^4 - x^2 - x + 2 - 3x^4 + 4x^2 - 3x$ Distributive Property

$= (7x^4 - 3x^4) + (-x^2 + 4x^2) + (-x - 3x) + 2$ Group like terms.

$= 4x^4 + 3x^2 - 4x + 2$ Combine like terms.

The simplest type of polynomial multiplication involves a monomial multiplier. The product is obtained by direct application of the Distributive Property.

EXAMPLE 7   **Finding a Product with a Monomial Multiplier**

Find the product of $4x^2$ and $-2x^3 + 3x + 1$.

SOLUTION

$$4x^2(-2x^3 + 3x + 1) = (4x^2)(-2x^3) + (4x^2)(3x) + (4x^2)(1)$$

$$= -8x^5 + 12x^3 + 4x^2$$

To multiply two binomials, use the *FOIL Method* illustrated below.

First
Outer

$$(ax + b)(cx + d) = ax(cx) + ax(d) + b(cx) + b(d)$$

Inner

F   O   I   L

Last

EXAMPLE 8   Using the FOIL Method

Use the FOIL Method to find the product of $x - 3$ and $x - 9$.

SOLUTION

$$
\begin{array}{cccc}
\text{F} & \text{O} & \text{I} & \text{L}
\end{array}
$$
$$(x - 3)(x - 9) = x^2 - 9x - 3x + 27$$
$$= x^2 - 12x + 27 \qquad \text{Combine like terms.}$$

EXAMPLE 9   Using the FOIL Method

Use the FOIL Method to find the product of $2x - 4$ and $x + 5$.

SOLUTION

$$
\begin{array}{cccc}
\text{F} & \text{O} & \text{I} & \text{L}
\end{array}
$$
$$(2x - 4)(x + 5) = 2x^2 + 10x - 4x - 20$$
$$= 2x^2 + 6x - 20 \qquad \text{Combine like terms.}$$

To multiply two polynomials that have three or more terms, you can use the same basic principle that you use when multiplying monomials and binomials. That is, each term of one polynomial must be multiplied by each term of the other polynomial. This can be done using either a vertical or a horizontal format.

EXAMPLE 10   Multiplying Polynomials (Vertical Format)

Multiply $x^2 - 2x + 2$ by $x^2 + 3x + 4$ using a vertical format.

SOLUTION

$$
\begin{array}{r}
x^2 - 2x + 2 \\
\times\ x^2 + 3x + 4 \\
\hline
4x^2 - 8x + 8 \\
3x^3 - 6x^2 + 6x \\
x^4 - 2x^3 + 2x^2 \\
\hline
x^4 + x^3 + 0x^2 - 2x + 8
\end{array}
$$

$\Longleftarrow \quad 4(x^2 - 2x + 2)$

$\Longleftarrow \quad 3x(x^2 - 2x + 2)$

$\Longleftarrow \quad x^2(x^2 - 2x + 2)$

Combine like terms.

So, $(x^2 - 2x + 2)(x^2 + 3x + 4) = x^4 + x^3 - 2x + 8$.

EXAMPLE 11   **Multiplying Polynomials (Horizontal Format)**

$(4x^2 - 3x - 1)(2x - 5)$

$\quad = 4x^2(2x - 5) - 3x(2x - 5) - 1(2x - 5)$ Distributive Property

$\quad = 8x^3 - 20x^2 - 6x^2 + 15x - 2x + 5$ Distributive Property

$\quad = 8x^3 - 26x^2 + 13x + 5$ Combine like terms.

Some binomial products have special forms that occur frequently in algebra. These special products are listed below.

1. Sum and Difference of Two Terms: $(a + b)(a - b) = a^2 - b^2$

2. Square of a Binomial: $(a + b)^2 = a^2 + 2ab + b^2$
$$(a - b)^2 = a^2 - 2ab + b^2$$

EXAMPLE 12   **Finding Special Products**

a. $(3x - 2)(3x + 2) = (3x)^2 - 2^2$ Special product

$\quad\qquad\qquad\qquad = 9x^2 - 4$ Simplify.

b. $(2x - 7)^2 = (2x)^2 - 2(2x)(7) + 7^2$ Special product

$\quad\qquad\qquad = 4x^2 - 28x + 49$ Simplify.

c. $(4a + 5b)^2 = (4a)^2 + 2(4a)(5b) + (5b)^2$ Special product

$\quad\qquad\qquad = 16a^2 + 40ab + 25b^2$ Simplify.

To divide a polynomial by a monomial, separate the original division problem into multiple division problems, each involving the division of a monomial by a monomial.

EXAMPLE 13   **Dividing a Polynomial by a Monomial**

Perform the division and simplify.

$$\frac{7x^3 - 12x^2 + 4x + 1}{4x}$$

SOLUTION

$\dfrac{7x^3 - 12x^2 + 4x + 1}{4x} = \dfrac{7x^3}{4x} - \dfrac{12x^2}{4x} + \dfrac{4x}{4x} + \dfrac{1}{4x}$ Divide each term separately.

$\qquad\qquad\qquad\qquad = \dfrac{7x^2}{4} - 3x + 1 + \dfrac{1}{4x}$ Use rules of exponents.

Polynomial division is similar to long division of integers. To use polynomial long division, write the dividend and divisor in descending powers of the variable, insert placeholders with zero coefficients for missing powers of the variable, and divide as you would with integers. Continue this process until the degree of the remainder is less than that of the divisor.

EXAMPLE 14 Long Division Algorithm for Polynomials

Use the long division algorithm to divide $x^2 + 2x + 4$ by $x - 1$.

SOLUTION

$$\text{Think } \frac{x^2}{x} = x.$$

$$\text{Think } \frac{3x}{x} = 3.$$

$$
\begin{array}{r}
x + 3 \\
x - 1 \overline{)x^2 + 2x + 4} \\
\underline{x^2 - x} \\
3x + 4 \\
\underline{3x - 3} \\
7
\end{array}
$$

Multiply x by $(x - 1)$.
Subtract and bring down 4.
Multiply 3 by $(x - 1)$.
Remainder

Considering the remainder as a fractional part of the divisor, the result is

$$
\underbrace{\frac{\overbrace{x^2 + 2x + 4}^{\text{Dividend}}}{\underbrace{x - 1}_{\text{Divisor}}}}_{} = \overbrace{x + 3}^{\text{Quotient}} + \underbrace{\frac{\overset{\text{Remainder}}{7}}{\underbrace{x - 1}_{\text{Divisor}}}}_{}.
$$

EXAMPLE 15 Accounting for Missing Powers of x

Divide $x^3 - 2$ by $x - 1$.

SOLUTION

Note how the missing x^2- and x-terms are accounted for.

$$
\begin{array}{r}
x^2 + x + 1 \\
x - 1 \overline{)x^3 + 0x^2 + 0x - 2} \\
\underline{x^3 - x^2} \\
x^2 + 0x \\
\underline{x^2 - x} \\
x - 2 \\
\underline{x - 1} \\
-1
\end{array}
$$

Insert $0x^2$ and $0x$.
Multiply x^2 by $(x - 1)$.
Subtract and bring down $0x$.
Multiply x by $(x - 1)$.
Subtract and bring down -2.
Multiply 1 by $(x - 1)$.
Remainder

So, you have $\dfrac{x^3 - 2}{x - 1} = x^2 + x + 1 - \dfrac{1}{x - 1}$.

Synthetic division is a nice shortcut for dividing by polynomials of the form $x - k$.

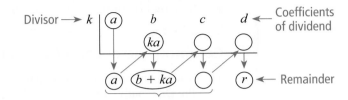

Synthetic Division of a Third-Degree Polynomial

Use synthetic division to divide $ax^3 + bx^2 + cx + d$ by $x - k$, as follows.

Vertical Pattern: Add terms.
Diagonal Pattern: Multiply by k.

EXAMPLE 16   **Using Synthetic Division**

Use synthetic division to divide $x^3 - 6x^2 + 4$ by $x - 3$.

SOLUTION

You should set up the array as follows. Note that a zero is included for the missing x-term in the dividend.

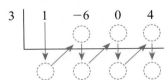

Then, use the synthetic division pattern by adding terms in columns and multiplying the results by 3.

Divisor: $x - 3$ Dividend: $x^3 - 6x^2 + 4$

$$
\begin{array}{r|rrrr}
3 & 1 & -6 & 0 & 4 \\
  &   & 3 & -9 & -27 \\
\hline
  & 1 & -3 & -9 & \boxed{-23} \\
\end{array}
$$

← Remainder: -23

Quotient: $x^2 - 3x - 9$

So, you have

$$\frac{x^3 - 6x^2 + 4}{x - 3} = x^2 - 3x - 9 - \frac{23}{x - 3}.$$

Study Tip

For more review on exponents and polynomials, refer to Chapter 5.

A.6 Factoring and Solving Equations

▶ Common Factors and Factoring by Grouping
▶ Factoring Trinomials
▶ Factoring Special Polynomial Forms
▶ Solving Polynomial Equations by Factoring

Common Factors and Factoring by Grouping

The process of writing a polynomial as a product is called *factoring*. Previously, you used the Distributive Property to *multiply* and *remove* parentheses. Now, you will use the Distributive Property in the reverse direction to *factor* and *create* parentheses.

Removing the common monomial factor is the first step in completely factoring a polynomial. When you use the Distributive Property to remove this factor from each term of the polynomial, you are factoring out the greatest common monomial factor.

EXAMPLE 1   **Common Monomial Factors**

Factor out the greatest common monomial factor from each polynomial.

a. $3x + 9$ **b.** $6x^3 - 4x$ **c.** $-4y^2 + 12y - 16$

SOLUTION

a. $3x + 9 = 3(x) + 3(3) = 3(x + 3)$

Greatest common monomial factor is 3.

b. $6x^3 - 4x = 2x(3x^2) - 2x(2) = 2x(3x^2 - 2)$

Greatest common monomial factor is $2x$.

c. $-4y^2 + 12y - 16 = -4(y^2) + (-4)(-3y) + (-4)(4)$

Greatest common monomial is -4.

$$= -4(y^2 - 3y + 4)$$

Factor -4 out of each term.

Some expressions have common factors that are not simple monomials. For instance, the expression $2x(x - 2) + 3(x - 2)$ has the common binomial factor $(x - 2)$. Factoring out this common factor produces

$$2x(x - 2) + 3(x - 2) = (x - 2)(2x + 3).$$

This type of factoring is called *factoring by grouping*.

EXAMPLE 2   **Common Binomial Factor**

Factor $7a(3a + 4b) + 2(3a + 4b)$.

SOLUTION

Each of the terms of this expression has a binomial factor of $(3a + 4b)$.

$$7a(3a + 4b) + 2(3a + 4b) = (3a + 4b)(7a + 2)$$

In Example 2, the expression was already grouped, and so it was easy to determine the common binomial factor. In practice, you will have to do the grouping *and* the factoring.

EXAMPLE 3   Factoring by Grouping

Factor $x^3 - 5x^2 + x - 5$.

SOLUTION

$$x^3 - 5x^2 + x - 5 = (x^3 - 5x^2) + (x - 5) \qquad \text{Group terms.}$$

$$= x^2(x - 5) + 1(x - 5) \qquad \text{Factoring out common monomial factor in each group.}$$

$$= (x - 5)(x^2 + 1) \qquad \text{Factored form}$$

Factoring Trinomials

To factor a trinomial $x^2 + bx + c$ into a product of two binomials, you must find two numbers m and n whose product is c and whose sum is b. If c is positive, then m and n have like signs that match the sign of b. If c is negative, then m and n have unlike signs. If $|b|$ is small relative to $|c|$, first try those factors of c that are close to each other in absolute value.

EXAMPLE 4   Factoring Trinomials

Factor each trinomial.

a. $x^2 - 7x + 12$

b. $x^2 - 2x - 8$

SOLUTION

a. You need to find two numbers whose product is 12 and whose sum is -7.

The product of -3 and -4 is 12.

$$x^2 - 7x + 12 = (x - 3)(x - 4)$$

The sum of -3 and -4 is -7.

b. You need to find two numbers whose product is -8 and whose sum is -2.

The product of -4 and 2 is -8.

$$x^2 - 2x - 8 = (x - 4)(x + 2)$$

The sum of -4 and 2 is -2.

Applications of algebra sometimes involve trinomials that have a common monomial factor. To factor such trinomials completely, first factor out the common monomial factor. Then try to factor the resulting trinomial by the methods given in this section.

EXAMPLE 5   **Factoring Completely**

Factor the trinomial $5x^3 + 20x^2 + 15x$ completely.

SOLUTION

$$5x^3 + 20x^2 + 15x = 5x(x^2 + 4x + 3)$$ Factor out common monomial factor $5x$.

$$= 5x(x + 1)(x + 3)$$ Factor trinomial.

To factor a trinomial whose leading coefficient is not 1, use the following pattern.

Factors of a

$$ax^2 + bx + c = \left(x + \right)\left(x + \right)$$

Factors of c

Use the following guidelines to help shorten the list of possible factorizations of a trinomial.

Guidelines for Factoring $ax^2 + bx + c$ ($a > 0$)

1. When the trinomial has a common monomial factor, you should factor out the common factor before trying to find the binomial factors.

2. Because the resulting trinomial has no common monomial factors, you do not have to test any binomial factors that have a common monomial factor.

3. Switch the signs of the factors of c when the middle term $(O + I)$ is correct except in sign.

EXAMPLE 6   **Factoring a Trinomial of the Form $ax^2 + bx + c$**

Factor the trinomial $6x^2 + 17x + 5$.

SOLUTION

First, observe that $6x^2 + 17x + 5$ has no common monomial factor. For this trinomial, $a = 6$, which factors as $(1)(6)$ or $(2)(3)$, and $c = 5$, which factors as $(1)(5)$.

$$(x + 1)(6x + 5) = 6x^2 + 11x + 5$$

$$(x + 5)(6x + 1) = 6x^2 + 31x + 5$$

$$(2x + 1)(3x + 5) = 6x^2 + 13x + 5$$

$$(2x + 5)(3x + 1) = 6x^2 + 17x + 5 \quad \Longleftarrow \quad \text{Correct factorization}$$

So, the correct factorization is $6x^2 + 17x + 5 = (2x + 5)(3x + 1)$.

EXAMPLE 7   **Factoring a Trinomial of the Form** $ax^2 + bx + c$

Factor the trinomial.

$$3x^2 - 16x - 35$$

SOLUTION

First, observe that $3x^2 - 16x - 35$ has no common monomial factor. For this trinomial,

$$a = 3 \text{ and } c = -35.$$

The possible factorizations of this trinomial are as follows.

$$(3x - 1)(x + 35) = 3x^2 + 104x - 35$$

$$(3x - 35)(x + 1) = 3x^2 - 32x - 35$$

$$(3x - 5)(x + 7) = 3x^2 + 16x - 35 \qquad \text{Middle term has opposite sign.}$$

$$(3x + 5)(x - 7) = 3x^2 - 16x - 35 \quad \Longleftarrow \quad \text{Correct factorization}$$

So, the correct factorization is

$$3x^2 - 16x - 35 = (3x + 5)(x - 7).$$

EXAMPLE 8   **Factoring Completely**

Factor the trinomial completely.

$$6x^2y + 16xy + 10y$$

SOLUTION

Begin by factoring out the common monomial factor $2y$.

$$6x^2y + 16xy + 10y = 2y(3x^2 + 8x + 5)$$

Now, for the new trinomial $3x^2 + 8x + 5$,

$$a = 3 \text{ and } c = 5.$$

The possible factorizations of this trinomial are as follows.

$$(3x + 1)(x + 5) = 3x^2 + 16x + 5$$

$$(3x + 5)(x + 1) = 3x^2 + 8x + 5 \quad \Longleftarrow \quad \text{Correct factorization}$$

So, the correct factorization is

$$6x^2y + 16xy + 10y = 2y(3x^2 + 8x + 5) = 2y(3x + 5)(x + 1).$$

Factoring a trinomial can involve quite a bit of trial and error. Some of this trial and error can be lessened by using factoring by grouping. The key to this method of factoring is knowing how to rewrite the middle term. In general, to factor a trinomial $ax^2 + bx + c$ by grouping, choose factors of the product ac that add up to b and use these factors to rewrite the middle term. This technique is illustrated in Example 9.

EXAMPLE 9   **Factoring a Trinomial by Grouping**

Use factoring by grouping to factor the trinomial $6y^2 + 5y - 4$.

SOLUTION

In the trinomial $6y^2 + 5y - 4$, $a = 6$ and $c = -4$, which implies that the product of ac is -24. Now, because -24 factors as $(8)(-3)$, and $8 - 3 = 5 = b$, you can rewrite the middle term as $5y = 8y - 3y$. This produces the following result.

$$6y^2 + 5y - 4 = 6y^2 + 8y - 3y - 4 \qquad \text{Rewrite middle term.}$$
$$= (6y^2 + 8y) - (3y + 4) \qquad \text{Group terms.}$$
$$= 2y(3y + 4) - (3y + 4) \qquad \text{Factor out common monomial factor in first group.}$$
$$= (3y + 4)(2y - 1) \qquad \text{Distributive Property}$$

So, the trinomial factors as $6y^2 + 5y - 4 = (3y + 4)(2y - 1)$.

Factoring Special Polynomial Forms

Some polynomials have special forms. You should learn to recognize these forms so that you can factor such polynomials easily.

Factoring Special Polynomial Forms

Let a and b be real numbers, variables, or algebraic expressions.
1. Difference of Two Squares: $a^2 - b^2 = (a + b)(a - b)$
2. Perfect Square Trinomial: $a^2 + 2ab + b^2 = (a + b)^2$
$$a^2 - 2ab + b^2 = (a - b)^2$$
3. Sum or Difference of Two Cubes: $a^3 + b^3 = (a + b)(a^2 - ab + b^2)$
$$a^3 - b^3 = (a - b)(a^2 + ab + b^2)$$

EXAMPLE 10   **Factoring the Difference of Two Squares**

Factor each polynomial.

a. $x^2 - 144$

b. $4a^2 - 9b^2$

SOLUTION

a. $x^2 - 144 = x^2 - 12^2$ Write as difference of two squares.

$= (x + 12)(x - 12)$ Factored form

b. $4a^2 - 9b^2 = (2a)^2 - (3b)^2$ Write as difference of two squares.

$= (2a + 3b)(2a - 3b)$ Factored form

To recognize perfect square terms, look for coefficients that are squares of integers and for variables raised to even powers.

EXAMPLE 11   Factoring Perfect Square Trinomials

Factor each trinomial.

a. $x^2 - 10x + 25$

b. $4y^2 + 4y + 1$

SOLUTION

a. $x^2 - 10x + 25 = x^2 - 2(5x) + 5^2$ Recognize the pattern.

$= (x - 5)^2$ Write in factored form.

b. $4y^2 + 4y + 1 = (2y)^2 + 2(2y)(1) + 1^2$ Recognize the pattern.

$= (2y + 1)^2$ Write in factored form.

EXAMPLE 12   Factoring Sum or Difference of Two Cubes

Factor each polynomial.

a. $x^3 + 1$

b. $27x^3 - 64y^3$

SOLUTION

a. $x^3 + 1 = x^3 + 1^3$ Write as sum of two cubes.

$= (x + 1)[x^2 - (x)(1) + 1^2]$ Factored form

$= (x + 1)(x^2 - x + 1)$ Simplify.

b. $27x^3 - 64y^3 = (3x)^3 - (4y)^3$ Write as difference of two cubes.

$= (3x - 4y)[(3x)^2 + (3x)(4y) + (4y)^2]$ Factored form

$= (3x - 4y)(9x^2 + 12xy + 16y^2)$ Simplify.

Solving Polynomial Equations by Factoring

A *quadratic equation* is an equation that can be written in the general form

$$ax^2 + bx + c = 0, \quad a \neq 0.$$

You can combine your factoring skills with the *Zero-Factor Property* to solve quadratic equations.

Zero-Factor Property

Let a and b be real numbers, variables, or algebraic expressions. If a and b are factors such that

$$ab = 0$$

then $a = 0$ or $b = 0$. This property also applies to three or more factors.

In order for the Zero-Factor Property to be used, a quadratic equation must be written in general form.

EXAMPLE 13 **Using Factoring to Solve a Quadratic Equation**

Solve the equation.

$$x^2 - x - 12 = 0$$

SOLUTION

First, check to see that the right side of the equation is zero. Next, factor the left side of the equation. Finally, apply the Zero-Factor Property to find the solutions.

$x^2 - x - 12 = 0$	Write original equation.
$(x + 3)(x - 4) = 0$	Factor left side of equation.
$x + 3 = 0 \quad \Rightarrow \quad x = -3$	Set 1st factor equal to 0 and solve for x.
$x - 4 = 0 \quad \Rightarrow \quad x = 4$	Set 2nd factor equal to 0 and solve for x.

So, the equation has two solutions: $x = -3$ and $x = 4$.

Remember to check your solutions in the original equation, as follows.

Check

First Solution

$x^2 - x - 12 = 0$	Write original equation.
$(-3)^2 - (-3) - 12 \stackrel{?}{=} 0$	Substitute -3 for x.
$9 + 3 - 12 \stackrel{?}{=} 0$	Simplify.
$0 = 0$	Solution checks. ✔

Second Solution

$x^2 - x - 12 = 0$	Write original equation.
$4^2 - 4 - 12 \stackrel{?}{=} 0$	Substitute 4 for x.
$16 - 4 - 12 \stackrel{?}{=} 0$	Simplify.
$0 = 0$	Solution checks. ✔

EXAMPLE 14   **Using Factoring to Solve a Quadratic Equation**

Solve $2x^2 - 3 = 7x + 1$.

SOLUTION

$$2x^2 - 3 = 7x + 1$$ Write original equation.

$$2x^2 - 7x - 4 = 0$$ Write in general form.

$$(2x + 1)(x - 4) = 0$$ Factor.

$$2x + 1 = 0 \implies x = -\tfrac{1}{2}$$ Set 1st factor equal to 0 and solve for x.

$$x - 4 = 0 \implies x = 4$$ Set 2nd factor equal to 0 and solve for x.

So, the equation has two solutions: $x = -\tfrac{1}{2}$ and $x = 4$. Check these in the original equation, as follows.

Check

First Solution

$$2\left(-\tfrac{1}{2}\right)^2 - 3 \overset{?}{=} 7\left(-\tfrac{1}{2}\right) + 1$$ Substitute $-\tfrac{1}{2}$ for x in original equation.

$$\tfrac{1}{2} - 3 \overset{?}{=} -\tfrac{7}{2} + 1$$ Simplify.

$$-\tfrac{5}{2} = -\tfrac{5}{2}$$ Solution checks. ✔

Second Solution

$$2(4)^2 - 3 \overset{?}{=} 7(4) + 1$$ Substitute 4 for x in original equation.

$$32 - 3 \overset{?}{=} 28 + 1$$ Simplify.

$$29 = 29$$ Solution checks. ✔

The Zero-Factor Property can be used to solve polynomial equations of degree 3 or higher. To do this, use the same strategy you used with quadratic equations.

Study Tip

The solution $x = -6$ in Example 15 is called a *repeated solution*.

Study Tip

For more review on factoring and solving equations, refer to Chapter 6.

EXAMPLE 15   **Solving a Polynomial Equation with Three Factors**

Solve $x^3 + 12x^2 + 36x = 0$.

SOLUTION

$$x^3 + 12x^2 + 36x = 0$$ Write original equation.

$$x(x^2 + 12x + 36) = 0$$ Factor out common monomial factor.

$$x(x + 6)(x + 6) = 0$$ Factor perfect square trinomial.

$$x = 0$$ Set 1st factor equal to 0.

$$x + 6 = 0 \implies x = -6$$ Set 2nd factor equal to 0 and solve for x.

$$x + 6 = 0 \implies x = -6$$ Set 3rd factor equal to 0 and solve for x.

Note that even though the left side of the equation has three factors, two of the factors are the same. So, you conclude that the solutions of the equation are $x = 0$ and $x = -6$. Check these in the original equation.

Answers to Odd-Numbered Exercises

CHAPTER 1

Section 1.1 *(pp. 2–9)*

1. (a) $20, \frac{9}{3}$ (b) $-3, 20, \frac{9}{3}$
   (c) $-3, 20, -\frac{3}{2}, \frac{9}{3}, 4.5$ (d) $\pi, -\sqrt{3}$

3. (a) none   (b) $-\sqrt{25}, -\frac{5}{1}, 0, -12$
   (c) $-\sqrt{25}, -\frac{5}{1}, 9.4, 0, -12, \frac{7}{14}$ (d) $\sqrt{7}$

5. 3   **7.** 0   **9.** 1.5 or $\frac{3}{2}$

11.

13.

15.

17. $>$

19. $<$

21. $<$

23. $>$

25. $<$ **27.** $>$

29. Your scores decreased as you played more rounds.

31. 2   **33.** 8   **35.** 10   **37.** 3   **39.** 3.4   **41.** $\frac{7}{2}$

43. -23.6 **45.** 0   **47.** $=$ **49.** $>$ **51.** $<$

53. 456 m

55. 49.12; As a change in the account balance, the amount is negative, as a payment, the amount is positive.

57. -2 is to the right of -2.5.

59. The fractions are converted to decimals and plotted on a number line to determine the order.

61.

63.

65. $-4, 20$ **67.** $15.3, 27.3$ **69.** $-5.5, 1.5$

71. *Sample answer:* $\sqrt{2}, \pi, -\sqrt{3}$

73. *Sample answer:* $-7, 1, 341$

75. *Sample answer:* $-\sqrt{7}, 0, -\frac{1}{3}$

77. $n \geq 0$

79. (a) Lo`ihi and Ruby   (b) Ruby   (c) Mauna Loa

81. The number 4 is plotted four units to the right of 0, and the number -4 is plotted four units to the left of 0.

83. The number $\frac{8}{4}$ is a natural number because it equals 2, and $\frac{7}{4}$ is not a natural number because it equals 1.75.

85. $3; \left|3 - (-4)\right| > \left|-10 - (-4)\right|$

87. False. $\left|0\right| = 0$ **89.** True. $\left|\frac{x}{y}\right| = \frac{x}{y}$

91. True. This is the definition of opposite.

Section 1.2 *(pp. 10–17)*

1.

3.

5.

7.

9.

11.

13. Yes   **15.** 23°C   **17.** 16   **19.** 0   **21.** 0

23. -27 **25.** -27 **27.** 6   **29.** 25   **31.** -5

33. 363   **35.** 726   **37.** 38   **39.** -10 **41.** -5

43. 300   **45.** -233 **47.** \$109   **49.** 67 ft   **51.** 3

53. 26   **55.** 5   **57.** 25   **59.** 36   **61.** -30

63. -24 **65.** -109 **67.** -9 **69.** -11 **71.** -21

73. 0   **75.** -6 **77.** -103 **79.** -610 **81.** -80

83. -12 **85.** 17   **87.** -2 **89.** 7000 ft

91. $3 \boxed{+/-} \boxed{-} 7 \boxed{=}$
   $\boxed{(-)} 3 \boxed{-} 7 \boxed{\text{ENTER}}$
   -10

93. $6 \boxed{+} 5 \boxed{-} \boxed{(} 7 \boxed{+/-} \boxed{)} \boxed{=}$
   $6 \boxed{+} 5 \boxed{-} \boxed{(} \boxed{(-)} 7 \boxed{)} \boxed{\text{ENTER}}$
   18

95. $6 \boxed{+/-} \boxed{+} \boxed{(} 2 \boxed{+/-} \boxed{)} \boxed{-} 5 \boxed{=}$
   $\boxed{(-)} 6 \boxed{+} \boxed{(} \boxed{(-)} 2 \boxed{)} \boxed{-} 5 \boxed{\text{ENTER}}$
   -13

97. To add two integers with unlike signs, you subtract the smaller absolute value from the larger absolute value; You attach the sign of the integer with the larger absolute value.

99. To add two integers with like signs, add their absolute values and attach the common sign to the result.

101. -15 **103.** -36 **105.** 1371 m

107. (a)

Day	Daily Gain or Loss
Tuesday	\$5
Wednesday	\$8
Thursday	$-\$5$
Friday	\$16

   (b) \$11
   (c) \$24; The stock gained \$24 in value during the week; Find the difference between the first bar (Monday) and the last bar (Friday).

109. (a) $3 + 2 = 5$
   (b) Adding two integers with like signs

111. To add two negative integers, add their absolute values and attach the negative sign.

113. No. To add two positive integers, add their absolute values and attach the common sign, which is always the positive sign.

Section 1.3 *(pp. 18–27)*

1. 35   **3.** 0   **5.** -32 **7.** -36 **9.** -690

11. 91   **13.** 1600   **15.** -90 **17.** 21   **19.** -30

21. 12   **23.** 90   **25.** 338   **27.** -4725 **29.** 260

31. 9009   **33.** 57,600 ft^2 **35.** 180 in.^{3}

37. 3   **39.** -6 **41.** -7 **43.** 7

45. Division by zero is undefined.   **47.** 0   **49.** 27

51. −7   **53.** 32   **55.** −160   **57.** −82   **59.** 331

61. 713   **63.** −1045   **65.** 5 mi/sec

67. (a) 82

     (b) 73   77   82   87   91

          72   76   80   84   88   92

     (c) 5, −9, −5, 9; Sum is 0; Explanations will vary.

69. Prime   **71.** Composite   **73.** Prime

75. Composite   **77.** Composite   **79.** Prime

81. 11 is prime.   **83.** $2 \cdot 2$ **85.** $2 \cdot 2 \cdot 2 \cdot 2$

87. 37 is prime.   **89.** $2 \cdot 2 \cdot 3$ **91.** $3 \cdot 11 \cdot 17$

93. $2 \cdot 3 \cdot 5 \cdot 7$ **95.** $2 \cdot 2 \cdot 2 \cdot 2 \cdot 2 \cdot 2 \cdot 3$

97. $3 \cdot 5 \cdot 13 \cdot 13$

99. 1 row and 24 columns, 2 rows and 12 columns,
3 rows and 8 columns, 4 rows and 6 columns,
6 rows and 4 columns, 8 rows and 3 columns,
12 rows and 2 columns, 24 rows and 1 column

101. 3   **103.** 0

105. Example: $1 \cdot (-4) = -4$; Algebraic description: If a is a real number, then $1 \cdot a = a$.

107. $-1(0) = 0$
$-2(0) = 0$
The product of an integer and zero is 0.

109. $|0| = 0$
$|-1| = 1$
$|-2| = 2$
$|a| = \begin{cases} a, & \text{if } a \geq 0 \\ -a, & \text{if } a < 0 \end{cases}$

111. Unlike signs

113. Positive; Because −6 and −4 have like (negative) signs, the product is positive.

115. −120   **117.** 40   **119.** 840   **121.** −2520

123. −192   **125.** 0   **127.** Yes   **129.** No   **131.** Yes

133. Yes   **135.** No   **137.** Yes   **139.** −24°

141. −$0.84   **143.** 21 ft

145. 2; It is divisible only by 1 and itself. Any other even number is divisible by 1, itself, and 2.

147. $(2m)n = 2(mn)$. The product of two odd integers is odd.

149. The product is negative because there is an odd number of negative factors.

151. 5 and 7; 11 and 13; 17 and 19; 29 and 31; 41 and 43; 59 and 61; 71 and 73

153. 114, 115, 116, 117, 118, 119, 120, 121, 122, 123

155. (a) 01 02 03 04 05 06 07 08 09 10
11 12 13 14 15 16 17 18 19 20
21 22 23 24 25 26 27 28 29 30
31 32 33 34 35 36 37 38 39 40
41 42 43 44 45 46 47 48 49 50
51 52 53 54 55 56 57 58 59 60
61 62 63 64 65 66 67 68 69 70
71 72 73 74 75 76 77 78 79 80
81 82 83 84 85 86 87 88 89 90
91 92 93 94 95 96 97 98 99 100

     (b) Prime numbers; The multiples of 2, 3, 5, and 7, other than the numbers themselves, cannot be prime because they have 2, 3, 5, and 7 as factors.

Mid-Chapter Quiz   *(p. 28)*

1. <

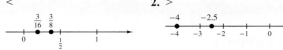

2. >

3. <   **4.** >

5. −0.75   **6.** $\frac{17}{19}$ **7.** =   **8.** >   **9.** −9   **10.** 28

11. 99   **12.** −53   **13.** −27   **14.** −25   **15.** −50

16. −62   **17.** 8   **18.** −5   **19.** −60   **20.** 91

21. 15   **22.** −4   **23.** Prime

24. Composite; $7 \cdot 13$ **25.** Composite; $3 \cdot 37$

26. Composite; $2 \cdot 2 \cdot 2 \cdot 2 \cdot 3 \cdot 3$

27. $450,450   **28.** 128 ft³   **29.** 15 ft   **30.** $367

Section 1.4   *(pp. 30–39)*

1. 5   **3.** 5   **5.** 45   **7.** 6   **9.** $\frac{1}{2}$ **11.** $\frac{4}{5}$

13. $\frac{5}{16}$ **15.** $\frac{2}{25}$ **17.** 6   **19.** 10

21. $\frac{2}{3}$; 0.6666…; No   **23.** $\frac{3}{5}$; 0.6; Yes   **25.** $\frac{8}{15}$

27. 4   **29.** $\frac{3}{8}$ **31.** $-\frac{1}{2}$ **33.** $\frac{5}{6}$ **35.** $\frac{9}{16}$

37. $\frac{1}{2}$ **39.** $\frac{7}{20}$ **41.** $\frac{55}{6}$ **43.** $-\frac{17}{16}$ **45.** $-\frac{21}{4}$

47. $-\frac{39}{4}$ **49.** $\frac{17}{48}$ **51.** $\frac{3}{10}$ **53.** $\frac{3}{8}$ **55.** $-\frac{10}{21}$

57. $-\frac{3}{8}$ **59.** $\frac{1}{3}$ **61.** $\frac{3}{16}$ **63.** $\frac{12}{5}$ **65.** $\frac{1}{7}$; $7 \cdot \frac{1}{7} = 1$

67. $\frac{7}{4}$; $\frac{4}{7} \cdot \frac{7}{4} = 1$ **69.** $\frac{1}{2}$ **71.** $-\frac{8}{27}$ **73.** 1

75. $\frac{3}{7}$ **77.** $\frac{25}{24}$ **79.** −90

81. 2150 cal; $716\frac{2}{3}$ cal/day

83. $\frac{3}{5}$ hr or 36 min

85. 27.09   **87.** 4.1302   **89.** 106.65   **91.** 6.123

93. −2.128   **95.** 2.27   **97.** −57.02   **99.** 4.30

101. 39.08   **103.** −0.51   **105.** $4.71   **107.** $1872.11

109. Use the LCD to rewrite the two fractions so that they have like denominators. Then add the numerators and write the result over the LCD.

111. Use the decimal points to line up the digits of the decimals according to place value. Then add vertically down each column starting from the right and carrying when needed.

113. $\frac{13}{60}$ **115.** −1   **117.** $\frac{2}{5}$ **119.** $-\frac{7}{24}$ **121.** $\frac{7}{5}$

123. $\frac{121}{12}$ **125.** $-\frac{51}{2}$ **127.** $\frac{27}{40}$ **129.** 1

131. Division by zero is undefined.   **133.** $\frac{5}{2}$ **135.** −1.90

137. −63.22   **139.** $1.15   **141.** No; $-\frac{1}{2} + \left(-\frac{3}{4}\right) = -\frac{5}{4}$

143. The product of two fractions with like signs is positive. The product of two fractions with different signs is negative.

145. 12; $3 \div \frac{1}{4} = 12$

147. 43.6; 42.12
The first method produces the more accurate answer because you round only the answer, while in the second method, you round each dimension before you multiply to get the answer.

149. True. The reciprocal of a rational number can always be written as a ratio of two integers.

151. False. $\frac{1}{2} \cdot \frac{1}{3} = \frac{1}{6}$

153. False. If $u = 1$ and $v = 2$, then $u - v = 1 - 2 = -1 \not> 0$.

155. $\frac{4}{5} + \frac{3}{6} = \frac{13}{10}$

Section 1.5 *(pp. 40–47)*

1. 2^6 **3.** 9 **5.** 64 **7.** $\frac{1}{64}$ **9.** -125

11. -16 **13.** -1.728 **15.** 512 boxes

17. Stage 6

Stage	Emails sent, as a power	Emails sent
1	5^1	5
2	5^2	25
3	5^3	125
4	5^4	625
5	5^5	3125
6	5^6	15,625

19. 8 **21.** 12 **23.** 27 **25.** 17 **27.** 9

29. $-\frac{11}{2}$ **31.** 36 **33.** 8 **35.** 68 **37.** 17

39. 33 **41.** $\frac{7}{3}$ **43.** 21 **45.** $\frac{7}{80}$ **47.** $\frac{5}{6}$ **49.** $-\frac{1}{8}$

51. 4 **53.** Division by zero is undefined. **55.** 0

57. 13 **59.** Associative Property of Addition

61. Commutative Property of Multiplication

63. Commutative Property of Addition

65. Distributive Property

67. Multiplicative Inverse Property

69. Additive Identity Property

71. $-3(10)$ **73.** $(18 + 12) + 9$

75. (a) $30(30 - 8)$ (b) $30(30) - 30(8)$ (c) 660 ft^2

77. (a) $2 \cdot 2 + 2 \cdot 3 = 4 + 6 = 10$
(b) $2 \cdot 5 = 10$
(c) Explanations will vary.

79. base: 2, exponent: 4

81. Added inside **P**arentheses, evaluated the **E**xponential expression, **M**ultiplied, and **A**dded.

83. -1 **85.** 10 **87.** 366.12 **89.** 10.69

91. (a) -50 (b) $\frac{1}{50}$

93. $8 + 2 + 6 \cdot 2 - 4 + 2 \cdot 3 = 24$

95. No. $-6^2 = -(6 \cdot 6) = -36$; $(-6)^2 = (-6)(-6) = 36$

97. $24^2 = (4 \cdot 6)^2 = 4^2 \cdot 6^2$ **99.** $4 - (6 - 2) = 4 - 6 + 2$

101. $100 \div 2 \times 50 = 50 \times 50 = 2500$

103. $5(7 + 3) = 5(7) + 5(3)$

105. Division by zero is undefined.

107. Fraction was simplified incorrectly.
$$-9 + \frac{9 + 20}{3(5)} - (-3) = -9 + \frac{29}{15} + 3$$
$$= -6 + \frac{29}{15}$$
$$= \frac{-90 + 29}{15}$$
$$= -\frac{61}{15}$$

109.

Expression	Value
$(6 + 2) \cdot (5 + 3)$	$= 64$
$(6 + 2) \cdot 5 + 3$	$= 43$
$6 + 2 \cdot 5 + 3$	$= 19$
$6 + 2 \cdot (5 + 3)$	$= 22$

111. $8 \cdot 15 - 8 \cdot 6 = 120 - 48 = 72$ **113.** No
$8(15 - 6) = 8(9) = 72$
Explanations will vary.

Review Exercises *(pp. 50–53)*

1. (a) $\sqrt{4}$ (b) $-1, \sqrt{4}$
(c) $-1, 4.5, \frac{2}{5}, -\frac{1}{7}, \sqrt{4}$ (d) $\sqrt{5}$

3. (a) $\frac{30}{2}, 2$ (b) $\frac{30}{2}, 2$
(c) $\frac{30}{2}, 2, 1.5, -\frac{10}{7}$ (d) $-\sqrt{3}, -\pi$

5.  **7.**

9.

11. $<$ **13.** $>$

15. $>$

17. 152 **19.** $\frac{7}{3}$ **21.** 8.5 **23.** 3.4

25. -6.2 **27.** $-\frac{8}{5}$ **29.** $=$ **31.** $>$

33. $>$ **35.** $-2, 12$ **37.** $-2.4, 7.6$

39. 7 **41.** -5

43. 11 **45.** -95 **47.** -89 **49.** 5 **51.** -29

53. $82,400 **55.** 21 **57.** -7 **59.** 33 **61.** -22

63. -9 **65.** \$765 **67.** 45 **69.** -72 **71.** -48

73. 45 **75.** -54 **77.** -40 **79.** \$3600 **81.** 9

83. -12 **85.** -15 **87.** 13 **89.** 0

91. Division by zero is undefined. **93.** 65 mi/hr

95. Prime **97.** Prime **99.** Composite

101. $2 \cdot 2 \cdot 2 \cdot 3 \cdot 11$ **103.** $2 \cdot 3 \cdot 3 \cdot 3 \cdot 7$

105. $2 \cdot 2 \cdot 13 \cdot 31$ **107.** -36 **109.** 7 **111.** 18

113. 1 **115.** 21 **117.** $\frac{1}{4}$ **119.** $\frac{5}{8}$ **121.** 10

123. 15 **125.** $\frac{2}{5}$ **127.** $\frac{3}{4}$ **129.** $\frac{7}{8}$ **131.** $\frac{7}{8}$

133. $-\frac{103}{96}$ **135.** $\frac{5}{4}$ **137.** $\frac{17}{8}$ **139.** $2\frac{3}{4}$ in.

141. $-\frac{1}{12}$ **143.** 1 **145.** $-\frac{1}{36}$ **147.** $\frac{2}{3}$ **149.** $\frac{6}{7}$

151. Division by zero is undefined. **153.** 0

155. $\frac{27}{32}$ in./hr **157.** 5.65 **159.** -1.38

161. -0.75 **163.** 21 **165.** \$947.75 **167.** 6^5

169. $\left(\frac{6}{7}\right)^4$ **171.** 16 **173.** $-\frac{27}{64}$ **175.** -49

177. 6 **179.** 21 **181.** 52 **183.** 160 **185.** 81

187. $\frac{37}{8}$ **189.** 140 **191.** -3 **193.** 7 **195.** 0

197. 796.11 **199.** 1841.74

201. (a) \$10,546.88 (b) \$14,453.12

203. Additive Inverse Property

205. Commutative Property of Multiplication

207. Multiplicative Identity Property

209. Distributive Property

211. -16 **213.** $1 + 24$

215. $6 \cdot 18 - 6 \cdot 5 = 108 - 30 = 78$
$6(18 - 5) = 6(13) = 78$
Explanations will vary.

Chapter Test   (p. 54)

1. (a) 4   (b) 4, -6, 0   (c) 4, -6, $\frac{1}{2}$, 0, $\frac{7}{9}$ (d) π

2. $>$ **3.** 13   **4.** -6.8 **5.** -4 **6.** 10   **7.** 10

8. 47   **9.** -160 **10.** 8   **11.** -30 **12.** 1

13. $\frac{17}{24}$ **14.** $\frac{2}{15}$ **15.** $\frac{7}{12}$ **16.** -27 **17.** -0.64

18. 33   **19.** 235   **20.** -2 **21.** Distributive Property

22. Multiplicative Inverse Property

23. Associative Property of Addition

24. Commutative Property of Multiplication

25. $\frac{2}{9}$ **26.** $2 \cdot 2 \cdot 2 \cdot 3 \cdot 3 \cdot 3$ **27.** 58 ft/sec

28. $6.43

CHAPTER 2

Section 2.1   (pp. 56–63)

1. x **3.** m, n **5.** $4x, 3$ **7.** $\frac{5}{3}, -3y^3$

9. $a^2, 4ab, b^2$ **11.** 14   **13.** $-\frac{1}{3}$ **15.** $\frac{2}{5}$

17. 2π **19.** $y \cdot y \cdot y \cdot y \cdot y$ **21.** $2 \cdot 2 \cdot x \cdot x \cdot x \cdot x$

23. $4 \cdot y \cdot y \cdot z \cdot z \cdot z$

25. $a^2 \cdot a^2 \cdot a^2 = a \cdot a \cdot a \cdot a \cdot a \cdot a$

27. $-4 \cdot x \cdot x \cdot x \cdot x \cdot x \cdot x \cdot x$ **29.** $-9 \cdot a \cdot a \cdot a \cdot b \cdot b \cdot b$

31. $(x + y)(x + y)$ **33.** $\left(\frac{a}{3s}\right)\left(\frac{a}{3s}\right)\left(\frac{a}{3s}\right)\left(\frac{a}{3s}\right)$

35. $2 \cdot 2 \cdot (a - b)(a - b)(a - b)(a - b)(a - b)$

37. (a) 0   (b) -9 **39.** (a) 3   (b) 13

41. (a) 6   (b) 4   **43.** (a) 3   (b) -20

45. (a) 33   (b) 112   **47.** (a) 5   (b) 14

49. (a) 0   (b) Division by zero is undefined.

51. (a) $-\frac{1}{5}$ (b) $\frac{3}{10}$ **53.** (a) 0   (b) 11

55. $646   **57.** $7.55w$ **59.** $3.79m$

61. $(n - 5)^2$; 9 square units   **63.** $a(a + b)$; 45 square units

65. No. $-3^2 = -9$ and $(-3)^2 = 9$. An exponent affects what is directly to its left. In the expression -3^2, the 3 is the only portion of the expression being squared.

67. addition, subtraction, multiplication, and division

69. k **71.** $3(x + 5)$, 10   **73.** $-2u^4$ **75.** $(-3)^3(x - y)^2$

77. (a) $\frac{15}{2}$ (b) 10   **79.** (a) 72   (b) 320

81. (a) $x + 6$ (b) 29 in.   (c) 26.5 in.

83. (a) $\dfrac{3(4)}{2} = 6 = 1 + 2 + 3$

(b) $\dfrac{6(7)}{2} = 21 = 1 + 2 + 3 + 4 + 5 + 6$

(c) $\dfrac{10(11)}{2} = 55 = 1 + 2 + 3 + 4 + 5 + 6 + 7$
$+ 8 + 9 + 10$

85. No. The term includes the minus sign and is $-3x$.

87. The product of an even number and an odd number $[n(n + 1)$, where $n \geq 1$, and $n(n - 3)$, where $n \geq 4]$ is even, so it divides evenly by 2. This will always yield a natural number.

89. 17   **91.** 10   **93.** 24   **95.** 12

97. Commutative Property of Multiplication

99. Distributive Property

Section 2.2   (pp. 64–73)

1. $2v$, Commutative Property of Multiplication

3. $t, -2$, Distributive Property   **5.** $32 + 16z$

7. $-10x + 5y$ **9.** $8x + 8$ **11.** $-36s^2 + 6s$

13. $ab; ac; a(b + c) = ab + ac$

15. $2a; 2(b - a); 2a + 2(b - a) = 2b$

17. $8(1) + 8(0.25) = 8 + 2 = 10$

19. $5(20) - 5(2) = 100 - 10 = 90$

21. $16t^3, 3t^3; 4t, -5t$ **23.** $4rs^2, 12rs^2; -5, 1$

25. $-2y$ **27.** $-2x + 5$ **29.** $11x + 4$

31. $3r + 7$ **33.** $12x$ **35.** $-4x$ **37.** $6x^2$

39. $-10z^3$ **41.** $9a$ **43.** $-\dfrac{x^3}{3}$ **45.** $-24x^4y^4$

47. $2x$ **49.** $13s - 2$ **51.** $-2m + 21$ **53.** $8x + 38$

55. $8x + 26$ **57.** $2x - 17$ **59.** $10x - 7x^2$

61. $3x^2 + 5x - 3$ **63.** $4t^2 - 11t$ **65.** $26t - 2t^2$

67. (a) $10x + 10$ (b) $4x^2 + 20x$ **69.** $x^2 + 50x$

71. In an algebraic expression, two terms are said to be like terms if they are both constant terms or if they have the same variable factor(s).
Like terms: $-3x^2y, x^2y$ Unlike terms: x^2y, x^2y^2

73. Beginning with the innermost parentheses, use the Distributive Property to remove nested symbols of grouping.

75. $x^2 - xy + 4$ **77.** $17z + 11$ **79.** $-21y^3$

81. $\dfrac{x^2}{5}$ **83.** $23x + 10$ **85.** $3x - 5$

87. $5 + (3x - 1) + (2x + 5) = 5x + 9$

89. (a) $2(3x) + 2(x + 7)$; $8x + 14$ (b) $(3x)(x + 7)$; $3x^2 + 21x$

91. (a) Answers will vary.   (b) 56 square units

93. The exponents of y are not the same.

95. $(6x)^4 = (6x)(6x)(6x)(6x) = 6 \cdot 6 \cdot 6 \cdot 6 \cdot x \cdot x \cdot x \cdot x$;
$6x^4 = 6 \cdot x \cdot x \cdot x \cdot x$

97. 12   **99.** -11 **101.** $\frac{1}{80}$

103. (a) 4   (b) -5

Mid-Chapter Quiz   (p. 74)

1. (a) 0   (b) 10

2. (a) Division by zero is undefined.   (b) 0

3. Terms: $4x^2, -2x$ **4.** Terms: $5x, 3y, -z$
Coefficients: $4, -2$ Coefficients: $5, 3, -1$

5. $(-3y)^4$ **6.** $2^3(x - 3)^2$

7. Associative Property of Multiplication

8. Distributive Property   **9.** Multiplicative Inverse Property

10. Commutative Property of Addition   **11.** $6x^2 - 2x$

12. $-12y - 18y^2 + 36$ **13.** $20y^2$ **14.** $-\dfrac{x^2}{5}$

15. $9y^5$ **16.** $\dfrac{10z^3}{21y}$ **17.** $y^2 + 4xy + y$

18. $7x - 4$ **19.** $8a - 7b$ **20.** $-8x - 66$

21. $8 + (x + 6) + (3x + 1) = 4x + 15$

22. (a) $\dfrac{x}{6}$ (b) 5 students

Section 2.3   (pp. 76–85)

1. Pay per hour $\cdot$ Number of hours

3. Original number of coupons $-$ Number of coupons used

5. Price per carton $\cdot$ Number of cartons

7. $0.10d$ **9.** $0.10d + 0.25q$ **11.** $x + 5$

13. $b - 25$ **15.** $g - 6$ **17.** $2h$ **19.** $\dfrac{w}{3}$

21. $\dfrac{x}{50}$ **23.** A number decreased by 10

25. The product of 3 and a number, increased by 2

27. One-half of a number, decreased by 6

29. Three times the difference of 2 and a number

31. The sum of a number and 1, all divided by 2

33. One-half decreased by the quotient of a number and 5

35. The square of a number, increased by 5

37. $0.06L$ **39.** $3m + 4v$ **41.** $12.50 + 0.75q$

43. $0.99a + 1.99r$ **45.** $t \approx 10.2$ yr

47. $t \approx 11.9$ yr   **49.** $t \approx 11$ yr

51.

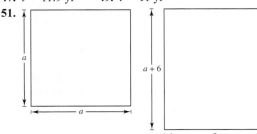

Perimeter of the square: $4a$ cm
Area of the square: a^2 cm^2
Perimeter of the rectangle: $4a + 12$ cm
Area of the rectangle: $a^2 + 6a$ cm^2

53. 5.75 ft

55. Guess, Check, and Revise; Make a Table; Look for a Pattern; Draw a Diagram; Solve a Simpler Problem

57. Answers will vary.

59. The product of a number and the sum of the same number and 16

61. 4 divided by the difference between a number and 2

63. $x(x + 3) = x^2 + 3x$ **65.** $x - (25 + x) = -25$

67. $x^2 - x(2x) = -x^2$ **69.** $0.25p + 0.1n$

71. One less than 2 times a number; 39

73. The start time is missing.   **75.** (a), (b), and (e)

77. $\dfrac{5}{3n}$, $\dfrac{5}{n} \cdot 3 = \dfrac{15}{n}$; The expression $\dfrac{3n}{5}$ is not a possible interpretation because the phrase "the quotient of 5 and a number" indicates that the variable is in the denominator.

79. 78   **81.** $\dfrac{3}{4}$ **83.** $\dfrac{23}{9}$

85. Commutative Property of Addition

87. Distributive Property

Section 2.4 *(pp. 86–93)*

1. (a) Not a solution   (b) Solution

3. (a) Not a solution   (b) Solution

5. (a) Not a solution   (b) Solution

7. (a) Not a solution   (b) Not a solution

9. (a) Solution   (b) Not a solution

11. Equation   **13.** Expression   **15.** Equation

17. Expression   **19.** Equation   **21.** Expression

23. Equation   **25.** Expression   **27.** $x = 13$

29. $x = 16$ **31.** $x = 10$

33. $x - 8 = 3$ Original equation
    $x - 8 + 8 = 3 + 8$ Add 8 to each side.
    $x = 11$ Solution
    Addition Property of Equality

35. $\dfrac{2}{3}x = 12$ Original equation
   $\dfrac{3}{2}\left(\dfrac{2}{3}x\right) = \dfrac{3}{2}(12)$ Multiply each side by $\dfrac{3}{2}$.
   $x = 18$ Solution
   Multiplication Property of Equality

37. $5x + 12 = 22$ Original equation
   $5x + 12 - 12 = 22 - 12$ Subtract 12 from each side
   $5x = 10$ Combine like terms.
   $\dfrac{5x}{5} = \dfrac{10}{5}$ Divide each side by 5.
   $x = 2$ Solution
   Addition & Multiplication Properties of Equality

39. $x + 6 = 94$ **41.** $\dfrac{x}{18} = 4.5$ **43.** $7x = 150 - 72$

45. Yes. Answers will vary.   **47.** Dividing by zero is undefined.

49. $\dfrac{x}{3} = x + 1$ Original equation
   $3\left(\dfrac{x}{3}\right) = 3(x + 1)$ Multiply each side by 3.
   $x = 3x + 3$ Distributive Property
   $x - 3x = 3x + 3 - 3x$ Subtract $3x$ from each side.
   $x - 3x = 3x - 3x + 3$ Group like terms.
   $-2x = 3$ Combine like terms.
   $\dfrac{-2x}{-2} = \dfrac{3}{-2}$ Divide each side by -2.
   $x = -\dfrac{3}{2}$ Solution
   Addition & Multiplication Properties of Equality

51. (a) Not a solution   (b) Not a Solution

53. (a) Not a solution   (b) Not a Solution

55. $0.25n + 7 = 8.75$

57. $10a + 6\left(\dfrac{3}{4}a\right) = 986$
      $\dfrac{29}{2}a = 986$

59. No, there is only one value of x, $\dfrac{b}{a}$, for which the equation is true.

61. *Sample answer:* The total cost of a shipment of bulbs is $840. Find the number of cases of bulbs when each case costs $35.

63. t^7 **65.** $15x$ **67.** $8b$ **69.** $x + 23$ **71.** $4y + 7$

Review Exercises *(pp. 96–99)*

1. x **3.** a, b

5. Terms: $12y, y^2$ **7.** Terms: $5x^2, -3xy, 10y^2$
Coefficients: 12, 1   Coefficients: 5, -3, 10

9. Terms: $\dfrac{2y}{3}, -\dfrac{4x}{y}$

Coefficients: $\dfrac{2}{3}, -4$

11. $(5z)^3$ **13.** $(-3x)^5$ **15.** $6^2(b - c)^2$

17. (a) 5 (b) 5   **19.** (a) 4 (b) -2

21. (a) 0 (b) -7 **23.** (a) -3 (b) 6

25. Multiplicative Inverse Property

27. Commutative Property of Multiplication

29. Associative Property of Addition   **31.** $-2a$

33. $11p - 3q$ **35.** $\dfrac{15}{4}s - 5t$ **37.** $\dfrac{19}{15}a + \dfrac{1}{6}b$

39. $3x - 3y + 3xy$ **41.** $6n + 3$ **43.** $48t$ **45.** $45x^2$

47. $-12x^3$ **49.** $8x$ **51.** (a) $6x + 12$ (b) $2x^2 + 12x$

53. $(4x)(16x) - x(6x) = 58x^2$ **55.** $5u - 10$ **57.** $5s - r$

59. $10z - 1$ **61.** $2z - 2$ **63.** $8x - 32$ **65.** $-2x + 4y$

67. *Verbal model:*

$$\begin{array}{c}\text{Base pay} \\ \text{per hour}\end{array} + \begin{array}{c}\text{Additional} \\ \text{pay per unit}\end{array} \cdot \begin{array}{c}\text{Number of units} \\ \text{produced per hour}\end{array}$$

Algebraic expression: $8.25 + 0.60x$

69. $\frac{2}{3}x + 5$ **71.** $2y - 10$ **73.** $50 + 7z$ **75.** $\frac{s + 10}{8}$

77. $g^2 + 64$ **79.** $0.05x$ **81.** $625n$

83. Four more than 3 times a number; 64

85. (a) Not a solution (b) Solution

87. (a) Not a solution (b) Solution

89. (a) Solution (b) Not a solution

91. (a) Not a solution (b) Solution

93. (a) Solution (b) Solution

95.

$-7x + 20 = -1$	Original equation
$-7x + 20 - 20 = -1 - 20$	Subtract 20 from each side.
$-7x = -21$	Combine like terms.
$\dfrac{-7x}{-7} = \dfrac{-21}{-7}$	Divide each side by -7.
$x = 3$	Solution

Addition & Multiplication Properties of Equality

97.

$x = -(x - 14)$	Original equation
$x = -x + 14$	Distributive Property
$x + x = -x + 14 + x$	Add x to each side.
$x + x = -x + x + 14$	Group like terms.
$2x = 14$	Combine like terms.
$\dfrac{2x}{2} = \dfrac{14}{2}$	Divide each side by 2.
$x = 7$	Solution

Addition & Multiplication Properties of Equality

99. $x + \dfrac{1}{x} = \dfrac{37}{6}$ **101.** $6x - \dfrac{1}{2}(6x) = 24$

Chapter Test *(p. 100)*

1. Terms: $2x^2, -7xy, 3y^3$; Coefficients: $2, -7, 3$

2. $x^3(x + y)^2$ **3.** Associative Property of Multiplication

4. Commutative Property of Addition

5. Additive Identity Property

6. Multiplicative Inverse Property **7.** $3x + 24$

8. $20r - 5s$ **9.** $-3y + 2y^2$ **10.** $-36 + 18x - 9x^2$

11. $-a - 7b$ **12.** $8u - 8v$ **13.** $4z - 4$ **14.** $18 - 2t$

15. 6 **16.** -28 **17.** Division by zero is undefined.

18. $\frac{1}{3}n - 4$

19. (a) Perimeter: $2w + 2(2w - 4) = 6w - 8$; Area: $w(2w - 4)$
$= 2w^2 - 4w$

(b) Perimeter: 34 units; Area: 70 square units

20. (a) $25m + 20n$ (b) \$110

21. (a) Not a solution (b) Solution

CHAPTER 3

Section 3.1 *(pp. 102–109)*

1. 9 **3.** -6 **5.** Subtraction

7. Multiplication **9.** Subtraction **11.** Addition

13.

$5x + 15 = 0$	Original equation
$5x + 15 - 15 = 0 - 15$	Subtract 15 from each side.
$5x = -15$	Combine like terms.
$\dfrac{5x}{5} = \dfrac{-15}{5}$	Divide each side by 5.
$x = -3$	Simplify

15. -1 **17.** 2 **19.** 6 **21.** -26 **23.** $\frac{1}{4}$

25. -28 **27.** 9 **29.** -3 **31.** $\frac{1}{3}$ **33.** 2 **35.** $\frac{1}{3}$

37. -2 **39.** No solution **41.** 3

43. Infinitely many solutions **45.** $\frac{2}{5}$ **47.** $\frac{9}{2}$ **49.** 0

51. 75 cm **53.** 20 in. $\times$ 40 in. **55.** 150 seats

57. Yes. Subtract the cost of parts from the total to find the cost of labor. Then divide by 44 to find the number of hours spent on labor (2.25 hours).

59. \$1430 **61.** 8 wk

63. Substitute the solution into the original equation and simplify each side.

$$\begin{aligned} 3x + 2 &= 11 \\ 3(3) + 2 &\overset{?}{=} 11 \\ 9 + 2 &\overset{?}{=} 11 \\ 11 &= 11 \checkmark \end{aligned}$$

65. You are trying to isolate the variable term on the left-hand side of the equation. To do this, you must eliminate the $+2$ by subtracting 2; add 2

67. 5 **69.** 30 **71.** 1 **73.** $\frac{2}{3}$

75. Infinitely many solutions **77.** No solution

79. $l = 80$ m, $w = 50$ m **81.** 12 units **83.** 12.5 hr

85. Yes; The Addition Property of Equality; 1 oz

87. False. Multiplying each side of the equation $3x = 9$ by 0 yields $0 = 0$. The equation $3x = 9$ has one solution, $x = 3$, and the result $0 = 0$ suggests that the equation has infinitely many solutions.

89. False. $(2m + 1) + 2n = 2m + 2n + 1 = 2(m + n) + 1$, which is odd.

91. 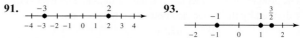 **93.**

95. (a) Solution (b) Not a solution

97. (a) Not a solution (b) Not a solution

Section 3.2 *(pp. 110–117)*

1. 4 **3.** 3 **5.** 2 **7.** -5 **9.** 2 **11.** -10

13. 9 **15.** 30 **17.** No solution **19.** -4

21. No solution **23.** No solution **25.** 1 **27.** $\frac{5}{6}$

29. 1 **31.** $\frac{5}{2}$ **33.** $-\frac{2}{5}$ **35.** $\frac{35}{2}$ **37.** $-\frac{10}{3}$ **39.** $\frac{3}{4}$

41. $\frac{1}{6}$ **43.** 4.8 hr **45.** 77 points **47.** 5

49. 6.18 **51.** 5 **53.** 7.71 **55.** 66.67

57. 0.42 **59.** 123 **61.** 3.51 **63.** 8.99 **65.** 2054

67. The least common multiple of the denominators of two or more fractions is the least number that is a common multiple of all of the denominators.

69. Multiplying each side of the equation by the least common multiple of the denominators clears the equation of fractions, making the equation easier to solve.

71. 6 **73.** 5 **75.** 6 **77.** No solution **79.** $\frac{4}{11}$

81. 10 **83.** 10 **85.** 25 qt

87. Because each brick is 8 inches long and there are n bricks, the width that is made up of bricks is represented by $8n$. Because there is $\frac{1}{2}$ inch of mortar between adjoining bricks and there are $n - 1$ widths of mortar, the width that is made up of mortar is represented by $\frac{1}{2}(n - 1)$. Because the width of the fireplace is 93 inches, the equation is $8n + \frac{1}{2}(n - 1) = 93$.

89. You could divide each side of the equation by 3.

91. Dividing by a variable assumes that is does not equal zero, which may yield a false solution.

93. $4x^6$ **95.** $5z^5$ **97.** $x - 4$ **99.** $-y^4 + 2y^2$

101. $\frac{17}{3}$ **103.** -9

Section 3.3   *(pp. 118–125)*

1. 62%   **3.** 7.5%   **5.** 80%   **7.** 125%   **9.** 0.12

11. 1.25   **13.** 0.085   **15.** 0.0075   **17.** $\frac{3}{10}$ **19.** $\frac{13}{10}$

21. $\frac{7}{500}$ **23.** $\frac{1}{200}$ **25.** 45   **27.** 0.42   **29.** $37,380

31. 2100   **33.** 132   **35.** 430 points   **37.** 72%

39. 2.75%   **41.** 9.5%

	Cost	Selling Price	Markup	Markup Rate
43.	$26.97	$49.95	$22.98	85.2%
45.	$40.98	$74.38	$33.40	81.5%
47.	$69.29	$125.98	$56.69	81.8%
49.	$13,250.00	$15,900.00	$2650.00	20%

	Original Price	Sale Price	Discount	Discount Rate
51.	$39.95	$29.95	$5.00	9.8%
53.	$315.00	$18.95	$126.00	20%
55.	$189.99	$10.95	$30.00	42.2%
57.	$119.96	$29.73	$55.22	50%
59.	$394.97	$695.00	$300.00	34.2%

61. Percent means per hundred or parts of 100.

63. $x = 0.25y$

	Percent	Parts out of 100	Decimal	Fraction
65.	40%	40	0.40	$\frac{2}{5}$
67.	7.5%	7.5	0.075	$\frac{3}{40}$
69.	63%	63	0.63	$\frac{63}{100}$
71.	15.5%	15.5	0.155	$\frac{31}{200}$
73.	60%	60	0.60	$\frac{3}{5}$

75. $37\frac{1}{2}\%$ **77.** $41\frac{2}{3}\%$ **79.** 4%   **81.** 0.107%

83. If $a > b$, the percent is greater than 100%.
If $a < b$, the percent is less than 100%.
If $a = b$, the percent is equal to 100%.

85. False. 1% $= 0.01 \neq 1$

87. False. Because 68% $= 0.68$, $a = 0.68(50)$.

89. 0   **91.** (a) 7   (b) 16   **93.** $8x - 20$

95. -3 **97.** -12

Section 3.4   *(pp. 126–133)*

1. $\frac{4}{1}$ **3.** $\frac{1}{2}$ **5.** $\frac{17}{4}$ **7.** $\frac{2}{3}$ **9.** $\frac{9}{1}$ **11.** $\frac{32}{53}$

13. $\frac{2}{1}$ **15.** $\frac{2}{3}$ **17.** $\frac{3}{50}$ **19.** $\frac{7}{15}$ **21.** $\frac{2}{1}$ **23.** $\frac{3}{8}$

25. $0.049/oz   **27.** $0.073/oz   **29.** a   **31.** b

33. 12   **35.** $\frac{10}{3}$ **37.** $\frac{175}{8}$ **39.** $\frac{3}{16}$ **41.** $\frac{1}{2}$

43. 27   **45.** $\frac{5}{2}$ **47.** 6   **49.** 16 gal

51. 22,691 votes   **53.** 250 blocks

55. *Sample answer:* You want to calculate the average miles per hour of a trip.

57. Cross-multiply and then solve for the variable.

59. $\frac{1}{4}$ **61.** $\frac{4}{5}$ **63.** $\frac{3}{10}$ **65.** a   **67.** $\frac{2}{3}$ **69.** 384 mi

71. $6\frac{2}{3}$ ft   **73.** $46\frac{2}{3}$ min   **75.** $12\frac{1}{2}$ lb   **77.** 20%

79. No. It is also necessary to know either the number of men in the class or the number of women in the class.

81. Answers will vary.   **83.** 13   **85.** 9,300,000

87. $\frac{77}{5}$ **89.** 62.5   **91.** 60   **93.** 62.5

Mid-Chapter Quiz   *(p. 134)*

1. 6   **2.** 8   **3.** $\frac{19}{2}$ **4.** 0   **5.** $-\frac{1}{3}$ **6.** $\frac{35}{12}$

7. 36   **8.** $\frac{11}{5}$ **9.** 5   **10.** -2 **11.** 2.06

12. 51.23   **13.** 15.5   **14.** 42   **15.** 200%   **16.** 455

17. 10 hr   **18.** 6 m², 12 m², 24 m²   **19.** 93

20. $11,550   **21.** 3 hr   **22.** $\frac{225}{64}$ **23.** 26.25 gal

Section 3.5   *(pp. 136–143)*

1. $49.59   **3.** 6%   **5.** 4 in.   **7.** 2.5 m

9. $2000 at 7%, $4000 at 9%   **11.** 35 mL   **13.** $1\frac{1}{5}$ hr

15. Answers will vary.   **17.** 125 mL/hr

19. 16 dozen roses, 8 dozen carnations   **21.** $V = \pi r^2 h$

23. Perimeter: linear units—inches and meters
Area: square units—square feet and square centimeters
Volume: cubic units—cubic inches and cubic feet

	Distance, d	Rate, r	Time, t
25.	48 m	4 m/min	12 min
27.	210 mi	50 mi/hr	4.2 hr

29. $h = \dfrac{2A}{b}$

31. Solution 2: 75 gal; Final solution: 100 gal

33. Solution 2: 5 qt; Final solution: 10 qt

35. 8.6%   **37.** 15 m, 15 m, 53 m   **39.** 28 mi

41. 3 hours on the first part, 1 hour and 15 minutes on the last part

43. Divide by 2 to obtain 90 miles per 2 hours. Divide by 2 again to obtain 45 miles per hour.

$$d = rt \Rightarrow \frac{d}{t} = r \Rightarrow \frac{180 \text{ miles}}{4 \text{ hours}} = 45 \text{ mi/hr}$$

45. Use $h = \dfrac{2A}{x + y}$ to find the height of a trapezoid.

Use $x = \dfrac{2A}{h} - y$ to find the base x of a trapezoid.

Use $y = \dfrac{2A}{h} - x$ to find the base y of a trapezoid.

47. The circumference would double; the area would quadruple.
Circumference: $C = 2\pi r$, Area: $A = \pi r^2$
If r is doubled, $C = 2\pi(2r) = 2(2\pi r)$ and $A = \pi(2r)^2 = 4\pi r^2$

49. (a) 7, 1   (b) 7, 1, -3
(c) 1.8, $\frac{1}{10}$, 7, -2.75, 1, -3 (d) None

51. 9   **53.** 6   **55.** 16

Section 3.6   *(pp. 144–151)*

1. 8   **3.** 11   **5.** 3

7.   **9.**

11.   **13.**

15. (number line: bracket at 1, parenthesis at 4; 0 1 2 3 4 5) **17.** (number line: parenthesis at 0, bracket at $\frac{3}{2}$; -1 0 1 2 3)

19. (number line: 3.5 to 4.5, parenthesis to bracket; 2 3 4 5)

21. Not equivalent **23.** Equivalent

25. $x \ge 4$ (number line: bracket at 4; 0 1 2 3 4 5 6) **27.** $x \le 2$ (number line: bracket at 2; 0 1 2 3)

29. $x < 4$ (number line: parenthesis at 4; -2 0 2 4 6) **31.** $x \le -4$ (number line: bracket at -4; -5 -4 -3 -2 -1 0)

33. $x > 8$ (number line: parenthesis at 8; 0 2 4 6 8 10) **35.** $x \ge 7$ (number line: bracket at 7; 5 6 7 8 9)

37. $x > 7.55$ (number line: parenthesis at 7.55; 5 6 7 8 9) **39.** $x > -\frac{2}{3}$ (number line: parenthesis at $-\frac{2}{3}$; -2 -1 0 1)

41. $x > \frac{9}{2}$ (number line: parenthesis at $\frac{9}{2}$; 0 2 4 6) **43.** $x > \frac{20}{11}$ (number line: parenthesis at $\frac{20}{11}$; 0 1 2 3)

45. $y \le -10$ (number line: bracket at -10; -15 -10 5 0) **47.** $\frac{5}{2} < x < 7$ (number line: parenthesis at $\frac{5}{2}$ and 7; 0 2 4 6 8)

49. $-3 \le x < -1$ (number line: bracket at -3, parenthesis at -1; -4 -3 -2 -1 0) **51.** $-5 < x < 5$ (number line: parenthesis at -5 and 5; -8 -4 0 4 8)

53. $-1 < x \le 4$ (number line: parenthesis at -1, bracket at 4; -2 -1 0 1 2 3 4 5) **55.** No solution

57. $x < -\frac{8}{3}$ or $x \ge \frac{5}{2}$ (number line: parenthesis at $-\frac{8}{3}$, bracket at $\frac{5}{2}$; -4 -3 -2 -1 0 1 2 3 4) **59.** $-\infty < x < \infty$ (number line; -2 -1 0 1 2)

61. \$2600 **63.** $23 \le x \le 38$

65. Parenthesis; A parenthesis is used when the endpoint is excluded.

67. Yes. By definition, dividing by a number is the same as multiplying by its reciprocal.

69. a **70.** e **71.** d **72.** b **73.** f **74.** c

75. $x > \frac{8}{3}$ (number line: parenthesis at $\frac{8}{3}$; 0 1 2 3) **77.** $x \le -8$ (number line: bracket at -8; -12 -10 -8 -6 -4 -2 0)

79. $x > -15$ (number line: parenthesis at -15; -20 -16 -12 -8 -4 0) **81.** $-5 < x \le 0$ (number line: parenthesis at -5, bracket at 0; -6 -4 -2 0 2)

83. $-\frac{3}{2} < x < \frac{9}{2}$ (number line: parenthesis at $-\frac{3}{2}$ and $\frac{9}{2}$; -2 0 2 4 6) **85.** $1 < x < 10$ (number line: parenthesis at 1 and 10; 0 3 6 9 12)

87. The average temperature in Miami is greater than the average temperature in New York City.

89. 26,000 mi

91. $12.50 < 8 + 0.75n;\ n > 6$

93. The multiplication and division properties differ. The inequality symbol is reversed if both sides of an inequality are multiplied or divided by a negative real number.

95. The solution set of a linear inequality is bounded if the solution is written as a double inequality. Otherwise, the solution set is unbounded.

97. $a < x < b$; A double inequality is always bounded.

99. $x > a$ or $x < b$; $x > a$ includes all real numbers between a and b and greater than or equal to b, all the way to ∞. $x < b$ adds all real numbers less than or equal to a, all the way to $-\infty$.

101. $<$ **103.** $=$ **105.** No; Yes **107.** Yes; No

109. $\frac{17}{2}$ **111.** $-\frac{1}{4}$

Section 3.7 *(pp. 152–159)*

1. Not a solution **3.** Solution **5.** 4, −4

7. No solution **9.** 0 **11.** 3, −3 **13.** 4, −6 **15.** $\frac{4}{3}$

17. 11, −14 **19.** No solution **21.** 2 **23.** 2, 3

25. −5, 1 **27.** 7, −3 **29.** $\frac{1}{3}$ **31.** $\frac{1}{2}$ **33.** Solution

35. Not a solution **37.** $-4 < y < 4$ **39.** $x \le -6$ or $x \ge 6$

41. $x < -16$ or $x > 4$ **43.** $-7 < x < 7$ **45.** $-1 \le y \le 1$

47. $x < -6$ or $x > 3$ **49.** $-3 \le x \le 4$ **51.** No solution

53. $a \le -38$ or $a \ge 26$ **55.** $-5 < x < 35$

57. (number line: 41.826 to 42.65, bracket to bracket; 40 41 42 43 44 45)
Fastest: 41.826 sec
Slowest: 42.65 sec

59. To solve an absolute value equation, first write the equation in standard form. Because the solutions of $|x| = a$ are $x = a$ and $x = -a$, rewrite the equation as two equivalent equations. Then, solve for each value of x. For example: $|x - 3| + 2 = 7$ in standard form is $|x - 3| = 5$, which means that $x - 3 = 5$ or $x - 3 = -5$. So, $x = 8$ or $x = -2$.

61. $|x| = a$ **63.** $-\frac{1}{8}, -\frac{3}{8}$ **65.** $|x - 4| = 9$

67. $|x| \le 2$ **69.** $|x - 19| > 2$ **71.** $|x - 4| \ge 2$

73. $-\infty < x < \infty$ **75.** $|x| < 3$ **77.** $|2x - 3| > 5$

79. (a) $|s - x| \le \frac{3}{16}$ (b) $4\frac{15}{16} \le x \le 5\frac{5}{16}$

81. $65.8\text{ in.} \le h \le 71.2\text{ in.}$

83. All real numbers that are within one unit of 4

85. 6 **87.** $4(n + 3)$ **89.** $x > 20$ **91.** $x \ge 4$

Review Exercises *(pp. 162–164)*

1. 5 **3.** $\frac{3}{5}$ **5.** 3 **7.** 5 **9.** 4 **11.** 20

13. 12 units **15.** 20 **17.** 6 **19.** 1 **21.** 7

23. 20 **25.** $-\frac{1}{7}$ **27.** 23.26 **29.** 224.31 **31.** 3 hr

Percent	Parts out of 100	Decimal	Fraction
33. 60%	60	0.60	$\frac{3}{5}$
35. 80%	80	0.80	$\frac{4}{5}$
37. 20%	20	0.20	$\frac{1}{5}$
39. 55%	55	0.55	$\frac{11}{20}$

41. 20 **43.** 400 **45.** 60% **47.** \$85.44 **49.** $\frac{1}{8}$

51. $\frac{4}{3}$ **53.** (b) 24-ounce container **55.** $\frac{7}{2}$

57. $-\frac{10}{3}$ **59.** 9 **61.** \$133.33

Distance, d	Rate, r	Time, t
63. 520 mi	65 mi/hr	8 hr
65. 855 m	5 m/min	171 min
67. 3000 mi	60 mi/hr	50 hr

69. 30 ft × 26 ft **71.** \$475

73. 13 dimes, 17 quarters **75.** $\frac{30}{11} \approx 2.7$ hr

77.  **79.**

81. $x \le 4$ **83.** $x > 4$

85. $x > 3$ **87.** $x \le 6$

89. $y > -\frac{70}{3}$ **91.** $x \le -3$

93. $-7 \le x < -2$ **95.** $-16 < x < -1$

97. $-3 < x < 2$

99. 125 min **101.** 6, −6 **103.** 4, $-\frac{4}{3}$ **105.** 0, $-\frac{8}{5}$

107. $\frac{1}{2}$, 3 **109.** $x < 1$ or $x > 7$ **111.** $-4 < x < 4$

113. $-4 < x < 11$ **115.** $b < -9$ or $b > 5$

117. $|x - 3| < 2$

119.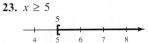

Maximum: 116.6°F

Minimum: 40°F

Chapter Test *(p. 165)*

1. −13 **2.** $\frac{21}{4}$ **3.** 7 **4.** 1 **5.** $-\frac{1}{3}$ **6.** −6

7. 5, −11 **8.** $\frac{2}{3}$, $-\frac{4}{3}$ **9.** 11.03 **10.** $2\frac{1}{2}$ hr

11. $31\frac{1}{4}$%, 0.3125 **12.** 1200 **13.** 36% **14.** 6

15. $\frac{5}{9}$; 2 yd = 6 ft = 72 in. **16.** $\frac{12}{7}$ **17.** 5 **18.** 66 mi/hr

19. $\frac{36}{7} \approx 5.1$ hr **20.** \$6250 **21.** $t \ge 8$

22. 25,000 miles

23. $x \ge 5$ **24.** $x > 1$

25. $-7 < x \le 1$ **26.** $-1 \le x < \frac{5}{4}$

27. $1 \le x \le 5$ **28.** $x < -\frac{9}{5}$ or $x > 3$

Cumulative Test *(p. 166)*

1. < **2.** 1200 **3.** $-\frac{11}{24}$ **4.** $-\frac{25}{12}$ **5.** 8

6. 14 **7.** 28 **8.** −30 **9.** $-\frac{11}{2}$ **10.** $3^3(x + y)^2$

11. $-2x^2 + 6x$ **12.** Associative Property of Addition

13. $15x^7$ **14.** $7x^2 - 6x - 2$ **15.** $-3x^2 + 18x$

16. 6 **17.** $\frac{52}{3}$ **18.** −7

19. $x \ge -7$;

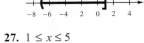

20. $\dfrac{15{,}000 \text{ miles}}{1 \text{ year}} \cdot \dfrac{1 \text{ gallon}}{30 \text{ miles}} \cdot \dfrac{\$4.00}{1 \text{ gallon}} \approx \$2000/\text{yr}$

21. $\frac{3}{4}$ **22.** \$920 **23.** \$57,000

CHAPTER 4

Section 4.1 *(pp. 168–175)*

1. **3.**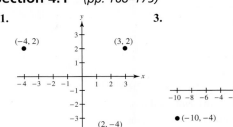

5. **7.**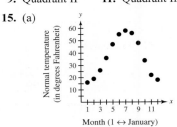

9. Quadrant II **11.** Quadrant III **13.** Quadrant III

15. (a) 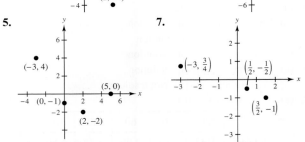 No, because there are only 12 months, but the temperature ranges from 16°F to 58°F.

(b) August

17.

x	−2	0	2	4	6
$y = 3x - 4$	−10	−4	2	8	14

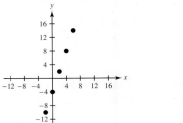

19.

x	−2	0	4	6	8
$y = -\frac{3}{2}x + 5$	8	5	−1	−4	−7

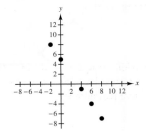

21.

x	−2	−1	0	1	2
y = −4x − 5	3	−1	−5	−9	−13

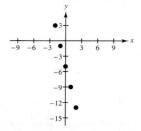

23. (a) Solution   (b) Not a solution
     (c) Not a solution   (d) Solution
25. (a) Solution   (b) Solution
     (c) Not a solution   (d) Solution
27. (a) Not solution   (b) Solution
     (c) Solution   (d) Not a solution
29. (a) Solution   (b) Not a solution
     (c) Not a solution   (d) Not a solution
31. (a) $-\frac{4}{3}$ (b) 16   (c) −2
33.

x	20	40	60	80	100
y = 0.066x	1.32	2.64	3.96	5.28	6.60

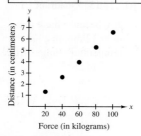

35. $y = 25x + 5000$;

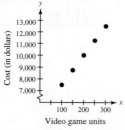

37. Yes. The scale on the vertical axis makes it appear that the changes in the surplus are dramatic.
39. (2, 2)   **41.** (−4, −1); (0, 1)
43. A: (5, 2), B: (−3, 4),   **45.** A: (−1, 3), B: (5, 0),
     C: (2, −5), D: (−2, −2)   C: (2, 1), D: (−1, −2)
47.   **49.**

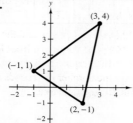 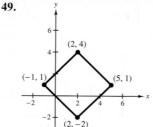

51.   **53.**

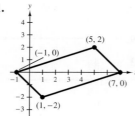 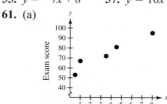

55. $y = -7x + 8$ **57.** $y = 10x - 2$ **59.** Quadrant I or IV
61. (a)

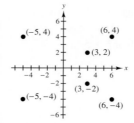

     (b) Scores increase with increased study time.
63. (a) and (b)   (c) Reflection in the x-axis

65. (6, 4)
67. No. *Sample answer:* If y is measuring revenue for a product and x is measuring time in years, then the scale on the y-axis may be in units of $100,000 and the scale on the x-axis may be in units of 1 year.
69. −10   **71.** 14   **73.** $x > -1$ **75.** $x < 4$

Section 4.2   *(pp. 176–183)*

1.

x	−2	−1	0	1	2
y	11	10	9	8	7

3.

x	−2	−1	0	1	2
y	7	4	3	4	7

5.

x	−3	−2	−1	0	1
y	2	1	0	1	2

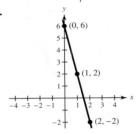

7.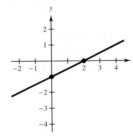

9.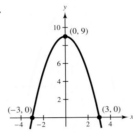

11.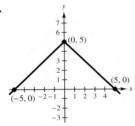

13. (−2, 0), (0, 4)   **15.** (−1, 0), (1, 0), (0, 1)

17. (2, 0), (0, −1)   **19.** (−1, 0), (0, −2)

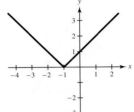 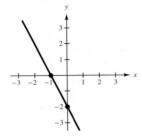

21. (0, 50)

23. (a) $y = 1120 - 80x$

(b)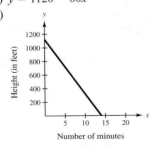

(c) (0, 1120); the initial height of the hot-air balloon

25. (a) 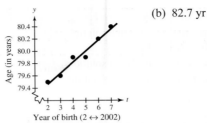   (b) 82.7 yr

27. (2, 1)   **29.** Let $x = 0$ and solve the equation for y.

31. 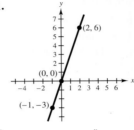   **33.**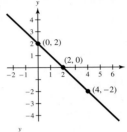

35. 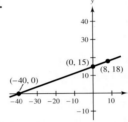   **37.**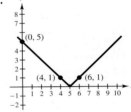

39. (6, 0), (0, 2)   **41.** (−4, 0), (4, 0), (0, 4)

43. No   **45.** Yes; Distributive Property

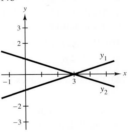 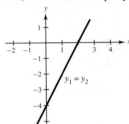

47. $y = 35t$;

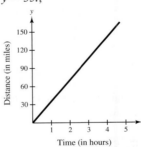

49. Yes. For any linear equation in two variables, x and y, there is a resulting value for y when $x = 0$. The corresponding point $(0, y)$ is the y-intercept of the graph of the equation.

51. 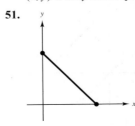   The distance between you and the tree decreases as you move from left to right on the graph. The x-intercept represents the number of seconds it takes you to reach the tree.

53. -13 **55.** 13 **57.** $-\frac{1}{6}$ **59.** $\frac{3}{10}$ **61.** $-\frac{36}{5}$

63. (a) Not a solution   (b) Not a solution
(c) Solution   (d) Solution

65. (a) Not a solution   (b) Not a solution
(c) Not a solution   (d) Not a solution

Section 4.3 *(pp. 184–191)*

1. Domain: $\{-4, 1, 2, 4\}$; Range: $\{-3, 2, 3, 5\}$

3. Domain: $\{-9, \frac{1}{2}, 2\}$; Range: $\{-10, 0, 16\}$

5. Domain: $\{-1, 1, 5, 8\}$; Range: $\{-7, -2, 3, 4\}$

7. Function   **9.** Function   **11.** Function

13. Function   **15.** Not a function   **17.** Not a function

19. (a) 6   (b) 6   (c) 66   (d) 11

21. (a) 1   (b) 15   (c) 0   (d) 0

23. $D = \{0, 1, 2, 3, 4\}$
$R = \{4, 3, 2, 1, 0\}$

25. $D = \{-8, -6, 2, 5, 12\}$
$R = \{-1, 0, 7, 10\}$

27. $D = \{-5, -4, -3, -2, -1\}$
$R = \{2\}$

29. Domain: The set of all real numbers r such that $r > 0$
Range: The set of all real numbers A such that $A > 0$

31. 2400

33. A relation is any set of ordered pairs. A function is a relation in which no two ordered pairs have the same first component and different second components.

35. Given the graph of a set of points on a rectangular coordinate system, if a vertical line intersects the graph at more than one point, the relation does not represent a function.

37. Not a function   **39.** (a) 4   (b) 0   (c) 12   (d) $\frac{1}{2}$

41. (a) -1 (b) 0   (c) 26   (d) $-\frac{7}{8}$

43. (a) $f(10) = 15, f(15) = 12.5$ (b) Demand decreases.

45. (a) 100 mi   (b) 200 mi   (c) 500 mi

47. $P = 4s$; P is a function of s. If you make a table of values where $s > 0$, no first component will have two different second components.

49. Yes; *Domain*   *Range*   **51.** Yes; *Domain*   *Range*

```
  1 ─────► 4            1 ─────► 6
  2 ─────► 5            2 ─────► 7
  3 ─────► 6            3 ─────► 8
                        4 ─────► 9
                        5
```

53. Not equivalent   **55.** Equivalent   **57.** $-8, 8$

59. $-6, 6$ **61.** $-9, 1$ **63.** No solution

Mid-Chapter Quiz *(p. 192)*

1.
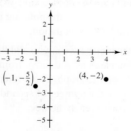

2. Quadrant I or IV

3. (a) Solution   (b) Not a solution
(c) Solution   (d) Not a solution

4. $(12, 0), (0, -4)$ **5.** $\left(\frac{2}{7}, 0\right), (0, 2)$

6.
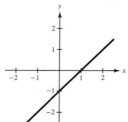

7.

8.
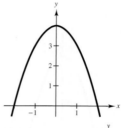

9.

10.
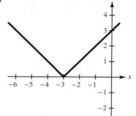

11.

12. Domain: $\{1, 2, 3\}$, Range: $\{0, 4, 6, 10, 14\}$

13. Domain: $\{-3, -2, -1, 0\}$, Range: $\{6\}$

14. Not a function

15. (a) 2   (b) -7

16. (a) 3   (b) -60

17. Domain: $\{10, 15, 20, 25\}$

18. 2006: \$120 million   2009: \$122 million
2007: \$143 million   2010: \$162 million
2008: \$133 million   2011: \$162 million

19. (a) $y = 2000 - 500t$
(b)
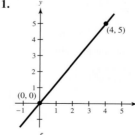

(c) $(0, 2000)$; the original value of the computer system

Section 4.4 *(pp. 194–201)*

1.

$m = \frac{5}{4}$

3.

$m = -\frac{1}{2}$

5.

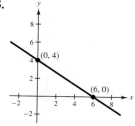

$$m = -\frac{2}{3}$$

7. $m = \frac{7}{6}$; The line rises.

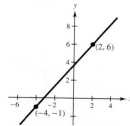

9. m is undefined; The line is vertical.

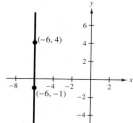

11. $m = 0$; The line is horizontal.   **13.** $m = -\frac{18}{17}$; The line falls.

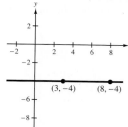   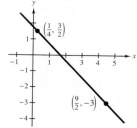

15. $-\frac{2}{3}$ **17.** $y = -\frac{1}{2}x + 2$; $m = -\frac{1}{2}$, $(0, 2)$

19. $y = -\frac{2}{3}x + 2$; $m = -\frac{2}{3}$, $(0, 2)$

21. $y = \frac{3}{4}x + \frac{1}{2}$; $m = \frac{3}{4}$, $\left(0, \frac{1}{2}\right)$

23. $y = 2x - 3$ **25.** $y = \frac{1}{2}x + 1$

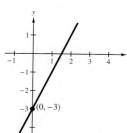   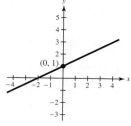

27. $y = \frac{1}{3}x - \frac{5}{2}$ **29.**

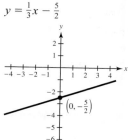   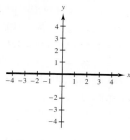

31.   **33.**

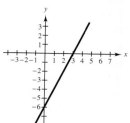   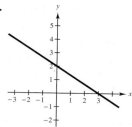

35. Perpendicular; The lines have slopes of -2 and $\frac{1}{2}$, which are negative reciprocals of each other.

37. Neither; The lines have slopes of $\frac{7}{8}$ and $\frac{5}{6}$, which are not the same and are not negative reciprocals of each other.

39. Parallel; The lines both have the same slope: 2.

41. Neither; The equations represent the same line, not 2 distinct lines.

43. Perpendicular; The lines have slopes of $-\frac{1}{3}$ and 3, which are negative reciprocals of each other.

45. (a) -2 (b) $\frac{1}{2}$ **47.** (a) 0   (b) Undefined

49. -2; 3   **51.** It falls from left to right.

53. (a) L_2 (b) L_3 (c) L_4 (d) L_1

55.   **57.**

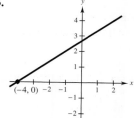   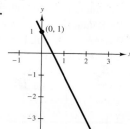

59. Perpendicular   **61.** Parallel

63. $\frac{2}{5}$ **65.** No; $\left|\frac{12}{60}\right| < \left|\frac{12}{50}\right|$

67. No. The slopes of nonvertical perpendicular lines have opposite signs. The slopes are the negative reciprocals of each other.

69. The slope

71. Yes. You are free to label either one of the points as (x_1, y_1) and the other as (x_2, y_2). However, once this is done, you must form the numerator and denominator using the same order of subtraction.

73. x^5 **75.** $-y^3$ **77.** $50x^5$ **79.** $x + 2$

81. $\left(\frac{1}{2}, 0\right)$, $(0, -3)$ **83.** $\left(-\frac{3}{2}, 0\right)$, $(0, -3)$

Section 4.5 *(pp. 202–209)*

1. $y = -\frac{1}{3}x + \frac{5}{3}$ **3.** $y = -\frac{1}{5}x - \frac{13}{5}$ **5.** $y = \frac{1}{3}x - \frac{1}{3}$

7. $y = -\frac{3}{5}x + \frac{8}{5}$ **9.** $y = -\frac{1}{2}x + \frac{13}{2}$

11. (a) $y = x - 1$ (b) $y = -x - 3$

13. (a) $y = -\frac{1}{2}x + \frac{5}{2}$ (b) $y = 2x - 5$

15. (a) $y = x - 1$ (b) $y = -x + 3$

17. (a) $y = -\frac{3}{4}x - 5$ (b) $y = \frac{4}{3}x + 20$ **19.** $x = -2$

21. $y = \frac{2}{3}$ **23.** $x = 4$ **25.** $y = -8$

27. $P = 146t + 735$; $1.027 billion

29. (a) $V = 25,000 - 2300t$ (b) $18,100

31. $-\frac{1}{2}$ mi/day; $y = -\frac{1}{2}x + 16$

33. The slope is needed to write an equation using the point-slope form.

35. The equation is written in slope-intercept form.

37. $y = \frac{2}{3}x + \frac{3}{2}$ **39.** $y = -0.8x + 5.6$ **41.** 5   **43.** $\frac{3}{2}$

45. $8x + 6y - 19 = 0$ **47.** $6x + 5y - 9 = 0$

49. $W = 2000 + 0.02S$

51. (a) f: $m = -10$; Loan decreases by $10 per week.

    (b) e: $m = 1.50$; Pay increases by $1.50 per unit.

    (c) g: $m = 0.32$; Amount increases by $0.32 per mile.

    (d) h: $m = -100$; Annual depreciation is $100.

53. Let $y = 0$ and solve for x.

$$y = mx + b$$
$$0 = mx + b$$
$$-b = mx$$
$$-\frac{b}{m} = x$$

55. The lines are parallel or they coincide.   **57.** $12 - 8x$

59. $x + 10$ **61.** $y = -3x + 4$ **63.** $y = \frac{4}{5}x + \frac{2}{5}$

65. 2   **67.** $-\frac{1}{7}$

Section 4.6   *(pp. 210–217)*

1. (a) Not a solution   (b) Solution

    (c) Not a solution   (d) Solution

3. (a) Solution   (b) Solution

    (c) Solution   (d) Not a solution

5. Dashed   **7.** Solid   **9.** c   **10.** d

11. b   **12.** a

13. 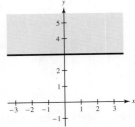   **15.**

17. c   **18.** a   **19.** b   **20.** d

21. 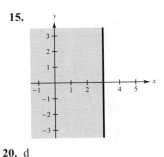   **23.**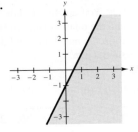

25. 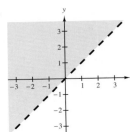   **27.**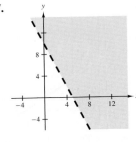

29. d   **30.** b   **31.** c   **32.** a

33. 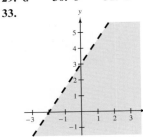   **35.**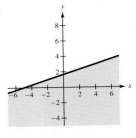

37.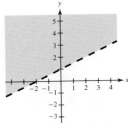

39. $9x + 12y \geq 210$, where x represents the number of hours at the grocery store and y represents the number of hours mowing lawns.

41.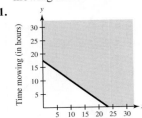

    Sample answer: $(2, 16)$, $(15, 7)$, $(20, 20)$

43. The inequality symbol is $<$.

45. No. Only points in the shaded region below the dashed line represent solutions. $(0, 1)$ is on the dashed line.

47. $y > 6x$ **49.** $x + y \geq 9$

   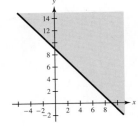

51. $y \le x + 3$

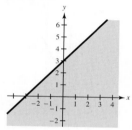

53. $y > -\frac{2}{3}x + 2$ **55.** $y < \frac{1}{2}x + 1$

57. (a) $y \le 0.35x$, where x represents the total calories consumed per day and y represents the fat calories consumed per day.

(b)

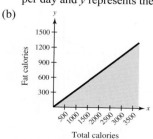

Total calories

Sample answer: $(1500, 500)$, $(2000, 700)$, $(2500, 800)$

59. (a) $t + \frac{3}{2}c \le 12$, where t represents the number of tables and c represents the number of chairs.

(b)

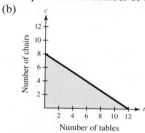

Number of tables

61. $y > 0$

63. Yes. When you divide each side of $2x < 2y$ by 2, you get $y > x$.

65. $\frac{9}{2}$ **67.** $\frac{16}{3}$ **69.** $\frac{18}{1}$ **71.** $y = -2x$ **73.** $y = \frac{2}{5}x - 2$

75. $y = \frac{2}{9}x - \frac{7}{3}$

Review Exercises *(pp. 220–223)*

1.

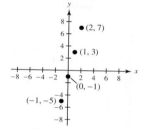

3. A: $(3, -2)$; B: $(0, 5)$; **5.** Quadrant II **7.** Quadrant III
C: $(-1, 3)$; D: $(-5, -2)$

9.

x	-1	0	1	2
$y = 4x - 1$	-5	-1	3	7

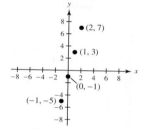

11. $y = -\frac{3}{4}x + 3$ **13.** $y = 3x - 4$

15. (a) Solution (b) Not a solution
(c) Not a solution (d) Not a solution

17. (a)

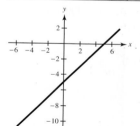

Wattage of 120-V light bulb

(b) The energy rate increases as the wattage increases.

19.

x	-2	-1	0	1	2
y	-7	-6	-5	-4	-3

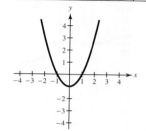

21.

x	-2	-1	0	1	2
y	3	0	-1	0	3

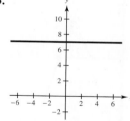

23.

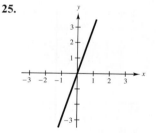

25.

27.

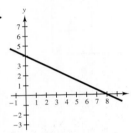

29.

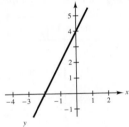

69. $\frac{2}{3}$

71. $y = -x + 6$

73. $y = 2x + 1$

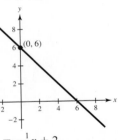

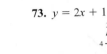

31.

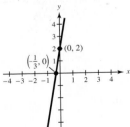

33.

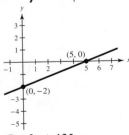

75. $y = -\frac{1}{2}x + 2$

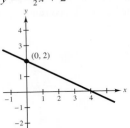

35.

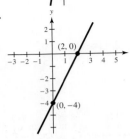

37. $C = 3x + 125$;

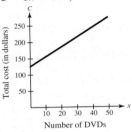

Number of DVDs

77. Parallel   **79.** Neither   **81.** $y = 2x - 9$

83. $y = -\frac{8}{3}x + \frac{1}{3}$ **85.** $x = 3$ **87.** $x + 2y + 4 = 0$

89. $y - 8 = 0$ **91.** $4x + 9y - 12 = 0$

93. (a) $y = x + 9$ (b) $y = -x - 3$

95. $y = 5$ **97.** $y = 3$ **99.** $W = 5500 + 0.07S$

39. Domain: $\{-2, 3, 5, 8\}$; Range: $\{1, 3, 7, 8\}$

41. Domain: $\{-2, 2, 4\}$; Range: $\{-3, 0, 3, 4\}$

43. Function   **45.** Function   **47.** Not a function

49. (a) $-\frac{3}{4}$ (b) 3   (c) $\frac{15}{2}$ (d) -1

51. (a) 64   (b) 63   (c) 48   (d) 0

53. (a) 3   (b) 13   (c) 5   (d) 0

55. $D = \{1, 2, 3, 4, 5\}$
$R = \{5, 10, 15, -10, -15\}$

57. $D = \{-2, 0, 3, 4, 7\}$
$R = \{-1, 6, -2, 0\}$

59. $\frac{1}{2}$

101. (a) Not a solution   (b) Not a solution
(c) Solution   (d) Solution

103. b   **104.** c   **105.** d   **106.** a

107.

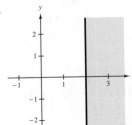

109.

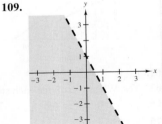

111.

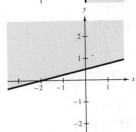

61.

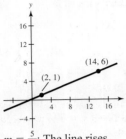

$m = \frac{5}{12}$; The line rises.

63.

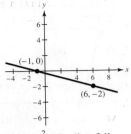

$m = -\frac{2}{7}$; The line falls.

113. (a) $2x + 3y \leq 120$, where x represents the number of DVD players and y represents the number of camcorders.

(b)

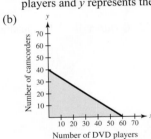

Number of DVD players

Sample answer: $(10, 15), (20, 20), (30, 20)$

65.

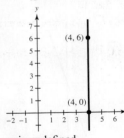

m is undefined.
The line is vertical.

67.

$m = -3$; The line falls.

Chapter Test *(p. 224)*

1.

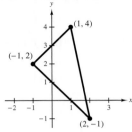

2. (a) Not a solution (b) Solution
   (c) Solution (d) Not a solution

3. 0   **4.** $(-2, 0), (0, 8)$

5.

x	-2	-1	0	1	2
y	2	-1	-4	-7	-10

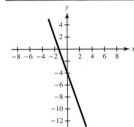

6.

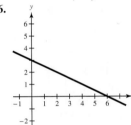

7.

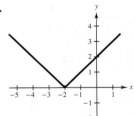

8.

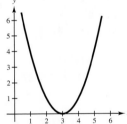

9. No, because some input values (0 and 1) have two different output values.

10. Yes, because it passes the Vertical Line Test.

11. (a) 0 (b) 0 (c) -16 (d) $-\frac{3}{8}$

12. $\frac{3}{14}; y = \frac{3}{14}x + \frac{15}{14}$

13.

      $(-2, 2), (-1, 0)$

14. $-\frac{8}{7}$ **15.** $y = -\frac{3}{8}x + 6$ **16.** $x = 3$

17. (a) Solution (b) Solution
   (c) Solution (d) Solution

18.

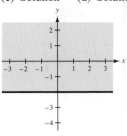

19.

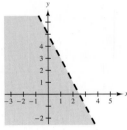

20.

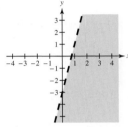

21. Sales are increasing at a rate of 230 units per year.

CHAPTER 5

Section 5.1 *(pp. 226–233)*

1. $2u^3v^3$ **3.** $-3x^6$ **5.** $-125z^6$ **7.** $64u^5$

9. $3m^4n^3$ **11.** $-\frac{4a^2}{9y^2}$ **13.** $x^{n-3}y^{2n-2}$ **15.** $-\frac{8x^4y}{9}$

17. 1   **19.** $\frac{1}{25}$ **21.** $\frac{3}{2}$ **23.** $-\frac{1}{1000}$ **25.** $\frac{7}{x^4}$

27. $\frac{1}{8x}$ **29.** $\frac{1}{64x^3}$ **31.** y^2 **33.** x^6 **35.** t^2

37. $25y^2$ **39.** $\frac{1}{625u^4}$ **41.** $\frac{10}{x}$ **43.** $\frac{1}{4x^4}$

45. $-\frac{y^9}{125x^3}$ **47.** $\frac{x^5}{2y^4}$ **49.** $\frac{81v^8}{u^6}$ **51.** $\frac{5}{y^2}$

53. 3.1×10^{-4} **55.** 3.6×10^6 **57.** 9.4608×10^{12} km

59. 720,000,000   **61.** 0.0000001359   **63.** 15,000,000°C

65. 9×10^{15} **67.** 1.6×10^{12} **69.** 4.50×10^{13}

71. 3.46×10^{10} **73.** 3.33×10^5 **75.** 2.4

77. Write the number in the form $c \times 10^{-n}$, where $1 \le c < 10$, by moving the decimal point n places to the right.

79. $\frac{25u^8v^2}{4}$ **81.** $x^{2n-1}y^{2n-1}$ **83.** $\frac{1}{2x^8y^3}$ **85.** $6u$

87. x^8y^{12} **89.** $\frac{2b^{11}}{25a^{12}}$ **91.** 6.8×10^5 **93.** 2.5×10^9

95. 6×10^6 **97.** 7×10^{22} stars

99. $\$1.35 \times 10^{13} = \$13,500,000,000,000$

101. Scientific notation makes it easier to multiply or divide very large or very small numbers because the properties of exponents make it more efficient.

103. False. $\frac{1}{3^{-3}} = 27$, which is greater than 1.

105. The product rule can be applied only to exponential expressions with the same base.

107. The power-to-power rule applied to this expression raises the base to the *product* of the exponents.

109. $6x$ **111.** $a^2 + 3ab + 3b^2$

113. 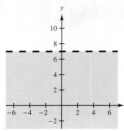   **115.**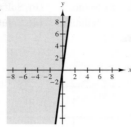

117.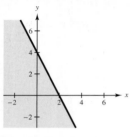

Section 5.2   *(pp. 234–241)*

1. Polynomial
3. Not a polynomial because the exponent in the first term is not an integer
5. Not a polynomial because the exponent is negative
7. Standard form: $-5x^2 + 7x + 10$; Degree: 2; Leading coefficient: -5
9. Standard form: $-3m^5 - m^2 + 6m + 12$; Degree: 5; Leading coefficient: -3
11. $20w - 4$ 13. $4z^2 - z - 2$ 15. $2b^3 - b^2$
17. $5x + 13$ 19. $-x - 28$ 21. $4x^2 + 2x + 2$
23. $10z + 4$
25. (a) $T = -0.026t^2 + 0.73t + 2.0$

   (b)
   Civilian Labor Force (65 years old or older)

27. $9x - 11$ 29. $x^2 - 2x + 2$ 31. $5z^3 - z - 4$
33. $x - 1$ 35. $-x^2 - 2x + 3$ 37. $-x^3 + 9x^2 - x - 10$
39. $-2x^4 - 5x^3 - 4x^2 + 6x - 10$ 41. $-2x - 20$
43. $3x^3 - 2x + 2$ 45. $2x^4 + 9x + 2$ 47. $12z + 8$
49. $y^3 - y^2 - 3y + 7$
51. $4t^2 + 20$ 53. $6v^2 + 90v + 30$ 55. $x^2 - 2x$
57. $2x^2 - 2x$ 59. $21x^2 - 8x$ 61. $3x^2$ and x^2, $2x$ and $-4x$
63. Lining up the like terms
65. $-3x^5 - 3x^4 + 2x^3 - 6x + 6$ 67. $\frac{3}{2}y^2 + \frac{5}{4}$
69. $1.6t^3 - 3.4t^2 - 7.3$ 71. $4b^2 - 3$ 73. $11x^2 - 3x$
75. (a) $P = -x^2 + 75x - 200, 0 \le x \le 60$
   (b) $1200   (c) $-$150
77. No, two binomials may have no like terms or like terms that differ only in sign.
79. Two terms are like terms if they are both constant or if they have the same variable factor(s); Numerical coefficients

81. Yes. A polynomial is an algebraic expression whose terms are all of the form ax^k, where a is any real number and k is a nonnegative integer.
83. 6   85. $-\frac{3}{5}$
87.    89. -5 91. 3.3714×10^{-15}

Mid-Chapter Quiz   *(p. 242)*

1. $81m^6$ 2. $-24x^7y$ 3. $-\dfrac{4}{3x^2y}$ 4. $\dfrac{t}{12}$
5. $\dfrac{5}{x^2y^3}$ 6. $\dfrac{3yz}{5x^2}$ 7. $\dfrac{a^6}{9b^4}$ 8. 1   9. 8.168×10^{12}
10. 0.005021
11. Because the exponent of the third term is negative.
12. Degree: 3; Leading coefficient: -4 13. $x^2 + 5x - 3$
14. $y^2 + 6y + 3$ 15. $-v^3 + v^2 + 6v - 5$ 16. $10s - 11$
17. $3x^2 + 5x - 4$ 18. $3x^3 - 2x + 2$
19. $x^3 - x^2 + 5x - 19$ 20. $x^2 - 2x + 6$
21. $5x^4 + 3x^3 - 4x^2 + 6$ 22. $10x + 36$

Section 5.3   *(pp. 244–253)*

1. $-x^3 + 4x$ 3. $-6x^3 - 15x$ 5. $24x^4 - 12x^3 - 12x$
7. $2x^3 - 4x^2 + 16x$ 9. $4x^4 - 3x^3 + x^2$
11. $4t^4 - 12t^3$ 13. $x^2 + 7x + 12$ 15. $3x^2 - 2x - 5$
17. x^2; $4x$; $2x$; 8   19. $2x^2 - 5xy + 2y^2$
21. $15x^2 + 23x + 6$ 23. $-8x^2 + 18x + 18$
25. $3x^2 - 5xy + 2y^2$ 27. $2x^2 + 4x = 2x(x + 2)$
29. $x^2 + 3x + 2 = (x + 1)(x + 2)$ 31. $4x^2 + 16x + 16$
33. $64x^2 + 32x + 4$ 35. $3x^3 + 6x^2 - 4x - 8$
37. $15s + 4$ 39. $-x^6 + 8x^3 + 32x^2 - 2x - 8$
41. $(x + 2)^2 = x^2 + 4x + 4$
43. $x^3 + 3x^2 + x - 1$ 45. $x^3 - 7x^2 + 11x - 5$
47. $5x^3 - 8x^2 - 8$ 49. $x^2 + x - 6$ 51. $8x^3 + 27$
53. $3x^5 + 4x^3 + 7x^2 + x + 7$ 55. $x^4 - x^2 + 4x - 4$
57. $-2x^4 - 9x^3 + 16x^2 + 11x - 12$
59. $x^3 - 6x^2 + 12x - 8$ 61. $x^2 - 9$ 63. $16t^2 - 36$
65. $16x^2 - y^2$ 67. $x^2 + 12x + 36$ 69. $t^2 - 6t + 9$
71. $64 - 48z + 9z^2$ 73. $16 + 56s^2 + 49s^4$
75. $4x^2 - 20xy + 25y^2$ 77. 4 ft $\times$ 4 ft
79. $x - y$ and $x + 2$
81. First: x, x; Outer: $2, x$; Inner: $x, -y$; Last: $2, -y$
83. $(x + 4)(x + 5) = x^2 + 9x + 20$
85. $8x$ 87. $x^3 - 12x - 16$
89. $x^2 + y^2 + 2xy + 2x + 2y + 1$
91. Yes; The right side is obtained by simplifying the left side.
93. $x^3 + 6x^2 + 12x + 8$ 95. $(x^2 + 10x)$ ft^2
97. False. $(3x)^2 = 3^2 \cdot x^2 = 9x^2 \ne 3x^2$
99. Multiplying a polynomial by a monomial is an application of the Distributive Property. Polynomial multiplication requires repeated use of this property. $4x(x - 2) = 4x^2 - 8x$

101. *mn.* Each term of the first polynomial must be multiplied by each term of the second polynomial.

103. $15x - 7$ **105.** $-4x + 17$ **107.** 11.25

109. $33\frac{1}{3}\%$ **111.** 4   **113.** $\frac{15}{4}$

Section 5.4   *(pp. 254–261)*

1. $7x^2 - 2x, x \neq 0$ **3.** $m^3 + 2m - \dfrac{7}{m}$ **5.** $-4x + 2, x \neq 0$

7. $-10z^2 - 6, z \neq 0$ **9.** $v^2 + \dfrac{5}{2}v - 2, v \neq 0$

11. $\dfrac{5}{2}x - 4 + \dfrac{7}{2}y, x \neq 0, y \neq 0$ **13.** $112 + \dfrac{5}{9}$ **15.** $215 + \dfrac{2}{3}$

17. $242\frac{1}{5}$ **19.** $x - 5, x \neq 3$ **21.** $x + 7, x \neq 3$

23. $3t - 4, t \neq \dfrac{3}{2}$ **25.** $3x^2 - 3x + 1 + \dfrac{2}{3x + 2}$

27. $x - 4 + \dfrac{32}{x + 4}$ **29.** $x^2 - 5x + 25, x \neq -5$

31. $x + 2 + \dfrac{1}{x^2 + 2x + 3}$ **33.** $4x^2 + 12x + 25 + \dfrac{52x - 55}{x^2 - 3x + 2}$

35. $x + 3, x \neq 2$ **37.** $x^2 - x + 4 - \dfrac{17}{x + 4}$

39. $5x^2 + 14x + 56 + \dfrac{232}{x - 4}$ **41.** $0.1x + 0.82 + \dfrac{1.164}{x - 0.2}$

43. $(x - 3)(x + 4)(x - 2)$ **45.** $(x - 1)(2x + 1)^2$

47. $(x + 3)^2(x - 3)(x + 4)$ **49.** $5\left(x - \dfrac{4}{5}\right)(3x + 2)$ **51.** -8

53. Dividend: 1253, Divisor: 12, Quotient: 104, Remainder: 5

55. A divisor divides evenly into a dividend when the remainder is zero.

57. $x^2 + 2x - 5 - \dfrac{10}{x - 2}$

59. $7uv, u \neq 0, v \neq 0$ **61.** $3x + 4, x \neq -2$

63. $x^{2n} + x^n + 4, x^n \neq -2$ **65.** $x^3 - 5x^2 - 5x - 10$

67. $x^2 + 2x + 1$ **69.** $2x + 8$

71. x is not a factor of the numerator.

$$\dfrac{6x + 5y}{x} = \dfrac{6x}{x} + \dfrac{5y}{x} = 6 + \dfrac{5y}{x}$$

73. $\dfrac{x^2 + 4}{x + 1} = x - 1 + \dfrac{5}{x + 1}$

Dividend: $x^2 + 4$, Divisor: $x + 1$,

Quotient: $x - 1$, Remainder: 5

75. $x < \dfrac{3}{2}$ **77.** $1 < x < 5$ **79.** $x \leq -8$ or $x \geq 16$

81. Quadrant II or III   **83.** Located on the y-axis

Review Exercises   *(pp. 264–267)*

1. x^9 **3.** u^6 **5.** $-8z^3$ **7.** $4u^7v^3$ **9.** $2z^3$

11. $\dfrac{5g^2}{16}$ **13.** $144x^4$ **15.** $\dfrac{1}{72}$ **17.** $\dfrac{64}{27}$ **19.** $12y$

21. $\dfrac{2}{x^3}$ **23.** 1   **25.** $\dfrac{a^7}{2b^6}$ **27.** $\dfrac{4x^6}{y^5}$ **29.** $\dfrac{405u^5}{v}$

31. 3.19×10^{-5} **33.** 1.735×10^7 **35.** 1,950,000

37. 0.0000205   **39.** 3.6×10^7 **41.** 500

43. Standard form: $-5x^3 + 10x - 4$; Degree: 3

Leading coefficient: -5

45. Standard form: $5x^4 + 4x^3 - 7x^2 - 2x$; Degree: 4

Leading coefficient: 5

47. Standard form: $7x^4 + 11x^2 - 1$; Degree: 4

Leading coefficient: 7

49. $9x$ **51.** $5y^3 + 5y^2 - 12y + 10$ **53.** $7u^2 + 8u + 5$

55. $x^3 + 6x + 2$ **57.** $2x^4 - 7x^2 + 3$ **59.** $(4x - 6)$ units

61. 4   **63.** $-6x^2 + 9x - 8$ **65.** $7y^2 - y + 6$

67. $3x^2 + 4x - 14$ **69.** $36x^2 - 10x$

71. $2x^2 + 8x$ **73.** $-12x^3 - 6x^2$ **75.** $x^2 + 12x + 20$

77. $14x^2 - 31x - 10$ **79.** $x^2 + 10x + 24$

81. $12x^2 + 7x - 12$ **83.** $-28x^2 + 45x - 18$

85. $x^2 + 4xy + 3y^2$ **87.** $x^3 - x^2 - 28x - 12$

89. $2t^3 - 7t^2 + 9t - 3$ **91.** $12x^3 - 11x^2 - 13x + 10$

93. $y^4 - 2y^3 - 6y^2 + 22y - 15$ **95.** $8x^3 + 12x^2 + 6x + 1$

97. $(6x^2 + 38x + 60)$ in.²   **99.** $x^2 + 6x + 9$

101. $\dfrac{1}{4}x^2 - 4x + 16$ **103.** $u^2 - 36$

105. $4r^2 - 20rt + 25t^2$ **107.** $4x^2 - 16y^2$

109. $2x^2 - \dfrac{1}{2}, x \neq 0$ **111.** $3xy - y + 1, x \neq 0, y \neq 0$

113. $5x - 8 + \dfrac{19}{x + 2}$ **115.** $x^2 - 2, x \neq \pm 1$

117. $x^2 - x - 3 - \dfrac{3x^2 - 2x - 3}{x^3 - 2x^2 + x - 1}$

119. $x + 2 + \dfrac{3}{x + 1}$ **121.** $x^2 + 5x - 7, x \neq -2$

123. $x^3 + 3x^2 + 6x + 18 + \dfrac{29}{x - 3}$

125. $(x - 2)(x + 1)(x + 3)$

Chapter Test   *(p. 268)*

1. x^9 **2.** $25y^{14}$ **3.** $\dfrac{2a}{3}$ **4.** $\dfrac{9y^6}{x^4}$

5. $-24u^9v^5$ **6.** $\dfrac{27x^6}{2y^4}$

7. (a) 6.9×10^8 (b) 0.0000472

8. Degree: 4; Leading coefficient: -3

9. $2z^2 - 3z + 15$ **10.** $7u^3 - 1$ **11.** $y^2 + 8y - 3$

12. $-6x^2 + 12x$ **13.** $x^2 - 14x + 49$ **14.** $4x^2 - 9$

15. $2z^3 + z^2 - z + 10$ **16.** $y^3 + 3y^2 - 4y - 12$

17. $2z^2 + \dfrac{1}{2}$ **18.** $-2x^2 + \dfrac{3}{2}$

19. $2x^3 + 6x^2 + 3x + 9 + \dfrac{20}{x - 3}$ **20.** $t^2 + 3 - \dfrac{6t - 6}{t^2 - 2}$

21. $4x^2 - x$ **22.** $2x^2 + 11x - 6$

23. $1500r^2 + 3000r + 1500$ **24.** $x - 3$

CHAPTER 6

Section 6.1   *(pp. 270–277)*

1. z^2 **3.** $2x$ **5.** u^2v **7.** $3y^5z^4$ **9.** 1

11. $14a^2$ **13.** $3(x + 1)$ **15.** $8(t - 2)$ **17.** $6(4y^2 - 3)$

19. $x(x + 1)$ **21.** $u(25u - 14)$ **23.** $2x^3(x + 3)$

25. No common factor   **27.** $4(3x^2 + 4x - 2)$

29. $25(4 + 3z + 2z^2)$ **31.** $3x^2(3x^2 + 2x + 6)$

33. $5u(2u + 1)$ **35.** $10ab(1 + a)$

37. $4xy(1 - 2x + 6x^3y^4)$ **39.** $-5(2x - 1)$

41. $-5(3x^2 - x - 2)$ **43.** $-2(x^2 - 6x - 2)$

45. $(x - 3)(x + 5)$ **47.** $(q - 5)(y - 1)$

49. $(x - 4)(x^3 - 2)$ **51.** $(x + 10)(x + 1)$

53. $(x + 3)(x + 4)$ **55.** $(x + 3)(x - 5)$

57. $(2x - 7)(2x + 7)$ **59.** $(2x + 1)(3x - 1)$

61. $(x + 4)(8x + 1)$ **63.** $x + 1$ **65.** $3x^2 - 2$

67. The terms were grouped so that the terms in each group have a common monomial factor: x^3 and x^2 have a common factor of x^2, and $2x$ and 2 have a common factor of 2.

69. $x + 1$ **71.** $x + 3$ **73.** $10y - 1$ **75.** $14x + 5y$

77. $(a + b)(2a - b)$ **79.** $(y - 4)(ky + 2)$

81. $6x^2$ **83.** $2\pi r(r + h)$ **85.** $kx(Q - x)$

87. There are no more common monomials that can be factored out.

89. *Sample answer:* $4x^3 - 24x^2 = 4x^2(x - 6)$

91. *Sample answer:* $3x^3 + 2x + 7$

93. (a) Not a solution (b) Solution

95. (a) Solution (b) Not a solution

97. $\dfrac{9y^2}{4x^6}$ **99.** $\dfrac{1}{z^4}$ **101.** $3mn$ **103.** $x + 1$

Section 6.2 *(pp. 278–285)*

1. $x + 1$ **3.** $(x + 4)(x + 2)$ **5.** $(x + 5)(x - 3)$

7. $(x + 1)(x + 11)$; $(x - 1)(x - 11)$

9. $(x + 14)(x + 1)$; $(x - 14)(x - 1)$;
$(x + 7)(x + 2)$;$(x - 7)(x - 2)$

11. $(x - 11)(x + 2)$ **13.** $(x - 7)(x - 2)$

15. $-(x - 5)(x + 3)$ **17.** $(x + 10)(x - 7)$

19. 21 square units **21.** 9 units **23.** $(x - 9z)(x + 2z)$

25. $(x - 2y)(x - 3y)$ **27.** $(x + 5y)(x + 3y)$

29. $(a + 5b)(a - 3b)$ **31.** $4(x - 5)(x - 3)$

33. $9(x^2 + 2x - 2)$ **35.** $x(x - 10)(x - 3)$

37. $3x(x + 4)(x + 2)$ **39.** $x^2(x - 2)(x - 3)$

41. $2x^2(x - 3)(x - 7)$

43. (a) $4x(x - 4)(x - 3)$; The model was found by multiplying
the length, width, and height of the box.
$V = lwh = (8 - 2x)(6 - 2x)(x) = 4x(x - 4)(x - 3)$
(b) About 1.1 ft

45. $c = -35$, which is negative.

47. A product of -35 and a sum of 2. **49.** Prime

51. $x(x + 2y)(x + 3y)$ **53.** Prime **55.** $2xy(x + 3y)(x - y)$

57. $\pm 9, \pm 11, \pm 19$ **59.** *Sample answer:* $2, -10$

61. $(x + 3)(x + 1)$

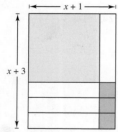

63. (a) $V = (y - 2x)(z - 2x)x$ (b) $4x(x - 4)(x - 2)$
(c) $y = 8$ ft, $z = 4$ ft

65. (a) and (d)
(a) Not completely factored; (d) Completely factored

67. Because c is positive, m and n have like signs that match
the sign of b.

69. 7 **71.** $-\frac{1}{3}$ **73.** 3 **75.** $-\frac{7}{2}$ **77.** $2x^3$

79. a^3b^4 **81.** $18r^3s^3$ **83.** $2u^2v^3$

Section 6.3 *(pp. 286–293)*

1. $x + 4$ **3.** $t + 3$ **5.** $(2x + 3)(x + 1)$

7. $(4y + 1)(y + 1)$ **9.** $(6y - 1)(y - 1)$

11. Prime **13.** $(x + 4)(4x - 3)$

15. $(3x - 2)(3x - 4)$ **17.** $2(v + 7)(v - 3)$

19. $3(z - 1)(3z - 5)$ **21.** $2s(4s - 1)(2s - 3)$

23. $-(2x - 9)(x + 1)$ **25.** $-(6x + 5)(x - 2)$

27. (a) $(2x - 1)$ ft (b) 168 ft^3

29. $(3x + 1)(x + 1)$ **31.** $(7x - 1)(x + 3)$

33. $(3x + 4)(2x - 1)$ **35.** $(5x - 2)(3x - 1)$

37. (a) $(x + 2)$ in. (b) 5865 in.^{3} **39.** No

41. The middle term of the trinomial represented by
$(3x - 1)(x + 16)$ would be $-13x$ rather than $47x$
if $(3x + 1)(x - 16)$ were the correct factorization.

43. $-x^2(5x + 4)(3x - 2)$ **45.** $6x(x - 4)(x + 8)$

47. $9u^2(2u^2 + 2u - 3)$ **49.** $\pm 11, \pm 13, \pm 17, \pm 31$

51. $\pm 1, \pm 4, \pm 11$

53. $\pm 22, \pm 23, \pm 26, \pm 29, \pm 34, \pm 43, \pm 62, \pm 121$

55. *Sample answer:* $-1, -7$ **57.** *Sample answer:* $-8, 3$

59. *Sample answer:* $-6, -1$ **61.** $2(x + 5) = 2x + 10$

63. Rewrite the middle term so that you can group the first two
terms and the last two terms to factor the trinomial.

65. The product of the last terms of the binomials is 15, not -15.

67. *Sample answer:* $2x^3 + 2x^2 + 2x$

69. Four $(ax + 1)(x + c)$, $(ax + c)(x + 1)$,
$(ax - 1)(x - c)$, $(ax - c)(x - 1)$

71. $2^2 \cdot 5^3$ **73.** $2^3 \cdot 3^2 \cdot 11$ **75.** $2x^2 + 9x - 35$

77. $49y^2 - 4$ **79.** $(x - 9)(x + 5)$ **81.** $(z + 2)(z + 20)$

Mid-Chapter Quiz *(p. 294)*

1. $2x - 1$ **2.** $x - y$ **3.** $y - 6$ **4.** $y + 3$

5. $3(3x^2 + 7)$ **6.** $5a^2(a - 5)$ **7.** $(x + 7)(x - 6)$

8. $(t - 3)(t^2 + 1)$ **9.** $(y + 6)(y + 5)$

10. $(u + 8)(u - 7)$ **11.** $x(x - 6)(x + 5)$

12. $2y(x + 8)(x - 4)$ **13.** $(2y - 9)(y + 3)$

14. $(3 + z)(2 - 5z)$ **15.** $(3x - 2)(4x + 1)$

16. $2s^2(5s^2 - 7s + 1)$

17. $\pm 7, \pm 8, \pm 13$; These integers are the sums of the factors
of 12.

18. 16, 21; Find a number c that has a pair of factors with a sum
of -10.

19. m and n are factors of 6.
$(3x + 1)(x + 6)$ $(3x - 1)(x - 6)$
$(3x + 6)(x + 1)$ $(3x - 6)(x - 1)$
$(3x + 2)(x + 3)$ $(3x - 2)(x - 3)$
$(3x + 3)(x + 2)$ $(3x - 3)(x - 2)$

20. $10(2x + 8) = 20x + 80$

Section 6.4 *(pp. 296–303)*

1. $(x + 3)(x - 3)$ **3.** $\left(u + \frac{1}{2}\right)\left(u - \frac{1}{2}\right)$

5. $(4y + 3)(4y - 3)$ **7.** $(x + 1)(x - 3)$

9. (a) $-(4t + 29)(4t - 29)$ (b) 1.5 sec

11. $2(x + 6)(x - 6)$ **13.** $x(2 + 5x)(2 - 5x)$

15. $2y(2y + 5)(2y - 5)$ **17.** $(y^2 + 9)(y + 3)(y - 3)$

19. $(1 + x^2)(1 + x)(1 - x)$ **21.** $2(x^2 + 9)(x + 3)(x - 3)$

23. Yes **25.** No **27.** Yes **29.** $x^2 + 2x + 1 = (x + 1)^2$

31. $(x - 4)^2$ **33.** $(x + 7)^2$ **35.** $(2t + 1)^2$

37. $(5y - 1)^2$ **39.** $(x - 3y)^2$ **41.** $(2y + 5z)^2$

43. x cm $\times$ $(x + 8)$ cm $\times$ $(x + 8)$ cm

45. $(x - 2)(x^2 + 2x + 4)$ **47.** $(y + 4)(y^2 - 4y + 16)$

49. $(1 + 2t)(1 - 2t + 4t^2)$ **51.** $(3u - 2)(9u^2 + 6u + 4)$

53. $(3x + 4y)(9x^2 - 12xy + 16y^2)$ **55.** No

57. $a = x, b = 2$ **59.** ± 2 **61.** ± 36 **63.** 9

65. 4 **67.** $y^2(y + 5)(y - 5)$ **69.** $(x - 1)^2$

71. $(t + 10)(t - 12)$ **73.** Prime

75. $(x^2 + 9)(x + 3)(x - 3)$ **77.** $(x + 1)(x - 1)(x - 4)$

79. 441   **81.** 3599   **83.** $(x + 3)$; 1; $(x + 4)(x + 2)$

85. Box 1: $(a - b)a^2$; Box 2: $(a - b)ab$; Box 3: $(a - b)b^2$
The sum of the volumes of boxes 1, 2, and 3 equals the volume of the large cube minus the volume of the small cube, which is the difference of two cubes.

87. False. $x(x + 2) - 2(x + 2) = (x + 2)(x - 2)$

89. False. $x^3 - 27 = (x - 3)(x^2 + 3x + 9)$, whereas $(x - 3)^3 = (x^2 - 6x + 9)(x - 3)$.

91. To identify the difference of two squares, look for coefficients that are squares of integers and for variables raised to even powers.
Let a and b be real numbers, variables, or algebraic expressions.
$a^2 - b^2 = (a + b)(a - b)$

93. 4   **95.** -1 **97.** $(2x + 1)(x + 3)$

99. $(3m - 4)(2m + 5)$

Section 6.5 *(pp. 304–311)*

1. No   **3.** No   **5.** Yes   **7.** 0, 4   **9.** $-10, 3$

11. $-4, 2$ **13.** $-\frac{5}{2}, -\frac{1}{3}$ **15.** $-4, -\frac{1}{2}, 3$ **17.** $-2, 0$

19. 0, 8   **21.** ± 5 **23.** $-2, 5$ **25.** 4, 6   **27.** $-5, \frac{5}{4}$

29. $-\frac{1}{2}, 7$ **31.** $-1, \frac{2}{3}$ **33.** 4, 9   **35.** 4   **37.** -8

39. $\frac{3}{2}$ **41.** 0, 7, 12   **43.** $-\frac{1}{3}, 0, \frac{1}{2}$ **45.** ± 2

47. $\pm 3, -2$ **49.** ± 3 **51.** $\pm 1, 0, 3$ **53.** $\pm 2, -\frac{3}{2}, 0$

55. 20 ft × 27 ft   **57.** Base: 6 in.; Height: 9 in.

59. 5 sec   **61.** About 1.9 sec

63. Yes. It has the form $ax^2 + bx + c = 0$ with $a \neq 0$.

65. The factors $2x$ and $x + 9$ are set equal to zero to form two linear equations.

67. $0, \frac{5}{8}$ **69.** $-4, 9$ **71.** $-2, 10$ **73.** $-6, 5$

75. $-13, 5$ **77.** $-2, 6$

79. $(-3, 0), (3, 0)$; The x-intercepts are solutions of the polynomial equation.

81. $(0, 0), (3, 0)$; The x-intercepts are solutions of the polynomial equation.

83. $x^2 - 4x - 12 = 0$ **85.** 15

87. 2 sec   **89.** 40 systems, 50 systems

91. Maximum number: n. The third-degree equation $(x + 1)^3 = 0$ has only one real solution: $x = -1$.

93. When a quadratic equation has a repeated solution, the graph of the equation has one x-intercept which is the vertex of the graph.

95. Many polynomial equations with irrational solutions cannot be factored.

97. \$0.0625/oz   **99.** \$0.071/oz

101. All real numbers x such that $x \neq -1$

103. All real numbers x such that $x \leq 3$

Review Exercises *(pp. 314–316)*

1. t^2 **3.** $3x^2$ **5.** $7x^2y^3$ **7.** $4xy$ **9.** $3(x - 2)$

11. $t(3 - t)$ **13.** $5x^2(1 + 2x)$ **15.** $4a^2(2 - 3a)$

17. $5x(x^2 + x - 1)$ **19.** $4(2y^2 + y + 3)$ **21.** $x(3x + 4)$

23. $(x + 1)(x - 3)$ **25.** $(u - 2)(2u + 5)$

27. $(y + 3)(y^2 + 2)$ **29.** $(x^2 + 1)(x + 2)$

31. $(x + 3)(x - 4)$ **33.** $(x - 7)(x + 4)$

35. $(u - 4)(u + 9)$ **37.** $(x - 6)(x + 4)$

39. $(y + 7)(y + 3)$ **41.** $(b + 15)(b - 2)$

43. $(w + 8)(w - 5)$ **45.** $\pm 6, \pm 10$ **47.** ± 12

49. $(x - y)(x + 10y)$ **51.** $(y + 3x)(y - 9x)$

53. $(x + 2y)(x - 4y)$ **55.** $4(x - 2)(x - 4)$

57. $x(x + 3)(x + 6)$ **59.** $3(x + 9)(x - 3)$

61. $4x(x + 2)(x + 7)$ **63.** $(3x + 5)(x - 1)$

65. $(2x - 7)(x + 1)$ **67.** $(3x + 2)(2x + 1)$

69. $(4y + 1)(y - 1)$ **71.** $(3x - 2)(x + 3)$

73. $(3x - 1)(x + 2)$ **75.** $(2x - 1)(x - 1)$

77. $\pm 2, \pm 5, \pm 10, \pm 23$ **79.** *Sample answer:* $2, -6$

81. $3(x + 6)(x + 5)$ **83.** $3(2y - 1)(y + 7)$

85. $3u(2u + 5)(u - 2)$ **87.** $4y(2y - 3)(y - 1)$

89. $2x(3x - 2)(x + 3)$ **91.** $(3x + 1)$ in.

93. $(2x - 7)(x - 3)$ **95.** $(4y - 3)(y + 1)$

97. $(3x - 2)(2x + 5)$ **99.** $(7x + 5)(2x + 1)$

101. $(a + 10)(a - 10)$ **103.** $(5 + 2y)(5 - 2y)$

105. $3(2x + 3)(2x - 3)$ **107.** $(u + 3)(u - 1)$

109. $-(z - 1)(z - 9)$ **111.** $3y(y + 5)(y - 5)$

113. $st(s + t)(s - t)$ **115.** $(x^2 + 9)(x + 3)(x - 3)$

117. $(x^2 + 4)(x - 2)$ **119.** No   **121.** Yes

123. $(x - 4)^2$ **125.** $(3s + 2)^2$ **127.** $(y + 2z)^2$

129. $\left(x + \frac{1}{3}\right)^2$ **131.** $(a + 1)(a^2 - a + 1)$

133. $(3 - 2t)(9 + 6t + 4t^2)$ **135.** $(2x + y)(4x^2 - 2xy + y^2)$

137. 0, 2   **139.** $-\frac{1}{2}, 3$ **141.** $-10, -\frac{9}{5}, \frac{1}{4}$

143. $-\frac{4}{3}, 2$ **145.** $-3, \frac{1}{2}$ **147.** $-4, 9$ **149.** ± 10

151. $-4, 0, 3$ **153.** 0, 2, 9   **155.** $\pm 1, 6$ **157.** $\pm 3, 0, 5$

159. 9, 11   **161.** 45 in. × 20 in.   **163.** 8 seconds

Chapter Test *(p. 317)*

1. $9x^2(1 - 7x^3)$ **2.** $(z + 17)(z - 10)$

3. $(t - 10)(t + 8)$ **4.** $(3x - 4)(2x - 1)$

5. $3y(y - 1)(y + 25)$ **6.** $(2 + 5v)(2 - 5v)$

7. $(x + 2)(x^2 - 2x + 4)$ **8.** $-(z + 1)(z + 21)$

9. $(x + 2)(x + 3)(x - 3)$ **10.** $(4 + z^2)(2 + z)(2 - z)$

11. $2x - 3$ **12.** ± 6 **13.** 36

14. $3x^2 - 3x - 6 = 3(x + 1)(x - 2)$ **15.** $-4, \frac{3}{2}$

16. $-3, \frac{2}{3}$ **17.** $-\frac{3}{2}, 2$ **18.** $-4, -1, 0$ **19.** $x + 4$

20. 7 in. × 12 in.   **21.** 8.875 sec   **22.** 24, 26

Cumulative Test *(p. 318)*

1. Quadrant II or Quadrant III

2. (a) Not a solution   (b) Solution
(c) Solution   (d) Not a solution

3. $(0, 3)$ **4.** $(6, 0), (0, 3)$

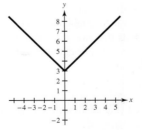   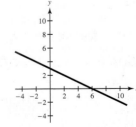

5. Not a function

6. *Sample answer:* $(-2, 2)$; There are infinitely many points on a line.

7. $y = \frac{2}{5}x - \frac{3}{2}$

8. Perpendicular **9.** Parallel

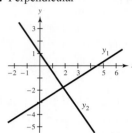 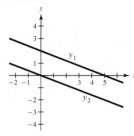

10. $-5x^2 + 5$ **11.** $-42z^4$ **12.** $3x^2 - 7x - 20$

13. $25x^2 - 9$ **14.** $25x^2 + 60x + 36$ **15.** $x + 12$

16. $x + 1 + \dfrac{2}{x - 4}$ **17.** $\dfrac{x}{54y^4}$ **18.** $2u(u - 3)$

19. $(x + 2)(x - 10)$ **20.** $x(x + 4)^2$ **21.** $(x + 2)^2(x - 2)$

22. $0, 12$ **23.** $-\dfrac{3}{5}, 3$ **24.** $\dfrac{4}{x^2}$

25. $C = 150 + 0.45x$; $\$181.50$

26. $x + y \le 10$; The second song can be up to 4 minutes long.

CHAPTER 7

Section 7.1 *(pp. 320–327)*

1. All real values of x such that $x \ne 3$

3. All real values of z such that $z \ne 0$ and $z \ne 4$

5. All real values of t such that $t \ne -4$ and $t \ne 4$

7. $x > 0$ **9.** $\{1, 2, 3, 4, \dots\}$ **11.** $\dfrac{x - 3}{4x}$

13. $x, x \ne 8, x \ne 0$ **15.** $u - 6, u \ne 6$ **17.** $\dfrac{y(y + 2)}{y + 6}, y \ne 2$

19. $-\dfrac{3x + 5}{x + 3}, x \ne 4$ **21.** $\dfrac{x + 8}{x - 3}, x \ne -\dfrac{3}{2}$

23. $\dfrac{3y^2}{y^2 + 1}, x \ne 0$ **25.** $\dfrac{y - 8x}{15}, y \ne -8x$

27. $\dfrac{5 + 3xy}{y^2}, x \ne 0$ **29.** $\dfrac{u - 2v}{u - v}, u \ne -2v$ **31.** $\dfrac{3(m - 2n)}{m + 2n}$

33. $\dfrac{x}{x + 3}, x > 0$ **35.** $\dfrac{1}{4}, x > 0$

37. The rational expression is in simplified form when the numerator and denominator have no factors in common (other than ± 1).

39. A change in signs is one additional step sometimes needed in order to find common factors.

41. $x(x - 2)$ **43.** $x + 3$

45. (a) $C = 60,000 + 6.50x$ (b) $\overline{C} = \dfrac{60,000 + 6.50x}{x}$

(c) $\{1, 2, 3, 4, \dots\}$ (d) $\$11.95$

47. $\pi, d > 0$

49. A rational expression is undefined when the denominator is equal to zero. To find the implied domain restriction of a rational function, find the values that make the denominator equal to zero (by setting the denominator equal to zero), and exclude those values from the domain.

51. $\dfrac{1}{x^2 + 1}$

53. The student incorrectly divided (the denominator may not be split up) and the domain is not restricted.

Correct solution: $\dfrac{x^2 + 7x}{x + 7} = \dfrac{x(x + 7)}{x + 7} = x, x \ne -7$

55. To write the polynomial $g(x)$, multiply $f(x)$ by $(x - 2)$ and divide by $(x - 2)$.

$$g(x) = \dfrac{f(x)(x - 2)}{(x - 2)} = f(x), x \ne 2$$

57. $\dfrac{3}{16}$ **59.** 1 **61.** $4a^4$ **63.** $-3b^3 + 9b^2 - 15b$

Section 7.2 *(pp. 328–335)*

1. x^2 **3.** $(x + 2)^2$ **5.** $\dfrac{7}{3}, x \ne 0$ **7.** $\dfrac{s^3}{6}, s \ne 0$

9. $24u^2, u \ne 0$ **11.** $24, x \ne -\dfrac{3}{4}$ **13.** $\dfrac{2uv(u + v)}{3(3u + v)}, u \ne 0$

15. $-1, r \ne 12$ **17.** $-\dfrac{x + 8}{x^2}, x \ne \dfrac{3}{2}$

19. $4(r + 2), r \ne 3, r \ne 2$ **21.** $2t + 5, t \ne 3, t \ne -2$

23. $-\dfrac{xy(x + 2y)}{x - 2y}$ **25.** $\dfrac{(x - y)^2}{x + y}, x \ne -3y$

27. $\dfrac{(x - 1)(2x + 1)}{(3x - 2)(x + 2)}, x \ne \pm 5, x \ne -1$

29. $-\dfrac{x^2(x + 3)(x - 3)(2x + 5)(3x - 1)}{2(2x + 1)(2x + 3)(2x - 3)}, x \ne 0, x \ne \dfrac{1}{2}$

31. $\dfrac{x(x + 1)}{3(x + 2)}, x \ne -1$ **33.** $\dfrac{4x}{3}, x \ne 0$ **35.** $\dfrac{1}{10u}$

37. $\dfrac{4}{3}, x \ne -2, x \ne 0, x \ne 1$ **39.** $\dfrac{3y^2}{2x^3}, y \ne 0$ **41.** $\dfrac{3}{2(a + b)}$

43. $\dfrac{2}{3}, x \ne y, x \ne -y$ **45.** $x^4 y(x + 2y), x \ne 0, y \ne 0, x \ne -2y$

47. $\dfrac{x - 4}{x - 5}, x \ne -6, x \ne -5, x \ne 3$

49. $Y = \dfrac{(6591.43t + 102,139.0)(0.029t + 1)}{0.184t + 4.3}, 4 \le t \le 9$

51. To multiply two rational expressions, let a, b, c, and d represent real numbers, variables, or algebraic expressions such that $b \ne 0$ and $d \ne 0$. Then the product of $\dfrac{a}{b}$ and $\dfrac{c}{d}$ is $\dfrac{a}{b} \cdot \dfrac{c}{d} = \dfrac{ac}{bd}$.

53. To divide rational expressions, multiply the first expression be the reciprocal of the second expression.

55. $\dfrac{x + 4}{3}, x \ne -2, x \ne 0$ **57.** $\dfrac{1}{4}, x \ne -1, x \ne 0, y \ne 0$

59. $\dfrac{(x + 1)(2x - 5)}{x}, x \ne -5, x \ne -1, x \ne -\dfrac{2}{3}$

61. $\dfrac{x^4}{(x^n + 1)^2}, x^n \ne -3, x^n \ne 3, x \ne 0$ **63.** $\dfrac{x}{4(2x + 1)}, x > 0$

65. $\dfrac{\pi x}{4(2x + 1)}, x > 0$

67. In simplifying a product of rational expressions, you divide the common factors out of the numerator and denominator.

69. The domain needs to be restricted, $x \ne a, x \ne b$.

71. The first expression needs to be multiplied by the reciprocal of the second expression (not the second by the reciprocal of the first), and the domain needs to be restricted.

$$\dfrac{x^2 - 4}{5x} \div \dfrac{x + 2}{x - 2} = \dfrac{x^2 - 4}{5x} \cdot \dfrac{x - 2}{x + 2}$$
$$= \dfrac{(x - 2)^2(x + 2)}{5x(x + 2)}$$
$$= \dfrac{(x - 2)^2}{5x}, x \ne \pm 2$$

73. $\dfrac{9}{8}$ **75.** $\dfrac{13}{15}$ **77.** $-3, 0$

Section 7.3 *(pp. 336–343)*

1. $\dfrac{3x}{2}$ **3.** $-\dfrac{3}{a}$ **5.** $-\dfrac{2}{9}$ **7.** $\dfrac{1}{x - 3}, x \ne 0$

9. $\dfrac{1}{c+4}, c \neq 1$　**11.** $20x^3$　**13.** $36y^3$　**15.** $15x^2(x+5)$

17. $126z^2(z+1)^4$　**19.** $56t(t+2)(t-2)$

21. $2y(y+1)(2y-1)$　**23.** $\dfrac{-12x+25}{20x}$　**25.** $\dfrac{7(a+2)}{a^2}$

27. $\dfrac{5(5x+22)}{x+4}$　**29.** $\dfrac{x^2-7x-15}{(x+3)(x-2)}$　**31.** $-\dfrac{2}{x+3}, x \neq 3$

33. $0, x \neq 4$　**35.** $\dfrac{3(x+2)}{x-8}$　**37.** $1, x \neq \dfrac{2}{3}$

39. $\dfrac{5(x+1)}{(x+5)(x-5)}$　**41.** $\dfrac{3(8x-3)}{5x(x-1)}$　**43.** $\dfrac{6}{(x-6)(x+5)}$

45. $\dfrac{4x}{(x-4)^2}$　**47.** $\dfrac{y-x}{xy}, x \neq -y$　**49.** $\dfrac{2(4x^2+5x-3)}{x^2(x+3)}$

51. $-\dfrac{u^2-uv-5u+2v}{(u-v)^2}$

53. $T = \dfrac{0.023587t^2 - 1.14494t + 13.24}{(-0.049t+1)(-0.055t+1)}, 5 \leq t \leq 10$

55. True

57. To add or subtract rational expressions with like denominators, simply add (or subtract) the terms in the numerators and keep the common denominator.

59. $\dfrac{4}{x^2(x^2+1)}$　**61.** $\dfrac{x}{x-1}, x \neq -6$　**63.** $\dfrac{5t}{12}$

65. $A = -4, B = 2, C = 2$

67. When the numerators are subtracted, the result should be $(x-1)-(4x-11) = x-1-4x+11.$

69. Yes. The LCD of $4x+1$ and $\dfrac{x}{x+2}$ is $x+2.$

71. $2v+4$　**73.** x^2-2x-3　**75.** $(x-3)(x-4)$

77. $(a-6)(2a+3)$

Mid-Chapter Quiz　*(p. 344)*

1. All real values of x such that $x \neq -1$ and $x \neq 0$

2. All real values of x such that $x \neq -2$ and $x \neq 2$

3. $\dfrac{3}{2}y, y \neq 0$　**4.** $\dfrac{2u^3}{5}, u \neq 0 \ v \neq 0$　**5.** $-\dfrac{2x+1}{x}, x \neq \dfrac{1}{2}$

6. $\dfrac{z+3}{2z-1}, x \neq -3$　**7.** $\dfrac{5a+3b^2}{ab}$　**8.** $\dfrac{n^2}{m+n}, 2m \neq n$

9. $\dfrac{t}{2}, t \neq 0$　**10.** $\dfrac{5x}{x-2}, x \neq -2$　**11.** $\dfrac{8x}{3(x+3)(x-1)^2}$

12. $\dfrac{2x^3y}{z}, x \neq 0, x \neq 0$　**13.** $\dfrac{(a+1)^2}{9(a+b)^2}, a \neq b, a \neq -1$

14. $\dfrac{4(u-v)^2}{5uv}, u \neq \pm v$　**15.** $\dfrac{7x-11}{x-2}$　**16.** $-\dfrac{4x^2-25x+36}{(x-3)(x+3)}$

17. $0, x \neq 2, x \neq -1$　**18.** $-\dfrac{3t^2}{2}, t \neq 0, t \neq 3$

19. $\dfrac{2(x+1)}{3x}, x \neq -2, x \neq -1$　**20.** $\dfrac{(3y-x)(x-y)}{xy}, x \neq y$

21. (a) $C = 25,000 + 144x$

　(b) $\overline{C} = \dfrac{25,000+144x}{x}$

　(c) $194

Section 7.4　*(pp. 346–353)*

1. $\dfrac{1}{4}$　**3.** $6xz^3, x \neq 0, y \neq 0, z \neq 0$　**5.** $\dfrac{2xy^2}{5}, x \neq 0, y \neq 0$

7. $-\dfrac{1}{y}, y \neq 3$　**9.** $-\dfrac{5x(x+1)}{2}, x \neq -1, x \neq 0, x \neq 5$

11. $\dfrac{x+5}{3(x+4)}, x \neq 2$　**13.** $\dfrac{2(x+3)}{x-2}, x \neq -3, x \neq 7$

15. $\dfrac{(2x-5)(3x+1)}{3x(x+1)}, x \neq \pm\dfrac{1}{3}$

17. $\dfrac{(x+3)(4x+1)}{(3x-1)(x-1)}, x \neq -3, x \neq -\dfrac{1}{4}$

19. $x+2, x \neq \pm2, x \neq -3$

21. $\dfrac{(x-5)(x+2)^2}{(x-2)^2(x+3)}, x \neq -2, x \neq 7$

23. $\dfrac{y+4}{y^2}$　**25.** $-\dfrac{3x+4}{3x-4}, x \neq 0$　**27.** $\dfrac{x^2}{2(2x+3)}, x \neq 0$

29. $\dfrac{3}{4}, x \neq 0, x \neq 3$　**31.** $\dfrac{1}{x}, x \neq -1$　**33.** $\dfrac{y(2y^2-1)}{10y^2-1}, y \neq 0$

35. $\dfrac{x^2(7x^3+2)}{x^4+5}, x \neq 0$　**37.** $\dfrac{R_1R_2}{R_1+R_2}$

39. A complex fraction is a fraction that has a fraction it its numerator or denominator, or both.

$$\dfrac{\dfrac{x-3}{x}}{\dfrac{3x-9}{4}}$$

41. To simplify a complex fraction, multiply the numerator and denominator by the least common denominator of all of the fractions in the numerator and denominator.

43. $y-x, x \neq 0, y \neq 0, x \neq -y$

45. $\dfrac{x(x+6)}{3x^3+10x-30}, x \neq 0, x \neq 3$　**47.** $\dfrac{20}{7}, x \neq -1$

49. $\dfrac{y+x}{y-x}, x \neq 0, y \neq 0$

51. $\dfrac{y-x}{x^2y^2(y+x)}$　**53.** $-\dfrac{1}{2(h+2)}, h \neq 0$

55. $\dfrac{11x}{60}$　**57.** $\dfrac{11x}{24}$　**59.** $\dfrac{b^2+5b+8}{8b}$　**61.** $\dfrac{x}{8}, \dfrac{5x}{36}, \dfrac{11x}{72}$

63. $\dfrac{R_1R_2R_3}{R_1R_2+R_1R_3+R_2R_3}$

65. No. A complex fraction can be written as the division of two rational expressions, so the simplified form will be a rational expression.

67. In the second step, the set of parentheses cannot be moved because division is not associative.

$$\dfrac{(a/b)}{b} = \dfrac{a}{b}\cdot\dfrac{1}{b} = \dfrac{a}{b^2}$$

69. $72y^5$　**71.** $(3x-1)(x+2)$　**73.** $\dfrac{x}{8}, x \neq 0$

75. $(x+1)(x+2)^2, x \neq -2, x \neq -1$

Section 7.5　*(pp. 354–361)*

1. (a) Not a solution　(b) Not a solution
　(c) Not a solution　(d) Solution

3. 10　**5.** 1　**7.** 0　**9.** 8　**11.** $-\dfrac{40}{9}$　**13.** $\dfrac{1}{5}, 1$

15. $-3, \dfrac{8}{3}$　**17.** 61　**19.** $\dfrac{18}{5}$　**21.** $-\dfrac{26}{5}$　**23.** 3

25. 3　**27.** $-\dfrac{11}{5}$　**29.** $\dfrac{4}{3}$　**31.** No solution

33. No solution　**35.** 20　**37.** $-3, -2$　**39.** -5

41. 8　**43.** $3, -1$　**45.** $-3, 4$　**47.** 5 hits

49. 6 mi/hr, 4.5 mi/hr

51. A rational equation is an equation containing one or more rational expressions.

53. The domain of a rational equation must be restricted if any value of any of the variables makes any of the denominators zero.

55. ±6　**57.** $\dfrac{43}{8}$　**59.** -12　**61.** $\dfrac{14x+57}{3(x+3)}$

63. (a) and (b) $(-2, 0)$　**65.** (a) and (b) $(-1, 0), (1, 0)$

67. $\frac{3}{20}$, 10   **69.** 8.75 hr = 8 hr, 45 min

71. An extraneous solution is a "trial solution" that does not satisfy the original equation. When you substitute this "trial solution" into the original equation, the result is false—the solution does not check, or is undefined.

73. When the equation is solved, the solution is $x = 0$. However, if $x = 0$, then there is division by zero, so the equation has no solution.

75. $(x + 9)(x - 9)$ **77.** $\left(2x - \frac{1}{2}\right)\left(2x + \frac{1}{2}\right)$

79. All real values of x

Section 7.6   *(pp. 362–369)*

1. $R = 9.65x$ **3.** (a) 2 in.   (b) 15 lb

5. 18 lb   **7.** 192 ft   **9.** 667 boxes

11. $6 per pizza; Answers will vary.   **13.** $22.50

15. The other variable also increases because when one side of the equation increases, so must the other side.

17. No. The equation $y = kx$ is not equivalent to $y = kx^2$.

19. $I = kV$ **21.** $p = k/d$ **23.** $A = klw$ **25.** $h = \frac{7}{3}r$

27. $F = \frac{25}{6}xy$ **29.** $d = 120x^2/r$ **31.** $y = k/x$ with $k = 4$

33. 1350 W

35. (a) $P = \dfrac{kWD^2}{L}$ (b) Unchanged

(c) Increases by a factor of 8   (d) Increases by a factor of 4
(e) Decreases by a factor of $\frac{1}{4}$ (f) 3125 lb

37. False. If x increases, then z and y do not both necessarily increase.

39. The variable y will be one-fourth as great. If $y = k/x^2$ and x is replaced with $2x$, the result is $y = \dfrac{k}{(2x)^2} = \dfrac{k}{4x^2}$.

41. 6^4 **43.** $\left(\frac{1}{5}\right)^5$ **45.** $x - 7, x \neq -2$
47. $4x^4 - 2x^3, x \neq 3$

Review Exercises   *(pp. 372–375)*

1. All real values of y such that $y \neq 8$
3. All real values of x.
5. All real values of u such that $u \neq 1$ and $u \neq 6$

7. $(0, \infty)$ **9.** $\dfrac{2x^3}{5}, x \neq 0, y \neq 0$ **11.** $\dfrac{b - 3}{6(b - 4)}$

13. $-9, x \neq y$ **15.** $\dfrac{x}{2(x + 5)}, x \neq 5$ **17.** $\frac{1}{2}, x > 0$

19. $\frac{x}{3}, x \neq 0$ **21.** $\dfrac{y}{8x}, y \neq 0$ **23.** $12z(z - 6), z \neq -6$

25. $-\frac{1}{4}, u \neq 0, u \neq 3$ **27.** $20x^3, x \neq 0$ **29.** $\dfrac{125y}{x}, y \neq 0$

31. $\dfrac{1}{3x - 2}, x \neq -2, x \neq -1$ **33.** $3x$ **35.** $\dfrac{4}{x}$

37. $\dfrac{x - 13}{4x}$ **39.** $\dfrac{5y + 11}{2y + 1}$ **41.** $\dfrac{7x - 16}{x + 2}$ **43.** $\dfrac{2x + 3}{5x^2}$

45. $\dfrac{4x + 3}{(x + 5)(x - 12)}$ **47.** $\dfrac{5x^3 - 5x^2 - 31x + 13}{(x - 3)(x + 2)}$

49. $\dfrac{2x + 17}{(x - 5)(x + 4)}$ **51.** $\dfrac{6(x - 9)}{(x + 3)^2(x - 3)}$ **53.** $3x^2, x \neq 0$

55. $-\frac{1}{2}, x \neq 0, x \neq 2$ **57.** $\dfrac{6(x + 5)}{x(x + 7)}, x \neq \pm 5$

59. $\dfrac{3t^2}{5t - 2}, t \neq 0$ **61.** $x - 1, x \neq 0, x \neq 2$

63. $\dfrac{-a^2 + a + 16}{(4a^2 + 16a + 1)(a - 4)}, a \neq -4, a \neq 0$

65. 6   **67.** -120 **69.** $2, -\frac{3}{2}$ **71.** $\frac{36}{23}$ **73.** 5

75. $-4, 6$ **77.** ± 4 **79.** $-4, -2$ **81.** No solution

83. $-2, 2$ **85.** $-\frac{9}{5}, 3$ **87.** 16 mi/hr, 10 mi/hr

89. 8 yr   **91.** 4 people   **93.** 150 lb   **95.** 2.44 hr

97. $61.49; Answers will vary.   **99.** $44.80

Chapter Test   *(p. 376)*

1. All real values of x such that $x \neq 1$ and $x \neq 5$

2. $-2, x \neq 2$ **3.** $\dfrac{2a + 3}{5}, a \neq 4$ **4.** $3x^3(x + 4)^2$

5. $\dfrac{5z}{3}, z \neq 0$ **6.** $\dfrac{4}{y + 4}, y \neq 2$ **7.** $\dfrac{14y^6}{15}, x \neq 0$

8. $(2x - 3)^2(x + 1), x \neq -\frac{3}{2}, x \neq -1, x \neq \frac{3}{2}$ **9.** $\dfrac{x + 1}{x - 3}$

10. $\dfrac{-2x^2 + 2x + 1}{x + 1}$ **11.** $\dfrac{5x^2 - 15x - 2}{(x - 3)(x + 2)}$

12. $\dfrac{5x^3 + x^2 - 7x - 5}{x^2(x + 1)^2}$ **13.** $\dfrac{x^3}{4}, x \neq -2, x \neq 0$

14. $-(3x + 1), x \neq 0, x \neq \frac{1}{3}$ **15.** $\dfrac{(3y + x^2)(x + y)}{x^2 y}, x \neq -y$

16. $3x - 2 + \dfrac{4}{x}$ **17.** 16   **18.** $-1, -\frac{15}{2}$

19. No solution   **20.** $v = \frac{1}{4}\sqrt{u}$ **21.** 240 m³

CHAPTER 8

Section 8.1   *(pp. 378–387)*

1. (a) Solution   (b) Not a solution
3. (a) Not a solution   (b) Solution
5. $(1, 2)$ **7.** $(2, 0)$ **9.** $(3, 0)$ **11.** One solution
13. Infinitely many solutions
15. $(2, 3)$ **17.** $(0, 0)$ **19.** $\left(\frac{1}{4}, -1\right)$ **21.** $(-4, 2)$
23. $(5, -2)$ **25.** Infinitely many solutions
27. No solution   **29.** Infinitely many solutions
31. 30 tons at $125 per ton, 70 tons at $75 per ton
33. 10,000 candy bars
35. • Solve one of the equations for one variable in terms of the other.
• Substitute the expression obtained in Step 1 into the other equation to obtain an equation in one variable.
• Solve the equation obtained in Step 2.
• Back-substitute the solution from Step 3 into the expression obtained in Step 1 to find the value of the other variable.
• Check the solution to see that it satisfies both of the original equations.
37. When you obtain a true result such as $15 = 15$, the system of linear equations has infinitely many solutions.
39. No solution   **41.** $(1, 1)$ **43.** Infinitely many solutions
45. $\left(\frac{5}{2}, \frac{3}{4}\right)$ **47.** $(8, 4)$ **49.** $b = 2$ **51.** $b = -\frac{1}{3}$
53. *Sample answer*:   **55.** 18, 25
$\begin{cases} x - 2y = 0 \\ x + y = 3 \end{cases}$
57. $4000 at 8.5%, $8000 at 10%
59. Because the slopes of the two lines are not equal, the lines intersect and the system has one solution: $(79,400, 398)$.
61. The substitution method yields exact solutions.
63. 1   **65.** $-\frac{1}{21}$ **67.** $-\frac{3}{2}$ **69.** $\frac{5}{11}$

71. 5   **73.** -13

Section 8.2 *(pp. 388–395)*

1. $(8, 4)$ **3.** $(-4, 4)$ **5.** $(2, 1)$ **7.** $\left(\frac{21}{4}, -\frac{3}{2}\right)$

9. $(-2, 2)$ **11.** $\left(\frac{3}{4}, \frac{1}{2}\right)$ **13.** $(-2, 3)$ **15.** $(4, -1)$

17. $\left(\frac{50}{7}, \frac{4}{7}\right)$ **19.** $(8, 7)$ **21.** $\left(\frac{15}{11}, \frac{15}{11}\right)$ **23.** $\left(\frac{1}{2}, 0\right)$

25. Infinitely many solutions   **27.** No solution

29. 12 hr   **31.** $(-1, 2)$

33. When you obtain a false result such as $0 = 11$, the system of linear equations has no solution.

35. $\left(\frac{4}{3}, \frac{4}{3}\right)$ **37.** $\left(1, -\frac{5}{4}\right)$ **39.** Consistent   **41.** Consistent

43. $(48, 34)$ **45.** $y = \frac{1}{3}x + \frac{2}{3}$

47. \$3.78 per gallon of regular unleaded, \$3.89 per gallon of premium unleaded

49. 24 liters of 40% solution, 6 liters of 70% solution

51. *Sample answer*:
$$\begin{cases} 0.02x - 0.03y = 0.12 \\ 0.5x + 0.3y = 0.9 \end{cases}$$
Multiply each side of the first equation by 100 and multiply each side of the second equation by 10.
$$\begin{cases} 2x - 3y = 12 \\ 5x + 3y = 9 \end{cases}$$

53. *Sample answer*:
$$\begin{cases} y = 3x + 4 \\ y = -x - 8 \end{cases}$$

55.
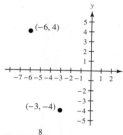
$m = -\frac{8}{3}$

57.
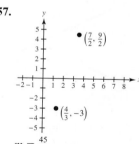
$m = \frac{45}{13}$

59.
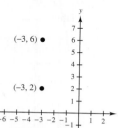
m is undefined.

61.
$x \le 3$

63.
$x < 1$

65. $(5, 5)$ **67.** $(3, -1)$

Section 8.3 *(pp. 396–403)*

1. $(22, -1, -5)$ **3.** $(14, 3, -1)$ **5.** $(1, 2, 3)$

7. $(1, 2, 3)$ **9.** $(2, -3, -2)$ **11.** $(5, 2, -4)$

13. $(5, -2, 0)$ **15.** $\left(\frac{3}{10}, \frac{2}{5}, 0\right)$ **17.** No solution

19. $(-4, 2, 3)$ **21.** $\left(-\frac{1}{2}a + \frac{1}{2}, \frac{3}{5}a + \frac{2}{5}, a\right)$

23. $\left(\frac{6}{13}a + \frac{10}{13}, \frac{5}{13}a + \frac{4}{13}, a\right)$ **25.** $(-3a + 10, 5a - 7, a)$

27. $s = -16t^2 + 144$

29. Gaussian elimination enables you to obtain an equivalent system in row-echelon form. The next steps involve the use of back-substitution to find the solution.

31. $\begin{cases} x + 2y - 3z = 13 \\ \quad y + 4z = -11 \\ \quad\quad z = -3 \end{cases}$

33. Yes. The first equation was multiplied by -2 and added to the second equation. Then the first equation was multiplied by -3 and added to the third equation.

35. $\begin{cases} x + 2y - z = -4 \\ \quad y + 2z = 1 \\ 3x + y + 3z = 15 \end{cases}$

37. $88°, 32°, 60°$

39. 2 pounds of vanilla, 4 pounds of hazelnut, 4 pounds of French roast

41. 50 students in strings, 20 students in winds, 8 students in percussion

43. The solution is apparent because the row-echelon form is
$\begin{cases} x = 1 \\ \quad y = -3. \\ \quad\quad z = 4 \end{cases}$

45. Three planes have no point in common when two of the planes are parallel and the third plane intersects the other two planes.

47. The graphs are three planes with three possible situations. If all three planes intersect in one point, there is one solution. If all three planes intersect in one line, there are an infinite number of solutions. If each pair of planes intersects in a line, but the three lines of intersection are all parallel, there is no solution.

49. $3x, 2; 3, 2$ **51.** $14t^5, -t, 25; 14, -1, 25$

53. $(4, 3)$ **55.** $(-2, 6)$

Mid-Chapter Quiz *(p. 404)*

1. Not a solution, because substituting $x = 4$ and $y = 2$ into $3x + 4y = 4$ yields $20 = 4$, which is a contradiction.

2. Solution, because substituting $x = 2$ and $y = -1$ into the equations yields true equalities.

3. $(3, 2)$ **4.** $(4, 1)$ **5.** $(4, -1)$ **6.** $(-3, 11)$

7. $(2, 2)$ **8.** $(-1, 4)$ **9.** $(6, 2)$ **10.** $(3, 3)$

11. $\left(-\frac{1}{2}, 6\right)$ **12.** $(8, 1)$ **13.** $(3, -1)$

14. $\left(\frac{1}{2}, -\frac{1}{2}, 1\right)$ **15.** $(5, -1, 3)$

16. $\begin{cases} x + y = -2 \\ 2x - y = 32 \end{cases}$

17. $\begin{cases} x + y + z = 7 \\ 2x - y = 9 \\ -2x + y + 3z = 21 \end{cases}$

18. $13\frac{1}{3}$ gallons of 20% solution, $6\frac{2}{3}$ gallons of 50% solution

19. $68°, 41°, 71°$

Section 8.4 *(pp. 406–415)*

1. 4×2 **3.** 4×1 **5.** 2×2 **7.** 1×1 **9.** 1×4

11. (a) $\begin{bmatrix} 4 & -5 \\ -1 & 8 \end{bmatrix}$ (b) $\begin{bmatrix} 4 & -5 & \vdots & -2 \\ -1 & 8 & \vdots & 10 \end{bmatrix}$

13. (a) $\begin{bmatrix} 1 & 1 & 0 \\ 5 & -2 & -2 \\ 2 & 4 & 1 \end{bmatrix}$ (b) $\begin{bmatrix} 1 & 1 & 0 & \vdots & 0 \\ 5 & -2 & -2 & \vdots & 12 \\ 2 & 4 & 1 & \vdots & 5 \end{bmatrix}$

15. $\begin{cases} 4x + 3y = 8 \\ x - 2y = 3 \end{cases}$

17. $\begin{cases} x \quad\quad + 2z = -10 \\ \quad 3y - z = 5 \\ 4x + 2y \quad\quad = 3 \end{cases}$

19. $\begin{bmatrix} 1 & 1 & -4 & 2 \\ 0 & 4 & 5 & 5 \\ 0 & 0 & 8 & 3 \end{bmatrix}$ **21.** $\begin{bmatrix} 1 & -2 & 3 \\ 3 & 4 & 5 \end{bmatrix}$ **23.** $\begin{bmatrix} 1 & 4 & 3 \\ 0 & 0 & 0 \end{bmatrix}$

25. $(2, -3, 2)$ **27.** $(1, 1)$ **29.** $(-2, 1)$ **31.** $(1, 2, -1)$
33. $(34, -4, -4)$ **35.** $(2a + 1, 3a + 2, a)$ **37.** No solution
39. \$800,000 at 8%, \$500,000 at 9%, \$200,000 at 12%
41. 1×4 (one row and four columns), 2×2 (two rows and two columns), 4×1 (four rows and one column)
43. Row operations are performed on systems of equations and elementary row operations are performed on matrices.
45. $(4, -2, -1)$ **47.** $\left(2, 5, \frac{5}{2}\right)$
49. Theater A: 300 tickets, theater B: 600 tickets, theater C: 600 tickets
51. 5, 8, 20
53. 15 computer chips, 10 resistors, 10 transistors
55. 3×5. There are 15 entries in the matrix, so the order is 3×5, 5×3, or 15×1. Because there are more columns than rows, the second number in the order must be larger than the first.
57. There will be a row in the matrix with all zero entries except in the last column.
59. The first entry in the first column is 1, and the other two are zero. In the second column, the first entry is a nonzero real number, the second number is 1, and the third number is zero. In the third column, the first two entries are nonzero real numbers and the third entry is 1.
61. -42 **63.** 26 **65.** $(4, -2, 1)$

Section 8.5 (pp. 416–425)

1. 5 **3.** 27 **5.** 0 **7.** 0 **9.** -24 **11.** -0.33
13. -24 **15.** -2 **17.** -30 **19.** 3 **21.** $(1, 2)$
23. $(2, -2)$ **25.** Not possible, $D = 0$ **27.** $(-1, 3, 2)$
29. $\left(\frac{1}{2}, -\frac{2}{3}, \frac{1}{4}\right)$ **31.** Not possible, $D = 0$ **33.** $\left(\frac{51}{16}, -\frac{7}{16}, -\frac{13}{16}\right)$
35. 16 **37.** 21 **39.** $\frac{31}{2}$ **41.** $\frac{33}{8}$ **43.** 42
45. 16 **47.** Collinear **49.** Not collinear
51. Not collinear **53.** $x - 2y = 0$ **55.** $7x - 6y - 28 = 0$
57. $9x + 10y + 3 = 0$ **59.** $16x - 15y + 22 = 0$
61. Find the difference of the products of the two diagonals of the matrix.
63. No. Only square matrices have determinants.
65. 248 **67.** 105.625 **69.** 4.32 **71.** 250 mi^2
73. $I_1 = 2, I_2 = 3, I_3 = 5$
75. (a) $\left(\dfrac{13}{2k + 6}, \dfrac{3k - 4}{-2k^2 - 6k}\right)$ (b) $0, -3$
77. A square matrix is a square array of numbers. The determinant of a square matrix is a real number.

79. The determinant is zero. Because two rows are identical, each term is zero when expanding by minors along the other row. Therefore, the sum is zero.

81. 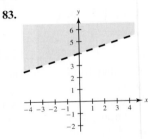 **83.**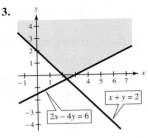

85. $x - 2y = 0$ **87.** $y = 2$
89. Function **91.** Not a function

Section 8.6 (pp. 426–433)

1. 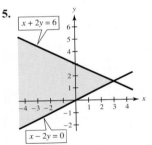 **3.**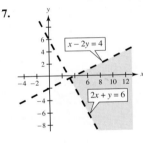

5. 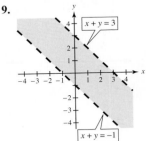 **7.**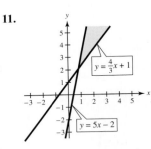

9. 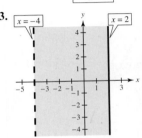 **11.**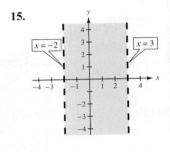

13. **15.**

17.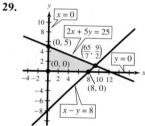

19.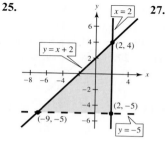

21.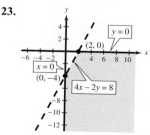

23.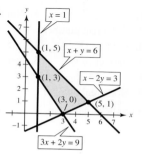

25.

27.

29.

31. $\begin{cases} x \geq 1 \\ y \geq x - 3 \\ y \leq -2x + 6 \end{cases}$

33. $\begin{cases} x \geq -2 \\ x \leq 2 \\ y \geq -\frac{1}{2}x - 4 \\ y \leq -\frac{1}{2}x + 2 \end{cases}$

35. $\begin{cases} x + \frac{3}{2}y \leq 12 \\ \frac{4}{3}x + \frac{3}{4}y \leq 16 \\ x \geq 0 \\ y \geq 0 \end{cases}$

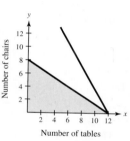

Number of chairs / Number of tables

37. A system of linear inequalities in two variables consists of two or more linear inequalities in two variables.

39. Not necessarily. Two boundary lines can intersect outside the solution region at a point that is not a vertex of the region.

41.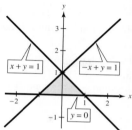

43. No solution

45.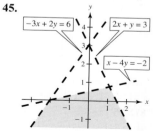

47. $\begin{cases} x + y \leq 25{,}000 \\ y \geq 4{,}000 \\ x \geq 3y \end{cases}$

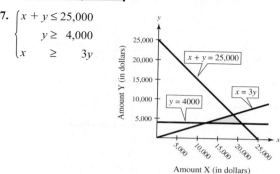

49. $\begin{cases} 30x + 20y \geq 75{,}000 \\ x + y \leq 3{,}000 \\ x \leq 2{,}000 \end{cases}$

51. $\begin{cases} y \leq 22 \\ y \geq 10 \\ y \geq 2x - 24 \\ y \geq -2x - 24 \end{cases}$

53. The point (x_1, y_1) is a solution if it satisfies each inequality in the system.

55. The solution set of a system of linear equations is either a point, a line, or a plane, whereas the solution set of a system of linear inequalities is an infinite number of solutions in a specific region of the plane.

57. $\left(\frac{8}{3}, 0\right)$, $(0, 8)$ **59.** $(4, 0)$, $(0, -2)$

61. $(-1, 0)$, $(3, 0)$, $(0, -1)$ **63.** (a) 12   (b) 2

65. (a) $\dfrac{11}{7}$ (b) $\dfrac{k + 1}{k - 3}$

Review Exercises *(pp. 436–439)*

1. (a) Solution   (b) Not a solution
3. (a) Not a solution   (b) Solution
5. $(5, 1)$ **7.** $(1, 1)$ **9.** No solution
11. Infinitely many solutions   **13.** $(-1, 1)$
15. $(4, 8)$ **17.** $(4, 2)$ **19.** $(4, -1)$ **21.** $\left(\frac{5}{2}, 3\right)$
23. $(5, -5)$ **25.** No solution
27. Infinitely many solutions   **29.** 5556 cameras

31. $(5, -2)$ **33.** $(3, 0)$ **35.** $(8, -3)$ **37.** $\left(\frac{4}{3}, \frac{5}{6}\right)$
39. $\left(\frac{3}{5}, \frac{1}{2}\right)$ **41.** No solution   **43.** Infinitely many solutions
45. 400 adult tickets, 100 children's tickets
47. 30 gallons of 100% solution, 20 gallons of 75% solution
49. $(3, 2, 5)$ **51.** $(0, 3, 6)$ **53.** $(2, -3, 3)$
55. $(0, 1, -2)$ **57.** $8000 at 7%, $5000 at 9%, $7000 at 11%

59. (a) $\begin{bmatrix} 7 & -5 \\ 1 & -1 \end{bmatrix}$ (b) $\begin{bmatrix} 7 & -5 & \vdots & 11 \\ 1 & -1 & \vdots & -5 \end{bmatrix}$

61. $\begin{cases} 4x - y = 2 \\ 6x + 3y + 2z = 1 \\ y + 4z = 0 \end{cases}$

63. $(10, -12)$ **65.** $(0.6, 0.5)$ **67.** $(3, -6, 7)$
69. $\left(\frac{1}{2}, -\frac{1}{3}, 1\right)$ **71.** 10   **73.** -51 **75.** 1
77. $(-3, 7)$ **79.** $(2, -3, 3)$ **81.** 16   **83.** 7
85. Not collinear   **87.** $x - 2y + 4 = 0$

89. 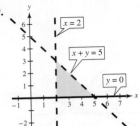   **91.**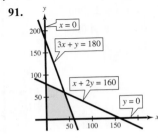

93. $\begin{cases} x + y \le 500 \\ x \ge 150 \\ y \ge 220 \end{cases}$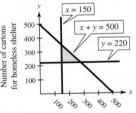

Chapter Test   *(p. 440)*

1. (a) Solution   (b) Not a solution   **2.** $(3, 2)$ **3.** $(1, 3)$
4. $(2, 3)$ **5.** $(-2, 2)$ **6.** $(2, 2a - 1, a)$ **7.** $(-1, 3, 3)$
8. $(2, 1, -2)$ **9.** $\left(4, \frac{1}{7}\right)$ **10.** $(5, 1, -1)$
11. -16 **12.** 12

13. 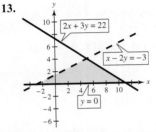   **14.** 50,000 mi

15. $13,000 at 4.5%, $9000 at 5%, $3000 at 8%

16. $\begin{cases} 30x + 40y \ge 300{,}000 \\ x \le 9{,}000 \\ y \le 4{,}000 \end{cases}$

where x is the number of reserved seat tickets and y is the number of floor seat tickets.

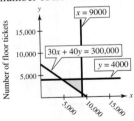

CHAPTER 9

Section 9.1   *(pp. 442–449)*

1. 8   **3.** -10 **5.** Not a real number
7. Not a real number   **9.** 3   **11.** -3
13. -1 **15.** 3   **17.** 25   **19.** $\frac{1}{16}$ **21.** Perfect square
23. Perfect cube   **25.** Neither   **27.** 11 in.   **29.** 6 ft
31. 8   **33.** 5   **35.** -7 **37.** 10
39. Not a real number   **41.** $-\frac{1}{5}$ **43.** -2 **45.** -7
47. $36^{1/2} = 6$ **49.** $\sqrt[4]{256^3} = 64$ **51.** $\sqrt[3]{x}$
53. $\left(\sqrt[5]{y}\right)^2$ or $\sqrt[5]{y^2}$ **55.** 9   **57.** 9   **59.** $\frac{1}{4}$ **61.** $\frac{4}{9}$
63. -6 **65.** $\frac{1}{9}$ **67.** x^3 **69.** $t^{-9/4} = \frac{1}{t^{9/4}}$
71. $x^{3/4}y^{1/4}$ **73.** $y^{1/8}$ **75.** $(x + y)^{1/2}$
77. (a) 3   (b) 5   (c) Not a real number   (d) 9
79. $-\infty < x < \infty$ **81.** $0 \le x < \infty$ **83.** $-\frac{9}{2} \le x < \infty$
85. $\sqrt[3]{x}, x^{1/3}$ **87.** The product rule of exponents
89. $\pm \frac{3}{4}$ **91.** ± 0.4 **93.** $\frac{1}{10}$ **95.** 0.1   **97.** $u^{7/3}$
99. $x^{-1} = \frac{1}{x}$ **101.** $y^{13/12}$ **103.** $y^{5/2}z^4$
105. $x^{3/8}$ **107.** $\frac{1}{(3u - 2v)^{5/6}}$ **109.** 0.128

111. 23 ft $\times$ 23 ft
113. The nth root of a number a is the number b for which $a = b^n$.
115. No. $\sqrt{2}$ is an irrational number. Its decimal representation is a nonterminating, nonrepeating decimal.
117. $0, 1, 4, 5, 6, 9$; Yes   **119.** 5   **121.** $-\frac{4}{5}$
123. $s = kt^2$ **125.** $a = kbc$

Section 9.2   *(pp. 450–457)*

1. $2\sqrt{7}$; Yes. For a square, $A = s^2$, so $\sqrt{A} = s$.
So, the figure shows that $\sqrt{A} = \sqrt{28} = s = 2\sqrt{7}$.
3. $2\sqrt{2}$ **5.** $3\sqrt{2}$ **7.** $3\sqrt{5}$ **9.** $4\sqrt{6}$ **11.** $3\sqrt{17}$
13. $13\sqrt{7}$ **15.** $2|y|$ **17.** $3x^2\sqrt{x}$ **19.** $4y^2\sqrt{3}$
21. $3\sqrt{13}|y^3|$ **23.** $2|x|y\sqrt{30y}$ **25.** $8a^2b^3\sqrt{3ab}$
27. $2\sqrt[3]{6}$ **29.** $2\sqrt[3]{14}$ **31.** $x\sqrt[4]{x^3}$ **33.** $|x|\sqrt[4]{x^2}$
35. $2x\sqrt[3]{5x^2}$ **37.** $3|y|\sqrt[4]{4y^2}$ **39.** $xy\sqrt[3]{x}$
41. $|xy|\sqrt[4]{4y^2}$ **43.** $2xy\sqrt[5]{y}$ **45.** $\frac{4}{3}$ **47.** $\frac{\sqrt[3]{35}}{4}$
49. $|y|\sqrt{13}$ **51.** $\frac{4a^2\sqrt{2}}{|b|}$ **53.** $\frac{2\sqrt[5]{x^2}}{y}$ **55.** $\frac{\sqrt{3}}{3}$

57. $\dfrac{\sqrt{7}}{7}$ **59.** $\dfrac{\sqrt[4]{20}}{2}$ **61.** $\dfrac{3\sqrt[3]{2}}{2}$ **63.** $\dfrac{\sqrt{y}}{y}$

65. $\dfrac{2\sqrt{x}}{x}$ **67.** $\dfrac{\sqrt{2}}{2x}$ **69.** $\dfrac{2\sqrt{3b}}{b^2}$ **71.** $\dfrac{\sqrt[3]{18xy^2}}{3y}$

73. $\dfrac{2xy\sqrt[3]{15yz^2}}{5z}$ **75.** $3\sqrt{5}$ **77.** 9

79. $2\sqrt{194} \approx 27.86$ ft

81. If the nth roots of a and b are real, then
$$\sqrt[n]{ab} = \sqrt[n]{a} \cdot \sqrt[n]{b}.$$
$$\sqrt[3]{108} = \sqrt[3]{27} \cdot \sqrt[3]{4} = 3\sqrt[3]{4}$$

83. • All possible nth-powered factors have been removed from each radical.
• No radical contains a fraction.
• No denominator of a fraction contains a radical.

85. 0.2   **87.** $0.06\sqrt{2}$ **89.** $2\sqrt{5}$ **91.** $\dfrac{\sqrt{13}}{5}$

93. $\dfrac{3a\sqrt[3]{2a}}{b^3}$ **95.** $\dfrac{1}{50}$ **97.** $20\sqrt[3]{300}$

99. 89.44 cycles/sec

101. (a) 17 ft   (b) 1360 ft^2

103. The perfect square factor 4 needs to be removed from the radical. $\sqrt{8} = \sqrt{4 \cdot 2} = \sqrt{4} \cdot \sqrt{2} = 2\sqrt{2}$

105. 1

107. Because $\sqrt[3]{u} = u^{1/3}$ and $\sqrt[4]{u} = u^{1/4}$, when $\sqrt[3]{u}$ and $\sqrt[4]{u}$ are multiplied, the rational exponents need to be added together. Therefore, $\sqrt[3]{u} \cdot \sqrt[4]{u} = u^{1/3} \cdot u^{1/4} = u^{7/12}$.

109. (3, 2)   **111.** No solution   **113.** (4, −8, 10)

Section 9.3   (pp. 458–465)

1. $2\sqrt{2}$ **3.** $13\sqrt[3]{y}$ **5.** $2\sqrt{7}$ **7.** $9\sqrt{2}$ **9.** $9\sqrt{2}$

11. $4\sqrt[4]{5} - 7\sqrt[4]{13}$ **13.** $13\sqrt[3]{7} + \sqrt{3}$ **15.** $8\sqrt[3]{9} + 4\sqrt{x}$

17. $-3\sqrt[4]{x} - 3\sqrt{x}$ **19.** $21\sqrt{3}$ **21.** $23\sqrt{5}$ **23.** $\sqrt{x}$

25. $12\sqrt{x}$ **27.** $27\sqrt{2y}$ **29.** $3x\sqrt{x}$ **31.** $\dfrac{12\sqrt{10}}{x^2}$

33. $7x^2\sqrt{3x}$ **35.** $30\sqrt[3]{2}$ **37.** $(x + 2)\sqrt[3]{6x}$

39. $(10 + 6u)\sqrt[3]{3u^2}$ **41.** $(3x + 3)\sqrt[3]{3x}$ **43.** $\dfrac{9}{2}\sqrt[3]{2x}$

45. $h = 3\sqrt[3]{x}$ ft   **47.** $\dfrac{2\sqrt{5}}{5}$ **49.** $\dfrac{9\sqrt{2}}{2}$ **51.** $\dfrac{5\sqrt{2y}}{2}$

53. $\dfrac{(3x + 2)\sqrt{3x}}{3x}$ **55.** $\dfrac{(7y^3 - 3)\sqrt{7y}}{7y^2}$ **57.** $12\sqrt{6x}$

59. $18\sqrt{7x}$ **61.** $19\sqrt{2x} + 8x\sqrt{2}$ **63.** $9x\sqrt{3} + 5\sqrt{3x}$

65. (a) $T = 2{,}245{,}671 - 2{,}127{,}778\sqrt{t} + 683{,}203t - 73{,}313\sqrt{t^3}$
(b) 31,713 people

67. $6\sqrt{6x} + \sqrt{6x} + 2x\sqrt{6} - 3\sqrt{6}$

69. $6\sqrt{6x}$ and $\sqrt{6x}$, $2x\sqrt{6}$ and $-3\sqrt{6}$

71. $13\sqrt{x + 1}$ **73.** $-t\sqrt[3]{2t}$ **75.** $(6a + 1)\sqrt{5a}$

77. $4\sqrt{x - 1}$ **79.** $(x + 2)\sqrt{x - 1}$ **81.** $5a\sqrt[3]{ab^2}$

83. $3r^3s^2\sqrt{rs}$ **85.** 0   **87.** >   **89.** <

91. (a) $5\sqrt{10} \approx 15.8$ ft   (b) $80 + 20\sqrt{10} \approx 143$ ft

93. No. $\sqrt{5} + (-\sqrt{5}) = 0$

95. Yes. $\sqrt{2x} + \sqrt{2x} = 2\sqrt{2x} = \sqrt{4} \cdot \sqrt{2x} = \sqrt{8x}$

97. (a) The student combined terms with unlike radicands and added the radicands. The radical expression can be simplified no further.
(b) The student combined terms with unlike indices. The radical expressions can be simplified no further.

99. $\dfrac{3(z - 1)}{2z}$ **101.** $-3, x \neq 3$ **103.** $\dfrac{2v + 3}{v - 5}$

105. $\dfrac{5}{2}$ **107.** $\dfrac{3(w + 1)}{w - 7}, w \neq -1, w \neq 3$

Mid-Chapter Quiz   (p. 466)

1. 15   **2.** $\dfrac{3}{2}$ **3.** 7   **4.** 9

5. (a) Not a real number   (b) 1   (c) 5

6. (a) 4   (b) 2   (c) 0

7. $-\infty < x < 0$ and $0 < x < \infty$ **8.** $-\dfrac{10}{3} \leq x < \infty$

9. $3|x|\sqrt{3}$ **10.** $2x^2\sqrt[4]{2}$ **11.** $\dfrac{2u\sqrt{u}}{3}$ **12.** $\dfrac{2\sqrt[3]{2}}{u^2}$

13. $5x|y|z^2\sqrt{5x}$ **14.** $4a^2b\sqrt[3]{2b^2}$ **15.** $4\sqrt{3}$

16. $3x\sqrt{7x}, x \neq 0$ **17.** $3\sqrt{3} - 4\sqrt{7}$ **18.** $4\sqrt{2y}$

19. $7\sqrt{3}$ **20.** $4\sqrt{x + 2}$ **21.** $6x\sqrt[3]{5x^2} + 4x\sqrt[3]{5x}$

22. $4xy^2z^2\sqrt{xz}$ **23.** $23 + 8\sqrt{2} \approx 34.3$ in.

Section 9.4   (pp. 468–475)

1. 4   **3.** $3\sqrt{5}$ **5.** $2\sqrt[3]{9}$ **7.** 2   **9.** $3\sqrt{7} - 7$

11. $2\sqrt{10} + 8\sqrt{2}$ **13.** $3\sqrt{2}$ **15.** $12 - 4\sqrt{15}$

17. $y + 4\sqrt{y}$ **19.** $4\sqrt{a} - a$ **21.** $\sqrt{15} - 5\sqrt{5} + 3\sqrt{3} - 15$

23. $8\sqrt{5} + 24$ **25.** $8 - 2\sqrt{15}$ **27.** $100 + 20\sqrt{2x} + 2x$

29. $45x - 17\sqrt{x} - 6$ **31.** $27 + \sqrt{3}$ **33.** $8 + 2\sqrt{x} - x$

35. $2 - \sqrt{5}, -1$ **37.** $\sqrt{11} + \sqrt{3}, 8$ **39.** $\sqrt{15} - 3, 6$

41. $\sqrt{x} + 3, x - 9$ **43.** $\sqrt{2u} + \sqrt{3}, 2u - 3$

45. $2\sqrt{2} - \sqrt{4}, 4$ **47.** $\sqrt{x} - \sqrt{y}, x - y$ **49.** 23

51. $\sqrt{5} - 2$ **53.** $-\sqrt{2} - \sqrt{3}$ **55.** $\dfrac{9 + \sqrt{6}}{15}$

57. $\dfrac{6(\sqrt{11} + 2)}{7}$ **59.** $\dfrac{7(5 - \sqrt{3})}{22}$ **61.** $\sqrt{30} + 5$

63. $2\sqrt{15} - 6$ **65.** $-\dfrac{\sqrt{x} + \sqrt{x + 3}}{3}$ **67.** $\dfrac{(\sqrt{15} + \sqrt{3})x}{4}$

69. $\dfrac{8\sqrt{x} + 24}{x - 9}$ **71.** $\dfrac{5\sqrt{t} + t\sqrt{5}}{5 - t}$ **73.** $4(\sqrt{3a} - \sqrt{a}), a \neq 0$

75. $\dfrac{4\sqrt{7} + 11}{3}$ **77.** $\dfrac{2x - 9\sqrt{x} - 5}{4x - 1}$ **79.** $\dfrac{20\sqrt{3} - 10\sqrt{5}}{7}$

81. The FOIL Method   **83.** They are conjugates.

85. $8x - 5$ **87.** $t + 5\sqrt[3]{t^2} + \sqrt[3]{t} - 3$

89. $4xy^3\left(x^4\sqrt[3]{x} + y\sqrt[3]{2x^2y^2}\right)$ **91.** (a) $2\sqrt{3} - 4$ (b) 0

93. $\dfrac{10}{\sqrt{30x}}$ **95.** $\dfrac{4}{5(\sqrt{7} - \sqrt{3})}$ **97.** $192\sqrt{2}$ in.^{2}

99. $\dfrac{\sqrt{4445}}{127}$

101. (a) If either a or b (or both) equal zero, the expression is zero and therefore rational.
(b) If the product of a and b is a perfect square, then the expression is rational.

103. $\sqrt{a} - \sqrt{b}; (\sqrt{a} + \sqrt{b})(\sqrt{a} - \sqrt{b}) = a - b;$
$\sqrt{b} - \sqrt{a}; (\sqrt{b} + \sqrt{a})(\sqrt{b} - \sqrt{a}) = b - a;$
When the order of the terms is changed, the conjugate and the product both change by a factor of -1.

105. 6 **107.** No solution **109.** $-12, 12$ **111.** $-5, 3$

113. $4|x|y^2\sqrt{2y}$ **115.** $2y\sqrt[4]{2x^2y}$

Section 9.5 *(pp. 476–483)*

1. (a) Not a solution (b) Not a solution
(c) Not a solution (d) Solution

3. (a) Not a solution (b) Solution
(c) Not a solution (d) Not a solution

5. 144 **7.** 49 **9.** 27 **11.** 49 **13.** 64

15. No solution **17.** 90 **19.** -27 **21.** 5

23. No solution **25.** 215 **27.** 4 **29.** 1

31. No solution **33.** 7 **35.** -15 **37.** $-\frac{9}{4}$ **39.** 8

41. 1, 3 **43.** $\frac{1}{4}$ **45.** $\frac{1}{2}$ **47.** 4 **49.** 7

51. 64 ft **53.** 80 ft **55.** 9 **57.** 12

59. $8\sqrt{6} \approx 19.6$ ft

61. Each side is squared to eliminate the radical.

63. To determine whether the solution is extraneous.

65. 216 **67.** $4, -12$ **69.** 25 **71.** 2, 6 **73.** $\frac{4}{9}$

75. 2, 6 **77.** 26,250 passengers

79. $h = \dfrac{\sqrt{S^2 - \pi^2 r^4}}{\pi r}$; 34 cm

81. $(\sqrt{x} + \sqrt{6})^2 \neq (\sqrt{x})^2 + (\sqrt{6})^2$

83. Substitute $x = 20$ into the equation, and then choose any value of a such that $a \leq 20$ and solve the resulting equation for b.

85. Parallel **87.** Perpendicular **89.** $(3, 2)$

91. $a^{4/5}$ **93.** $x^{3/2}, x \neq 0$

Section 9.6 *(pp. 484–491)*

1. $2i$ **3.** $-12i$ **5.** $\frac{2}{5}i$ **7.** $-\frac{6}{11}i$ **9.** $2\sqrt{2}i$

11. $\sqrt{7}i$ **13.** 2 **15.** $\frac{3\sqrt{2}}{5}i$ **17.** $0.3i$ **19.** $10i$

21. $3\sqrt{2}i$ **23.** $-2\sqrt{6}$ **25.** $-3\sqrt{6}$ **27.** -0.44

29. $-3 - 2\sqrt{3}$ **31.** $4 + 3\sqrt{2}i$ **33.** -16 **35.** Equal

37. Not equal **39.** $a = 3, b = -4$ **41.** $a = 2, b = -3$

43. $a = -4, b = -2\sqrt{2}$ **45.** $a = 64, b = 7$

47. $-14 - 40i$ **49.** $-14 + 20i$ **51.** $3 + 6i$

53. $-3 + 49i$ **55.** 24 **57.** $4 + 18i$ **59.** $-40 - 5i$

61. 45 **63.** 68 **65.** 5 **67.** 100 **69.** 144

71. $-\frac{1}{5} + \frac{2}{5}i$ **73.** $-\frac{24}{53} + \frac{84}{53}i$ **75.** $\frac{12}{13} - \frac{5}{13}i$

77. $\frac{11}{10} - \frac{7}{10}i$ **79.** $-\frac{23}{58} + \frac{43}{58}i$ **81.** $\sqrt{2}i$

83. The FOIL Method **85.** $3\sqrt{3}i$ **87.** $-35i$ **89.** $27i$

91. 9 **93.** $-21 + 20i$ **95.** $18 + 26i$ **97.** i **99.** -1

101. $-i$ **103.** -1 **105.** $\frac{9}{5} - \frac{2}{5}i$ **107.** $\frac{47}{26} + \frac{27}{26}i$

109. $2a$ **111.** $2bi$

113. (a) $\left(\dfrac{-5 + 5\sqrt{3}i}{2}\right)^3 = 125$ (b) $\left(\dfrac{-5 - 5\sqrt{3}i}{2}\right)^3 = 125$

115. (a) $1, \dfrac{-1 + \sqrt{3}i}{2}, \dfrac{-1 - \sqrt{3}i}{2}$
(b) $2, -1 + \sqrt{3}i, -1 - \sqrt{3}i$
(c) $4, -2 + 2\sqrt{3}i, -2 - 2\sqrt{3}i$

117. Exercise 109: The sum of complex conjugates of the form $a + bi$ and $a - bi$ is twice the real number of a, or $2a$.
Exercise 110: The product of complex conjugates of the form $a + bi$ and $a - bi$ is the sum of the squares of a and b, or $a^2 + b^2$.
Exercise 111: The difference of complex conjugates of the form $a + bi$ and $a - bi$ is twice the imaginary number bi, or $2bi$.
Exercise 112: The sum of the squares of complex conjugates of the form $a + bi$ and $a - bi$ is the difference of twice the squares of a and b, or $2a^2 - 2b^2$.

119. The numbers must be written in i-form first.
$\sqrt{-3}\sqrt{-3} = (\sqrt{3}i)(\sqrt{3}i) = 3i^2 = -3$

121. To simplify the quotient, multiply the numerator and the denominator by $-bi$. This will yield a positive real number in the denominator. The number i can also be used to simplify the quotient. The denominator will be the opposite of b, but the resulting number will be the same.

123. $-7, 5$ **125.** $-4, 0, 3$ **127.** 81 **129.** 25

Review Exercises *(pp. 494–497)*

1. -9 **3.** -4 **5.** $-\frac{3}{4}$ **7.** $-\frac{1}{5}$

9. Not a real number **11.** $\frac{4}{9}$ cm **13.** $27^{1/3} = 3$

15. $\sqrt[3]{216} = 6$ **17.** 81 **19.** Not a real number **21.** $\frac{1}{16}$

23. $x^{7/12}$ **25.** $z^{5/3}$ **27.** $\frac{1}{x^{5/4}}$ **29.** $ab^{2/3}$ **31.** $x^{1/8}$

33. $(3x + 2)^{1/3}, x \neq -\frac{2}{3}$ **35.** (a) Not a real number (b) 7

37. (a) -1 (b) 3 **39.** $-\infty < x < \infty$ **41.** $0 \leq x < \infty$

43. $-\infty < x \leq \frac{9}{2}$ **45.** $3\sqrt{7}$ **47.** $11\sqrt{2}$ **49.** $6u^2|v|\sqrt{u}$

51. $0.5x^2\sqrt{y}$ **53.** $2ab\sqrt[3]{6b}$ **55.** $\dfrac{\sqrt{30}}{6}$

57. $\dfrac{\sqrt[3]{4x^2}}{x}$ **59.** $\sqrt{145}$ **61.** 17 **63.** $8\sqrt{6}$

65. $14\sqrt{x} - 9\sqrt[3]{x}$ **67.** $7\sqrt[4]{y + 3}$ **69.** $3x\sqrt[3]{3x^2y}$

71. $6\sqrt{x} + 2\sqrt{3x}$ feet **73.** $10\sqrt{3}$ **75.** $2\sqrt{5} + 5\sqrt{2}$

77. $3 - x$ **79.** $3 + \sqrt{7}; 2$ **81.** $\sqrt{x} - 20; x - 400$

83. $-\dfrac{\sqrt{6} + 4\sqrt{2} - \sqrt{3} - 4}{13}$ **85.** $\dfrac{x + 20\sqrt{x} + 100}{x - 100}$

87. 32 **89.** No real solution **91.** 5 **93.** $-5, -3$

95. $\frac{3}{32}$ **97.** $\sqrt{409.76} \approx 20.24$ in. **99.** 2.93 ft

101. 64 ft **103.** $4\sqrt{3}i$ **105.** $10 - 9\sqrt{3}i$

107. $\frac{3}{4} - \sqrt{3}i$ **109.** $8.4 + 18i$ **111.** $2i$ **113.** $15i$

115. $\sqrt{70} - 2\sqrt{10}$ **117.** $a = 10, b = -4$

119. $a = 4, b = 7$ **121.** $8 - 3i$ **123.** 25

125. $11 - 60i$ **127.** $-\frac{7}{3}i$ **129.** $\frac{9}{26} - \frac{3}{13}i$ **131.** $\frac{13}{37} - \frac{33}{37}i$

Chapter Test *(p. 498)*

1. (a) 64 (b) 10 **2.** (a) $\frac{1}{25}$ (b) 6

3. $f(-8) = 7, f(0) = 3$ **4.** $\left[\frac{3}{7}, \infty\right)$

5. (a) $x^{1/3}, x \neq 0$ (b) 25 **6.** (a) $\dfrac{4\sqrt{2}}{3}$ (b) $2\sqrt[3]{3}$

7. (a) $2x\sqrt{6x}$ (b) $2xy^2\sqrt[4]{x}$ **8.** $\dfrac{2\sqrt[3]{3y^2}}{3y}$

9. $\dfrac{5(\sqrt{6} + \sqrt{2})}{2}$ **10.** $6\sqrt{2x}$ **11.** $5\sqrt{3x} + 3\sqrt{5}$

12. $16 - 8\sqrt{2x} + 2x$ **13.** $3 + 4y$ **14.** 24

15. No solution **16.** 9 **17.** $2 - 2i$ **18.** $-16 - 30i$

19. $-8 + 4i$ **20.** $13 + 13i$ **21.** $\frac{13}{10} - \frac{11}{10}i$ **22.** 144 ft

Cumulative Test *(pp. 499–500)*

1. $\dfrac{x(x+2)(x+4)}{9(x-4)}, x \neq -4, x \neq 0$

2. $\dfrac{x}{(2x-1)(x-4)}, x \neq -4, x \neq 3$ **3.** $\dfrac{5x^2 - 15x - 2}{(x+2)(x-3)}$

4. $\dfrac{3x+5}{x(x+3)}$ **5.** $\dfrac{x^3}{4}, x \neq -2, x \neq 0$

6. $x + y, x \neq 0, y \neq 0, x \neq y$ **7.** 2, 5 **8.** 2, 9

9. All real values of x such that $x \neq \frac{3}{8}$

10. (a) Solution (b) Not a solution

11. $(2, 1)$ **12.** $(3, -2)$ **13.** $(5, 4)$ **14.** $(0, 1, -2)$

15. $(1, -5, 5)$ **16.** $\left(-\dfrac{1}{5}, -\dfrac{22}{5}\right)$

17.

18. $2|x|y\sqrt{6y}$ **19.** $2a^5b^2\sqrt[3]{10b^2}$ **20.** $\dfrac{2|b^3|\sqrt{3}}{a^2}$

21. $\sqrt{t}, t \neq 0$ **22.** $35\sqrt{5x}$ **23.** $2x - 6\sqrt{2x} + 9$

24. $\dfrac{3(\sqrt{10} + \sqrt{x})}{10 - x}$ **25.** 41 **26.** 2 **27.** 11 **28.** 29

29. $-4 + 3\sqrt{2}i$ **30.** $-7 + 16i$ **31.** $21 + 20i$

32. $\frac{3}{20} + \frac{11}{20}i$ **33.** $\dfrac{11t}{14}$

34. (a) 2.8 cm (b) 30 lb

35. \$20,000 at 8% and \$30,000 at 8.5%

36. $16(1 + \sqrt{2}) \approx 38.6$ in

37. $90\sqrt{5} \approx 201.25$ ft **38.** 400 ft

CHAPTER 10

Section 10.1 *(pp. 502–509)*

1. 6, 9 **3.** $-5, 6$ **5.** $-9, 5$ **7.** 8 **9.** $-\frac{8}{9}, 2$

11. 9, 12 **13.** ± 3 **15.** $\pm\frac{4}{5}$ **17.** ± 14 **19.** $\pm\frac{5}{2}$

21. $-12, 4$ **23.** 2.5, 3.5 **25.** $2 \pm \sqrt{7}$ **27.** $-\frac{1}{2} \pm \frac{5\sqrt{2}}{2}$

29. $\pm 6i$ **31.** $\pm 2i$ **33.** $\pm\dfrac{\sqrt{17}}{3}i$ **35.** $3 \pm 5i$

37. $-\frac{4}{3} \pm 4i$ **39.** $-\frac{1}{4} \pm \sqrt{5}i$ **41.** $-3 \pm \frac{5}{3}i$

43. $\dfrac{15}{\pi} \approx 4.77$ in. **45.** $\pm 1, \pm 2$ **47.** $\pm\sqrt{2}, \pm\sqrt{3}$

49. $\pm 1, \pm\sqrt{5}$ **51.** 16 **53.** 4, 25 **55.** $-8, 27$

57. $1, \frac{125}{8}$ **59.** 1, 32 **61.** $\frac{1}{32}, 243$ **63.** 729

65. Write the equation in the form $u^2 = d$, where u is an algebraic expression and d is a positive constant. Take the square root of each side of the equation to obtain the solutions $u = \pm\sqrt{d}$.

67. Two complex solutions. When the squared expression is isolated on one side of the equation, the other side is negative. When the square root of each side is taken, the square root of the negative number is imaginary.

69. $\pm 30i$ **71.** ± 3 **73.** $2 \pm 6\sqrt{3}$ **75.** $2 \pm 6\sqrt{3}i$

77. $-2 \pm 3\sqrt{2}i$ **79.** 1, 16 **81.** $\frac{1}{2}, 1$ **83.** $-1, \frac{4}{5}$

85. $\pm 1, 2, 4$ **87.** $\frac{12}{5}$ **89.** 4 sec

91. $2\sqrt{2} \approx 2.83$ sec **93.** 9 sec **95.** 6%

97. A quadratic equation is of degree 2, so it must have an x^2-term, but it does not need an x-term or a constant.

99. No. Complex solutions always occur in complex conjugate pairs.

101. $x > 4$ **103.** $x \leq 5$

105. $(3, 2, 1)$ **107.** $3\sqrt{3}$ **109.** $16\sqrt[3]{y} - 9\sqrt[3]{x}$

111. $(4n + 1)m^2\sqrt{n}$

Section 10.2 *(pp. 510–517)*

1. 16 **3.** 100 **5.** 49 **7.** $\frac{25}{4}$ **9.** $\frac{81}{4}$ **11.** $\frac{1}{36}$

13. $\frac{16}{25}$ **15.** 0.04 **17.** 0, 20 **19.** $-6, 0$ **21.** 0, 5

23. 1, 7 **25.** $-4, -3$ **27.** $-3, 6$ **29.** $-7, -1$

31. 3, 7

33. $2 + \sqrt{7} \approx 4.65$ | **35.** $-2 + \sqrt{7} \approx 0.65$
  $2 - \sqrt{7} \approx -0.65$ | $\quad -2 - \sqrt{7} \approx -4.65$

37. $-\frac{3}{2} + \frac{\sqrt{21}}{6} \approx -0.74$ | **39.** $-2 + \frac{\sqrt{10}}{2} \approx -0.42$
  $-\frac{3}{2} - \frac{\sqrt{21}}{6} \approx -2.26$ | $\quad -2 - \frac{\sqrt{10}}{2} \approx -3.58$

41. $-\frac{1}{2} + \frac{\sqrt{10}}{2} \approx 1.08$ | **43.** $\frac{1}{3} + \frac{\sqrt{127}}{3} \approx 4.09$
  $-\frac{1}{2} - \frac{\sqrt{10}}{2} \approx -2.08$ | $\quad \frac{1}{3} - \frac{\sqrt{127}}{3} \approx -3.42$

45. $-\frac{5}{2} + \frac{\sqrt{17}}{2} \approx -0.44$ | **47.** $-2 \pm 3i$
  $-\frac{5}{2} - \frac{\sqrt{17}}{2} \approx -4.56$

49. $2 + \sqrt{5}i \approx 2 + 2.24i$ | **51.** $\frac{1}{2} + \frac{\sqrt{3}}{2}i \approx 0.5 + 0.87i$
  $2 - \sqrt{5}i \approx 2 - 2.24i$ | $\quad \frac{1}{2} - \frac{\sqrt{3}}{2}i \approx 0.5 - 0.87i$

53. $\frac{3}{8} + \frac{\sqrt{23}}{8}i \approx 0.38 + 0.60i$
  $\frac{3}{8} - \frac{\sqrt{23}}{8}i \approx 0.38 + 0.60i$

55. (a) $x^2 + 8x$ (b) $x^2 + 8x + 16$ (c) $(x + 4)^2$

57. 6 in. $\times$ 10 in. $\times$ 14 in.

59. A perfect square trinomial is one that can be written as $(x + k)^2$.

61. Each side of the equation must be divided by 2 to obtain a leading coefficient of 1. The resulting equation is the same by the Multiplication Property of Equality.

63. $-\frac{1}{4} + \frac{\sqrt{39}}{4}i \approx -0.25 + 1.56i$
  $-\frac{1}{4} - \frac{\sqrt{39}}{4}i \approx -0.25 - 1.56i$

65. $1 \pm \sqrt{3}$ **67.** $-2, 6$ **69.** $4 \pm 2\sqrt{2}$

71. 15 m $\times$ $46\frac{2}{3}$ m or 20 m $\times$ 35 m

73. 50 pairs, 110 pairs

75. Use the method of completing the square to write the quadratic equation in the form $u^2 = d$. Then use the Square Root Property to simplify.

77. (a) $d = 0$ (b) d is positive and is a perfect square.
   (c) d is positive and is not a perfect square.   (d) $d < 0$

79. 150   **81.** 7   **83.** $11 + 6\sqrt{2}$ **85.** $\dfrac{4\sqrt{10}}{5}$

87. $\sqrt{a} \cdot \sqrt{b}$

Section 10.3   (pp. 518–525)

1. $2x^2 + 2x - 7 = 0$ **3.** $-x^2 + 10x - 5 = 0$ **5.** 4, 7

7. $-2, -4$ **9.** $1 \pm \sqrt{5}$ **11.** $-2 \pm \sqrt{3}$ **13.** $5 \pm \sqrt{2}$

15. $-\dfrac{1}{4}$ **17.** $-\dfrac{3}{4} \pm \dfrac{\sqrt{15}}{4}i$

19. Two distinct complex solutions
21. Two distinct rational solutions
23. One repeated rational solution
25. Two distinct complex solutions

27. ± 13 **29.** $-3, 0$ **31.** $\dfrac{9}{5}, \dfrac{21}{5}$ **33.** $-\dfrac{3}{2}, 18$

35. $-4 \pm 3i$ **37.** $\dfrac{13}{6} \pm \dfrac{13\sqrt{11}}{6}i$ **39.** $-\dfrac{8}{5} \pm \dfrac{\sqrt{3}}{5}$

41. $-\dfrac{11}{8} \pm \dfrac{\sqrt{41}}{8}$

43. $(1, 0), (3, 0)$ **45.** $(0.20, 0), (66.47, 0)$
   The result is the same.   The result is the same.

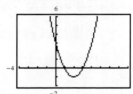 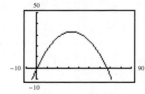

47. $x^2 - 3x - 10 = 0$ **49.** $x^2 - 8x + 7 = 0$
51. $x^2 - 2x - 1 = 0$ **53.** $x^2 + 25 = 0$
55. $x^2 - 24x + 144 = 0$ **57.** $x^2 - x + \dfrac{1}{4} = 0$

59. The opposite of b, plus or minus the square root of b squared minus $4ac$, all divided by $2a$.

61. The equation in general form is $3x^2 + x - 3 = 0$, so $a = 3$, $b = 1$, and $c = -3$.

63. $\dfrac{4}{3} \pm \dfrac{2\sqrt{13}}{3}$ **65.** $\dfrac{3}{2} + \dfrac{\sqrt{17}}{2}$ **67.** 5.1 in. $\times$ 11.4 in.

69. 2.15 or 4.65 hr

71. (a)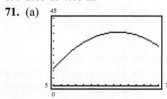

   (b) 31.3 or 64.8 mi/hr

73. The Square Root Property would be convenient because the equation is of the form $u^2 = d$.

75. The Quadratic Formula would be convenient because the equation is already in general form, the expression cannot be factored, and the leading coefficient is not 1.

77. When the Quadratic Formula is applied to $ax^2 + bx + c = 0$, the square root of the discriminant is evaluated. When the discriminant is positive, the square root of the discriminant is positive and will yield two real solutions (or x-intercepts). When the discriminant is zero, the equation has one real solution (or x-intercept). When the discriminant is negative, the square root of the discriminant is negative and will yield two complex solutions (no x-intercepts).

79. $3\sqrt{10}$ **81.** $\sqrt{13}$

83.   **85.**

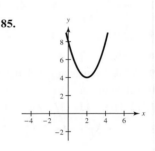

Mid-Chapter Quiz   (p. 526)

1. ± 6 **2.** $-4, \dfrac{5}{2}$ **3.** $\pm 2\sqrt{3}$ **4.** $-1, 7$

5. $-\dfrac{7}{2} \pm \dfrac{\sqrt{41}}{2}$ **6.** $-\dfrac{3}{2} \pm \dfrac{\sqrt{19}}{2}$ **7.** $-2 \pm \sqrt{10}$

8. $\dfrac{1}{4} \pm \dfrac{\sqrt{105}}{12}$ **9.** $-\dfrac{5}{2} \pm \dfrac{\sqrt{3}}{2}i$ **10.** $-2, 10$ **11.** $-3, 10$

12. $-2, 5$ **13.** $\dfrac{3}{2}$ **14.** $-\dfrac{5}{3} \pm \dfrac{\sqrt{10}}{3}$ **15.** 49

16. $\pm 2i, \pm \sqrt{3}i$ **17.** 1, 9   **18.** $\pm \sqrt{2}, \pm 2\sqrt{3}$

19. 60 video games   **20.** 35 m $\times$ 65 m   **21.** 392.6 ft

Section 10.4   (pp. 528–535)

1. $y = (x - 1)^2 - 1, (1, -1)$ **3.** $y = (x - 2)^2 + 3, (2, 3)$
5. $y = (x + 3)^2 - 4, (-3, -4)$
7. $y = -(x - 3)^2 - 1, (3, -1)$
9. $y = 2\left(x + \dfrac{3}{2}\right)^2 - \dfrac{5}{2}, \left(-\dfrac{3}{2}, -\dfrac{5}{2}\right)$ **11.** $(4, -1)$
13. $(-1, 2)$ **15.** $\left(-\dfrac{1}{2}, 3\right)$

17.   **19.**

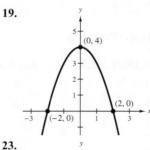

21.   **23.**

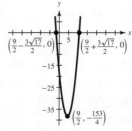

25. $y = (x - 2)^2$ **27.** $y = -\dfrac{1}{2}(x - 2)^2 + 6$
29. $y = (x + 1)^2 - 3$ **31.** $y = (x - 2)^2 + 1$

33. $y = (x - 2)^2 - 4$ **35.** $y = (x + 2)^2 - 1$

37. $y = \frac{2}{3}(x + 1)^2 + 1$ **39.** $y = \frac{2}{5}(x - 2)^2 - 2$

41. $y = \frac{1}{60}x^2$

43. A parabola that opens upward when $a > 0$ and downward when $a < 0$.

45. To find any x-intercepts, set $y = 0$ and solve the resulting equation for x. To find any y-intercepts, set $x = 0$ and solve the resulting equation for y.

47. Vertical shift   **49.** Horizontal shift

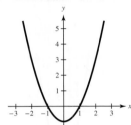   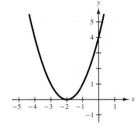

51. Horizontal shift and reflection in the x-axis

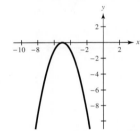

53. Horizontal and vertical shifts, reflection in the x-axis

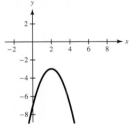

55. (a) 4 ft   (b) 16 ft   (c) $12 + 8\sqrt{3} \approx 25.9$ ft

57. (a) 0 yd   (b) 30 yd   (c) 240 yd

59. 14 ft

61. If the discriminant is positive, the parabola has two x-intercepts; if it is zero, the parabola has one x-intercept; and if it is negative, the parabola has no x-intercepts.

63. Find the y-coordinate of the vertex of the graph of the function.

65. $y = -\frac{1}{2}x$ **67.** $y = 2x$ **69.** $y = -\frac{11}{8}x + \frac{161}{16}$

71. $y = 8$ **73.** $8i$ **75.** $0.09i$

Section 10.5   *(pp. 536–543)*

1. 18 dozen, \$1.20/dozen   **3.** 12 in. × 16 in.

5. 9.5%   **7.** 7%   **9.** 15 people

11. 10.7 min, 13.7 min   **13.** 23.5 yd, 76.5 yd   **15.** 1.2 sec

17. The Pythagorean Theorem can be used to set the sum of the square of a missing length and the square of the difference of the sum of the lengths and the missing length equal to the square of the length of the hypotenuse.

19. The quotient of the investment amount and the number of units is equal to the quotient of the investment amount and the number of units sold, plus the profit per unit sold.

21. 177.19 in.^{2} **23.** 70 ft   **25.** 180 km^2

27. 440 m   **29.** 10 DVDs at \$5 per DVD

31. 13, 14   **33.** Height: 12 in.; Width: 24 in.

35. 50 mi/hr   **37.** $(-6, 9), (10, 9)$

39. To solve a rational equation, each side of the equation is multiplied by the LCD. The resulting equations in this section are quadratic equations.

41. No. For each additional person, the cost-per-person decrease gets smaller because the discount is distributed to more people.

43. $x < -4$

45. 0, 8

Section 10.6   *(pp. 544–551)*

1. Negative: $(-\infty, 4)$; Positive: $(4, \infty)$

3. Negative: $(6, \infty)$; Positive: $(-\infty, 6)$

5. Negative: $(0, 5)$; Positive: $(-\infty, 0)$ and $(5, \infty)$

7. Negative: $(-\infty, -2)$ and $(2, \infty)$; Positive: $(-2, 2)$

9. $0 < x < 2$ **11.** $0 \le x \le 2$

13. $x \le -4$ or $x > 0$ **15.** $x \le -2$ or $x \ge 5$

17. $x \le -2$ or $x > 2$ **19.** $-9 \le x \le 4$

21. $x < -3$ or $x > 1$ **23.** No solution

25. $-\infty < x < \infty$ **27.** $x = -8$

29. $x > 3$ **31.** $x < 3$

33. $x < -4$ or $x > 2$ **35.** $-2 \le x < 1$

37. $(3, 5)$ **39.** The critical numbers are -1 and 3.

41. No. $x = 4$ is not in the solution set $0 < x < 4$.

43. $0, \frac{5}{2}$ **45.** $\pm\frac{9}{2}$

47. $x < 2 - \sqrt{6}$ or $x > 2 + \sqrt{6}$

49. $x \le -9$ or $x \ge -1$ **51.** $\frac{5}{3} < x \le 6$

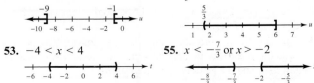

53. $-4 < x < 4$ **55.** $x < -\frac{7}{3}$ or $x > -2$

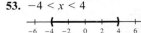   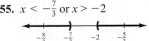

57. $-20 < x < 4$

59. $r > 7.24\%$ **61.** $(12, 20)$ **63.** $72{,}589 < x < 170{,}411$
65. The critical numbers of a polynomial are its zeros, so the value of the polynomial is zero at its critical numbers.
67. No solution. The value of the polynomial is positive for every real value of x, so there are no values that would make the polynomial negative.
69. $\frac{y^3}{2x^2}, y \neq 0$ **71.** $\frac{(x-3)(x+1)}{4x^3(x+3)}, x \neq -2$
73. $\frac{(x+4)^2}{3x(x-6)(x-8)}$ **75.** $\frac{1}{9}$ **77.** 79.21

Review Exercises   *(pp. 554–557)*

1. $-12, 0$ **3.** ± 3 **5.** $-\frac{5}{2}$ **7.** $-9, 10$
9. $-\frac{3}{2}, 6$ **11.** ± 12 **13.** $\pm 2\sqrt{3}$ **15.** $-4, 36$
17. $\pm 11i$ **19.** $\pm 5\sqrt{2}i$ **21.** $-4 \pm 3\sqrt{2}i$ **23.** $\pm\sqrt{5}, \pm i$
25. $1, 9$ **27.** $1, 1 \pm \sqrt{6}$ **29.** $-343, 64$ **31.** 81
33. $\frac{225}{4}$ **35.** $\frac{1}{25}$ **37.** $3 + 2\sqrt{3} \approx 6.46; 3 - 2\sqrt{3} \approx -0.46$
39. $-4, -1$
41. $\frac{1}{3} + \frac{\sqrt{17}}{3}i \approx 0.33 + 1.37i; \frac{1}{3} - \frac{\sqrt{17}}{3}i \approx 0.33 - 1.37i$
43. $-7, 6$ **45.** $\frac{8}{5} \pm \frac{3\sqrt{6}}{5}$ **47.** $\frac{3}{8} \pm \frac{\sqrt{7}}{8}i$
49. One repeated rational solution
51. Two distinct rational solutions
53. Two distinct irrational solutions
55. Two distinct complex solutions
57. $x^2 + 4x - 21 = 0$ **59.** $x^2 - 10x + 18 = 0$
61. $x^2 - 12x + 40 = 0$
63. $y = (x-4)^2 - 13$; Vertex: $(4, -13)$
65. $y = 2\left(x - \frac{1}{4}\right)^2 + \frac{23}{8}$; Vertex: $\left(\frac{1}{4}, \frac{23}{8}\right)$
67.

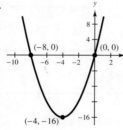

69.

71. $y = 2(x-2)^2 - 5$ **73.** $y = \frac{1}{16}(x-5)^2$
75. (a) 6 ft   (b) 28.5 ft   (c) 31.9 ft
77. 16 cars; $5000   **79.** 5 in. $\times$ 17 in.
81. 15 people   **83.** 4.7 mi or 15.3 mi
85. $9 + \sqrt{101} \approx 19$ hr, $11 + \sqrt{101} \approx 21$ hr
87. $-7, 0$ **89.** $-3, 9$
91. $0 < x < 7$

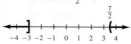

93. $x \leq -2$ or $x \geq 6$

95. $-4 < x < \frac{5}{2}$

97. $x \leq -3$ or $x \geq \frac{7}{2}$

99. $-4 < x < 1$

101. $(5.3, 14.2)$

Chapter Test   *(p. 558)*

1. $3, 10$ **2.** $-\frac{1}{3}, 6$ **3.** $1.7, 2.3$ **4.** $-4 \pm 10i$
5. $\frac{3}{2} \pm \frac{\sqrt{3}}{2}$ **6.** $1 \pm \frac{3\sqrt{2}}{2}$ **7.** $\frac{3}{4} \pm \frac{\sqrt{5}}{4}$ **8.** $1, 512$
9. -56; A negative discriminant tells us the equation has two imaginary solutions.
10. $x^2 + 10x + 21 = 0$
11.

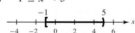

12.

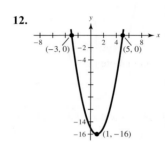

13. $x \leq -2$ or $x \geq 6$

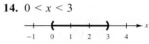

14. $0 < x < 3$

15. $-1 \leq x < 5$

16. 8 ft $\times$ 30 ft   **17.** 40 members
18. $\frac{\sqrt{10}}{2} \approx 1.58$ sec   **19.** 35 ft, 120 ft

CHAPTER 11

Section 11.1   *(pp. 560–569)*

1. 0.203 **3.** 9.739 **5.** 1.817
7. (a) $\frac{1}{9}$ (b) 1   (c) 3
9. (a) 0.455 (b) 0.094 (c) 0.145
11. (a) 500   (b) 250   (c) 56.657
13. (a) 1000   (b) 1628.895 (c) 2653.298
15. (a) 486.111 (b) 47.261 (c) 0.447
17. d **18.** c **19.** a **20.** b
21.

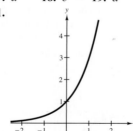

Domain: $-\infty < x < \infty$
Range: $x > 0$

23.

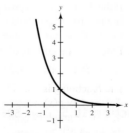

Domain: $-\infty < x < \infty$
Range: $x > 0$

25. The function g is related to $f(x) = 3^x$ by $g(x) = f(x) - 1$. To sketch the graph of g, shift the graph of f one unit downward.
y-intercept: $(0, 0)$
Asymptote: $y = -1$

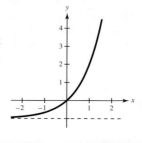

27. The function g is related to $f(x) = 5^x$ by $g(x) = f(x - 1)$. To sketch the graph of g, shift the graph of f one unit to the right.
y-intercept: $\left(0, \frac{1}{5}\right)$
Asymptote: x-axis

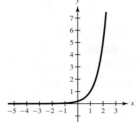

29. The function g is related to $f(x) = 2^x$ by $g(x) = f(x) + 3$. To sketch the graph of g, shift the graph of f three units upward.
y-intercept: $(0, 4)$
Asymptote: $y = 3$

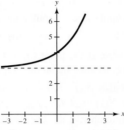

31. The function g is related to $f(x) = 2^x$ by $g(x) = f(x - 4)$. To sketch the graph of g, shift the graph of f four units to the right.
y-intercept: $\left(0, \frac{1}{16}\right)$
Asymptote: x-axis

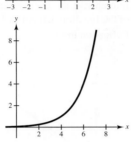

33. b   **34.** d   **35.** a   **36.** c

37. The function g is related to $f(x) = 4^x$ by $g(x) = -f(x)$. To sketch the graph of g, reflect the graph of f in the x-axis.

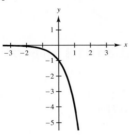

39. The function g is related to $f(x) = 5^x$ by $g(x) = f(-x)$. To sketch the graph of g, reflect the graph of f in the y-axis.

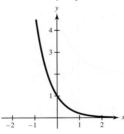

41. 1.396   **43.** 107.561
45. (a) 73.891   (b) 1.353   (c) 0.183

47. The function g is related to $f(x) = e^x$ by $g(x) = -f(x)$. To sketch the graph of g, reflect the graph of f in the x-axis.

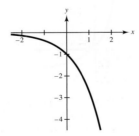

49. The function g is related to $f(x) = e^x$ by $g(x) = f(x) + 1$. To sketch the graph of g, shift the graph of f one unit upward.

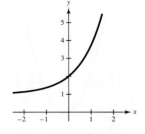

51. 2.520 g

53.

n	1	4	12	365	Continuous
A	\$275.90	\$283.18	\$284.89	\$285.74	\$285.77

55.

n	1	4	12	365	Continuous
A	\$4956.46	\$5114.30	\$5152.11	\$5170.78	\$5171.42

57. Reflect the graph of f in the x-axis.

59. The behavior is reversed. When the graph of the original function is increasing, the reflected graph is decreasing. When the graph of the original function is decreasing, the reflected graph is increasing.

61. $3^{2x + 2}$ **63.** $\dfrac{3}{e^{2x}}$ **65.** e^2 **67.** $-2e^x$

69.

n	1	4	12	365	Continuous
P	\$2541.75	\$2498.00	\$2487.98	\$2483.09	\$2482.93

71. (a) \$22.04   (b) \$20.13
73. (a) \$80,634.95   (b) \$161,269.89
75. $f(t) = 2^{t - 1}$; $f(30) = 536{,}870{,}912$ pennies
77. By definition, the base of an exponential function must be positive and not equal to 1. The function $y = 1x$ simplifies to the constant function $y = 1$.
79. No. e is an irrational number.
81. Because $1 < \sqrt{2} < 2$ and $2 > 0$, $2^1 < 2^{\sqrt{2}} < 2^2$.
So, $2 < 2^{\sqrt{2}} < 4$.
83. When $k > 1$, the values of f will increase. When $0 < k < 1$, the values of f will decrease. When $k = 1$, the values of f will remain constant.
85. $x \geq 4$
87.

y is not a function of x.

Section 11.2   (pp. 570–577)

1. (a) $2x - 9$ (b) $2x - 3$ (c) -1 (d) 11
3. (a) $x^2 + 4x + 7$ (b) $x^2 + 5$ (c) 19   (d) 14
5. (a) $(f \circ g)(x) = \sqrt{x - 2}$
Domain: $x \geq 2$
(b) $(g \circ f)(x) = \sqrt{x + 2} - 4$
Domain: $x \geq -2$

7.

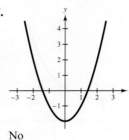

No

9.

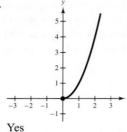

Yes

11.

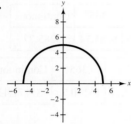

No

13. $f^{-1}(x) = \dfrac{3-x}{4}$ **15.** $f^{-1}(t) = \sqrt[3]{t+1}$

17.

19.

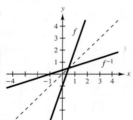

21.

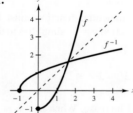

23. b **24.** c **25.** d **26.** a

27.

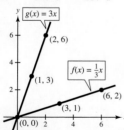

Points: $(0, f(0)), (3, f(3))$, and $(6, f(6))$

29.

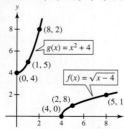

Points: $(4, f(4)), (5, f(5))$, and $(8, f(8))$

31. $y = 2x + 6$ **33.** $f^{-1}(x) = \dfrac{x-6}{2}$ **35.** -1 **37.** 1

39. $f(g(x)) = f\left[\frac{1}{2}(1-x)\right] = 1 - 2\left[\frac{1}{2}(1-x)\right] = 1 - (1-x) = x$

$g(f(x)) = g(1-2x) = \frac{1}{2}[1-(1-2x)] = \frac{1}{2}(2x) = x$

41. Inverse does not exist. **43.** $h^{-1}(x) = x^2, x \geq 0$

45. Domain of f: $x \geq 2$

$f^{-1}(x) = \sqrt{x} + 2$

Domain of f^{-1}: $x \geq 0$

47. $A(r(t)) = 0.36\pi t^2$;

Input: time, Output: area; $A(r(3)) = 10.2$ ft^2

49. (a) $y = \frac{20}{13}(x-9)$ (b) 8 units

51. True. If the point (a, b) lies on the graph of f, then the point (b, a) must lie on the graph of f^{-1}, and vice versa.

53. False. $f(x) = \sqrt{x-1}$, Domain: $x \geq 1$

$f^{-1}(x) = x^2 + 1, x \geq 0$, Domain: $x \geq 0$

55. Interchange the coordinates of each ordered pair. The inverse of the function defined by $\{(3,6), (5,-2)\}$ is $\{(6,3), (-2,5)\}$.

57. Reflection in the x-axis **59.** Horizontal and vertical shifts

61. $(6+y)(2-y)$ **63.** $-(u^2+1)(u-5)$

65.

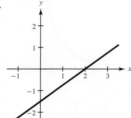

67.

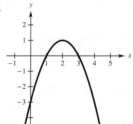

Section 11.3 *(pp. 578–585)*

1. 3 **3.** $\frac{1}{2}$ **5.** $\frac{3}{4}$ **7.** 0 **9.** -4 **11.** Undefined

13. 3 **15.** -1 **17.** -4 **19.** 1.6232 **21.** -1.6383

23. 0.7190

25.

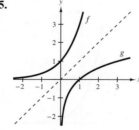

27.

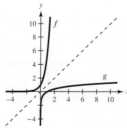

29.

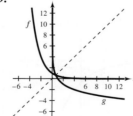

31. c **32.** b **33.** a **34.** d

35. Vertical shift two units upward; $x = 0$; $(0, \infty)$

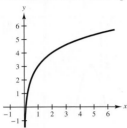

37. Horizontal shift two units to the right; $x = 2$; $(2, \infty)$

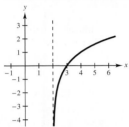

39. Reflection in the y-axis; $x = 0$; $(-\infty, 0)$

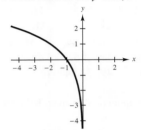

41. 3   **43.** -2

45. 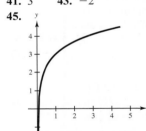   **47.**

49. 1.6309   **51.** 1.6397   **53.** -0.4739 **55.** 2.6332

57. -2 **59.** 1.3481   **61.** 1.8946

63. $y = \log_a x$ or $x = a^y$ **65.** e **67.** $7^2 = 49$

69. $2^{-5} = \frac{1}{32}$ **71.** $36^{1/2} = 6$ **73.** $8^{2/3} = 4$

75. $\log_6 36 = 2$ **77.** $\log_8 4 = \frac{2}{3}$

79. 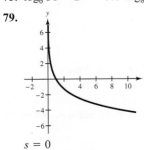   **81.**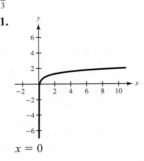

$s = 0$ $x = 0$

83. 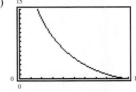   **85.**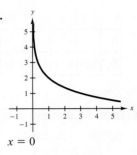

$x = 1$ $x = 0$

87. 53.4 in.

89. (a)

Domain: $0 < x \le 10$

(b) $x = 0$ (c) $(0, 22.9)$

91. $(0, \infty)$ **93.** $3 \le f(x) \le 4$ **95.** A factor of 10

97. False. $8 = 2^3$ is equivalent to $3 = \log_2 8$.

99. Logarithmic functions with base 10 are common logarithms. Logarithmic functions with base e are natural logarithms.

101. A vertical shift or reflection in the x-axis of a logarithmic graph does not affect the domain or range. A horizontal shift or reflection in the y-axis of a logarithmic graph affects the domain, but the range stays the same.

103. $-m^{10}n^4$ **105.** $\dfrac{9x^3}{2y^2}$, $x \ne 0$

107. $19\sqrt{3x}$ **109.** $\sqrt{5u}$

Mid-Chapter Quiz   *(p. 586)*

1. (a) $\frac{16}{9}$ (b) 1   (c) $\frac{3}{4}$ (d) 1.540

2. Horizontal asymptote: $y = 0$

3. b   **4.** d   **5.** a   **6.** c

7. (a) $2x^3 - 3$

(b) $(2x - 3)^3 = 8x^3 - 36x^2 + 54x - 27$

(c) -19

(d) 125

8. $f(g(x)) = 5 - 2\left[\frac{1}{2}(5 - x)\right]$

$= 5 - 5 + x = x$

$g(f(x)) = \frac{1}{2}\left[5 - (5 - 2x)\right]$

$= \frac{1}{2}(2x) = x$

9. $f^{-1}(x) = \dfrac{x - 3}{10}$ **10.** $f^{-1}(t) = \sqrt[3]{2t - 4}$

11. $9^{-2} = \frac{1}{81}$ **12.** $\log_2 64 = 6$ **13.** 3

14. 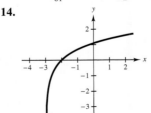   **15.**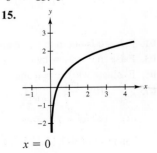

$t = -3$ $x = 0$

16. $h = 2, k = 1$ **17.** 6.0639

18.

n	1	4	12	365	Continuous
A	\$2979.31	\$3042.18	\$3056.86	\$3064.06	\$3064.31

19. 1.60 g

Section 11.4   *(pp. 588–595)*

1. 0.5108   **3.** 2.1972   **5.** 4.3174   **7.** 1.9033

9. $-3 \log_4 2 = \log_4 2^{-3} = \log_4 \frac{1}{8}$

11. $-3 \log_{10} 3 + \log_{10} \frac{3}{2} = \log_{10} 3^{-3} + \log_{10} \frac{3}{2}$

$$= \log_{10} \frac{1}{27} + \log_{10} \frac{3}{2}$$

$$= \log_{10} \left(\frac{1}{27} \cdot \frac{3}{2} \right)$$

$$= \log_{10} \frac{1}{18}$$

13. $-\ln \frac{1}{7} = \ln\left(\frac{1}{7}\right)^{-1} = \ln 7 = \ln \frac{56}{8} = \ln 56 - \ln 8$

15. 3   **17.** -4 **19.** $\frac{1}{3}$ **21.** 0   **23.** -9

25. 2   **27.** 3   **29.** -3 **31.** 12   **33.** 1

35. $\log_3 11 + \log_3 x$ **37.** $\ln 3 + \ln y$ **39.** $2 \log_7 x$

41. $-3 \log_4 x$ **43.** $\frac{1}{2}(\log_4 3 + \log_4 x)$

45. $\log_2 z - \log_2 17$ **47.** $\frac{1}{2} \log_9 x - \log_9 12$

49. $2 \ln x + \ln(y + 2)$ **51.** $6 \log_4 x + 2 \log_4(x + 7)$

53. $\frac{1}{3} \log_3(x + 1)$ **55.** $\frac{1}{2}[\ln x + \ln(x + 2)]$

57. $\ln x + 2 \ln y - 3 \ln z$ **59.** $\log_{12} \frac{x}{3}$

61. $\log_3 5x$ **63.** $\log_2 x^7 z^3$ **65.** $\ln x^4 y^4$

67. $\log_4 \frac{x + 8}{x^3}$

69. $f(t) = 80 - 12 \log_{10}(t + 1)$;

$f(2) \approx 74.27; f(8) \approx 68.55$

71. $B = 10 \log_{10} I + 120$; 110 dB

73. Condensing   **75.** The Power Property of Logarithms

77. $\ln \frac{x^3 y}{z^2}$ **79.** $\ln\left(\frac{x}{x + 1}\right)^2$ **81.** $\log_5 \frac{\sqrt[3]{x + 3}}{x - 6}$

83. $\log_6 \frac{(c + d)^5}{\sqrt{m - n}}$ **85.** $\log_2 \sqrt[5]{\frac{x^3}{y^4}}$ **87.** $2 + \ln 3$

89. $1 + \frac{1}{2} \log_5 2$ **91.** $1 - 3 \log_8 x$ **93.** $E = \log_{10}\left(\frac{C_2}{C_1}\right)^{1.4}$

95. True. $\log_2 8x = \log_2 8 + \log_2 x = 3 + \log_2 x$

97. False. $\log_3(u + v)$ does not simplify.

99. True. $f(ax) = \log_a ax$

$$= \log_a a + \log_a x$$

$$= 1 + \log_a x$$

$$= 1 + f(x)$$

101. False. 0 is not in the domain of f.

103. False. $f(x - 3) = \ln(x - 3)$

105. Evaluate when $x = e$ and $y = e$.

107. 2   **109.** $-\frac{1}{2}$ **111.** 1, 7

113.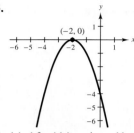

115.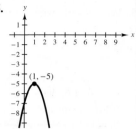

117. (a) $(f \circ g)(x) = 4x - 11$

Domain: $-\infty < x < \infty$

(b) $(g \circ f)(x) = 4x + 4$

Domain: $-\infty < x < \infty$

119. (a) $(f \circ g)(x) = \frac{1}{x + 2}$

Domain: $x < -2$ or $x > -2$

(b) $(g \circ f)(x) = \frac{1}{x} + 2 = \frac{2x + 1}{x}$

Domain: $x < 0$ or $x > 0$

Section 11.5   *(pp. 596–603)*

1. 3   **3.** -3 **5.** -2 **7.** 1   **9.** $\frac{22}{5}$ **11.** -7

13. -1 **15.** 4.11   **17.** 1.31   **19.** 0.11   **21.** -3.60

23. 3   **25.** -1.79 **27.** -1.04 **29.** 0.90   **31.** 4.39

33. 2.48   **35.** 8.99   **37.** 0.1   **39.** 2187   **41.** 9

43. 27   **45.** 3   **47.** 100,000   **49.** 37.94

51. 5   **53.** 5%   **55.** 105°

57. Each side of the equation is in exponential form with the same base.

59. Exponentiate each side of the equation using the base 6, and then rewrite the left side of the equation as $3x$.

61. -3 **63.** -6 **65.** 6   **67.** No solution

69. 3.28   **71.** No solution   **73.** 0.75   **75.** 2.46

77. 3.33   **79.** 6.00   **81.** 7.7 yr

83. 10^{-8} W/cm^2

85. (a) -0.144 (b) 6:24 P.M.   (c) 90.2°F

87. $2^{x-1} = 32$, because you can write 32 as 2^5 and then apply the one-to-one property of exponential equations.

89. To solve an exponential equation, first isolate the exponential expression, then take the logarithm of each side of the equation and solve for the variable.

To solve a logarithmic equation, first isolate the logarithmic expression, then exponentiate each side of the equation and solve for the variable.

91. $\pm 5i$ **93.** $\pm \frac{4}{3}$ **95.** $\pm 2, \pm 3$ **97.** 90 in.^{2}

99. 56 km

Section 11.6   *(pp. 604–611)*

1. 7%   **3.** 8.66 yr   **5.** 10.27 yr   **7.** Yearly

9. 8.33%   **11.** 7.23%   **13.** 6.14%   **15.** 5.39%

17. No. The effective yield is the ratio of the year's interest to the amount invested. The ratio will remain the same regardless of the amount invested.

19. (a) $y = 73e^{0.4763t}$ (b) 9 hr   **21.** 3.3 g

23. The earthquake of 1960 was about 501 times as intense.

25. The earthquake on July 6 was about 40 times as intense.

27. It is the formula for continuously compounded interest.

29. The variable r represents the interest rate in decimal form.

31. 9.27 yr   **33.** 12.14 yr   **35.** $k = \frac{1}{2} \ln \frac{8}{3} \approx 0.4904$

37. $k = \frac{1}{3} \ln \frac{1}{2} \approx -0.2310$ **39.** 7.04   **41.** 10^7 times

43. 7.761 billion people

45. (a) 1000 rabbits   (b) 2642 rabbits   (c) 5.88 yr

47. When $k > 0$, the model represents exponential growth, and when $k < 0$, the model represents exponential decay.

49. When the investment is compounded more than once in a year (quarterly, monthly, daily, continuously), the effective yield is greater than the interest rate.

51. $\frac{7}{2} \pm \frac{\sqrt{69}}{2}$ **53.** $-\frac{3}{2} \pm \frac{\sqrt{33}}{6}$

55. $x > 4$ **57.** $x < -3$ or $x > 3$

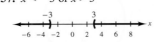

Review Exercises   *(pp. 614–617)*

1. (a) $\frac{1}{64}$ (b) 4   (c) 16

3. (a) 5   (b) 0.185   (c) $\frac{1}{25}$

5.
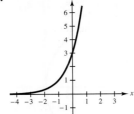

7.
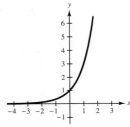

9.
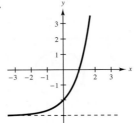

11. (a) 0.007   (b) 3   (c) 9.557×10^{16}

13. The function y is related to $f(x) = e^x$ by $y = f(-x) + 1$. To sketch y, reflect the graph of f in the y-axis, and shift the graph one unit upward.

y-intercept: $(0, 2)$

Asymptote: $y = 1$

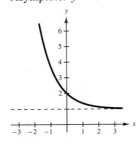

15. The function g is related to $f(x) = e^x$ by $y = f(x + 2)$. To sketch g, shift the graph of f two units to the left.

y-intercept: $(0, e^2)$

Asymptote: $y = 0$

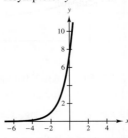

17.

n	1	4	12
A	\$226,296.28	\$259,889.34	\$268,503.32

n	365	Continuous
A	\$272,841.23	\$272,990.75

19. 4.21 g   **21.** (a) 6   (b) 1   **23.** (a) 5   (b) -1

25. (a) $(f \circ g)(x) = \sqrt{2x + 6}$

  Domain: $[-3, \infty)$

  (b) $(g \circ f)(x) = 2\sqrt{x + 6}$

  Domain: $[-6, \infty)$

27. No   **29.** $f^{-1}(x) = \frac{1}{3}(x - 4)$ **31.** $h^{-1}(x) = \frac{1}{5}x^2,\ x \geq 0$

33. $f^{-1}(t) = \sqrt[3]{t - 4}$

35.
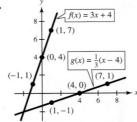

Points: $(-1, f(-1)), (0, f(0)), (1, f(1))$

37. 3   **39.** -2 **41.** 6   **43.** 0

45.
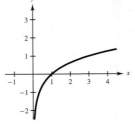

$x = 0$

47.
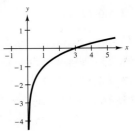

$x = 0$

49.
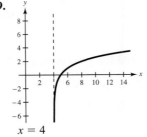

$x = 4$

51. 7

53. 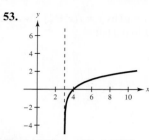   **55.**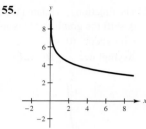

57. 1.5850   **59.** 2.4406   **61.** 1.7959   **63.** −0.4307
65. 1.0293   **67.** $\log_4 6 + 4 \log_4 x$ **69.** $\frac{1}{2}\log_5(x + 2)$

71. $\ln(x + 2) - \ln(x + 3)$ **73.** $\log_2 y^5$ **75.** $\log_8 32x^3$

77. $\ln \frac{9}{4x^2}, x > 0$ **79.** $\log_2\left(\frac{k}{k - t}\right)^4, k > t, k > 0$

81. False. $\log_2 4x = \log_2 4 + \log_2 x = 2 + \log_2 x$
83. True   **85.** True   **87.** 80 dB   **89.** 6   **91.** 1
93. 6   **95.** 5.66   **97.** 6.23   **99.** No solution
101. 1408.10   **103.** 31.47   **105.** 4   **107.** 64
109. 11.57   **111.** 7%   **113.** 5%   **115.** 7.5%
117. 6.5%   **119.** 5.65%   **121.** 7.71%
123. 7.79%   **125.** 2.282
127. The earthquake of 1906 was about 19,953 times as intense.

Chapter Test   *(p. 618)*

1. $f(-1) = 81$
   $f(0) = 54$
   $f\left(\frac{1}{2}\right) = 18\sqrt{6} \approx 44.09$
   $f(2) = 24$

2.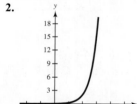
   $y = 0$

3. (a) $(f \circ g)(x) = 18x^2 - 63x + 55$
      Domain: $(-\infty, \infty)$
   (b) $(g \circ f)(x) = -6x^2 - 3x + 5$
      Domain: $(-\infty, \infty)$

4. $f^{-1}(x) = \frac{1}{9}(x + 4)$

5. $(f \circ g)(x) = -\frac{1}{2}(-2x + 6) + 3$
      $= (x - 3) + 3$
      $= x$
   $(g \circ f)(x) = -2\left(-\frac{1}{2}x + 3\right) + 6$
      $= (x - 6) + 6$
      $= x$

6. −4
7. The graph of g is a reflection in the line $y = x$ of the graph of f.

8. $\frac{2}{3} + \frac{1}{2}\log_8 x - 4 \log_8 y$ **9.** $\ln \frac{x}{y^4}, y > 0$ **10.** 32

11. 1.18   **12.** 13.73   **13.** 15.52   **14.** 5
15. 8   **16.** 109.20   **17.** 0
18. (a) $8012.78   (b) $8110.40

19. $10,806.08   **20.** 7%   **21.** $4746.09
22. 600 foxes   **23.** 1141 foxes   **24.** 4.4 yr

CHAPTER 12

Section 12.1   *(pp. 620–627)*

1. parabola   **3.** circle   **5.** $x^2 + y^2 = 25$
7. $x^2 + y^2 = \frac{4}{9}$

9. Center: $(0, 0), r = 4$ **11.** Center: $(0, 0), r = 6$

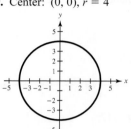   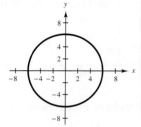

13. $(x - 4)^2 + (y - 3)^2 = 100$ **15.** $(x - 6)^2 + (y + 5)^2 = 9$
17. $(x + 2)^2 + (y - 1)^2 = 4$
19. Center: $(-1, -3), r = 2$ **21.** $y = \sqrt{1600 - x^2}$

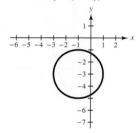

23. $x^2 = -6y$ **25.** $y^2 = -8x$ **27.** $(y - 2)^2 = -8(x - 3)$

29. 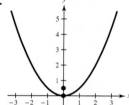   **31.**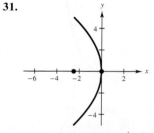

Vertex: $(0, 0)$, Focus: $\left(0, \frac{1}{2}\right)$ Vertex: $(0, 0)$, Focus: $\left(-\frac{5}{2}, 0\right)$

33.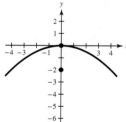

Vertex: $(0, 0)$, Focus: $(0, -2)$

35.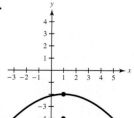

Vertex: $(1, -2)$, Focus: $(1, -4)$

37.

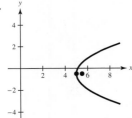

Vertex: $\left(5, -\frac{1}{2}\right)$, Focus: $\left(\frac{11}{2}, -\frac{1}{2}\right)$

39.

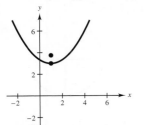

Vertex: $(1, 3)$, Focus: $\left(1, \frac{15}{4}\right)$

41.

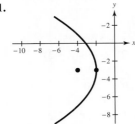

Vertex: $(-2, -3)$, Focus: $(-4, -3)$

43. First write the equation in standard form: $x^2 + y^2 = 6^2$. Because $h = 0$ and $k = 0$, the center is the origin. Because $r^2 = 6^2$, the radius is 6 units.

45. The equation $(x - h)^2 = 4p(y - k)$, $p \neq 0$, should be used because the vertex and the focus lie on the same vertical axis.

47. Center: $\left(-\frac{9}{4}, 4\right)$, $r = 4$ **49.** Center: $(-5, 2)$, $r = 6$

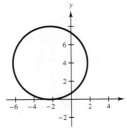
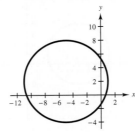

51. $x^2 + y^2 = 75^2$
53. (a) $x^2 = 180y$

(b)

x	0	20	40	60
y	0	$2\frac{2}{9}$	$8\frac{8}{9}$	20

55. No. The equation of the circle is $(x + 1)^2 + (y - 1)^2 = 13$, and the point $(3, 2)$ does not satisfy the equation.

57. No. For each $x > 0$, there correspond two values of y.

59. Yes. The directrix of a parabola is perpendicular to the line through the vertex and focus.

61. $-3 \pm \sqrt{5}$ **63.** $\frac{3}{2} \pm \frac{\sqrt{19}}{2}$ **65.** $\frac{2}{3} \pm \sqrt{2}$

67. $\frac{1}{2}\left(\log_{10} x + 3 \log_{10} y\right)$ **69.** $\ln x - 4 \ln y$

71. $\log_3 \dfrac{x^2}{y}$ **73.** $\ln \dfrac{x^4 y^4}{x^4 + y^4}$

Section 12.2 *(pp. 628–635)*

1. $\dfrac{x^2}{b^2} + \dfrac{y^2}{a^2} = 1$ **3.** $\dfrac{x^2}{a^2} + \dfrac{y^2}{b^2} = 1$ **5.** $\dfrac{x^2}{16} + \dfrac{y^2}{9} = 1$

7. $\dfrac{x^2}{9} + \dfrac{y^2}{36} = 1$

9. **11.**

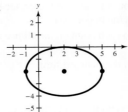
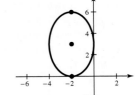

Vertices: $(\pm 4, 0)$ Vertices: $(0, \pm 2)$
Co-vertices: $(0, \pm 2)$ Co-vertices: $(\pm 1, 0)$

13. $\dfrac{(x - h)^2}{b^2} + \dfrac{(y - k)^2}{a^2} = 1$ **15.** $\dfrac{(x - h)^2}{a^2} + \dfrac{(y - k)^2}{b^2} = 1$

17. $\dfrac{(x - 2)^2}{9} + \dfrac{(y - 2)^2}{4} = 1$ **19.** $\dfrac{(x - 4)^2}{9} + \dfrac{y^2}{16} = 1$

21. **23.**

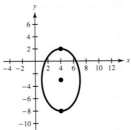

Center: $(2, -2)$ Center: $(-2, 3)$
Vertices: $(-1, -2), (5, -2)$ Vertices: $(-2, 6), (-2, 0)$

25.

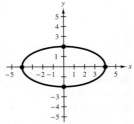

Center: $(4, -3)$
Vertices: $(4, -8), (4, 2)$

27. $\dfrac{x^2}{615^2} + \dfrac{y^2}{290^2} = 1$ or $\dfrac{x^2}{290^2} + \dfrac{y^2}{615^2} = 1$

29. An ellipse is the set of all points (x, y) such that the sum of the distances between (x, y) and two distinct fixed points is a constant.

$\dfrac{x^2}{a^2} + \dfrac{y^2}{b^2} = 1$ or $\dfrac{x^2}{b^2} + \dfrac{y^2}{a^2} = 1$

31. The length of the major axis is $2a$, and the length of the minor axis is $2b$.

33. $\dfrac{x^2}{9} + \dfrac{y^2}{25} = 1$

35.

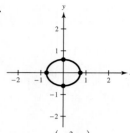

Vertices: $\left(\pm\frac{3}{4}, 0\right)$

Co-vertices: $\left(0, \pm\frac{3}{5}\right)$

37. 36 ft, 28 ft

39. (a) Every point on the ellipse represents the maximum distance (800 miles) that the plane can safely fly with enough fuel to get from airport A to airport B.

(b) Airport A: $(-250, 0)$; Airport B: $(250, 0)$

(c) 800 mi; Vertices: $(\pm 400, 0)$

(d) $\dfrac{x^2}{400^2} + \dfrac{y^2}{\left(50\sqrt{39}\,\right)^2} = 1$

(e) $20{,}000\sqrt{39}\,\pi \approx 392{,}385$ mi^2

41. A circle is an ellipse in which the major axis and the minor axis have the same length. Both circles and ellipses have foci. However, a circle has a single focus located at the center and an ellipse has two foci that lie on the major axis.

43. The sum of the distances between each point on the ellipse and the two foci is a constant.

45. The graph of an ellipse written in the standard form

$$\dfrac{(x-h)^2}{a^2} + \dfrac{(y-k)^2}{b^2} = 1$$

intersects the y-axis when $|h| > a$ and intersects the x-axis when $|k| > b$. Similarly, the graph of

$$\dfrac{(x-h)^2}{b^2} + \dfrac{(y-k)^2}{a^2} = 1$$

intersects the y-axis when $|h| > b$ and intersects the x-axis when $|k| > a$.

47. (a) $f(-2) = 9$ (b) $f(2) = \frac{1}{9}$

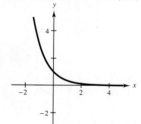

49. (a) $g(-1) \approx 3.639$ (b) $g(2) \approx 16.310$

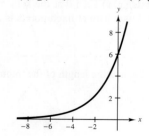

51. (a) $h(4) = 1$ (b) $h(64) = 2$

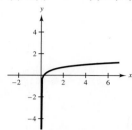

53. (a) $f(3)$ does not exist.   (b) $f(35) = \frac{5}{2}$

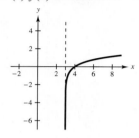

Mid-Chapter Quiz   *(p. 636)*

1. $x^2 + y^2 = 25$ **2.** $(x-3)^2 + (y+5)^2 = 25$

3. $(y-1)^2 = 8(x+2)$ **4.** $(x-2)^2 = -8(y-3)$

5. $\dfrac{(x+2)^2}{16} + \dfrac{(y+1)^2}{4} = 1$ **6.** $\dfrac{x^2}{36} + \dfrac{y^2}{100} = 1$

7. $x^2 + (y+3)^2 = 16$; Center: $(0, -3)$, $r = 4$

8. $(x+1)^2 + (y-2)^2 = 1$; Center $(-1, 2)$, $r = 1$

9. $(y-3)^2 = x + 16$; Vertex: $(-16, 3)$, Focus: $\left(-\frac{63}{4}, 3\right)$

10. $(x-4)^2 = -(y-4)$; Vertex: $(4, 4)$, Focus: $\left(4, \frac{15}{4}\right)$

11. $\dfrac{(x-2)^2}{9} + \dfrac{y^2}{36} = 1$ **12.** $\dfrac{(x-6)^2}{9} + \dfrac{(y+2)^2}{4}$

Center: $(2, 0)$ Center: $(6, -2)$

Vertices: $(2, -6), (2, 6)$ Vertices: $(3, -2), (9, -2)$

13.

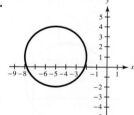

14.

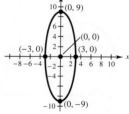

15.

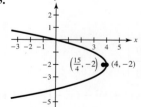

16.

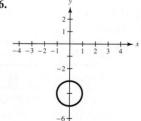

17.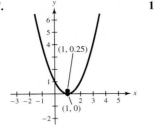

18.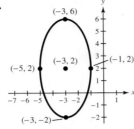

Section 12.3 *(pp. 638–645)*

1. $\dfrac{y^2}{a^2} - \dfrac{x^2}{b^2} = 1$ **3.** $\dfrac{x^2}{a^2} - \dfrac{y^2}{b^2} = 1$

5. Vertices: $(\pm 3, 0)$ **7.** Vertices: $(0, \pm 3)$

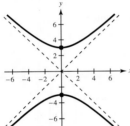 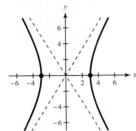

9. Vertices: $(0, \pm 3)$

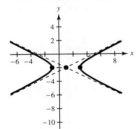

11. $\dfrac{x^2}{16} - \dfrac{y^2}{64} = 1$ **13.** $\dfrac{y^2}{16} - \dfrac{x^2}{64} = 1$ **15.** $\dfrac{x^2}{81} - \dfrac{y^2}{36} = 1$

17. Center: $(1, -2)$ **19.** Center: $(3, -4)$
Vertices: $(-1, -2), (3, -2)$ Vertices: $(3, 1), (3, -9)$

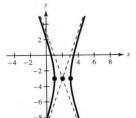 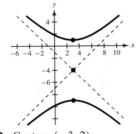

21. Center: $(2, -3)$ **23.** Center: $(-3, 2)$
Vertices: $(1, -3), (3, -3)$ Vertices: $(-4, 2), (-2, 2)$

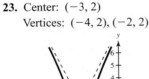 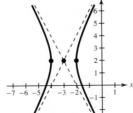

25. $\left(24 + \sqrt{220}, 0 \right) \approx (38.8, 0)$

27. The sides of the central rectangle will pass through the points $(\pm a, 0)$ and $(0, \pm b)$. To sketch the asymptotes, draw and extend the diagonals of the central rectangle.

29. Central rectangle dimensions: $2a \times 2b$, Center: (h, k)

31. $\dfrac{y^2}{9} - \dfrac{x^2}{9/4} = 1$ **33.** $\dfrac{(x-3)^2}{4} - \dfrac{(y-2)^2}{16/5} = 1$

35. Vertices: $(\pm 1, 0)$ **37.** Vertices: $(\pm 4, 0)$
Asymptotes: $y = \pm \frac{3}{2}x$ Asymptotes: $y = \pm \frac{1}{2}x$

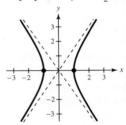 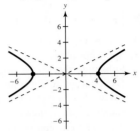

39. Ellipse   **41.** Hyperbola   **43.** Hyperbola
45. 10 mi
47. Infinitely many. The constant difference of the distances can be different for an infinite number of hyperbolas that have the same set of foci.
49. Answers will vary.   **51.** No solution   **53.** $(3, -3)$

Section 12.4 *(pp. 646–653)*

1. 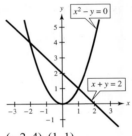   **3.**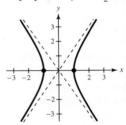

$(-2, 4), (1, 1)$ $(2, 5), (-3, 0)$

5. 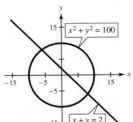   **7.**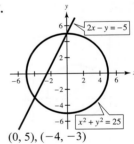

$(-6, 8), (8, -6)$ $(0, 5), (-4, -3)$

9. 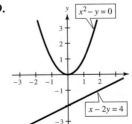   **11.**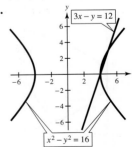

No real solution $(5, 3), (4, 0)$

13. $(1, 2), (2, 8)$ **15.** $(0, 2), (2, 0)$ **17.** No real solution

19. $\left(\pm\sqrt{5}, 2\right), (0, -3)$ **21.** $(0, 4), (3, 0)$

23. No real solution   **25.** $\left(\pm\sqrt{3}, -1\right)$

27. $\left(2, \pm2\sqrt{3}\right), (-1, \pm3)$

29. Between points $\left(-\frac{3}{5}, -\frac{4}{5}\right)$ and $\left(\frac{4}{5}, -\frac{3}{5}\right)$

31. In addition to zero, one, or infinitely many solutions, a system of nonlinear equations can have two or more solutions.

33. Substitution   **35.** $(-4, 11), \left(\frac{5}{2}, \frac{5}{4}\right)$ **37.** $(0, 2), (3, 1)$

39. $\left(\pm\dfrac{2\sqrt{5}}{5}, \pm\dfrac{2\sqrt{5}}{5}\right)$ **41.** $\left(\pm\sqrt{3}, \pm\sqrt{13}\right)$

43. $9 \text{ m} \times 12 \text{ m}$ **45.** $(3.633, 2.733)$

47. Solve one of the equations for one variable in terms of the other. Substitute that expression into the other equation and solve. Back-substitute the solution into the first equation to find the value of the other variable. Check the solution to see that it satisfies both of the original equations.

49. Two. The line can intersect a branch of the hyperbola at most twice, and it can intersect only one point on each branch at the same time.

51. -5 **53.** 9 **55.** 5 **57.** 4.564 **59.** 1.023

61. -0.632

Review Exercises   *(pp. 656–659)*

1. Ellipse   **3.** Circle   **5.** Hyperbola   **7.** Circle

9. Parabola   **11.** $x^2 + y^2 = 36$

13. Center: $(0, 0), r = 8$ **15.** $(x - 2)^2 + (y - 6)^2 = 9$

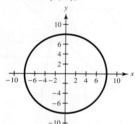

17. Center: $(-3, -4), r = 2$ **19.** $y^2 = 24x$

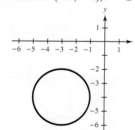

21. $(y - 5)^2 = 8x$

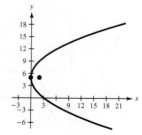

23. Vertex: $(8, -25)$, Focus: $\left(8, -\frac{49}{2}\right)$

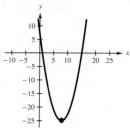

25. $\dfrac{x^2}{4} + \dfrac{y^2}{25} = 1$ **27.** $\dfrac{x^2}{4} + \dfrac{y^2}{9} = 1$

29.

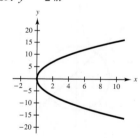

Vertices: $(\pm8, 0)$
Co-vertices: $(0, \pm4)$

31.

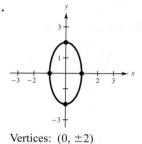

Vertices: $(0, \pm2)$
Co-vertices: $(\pm1, 0)$

33. $\dfrac{(x - 3)^2}{25} + \dfrac{(y - 4)^2}{16} = 1$ **35.** $\dfrac{x^2}{9} + \dfrac{(y - 4)^2}{16} = 1$

37. Center: $(-1, 2)$ **39.** Center: $(0, -3)$
Vertices: $(-1, -4), (-1, 8)$ Vertices: $(0, -7), (0, 1)$

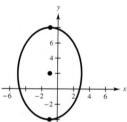   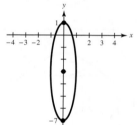

41. Vertices: $(\pm5, 0)$ **43.** Vertices: $(0, \pm5)$
Asymptotes: $y = \pm x$ Asymptotes: $y = \pm\frac{5}{2}x$

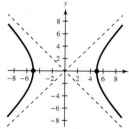   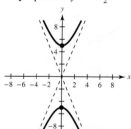

45. $\dfrac{x^2}{4} - \dfrac{y^2}{9} = 1$ **47.** $\dfrac{y^2}{64} - \dfrac{x^2}{100} = 1$

49. Center: $(3, -1)$ **51.** Center: $(4, -3)$
Vertices: $(0, -1), (6, -1)$ Vertices: $(4, -1), (4, -5)$

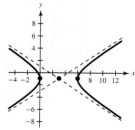   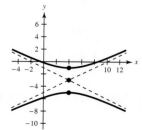

53. $\dfrac{(x+4)^2}{4} - \dfrac{(y-6)^2}{12} = 1$

55.

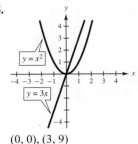

57.

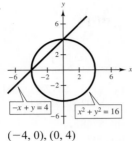

$(0, 0), (3, 9)$ $(-4, 0), (0, 4)$

59. $(-1, 5), (-2, 20)$ **61.** $(-1, 0), (0, -1)$

63. $\left(-\dfrac{17}{2}, -6\right), \left(-\dfrac{7}{2}, 4\right)$ **65.** $\left(-\dfrac{9}{5}, \dfrac{12}{5}\right), (3, 0)$

67. $(\pm 2, \pm 3)$ **69.** $\left(\pm 2, \pm \sqrt{3}\right)$ **71.** $(\pm 2, 0)$

73. $(\pm 3, \pm 2)$ **75.** $(\pm 3, 0)$

77. 6 cm × 8 cm   **79.** 40 ft × 75 ft

81. Piece 1: 38.48 in.
Piece 2: 61.52 in.

Chapter Test   *(p. 660)*

1. $(x+2)^2 + (y+3)^2 = 16$

2. $(x-1)^2 + (y-3)^2 = 9$ **3.** $(x+2)^2 + (y-3)^2 = 9$

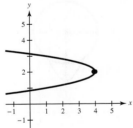

4. Vertex: $(4, 2)$; Focus: $\left(\dfrac{47}{12}, 2\right)$

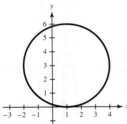

5. $(x-7)^2 = 8(y+2)$ **6.** $\dfrac{(x-2)^2}{25} + \dfrac{y^2}{9} = 1$

7. Center: $(0, 0)$ **8.** Center: $(1, 3)$
Vertices: $(0, \pm 4)$ Vertices: $(1, -2), (1, 8)$

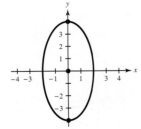
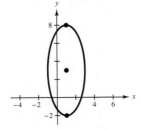

9. $\dfrac{x^2}{9} - \dfrac{y^2}{4} = 1$ **10.** $\dfrac{y^2}{25} - \dfrac{x^2}{4} = 1$

11. Center: $(3, 0)$
Vertices: $(1, 0), (5, 0)$

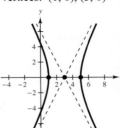

12. Center: $(4, -2)$
Vertices: $(4, -7); (4, 3)$

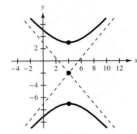

13. $(0, 3), (4, 0)$ **14.** $(\pm 4, 0)$

15. $\left(\sqrt{6}, 2\right), \left(\sqrt{6}, -2\right),$
$\left(-\sqrt{6}, 2\right), \left(-\sqrt{6}, -2\right)$

16. $x^2 + y^2 = 5000^2$ **17.** 16 in. × 12 in.

Cumulative Test   *(pp. 661–662)*

1. $-\dfrac{3}{4}, 3$ **2.** $-3, 13$ **3.** $5 \pm 5\sqrt{2}$

4. $-1 \pm \dfrac{\sqrt{3}}{3}$ **5.** $\pm\sqrt{3}, \pm\sqrt{5}$

6. $\left[-3, \dfrac{1}{3}\right]$ **7.** $\left(-\dfrac{4}{3}, \dfrac{1}{2}\right)$

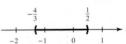

8. $x^2 - 4x - 12 = 0$

9. (a) $(f \circ g)(x) = 50x^2 - 20x - 1$; Domain: $(-\infty, \infty)$
(b) $(g \circ f)(x) = 10x^2 - 16$; Domain: $(-\infty, \infty)$

10. $f^{-1}(x) = -\dfrac{4}{3}x + \dfrac{5}{3}$

11. $f(1) = \dfrac{15}{2}, f(0.5) \approx 7.707, f(3) = \dfrac{57}{8}$

12.

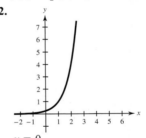

$y = 0$

13. The graphs are reflections of each other in the line $y = x$.

14.

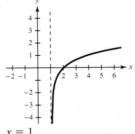

$x = 1$

15. -2 **16.** $\log_2 \dfrac{x^3 y^3}{z}$ **17.** $\dfrac{1}{2}\log_{10}(x+1) - 4\log_{10} x$

18. 3   **19.** 12.182   **20.** 18.013   **21.** 0.867

22. \$34.38   **23.** 8.33%   **24.** 19.8 yr

25. $(x - 3)^2 + (y + 7)^2 = 64$

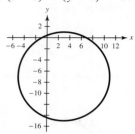

26. Vertex: $(5, -45)$

Focus: $5, \left(-\frac{359}{8}\right)$

27. $\dfrac{(x + 3)^2}{4} + \dfrac{(y - 2)^2}{25} = 1$

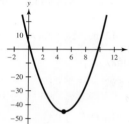

28. Center: $(0, 0)$

Vertices: $(0, \pm 2)$

29. $\dfrac{y^2}{9} - x^2 = 1$

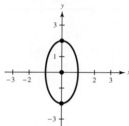

30. Center: $(0, 1)$

Vertices: $(\pm 12, 1)$

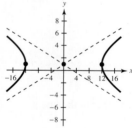

31. $(1, -1), (3, 5)$ **32.** $(-4, \pm 1), (-1, \pm 2)$
33. $8 \text{ ft} \times 4 \text{ ft}$ **34.** $7 \text{ ft} \times 3 \text{ ft}$

CHAPTER 13

Section 13.1   *(pp. 664–671)*

1. $2, 4, 6, 8, 10$ **3.** $\frac{1}{4}, \frac{1}{16}, \frac{1}{64}, \frac{1}{256}, \frac{1}{1024}$ **5.** $3, 8, 13, 18, 23$
7. $1, \frac{4}{5}, \frac{2}{3}, \frac{4}{7}, \frac{1}{2}$ **9.** $\frac{3}{4}, \frac{2}{3}, \frac{9}{14}, \frac{12}{19}, \frac{5}{8}$ **11.** $-2, 4, -6, 8, -10$
13. $\frac{1}{4}, -\frac{1}{8}, \frac{1}{16}, -\frac{1}{32}, \frac{1}{64}$ **15.** $-1, \frac{1}{4}, -\frac{1}{9}, \frac{1}{16}, -\frac{1}{25}$
17. $1 \cdot 2 \cdot 3 \cdot 4 \cdot 5 \cdot 6; 720$
19. $1 \cdot 2 \cdot 3 \cdot 4 \cdot 5 \cdot 6 \cdot 7 \cdot 8 \cdot 9; 362{,}880$
21. $1, 1, 2, 6, 24$ **23.** $2, 3, 4, 5, 6$ **25.** 5 **27.** $53{,}130$
29. $\dfrac{1}{n + 1}$ **31.** $n(n + 1)$ **33.** b   **34.** d   **35.** c

36. a   **37.** $a_n = 2n - 1$ **39.** $a_n = n^2 - 1$
41. $a_n = \left(-\frac{1}{5}\right)^n$ **43.** $S_1 = 7; S_2 = 16; S_6 = 72$
45. $S_2 = \frac{3}{2}; S_3 = \frac{11}{6}; S_9 = \frac{7129}{2520}$ **47.** 30 **49.** 77
51. $\frac{9}{5}$ **53.** $\frac{182}{243}$ **55.** $\displaystyle\sum_{k=1}^{5} k$ **57.** $\displaystyle\sum_{k=1}^{11} \frac{k}{k+1}$ **59.** $\frac{2}{9}$

61. Substitute 6 for n in a_n: $a_6 = \dfrac{2}{3(6)} = \dfrac{1}{9}$

63. (a) \$535, \$572.45, \$612.52, \$655.40, \$701.28, \$750.37, \$802.89, \$859.09

(b) \$7487.23

(c)

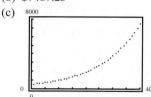

(d) No; For increasing values of N, the values of the terms A_N increase by greater amounts.

65. $36°, 60°, 77.1°, 90°, 100°, 108°$

67. A sequence is a function because there is only one value for each term of the sequence.

69. $a_n = 4n! = 4[1 \cdot 2 \cdot 3 \cdot 4 \cdot \cdots \cdot (n-1) \cdot n]$
$a_n = (4n)! = 1 \cdot 2 \cdot 3 \cdot 4 \cdot \cdots \cdot (4n - 1) \cdot (4n)$

71. True

$$\sum_{k=1}^{4} 3k = 3 + 6 + 9 + 12$$
$$= 3(1 + 2 + 3 + 4) = 3\sum_{k=1}^{4} k$$

73. 9 **75.** -11
77. Center: $(0, 0), r = 6$ **79.** Center: $(-2, 0), r = 4$

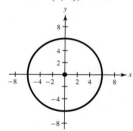   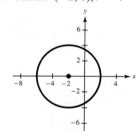

81. Vertex: $(0, 0)$, Focus: $\left(0, \frac{3}{2}\right)$

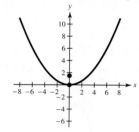

83. Vertex: $(0, -4)$, Focus: $(0, -6)$

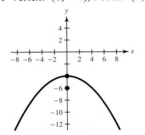

Section 13.2 *(pp. 672–679)*

1. 3   **3.** -6 **5.** $\frac{1}{2}$ **7.** 4   **9.** -3 **11.** $\frac{1}{2}$

13. $a_n = 3n + 1$ **15.** $a_n = \frac{3}{2}n - 1$ **17.** $a_n = -5n + 105$

19. $a_n = \frac{3}{2}n + \frac{3}{2}$ **21.** $a_n = \frac{5}{2}n + \frac{5}{2}$ **23.** 59

25. 14, 20, 26, 32, 38   **27.** 23, 18, 13, 8, 3

29. $a_1 = 5, a_5 = 29$ **31.** 210   **33.** 1425   **35.** 255

37. 62,625   **39.** 35   **41.** 522   **43.** 12,200   **45.** 243

47. 23   **49.** 2850   **51.** \$369,000   **53.** 632 bales

55. Add 3 to the value of the term.

57. The nth partial sum can be found by multiplying the number of terms by the average of the first term and the nth term.

59. 7, 12, 17, 22, 27   **61.** 11, 15, 19, 23, 27

63. 7, 10, 13, 16, 19   **65.** $\frac{3}{2}, 4, \frac{13}{2}, 9, \frac{23}{2}$ **67.** d   **68.** c

69. a   **70.** b   **71.** 1024 ft   **73.** 114 times

75. A recursion formula gives the relationship between the terms a_{n+1} and a_n.

77. Yes. Because a_{2n} is n terms away from a_n, add n times the difference d to a_n.
$$a_{2n} = a_n + nd$$

79. (a) 4, 9, 16, 25, 36

    (b) No. There is no common difference between consecutive terms of the sequence.

    (c) $49, \displaystyle\sum_{k=1}^{n}(2k-1) = n^2$

81. Center: $(-2, 8)$, Vertices: $(-4, 8), (0, 8)$

83. Center: $(4, 0)$, Vertices: $(2, 0), (6, 0)$

85. $\displaystyle\sum_{k=1}^{5} 3k$ **87.** $\displaystyle\sum_{k=1}^{5} 2^k$

Mid-Chapter Quiz *(p. 680)*

1. 4, 8, 12, 16, 20   **2.** 7, 9, 11, 13, 15   **3.** $32, 8, 2, \frac{1}{2}, \frac{1}{8}$

4. $-\frac{3}{5}, 3, -\frac{81}{7}, \frac{81}{2}, -135$ **5.** 100   **6.** 40   **7.** 87

8. 25   **9.** 40   **10.** 26   **11.** $\displaystyle\sum_{k=1}^{20}\frac{2}{3k}$ **12.** $\displaystyle\sum_{k=1}^{25}\frac{(-1)^{k+1}}{k^3}$

13. $\displaystyle\sum_{k=1}^{20}\frac{k-1}{k}$ **14.** $\displaystyle\sum_{k=1}^{10}\frac{k^2}{2}$ **15.** $\frac{1}{2}$ **16.** -6

17. $a_n = -3n + 23$ **18.** $a_n = -4n + 36$

19. 40,200   **20.** \$33,397.50

Section 13.3 *(pp. 682–689)*

1. Geometric, 2   **3.** Geometric, π **5.** Not geometric

7. Geometric, $\frac{1}{2}$ **9.** Geometric, $-\frac{1}{2}$

11. $a_n = 2^{n-1}; a_7 = 64$ **13.** $a_n = 9\left(\frac{2}{3}\right)^{n-1}; a_7 = \frac{64}{81}$

15. $a_n = 2(2)^{n-1}; a_7 = 128$ **17.** $a_n = 8\left(\frac{1}{4}\right)^{n-1}; a_7 = \frac{1}{512}$

19. 13,120   **21.** $-14,762$ **23.** 16.00   **25.** 2.47

27. 1023   **29.** 772.48   **31.** 32   **33.** 6   **35.** $\frac{7}{10}$

37. 2   **39.** $\frac{2}{3}$ **41.** $\frac{10}{9}$ **43.** \$3,623,993.23

45. (a) \$5,368,709.11   (b) \$10,737,418.23

47. \$9748.28   **49.** \$105,428.44   **51.** \$100,953.76

53. The sum of the first 6 terms of the geometric sequence whose nth term is $a_n = 3(2)^{i-1}$

55. $\frac{4}{5}$ **57.** 5, -10, 20, -40, 80

59. $-4, 2, -1, \frac{1}{2}, -\frac{1}{4}$ **61.** 32   **63.** $\frac{37}{64} \approx 0.578$ square unit

65. $a_n = 70(0.8)^n; 18.4°F$ **67.** $a_n = a_1 r^{n-1}$

69. An arithmetic sequence has a common difference between consecutive terms whereas a geometric sequence has a common ratio between consecutive terms.

71. When a positive number is multiplied by a number between 0 and 1, the result is a smaller positive number, so the terms of the sequence decrease.

73. $(-1, 2), (2, 8)$ **75.** 8%   **77.** 7.5%

79.

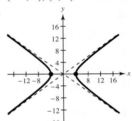

Vertices: $(\pm 4, 0)$

Asymptotes: $\pm\frac{3}{4}x$

Section 13.4 *(pp. 690–697)*

1. 15   **3.** 252   **5.** 1   **7.** 38,760   **9.** 38,760

11. 1, 8, 28, 56, 70, 56, 28, 8, 1   **13.** 15   **15.** 35   **17.** 70

19. 10   **21.** 35   **23.** $t^3 + 15t^2 + 75t + 125$

25. $m^5 - 5m^4n + 10m^3n^2 - 10m^2n^3 + 5mn^4 - n^5$

27. $x^6 + 18x^5 + 135x^4 + 540x^3 + 1215x^2 + 1458x + 729$

29. $u^3 - 3u^2v + 3uv^2 - v^3$

31. $243a^5 - 405a^4 + 270a^3 - 90a^2 + 15a - 1$

33. $64y^6 + 192y^5z + 240y^4z^2 + 160y^3z^3 + 60y^2z^4 + 12yz^5 + z^6$

35. $x^8 + 8x^6 + 24x^4 + 32x^2 + 16$

37. $81a^4 + 216a^3b + 216a^2b^2 + 96ab^3 + 16b^4$

39. $x^4 + \dfrac{8x^3}{y} + \dfrac{24x^2}{y^2} + \dfrac{32x}{y^3} + \dfrac{16}{y^4}$

41. $32x^{10} - 80x^8y + 80x^6y^2 - 40x^4y^3 + 10x^2y^4 - y^5$

43. $120x^7y^3$ **45.** $163,296a^5b^4$ **47.** $1,259,712x^2y^7$

49. 120   **51.** 5940

53. The first and last numbers in each row are 1. Every other number in the row is formed by adding the two numbers immediately above the number.

55. The signs of the terms alternate in the expansion of $(x - y)^n$.

57. $x^{3/2} + 15x + 75x^{1/2} + 125$

59. $x^2 - 3x^{4/3}y^{1/3} + 3x^{2/3}y^{2/3} - y$

61. $81t^2 + 108t^{7/4} + 54t^{3/2} + 12t^{5/4} + t$

63. 593,775 **65.** 2,598,960 **67.** 85,013,600

69. -4 **71.** 1.172 **73.** 510,568.785

75. $\frac{1}{32} + \frac{5}{32} + \frac{10}{32} + \frac{10}{32} + \frac{5}{32} + \frac{1}{32}$

77. $\frac{1}{256} + \frac{12}{256} + \frac{54}{256} + \frac{108}{256} + \frac{81}{256}$

79. The sum of the numbers in each row is a power of 2. Because the sum of the numbers in Row 2 is $1 + 2 + 1 = 4 = 2^2$, the sum of the numbers in Row n is 2^n.

81. No. The coefficient of x^4y^6 is $_{10}C_6$ because $n = 10$ and $r = 6$.

83. The value for r is 6 because $6 + 1 = 7$.

85. 390 **87.** 97,656

Review Exercises *(pp. 700–702)*

1. 8, 11, 14, 17, 20 **3.** $\frac{1}{2}, \frac{2}{5}, \frac{3}{8}, \frac{4}{11}, \frac{5}{14}$ **5.** 2, 6, 24, 120, 720

7. $\frac{1}{2}, \frac{1}{2}, 1, 3, 12$ **9.** $a_n = 3n + 1$ **11.** $a_n = \dfrac{1}{n^2 + 1}$

13. $a_n = -2n + 5$ **15.** $a_n = \dfrac{3n^2}{n^2 + 1}$ **17.** 28 **19.** $\frac{13}{20}$

21. $\sum\limits_{k=1}^{4} (5k - 3)$ **23.** $\sum\limits_{k=1}^{6} \dfrac{1}{3k}$ **25.** -5.5

27. 127, 122, 117, 112, 107 **29.** $2, \frac{7}{3}, \frac{8}{3}, 3, \frac{10}{3}$

31. $80, \frac{155}{2}, 75, \frac{145}{2}, 70$ **33.** $a_n = 4n + 6$

35. $a_n = -50n + 1050$ **37.** 486 **39.** 1935

41. 2527.5 **43.** 5100 **45.** 462 seats **47.** Geometric, $\frac{5}{2}$

49. 10, 30, 90, 270, 810 **51.** $100, -50, 25, -\frac{25}{2}, \frac{25}{4}$

53. $4, 6, 9, \frac{27}{2}, \frac{81}{4}$ **55.** $a_n = \left(-\frac{2}{3}\right)^{n-1}$ **57.** $a_n = 24(3)^{n-1}$

59. $a_n = 12\left(-\frac{1}{2}\right)^{n-1}$ **61.** 27,846.96 **63.** -637.85

65. 8190 **67.** -1.928 **69.** 8 **71.** 12

73. (a) $a_n = 120,000(0.70)^n$ (b) \$20,168.40

75. 321,222,672 visits **77.** 56 **79.** 1365 **81.** 91,390

83. 177,100 **85.** 6 **87.** 120

89. $x^4 - 20x^3 + 150x^2 - 500x + 625$

91. $125x^3 + 150x^2 + 60x + 8$

93. $x^{10} + 10x^9 + 45x^8 + 120x^7 + 210x^6 + 252x^5 + 210x^4$
$\quad + 120x^3 + 45x^2 + 10x + 1$

95. $81x^4 - 216x^3y + 216x^2y^2 - 96xy^3 + 16y^4$

97. $u^{10} + 5u^8v^3 + 10u^6v^6 + 10u^4v^9 + 5u^2v^{12} + v^{15}$

99. $13,440x^4$ **101.** $-61,236$

Chapter Test *(p. 703)*

1. $1, -\frac{3}{5}, \frac{9}{25}, -\frac{27}{125}, \frac{81}{625}$ **2.** 2, 10, 24, 44, 70 **3.** 60

4. 45 **5.** -45 **6.** $\sum\limits_{k=1}^{12} \dfrac{2}{3k+1}$ **7.** $\sum\limits_{k=1}^{6} \left(\dfrac{1}{2}\right)^{2k-2}$

8. 12, 16, 20, 24, 28 **9.** $a_n = -100n + 5100$

10. 3825 **11.** $-\frac{3}{4}$ **12.** $a_n = 4\left(\frac{1}{2}\right)^{n-1}$ **13.** 1020

14. $\frac{3069}{1024}$ **15.** 1 **16.** $\frac{50}{3}$ **17.** 1140

18. $x^5 - 10x^4 + 40x^3 - 80x^2 + 80x - 32$

19. 56 **20.** 490 m **21.** \$153,287.87

Index of Applications

Biology and Life Sciences

Antler spread of an elk, 584
Bamboo plant growth, 201
Blood concentration of an antibiotic, 551
Calories burned
    playing basketball, 35
    playing tennis, 35
    racewalking, 35
Environment: oil spill, 369
Fitness trail, 180
Human memory model, 592, 616
Killer whales age and weight, 483
Life expectancy, 181
Nutrition, 217, 431
Nutritional supplement for dogs, 432
Population, 611, 702
    United States, 115
    wildlife, 375, 611, 618
Range of human heights, 159
Range of normal body temperatures, 159
Rattlesnake pit organ, 157
Weight loss, 113

Business

Advertising effect, 611
Apartment rental, 223
Average cost per unit, 321, 326, 344, 372, 375, 557
Break-even analysis, 311, 385, 437
Cost
    of daily expenses for a sales employee, 209, 318
    of monthly flights for an airline, 483
    of printing a book, 183
Defective units, expected number, 131
Demand and advertising, 366, 375
Demand and price, 365, 375, 568
Demand for a product, 191
Depreciation of equipment, 91, 172, 702
Depreciation of a vehicle, 449
Interest rate in terms of monthly payment, 351
Inventory of air conditioning units, 214, 215
Inventory costs, 321, 439
Manufacturing, 217, 223
Manufacturing process time study, 157
Operating costs of fleet vehicles, 151, 165
Petty cash, 217
Production
    of baked goods, 415
    of computer parts, 415
    cost, 172, 577
    of electronics, 431
    of furniture, 431
    of an oil refinery, 157
Profit, 5, 15, 17, 27, 28, 50, 173, 241, 551
    and advertising, 509

of Coach, 206
of Hewlett-Packard, 169
Real estate commission, 121
Retail discount, 123
Retail markup, 122
Revenue, 362, 517, 526
Sales, 677
    of Aaron's, 206
    of AutoZone, 206
    of Coach, 115
Selling price, 536, 542, 556
Shared cost, 375
Stock
    total cost, 37
    value, 17, 21
Ticket pricing for an event, 677
Total salaries of the Boston Red Sox, 192
Transporting capacity, 41
Unit sales, 224

Chemistry and Physics

Acidity, 610
Ball rolling down an incline plane, 364
Boyle's Law for the volume of a gas, 376
Car
    road handling, 364
    stopping distance, 175, 364, 375
Chemical reaction, 277
Computer operating temperatures, 164
Cooling, 689
Earthquake intensity, 609, 617
Electrical networks, 425
Electrical resistance, 351, 353
Fluid rate problem, 207
Force to move a steel block, 475
Force on a spring, 131, 172
Free-falling object, 297, 316, 525
    distance fallen at time t, 679, 703
    height at time t, 541
    height when dropped, 480, 497, 498, 500
    time to reach a given height, 309, 311, 509, 558
Frequencies of piano notes, 23
Frequency of a vibrating string, 457
Friction, 601
Fuel economy, 525
Height of a projectile, 549, 557
Hooke's Law, 363, 375, 500
Light bulb wattage and lumens, 220
Light year, 230, 231
Mass of Earth and the Sun, 231
Metal expansion, 233
Meteorology
    average temperatures of cities, 151
    tornado wind speed, 585
Molecular transport, 595
Newton's Law of Cooling, 603

Pendulum length and period, 457, 496
Power used by an electric heater, 480
Radioactive decay, 566, 586, 608, 614, 617
Safe load for a wooden beam, 191, 369
Salt water mixture, 133
Satellite orbit, 660
Solution mixture, 117, 139, 142, 395, 404, 437
Sonic boom, 645
Sound intensity, 593, 603, 616, 617
Sunlight by day of the year, 189
Temperature
    Anchorage, Alaska, 169
    conversion, 60
    as measured by a weather balloon, 27
    of the Sun, 230
    of thawing meat, 5
Thickness of a soap bubble, 230
Time for a pump to fill a tank, 133
Tractrix, 703
Vertical motion, 401, 438
Width of an air molecule, 230
Wind power generation, 369

Construction

Acceptable wood lengths for a project, 159
Beam deflection, 627
Bike path, 207
Cement block wall, 131
Cutting a beam from a log, 475
Dry mix mortar, 133
Exercise area for a Border Collie, 105
Fireplace, 117
Highway design, 533
Length of a guy wire, 500
Mechanical drawing, 623
Open conduit, 543
Roller coaster track, 533
Roof pitch, 201
Semielliptical archway, 633
Shipping carton design, 23
Skateboarding ramp, 201
Suspension bridge, 533, 627
Tracking progress, 33
Wooden box with a square base, 375

Consumer

Account, 7, 13, 28, 51
Amount owed on a loan, 209
Bird seed mixture, 163
Calling card time available, 164
Car rental, 179
Change due, 37, 54
College expenses, 463
Commissions,
    production, 107, 108, 162, 172

sales, 98, 107, 121, 223
Cost(s)
  of admission, 81, 90, 100
  of an automobile, 81, 440
  of a band performance, 163
  of camping, 81, 98
  of car repair, 106
  of a cell phone, 37, 81
  of an engagement ring, 52
  of fruit, 77
  of fuel for a car, annually, 37, 166
  of fuel for a trip, 131
  of labor to build a deck, 108
  of living raise, 119
  of multiple units, 66, 85
  of an oil change, 662
  of paint, 85
  percent increase in, 125
  of a plasma television, 52
  to produce a DVD, 221
  of renting movies and videos, 80
  of repairing an oven, 165
  of a sweater, 45
Depreciation
  of a boat, 207
  of a car, 53, 180, 569, 618
  of a computer system, 193
  of a pizza oven, 207
  of a television, 209
  of a truck, 614
Earnings, 164
Exponential savings plan, 569
Fuel consumption, 52, 134
Hourly wage, 577
Housing budget, 134
Price(s)
  of bottled soda, 39
  of a camcorder, 166
  of electronics, 163
  of gasoline, 37, 51
  of gold, 201
  of golf equipment, 163
  of a phone, 98
Property assessment, 137
Property value, 568
Reduced rates, 538, 556, 558
Rent for an apartment, 98
Salary, 107, 119, 149, 677, 686
Savings, 27, 680
Staying within budget, 149
Tax
  income, 80, 132, 577
  sales, 80
Unit price, 51, 128, 132, 163, 277
Wages, 66, 81, 90, 98, 209, 220, 677, 686, 701
Working two jobs, 109, 134, 214, 215

Geometry
Angle measures of a triangle, 402, 404
Area,
  of an annulus, 297

of an apartment, 83
of the base of a rectangular solid, 261
circular, 124, 125, 134, 188
comparing a portion of a region to the whole, 325
divided into subregions, 134
of a property, 543
of a rectangle, 19, 45, 61, 65, 73, 83, 97, 124, 266, 267, 469, 470, 483
of a rectangle of fixed perimeter, 109
of a region, 239, 240, 265, 268, 276, 280, 293, 294, 314, 315, 372, 421
of a roof, 457
of a square, 188
of subdivided squares, 689
of subdivided triangles, 688
trapezoidal, 71
of a triangle, 97, 253, 268, 420
of a triangular region, 424
Cereal display, 41
Chorus platform, 433
Circle, doubling the radius, 143
Circular mirror, 626
Circumference of the Sun, 264
Comparing measurements, 127, 132, 163, 166
Cross section of a swimming area, 433
Diagonal length of a board, 455
Diagonals of a polygon, 63
Diameter
  of a softball, 505
  of the Unisphere, 505
Dimension(s)
  of a base drum, 137
  of a box, 315, 316
  of a building, 300
  of a cell phone display, 659
  of a ceramic tile, 659
  of a circuit board, 659
  of a closed box, 317
  of a cone, 483
  of a corral, 516, 659
  of a cube, 443, 460, 494
  of a dog park, 652
  of a golf tee area, 251
  of an ice rink, 659
  of an iPhone, 515
  of a Jamaican flag, 105
  of a kennel, 517
  of an open box, 515
  of a picture frame, 537
  of a rectangle, 308, 317, 473, 524, 526, 551, 556, 558, 660, 662
  of a rectangular prism, 261
  of a right triangle, 454, 481, 495
  of a room, 251
  of a sandbox, 251, 289
  of a softball diamond, 455
  of a square, 443, 449, 582
  of a suitcase, 317
  of a swimming pool, 137, 163
  of a television, 496

of a triangle, 137, 143, 163, 308, 537, 556
of a triangular prism, 261
of a vaccine cooler, 300
of a window, 316
Ellipse
  airplane range, 635
  bicycle chain wheel, 633
  sprint car dirt track, 633
  wading pool, 634
  White House Oval Office, 634
Geometric model
  for completing the square, 515
  for a difference of two cubes, 303
  for factoring a trinomial, 284, 299
  for a polynomial product, 246, 247, 252
Geometric probability, 335
Golden section, 475
Height of a prism, 27
Hyperbolic
  mirror, 643, 653
  radio navigation, 643
  sculpture, 643
Length
  of the cut pieces of a board, 105, 108, 162
  of a ladder, 455
  of a rectangle, 99, 149, 275
  of rope segments, 28
  of the sides of a sign, 105
  of two wires, 659
Maximum width of a package, 149
Measurement error, 159
Miniature golf, 653
Perimeter, 97, 235, 242
  of a dining hall, 496
  of a figure, 465, 466, 500
  of a rectangle, 70, 73, 80, 97, 108, 116, 265, 321
    of given area, 372
  of a roof, 465
  of a square, 191
  of a triangle, 70, 72, 74, 116, 462
Pythagorean Theorem, 540, 556, 558
Radius of a dime, 137
Resizing a rectangular picture, 133, 163
Ripples in water, 577
Sail of a sailboat, 652
Semicircular boundary of a dog leash, 623
Stars, 671
Surface area
  of a basketball, 505
  of a cylinder, 277
Testing for collinear points, 422
Triangle, doubling the height, 143
Using similar triangles, 130, 133, 165
Volume
  of a box, 19, 283, 285, 289
  of a cord of wood, 28
  of a cube, 191
  of a hot tub, 53

of a shower stall, 289
of two swimming pools, 327
Width of a tennis court, 105

Interest Rate

Compound interest, 253, 268, 509, 537, 618, 670
   comparing types of compounding, 567, 586
Continuous compounding, 601
Doubling time, 603, 605, 610
Effective yield, 606, 617, 662
   compound interest, 551, 556, 604, 617
Finding the interest rate, 142, 163
Finding the principal, 568, 618
Finding the time invested, 142
Finding the type of compounding, 605
Increasing annuity, 687, 702, 703
Investment mixture, 138, 227, 387, 413, 414, 432, 438
Length of a home mortgage, 585
Simple interest, 60, 136, 163, 165, 277, 367, 375
Time to double an investment, 82
Time to quadruple an investment, 662

Miscellaneous

Album downloads, 607
Auditorium seating capacity, 701
Avalanche rescue system, 651
Babylonian number system, 31
Baling hay, 677
Busing boundary, 651
Clock chimes, 679
Computer virus, 607
Cookie recipe, 135
Cooking, 39
Cost to seize an illegal drug, 321
Course grade, 120
Depth of a river, 524
Election poll, 131
Exam scores, 21, 90, 113, 134, 162, 175
Exponential communication, 41
Flower order, 141
Forensic archaeology, 13
Fund drive, 33
Inventory arrangement, 23
Jazz band audition, 318
Joint time to complete a task, 113, 134, 162, 165
Miniature golf scores, 5
Mixture problem, 384, 393, 395, 402, 403, 404, 413, 437
Number attending an event, 93
Number problem, 311, 316, 317, 353, 361, 387, 394, 415, 543, 701
Number of visible stars, 233
Observation wheel, 626
Public TV station membership, 125
Rainfall, 52, 90
SAT score and grade-point average, 189
Search team, 651

Sieve of Eratosthenes, 27
Slope of a ladder, 195
Slope of a loading ramp, 222
Snowfall, 52
Soup distribution, 439
Sports
   baseball batting average, 359
   basketball, ball and hoop diameters, 475
   bungee jumping, 689
   hockey save percentage, 359
   soccer ball, 670
   speed skating, 157
   Super Bowl scores, 169
Team score in a game, 11
Temperature change, 11, 15
Ticket sales
   for a concert, 91, 106, 433, 440
   for a dinner, 437
   for a drumline competition, 106
   for an ice show, 93, 362
   for a play, 93, 106, 141
   at three theaters, 414
Time spent in class and studying, 132
Value of coins or bills, 77, 80, 93, 163
Volunteer services, 91
Website growth, 607, 702
Work-rate problem, 140, 163, 342, 361, 375, 500, 539, 557

Time and Distance

Altitude of an airplane, 15
Average speed, 374, 375, 543
Depth of a river, 524
Distance
   between two cars, 143
   between two planes, 143
   between two vehicles, 327
   jogging, 142
   on a line, 543
   traveled by a bicycle, 80
   traveled by a car, 60, 63, 83, 85, 98, 99, 183, 191
   traveled by a train, 83, 221
Elevation
   at Death Valley, California, 7
   of a falcon, 7
   at the Grand Canyon, 16
   of a hot-air balloon, 179, 180
   of a rock climber, 207
   of scuba divers, 9
   at the summits of volcanoes, 9
   of a whale, 7
Fishing depth, 13
Flight path of an aircraft, 222
Jumping height with new shoes, 63
Map scale, 132
Path
   of a ball, 526, 534, 535, 555
   of a diver, 535
   of a fishing lure, 556
   of a toy rocket, 534

Speed
   determined by the cost of fuel, 543
   highway driving, 11, 91
   jogging, 21
   of sound, 191
   of a space shuttle, 21
   of a train, 51, 54
   of travel, average, 165
   of a truck, 51
   of two runners, 359
   up and down a trail, 470
Time
   to cross-country ski, 35
   to overtake a slower jogger, 143
   of travel, 81
   to travel between two cities, 375
   traveled at each rate on a trip, 143
   to walk to the subway, 35
Traveling upwind and downwind, 361
Yards gained or lost in football, 11, 21, 90

U.S. Demographics

Civilian labor force, 236
Federal debt, 233
Fishery products, 233
Government surplus, 173
Immigrants, 463
Marital status, 181, 341
Online retail sales, 601
Per capita income (Montana), 333
Personal income (Alabama), 333
Postsecondary school enrollment, 115
Public and private college enrollments, 236
Spending on meals and beverages, 333
Undergraduate enrollment, 341

Index

A

Absolute value, 6, 6*
  equation, 152, 152*
    solving, 152
    standard form of, 153*
  inequality, solving, 155, 155*, A13
Abundant number, 27
Add two integers, 12
Adding rational expressions
  with like denominators, 336
  with unlike denominators, 337
Addition
  Associative Property of, 43, 64*, A5
  Commutative Property of, 43, 64*, A5
  of fractions, 32
    alternative rule, 32
  of integers, 12
  Property of Inequalities, 146
Additional problem-solving strategies,
    summary of, 82
Additive
  Identity Property, 43, 64*, A5
  inverse, 11*
  Inverse Property, 43, 64*, A5
Algebra, properties of, 64*
Algebraic
  equation, 86
  expression, 56, 56*, 87
    evaluating, 58
    expanding, 64*
    simplifying, 68, 68*
    terms of, 56
    translating phrases into, 78, 78*
  inequalities, 144*
Algorithm
  borrowing, 14*
  carrying, 12*
  long division, 20
  vertical multiplication, 18
Alternative rule
  for adding two fractions, 32*
  for subtracting two fractions, 32*
Approximately equal to, 36*
Area
  formulas for, 137
  of a triangle, 420
Arithmetic sequence, 672
  common difference of, 672
  nth partial sum of, 675
  nth term of, 673
Arithmetic summary, 24
Associative Property
  of Addition, 43, 64*, A5
  of Multiplication, 43, 64*, A5
Asymptote(s), 563
  horizontal, 563*
  of a hyperbola, 639

Augmented matrix, 406
Average, 21*
Axis of a parabola, 528, 624

B

Back-substitute, 381
Base, 40
  natural, 565
Binomial, 234, 690*
  coefficients, 690*
  expanding a, 693*
  square of a, 250, 250*
Binomial Theorem, 690
Borrowing algorithm, 14*
Bounded intervals, 144, 144*
Branch of a hyperbola, 639
Break-even point, 385*

C

Carrying algorithm, 12*
Cartesian plane, 168*
Center
  of a circle, 621, 622
  of an ellipse, 628
Central rectangle of a hyperbola, 639
Change-of-base formula, 583
Check a solution of an equation, 86, 86*
Circle, 621
  center of, 621, 622
  radius of, 621, 622
  standard form of the equation of
    center at (h, k), 622
    center at origin, 621
Clearing an equation of fractions, 112,
    112*
Coefficient(s), 56, 234*, 690*
  binomial, 690*
  matrix, 406
Collinear points, test for, 422
Combined variation, 366*
Common
  difference, 672
  formulas, 137
    miscellaneous, 136
  logarithmic function, 579
  ratio, 682
Commutative Property
  of Addition, 43, 64*, A5
  of Multiplication, 43, 64*, A5
Completing the square, 510, 522
Complex
  conjugates, 488
  fraction, 346
  number(s), 486
    imaginary part of, 486
    real part of, 486
    standard form of, 486

Composite function, 570
Composite number, 22
Composition of two functions, 570
Compound inequality, 148
Compound interest, 567
  continuous, 567
  formulas for, 567
Condensing a logarithmic expression, 591
Conic, 620
  section, 620
Conjugates, 470
  complex, 488
Conjunctive inequality, 148*
Consistent system, 380, 380*
Constant, 56, 56*
  of proportionality, 362
  term, 234
Continuous compounding formula, 567
Coordinate(s), 168
  x-coordinate, 168*
  y-coordinate, 168*
Cost, 122*
Co-vertices of an ellipse, 628
Cramer's Rule, 418
Critical numbers
  of a polynomial, 544
  of a rational inequality, 548
Cross-multiplication, 129, 358
Cube root, 442
Cubes
  difference of two, 301
  sum of two, 301

D

Decimal
  repeating, 36
  rounding a, 36, 36*
  terminating, 36
Decision digit, 36
Declining balances method, 449
Degree
  of a polynomial, 234, 234*
  of a term, 234*
Denominator, 20
  least common, 337
  rationalizing the, 453*
Dependent
  system, 380, 380*
  variable, 186
Determinant, 416*
  expanding by minors, 417
  of a 2×2 matrix, 416
Difference, 14
  common, 672
  of two cubes, 301
  of two squares, 296, 296*
Direct variation, 362

as the nth power, 364
Directly proportional, 362*
 to the nth power, 364*
Directrix of a parabola, 624
Discount, 123
 rate, 123
Discriminant, 521
 using the, 521
Disjunctive inequality, 148*
Distance-rate-time formula, 136
Distributive Property, 43, 64*, A5
Divide evenly, 257
Dividend, 20, 255
Dividing
 integers, rules for, 20
 a polynomial by a monomial, 254
 rational expressions, 331
Divisibility tests, 22*
Divisible, 22*
Division
 of fractions, 34
 of integers, 20
 long, of polynomials, 255
 Property of Inequalities, 146
 synthetic, 258
  of a third-degree polynomial, A31
Divisor, 20, 24, 255
Domain
 of a function, 187
 of a radical function, 447
 of a rational expression, 320
 of a rational function, 320
 of a relation, 184
Double inequality, 148
Double solution, 502

E

e, 565
Effective yield, 606
Elementary row operations, 408
Elimination
 Gaussian, 397
  with back-substitution, 410
 method of, 388, 388*
Ellipse, 628
 center of, 628
 co-vertices of, 628
 focus of, 628
 major axis of, 628
 minor axis of, 628
 standard form of the equation of
  center at (h, k), 630
  center at origin, 628
 vertices of, 628
Endpoints of an interval, 144
Entry of a matrix, 406*
 minor of, 417*
Equation(s), 152
 absolute value, 152, 152*
  solving 152
  standard form of, 153*
 clearing of fractions, 112, 112*

equivalent, 88, 88*
 first-degree, 102*
 graph of an, 176, 176*
 of a line
  general form, 202, 205*
  point-slope form, 202
  slope-intercept form, 196
  summary, 207*
  two-point form, 203*, 423
 linear
  forms of, A20
  in one variable, 102, 102*
  in two variables, 176*
 operations that yield equivalent, A9
 percent, 119, 119*
 position, 401
 quadratic, 305
  guidelines for solving, 305
 of quadratic form, 506
 raising each side to the nth power, 476
 solution of, 170*
 solving, 86
 standard form
  absolute value, 153*
  circle, 621, 622
  ellipse, 628, 630
  hyperbola, 638, 641
  parabola, 624
 systems of, 378
  row-echelon form of, 396
  solution of, 378
  solving, 378
Equivalent
 equations, 88, 88*
  operations that yield, A9
 fractions, 30
 inequalities, 146*
 systems, 397
  operations that produce, 397
Evaluating
 an expression, 6*, 58
 a function, 187
Expanding
 an algebraic expression, 64*
 a binomial, 693*
 a logarithmic expression, 591
 by minors, 417
Exponent(s), 40, 40*
 Inverse Property of, 597
 negative, 227
 rational, 445
 Rules of, 226, 227
  Negative, 227, A24
  Power, 227
  Power-to-Power, 226, A24
  Product, 226, 227, A24
  Product-to-Power, 226, A24
  Quotient, 226, 227, A24
  Quotient-to-Power, 226, A24
  Summary of, 445, A24
  Zero, 227, A24
 zero, 227

Exponential
 decay, 607
 equations
  One-to-One Property of, 596
  solving, 597
 form, 40, 40*, 57, 57*, 226*
 functions(s), 560
  graphing, 562
  natural, 565
  natural base of, 565
  Rules of, 560
  with base a, 560
 growth, 607
  model, 607
Exponentiate each side of an equation, 599
Expression, 6*
 algebraic, 56
 radical, 412*
 rational, 320
  dividing, 331
  domain of, 320
  least common denominator of, 337
  logarithmic
   condensing, 591
   expanding, 591
  multiplying, 328
  reciprocal of, 331*
  reduced form of, 322*
  simplified form of, 322*
  simplifying, 322*
Extracting square roots, 503
Extraneous solutions, 357*, 477

F

Factor, 22
 greatest common, 30*, 270*
  monomial, 271, 271*
 proper, 27
Factorial, 665
Factoring, 270, 522
 $ax^2 + bx + c$, 286, 286*
  by grouping, guidelines for, 290
 guidelines for, 287
 by grouping, 273, 273*
 completely, 282
 out, 271*
 polynomials, guidelines for, 301*
 special polynomial forms, A36
 $x^2 + bx + c$, guidelines for, 278, 278*
Factors, variable, 67, 67*
Finding an inverse function algebraically, 673
Finding test intervals for a polynomial, 544
Finite sequence, 664
First-degree equation, 102*
Focus
 of an ellipse, 628
 of a hyperbola, 638
 of a parabola, 624
FOIL Method, 245, 245*

Forming equivalent equations, 88, 88*
Forms of linear equations, A20
Formula(s)
  area, 137
  change-of-base, 583
  common, 137
  compound interest, 567
  distance-rate-time, 136
  miscellaneous, 136
  perimeter, 137
  Quadratic, 518
  recursion, 674
  simple interest, 136
  temperature, 136
  volume, 137
Fractions, 2*, 30*
  addition of, 32
    alternative rule, 32*
  clearing an equation of, 112, 112*
  complex, 346
  division of, 34
  equivalent, 30
  multiplication of, 34
  rules of signs, 30*
  subtraction of, 32
    alternative rule, 32*
  summary of rules of, 35*
  writing in simplest form, 30*
Function(s), 185
  composite, 482
  composition of, 482
  domain, 187
  evaluating, 187
  exponential, 570
    natural, 565
    Rules of, 460
    with base a, 460
  inverse, 572
    finding algebraically, 573
    Horizontal Line Test for, 572
  logarithmic, 578
    common, 579
    natural, 582
    with base a, 578
  name, 187
  notation, 187, 187*
  one-to-one, 572
  quadratic
    graph of, 528
    standard form of, 528
  radical, 447*
    domain of, 447
  range, 187
  rational, 320
    domain of, 320

G

Gaussian elimination, 397
  with back-substitution, 410
General form
  of the equation of a line, 202, 205*
  of a polynomial equation, 305*
  of a quadratic equation, 305

of a quadratic inequality, 546*
Geometric sequence, 682
  common ratio of, 682
  nth partial sum of, 684
  nth term of, 683
Geometric series, 685*
  infinite, 685*
  sum of an infinite, 685
Golden section, 475
Graph
  of an equation, 176, 176*
  of an exponential function, 562
  of an inequality, 144
  of an inverse function, 574
  of a linear inequality in two variables,
    sketching, 211, 211*
  of a logarithmic function, 580
  of a parabola, 177*
  of a quadratic function, 528
Graphing
  point-plotting method, 176
  solution by, 379*
  a system of linear inequalities, 427
Greater than, 4
  or equal to, 4*
Greatest common factor (GCF), 30*,
    270*
Greatest common monomial factor, 271,
    271*
Grouping, factoring by, 273, 273*
Guidelines
  for factoring
    $ax^2 + bx + c$, 287, A34
      by grouping, 290
    polynomials, 285*
    $x^2 + bx + c$, 278
  for solving
    a linear equation containing symbols
      of grouping, 110
    quadratic equations, 305
    word problems, A14
  for verifying solutions, 171

H

Half-life, 608*
Half-planes, 211*
Horizontal asymptote, 563*
Horizontal Line Test, 572
Human memory model, 592
Hyperbola, 638
  asymptotes of, 639
  branch of, 639
  central rectangle of, 639
  focus of, 638
  standard form of equation of
    center at (h, k), 641
    center at origin, 638
  transverse axis of, 638
  vertex of, 638

I

Identity Property
  Additive, 43, A5
  Multiplicative, 43, A5

If and only if, 198
i-form, 484
Imaginary number(s), 486
  pure, 486
Imaginary part of a complex number, 486
Imaginary unit i, 484*
Implied domain restrictions, 321
Inconsistent system, 380, 380*
Increasing annuity, 687
Independent variable, 186
Index
  of a radical, 442
  of summation, 689
Inequality (inequalities)
  absolute value, solving, 155, 155*, A13
  algebraic, 144*
  compound, 148
  conjunctive, 148*
  disjunctive, 148*
  double, 148
  equivalent, 146*
  graph of, 144
  linear (one variable), 146*
  linear, in two variables, 210
    graph of, 211, 211*
    solution of, 210
    systems of, 426*
  properties of, 146, 146*
  quadratic, general form of, 546
  rational, critical numbers of, 548
  solution set of, 144
  solutions of, 144
  solve an, 144
  symbol, 4
  systems of linear, 426*
Infinite
  geometric series, 685*
    sum of, 685
  interval, 144*
  sequence, 664
  series, 668
Infinity
  negative, 145
  positive, 145
Integer(s), 2*
  addition of, 12
  negative, 2*
  positive, 2*
  rules for
    dividing, 20
    multiplying, 18
  subtraction of, 14
Intensity model, 609
Intercepts, 178
  x-intercept, 178
  y-intercept, 178
Interest formulas
  compound, 567
Intersection symbol, 148*

---

*Terms that appear in the Math Help feature at AlgebraWithinReach.com

Intervals on the real number line
  bounded, 144, 144*
  closed, 144
  endpoints, 144
  infinite, 144*
  length of, 144
  open, 144
  unbounded, 145, 144*
Inverse
  additive, 11*
  function, 572
    finding algebraically, 573
    Horizontal Line Test for, 572
  multiplicative, 34*
  Properties
    of Exponents and Logarithms, 597
    of nth Powers and nth Roots, 444
  variation, 365
Inversely proportional, 365*
Irrational number, 2, 2*

J

Joint variation, 367
Jointly proportional, 367*

L

Leading 1, 409
Leading coefficient of a polynomial, 234, 234*
Least common denominator (LCD), 337
Least common multiple (LCM), 32, 322*, 337
Length of an interval, 144
Less than, 4
  or equal to, 4*
Like denominators, combining rational expressions, 32, 336
Like radicals, 458
Like terms, 67, 67*
Line(s)
  parallel, 198
  perpendicular, 198
  slope of, A20
  summary of equations of, 207*
Linear
  equation(s), 102, 102*
    containing symbols of grouping, 110
    forms of, A20
    in one variable, 102, 102*
    in two variables, 176*
  extrapolation, 206*
  inequality (inequalities), 146, 210
    graph of, 211, 211*
    in one variable, 146*
    solution of, 210
    in two variables, 210
      system of, 426*
        graphing, 427
        solution of, 426*
        solution set of, 426*
  interpolation, 206*
  system, number of solutions of, 399

Logarithm(s)
  Inverse Properties of, 597
  Natural, Properties of, 582, 588
    Power, 588
    Product, 588
    Quotient, 588
  Properties of, 579, 588
    Power, 588
    Product, 588
    Quotient, 588
  of x with base a, 578
Logarithmic equations
  One-to-One Property of, 596
  solving, 599
Logarithmic expression
  condensing, 591
  expanding, 591
Logarithmic function, 578
  common, 579
  natural, 582
  with base a, 578
Long division algorithm, 20
Long division of polynomials, 255
Lower limit of summation, 669

M

Major axis of an ellipse, 628
Markup, 122
  rate, 122
Mathematical model, 90
  verbal, 76
Matrix (matrices), 406
  augmented, 406
  coefficient, 406
  determinant of, 416*
    expanding by minors, 417
  determinant of a 2×2, 416
  elementary row operations on, 408
  entry of, 406*
    minor of, 417*
  leading 1, 409
  order of, 406
  row-echelon form of, 409
  row-equivalent, 408
  square, 406
Method of elimination, 388, 388*
Method of substitution, 381, 381*, 648
Minor axis of an ellipse, 628
Minor of a matrix entry, 417*
Miscellaneous common formulas, 136
Mixture problem, 138
Model
  mathematical, 90
  verbal, 90
  verbal mathematical, 76
Monomial, 234
Multiple, least common, 337
Multiplication
  Associative Property of, 43, 64*, A5
  Commutative Property of, 43, 64*, A5
  of fractions, 34

of integers, 18
  Property of Inequalities, 146
Multiplicative
  Identity Property, 43, 64*, A5
  inverse, 34*
  Inverse Property, 43, 64*, A5
Multiplying
  integers, rules for, 18
  rational expressions, 328

N

Name of a function, 187
Natural
  base, 565
  exponential function, 565
  logarithmic function, 582
  Logarithms, Properties of, 582, 588
    Inverse, 597
    Power, 588
    Product, 588
    Quotient, 588
  number, 2, 2*
Negative, 3*
  exponent(s), 227
    Rules for, 227, 445, A24
  infinity, 145
  integer, 2*
  number, square root of, 484
Nonlinear system of equations, 646
  solving by elimination, 650
  solving graphically, 646
  solving by method of substitution, 648
Nonnegative, 3*
Notation, function, 187, 187*
Notation, sigma, 669*
nth partial sum
  of an arithmetic sequence, 675
  of a geometric sequence, 684
nth power(s)
  Inverse Properties of, 444
  Raising each side of an equation to, 476
nth root
  Inverse Properties of, 444
  of a number, 442
    principal, 442
  Properties of, 443
nth term
  of an arithmetic sequence, 673
  of a geometric sequence, 683
Number(s)
  abundant, 27
  complex, 486
    imaginary part of, 486
    real part of, 486
    standard form of, 486
  composite, 22
  critical, 544, 548
  fraction, 2*
  imaginary, 486
    pure, 486

integer, 2, 2*
irrational, 2, 2*
natural, 2*
negative, 3*
nonnegative, 3*
perfect, 27
positive, 3*
prime, 22
rational, 2, 2*
real, 2
of solutions of a linear system, 399
whole, 2, 2*
Numerator, 20

O

One-to-one
function, 572
Properties of Exponential and
Logarithmic Equations, 596
Operations that produce equivalent
systems, 397
Operations that yield equivalent
equations, A9
Opposite of a number, 6, 6*, 11*
Order, 4*
of a matrix, 406
of operations, 42, 42*, 59*, A4
Ordered pair, 168
Ordered triple, 396
Origin, 3*, 168

P

Parabola, 177*, 528, 624
axis of, 528, 624
directrix of, 624
focus of, 624
graph of, 177*
sketching, 530
standard form of equation, 528, 624
vertex of, 528, 624
Parallel lines, 198
Partial sum, 668
nth, of an arithmetic sequence, 675
nth, of a geometric sequence, 684
Pascal's Triangle, 691
Percent, 118, 118*
Percent equation, 119, 119*
Perfect
cube, 443*
number, 27
square, 443*
square trinomial, 299, 299*
Perimeter formulas, 137
Perpendicular lines, 198, 198*
Plotting, 3*
points, 168, 168*
Point-plotting method of sketching a
graph, 175
Point-slope form of the equation of a line,
202, 202*
Polynomial(s), 234, 234*
constant term, 234

critical numbers of, 544
degree, 234, 234*
dividing by a monomial, 254
equation, general form of, 305*
forms, factoring special, A31
finding test intervals for, 544
guidelines for factoring, 285*
leading coefficient, 234, 234*
long division of, 255,
in one variable, 234*
prime, 279
standard form, 234*
synthetic division of a third degree,
258, A31
in x, 234
zeros of, 544
Position equation, 401
Positive, 3*
infinity, 145
integer, 2*
Power, 40, 40*
Property
of Logarithms, 588
of Natural Logarithms, 588
Rules of Exponents, 227
Power-to-Power Rule of Exponents, 226,
227, A24
Price, 122*
unit, 128, 128*
Prime
factorization, 452
number, 22
polynomials, 279
Principal nth root, 442
Problem-solving strategies, 82
Product, 18
Property of Logarithms, 588
Property of Natural Logarithms, 588
Rule of Exponents, 226, 227, 445, A24
Rule for Radicals, 450
of the sum and difference of two terms,
250, 250*
Product-to-Power Rule of Exponents,
226, A24
Proper factor, 27
Properties
Additive Identity, 43, 64*, A5
Additive Inverse, 43, 64*, A5
of Algebra, 64*
Associative Property of Addition, 43,
64*, A5
Associative Property of Multiplication,
43, 64*, A5
Commutative Property of Addition, 43,
64*, A5
Commutative Property of
Multiplication, 43, 64*, A5
Distributive, 43, 64*, A5
of equality, 88
of Inequalities, 146, 146*
Addition and Subtraction, 146
Multiplication and Division, 146

Negative Quantities, 146
Positive Quantities, 146
Transitive, 146
of Logarithms, 579, 588
Inverse, 597
Power, 588
Product, 588
Quotient, 588
Multiplicative Identity, 43, 64*, A5
Multiplicative Inverse, 43, 64*, A5
of Natural Logarithms, 582, 588
Inverse, 597
Power, 588
Product, 588
Quotient, 588
of nth Roots, 443
of real numbers, 43, A5
Square Root, 503, 522
Complex Square Root, 504
Zero-Factor, 304, A38
Reverse of, 523
Proportion, 129
solving, 129
Proportional
directly, 364
inversely, 365
jointly, 367
Pure imaginary number, 486
Pythagorean Theorem, 454

Q

Quadrant, 168
Quadratic
equation(s), 305
general form of, 305*
guidelines for solving, 305
solving
completing the square, 510, 522
extracting square roots, 503
factoring, 522
Quadratic Formula, 518, 522
Square Root Property, 503, 522
Complex Square Root, 504
summary of methods, 522
form, 506
Formula, 518, 522
discriminant, 521
function
graph of, 528
standard form of, 528
inequality, general form, 546
Quotient, 20, 255
Property of Logarithms, 588
Property of Natural Logarithms, 588
Rule of Exponents, 226, 227, A24
Rule for Radicals, 450
Quotient-to-Power Rule of Exponents,
226

R

Radical(s), 442
expressions, simplifying, 453

function, 447*
  domain of, 447
index of, 442
like, 458
Product Rule for, 450
Quotient Rule for, 450
removing perfect square factors from, 450*
Radicand, 442
Radioactive decay, 566*
Radius of a circle, 621
Raising each side of an equation to the nth power, 476
Range
  of a function, 187
  of a relation, 184
Rate
  discount, 123
  markup, 122
Ratio, 126
  common, 682
Rational
  equation, 354
  exponent, 445
  expression(s), 320
    adding
      with like denominators, 336
      with unlike denominators, 337
    complex fraction, 346
    dividing, 331
    domain of, 320
    least common denominator of, 337
    multiplying, 328
    reciprocal of, 331*
    reduced form of, 322*
    simplified form of, 322*
    simplifying, 322*
    subtracting
      with like denominators, 336
      with unlike denominators, 337
  function, 320
    domain of, 320
  inequality, critical numbers of, 548
  number, 2, 2*
Rationalizing the denominator, 453*
Rationalizing factor, 453*
Real numbers, 2
  properties of, 43, A5
Real number line, 3
  bounded intervals on, 144
  unbounded intervals on, 145
Real part of a complex number, 486
Reciprocal, 34*, 331*
Rectangular coordinate system, 168, 169*
Reduced form of a rational expression, 322*
Relation, 184, 184*
  domain of, 184
  range of, 184
Remainder, 255
Removing perfect square factors from a radical, 450*

Repeated solution, 306, 502
Repeating decimal, 36
Reverse of Zero-Factor Property, 523
Rewriting an exponential equation in logarithmic form, 597*
Richter scale, 609*
Root(s)
  cube, 442
  nth, 442
    Inverse Properties of, 444
    principal, 442
    Properties of, 443
  square, 442
    extracting, 503
Rounding a decimal, 36, 36*
Rounding digit, 36
Row operations, 397
Row-echelon form
  of a matrix, 409
  of a system of equations, 396
Row-equivalent matrices, 408
Rules
  for dividing integers, 20
  of Exponents, 226, 227, A24
    Negative, 227, A24
    Power, 227
    Power-to-Power, 226, A24
    Product, 226, 227, A24
    Product-to-Power, 226, A24
    Quotient, 226, 227, A24
    Quotient-to-Power, 226, A24
    Summary of, 227, A24
    Zero, 227, A24
  of fractions, summary of, 35*
  for multiplying integers, 18
  of signs for fractions, 30

S

Satisfy
  an equation, 86, 171*
  an inequality, 144
  solutions of an equation, 86
Scientific notation, 230
Sequence, 664
  arithmetic, 672
  geometric, 682
  finite, 596
  infinite, 596
  term of, 596
Series, 596
  geometric, 685*
    infinite, 685*
  infinite, 668
Set, 2*
Sigma notation, 669
Simple interest formula, 136
Simplest form, 30*
Simplified form of a rational expression, 322*
Simplify an algebraic expression, 68, 68*
Simplifying

radical expressions, 453
rational expressions, 322
Sketching
  the graph of a linear inequality in two variables, 211, 211*
  a graph, point-plotting method of, 176
  a parabola, 530
Slope of a line, 194, 195, A20
Slope-intercept form, 196
Solution(s)
  checking, of an equation, 86, 86*
  double, 502
  of an equation, 86, 170*
  extraneous, 357*, 477
  guidelines for verifying, 171
  of an inequality, 144
  of a linear inequality, 210
  point, 170*
  repeated, 306, 502
  satisfy an equation, 86
  set
    of an inequality, 144
    of a system of linear inequalities, 426*
  steps, 88, 102*
  of a system of linear equations, 354
    by graphing, 379*
  of a system of linear inequalities, 426*
Solve an inequality, 144
Solving
  an absolute value equation, 152
  an absolute value inequality, 155, 155*, A13
  an equation, 152
  exponential equations, 597
  an inequality, 144
  a linear equation containing symbols of grouping, 110
  a linear equation in nonstandard form, 104*
  logarithmic equations, 598
  a nonlinear system graphically, 646
  a proportion, 129
  quadratic equations,
    by completing the square, 510, 522
    by extracting square roots, 503
    by factoring, 522
    guidelines for, 305
    summary of methods for, 522
    using the Quadratic Formula, 518, 522
    using the Square Root Property, 503, 522
      Complex Square Root, 504
  a system of equations, 378
    by Gaussian elimination, 397
      with back-substitution, 198
    using Cramer's Rule, 418
  word problems, guidelines for, A14
Special polynomial forms, factoring, A36
Special products, 250, 250*

Square
  of a binomial, 250, 250*
  completing the, 510, 522
  matrix, 406
  root, 442
    extracting, 503
    of a negative number, 484
  Root Property, 503, 522
    Complex Square Root, 504
Standard form
  of an absolute value equation, 153*
  of a complex number, 486
  of the equation of a circle
    center at (h, k), 622
    center at origin, 621
  of the equation of an ellipse
    center at (h, k), 630
    center at origin, 628
  of the equation of a hyperbola
    center at (h, k), 641
    center at origin, 638
  of the equation of a parabola, 624
  of a polynomial, 234*
  of a quadratic function, 528
Steps of a solution, 88, 102*
Subset, 2*
Substitution, method of, 381, 381*, 648
Subtract one integer from another, 14
Subtracting rational expressions
  with like denominators, 336
  with unlike denominators, 337
Subtraction
  of fractions, 32
    alternative rule, 32*
  of integers, 14
  Property of Inequalities, 146
Sum, 10
  of an infinite geometric series, 685
  of two cubes, 301
  of two squares, 298
Summary
  of additional problem-solving
    strategies, 82
  arithmetic, 24
  of equations of lines, 207*
  of methods for solving quadratic
    equations, 522
  of Rules of Exponents, 227, 445, A24
  of Rules of Fractions, 35
Summation
  index of, 669
  lower limit of, 669
  upper limit of, 669
Symbols
  of grouping, 42, 42*
  intersection, 148*
  union, 148*

Symmetric (about an axis), 528
Synthetic division, 358
  of a third-degree polynomial, 358, A31
Systems of equations, 378
  consistent, 380, 380*
  dependent, 380, 380*
  equivalent, 397
    operations that produce, 397
  inconsistent, 380, 380*
  nonlinear, 646
  row-echelon form of, 396
  solution of, 378
  solving, 378
    by Gaussian elimination, 397
      with back-substitution, 410
    by the method of elimination, 388
    by the method of substitution, 381
    using Cramer's Rule, 418
Systems of linear inequalities, 426*
  graphing, 427
  solution of, 426*
  solution set of, 426*

T

Table of values, 170*
Taking the logarithm of each side of an
    equation, 597
  of a sequence, 664
Temperature formula, 136
Term(s)
  of an algebraic expression, 56
  like, 67, 67*
  of a sequence, 664
Terminating decimal, 36
Test for collinear points, 422
Test intervals, 544
  finding, 544
Tests, divisibility, 22*
Three approaches to problem solving, 170
Transitive Property of Inequalities, 146
Translating phrases into algebraic
    expressions, 78, 78*, 90*
Transverse axis of a hyperbola, 638
Triangle, area of, 137, 420
Trinomial, 234
  perfect square, 299, 299*
Two-point form of the equation of a line,
    203*, 423

U

Unbounded, 144*, 145
Unbounded intervals, 144*, 145
  on the real number line, 145
Undefined, 20*
Union symbol, 148*
Unit price, 128, 128*
Unlike denominators, combining rational
    expressions, 32
Upper limit of summation, 669
Using the discriminant, 521

V

Variable, 56, 56*
  dependent, 186
  factors, 67
  independent, 186
Variation
  combined, 366*
  direct, 362
    as nth power, 364*
  inverse, 365
  joint, 367
Varies
  directly, 362*
    as the nth power, 362*
  inversely, 365*
  jointly, 367*
Verbal mathematical model, 76
Verbal model, 76, 90
Verifying solutions, guidelines for, 171
Vertex
  of an ellipse, 628
  of a hyperbola, 638
  of a parabola, 528, 624
  of a region, 427*
Vertical Line Test, 186
Vertical multiplication algorithm, 18
Volume formulas, 137

W

Whole number, 2, 2*
Word problems, guidelines for solving,
    A14
Work-rate problem, 140
Writing a fraction in simplest form, 30*

X

x-axis, 168
x-coordinate, 168*
x-intercept, 178

Y

y-axis, 168
y-coordinate, 168*
y-intercept, 178

Z

Zero exponent, 227
Zero Exponent Rule, 227, A24
Zero-Factor Property, 304, A38
  Reverse of, 523
Zeros of a polynomial, 544